T0188947

Graduate Texts in Physics

Series Editors

Kurt H. Becker, NYU Polytechnic School of Engineering, Brooklyn, NY, USA

Jean-Marc Di Meglio, Matière et Systèmes Complexes, Bâtiment Condorcet, Université Paris Diderot, Paris, France

Sadri Hassani, Department of Physics, Illinois State University, Normal, IL, USA

Morten Hjorth-Jensen, Department of Physics, Blindern, University of Oslo, Oslo, Norway

Bill Munro, NTT Basic Research Laboratories, Atsugi, Japan

Richard Needs, Cavendish Laboratory, University of Cambridge, Cambridge, UK

William T. Rhodes, Department of Computer and Electrical Engineering and Computer Science, Florida Atlantic University, Boca Raton, FL, USA

Susan Scott, Australian National University, Acton, Australia

H. Eugene Stanley, Center for Polymer Studies, Physics Department, Boston University, Boston, MA, USA

Martin Stutzmann, Walter Schottky Institute, Technical University of Munich, Garching, Germany

Andreas Wipf, Institute of Theoretical Physics, Friedrich-Schiller-University Jena, Jena, Germany

Graduate Texts in Physics publishes core learning/teaching material for graduate- and advanced-level undergraduate courses on topics of current and emerging fields within physics, both pure and applied. These textbooks serve students at the MS- or PhD-level and their instructors as comprehensive sources of principles, definitions, derivations, experiments and applications (as relevant) for their mastery and teaching, respectively. International in scope and relevance, the textbooks correspond to course syllabi sufficiently to serve as required reading. Their didactic style, comprehensiveness and coverage of fundamental material also make them suitable as introductions or references for scientists entering, or requiring timely knowledge of, a research field.

More information about this series at https://link.springer.com/bookseries/8431

Ursula Keller

Ultrafast Lasers

A Comprehensive Introduction to Fundamental Principles with Practical Applications

Ursula Keller
Physics Department,
Ultrafast Laser Physics
ETH Zürich
Zürich, Switzerland

ISSN 1868-4513 ISSN 1868-4521 (electronic)
Graduate Texts in Physics
ISBN 978-3-030-82534-8 ISBN 978-3-030-82532-4 (eBook)
https://doi.org/10.1007/978-3-030-82532-4

© The Editor(s) (if applicable) and The Author(s), under exclusive licence to Springer Nature Switzerland AG 2021
This work is subject to copyright. All rights are solely and exclusively licensed by the Publisher, whether the whole or part of the material is concerned, specifically the rights of translation, reprinting, reuse of illustrations, recitation, broadcasting, reproduction on microfilms or in any other physical way, and transmission or information storage and retrieval, electronic adaptation, computer software, or by similar or dissimilar methodology now known or hereafter developed.
The use of general descriptive names, registered names, trademarks, service marks, etc. in this publication does not imply, even in the absence of a specific statement, that such names are exempt from the relevant protective laws and regulations and therefore free for general use.
The publisher, the authors and the editors are safe to assume that the advice and information in this book are believed to be true and accurate at the date of publication. Neither the publisher nor the authors or the editors give a warranty, expressed or implied, with respect to the material contained herein or for any errors or omissions that may have been made. The publisher remains neutral with regard to jurisdictional claims in published maps and institutional affiliations.

This Springer imprint is published by the registered company Springer Nature Switzerland AG
The registered company address is: Gewerbestrasse 11, 6330 Cham, Switzerland

Preface

This graduate-level textbook starts with Maxwell's equations. If the reader finds this reasonable, this book should be helpful. Focusing on ultrafast solid-state laser oscillators, the book provides detailed derivations of most key equations. Subjects such as nonlinear pulse propagation and frequency metrology are discussed, but only to a level that should help the researcher to get started in this field. Certain topics, such as fiber lasers, are left for other more qualified colleagues to author. In general, I would like to encourage more such textbooks, with detailed derivations critical for new research fields, as this would greatly assist the current and new generations of researchers.

This book is the product of nearly 30 years of research and teaching in ultrafast laser physics at ETH Zurich, Switzerland. I started my academic career in March 1993 as an associate tenured professor in physics at ETH Zurich, and taught my first course one year later, in 1994. In addition to educating new graduate students, one re-discovers, and must even re-learn, much of the subject matter again. I found that these re-learned basics ultimately strengthened my own research efforts, in addition to engaging excellent students and postdocs.

Many students of this course went on to become members of my research group, and many have contributed to this textbook over the years. As this book goes to press, I have graduated 87 Ph.D. students, and expect to have graduated 100 by my mandatory retirement from ETH in 2024. Scattered across industry, from multinational companies to start-ups, and academia including a number of professorships, they have strongly encouraged me to complete this textbook and will hopefully use it to continue teaching the next generations in ultrafast lasers!

I have been active in ultrafast lasers for more than 30 years and have added, on occasion, personal insights into how some of our progress was made. These historical perspectives, based solely on my own experiences, attempt to show how innovation and new discoveries come about. In any event, I encourage new researchers to simply take action and push the knowledge horizon. It has been a great experience for me.

Starting and finishing a textbook is always challenging. In this case, the covid pandemic gets credit for providing the extra required push. On March 16, 2020, all labs at ETH Zurich were shut down. Now forced to work from home, I finalized a plan for the textbook by April 6, and signed the book contract with Springer on May 4, 2020. I also overcame my career-long resistance to learning LaTeX (such a textbook filled with equations goes well beyond the capabilities of normal word processors I discovered), so with very good support from Springer, my old word files were converted into LaTeX files, and we started this more than one-year project.

My first taste of ultrafast lasers started during my Ph.D. at Stanford University. My ETH Diplom (master's equivalent) in physics had me well-prepared with respect to the basic physics and math. Key coursework at Stanford was often taught by the authors themselves, and laid the necessary groundwork for lasers and other essential topics. A standout in my memory was Prof. Anthony Siegman with his laser textbook (published 1986), the go-to reference for most laser challenges and issues. There were many other excellent professors and courses in lasers and optics at Stanford. For example, we learned about Fourier transforms from Ronald Bracewell, nonlinear optics from Steve Harris and Robert Byer, diode-pumped solid-state lasers from Byer, and microwave measurement techniques and laser noise characterization from my Ph.D. advisor Dave Bloom. A visiting professor Geraldine Kenney-Wallace—the first woman professor I had ever met in my education—encouraged me to join this very young and energetic professor in electrical engineering. Between my budding English skills and Dave's electrical-engineering talk, using foreign-sounding terms such as dBm, noise floor, and sampling, I barely understood what his research was about. Nonetheless, his positive energy and maybe some further encouragement from Geraldine, and my strong physics background from ETH, convinced him to offer me a Ph.D. student position, and me to take it. A decision I never regretted. All of the teaching of these great mentors at Stanford, in one way or another, is integrated into this textbook.

For those new to photonics, this Stanford-Ginzton Lab thread runs all the way back to the origin of lasers. Anthony Siegman was the Ph.D. advisor of Steve Harris. Harris supervised Byer and Bloom, my supervisor. Siegman is my academic great-grandfather, and the academic great-great grandfather for my Ph.D. students. So if you happen to be studying under one of my former students, Siegman is your great-cubed (great^3)-grandfather. Theodore H. Maiman, who built and demonstrated the very first laser in 1960, learned key skills as a Ph.D. at Stanford. There has been a long tradition in understanding and building better lasers, cultivated in Ginzton Labs for many decades.

After my Ph.D., I joined AT&T Bell Labs (Homdel, NJ) as a Member of Technical Staff (MTS), where I continued working with excellent scientists. One of the last permanent hires for many years due to an economic recession, I was empowered to pursue my own independent research efforts, with the guideline: Do something different from everybody else, but it better be good. This is where I invented the SESAM for passive modelocking, and how this happened is explained in this textbook. A key asset was the state-of-the-art semiconductor epitaxial growth expertise at Bell Labs. Some of my great role models there were Chuck Shank (who hired me, but had left for an academic position before I arrived), Daniel Chemla, Dave Auston, David A. B. Miller, Rick Freeman (who offered me an MTS position in Murray Hill, based on a talk at my first Gordon Conference during my Ph.D.), some great office neighbors Ted Woodward and Keith Goossen, and my first postdoc Gloria Jacobovitz Veselka, still a good friend to this day.

My work at Bell Labs, combined with growing political pressure to hire more women professors at ETH Zurich, resulted in an unexpected call asking me to consider a position in the physics department at ETH Zurich. At that time, many professors in Switzerland had first been in the U.S., very often with a postdoc or an MTS position at Bell Labs or at IBM Research Labs, in photonics, quantum electronics, or lasers. Following this pattern, my nearly four years as MTS at Bell Labs had prepared me well, and resulted in an offer for a tenured position. The professorship meant I could multiply my work with many excellent Ph.D. students and postdocs, and I knew exactly what I wanted to do. At first lacking sufficient cleanroom and semiconductor growth facilities at ETH Zurich, we eventually achieved a critical mass of new professors and could fund and launch a shared ETH technology platform, the FIRST lab. FIRST stands for Frontier In Research: Space and Time, and became a key building block for my research work here.

I came to greatly appreciate the warm welcome after arriving in Switzerland from leaders in the European laser community, including Profs. Günter Huber, Orazio Svelto, David Hanna, and Gerd Leuchs. Peer review for my Swiss proposals is mostly based on international experts and fortunately was often highly ranked (in Switzerland and at ETH Zurich, however, there was, too often, a surprisingly less-than-welcome feedback, which may be the topic of a future book). In the end, the resources at ETH were significantly better than at many other places, and I always focused on excellent results. As an experimentalist, most of my laser results could be measured, and this helped me to continue to push the frontier in ultrafast laser physics by always working on better lasers that had useful applications.

In the meantime, the leadership in ultrafast lasers, and indeed many other types of lasers, has shifted from mostly in the U.S. to Europe, with key contributions from many European universities and institutions, and innovative products from many European start-ups. This has been a great trend to witness, and I can only hope to continue to see it strengthen in the coming decades. Thinking back to my start in lasers at Stanford in mid-80s, and looking at the technical performance routinely achieved then in research labs and many commercial products, the scope of today's laser performance in output power, pulse energy, pulse width, repetition rate, timing, and even optical phase stability, is jaw-droppingly stunning. Many of these features were considered impossible, even by some of the leaders in the field of the time. In the meantime, we, and by this, I mean contributions of thousands of scientists and engineers worldwide, have made it happen.

I hope that this textbook, and the many students and researchers I have had the pleasure and opportunity to teach and learn from, continue to contribute. I am sure that ultrafast lasers still hold many useful developments, applications, and positive surprises for the future.

Zürich, Switzerland Ursula Keller

Acknowledgements

I would like to acknowledge the many contributions from my current and former Ph.D. students and postdocs. The list of Ph.D. students is summarized below. Three new Ph.D. students starting in 2020 (i.e., Sandro Camenzind, Erik de Vos, and Alexander Nussbaum-Lapping) had, as their first assignment, a last check on the current textbook while they were also learning the basics. I am very grateful for their careful reading. Many postdocs over the years significantly contributed to this textbook such as Franz Kärtner (1993–1997), Luigi Brovelli (1993–1995), Rüdiger Paschotta (1997–2005), Günter Steinmeyer (1998–2002), Markus Haiml (2002–2005), Thomas Südmeyer (2005–2011), Clara Saraceno (2013–2016), Chris Phillips (since 2012), and Lukas Gallmann (since 2006). Professor Lukas Gallmann also took over the lecturing for many years during my additional leadership duties within the NCCR MUST. Special thanks also goes to my MBE growers in the FIRST lab, especially for the great long-lasting support from Dr. Matthias Golling since 2002. I also would like to thank Stephen Lyle for his support in LaTeX for the final preparation of this textbook and my former Ph.D. student Birgit Gallmann-Schenkel for help in improving the growing lecture notes with temporary assignments between 2006 and 2009.

For our new effort in frequency metrology (Chap. 12) I would like to acknowledge the great collaborations with Harald R. Telle (PTB Braunscheig, Germany), and Prof. Pierre Thomann and his senior group leader Dr. Stephane Schilt (University Neuchatel, Switzerland). Especially the great tutorials from the group of Prof. Thomann have been used in Chap. 12. In addition we had many other international collaborations which resulted in joint publications referenced throughout this textbook.

A special thank you goes to my graduate students as follows (with year of Ph.D. exam): Daniel Kopf (1996), Bernd Braun (1996), Isabella Jung (1997), Regula Fluck (1998), Clemens Hönninger (1998), Nicolai Matuschek (1999), Sebastian Arlt (1999), Dirk Sutter (2000), Gabriel Spühler (2000), Jürg Aus der Au (2000), Jens Kunde (2000), Markus Haiml (2001), Marc Achermann (2001), Reto Häring (2001), Lukas Gallmann (2001), Lukas Krainer (2002), Thomas Südmeyer (2003), Felix Brunner (2004), Birgit Schenkel (2004), Alex Aschwanden (2004), Florian Helbing (2004), Steve Lecomte (2005), Wouter Kornelis (2005),

Christoph Hauri (2005), Edith Innerhofer (2005), Dirk Lorenser (2005), Anastassia Gosteva (2005), Rachel Grange (2006), Arne Heinrich (2006), Adrian Schlatter (2006), Simon Zeller (2006), Valeria Liverini (2007), Sergio Marchese (2007), Andreas Rutz (2008), Christian Erny (2008), Petrissa Eckle (2009), Deran Maas (2008), Aude-Reine Bellancourt (2009), Florian Schapper (2009), Max Stumpf (2009), Andreas Oehler (2009), Mirko Holler (2009), Benjamin Rudin (2010), Cyrill Bär (2010), Adrian Pfeiffer (2011), Ekatarina Vorobeva (2011), Martin Hoffmann (2011), Oliver Heckl (2011), Selina Pekarek (2011), Clemens Heese (2012), Valentin Wittwer (2012), Matthias Weger (2012), Clara Saraceno (2012), Oliver Sieber (2013), Wolfgang Pallmann (2013), Reto Locher (2013), Mazyar Sabbar (2014), Christian Zaugg (2014), Jens Herrmann (2014), Robert Boge (2014), Alexander Klenner (2015), Cinia Schriber (2015), Mario Mangold (2015), Florian Emaury (2015), Sebastian Heuser (2015), Benedikt Mayer (2015), Andre Ludwig (2016), Cornelia Hofmann (2016), Sandro Link (2017), Lamia Kasmi (2017), Andreas Diebold (2018), Aline Mayer (2018), Dominik Waldburger (2018), Jannie Vos (2018), Nicolas Bigler (2018), Cesare Alfieri (2018), Mikhail Volkov (2018), Fabian Schläpfer (2019), Benjamin Willenberg (2019), Justinas Pupeikis (2019), Ivan Graumann (2019), Francesco Saltarelli (2020), Luca Pedrelli (2020), Nadja Hartmann (2020), Stefan Hrisafov (2020), Jacob Nürnberg (2020), Jaco Fuchs (2021). My current Ph.D. students: Fabian Brunner, Leonard Krüger, Arthur Niedermayr, Lukas Lang, Pierre-Alexis Chevreuil, Jonas Heidrich, Marco Gaulke, Carolin Bauer, Erik de Vos, Alexander Nussbaum-Lapping, Sandro Camenzind, Moritz Seidel.

I am particularly grateful for the excellent funding support in Switzerland, within ETH Zurich, the Swiss National Science Foundation, the Innosuisse, and from special programs such as NCCR quantum photonics and NCCR MUST. For MUST I was also serving as a director. Additional support is also acknowledged from EU funding programs, especially the two ERC advanced grants. My efforts at ETH Zurich for a joint new technology platform (i.e., FastLab) based on ultrafast lasers is still an ongoing effort and I hope that this FastLab will become possible soon to better empower the next generation of researchers in ultrafast science.

Last but not least, a very special thank you goes to my family, my husband Kurt Weingarten, and our two sons Matthew and Christopher. My husband and I focused on combining both our careers and raising our two children together. And many thanks to our greater family and friends support network, as they were often crucial when we had to ask for additional working time under difficult or tight deadlines. In the end, our sons flourished and are currently completing degrees at ETH Zurich.

Zürich, Switzerland
April 2021

Ursula Keller

Contents

1 Plane Wave Propagation in Dispersive Media

1.1 Maxwell's Equations in SI Units

$$\text{rot}\,\mathbf{H} = \mathbf{j}_{\text{free}} + \frac{\partial}{\partial t}\mathbf{D}\,, \tag{1.1}$$

$$\text{rot}\,\mathbf{E} = -\frac{\partial}{\partial t}\mathbf{B}\,, \tag{1.2}$$

$$\text{div}\,\mathbf{D} = \rho_{\text{free}}\,, \tag{1.3}$$

$$\text{div}\,\mathbf{B} = 0\,. \tag{1.4}$$

Here we use Maxwell's equations in matter, expressed in SI units. This has the advantage that the influence of the bound charge and bound current is incorporated into the displacement field $\mathbf{D}$ and the magnetizing field $\mathbf{H}$, and the equations only depend on the free charge density ρ_{free} and the free current density $\mathbf{j}_{\text{free}}$. By free charges we mean for example the electrons in the conduction band of a metal or a semiconductor, in contrast to the core or bonding electrons which are localized within the atomic structure and are therefore considered not free, but bound. The bonding electron is an electron involved in chemical bonding, localized between atoms, ions, or molecules.

Vectors are represented by bold symbols. In isotropic media it is sufficient for many cases to use a scalar (i.e., one-dimensional) notation, e.g., $\mathbf{E} \rightarrow E$. The notation is as follows, where each quantity generally has a time and a position dependence:

$\mathbf{E}$, $\mathbf{B}$ electric and magnetic field, respectively,
$\mathbf{j}$ electric current density,
ρ charge density.

© The Author(s), under exclusive licence to Springer Nature Switzerland AG 2021

U. Keller, *Ultrafast Lasers*, Graduate Texts in Physics,
https://doi.org/10.1007/978-3-030-82532-4_1

1.2 Material Equations

The material equations describe how the bound charges respond under the influence of the electromagnetic (e.m.) wave, driven by the Coulomb and Lorentz force:

$$\mathbf{F} = q\left(\mathbf{E} + \upsilon \times \mathbf{B}\right) , \tag{1.5}$$

where q is a point charge, e.g., an electron, and υ is its velocity. As long as the force is not too large, the response of each atom or molecule can be described by a harmonic oscillator. In this oscillator a charge, e.g., from the electron, oscillates with respect to the positively charged core. Under the influence of an e.m. wave the electron is brought to forced oscillations induced by the incident e.m. wave. In the linear regime, the oscillating electron then emits radiation with the same frequency, but generally a different frequency-dependent phase. Therefore, a superposition of the incident wave and the phase-shifted wave emitted by the electrons is formed. In the visible and near-infrared spectrum, this superposition of waves propagates more slowly through a transparent medium than the original incident wave.

In the following material equations, the medium is treated as a continuum, which is a good approximation as long as the wavelength is long compared to atomic dimensions. The interaction with a material with bound electrons is described by the following material equations:

$$\mathbf{D} = \varepsilon_0\mathbf{E} + \mathbf{P} , \tag{1.6}$$

where $\mathbf{P}$ is the induced electric polarization of the material, caused by the deflection of the bound electrons relative to the lattice ion, which satisfies

$$\operatorname{div} \mathbf{P} = -\rho_{\text{bound}} , \tag{1.7}$$

and ϵ_0 the vacuum permittivity, which has value

$$\varepsilon_0 = 8.8542 \times 10^{-12}\frac{\text{As}}{\text{Vm}} . \tag{1.8}$$

From (1.3) we then obtain Maxwell's equation $\operatorname{div} \mathbf{E} = \rho/\epsilon_0$ with the total charge density $\rho = \rho_{\text{free}} + \rho_{\text{bound}}$.

For small electric fields, i.e., linear optics, the instantaneous polarization is directly proportional to the applied electric field, that is,

$$\mathbf{P}(t) = \chi\varepsilon_0\mathbf{E}(t) , \tag{1.9}$$

where χ is the electric susceptibility. If χ is frequency dependent, the simple proportionality is only valid in the frequency domain, so that

$$\tilde{\mathbf{P}}(\omega) = \chi(\omega)\,\varepsilon_0\tilde{\mathbf{E}}(\omega) , \tag{1.10}$$

where the polarization and the electric field are no longer in phase, i.e., χ generally has a complex value.

For the magnetic field the material equation is

$$\mathbf{H} = \frac{1}{\mu_0}\mathbf{B} - \mathbf{M}\,, \tag{1.11}$$

where $\mathbf{M}$ is the induced magnetization field of the material and μ_0 the vacuum permeability with value

$$\mu_0 = 1.2566 \times 10^{-6}\frac{\text{Vs}}{\text{Am}}\,. \tag{1.12}$$

Using Maxwell's equation

$$\text{rot}\,\mathbf{B} = \mu_0\left(\mathbf{j} + \epsilon_0\frac{\partial}{\partial t}\mathbf{E}\right)$$

with the total current density $\mathbf{j} = \mathbf{j}_{\text{free}} + \mathbf{j}_{\text{bound}}$, the magnetization $\mathbf{M}$ is then given by

$$\text{rot}\,\mathbf{M} = \mathbf{j}_{\text{bound}} - \frac{\partial}{\partial t}\mathbf{P}\,. \tag{1.13}$$

For small fields, we can again assume that the magnetic polarization is directly proportional to the applied magnetic field:

$$\mathbf{M} = \chi_{\text{m}}\frac{1}{\mu_0}\mathbf{B}\,, \tag{1.14}$$

where χ_{m} is the magnetic susceptibility.

In linear optics it is useful to introduce the dielectric constant ϵ and the magnetic permeability μ. These are defined by

$$\mathbf{D} \equiv \varepsilon\varepsilon_0\mathbf{E}\,, \tag{1.15}$$

$$\mathbf{H} \equiv \frac{1}{\mu\mu_0}\mathbf{B}\,, \tag{1.16}$$

analogous to the vacuum relation $\mathbf{D} = \varepsilon_0\mathbf{E}$ with $\mathbf{P} = 0$ in (1.6) and $\mathbf{H} = \mathbf{B}/\mu_0$ with $\mathbf{M} = 0$ in (1.11). Since $\mathbf{P} = \chi\varepsilon_0\mathbf{E}$ from (1.9) and $\mathbf{M} = \chi_{\text{m}}\mathbf{B}/\mu_0$ from (1.14), this implies that

$$\chi = \varepsilon - 1\,, \tag{1.17}$$

$$\chi_{\text{m}} = \mu - 1\,. \tag{1.18}$$

In addition, we define a refractive index n by

$$n \equiv \sqrt{\varepsilon\mu}\,. \tag{1.19}$$

Note that some literature also uses the following notation:

$$\varepsilon\varepsilon_0 \to \varepsilon\,, \qquad \mu\mu_0 \to \mu\,.$$

In this case $\varepsilon/\varepsilon_0$ is the relative dielectric constant and μ/μ_0 is the relative magnetic permeability. A look at the units quickly shows which definition is adopted. We will use the notation introduced in (1.15) and (1.16), where ε and μ are dimensionless.

In general, in this book we will not discuss magnetic materials such as ferromagnets, which have a magnetic field **M** even when **H** is zero. The discussion will focus more on dielectric materials, that is, materials that can be polarized by an applied electric field. For such a dielectric material, we can neglect the magnetic interaction with the material compared to the electric interaction at optical frequencies. Therefore, we may assume

$$\mu \approx 1 \,. \tag{1.20}$$

This also means that we neglect any frequency dependence of the magnetic permeability. This will not be the case for the induced polarization, for which the frequency dependence will become very important in the context of ultrafast lasers. Note that, in the microwave regime, the condition in (1.20) is no longer valid.

1.3 Wave Equation with Refractive Index

1.3.1 Derivation of the Wave Equation

(a) Wave Equation for the Electric Field

The derivation of the wave equation in a material with bound electrons, i.e., without free charges, follows directly from Maxwell's equations and the material equations:

$$\mathrm{rot}\,(\mathrm{rot}\,\mathbf{E}) = \mathrm{grad}\,(\mathrm{div}\,\mathbf{E}) - \Delta\mathbf{E} = -\Delta\mathbf{E}$$

$$\mathrm{rot}\,(\mathrm{rot}\,\mathbf{E}) = -\frac{\partial}{\partial t}\mathrm{rot}\,\mathbf{B}$$

$$\Longrightarrow \Delta\mathbf{E} = \frac{\partial}{\partial t}\mathrm{rot}\,\mathbf{B}\,, \tag{1.21}$$

where

$$\Delta = \frac{\partial^2}{\partial x^2} + \frac{\partial^2}{\partial y^2} + \frac{\partial^2}{\partial z^2}\,.$$

In (1.21), we assume that there are no free charges, i.e.,

$$\mathrm{div}\,\mathbf{D} = \rho_{\mathrm{free}} = 0\,.$$

The influence of the bound electrons was already considered with $\mathbf{D} = \varepsilon\varepsilon_0\mathbf{E}$. This implies that, for a homogeneous material, i.e., when $\varepsilon(\mathbf{r}) = \varepsilon = \mathrm{const.}$, we also have $\mathrm{div}\,\mathbf{E} = 0$:

$$\mathrm{div}\,\mathbf{D} = \mathrm{div}\,(\varepsilon\varepsilon_0\mathbf{E}) = \varepsilon\varepsilon_0\mathrm{div}\,\mathbf{E} = 0\,, \quad \text{with } \varepsilon(\mathbf{r}) = \mathrm{const.}$$

With Maxwell's equation

$$\text{rot}\,\mathbf{H} = \mathbf{j}_{\text{free}} + \frac{\partial}{\partial t}\mathbf{D}$$

and the material equation $\mathbf{B} = \mu\mu_0\mathbf{H}$, and using (1.21) for a homogeneous material, i.e., $\mu(\mathbf{r}) = \mu = 1$, without electric currents, i.e., $\mathbf{j}_{\text{free}} = 0$, it follows that

$$\Delta\mathbf{E} = \frac{\partial}{\partial t}\text{rot}\,\mathbf{B} = \mu\mu_0\frac{\partial}{\partial t}\text{rot}\,\mathbf{H} = \mu\mu_0\left(\frac{\partial}{\partial t}\mathbf{j}_{\text{free}} + \frac{\partial^2}{\partial t^2}\mathbf{D}\right) = \mu\mu_0\frac{\partial^2}{\partial t^2}\mathbf{D}\,. \quad (1.22)$$

Using the material equation (1.6)

$$\mathbf{D} = \varepsilon_0\mathbf{E} + \mathbf{P}\,,$$

we obtain from (1.22) the wave equation for a homogenous dielectric material without free charges and electric currents:

$$\boxed{\Delta\mathbf{E} - \frac{1}{c^2}\frac{\partial^2}{\partial t^2}\mathbf{E} = \mu_0\frac{\partial^2}{\partial t^2}\mathbf{P}}\,, \quad (1.23)$$

where c is the speed of light in vacuum given by

$$c = \frac{1}{\sqrt{\varepsilon_0\mu_0}} = 3\times 10^8\,\frac{\text{m}}{\text{s}} \quad (1.24)$$

and

$$\mu \approx 1$$

in (1.23). This is the wave equation for a homogeneous, isotropic, nonmagnetic medium with no free charge and no free currents. This is the relevant wave equation for dielectric dispersive materials, for which the induced polarization is frequency-dependent and the electric susceptibility is complex. Without polarization of the dielectric material, (1.23) reduces to the wave equation in vacuum, and the solution is an electromagnetic wave which propagates with the speed of light in vacuum c. The influence of the dielectric medium on the propagation of an e.m. wave $\mathbf{E}(\mathbf{r}, t)$ shows up in the polarization term or source term in the wave equation.

The wave equation (1.23) can be further simplified in a dielectric material free of dispersion. In this case, we can assume that $\chi(\omega) = \chi$ is constant. Using (1.9), (1.17), and (1.23), it follows that

$$\Delta\mathbf{E} - \frac{1}{c^2}(1+\chi)\frac{\partial^2}{\partial t^2}\mathbf{E} = \Delta\mathbf{E} - \frac{1}{c^2}n^2\frac{\partial^2}{\partial t^2}\mathbf{E} = 0\,,$$

and finally,

$$\boxed{\Delta\mathbf{E} - \frac{1}{v_{\text{p}}^2}\frac{\partial^2}{\partial t^2}\mathbf{E} = 0}\,. \quad (1.25)$$

The solutions of the wave equation (1.25) in a linear dielectric medium with constant susceptibility are therefore e.m. waves that propagate with the velocity υ_p. The propagation velocity υ_p of the phase fronts of the e.m. plane wave is called the phase velocity. It is given by the refractive index n:

$$\upsilon_\mathrm{p} = \frac{c}{n} \equiv c_n , \tag{1.26}$$

where c is the speed of light in vacuum of (1.24) and c_n is the speed of light in a dielectric material. In the following, we will use both υ_p and c_n as symbols for the phase velocity in a dielectric medium.

The macroscopic change in the phase velocity of a plane e.m. wave in a material is given by the refractive index n, which is determined by the material's atomic structure. The dispersion relation $n(\omega)$ describes an electromagnetic wave's macroscopic propagation through a material.

Nonlinear optical processes like frequency doubling could also be included in the source term of the wave equation. For high field intensities, the material equations become nonlinear because of the anharmonic binding potential. The nonlinear polarization in the perturbation regime is usually expanded in a Taylor series as a function of the field **E**:

$$P_{i,\mathrm{NL}} = \varepsilon_0 \chi^{(2)}_{ijk} E_j E_k + \varepsilon_0 \chi^{(3)}_{ijkl} E_j E_k E_l + \varepsilon_0 \chi^{(4)}_{ijklm} E_j E_k E_l E_\mathrm{m} + \cdots ,$$

through which the wave equation generalizes to

$$\boxed{\Delta \mathbf{E} - \frac{1}{c^2}\frac{\partial^2}{\partial t^2}\mathbf{E} = \mu_0 \frac{\partial^2}{\partial t^2}\mathbf{P}_\mathrm{L} + \mu_0 \frac{\partial^2}{\partial t^2}\mathbf{P}_\mathrm{NL}} , \tag{1.27}$$

where $\mathbf{P}_L$ only contains the linear part of the polarization. In the first three chapters of the book, we will neglect the nonlinear terms in the wave equation.

(b) Wave Equation for the Magnetic Field

Analogously to the electric field, it is possible to derive the wave equation for the field **H** directly from Maxwell's equations:

$$\boxed{\Delta \mathbf{E} - \frac{1}{\upsilon_\mathrm{p}^2}\frac{\partial^2 \mathbf{E}}{\partial t^2} = 0 , \qquad \Delta \mathbf{H} - \frac{1}{\upsilon_\mathrm{p}^2}\frac{\partial^2 \mathbf{H}}{\partial t^2} = 0} . \tag{1.28}$$

The fields **H** and **E** bear a fixed relation to each other. It is therefore common to restrict discussion to the electric field. But one should not forget that both fields are always necessary for the propagation of an electromagnetic wave.

In many cases, it is possible to choose the coordinate system so that the propagation of the plane wave is along one coordinate axis, in this book usually the z-axis. In this case, the wave equations (1.28) reduce to

$$\boxed{\frac{\partial^2}{\partial z^2}\mathbf{E} - \frac{1}{\upsilon_\mathrm{p}^2}\frac{\partial^2}{\partial t^2}\mathbf{E} = 0 , \qquad \frac{\partial^2}{\partial z^2}\mathbf{H} - \frac{1}{\upsilon_\mathrm{p}^2}\frac{\partial^2}{\partial t^2}\mathbf{H} = 0} . \tag{1.29}$$

1.3.2 Solution of the Wave Equation: Plane Wave

The plane monochromatic wave is an important special solution of the wave equation. Later, we will see that any wave can be expanded in terms of monochromatic plane waves (Fourier transform, Appendix A). Therefore, we will only discuss the solution of the plane wave here. We will address the general case later, with wave packets limited in time such as modelocked laser pulses.

(a) Real Notation

The following applies for the real solution of a plane wave propagating in the positive z-direction (to the right):

$$\mathbf{E}(z,t) = \left|\mathbf{E}_0^+\right| \cos(\omega t - kz + \varphi) \ , \tag{1.30}$$

where $\mathbf{E}_0^+$ is the complex amplitude $\mathbf{E}_0^+ = \left|\mathbf{E}_0^+\right| \mathrm{e}^{\mathrm{i}\varphi}$, ω the angular frequency, and k the wave number. We use the plus sign here because the wave propagates in the positive z-direction. In the following we will very often omit the plus or minus sign for the amplitude. For a plane wave propagating in the negative z-direction (to the left), we have analogously:

$$\mathbf{E}(z,t) = \left|\mathbf{E}_0^-\right| \cos(\omega t + kz + \varphi) \ . \tag{1.31}$$

Likewise, for the field $\mathbf{H}$, we have

$$\mathbf{H}(z,t) = \left|\mathbf{H}_0^+\right| \cos\left(\omega t - kz + \varphi^+\right) + \left|\mathbf{H}_0^-\right| \cos\left(\omega t + kz + \varphi^-\right) \ ,$$

for the superposition of plane waves propagating in the positive and negative z-directions.

The period T (see Fig. 1.1) is defined by

$$T = \frac{2\pi}{\omega} = \frac{1}{\nu} \ , \tag{1.32}$$

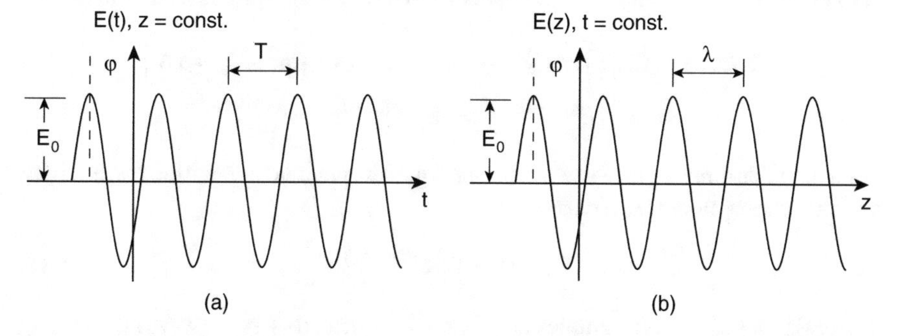

Fig. 1.1 **a** Depiction of a plane wave for a constant position as a function of time $E(t) = |E_0| \cos(\omega t + \varphi)$. **b** Depiction of a plane wave for a constant time as a function of position $E(z) = |E_0| \cos(-kz + \varphi)$

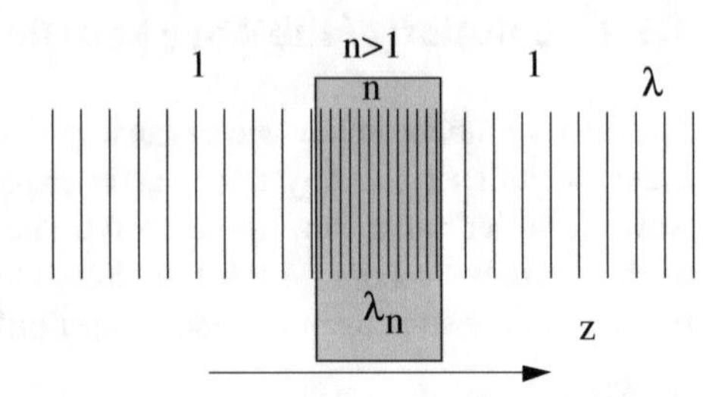

Fig. 1.2 Shorter wavelength in an optically denser medium, i.e., a medium with higher refractive index. Example: transmission of a plane wave through a transparent plate

where ν is the frequency. The plane wave solution is a function of two variables in space and time. If we want to depict this solution in a 2D plot, we show the solution either as a function of time at a fixed position in space (see Fig. 1.1a) or as a function of space at a fixed position in time (see Fig. 1.1b). The period is T in the time domain and λ in space (in vacuum). In the linear regime, when a monochromatic plane wave propagates from vacuum into a medium, the optical frequency does not change. This follows directly from Maxwell's equations (not proven here).

A plane wave in vacuum propagates at the speed of light c over a distance λ in a time T. Therefore, the wavelength of a plane wave is different in vacuum, where it is given by λ, from the wavelength in a dispersive medium, where it is given by λ_n. This is also depicted in Fig. 1.2. This is required to be consistent with the solution of the wave equation in a dispersive medium. In my experience, young students tend to be confused about this fact, at least initially, because the frequency is the same in vacuum and in a dispersive medium, but the wavelength is different!

Therefore, to summarize, over the duration of a period T, the maximum of a wavefront travels the distance of a wavelength, so

$$\boxed{\text{vacuum} \quad c = \frac{\lambda}{T}, \qquad \text{dispersive medium} \quad c_n = \frac{\lambda_n}{T}} . \tag{1.33}$$

(b) Complex Notation

In linear optics the wave equation is linear. Therefore, if $\cos(\omega t - kz)$ from (1.30) and $\sin(\omega t - kz)$ from (1.31) with $\varphi = 90°$ are solutions, then so also is any superposition of these. For example, the complex form of the plane wave is again a solution:

$$\begin{aligned} \mathbf{E}(z,t) &= \left|\mathbf{E}_0^+\right| \left[\cos(\omega t - kz + \varphi) + \mathrm{i}\sin(\omega t - kz + \varphi)\right] \\ &= \left|\mathbf{E}_0^+\right| \mathrm{e}^{\mathrm{i}(\omega t - kz + \varphi)} = \mathbf{E}_0^+ \mathrm{e}^{\mathrm{i}(\omega t - kz)} . \end{aligned}$$

For a monochromatic plane wave propagating in the positive z-direction (to the right) and in complex notation, we have

$$\mathbf{E}(z,t) = \mathbf{E}_0^+ \mathrm{e}^{\mathrm{i}(\omega t - kz)} , \tag{1.34}$$

where $\mathbf{E}_0^+ = |\mathbf{E}_0^+| \mathrm{e}^{\mathrm{i}\phi}$ is a complex amplitude. Likewise, for a plane wave propagating in the negative z-direction (to the left), we have

$$\mathbf{E}(z,t) = \mathbf{E}_0^- \mathrm{e}^{\mathrm{i}(\omega t + kz)} . \tag{1.35}$$

For the field **H**, we have analogously

$$\mathbf{H}(z,t) = \mathbf{H}_0^+ \mathrm{e}^{\mathrm{i}(\omega t - kz)} + \mathbf{H}_0^- \mathrm{e}^{\mathrm{i}(\omega t + kz)} , \tag{1.36}$$

for the superposition of plane waves traveling in the positive and negative z-directions.

The real solution can be written as a superposition of two complex plane waves:

$$\mathbf{E}_0^+ \cos(\omega t - kz) = \frac{1}{2}\left[\mathbf{E}_0^+ \mathrm{e}^{\mathrm{i}(\omega t - kz)} + \text{c.c.}\right] ,$$

where c.c. denotes the complex conjugate $\left(\mathbf{E}_0^+\right)^* \mathrm{e}^{-\mathrm{i}(\omega t - kz)}$. With this, the complex treatment is completely equivalent to the real one. But the mathematics is much easier using the complex notation.

Note. Care must be taken when multiplying fields! For two complex numbers a and b with nonzero imaginary parts, we have

$$\mathrm{Re}\{a\} \cdot \mathrm{Re}\{b\} \neq \mathrm{Re}\{a \cdot b\} .$$

But in contrast to quantum mechanics, the correct physical value is the real part of the particular field, i.e., $\mathrm{Re}\{a\} \cdot \mathrm{Re}\{b\}$. This problem arises, for example, with energy density, energy flow, and nonlinear processes.

(c) Wave Number, Angular Frequency, and Wavelength

When inserting (1.34) into the wave equation (1.29) with $\upsilon_\mathrm{p} = c$, we obtain for the wave number k in vacuum

$$(-\mathrm{i}k)^2 - \frac{1}{c^2}(\mathrm{i}\omega)^2 = 0 .$$

Then,

$$\boxed{\text{wave number in vacuum } k = \frac{\omega}{c}} . \tag{1.37}$$

For a dispersive medium, it follows with $\upsilon_\mathrm{p} = c_n$ (1.26) that

$$\mathbf{E}(z,t) = \mathbf{E}_0^+ \mathrm{e}^{\mathrm{i}(\omega t - k_n z)} , \tag{1.38}$$

which is analogous to the vacuum situation but with a modified wave number k_n. As before, it follows directly from the wave equation that, for a dispersive medium with refractive index n,

$$(-\mathrm{i}k_n)^2 - \frac{1}{\upsilon_\mathrm{p}^2}(\mathrm{i}\omega)^2 = 0 ,$$

and therefore

$$\boxed{\text{wave number in dispersive medium } k_n = \frac{\omega}{v_\mathrm{p}} = \frac{\omega}{c} n \implies k_n = kn} \,. \tag{1.39}$$

This directly implies the relationship between wavelength and wave number:

$$\boxed{\text{vacuum } k = \frac{2\pi}{\lambda}, \quad \text{dispersive medium } k_n = \frac{2\pi}{\lambda_n}} \,. \tag{1.40}$$

Proof. In vacuum, the plane wave propagates with the speed of light c over a distance λ during the time of one period T (1.32):

$$\lambda = cT = \frac{c}{\nu} = 2\pi \frac{c}{\omega} = 2\pi \frac{1}{k} \,.$$

In a dispersive medium, the plane wave propagates with the speed c_n over a distance λ_n during the time of one period T (1.32):

$$\lambda_n = c_n T = \frac{c_n}{\nu} = 2\pi \frac{c_n}{\omega} = 2\pi \frac{c}{\omega n} = 2\pi \frac{1}{k_n} \,.$$

1.3.3 Summary of the Notation in Vacuum and in a Dispersive Medium

Let us assume we have a monochromatic source for e.m. waves in vacuum and in a dielectric material. The source emits with the same frequency in both cases. Table 1.1 shows a summary of how the physical parameters compare. There are some useful conversion rules to translate photon energy into wavelength, reduced wave number into frequency, and wave number into angular frequency. These are summarized in Table 1.2.

1.3.4 TEM Wave and Impedance

Maxwell's equations imply

$$\boxed{\mathbf{E} \perp \mathbf{H} \,, \quad \mathbf{k} \perp \mathbf{E} \,, \quad \mathbf{k} \perp \mathbf{H}} \,. \tag{1.41}$$

This wave is called a transverse electromagnetic wave or TEM wave, because both electric and magnetic field are perpendicular to the wave vector. For a plane harmonic wave, the fields **E** and **B** are in phase, as shown in Fig. 1.3.

Table 1.1 Physical parameters for a monochromatic e.m. wave, where ν is the frequency, $T = 1/\nu$ is the period, $\omega = 2\pi\nu$ is the angular frequency, λ is the wavelength in vacuum, λ_n is the wavelength in a material, c is the speed of light in vacuum, c_n is the speed of light in a material, $k = 2\pi/\lambda$ is the wave number in vacuum, and $k_n = 2\pi/\lambda_n$ is the wave number in a material

Vacuum	Dispersive medium
ν	ν
$T = 1/\nu, \omega = 2\pi\nu$	$T = 1/\nu, \omega = 2\pi\nu$
$\upsilon_p = c$	$\upsilon_p = c_n = c/n$
$k = \frac{\omega}{c}$	$k_n = \frac{\omega}{\upsilon_p} = \frac{\omega}{c}n = kn$
$k = \frac{2\pi}{\lambda}$	$k_n = \frac{2\pi}{\lambda_n} = kn$
λ	$\lambda_n = \frac{\lambda}{n}$

Table 1.2 Useful conversion equations from energy to wavelength and from frequency to wave numbers

Energy $E = \hbar\omega \iff$ wavelength λ	$E\,[\text{eV}] = \frac{1.24}{\lambda\,[\mu\text{m}]}$
Frequency $\nu \iff$ reduced wave number $\frac{1}{\lambda}$	$\nu\,[\text{GHz}] = 30 \cdot \frac{1}{\lambda}\left[\text{cm}^{-1}\right]$
Angular frequency $\omega \iff$ wave number k	$\omega\,[\text{GHz}] = 30 \cdot k\left[\text{cm}^{-1}\right]$

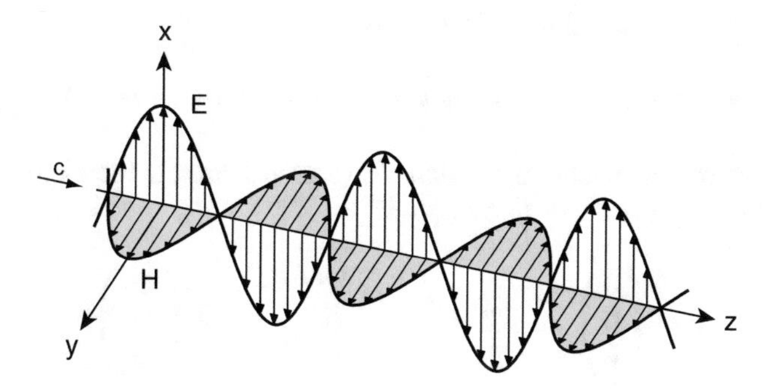

Fig. 1.3 TEM wave

Maxwell's equations also imply that the amplitudes of the electric and magnetic fields of a monochromatic plane wave bear a fixed relation to each other:

$$\boxed{H_0^{\pm} = \pm\frac{1}{Z}E_0^{\pm}}\,, \tag{1.42}$$

where Z is the impedance, defined by

$$\boxed{\begin{array}{l} \text{in vacuum } Z \equiv Z_0 \equiv \sqrt{\dfrac{\mu_0}{\varepsilon_0}} = 377\ \Omega \\ \text{in a medium } Z \equiv \sqrt{\dfrac{\mu\mu_0}{\varepsilon\varepsilon_0}} \overset{\mu=1}{=} \dfrac{Z_0}{n} \end{array}} \tag{1.43}$$

with n the refractive index. Z_0 is also called the characteristic vacuum impedance, and is a natural constant like the speed of light $c \equiv \sqrt{1/\varepsilon_0\mu_0}$. The relationship with the speed of light is

$$\boxed{\begin{array}{l} \text{in vacuum } Z_0 = \mu_0 c \\ \text{in a dispersive medium } Z = \mu\mu_0 c_n \overset{\mu=1}{=} \mu_0 c_n \end{array}} \tag{1.44}$$

This means that the phase velocity in a medium with $\mu = 1$ is directly proportional to the impedance.

Proof of (1.41) and (1.42). We assume that there are no free charges. Using $\mathbf{D} = \varepsilon\varepsilon_0\mathbf{E}$ and Maxwell's equation

$$\operatorname{div}\mathbf{E} = \rho_{\text{free}}/\varepsilon\varepsilon_0 = 0\,,$$

it follows that for a monochromatic plane wave

$$\operatorname{div}\mathbf{E} = \mp\mathrm{i}\left(k_x E_x + k_y E_y + k_z E_z\right) = \mp\mathrm{i}\mathbf{k}\cdot\mathbf{E} = 0 \implies \mathbf{k}\perp\mathbf{E}\,.$$

This is therefore a monochromatic plane wave propagating in the z-direction, with electric field along the x-axis (see Fig. 1.3):

$$\mathbf{k} = \begin{pmatrix} 0 \\ 0 \\ k \end{pmatrix}, \quad \mathbf{E} = \begin{pmatrix} E_0^{\pm}\mathrm{e}^{\mathrm{i}(\omega t \mp k_n z)} \\ 0 \\ 0 \end{pmatrix} = \begin{pmatrix} E_x \\ 0 \\ 0 \end{pmatrix}.$$

The field $\mathbf{H}$ follows directly from $\mathbf{B} = \mu\mu_0\mathbf{H}$ and Maxwell's equation

$$\operatorname{rot}\mathbf{E} = -\mu\mu_0\frac{\partial}{\partial t}\mathbf{H}\,.$$

Therefore, rot **E** is given by

$$\text{rot } \mathbf{E} = \begin{vmatrix} e_x & \partial_x & E_x \\ e_y & \partial_y & 0 \\ e_z & \partial_z & 0 \end{vmatrix} = \begin{pmatrix} 0 \\ \partial_z E_x \\ -\partial_y E_x \end{pmatrix} \overset{\partial_y E_x = 0}{=} \begin{pmatrix} 0 \\ \partial_z E_x \\ 0 \end{pmatrix} = \mp \mathrm{i} k_n E_x \begin{pmatrix} 0 \\ 1 \\ 0 \end{pmatrix} \implies \mathbf{E} \perp \mathbf{H} .$$

This shows that the e.m. wave is a TEM wave, i.e., $\mathbf{E}\perp\mathbf{H}$, $\mathbf{k}\perp\mathbf{E}$, and $\mathbf{k}\perp\mathbf{H}$, as required in (1.41).

H is also a monochromatic plane wave proportional to $\mathrm{e}^{\mathrm{i}(\omega t \mp kz)}$, and we have

$$\frac{\partial}{\partial t}\mathbf{H} = \mathrm{i}\omega\mathbf{H} .$$

Maxwell's equation

$$\text{rot } \mathbf{E} = -\mu\mu_0 \frac{\partial}{\partial t}\mathbf{H}$$

thus implies

$$\mp \mathrm{i} k_n E_x = -\mu\mu_0 \mathrm{i}\omega H_y \implies \mp \mathrm{i} k_n E_0^{\pm} = -\mu\mu_0 \mathrm{i}\omega H_0^{\pm} .$$

Using $k_n = \omega/c_n$, it follows that

$$E_0^{\pm} = \pm\mu\mu_0 c_n H_0^{\pm} = \pm Z H_0^{\pm} .$$

This proves (1.42).

The fields **E** and **B** oscillate in phase as shown in Fig. 1.3 because, by Maxwell's equations,

$$\text{rot } \mathbf{E} = -\frac{\partial}{\partial t}\mathbf{B} .$$

Thus, for a harmonic plane wave, it follows that $-\mathrm{i}\mathbf{k} \times \mathbf{E} = -\mathrm{i}\omega\mathbf{B}$ and therefore $\mathbf{k} \times \mathbf{E} = \omega\mathbf{B}$.

1.3.5 Polarization

The electromagnetic wave discussed so far is a linearly polarized wave, because the electric field oscillates in one fixed direction, and likewise for the magnetic field. The direction in which the electric field oscillates is called the polarization direction.

$$\text{linearly polarized plane wave (x-polarization)} \quad \mathbf{E} = \begin{pmatrix} E_0 \\ 0 \\ 0 \end{pmatrix} \mathrm{e}^{\mathrm{i}(\omega t - k_n z)} . \tag{1.45}$$

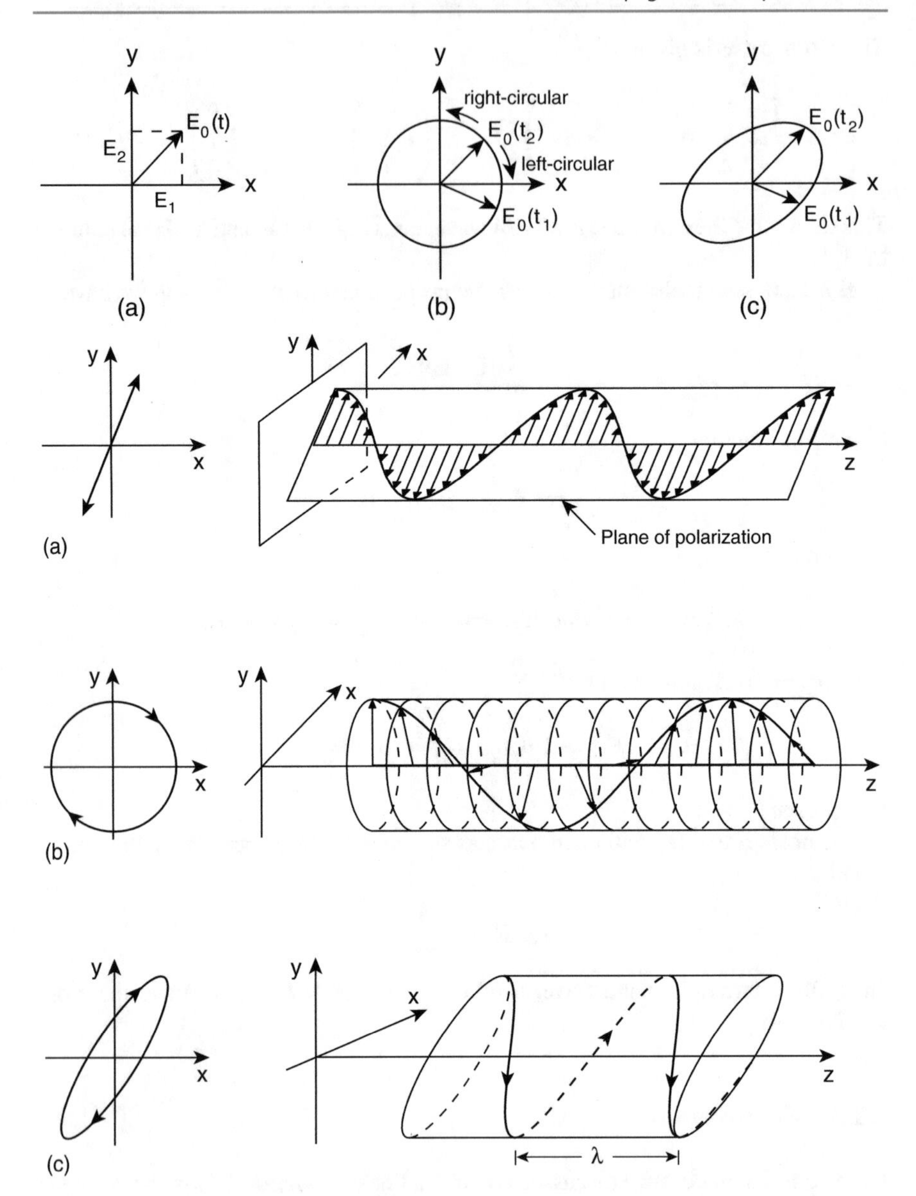

Fig. 1.4 Polarization states of a TEM wave. **a** Linearly polarized wave. **b** Circularly polarized wave. Convention for right-circular polarization: we use the right-hand rule, i.e., with the thumb in the propagation direction, the electric field vector rotates counter-clockwise at a given position in space. For left-circular polarization, we use the left-hand rule. **c** Elliptically polarized wave

The superposition of two linearly polarized waves with a phase difference $\Delta\varphi$ and polarizations perpendicular to each other (e.g., x- and y-axis for a plane wave propagating in the z-direction) generally results in:

$$\text{elliptically polarized plane wave} \quad \mathbf{E} = \begin{pmatrix} E_1 \\ E_2 e^{i\Delta\varphi} \\ 0 \end{pmatrix} e^{i(\omega t - k_n z)} . \tag{1.46}$$

The special case $\Delta\varphi = 0$ results in a linearly polarized wave with a polarization direction $\mathbf{E}_0$ in the xy-plane (see Fig. 1.4a) which is constant in time. A linearly polarized wave with the electric field along the x-axis is the special case $E_1 = E_0$, $E_2 = 0$. The special case $\Delta\varphi = \pm\pi/2$ and $E_1 = E_2$ results in a circularly polarized wave, where the polarization vector moves around a circle with frequency ω (Fig. 1.4b). In the general case of an elliptically polarized wave, the polarization vector moves around an ellipse during propagation (Fig. 1.4c).

1.3.6 Energy Density, Poynting Vector, and Intensity

For optically isotropic materials, both the Poynting vector and the wave number vector are perpendicular to the wave fronts. This is not necessarily the case for anisotropic materials. The intensity I is typically given in units W/cm^2 (Tables 1.3 and 1.4).

We may derive

$$\mathbf{S}(\mathbf{r}, t) = \frac{1}{Z}\mathbf{E}^2\mathbf{k}_0$$

from the definition

$$\mathbf{S} = \mathbf{E} \times \mathbf{H} = \mathbf{E}(\mathbf{r}, t) \times \left[\frac{\mathbf{k}_0}{Z} \times \mathbf{E}(\mathbf{r}, t)\right] ,$$

using the vector identity

$$\mathbf{A} \times (\mathbf{B} \times \mathbf{C}) = (\mathbf{A} \cdot \mathbf{C})\,\mathbf{B} - (\mathbf{A} \cdot \mathbf{B})\,\mathbf{C}$$

Table 1.3 Energy density, Poynting vector, and intensity

Energy density (instantaneous)	$w = \frac{1}{2}(\mathbf{E} \cdot \mathbf{D} + \mathbf{H} \cdot \mathbf{B}) = \frac{1}{2}\varepsilon\varepsilon_0\mathbf{E}^2 + \frac{1}{2}\mu\mu_0\mathbf{H}^2$
Poynting vector (instantaneous)	$\mathbf{S} = \mathbf{E} \times \mathbf{H}$ (for real fields)
Poynting vector (averaged over time)	$\overline{\mathbf{S}} = \frac{1}{2}\mathrm{Re}\left(\mathbf{E}_0 \times \mathbf{H}_0^*\right)$ (for complex fields)
Intensity	$I = \lvert\overline{\mathbf{S}}\rvert = c_n\overline{w}$

Table 1.4 The special case of a monochromatic plane wave with amplitudes $\mathbf{E}_0$ and $\mathbf{H}_0$

Magnetic field	$\mathbf{H}(\mathbf{r},t)=\dfrac{\mathbf{k}_0}{Z}\times\mathbf{E}(\mathbf{r},t),\ \mathbf{k}_0\equiv\dfrac{\mathbf{k}}{\lvert\mathbf{k}\rvert}$
Energy density (averaged over time)	$\overline{w}=\dfrac{1}{2}\varepsilon\varepsilon_0\,\lvert\mathbf{E}_0\rvert^2$
Poynting vector (instantaneous)	$\mathbf{S}(\mathbf{r},t)=\dfrac{1}{Z}\mathbf{E}^2\mathbf{k}_0,\ \mathbf{k}_0\equiv\dfrac{\mathbf{k}}{\lvert\mathbf{k}\rvert}$
Poynting vector (averaged over time)	$\overline{\mathbf{S}}=\dfrac{1}{2Z}\,\lvert\mathbf{E}_0\rvert^2\,\mathbf{k}_0,\ \mathbf{k}_0\equiv\dfrac{\mathbf{k}}{\lvert\mathbf{k}\rvert}$
Intensity	$I=\dfrac{1}{2Z}\,\lvert\mathbf{E}_0\rvert^2=\dfrac{1}{2}Z\,\lvert\mathbf{H}_0\rvert^2$

and the condition for a transverse wave,

$$\mathbf{k}_0\cdot\mathbf{E}=0\,,$$

whence the desired equations follow.

1.4 Dispersion

1.4.1 Dispersion for Electromagnetic Waves

Due to the fact that the refractive index $n(\omega)$ is frequency dependent, the wave number $k_n(\omega)=(\omega/c)n(\omega)$ in a dispersive medium, as given in (1.39), is in general no longer directly proportional to the frequency, as it is in vacuum, where $k(\omega)=\omega/c$ according to (1.37). This 'complicated' frequency dependence is called dispersion. It is very important to have exact knowledge of the frequency dependence of the refractive index $n(\omega)$. For example, dispersive materials refract waves with different frequencies at different angles (see Fig. 1.5). Another important example is the propagation of a light pulse through a dispersive medium (see Fig. 1.6). This effect will be discussed in detail in the following chapters.

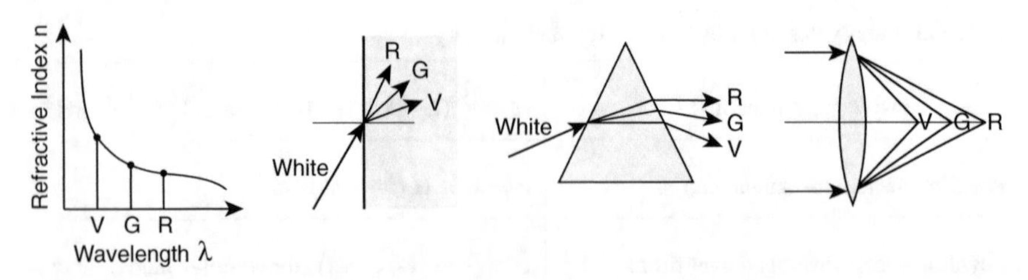

Fig. 1.5 One important consequence of dispersion on wave propagation is refraction. Adapted from [91Sal2]. Here, we assume a dispersive medium with $\lambda_R=\lambda$ (red) $>\lambda_G=\lambda$ (green) $>\lambda_V=\lambda$ (violet)

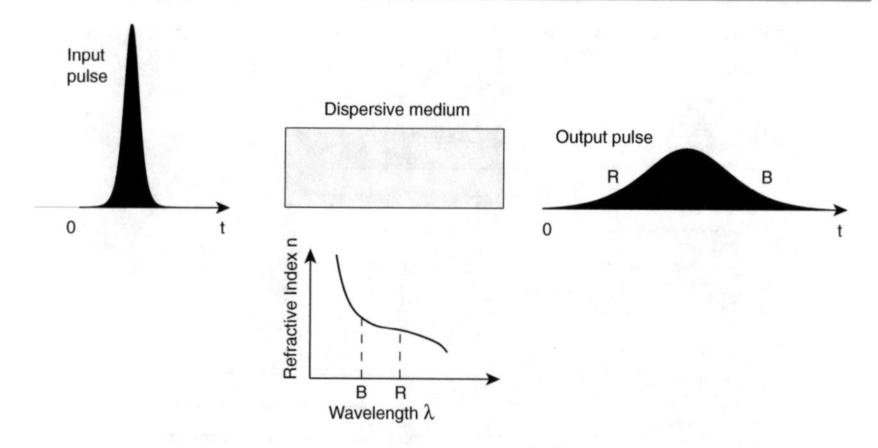

Fig. 1.6 Another important consequence of dispersion on wave propagation is the effect on pulse propagation. Adapted from [91Sal2]. Here, we assume a normally dispersive medium, i.e., $\mathrm{d}^2n/\mathrm{d}\omega^2 > 0$, or equivalently, $\mathrm{d}\upsilon_g/\mathrm{d}\lambda > 0$ which means the longer wavelength plane wave (i.e., red or R) propagates faster than the shorter wavelength (i.e., blue or B)

Typically, the dispersion is classified into three types: normal or positive dispersion for $\mathrm{d}^2n/\mathrm{d}\omega^2 > 0$, anomalous or negative dispersion for $\mathrm{d}^2n/\mathrm{d}\omega^2 < 0$, and no dispersion for $\mathrm{d}^2n/\mathrm{d}\omega^2 = 0$.

1.4.2 Sellmeier Equation in the Visible and Near-Infrared

In many cases we will discuss dielectric materials which are transparent, and therefore have negligible absorption. Glass is transparent in the visible spectrum, but absorbs in the ultraviolet and infrared spectral regions. The extent of the spectral window, i.e., the wavelength region where a material is transparent, is shown in Fig. 1.7 for different materials. In this spectral window, one is typically in the normal or positive dispersion region (Fig. 1.8), because $\mathrm{d}^2n/\mathrm{d}\omega^2 > 0$. Exact knowledge of the dispersion relation in the spectral window is very important for many applications. The Sellmeier equations typically provide this dispersion relation.

Far away from absorption, the dispersion $n(\omega)$ can be described by the Sellmeier equations. Far away from an absorption line, the frequency-dependent absorption coefficient $\alpha(\omega)$ can be replaced by an infinitely sharp absorption line at ω_0, given by a delta function $\delta(\omega - \omega_0)$. We use the Kramers–Kronig relation

$$\boxed{\chi_r(\omega) = \frac{2}{\pi}\int_0^\infty \frac{\omega' \chi_i(\omega')}{\omega'^2 - \omega^2}\mathrm{d}\omega', \qquad \chi_i(\omega) = \frac{2}{\pi}\int_0^\infty \frac{\omega' \chi_r(\omega')}{\omega^2 - \omega'^2}\mathrm{d}\omega'}$$

where the real part of the complex susceptibility χ_r determines the refractive index and the imaginary part χ_i determines the absorption of the medium. This simplifies

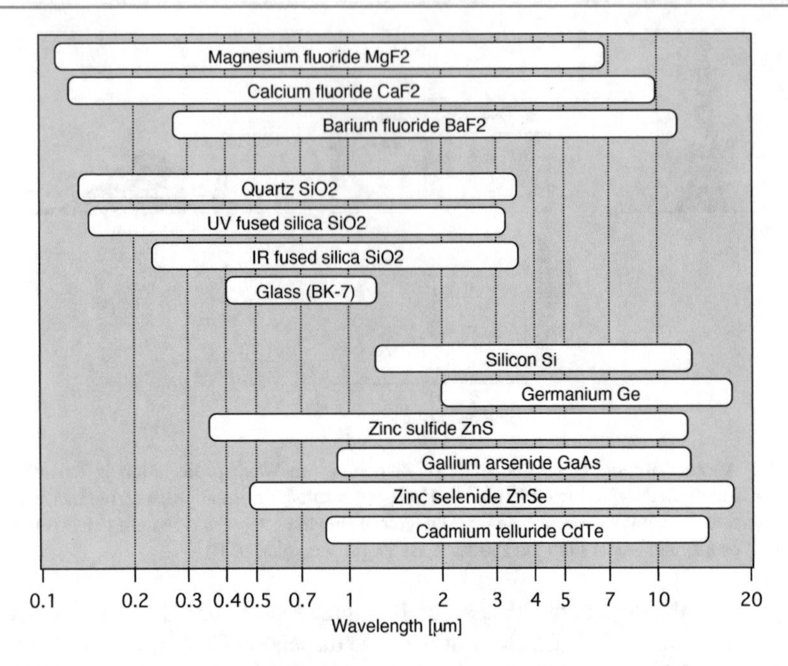

Fig. 1.7 Spectral windows, i.e., transparent spectral regions, for different materials. According to [91Sal2]

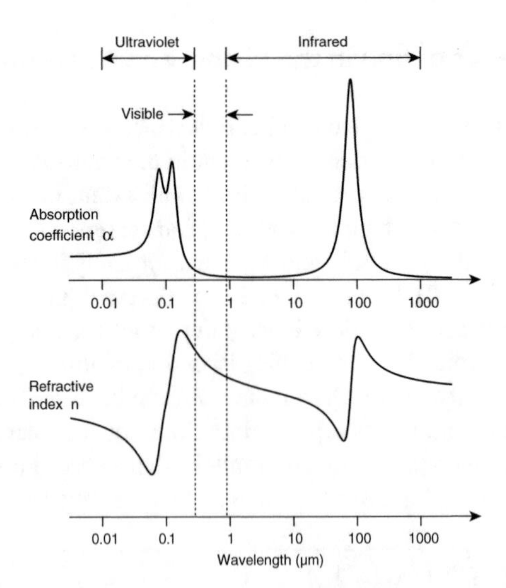

Fig. 1.8 Typical wavelength dependence of the absorption coefficient and refractive index for dielectric materials having strong absorptions in the ultraviolet and infrared, and negligibly small absorptions in the visible spectral region. The physical origin of the infrared resonances is scattering on atomic nuclei, which have a higher mass than electrons, and therefore produce lower resonances. Quantum mechanically, this corresponds to phonon absorption. The resonances in the UV and higher originate from scattering by bound electrons

in the present case to yield

$$n^2 = 1 + \frac{A\lambda^2}{\lambda^2 - \lambda_0^2} , \tag{1.47}$$

where A is a constant. This relation was already obtained by Sellmeier in 1871. It can be extended to more absorption lines by inserting

$$\alpha(\omega) = \delta(\omega - \omega_1) + \delta(\omega - \omega_2) + \cdots .$$

Derivation of (1.47)
Using 1.17 and 1.19 with $\mu = 1$ yields

$$\tilde{n}(\omega)^2 = 1 + \tilde{\chi}(\omega) = 1 + \chi_r(\omega) - i\chi_i(\omega) = n(\omega)^2 - in_i(\omega)^2 , \tag{1.48}$$

where the tilde represents a complex quantity and $n(\omega)^2$ and $n_i(\omega)^2$ correspond to the real and imaginary parts of the square of the refractive index, respectively. Furthermore, we define

$$n(\omega) = \eta(\omega) - i\alpha(\omega) , \tag{1.49}$$

where $\eta(\omega)$ and $\alpha(\omega)$ are the real and imaginary parts of the refractive index. By squaring (1.49) and comparing with (1.48), we obtain

$$n(\omega)^2 = 1 + \chi_r(\omega) = \eta(\omega)^2 - \alpha(\omega)^2 , \tag{1.50}$$

$$n_i(\omega)^2 = \chi_i(\omega) = 2\alpha(\omega)\eta(\omega) , \tag{1.51}$$

so using the above approximation of an infinitely sharp absorption line we can substitute $\alpha(\omega) = \delta(\omega - \omega_0)$. We can now substitute (1.51) into the real part of the Kramers–Kronig relation to get

$$\chi_r(\omega) = \frac{4}{\pi} \int_0^\infty \frac{\omega'\eta(\omega')\delta(\omega' - \omega_0)}{\omega'^2 - \omega^2} d\omega' . \tag{1.52}$$

This integral simplifies to

$$\chi_r = \frac{4\omega_0\eta(\omega_0)}{\pi} \frac{1}{\omega_0^2 - \omega^2} = \frac{2\lambda_0\eta(\lambda_0)}{\pi^2 c} \frac{\lambda^2}{\lambda^2 - \lambda_0^2} , \tag{1.53}$$

by using the definition of the delta function and the fact that $\omega = 2\pi c/\lambda$. As the first fraction is a constant, we can group this into a constant A, substitute into (1.50), and thereby obtain the Sellmeier equation

$$n^2 = 1 + \frac{A\lambda^2}{\lambda^2 - \lambda_0^2} . \tag{1.54}$$

Note that in this textbook we shall normally use n for the real part of the refractive index, and we only temporarily introduced η in (1.49) for the purposes of this derivation.

Examples of Materials

(a) Fused Silica. The Sellmeier equation for fused silica is

$$n^2 = 1 + \frac{A\lambda^2}{\lambda^2 - \lambda_1^2} + \frac{B\lambda^2}{\lambda^2 - \lambda_2^2} + \frac{C\lambda^2}{\lambda^2 - \lambda_3^2} ,$$

where the wavelength λ is given in units of μm. There is absorption in the UV region (at λ_1 and λ_2), and in the far-infrared (at λ_3):

$$\begin{aligned} A &= 0.6961663 , \quad \lambda_1 = 0.0684043 , \\ B &= 0.4079426 , \quad \lambda_2 = 0.1162414 , \\ C &= 0.8974794 , \quad \lambda_3 = 9.896161 . \end{aligned}$$

The spectral region of validity is very broad, from 0.21 to 6.7 μm [65Mal,98Tan].

(b) SF10 Glass (e.g., Schott N-SF10). The Sellmeier equation is

$$n^2 = 1 + \frac{A\lambda^2}{\lambda^2 - \lambda_1^2} + \frac{B\lambda^2}{\lambda^2 - \lambda_2^2} + \frac{C\lambda^2}{\lambda^2 - \lambda_3^2} ,$$

where the wavelength λ is given in units of μm. There is absorption in the UV region (at λ_1 and λ_2), and in the far-infrared (at λ_3):

$$\begin{aligned} A &= 1.62153902 , \quad \lambda_1 = 0.110562859 , \\ B &= 0.256287842 , \quad \lambda_2 = 0.244077196 , \\ C &= 1.64447552 , \quad \lambda_3 = 12.14367296 . \end{aligned}$$

The spectral region of validity is very broad, from 0.38 to 2.5 μm.

(b) Sapphire (Al_2O_3, Ordinary Ray o). The Sellmeier equation is

$$n^2 = 1 + \frac{A\lambda^2}{\lambda^2 - \lambda_1^2} + \frac{B\lambda^2}{\lambda^2 - \lambda_2^2} + \frac{C\lambda^2}{\lambda^2 - \lambda_3^2} ,$$

where the wavelength λ is given in units of μm. There is absorption in the UV region (at λ_1 and λ_2), and in the far-infrared (at λ_3):

$$\begin{aligned} A &= 1.4313493 , \quad \lambda_1 = 0.0726631 , \\ B &= 0.65054713 , \quad \lambda_2 = 0.1193242 , \\ C &= 5.3414021 , \quad \lambda_3 = 18.028251 . \end{aligned}$$

The spectral region of validity is very broad, from 0.2 to 5 μm (Fig. 1.9) [86Dod].

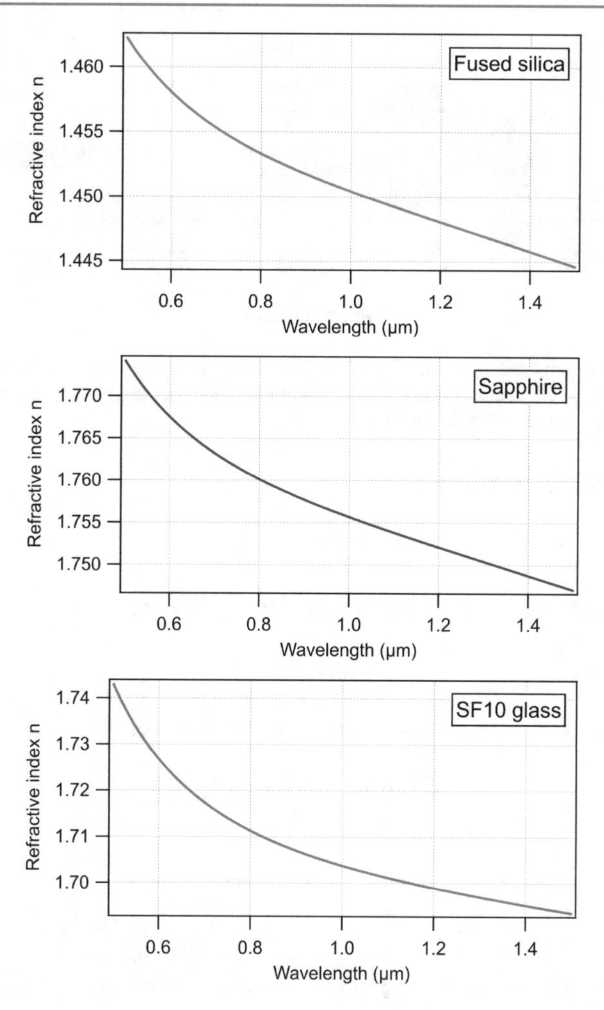

Fig. 1.9 Normal dispersion of fused quartz, SF10 glass, and sapphire in the spectral region from 500 nm to 1.5 μm, plotted as refractive index versus wavelength

1.4.3 Refractive Index from the VUV to the X-Ray Region

Typical optical materials are based on fused quartz, which transmits well to a lower wavelength of approximately 200 nm. But below 200 nm we reach the vacuum ultraviolet (VUV) region, where air and all other materials absorb very strongly. The strong atomic resonances and the associated strong absorptions make optical access to the extreme UV (EUV) and the soft X-ray regions much more difficult (see Fig. 1.7).

Above all resonances, the refractive index in the EUV and soft X-ray region differs only very slightly from unity and is less than unity. Typically, the refractive index in

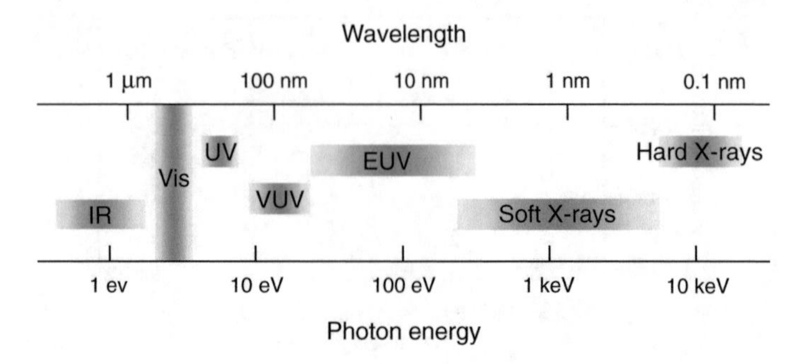

Fig. 1.10 Electromagnetic spectrum from the infrared to the X-ray region: infrared (IR), visible (Vis), ultraviolet (UV), vacuum UV, i.e., absorbed by air (VUV), extreme (EUV), sometimes also referred to as XUV. According to [99Att]

this spectral region is written as

$$\boxed{n = 1 - \delta - \mathrm{i}\beta}\ , \tag{1.55}$$

where β is responsible for the exponential decay of the intensity

$$I = I_0 \mathrm{e}^{-(4\pi\beta/\lambda)r}\ . \tag{1.56}$$

The relationship with the complex atomic scattering factor

$$f^0(\omega) = f_1^0(\omega) + \mathrm{i} f_2^0(\omega)\ , \tag{1.57}$$

is given by

$$\boxed{\delta = \frac{n_\mathrm{a} r_\mathrm{e} \lambda^2}{2\pi} f_1^0(\omega)\ , \qquad \beta = \frac{n_\mathrm{a} r_\mathrm{e} \lambda^2}{2\pi} f_2^0(\omega)}\ , \tag{1.58}$$

where n_a is the number of atoms per volume, and r_e is the classical electron radius

$$r_\mathrm{e} = \frac{e^2}{4\pi\epsilon_0 m_\mathrm{e} c^2}\ ,$$

with m_e the electron mass.

The complex atomic scattering factor can be found at the website of the National Institute of Standards and Technology[1] for many different materials. Here are some examples.

Carbon. Wavelength 4 Å (with an energy of approximately 3.1 keV) [99Att]:

$$n_\mathrm{a} = 1.13 \times 10^{23}\ \text{atoms/cm}^3\ ,$$

[1] http://physics.nist.gov/PhysRefData/FFast/Text/cover.html

$$f_1^0\left(\lambda = 4\,\text{Å}\right) = 6.09\,,$$

$$f_2^0\left(\lambda = 4\,\text{Å}\right) = 0.071\,.$$

Thus we obtain

$$\delta = 4.90 \times 10^{-5}\,, \qquad \beta = 5.71 \times 10^{-7}\,.$$

Note that materials with high Z values like gold still have very small δ and β values (in the region of 10^{-2}) in the soft X-ray region.

We can introduce an absorption length

$$I = I_0 \mathrm{e}^{-(4\pi\beta/\lambda)r} = I_0 \mathrm{e}^{-r/L_\mathrm{a}}\,, \tag{1.59}$$

with which

$$\boxed{L_\mathrm{a} = \frac{\lambda}{4\pi\beta} = \frac{1}{2n_\mathrm{a} r_\mathrm{e} \lambda f_2^0(\omega)}}\,. \tag{1.60}$$

Typically, in this wavelength region, we may define an absorption coefficient μ such that

$$I = I_0 \mathrm{e}^{-\rho_\mathrm{m}\mu r}\,, \tag{1.61}$$

where I is the intensity and ρ_m is the mass density. It follows that

$$\boxed{\mu = \frac{2n_\mathrm{a} r_\mathrm{e} \lambda}{\rho_\mathrm{m}} f_2^0(\omega)}\,. \tag{1.62}$$

Aluminum (Z = 13) (see Table 1.5). In this case,

$$n_\mathrm{a} = 6.02 \times 10^{22}\mathrm{cm}^{-3}\,.$$

Ionization energies are

$$K,\ 1s:\ 1559.6\,\mathrm{eV}\,,$$

and

$$L,\ 2s:\ 117.8\,\mathrm{eV}\,,\quad L,\ 2p_{1/2}:\ 72.9\,\mathrm{eV}\,,\quad L,\ 2p_{3/2}:\ 72.5\,\mathrm{eV}\,.$$

For more details on X-ray optics the interested reader is referred to the book by D. Attwood [99Att].

Table 1.5 Values of constants for aluminum

E [eV]	λ [nm]	f_1^0	f_2^0	$\mu\ [\text{cm}^2/\text{g}]$	L_a [μm]
30	41.3	2.320	1.737×10^{-1}	9.030×10^3	0.41
70	17.7	−1.449	2.495×10^{-1}	5.559×10^3	0.667
100	12.4	0.059	6.908	1.077×10^5	0.069
1000	1.24	11.65	7.626×10^{-1}	1.189×10^3	6.23
10000	0.124	13.16	1.581×10^{-1}	2.466×10^1	300

Linear Pulse Propagation

2

2.1 Motivation

A monochromatic plane wave as described in Chap. 1 extends infinitely in space and time. We can only obtain a limited wave packet in time (a laser pulse) when we superpose different plane waves with different wavelengths (see Fig. 2.1). In the visible spectrum different wavelengths correspond to different colors, and therefore, a visible laser pulse can never be single-colored, and must necessarily differ from the monochromatic light of a continuous-wave (cw) laser. The more varied the color (i.e., frequency) components forming the pulse, the broader its spectrum and the shorter the pulse can be. We will show that the product of the pulse duration and the frequency bandwidth of a pulse cannot fall below a minimum value and will depend on the pulse shape.

Figure 2.2 shows various items and events on logarithmic time scales. Our eye can only distinguish movements on timescales longer than about 100 milliseconds (i.e., 100 ms). A good example is a filmstrip, where there is a delay of about 20 ms before each new picture is shown, and yet we perceive a continuous sequence. Mechanical shutters that determine the exposure time in photography are limited to about 1 ms. Any movement that occurs faster than that will smear out on the photo and will result in a blurry picture. Using strobe photography it is possible to achieve exposure times as short as 1 μs and hence to picture a bullet during its flight, for example. A stroboscope generates repeated short bursts of light which allow an observer to view quickly moving objects in a series of static, frozen images, rather than a single continuous blur. Synchronizing strobe flashes with the fast motion (for example, the fast rotation of a turbine rotor), then allows one to take a series of photos through an open shutter at a much slower rate. The frequency offset of the strobe flashes and the frequency of the fast repetitive motion then determines the sampling rate, which can in principle be set arbitrarily low. Of course, this technique can also be used to take a single ultra-high speed picture, as was pioneered by H.E. Edgerton at MIT.

© The Author(s), under exclusive licence to Springer Nature Switzerland AG 2021

U. Keller, *Ultrafast Lasers*, Graduate Texts in Physics,
https://doi.org/10.1007/978-3-030-82532-4_2

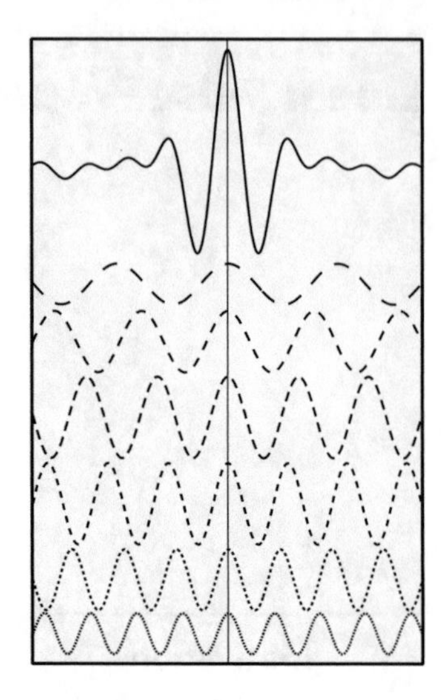

Fig. 2.1 Only the superposition of several differently colored wavelengths results in a short light pulse. Like an acoustic "bang", where it is no longer possible to assign a fixed pitch, such short light pulses already contain many frequency components. In this sense, they are "colored"

His "coronet" milk drop photo (see Fig. 2.2 at 1 μs) was featured in the New York Museum of Modern Art's first photography exhibition in 1937.

With strobe photography, short light flashes started to become important for time resolved measurements. Lasers are relevant here because we can generate much shorter flashes of light with lasers than with electrically switched light bulbs. Lasers are highly nonlinear systems for which only a minor loss modulation (e.g., even less than 1%) inside the laser cavity can produce short light pulses with high contrast and with very accurate pulse repetition rates. Today we can generate pulses as short as a few femtoseconds in the visible and near-infrared wavelength regimes directly from a laser. This is much faster than any electronic switching speed. For example, in the nanosecond region, we find the typical computer switching elements and the fastest commercial microwave sampling scope have a time resolution of around a picosecond.

Molecular vibration is a periodic motion of atoms relative to each other such that the center of mass of the molecule remains unchanged. Typical molecular vibration periods are in the range of 100–10 fs, corresponding to vibrational frequencies of 10–100 THz. For example, chemical reactions can take place in just 100 fs. In femtochemistry, ultrashort laser pulses are used to study these chemical reactions on their natural timescales. Femtochemistry resulted in a Nobel Prize in Chemistry in 1999. If we want to observe the underlying electron dynamics within the molecular

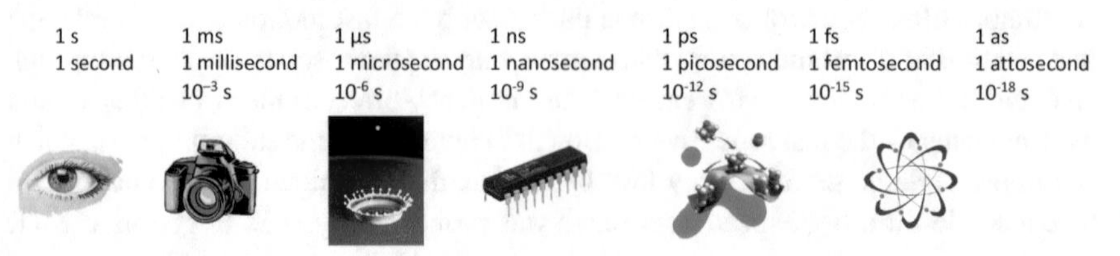

Fig. 2.2 Time scales over 18 orders of magnitude

structure before the electrons adapt to the moving atomic structure, we need to resolve their dynamics with attosecond resolution. For example, in attosecond science we recently managed to measure ionization dynamics during which electrons are liberated from their parent ion through tunnel ionization, multi-photon ionization, and high energy single photon ionization. With regard to condensed matter for example electron relaxation processes in optically excited semiconductors occur on different time scales between femtoseconds and nanoseconds and energy or charge transfer can even occur in the attosecond regime. What makes ultrafast laser physics so exciting is that we can observe 18 orders of magnitude in time—from 1 s to 1 attosecond—on a laser table in a laboratory.

Ultrafast lasers can easily generate femtosecond pulses in the near infrared and nonlinear processes can move the center wavelength into the terahertz and X-ray regime. Attosecond pulses have been generated with a highly nonlinear process, beyond the perturbation theory of nonlinear optics, with high harmonic generation using intense ultrafast laser pulses in the near-infrared. For sub-femtosecond pulses, however, we have to move into vacuum systems because the center wavelength has to be shifted into the VUV and soft-X-ray regime to support such short pulses. In this wavelength regime, everything becomes highly absorptive.

Today, short laser pulses have many applications. These applications can be grouped according to the following properties of a short pulse:

- **Short pulse duration.** With short-time spectroscopy using ultrashort laser pulses, we can examine fast processes in nature and technology. This is typically done with so-called pump–probe measurements, where for example a stronger pump pulse starts a fast process and a delayed probe pulse probes the change in time. The two pulses are synchronized (ideally coming from the same laser source), so a femtosecond delay between the pulses can be adjusted using a simple mechanical delay line with which we can control the time of flight of the laser pulses, i.e., approximately 0.3 μm in 1 fs. Piezomotors can adjust the optical path length with (sub-)nanometer precision, which corresponds to a delay that can be controlled with a resolution of a few attoseconds. The pulse durations of the pump and probe pulse are usually at least one order of magnitude longer. This means that the pulse duration is often the limiting factor on the timescales at which the dynamics of a fast process can be resolved.
- **Short pulses and high pulse repetition rates.** The pulse repetition rate of ultrafast lasers is ideally set by the roundtrip time of the pulses inside the laser cavity. Every time the pulse hits the output coupler, a smaller part of the pulse is transmitted to the outside. Therefore, we obtain a pulse train with a pulse repetition rate given by the roundtrip frequency of the intracavity pulses. For example, a linear cavity of length 3 mm results in a pulse repetition rate of 50 GHz. Today, long distance optical communication at a single carrier frequency typically occurs at a pulse repetition rate of 5–10 GHz. In computers, the clock rate has continuously increased, and typical personal computers have a clock rate above 1 GHz. In the future it is very likely that light pulses will support high speed interconnects and clocks for multi-core microprocessors.

- **Broad spectrum.** A short pulse also has a broad coherent spectrum. Applications include for example optical coherence tomography (OCT) or frequency metrology. For frequency metrology, we can benefit from the extremely stable frequency combs generated by modelocked lasers. Modelocking is the technique generally used to produce ultrashort pulses from lasers. It will be possible in the future to build optical clocks that will be much more accurate than today's best atomic clocks. We hope that such clocks will be able to measure small changes in the physical constants over time—if they really exist.
- **High intensity.** In a short pulse, all energy is confined within a very short time duration. Thus, very high peak intensities can be achieved. This is important for many applications, such as nonlinear optics, material processing (for machining and medicine), high-field physics, fusion, soft X-ray generation, and attosecond pulse generation.

2.2 Wave Equation in the Spectral Domain: Helmholtz Equation

2.2.1 Fourier Transform

From mathematics, we know, that every function can be expanded in terms of a complete set of orthogonal functions. Examples of such developments are the Fourier series for periodic functions and the Fourier integrals for arbitrary functions. A Fourier transform decomposes any arbitrary function of time into its harmonic frequency components [see (2.2)]. An inverse Fourier transform is a mathematical transform that expresses for example an arbitrary function of time as a superposition of harmonic functions of time with different frequencies, [see Fig. 2.3 and (2.3)]. This is a very useful mathematical tool to describe ultrashort laser pulses because the harmonic components have the physical meaning of plane waves with different frequencies. We know from Chap. 1 how plane waves propagate in a dispersive medium and from the superposition of plane waves in linear optics we can also learn how light pulses propagate. This will be the focus of this chapter.

Figure 2.4 shows how a photon wave packet (i.e., a light pulse) is formed by a superposition of plane waves with different discrete frequencies at a fixed position in space. The oscillations far away from the pulse maximum decrease as more frequency components are used in the superposition. Figure 2.4 shows an example of a superposition of plane waves with a constant frequency spacing $\Delta\omega$. Due to these discrete frequencies, the final solution is not just one pulse, but a periodic pulse train

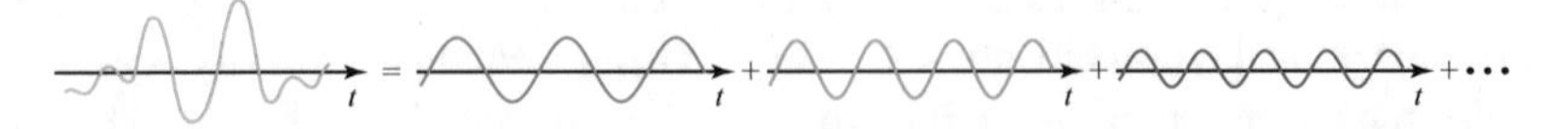

Fig. 2.3 An inverse Fourier transform describes an arbitrary function in time as a superposition of harmonic functions of time with different frequencies [see (2.3)]

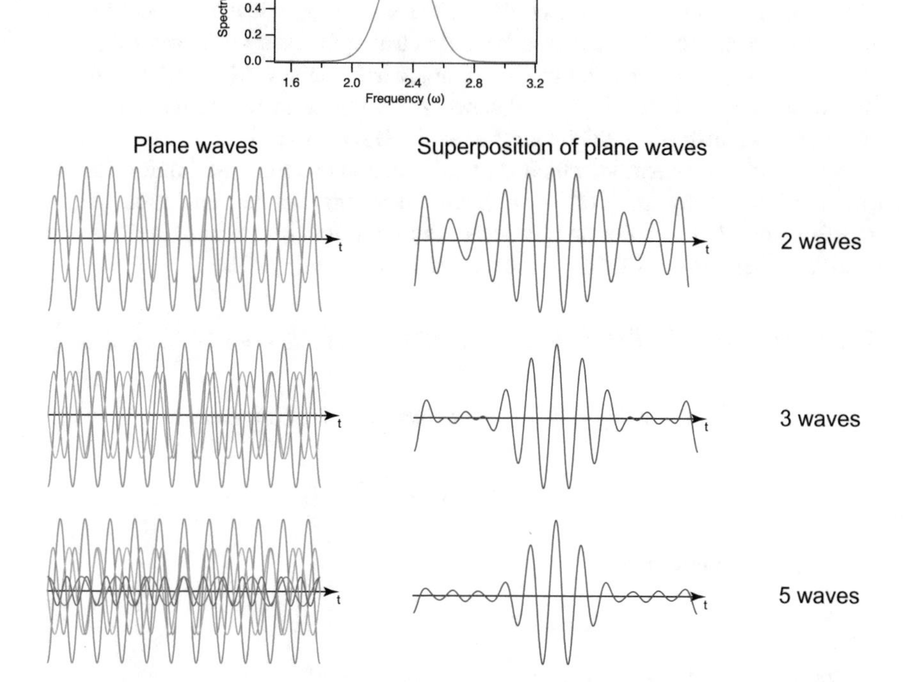

Fig. 2.4 Discrete superposition of plane waves. Numerical example: Gaussian with $\omega_0 = 20$ and $\tilde{A}(\Delta\omega) = \mathrm{e}^{-a\Delta\omega^2}$ with $a = 0.5$ in frequency steps of $\Delta\omega = 1$. Amplitudes of the plane waves $\tilde{A}(\Delta\omega)$. Superposition of two waves ω_0 and $\omega_0 + 1$. Superposition of three waves ω_0 and $\omega_0 \pm 1$. Superposition of five waves ω_0, $\omega_0 \pm 1$, and $\omega_0 \pm 2$

with period $T_{\mathrm{rep}} = 2\pi/\Delta\omega$. In contrast, a continuous spectrum results in only one pulse as will be discussed in more detail later.

We follow the notation which is generally used in electrical engineering and laser physics, and write the complex scalar electric field of an electromagnetic (e.m.) plane wave in the form

$$E(z,t) = E_0 \mathrm{e}^{\mathrm{i}(\omega t - kz)} \,. \tag{2.1}$$

With this choice of the complex notation for a plane wave $E \propto \mathrm{e}^{+\mathrm{i}\omega t}$, the inverse Fourier transform with the physical meaning of a superposition of plane waves is given by (2.3) below. We will therefore use the Fourier transform $F\{\}$ and the inverse Fourier tranform $F^{-1}\{\}$ with the following convention (Appendix A):

$$\tilde{E}(z,\omega) \equiv F\left\{E(z,t)\right\} = \int E(z,t)\mathrm{e}^{-\mathrm{i}\omega t}\mathrm{d}t \,, \tag{2.2}$$

$$E(z,t) = F^{-1}\left\{\tilde{E}(z,\omega)\right\} = \frac{1}{2\pi}\int \tilde{E}(z,\omega)\mathrm{e}^{+\mathrm{i}\omega t}\mathrm{d}\omega \,. \tag{2.3}$$

Furthermore, in the following the integral sign without indicated limits will always mean integration from $-\infty$ to $+\infty$, i.e., $\int_{-\infty}^{+\infty}$. Note also that, regarding the constant $1/2\pi$, we have chosen an asymmetric Fourier transform, which is typical in quantum electronics, because we measure the spectrum of a laser pulse as a function of frequency $\tilde{E}(z, \nu)$ and not as function of angular frequency $\tilde{E}(z, \omega)$. Therefore, in this convention the factor $1/2\pi$ will cancel out with the extra multiplicative factor of 2π after a transform of the integration variable $\mathrm{d}\omega \to 2\pi \mathrm{d}\nu$.

For the other convention, which is mainly used in quantum mechanics, the electric field is given by $\exp[\mathrm{i}(kz - \omega t)]$. But all results can be transferred with the transformation $\mathrm{i} \to -\mathrm{i}$. For our purposes later on, it will be useful to derive several relations. First, the derivative with respect to time of (2.3) yields

$$\frac{\partial}{\partial t}E(z,t) = \frac{\partial}{\partial t}F^{-1}\left\{\tilde{E}(z,\omega)\right\} = \frac{1}{2\pi}\int \tilde{E}(z,\omega)\mathrm{i}\omega \mathrm{e}^{\mathrm{i}\omega t}\mathrm{d}\omega = F^{-1}\left\{\mathrm{i}\omega\tilde{E}(z,\omega)\right\} .$$

This results in the following relationship between the spectral and time domains:

$$F^{-1}\left\{\mathrm{i}\omega\tilde{E}(z,\omega)\right\} = \frac{\partial}{\partial t}E(z,t) ,$$

which can be abbreviated as

$$\omega \Leftrightarrow -\mathrm{i}\frac{\partial}{\partial t} .$$

For the second and third derivatives, we obtain

$$\boxed{\omega \Longleftrightarrow -\mathrm{i}\frac{\partial}{\partial t} , \quad \omega^2 \Longleftrightarrow -\frac{\partial^2}{\partial t^2} = \left(-\mathrm{i}\frac{\partial}{\partial t}\right)^2 , \quad \omega^3 \Longleftrightarrow \mathrm{i}\frac{\partial^3}{\partial t^3} = \left(-\mathrm{i}\frac{\partial}{\partial t}\right)^3} . \tag{2.4}$$

2.2.2 Derivation of the Helmholtz Equation

If it is no longer possible to replace the frequency dependence of the susceptibility by a constant, the wave equation has to be examined in the spectral domain. To determine the wave equation (1.23) in the spectral domain, we use the relationships of (2.4):

$$F^{-1}\left\{\omega^2\tilde{E}(z,\omega)\right\} = -\frac{\partial^2}{\partial t^2}E(z,t) . \tag{2.5}$$

Using the wave equation (1.23)

$$\Delta\mathbf{E} - \frac{1}{c^2}\frac{\partial^2}{\partial t^2}\mathbf{E} = \mu_0\frac{\partial^2}{\partial t^2}\mathbf{P}$$

and (2.5) we obtain the wave equation in the spectral domain:

$$\frac{\partial^2 \tilde{E}(z,\omega)}{\partial z^2} + \frac{\omega^2}{c^2}\tilde{E}(z,\omega) = -\mu_0 \omega^2 \tilde{P}(z,\omega) \,, \tag{2.6}$$

in the case of propagation along the z-axis. Using the material equations in Chap. 1, it follows that

$$\tilde{P}(z,\omega) = \chi(\omega)\varepsilon_0 \tilde{E}(z,\omega) = [\varepsilon(\omega) - 1]\,\varepsilon_0 \tilde{E}(z,\omega) \tag{2.7}$$

and

$$k_n(\omega) = \frac{\omega}{c} n(\omega) \,. \tag{2.8}$$

Since $c = 1/\sqrt{\varepsilon_0 \mu_0}$ and $n(\omega) = \sqrt{\varepsilon(\omega)}$, (2.6) implies

$$\boxed{\frac{\partial^2 \tilde{E}(z,\omega)}{\partial z^2} + [k_n(\omega)]^2\, \tilde{E}(z,\omega) = 0} \,. \tag{2.9}$$

Equation (2.9) is the wave equation in the spectral domain for a dispersive linear medium with an arbitrary dispersion relation $n(\omega)$ and $\mu = 1$. This equation is also called the one-dimensional Helmholtz equation.

The solution of the Helmholtz equation is very simple, as can be checked by inserting the following:

$$\boxed{\tilde{E}(z,\omega) = \tilde{E}_0^+(\omega)\mathrm{e}^{-\mathrm{i}k_n(\omega)z} + \tilde{E}_0^-(\omega)\mathrm{e}^{\mathrm{i}k_n(\omega)z}} \,. \tag{2.10}$$

Note that (2.9) is analogous to the time-independent Schrödinger equation for a free particle,

$$\frac{\partial^2 \psi}{\partial x^2} + k^2 \psi = 0 \,, \tag{2.11}$$

while, for a particle in a potential energy landscape $E_\mathrm{p}(x)$, we have

$$\frac{\partial^2 \psi}{\partial x^2} + \frac{2m}{\hbar^2}\left[E - E_\mathrm{p}(x)\right]\psi = 0 \,. \tag{2.12}$$

Hence, the discussion on laser pulse propagation also applies to any quantum mechanical wave packet. Appendix B discusses the dispersion relations for various particles such as free electrons, electrons in semiconductor materials, and phonons. For these cases, we can predict wave packet spreading and wave packet propagation in a very similar way as for the discussion for light pulses. The reason why the dispersion relation is typically described by $\omega(k)$ relies on the wave–particle duality of the e.m. wave in quantum mechanics. For a photon, the dispersion relation $\omega(k)$ together with the de Broglie relation results in an expression for the energy $E = \hbar\omega$ as a function of the momentum $p = \hbar k$:

$$\text{photon in vacuum} \quad E = cp \iff \omega(k) = ck \ , \tag{2.13}$$

$$\text{photon in medium} \quad E = c_n p_n \iff \omega(k) = c_n k_n \ . \tag{2.14}$$

2.3 Linear Versus Nonlinear Wave Propagation

2.3.1 Superposition Principle

In linear optics, the wave equation is linear, i.e., the electric and magnetic fields and their derivatives only appear to first order. Therefore, the superposition principle is valid in this linear system. This means, that if E_i and H_i are solutions of the wave equation, then any linear combination of them will also be a solution of the wave equation:

$$\boxed{E_{\text{tot}} = \sum_i a_i E_i \ , \qquad H_{\text{tot}} = \sum_i b_i H_i} \ . \tag{2.15}$$

Because $(E_1 + E_2)^2 \neq E_1^2 + E_2^2$, the superposition principle is not generally valid for intensities:

$$I_{\text{tot}} \neq \sum_i I_i \ . \tag{2.16}$$

But the superposition principle is valid for intensities, too, if the e.m. fields do not satisfy the conditions for interference. An example occurs when the e.m. waves are polarized perpendicularly to each other or when the two waves are fully incoherent:

$$I_{\text{tot}} = \sum_i I_i \quad \text{if no interference is possible} \ . \tag{2.17}$$

2.3.2 Linear System Theory

In a linear system, it is generally much easier to examine the system dynamics directly in the spectral domain. For a linear system we can directly connect the input signal with the output signal as follows

$$E_{\text{in}\,1} \overset{\text{linear system}}{\longrightarrow} E_{\text{out}}\,1 \implies a E_{\text{in}}\,1 \overset{\text{linear system}}{\longrightarrow} a E_{\text{out}\,1} \ , \tag{2.18}$$

where a is a constant. The superposition principle is valid and yields:

$$\left.\begin{array}{l} E_{\text{in}\,1} \overset{\text{linear system}}{\longrightarrow} E_{\text{out}\,1} \\ E_{\text{in}\,2} \overset{\text{linear system}}{\longrightarrow} E_{\text{out}\,2} \end{array}\right\} \Longrightarrow \quad E_{\text{in}\,1} + E_{\text{in}\,2} \overset{\text{linear system}}{\longrightarrow} E_{\text{out}\,1} + E_{\text{out}\,2} \,. \tag{2.19}$$

Examples of linear systems are dispersive media, photodetectors, where the output and input signal relates the photocurrent to the incoming light power, active light modulators, lens systems, and stochastic processes like amplitude and phase noise, which are considered as linearized perturbations added to a much stronger signal.

In a linear system, one can define an impulse response $h(t)$, which describes the complete dynamic evolution $E_{\text{in}}(t) \rightarrow E_{\text{out}}(t)$, where E_{in} is the input signal and E_{out} is the output signal. The system is linear if the impulse response $h(t)$ does not depend on the strength of the input signal E_{in}:

$$E_{\text{out}}(t) = \int h\left(t'\right) E_{\text{in}}\left(t - t'\right) \mathrm{d}t' \equiv h(t) * E_{\text{in}}(t) \,. \tag{2.20}$$

Here, $h(t)$ is called the impulse response of the linear system because, for any short input pulse $E_{\text{in}}(t) = \delta(t)$ (i.e., a δ-pulse, see Appendix C), the output impulse $E_{\text{out}}(t)$ is given by $h(t)$:

$$h(t) = \int h\left(t'\right) \delta\left(t - t'\right) \mathrm{d}t' \,. \tag{2.21}$$

The integral in (2.20) describes a convolution abbreviated by an asterisk $*$.

In the spectral domain, the dynamics of a linear system are described by the transfer function $\tilde{h}(\omega)$, which is the Fourier transform of $h(t)$. In the spectral domain, the linear system response is then given by a simple multiplication:

$$\boxed{\tilde{E}_{\text{out}}(\omega) = \tilde{h}(\omega) \tilde{E}_{\text{in}}(\omega)} \,. \tag{2.22}$$

Equation (2.22) follows from basic Fourier transform theory:

$$\begin{aligned} \tilde{E}_{\text{out}}(\omega) &= F\left\{\int h(t') E_{\text{in}}(t - t') \mathrm{d}t'\right\} \\ &= \int\int h(t') E_{\text{in}}(t'' - t') \mathrm{e}^{-\mathrm{i}\omega t''} \mathrm{d}t' \mathrm{d}t'' \\ &= \int\int h(t') E_{\text{in}}(t'' - t') \mathrm{e}^{-\mathrm{i}\omega (t'' - t')} \mathrm{e}^{-\mathrm{i}\omega t'} \mathrm{d}t' \mathrm{d}t'' \\ &= \int h(t') \mathrm{e}^{-\mathrm{i}\omega t'} \mathrm{d}t' \int E_{\text{in}}(t''') \mathrm{e}^{-\mathrm{i}\omega t'''} \mathrm{d}t''' \\ &= F\{h(t)\}\, F\{E_{\text{in}}(t)\} = \tilde{h}(\omega) \tilde{E}_{\text{in}}(\omega) \,, \end{aligned}$$

where the integration variable was changed to $t''' = t'' - t'$ and the integrals were separated.

The power spectral density $P(\omega)$ of $E(t)$ is given by

$$P(\omega) \equiv \left|\tilde{E}(\omega)\right|^2 \implies P_{\text{in}}(\omega) = \left|\tilde{E}_{\text{in}}(\omega)\right|^2, \quad P_{\text{out}}(\omega) = \left|\tilde{E}_{\text{out}}(\omega)\right|^2, \tag{2.23}$$

and analogously for the power spectral density $S(\omega)$ of the impulse response $h(t)$:

$$S(\omega) \equiv \left|\tilde{h}(\omega)\right|^2. \tag{2.24}$$

For a linear system, the much easier relation in the spectral domain is valid once again:

$$\boxed{P_{\text{out}}(\omega) = S(\omega) P_{\text{in}}(\omega)}. \tag{2.25}$$

Equations (2.20), (2.22), and (2.25) clearly show the advantages of the spectral domain. It is much easier to examine the dynamics of a linear system in the spectral domain, because the complicated convolution (2.20) is turned into an easy multiplication (2.22). A simple inverse Fourier tranform of the output signal $\tilde{E}_{\text{out}}(\omega)$ after the interaction with a linear system brings the dynamics back into the time domain.

Example: Dispersion

For a dispersive medium with length L_{d} the system function is given by the solution of the Helmholtz equation (2.10):

$$\tilde{h}(\omega) = \mathrm{e}^{-\mathrm{i}k_n(\omega)L_{\text{d}}} = \mathrm{e}^{-\mathrm{i}\omega L_{\text{d}} n(\omega)/c}. \tag{2.26}$$

For comparison, the system function in vacuum is given by

$$\tilde{h}(\omega) = \mathrm{e}^{-\mathrm{i}k(\omega)L_{\text{d}}} = \mathrm{e}^{-\mathrm{i}\omega L_{\text{d}}/c}. \tag{2.27}$$

In vacuum it is very easy to calculate the impulse response. It follows from (2.27):

$$h(t) = F^{-1}\left\{\tilde{h}(\omega)\right\} = F^{-1}\left\{\mathrm{e}^{-\mathrm{i}\omega L_{\text{d}}/c}\right\} = \frac{1}{2\pi}\int \mathrm{e}^{\mathrm{i}\omega(t-L_{\text{d}}/c)}\mathrm{d}\omega = \delta\left(t - \frac{L_{\text{d}}}{c}\right), \tag{2.28}$$

where the last step follows from the definition of the Dirac delta function in integral form (Appendix C). Thus, in vacuum, a pulse will propagate as follows (2.20):

$$E_{\text{out}}(t) = \int \delta\left(t' - \frac{L_{\text{d}}}{c}\right) E_{\text{in}}(t - t')\,\mathrm{d}t' = E_{\text{in}}\left(t - \frac{L_{\text{d}}}{c}\right). \tag{2.29}$$

Equation (2.29) shows that an arbitrary pulse $E_{\text{in}}(t)$ can propagate in vacuum without changing its pulse shape. $E_{\text{in}}(t)$ is only shifted in time with the speed of light in vacuum c. This is no longer the case in a dispersive material, because the impulse response of (2.26) no longer results in a delta function for an arbitrary dispersion function $n(\omega)$. We will show later that, in a dispersive material with a frequency-dependent refractive index the pulse shape can change considerably during linear propagation.

However, in the spectral domain, linear system theory results in a very important general condition for linear pulse propagation in a dispersive medium. Equations (2.24)–(2.26) imply

$$S(\omega) = \left| e^{-ik_n L_d} \right|^2 = 1 , \tag{2.30}$$

for the power spectral density of the impulse response in a dispersive medium. Thus, the pulse power spectrum never changes during linear propagation in a dispersive medium (2.25). It is important to remember that this is in contrast to nonlinear pulse propagation. This is very useful to remember for experimental work in a laser lab where we can often very easily measure the optical spectrum of a laser pulse. If this spectrum changes within an experimental setup, it means that some nonlinear effects or spectral filtering are taking place.

2.3.3 Nonlinear Systems

In a nonlinear system, it is no longer possible to examine the system dynamics so easily in the spectral domain. The explanation is very simple: in a nonlinear system, the impulse response depends on the incoming intensity. If the system is examined in the spectral domain, the result does not depend on whether the individual spectral components are in phase or out of phase with each other. If the spectral components are in phase and interfere constructively, we obtain a pulse with a high intensity. In contrast, if the spectral components are out of phase with each other and interfere destructively, we obtain a pulse with a lower intensity. The spectral analysis would then lead to the same result for the dynamic evolution in both cases, which is not the case for nonlinear processes. In nonlinear systems the output signal depends on the intensity of the input signal.

Examples of nonlinear systems are a saturable absorber, the Kerr effect, self-phase modulation, and frequency doubling. In nonlinear systems, it is therefore advantageous to examine the dynamic evolution in the time domain. But of course it is possible to determine the accompanying spectrum at any moment in time with a Fourier transform. In this case, however, the spectrum will change with time.

2.4 Ultrafast Pulses

2.4.1 Pulse Wave Packet and Pulse Envelope

A coherent light pulse forms a photon wave packet (see, for example, Fig. 2.4) and can be described by a superposition of plane waves with different frequencies and phases at a fixed position in space. The shortest pulse for a given spectrum is obtained

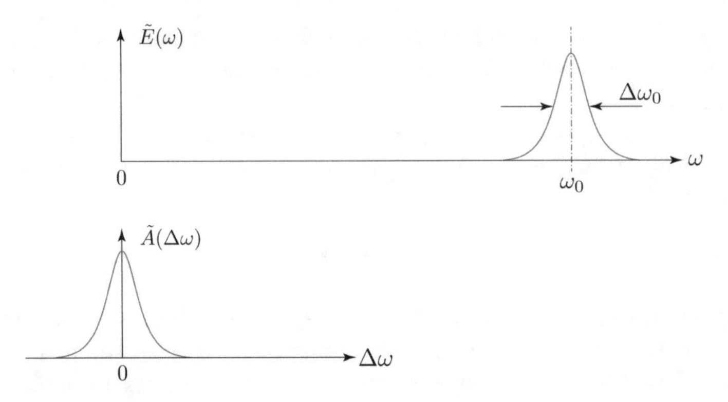

Fig. 2.5 Spectrum of the light pulse given by $\tilde{E}(\omega)$ and $\tilde{A}(\Delta\omega)$

for a constant phase, as shown later. For a light or laser pulse observed at a fixed position, e.g., $z = 0$, as a function of time, we find

$$\boxed{E(t) = \frac{1}{2\pi}\int \tilde{E}(\omega)\mathrm{e}^{\mathrm{i}\omega t}\mathrm{d}\omega} \,. \tag{2.31}$$

We have chosen the factor $1/2\pi$ in (2.31), because this integral just corresponds to an inverse Fourier transform, as in (2.3). The function $\tilde{E}(\omega)$ is the complex amplitude of the plane wave with frequency ω, which is given by the Fourier transform of $E(t)$. The function $\tilde{E}(\omega)$ determines the spectral amplitude and phase of the light pulse. The spectrum is typically a function with a finite width $\Delta\omega_0$ and center frequency ω_0, i.e., typically $\omega_0 \gg \Delta\omega_0$ for pulse durations that are significantly longer than one optical period (see Fig. 2.5). For example for a center wavelength of 800 nm the optical period (cycle) is 2.7 fs. The exact shape and phase of the spectrum determines the shape of the light pulse.

It is useful to rewrite (2.31), replacing ω by $\omega_0 + \Delta\omega$ and carrying out the integral over $\Delta\omega$:

$$E(t) = \frac{1}{2\pi}\int \tilde{E}\left(\omega_0 + \Delta\omega\right)\mathrm{e}^{\mathrm{i}(\omega_0+\Delta\omega)t}\mathrm{d}\Delta\omega = \frac{1}{2\pi}\mathrm{e}^{\mathrm{i}\omega_0 t}\int \tilde{A}(\Delta\omega)\mathrm{e}^{\mathrm{i}\Delta\omega t}\mathrm{d}\Delta\omega \,. \tag{2.32}$$

Then $\tilde{A}(\Delta\omega)$ is the frequency-shifted spectrum (see Fig. 2.5):

$$\boxed{\tilde{A}(\Delta\omega) = \tilde{E}\left(\omega_0 + \Delta\omega\right)} \,. \tag{2.33}$$

Equation (2.32) then implies

$$\boxed{E(t) = A(t)\mathrm{e}^{\mathrm{i}\omega_0 t} \,, \quad \text{where } A(t) = \frac{1}{2\pi}\int \tilde{A}(\Delta\omega)\mathrm{e}^{\mathrm{i}\Delta\omega t}\mathrm{d}\Delta\omega} \,. \tag{2.34}$$

The function $A(t)$ is called the pulse envelope (see Fig. 2.6).

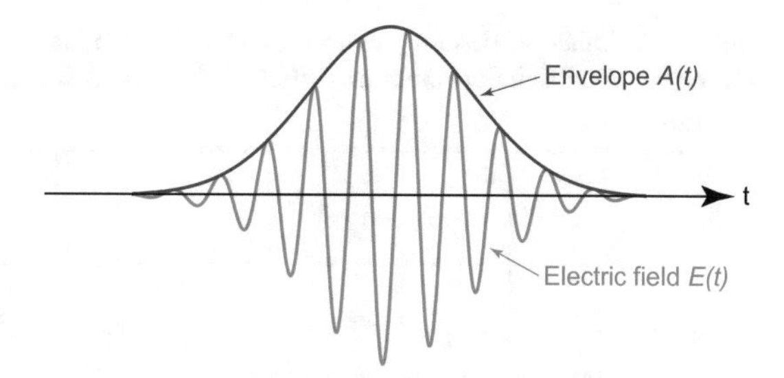

Fig. 2.6 Electromagnetic pulse in the time domain, given by the electric field $E(t)$ and the pulse envelope $A(t)$

2.4.2 Time–Bandwidth Product. Analogy with Heisenberg's Uncertainty Relation

The laser pulse duration τ_p in laser physics is normally defined by the full width at half maximum (FWHM) of the time-dependent pulse intensity. Analogously, the laser pulse spectral width $\Delta\nu_\mathrm{p}$ in laser physics is normally defined by the FWHM of the spectral intensity. The shorter the laser pulse, i.e., the smaller the value of τ_p, the broader its spectrum $\Delta\nu_\mathrm{p}$. With the Fourier transform defined above, we summarize the time–bandwidth products $\Delta\nu_\mathrm{p} \times \tau_\mathrm{p}$ of some common pulse shapes in Table 2.1. Depending on the pulse shape the time–bandwidth product is slightly different. The example of a Gaussian pulse is calculated explicitly as follows.

Consider an example of a chirped Gaussian pulse (see Fig. 2.7):

$$E(t) = A(t)\exp(\mathrm{i}\omega_0 t) = \exp\left(-\Gamma t^2\right)\exp(\mathrm{i}\omega_0 t) \,, \tag{2.35}$$

where ω_0 is the center angular frequency and Γ is the complex Gauss parameter, defined by

$$\Gamma \equiv \Gamma_1 - \mathrm{i}\Gamma_2 \,. \tag{2.36}$$

The real part Γ_1 of the complex Gauss parameter determines the pulse length, and the imaginary part Γ_2 of the complex Gauss parameter determines the chirp of the Gaussian pulse (see Fig. 2.7). If a pulse is chirped, its instantaneous frequency is time dependent.

Equations (2.35) and (2.36) imply

$$\phi_{\mathrm{tot}}(t) \equiv \omega_0 t + \Gamma_2 t^2 \,, \tag{2.37}$$

whence we have

$$\omega(t) \equiv \frac{\mathrm{d}\phi_{\mathrm{tot}}(t)}{\mathrm{d}t} = \omega_0 + 2\Gamma_2 t \,. \tag{2.38}$$

From this expression we can see that the instantaneous frequency will only vary with time for $\Gamma_2 \neq 0$. The pulse length τ_p of the laser pulse is defined by the FWHM of the intensity $I(t)$:

$$I(t) \propto |E(t)|^2 = \exp\left(-2\Gamma_1 t^2\right) = \exp\left[-(4\ln 2)\left(\frac{t}{\tau_\mathrm{p}}\right)^2\right] \,, \tag{2.39}$$

Table 2.1 Time–bandwidth product for different pulse shapes [80Sal]. τ_p is the pulse length, given by the full width at half maximum (FWHM) of $I(t)$, and $\Delta\nu_p$ is the bandwidth, given by the FWHM of $\tilde{I}(\omega)$

	$I(t)\ (x \equiv t/\tau)$	τ_p/τ	$\Delta\nu_p \times \tau_p$
Gaussian	$I(t) = e^{-x^2}$	$2\sqrt{\ln 2}$	0.4413
Hyperbolic secant (soliton pulse)	$I(t) = \mathrm{sech}^2 x$	1.7627	0.3148
Rectangle	$I(t) = \begin{cases} 1, & \lvert t\rvert \le \tau/2 \\ 0, & \lvert t\rvert > \tau/2 \end{cases}$	1	0.8859
Parabolic	$I(t)= \begin{cases} 1-x^2, & \lvert t\rvert \le \tau \\ 0, & \lvert t\rvert > \tau \end{cases}$	$\sqrt{2}$	0.7276
Lorentzian	$I(t) = \dfrac{1}{1+x^2}$	2	0.2206
Symmetric two-sided exponent	$I(t) = e^{-2\lvert x\rvert}$	$\ln 2$	0.1420

where the last expression is simply the definition of a Gaussian function, rewritten in terms of its FWHM (i.e., its τ_p). Hence, the pulse length τ_p is given by

$$\boxed{\tau_p = \sqrt{\frac{2\ln 2}{\Gamma_1}}}\,. \tag{2.40}$$

By doing a Fourier transform and completing the square in the exponent, the continuous frequency spectrum is found to be

$$\tilde{E}(\omega) \propto \exp\left[-\frac{(\omega-\omega_0)^2}{4\Gamma}\right]. \tag{2.41}$$

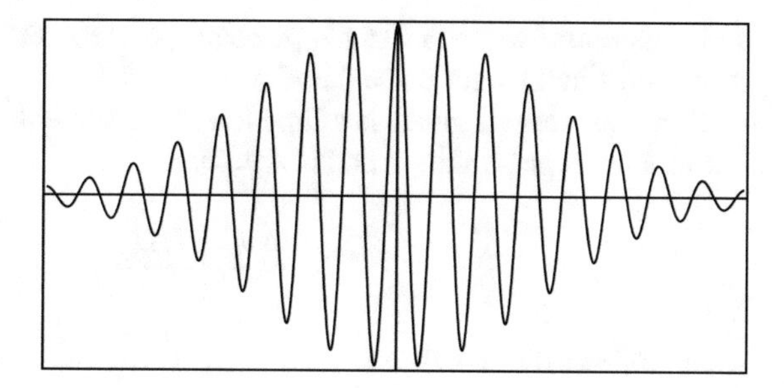

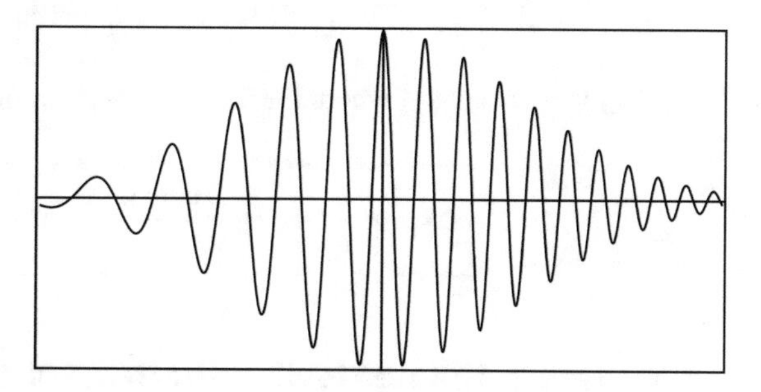

Fig. 2.7 The electric field $E(t)$ of a Gaussian given by (2.35) and (2.36). *Upper*: Unchirped Gaussian ($\Gamma_1 = 3$, $\Gamma_2 = 0$, $\omega_0 = 50$). *Lower*: Chirped Gaussian ($\Gamma_1 = 3$, $\Gamma_2 = 15$, $\omega_0 = 50$)

The intensity spectrum or intensity spectral density $\tilde{I}(\omega)$ is

$$\tilde{I}(\omega) \propto \left|\tilde{E}(\omega)\right|^2 \propto \exp\left[-(4\ln 2)\frac{(\omega-\omega_0)^2}{\Delta\omega_p^2}\right], \tag{2.42}$$

where $\Delta\omega_p$, the FWHM spectral width of $\tilde{I}(\omega)$, is given by

$$\Delta\nu_p \equiv \frac{\Delta\omega_p}{2\pi} = \frac{\sqrt{2\ln 2}}{\pi}\sqrt{\Gamma_1\left[1+\left(\frac{\Gamma_2}{\Gamma_1}\right)^2\right]}. \tag{2.43}$$

Equation (2.43) shows that a chirped pulse ($\Gamma_2 \neq 0$) for a given pulse duration, i.e., fixed Γ_1, always has a broader spectral bandwidth than an unchirped pulse ($\Gamma_2 = 0$). The time–bandwidth product for a Gaussian pulse is then given by (2.40) and (2.43):

$$\boxed{\tau_p\Delta\nu_p = \frac{2\ln 2}{\pi}\sqrt{1+\left(\frac{\Gamma_2}{\Gamma_1}\right)^2} = 0.4413\times\sqrt{1+\left(\frac{\Gamma_2}{\Gamma_1}\right)^2}}\,. \tag{2.44}$$

Without a chirp, i.e., $\Gamma_2 = 0$, the time–bandwidth product is 0.4413 (see Table 2.1). Such a light pulse, corresponding to this minimal time–bandwidth product, is said to be a time–bandwidth-limited, bandwidth-limited, or transform-limited pulse. For a given spectrum, a transform-limited pulse has the shortest possible pulse duration for a given spectral width and given pulse shape. Any additional chirp in the spectrum then results in a longer pulse duration.

If the spectrum is given as a function of wavelength, the following conversion is useful. For all pulse shapes and if $\Delta\nu_p \ll \nu$,

$$\frac{\Delta\nu_p}{\nu} = \frac{\Delta\lambda_p}{\lambda} , \tag{2.45}$$

which follows directly from

$$\nu = \frac{c}{\lambda} \Longrightarrow \frac{d\nu}{d\lambda} = -\frac{c}{\lambda^2} = -\frac{\nu}{\lambda} \Longrightarrow \frac{d\nu}{\nu} = -\frac{d\lambda}{\lambda} .$$

With $\Delta\nu_p \times \tau_p = \text{const.}$, where the constant is given in Table 2.1, we obtain

$$\boxed{\Delta\lambda_p = \frac{\lambda}{\nu}\Delta\nu_p = \frac{\lambda}{\nu}\frac{\text{const.}}{\tau_p} = \text{const.}\frac{\lambda^2}{c}\frac{1}{\tau_p}} . \tag{2.46}$$

2.4.3 Spectral Phase Yielding Shortest Pulse

Continuous progress in the field of ultrashort pulse generation has led to pulse durations below 6 fs in the visible and near-infrared spectral range [99Ste]. See Sect. 9.10.1 for more details. Such ultrashort pulse durations have been independently generated using different techniques, namely direct generation in Ti:sapphire laser oscillators [99Sut,01Ell], optical parametric amplification [99Shi], and compression in either single-mode fibers [97Bal], microstructured fibers [05Sch4], hollow optical waveguides [97Nis,03Sch], or filamentation [04Hau,07Zai]. All these pulses exhibit broad and complex spectra which do not agree with an ideal pulse shape as discussed so far. Representative examples are shown in Fig. 2.8 for a dual stage filament compressor [07Zai].

In such cases it becomes increasingly questionable to simply specify the pulse duration by an FWHM value. A more appropriate pulse width measurement is given by an rms pulse duration τ_{rms}:

$$\tau_{rms}^2 = \left\langle (t - \langle t \rangle)^2 \right\rangle = \int_{-\infty}^{\infty} (t - \langle t \rangle)^2 \, I(t) dt , \tag{2.47}$$

$$\tau_{rms}^2 = \int_{-\infty}^{\infty} \left| (t - \langle t \rangle) \, E(t) \right|^2 dt , \tag{2.48}$$

whence $\langle t \rangle$ corresponds to the center of gravity of the temporal pulse profile. It can be shown that, for a given power spectrum $\left|\tilde{E}(\omega)\right|^2$, the shortest rms pulse duration is obtained with a linear spectral phase $\varphi(\omega)$ [02Gal2,01Wal].

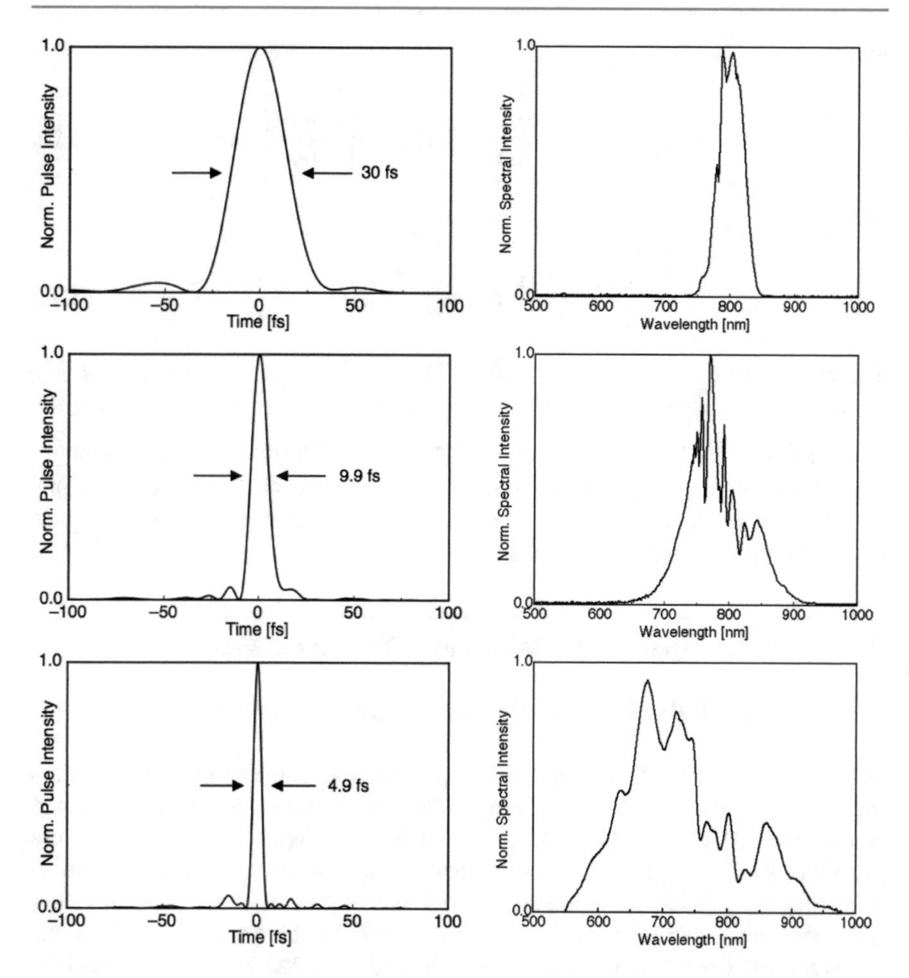

Fig. 2.8 Pulse shapes and associated spectra according to [07Zai] of (*top row*) the input pulse (30 fs), (*center row*) after the first filament cell (9.9 fs), and (*bottom row*) after the second filament cell (4.9 fs). The optical spectrum is measured directly and the temporal pulse shape is determined through a SPIDER measurement. See Chap. 10 for more detail

Using Parseval's theorem in Fourier analysis (Appendix A), (2.48) can be rewritten as

$$\tau_{\mathrm{rms}}^2 = \int_{-\infty}^{\infty} \left| \frac{\partial \left[\tilde{E}(\omega) \exp\left(-\mathrm{i}\omega \langle t \rangle\right) \right]}{\partial \omega} \right|^2 \mathrm{d}\omega \,. \tag{2.49}$$

We define $\phi(\omega) := \varphi(\omega) - \omega \langle t \rangle$, which is thus the spectral phase without its linear term. From this definition and the relation $\tilde{E}(\omega) = \left|\tilde{E}(\omega)\right| \exp(\mathrm{i}\varphi(\omega))$, it follows that

$$\tau_{\mathrm{rms}}^{2}=\int_{-\infty}^{\infty}\left|\frac{\partial\left|\tilde{E}(\omega)\right|}{\partial\omega}+\mathrm{i}\left|\tilde{E}(\omega)\right|\frac{\partial\phi(\omega)}{\partial\omega}\right|^{2}\mathrm{d}\omega \tag{2.50}$$

and

$$\tau_{\mathrm{rms}}^{2}=\int_{-\infty}^{\infty}\left|\frac{\partial\left|\tilde{E}(\omega)\right|}{\partial\omega}\right|^{2}\mathrm{d}\omega+\int_{-\infty}^{\infty}\left|\left|\tilde{E}(\omega)\right|\frac{\partial\phi(\omega)}{\partial\omega}\right|^{2}\mathrm{d}\omega\,. \tag{2.51}$$

The second term on the right-hand side of this last equation is always greater than or equal to zero. Therefore, the shortest rms duration of an optical pulse for a given pulse spectrum $\left|\tilde{E}(\omega)\right|^{2}$ occurs when the phase $\phi(\omega)$ is a constant. From the above definitions, one may directly conclude that in this case the spectral phase $\phi(\omega)$ only consists of a constant and a linear term. All other spectral phase functions yield longer rms pulse durations.

2.5 Linear Pulse Propagation in a Dispersive Material

2.5.1 Linear Pulse Propagation Versus Linear Dispersion

We have shown in Sect. 2.3.2 that for linear pulse propagation in a dispersive medium the frequency spectrum does not change. This is the case for any arbitrary dispersion relation which can exhibit a nonlinear frequency dependence (see, for example, Fig. 1.8). Linear pulse propagation means that no intensity-dependent material properties are considered. However, the refractive index $n(\omega)$ can have any nonlinear frequency dependence. We will show that the linear pulse propagation can be completely described by the frequency-dependent refractive index $n(\omega)$ and the absorption coefficient $\alpha(\omega)$. Dispersion and absorption are linear effects in pulse propagation. On the other hand, in nonlinear pulse propagation, one also considers intensity-dependent effects such as the saturation of absorption or amplification, self-focusing, solitons, etc.

2.5.2 Slowly-Varying-Envelope Approximation

For the examination of linear pulse propagation it makes sense to separate the fast oscillating part of the complex electric field, and to consider only the dynamics of the field's envelope $A(z,t)$ (see Fig. 2.6):

$$E(z,t)=A(z,t)\mathrm{e}^{\mathrm{i}[\omega_0 t-k_n(\omega_0)z]}\,, \tag{2.52}$$

where $k_n(\omega_0)=n\omega_0/c$, and c is the velocity of light in vacuum. Depending on the case, one can choose an additional scaling of $A(z,t)$ such that A determines the intensity or the power directly:

$$I(z, t) = |A(z, t)|^2 \ , \quad P(z, t) = |A(z, t)|^2 \ . \tag{2.53}$$

In the following, we will skip such scale factors.

The Fourier transforms $\tilde{E}(z, \omega)$ and $\tilde{A}(z, \Delta\omega)$ satisfy the relation

$$\tilde{E}(z, \omega_0 + \Delta\omega) = \tilde{A}(z, \Delta\omega)\, \mathrm{e}^{-\mathrm{i}k_n(\omega_0)z} \ , \tag{2.54}$$

where

$$\Delta\omega \equiv \omega - \omega_0 \tag{2.55}$$

describes the frequency deviation from ω_0, and $\tilde{A}(z, \Delta\omega)$ is given by Fourier transform [analogously to (2.34)] as

$$\tilde{A}(z, \Delta\omega) = \int A(z, t)\mathrm{e}^{-\mathrm{i}\Delta\omega t}\mathrm{d}t \ . \tag{2.56}$$

Using (2.54) with the wave equation in the spectral domain as given by (2.9),

$$\frac{\partial^2 \tilde{E}(z, \omega)}{\partial z^2} + [k_n(\omega)]^2\, \tilde{E}(z, \omega) = 0 \ ,$$

in the differential equation for $\tilde{A}(z, \Delta\omega)$, we find that

$$\begin{aligned}\frac{\partial^2}{\partial z^2}\tilde{A}(z, \Delta\omega) - 2\mathrm{i}k_n(\omega_0)\frac{\partial}{\partial z}\tilde{A}(z, \Delta\omega) - [k_n(\omega_0)]^2\, \tilde{A}(z, \Delta\omega)& \\ + [k_n(\omega_0 + \Delta\omega)]^2\, \tilde{A}(z, \Delta\omega) = 0 \ .&\end{aligned}$$

In addition, we make the slowly-varying-envelope approximation (SVEA):

$$\boxed{\left|\frac{\partial^2 \tilde{A}}{\partial z^2}\right| \ll \left|k_n(\omega_0)\frac{\partial \tilde{A}}{\partial z}\right|} \ . \tag{2.57}$$

Equation (2.57) means that the envelope does not change much over the distance of a wavelength $\lambda_{n0} = \lambda_0/n$. Numerical simulations show that this approximation is surprisingly well satisfied even if only a couple of cycles of the e.m. wave are underneath the envelope. Note that, at 800 nm, an optical cycle is 2.7 fs.

With this approximation, we can neglect higher-order terms like $\partial^2\tilde{A}/\partial z^2$. Therefore, we obtain

$$\frac{\partial}{\partial z}\tilde{A}(z, \Delta\omega) + \mathrm{i}\frac{1}{2}\left[\frac{k_n^2(\omega_0 + \Delta\omega)}{k_n(\omega_0)} - k_n(\omega_0)\right]\tilde{A}(z, \Delta\omega) = 0 \ . \tag{2.58}$$

Expanding k_n around the center frequency

$$k_n(\omega_0 + \Delta\omega) \approx k_n(\omega_0) + \Delta k_n \ , \tag{2.59}$$

gives

$$\begin{aligned}\frac{1}{2}\left[\frac{k_n^2(\omega_0 + \Delta\omega)}{k_n(\omega_0)} - k_n(\omega_0)\right] &= \frac{1}{2}\left[\frac{k_n^2(\omega_0) + 2\Delta k_n k_n(\omega_0) + \Delta k_n^2 - k_n^2(\omega_0)}{k_n(\omega_0)}\right] \\ &= \frac{1}{2}\left[\frac{2\Delta k_n k_n(\omega_0) + \Delta k_n^2}{k_n(\omega_0)}\right] ,\end{aligned}$$

with $\Delta k_n \ll k_n$, and (2.58) implies

$$\boxed{\frac{\partial}{\partial z}\tilde{A}(z, \Delta\omega) + \mathrm{i}\Delta k_n \tilde{A}(z, \Delta\omega) = 0} \ . \tag{2.60}$$

Equation (2.60) is then the equivalent wave equation for the pulse envelope. The solution of this differential equation is trivial:

$$\boxed{\tilde{A}(z, \Delta\omega) = \tilde{A}(0, \Delta\omega)\, \mathrm{e}^{-\mathrm{i}\Delta k_n z} = \tilde{A}(0, \Delta\omega)\, \mathrm{e}^{-\mathrm{i}[k_n(\omega_0 + \Delta\omega) - k_n(\omega_0)]z}} \tag{2.61}$$

This can be checked by inserting (2.61) into (2.60).

2.5.3 First and Second Order Dispersion

An optical pulse consists of a continuous superposition of monochromatic plane waves, the components of its Fourier spectrum, which are grouped around a center frequency ω_0. Each part of the spectrum has its own phase velocity for a given medium, and this leads to phase shifts between individual spectral components during propagation. Note that the dispersion of the group velocity and not the phase velocity is responsible for the pulse broadening. This is initially counterintuitive and will be shown in the following.

The shorter the pulse, the broader its spectral bandwidth, and the stronger the effects of dispersion on the pulse. If the spectral width of the pulse is much smaller than its carrier frequency, which is satisfied in the optical region even for ultrashort light pulses due to the enormous carrier frequency, we only need to expand the wave number $k_n(\omega)$ up to second order around the carrier frequency ω_0:

$$k_n(\omega) \approx k_n(\omega_0) + k_n' \Delta\omega + \frac{1}{2} k_n'' \Delta\omega^2 \ , \tag{2.62}$$

where $k_n' = \mathrm{d}k_n/\mathrm{d}\omega$ is called the first order dispersion and $k_n'' = \mathrm{d}^2 k_n/\mathrm{d}\omega^2$ the second order dispersion, determined by the curvature of the parabola in (2.62).

2.5.4 Phase Velocity and Group Velocity

Due to the fact that a dispersive medium is a linear system, one can solve the temporal evolution without difficulty in the spectral domain and then use the inverse Fourier transform to map it back into the time domain. For a Gaussian pulse, linear pulse propagation can be solved analytically in a dispersive medium taking into account up to second order dispersion. We benefit from the fact that the Fourier transform of a Gauss function is a Gauss function again (Appendix A).

A Gaussian pulse envelope is given by

$$A\,(z=0,t)=\exp\left[-\Gamma(0)t^2\right]\,, \tag{2.63}$$

$$E\,(z=0,t)\propto\exp\left[-\Gamma(0)t^2\right]\mathrm{e}^{\mathrm{i}[\omega_0 t-k_n(\omega_0)z]}\,, \tag{2.64}$$

and the Fourier transform $\tilde{A}\,(z=0,\Delta\omega)$ by (2.56):

$$\tilde{A}\,(z=0,\Delta\omega)=\int A\,(0,t)\,\mathrm{e}^{-\mathrm{i}\Delta\omega t}\mathrm{d}t=\sqrt{\frac{\pi}{\Gamma(0)}}\exp\left[-\frac{(\omega-\omega_0)^2}{4\Gamma(0)}\right]\,, \tag{2.65}$$

where $\Delta\omega=\omega-\omega_0$. The frequency spectrum of a pulse which has propagated through a dispersive medium of length L_d is given by (2.61):

$$\tilde{A}\,(L_\mathrm{d},\omega)=\tilde{A}\,(0,\omega)\,\mathrm{e}^{-\mathrm{i}[k_n(\omega)-k_n(\omega_0)]L_\mathrm{d}}\,. \tag{2.66}$$

Considering up to second order dispersion, (2.62) implies

$$\tilde{A}\,(L_\mathrm{d},\omega)=\sqrt{\frac{\pi}{\Gamma(0)}}\exp\left[-\frac{(\omega-\omega_0)^2}{4\Gamma(0)}\right]\exp\left[-\mathrm{i}k_n'L_\mathrm{d}\,(\omega-\omega_0)-\mathrm{i}\frac{k_n''L_\mathrm{d}}{2}\,(\omega-\omega_0)^2\right]\,. \tag{2.67}$$

The output pulse $A\,(L_\mathrm{d},t)$ after the dispersive medium is determined by an inverse Fourier transform:

$$A\,(L_\mathrm{d},t)=\frac{1}{2\pi}\sqrt{\frac{\pi}{\Gamma(0)}}\int\exp\left[-\frac{(\omega-\omega_0)^2}{4\Gamma(L_\mathrm{d})}+\mathrm{i}\,(\omega-\omega_0)\left(t-k_n'L_\mathrm{d}\right)\right]\mathrm{d}\,(\omega-\omega_0)\,. \tag{2.68}$$

Using this and dropping the prefactor, we obtain

$$E\,(L_\mathrm{d},t)=\exp\left[\mathrm{i}\omega t-\mathrm{i}k_n(\omega_0)L_\mathrm{d}\right]\times\int\exp\left[-\frac{(\omega-\omega_0)^2}{4\Gamma(L_\mathrm{d})}+\mathrm{i}\,(\omega-\omega_0)\left(t-k_n'L_\mathrm{d}\right)\right]\mathrm{d}\,(\omega-\omega_0)\,. \tag{2.69}$$

Equation (2.69) corresponds to the Fourier transform of a Gaussian pulse $\exp(\Gamma t^2)$ with a temporal shift of $(t-k_n'L_\mathrm{d})$. In this equation we can identify the complex Gauss parameter after propagation $\Gamma(L_\mathrm{d})$ as

$$\boxed{\frac{1}{\Gamma(L_\mathrm{d})}=\frac{1}{\Gamma(0)}+2\mathrm{i}k_n''L_\mathrm{d}}\,, \tag{2.70}$$

where we see that only the term for second order dispersion (k_n'') affects the complex Gauss parameter. This is the dispersion of the group velocity and not the phase velocity, which is initially counterintuitive. This is indicative of the fact that only second and higher order dispersion is responsible for pulse broadening, as will be further elaborated in this section. To calculate the integral in (2.69), the following relation is useful:

$$\int \mathrm{e}^{-Ay^2-2By}\mathrm{d}y = \sqrt{\frac{\pi}{A}}\mathrm{e}^{B^2/A}\,, \quad \mathrm{Re}(A) > 0\,. \tag{2.71}$$

It follows that

$$E\left(L_\mathrm{d}, t\right) \propto \exp\left[\mathrm{i}\omega_0\left(t - \frac{L_\mathrm{d}}{\upsilon_\mathrm{p}(\omega_0)}\right)\right]\exp\left[-\Gamma(L_\mathrm{d})\left(t - \frac{L_\mathrm{d}}{\upsilon_\mathrm{g}(\omega_0)}\right)^2\right]\,, \tag{2.72}$$

where υ_p is the phase velocity, given by

$$\boxed{\upsilon_\mathrm{p}(\omega_0) \equiv c_n = \left.\frac{\omega}{k_n}\right|_{\omega=\omega_0}}\,, \tag{2.73}$$

and υ_g is the group velocity, given by

$$\boxed{\upsilon_\mathrm{g}(\omega_0) \equiv \frac{1}{k_n'(\omega_0)} = \frac{1}{\left.\frac{\mathrm{d}k_n}{\mathrm{d}\omega}\right|_{\omega=\omega_0}} = \left.\frac{\mathrm{d}\omega}{\mathrm{d}k_n}\right|_{\omega=\omega_0}}\,. \tag{2.74}$$

The first exponent in (2.72) has the form of a phase factor of a monochromatic wave with frequency ω_0 shifted in time by $L_\mathrm{d}/\upsilon_\mathrm{p}(\omega_0)$. This is why we identify $\upsilon_\mathrm{p}(\omega_0)$ as the phase velocity, i.e., the propagation velocity of the carrier wave with frequency ω_0. The second exponent has the form of a Gaussian envelope, shifted in time by $L_\mathrm{d}/\upsilon_\mathrm{g}(\omega_0)$. Therefore $\upsilon_\mathrm{g}(\omega_0)$ corresponds to the group velocity, i.e., the propagation speed of the envelope of the pulse (see Fig. 2.9). The different propagation velocities of the wave packet and the wave fronts can be very nicely observed for example in water waves. In vacuum, the dispersion relation $k(\omega) = \omega/c$ applies. This is why the group velocity of (2.74) and the phase velocity of (2.73) are identical in vacuum.

The first order dispersion determines the temporal shift of the pulse and is called the group delay, denoted T_g. This is given by

$$\boxed{T_\mathrm{g} \equiv \frac{z}{\upsilon_\mathrm{g}} = k_n' z}\,. \tag{2.75}$$

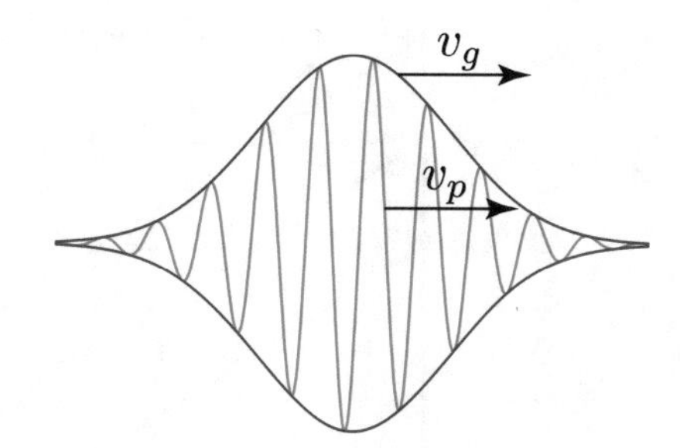

Fig. 2.9 Group velocity $v_g(\omega_0)$ and phase velocity $v_p(\omega_0)$. The phase front propagates with the phase velocity and the peak of the pulse envelope with the group velocity

2.5.5 Dispersive Pulse Broadening

A pulse is broadened by propagation through a dispersive medium (see Fig. 2.10). How much a pulse is broadened will be discussed in this section. The FWHM $\tau_p(L_d)$ of a Gaussian pulse is given by (2.40) and (2.70):

$$\tau_p(L_d) = \sqrt{\frac{2\ln 2}{\mathrm{Re}\left[\Gamma(L_d)\right]}} . \tag{2.76}$$

Equations (2.70) and (2.76) clearly show that linear dispersion alone does not contribute to pulse broadening, but only second and higher order dispersion. For the reasons previously discussed, we have neglected third order and higher order dispersion. Higher order dispersion becomes important for very short pulses, or negligibly small lower order dispersion.

For the special case of a transform-limited input pulse where no chirp is present, (2.70) reduces to

$$\frac{1}{\Gamma(L_d)} = \frac{\tau_p^2(0)}{2\ln 2} + 2\mathrm{i}k_n'' L_d . \tag{2.77}$$

Using (2.76), it then follows that

$$\tau_p(L_d) = \frac{\sqrt{\tau_p^4(0) + \left(4\ln 2 k_n'' L_d\right)^2}}{\tau_p(0)} , \tag{2.78}$$

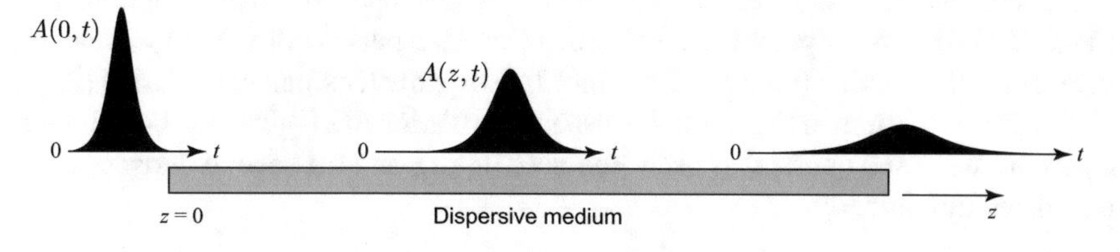

Fig. 2.10 Dispersive pulse broadening for the pulse envelope $A(z, t)$

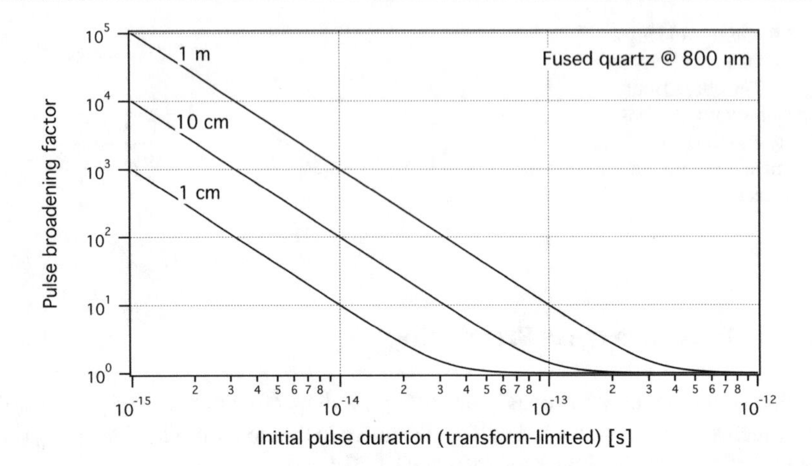

Fig. 2.11 Pulse broadening factor by second order dispersion in fused quartz vs. input pulse duration for thicknesses of 1 cm, 10 cm, and 1 m at a wavelength of 800 nm (2.79). Please note that the initial pulse is Gaussian and transform-limited. The initial pulse defines the spectral width, which does not change (2.30). In the linear regime the pulse broadening factor is then directly proportional to the spectral width of the pulse. This is shown with the approximation for strong pulse broadening (2.80)

whence

$$\boxed{\frac{\tau_\mathrm{p}\left(L_\mathrm{d}\right)}{\tau_\mathrm{p}(0)} = \sqrt{1 + \frac{\left(4\ln 2 k_n'' L_\mathrm{d}\right)^2}{\tau_\mathrm{p}^4(0)}} = \sqrt{1 + \left(\frac{4\ln 2 \frac{\mathrm{d}^2\phi}{\mathrm{d}\omega^2}}{\tau_\mathrm{p}^2(0)}\right)^2}}\,, \tag{2.79}$$

where $\phi = k_n L_\mathrm{d}$ is the accumulated phase shift of a plane wave in a dispersive medium of thickness L_d. In the literature it is actually quite common to describe dispersion in terms of this phase shift rather than the frequency-dependent refractive index or the wave number. Equation (2.79) describes pulse broadening by second order dispersion for a transform-limited input pulse duration $\tau_\mathrm{p}(0)$. The shorter the original pulse, the more the pulse is stretched by the dispersive medium. Here, the pulse remains Gaussian. But in general the pulse shape can change significantly. Figure 2.11 depicts the pulse broadening of a Gaussian pulse with a center wavelength of 800 nm after passing through a 1 cm thick fused quartz lens, calculated using (2.79).

For the calculation in Fig. 2.11, the dispersion relation $n(\lambda)$ (given by the Sellmeier equation) for fused quartz was used (see Sect. 1.4.2), and the second derivative was calculated numerically:

$$\left.\frac{\mathrm{d}^2 n}{\mathrm{d}\lambda^2}\right|_{\lambda_0=800\ \mathrm{nm}} = 0.04 \frac{1}{\mu\mathrm{m}^2}\,.$$

The second order dispersion was then calculated (Table 2.3):

$$\frac{\mathrm{d}^2\phi}{\mathrm{d}\omega^2} = \frac{\lambda^3 L_\mathrm{d}}{2\pi c^2}\frac{\mathrm{d}^2 n}{\mathrm{d}\lambda^2} .$$

For a 1 cm thick fused quartz lens, we obtain

$$\left.\frac{\mathrm{d}^2\phi}{\mathrm{d}\omega^2}\right|_{\lambda_0=800\ \mathrm{nm}} = 3.61 \times 10^{-26}\frac{\mathrm{s}^2}{\mathrm{m}} \times 1\ \mathrm{cm} = 3.61 \times 10^{-28}\mathrm{s}^2 = 361\ \mathrm{fs}^2 .$$

For the case of an originally unchirped, i.e., transform-limited, Gaussian pulse we can then insert this value in (2.79). Figure 2.11 shows the resulting pulse broadening in a dispersive medium with positive dispersion. Pulse broadening by optical elements like lenses, beam-splitters, etc., becomes important for pulse lengths below 100 fs. In measurements where a short time resolution is required, one therefore has to compensate for all lenses and optical elements or one has to replace them with reflective optical elements.

Approximation for Strong Pulse Broadening

In the case where

$$\frac{\mathrm{d}^2\phi}{\mathrm{d}\omega^2} \gg \tau_\mathrm{p}^2(0) ,$$

equation (2.79) can be greatly simplified:

$$\frac{\tau_\mathrm{p}\left(L_\mathrm{d}\right)}{\tau_\mathrm{p}(0)} = \sqrt{1 + \left(\frac{4\ln 2\dfrac{\mathrm{d}^2\phi}{\mathrm{d}\omega^2}}{\tau_\mathrm{p}^2(0)}\right)^2} \approx \frac{4\ln 2\dfrac{\mathrm{d}^2\phi}{\mathrm{d}\omega^2}}{\tau_\mathrm{p}^2(0)} , \quad \text{for } \frac{\mathrm{d}^2\phi}{\mathrm{d}\omega^2} \gg \tau_\mathrm{p}^2(0) .$$

Using (2.44) and assuming that $\tau_\mathrm{p}(0)$ is unchirped, we have

$$\tau_\mathrm{p}(0)\Delta\nu_\mathrm{p} = \frac{2\ln 2}{\pi} ,$$

and hence

$$\tau_\mathrm{p}\left(L_\mathrm{d}\right) \approx \frac{4\ln 2\dfrac{\mathrm{d}^2\phi}{\mathrm{d}\omega^2}}{\tau_\mathrm{p}(0)} = \frac{\Delta\nu_\mathrm{p}\pi}{2\ln 2}4\ln 2\frac{\mathrm{d}^2\phi}{\mathrm{d}\omega^2} = \frac{\mathrm{d}^2\phi}{\mathrm{d}\omega^2}\Delta\omega_\mathrm{p} .$$

Therefore, for strong pulse broadening, (2.79) reduces to

$$\boxed{\frac{\mathrm{d}^2\phi}{\mathrm{d}\omega^2} \gg \tau_\mathrm{p}^2(0) \implies \tau_\mathrm{p}\left(L_\mathrm{d}\right) \approx \frac{\mathrm{d}^2\phi}{\mathrm{d}\omega^2}\Delta\omega_\mathrm{p}} . \tag{2.80}$$

Table 2.2 Examples of numerically calculated dispersion values from the corresponding Sellmeier equations. The propagation constants have been determined using Table 2.3

Material	Refractive index $n(\lambda)$ at 800 nm	Propagation constant $k_n(\omega)$ at 800 nm
Fused quartz	n (800 nm) = 1.45332	
	$\left.\frac{\partial n}{\partial \lambda}\right\|_{800\text{ nm}} = -0.017 \frac{1}{\mu\text{m}}$	$\left.\frac{\partial k_n}{\partial \omega}\right\|_{800\text{ nm}} = 4.84 \times 10^{-9} \frac{\text{s}}{\text{m}}$
	$\left.\frac{\partial^2 n}{\partial \lambda^2}\right\|_{800\text{ nm}} = 0.04 \frac{1}{\mu\text{m}^2}$	$\left.\frac{\partial^2 k_n}{\partial \omega^2}\right\|_{800\text{ nm}} = 3.61 \times 10^{-26} \frac{\text{s}^2}{\text{m}} = 36.1 \frac{\text{fs}^2}{\text{mm}}$
	$\left.\frac{\partial^3 n}{\partial \lambda^3}\right\|_{800\text{ nm}} = -0.24 \frac{1}{\mu\text{m}^3}$	$\left.\frac{\partial^3 k_n}{\partial \omega^3}\right\|_{800\text{ nm}} = 2.74 \times 10^{-41} \frac{\text{s}^3}{\text{m}} = 27.4 \frac{\text{fs}^3}{\text{mm}}$
SF10 glass	n (800 nm) = 1.71125	
	$\left.\frac{\partial n}{\partial \lambda}\right\|_{800\text{ nm}} = -0.0496 \frac{1}{\mu\text{m}}$	$\left.\frac{\partial k_n}{\partial \omega}\right\|_{800\text{ nm}} = 5.70 \times 10^{-9} \frac{\text{s}}{\text{m}}$
	$\left.\frac{\partial^2 n}{\partial \lambda^2}\right\|_{800\text{ nm}} = 0.176 \frac{1}{\mu\text{m}^2}$	$\left.\frac{\partial^2 k_n}{\partial \omega^2}\right\|_{800\text{ nm}} = 1.59 \times 10^{-25} \frac{\text{s}^2}{\text{m}} = 159 \frac{\text{fs}^2}{\text{mm}}$
	$\left.\frac{\partial^3 n}{\partial \lambda^3}\right\|_{800\text{ nm}} = -0.997 \frac{1}{\mu\text{m}^3}$	$\left.\frac{\partial^3 k_n}{\partial \omega^3}\right\|_{800\text{ nm}} = 1.04 \times 10^{-40} \frac{\text{s}^3}{\text{m}} = 104 \frac{\text{fs}^3}{\text{mm}}$
Sapphire	n (800 nm) = 1.76019	
	$\left.\frac{\partial n}{\partial \lambda}\right\|_{800\text{ nm}} = -0.0268 \frac{1}{\mu\text{m}}$	$\left.\frac{\partial k_n}{\partial \omega}\right\|_{800\text{ nm}} = 5.87 \times 10^{-9} \frac{\text{s}}{\text{m}}$
	$\left.\frac{\partial^2 n}{\partial \lambda^2}\right\|_{800\text{ nm}} = 0.064 \frac{1}{\mu\text{m}^2}$	$\left.\frac{\partial^2 k_n}{\partial \omega^2}\right\|_{800\text{ nm}} = 5.80 \times 10^{-26} \frac{\text{s}^2}{\text{m}} = 58 \frac{\text{fs}^2}{\text{mm}}$
	$\left.\frac{\partial^3 n}{\partial \lambda^3}\right\|_{800\text{ nm}} = -0.377 \frac{1}{\mu\text{m}^3}$	$\left.\frac{\partial^3 k_n}{\partial \omega^3}\right\|_{800\text{ nm}} = 4.21 \times 10^{-41} \frac{\text{s}^3}{\text{m}} = 42.1 \frac{\text{fs}^3}{\text{mm}}$

Table 2.2 lists the first to third order dispersion at a center wavelength of 800 nm for different materials. The values are based on the Sellmeier equations discussed and given in Sect. 1.4.2. These values are used for the following examples.

Example. A 10 fs bandwidth-limited Gaussian pulse with center wavelength 800 nm propagates through a 1 cm thick fused quartz material. Hence,

$$\frac{\mathrm{d}^2\phi}{\mathrm{d}\omega^2} = L_\mathrm{d}\frac{\mathrm{d}^2 k_n}{\mathrm{d}\omega^2} = 361\ \mathrm{fs}^2\ .$$

With $\Delta\nu_\mathrm{p}\tau_\mathrm{p} = 0.4413$ (see Table 2.1), we have $\Delta\omega_\mathrm{p} = 0.277\ \mathrm{fs}^{-1}$. Using (2.80), it turns out that the broadened pulse length after passing through 1 cm fused quartz is 100 fs, which corresponds to pulse broadening by a factor of 10. This also corresponds to the exactly calculated result in Fig. 2.11.

Pulse broadening under positive dispersion is shown schematically in Fig. 2.12. The pulse becomes longer and chirped, which means that the spectral components with a longer wavelength propagate faster than the components with shorter wavelengths, i.e., $\lambda_1 > \lambda_2 \Rightarrow v_\mathrm{g}(\lambda_1) > v_\mathrm{g}(\lambda_2)$. The broadened pulse is chirped by the dispersion. But it is important to remember that the pulse spectrum does not change during propagation (2.30). This property is shown again schematically in Fig. 2.13.

Equations (2.79) and (2.80) show that the pulse broadens only due to second or higher order dispersion, i.e., first order dispersion does not change the pulse shape. In this chapter this was explicitly derived for a Gaussian pulse, but this statement can be generalized to arbitrary pulse shapes. For a material whose higher order dispersion is negligible in a certain spectral region, i.e., $\mathrm{d}^j\phi/\mathrm{d}\omega^j \approx 0$ for $j \geq 2$, a pulse with a spectrum within this spectral region will not be broadened.

It is important to learn to accept that the linear dispersive pulse broadening only occurs for second and higher order dispersion. This means that it is not the frequency-dependent difference in the phase velocity but the group velocity that is relevant. We

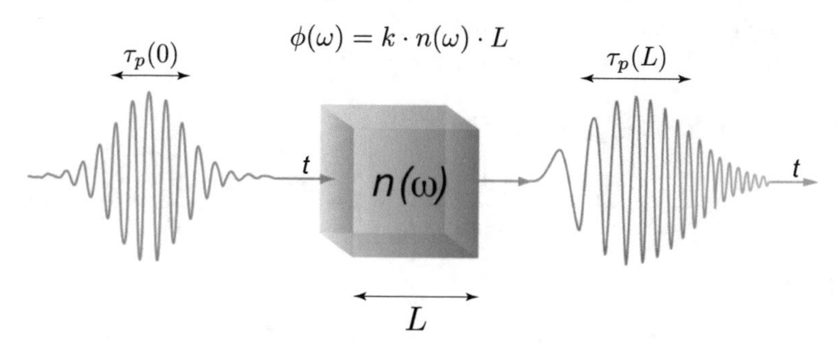

Fig. 2.12 Pulse broadening in a dispersive medium of thickness L with positive dispersion for $E(z, t)$ according to (2.72). Note the horizontal axis for the electric field in this figure represents time. Therefore, the blue wavelengths of the pulse after the dispersive medium are in fact trailing the red wavelengths

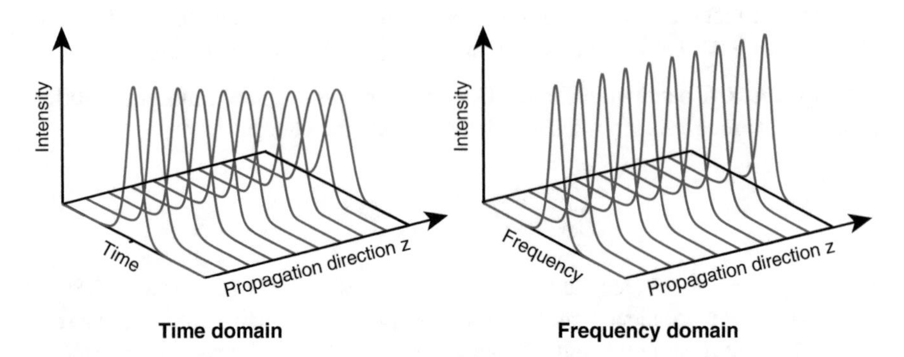

Fig. 2.13 Linear pulse propagation in a dispersive medium in the time and spectral domain. The pulse broadens in the time domain (2.79), while its power spectrum (2.30) remains unchanged

can show this more schematically as follows. For a descriptive observation of pulse broadening, one can split a short pulse into longer sub-pulses with slightly shifted center wavelengths (Fig. 2.14). Superposition of the longer sub-pulses results in a shorter pulse, because the electric fields in the shoulders of the sub-pulse interfere destructively. This is completely analogous to a wave packet formation (see Figs. 2.1 and 2.4). One assumes that all sub-pulses have a sufficiently small spectral width to justify neglecting higher order dispersion. The individual sub-pulses will not broaden, but they will have a different group delay because $\mathrm{d}\phi/\mathrm{d}\omega$ is not the same for the different center wavelength (Table 2.3). This different group delay then leads to pulse broadening, shown schematically in Fig. 2.14.

2.5.6 Dispersion as a Function of Frequency and Wavelength

It is useful to express the dispersion as a function of frequency ω or wavelength λ (in vacuum). We will derive this dependence explicitly for the example of the group velocity υ_g and the second and third order dispersions. All other relations follow analogously with a little more algebraic effort. The results are summarized in Table 2.3.

The group velocity is given by (2.74):

$$\frac{1}{\upsilon_\mathrm{g}} = \frac{\mathrm{d}k_n(\omega)}{\mathrm{d}\omega} = \frac{\mathrm{d}}{\mathrm{d}\omega}\left[\frac{n(\omega)\omega}{c}\right] = \frac{\mathrm{d}n(\omega)}{\mathrm{d}\omega}\frac{\omega}{c} + \frac{n}{c} \,.$$

Using the relation

$$\nu = \frac{c}{\lambda} \implies \frac{\mathrm{d}\nu}{\mathrm{d}\lambda} = -\frac{c}{\lambda^2} = -\frac{\nu}{\lambda} \implies \frac{\mathrm{d}\nu}{\nu} = -\frac{\mathrm{d}\lambda}{\lambda} \,,$$

it follows that

$$\frac{\mathrm{d}\omega}{\omega} = -\frac{\mathrm{d}\lambda}{\lambda} \,,$$

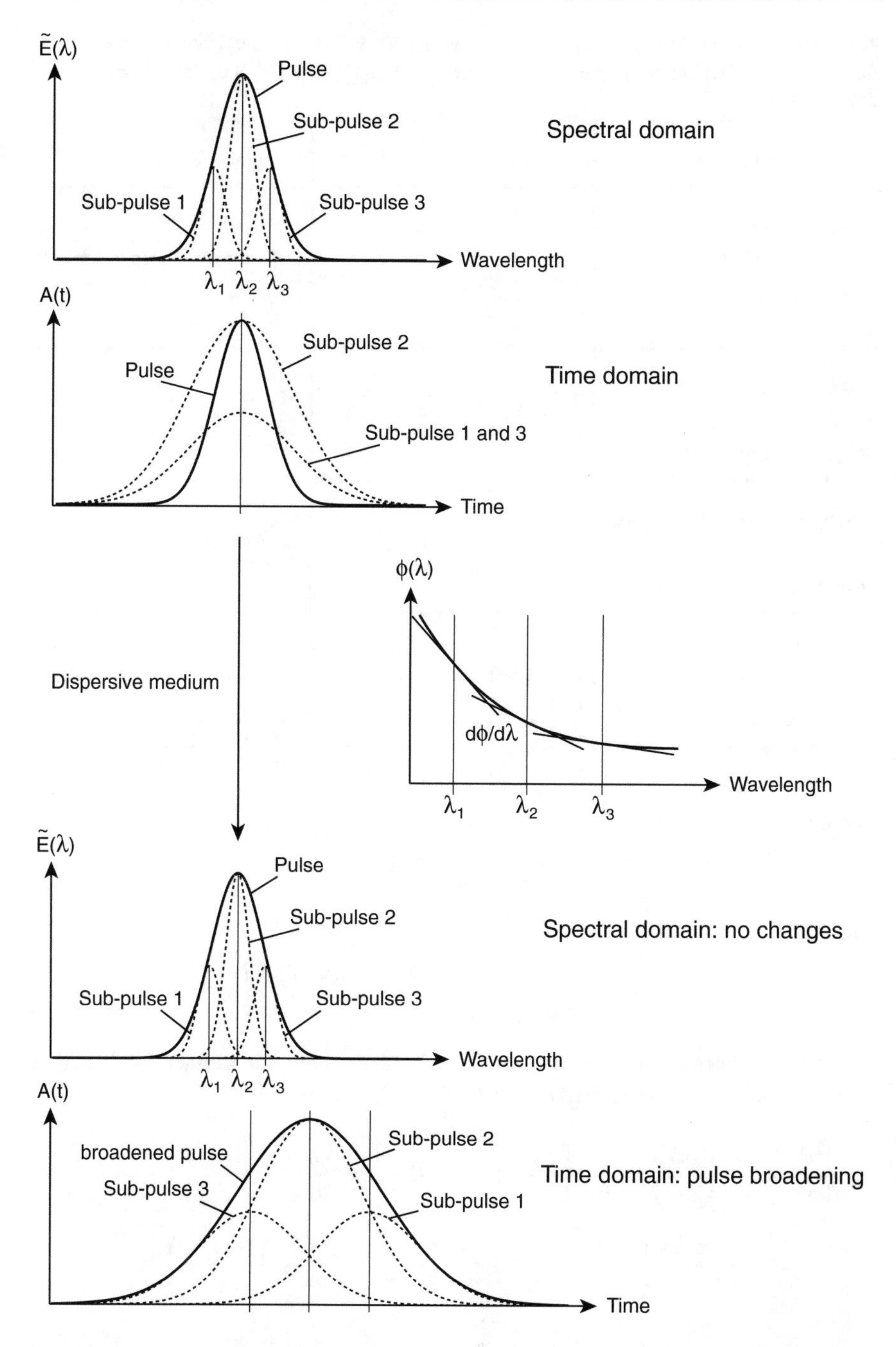

Fig. 2.14 Pulse broadening depicted schematically for different delays of the longer sub-pulses, with shifted center wavelengths in the spectral domain and in the time domain. Note that in the time and spectral domain the individual sub-pulses have not broadened. Rather, the pulse broadening of the resultant pulse comes about due to the shifted locations of the sub-pulses in the time domain after dispersion

Table 2.3 Phase velocity v_p, group velocity v_g, group delay $T_\mathrm{g} \equiv z/v_\mathrm{g}$, and first and second order dispersions as a function of frequency ω and wavelength λ. Note that $n' \equiv \mathrm{d}n/\mathrm{d}\lambda$, $n'' \equiv \mathrm{d}^2n/\mathrm{d}\lambda^2$, and $\phi \equiv k_n L_\mathrm{d}$

Dispersion parameter	Definition	Calculation from $n(\lambda)$
Phase velocity v_p	ω/k_n	$c_n = c/n$
Group velocity v_g	$\frac{\mathrm{d}\omega}{\mathrm{d}k_n} = \left(\frac{\mathrm{d}k_n}{\mathrm{d}\omega}\right)^{-1}$	$\frac{c}{n}\frac{1}{1-\frac{n'}{n}\lambda}$
Group delay T_g (first order dispersion)	$\frac{\mathrm{d}\phi}{\mathrm{d}\omega} = L_\mathrm{d}\frac{\mathrm{d}k_n}{\mathrm{d}\omega} = \frac{L_\mathrm{d}}{v_\mathrm{g}}$	$\frac{nL_\mathrm{d}}{c}\left(1-\frac{n'}{n}\lambda\right)$
Group delay dispersion $\mathrm{d}T_\mathrm{g}/\mathrm{d}\omega$ (second order dispersion)	$\frac{\mathrm{d}^2\phi}{\mathrm{d}\omega^2}$	$\frac{\lambda^3 L_\mathrm{d}}{2\pi c^2}n''$ [fs^2]
$\frac{\mathrm{d}T_\mathrm{g}}{\mathrm{d}\lambda}$	$\mathrm{d}\lambda = -\frac{\lambda^2}{2\pi c}\mathrm{d}\omega$	$-\frac{\lambda L_\mathrm{d}}{c}n''$ [fs/nm]
Third order dispersion	$\frac{\mathrm{d}^3\phi}{\mathrm{d}\omega^3}$	$-\frac{\lambda^4 L_\mathrm{d}}{4\pi^2 c^3}(3n'' + \lambda n''')$

and hence,

$$\frac{\mathrm{d}n(\omega)}{\mathrm{d}\omega}\omega = -\frac{\mathrm{d}n(\lambda)}{\mathrm{d}\lambda}\lambda\,.$$

Therefore,

$$\frac{1}{v_\mathrm{g}} = -\frac{\mathrm{d}n(\lambda)}{\mathrm{d}\lambda}\frac{\lambda}{c} + \frac{n}{c} = \frac{n}{c}\left[1 - \frac{\mathrm{d}n(\lambda)}{\mathrm{d}\lambda}\frac{\lambda}{n}\right].$$

Analogous relations for the second and third order dispersion, defined by $\mathrm{d}^2\phi/\mathrm{d}\omega^2$ and $\mathrm{d}^3\phi/\mathrm{d}\omega^3$, can be derived as follows:

$$\begin{aligned}
\frac{\mathrm{d}^2\phi}{\mathrm{d}\omega^2} &= \frac{\mathrm{d}}{\mathrm{d}\omega}\left(\frac{\mathrm{d}\phi}{\mathrm{d}\omega}\right) = \frac{\mathrm{d}}{\mathrm{d}\omega}\left[\frac{nL_\mathrm{d}}{c}\left(1-\frac{n'}{n}\lambda\right)\right] = \frac{\mathrm{d}\lambda}{\mathrm{d}\omega}\frac{\mathrm{d}}{\mathrm{d}\lambda}\left[\frac{nL_\mathrm{d}}{c}\left(1-\frac{n'}{n}\lambda\right)\right] \\
&= -\frac{2\pi c}{\omega^2}\frac{L_\mathrm{d}}{c}\left[n'\left(1-\frac{n'}{n}\lambda\right) + n\left(-\frac{n''}{n}\lambda - \frac{n'}{n} + \left(\frac{n'}{n}\right)^2\lambda\right)\right] \\
&= -\frac{\lambda^2}{2\pi c}\frac{L_\mathrm{d}}{c}\left(n' - \frac{n'^2}{n}\lambda - n''\lambda - n' + \frac{n'^2}{n}\lambda\right) = \frac{\lambda^3 L_\mathrm{d}}{2\pi c^2}n''\,.
\end{aligned}$$

and

$$\frac{d^3\phi}{d\omega^3} = \frac{d}{d\omega}\left(\frac{d^2\phi}{d\omega^2}\right) = \frac{d\lambda}{d\omega}\frac{d}{d\lambda}\left(\frac{\lambda^3 L_d}{2\pi c^2}n''\right)$$
$$= -\frac{\lambda^2}{2\pi c}\frac{L_d}{2\pi c^2}\left(3\lambda^2 n'' + \lambda^3 n'''\right) = -\frac{\lambda^4 L_d}{4\pi^2 c^3}\left(3n'' + \lambda n'''\right) .$$

The second order dispersion (see Table 2.3 for definitions), i.e., group velocity dispersion (GVD) or group delay dispersion (GDD) is given in different forms, and depending on the application, different units are usually used:

$$\text{GVD}: \quad \frac{d\upsilon_g}{d\omega} , \tag{2.81}$$

$$\text{GDD}: \quad \frac{dT_g}{d\omega} = \frac{d^2\phi}{d\omega^2} \quad [\text{fs}^2] , \tag{2.82}$$

$$\text{GDD}: \quad \frac{dT_g}{d\lambda} \quad \left[\frac{\text{fs}}{\text{nm}}\right] . \tag{2.83}$$

The relationship in the table follows from

$$\frac{dT_g}{d\lambda} = \frac{dT_g}{d\omega}\frac{d\omega}{d\lambda}$$
$$= -\frac{2\pi c}{\lambda^2}\frac{dT_g}{d\omega} = -\frac{2\pi c}{\lambda^2}\frac{\lambda^3 L_d}{2\pi c^2}n''$$
$$= -\frac{\lambda L_d}{c}n'' .$$

Note that (2.82) and (2.83) also result in different algebraic signs for the value of the GDD (apart from the difference in the units). For fibers, the following notation is typically used:

$$\text{GDD}: \quad \frac{d^2 k_n}{d\omega^2} \quad \left[\frac{\text{ps}^2}{\text{km}}\right] . \tag{2.84}$$

For a given fiber length L_F, this yields once again

$$\frac{d^2 k_n}{d\omega^2} L_F = \frac{d^2\phi}{d\omega^2} = \frac{dT_g}{d\omega} .$$

Additionally, in optical communications one finds yet another description of the second order dispersion:

$$\text{GDD}: \quad D_\lambda \equiv -\frac{1}{L_F}\frac{dT_g}{d\lambda} = \frac{\omega^2}{2\pi c L_F}\frac{dT_g}{d\omega} \quad \left[\frac{\text{ps}}{\text{km}\cdot\text{nm}}\right] . \tag{2.85}$$

We used (2.80) to derive a simple relation for calculating the dispersive pulse broadening of a Gaussian pulse in the approximation of strong pulse broadening, viz.,

$$\tau_p(L_d) \approx \frac{d^2\phi}{d\omega^2}\Delta\omega_p .$$

We will now derive the equivalent relation in the wavelength region, using the dispersion coefficient D_λ.

First we derive the relation between $d^2\phi/d\omega^2$ and D_λ. Since

$$D_\lambda := -\frac{1}{L_d}\frac{dT_g}{d\lambda} = \frac{\omega^2}{2\pi c L_d}\frac{dT_g}{d\omega}, \qquad T_g := \frac{d\phi}{d\omega},$$

we obtain

$$D_\lambda = \frac{\omega^2}{2\pi c L_d}\frac{d^2\phi}{d\omega^2},$$

and hence,

$$\tau_p(L_d) \approx \frac{d^2\phi}{d\omega^2}\Delta\omega = \frac{2\pi c L_d D_\lambda}{\omega^2}\Delta\omega = D_\lambda L_d \Delta\lambda . \tag{2.86}$$

In the last step, we used

$$\frac{\Delta\omega}{\omega} = \frac{\Delta\lambda}{\lambda},$$

hence

$$\Delta\omega = \frac{\omega}{\lambda}\Delta\lambda = \frac{2\pi c}{\lambda^2}\Delta\lambda .$$

Thus, the pulse duration at the end of the fiber is the product of the dispersion coefficient D_λ, the fiber length, and the spectral bandwidth of the pulse. Optical telecommunication usually uses picosecond pulses over kilometer-long fibers. Since the center wavelength for fiber communication is usually about 1.3 μm or 1.55 μm, the corresponding pulse bandwidth is in the nm region. Therefore, for practical reasons, it makes a lot of sense to mix these different length scales in the units of D_λ: [ps/km · nm].

2.5.7 Optical Communication

Most long-distance telephone calls are transmitted via optical glass fibers. Glass fibers have a high refractive index in the core and a low refractive index in the cladding (see Fig. 2.15). Light can propagate through the fiber core with low loss, because it stays in the core due to total internal reflection on the cladding. If only one possible transverse field distribution can be guided in the fiber, one speaks of a single-mode fiber.

The transmission of information can be digital, i.e., the information is transmitted in bits at a certain bit rate. For example, in a given time frame, bit 1 corresponds to a pulse and bit 0 corresponds to no pulse. For a 5 GHz bit rate, optical pulses with a

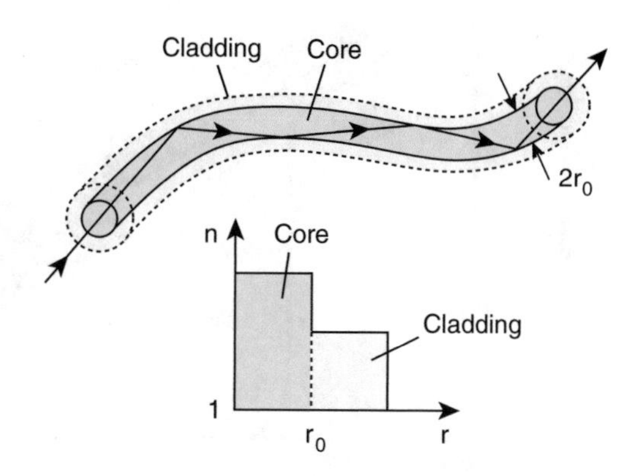

Fig. 2.15 Light-guiding fiber made of glass. The light is guided by total internal reflection on the cladding inside the core

spacing as short as 200 ps are sent through the glass fiber. Because the transmission distance can exceed 1000 km, it is advantageous to choose an optical carrier frequency at which glass fibers have minimal loss. This is the case at a wavelength of 1.55 μm (see Fig. 2.16). Nevertheless, after a certain distance, additional optical amplifiers are necessary. At a wavelength of 1.55 μm, glass fibers have second order dispersion (see Fig. 2.16). The dispersion in the waveguide depends on the material, but also on the geometry of the waveguide.

Table 2.4 summarizes loss and second order dispersion for certain wavelengths. Consider the special case in which we send 20 ps pulses with a center wavelength of $\lambda_0 = 1.55$ μm and a pulse repetition rate of 5 GHz through a fiber of length $L_F = 5000$ km. We may now ask how the pulses will have changed after this distance and whether the digital information is conserved. For the moment we neglect loss and all nonlinear processes, and consider only the influence of dispersion. We will see that, to calculate pulse broadening, the approximation for strong pulse broadening in (2.80) is very good. The spectral width $\Delta\lambda_p$ of a 20 ps pulse with center wavelength $\lambda_0 = 1.55$ μm is given by the time–bandwidth product of

$$\Delta\nu_p\tau_p = 0.315 ,$$

assuming the pulse shape corresponds to a hyperbolic secant (see Table 2.1). For $\Delta\lambda_p$ and using (2.45),

$$\frac{\Delta\lambda_p}{\lambda_0} = \frac{\Delta\nu_p}{\nu_0} ,$$

which yields

$$\Delta\lambda_p = \frac{\lambda_0}{\nu_0}\Delta\nu_p = \frac{\lambda_0}{\nu_0}\frac{0.315}{\tau_p} = 0.315\frac{\lambda_0^2}{c}\frac{1}{\tau_p} = 0.13 \text{ nm} .$$

Using (2.86), the pulse broadening in the strong pulse broadening approximation is

$$\tau_p\,(L_F = 5000 \text{ km}) \approx 17\frac{\text{ps}}{\text{km}\cdot\text{nm}} \times 5000 \text{ km} \times 0.13 \text{ nm} \approx 10.6 \text{ ns} .$$

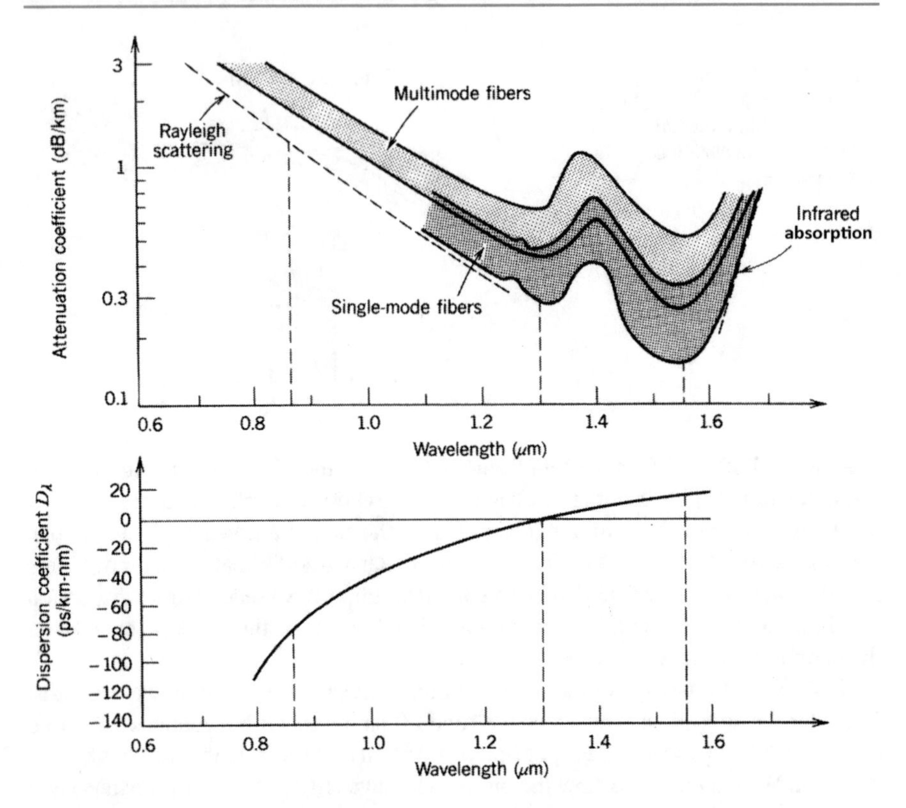

Fig. 2.16 Dispersion and absorption in a glass fiber. The second order dispersion is given here by D_λ in units of ps/km · nm as in (2.85) [91Sal2]

Table 2.4 Losses in dB/km and second order dispersion D_λ in ps/km·nm as in (2.85) for different wavelengths (from Fig. 2.16)

λ_0 [μm]	Losses [dB/km]	Second order dispersion [ps/(km·nm)]
0.87	1.5	−80
1.312	0.3	0
1.55	0.16	+17

This means that, after 5000 km, the pulse broadens to a pulse length of much more than 200 ps, which is longer than the pulse spacing, i.e., the digital information will be completely lost.

Until now we have neglected loss and all nonlinear processes. The nonlinear processes in particular are extremely important because the effects also accumulate at low intensities over long distances. Nonlinear effects cause spectral broadening, fluctuations in the pulse group delay, and much more, and this distorts the pulses much faster than linear dispersive pulse propagation. In reality, after a distance of about 50–100 km, the optical signals are typically electrically detected in so-called repeaters and are then newly emitted by laser diodes, i.e., the optical information is regenerated. For long-distance optical communication (especially with multiple carrier frequencies) it is better to avoid such repeaters because they are complex, expensive, and degrade with time. Moreover, the bit rate is limited by these repeaters, while the optical communication requirements will continue to increase in the future, and higher bit rates will be required. Soliton pulses have been discussed for a possible solution in optical communications, but there have been some fundamental issues with this approach (see Sect. 4.3). Scaling up time-division multiplexing (TDM) also becomes increasingly more challenging because the shorter pulses at higher data rates require a narrow receiver time window and a large optical bandwidth.

Therefore, wavelength-division multiplexing (WDM) with longer pulses and coherent communication has become more interesting for high bit rate transmission over optical fibers. In WDM, several carrier wavelengths transport several data streams in parallel while keeping pulse durations longer and optical bandwidths more moderate. In 2002 all-optical wavelength converters separated by 400 km were used for a 40 Gbit/s optical transmission demonstration over 1 000 000 km to potentially replace the optical-to-electrical-to-optical repeaters [02Leu]. This was possible because the signal was based on phase rather than amplitude modulation, which is also referred to as coherent optical communication. Currently, coherent optical communication has made a significant comeback using erbium-doped fiber amplifiers with improved coherent receivers and digital signal processing [10Ip, 10Sav, 10Lev, 10Spi].

2.5.8 Can a Pulse Propagate Faster Than the Speed of Light in Vacuum?

Here we want to show that the group velocity can be larger than the speed of light in vacuum. An electromagnetic field propagating in one dimension and in a medium which only has an induced electric polarization is described by the wave equation (1.23)

$$\left(\frac{\partial}{\partial z^2} - \frac{1}{c^2}\frac{\partial}{\partial t^2}\right) E(z,t) = \mu_0 \frac{\partial}{\partial t^2} P(z,t)\,, \tag{2.87}$$

where $c^2 = 1/\varepsilon_0\mu_0$ is the square of the velocity of light in vacuum. At the same time, the polarization induced in the material generally depends on the electric field itself. In a causal, local, and time-invariant medium, the polarization in the medium can be described as in (1.9),

$$\tilde{P}(z,\omega) = \tilde{\chi}(\omega)\varepsilon_0\tilde{E}(z,\omega)\,.$$

A multiplication becomes a convolution after Fourier transform (Appendix A):

$$P(z,t) = \int_{-\infty}^{t} \chi\left(t-t'\right)\varepsilon_0 E\left(t'\right)\mathrm{d}t'\,, \tag{2.88}$$

where the dielectric susceptibility $\chi(t)$ is a causal function, i.e., $\chi(t) = 0$ for $t < 0$. We will limit the following discussion to this case.

In such media, group velocities greater than the velocity of light in vacuum are already possible. This fact is very easy to understand. In particular, we will see that there is no contradiction with the theory of relativity, i.e., there is no new physics here.

We solve (2.87) by Fourier transform, turning it into the one-dimensional Helmholtz equation, as discussed in Sect. 2.2.2:

$$\tilde{E}(z,\omega) \equiv F\{E(z,t)\} = \int E(z,t)\mathrm{e}^{-\mathrm{i}\omega t}\,\mathrm{d}t \ , \tag{2.89}$$

and use the complex susceptibility

$$\tilde{\chi}(\omega) = \int_0^\infty \chi(t)\mathrm{e}^{-\mathrm{i}\omega t}\,\mathrm{d}t \tag{2.90}$$

in the frequency domain with the wave number

$$k_n(\omega) = \pm\frac{\omega}{c}n(\omega) \ , \tag{2.91}$$

and the complex refractive index $\tilde{n}(\omega)$ which also includes absorption

$$\tilde{n}(\omega) = \sqrt{1+\tilde{\chi}(\omega)} \ , \quad \text{where } \tilde{n}(\omega) \equiv \eta(\omega) - i\alpha(\omega) \ . \tag{2.92}$$

Here we have temporarily used the definition in Sect. 1.4.2 [see (1.49)], where $\eta(\omega)$ and $\alpha(\omega)$ are the real and imaginary parts of the refractive index.

In this case, the solution to the Helmholtz equation is

$$\tilde{E}(z,\omega) = \tilde{E}(0,\omega)\mathrm{e}^{-\mathrm{i}k\tilde{n}(\omega)z} = \tilde{E}(0,\omega)\mathrm{e}^{-\mathrm{i}k\eta(\omega)z}\mathrm{e}^{-k\alpha(\omega)z} \ . \tag{2.93}$$

Before we continue calculating, we should think about the kind of problem we would actually like to solve (see Fig. 2.17).

Whether one has to solve (2.87) as a boundary value problem or as an initial value problem depends on the considered experimental situation one wants to describe.

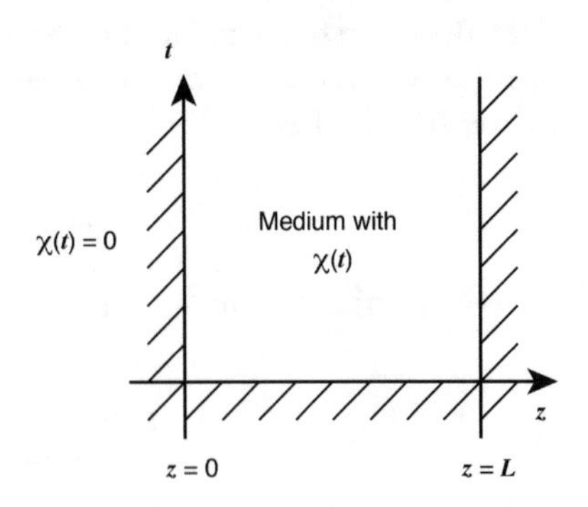

Fig. 2.17 Sketch of a common experimental configuration for pulse propagation along the z-axis through a medium with complex susceptibility $\chi(t)$

Very often the following configuration is realized. An electromagnetic pulse is sent from the half space $z < 0$ onto a medium with thickness L. One part of this pulse is reflected at the first vacuum–medium interface. One part penetrates into the medium and crosses the medium until it reaches the second medium–vacuum interface. There one part is reflected once again and one part finally leaves the medium. Due to the fact that we are dealing with a linear medium, we could of course determine the transfer function of the whole configuration for each frequency component and calculate the transmitted pulse. This transfer function would then take into account all multiple reflections within the structure. Here, we would like to consider a simplified problem, where we can neglect the multiple reflections, but which provides interesting results nonetheless. We consider the case where the refractive index change due to the medium is small. Then we can neglect the reflections on the boundaries. With this approximation, we can simplify the complex refractive index (2.92):

$$|\tilde{\chi}(\omega)| \ll 1 \implies \tilde{n}(\omega) \approx 1 + \frac{\tilde{\chi}(\omega)}{2} . \tag{2.94}$$

As we have seen, the maximum of a Gaussian pulse propagates with the group velocity

$$\frac{1}{v_g} = \frac{dk_{\tilde{n}}}{d\omega} = \frac{d}{d\omega}\left[\frac{\omega}{c}\tilde{n}(\omega)\right] . \tag{2.95}$$

From classical dispersion theory, we also know the typical curve of the complex susceptibility for positive frequencies in the vicinity of an absorption line:

$$\tilde{\chi}(\omega) = -2i\alpha \frac{1}{1 + i(\omega - \omega_0)/\Delta\omega_a} . \tag{2.96}$$

where α is directly proportional to the amplitude attenuation per wave number at resonance [see (2.93) and Fig. 2.19], and $\Delta\omega_a$ is the half FWHM of the absorption line. Figures 2.18 and 2.19 show the curves of the real and imaginary parts of the susceptibility as a function of frequency.

Figure 2.18 depicts the real part of the complex susceptibility χ_r for a given absorption linewidth. With the approximation of (2.94) the real part of the refractive index is determined by $n(\omega) \approx 1 + \chi_r(\omega)/2$ and therefore the inverse group velocity is determined by the derivative of $\chi_r(\omega)$ (2.95). Figure 2.20 then clearly shows that we can easily obtain every arbitrary value for the group velocity without violating any approximations we have made. The group velocity can even become negative.

The dashed line in Figs. 2.18 and 2.19 shows the spectrum of a Gaussian pulse with a chosen center frequency that just corresponds to the absorption line. We choose the pulse long enough, i.e., the spectrum narrow enough, so that we can approximate the real part of the susceptibility as linear in the frequency and treat the imaginary part as constant. With this approximation and with $\omega - \omega_0 \ll \Delta\omega_a$, the susceptibility of (2.96) is

$$\boxed{\tilde{\chi}(\omega) \approx -2i\alpha\left(1 - i\frac{\omega - \omega_0}{\Delta\omega_a}\right) = -2\alpha\frac{\omega - \omega_0}{\Delta\omega_a} - 2i\alpha} , \tag{2.97}$$

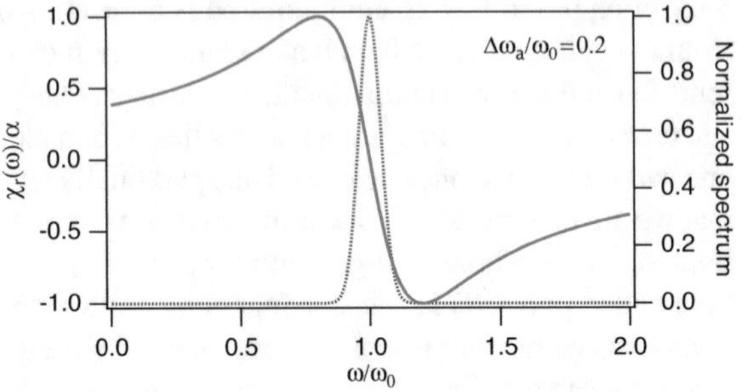

Fig. 2.18 Typical curve of the real part of the complex susceptibility χ_r near an absorption line (*solid line*), and the spectrum of a pulse (*dashed line*)

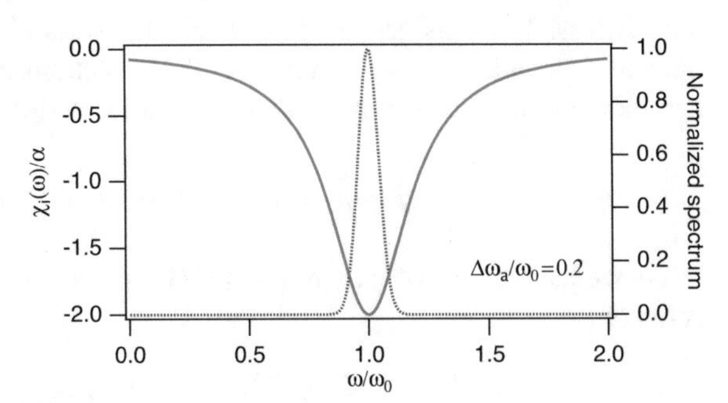

Fig. 2.19 Typical curve of the imaginary part of the complex susceptibility χ_i near an absorption line (*solid line*), and the spectrum of a pulse (*dashed line*)

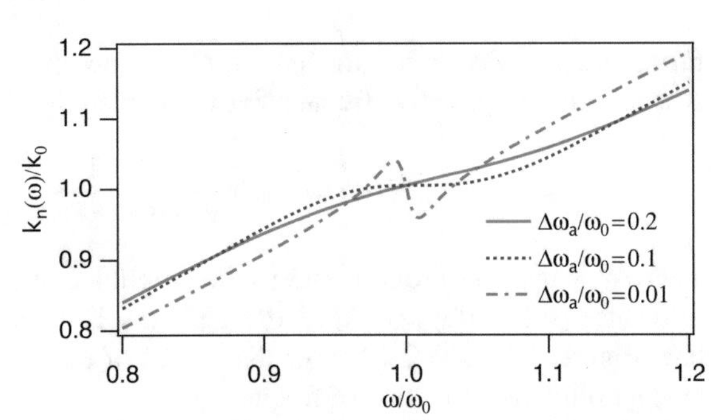

Fig. 2.20 Dispersion relation with $\alpha = 0.1$ for the group velocity $\upsilon_g^{-1} = \mathrm{d}k_n/\mathrm{d}\omega$

and using (2.94),

$$\tilde{n}(\omega) \approx 1 - \alpha \left(\frac{\omega - \omega_0}{\Delta\omega_a} \right) - \mathrm{i}\alpha \, . \tag{2.98}$$

Then with (2.95) the group velocity at the center frequency of the pulse is given by

$$\frac{1}{\upsilon_g(\omega_0)} = \frac{\tilde{n}(\omega_0)}{c} + k \left. \frac{\mathrm{d}\tilde{n}}{\mathrm{d}\omega} \right|_{\omega=\omega_0} = \frac{1}{c} \left[\tilde{n}(\omega_0) - \alpha \frac{\omega_0}{\Delta\omega_a} \right] \approx \frac{1}{c} \left(1 - \alpha \frac{\omega_0}{\Delta\omega_a} \right) ,$$

from which we obtain

$$\upsilon_g(\omega_0) \approx \frac{c}{1 - \alpha\omega_0/\Delta\omega_a} \, . \tag{2.99}$$

For certain parameters in (2.99), we can obtain $|\upsilon_g(\omega_0)| > c$. For example, with $\omega_0/\Delta\omega_a = 5$ and $\alpha = 0.1$, we obtain $\upsilon_g(\omega_0) = 2c$ (see Fig. 2.20).

Does this imply a violation of the theory of relativity? Can information now be transmitted with superluminal velocity? The answer is negative, because the Gaussian pulse is an analytical signal, i.e., the information about the Gaussian pulse already exists at all times everywhere in space, and it can already be completely reconstructed everywhere in space by Taylor series expansion. During propagation, only the pulse maximum is shifted. To treat the transmission velocity more realistically, one has to consider the propagation of a causal electromagnetic pulse with the initial boundary values $E(z=0,t)=0$ for $t<0$ and otherwise arbitrary behaviour. This pulse edge cannot propagate with superluminal velocity. The mathematical treatment of this problem is more involved and not shown here.

2.5.9 Definition of the Group Index

The discussion in Sect. 2.5.8 shows that it can be useful to define a group index n_g with the group velocity $\upsilon_g(\omega)$ given by

$$\upsilon_g(\omega) \equiv \frac{c}{n_g} , \tag{2.100}$$

by analogy with the phase velocity $\upsilon_p(\omega) = c/n(\omega)$, where

$$\boxed{n_g = n + \omega\frac{dn}{d\omega}} . \tag{2.101}$$

The proof is very simple. Using (2.95) and (2.100) it follows that

$$\frac{n_g}{c} = \frac{1}{\upsilon_g} = \frac{dk_n}{d\omega} = \frac{d}{d\omega}\left(\frac{\omega}{c}n(\omega)\right) = \frac{n(\omega)}{c} + \frac{\omega}{c}\frac{d}{d\omega}n(\omega) .$$

From the definition of the group index and Fig. 2.18 it follows that the group velocity around absorption resonances can become strongly frequency-dependent, positive or negative, and larger or smaller than the speed of light. This allows for group-velocity engineering with the trade-off of loss during propagation. Replacing the absorption resonance with a gain resonance solves this issue. This example shows that we can slow the pulse propagation down to very low speed.

Slow light has become a hot topic with regard to computer and quantum information applications. Lene Hau demonstrated how to stop light completely, and developing methods for stopping light and later restarting propagation again [01Liu2].

2.5.10 Higher Order Dispersion

The production of ultrashort pulses requires precise control of the dispersion and the spectral bandwidth in the laser resonators as well as in the subsequent systems.

While pulses with a length of 10–30 fs can be generated straightforwardly, increasing difficulties arise for shorter pulses. Precise knowledge of higher order dispersion becomes more important. Therefore, higher orders in the Taylor expansion must be considered.

To characterize higher order dispersion, it can be useful to introduce the dispersion lengths L_D and L'_D [95Agr]:

$$L_D \equiv \frac{\tau^2}{|k''_n|}\,, \qquad L'_D \equiv \frac{\tau^3}{|k'''_n|}\,, \tag{2.102}$$

where $k''_n = \partial^2 k_n/\partial\omega^2$ and $k'''_n = \partial^3 k_n/\partial\omega^3$. Dispersion lengths for even higher order dispersion can be defined in a similar way.

For Gaussian pulses, τ is given by

$$E(t) \propto \exp\left(-\frac{t^2}{2\tau^2}\right) \implies \tau_p = 2\sqrt{\ln 2}\,\tau = 1.665\tau\,, \tag{2.103}$$

and for sech pulses by

$$E(t) \propto \operatorname{sech}\left(\frac{t}{\tau}\right) \implies \tau_p = 2\ln\left(1+\sqrt{2}\right)\tau = 1.763\tau\,, \tag{2.104}$$

where $\tau_p{=}\tau_p(0)$ is the transform-limited FWHM pulse length at the beginning of the dispersive propagation.

If we compare the pulse broadening of a Gaussian pulse due to second order dispersion as in (2.79):

$$\frac{\tau_p(z)}{\tau_p(0)} = \sqrt{1+\left(\frac{z}{L_D}\right)^2}\,, \tag{2.105}$$

we see that, after propagation through the dispersion length L_D, a bandwidth-limited Gaussian pulse will be broadened by a factor of $\sqrt{2}$.

In most cases, third order dispersion is negligible compared to the second order, i.e., the higher order terms in the Taylor expansion in $k_n(\omega)$ can be neglected. Third order dispersion only plays a role if $L'_D \le L_D$ or if L'_D becomes comparable to L_D. Similar arguments apply for even higher order dispersion. We will show later that, even for pulse generation of ≤ 30 fs, third order dispersion inside the laser resonators typically starts to play a non-negligible role and becomes the main limitation for sub-10-femtosecond pulse generation in the near infrared spectral regime. Third order dispersion is also important in fibers, for example, when light with a center wavelength of 1.32 μm has vanishing second order dispersion, i.e., $|k''_n| \approx 0$.

For $z \gg L_D$ the following approximation is valid for a Gaussian pulse [95Agr]:

$$\frac{\tau_p(z)}{\tau_p(0)} \approx z\sqrt{1+\frac{1}{L_D^2}+\frac{1}{4L'^{\,2}_D}}\,. \tag{2.106}$$

This describes pulse broadening due to second and third order dispersion. To calculate the third order dispersion, we may use the dispersion relation $n(\lambda)$ to calculate the third derivative numerically (see Table 2.3). The third order dispersion is found to be

$$\frac{\mathrm{d}^3\phi}{\mathrm{d}\omega^3} = -\frac{\lambda^4 z}{4\pi^2 c^3}\left(3\frac{\mathrm{d}^2 n}{\mathrm{d}\lambda^2} + \lambda\frac{\mathrm{d}^3 n}{\mathrm{d}\lambda^3}\right) . \tag{2.107}$$

For a 1 cm thick fused quartz lens, we then find

$$\left.\frac{\mathrm{d}^3\phi}{\mathrm{d}\omega^3}\right|_{\lambda_0=800\ \mathrm{nm}} = 2.74 \times 10^{-43}\ \mathrm{s}^3 = 274\ \mathrm{fs}^3 .$$

2.5.11 Slowly-Evolving-Wave Approximation

In Sect. 2.5.2 we showed that it is useful in linear pulse propagation to consider only the dynamics of the pulse envelope $A(z, t)$. Using the slowly-varying-envelope approximation (2.57), we derived a wave equation for the envelope which contains only the first derivative with respect to the spatial coordinate, and is therefore very easy to solve [see (2.60)]. The slowly-varying-envelope approximation is very often still valid in the few-cycle regime, where the pulse contains only a few periods of the e.m. wave under the pulse envelope. But what happens for pulses with durations in the regime of one or two optical cycles? Is this first-order envelope equation (2.60) still valid for such short pulses?

In this section we will see, that the envelope equation can be extended to pulse durations equal to the carrier oscillation period T, and that the validity of a modified first-order envelope equation also extends down to pulse durations as short as one optical cycle T. For the derivation of the envelope equation, we follow very closely [97Bra2], including the notation.

This section is more advanced for few-cycle pulse propagation and is not required for further reading in this textbook. So in principle it can be skipped in a first reading.

Note. We had some difficulties with the derivation of (4) in the Brabec paper [97Bra2]. We developed a new step-by-step derivation which also provides a stronger criterion for the slowly-evolving-wave equation. Therefore, we decided to use this new derivation here. This is the reason why our (2.121) is not identical with (4) in Brabec's paper, although it is the corresponding equation.

The three-dimensional wave equation is given as

$$\left(\partial_z^2 + \nabla_\perp^2\right) E(r,t) - \frac{1}{c^2}\partial_t^2 \int_{-\infty}^{t} \mathrm{d}t' \varepsilon(t-t') E(r,t') = \frac{4\pi}{c^2}\partial_t^2 P_{\mathrm{nl}}(r,t) . \tag{2.108}$$

Table 2.5 Examples of numerically calculated dispersion values from the corresponding Sellmeier equations. The propagation constants have been determined using Table 2.3

Material	Refractive index $n(\lambda)$ at 800 nm	Propagation constant $k_n(\omega)$ at 800 nm
Fused quartz	n (800 nm) = 1.45332	
	$\left.\frac{\partial n}{\partial \lambda}\right\|_{800\text{ nm}} = -0.017\frac{1}{\mu\text{m}}$	$\left.\frac{\partial k_n}{\partial \omega}\right\|_{800\text{ nm}} = 4.84\times10^{-9}\frac{\text{s}}{\text{m}}$
	$\left.\frac{\partial^2 n}{\partial \lambda^2}\right\|_{800\text{ nm}} = 0.04\frac{1}{\mu\text{m}^2}$	$\left.\frac{\partial^2 k_n}{\partial \omega^2}\right\|_{800\text{ nm}} = 3.61\times10^{-26}\frac{\text{s}^2}{\text{m}} = 36.1\frac{\text{fs}^2}{\text{mm}}$
	$\left.\frac{\partial^3 n}{\partial \lambda^3}\right\|_{800\text{ nm}} = -0.24\frac{1}{\mu\text{m}^3}$	$\left.\frac{\partial^3 k_n}{\partial \omega^3}\right\|_{800\text{ nm}} = 2.74\times10^{-41}\frac{\text{s}^3}{\text{m}} = 27.4\frac{\text{fs}^3}{\text{mm}}$
SF10 glass	n (800 nm) = 1.71125	
	$\left.\frac{\partial n}{\partial \lambda}\right\|_{800\text{ nm}} = -0.0496\frac{1}{\mu\text{m}}$	$\left.\frac{\partial k_n}{\partial \omega}\right\|_{800\text{ nm}} = 5.70\times10^{-9}\frac{\text{s}}{\text{m}}$
	$\left.\frac{\partial^2 n}{\partial \lambda^2}\right\|_{800\text{ nm}} = 0.176\frac{1}{\mu\text{m}^2}$	$\left.\frac{\partial^2 k_n}{\partial \omega^2}\right\|_{800\text{ nm}} = 1.59\times10^{-25}\frac{\text{s}^2}{\text{m}} = 159\frac{\text{fs}^2}{\text{mm}}$
	$\left.\frac{\partial^3 n}{\partial \lambda^3}\right\|_{800\text{ nm}} = -0.997\frac{1}{\mu\text{m}^3}$	$\left.\frac{\partial^3 k_n}{\partial \omega^3}\right\|_{800\text{ nm}} = 1.04\times10^{-40}\frac{\text{s}^3}{\text{m}} = 104\frac{\text{fs}^3}{\text{mm}}$
Sapphire	n (800 nm) = 1.76019	
	$\left.\frac{\partial n}{\partial \lambda}\right\|_{800\text{ nm}} = -0.0268\frac{1}{\mu\text{m}}$	$\left.\frac{\partial k_n}{\partial \omega}\right\|_{800\text{ nm}} = 5.87\times10^{-9}\frac{\text{s}}{\text{m}}$
	$\left.\frac{\partial^2 n}{\partial \lambda^2}\right\|_{800\text{ nm}} = 0.064\frac{1}{\mu\text{m}^2}$	$\left.\frac{\partial^2 k_n}{\partial \omega^2}\right\|_{800\text{ nm}} = 5.80\times10^{-26}\frac{\text{s}^2}{\text{m}} = 58\frac{\text{fs}^2}{\text{mm}}$
	$\left.\frac{\partial^3 n}{\partial \lambda^3}\right\|_{800\text{ nm}} = -0.377\frac{1}{\mu\text{m}^3}$	$\left.\frac{\partial^3 k_n}{\partial \omega^3}\right\|_{800\text{ nm}} = 4.21\times10^{-41}\frac{\text{s}^3}{\text{m}} = 42.1\frac{\text{fs}^3}{\text{mm}}$

Table 2.6 Fused quartz. Refractive index n, first and second derivatives of the refractive index n' and n'', second order dispersion $\mathrm{d}T_g/\mathrm{d}\omega$ in units of fs^2 and $\mathrm{d}T_g/\mathrm{d}\lambda$ in units of fs/nm as a function of the wavelength λ in μm for 1 mm thickness of fused quartz

Wavelength λ [μm]	n	n'	n''	$\frac{\mathrm{d}T_g}{\mathrm{d}\omega}$ [fs^2]	$\frac{\mathrm{d}T_g}{\mathrm{d}\lambda}$ [fs/nm]
0.64	1.45681				
0.66	1.45627	−0.0262			
0.68	1.45576	−0.0244	0.0848	47.2	0.192
0.7	1.45529	−0.0228	0.0744	45.2	0.174
0.72	1.45485	−0.0214	0.0655	43.2	0.157
0.74	1.45444	−0.0202	0.0578	41.4	0.143
0.76	1.45404	−0.0191	0.0511	39.7	0.129
0.78	1.45367	−0.0182	0.0452	38.0	0.118
0.8	1.45332	−0.0173	0.0401	36.3	0.107
0.82	1.45298	−0.0165	0.0356	34.7	0.097
0.84	1.45266	−0.0159	0.0316	33.1	0.088
0.86	1.45234	−0.0153	0.0281	31.6	0.080
0.88	1.45204	−0.0148	0.0249	30.0	0.073
0.9	1.45175	−0.0143	0.0221	28.6	0.066
0.92	1.45147	−0.0139	0.0197	27.1	0.060
0.94	1.45120	−0.0135	0.0174	25.6	0.055
0.96	1.45093	−0.0132	0.0154	42.1	0.049
0.98	1.45067	−0.0129	0.0136	22.7	0.045
1	1.45042	−0.0126	0.0120	21.2	0.040
1.02	1.45017	−0.0124	0.0105	19.8	0.036
1.04	1.44992	−0.0122	0.0092	18.3	0.032
1.06	1.44968	−0.0120	0.0080	16.8	0.028
1.08	1.44944	−0.0119			
1.1	1.44920				

We make the ansatz

$$E(r,t) = A(r_\perp, z, t)\mathrm{e}^{\mathrm{i}(\beta_0 z - \omega_0 t + \psi_0)} + \text{c.c.}\,, \tag{2.109}$$

and for the nonlinear polarization we write

$$P_{\mathrm{nl}}(r,t) = B(r_\perp, z, t, A)\mathrm{e}^{\mathrm{i}(\beta_0 z - \omega_0 t + \psi_0)} + \text{c.c.}\,, \tag{2.110}$$

with $\beta_0 = \mathrm{Re}\,[k(\omega_0)] = (\omega_0/c)n_0$ and $k(\omega) = (\omega/c)\sqrt{\varepsilon(\omega)}$. For simplicity and easier readability, we won't write the complex conjugate in the following. Indeed, including the complex conjugate throughout this derivation only leads to a second, redundant differential equation which is the complex conjugate of the one we derive here.

We substitute the ansatz into (2.108) and thereby obtain for its first term

$$
\begin{aligned}
\partial_z^2 + \nabla_\perp^2 & \left[A(r_\perp, z, t)\mathrm{e}^{\mathrm{i}(\beta_0 z-\omega_0 t+\psi_0)}\right] \\
&= \nabla_\perp^2 A(r_\perp, z, t)\mathrm{e}^{\mathrm{i}(\beta_0 z-\omega_0 t+\psi_0)} + \partial_z\left[\partial_z\left(A(r_\perp, z, t)\mathrm{e}^{\mathrm{i}(\beta_0 z-\omega_0 t+\psi_0)}\right)\right] \\
&= \nabla_\perp^2 A(r_\perp, z, t)\mathrm{e}^{\mathrm{i}(\beta_0 z-\omega_0 t+\psi_0)} \\
&\qquad + \partial_z\left[\partial_z A(r_\perp, z, t)\mathrm{e}^{\mathrm{i}(\beta_0 z-\omega_0 t+\psi_0)} + A(r_\perp, z, t)\mathrm{i}\beta_0\mathrm{e}^{\mathrm{i}(\beta_0 z-\omega_0 t+\psi_0)}\right] \\
&= \nabla_\perp^2 A(r_\perp, z, t)\mathrm{e}^{\mathrm{i}(\beta_0 z-\omega_0 t+\psi_0)} + \partial_z^2 A(r_\perp, z, t)\mathrm{e}^{\mathrm{i}(\beta_0 z-\omega_0 t+\psi_0)} \\
&\qquad + 2\partial_z A(r_\perp, z, t)\mathrm{i}\beta_0\mathrm{e}^{\mathrm{i}(\beta_0 z-\omega_0 t+\psi_0)} - A(r_\perp, z, t)\beta_0^2\mathrm{e}^{\mathrm{i}(\beta_0 z-\omega_0 t+\psi_0)} \\
&= \left[\left(-\beta_0^2 + 2\mathrm{i}\beta_0\partial_z + \partial_z^2 + \nabla_\perp^2\right) A(r_\perp, z, t)\right]\mathrm{e}^{\mathrm{i}(\beta_0 z-\omega_0 t-\psi_0)} \, . \qquad (2.111)
\end{aligned}
$$

For the next term of (2.108), we make extensive use of the following property of the Fourier transform:

$$
\partial_t^m F^{-1}\{f(\omega)\} = F^{-1}\left\{(-\mathrm{i}\omega)^m f(\omega)\right\} . \qquad (2.112)
$$

First, we use the convolution theorem to rewrite the second term of (1) as

$$
\begin{aligned}
-\frac{1}{c^2}\partial_t^2 \int_{-\infty}^{t} & \mathrm{d}t'\varepsilon(t-t')A(r_\perp, z, t')\mathrm{e}^{\mathrm{i}(\beta_0 z-\omega_0 t'+\psi_0)} \\
&= -\frac{1}{c^2}\partial_t^2 F^{-1}\left\{\varepsilon(\omega)A(r_\perp, z, \omega-\omega_0)\mathrm{e}^{\mathrm{i}(\beta_0 z+\psi_0)}\right\} \\
&= F^{-1}\left\{\frac{\omega^2}{c^2}\varepsilon(\omega)A(r_\perp, z, \omega-\omega_0)\mathrm{e}^{\mathrm{i}(\beta_0 z+\psi_0)}\right\} . \qquad (2.113)
\end{aligned}
$$

We now expand $k(\omega)$ in a Taylor series about ω_0:

$$
\varepsilon(\omega) = \frac{c^2}{\omega^2}k^2(\omega) = \frac{c^2}{\omega^2}\left[\sum_{m=0}^{\infty}\frac{1}{m!}\left.\frac{\partial^m k}{\partial\omega^m}\right|_{\omega_0}(\omega-\omega_0)^m\right]^2 . \qquad (2.114)
$$

Defining

$$
\beta_m = \mathrm{Re}\left.\frac{\partial^m k}{\partial\omega^m}\right|_{\omega_0} , \qquad \alpha_m = 2\mathrm{Im}\left.\frac{\partial^m k}{\partial\omega^m}\right|_{\omega_0} ,
$$

we can further expand (2.113) and move back to the time domain:

$$
\begin{aligned}
F^{-1} & \left\{\left[\sum_{m=0}^{\infty}\frac{1}{m!}\left(\beta_m + \mathrm{i}\frac{\alpha_m}{2}\right)(\omega-\omega_0)^m\right]^2 A(r_\perp, z, \omega-\omega_0)\mathrm{e}^{\mathrm{i}(\beta_0 z+\psi_0)}\right\} \\
&= \left[\sum_{m=0}^{\infty}\frac{1}{m!}\left(\beta_m + \mathrm{i}\frac{\alpha_m}{2}\right)(\mathrm{i}\partial_t)^m\right]^2\left[A(r_\perp, z, t)\mathrm{e}^{\mathrm{i}(\beta_0 z-\omega_0 t+\psi_0)}\right] \\
&= \left(\beta_0 + \mathrm{i}\frac{\alpha_0}{2} + \mathrm{i}\beta_1\partial_t + \hat{D}\right)^2\left[A(r_\perp, z, t)\mathrm{e}^{\mathrm{i}(\beta_0 z-\omega_0 t+\psi_0)}\right] . \qquad (2.115)
\end{aligned}
$$

The dispersion operator $\hat{D}$ is defined by

$$\hat{D} = -\frac{\alpha_1}{2}\partial_t + \sum_{m=2}^{\infty} \frac{1}{m!}\left(\beta_m + \mathrm{i}\frac{\alpha_m}{2}\right)(\mathrm{i}\partial_t)^m . \tag{2.116}$$

Finally, we insert the ansatz (2.110) into the right-hand side of (2.108):

$$\begin{aligned}
\frac{4\pi}{c^2}\partial_t^2 P_{\mathrm{nl}}(r,t) &= \frac{4\pi}{c^2}\partial_t^2\left[B(r_\perp,z,t,A)\mathrm{e}^{\mathrm{i}(\beta_0 z-\omega_0 t+\psi_0)}\right] \\
&= \frac{4\pi}{c^2}\partial_t\left[\partial_t B(r_\perp,z,t,A)\mathrm{e}^{\mathrm{i}(\beta_0 z-\omega_0 t+\psi_0)} - B(r_\perp,z,t,A)\mathrm{i}\omega_0\mathrm{e}^{\mathrm{i}(\beta_0 z-\omega_0 t+\psi_0)}\right] \\
&= \frac{4\pi}{c^2}\Big[\partial_t^2 B(r_\perp,z,t,A)\mathrm{e}^{\mathrm{i}(\beta_0 z-\omega_0 t+\psi_0)} - 2\partial_t B(r_\perp,z,t,A)\mathrm{i}\omega_0\mathrm{e}^{\mathrm{i}(\beta_0 z-\omega_0 t+\psi_0)} \\
&\qquad - B(r_\perp,z,t,A)\omega_0^2\mathrm{e}^{\mathrm{i}(\beta_0 z-\omega_0 t+\psi_0)}\Big] \\
&= -\frac{4\pi\omega_0^2}{c^2}\left(1+\frac{\mathrm{i}}{\omega_0}\partial_t\right)^2 B(r_\perp,z,t,A)\mathrm{e}^{\mathrm{i}(\beta_0 z-\omega_0 t+\psi_0)} .
\end{aligned} \tag{2.117}$$

Putting the results (2.111), (2.115), and (2.117) back into (2.108) and dividing by $\exp[\mathrm{i}(\beta_0 z - \omega_0 t + \psi_0)]$, we obtain

$$\begin{aligned}
&\left(-\beta_0^2 + 2\mathrm{i}\beta_0\partial_z + \partial_z^2 + \nabla_\perp^2\right)A + \left(\beta_0 + \mathrm{i}\frac{\alpha_0}{2} + \mathrm{i}\beta_1\partial_t + \hat{D}\right)^2 A \\
&\qquad = -\frac{4\pi\omega_0^2}{c^2}\left(1+\frac{\mathrm{i}}{\omega_0}\partial_t\right)^2 B .
\end{aligned} \tag{2.118}$$

This corresponds to (2) in [97Bra2]. This expression can be expanded to give

$$\begin{aligned}
&\left(-\beta_0^2 + 2\mathrm{i}\beta_0\partial_z + \partial_z^2 + \nabla_\perp^2\right)A \\
&\quad + \Big(\beta_0^2 + 2\mathrm{i}\frac{\alpha_0}{2}\beta_0 + 2\mathrm{i}\beta_1\beta_0\partial_t + 2\beta_0\hat{D} + \mathrm{i}^2\frac{\alpha_0^2}{4} + 2\mathrm{i}^2\frac{\alpha_0}{2}\beta_1\partial_t \\
&\qquad + 2\mathrm{i}\frac{\alpha_0}{2}\hat{D} + \mathrm{i}^2\beta_1^2\partial_t^2 + 2\mathrm{i}\beta_1\partial_t\hat{D} + \hat{D}^2\Big)A \\
&\quad = -\frac{4\pi\omega_0^2}{c^2}\left(1+\frac{\mathrm{i}}{\omega_0}\partial_t\right)^2 B .
\end{aligned}$$

Some reordering yields

$$\begin{aligned}
&\left(-\beta_0^2 + 2\mathrm{i}\beta_0\partial_z + \partial_z^2\right)A + \nabla_\perp^2 A + \beta_0^2 A + 2\beta_0\left(\mathrm{i}\frac{\alpha_0}{2} + \mathrm{i}\beta_1\partial_t + \hat{D}\right)A \\
&\quad + \left(\mathrm{i}\frac{\alpha_0}{2} + \mathrm{i}\beta_1\partial_t + \hat{D}\right)^2 A = -\frac{4\pi\omega_0^2}{c^2}\left(1+\frac{\mathrm{i}}{\omega_0}\partial_t\right)^2 B .
\end{aligned}$$

Further reordering leads to

$$2\mathrm{i}\beta_0\partial_z A + 2\mathrm{i}\beta_0\beta_1\partial_t A + 2\mathrm{i}\beta_0\left(\frac{\alpha_0}{2} - \mathrm{i}\hat{D}\right)A + \nabla_\perp^2 A + \partial_z^2 A - \beta_1^2\partial_t^2 A$$
$$+2\mathrm{i}\beta_1\partial_t\left(\mathrm{i}\frac{\alpha_0}{2} + \hat{D}\right)A + \left(\mathrm{i}\frac{\alpha_0}{2} + \hat{D}\right)^2 A = -\frac{4\pi\omega_0^2}{c^2}\left(1 + \frac{\mathrm{i}}{\omega_0}\partial_t\right)^2 B\,.$$

We now insert

$$\Big[-2\beta_1\partial_t\left(\partial_z + \beta_1\partial_t\right) + 2\beta_1\partial_t\left(\partial_z + \beta_1\partial_t\right)\Big]A = 0$$

to obtain

$$2\mathrm{i}\beta_0\left(\partial_z + \beta_1\partial_t + \frac{\alpha_0}{2} - \mathrm{i}\hat{D}\right)A + \nabla_\perp^2 A + \partial_z^2 A - \beta_1^2\partial_t^2 A - 2\beta_1\partial_t\left(\frac{\alpha_0}{2} - \mathrm{i}\hat{D}\right)A$$
$$+\Big[-2\beta_1\partial_t\left(\partial_z + \beta_1\partial_t\right) + 2\beta_1\partial_t\left(\partial_z + \beta_1\partial_t\right)\Big]A + \left(\mathrm{i}\frac{\alpha_0}{2} + \hat{D}\right)^2 A$$
$$= -\frac{4\pi\omega_0^2}{c^2}\left(1 + \frac{\mathrm{i}}{\omega_0}\partial_t\right)^2 B\,.$$

This then yields

$$2\mathrm{i}\beta_0\left(\partial_z + \beta_1\partial_t + \frac{\alpha_0}{2} - \mathrm{i}\hat{D}\right)A - 2\beta_1\partial_t\left(\partial_z + \beta_1\partial_t + \frac{\alpha_0}{2} - \mathrm{i}\hat{D}\right)A$$
$$+\nabla_\perp^2 A + \frac{4\pi\omega_0^2}{c^2}\left(1 + \frac{\mathrm{i}}{\omega_0}\partial_t\right)^2 B$$
$$= -\partial_z^2 A + \beta_1^2\partial_t^2 A - 2\beta_1\partial_t\left(\partial_z + \beta_1\partial_t\right)A - \left(\mathrm{i}\frac{\alpha_0}{2} + \hat{D}\right)^2 A\,,$$

which simplifies to

$$2\mathrm{i}\beta_0\left(1 + \frac{\mathrm{i}\beta_1}{\beta_0}\partial_t\right)\left(\partial_z + \beta_1\partial_t + \frac{\alpha_0}{2} - \mathrm{i}\hat{D}\right)A + \nabla_\perp^2 A + \frac{4\pi\omega_0^2}{c^2}\left(1 + \frac{\mathrm{i}}{\omega_0}\partial_t\right)^2 B$$
$$= \left(-\partial_z^2 - 2\beta_1\partial_z\partial_t - \beta_1^2\partial_t^2\right)A - \left(\mathrm{i}\frac{\alpha_0}{2} + \hat{D}\right)^2 A\,. \quad (2.119)$$

We perform a coordinate transformation to a reference frame moving with the pulse envelope, viz., $\tau = t - \beta_1 z$ and $\xi = z$. Thus, in this new reference frame, with

$$\partial_z + \beta_1\partial_t = \partial_\xi\,, \qquad \left(\partial_z + \beta_1\partial_t\right)^2 = \partial_\xi^2\,,$$

equation (2.119) becomes

$$2\mathrm{i}\beta_0\left(1 + \frac{\mathrm{i}\beta_1}{\beta_0}\partial_\tau\right)\left(\partial_\xi + \frac{\alpha_0}{2} - \mathrm{i}\hat{D}\right)A + \nabla_\perp^2 A + \frac{4\pi\omega_0^2}{c^2}\left(1 + \frac{\mathrm{i}}{\omega_0}\partial_\tau\right)^2 B$$
$$= -\left[\partial_\xi^2 + \left(\mathrm{i}\frac{\alpha_0}{2} + \hat{D}\right)^2\right]A\,. \quad (2.120)$$

Using $\beta_0/\beta_1 \approx \omega_0$ and $\beta_0 = \omega_0 n_0/c$, (2.120) can be rewritten as

$$\left(1+\frac{\mathrm{i}}{\omega_0}\partial_\tau\right)\left(\partial_\xi + \frac{\alpha_0}{2} - \mathrm{i}\hat{D}\right)A + \frac{1}{2\mathrm{i}\beta_0}\nabla_\perp^2 A + \frac{2\pi\beta_0}{in_0^2}\left(1+\frac{\mathrm{i}}{\omega_0}\partial_\tau\right)^2 B$$
$$\approx -\frac{1}{2\mathrm{i}\beta_0}\left[\partial_\xi^2 + \left(\mathrm{i}\frac{\alpha_0}{2} + \hat{D}\right)^2\right]A\,.$$

This corresponds to (4) in [97Bra2]. The terms on the right-hand side of this equation can be neglected when compared with those on the left if the following requirements are fulfilled:

$$\partial_\xi \ll \beta_0 \tag{2.121}$$

and

$$\mathrm{i}\frac{\alpha_0}{2} + \hat{D} \ll \beta_0\,. \tag{2.122}$$

Note that the latter requirement is always fulfilled when a Taylor expansion is used. In this case, (2.121) simplifies to

$$\partial_\xi A = -\frac{\alpha_0}{2}A + \mathrm{i}\hat{D}A + \frac{\mathrm{i}}{2\beta_0}\left(1+\frac{\mathrm{i}}{\omega_0}\partial_\tau\right)^{-1}\nabla_\perp^2 A + \mathrm{i}\frac{2\pi\beta_0}{n_0^2}\left(1+\frac{\mathrm{i}}{\omega_0}\partial_\tau\right)B\,. \tag{2.123}$$

This nonlinear envelope equation corresponds exactly to (6) in [97Bra2].

From the solution $A(r_\perp, \xi, \tau)$ of (2.123), the electric field can be reconstructed as

$$E(r_\perp, \xi, \tau) = A\mathrm{e}^{-\mathrm{i}\omega_0\tau + \mathrm{i}\psi(\xi)} + \text{c.c.}\,, \quad \psi(\xi) = \psi_0 + (\beta_0 - \omega_0\beta_1)\,\xi\,. \tag{2.124}$$

These equations describe the evolution of light wave packets in terms of a fixed carrier frequency ω_0 defined at the entrance to the propagation medium $\xi = 0$, and an evolving complex envelope $A(r_\perp, \xi, \tau)$ and phase $\psi(\xi)$, which determines the "position" of the carrier wave relative to the envelope. The complex envelope $A(r_\perp, \xi, \tau)$ evolves due to absorption (or gain), dispersion, diffraction, and nonlinearities, whereas $\psi(\xi)$ evolves due to a difference between the group velocity β_1^{-1} and the phase velocity in the propagation medium.

The mathematical condition

$$\left|\partial_\xi E\right| \ll \beta_0\,|E| \tag{2.125}$$

is referred to as the slowly-evolving-wave approximation. This approximation requires more from the propagation medium than the slowly-varying-envelope approximation: not only the envelope A, but also the relative carrier phase ψ must not significantly vary as the pulse covers a distance equal to the wavelength $\lambda_0 = 2\pi c/\omega_0$. In return, it does not explicitly impose a limitation on the pulse width. Therefore, in the framework of the slowly-evolving-wave approximation, the nonlinear envelope equation accurately describes light pulse propagation down to the single cycle regime.

3 Dispersion Compensation

3.1 Introduction and Motivation

Generating the shortest laser pulses is an important goal in research and development of ultrafast laser oscillators. This fast-moving frontier with ever shorter pulse durations has led to new and even surprising breakthroughs in science and technology and has been a crucial contributor to several Nobel prizes in the past, in femtosecond chemistry, frequency metrology, and chirped pulse amplification. Those who first got access to the cutting-edge laser technology have very often been able to achieve groundbreaking results. In the visible and near-infrared regime, we can obtain pulses of about 5 fs directly from a passively modelocked oscillator. Even shorter pulses in the range of 3 fs can be achieved by external pulse compression. Such ultrashort pulses have a very broad spectral bandwidth and experience strong dispersive broadening during propagation.

In this case, the dispersive pulse broadening can be estimated as follows taking into account only second order dispersion (see Chap. 2):

$$\frac{\mathrm{d}^2\phi}{\mathrm{d}\omega^2} \gg \tau_\mathrm{p}^2(0) \implies \tau_\mathrm{p}\left(L_\mathrm{d}\right) \approx \frac{\mathrm{d}^2\phi}{\mathrm{d}\omega^2}\Delta\omega_\mathrm{p}\ . \tag{3.1}$$

Based on the discussion in Chap. 2, we can make a quick estimate of what happens when a 5 fs pulse propagates through a dispersive material. A bandwidth-limited Gaussian pulse with a pulse duration of 5 fs at a center wavelength of 800 nm has a spectral frequency bandwidth of

$$\Delta\nu_\mathrm{p}\tau_\mathrm{p} = 0.4413 \implies \Delta\omega_\mathrm{p} = 0.554\ \mathrm{fs}^{-1}\ . \tag{3.2}$$

© The Author(s), under exclusive licence to Springer Nature Switzerland AG 2021
U. Keller, *Ultrafast Lasers*, Graduate Texts in Physics,
https://doi.org/10.1007/978-3-030-82532-4_3

For example, according to Table 2.5, the dispersion in fused quartz is

$$\left.\frac{\partial^2 k_n}{\partial \omega^2}\right|_{800\,\text{nm}} = 3.61 \times 10^{-26}\frac{\text{s}^2}{\text{m}} = 36.1\frac{\text{fs}^2}{\text{mm}} \,. \tag{3.3}$$

Therefore, just 1 mm of fused quartz broadens the 5 fs pulses to about 20 fs. Even propagation in air becomes an issue that needs to be taken into account. Therefore, we need to compensate for this pulse broadening.

We have seen, that dispersive materials always broaden transform-limited pulses. Due to the fact that this process is linear, one can subsequently revoke this pulse broadening again by propagating through a "material" that compensates for the corresponding dispersion: the frequency components that have been shifted in time need to be shifted back by the right time delay. But we have seen in Chap. 2 that materials in the transparent region have positive dispersion. If we want to avoid big losses, we have to find other ways to obtain negative dispersion. This is possible with "geometric structures". Some use angular dispersion produced in refraction or diffraction to make certain frequency components propagate along a longer optical path length, such as grating and prism pairs. The condition is that, at the end, we need to obtain a spatially coherent beam. Other geometric structures use interference effects that exhibit strong wavelength dependence such as a Gires–Tournois interferometer (GTI) and dispersive mirrors. Examples will be discussed in this chapter.

For correct compensation, we also need to know the dispersion of different materials very well. The refractive index n of transparent substances of high density typically satisfies a Sellmeier equation with properly chosen coefficients (see also Sect. 1.4.2 for an introduction to Sellmeier equations):

$$n^2 - 1 = \sum_k \frac{B_k \lambda^2}{\lambda^2 - C_k} \,. \tag{3.4}$$

The derivatives of the refractive index are then given by

$$\frac{\mathrm{d}n}{\mathrm{d}\lambda} = -\frac{\lambda}{n}\sum_k \frac{B_k C_k}{\left(\lambda^2 - C_k\right)^2}\,, \qquad \frac{\mathrm{d}^2 n}{\mathrm{d}\lambda^2} = \frac{1}{n}\left[\sum_k \frac{B_k C_k\left(3\lambda^2 + C_k\right)}{\left(\lambda^2 - C_k\right)^3} - \left(\frac{\mathrm{d}n}{\mathrm{d}\lambda}\right)^2\right], \tag{3.5}$$

and

$$\frac{\mathrm{d}^3 n}{\mathrm{d}\lambda^3} = -\frac{1}{n}\left[12\lambda \sum_k \frac{B_k C_k\left(\lambda^2 + C_k\right)}{\left(\lambda^2 - C_k\right)^4} + 3\frac{\mathrm{d}n}{\mathrm{d}\lambda}\frac{\mathrm{d}^2 n}{\mathrm{d}\lambda^2}\right]. \tag{3.6}$$

We usually speak of positive (negative) dispersion when GDD or, likewise, $\mathrm{d}^2 n/\mathrm{d}\omega^2$ or $\mathrm{d}^2 n/\mathrm{d}\lambda^2$ are positive (negative). Within the wavelength range from 600 nm to 1.1 μm, glass, sapphire and other transparent media like quartz, CaF_2, or BaF_2 show positive dispersion. Signals at shorter wavelengths have a lower group velocity (red is faster than blue), and originally unchirped pulses acquire a positive chirp during propagation, i.e., they become "up-chirped". By an up-chirp or down-chirp we refer to the changes in the frequency components as a function of time. An up-chirp then means that $\mathrm{d}\omega/\mathrm{d}t > 0$.

According to Chap. 2, we obtain for an initially transform-limited Gaussian pulse the following chirp after second order dispersion (2.38):

$$\omega(t) \equiv \frac{\mathrm{d}\phi_{\mathrm{tot}}(t)}{\mathrm{d}t} = \omega_0 + 2\Gamma_2 t\ , \tag{3.7}$$

where (2.70)

$$\frac{1}{\Gamma(L_\mathrm{d})} = \frac{1}{\Gamma(0)} + 2\mathrm{i}k_n'' L_\mathrm{d}\ , \qquad \Gamma \equiv \Gamma_1 - \mathrm{i}\Gamma_2\ , \tag{3.8}$$

and therefore

$$\Gamma(L_\mathrm{d}) = \frac{1}{\dfrac{1}{\Gamma(0)} + 2\mathrm{i}k_n'' L_\mathrm{d}} = \frac{\dfrac{1}{\Gamma(0)} - 2\mathrm{i}k_n'' L_\mathrm{d}}{\left[\dfrac{1}{\Gamma(0)}\right]^2 + \left(2k_n'' L_\mathrm{d}\right)^2} \Longrightarrow \Gamma_2 \propto k_n'' L_\mathrm{d}\ . \tag{3.9}$$

Therefore, for positive dispersion, we have $k_n'' > 0$ and, using (3.9), $\Gamma_2 > 0$. Hence, this is an up-chirp with $\mathrm{d}\omega/\mathrm{d}t > 0$, according to (3.7).

Table 3.1 lists Sellmeier coefficients for some materials used in this work. Figure 3.1 shows refractive indices calculated using these coefficients. Figure 3.2

Table 3.1 Sellmeier coefficients given by (3.4) for selected materials, as reported by different commercial vendors. The last column gives the refractive index n at 800 nm. BK7, FK54, and SF10 are Schott glasses. The last two rows represent exemplary fits to the refractive index of ion-beam sputtered amorphous silicon dioxide (SiO_2) and titanium dioxide (TiO_2), but depend on the deposition parameters. The dispersion of dry air [99Cid] can be fitted with two absorption lines: $B_1 = 0.00048673$, $C_1 = (65\ \mathrm{nm})^2$, $B_2 = 0.000058583$, and $C_2 = (132\ \mathrm{nm})^2$

Material	B_1	$\sqrt{C_1}$ (nm)	B_2	$\sqrt{C_2}$ (nm)	B_3	$\sqrt{C_3}$ (μm)	n(800 nm)
Sapphire (e)	1.504	74	0.5507	122	6.5927	20.072	1.752
Sapphire (o)	1.4314	73	0.6505	119	5.3414	18.028	1.760
Fused silica	0.9540	68	0.1505	166	1.2777	11.673	1.453
BK7	1.0396	77	0.2318	141	1.0105	10.176	1.511
BaF_2	0.6336	58	0.5068	110	3.8261	46.386	1.467
CaF_2	0.5676	51	0.4711	100	3.8485	34.649	1.431
FK54	0.7453	60	0.3023	111	1.0189	14.595	1.433
SF10	1.6163	113	0.2592	241	1.0776	10.799	1.711
SiO_2	1.2176	90	0	0	0.8035	9.598	1.492
TiO_2	3.5193	171	0	0	2.2799	10.954	2.162

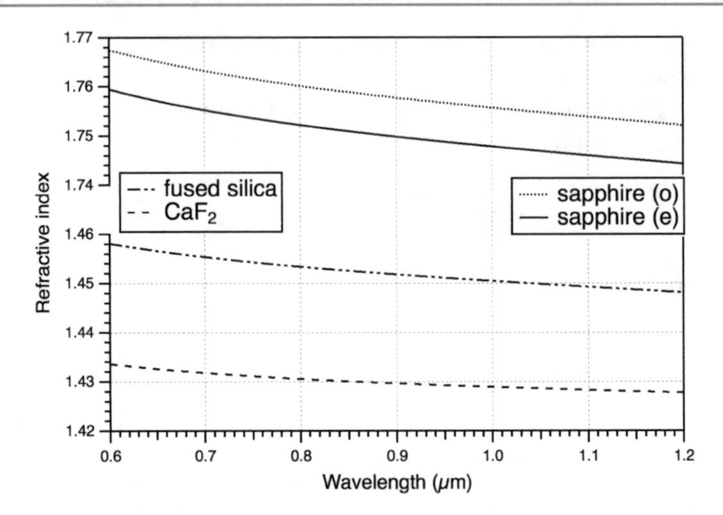

Fig. 3.1 Refractive index for different materials calculated using the Sellmeier coefficients from Table 3.1. The laser beam propagates perpendicularly to the optical axis as either the ordinary (o) or the extraordinary (e) wave

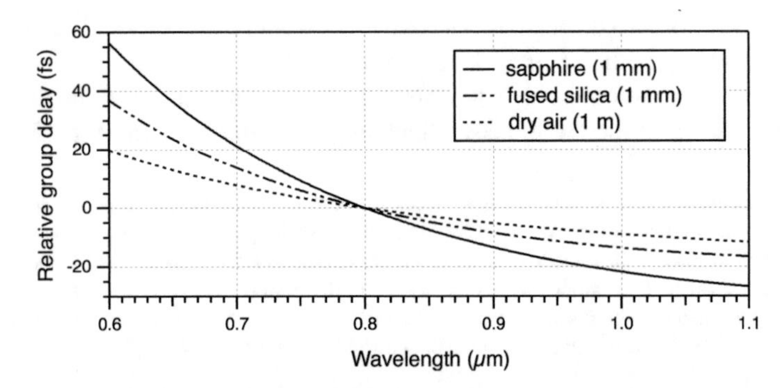

Fig. 3.2 Relative group delay with respect to the absolute group delay at 0.8 μm, which is about 5.9 ps for both the ordinary and the extraordinary beam in sapphire and 4.9 ps for fused silica for a material thickness of 1 mm, and about 3.3 ns for 1 m of air, calculated using the Sellmeier coefficients from Table 3.1

displays the group delay, and Fig. 3.3 the group delay dispersion (GDD) for propagation in sapphire, fused silica, and air. The importance of proper intracavity dispersion compensation is illustrated by the following consideration. For a roundtrip inside a Ti:sapphire laser resonator, to be described later, the laser pulse passes twice through a thin Ti:sapphire laser crystal with a roundtrip physical thickness $L = 5.2$ mm. For a single cavity roundtrip without negative dispersion compensation a broadband pulse with a pulse duration of 5 fs would be strongly broadened, because for example the short wavelength components at 700 nm would lag behind those at 1 μm by an accumulated relative group delay of 220 fs (see Fig. 3.2).

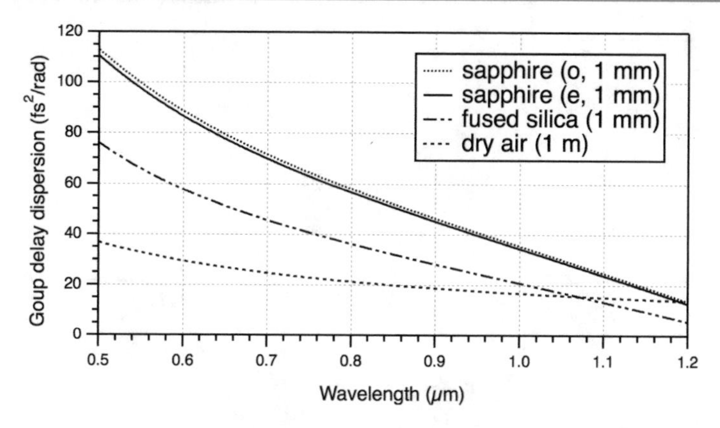

Fig. 3.3 Group delay dispersion per physical path length for 1 mm of sapphire (ordinary and extraordinary beam), or fused silica, and for 1 m of air, calculated using the Sellmeier coefficients from Table 3.1

Figure 3.3 shows a comparison of the GDD of 1 mm of sapphire or fused silica with 1 m of air. Though air is much less dispersive, a few meters in air will lead to a dispersion comparable to the dispersion of a laser crystal. This needs to be considered in the dispersion compensation for few-cycle laser oscillators.

3.2 Prism Compressor

3.2.1 Second Order Dispersion of the Four-Prism Compressor

Especially for wavelengths in the visible region or in the near infrared, most materials show positive (normal) dispersion. To compensate for this, we need sources of negative (anomalous) dispersion. Prism or grating pair configurations offer one possibility to achieve this. Here, one takes advantage of the so-called "geometric dispersion" or "angular dispersion" of such configurations: the wavelength-dependent refractive index in Snell's law results in an angle-dependent refraction and therefore an angle dispersion of the refracted light. Therefore, different frequency components travel along different geometric paths. The resulting frequency dependence of the optical path length can give rise to negative dispersion.

For dispersion compensation with prisms, one usually does not use a single prism, but a prism pair (see Fig. 3.4). After the first prism, frequency components with different wavelengths travel in different directions: the short wavelength part (i.e., the "blue" part) is refracted more strongly than the long wavelength part (i.e., the "red" part). After the second prism, these beams travel parallel again, but are spatially separated in the transverse direction. This spatial separation can be reversed either by retro-reflecting the beams or by inserting a second prism pair (see Fig. 3.4). In both cases, the resulting path length differences double and so does the dispersion. After a double path through the prism pair, we can obtain negative dispersion, i.e., "blue is faster than red".

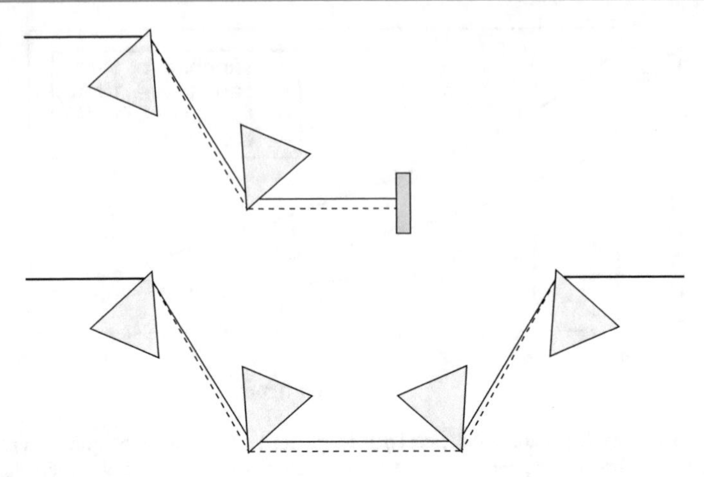

Fig. 3.4 *Top*: Prism pair with wavelength-dependent beam paths assuming normal positive dispersion. The *solid line* corresponds to the long (red) wavelength λ_2 and the *dashed line* to the short (blue) wavelength λ_1 part of the pulse spectrum (i.e., $\lambda_2 > \lambda_1$). Through additional retro-reflection on a mirror, the beams travel back along the same path. The incident and reflected beams from this prism pair arrangement can be spatially separated when the beam in the retro-reflector is also shifted up in space. The double path through the prism pair doubles the net effect of geometric dispersion. *Bottom*: Equivalent to a double pass through a prism pair is a four-prism sequence. This is required to obtain a spatially coherent beam profile

The prisms are cut in such a way that the angle of incidence is the Brewster angle θ_B for the center wavelength of the laser pulse. Therefore the reflection losses for p-polarized light are reduced to a minimum. This also results in negligible loss for any wavelength slightly different from the center wavelength within the spectral bandwidth of the ultrashort pulses. In addition, the apex angle α is chosen for a symmetric beam path with equal incidence angle θ_1 and exit angle θ_2 (see Fig. 3.5). A symmetric beam path with equal incidence and exit angles always results in a minimum deviation angle. The prism rotation required for a symmetric path can therefore be aligned easily.

The total GDD of the prism configuration can be continuously adjusted by moving the prisms without spatially shifting the beam (see Fig. 3.5). In the bottom image of Fig. 3.5 this would amount to shifting the first or second prism upwards or downwards. By moving a prism into the beam, positive dispersion is continuously added because the beam has to propagate through a longer stretch of prism material with positive dispersion. We will show, that the negative dispersion part is determined by the distance between the two prisms. In the following we will neglect the dispersion in air between the prisms.

The exit angle after the prism depends on the wavelength due to the dispersion of the prism material. According to Snell's law and the notation introduced in the top-left image of Fig. 3.5, we obtain

$$n(\lambda) \sin \theta_1' = \sin \theta_1 , \tag{3.10}$$

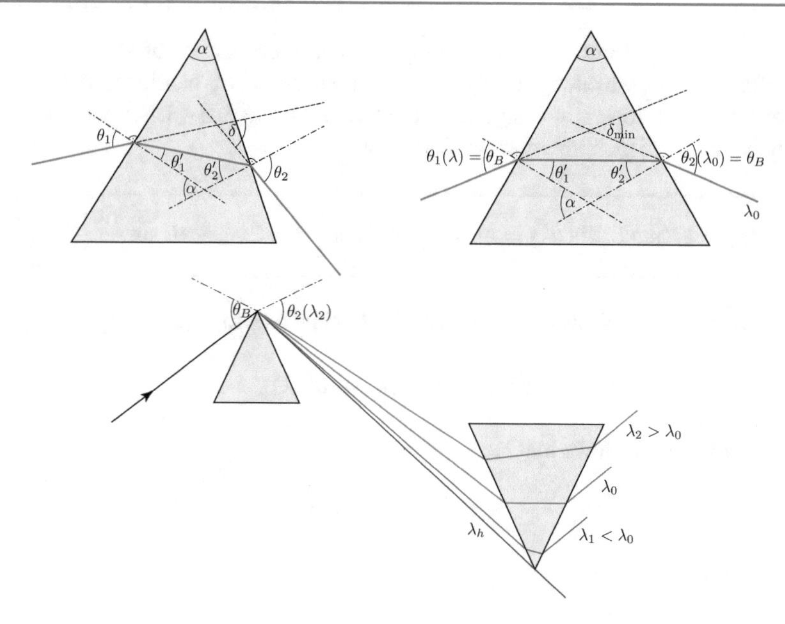

Fig. 3.5 Beam path through (*top left*) an arbitrary prism, where δ is the deviation angle and (*top right*) a prism with a minimal deviation angle δ_{min}, for which the entrance and exit angles are equal for a center wavelength λ_0. In addition for minimal loss the entrance angle is the Brewster angle (3.13). *Bottom*: Angular dispersion $\theta_2(\lambda)$ after the first prism according to (3.15). Here we assume $\lambda_1 < \lambda_0 < \lambda_2$. The horizon wavelength λ_{h} is the shortest wavelength that still passes through both prisms

$$n(\lambda) \sin \theta_2' = \sin \theta_2 \ , \tag{3.11}$$

where n is the refractive index of the prism material. For an arbitrary prism (see top images in Fig. 3.5), we also have

$$\alpha = 180^\circ - \left(90^\circ - \theta_1'\right) - \left(90^\circ - \theta_2'\right) = \theta_1' + \theta_2' \ . \tag{3.12}$$

We can reduce the reflection losses through the prisms if we choose the entrance angle as the Brewster angle for the center wavelength λ_0 of the pulse with p-polarization with a reflectivity $R_{\mathrm{p}}\left(\theta_{\mathrm{B}}\right) = 0$ and a prism arrangement for a minimal deviation angle (see top-right image of Fig. 3.5). For a minimal deviation angle δ_{min} the beam path through the prism is symmetric with the same entrance and exit angles i.e., $\theta_1(\lambda_0) = \theta_{\mathrm{B}} = \theta_2(\lambda_0)$, where θ_{B} corresponds to the Brewster angle at the center wavelength λ_0:

$$\tan \theta_{\mathrm{B}} = n(\lambda_0) \ , \qquad \theta_{\mathrm{B}} + \theta_{\mathrm{B}}' = 90^\circ \ . \tag{3.13}$$

Therefore we have minimal losses after a double path through a prism pair. At the minimum deviation angle configuration with $\theta_1' = \theta_2'$ (see Fig. 3.5 top right), we then obtain, using (3.12) and (3.13),

$$\alpha = 2(90^\circ - \theta_{\mathrm{B}}) = 180^\circ - 2\theta_{\mathrm{B}} = \pi - 2\theta_{\mathrm{B}} \ . \tag{3.14}$$

The entrance angle into the prism compressor is independent of the wavelength. Therefore for all wavelengths, we have a fixed angle of incidence $\theta_1(\lambda) = \theta_B$. According to the top-right image of Fig. 3.5 and with (3.10)–(3.14), we obtain the following angle dispersion after the first prism:

$$\boxed{\theta_2(\lambda) = \arcsin\left(n \sin\theta_2'\right) = \arcsin\left[n(\lambda)\sin\left(\pi - 2\theta_B - \arcsin\frac{\sin\theta_B}{n(\lambda)}\right)\right]} . \tag{3.15}$$

The proof of (3.15) follows directly from Snell's law (see top-right image of Fig. 3.5)

$$\sin\theta_2(\lambda) = n(\lambda)\sin\theta_2'(\lambda) ,$$

and with (3.12) it follows that

$$\theta_2'(\lambda) = \alpha - \theta_1'(\lambda) ,$$

and therefore

$$\theta_2(\lambda) = \arcsin\left[n(\lambda)\sin\theta_2'(\lambda)\right] = \arcsin\left[n(\lambda)\sin\left(\alpha - \theta_1'(\lambda)\right)\right] .$$

Using Snell's law for the first interface we obtain

$$\sin\theta_B = n(\lambda)\sin\theta_1'(\lambda) ,$$

given that $\theta_1(\lambda) = \theta_B$ for all wavelengths and therefore

$$\theta_2(\lambda) = \arcsin\left[n(\lambda)\sin\left(\alpha - \arcsin\left(\frac{\sin\theta_B}{n(\lambda)}\right)\right)\right] .$$

Equation (3.14) then implies (3.15).

In principle, one can now calculate the frequency-dependent optical path length in the prism pair by simply adding up all path lengths as a function of the wavelength. However, the necessary geometric calculations get rather involved if one does not proceed cleverly. A very elegant calculation stems from Fork [84For]. He relies on the fact that the wave fronts (points with the same phase) in optical isotropic materials are always perpendicular to the propagation direction.

First, we consider the simple special case in which the beam travels exactly over the apex of the two prisms for a certain reference wavelength. In this case we have pure geometric dispersion, and can initially neglect possible material dispersion in both prisms. This case is called the horizon of the prism configuration, because at the same time this defines the shortest wavelength, also referred to as the horizon wavelength λ_h, that can actually travel through the prism sequence (see bottom image of Fig. 3.5).

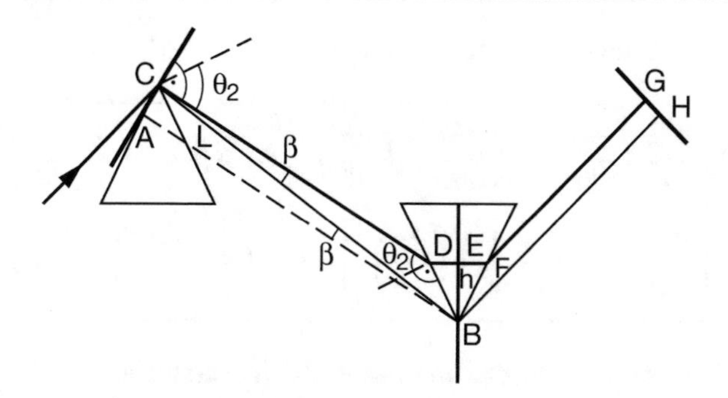

Fig. 3.6 Wavelength-dependent path length through a prism pair, aligned according to minimal deviation as shown in the top-right panel of Fig. 3.5. The distance $L = \overline{CB}$ is the distance between the two prism apexes and is defined by the position of the two prisms. The wavelength-dependent optical path $\overline{CDEG}_{\text{optical}}$ is defined by the angle dispersion $\theta_2(\lambda)$ (3.15), which also determines the angle β

The distance $\overline{CB}$ (see Fig. 3.6) is defined by the positions of the two prisms and is the reference line for the angle β. Thereby, the angle β defines the beam for a longer wavelength than the horizon wavelength, which propagates through both prisms. Three phase fronts for this beam are given by the direction of the distances $\overline{AC}$, $\overline{BE}$, and $\overline{GH}$. The crucial trick now is not to painstakingly calculate the optical path length $\overline{CDE}\big|_{\text{opt}}$, but to make use of the fact that, due to the phase fronts, this corresponds to the path $\overline{AB}$. By analogy, the optical path length $\overline{EFG}\big|_{\text{opt}}$ corresponds to the path $\overline{BH}$. Thus, on this section, no wavelength-dependent phase difference appears. Therefore, we only have to consider the first section. Thus, after a double pass through the prism pair, we find that the total wavelength-dependent optical path length is simply twice the value of $\overline{CDE}\big|_{\text{opt}}$ resp. $\overline{AB}$:

$$P = 2\,\overline{CDE}\big|_{\text{opt}} = 2\overline{AB} = 2L\cos\beta\ , \tag{3.16}$$

where $L = \overline{CB}$ is the distance between the prism apexes. The corresponding phase shift is given by

$$\phi_{\text{P}} = kP(\lambda) = \frac{2\pi}{\lambda}P(\lambda)\ . \tag{3.17}$$

Because we neglect dispersion in air and now the path length is dependent on the wavelength, we can use Table 2.3 to obtain the result for the second order dispersion of the four-prism compressor:

$$\boxed{\frac{\mathrm{d}^2\phi_{\text{P}}}{\mathrm{d}\omega^2} = \frac{\lambda^3}{2\pi c^2}\frac{\mathrm{d}^2P}{\mathrm{d}\lambda^2}}\ . \tag{3.18}$$

As we will show below,

$$\boxed{\begin{aligned}\frac{\mathrm{d}^2P}{\mathrm{d}\lambda^2} &= 2\left[\frac{\partial^2 n}{\partial\lambda^2}\left(\frac{\partial\theta_2}{\partial n}\right)+\left(\frac{\partial^2\theta_2}{\partial n^2}\right)\left(\frac{\partial n}{\partial\lambda}\right)^2\right]L\sin\beta-2\left(\frac{\partial\theta_2}{\partial n}\frac{\partial n}{\partial\lambda}\right)^2 L\cos\beta \\ &\approx 4\left[\frac{\mathrm{d}^2 n}{\mathrm{d}\lambda^2}+\left(2n-\frac{1}{n^3}\right)\left(\frac{\mathrm{d}n}{\mathrm{d}\lambda}\right)^2\right]L\sin\beta-8\left(\frac{\mathrm{d}n}{\mathrm{d}\lambda}\right)^2 L\cos\beta\end{aligned}} \tag{3.19}$$

In this equation we assume that the beam hits the first prism at the apex. Most of the time, the prism compressor is operated for small angles β, and we obtain approximately

$$\beta = 0 \implies \frac{\mathrm{d}^2P}{\mathrm{d}\lambda^2} = -2\left(\frac{\partial\theta_2}{\partial n}\frac{\partial n}{\partial\lambda}\right)^2 L \approx -8\left(\frac{\mathrm{d}n}{\mathrm{d}\lambda}\right)^2 L < 0\,. \tag{3.20}$$

We thus obtain negative dispersion, and it is in fact independent of the sign of the material dispersion. The amount is proportional to the distance between the prisms and depends on the material dispersion squared (see Fig.3.7).

Derivation of (3.19)
The second derivative of the optical path length P (3.16) with $P = P\big(\beta\,(n(\lambda))\big)$ is given by

$$\frac{\mathrm{d}P}{\mathrm{d}\lambda} = \frac{\partial P}{\partial\beta}\frac{\partial\beta}{\partial n}\frac{\partial n}{\partial\lambda} \tag{3.21}$$

and

$$\frac{\mathrm{d}^2P}{\mathrm{d}\lambda^2} = \frac{\partial P}{\partial\beta}\left[\frac{\partial^2 n}{\partial\lambda^2}\frac{\partial\beta}{\partial n}+\frac{\partial^2\beta}{\partial n^2}\left(\frac{\partial n}{\partial\lambda}\right)^2\right]+\left(\frac{\partial\beta}{\partial n}\frac{\partial n}{\partial\lambda}\right)^2\frac{\mathrm{d}^2P}{\mathrm{d}\beta^2}\,. \tag{3.22}$$

Equation (3.21) follows directly from the chain rule:

$$\frac{\mathrm{d}}{\mathrm{d}\lambda}P\big(\beta\,(n(\lambda))\big) = \frac{\partial P\big(\beta\,(n(\lambda))\big)}{\partial\beta}\frac{\partial\beta\,(n(\lambda))}{\partial n}\frac{\partial n(\lambda)}{\partial\lambda}\,.$$

To obtain (3.22), we use the product rule:

$$\frac{\mathrm{d}^2P}{\mathrm{d}\lambda^2} = \frac{\mathrm{d}}{\mathrm{d}\lambda}\left[\frac{\partial P}{\partial\beta}\left(\frac{\partial\beta}{\partial n}\frac{\partial n}{\partial\lambda}\right)\right].$$

Hence,

$$\frac{\mathrm{d}^2P}{\mathrm{d}\lambda^2} = \frac{\partial P}{\partial\beta}\frac{\mathrm{d}}{\mathrm{d}\lambda}\left(\frac{\partial\beta}{\partial n}\frac{\partial n}{\partial\lambda}\right)+\left(\frac{\mathrm{d}}{\mathrm{d}\lambda}\frac{\partial P}{\partial\beta}\right)\left(\frac{\partial\beta}{\partial n}\frac{\partial n}{\partial\lambda}\right).$$

Using the product rule again, we obtain

$$\frac{\mathrm{d}^2P}{\mathrm{d}\lambda^2} = \frac{\partial P}{\partial\beta}\left[\left(\frac{\mathrm{d}}{\mathrm{d}\lambda}\frac{\partial n}{\partial\lambda}\right)\frac{\partial\beta}{\partial n}+\left(\frac{\mathrm{d}}{\mathrm{d}\lambda}\frac{\partial\beta}{\partial n}\right)\frac{\partial n}{\partial\lambda}\right]+\left(\frac{\mathrm{d}}{\mathrm{d}\lambda}\frac{\partial P}{\partial\beta}\right)\left(\frac{\partial\beta}{\partial n}\frac{\partial n}{\partial\lambda}\right).$$

The chain rule then implies

$$\frac{\mathrm{d}^2 P}{\mathrm{d}\lambda^2} = \frac{\partial P}{\partial \beta}\left(\frac{\partial^2 n}{\partial \lambda^2}\frac{\partial \beta}{\partial n} + \frac{\partial^2 \beta}{\partial n^2}\frac{\partial n}{\partial \lambda}\frac{\partial n}{\partial \lambda}\right) + \frac{\partial^2 P}{\partial \beta^2}\frac{\partial \beta}{\partial n}\frac{\partial n}{\partial \lambda}\left(\frac{\partial \beta}{\partial n}\frac{\partial n}{\partial \lambda}\right) .$$

Equation (3.16) gives

$$\frac{\partial P}{\partial \beta} = -2L \sin\beta , \qquad \frac{\mathrm{d}^2 P}{\mathrm{d}\beta^2} = -2L\cos\beta . \tag{3.23}$$

According to Fig. 3.6 we keep the distance between the two prism apexes constant, i.e., $L = \overline{CB} = \text{constant}$. An increase in $\theta_2(\lambda)$ gives a decrease in β. We then obtain

$$\frac{\partial \beta}{\partial n} = -\frac{\partial \theta_2}{\partial n} , \qquad \frac{\partial^2 \beta}{\partial n^2} = -\frac{\partial^2 \theta_2}{\partial n^2} , \tag{3.24}$$

where $\theta_2(n)$ is given by (3.15). However, it is easier to calculate the first and second derivatives in (3.24) using (3.10)–(3.12). The entrance angle into the prism pair θ_1 is fixed for all wavelengths and prism materials, whence

$$\frac{\partial}{\partial n}\sin\theta_1 = 0 .$$

Equation (3.10) then implies

$$0 = \frac{\partial}{\partial n}(n \sin\theta_1') = \sin\theta_1' + n\cos\theta_1' \frac{\partial \theta_1'}{\partial n} \implies \tan\theta_1' = n\frac{\partial \theta_1'}{\partial n} .$$

Equation (3.12) implies

$$\alpha = \theta_1' + \theta_2' \implies \frac{\partial \theta_2'}{\partial n} = \frac{\partial}{\partial n}(\alpha - \theta_1') = -\frac{\partial \theta_1'}{\partial n} ,$$

and using (3.11), it follows that

$$\sin\theta_2 = n\sin\theta_2' \implies \cos\theta_2 \frac{\partial \theta_2}{\partial n} = \sin\theta_2' + n\cos\theta_2'\frac{\partial \theta_2'}{\partial n} .$$

Combining all equations, this then results in

$$\cos\theta_2 \frac{\partial \theta_2}{\partial n} = \sin\theta_2' + \tan\theta_1' \cos\theta_2' ,$$

and therefore

$$\frac{\partial \theta_2}{\partial n} = \frac{1}{\cos\theta_2}\left(\sin\theta_2' + \tan\theta_1'\cos\theta_2'\right) , \tag{3.25}$$

and

$$\frac{\partial^2\theta_2}{\partial^2 n} = \tan\theta_2\left(\frac{\partial\theta_2}{\partial n}\right)^2 - \frac{\tan^2\theta_1'}{n}\left(\frac{\partial\theta_2}{\partial n}\right) . \tag{3.26}$$

If we assume an approximately symmetric beam path for all wavelengths, i.e., $\theta_1'(\lambda) \approx \theta_2'(\lambda)$ and Brewster angle condition $\tan\theta_2 = n$, the derivatives in (3.25) and (3.26) reduce to

$$\frac{\partial\beta}{\partial n} = -\frac{\partial\theta_2}{\partial n} \approx -2 , \qquad \frac{\partial^2\beta}{\partial n^2} = -\frac{\partial^2\theta_2}{\partial n^2} \approx -4n + \frac{2}{n^3} . \tag{3.27}$$

This approximation is satisfied very well, but for ultrashort pulse generation below 20 fs, the exact solution should be used [the first line in (3.19)].

Note. Up to now the calculation was valid for the case in which the reference beam travels through the apex of the prisms. In practice, of course, all beams have to maintain a certain distance to the apex to minimize diffraction loss. In fact, h in Fig. 3.6 has to be at least 1.5 times the spectrally broadened beam radius, so that roughly 99% of the power in the Gaussian beam gets through the prism (see the following note). A Gaussian beam propagating in the z-direction can be approximated by a parabolic wave in the paraxial approximation. Then its intensity is given by

$$I(\rho, z) = I_0\left[\frac{w_0}{w(z)}\right]^2 \exp\left[-\frac{2\rho^2}{w^2(z)}\right] ,$$

where $\rho^2 = x^2 + y^2$, w_0 is the beam waist, and $w(z)$ is the beam radius. Its power is given by

$$P = \int_0^\infty I(\rho, z)\, 2\pi\rho \mathrm{d}\rho = \frac{1}{2}\pi w_0^2 I_0 \implies I_0 = \frac{2P}{\pi w_0^2} .$$

This yields

$$I(\rho, z) = \frac{2P}{\pi w^2(z)} \exp\left[-\frac{2\rho^2}{w^2(z)}\right] .$$

The ratio of the power within a circular area with radius ρ_0 to the total power is given by

$$\frac{P(\rho \le \rho_0)}{P} = \frac{1}{P}\int_0^{\rho_0} I(\rho, z)\, 2\pi\rho \mathrm{d}\rho = 1 - \exp\left[-\frac{2\rho_0^2}{w^2(z)}\right] ,$$

which implies

$$\frac{P\big(\rho \le 1.5w(z)\big)}{P} \approx 99\% ,$$

as claimed above.

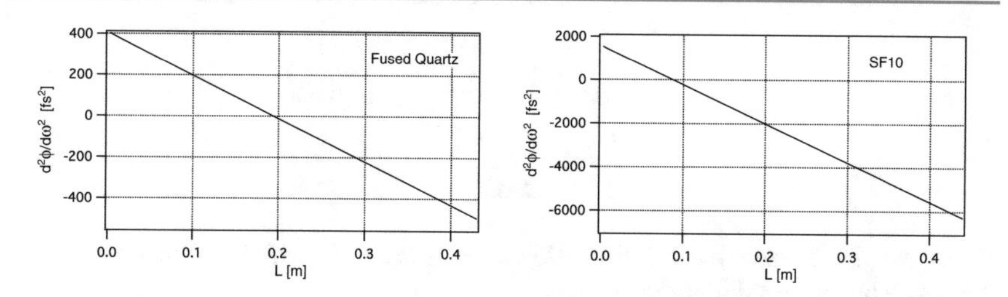

Fig. 3.7 Second order dispersion of a four-prism compressor given by (3.18) and (3.19) with fused quartz and N-SF10 glass prisms as a function of the prism separation L for a center wavelength of 800 nm and $h = 4$ mm

3.2.2 Third Order Dispersion of the Four-Prism Compressor

For the third order dispersion of the four-prism compressor, derived below, we obtain

$$\boxed{\frac{\mathrm{d}^3\phi_\mathrm{P}}{\mathrm{d}\omega^3} = \frac{-\lambda^4}{4\pi^2c^3}\left(3\frac{\mathrm{d}^2P}{\mathrm{d}\lambda^2} + \lambda\frac{\mathrm{d}^3P}{\mathrm{d}\lambda^3}\right)}\,, \tag{3.28}$$

and by analogy with the derivation discussed above,

$$\boxed{\frac{\mathrm{d}^3P}{\mathrm{d}\lambda^3} \approx 4\frac{\mathrm{d}^3n}{\mathrm{d}\lambda^3}L\sin\beta - 24\frac{\mathrm{d}n}{\mathrm{d}\lambda}\frac{\mathrm{d}^2n}{\mathrm{d}\lambda^2}L\cos\beta}\,. \tag{3.29}$$

In (3.29), we once again make the approximation of a symmetric beam path for all wavelengths, as in (3.27) (see Fig. 3.8).

Derivation of (3.28) and (3.29)
We start with the first derivative of $\phi_\mathrm{P} = (2\pi/\lambda)P(\lambda)$:

$$\begin{aligned}\frac{\mathrm{d}\phi_\mathrm{P}}{\mathrm{d}\omega} &= \frac{\mathrm{d}\lambda}{\mathrm{d}\omega}\frac{\mathrm{d}}{\mathrm{d}\lambda}\phi_\mathrm{P}\\ &= -\frac{2\pi c}{\omega^2}\frac{\mathrm{d}}{\mathrm{d}\lambda}\left(\frac{2\pi}{\lambda}P\right)\end{aligned}$$

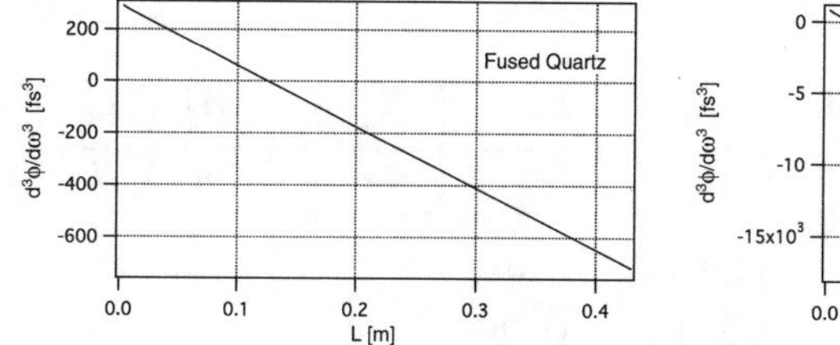

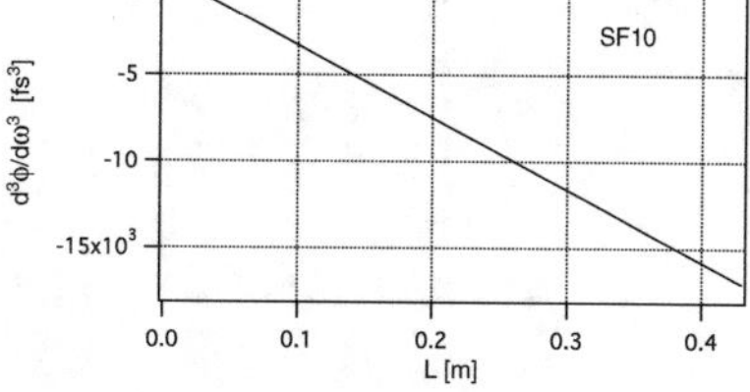

Fig. 3.8 Third order dispersion of a four-prism compressor given by (3.28) and (3.29) with fused quartz and N-SF10-glass prisms as a function of the prism separation L for a center wavelength of 800 nm and $h = 4$ mm

$$= -\frac{\lambda^2}{2\pi c}\left(-\frac{2\pi}{\lambda^2}P + \frac{2\pi}{\lambda}\frac{\mathrm{d}P}{\mathrm{d}\lambda}\right)$$
$$= \frac{1}{c}P - \frac{\lambda}{c}\frac{\mathrm{d}P}{\mathrm{d}\lambda},$$

where we have used the relation $\mathrm{d}\lambda/\mathrm{d}\omega = -2\pi c/\omega^2$, which follows from $\lambda = 2\pi c/\omega$. The second derivative is then

$$\frac{\mathrm{d}^2\phi_\mathrm{P}}{\mathrm{d}\omega^2} = \frac{\mathrm{d}}{\mathrm{d}\omega}\left(\frac{\mathrm{d}\phi_\mathrm{P}}{\mathrm{d}\omega}\right)$$
$$= \frac{\mathrm{d}\lambda}{\mathrm{d}\omega}\frac{\mathrm{d}}{\mathrm{d}\lambda}\left(\frac{1}{c}P - \frac{\lambda}{c}\frac{\mathrm{d}P}{\mathrm{d}\lambda}\right)$$
$$= -\frac{\lambda^2}{2\pi c}\left[\frac{1}{c}\frac{\mathrm{d}P}{\mathrm{d}\lambda} - \left(\frac{1}{c}\frac{\mathrm{d}P}{\mathrm{d}\lambda} + \frac{\lambda}{c}\frac{\mathrm{d}^2P}{\mathrm{d}\lambda^2}\right)\right]$$
$$= \frac{\lambda^3}{2\pi c^2}\frac{\mathrm{d}^2P}{\mathrm{d}\lambda^2}.$$

And finally the third derivative is

$$\frac{\mathrm{d}^3\phi_\mathrm{P}}{\mathrm{d}\omega^3} = \frac{\mathrm{d}}{\mathrm{d}\omega}\left(\frac{\mathrm{d}^2\phi_\mathrm{P}}{\mathrm{d}\omega^2}\right)$$
$$= \frac{\mathrm{d}\lambda}{\mathrm{d}\omega}\frac{\mathrm{d}}{\mathrm{d}\lambda}\left(\frac{\lambda^3}{2\pi c^2}\frac{\mathrm{d}^2P}{\mathrm{d}\lambda^2}\right)$$
$$= -\frac{\lambda^2}{2\pi c}\left(\frac{3\lambda^2}{2\pi c^2}\frac{\mathrm{d}^2P}{\mathrm{d}\lambda^2} + \frac{\lambda^3}{2\pi c^2}\frac{\mathrm{d}^3P}{\mathrm{d}\lambda^3}\right)$$
$$= -\frac{\lambda^4}{4\pi^2c^3}\left(3\frac{\mathrm{d}^2P}{\mathrm{d}\lambda^2} + \lambda\frac{\mathrm{d}^3P}{\mathrm{d}\lambda^3}\right),$$

which is precisely (3.28).

Starting from (3.22), the third derivative of the optical path length P with respect to λ is given by

$$\frac{\mathrm{d}^3P}{\mathrm{d}\lambda^3} = \frac{\mathrm{d}}{\mathrm{d}\lambda}\frac{\partial P}{\partial\beta}\left[\frac{\partial^2 n}{\partial\lambda^2}\frac{\partial\beta}{\partial n} + \frac{\partial^2\beta}{\partial n^2}\left(\frac{\partial n}{\partial\lambda}\right)^2\right] + \frac{\partial P}{\partial\beta}\frac{\mathrm{d}}{\mathrm{d}\lambda}\left[\frac{\partial^2 n}{\partial\lambda^2}\frac{\partial\beta}{\partial n} + \frac{\partial^2\beta}{\partial n^2}\left(\frac{\partial n}{\partial\lambda}\right)^2\right]$$
$$+\frac{\mathrm{d}}{\mathrm{d}\lambda}\left(\frac{\partial\beta}{\partial n}\frac{\partial n}{\partial\lambda}\right)^2\frac{\partial^2P}{\partial\beta^2} + \left(\frac{\partial\beta}{\partial n}\frac{\partial n}{\partial\lambda}\right)^2\frac{\mathrm{d}}{\mathrm{d}\lambda}\frac{\partial^2P}{\partial\beta^2}.$$

This yields

$$\begin{aligned}\frac{\mathrm{d}^3 P}{\mathrm{d}\lambda^3} &= \frac{\partial^2 P}{\partial \beta^2}\frac{\partial \beta}{\partial n}\frac{\partial n}{\partial \lambda}\left[\frac{\partial^2 n}{\partial \lambda^2}\frac{\partial \beta}{\partial n} + \frac{\partial^2 \beta}{\partial n^2}\left(\frac{\partial n}{\partial \lambda}\right)^2\right] \\ &+\frac{\partial P}{\partial \beta}\left[\frac{\partial^3 n}{\partial \lambda^3}\frac{\partial \beta}{\partial n} + \frac{\partial^2 n}{\partial \lambda^2}\frac{\partial^2 \beta}{\partial n^2}\frac{\partial n}{\partial \lambda} + \frac{\partial^3 \beta}{\partial n^3}\left(\frac{\partial n}{\partial \lambda}\right)^2 + \frac{\partial^2 \beta}{\partial n^2}2\frac{\partial n}{\partial \lambda}\frac{\partial^2 n}{\partial \lambda^2}\right] \\ &+2\left(\frac{\partial \beta}{\partial n}\frac{\partial n}{\partial \lambda}\right)\left(\frac{\partial^2 \beta}{\partial n^2}\frac{\partial n}{\partial \lambda}\frac{\partial n}{\partial \lambda} + \frac{\partial \beta}{\partial n}\frac{\partial^2 n}{\partial \lambda^2}\right)\frac{\partial^2 P}{\partial \beta^2} \\ &+\left(\frac{\partial \beta}{\partial n}\frac{\partial n}{\partial \lambda}\right)^2\frac{\partial^3 P}{\partial \beta^3}\frac{\partial \beta}{\partial n}\frac{\partial n}{\partial \lambda}\,.\end{aligned}$$

Then, finally,

$$\begin{aligned}\frac{\mathrm{d}^3 P}{\mathrm{d}\lambda^3} &= \frac{\partial P}{\partial \beta}\left[\frac{\partial^3 n}{\partial \lambda^3}\frac{\partial \beta}{\partial n} + 3\frac{\partial^2 n}{\partial \lambda^2}\frac{\partial^2 \beta}{\partial n^2}\frac{\partial n}{\partial \lambda} + \frac{\partial^3 \beta}{\partial n^3}\left(\frac{\partial n}{\partial \lambda}\right)^2\right] \\ &+3\frac{\partial^2 P}{\partial \beta^2}\left(\frac{\partial \beta}{\partial n}\frac{\partial n}{\partial \lambda}\right)\left[\frac{\partial^2 n}{\partial \lambda^2}\frac{\partial \beta}{\partial n} + \frac{\partial^2 \beta}{\partial n^2}\left(\frac{\partial n}{\partial \lambda}\right)^2\right] + \left(\frac{\partial \beta}{\partial n}\frac{\partial n}{\partial \lambda}\right)^3\frac{\partial^3 P}{\partial \beta^3}\,.\end{aligned}$$

We now make use of (3.23)–(3.27), inserting them into the previous equation:

$$\begin{aligned}\frac{\mathrm{d}^3 P}{\mathrm{d}\lambda^3} \approx -2L\sin\beta\Bigg[&-\frac{\partial^3 n}{\partial \lambda^3}2 + 3\frac{\partial^2 n}{\partial \lambda^2}\frac{\partial n}{\partial \lambda}\left(-4n + \frac{2}{n^3}\right) \\ &-3\frac{\partial^2 n}{\partial \lambda^2}\left(\frac{\partial n}{\partial \lambda}\right)^{-1}\left(-4n + \frac{2}{n^3}\right)\left(\frac{\partial n}{\partial \lambda}\right)^2\Bigg] \\ -3\times 2L\cos\beta&\left(-2\frac{\partial n}{\partial \lambda}\right)\left[-\frac{\partial^2 n}{\partial \lambda^2}2 + \left(-4n + \frac{2}{n^3}\right)\left(\frac{\partial n}{\partial \lambda}\right)^2\right] \\ +\left(-2\frac{\partial n}{\partial \lambda}\right)^3&2L\sin\beta\,.\end{aligned}$$

This equation simplifies to

$$\begin{aligned}\frac{\mathrm{d}^3 P}{\mathrm{d}\lambda^3} &\approx 4L\sin\beta\frac{\partial^3 n}{\partial \lambda^3} - 6L\sin\beta\frac{\partial^2 n}{\partial \lambda^2}\frac{\partial n}{\partial \lambda}\left(-4n + \frac{2}{n^3}\right) \\ &+6L\sin\beta\frac{\partial^2 n}{\partial \lambda^2}\frac{\partial n}{\partial \lambda}\left(-4n + \frac{2}{n^3}\right) - 24L\cos\beta\frac{\partial n}{\partial \lambda}\frac{\partial^2 n}{\partial \lambda^2} \\ &-12L\cos\beta\left(-4n + \frac{2}{n^3}\right)\left(\frac{\partial n}{\partial \lambda}\right)^3 - 16L\sin\beta\left(\frac{\partial n}{\partial \lambda}\right)^3\,.\end{aligned}$$

Since we can neglect terms in $(\partial n/\partial \lambda)^3$, we finally obtain the desired (3.29):

$$\frac{\mathrm{d}^3 P}{\mathrm{d}\lambda^3} \approx 4\frac{\mathrm{d}^3 n}{\mathrm{d}\lambda^3} L \sin\beta - 24\frac{\mathrm{d}n}{\mathrm{d}\lambda}\frac{\mathrm{d}^2 n}{\mathrm{d}\lambda^2} L \cos\beta ,$$

as required.

3.2.3 Continuous Adjustment of Dispersion

The big advantage of prism compressors is that the dispersion can be adjusted continuously, and in fact without introducing a spatial offset of the beam after the prism pair. To do this, the prism is moved along the x-axis by a distance Δx away from the apex, which determines the beam path through the prism, as shown in Fig. 3.9. From Fig. 3.6 with a prism configuration at minimum deviation we can obtain a bigger "insertion" into the second prism (i.e., a larger distance $h = \overline{BE}$) without changing the angles, when we translate the second prism perpendicular to the line $\overline{DE}$. Therefore one can add a certain amount of additional positive material dispersion, and this can be adjusted via this prism insertion (see derivation below):

$$\boxed{h(\Delta x) = 2\Delta x \sin^2\theta_\mathrm{B}} . \tag{3.30}$$

This also applies to the first prism. But in doing this, the distance between the prisms is also changed. However, this can usually be neglected, because the prism separation is typically greater than 10 cm, and the prisms are only moved by millimeters.

In contrast to the angle β, the prism insertion h is much more easily accessible to experiment with (3.30). Therefore, it is useful to determine the angle β in terms of h. From Fig. 3.6, it follows that

$$\cos\frac{\beta}{2} = \frac{h}{s} , \tag{3.31}$$

where s is the distance $\overline{DB}$ in Fig. 3.6. In addition we can use the law of sines in Fig. 3.9, as follows:

$$\frac{L}{\sin(\theta_2 + \pi/2)} = \frac{L}{\cos\theta_2} = \frac{s}{\sin\beta} \Longrightarrow s = L\frac{\sin\beta}{\cos\theta_2} . \tag{3.32}$$

Using (3.31) and (3.32), we can eliminate the unkown distance s to obtain

$$\boxed{\sin\beta = \frac{h}{L}\frac{\cos\theta_2}{\cos(\alpha/2)}} . \tag{3.33}$$

Thus, using (3.30) and (3.33) the second order dispersion, i.e., (3.18) and (3.19), and the third order dispersion, i.e., (3.28) and (3.29), can be determined as a function of Δx.

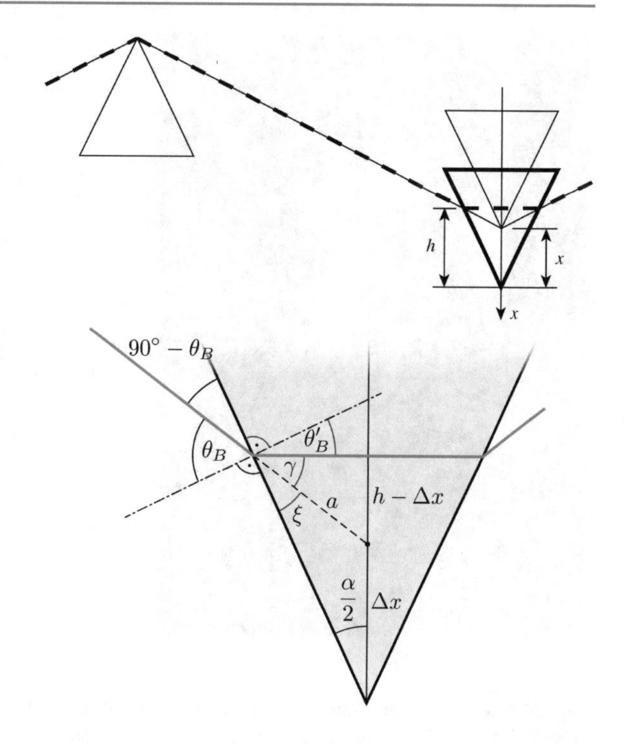

Fig. 3.9 Shifting the prisms along the x-axis changes the amount of positive dispersion. Thus the final dispersion can be continuously adjusted. With this arrangement, the beam direction after the prism is not affected. The *bottom figure* is a zoom to define the notation used for the proof of (3.30)

Derivation of (3.30)

We use the notation given in Fig. 3.9 and it follows that $\xi = 90° - \theta_B$ and from (3.14) it follows that

$$\frac{\alpha}{2} = 90° - \theta_B = \xi \implies a = \Delta x \; .$$

From Fig. 3.9, it follows that γ is given by $90° = \theta'_B + \gamma + \xi$, and with

$$\xi = 90° - \theta_B \; ,$$

it follows that

$$\gamma = 2\theta_B - 90° \; .$$

From Fig. 3.9, we then have

$$\sin\gamma = \frac{h - \Delta x}{a} \; .$$

With some trigonometric identities, it follows that

$$\sin\gamma = -\cos(2\theta_B) = -1 + 2\sin^2\theta_B \; ,$$

and therefore with $a = \Delta x$,

$$-\Delta x + 2\Delta x \sin^2\theta_B = h - \Delta x \; ,$$

which then results in (3.30).

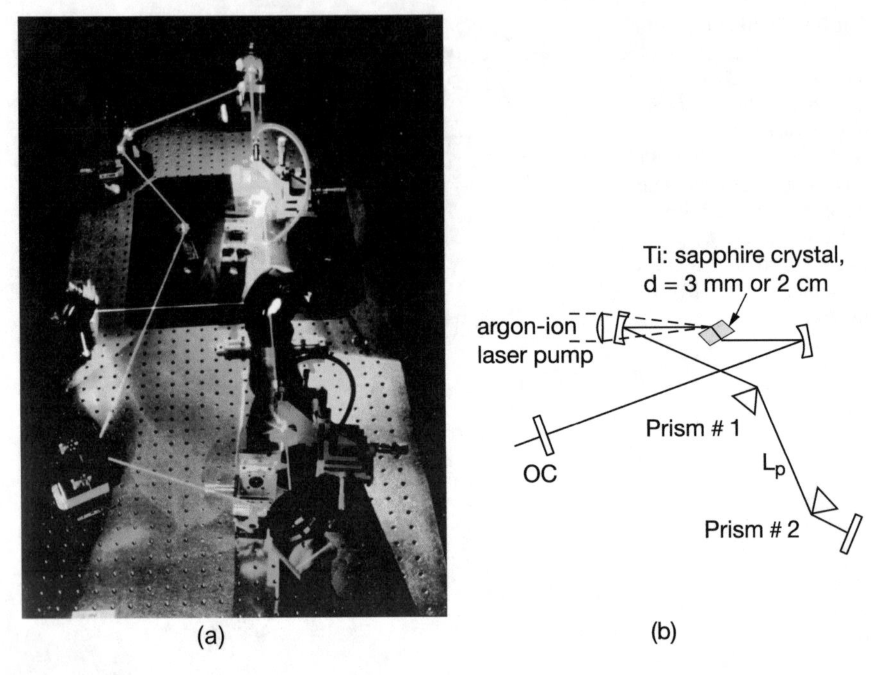

Fig. 3.10 **a** Dispersion compensation in a CPM dye laser. **b** Dispersion compensation in a Ti:sapphire laser using a prism pair. OC is the output coupler

In ultrafast lasers, prism compressors are often used for dispersion compensation. In a ring cavity, a four-prism compressor can be used as shown in Fig. 3.10a for a colliding pulse modelocked (CPM) dye laser (see Chap. 9 for more details). In a linear cavity, a two prism compressor is typically used, and if possible the output coupler is chosen at the opposite end of the linear cavity. Otherwise an additional prism pair sequence has to be used to obtain a spatially coherent beam. Figure 3.10b shows such a laser, where prisms are used to compensate for the positive dispersion in the Ti:sapphire laser crystal. To minimize the influence of third order dispersion at a center wavelength of 800 nm, one typically uses fused quartz as the prism material. For pulse generation of pulses longer than 20 to 30 fs, one typically uses SF10-glass, because this glass has a stronger dispersion and the prism separation can be chosen smaller. This results in a more compact laser cavity. We will see later, that in principle one always works with a small negative second order dispersion to satisfy the condition of a stable soliton pulse.

One can usually only compensate for one order of dispersion, e.g., the second order. For the generation of very short pulses, this limits the possible pulse length, because the uncompensated higher order dispersions contribute dominantly to pulse broadening. Figure 3.11 shows the GDD for a fused-silica Brewster pair at 50 cm apex separation with a different second prism insertion h (as shown in Fig. 3.9). It is assumed that the prisms are cut and the incidence angle is chosen according to the Brewster condition for the design wavelength $\lambda_B = 800$ nm. The location of the apexes is determined for an infinitesimally thin ray of a specific horizon wavelength

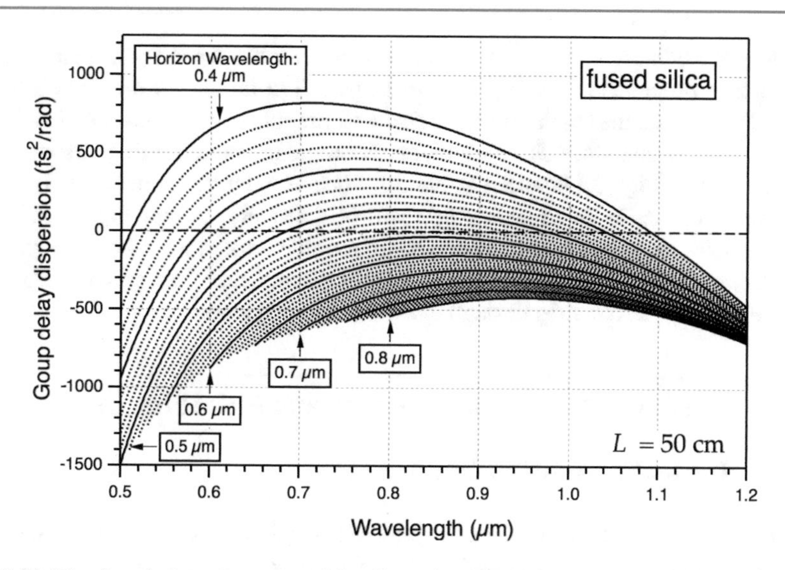

Fig. 3.11 Wavelength-dependent group delay dispersion (GDD) for a single pass through a fused silica prism pair for different horizon wavelengths [00Sut]. The apex angle α is cut for Brewster's incidence at 0.8 μm according to (3.13) and (3.14). The same entrance angle, determined by Brewster's condition for 0.8 μm, is assumed for all wavelengths. The GDD scales with the apex separation L (3.16)–(3.18), which is chosen to be $L = 50$ cm for this calculation [00Sut2]

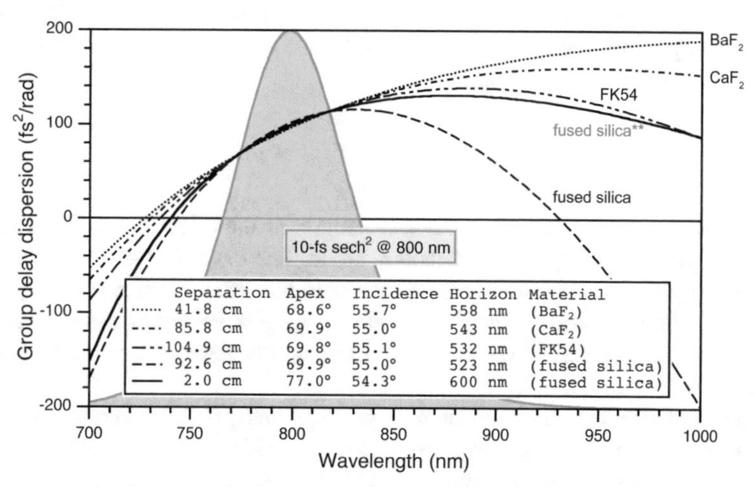

** non-Brewster geometry: enhanced dispersion, smaller apex separation (2 cm instead of ≈90 cm), but more sensitive to misalignment

Fig. 3.12 Dispersion for Brewster prism pairs of different materials (BaF_2, CaF_2, FK54, and fused silica) and for a fused silica prism pair with larger apex angle (*solid line*). The prism separation and horizon wavelengths have been fitted for minimum squared deviation using the spectrum shown as a weight function. Due to enhanced angular dispersion, the non-Brewster geometry gives similar dispersion at a greatly reduced apex separation, but it is also much more sensitive to misalignment [00Sut2,00Sut]

λ_h. Any wavelength $\lambda < \lambda_h$ will not be refracted by the second prism and will be lost during propagation through the prism pair. The larger the insertion into the second prism, i.e., the larger h in Fig. 3.9, the shorter the horizon wavelength λh. But the trade-off is that, with shorter horizon wavelengths, more positive dispersion is introduced (see Figs. 3.11 and 3.12). Note that beams of larger diameters will already be clipped for $\lambda > \lambda_h$ and this must normally be taken into account. Figures 3.11 and 3.12 show the wavelength-dependent GDD for a single pass through a prism pair for different materials. The dispersion doubles for a symmetric four-prism compressor or, equivalently, a roundtrip through the two-prism sequence.

3.3 Grating Compressor, Stretcher, and Pulse Shaper

3.3.1 Diffraction Grating Compressor

Diffraction gratings also introduce an angular dispersion. Figure 3.13 shows the simplest grating configuration, where long-wavelength components travel a greater distance than short-wavelength components thus introducing negative GDD. This setup is normally used as a compressor. The dispersion is easily tuned by changing the distance between the two gratings.

An optical pulse compressor with a diffraction grating pair was developed very early [69Tre,84Tom]. Treacy's publication [69Tre] has become a classic. For dispersion compensation in a diffraction grating pair, we can benefit from the angular dispersion in diffraction from gratings:

$$\nu_{x,m} = m\frac{1}{\Lambda} \iff \theta_{x,m} \approx m\frac{\lambda}{\Lambda}, \qquad (3.34)$$

where Λ is the grating period, λ the wavelength, m the diffraction order, $\nu_{x,m}$ the spatial frequency, and $\theta_{x,m}$ the diffraction angle. The long-wavelength component, i.e., "red" light, is more strongly diffracted than the short-wavelength component, i.e., "blue" light. Therefore, the path length for long wavelengths can be increased in such a way that negative second order dispersion is generated, i.e., blue is faster than red (see Fig. 3.13).

Angular dispersive elements such as gratings and prisms generate an angular dispersion, for which the spectral components travel in different directions. When we propagate a short pulse through such an element, we also introduce a tilt in the pulse front (see Fig. 3.14). Therefore the pulse front rather than the phase front is the contour of constant intensity at the peak of the pulse. Note that the phase front is still perpendicular to the pulse propagation direction. This pulse front tilt is given [99Pre]

$$\tan\alpha_t = \lambda_0 \left.\frac{d\varphi}{d\lambda}\right|_{\lambda_0}. \qquad (3.35)$$

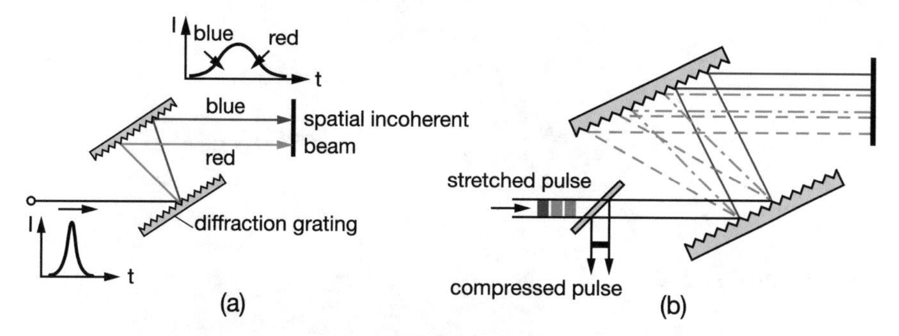

Fig. 3.13 **a** Negative dispersion through a diffraction grating pair. After two gratings the beam is spatially incoherent and has a tilted pulse intensity front and negative GDD (blue is faster than red). **b** Schematic layout of a grating-based compressor with negative dispersion, i.e., the short wavelengths (*in blue*) traverse a shorter distance than longer wavelengths

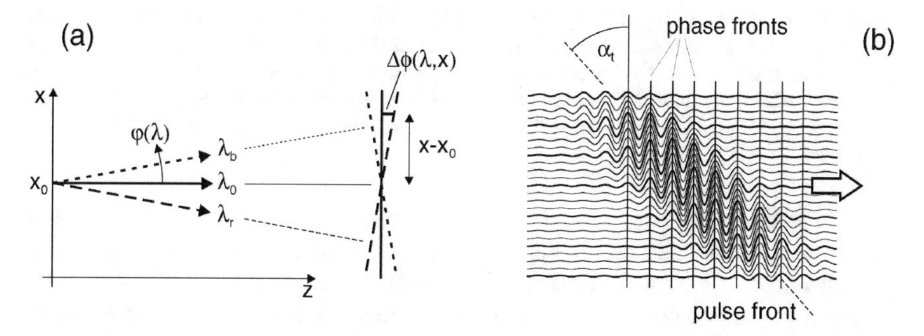

Fig. 3.14 Pulse front tilt: **a** An angular dispersion is introduced because the different spectral components of a laser pulse propagate at an angle of $\varphi(\lambda)$ relative to the pulse propagation direction z. The central (λ_0, *solid line*), a red (λ_r, *dashed line*), and a blue (λ_b, *short dashed line*) components of the pulse are indicated, as well as the (virtual!) phase fronts of these spectral components. The angular chirp leads to a spectral phase shift $\Delta\phi(\lambda, x)$ varying with the x coordinate. **b** Resulting pulse front tilt shown where the phase fronts of the pulse are always perpendicular to the pulse propagation direction. The pulse front (i.e., the plane of the maximum of the pulse envelope), however, is tilted by an angle α_t. According to [99Pre, Fig. 1]

In the Littrow configuration, where the diffracted beam propagates along the same direction as the incident beam, the grating can be designed for a maximum diffraction efficiency into the first order (for horizontal polarization). Thus the grating equation for the first diffraction order simplifies to

$$\sin\varphi = \frac{\lambda}{\Lambda} ,$$

where $\varphi = \theta$ is the diffraction angle as a function of the wavelength λ (3.34), using the notation in [99Pre] and Fig. 3.14. This pulse front tilt can be very useful for group velocity matching in nonlinear frequency conversion. Using such tilted pulse fronts in combination with a noncollinear geometry in sum-frequency generation resulted in new world-record deep-UV generation for example [20Wil].

A grating compressor has an additional position-dependent contribution to the phase. This contribution can be neglected on the first grating, because here the beams

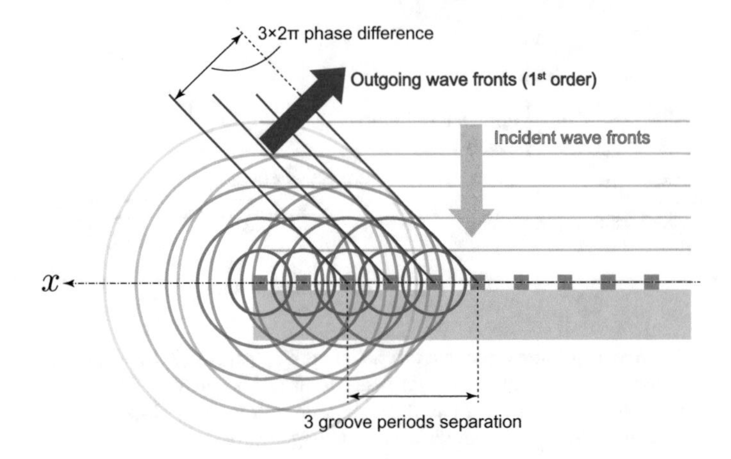

Fig. 3.15 Grating phase, explained with the first-order diffraction on a reflective grating. Here, the incoming wave front is assumed to be perpendicular to the surface. The wave diffracted in first order results from constructive interference of the secondary wave fronts from different grating lines. Each additional grating line introduces an additional phase shift of 2π. Thus, the relative grating phase amounts to $2\pi m$/grating line, where m is the diffraction order in which the grating is operated

are not yet spectrally dispersed, i.e., all frequency components experience the same phase shift, which can thus be neglected regarding GDD. The situation is different at the second grating, because at this point the beams are spectrally dispersed. Different Fourier components of the pulse are now diffracted at different positions of the grating, which results in a position-dependent grating phase and therefore a frequency-dependent phase, which influences GDD.

The easiest way to understand this grating phase is to illustrate the construction of the diffracted wave fronts according to Huygens' principle (see Fig. 3.15). The Huygens principle assumes that each point of an advancing wave front is in fact the center of a fresh disturbance, and the source of a new train of secondary waves that all interfere constructively to form the new wave front. Each point source creates an ideal circular wave front—the secondary wave. In Fig. 3.15, we show the easiest case where the incoming wave fronts arrive parallel to the grating surface: the first grating-order wave front results from constructive interference of all secondary waves coming from each grating line. We obtain constructive interference with a grating line distance that results in a 2π-shifted secondary wave front compared to the neighboring grating lines. This grating phase has to be taken into account when the two-grating configuration is evaluated in the phase picture.

There is one way to avoid this complication. The grating phase has no influence on the group delay in the grating sequence. If one interprets the optical path length in the grating compressor, not as a phase delay, but as a group delay, then the relevant second order dispersion can be much more easily calculated using only the first derivative in ω. Towards the end of this section, we will show that the two approaches result in the same solution for the second order dispersion of the grating compressor.

The incoming beam has an angle of incidence θ_i and is diffracted by the grating with a grating period Λ. The resulting reflected and diffracted beam (of order m) is determined by the angle θ_m as shown in Fig. 3.17 and using (3.34):

$$\sin\theta_m = m\frac{\lambda}{\Lambda} + \sin\theta_i\,, \quad \text{where } m = 0\,,\ \pm 1\,,\ \pm 2\,,\ldots\,. \tag{3.36}$$

Inserting $m = 0$ in (3.36), it follows that $\theta_m = \theta_i$, i.e., (3.36) implies a simple reflection on a plane interface for which θ_m is the same for all wavelengths. We define the diffraction angle θ_m to be negative if it is located on the same side of the z-axis as the angle of incidence θ_i.

For the second order dispersion, it makes no difference at all whether one operates the grating in the positive or negative first order, but in practice the negative first order is almost always used. The reason for this is that it is possible to work with much higher diffraction angles, which permits the use of gratings with larger line numbers. Figure 3.13 shows schematically that by choosing θ_i at 45°, for example, the biggest possible diffraction angle in positive first order is also only 45°. The diffracted beam is then tangential. However, in the back-direction, an angle of 135° is possible before the beam becomes tangential.

Compressor gratings are generally used in a near-Littrow configuration. This means, that the beam is diffracted back into the direction it came from. In this configuration, very efficient gratings can be made, and in general blazed gratings are optimized precisely for this case. To keep the losses in a diffraction grating compressor at a minimum, one typically uses a blazed grating, because in such a grating, the diffraction efficiency into one diffraction order can be optimized. According to the above discussion, one typically chooses a blazed grating for the diffraction order $m = -1$. Low diffraction loss is essential because the light has to be diffracted four times off the diffraction grating to obtain a wavelength-independent beam profile again, i.e., to obtain a spatially coherent beam (see Fig. 3.16). The input and output beams in Fig. 3.16a can be spatially separated using a prism mirror configuration with two mirrors mounted perpendicular to each other to translate the reflected beam into a parallel image plane.

It is also important to take into account the fact that the diffraction efficiency is frequency-dependent. However, at this point we only derive the frequency-dependent phase shift due to a grating pair. This phase shift has two parts: first, a grating phase $\phi_g(x)$ as in (3.38) and second, the geometric dispersion given by (3.39):

1. According to Fig. 3.17, the path of the light through the grating pair is given by

$$L = \overline{PABQ} = \overline{P'A} + b = b\big[1 + \cos(\theta_m + \theta_i)\big] = \frac{L_g}{\cos\theta_m}\big[1 + \cos(\theta_m + \theta_i)\big]\,. \tag{3.37}$$

   where θ_m is the wavelength dependent diffraction angle (3.36).
2. An x- and λ-dependent phase shift $\phi_g(x)$ occurs for diffraction off the second grating in the plane $z = L_g$ (see Fig. 3.17):

$$\phi_g(x) = \pi - m\frac{2\pi}{\Lambda}x\,, \tag{3.38}$$

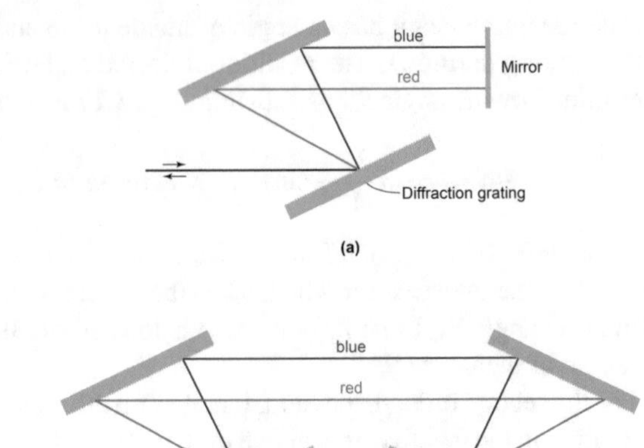

Fig. 3.16 Four-grating compressor. Generation of a spatially coherent beam by four diffractions on a grating. Configuration with **a** a double-pass through a grating pair and **b** a single-pass through two symmetric grating pairs

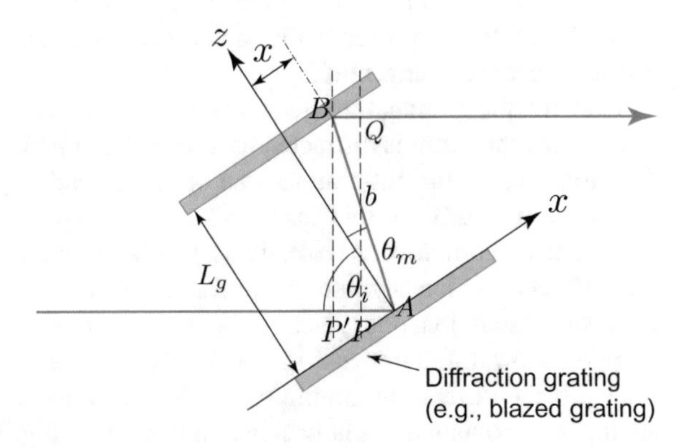

Fig. 3.17 Notation for the derivation of the frequency-dependent phase shift $\phi(\omega)$ of a diffraction grating pair

where x is located in the grating plane according to Fig. 3.17 and is λ-dependent because of the λ-dependent diffraction angle θ_m (3.36).

This two parts contribute to the total wavelength-dependent phase shift at the grating pair:

$$\phi = \frac{\omega}{c}L + \phi_{\mathrm{g}}(x) , \qquad x = L_{\mathrm{g}} \tan\theta_m , \tag{3.39}$$

where we neglected any material dispersion between the prisms, i.e., the group velocity is equal to the speed of light in vacuum c. This is normally justified because these gratings are operated in air or even in vacuum. For the second order dispersion of the four-grating compressor, one then has, with $m = -1$,

$$\boxed{\frac{\mathrm{d}^2\phi}{\mathrm{d}\omega^2} = -\frac{\lambda^3 L_{\mathrm{g}}}{\pi c^2 \Lambda^2}\left[1 - \left(\frac{\lambda}{\Lambda} - \sin\theta_{\mathrm{i}}\right)^2\right]^{-3/2}} . \tag{3.40}$$

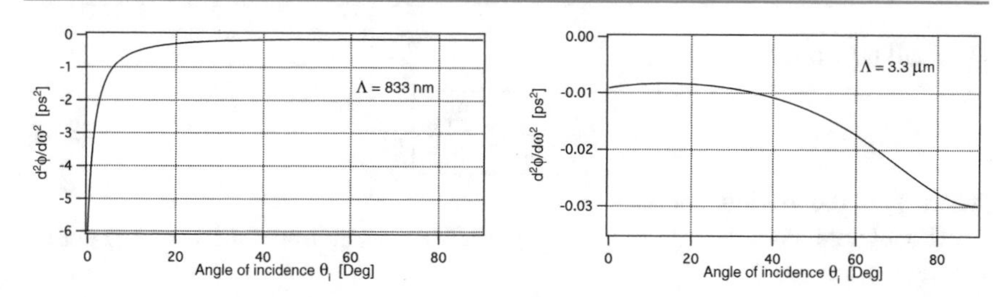

Fig. 3.18 Second order dispersion of a four-grating compressor as a function of the angle of incidence θ_i. Example for identical gratings with **a** 1200 lines/mm, i.e., $\Lambda \approx 833$ nm, and **b** 300 lines/mm, i.e., $\Lambda \approx 3.3$ μm, at a wavelength of 800 nm and a grating distance L_g of 5 cm, according to (3.40)

This is derived at the end of this section. If the pulse has originally been broadened by a dispersive material (GDD > 0), the four-grating compressor can compress the pulse again. The dispersion can be continuously adjusted through the distance L_g between the two parallel gratings as described in (3.40) (see Fig. 3.18):

$$\frac{\mathrm{d}^2\phi}{\mathrm{d}\omega^2} \propto L_\mathrm{g} \,. \tag{3.41}$$

For the third order dispersion of the four-grating compressor, one then has, with $m = -1$,

$$\boxed{\frac{\mathrm{d}^3\phi}{\mathrm{d}\omega^3} = -\frac{\mathrm{d}^2\phi}{\mathrm{d}\omega^2}\frac{3\lambda}{\pi c}\frac{1 + \dfrac{\lambda}{\Lambda}\sin\theta_\mathrm{i} - \sin^2\theta_\mathrm{i}}{1 - \left(\dfrac{\lambda}{\Lambda} - \sin\theta_\mathrm{i}\right)^2}} \,. \tag{3.42}$$

This is derived at the end of the present section. Because of (3.41)

$$\frac{\mathrm{d}^3\phi}{\mathrm{d}\omega^3} \propto L_\mathrm{g} \,. \tag{3.43}$$

Useful relationship between group delay dispersion of optical path length L and grating phase ϕ_g in a grating pair compressor (3.44):
The group delay is defined as

$$T_\mathrm{g} \equiv \frac{L}{c} \,,$$

where L is the optical path length. Thus, the group delay dispersion D_2 (i.e., the second order dispersion) is given by

$$D_2 = \frac{\mathrm{d}T_\mathrm{g}}{\mathrm{d}\omega} = \frac{\mathrm{d}}{\mathrm{d}\omega}\left(\frac{L}{c}\right) = \frac{1}{c}\frac{\mathrm{d}L}{\mathrm{d}\omega} \,.$$

We shall now prove that

$$\boxed{-\frac{\omega}{c}\frac{\mathrm{d}L}{\mathrm{d}\omega} = \frac{\mathrm{d}\phi_\mathrm{g}}{\mathrm{d}\omega}}\,, \tag{3.44}$$

where ϕ_g is the grating phase.

The optical path length is given by (3.37) and, noting that $b = L_\mathrm{g}/\cos\theta_m$ (Fig. 3.17), we see that

$$\frac{\mathrm{d}}{\mathrm{d}\omega}b = \frac{L_\mathrm{g}}{\cos^2\theta_m}\sin\theta_m\frac{\mathrm{d}}{\mathrm{d}\omega}\theta_m = b\tan\theta_m\frac{\mathrm{d}}{\mathrm{d}\omega}\theta_m\,.$$

Thus, we obtain for the derivation of the optical path length:

$$\begin{aligned}\frac{\mathrm{d}}{\mathrm{d}\omega}L &= \frac{\mathrm{d}b}{\mathrm{d}\omega}\big[1+\cos(\theta_m+\theta_\mathrm{i})\big] - b\left[\sin(\theta_m+\theta_\mathrm{i})\frac{\mathrm{d}\theta_m}{\mathrm{d}\omega}\right]\\ &= b\tan\theta_m\frac{\mathrm{d}\theta_m}{\mathrm{d}\omega}\big[1+\cos(\theta_m+\theta_\mathrm{i})\big] - b\sin(\theta_m+\theta_\mathrm{i})\frac{\mathrm{d}\theta_m}{\mathrm{d}\omega}\\ &= b\frac{\mathrm{d}\theta_m}{\mathrm{d}\omega}\Big[\tan\theta_m\big[1+\cos(\theta_m+\theta_\mathrm{i})\big] - \sin(\theta_m+\theta_\mathrm{i})\Big]\,.\end{aligned} \tag{3.45}$$

From the grating (3.36), and using $\lambda = 2\pi c/\omega$, we have

$$\frac{\mathrm{d}}{\mathrm{d}\omega}(\sin\theta_m - \sin\theta_\mathrm{i}) = \frac{\mathrm{d}}{\mathrm{d}\omega}\left(\frac{2\pi mc}{\omega\Lambda}\right)\,,$$

and hence,

$$\frac{\mathrm{d}}{\mathrm{d}\omega}\sin\theta_m = \cos\theta_m\frac{\mathrm{d}\theta_m}{\mathrm{d}\omega} = -\frac{2\pi mc}{\omega^2\Lambda}\,.$$

Finally,

$$\frac{\mathrm{d}\theta_m}{\mathrm{d}\omega} = -\frac{2\pi mc}{\omega^2\Lambda\cos\theta_m}\,. \tag{3.46}$$

Inserting (3.46) into (3.45), and using sum-to-product trigonometric identities yields

$$\begin{aligned}\frac{\mathrm{d}}{\mathrm{d}\omega}L &= -\frac{2\pi mc}{\omega^2\Lambda\cos\theta_m}b\left[\frac{\sin\theta_m}{\cos\theta_m}\big[1+(\cos\theta_m\cos\theta_\mathrm{i} - \sin\theta_m\sin\theta_\mathrm{i})\big]\right.\\ &\qquad\left. -(\sin\theta_\mathrm{i}\cos\theta_m + \sin\theta_m\cos\theta_\mathrm{i})\right]\\ &= -\frac{2\pi mc}{\omega^2\Lambda\cos\theta_m}b\frac{\sin\theta_m - \sin^2\theta_m\sin\theta_\mathrm{i} - \sin\theta_\mathrm{i}\cos^2\theta_m}{\cos\theta_m}\\ &= -\frac{2\pi mc}{\omega^2\Lambda\cos^2\theta_m}b(\sin\theta_m - \sin\theta_\mathrm{i})\\ &= -\frac{2\pi mc}{\omega^2\Lambda\cos^2\theta_m}\frac{L_\mathrm{g}}{\cos\theta_m}\frac{2\pi mc}{\omega\Lambda} = -\frac{(2\pi mc)^2}{\omega^3\Lambda^2\cos^3\theta_m}L_\mathrm{g}\,.\end{aligned} \tag{3.47}$$

The grating phase ϕ_g is given by

$$\phi_g = \pi - 2\pi Nm = \pi - 2\pi \frac{x}{\Lambda} m = \pi - 2\pi m \frac{L_g}{\Lambda} \tan\theta_m \, .$$

Its derivative is therefore

$$\frac{d}{d\omega}\phi_g = -\frac{2\pi m}{\Lambda \cos^2\theta_m} L_g \frac{d\theta_m}{d\omega} \, .$$

Inserting (3.46) yields

$$\frac{d\phi_g}{d\omega} = \frac{1}{c} \frac{(2\pi mc)^2}{\omega^2 \Lambda^2 \cos^3\theta_m} L_g \, . \qquad (3.48)$$

Comparison with (3.47) shows that

$$-\frac{\omega}{c}\frac{dL}{d\omega} = \frac{d\phi_g}{d\omega} \, .$$

This shows that the resulting second order dispersion is the same with both approaches.

Derivation of (3.40) **for the second order dispersion of a four-grating compressor:**
The optical phase ϕ is given by

$$\phi = \frac{\omega}{c} L + \phi_g \, .$$

Its first derivative is

$$\frac{d}{d\omega}\phi = \frac{1}{c} L + \frac{\omega}{c}\frac{dL}{d\omega} + \frac{d\phi_g}{d\omega} \, . \qquad (3.49)$$

We could use (3.44) to simplify the derivation. However, for now we want to continue without this equation (see at the end the alternative derivation). The second derivative is

$$\frac{d^2}{d\omega^2}\phi = \frac{2}{c}\frac{dL}{d\omega} + \frac{\omega}{c}\frac{d^2 L}{d\omega^2} + \frac{d^2\phi_g}{d\omega^2} \, . \qquad (3.50)$$

We first determine the derivatives in this equation. Using (3.47), we have

$$\frac{d^2}{d\omega^2} L = \frac{3\,(2\pi mc)^2 L_g}{\omega^4 \Lambda^2 \cos^3\theta_m} - \frac{3\,(2\pi mc)^2 L_g}{\omega^3 \Lambda^2 \cos^4\theta_m} \sin\theta_m \frac{d\theta_m}{d\omega} \, . \qquad (3.51)$$

Equation (3.48) implies

$$\frac{d^2}{d\omega^2}\phi_g = -\frac{1}{c}\frac{2\,(2\pi mc)^2 L_g}{\omega^3 \Lambda^2 \cos^3\theta_m} + \frac{1}{c}\frac{3\,(2\pi mc)^2 L_g}{\omega^2 \Lambda^2 \cos^4\theta_m} \sin\theta_m \frac{d\theta_m}{d\omega} \, . \qquad (3.52)$$

Inserting (3.47), (3.51), and (3.52) into (3.50) finally yields

$$
\begin{aligned}
\frac{\mathrm{d}^2}{\mathrm{d}\omega^2}\phi &= \frac{2}{c}\left[-\frac{(2\pi mc)^2 L_{\mathrm{g}}}{\omega^3 \Lambda^2 \cos^3\theta_m}\right] + \frac{\omega}{c}\left[\frac{3\,(2\pi mc)^2 L_{\mathrm{g}}}{\omega^4 \Lambda^2 \cos^3\theta_m} - \frac{3\,(2\pi mc)^2 L_{\mathrm{g}}}{\omega^3 \Lambda^2 \cos^4\theta_m}\sin\theta_m \frac{\mathrm{d}\theta_m}{\mathrm{d}\omega}\right] \\
&\quad - \frac{1}{c}\frac{2\,(2\pi mc)^2 L_{\mathrm{g}}}{\omega^3 \Lambda^2 \cos^3\theta_m} + \frac{1}{c}\frac{3\,(2\pi mc)^2 L_{\mathrm{g}}}{\omega^2 \Lambda^2 \cos^4\theta_m}\sin\theta_m \frac{\mathrm{d}\theta_m}{\mathrm{d}\omega} \\
&= -\frac{1}{c}\frac{(2\pi mc)^2 L_{\mathrm{g}}}{\omega^3 \Lambda^2 \cos^3\theta_m} \;. \qquad (3.53)
\end{aligned}
$$

If we now replace ω by λ, using the relation $\omega = 2\pi c/\lambda$, and express θ_m as a function of the angle of incidence θ_{i}, using the grating equation (3.34) and the well known identity $\cos^2\theta_m + \sin^2\theta_m = 1$, so that

$$
\cos\theta_m = \left[1 - \left(-m\frac{\lambda}{\Lambda} - \sin\theta_{\mathrm{i}}\right)^2\right]^{1/2} ,
$$

we obtain

$$
\frac{\mathrm{d}^2\phi}{\mathrm{d}\omega^2} = -\frac{m^2\lambda^3 L_{\mathrm{g}}}{2\pi c^2 \Lambda^2}\left[1 - \left(-m\frac{\lambda}{\Lambda} - \sin\theta_{\mathrm{i}}\right)^2\right]^{-3/2} . \qquad (3.54)
$$

Since this is the result for a single pass through a single grating pair, we have to multiply it by a factor of 2 to obtain the result for a four-grating compressor. We thus arrive at (3.40).

A faster, alternative derivation can be obtained using (3.44) in (3.49):

$$
\frac{\mathrm{d}\phi}{\mathrm{d}\omega} = \frac{L}{c}, \qquad \frac{\mathrm{d}^2\phi}{\mathrm{d}\omega^2} = \frac{1}{c}\frac{\mathrm{d}L}{\mathrm{d}\omega} . \qquad (3.55)
$$

Using (3.47) we then find

$$
\frac{\mathrm{d}^2\phi}{\mathrm{d}\omega^2} = -\frac{1}{c}\frac{(2\pi mc)^2 L_{\mathrm{g}}}{\omega^3 \Lambda^2 \cos^3\theta_m} ,
$$

which is the same as (3.53).

Derivation of (3.42) **for the third order dispersion of a four-grating compressor:** We can use (3.44) to simplify the derivation. Equation (3.55) implies

$$
\frac{\mathrm{d}^3\phi}{\mathrm{d}\omega^3} = \frac{1}{c}\frac{\mathrm{d}^2 L}{\mathrm{d}\omega^2} .
$$

However, for now we want to continue without this simplification and go on with (3.50), which implies

$$\frac{d^3}{d\omega^3}\phi = \frac{3}{c}\frac{d^2 L}{d\omega^2} + \frac{\omega}{c}\frac{d^3 L}{d\omega^3} + \frac{d^3\phi_g}{d\omega^3}\,, \tag{3.56}$$

for the third order dispersion. Inserting (3.46) into (3.51) yields

$$\frac{d^2}{d\omega^2}L = \frac{3\,(2\pi mc)^2\,L_g}{\omega^4\Lambda^2\cos^3\theta_m} + \frac{3\,(2\pi mc)^3\,L_g}{\omega^5\Lambda^3\cos^5\theta_m}\sin\theta_m\,, \tag{3.57}$$

for the second derivative of the optical path length. Hence, for the third derivative

$$\begin{aligned}\frac{d^3}{d\omega^3}L = &-\frac{12\,(2\pi mc)^2\,L_g}{\omega^5\Lambda^2\cos^3\theta_m} + \frac{9\,(2\pi mc)^2\,L_g}{\omega^4\Lambda^2\cos^4\theta_m}\sin\theta_m\frac{d\theta_m}{d\omega} - \frac{15\,(2\pi mc)^3\,L_g}{\omega^6\Lambda^3\cos^5\theta_m}\\ &+\frac{15\,(2\pi mc)^3\,L_g}{\omega^5\Lambda^3\cos^6\theta_m}\sin^2\theta_m\frac{d\theta_m}{d\omega} + \frac{3\,(2\pi mc)^3\,L_g}{\omega^5\Lambda^3\cos^4\theta_m}\frac{d\theta_m}{d\omega}\,.\end{aligned}$$

Inserting (3.46) yields

$$\begin{aligned}\frac{d^3}{d\omega^3}L = &-\frac{12\,(2\pi mc)^2\,L_g}{\omega^5\Lambda^2\cos^3\theta_m} - \frac{24\,(2\pi mc)^3\,L_g}{\omega^6\Lambda^3\cos^5\theta_m}\sin\theta_m\\ &-\frac{15\,(2\pi mc)^4\,L_g}{\omega^7\Lambda^4\cos^7\theta_m}\sin^2\theta_m - \frac{3\,(2\pi mc)^4\,L_g}{\omega^7\Lambda^4\cos^5\theta_m}\,.\end{aligned} \tag{3.58}$$

Inserting (3.46) into (3.52) yields

$$\frac{d^2}{d\omega^2}\phi_g = -\frac{2}{c}\frac{(2\pi mc)^2\,L_g}{\omega^3\Lambda^2\cos^3\theta_m} - \frac{3}{c}\frac{(2\pi mc)^3\,L_g}{\omega^4\Lambda^3\cos^5\theta_m}\sin\theta_m$$

for the second derivative of the grating phase. The third derivative of the grating phase is thus

$$\begin{aligned}\frac{d^3}{d\omega^3}\phi_g = &\frac{6}{c}\frac{(2\pi mc)^2\,L_g}{\omega^4\Lambda^2\cos^3\theta_m} - \frac{6}{c}\frac{(2\pi mc)^2\,L_g}{\omega^3\Lambda^2\cos^4\theta_m}\sin\theta_m\frac{d\theta_m}{d\omega}\\ &-\Bigg[-\frac{12}{c}\frac{(2\pi mc)^3\,L_g}{\omega^5\Lambda^3\cos^5\theta_m}\sin\theta_m + \frac{15}{c}\frac{(2\pi mc)^3\,L_g}{\omega^4\Lambda^3\cos^6\theta_m}\sin^2\theta_m\frac{d\theta_m}{d\omega}\\ &+\frac{3}{c}\frac{(2\pi mc)^3\,L_g}{\omega^4\Lambda^3\cos^4\theta_m}\frac{d\theta_m}{d\omega}\Bigg]\,.\end{aligned}$$

Inserting (3.46) yet again yields

$$\frac{\mathrm{d}^3}{\mathrm{d}\omega^3}\phi_\mathrm{g} = \frac{6}{c}\frac{(2\pi mc)^2 L_\mathrm{g}}{\omega^4 \Lambda^2 \cos^3\theta_m} + \frac{18}{c}\frac{(2\pi mc)^3 L_\mathrm{g}}{\omega^5 \Lambda^3 \cos^5\theta_m}\sin\theta_m$$
$$+\frac{15}{c}\frac{(2\pi mc)^4 L_\mathrm{g}}{\omega^6 \Lambda^4 \cos^7\theta_m}\sin^2\theta_m + \frac{3}{c}\frac{(2\pi mc)^4 L_\mathrm{g}}{\omega^6 \Lambda^4 \cos^5\theta_m}. \tag{3.59}$$

Inserting (3.57)–(3.59) into (3.56) yields

$$\frac{\mathrm{d}^3\phi}{\mathrm{d}\omega^3} = \frac{1}{c}\left[\frac{9\,(2\pi mc)^2 L_\mathrm{g}}{\omega^4 \Lambda^2 \cos^3\theta_m} + \frac{9\,(2\pi mc)^3 L_\mathrm{g}}{\omega^5 \Lambda^3 \cos^5\theta_m}\sin\theta_m\right]$$
$$+\frac{1}{c}\left[-\frac{12\,(2\pi mc)^2 L_\mathrm{g}}{\omega^4 \Lambda^2 \cos^3\theta_m} - \frac{24\,(2\pi mc)^3 L_\mathrm{g}}{\omega^5 \Lambda^3 \cos^5\theta_m}\sin\theta_m\right.$$
$$\left.-\frac{15\,(2\pi mc)^4 L_\mathrm{g}}{\omega^6 \Lambda^4 \cos^7\theta_m}\sin^2\theta_m - \frac{3\,(2\pi mc)^4 L_\mathrm{g}}{\omega^6 \Lambda^4 \cos^5\theta_m}\right]$$
$$+\frac{1}{c}\left[\frac{6\,(2\pi mc)^2 L_\mathrm{g}}{\omega^4 \Lambda^2 \cos^3\theta_m} + \frac{18\,(2\pi mc)^3 L_\mathrm{g}}{\omega^5 \Lambda^3 \cos^5\theta_m}\sin\theta_m\right.$$
$$\left.+\frac{15\,(2\pi mc)^4 L_\mathrm{g}}{\omega^6 \Lambda^4 \cos^7\theta_m}\sin^2\theta_m + \frac{3\,(2\pi mc)^4 L_\mathrm{g}}{\omega^6 \Lambda^4 \cos^5\theta_m}\right]$$

for the third order dispersion. Thus,

$$\frac{\mathrm{d}^3\phi}{\mathrm{d}\omega^3} = \frac{1}{c}\left[\frac{3\,(2\pi mc)^2 L_\mathrm{g}}{\omega^4 \Lambda^2 \cos^3\theta_m} + \frac{3\,(2\pi mc)^3 L_\mathrm{g}}{\omega^5 \Lambda^3 \cos^5\theta_m}\sin\theta_m\right]. \tag{3.60}$$

Using (3.53), we can transform this to

$$\frac{\mathrm{d}^3\phi}{\mathrm{d}\omega^3} = -\frac{\mathrm{d}^2\phi}{\mathrm{d}\omega^2}\frac{3}{\omega}\left[1 + \frac{(2\pi mc)\sin\theta_m}{\omega\Lambda\cos^2\theta_m}\right].$$

Replacing ω by λ then yields

$$\frac{\mathrm{d}^3\phi}{\mathrm{d}\omega^3} = -\frac{\mathrm{d}^2\phi}{\mathrm{d}\omega^2}\frac{3\lambda}{2\pi c}\left(1 + \frac{m\lambda\sin\theta_m}{\Lambda\cos^2\theta_m}\right)$$
$$= -\frac{\mathrm{d}^2\phi}{\mathrm{d}\omega^2}\frac{3\lambda}{2\pi c}\frac{\cos^2\theta_m + m\dfrac{\lambda}{\Lambda}\sin\theta_m}{\cos^2\theta_m}.$$

To express θ_m as a function of θ_i, we again use the grating (3.34), and insert

$$\sin\theta_m = m\frac{\lambda}{\Lambda} + \sin\theta_\mathrm{i}, \qquad \cos\theta_m = \left[1 - \left(-m\frac{\lambda}{\Lambda} - \sin\theta_\mathrm{i}\right)^2\right]^{1/2},$$

into the above equation:

$$\frac{\mathrm{d}^3\phi}{\mathrm{d}\omega^3} = -\frac{\mathrm{d}^2\phi}{\mathrm{d}\omega^2}\frac{3\lambda}{2\pi c}\frac{1-\left(-m\dfrac{\lambda}{\Lambda}-\sin\theta_\mathrm{i}\right)^2+m\dfrac{\lambda}{\Lambda}\left(m\dfrac{\lambda}{\Lambda}+\sin\theta_\mathrm{i}\right)}{1-\left(-m\dfrac{\lambda}{\Lambda}-\sin\theta_\mathrm{i}\right)^2},$$

which leads finally to

$$\frac{\mathrm{d}^3\phi}{\mathrm{d}\omega^3} = -\frac{\mathrm{d}^2\phi}{\mathrm{d}\omega^2}\frac{3\lambda}{2\pi c}\frac{1-m\dfrac{\lambda}{\Lambda}\sin\theta_\mathrm{i}-\sin^2\theta_\mathrm{i}}{1-\left(-m\dfrac{\lambda}{\Lambda}-\sin\theta_\mathrm{i}\right)^2}.$$

This is in perfect agreement with (3.42).

3.3.2 Turning a Grating Compressor into a Stretcher

Figure 3.19 shows a more complicated grating configuration that involves focusing elements, here depicted as lenses. The lenses are placed at a distance $2f$ from each other (they act as a 1:1 telescope), and at a distance L from the gratings. If $L < f$, the setup acts as a positive-dispersion stretcher and if $L > f$, it is a negative-dispersion stretcher. The case $L = f$ is used in the pulse shaper (see Fig. 3.21 and discussion in next section) [92Wei2]. For ultrashort pulses with a broad spectrum, the focusing element is a spherical or cylindrical mirror rather than a lens, in order to avoid introducing additional positive dispersion.

The GDD of a four-grating compressor is given by (3.40) as

$$\frac{\mathrm{d}^2\phi}{\mathrm{d}\omega^2} = -\frac{\lambda^3 L_\mathrm{g}}{\pi c^2\Lambda^2}\left[1-\left(\frac{\lambda}{\Lambda}-\sin\theta_\mathrm{i}\right)^2\right]^{-3/2}.$$

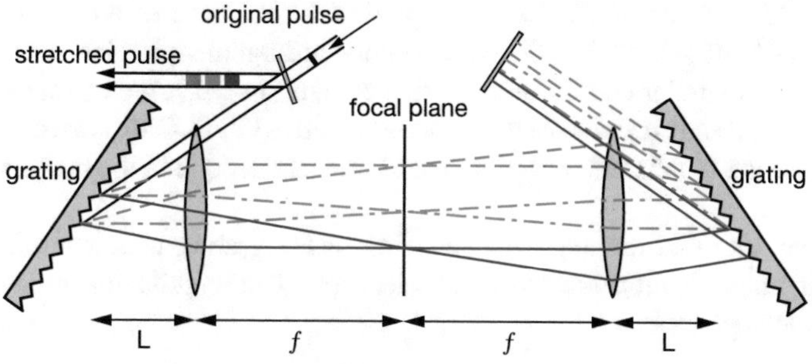

Fig. 3.19 Schematic layout of a grating-based stretcher. In this case, $L < f$, which leads to a positive GDD, i.e., the long wavelengths (*red*) come first. Positive GDD according to (3.63) with $M = 1$

Let us inspect the quantities on the right-hand side of this expression. The wavelength λ is always a positive quantity, hence λ^3 is positive, and the same is obviously true of $\pi c^2 \Lambda^2$. By rewriting (3.36) we found in the previous section that

$$\left[1 - \left(\frac{\lambda}{\Lambda} - \sin\theta_{\mathrm{i}}\right)^2\right]^{1/2} = \cos\theta_m \,, \qquad \theta_m \in [-\pi/2, \pi/2] \,.$$

Because the reflected beam at the grating needs to lie in the incident half-space of the grating (i.e., $\theta_m \in [-\pi/2, \pi/2]$), this part of the equation is also positive. As a result, with a positive grating separation L_{g}, the overall expression is always negative, regardless of the choice of the different parameters.

We will see next how one can achieve a 'negative' grating separation and thus flip the sign of the GDD. This is especially important for chirped-pulse amplification [85Str], where a large amount of adjustable positive dispersion is required to broaden the pulse before amplification to prevent any material damage. Here, we will find out how to obtain a positive GDD from a grating arrangement.

We refer to the paper by Martinez [84Mar2] who derived the GDD of a grating compressor taking into account diffraction and refraction. We only have to do the calculation for a two-grating compressor arrangement because the full GDD of a four-grating compressor is twice the result. Equation (5) of [84Mar2] ($n = n_2$ here) is given by

$$\frac{\mathrm{d}^2 P}{\mathrm{d}\lambda^2} = \left[\frac{\mathrm{d}^2 n}{\mathrm{d}\lambda^2} - n\left(\frac{\mathrm{d}\theta}{\mathrm{d}\lambda}\right)^2\right] L \,.$$

Note that, when looking at Fig. 3.17, L now indicates $\overline{AB}$ rather than $\overline{PABQ}$, as it did in the previous section. For the case of this grating arrangement, the above equation simplifies to

$$\frac{\mathrm{d}^2 P}{\mathrm{d}\lambda^2} = -\left(\frac{\mathrm{d}\theta}{\mathrm{d}\lambda}\right)^2 L \,,$$

because no material dispersion is present in this system based on diffraction in air or vacuum. This then results in (3.40).

We would now like to modify the equation for the case of an arrangement as shown in [84Mar2, Fig. 3.2], where a telescope arrangement is placed between the two gratings. A similar configuration is depicted in Fig. 3.19. We can show that this modified optical setup can be used to produce positive GDD. Only two modifications have to be made to take into account an additional telescope in the beam path:

- First, we note that the angular dispersion of the grating gets magnified by the angular magnification factor of the telescope. The angular magnification of a telescope is given by

$$M = f_1/f_2 \,, \tag{3.61}$$

where f_1 and f_2 are the focal lengths of the first and second lens of the telescope, respectively.

- Second, because the optical path lengths of all optical rays are identical between the two focal planes of a lens, this part of the geometrical path through the grating compressor does not contribute to the overall GDD. This leads to a corrected effective path length through the compressor given by

$$L_{\text{eff}} = L - 2(f_1 + f_2)\,. \tag{3.62}$$

Now using (8) of [84Mar2], we can replace L_g in (3.40) with $L_{\text{eff}} M^2$. With these modifications, we obtain

$$\frac{\mathrm{d}^2\phi}{\mathrm{d}\omega^2} = -\frac{m^2\lambda^3 M^2 L_{\text{eff}}}{2\pi c^2 \Lambda^2}\left[1-\left(-m\frac{\lambda}{\Lambda} - \sin\theta_{\mathrm{i}}\right)^2\right]^{-3/2} \tag{3.63}$$

for the GDD of a two-grating sequence with a telescope. With L_{eff}, we now have a parameter that allows us to flip the sign of the GDD. By choosing the distance L smaller than the distance between the input and output focal planes of the telescope, we obtain a positive GDD from this arrangement. The magnification of the telescope 'amplifies' the dispersion provided by the grating. Figure 3.19 shows an example for positive GDD with a magnification $M = 1$ (i.e., $f_1 = f_2$) and $L < f$.

Two questions arise: how would one choose the magnification factor in a real-world implementation of such a pulse stretcher and where would one put the grating in order to implement an optical pulse shaper? Although this 'magnification' would be a nice feature for building a compact setup, one practical problem is that the second grating would need to be chosen in such a way that, after it, all different spectral components would propagate parallel. Thus, the number of grooves per millimeter of the second grating also has to be corrected by the magnification factor. First, in practice, it would be very difficult to get such a perfectly matched second grating (it is much easier to manufacture two identical gratings). Second, to get a compact setup, a magnification factor greater than 1 would be used, which would mean that the second grating would need more grooves per millimeter than the first. So, because it is much easier to produce two identical gratings, why not also use an identical grating with double the grooves per millimeter as a first grating? This would then directly provide greater angular dispersion in the first place.

Furthermore, at the output of the grating compressor, we want all spectral components to overlap spatially once more. That is why we would use the two-grating setup in double-pass (to get a four-grating compressor). If we get a magnification factor M at the first pass, the magnification on the way back will be $1/M$, thereby completely eliminating the magnification effect from the first pass. As a result, one chooses a magnification factor of $M = 1$ (i.e., the two lenses have identical focal length) in practice.

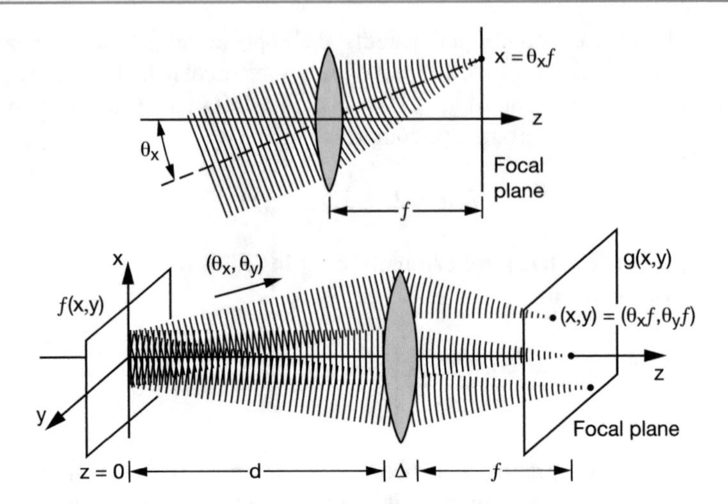

Fig. 3.20 A focusing lens carries out a Fourier transform, according to [91Sal2]

In an optical pulse shaper, one wants zero net GDD from the grating arrangement. Therefore, one places the gratings in the focal planes of the telescope ($L_{\mathrm{eff}} = 0$) [92Wei2].

3.3.3 Grating-Based Pulse Shaper

The basic principle of a pulse shaper is simple to understand. It is important to realize that a lens makes a Fourier transform of the incoming wave in the focal plane (see Fig. 3.20). The grating introduces angular dispersion, which means that each frequency component of the incoming beam is diffracted into a fixed direction in space. A lens images each incoming plane wave with a specific angle of incidence to a fixed position in its focal plane separated by a distance f from the lens. Therefore, we can address each individual frequency component separately with a special phase and amplitude mask in the focal plan of the lens. This is the basic principle of the diffraction grating based pulse shaper as shown in Fig. 3.21. Examples of different pulse shapes are given in Fig. 3.22.

Grating-based pulse shapers, compressors, and stretchers are very important tools in ultrafast laser physics. They have also been used to push forward the frontier in pulse amplification. The most important breakthrough was the invention of chirped pulse amplification (CPA) [85Str] (see Fig. 3.23), where damage in amplifier materials can be prevented by first stretching the pulse and reducing the peak intensity, thereby also reducing optical damage. After amplification, the different frequency components are regrouped in time with another grating pulse compressor. The CPA technique enables the scaling of the peak power by many orders of magnitude, which was also recognized by the Nobel Prize in Physics in 2018.

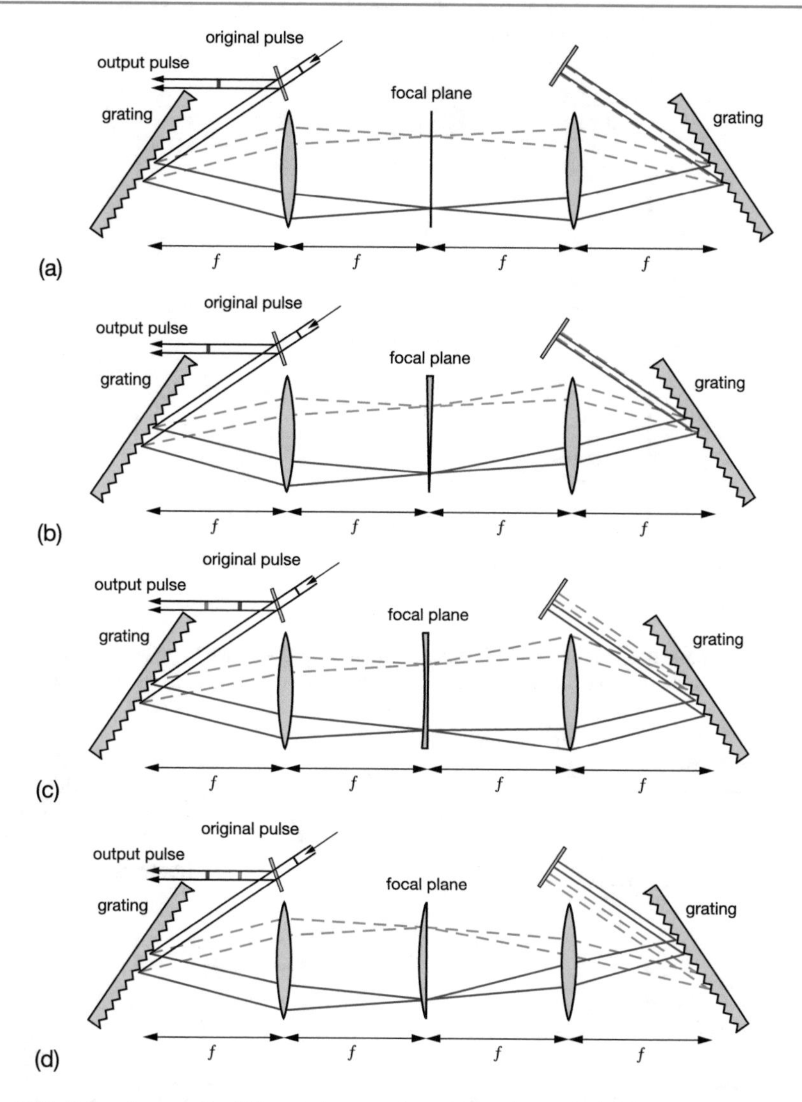

Fig. 3.21 Pulse shaper with different phase masks. **a** Pulse shaper in $4f$ design. This configuration does not affect the pulses that travel through. **b** Pulse shaper with a linear phase shift. This configuration changes the transit time of the pulse through the setup, but does not affect the pulse shape. **c** Pulse shaper with a quadratic phase shift, generating a pulse with a positive chirp, i.e., lower frequency components come out first. **d** Pulse shaper with a quadratic phase shift, generating a pulse with a negative chirp, i.e., higher frequency components come out first. The examples above are with static filters. With liquid crystal technology, it is possible to create a filter that can have an arbitrary computer-controlled phase and amplitude spectrum

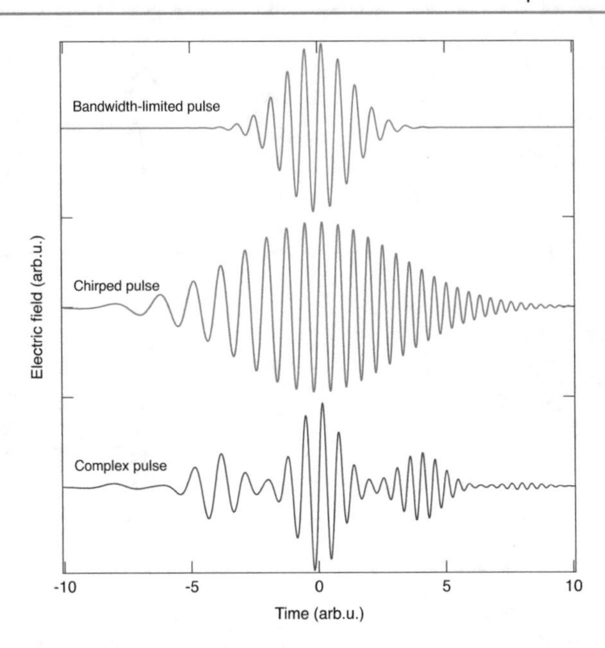

Fig. 3.22 Example of a bandwidth-limited pulse, a chirped pulse which has a longer duration, and a complex pulse shape. With a pulse shaper we can obtain pulse shapes in many different forms

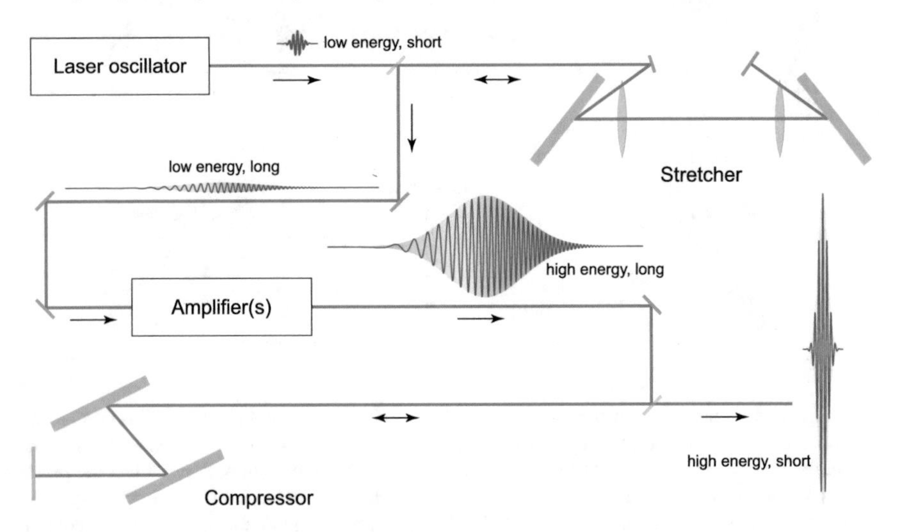

Fig. 3.23 Schematic set-up of chirped pulse amplification (CPA) [85Str]. This technique received the Nobel Prize in Physics in 2018

3.4 Gires–Tournois Interferometer (GTI)

A Gires–Tournois etalon or interferometer is a non-absorbing reflector with two reflecting surfaces, one of which has very high reflectivity (ideally 100%). The two reflecting surfaces then form a Fabry–Pérot interferometer, which ideally reflects 100% of the incoming light, but introduces a wavelength-dependent phase shift due to multi-beam interference that give rise to strong resonances in the phase alone if a 100% back reflector is used, while no other losses occur. Normally, we can neglect all material dispersion and can describe a GTI as shown in Fig. 3.24.

The full intensity is reflected when the left mirror of the GTI (R_b in Fig. 3.24) is a 100% reflector and there is no absorption inside the GTI. Then the GTI reflects 100% and introduces no loss:

$$|R_\mathrm{GTI}| = 1 , \qquad R_\mathrm{GTI} = \exp\left(\mathrm{i}2\phi_\mathrm{GTI}\right) . \tag{3.64}$$

The only parameter that changes is the phase ϕ_GTI, which becomes strongly frequency dependent due to the Fabry–Pérot structure. This frequency dependence is much stronger than the material dispersion of the layer thickness d, i.e., if no air gap is used, or of the substrate. Usually, one can neglect the material dispersion in the GTI. The distance between the two mirrors can be adjusted using piezo-elements, so a GTI can also be used to tune the dispersion. A fully integrated dielectric GTI mirror is discussed in Sect. 3.6.2.

According to Fig. 3.24, the reflectivity of this GTI is given by (see, e.g., [91Sal2])

$$r_\mathrm{GTI} \equiv \exp\left(-\mathrm{i}\phi_\mathrm{GTI}\right) = \frac{r_{12} + r_{23}\mathrm{e}^{-\mathrm{i}2\varphi}}{1 + r_{12}r_{23}\mathrm{e}^{-\mathrm{i}2\varphi}} = \frac{\sqrt{R_\mathrm{t}} - \mathrm{e}^{-\mathrm{i}2\varphi}}{1 - \sqrt{R_\mathrm{t}}\mathrm{e}^{-\mathrm{i}2\varphi}} , \tag{3.65}$$

where $r_{23} = -1$ and $r_{12} = \sqrt{R_\mathrm{t}}$, and we assume that R_t is a positive real number, i.e., $R_\mathrm{t} = |R_\mathrm{t}|$. The phase shift φ is given by

$$\varphi = nkd . \tag{3.66}$$

The group delay T_g of the GTI is then given by

$$T_\mathrm{g} = \frac{\mathrm{d}\phi_\mathrm{GTI}}{\mathrm{d}\omega} , \tag{3.67}$$

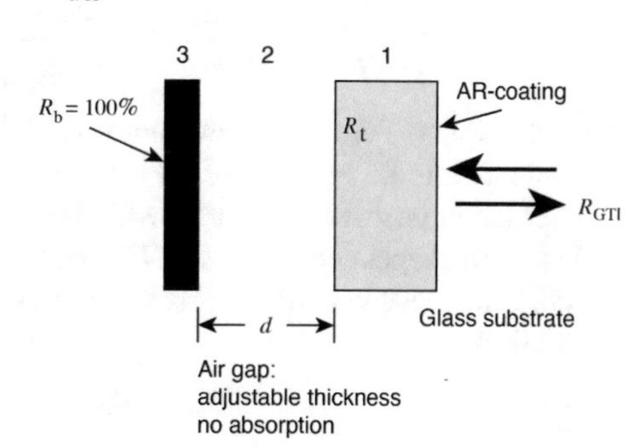

Fig. 3.24 A GTI [64Gir] is a Fabry–Pérot reflector with a 100% reflecting left mirror R_b, an air gap or a non-absorbing material of thickness d, and a right mirror R_t on a glass substrate with an AR-coating. Typically, a GTI has a low finesse, i.e., $R_\mathrm{t} < 10\%$

and the second order dispersion by

$$\frac{\mathrm{d}T_\mathrm{g}}{\mathrm{d}\omega} = \frac{\mathrm{d}^2\phi_\mathrm{GTI}}{\mathrm{d}\omega^2}, \tag{3.68}$$

where the phase is given by (3.65):

$$\tan\phi_\mathrm{GTI} = -\frac{\mathrm{Im}\,(r_\mathrm{GTI})}{\mathrm{Re}\,(r_\mathrm{GTI})} = -\frac{(1-R_\mathrm{t})\sin 2\varphi}{2\sqrt{R_\mathrm{t}} - (1+R_\mathrm{t})\cos 2\varphi}. \tag{3.69}$$

For the borderline case $R_\mathrm{t} = 0$ the phase ϕ_GTI is given by 2φ, which corresponds to the roundtrip phase through the mirror spacing d.

To calculate the dispersion, it is useful to rewrite the phase φ of (3.66) in the form

$$\varphi = nkd = n\frac{\omega}{c}d \equiv \frac{\omega t_0}{2}, \tag{3.70}$$

with

$$t_0 = \frac{2nd}{c}, \tag{3.71}$$

the roundtrip time in the Fabry–Pérot. In the following, we neglect the material dispersion in the GTI, i.e.,

$$\frac{\mathrm{d}\varphi}{\mathrm{d}\omega} \approx \frac{t_0}{2}. \tag{3.72}$$

Then using (3.67) and (3.69) and the relation

$$\frac{\mathrm{d}}{\mathrm{d}\omega}\tan\phi = \left(1+\tan^2\phi\right)\frac{\mathrm{d}\phi}{\mathrm{d}\omega}$$

and carrying out a few transformations

$$T_\mathrm{g} = \frac{\mathrm{d}\phi_\mathrm{GTI}}{\mathrm{d}\omega} = \frac{t_0\,(1-R_\mathrm{t})}{1+R_\mathrm{t}-2\sqrt{R_\mathrm{t}}\cos\omega t_0}. \tag{3.73}$$

Hence,

$$\frac{\mathrm{d}T_\mathrm{g}}{\mathrm{d}\omega} = \frac{\mathrm{d}^2\phi_\mathrm{GTI}}{\mathrm{d}\omega^2} = -\frac{2t_0^2\,(1-R_\mathrm{t})\sqrt{R_\mathrm{t}}\sin\omega t_0}{\left(1+R_\mathrm{t}-2\sqrt{R_\mathrm{t}}\cos\omega t_0\right)^2}. \tag{3.74}$$

Thus the Fabry–Pérot structure (not material dispersion) generates a wavelength-dependent periodic variation of the group delay and the second order dispersion with an alternating positive and negative GDD (see Figs. 3.25 and 3.26).

Due to the dependence on t_0, (3.73) and (3.74) imply that both the group delay and the GDD increase with increasing layer thickness d. For the GDD, it even increases quadratically:

$$\frac{\mathrm{d}T_\mathrm{g}}{\mathrm{d}\omega} \propto \mathrm{d}^2. \tag{3.75}$$

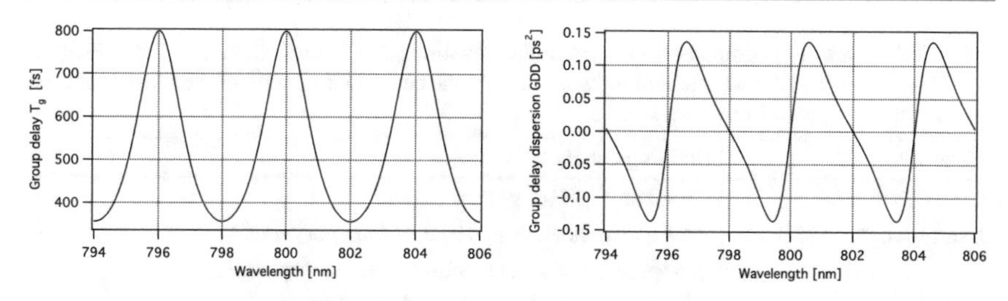

Fig. 3.25 Group delay and GDD as a function of wavelength in μm for a GTI with an 80 μm thick air gap, and $R_t = 4\%$, according to (3.73) and (3.74)

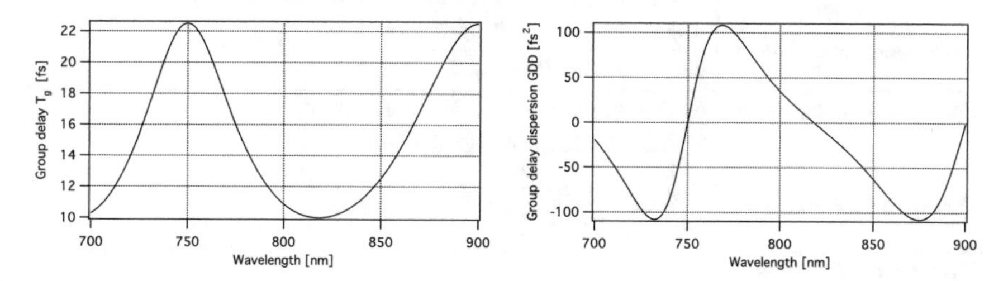

Fig. 3.26 Group delay and GDD as a function of wavelength in μm for a GTI with an 2.25 μm thick air gap, and $R_t = 4\%$, according to (3.73) and (3.74)

The trade-off is that the bandwidth of the GTI decreases with increasing layer thickness d:

$$\text{bandwidth of } \frac{\mathrm{d}T_g}{\mathrm{d}\omega} \propto \frac{1}{d} \,. \tag{3.76}$$

For a large negative dispersion, d has to be chosen sufficiently large [see (3.75)], but this constricts the usable bandwidth so much [see (3.76)] that short pulses cannot be supported. Usually, air-spaced GTIs are not used for pulses shorter than about 100 fs.

For example, for a Ti:sapphire laser with a center wavelength of about 800 nm, one can use a GTI with $d = 80$ μm and $R_t = 4\%$, whereby the usable bandwidth of the GTI is limited to about 0.5 nm and thus the minimal pulse length to about 1 ps (compare with Fig. 3.25). The usable bandwidth of the GTI is limited by the quickly growing higher order dispersion. A much thinner GTI increases the usable bandwidth, but the negative GDD is no longer sufficient to compensate for, e.g., the positive dispersion of 1160 fs^2, which is produced in a 2 cm long Ti:sapphire crystal (see Table 2.5). The only solution for this crystal length is to use a thinner GTI, but then one has to introduce several reflections within one resonator roundtrip. For example, using this approach in a pulsed Ti:sapphire laser, pulses as short as 150 fs have been generated with a GTI with $d = 2.25$ μm and 11 reflections per resonator roundtrip (Fig. 3.26). However, in this case, the laser resonator setup had become very complicated and the laser was no longer tunable [92Kaf]. The prism compressor, which we already discussed, constitutes a much better solution in the case of ultrashort pulse generation.

Table 3.2 Dispersion compensation, its defining equations and figures. Here, c is the velocity of light in vacuum, λ is the wavelength in vacuum, λ_0 is the center wavelength of the pulse spectrum, and ω is the frequency in radians/second

Quantity	Defining equation
Gires–Tournois Interferometer (GTI)	d thickness of Fabry–Pérot, n refractive index of material inside Fabry–Pérot (air-spaced $n = 1$), neglecting material dispersion, $t_0 = 2nd/c$ roundtrip time of Fabry–Pérot, R_t intensity reflectivity of top reflector of Fabry–Pérot (bottom reflector assumed to have 100% reflectivity)
Dispersion: second order	$\frac{d^2\phi}{d\omega^2} = -\frac{2t_0^2(1-R_t)\sqrt{R_t}\sin\omega t_0}{(1+R_t-2\sqrt{R_t}\cos\omega t_0)^2}$
Four-grating compressor	L_g grating pair spacing, Λ grating period, θ_i angle of incidence at grating
Dispersion: second order	$\frac{d^2\phi}{d\omega^2} = -\frac{\lambda^3 L_g}{\pi c^2 \Lambda^2}\left[1-\left(\frac{\lambda}{\Lambda}-\sin\theta_i\right)^2\right]^{-3/2}$
Dispersion: third order	$\frac{d^3\phi}{d\omega^3} = -\frac{d^2\phi}{d\omega^2}\frac{6\pi\lambda}{c}\frac{1+\frac{\lambda}{\Lambda}\sin\theta_i-\sin^2\theta_i}{1-\left(\frac{\lambda}{\Lambda}-\sin\theta_i\right)^2}$
Four-prism compressor	n refractive index of prisms, angle of incidence at prism is the Brewster angle $\theta_B = \arctan\left[n(\lambda_0)\right]$, apex angle of prism $\alpha = \pi - 2\theta_B$, $\theta_2(\lambda) = \arcsin\left[n(\lambda)\sin\left(\pi - 2\theta_B - \arcsin\frac{\sin\theta_B}{n(\lambda)}\right)\right]$, L apex-to-apex prism distance, h beam insertion into second prism, $\sin\beta = \frac{h}{L}\frac{\cos\theta_2}{\cos(\alpha/2)}$,
Dispersion: second order	$\frac{d^2\phi}{d\omega^2} = \frac{\lambda^3}{2\pi c^2}\frac{d^2P}{d\lambda^2}$, $\frac{d^2P}{d\lambda^2} = 2\left[\frac{\partial^2 n}{\partial\lambda^2}\frac{\partial\theta_2}{\partial n} + \frac{\partial^2\theta_2}{\partial n^2}\left(\frac{\partial n}{\partial\lambda}\right)^2\right]L\sin\beta - 2\left(\frac{\partial\theta_2}{\partial n}\frac{\partial n}{\partial\lambda}\right)^2 L\cos\beta$ $\approx 4\left[\frac{\partial^2 n}{\partial\lambda^2} + \left(2n-\frac{1}{n^3}\right)\left(\frac{\partial n}{\partial\lambda}\right)^2\right]L\sin\beta - 8\left(\frac{\partial n}{\partial\lambda}\right)^2 L\cos\beta$,
Dispersion: third order	$\frac{d^3\phi}{d\omega^3} = -\frac{\lambda^4}{4\pi^2c^3}\left(3\frac{d^2P}{d\lambda^2}+\lambda\frac{d^3P}{d\lambda^3}\right)$, $\frac{d^3P}{d\lambda^3} \approx 4\frac{d^3n}{d\lambda^3}L\sin\beta - 24\frac{dn}{d\lambda}\frac{d^2n}{d\lambda^2}L\cos\beta$.

3.5 Summary of Dispersion Compensation with Angular Dispersion and GTI

Table 3.2 summarizes the results for dispersion compensation with angular dispersion (i.e., prism and grating compressors) and with GTI mirrors.

3.6 Mirrors with Controlled Phase Properties

3.6.1 Bragg Mirror

A standard high-reflectance laser mirror is a Bragg mirror which consists of a stack of quarter-wave layers with alternating high and low refractive index (i.e., n_{H} and n_{L}), as shown in Fig. 3.27. The complex amplitude reflectivity of such dielectric multilayer structures can be calculated using different techniques. This will be discussed in more detail in Sect. 7.5.4. Here we summarize the most important information about a Bragg mirror, so that we understand the motivation for chirped mirrors.

A Bragg mirror shows maximum reflectivity at the Bragg wavelength λ_{B}, which is twice the optical thickness of the unit cell defining the quarter-wave structure, i.e., each individual layer has a thickness of a quarter of the Bragg wavelength $\lambda_B/4$. Its reflectivity spectrum shows high reflectivity for a range of wavelengths (called the stop band), that sharply drops off at the edge (see upper graph of Fig. 3.28). In the stop band around the angular frequency $\omega_{\mathrm{B}} = 2\pi c/\lambda_{\mathrm{B}}$, one has the condition $(\cos\phi - r^2)/(1 - r^2) \leq -1$, where $\phi = \pi\omega/\omega_{\mathrm{B}}$ is the phase acquired after propagation through a pair of layers and Fresnel's reflectance for the index step is given by

$$r = \frac{n_{\mathrm{H}} - n_{\mathrm{L}}}{n_{\mathrm{H}} + n_{\mathrm{L}}} . \tag{3.77}$$

If we neglect the wavelength dependence of the refractive index, the stop band covers the wavelength range between $\lambda_{\mathrm{S1}} = \pi\lambda_{\mathrm{B}}/(2\pi - 2\arccos r)$ and $\lambda_{\mathrm{S2}} = \pi\lambda_{\mathrm{B}}/2\arccos r$. Its fractional width $\Delta\omega/\omega_{\mathrm{B}}$ therefore depends only on the index ratio between the high- and the low-index material:

$$\frac{\Delta\omega}{\omega_{\mathrm{B}}} = \frac{4}{\pi}\arcsin r . \tag{3.78}$$

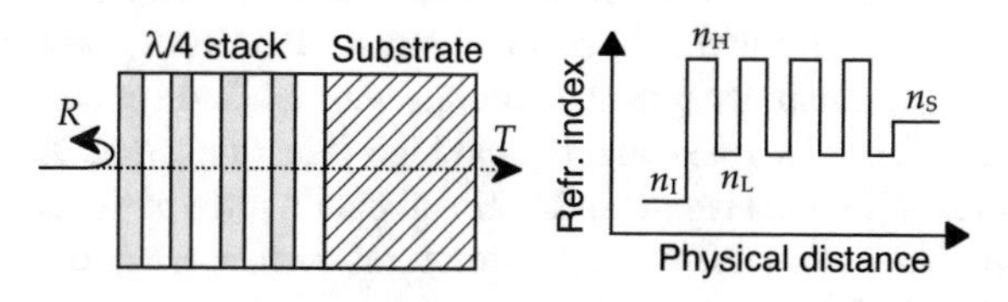

Fig. 3.27 Bragg mirror. Quarter-wave mirror consisting of four pairs ($m = 8$ layers) of alternating high- and low-index layers deposited on a substrate (*left*), and the refractive index profile (*right*). The optical thickness of the individual layers (i.e., $d_{\mathrm{H}}n_{\mathrm{H}}$ and $d_{\mathrm{L}}n_{\mathrm{L}}$) is given by $\lambda_{\mathrm{B}}/4$ where λ_{B} is the Bragg wavelength (in vacuum), for which maximum reflectance is obtained. R indicates the reflected beam and T the transmitted beam

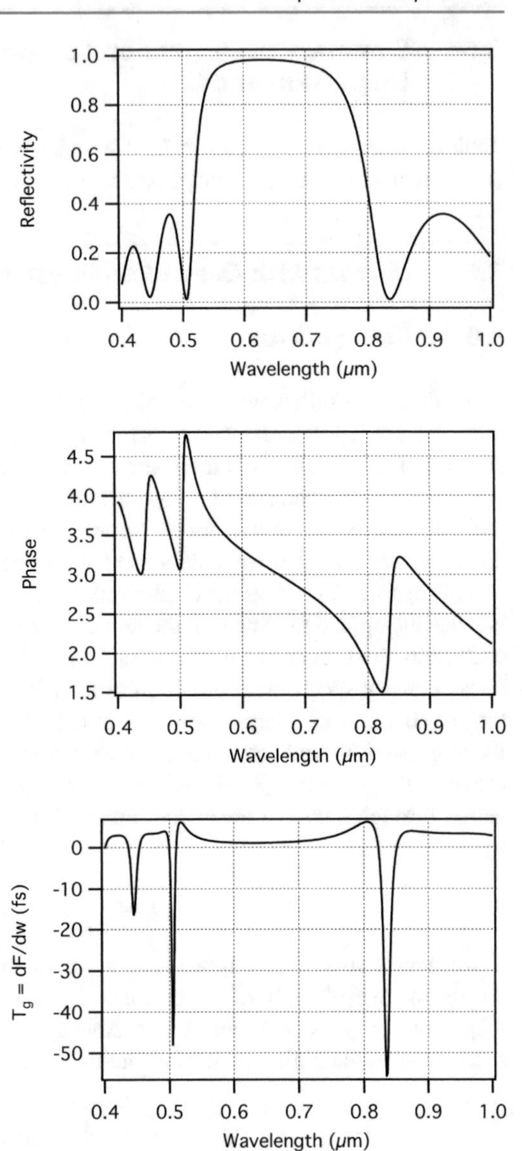

Fig. 3.28 Calculated reflectivity, phase, and group delay of a dielectric Bragg mirror. Bragg wavelength $\lambda = 630$ nm, material $5 \times SiO_2/TiO_2$ layers

For epitaxially grown semiconductor quarter-wave mirrors, all refractive indices are typically between 3.0 and 3.6, so their largest fractional bandwidth is about 10%. For amorphous coatings, typical materials with low indices are between MgF_2 ($n = 1.37$) and SiO_2 ($n = 1.45$), and typical high-index materials Ta_2O_5 ($n = 2.1$) and TiO_2 ($n = 2.35$), which results in fractional bandwidths of about 30%. Epitaxially grown $CaF_2/Al_{1-x}Ga_xAs$ multilayer structures even have a fractional bandwidth as high as 55% [02Sch1].

The maximum reflectance of a quarter-wave mirror with the structure shown in Fig. 3.27 can be calculated exactly for a lossless quarter-wave mirror, and it is given by

$$R_0(\lambda_B) = \left(\frac{1 - aqp^{m-1}}{1 + aqp^{m-1}}\right)^2 , \tag{3.79}$$

where $m = 2N$ is the number of layers for N layer pairs, and the parameters a, p, and q are the refractive index ratios at the three types of interface (see Fig. 3.27):

$$p = \frac{n_1}{n_H} , \quad q = \frac{n_L}{n_H} , \quad a = \frac{n_L}{n_S} . \tag{3.80}$$

The wavelength dependent reflectivity of a Bragg mirror is shown in Fig. 3.28.

Well within the stop band of the Bragg mirror, i.e., the high reflectivity bandwidth, we have a linear phase shift and therefore no second order dispersion. This means that, within the stop band, a pulse will be reflected without any distortion. However, this changes drastically as soon as the bandwidth of the ultrashort pulse extends beyond the highly reflective stop band.

For the first sub-100 fs dye lasers, prism compressors were not yet available for dispersion compensation. To obtain negative dispersion, a highly reflecting Bragg mirror with a shorter Bragg wavelength than the center wavelength of the laser was inserted into the laser. An example of a Bragg mirror with a Bragg wavelength of 630 nm is depicted in Fig. 3.28: for a longer wavelength of about 750 nm, the mirror generates negative dispersion, i.e., $\mathrm{d}T_g/\mathrm{d}\lambda > 0$ and thus $\mathrm{d}T_g/\mathrm{d}\omega < 0$, without introducing much loss.

3.6.2 Dielectric GTI-Type Mirrors

Large amounts of negative group delay dispersion (GDD) are required to achieve soliton modelocking at high pulse energy [15Sar]. Therefore, dielectric GTI-type mirrors with significant field enhancement in the multilayer structure are typically used [00Szi]. Instead of an air-spaced Fabry–Pérot (see Sect. 3.4), a dielectric multilayer coating is used with the layer thickness adjusted to obtain a Fabry–Pérot field enhancement. In this way, relatively large amounts of GDD can be achieved, but typically with a limited bandwidth [12Per]. However, as long as the pulse durations are sufficiently long (i.e., > 500 fs at a center wavelength of 1.030 μm), the bandwidth requirements for a flat negative GDD can be fulfilled by commercially available dispersive GTI-type mirrors with high dispersion per bounce (typically around $-500\ \mathrm{fs}^2$ per bounce). The use of low-loss chirped mirrors (see Sect. 3.6.3) can also be considered, but the low maximum achievable dispersion per bounce still remains impractical for the total overall dispersion required, so GTI-type mirrors still remain the preferred choice for such laser oscillators.

The large field enhancement in dielectric GTI-type mirror structures can directly result in a higher parasitic nonlinear absorption compared to a standard highly-reflective multi-layer stack, which deteriorates thermal properties, introduces a lower damage threshold, and increases losses. In high-power ultrafast solid-state lasers, this can become a significant issue [15Sar].

Semiconductor based GTIs offer another opportunity. For example, using a multi-layer semiconductor GTI mirror, which contains a saturable absorber in addition to the dispersion compensation, 160 fs pulses have been generated with a diode-pumped Cr:LiSAF laser [96Kop2].

3.6.3 Chirped Mirrors

In the area of ultrashort laser pulse generation, the 1990s can be regarded as the decade of the Ti:sapphire (Ti:Al_2O_3) laser. Today, one can obtain pulse durations in the range of one to two optical cycles ($\tau_p \approx 5$ fs) at a center wavelength of around 800 nm directly from a Ti:sapphire laser. The generation of such short pulses requires dispersion compensation techniques over an enormous spectral bandwidth. The prism pair, which is the standard component for dispersion compensation, has the disadvantage of generating a considerable amount of higher order dispersion (Figs. 3.11 and 3.12) (see Sect. 3.2). This limitation has been overcome with chirped mirrors, which were invented in 1994 [94Szi].

Very broadband mirrors that cover a stop band beyond the simple standard TiO_2/SiO_2 Bragg mirror involve several consecutive quarter-wave sections with overlapping stop bands. However, this introduces a strongly wavelength-dependent GDD with very large phase variations and it cannot be used for femtosecond applications. It was the invention of a "chirped mirror" by Szipöcs et al. [94Szi] that began a new era in ultrashort pulse generation. In chirped mirrors, the Bragg wavelength is gradually decreased during deposition (at least in the starting design, which had to be altered by subsequent computer optimization to prevent parasitic GDD oscillations). Therefore, longer wavelengths can penetrate deeper into the coating, as sketched in Fig. 3.29. This broadens the high-reflectance range as compared to quarter-wave mirrors, and simultaneously leads to a negative GDD (Fig. 3.29b).

In a chirped mirror the local Bragg wavelength is varied along the mirror structure, which allows for the generation of a controlled GDD. In addition, the region of high reflectivity compared to conventional dielectric quarter-wave Bragg mirrors can be expanded due to the chirp in the Bragg wavelength. Therefore, chirped mirrors are the ideal compact components for broadband dispersion compensation. The total dispersion that can be obtained from one reflection from a chirped mirror is not very large, so these mirrors are only used in the sub-10 fs regime in the near-infrared. Even in that regime, multiple reflections are typically used in chirped mirror compressors. A simple estimate for the maximum negative dispersion can be obtained by calculating the beam path difference in the mirror material, as shown in Fig. 3.29a. Simply from the total thickness of the multilayer mirror structure, it becomes clear

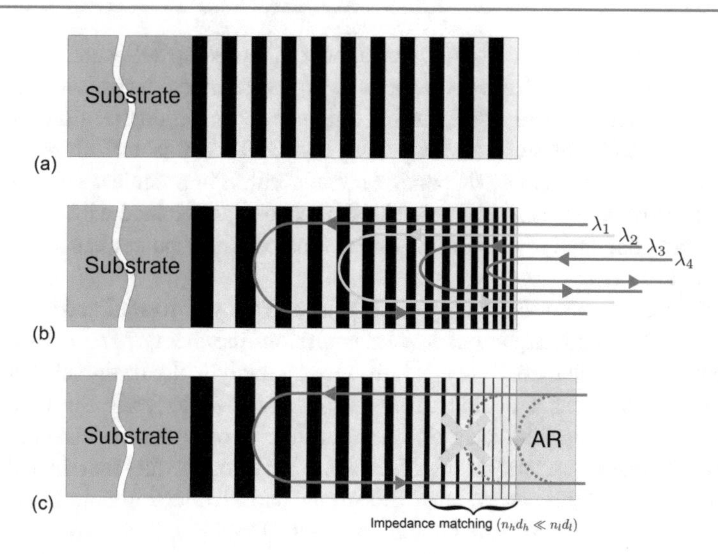

Fig. 3.29 Schematic comparison between a Bragg mirror (**a**), a single-chirped mirror (**b**), and a double-chirped mirror (**c**). In (**b**), the Bragg wavelength λ_B is gradually increased from the surface to the substrate. Therefore, shorter wavelengths (λ_4) are reflected closer to the surface, while longer wavelengths (λ_1) penetrate deeper into the coating, with $\lambda_1 > \lambda_2 > \lambda_3 > \lambda_4$. In (**c**), the ratio of the high- versus low-index layer (TiO_2) is also increased. This leads to the term "double-chirped mirror" (DCM) [97Kar,97Mat]. Both the additional chirping and the anti-reflection (AR) coating can be understood as a means of impedance matching from air into the coating which strongly reduces parasitic GDD oscillations

that we need many reflections to compensate the dispersion for laser crystals that are at least several mm long.

Chirped mirrors have one important disadvantage: the calculated dispersion properties show unwanted oscillations around the desired target functions for ideal dispersion compensation (Fig. 3.30). In 1994, no satisfying theory existed for the correct description of the spectral properties of such mirrors. In particular, no analytic expla-

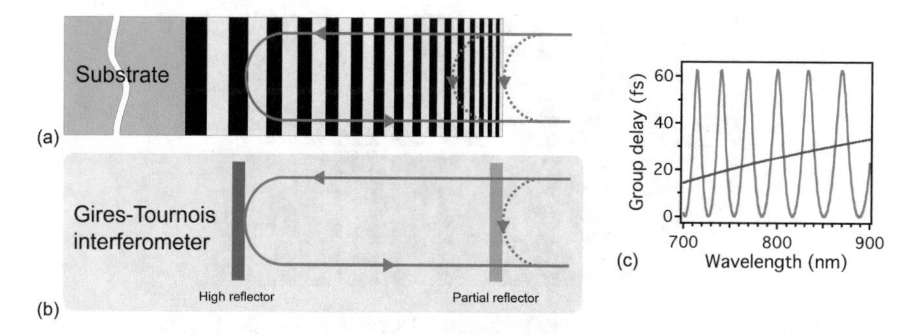

Fig. 3.30 Strong oscillations in the group delay and GDD of chirped mirrors occur due to a GTI effect (see Figs. 3.25 and 3.26) of different reflections from different sections inside the mirror structure. Due to this effect, longer wavelengths with their deeper penetration into the mirror stack (shown schematically here in *red*) typically have stronger GTI-type group delay oscillations

nation existed for the oscillations, which appear in the group delay and in the GDD. In spite of this deviation from the desired smooth dispersion curve, such structures were initially used as starting designs for computer optimization that aimed at reducing the unwanted oscillations in group delay and GDD. In this way, chirped mirrors were designed for intracavity dispersion compensation in prismless resonators and for hybrid mirror and prism schemes. The shortest pulses obtained with such chirped mirrors were generated in a prismless ring resonator. These pulses had a duration of about 7.5 fs [96Xu2].

In 1997 a comprehensive theory of chirped mirrors was formulated on the basis of exact coupled-mode equations and transmission-line theory [97Kar,97Mat]. As a consequence of this theory we can introduce an analytically designed multi-layer mirror stack known as a double-chirped mirror (DCM) to avoid the undesirable oscillations. For a detailed derivation I would like to refer the interested reader to [97Mat]. The simple understanding of DCMs is based on the fact that, in addition to the chirp of the local Bragg wavelength, the initial partial reflection at the beginning of the mirror has to be reduced to avoid the strong GDD-type oscillations (see Fig. 3.30). This can be solved with an impedance-matching approach, where a chirp in the thickness of the high-index layer slowly tapers the impedance and therefore reduces the partial reflection (see Figs. 3.29c and 3.31). We therefore end up with a chirp in the Bragg wavelength and a chirp in the ratio of the high- to low-index layer thickness, which slowly increases the strength of the reflection and therefore acts as an impedance-matching section for the longer wavelength that penetrates more deeply into the mirror. The impedance matching reduces the partial reflection within the first part of the mirror layers. We therefore called this mirror structure a double-chirped mirror. The analytical theory permits the design of DCMs with tailored dispersion characteristics over a broad region with high reflectivity [99Mat].

The impedance matching approach reduces the partial reflection for the initial part of the chirped mirror structure (Fig. 3.31 top), which is especially important for the long wavelength components. An additional broadband anti-reflection (AR) coating resolves the impedance-matching problem for the first interface coming from

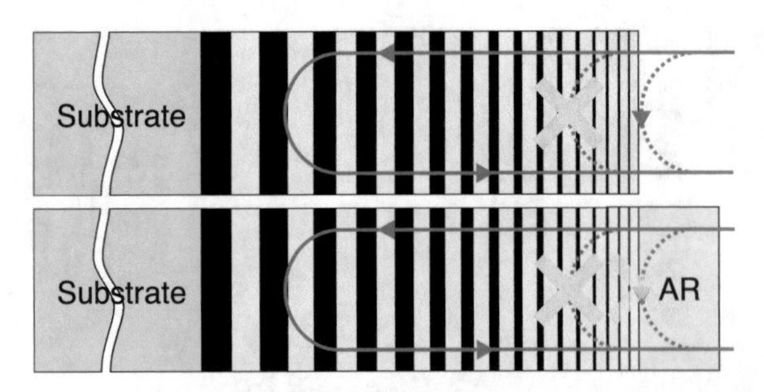

Fig. 3.31 Basic principle of a double-chirped mirror (DCM), for which the dispersion oscillations can be reduced with an analytical design to reduce the partial reflections at the beginning of the multilayer structure. Only simple standard numerical optimization is required to subsequently further minimize the residual GDD oscillations. The ultimate challenge, however, is the manufacture of such DCMs, because it requires very accurate control of the multi-layer deposition [00Mat] (see Fig. 3.36). Ion beam sputtering is typically used for their production

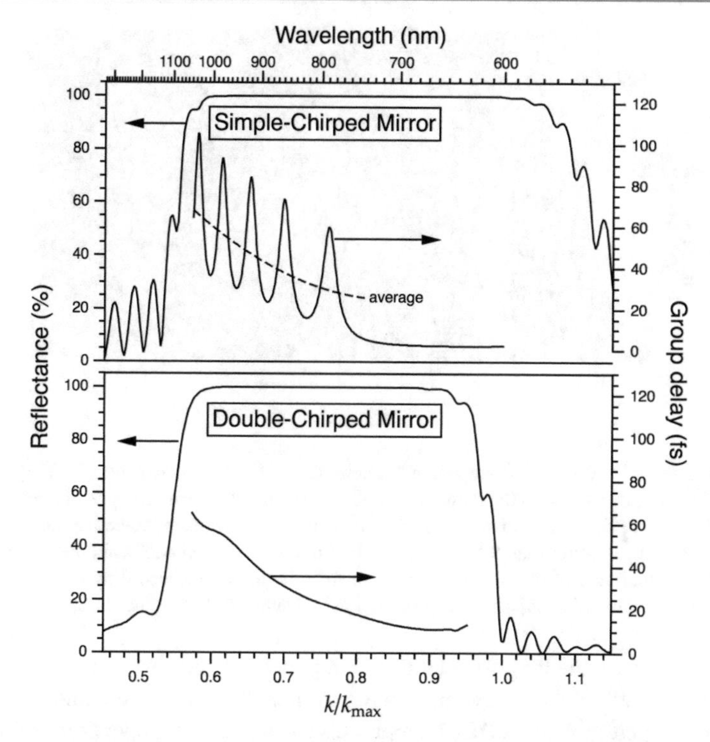

Fig. 3.32 Comparison of two different starting designs for a computer optimization, calculated using the exact transfer matrix method. The *upper plot* shows the reflectance and group delay for a single-chirped mirror consisting of 25 layer pairs with $n_H = 2.5$ and $n_L = 1.5$. The *dashed line* shows the average of the group delay for the single-chirped mirror, without the dispersion oscillations which are clearly visible for the solid line. The *lower plot* shows the reflectance and group delay of a DCM, whose thickness of the high-index layer is chirped quadratically over the first 12 index steps. In comparison to the single-chirped mirror, the group delay curve becomes very smooth, at the expense of a reduced high reflectance bandwidth on the short wavelength side. In both cases, the Bragg wave number is linearly chirped from $k_{B,max} = 2\pi/(600\,\text{nm})$ to $2\pi/(900\,\text{nm})$ over the first 20 index steps and then kept constant. Additionally, a perfect AR coating is assumed to match the impedance from air to the first (low-index) layer

air into the chirped mirror structure (Fig. 3.31c bottom). We typically need to have the residual reflectivity of this AR coating reduced to better than 10^{-4}. However, this limits the bandwidth that can be obtained with the DCMs. Unfortunately, we cannot obtain both an arbitrarily low reflectivity and an arbitrarily broad bandwidth [96Dob]. At a center wavelength of 800 nm, we can typically obtain a reflectivity bandwidth of about 230 nm with a residual reflectivity of less than 10^{-4} [00Mat].

The reduced dispersion oscillations are shown in Fig. 3.32. This compares the reflectance and the group delay for a single-chirped mirror and a double-chirped mirror. For the single-chirped mirror in the range between 1050 and 700 nm, the slope of the average group delay is negative. However, the strong group delay oscillations of ± 25 fs would dramatically distort any ultrashort pulse, and this prevents the use of such single-chirped mirrors for ultrashort pulse generation. These oscilla-

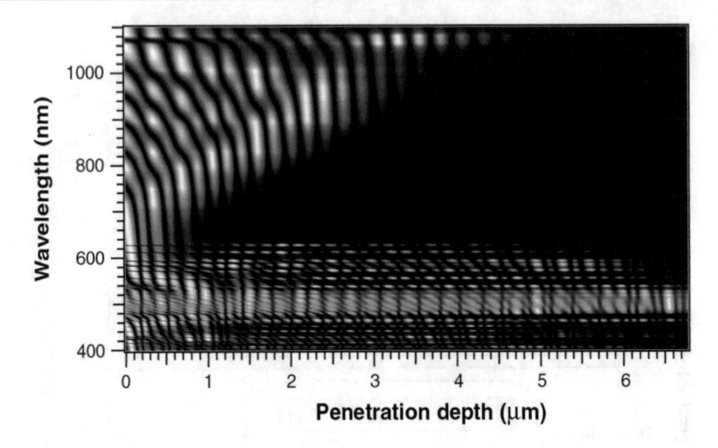

Fig. 3.33 Calculated wavelength-dependent penetration of the electromagnetic field into a broadband double-chirped mirror. In the range from 630 to 1030 nm, the penetration depth increases with wavelength, giving rise to negative group delay dispersion. The transmittance for the green pump light (i.e., 500 nm) is better than 90% and could be further optimized with a simple anti-reflection coating on the rear side of the substrate. A set of mirrors fabricated according to this design has been successfully used for dispersion compensation in a sub-6 fs Ti:sapphire laser

tions arise because the long wavelengths that penetrate deep into the mirror are also partially reflected on their way through the front sections of the coating. In contrast to single-chirped mirrors, a DCM suppresses the undesired dispersion oscillations, as explained above. Figure 3.32 further shows that the high-reflectance bandwidth of the DCM is reduced on the short wavelength side compared with the single-chirped mirror. The additional bandwidth of the single-chirped mirror, however, does not show the desired negative dispersion, so no bandwidth is lost in the dispersion compensation range. Figure 3.33 shows the wavelength-dependent penetration into the DCM. This DCM was also designed to have a high transmission for the green pump light in a Ti:sapphire laser.

The bandwidth limitation of the AR coating can be resolved with a further extension in the design of DCMs. The restrictions with regard to the maximal acceptable residual reflection from the AR coating can be substantially relaxed when these residual reflections do not interfere with the multi-layer DCM stack. This can be obtained with a wedged substrate, as shown in Fig. 3.34, where the front reflection from the substrate does not interfere with the rest of the mirror stack. The impedance matching from within the substrate into the DCM structure can be obtained perfectly if we choose the refractive index of the substrate $n_{\text{substrate}}$ to be equal to the refractive index of the first layer of the DCM section $n_{\text{first layer}}$. In this case, the AR coating on the substrate only needs to reduce the loss, which typically should be a fraction of

Fig. 3.34 Back-side coated (BASIC) double-chirped mirrors (BASIC DCM) [00Mat]

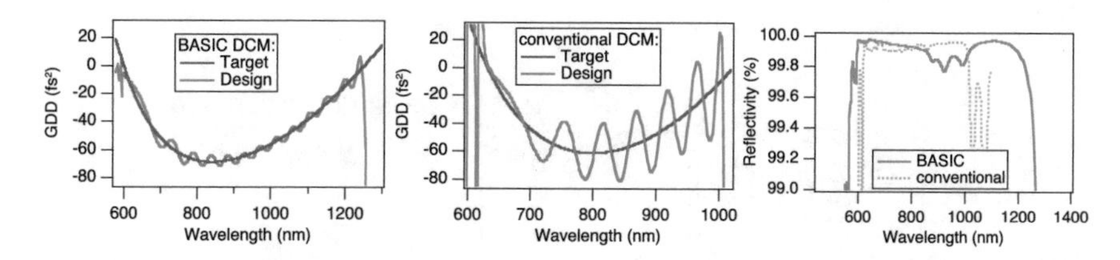

Fig. 3.35 Octave spanning BASIC DCM design compared to a standard DCM design. For the BASIC DCM, we can obtain a reflectivity greater than 99.7% over a wavelength range of 600 to 1240 nm. Over this range, the group delay varies less than 0.3 fs rms and the GDD 3.8 fs^2 rms. This covers a frequency bandwidth of 260 THz [04San]

1%. Therefore, a much broader AR coating can be produced. This concept is referred to as a back-side coated double-chirped mirror (BASIC DCM) [00Mat].

With such a BASIC DCM, the reflectivity bandwidth can be substantially extended while keeping the residual dispersion oscillation to less than 3.8 fs^2 rms (Fig. 3.35) [00Mat]. The disadvantage of the BASIC DCM design is that we have to introduce additional positive dispersion because the pulse now has to propagate through the substrate. However, this can be compensated for with a higher number of reflections for the chirped mirrors. The more serious disadvantage is that the beams are no longer spatially coherent because of the angle dispersion introduced by the wavelength-dependent refraction at the wedged substrate. This has to be compensated for with another matched reflection. For curved mirrors this is not a problem because then we can choose a flat substrate at the first interface, whereas the DCM structure will experience the radius of curvature at the second interface of the substrate.

Currently, the main limitation in chirped mirror dispersion compensation is the fabrication of such mirrors. We can obtain designs that are good enough for the one optical cycle regime, but the layer deposition errors during the fabrication process are normally too large and introduce GDD oscillation again. A layer deposition error of only 0.2 nm rms is sufficient for an error in the GDD of up to 50 fs^2 peak (Fig. 3.36). Therefore, we typically use many different chirped mirrors manufactured at different times and carefully selected to compensate for their dispersion oscillations. Thus by using a group of different chirped mirrors, the induced layer deposition error can be averaged out and thereby these configurations ultimately demonstrate the shortest pulses from a laser oscillator. However, it is clear that a more accurate and affordable coating deposition technique would be highly beneficial to the ultrafast laser community.

3.6.4 Design of Chirped Mirrors

For non-periodic structures, the analytical form of the complete transfer matrix becomes extremely complicated. Therefore, it normally needs to be calculated numerically. New physical insight into the origin of group delay oscillations of single-chirped mirrors was found soon after an exact coupled-mode theory for multilayer coatings with arbitrarily large index differences was established [97Mat]. Standard

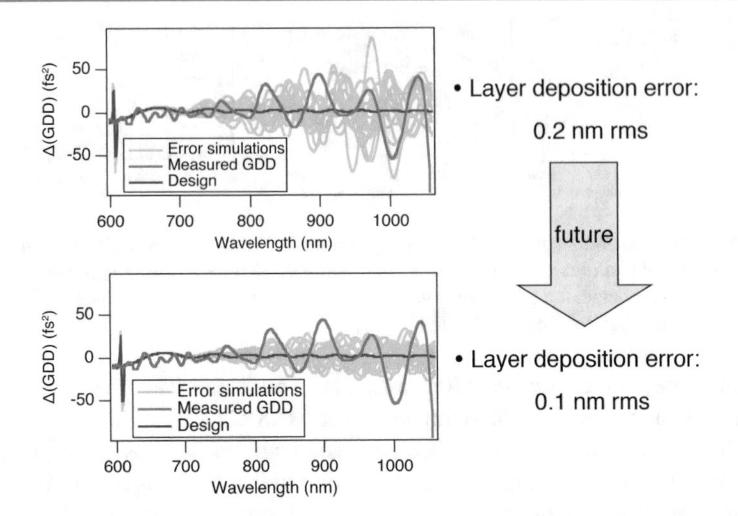

Fig. 3.36 Coating layer deposition errors and their impact on error-induced dispersion oscillations [00Mat]

coupled-mode theory [54Pie] applies for small index steps, and has typically been used for the calculation of fiber gratings [75Mat, 84Hau].

It was recognized that concepts such as impedance matching, well known from high-frequency transmission lines, can be fully transferred to coating design, and not only for structures of small index modulation. Coupled-mode theory connects the slowly varying field amplitudes of the backward and forward propagating modes, A and B, according to a coupled first-order differential equation

$$\frac{\mathrm{d}}{\mathrm{d}m}\begin{pmatrix} A \\ B \end{pmatrix} = \mathrm{i}\begin{bmatrix} -\delta & -\kappa \\ \kappa & \delta \end{bmatrix}\begin{pmatrix} A \\ B \end{pmatrix} , \tag{3.81}$$

where the detuning coefficient δ and the coupling coefficient κ are also functions of the normalized propagation distance m (see Fig. 3.37). Here m is a continuous variable that counts the index steps and is defined by

$$m \equiv \int_0^m \mathrm{d}\bar{m} \equiv \int_0^z \frac{\mathrm{d}\bar{z}}{d_{\mathrm{H}}(\bar{z}) + d_{\mathrm{L}}(\bar{z})} . \tag{3.82}$$

By suitable substitution [97Mat], the coupled-mode equations decouple, leading to two independent wave equations. With the exact coupled-mode equations, it is possible to show that the unwanted dispersion oscillation of the single-chirped structure results from an impedance mismatch. The (complex) impedance depends on the normalized position m via the coupling and detuning coefficients. The mismatch can be avoided by chirping not only the Bragg wavelength λ_{B} but also the duty cycle (the portion of the high index material that contributes to the Bragg wavelength, see Fig. 3.29). The difference between the phases due to propagation in the high- and low-index parts of a unit cell influences the coupling coefficient κ, and a slow

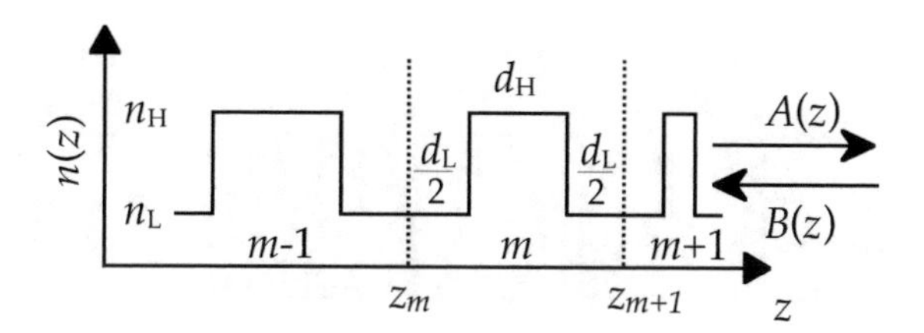

Fig. 3.37 Refractive index profile of a binary chirped mirror with a symmetrically defined local Bragg cell, as used in [99Mat2]. The refractive indices of the high- and low-index layers are n_H and n_L. The respective thicknesses are d_H and d_L may be different for the different Bragg cells. By convention, m is a negative number that counts the Bragg cells starting with $m = 0$ at the surface of the coating. $A(z)$ and $B(z)$ are the slowly varying field amplitudes of the rightward propagating and leftward propagating waves, while z is the physical position inside the coating, also negative inside the coating by convention. For light incident from the left at $z = 0$, $A(0)$ is the reflected field and $B(L)$ the transmitted field

ramp-up of the duty cycle reduces internal reflections at impedance discontinuities. We call chirped mirrors where both λ_B and κ vary systematically double-chirped mirrors (DCMs) [97Kar]. For optimal impedance matching, the coupling coefficient is ramped up from zero to its maximum value according to a power law [97Mat].

One of the important practical features of chirped-mirror theory is that it provides an approximate analytic formula showing how to chirp the Bragg wavelength in order to adjust the dispersive coating properties [99Mat]. The analytic formula for the chirp law given below relies on a simplified WKB solution that uses the standard detuning and constant coupling coefficients and thereby neglects any double-chirping, as explained in full detail in [99Mat,99Mat2]. It allows us to calculate the position m of the unit cell at which the Bragg wavenumber has the value $k_B \equiv 2\pi/\lambda_B$ in order to achieve a target GDD given as a function of k_B. The optical thickness of the m th unit cell is related to the Bragg wavelength by $d_{opt} = n_H d_H + n_L d_L = \pi/k_B$. With m defined as in (3.82), the inverse chirp law is

$$m\,(k_B) = \frac{c^2}{2\pi}\left(1 - \frac{2r}{\pi}\right)\int_{k_B}^{k_B^{max}} \bar{k}_B\,\mathrm{GDD}\left(\bar{k}_B\right)\mathrm{d}\bar{k}_B\ . \tag{3.83}$$

Solving this equation is simple if the desired GDD obeys a power law. For example, it follows that, to achieve a constant negative GDD, one has to chirp k_B according to a square-root law. If instead we decrease k_B linearly during deposition, $k_B(m) = k_B^{max} - a|m|$, $a > 0$, the resulting GDD will be proportional to λ, and therefore the TOD will be positive (see Fig. 3.32). One remarkable result of the DCM theory is that it is easier to design broadband chirped mirrors with negative GDD and positive TOD than with both negative GDD and negative TOD.

The analytical chirp law given in (3.83) is of great benefit also for numerical refinement: typical chirped mirrors consist of about 50 layers, so that computer optimization programs will have to search a parameter space of 50 dimensions to find a good minimum of a merit function (that depends on the GDD oscillations and high reflective stop band). The existence of analytic starting designs becomes even more advantageous if the number of layers is further increased, which is the case for chirped

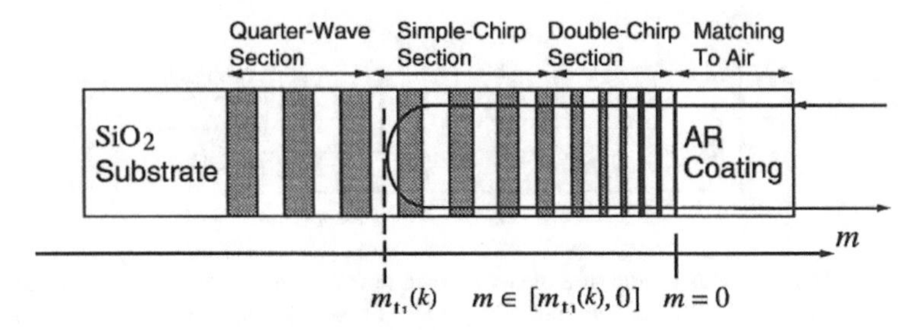

Fig. 3.38 Schematic growth of a DCM, comprising four multilayer subsections. The position inside the mirror is determined by the variable m, where $m = 0$ defines the beginning of the theoretical DCM structure directly after the AR coating. Light with wavenumber k is reflected at the wavenumber-dependent classical turning point $m_{t_1}(k)$ [99Mat2]

semiconductor mirrors [99Pas], where the relative refractive index difference is much smaller. Practically speaking, the analytical chirp law gives information on how to change the Bragg wavelength, but the double chirping cannot be included in a closed form of solution. In the design of DCMs (Fig. 3.38), we include double chirping in the first section of the coating. Like this double-chirped section, the subsequent single-chirped section obeys the analytical chirp law. An optional unchirped quarter-wave section may then be added, in order to further extend the high-reflectance range on the long-wavelength side. Finally, a separately designed AR coating is added at the top of the mirror to provide impedance matching to air. This structure does need further computer optimization due to the approximations made, the dispersion of the refractive index, the finite number of layers, the minimum thickness required for the layers, and the imperfect AR coating. Nevertheless, DCM theory leads to starting designs much closer to the desired target than any other approach. Details of the computer optimization go far beyond the scope of this chapter, and the reader is referred to [99Mat2] for a more detailed account of this issue.

3.7 Dazzlers

In ultrafast laser physics spectral phase control is important for the generation of short pulses with no pedestals and a given temporal shape. Therefore, programmable devices capable of adaptively compensating large amounts of dispersion over large spectral bandwidths have been developed. Although spatial light modulators (SLM) and deformable mirrors have also been successfully employed [92Wei2, 98Efi, 99Zee], we will limit our discussion in this section to the acousto-optic programmable dispersive filter (AOPDF or "dazzler"). In contrast to both SLMs and deformable mirrors, the dazzler does not require high-quality optical elements. Furthermore, it is compact and easy to implement in existing laser amplifier systems.

The AOPDF was proposed in 1997 by Tournois, who demonstrated in a first experiment the compression of Ti:sapphire laser pulses in the 100 fs range with a tellurium dioxide (TeO_2) acousto-optic modulator [97Tou]. Later, the AOPDF was

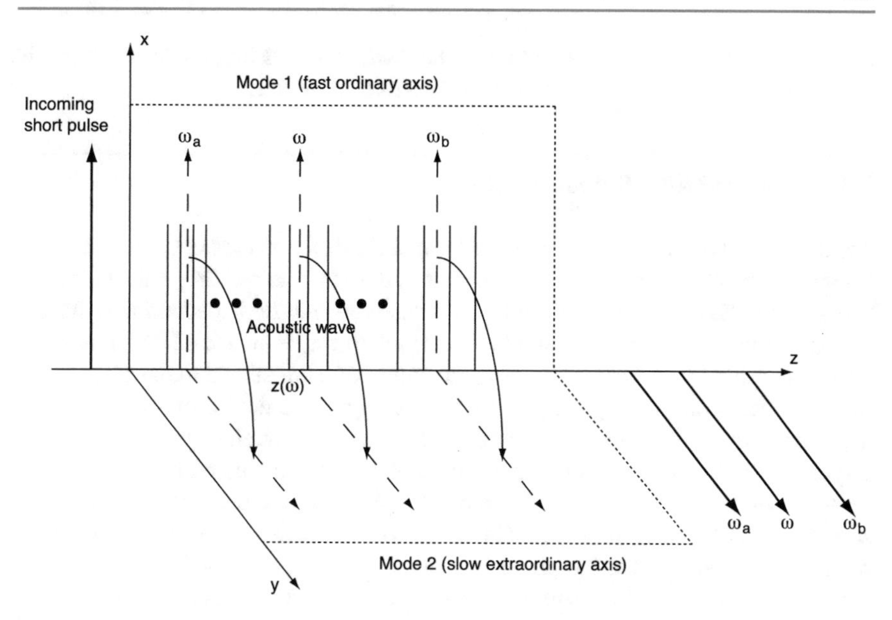

Fig. 3.39 Schematic of the acousto-optic programmable dispersive filter (AOPDF or "dazzler") principle [00Ver2]

implemented into a CPA laser system and successfully compensated for gain narrowing and residual phase errors resulting in the generation of 17 fs pulses [00Ver1].

Figure 3.39 shows a schematic of the AOPDF. An acoustic wave is launched into an acousto-optic device using a transducer excited by an RF temporal signal. The acoustic wave propagates with a velocity υ along the z-axis, in this way reproducing spatially the temporal shape of the RF signal. Two optical modes can only be coupled efficiently by acousto-optic interaction if the two optical modes and the acoustic wave are phase-matched. Since locally there is only one spatial frequency in the acoustic grating, only one optical frequency can be diffracted at a position z. The incoming optical pulse is initially in mode 1. Every frequency ω travels a certain distance before it encounters a phase-matched spatial frequency in the acoustic grating. At this position $z(\omega)$, part of the energy is diffracted into mode 2 and subsequently travels on that mode. The pulse leaving the device in mode 2 will be made of all the spectral components that have been diffracted at various positions. If the velocities of the two modes are different, then each frequency will experience a different time delay. This is usually accomplished by propagating the acoustic wave and the light pulse through a birefringent crystal. The light is initially polarized along the fast ordinary axis, but gets diffracted into the slow extraordinary axis. The diffraction efficiency, i.e., the amplitude of the output pulse, is controlled by the acoustic power at position $z(\omega)$.

The optical output of the AOPDF is a function of the optical input and of the electric signal. Thus, by properly choosing the temporal shape of the RF signal, and hence the spatial shape of the acoustic wave, an almost arbitrary group-delay distribution as a function of frequency can be created. The analytical expression relating the group delay at the output of the AOPDF to the input acoustic signal is obtained by coupled-wave theory and was derived in [00Ver2].

3.8 Dispersion Measurements

To analyze and balance the dispersive effects of the different elements of an ultrafast laser experiment, an accurate measurement system is required. In particular this is true for the characterization of multi-layer coatings whose dispersion may have a strong dependence on the wavelength. The standard way in which GDD is measured is via white-light interferometry. Here, the output of a broadband light source is fed into the interferometer, typically of Michelson type. The device under test (DUT) is placed in the sample arm, whereas the reference arm contains optics with known dispersion. The resulting interference pattern is monitored either with a photodiode (time domain sampling, spectrally integrated detection) [90Nag,96Did] or using spectral interferometry [00Dor,05Gos]. The obvious advantage of the spectrally integrated method is that the phase is evaluated simultaneously for all frequency components, so even a light source with low power spectral density can be used. In comparison with the spectrally resolved detection, the spectrally integrated method does not require monochromators, spectrometers, or detector arrays; therefore there is no need for an otherwise necessary calibration of these devices. Low equipment cost and the possibility of real-time alignment of tilt and zero delay monitoring make the temporal detection methods particularly attractive.

Here we will discuss two white-light Michelson interferometers to measure the wavelength-dependent phase difference between a sample arm, which contains the component to be tested, and a reference arm of known dispersion. The first scheme, shown in Fig. 3.40, utilizes a motorized mirror stage in one interferometer arm and a HeNe laser based distance calibration [90Nag]. An example of an experimental trace is shown in Fig. 3.41. The fringe term in the cross correlation is expressed as

$$S(\Delta t) = \mathrm{Re} \int_{-\infty}^{+\infty} E_{\mathrm{S}}^{*}(t) E_{\mathrm{R}}(t + \Delta t)\,\mathrm{d}t\,, \tag{3.84}$$

where E_{S} and E_{R} are the fields from the sample and reference arms, respectively, and we assume a quasi-infinite integration time, much longer than the coherence time of the white light. The fringes are sampled every time the path difference changes by half the wavelength of the monochromatic HeNe trigger laser. Fast Fourier transform of the interferogram (Fig. 3.41) yields the product of the amplitude spectra from the sample and reference arm. The argument of the Fourier transformed signal is the phase difference between the two arms, while its derivative with respect to frequency gives the relative group delay.

The relative dispersion of the two arms is first characterized without the sample. The spectral phase φ due to the sample is then measured as a function of frequency. From the phase we calculate the group delay numerically and average over many individual interference traces. The spectral resolution of the measurement is given

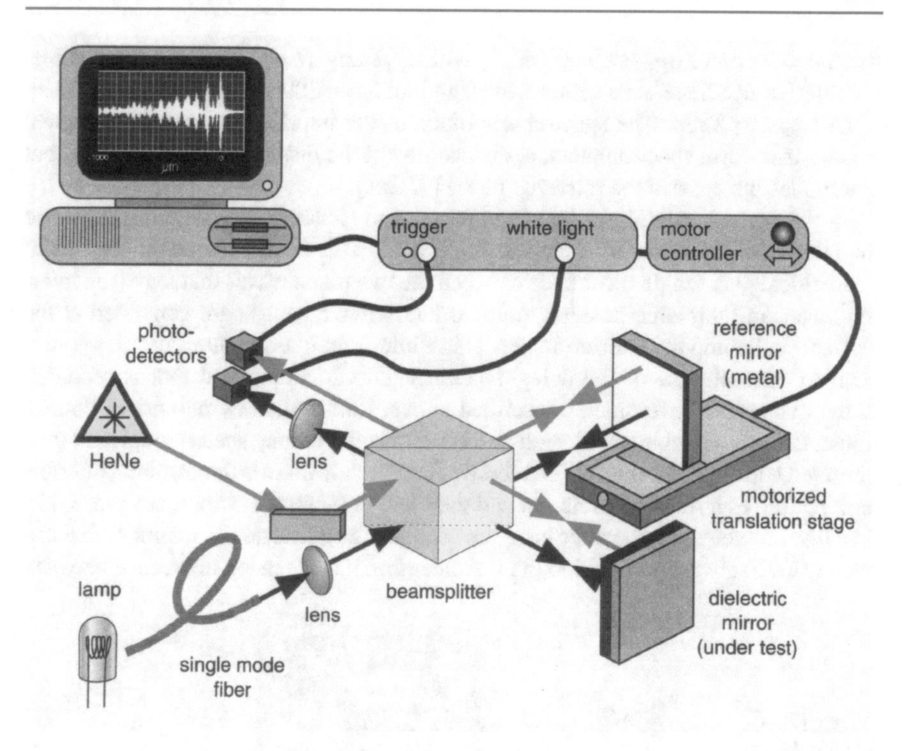

Fig. 3.40 Dynamic white light interferometer for dispersion measurement. While the reference arm length is scanned, the white light interference between the light from the reference mirror and the light from the dielectric mirror is recorded at integer multiples of half the HeNe laser wavelength $\lambda_{\mathrm{HeNe}}/2 = 316.4$ nm. The phase difference between the two Michelson arms is calculated from the recorded signal by a fast Fourier transform

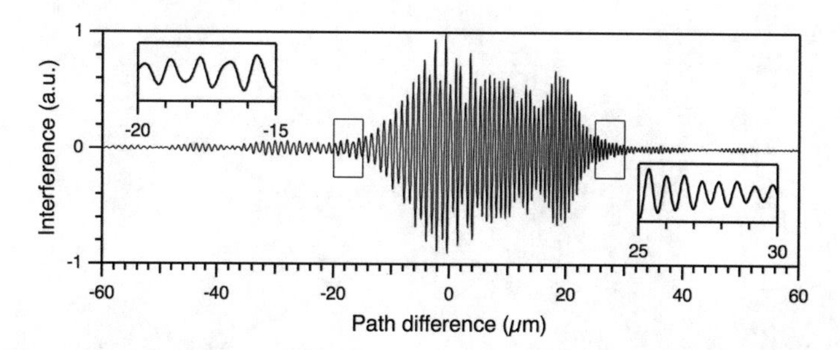

Fig. 3.41 Interference fringes (spectrally integrated) with a broadband double-chirped mirror in the sample arm. The path difference is the difference between the physical lengths of the sample arm and the reference arm. For longer wavelengths, the path needs to be decreased for interference as they penetrate more deeply into the coating (*left inset*)

by $\Delta\omega = 2\pi c/N\lambda_{\mathrm{HeNe}} \approx 0.006\ \mathrm{fs}^{-1}$, with typically $N = 512$ fringe data points per interference trace. At a center wavelength of $\lambda_c \approx 800$ nm, this results in $\Delta\lambda = \lambda^2/N\lambda_{\mathrm{HeNe}} \approx 2$ nm. The spectral amplitude of the interferogram, used in typical Fourier transform spectrometers, is disregarded in the dispersion measurement, but determines the error in the retrieved phase [05Gos].

In the second white-light interferometer setup [95Kov], we slightly tilt one of the interferometer arms, as sketched in Fig. 3.42. This leads to a height-dependent geometrical path length difference between the two plane waves that leave the interferometer. In their superposition, colored interference fringes are generated at the vertical slit behind the interferometer. This white-light interference pattern is equivalent to a small slice of the delay-dependent interference signal that is recorded in the dynamic interferometer described above. Rather than by numerical Fourier transform, this signal is wavelength-resolved with an imaging spectrograph and then observed with a CCD camera. The recorded image shows a two-dimensional interference pattern as a function of height and wavelength (Müller's stripes, see Fig. 3.43). At a fixed wavelength, the interferogram oscillates as a function of height with a frequency inversely proportional to the wavelength. The dispersion difference between

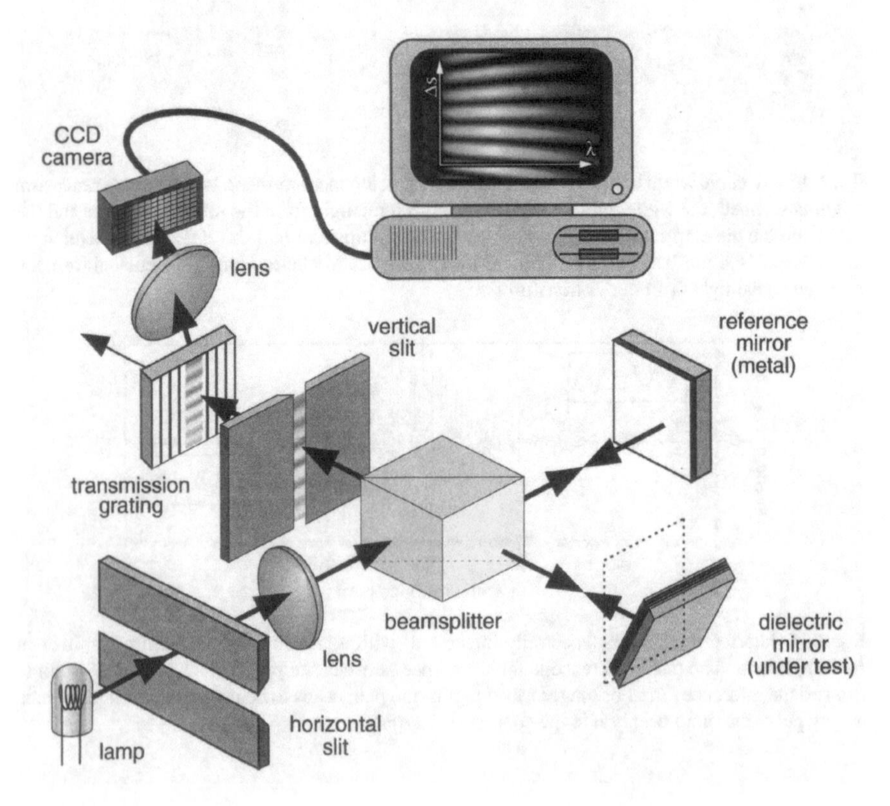

Fig. 3.42 Static white light interferometer for dispersion measurement. The tilt of the tested mirror results in a geometric path length difference at different heights of the plane wave that enters the interferometer

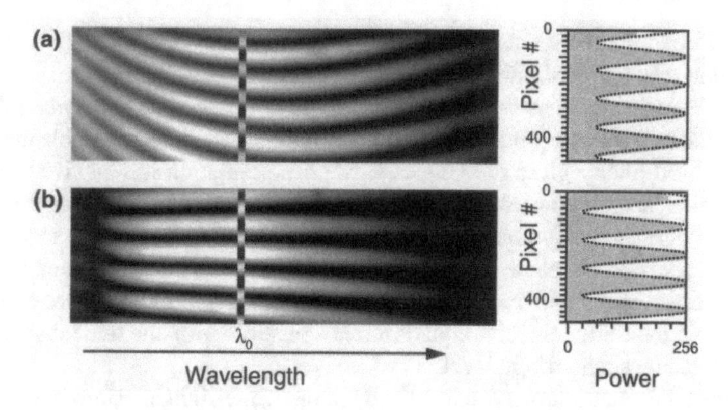

Fig. 3.43 Examples of recorded CCD images (480 × 640 pixels, 5 times smoothed). **a** With five thin fused silica microscope pellicles in the sample arm, the resulting phase difference causes a curvature of Müller's stripes. **b** With an output coupling mirror in the sample arm of the interferometer, the phase is almost constant with wavelength, except for a transmission resonance on the left-hand short wavelength side where the phase changes by π and where the fringe contrast is lower due to decreased mirror reflectance. The *right-hand plots* show cross-sections of the fringe pattern taken at λ_0, which is inverted in the image for display purposes [00Sut]

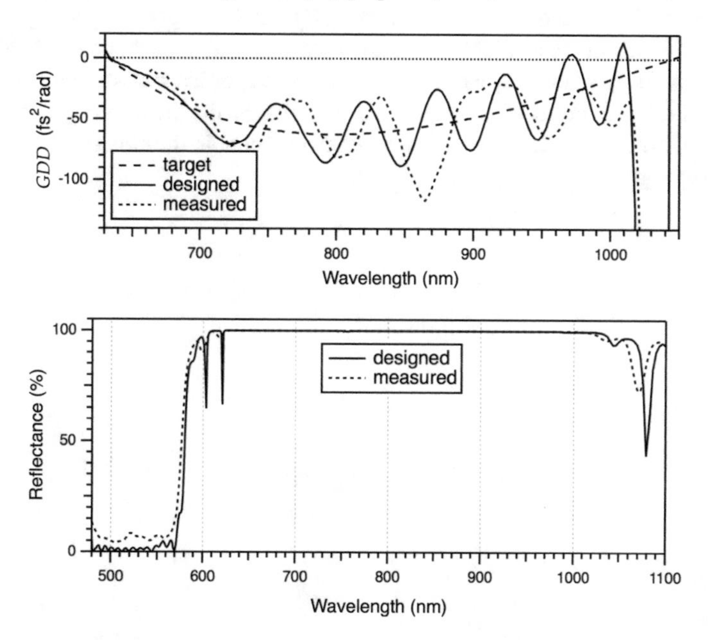

Fig. 3.44 GDD measurements using the dynamic white-light interferometer. Design and characterization of a broadband double-chirped mirror. **a** The measured group delay dispersion (GDD) follows the GDD expected from the design, but shows larger residual deviations from the target function. **b** The reflectance of the fabricated mirror is greater than 99.8% between 630 and 1020 nm [00Sut]

the sample and the reference arm causes a phase shift φ of this oscillation, which is recorded as a function of optical frequency.

In each interferometer, we measure the relative phase shift φ between the two interferometer arms as a function of frequency ω or wavelength λ. Double numerical differentiation finally gives the GDD. As the differentiation is sensitive to experimental noise in the retrieved phase, we usually smooth the resulting curves with a low-pass Fourier filter (binomial smoothing). The disadvantage of the second technique is that it relies on plane waves that enter and leave the interferometer. Measurement of curved mirrors is therefore complicated. This requirement is not crucial for the dynamic measurement, although different spot sizes from the two interferometer arms will decrease the fringe level.

For the GDD measurements using the dynamic white-light interferometer to characterize chirped mirrors at a center wavelength of around 800 nm typically an accuracy better than $\pm 5\ \mathrm{fs}^2$ is achieved for the GDD and better than ± 0.1 fs for the group delay, at a wavelength resolution of about ± 5 nm (Fig. 3.44). This is sufficient to resolve the dispersion variation of chirped mirrors and more than sufficient for components with smoother dispersion features. In case a better signal-to-noise ratio is required, we could replace the white-light source with a supercontinuum generated by femtosecond lasers, this being by far the brightest spatially coherent broadband light source available. To extend the spectral range that can be characterized to shorter wavelengths, we can either use a trigger laser of shorter wavelength or we can fold the beam of the trigger laser so that it hits the moving mirror twice. For narrow folding angles, this will approximately double the effective frequency of the trigger signal.

4 Nonlinear Pulse Propagation

In this chapter, we will discuss nonlinear pulse propagation as far as it is relevant in the context of ultrafast lasers discussed in this textbook. This is still an active field of research and there are other books dedicated to this topic. Thus this chapter should only be considered as an introduction to this field.

4.1 Self-Phase Modulation (SPM)

4.1.1 Kerr Effect and SPM

The intensity-dependent refractive index can be described by

$$n(I) = n + n_2 I \ , \tag{4.1}$$

where n is the refractive index, n_2 is the nonlinear refractive index, and I is the intensity. This is referred to as the (optical) Kerr effect, in analogy to the electrooptic Kerr effect, where the refractive index changes in proportion to the absolute square of the applied static electric field strength. In the literature, n_2 is given in cgs or SI units. The conversion from cgs to SI units is given by

$$n_2 \left[\frac{\mathrm{cm}^2}{\mathrm{W}}\right] = 4.19 \times 10^{-3} \frac{n_2\,[\mathrm{esu}]}{n} \ . \tag{4.2}$$

The nonlinear refractive index n_2 for different materials is summarized in Table 4.1. The Kerr effect yields a positive nonlinear refractive index n_2 when the laser wavelength is far enough from the bandgap to avoid multiphoton absorption. Using wide bandgap materials to satisfy this constraint means only moderate values of n_2 are

© The Author(s), under exclusive licence to Springer Nature Switzerland AG 2021

U. Keller, *Ultrafast Lasers*, Graduate Texts in Physics,
https://doi.org/10.1007/978-3-030-82532-4_4

Table 4.1 Nonlinear refractive index for different materials

Material	Refractive index n	n_2 (cm^2/W)	Reference
Sapphire (Al_2O_3)	1.76 @ 850 nm	3.1×10^{-16}	[04Maj]
Fused quartz	1.45 @ 1.06 μm	2.74×10^{-16}	[98Mil]
Glass (Schott BK7)	1.507 @ 1.06 μm	3.62×10^{-16}	[87Ada2]
Glass (Schott SF6)	1.774 @ 1.06 μm	18.9×10^{-16}	[87Ada2]
YAG ($Y_3Al_5O_{12}$)	1.815 @ 1.064 μm	6.24×10^{-16}	[89Ada]
YLF ($LiYF_4$)	$n_e = 1.47$ @ 1.047 μm	1.72×10^{-16}	[77Mil]
CALGO ($CaGdAlO_4$)	1.916 @ 1.06 μm	9.01×10^{-16}	[07Bou]
ZnS	2.289 @ 1.06 μm	32.6×10^{-16}	[94Kra]
ZnSe	2.483 @ 1.06 μm	63.0×10^{-16}	[94Kra]
Air (1 bar / 300 K)	$1 + 0.274 \times 10^{-3}$ @ 1.06 μm	4×10^{-19}	[97Nib]

typically accessible, since n_2 scales inversely with the fourth power of the bandgap [10Chr].

As can be seen, the order of magnitude in the near-infrared spectral range is around 10^{-16} cm^2/W for many different solid-state materials, whereas the nonlinear refractive index is reduced by about three orders of magnitude in air. So, we can normally neglect the nonlinear refractive index in air, but at higher intensities and inside ultrafast laser cavities, this can become significant and can no longer be neglected [08Mar] (Sect. 9.10.2).

The dispersion of the n_2 can be neglected for many cases. For example, for sapphire in the wavelength range 550–1550 nm the nonlinear refractive index decreases monotonically from around 3.3×10^{-16} cm^2/W to 2.8×10^{-16} cm^2/W [04Maj].

If a laser beam with constant beam area A_{eff} and thus with constant peak intensity propagates through a Kerr material of length L_K, a nonlinear phase shift is generated:

$$\varphi(t) = -kn(I)L_K = -k\left[n + n_2 I(t)\right] L_K \ , \tag{4.3}$$

where we have used (4.1). The nonlinear phase shift $\varphi_2(t)$ is given by

$$\varphi_2(t) = -kn_2 I(t) L_K = -kn_2 L_K \left|A(t)\right|^2 \equiv -\delta \left|A(t)\right|^2 \ , \tag{4.4}$$

where $|A(t)|$ is the amplitude of the pulse envelope, normalized such that $|A(t)| = I(t)$, and δ is the self-phase-modulation (SPM) coefficient defined by

$$\boxed{\delta \equiv kn_2 L_K} \ . \tag{4.5}$$

According to (4.5), the physical units of δ are the same as those of n_2.

The pulse envelope has already been introduced in Chap. 2, and we use the normalization convention that $|A(t)|^2$ is the intensity [see (2.52) and (2.53)]. Note that it is sometimes assumed in the literature and also in this textbook that $|A(t)|^2$ corresponds to the power, i.e., $|A(t)|^2 = A_{\text{eff}} I(t)$.

If a pulse $E(z,t)$ propagates through a Kerr material of length L_K and the dispersion can be neglected, $E(z,t)$ will change in the following way due to the Kerr effect:

$$E(L_K, t) = A(0,t)\exp\left[i\omega_0 t + i\varphi(t)\right] = A(0,t)\exp\left[i\omega_0 t - ik_n(\omega_0)L_K - i\delta|A(t)|^2\right] . \tag{4.6}$$

For small SPM we can linearize (4.6) as follows

$$\boxed{A(L_K, t) = e^{-i\delta|A|^2} A(0,t)e^{-ik_n(\omega_0)L_K} \overset{\delta|A|^2 \ll 1}{\longrightarrow} \approx \left[1 - i\delta|A(t)|^2\right] A(0,t)e^{-ik_n(\omega_0)L_K}} \tag{4.7}$$

The second term in the square brackets in (4.7) can then be defined as an SPM operator (Appendix F). This can be used to obtain analytical solutions for the theoretical treatment of nonlinear pulse propagation.

Self-phase modulation broadens the bandwidth of a pulse, because the instantaneous frequency is given by

$$\omega_2(t) = \frac{d\varphi_2(t)}{dt} = -\delta\frac{dI(t)}{dt} . \tag{4.8}$$

According to (4.8), the influence of SPM on the time-dependent frequency of a bandwidth-limited pulse can be inferred. In the leading edge of the pulse, where $dI(t)/dt > 0$, there is a reduction in the frequency, i.e., $\omega_2(t) < 0$. In the trailing edge of the pulse, where $dI(t)/dt < 0$, there is an increase in the frequency, i.e., $\omega_2(t) > 0$. This spectral broadening is shown in Fig. 4.1.

SPM-induced spectral broadening is accompanied by an oscillatory structure in the spectrum, as shown in Fig. 4.2. The number of oscillation peaks depends on the maximum nonlinear phase shift $\varphi_{2,\max}$ [70Cub]:

$$\left|\varphi_{2,\max}\right| = kn_2 I_p L_K \approx \left(M - \frac{1}{2}\right)\pi , \tag{4.9}$$

where I_p is the peak intensity and M is the number of peaks in the SPM-broadened spectrum. Therefore, counting the oscillation peaks, the SPM-broadened spectrum gives a quick estimate for the maximum nonlinear phase shift (Fig. 4.2).

4.1.2 Pulse Compressors Such as Fiber-Grating, Fiber-Prism, and Fiber-Chirped-Mirror Compressors

Spectral broadening produced by SPM can be used to compress bandwidth-limited pulses. Additional GDD can improve the quality of the compressed pulses. Depending on the available peak power, different gases, bulk materials, fibers, and waveg-

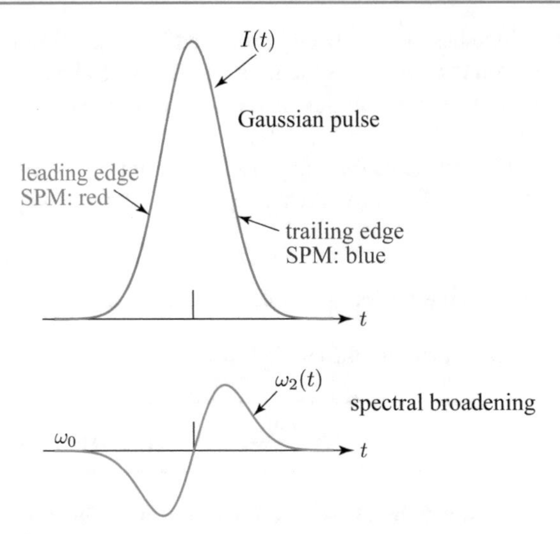

Fig. 4.1 Spectral broadening of a transform-limited pulse by SPM assuming $n_2 > 0$. *Top*: Time-dependent intensity with the leading and trailing edge of the pulse. *Bottom*: Instantaneous frequency versus time caused by SPM. SPM with $n_2 > 0$ will produce a spectral broadening generating lower frequencies (i.e., "red" frequency components) in the leading edge of the pulse and higher frequencies (i.e., "blue" frequency components) in the trailing edge of the pulse ("red is earlier than blue")

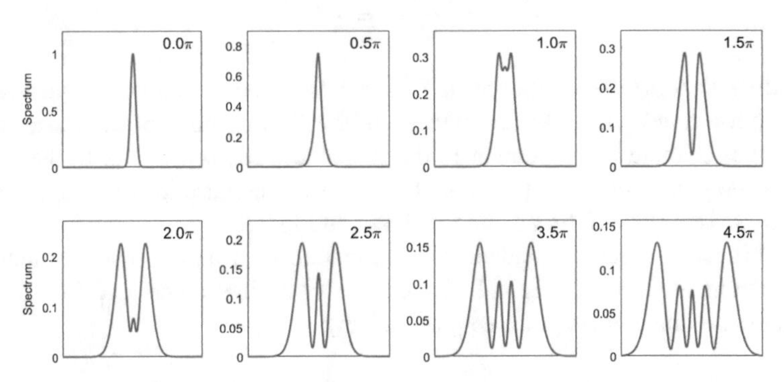

Fig. 4.2 Calculated SPM-broadened spectra for an initially unchirped 1 ps Gaussian pulse. Spectra are labeled by the maximum nonlinear phase shift $\varphi_{2,\max}$ added to the initial pulse, following the discussion given in [78Sto]

uides have been used for both SPM and GDD. With regard to fibers, standard single-mode fibers, microstructured or photonic crystal fibers, and hollow-core fibers have been used. With regard to waveguides, many different materials enable chip-based nonlinear photonics [19Gae]. Examples will be discussed in this section.

Mollenauer [80Mol] was the first to experimentally examine pulse compression with fibers using the Kerr effect. In this paper, the spectral width of the incoming bandwidth-limited pulse spectra was SPM-broadened and then compressed by the negative dispersion in the fiber. Using this approach, for example, a 7 ps pulse was compressed to 0.26 ps with a 100 m long fiber. This corresponds to shortening by a

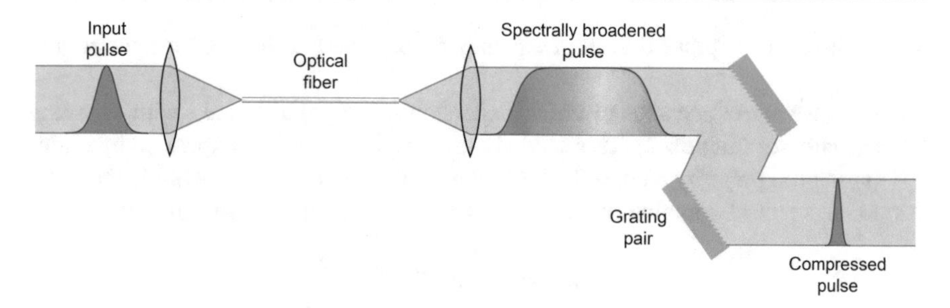

Fig. 4.3 Fiber-grating pulse compressor for femtosecond pulses

factor of 27. In the negative dispersion regime, one obtains so-called solitons, which we will discuss in more detail in Sect. 4.3.

SPM spectral broadening in the positive dispersion regime of the fiber can also be used to further shorten a bandwidth-limited pulse. In this case additional negative dispersion can be obtained to compress the pulses after the fiber using grating, prism, or chirped mirror pairs, as introduced in Chap. 3. A fiber-grating pulse compressor was pioneered by Grischkowsky (see Fig. 4.3) [82Gri, 83Nik]: first SPM spectrally broadens the pulse during propagation through the fiber, with positive dispersion which more strongly spreads the spectral components in time. Then in a second step, a pair of diffraction gratings provides the negative dispersion to compress the new frequency components in time [69Tre]. Optical pulse compression owes its heritage to radar development during World War II [60Kla].

Because of the exponential time function of the SPM operator [see (4.7)], the fiber-grating pulse compressor needs to be solved and optimized numerically. The combination of SPM and fiber dispersion generates a complex spectrum with a frequency-dependent phase after the fiber of length L, which can be expanded in a power series around a center frequency:

$$\phi(L,\omega) = k_n L = \alpha_1 + \alpha_1(\omega-\omega_0) + \alpha_2(\omega-\omega_0)^2 + \cdots \,. \tag{4.10}$$

Because the diffraction grating pair is a linear system, we can multiply the pulse spectrum by the grating transfer function (see Chap. 3, Table 3.2). When the linear chirp in the pulse (controlled by α_2) is equal and opposite to the chirp produced by the gratings, the quadratic phase term is eliminated and the pulse will be compressed. However, this neglects any higher order dispersion compensation, which can lead to some distortion of the pulse if it becomes significant.

Example. Fiber-Grating Pulse Compression. A Gaussian pulse, which propagates through a fiber, experiences a self-phase modulation (SPM) and gets chirped. Hence, the phase of the pulse is frequency dependent. We can write the Gaussian pulse as (2.35)

$$A(t) = \mathrm{e}^{-\Gamma_1 t^2}\mathrm{e}^{\mathrm{i}\Gamma_2 t^2}\mathrm{e}^{\mathrm{i}\omega_0 t} = \mathrm{e}^{-\Gamma t^2}\mathrm{e}^{\mathrm{i}\omega_0 t} \,,$$

where $\Gamma \equiv \Gamma_1 - \mathrm{i}\Gamma_2$. Here, Γ_1 determines the duration and Γ_2 the chirp of the pulse:

(a) Transform the pulse into the frequency domain and normalise the spectrum to unity.
(b) Now we introduce second order dispersion into the pulse, e.g., with a grating compressor (Table 3.2). This second order dispersion can be expressed by multiplying the pulse by a phase factor in the frequency domain. If $A(\omega)$ is the pulse in the spectral domain obtained in part (a) of this exercise, we can write

$$A'(\omega) = A(\omega)\mathrm{e}^{-\mathrm{i}D_2(\omega_0-\omega)^2} .$$

Express the new Γ of the pulse as a function of Γ_1, Γ_2, and D_2.
(c) Introduce $d = D_2/\tau_0^2$ and $\gamma = \Gamma_2\tau_0^2$. Determine and plot $\tau_{\mathrm{new}}^2/\tau_0^2$ as a function of d and γ. With which combination of parameters would you expect a pulse compression?

Solution

(a) Self-phase modulation (SPM) can be described by a nonlinear phase shift

$$\varphi_2(t) = -\delta|A(t)|^2 ,$$

where δ is the SPM coefficient given by (4.5). The pulse is described by

$$A(t) = \mathrm{e}^{-\Gamma_1 t^2}\mathrm{e}^{\mathrm{i}\Gamma_2 t^2}\mathrm{e}^{\mathrm{i}\omega_0 t} = \mathrm{e}^{-\Gamma t^2}\mathrm{e}^{\mathrm{i}\omega_0 t} ,$$

with $\Gamma \equiv \Gamma_1 - \mathrm{i}\Gamma_2$. The Fourier transform of the pulse into the spectral domain reads

$$\begin{aligned}
A(\omega) &= C\int_{-\infty}^{+\infty} \mathrm{e}^{-\Gamma t^2}\mathrm{e}^{\mathrm{i}\omega_0 t}\mathrm{e}^{-\mathrm{i}\omega t}\,\mathrm{d}t \\
&= C\int_{-\infty}^{+\infty} \exp\left\{-\Gamma\left[t^2 - \mathrm{i}\frac{\omega_0-\omega}{\Gamma}t + \left(\mathrm{i}\frac{\omega_0-\omega}{2\Gamma}\right)^2 - \left(\mathrm{i}\frac{\omega_0-\omega}{2\Gamma}\right)^2\right]\right\}\mathrm{d}t \\
&= C\times\exp\left[-\frac{(\omega_0-\omega)^2}{4\Gamma}\right]\int_{-\infty}^{+\infty}\exp\left[-\Gamma\left(t-\mathrm{i}\frac{\omega_0-\omega}{2\Gamma}\right)^2\right]\mathrm{d}t \\
&= C\times\exp\left[-\frac{(\omega_0-\omega)^2}{4\Gamma}\right]\sqrt{\frac{\pi}{\Gamma}} ,
\end{aligned}$$

where C is a constant to be determined. Imposing the normalisation condition $A(\omega=\omega_0)=1$, we obtain

$$A(\omega_0) = C\sqrt{\frac{\pi}{\Gamma}} = 1 ,$$

whence the constant in the above is $C = \sqrt{\Gamma/\pi}$ and $A(\omega) = \mathrm{e}^{-(\omega_0-\omega)^2/4\Gamma}$.

(b) We have

$$A'(\omega) = A(\omega)\mathrm{e}^{-\mathrm{i}D_2(\omega_0-\omega)^2}\mathrm{e}^{-(\omega_0-\omega)^2}\mathrm{e}^{\mathrm{i}D_2+1/4\Gamma} \; .$$

Hence, Γ' can be found from

$$\frac{1}{4\Gamma'} = \frac{1}{4\Gamma} + \mathrm{i}D_2 \; ,$$

which leads to

$$\Gamma' = \frac{\Gamma}{1+\mathrm{i}4\Gamma D_2} = \frac{\Gamma_1 - \mathrm{i}\Gamma_2}{1+\mathrm{i}4\Gamma_1 D_2 + 4\Gamma_2 D_2} \; .$$

(c) The real part of Γ, denoted Γ_1 here, defines the pulse duration through (2.40)

$$\tau_0^2 = \frac{2\ln 2}{\Gamma_1} \; .$$

Accordingly, the new time duration is related to the real part of Γ', denoted Γ_1' here, by the expression

$$\Gamma_1' = \frac{\Gamma_1}{(1+4\Gamma_2 D_2)^2 + 16\Gamma_1^2 D_2^2} \; .$$

Therefore,

$$\frac{\tau_{\mathrm{new}}^2}{\tau_0^2} = (1+4\Gamma_2 D_2)^2 + 16\Gamma_1^2 D_2^2 = (1+4\gamma d)^2 + (8d\ln 2)^2 \; ,$$

where we have used $D_2 = d\tau_0^2$ and $\Gamma_2 = \gamma/\tau_0^2$. For certain parameters, we can achieve pulse compression, with $\tau_{\mathrm{new}}^2/\tau_0^2 < 1$. For example, for $\gamma = -3$ and $d \approx 0.1$.

SPM generates a chirp that is only linear in the center of the pulse (Fig. 4.4). Only this part can be compressed by the grating pulse compressor. The remaining energy in the pulse, which is not properly compressed, will cause a broadening of the pulse from the minimum width expected from the spectral broadening ratio, and it introduces longer wings in the pulse. This degrades the quality of the compressed pulse. But the combination of GDD and SPM can be optimized to more strongly linearize the chirp in the pulse and thereby make the compression process more efficient, with higher quality pulses and less energy in the wings. This enhanced frequency chirping substantially improves the quality of the compressed pulses with an optimal length of the fiber [82Gri, 84Tom] (Fig. 4.4).

A compact and efficient pulse compressor for flashlamp-pumped actively modelocked 80 ps Nd:YAG lasers at a center wavelength of 1.06 μm was achieved for the first time using the grating compressor at near grazing incidence [84Kaf].

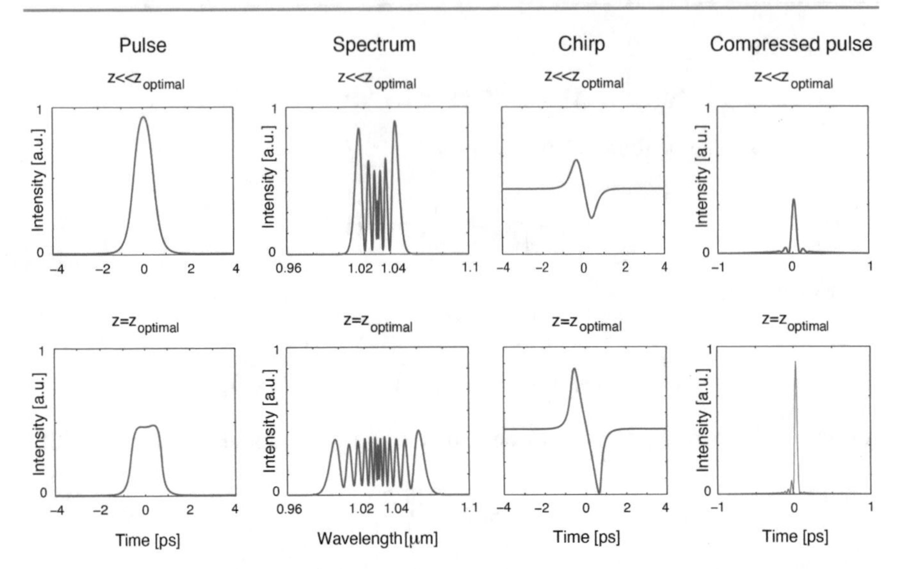

Fig. 4.4 Fiber-grating pulse compressor for higher quality pulses with an optimal length of fiber z_{optimal}. Pulse shape before and after compression, and spectrum and chirp for two different fiber lengths z. For $z \ll z_{\text{optimal}}$ the fiber length is too short to generate a sufficiently strong linear chirp through positive dispersion. Therefore, a large part of the energy cannot be compressed into a short pulse shape by a grating compressor. The figures at the bottom show the situation for a more optimum fiber length $z = z_{\text{optimal}}$

At that time, the pulse duration was rather long, typically 80 ps, which required a 300-m long single-mode fiber to sufficiently broaden the spectrum of the input pulse with SPM from 0.03 nm to around 3.5 nm, for an average input power of 6 W. With a 1800 lines/mm diffraction grating pair at near grazing incidence, the pulse was compressed to 1.8 ps. However, because of the limited GDD in the fiber at 1.06 μm, the compressed pulse still had strong wings. The final pulse compression ratio of about 45 agreed well with the theoretical prediction with no GDD.

To achieve pulse compression below 10 fs, the third order dispersion also had to be compensated for. Hence, a combination of a grating and a prism compressor was used for dispersion compensation of the second and third order (Fig. 4.5). This resulted in the long-standing world record pulse duration of 6 fs in 1987, at a pulse repetition rate of 8 kHz [87For]. This world record pulse duration was surpassed for the first time at ETH Zurich in 1999—12 years later—with a SESAM-assisted Kerr lens modelocked Ti:sapphire laser using double chirped mirrors (DCMs) and prism pairs for external higher order dispersion compensation (see Fig. 4.6) [99Sut]. This time, however, these ultrashort pulses were obtained directly from a modelocked oscillator at an 85 MHz pulse repetition rate without further spectral broadening in a fiber and without any external amplification. A higher pulse repetition rate provides a much higher signal-to-noise ratio in ultrafast pump–probe measurements because, for a given pulse energy, we have a much higher average photon flux, as discussed for example in [03Kel,08Sud].

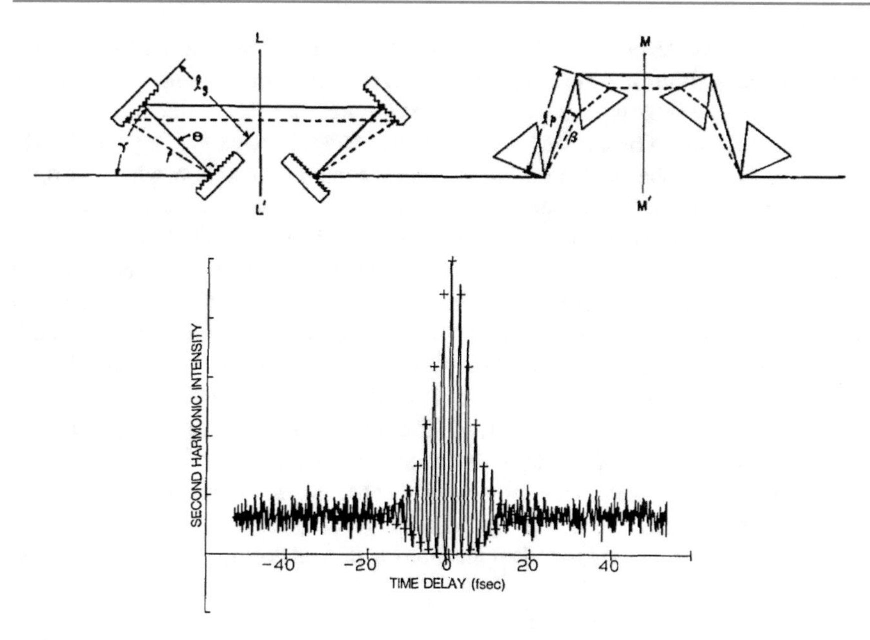

Fig. 4.5 World-record pulse duration in 1987. *Upper*: Fiber-grating-prism pulse compressor for the long-standing world record pulse compression from 50 fs to 6 fs: second and third order dispersion had to be compensated for by a combination of grating and prism compressors. The spectrum was SPM-broadened in a polarization-preserving quartz fiber with core diameter about 4 μm and length 0.9 cm. The optical intensity inside the fiber was 1 to 2 × 10^{12} W/cm^2. *Lower*: Measured interferometric autocorrelation trace, a pulse duration measurement technique that will be described in Chap. 10, at a center wavelength of 620 nm and 8 kHz repetition rate. After [87For]

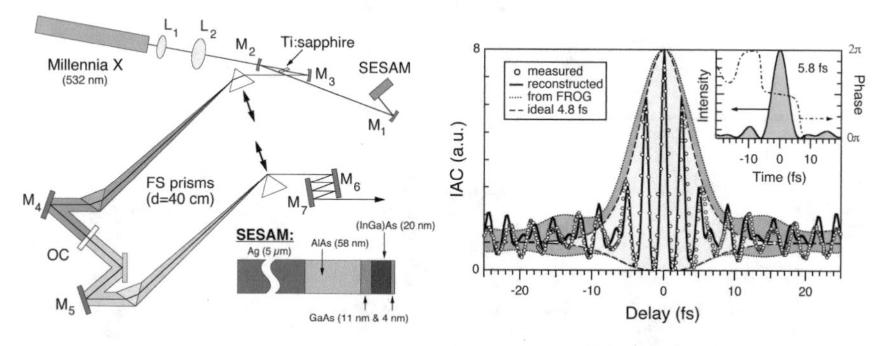

Fig. 4.6 World record pulse duration in 1999 generating 5.8 fs pulses—surpassing for the first time the long-standing world record of 6 fs in 1987. The center wavelength is shifted to about 800 nm because the passively modelocked laser is a Ti:sapphire laser. The different pulse duration measurement techniques will be described in Chap. 10. After [99Sut]

Because of the development of microstructure fibers [03Rus], low-energy pulses coming directly from an ultrafast laser oscillator could be further compressed into the few-cycle regime by using fiber-based pulse compression. These fibers consist of a solid core surrounded by an array of air holes and have already found applications in many fields. One of the most important is supercontinuum generation, where the optical spectrum is strongly broadened, exploiting many different nonlinear processes in addition to SPM, as discussed for example in [06Dud,02Gae]. A more detailed discussion of supercontinuum generation goes beyond the scope of this book. Small mode area fibers, as used in Fig. 4.7 for supercontinuum generation, benefit not only from the small mode area and therefore a high peak intensity, but also from the modified dispersion characteristics of these fibers. With these fibers, supercontinuum generation has become possible even with nanojoule pulses directly generated by a modelocked Ti:sapphire laser [00Ran]. The ultrabroad spectra generated in these fibers have some remarkable properties: they can span a spectral bandwidth covering a full optical octave, they are spatially coherent, and they have a brightness which exceeds the brightness of a light bulb by at least 5 orders of magnitude. These supercontinua have already found applications in optical coherence tomography and

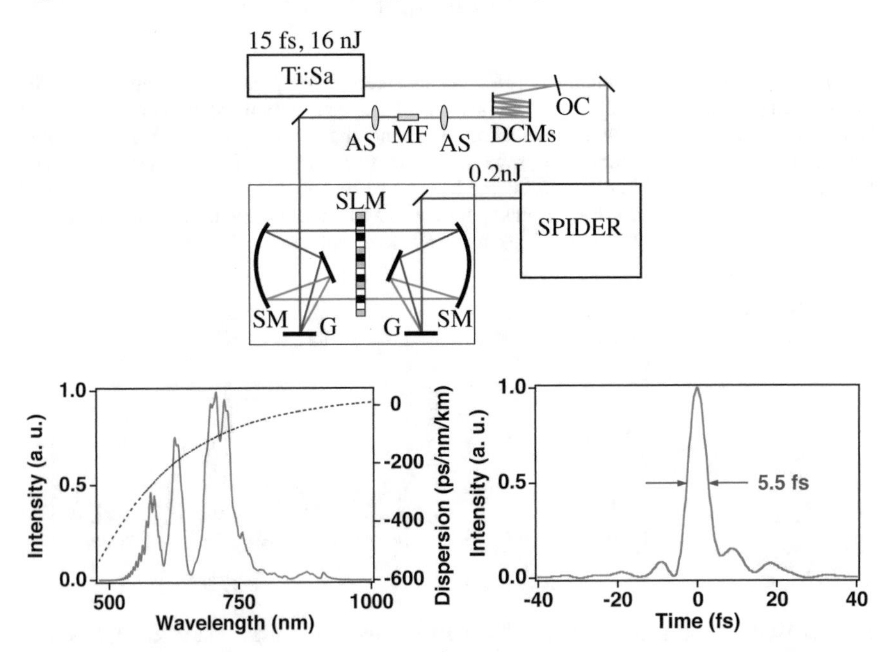

Fig. 4.7 Fiber-prism pulse compressor using a 5 mm long microstructure fiber (MF) (i.e., photonic crystal fiber), after [05Sch4]. *Upper*: Experimental setup: Ti:sapphire oscillator generating 15 fs pulses with pulse energy 16 nJ, center wavelength about 790 nm, and repetition rate 19 MHz. Output coupler OC, double chirped mirrors DCMs, aspherical lens AS, spherical mirror SM, grating G, spatial light modulator SLM. *Lower left*: The *solid curve* is the experimentally measured output spectrum of the fiber with core diameter 2.6 μm. The *dotted curve* shows the dispersion profile of this fiber. *Lower right*: Reconstructed temporal pulse profile showing a FWHM pulse duration of 5.5 fs (i.e., $\approx$2.1 optical cycles)

in frequency metrology, where they deliver octave-spanning frequency combs (see Chap. 12 for more details on optical frequency combs).

The broad coherent bandwidth also suggests that, with appropriate dispersive compression, such supercontinua should allow the generation of very short femtosecond pulses in the few-cycle regime. Indeed, in 2005, 5.5 fs pulses at a center wavelength of about 790 nm (i.e., about 2.1 optical cycles) were generated by dispersive compression of a supercontinuum generated with 15 fs input pulses from a Ti:sapphire laser into a 5 mm long microstructure fiber [05Sch4] (Fig. 4.7). Furthermore, by combining the output of two PCFs, more than an octave bandwidth was produced in 2010, which could be compressed to a single cycle pulse with a pulse duration of 4.3 fs at a center wavelength of about 1.5 μm (i.e., about 0.9 optical cycles). This requires a coherent superposition of two coherent ultrabroadband spectra generated in each of the two PCFs. Note that in this few-cycle pulse duration regime the pulse spectrum becomes more complex and there is typically significant energy in longer pulse wings or shoulders of the compressed pulses. A simple pulse duration based on FWHM in this case can be somewhat misleading (see Chap. 10 for more details).

The spectral coherence of supercontinua generated in highly nonlinear processes can be imperfect [02Dud], and this can limit the compression process, because a dispersive compressor can be adjusted only relatively slowly and so cannot be optimized for each pulse separately when the spectral phase undergoes strong fluctuations. Particularly for very broad spectra, this problem can be severe. However, this can be resolved with the right choice of fiber parameters (dispersion, length, etc.) and of the initial pulses. In addition, higher order dispersion compensation was required for the generation of 5.5 fs pulses in our group. For example, this can be obtained with a liquid crystal spatial light modulator (SLM) [00Wei]. The overall system can be optimized by numerical simulation, which ultimately allowed for pulse compression from 15 to 5.5 fs at a center wavelength of around 790 nm, a world record at that time. The pulse duration was measured using the SPIDER technique [98Iac], which will be discussed in Chap. 10.

The scaling of fiber-based pulse compression from Ti:sapphire laser amplifier systems towards both shorter and higher energy pulses became possible for the first time with the invention of the hollow fiber compressor [97Nis]. Scaling towards higher pulse energies in a normal or even microstructured fiber (Fig. 4.12) would ultimately damage the material. We can prevent damage if the hollow fiber is filled with a neutral gas and SPM in the gas leads to a large spectral broadening. Using a single-stage hollow-fiber-chirped-mirror compressor, pulses as short as 4.5 fs have been generated [97Nis], and 3.8 fs with a dual stage compressor [03Sch] (Fig. 4.8). The spectrally broadened pulses after the hollow fiber are typically compressed with many reflections from a chirped mirror pair to support a sufficient amount of second and third order dispersion compensation over a very large bandwidth. However, for the 3.8 fs pulses, an additional spatial light modulator (SLM) had to be used for optimized pulse shaping (Fig. 4.8). Note that an optical cycle at 800 nm corresponds to 2.7 fs. The pulse shaper is discussed in Sect. 3.3.3. The advantage is that, with such a pulse shaper, we can determine the required dispersion compensation and can in principle design DCMs that provide such dispersion. Therefore, we can assume that,

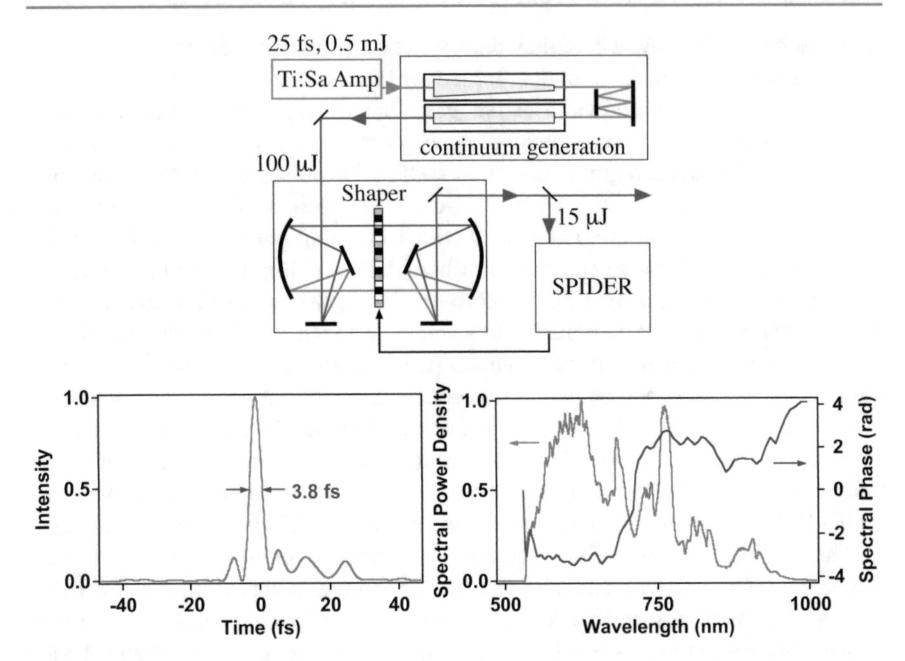

Fig. 4.8 Dual stage hollow fiber pulse compressor of pulses from a Ti:sapphire amplifier system (Ti:Sa Amp) that produces 25 fs pulses with 0.6 mJ pulse energy. Pulses as short as 3.8 fs are generated with energies of up to 100 μJ before and 15 μJ after the spatial light modulator (SLM) for adaptive dispersion compensation. Spectral broadening is achieved in two cascaded hollow fibers, dominated by SPM. Adaptive ultra-broadband dispersion compensation was achieved through a closed-loop combination of an SLM pulse shaper and a full pulse characterization for amplitude and phase using spectral phase interferometry for direct electric field reconstruction (SPIDER) as the feedback signal. After [03Sch]

with better DCMs, we could obtain 3.8 fs pulses at the full pulse energy of 100 μJ without the large loss introduced by an SLM. The pulse energies ideally need to be in the 100 μJ regime for high field physics experiments—generally, the higher the pulse energy and the shorter the pulse, the better it is for these applications. Initially, the hollow fiber compressor was the key enabling laser technology for single attosecond pulse generation through high harmonic generation [01Hen]. Through the work with hollow fiber compressors, filament compressors in the few-cycle regime were for the first time demonstrated in our group [04Hau]. Filament formation will be discussed in Sect. 4.2.

The dual stage hollow fiber pulse compressor shown in Fig. 4.8 [03Sch] is based on a 1 kHz Ti:sapphire laser-amplifier system, delivering 25 fs pulses with pulse energies of 0.5 mJ. This output is focused into the first argon-filled capillary consisting of a 60 cm long fiber with an inner diameter of 0.5 mm at the entrance side and 0.3 mm at the exit side. The emerging pulses are then compressed by chirped mirrors (5 bounces, 30 fs^2 each). The gas pressure in the first fiber (0.3 bar) was not optimized for maximum spectral broadening, but chosen such that, after compression, 10 fs pulses with negligible wings are obtained. Sending these pulses into a second Ar-filled hollow fiber (constant inner diameter 0.3 mm, length 60 cm) finally leads to a high-energy supercontinuum (100 μJ, 500 THz bandwidth) with excellent spatial characteristics. Note that, in comparison to a single hollow fiber, the use of two cascaded hollow fibers with an intermediate compression stage allows for the generation of a supercontinuum with more spectral energy in the blue and green part of the spectrum. The emerging beam was then collimated and sent into a pulse shaper [00Wei] consisting of a 640 pixel liquid crystal SLM (Jenoptik, each pixel of the SLM is 97 μm wide and separated by a 3 μm gap), two 300 l/mm gratings and two 300 mm focal length spherical mirrors (4-f setup). Alignment for normal incidence on the spherical mirrors proved to be critical in order to preserve good beam quality. This was necessary because of the very large beam diameter of 6.4 cm caused by the divergence of the gratings. The gratings introduced considerable losses: the 100 μJ pulse energy at the entrance of the pulse shaper was reduced to 15 μJ at the exit.

Note that in the few-cycle pulse duration regime the optical spectrum in most cases is complex and does not for example correspond to a simple sech^2 shape. In this case a FWHM pulse duration can be somewhat misleading and more information should be provided. For example, as shown in Sect. 2.4.3, an rms pulse duration τ_{rms} (2.47) was introduced for this reason. It also may be useful to calculate how much pulse energy is contained in an ideal pulse shape.

4.1.3 Nonlinear Optical Pulse Cleaner

Very pronounced pulse shoulders, which originate for example from non-ideal pulse compression or from higher-order solitons (as discussed in Sect. 4.3) can be suppressed by an optical pulse cleaner [92Tap, 92Bea], as shown in Fig. 4.9. One uses a birefringent material such as, for example in 1992, a slightly birefringent single-mode fiber. The input polarization is not along one of the birefringent principal axes. Thus, the incoming pulse is split into two polarization states which are perpendicular to each other and propagate with different phase velocities. The high-intensity part of the pulse will change the refractive index due to the Kerr effect (4.1). The

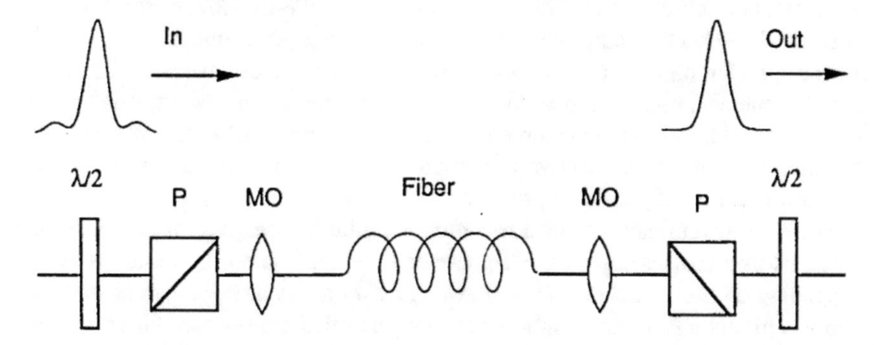

Fig. 4.9 Experimental setup of an optical pulse cleaner, based on nonlinear birefringence. After [92Tap]

birefringence as well as the rotation of the polarization direction of the pulse will thereby become intensity dependent. Depending on the alignment of the polarizer after the birefringent material, the weak intensity part of the pulse can therefore be separated.

This technique was initially used in 1992 to clean up the pulses after a double-stage fiber-grating pulse compressor to reduce the pulses from an actively modelocked Nd:YLF laser to around 1 ps. Note that only three years later, in 1995, pulses with an even better quality were generated directly from a compact diode-pumped Nd:glass laser using novel modelocking techniques, as discussed in Chap. 9 [95Kop1].

4.1.4 Average Power Scaling of Pulse Compressors

The continued advancement of laser technology opens up more and more possibilities for experiments, as peak intensities and repetition rates increase and new wavelengths become available. A higher pulse repetition rate also provides a much higher signal-to-noise ratio in ultrafast pump–probe measurements because, for a given pulse energy, we have a much higher average photon flux [03Kel, 08Sud]. Such short pulses were predicted to be useful for high field physics experiments because the output beam can be focused to a peak intensity of 10^{14} W/cm^2, a regime where high field laser physics such as high harmonic generation and laser plasma generated X-rays are possible at more than 10 MHz pulse repetition rate. This improves the signal-to-noise ratio in measurements by 4 orders of magnitude compared to the standard sources at kHz repetition rates. This would be important for low-power applications such as X-ray imaging and microscopy, femtosecond EUV and soft-X-ray photoelectron spectroscopy, and ultrafast X-ray diffraction.

Average power scaling of ultrafast laser systems is usually based on Yb-doped gain materials. See Sect. 9.10.2 for more detail. My own research work has focused on average-power scaling of ultrafast laser oscillators. The diode-pumped Yb-doped thin-disk laser [94Gie] is ideally suited for this application [15Sar]. In 2019, the average output power was scaled up to as high as 350 W with passively modelocked diode-pumped Yb-doped thin disk laser oscillators [19Sal]. Coherently combined ultrafast Yb-doped fiber amplifiers resulted in an average power above 3 kW [18Mul]. However, so far, high average power Yb-doped laser systems have not been able to directly generate pulse durations in the few-cycle regime without external pulse compression. In addition, external chirped-pulse amplification (CPA) generally increases the pulse duration due to effective gain narrowing. Therefore there is a need for average power scaling of pulse compression.

There are several approaches for compressing high average power femtosecond and picosecond pulses based on self-phase modulation. Table 4.2 gives an overview. Depending on the available peak and average power, compression can be obtained with gas-filled multi-pass cells, bulk crystals, gas-filled hollow-core fibers or capillaries, large mode area fibers, or in free space via filamentation. With more moderate average powers, compression in fibers and capillaries can offer excellent beam quality and low spatiotemporal coupling due to the guided wave nature of the device.

Table 4.2 Overview for power scaling of nonlinear pulse compression. Multi-pass cell (MPC), hollow-core fiber (HCF), hollow-core photonic cystal fibers (HC-PCF). The MPC is typically a Herriott-type MPC. A plus sign is considered a benefit and a minus a challenge. The best approach depends on the available peak power and average power to be compressed

Technology	Remarks	Typical compression factor
Capillary	+ Pioneering results for high energy	10 up to ∼ 20
Kagome HC-PCF	+ Guided mode, highly efficient	10 up to ∼ 20
Filamentation	− Spatially non-uniform	∼ 5
Bulk—single piece	− Spatially non-uniform	∼ 3–4
Bulk—cascaded χ^2	+ Prevents Kerr self-focusing	∼ 3–4
Bulk—multiple plates (MPC)	+ Reduced spatial non-uniformity	5 (first demonstration)
Free-space gas-filled MPC	+ For more power scaling	6 (at 1 mJ, 1 kW)

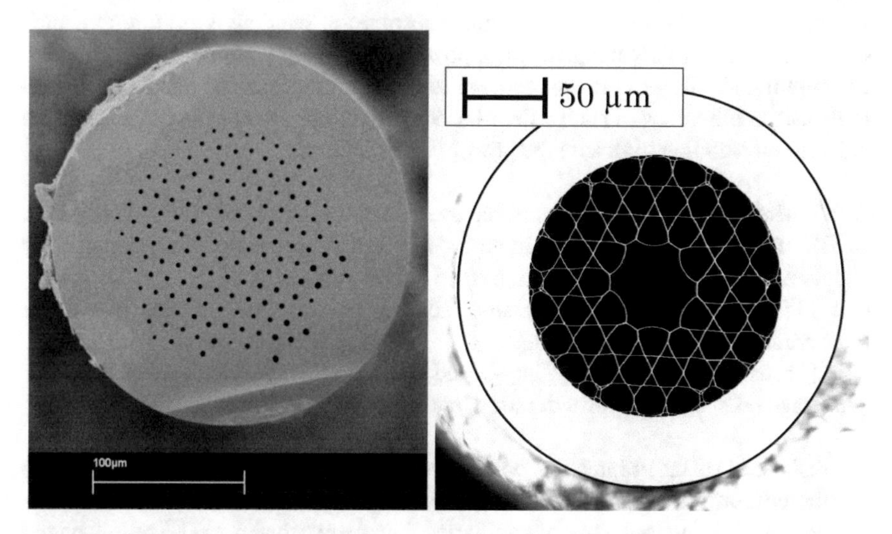

Fig. 4.10 Novel fibers for high average power pulse compression. *Left*: Scanning electron microscope image of a large mode area (LMA) photonic crystal fiber (PCF) with an effective mode area of about 200 μm^2, a hole spacing of about 11 μm, and a hole diameter of 2.7 μm [03Sud]. *Right*: A Kagome-type hollow-core photonic crystal fiber (HC-PCF) [14Ema]

For example, the large mode area fiber shown in Fig. 4.10 (left) has been used to compress 810 fs pulses to 33 fs with an average power of 18 W at a 34 MHz pulse repetition rate [03Sud]. The average power could be increased to 250 W, but only with strongly chirped input pulses which substantially increased the complexity of the system. The input peak power was limited to about 4.6 MW before damage occurred.

Gas-filled hollow-core photonic crystal fibers (HC-PCF) [11Ben] with a Kagome-type structure (Fig. 4.10 right) became important for high average power pulse com-

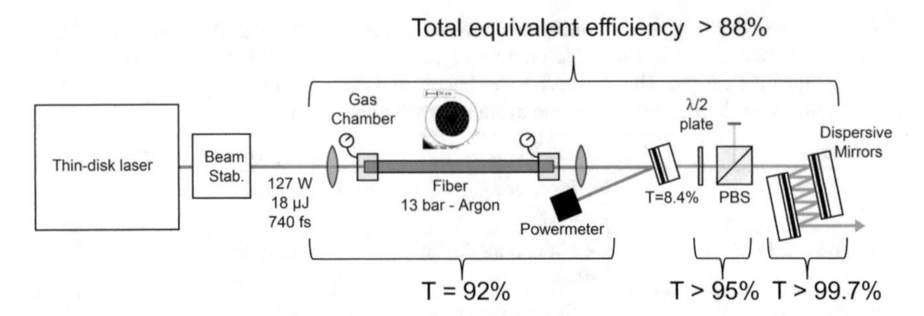

Fig. 4.11 Experimental setup for high average power pulse compression with a Kagome-type HC-PCF. The output of the thin-disk laser (TDL) is spatially stabilized and sent directly into the fiber. The 66 cm long hypocycloid-core Kagome HC-PCF is held in a gas chamber with 13 bar of argon. The transmission through the fiber alone reaches up to 118 W of output average power (93% of transmitted power). Power transmission T [14Ema]

pression in 2011 [11Hec]. These fibers further improved both the high damage threshold and the ultra-low losses over a wide transmission window. Working with such Kagome-type HC-PCFs at high average power requires input laser beams with excellent pointing stability to prevent damage to the input facet of the fiber. The overall transmission efficiency of such a fiber-based compressor system can be greater than 80% and outstanding high average power results include:

- In [14Ema] (Fig. 4.11), a transmitted average power of 112 W was obtained with 88% efficiency through a Kagome-type HC-PCF spectrally broadening 740 fs pulses 16×, and supporting compressed pulses as short as 88 fs.
- In [15Bal], the highest compression factor of 18 was demonstrated in a single Kagome fiber compression stage.
- In [16Had], a two-stage capillary-based compression scheme resulted in 216 W average power, 6.3 fs pulse duration, and 17 GW peak power.

The result of [16Had] represents one of the highest average powers for few-cycle fiber-based compression. Nonetheless, since the overall power transmission efficiency was 33% due to the use of capillaries, together with about 60% of this power being in the main compressed pulse, the absolute efficiency (energy in the main pulse) is about 20%. Hence, there are still ongoing efforts for significant further improvement. Indeed, this reduced energy in the main pulse is an issue common to many nonlinear compression methods based on self-phase modulation (SPM) followed by group delay dispersion (GDD), and mitigating this issue will be extremely beneficial. Filamentation is an alternative to capillaries and will be discussed in Sect. 4.2.

It is beyond the scope of this textbook to review all the high average power pulse compression results. I will therefore discuss only one earlier result in more detail, achieved in my own lab in 2003 using a high average power Yb:YAG thin disk laser generating 80 W of average power, with a pulse duration of 705 fs, at 57 MHz pulse repetition rate, and at a center wavelength of 1030 nm [03Inn,04Bru]. To further reduce the pulse duration of the laser output, this was the first application

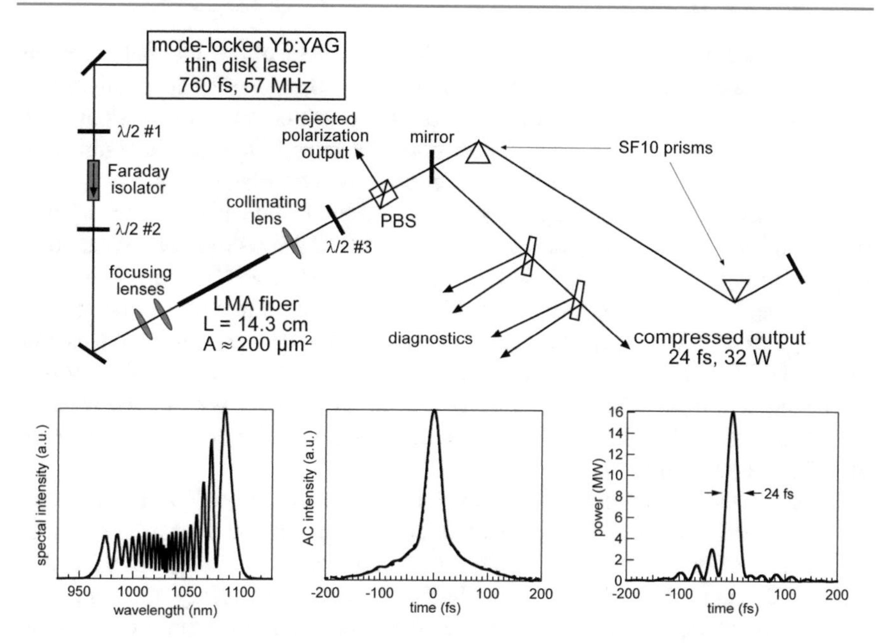

Fig. 4.12 Fiber-prism pulse compression in 2005 using a large-mode-area (LMA) fiber shown in Fig. 4.10 left. *Upper*: Experimental setup. For an incident average power of 60 W and a pulse duration of 760 fs, we obtained 42 W after the fiber and 32 W after the prism pair compressor, with a pulse duration of 24 fs. *Lower*: Measured SPM broadened pulse spectrum, autocorrelation (AC), and retrieved pulse with 24 fs FWHM pulse duration. The strong wings in the pulse reflect the fact that the pulses have not been compressed to an ideal pulse shape. After [05Inn]

of the well-known pulse compression technique using a nonlinear fiber and a prism compressor at such high average power levels [03Sud]. The use of a large-mode-area microstructured fiber (Fig. 4.10 left) allowed us to overcome limitations arising from the high peak intensity when operating at such power levels. Figure 4.12 (lower) shows the world-leading result at that time with a fiber-prism pulse compressor using input pulses with 60 W and 760 fs and compressing them to 24 fs with 32 W average output power [05Inn].

We used a large-mode-area microstructured fiber with a mode area of about 200 μm^2 (Fig. 4.10 left) and a length of 14.3 cm for the spectral broadening, while two SF10 prisms were used in a double-pass configuration to remove the resulting chirp and compress the pulses. Figure 4.12 (upper) shows the experimental setup. After the passively modelocked thin-disk laser, we used a set of polarizing optics to adjust the power incident on the fiber and to control the input polarization. A Faraday isolator was used to avoid residual reflections from the fiber face back into the laser cavity. The first half-wave plate ($\lambda/2$ #1) was used for the power adjustment, while the second half-wave plate ($\lambda/2$ #2) allowed us to change the polarization incident on the fiber. We used two focusing lenses to control the waist size and the position of the focus. The large-mode-area fiber used had a mode radius of about 8 μm and thus an effective mode area of 200 μm^2. After the fiber output, a third lens was used to collimate the beam, while the following half-wave plate ($\lambda/2$ #3) and polarizing beam splitter defined a linear polarization state of the fiber output. A pair of Brewster-angled SF10 prisms was used to linearly compress the spectrally broadened pulses. The pulses were extracted from the prism compressor on a mirror by introducing a small vertical offset for the beam traveling in the backward direction. The two wedged glass plates allowed beam extraction for diagnostics.

We measured 32 W of average power in the output using launch powers of up to 60 W and launch efficiencies of around 70%, while 10 W was rejected by the polarizing beam splitter after the third half-wave plate ($\lambda/2$ #3). Thus a polarization-maintaining fiber would be more efficient. The optical spectrum of the output spans the range from 970 to 1090 nm (Fig. 4.12 lower left) and clearly shows the oscillation in the SPM-broadened spectrum. The asymmetry results from additional self-steepening (Fig. 4.29). Assuming a flat phase, the calculated FWHM pulse duration of this spectrum would be 20 fs. However, the imperfect pulse compression resulted in a FWHM pulse duration of 24 fs.

Schemes using bulk crystals can exhibit reduced beam quality and increased spatiotemporal coupling. The presence of spatiotemporal coupling means that the dependence of the electric field on the spatial coordinates is no longer independent from its dependence on the time coordinate. We have seen plenty of examples in previous chapters, such as angular dispersion, where the angle and thereby often the path length differs for different spectral components. These spectral components are then temporally spread out across the spatial profile of the beam, thus inducing spatiotemporal coupling. The fact that bulk crystals can exhibit a reduced beam quality and spatiotemporal coupling means that they require spatial filtering and suffer from corresponding losses. Specifying the actual efficiency (energy in main compressed pulse vs input energy) is therefore challenging. Spatiotemporal coupling can be somewhat mitigated by using multiple thin crystal plates. The number of plates, material, and thickness can be numerically optimized. Self-focusing (see Sect. 4.2) can be avoided by using cascaded quadratic nonlinearities. A more compact procedure for compression in bulk crystals uses a multipass cell (MPC) [16Sch], allowing for high throughput and quite good beam quality. Results with SPM generation in bulk crystals and free-space gas-filled MPC demonstrating efficient compression at high average power include:

- In [16Sei1], 50 W average power bulk compression from 250 to 43 fs (compression factor 5.8) with a power transmission of about 60%.
- In [16Sei2], cascaded quadratic nonlinearities were used in a multi-stage scheme. Compression to 30 fs (compression factor about 6 over 3 stages) was demonstrated with a transmitted power of 70 W.
- In [16Sch], a Herriott cell was used for multipass bulk compression, to compress to 170 fs (compression factor 5), with 90% transmission efficiency and 375 W average output power.
- In [19Tsa], dual-stage compression from an ultrafast thin-disk laser oscillator compressing pulses from 534 to 27 fs with an average power transmission of 98 W and an 80% overall average power compression efficiency using a multipass cell followed by a multiple plate stage.

- In [20Gre], kW average power pulse compression with mJ pulse energy in a free-space gas-filled multi-pass cell (MPC). In this case using the output from a coherently combined Yb:fiber laser system with an argon-filled Herriott-type MPC with 96% efficiency resulted in 1 kW average power, 31 fs pulse duration, and a 1 mJ pulse energy.

4.2 Self-Focusing and Filamentation Compressor

4.2.1 Self-Focusing via a Kerr Lens

Due to the Kerr effect, a laser beam produces a focusing lens effect (Fig. 4.13). For a sufficiently thin Kerr material with length L_K, one can assume that the beam radius of the laser w is constant and one can make a parabolic approximation for the beam profile:

$$I(x, y) = I_p \exp\left(-2\frac{x^2+y^2}{w^2}\right) \overset{(x^2+y^2)\ll w^2}{\longrightarrow} \approx I_p\left(1 - 2\frac{x^2+y^2}{w^2}\right) , \tag{4.11}$$

where I_p is the peak intensity. With the Kerr effect (4.1) and (4.11), we obtain a parabolic refractive index variation:

$$n(x, y) = n + n_2 I(x, y) \approx n_p - 2\Delta n_p \frac{x^2+y^2}{w^2} , \tag{4.12}$$

where $\Delta n_p = n_2 I_p$ and $n_p = n + \Delta n_p$. This refractive index variation corresponds to a focusing lens with a focusing length f, given by [91She2]:

$$\boxed{f = a\frac{w^2}{4\Delta n_p L_K}} , \tag{4.13}$$

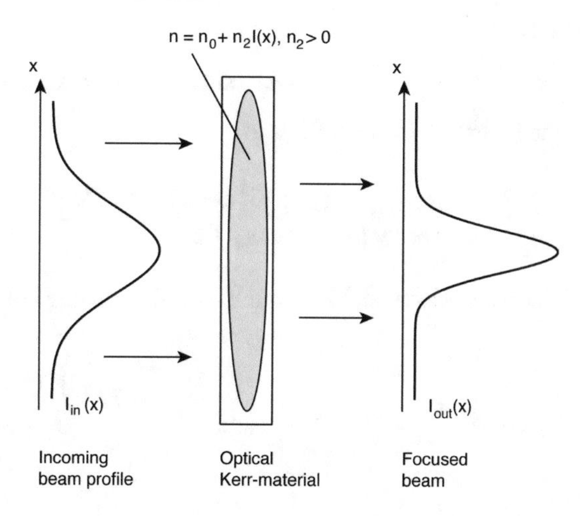

Fig. 4.13 Self-focusing through the Kerr effect. The incoming beam profile hits the optical Kerr material and emerges as a focused beam

where a is a scaling factor to take into account higher order parts of the phase profile which have been neglected in the expansion of (4.12). In the parabolic approximation $a = 1$ (proof follows). According to [91She2] the scaling factor a may take values between 3.77 and 6.4 depending on geometry and power. This scaling factor was also evaluated in the context of Kerr lens modelocking (to be discussed in Sect. 9.4.2) [94Her]. In this case it was found that (after (27) in [94Her]) "the parabolic approximation gives a four-times-larger Kerr-lens effect and the gain-guiding effect is independent of the laser-beam spot sizes."

If the Kerr material is not sufficiently thin, it is better to describe the self-focusing in the parabolic approximation by a waveguide with a Gaussian index profile, i.e., a Gaussian duct [86Sie]. The Gaussian index profile can be approximated by a parabolic index profile (4.12). In this parabolic approximation, the ABCD matrix of the Gaussian duct is given by [86Sie]

$$\begin{pmatrix} \cos\gamma L_K & (n_p\gamma)^{-1}\sin\gamma L_K \\ -n_p\gamma\sin\gamma L_K & \cos\gamma L_K \end{pmatrix}, \tag{4.14}$$

where γ is given by

$$\gamma = \sqrt{\frac{4\Delta n_p}{w^2 n_p}}\,. \tag{4.15}$$

However, it should be noted that the beam diameter changes in a long Kerr material with strong self-focusing. In this case, one has to divide the Kerr material numerically into thin slices and adjust the self-focusing to the new beam radius over and over again in consecutive slices.

In the borderline case of a very short Gaussian duct, i.e., $\gamma L_K \ll 1$, it follows that $\sin\gamma L_K \approx \gamma L_K$ and $\cos\gamma L_K \approx 1$. Hence, using

$$(n_p\gamma)^{-1}\sin\gamma L_K \approx L_K/n_p$$

and

$$-n_p\gamma\sin\gamma L_K \approx -4\Delta n_p L_K/w^2\,,$$

the matrix (4.14) simplifies to

$$\begin{pmatrix} \cos\gamma L_K & (n_p\gamma)^{-1}\sin\gamma L_K \\ -n_p\gamma\sin\gamma L_K & \cos\gamma L_K \end{pmatrix} \overset{\gamma L_K \ll 1}{\approx} \begin{pmatrix} 1 & L_K/n_p \\ -4\Delta n_p L_K/w^2 & 1 \end{pmatrix}. \tag{4.16}$$

For comparison, an ideal thin lens with focusing length f has an ABCD matrix

$$\begin{pmatrix} 1 & 0 \\ -1/f & 1 \end{pmatrix}, \tag{4.17}$$

and a material with length L_K with a refractive index n_p has an ABCD matrix

$$\begin{pmatrix} 1 & L_K/n_p \\ 0 & 1 \end{pmatrix} . \tag{4.18}$$

From this it becomes clear that we can write the matrix in (4.16) in the form

$$\begin{pmatrix} 1 & L_K/n_p \\ -4\Delta n_p L_K/w^2 & 1 \end{pmatrix} = \begin{pmatrix} 1 & 0 \\ -1/f & 1 \end{pmatrix} \begin{pmatrix} 1 & L_K/n_p \\ 0 & 1 \end{pmatrix} \overset{f \ll \frac{L_K}{n_p}}{\approx} \begin{pmatrix} 1 & L_K/n_p \\ -1/f & 1 \end{pmatrix} . \tag{4.19}$$

This yields (4.13) for the focusing length with $a = 1$.

A criterion for the strength of SPM is the so-called B-integral which is given by

$$B \equiv \frac{2\pi}{\lambda} \int_0^L n_2 I(z) \mathrm{d}z . \tag{4.20}$$

Note that the B-integral is a dimensionless quantity because the nonlinear refractive index has SI units [m^2/W]. To avoid serious material damage through self-focusing, B should be smaller than 3 to 5.

Example A pulse train with pulse durations $\tau_p = 100$ fs, a pulse repetition rate $f_{rep} = 100$ MHz, an average power P_{av} of 1 W, and a focused beam radius w of 10 μm is focused by the Kerr effect in a thin sapphire plate with length $L_K = 1$ mm. The peak intensity I_p is given by

$$I_p = \frac{P_{av}}{\pi w^2 \tau_p f_{rep}} = 3.2 \times 10^{10} \frac{\mathrm{W}}{\mathrm{cm}^2} .$$

With a nonlinear index coefficient of $n_2 = 3 \times 10^{-16}$ cm²/W and (4.13) the sapphire Kerr lens has a focusing length f of 2.6 mm in the parabolic approximation i.e., $a = 1$ in (4.13). For comparison, a focusing length of 2.9 mm corresponds to a 60× microscope lens.

4.2.2 Filament Formation

Filamentation was discovered in 1995 [95Bra3] when an infrared laser beam with femtosecond duration and gigawatt (GW) power was sent along a long corridor in air. The laser beam became confined to a long-lived, self-confined tube of light covering several tens of metres, without significant diffraction. The basic mechanism supporting this femtosecond filament results from the balance between Kerr focusing, which increases the local optical refractive index with the pulse intensity, and self-induced ionization which decreases the refractive index again.

There is a critical power associated with strong self-focusing that ultimately results in material damage, or in multiphoton ionization in gases, which then counteracts

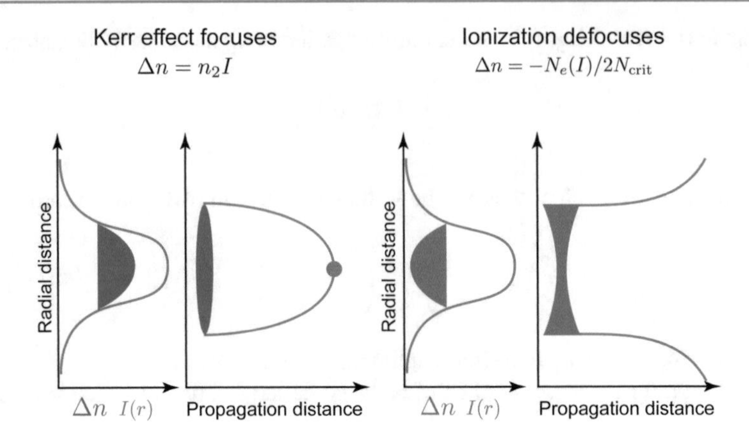

Fig. 4.14 Basic principle of filament formation, which is a balance between self-focusing due to the Kerr effect und defocusing due to the plasma effect. Free carriers are generated in the gas through multiphoton ionization at the high intensity that is generated when the beam starts to collapse

the self-focusing (see Fig. 4.14). This is the basic mechanism for filament formation. It can be understood as follows. Above a critical power, the Kerr effect becomes so strong that self-focusing becomes dominant, and this would ultimately collapse the beam. However, during this beam collapse, the peak intensity becomes so large that multiphoton ionization of the gas then counteracts the effective focusing lens of the Kerr effect. The free carriers (i.e., plasma) provide an effective defocusing lens. We then obtain a balance between these two effects which leads to filamentation, i.e., the beam maintains a narrow beam diameter well beyond the normal Rayleigh range of Gaussian beam propagation. The high intensity broadens the spectrum via SPM so much that white light is emitted by the filaments and transforms infrared lasers into 'white light lasers'. However, the details are much more complicated because this is a highly nonlinear process that cannot be explained by SPM and plasma dispersion alone [07Ber, 07Cou] and goes beyond the scope of this book. In addition, it is even possible to obtain self-compression of the pulses in filaments with few-cycle pulse generation [04Hau, 06Cou].

The following discussion gives an order of magnitude of these effects. Consider a short pulse with a center wavelength λ_0 and an input power $P > P_{\text{cr}}$, propagating in a transparent medium, which can be a gas, a liquid, or a solid-state material. P_{cr} is given by [00Fib]

$$P_{\text{cr}} = \alpha \frac{\lambda_0^2}{4\pi n_0 n_2} = 1.896 \frac{\lambda_0^2}{4\pi n_0 n_2} , \tag{4.21}$$

where n_0 is the index of refraction at the laser wavelength λ_0 and n_2 is the nonlinear refractive index at λ_0. The constant α does not depend on the material but on the beam shape. There is no analytical solution for α, but numerical solutions have been derived for different beam shapes [75Mar]. For a Gaussian beam it is typically assumed to be $\alpha = 1.896$. For air with $n_0 \approx 1$ and $n_2 = 4 \times 10^{-19}$ cm^2/W at 1.06 μm (Table 4.1) we obtain $P_{cr} = 4.2$ GW. For argon at atmospheric pressure, $n_0 \approx 1$ and $n_2 = 3 \times 10^{-19}$ cm^2/W at 800 nm we obtain $P_{cr} = 3.2$ GW. For fused quartz with $n_0 = 1.45$ and $n_2 = 2.74 \times 10^{-16}$ cm^2/W at 1.06 μm (Table 4.1) we obtain $P_{cr} = 4.2$ MW. Assuming a pulse duration $\tau_p = 100$ fs and a critical power of 1 GW (resp. 1 MW) we have a pulse energy $E_p \approx P_{cr}\tau_p$ of 0.1 mJ (resp. 0.1 μJ).

The condition $P > P_{cr}$ translates the fact that beam self-focusing always prevails over the defocusing effect due to diffraction. As a consequence, beam collapse is expected to occur after a propagation distance given by [75Mar]

$$L_c = \frac{0.376 L_{DF}}{\sqrt{\left[(P_{in}/P_{cr})^{1/2} - 0.852\right]^2 - 0.0219}}, \tag{4.22}$$

where

$$L_{DF} = \frac{\pi n_0 w_0^2}{\lambda_0} \tag{4.23}$$

is the Rayleigh length of the beam waist w_0.

In gases before beam collapse, the intensity has increased to the point where multiphoton ionization of the gas occurs. For example, in argon at 800 nm center wavelength, multiphoton ionization involves the simultaneous absorption of 11 photons. Beam collapse is therefore prevented by multiphoton absorption and the defocusing effect of the plasma. The complex interplay between diffraction, self-focusing by the optical Kerr effect, defocusing and absorption by the plasma, and other effects such as pulse self-steepening leads to the formation of a filament, i.e., a pulse maintaining a narrow beam diameter (about 50 μm) while keeping a high intensity (about 5×10^{13} W/cm^2) over long distances $L > L_{DF}$.

4.2.3 Filament Pulse Compression

In 2004, filament compression in the few-cycle regime was demonstrated for the first time at ETH Zurich [04Hau]. It was recognized during the tedious alignment of two successive hollow fiber compression stages as shown in Fig. 4.14 that similar or even better results were obtained when the two hollow fibers inside the gas chambers were simply removed. Without the hollow fibers filamentation occurred during propagation in the argon inside the larger chambers. Filamentation was first discovered in 1995 using 200 fs pulses in air [95Bra3] and it was assumed that pulse compression into the few-cycle regime was not possible because of the strong dispersion of the plasma. However, the unexpected discovery nearly 10 years later showed that filament compression in the few-cycle regime works and provides improved stability against input pulse fluctuations [06Gua]. Pulses as short as 4.9 fs have been generated in the near-infrared, i.e., around 800 nm center wavelength [07Zai], and self-compression to 7.8 fs has been obtained even at millijoule pulse energies [06Sti].

4.3 Solitons

4.3.1 Discovery of the Soliton

Wave packets propagating in linear media disperse due to the different propagation velocities of the spectral components inside the pulse (see Chap. 2). This is a linear effect. But if the propagation velocity of the wave in the medium also depends on the intensity, the two effects may cancel each other out for certain pulse shapes. Such pulses, also called soliton pulses, can propagate undisturbed through the medium.

The first soliton pulse was observed in August 1834 by the Scottish shipbuilding engineer and scientist J. Scott Russell on the Edinburgh–Glasgow canal. He observed that after stopping a barque pulled by horses a big soliton bump, as he called the phenomenon himself, disengaged from the boat. Because he was on horseback, he was able to follow the soliton for several kilometers. He was so impressed by this, that he himself conducted experimental studies on the propagation of water waves, and in 1844 finally gave a lecture on this to the British Association for the Advancement of Science. From his report, one can see that he studied the interaction of solitary pulses experimentally. In his drawings, he already correctly illustrated the way solitons propagate through each other in a collision and maintain their original shape after the collision. They possess particle-like character. But the equation of motion for water waves in shallow water was not derived until 60 years later in 1895 by Korteweg and de Vries. For this reason, it is also called the Korteweg–de Vries (KdV) equation. On the other hand, the propagation of optical solitons is described by the nonlinear Schrödinger equation.

Nonlinear Schrödinger Equation (NSE)

Optical solitons and their remarkable physical properties are important for the generation and propagation of ultrashort electromagnetic pulses. Therefore, they have been investigated for applications in optical communication, laser physics, and quantum optics. For the theoretical treatment of active and passive modelocking we will use the linearized operators that change the pulse envelope due to SPM and dispersion, introduced by Haus for his master equations (see Appendix F). We will therefore use these operators to derive the nonlinear Schrödinger equation in this section.

Solitons are pulses, for which the effects of SPM and negative GDD neutralize each other, so that they can propagate without pulse broadening in a dispersive material. Additionally, if the dynamics of the observed system allows several such pulses to collide, the pulses emerge from these collisions unchanged, except for position and phase changes. Regarding such collisions, solitons thus possess particle-like character.

If we also want to consider SPM in pulse propagation, we have to calculate the dynamics in the time domain, because in this case we are dealing with nonlinear pulse propagation (see the discussion in Chap. 2). To do this, we first have to derive the equation of motion in the time domain.

We consider the equation of motion of the envelope $\tilde{A}(z, \Delta\omega)$ for a dispersive medium in the slowly varying envelope approximation (2.61):

$$\tilde{A}(L_d, \Delta\omega) = e^{-i[k_n(\omega_0+\Delta\omega)-k_n(\omega_0)]L_d}\tilde{A}(0, \Delta\omega) , \tag{4.24}$$

where L_d is the length of the dispersive material. The dispersion can be expanded as follows:

$$k_n(\omega) \approx k_n(\omega_0) + k_n' \Delta\omega + \frac{1}{2} k_n'' \Delta\omega^2 , \tag{4.25}$$

where

$$k_n' = \left.\frac{\partial k_n}{\partial \omega}\right|_{\omega_0} , \quad k_n'' = \left.\frac{\partial^2 k_n}{\partial \omega^2}\right|_{\omega_0} .$$

Equations (4.24) and (4.25) yield

$$\tilde{A}(L_d, \Delta\omega) = \exp\left[-i\left(k_n' \Delta\omega + \frac{1}{2} k_n'' \Delta\omega^2\right) L_d\right] \tilde{A}(0, \Delta\omega) . \tag{4.26}$$

First Derivative k_n'

For

$$k_n' \Delta\omega L_d \ll 1 , \tag{4.27}$$

the term with the first derivative in (4.26) can be linearized:

$$\begin{aligned} \tilde{A}(L_d, \Delta\omega) &= e^{-ik_n'\Delta\omega L_d} e^{-ik_n''\Delta\omega^2 L_d/2} \tilde{A}(0, \Delta\omega) & (4.28) \\ &\approx (1 - ik_n'\Delta\omega L_d) e^{-ik_n''\Delta\omega^2 L_d/2} \tilde{A}(0, \Delta\omega). & (4.29) \end{aligned}$$

We can now use (2.4) to Fourier transform the linearized term into the time-domain, viz.,

$$F^{-1}\{\Delta\omega \tilde{A}(z, \Delta\omega)\} = -i\frac{\partial}{\partial t} A(z, t) \tag{4.30}$$

and obtain

$$A(L_d, t) \approx \left(1 - k_n' L_d \frac{\partial}{\partial t}\right) e^{-ik_n''\Delta\omega^2 L_d/2} A(0, t) . \tag{4.31}$$

So, with (4.31) and $k_n'' = 0$, we obtain

$$\boxed{A(L_d, t) \approx \left(1 - k_n' L_d \frac{\partial}{\partial t}\right) A(0, t) , \quad \text{for } k_n' \Delta\omega L_d \ll 1} . \tag{4.32}$$

We thus get a temporal shift that is determined by the group delay (2.75):

$$T_g = k_n' L_d = \frac{L_d}{v_g} ,$$

where υ_g is the group velocity (2.74).

Second Derivative K''_n

We can now further linearize the term with the second derivative. For

$$k''_n \Delta\omega^2 L_d \ll 1 , \tag{4.33}$$

we obtain in the frequency domain from (4.31)

$$\tilde{A}(L_d, \Delta\omega) \approx (1 - ik'_n \Delta\omega L_d) e^{-ik''_n \Delta\omega^2 L_d/2} \tilde{A}(0, \Delta\omega) \tag{4.34}$$

$$\approx (1 - ik'_n \Delta\omega L_d) \left(1 - \frac{i}{2} k''_n \Delta\omega^2 L_d\right) \tilde{A}(0, \Delta\omega) . \tag{4.35}$$

As before, we can transform this into the time domain with

$$F^{-1}\{\Delta\omega^2 \tilde{A}(z, \Delta\omega)\} = -\frac{\partial^2}{\partial t^2} A(z, t) . \tag{4.36}$$

This then yields

$$A(L_d, t) \approx \left(1 - k'_n L_d \frac{\partial}{\partial t}\right) \left(1 + \frac{i}{2} k''_n L_d \frac{\partial^2}{\partial t^2}\right) A(0, t) \tag{4.37}$$

$$\equiv \left(1 - k'_n L_d \frac{\partial}{\partial t}\right) \left(1 + iD \frac{\partial^2}{\partial t^2}\right) A(0, t) , \tag{4.38}$$

where the dispersion parameter D is defined by

$$\boxed{D \equiv \frac{1}{2} k''_n L_d} . \tag{4.39}$$

The dispersion parameter D describes the curvature of the dispersion curve $n(\omega)$. According to Table 2.3, the group delay dispersion (GDD) is given by the second order dispersion with $\phi = k_n L_d$

$$\text{GDD} = \frac{d^2\phi}{d\omega^2} = \frac{d^2 k_n}{d\omega^2} L_d = 2D .$$

This means that the dispersion parameter is half of the GDD.

So, with (4.31) and $k'_n = 0$, we obtain

$$\boxed{A\,(L_d, t) \approx \left(1 + i\frac{1}{2} k''_n L_d \frac{\partial^2}{\partial t^2}\right) A(0, t) \equiv \left(1 + iD \frac{\partial^2}{\partial t^2}\right) A(0, t) , \quad \text{for } k''_n \Delta\omega^2 L_d \ll 1} . \tag{4.40}$$

The operator in (4.40) represents the group delay dispersion, which causes dispersive pulse broadening (see Sect. 2.5).

Now we use the SVEA to neglect the cross-term including both temporal derivatives in (4.37) and use the definition of the spatial derivative

$$\frac{\partial}{\partial z}A(z,t) = \lim_{L_\mathrm{d}\to 0}\frac{A(L_\mathrm{d},t) - A(0,t)}{L_\mathrm{d}} \tag{4.41}$$

to obtain

$$\frac{\partial}{\partial z}A(z,t) \approx -k_n'\frac{\partial}{\partial t}A(z,t) + \frac{\mathrm{i}}{2}k_n''\frac{\partial^2}{\partial t^2}A(z,t)\,. \tag{4.42}$$

Rewriting in terms of the group velocity $\upsilon_\mathrm{g} = 1/k_n'$ then gives

$$\boxed{\frac{\partial}{\partial z}A(z,t) + \frac{1}{\upsilon_\mathrm{g}}\frac{\partial}{\partial t}A(z,t) = \mathrm{i}\frac{k_n''}{2}\frac{\partial^2}{\partial t^2}A(z,t)}\,. \tag{4.43}$$

We introduce a retarded time, such that we now only consider the temporal change of the envelope and move as an observer with the pulse:

$$t' = t - \frac{z}{\upsilon_\mathrm{g}}\,. \tag{4.44}$$

With this retarded time, (4.43) reduces to

$$\boxed{\frac{\partial}{\partial z}A\left(z,t'\right) = \mathrm{i}\frac{k_n''}{2}\frac{\partial^2}{\partial t'^2}A\left(z,t'\right)}\,. \tag{4.45}$$

If we also want to consider SPM, we can, by analogy with the dispersion, use the linearized SPM operators (4.7) with (4.37):

$$\begin{aligned}\frac{\partial}{\partial z}A(z,t) &= \lim_{L_\mathrm{d}\to 0}\frac{A\left(L_\mathrm{d},t\right) - A(0,t)}{L_\mathrm{d}}\\ &\approx -k_n'\frac{\partial}{\partial t}A(z,t) + \mathrm{i}\frac{1}{2}k_n''\frac{\partial^2}{\partial t^2}A(z,t) - \mathrm{i}kn_2\left|A(t)\right|^2A(z,t)\,.\end{aligned}$$

Then (4.43) and (4.45) imply

$$\boxed{\frac{\partial}{\partial z}A\left(z,t'\right) = \mathrm{i}\frac{k_n''}{2}\frac{\partial^2}{\partial t'^2}A\left(z,t'\right) - \mathrm{i}kn_2\left|A\left(z,t'\right)\right|^2A\left(z,t'\right)}\,. \tag{4.46}$$

Because we assumed that n_2 was not frequency dependent, we can simply add the effect of the intensity-dependent refractive index in the temporal domain. This means that the nonlinear refractive index variation depends instantaneously on the intensity of the pulse envelope. Equation (4.46) is the nonlinear Schrödinger equation (NSE).

This name arises from the analogy with the quantum-mechanical Schrödinger equation, where in the case of the NSE the potential depends on the absolute square of the pulse envelope.

One has to remember that the nonlinear Schrödinger equation contains the assumption that the dispersion relation can be approximated by a parabola. Therefore, terms higher than second order are neglected.

4.3.2 Solution of the NSE: The Fundamental Soliton

(a) $k''_n \neq 0$ and $n_2 = 0$ (i.e., GDD but no SPM)

Without SPM the NSE reduces to (4.45), i.e., to linear pulse propagation. An originally bandwidth-limited pulse is temporally broadened for positive as well as negative GDD, while the spectral width does not change (Fig. 4.15). But the phase of the spectrum does change.

(b) $k''_n = 0$ and $n_2 \neq 0$ (i.e., no GDD but SPM)

With SPM but without GDD, a bandwidth-limited pulse will propagate with unchanged intensity in the time domain, but its spectrum will be broadened more and more through SPM (Fig. 4.16).

(c) $k''_n > 0$ and $n_2 > 0$ (i.e., positive GDD and positive SPM)

Positive SPM will produce a spectral broadening generating lower frequencies (i.e., red frequency components) in the leading edge of the pulse and higher frequencies (i.e., blue frequency components) in the trailing edge of the pulse (see Fig. 4.1) ("red comes earlier than blue"). Positive dispersion makes "red faster than blue", i.e., the red components in the pulse travel with a higher group velocity than the blue components. Hence, the lower frequencies ("red") at the leading edge of the pulse will shift even further ahead, away from the pulse center, while the higher frequencies ("blue") at the trailing edge of the pulse will be delayed even further. Thus, the

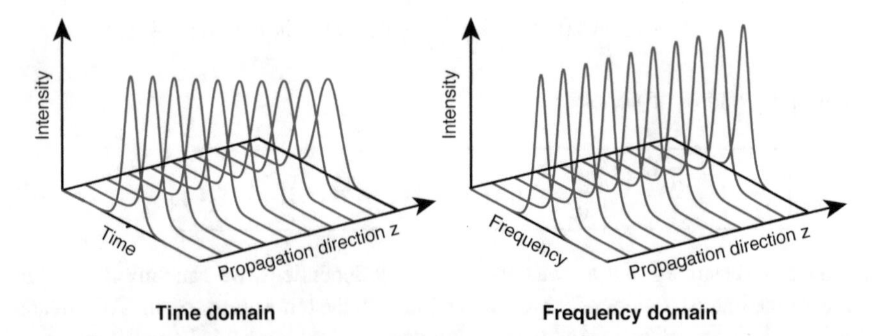

Fig. 4.15 Linear pulse propagation in a dispersive material in the time domain and in the frequency domain. The pulse broadens in the time domain (2.79), while its power spectrum (2.30) remains unchanged

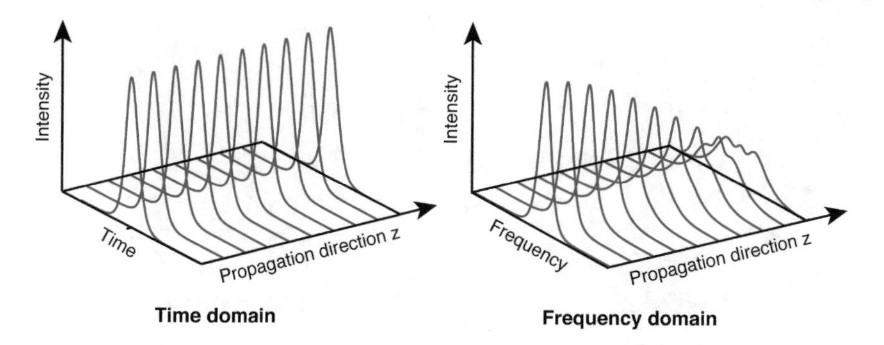

Fig. 4.16 Nonlinear pulse propagation in a material without dispersion in the time domain and in the frequency domain. The pulse remains unchanged in the time domain, while its power spectrum changes

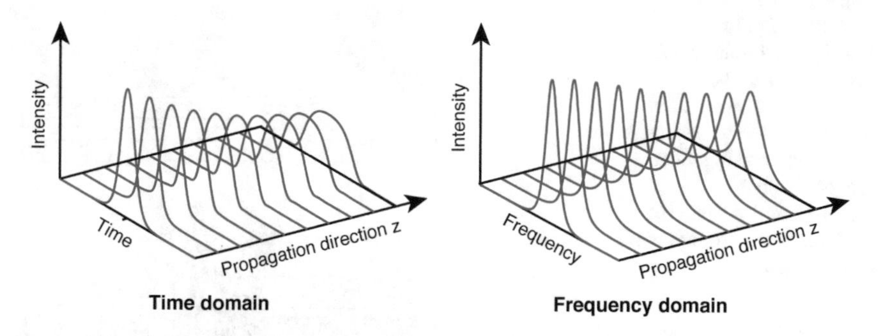

Fig. 4.17 Nonlinear pulse propagation with positive GDD and positive SPM. The pulse broadens even faster than with dispersion alone.

combination of SPM and positive dispersion (i.e., GDD) will broaden an originally bandwidth-limited pulse even faster than with positive GDD alone (Fig. 4.17).

(d) $k_n'' < 0$ and $n_2 > 0$ (i.e., negative GDD and positive SPM)

With positive SPM and negative GDD, the negative dispersion will reduce and ideally even compensate for the chirp produced by SPM. This means that a balance between SPM and negative GDD can cause a situation in which a pulse can propagate unchanged in the temporal and spectral domain, i.e., a soliton has formed (Fig. 4.18).

Positive dispersion and positive Kerr effect lead to very fast pulse broadening (Fig. 4.17), whereas negative dispersion and positive Kerr effect can cancel each other out for certain pulse shapes (Fig. 4.18). Thereby, a soliton pulse develops. The nonlinear Schrödinger equation shows that such pulses exist. According to Fig. 4.18, the intensity of a soliton pulse does not change temporally, i.e., it is in a steady state. Thus, we have to look for a steady-state solution of the Schrödinger equation, whose absolute value squared provides a potential for the solution. The solution to the NSE in this case is given by the hyperbolic secant function. This can be

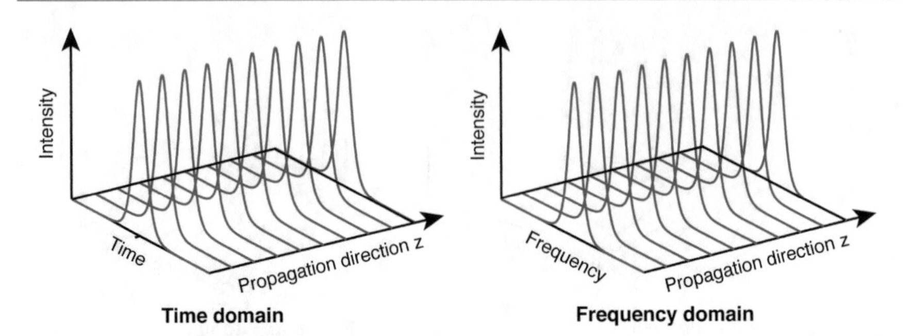

Fig. 4.18 Nonlinear pulse propagation with negative dispersion and positive SPM. The pulse, a fundamental soliton, remains unchanged in the time domain and in the frequency domain

Fig. 4.19 Schematic depiction of pulse broadening through linear dispersion, and the unchanged pulse duration of the soliton (sub-wavepackets are divided into individual 'people' with a 'self-induced' potential). With the kind permission of Linn Mollenauer

shown by substituting the following envelope function into the NSE in the case of $n_2 > 0, k_n'' < 0$ (Fig. 4.18):

$$\boxed{A_s\left(z, t'\right) = A_0 \operatorname{sech}\left(\frac{t'}{\tau}\right) \mathrm{e}^{-\mathrm{i}\phi_0(z)}} . \tag{4.47}$$

Thus the envelope of soliton pulses can be described by the hyperbolic secant function. In order to perform this substitution and other calculations relating to solitons, it is useful to work with several hyperbolic trigonometric identities, which are summarized in Appendix E.

Note that a stable soliton propagates without changing its pulse shape (but changes its phase). The FWHM pulse duration τ_p of the soliton is given by (see Table 2.1)

$$\boxed{\tau_p = 1.7627\tau} , \tag{4.48}$$

and the time–bandwidth product by

$$\boxed{\Delta\nu_p \tau_p = 0.3148} . \tag{4.49}$$

Substituting into the NSE (4.46), we obtain sech and sech^3 terms. Comparing coefficients, we deduce

$$\boxed{\phi_0(z) = \frac{|k_n''|}{2\tau^2} z = \frac{|D|}{\tau^2}}\,, \tag{4.50}$$

$$\boxed{kn_2 |A_0|^2 = \frac{|k_n''|}{\tau^2}}\,, \tag{4.51}$$

where D is the dispersion parameter, defined by (4.39). Combining the two conditions, we obtain

$$\phi_0(z) = \frac{|D|}{\tau^2} = \frac{1}{2}\delta I_\mathrm{p} = \frac{\delta F_\mathrm{p}}{4\tau} = \frac{kn_2 F_\mathrm{p}}{4\tau} z\,, \tag{4.52}$$

where z is the propagation distance, δ is the SPM coefficient (see Sect. 4.1.1) given by $\delta = kn_2 z$, and I_p is the peak intensity

$$I_\mathrm{p} = |A_0|^2\,, \tag{4.53}$$

while F_p is the pulse fluence (i.e., pulse energy density). The latter is given by (see Appendix E for the calculation of the integral)

$$F_\mathrm{p} = \frac{E_\mathrm{p}}{A_\mathrm{eff}} = \int I\left(z, t'\right) \mathrm{d}t' = \int \left|A_\mathrm{s}\left(z, t'\right)\right|^2 \mathrm{d}t' = 2\,|A_0|^2\,\tau\,, \tag{4.54}$$

where E_p is the pulse energy and A_eff is the area of the beam profile.

The eigenvalue of the steady-state solution corresponds to a phase $\phi_0(z)$, which increases continuously during propagation (4.50). The pulse as a whole experiences a homogeneous phase shift, in contrast to pure SPM, where the phase shift depends on the time-dependent intensity.

If we compare (4.51) with the nonlinear phase shift (4.4) we can rewrite (4.51) in the form

$$\varphi_{2\,\mathrm{max}}(z) = kn_2 A_0^2 z = \frac{|k_n''|}{\tau^2} z = 2\phi_0(z)\,. \tag{4.55}$$

This means that the soliton phase ϕ_0 corresponds to exactly half of the maximum nonlinear phase shift:

$$\boxed{\phi_0(z) = \frac{\varphi_{2\,\mathrm{max}}(z)}{2}}\,. \tag{4.56}$$

Equation (4.52) can be solved for τ, and with (4.48) we then obtain

$$\boxed{\tau_\mathrm{p} = 1.7627 \times \frac{4\,|D|}{\delta F_\mathrm{p}} = 1.7627 \times \frac{2\,|k_n''|}{kn_2 F_\mathrm{p}}}\,. \tag{4.57}$$

Because both D (4.39) and δ (4.5) scale with the length of the material, the pulse duration given in (4.57) becomes independent of the material length. This means that the soliton pulse duration is independent of the material length and is not broadened during propagation through a dispersive material, because the negative GDD and positive SPM exactly cancel each other out [compare with (4.50) and (4.51)].

Equation (4.57) shows that, for a fixed pulse fluence F_p and known material parameters, the pulse duration of a soliton is already given. For a higher pulse fluence and the same material conditions, the temporal pulse duration of the soliton becomes shorter:

$$\boxed{\tau_\mathrm{p} \propto \frac{1}{F_\mathrm{p}} \quad \text{for constant material parameters}} \,. \tag{4.58}$$

Moreover, we have the area theorem which says that solitons have constant areas. If we consider $A(t)$, then the soliton area is given by (see Appendix E)

$$\boxed{\text{Soliton area} = \int A_0 \operatorname{sech}\left(\frac{t}{\tau}\right) \mathrm{d}t = \pi A_0 \tau} \,. \tag{4.59}$$

If we keep the pulse fluence constant in soliton pulse propagation and change the dispersion, then the pulse length scales linearly with the dispersion according to (4.57):

$$\boxed{\tau_\mathrm{p} \propto |D| \quad \text{for constant pulse fluence}} \,. \tag{4.60}$$

The nonlinear Schrödinger equation contains the assumption that the dispersion relation can be approximated by a parabola. Higher-order terms are neglected. But then the GDD does not depend on the choice of the carrier frequency. Only the group velocity $\upsilon_\mathrm{g} = 1/k_n'$ itself, with which the pulse propagates, depends on the carrier frequency. Thus, for negative GDD, a soliton pulse with a high carrier frequency moves faster, i.e., "a blue soliton is faster than a red soliton" of GDD < 0.

Not only do soliton pulses have the property that their shape does not change. They also possess particle character if they are allowed to collide. If we place a pulse with a low carrier frequency in front of one with a high carrier frequency, they will necessarily collide during propagation. The result is shown in Fig. 4.20. The two pulses emerge undamaged from the collision. But if we take a closer look, we can see that the faster pulse is shifted ahead temporally by the interaction. Mathematical analysis also shows that the respective phases of the pulses change during the collision.

4.3.3 Solution of the NSE: Higher-Order Solitons

Additional solutions of the NSE are higher-order solitons. Wave fields composed of several fundamental solitons are called higher-order solitons. Superposition of two solitary pulses results in a second order soliton. Figure 4.20 indicates that the

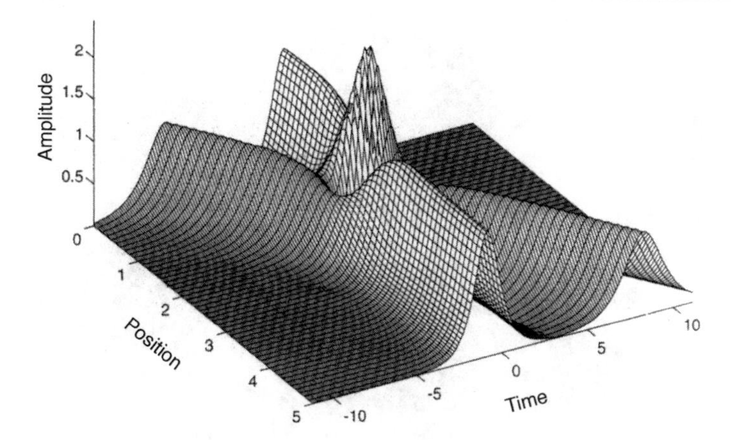

Fig. 4.20 A soliton with a high carrier frequency overtakes a soliton with a lower carrier frequency. Ideally, both pulses emerge from the collision completely undisturbed, and only the positions and the phases of the two solitons are shifted by the collision

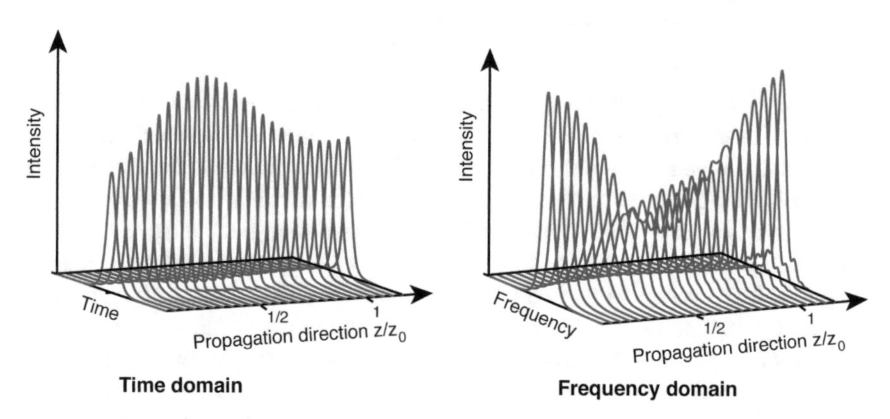

Fig. 4.21 Second order soliton

second order soliton solution is not simply a linear superposition of two first order solitons. Only if they are infinitely far away from each other can the higher-order solution be described by a linear superposition of fundamental solitons. However, in the interaction region, the overall field is strongly deformed.

In contrast to the fundamental soliton (Fig. 4.18), a higher-order soliton changes periodically in the temporal and spectral domain during propagation (see Fig. 4.21). The soliton period z_0 is defined as the distance after which the soliton phase change has reached the value $\pi/4$:

$$\boxed{\phi_0\left(z = z_0\right) = \frac{\pi}{4} \implies z_0 = \frac{\pi}{2}\frac{\tau^2}{\left|k_n''\right|}}\,. \tag{4.61}$$

After half of the soliton period, an originally bandwidth-limited pulse is reduced to a second order soliton with a minimal pulse duration and small shoulders, before the

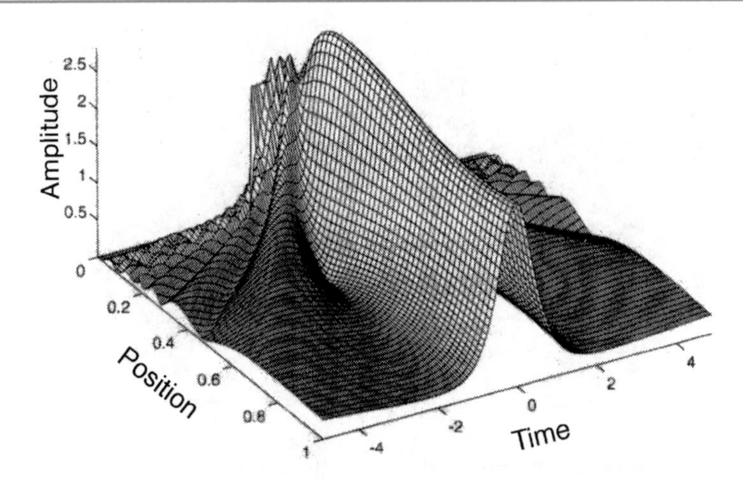

Fig. 4.22 Solution of the nonlinear Schrödinger equation for a rectangular-shaped initial pulse. The rectangular shape at the input can be seen at the time/position origin of the graph. During pulse propagation in the soliton regime, i.e., $k'' < 0$ and $n_2 > 0$, an originally arbitrary pulse reduces to a soliton

pulse resumes its original shape again after the whole soliton period. Therefore, if one chooses the fiber length to be exactly equal to half of the soliton period, one has produced a simple soliton pulse compressor. The relatively weak shoulders of the shortened pulse can be strongly reduced by additional filtering (see Sect. 4.1.3)

A soliton is a very stable formation. For a rectangular original pulse, the solution of the NSE is depicted in Fig. 4.22. One can clearly see the dispersive wave leaving the pulse, leaving behind a single soliton. Whether a first order or higher order soliton stays behind can be determined by the height of the original pulse, because the solution is given by the pulse energy. After a longer propagation distance, the remaining part of the original pulse energy, which is not used by the soliton, is reduced to a continuous weak background radiation (the so-called continuum). The continuum is determined by the linear properties of the equation of motion, i.e., by the linear dispersion of the medium, and diminishes over time. After a long time the dispersive part has spread over the whole space. Only a soliton is left to be observed.

4.3.4 Optical Communication with Solitons

In Sect. 2.5.7 we discussed the influence of dispersion for optical communication. Even with relatively low pulse energies, nonlinear effects can become important in optical fibers.

For example solitons were intially considered for optical communication over long distances. Here is a short explanation of why soliton-based optical communication never made the grade. Nevertheless a lot of interesting physical understanding of nonlinear pulse propagation and new mathematical approaches (e.g., soliton perturbation theory) were developed with this motivation. This has become extremely beneficial for ultrafast lasers.

In 1973, Hasegawa and Tappert showed [73Has] that the envelope of an optical pulse guided in a dielectric waveguide can be described by the NSE. They therefore concluded that it should be possible to demonstrate the existence of optical solitons experimentally in the negative dispersion region of optical fibers. For a glass fiber we have a high refractive index in the core and a low refractive index in the cladding (Fig. 2.15). For a single-mode fiber and a given wavelength the core diameter of just a few wavelengths guides only one specific transverse field distribution or mode in the fiber.

The material dispersion of glass fibers made of silica is positive in the visible wavelength region, while it becomes negative in the infrared for wavelengths longer than 1.3 μm (see Fig. 2.16). Because the material dispersion is dominant in common glass fibers, conventional single-mode fibers show negative dispersion for this wavelength region. In addition to that, at 1.5 μm they have their attenuation minimum (see Fig. 2.16). Due to the nonlinear refractive index of fused silica, the amplitude of the guided mode satisfies the NSE. Linn Mollenauer working at AT&T Bell Laboratories was the first to develop a color-center laser, which possessed a sufficiently high gain bandwidth to generate picosecond pulses in this wavelength region. Optical solitons were produced experimentally in glass fibers for the first time in 1980 [80Mol].

Immediately after the prediction of the existence of optical solitons in glass fibers, Hasegawa suggested using them for the transmission of optical signals over long distances. The motivation was to bridge transoceanic distances from America to Japan and Europe without "repeaters". With conventional linear transmission of signals in optical fibers, the signal is significantly blurred through dispersion, nonlinear effects, and attenuation. Therefore, despite the low attenuation, the signal has to be completely regenerated every 50–100 km, because otherwise it would disappear in the noise of the amplifiers. In conventional optical transmission lines based on intensity modulation, this happens in the following way. The information is digitally coded on the optical signal. As a simplified mechanism of coding the propagation of pulses in the fiber into binary information, we can imagine the following system: in a given time window a 1 corresponds to a pulse and a 0 corresponds to no pulse. In a "repeater", the optical signal is detected, i.e., transformed into an electrical signal. This is amplified, regenerated, synchronized to a clock signal, which is also extracted from the optical signal, then transformed back into an optical signal using a laser diode, and finally fed back into the glass fiber.

The expectation was that, by using solitons for transmission, there would be no signal blurring, because we are dealing with solitons which do not change their pulse shape and duration during propagation. Therefore, for such transmission, no "repeater" would be necessary to deal with dispersion, even for transatlantic distances. The only expected issue was to compensate for the inevitable fiber losses with a corresponding optical amplifier (Fig. 4.23).

The invention of the erbium-doped fiber amplifier in 1985 helped to get this idea more widely accepted. This fiber amplifier is a glass fiber doped with erbium as active material, into which a diode laser is coupled to optically pump the erbium atoms.

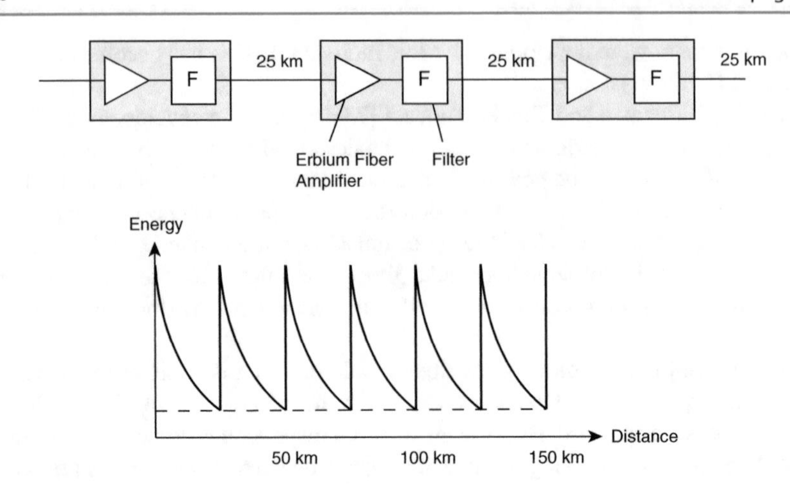

Fig. 4.23 Optical information transmission with periodically aligned erbium-doped fiber amplifiers

However, subsequent transatlantic cables have not yet used solitons for information transmission.

One of the reasons why solitons have not been used is the Gordon–Haus effect [86Gor]. If the solitons pass through an optical amplifier, they are not only amplified by induced emission, but noise is impressed on the signal by inevitable spontaneous emission. This is further amplified in subsequent amplifiers. But as we have seen above, the propagation of solitons is a nonlinear process. This means that the additional intensity noise affects the soliton propagation and leads to a random movement of the whole pulse, similar to the Brownian motion of a particle. This can result in an error in the information transmission whenever a soliton wanders from its assigned time interval into a neighboring time interval.

Naively, one would initially believe that additive noise should lead to the destruction of the soliton, because energy portions are randomly added to the system described by the NSE. Why does it "only" result in a stochastic motion of the whole pulse? The perturbation theory for solitons based on inverse scattering theory sheds light on this problem. If we add a random, but small signal to a soliton and consider the sum of soliton plus perturbation as a new initial value for the solution of the NSE, then we can conclude that the new solution can be written as a new soliton with new amplitude, phase, carrier frequency, position, and a part of the continuum. This works as long as the perturbation is not too big to excite a higher-order soliton. Once generated, the continuum no longer disturbs the soliton, because the two parts of the solution are decoupled. This is the reason for the remarkable stability of solitons against disturbance from outside.

During soliton propagation over distances of several thousand kilometers, the small fluctuations in the carrier frequency caused by noise can lead to large time shifts of the pulse due to dispersion, i.e., the Gordon–Haus effect. Figure 4.24 shows the maximum transmission distance for a bit error ratio of 10^{-9}, which results from the Gordon–Haus effect and the increase in the amplified spontaneous emission, for dif-

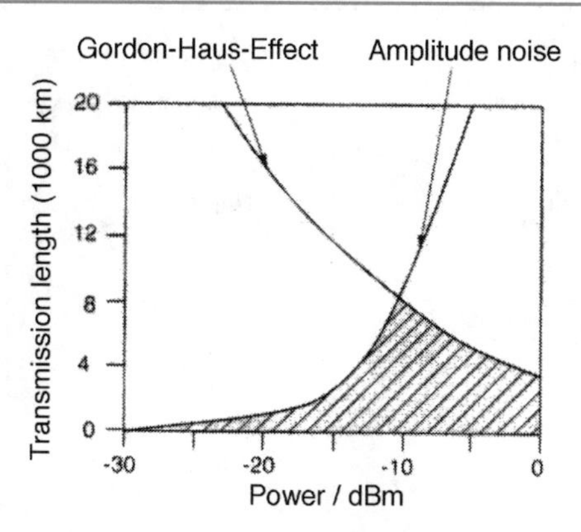

Fig. 4.24 Maximum transmission length due to the Gordon–Haus effect and amplified spontaneous emission for a bit error rate of 10^{-9}, after [93Hau]. If the signal power is too low, the signal-to-noise ratio is too small to ensure a lower bit error rate. If the signal power is too high, the Gordon–Haus effect dominates. Note that 1 dBm $= 10\log(P/1\text{ mW})$, where P is the power in watt, so 0 dBm corresponds to 1 mW, -10 dBm to 0.1 mW, -20 dBm to 0.01 mW, and so on

ferent medium signal powers in units of dBm [1 dBm $= 10\log(P/1\text{ mW})$, where P is the power in watt]. The signal power is adjusted with appropriate amplifier design. For a transatlantic line of length approximately 6000 km, these requirements lead to a window for the allowed signal power of only -3 dBm, i.e., 0.5 mW (Fig. 4.24). The amplifiers have to be designed so that, over the lifetime of a transatlantic line, they can keep the average signal power constant to a factor of about 2. However, this becomes more difficult to achieve due to aging of the pump diodes and other components. Therefore, transmission lines for the year 1995 were planned without the use of optical solitons.

However, in 1992, Hasegawa and Kodama [92Kod] found a method to greatly reduce the Gordon–Haus effect [92Mar]. As we already know from the discussion above, solitons are an extremely stable phenomenon. Therefore, using a filter, it is possible to predefine the carrier frequency of solitons (Fig. 4.25). Thus, fluctuations in the carrier frequency can no longer arbitrarily build themselves up. The carrier frequency ω_0 of the soliton is pulled again and again toward the center frequency of the filter ω_f, because the spectral parts far away from the center frequency are more strongly attenuated than the ones near the center frequency. The losses introduced by the filter are compensated for the soliton by the already present amplifiers. However, from the discussed configuration, it is clear that a monochromatic signal with the center frequency of the filter experiences a greater total amplification than the soliton over the whole distance.

Mollenauer inserted the filter technique into his existing setup und further developed it into the so-called "sliding-frequency filter" (Fig. 4.26). The filters along the transmission line do not all have the same center frequency. The center frequency is slowly tuned from filter to filter. Due to their nonlinear properties, the solitons are able to adjust themselves to the course of the filter while the amplifier noise is absorbed by them. Also transmission with several wavelength channels should be possible with a Fabry–Perot filter which provides an equally spaced wavelength filter

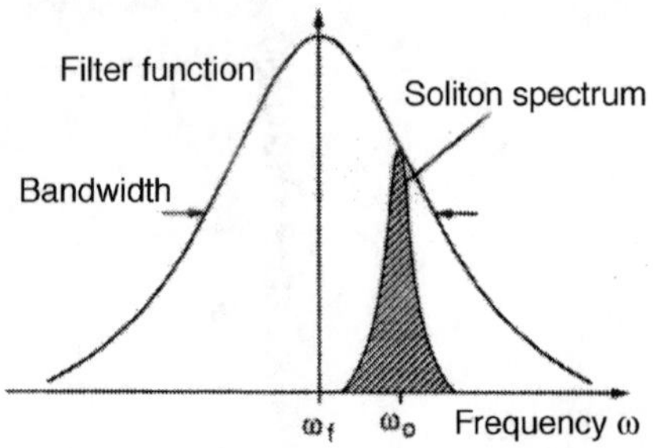

Fig. 4.25 A filter with a certain bandwidth inhibits the random walk of the carrier frequency of the soliton in the spectral region. After [93Hau]

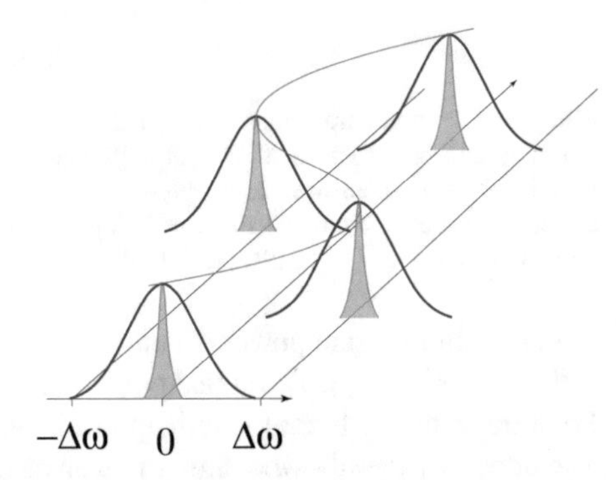

Fig. 4.26 "Sliding-frequency filter" to suppress the limitations of the Gordon–Haus effect

that can be used to build a wavelength multiplex system. In this way several channels with different carrier frequencies and sufficient frequency spacing can be transmitted along the same glass fiber.

In the end, soliton communication did not make it in reality because the notion of installing different fiber cables with slighty shifted filters in the correct order over long distances was considered too great a risk. There was the risk that the wrong cable segment might accidentally be used during installation.

Currently, coherent optical communication has made a significant comeback using erbium-doped fiber amplifiers with improved coherent receivers and digital signal processing [10Ip, 10Sav, 10Lev, 10Spi]. Coherent optical communication systems were the subject of intense research in the 1980s and early 1990s, but technical difficulties inhibited rapid transition into commercial systems. The combination of dense wavelength division multiplexing (DWDM) with longer pulses and coherent communication has become important for transmission over optical fibers with multi-terahertz bit rates.

4.3.5 Periodic Perturbations of Solitons

Periodic perturbations of solitons are considered important not only in optical communications but also in short-pulsed lasers. In optical communications, the soliton is amplified again and again after a certain distance. Therefore, the average power fluctuates periodically (Fig. 4.23). Solitons can also be formed in short-pulsed lasers, which we will discuss later in more detail. Within a laser, a soliton pulse is perturbed again and again, e.g., through losses at the output coupler or through amplification in the laser material. We therefore examine periodic perturbations of solitons in more detail in this section.

We consider a periodic perturbation with period z_a:

$$\mathrm{i}\xi \sum_{m=-\infty}^{\infty} \delta\left(z - m z_\mathrm{a}\right) A\left(z, t'\right) , \tag{4.62}$$

with the perturbation coefficient $\xi \ll 1$ (see Appendix D for the definition of the δ-comb). Examples of this are the periodic amplification stretches inside a communication cable (Fig. 4.23) or the periodic amplifications and losses inside a laser oscillator.

From the NSE (4.46), the periodic perturbation (4.62) yields

$$\frac{\partial}{\partial z} A\left(z, t'\right) = \mathrm{i}\frac{k_n''}{2}\frac{\partial^2}{\partial t'^2} A\left(z, t'\right) - \mathrm{i}k n_2 \left|A\left(z, t'\right)\right|^2 A\left(z, t'\right) + \mathrm{i}\xi \sum_{m=-\infty}^{\infty} \delta\left(z - m z_\mathrm{a}\right) A\left(z, t'\right) . \tag{4.63}$$

For the perturbed NSE, we make the ansatz

$$A\left(z, t'\right) = A_\mathrm{s}\left(z, t'\right) + u\left(z, t'\right) , \tag{4.64}$$

where, according to (4.47),

$$A_\mathrm{s}\left(z, t'\right) = A_0 \,\mathrm{sech}\left(\frac{t'}{\tau}\right) \mathrm{e}^{-\mathrm{i}\phi_0} \tag{4.65}$$

is the unperturbed soliton. Inserting this ansatz into (4.63) yields

$$\begin{aligned} \frac{\partial}{\partial z} A_\mathrm{s}\left(z, t'\right) + \frac{\partial}{\partial z} u\left(z, t'\right) = {} & \mathrm{i}\frac{k_n''}{2}\frac{\partial^2}{\partial t'^2}\left[A_\mathrm{s}\left(z, t'\right) + u\left(z, t'\right)\right] \\ & - \mathrm{i}k n_2 \left|A_\mathrm{s}\left(z, t'\right) + u\left(z, t'\right)\right|^2 \left[A_\mathrm{s}\left(z, t'\right) + u\left(z, t'\right)\right] \\ & + \mathrm{i}\xi \sum_{m=-\infty}^{\infty} \delta\left(z - m z_\mathrm{a}\right)\left[A_\mathrm{s}\left(z, t'\right) + u\left(z, t'\right)\right] . \end{aligned} \tag{4.66}$$

For $\left|u\left(z, t'\right)\right| \ll \left|A_\mathrm{s}\left(z, t'\right)\right|$, it follows that

$$\frac{\partial}{\partial z} u\left(z, t'\right) \approx \mathrm{i}\frac{k_n''}{2}\frac{\partial^2}{\partial t'^2} u\left(z, t'\right) + \mathrm{i}\xi \sum_{m=-\infty}^{\infty} \delta\left(z - m z_\mathrm{a}\right) A_\mathrm{s}\left(z, t'\right) , \tag{4.67}$$

where we can neglect second order terms like, e.g., $\xi u\left(z, t'\right)$ and $u^2\left(z, t'\right)$ and we have used the unperturbed NSE in order to get rid of the spatial derivative of the envelope. But it is not so obvious that we can neglect terms proportional to

$\left|A_{\mathrm{s}}\left(z, t'\right)\right|^2 u\left(z, t'\right)$ as well. Only with a more extensive treatment, e.g., using a soliton perturbation calculation or inverse scattering theory, is it possible to prove the validity of (4.67) in a clean mathematical way.

The dynamics of the perturbation are completely described by (4.67). According to Appendix D,

$$\sum_{m=-\infty}^{\infty} \delta\left(z - m z_{\mathrm{a}}\right) = \frac{1}{z_{\mathrm{a}}} \sum_{m=-\infty}^{\infty} \mathrm{e}^{\mathrm{i} m k_{\mathrm{a}} z}\,, \quad k_{\mathrm{a}} \equiv \frac{2\pi}{z_{\mathrm{a}}}\,. \tag{4.68}$$

The Fourier transform of (4.67) is

$$\frac{\partial}{\partial z}\tilde{u}(z, \omega) \approx -\mathrm{i}\frac{k_n''}{2}\omega^2 \tilde{u}(z, \omega) + \mathrm{i}\xi \sum_{m=-\infty}^{\infty} \frac{1}{z_{\mathrm{a}}} A_0 \pi \tau \,\mathrm{sech}\left(\frac{\pi}{2}\tau\omega\right) \exp(\mathrm{i} m k_{\mathrm{a}} z - \mathrm{i}\phi_0(z))\,. \tag{4.69}$$

One should note here that, in the negative dispersion region $k_n'' < 0$, and because $\phi_0 \propto \left|k_n''\right|$, it follows that $\phi_0 > 0$. The solution ansatz for (4.69) is, using (4.50),

$$\tilde{u}(z, \omega) = \sum_{m=-\infty}^{\infty} \tilde{u}_m \mathrm{e}^{\mathrm{i}(m k_{\mathrm{a}} z - \phi_0)} = \sum_{m=-\infty}^{\infty} \tilde{u}_m \mathrm{e}^{\mathrm{i}\left(m k_{\mathrm{a}} - \left|k_n''\right|/2\tau^2\right) z}\,. \tag{4.70}$$

Inserting into (4.69) yields

$$\mathrm{i}\left(m k_{\mathrm{a}} - \frac{1}{2\tau^2}\left|k_n''\right| - \frac{\left|k_n''\right|}{2}\omega^2\right)\tilde{u}_m = \mathrm{i}\xi \frac{1}{z_{\mathrm{a}}} A_0 \pi \tau \,\mathrm{sech}\left(\frac{\pi}{2}\tau\omega\right)\,, \tag{4.71}$$

whence

$$\tilde{u}(z, \omega) = \sum_{m=-\infty}^{\infty} \frac{\frac{1}{z_{\mathrm{a}}}\xi A_0 \pi \tau \,\mathrm{sech}\left(\frac{\pi}{2}\tau\omega\right)}{m k_{\mathrm{a}} - \frac{\left|k_n''\right|}{2}\left(\frac{1}{\tau^2} + \omega^2\right)} \mathrm{e}^{\mathrm{i}\left(m k_{\mathrm{a}} - \left|k_n''\right|/2\tau^2\right) z}\,. \tag{4.72}$$

In general, for $\xi \ll 1$, the perturbation is negligibly small compared to the soliton. Additionally, according to (4.67), the dynamics of $u\left(z, t'\right)$ is no longer influenced by SPM, because the intensity of the perturbation is too small. Therefore, SPM can no longer compensate for the dispersion. Thus, the pulse length of the perturbation is very quickly broadened by dispersion (see Fig. 4.15).

But the denominator in (4.72) can lead to a condition of resonance, which can arbitrarily increase the perturbation. It is clear that in this case the approximations which have been used in the derivation of (4.72) are no longer valid. We may expect that, under this resonance condition, the perturbation can no longer be neglected. The resonance condition is satisfied for

$$m k_{\mathrm{a}} = \frac{\left|k_n''\right|}{2}\left(\frac{1}{\tau^2} + \omega_m^2\right)\,,$$

which implies

$$\omega_m = \pm\sqrt{\frac{2mk_a}{|k_n''|} - \frac{1}{\tau^2}} = \pm\frac{1}{\tau}\sqrt{\frac{mk_a z}{\phi_0} - 1}\,, \tag{4.73}$$

and with the soliton period (4.61), it follows that

$$\omega_m = \pm\frac{1}{\tau}\sqrt{8m\frac{z_0}{z_a} - 1}\,. \tag{4.74}$$

This resonance has no influence on the soliton as long as the resonances are far outside $1/\tau$, whereby sech $(\pi\tau\omega/2) \approx 0$ in (4.72). In this regime, the perturbation calculation is still valid. With (4.74), it thus follows that

$$8\frac{z_0}{z_a} \gg 1\,.$$

In summary, we have

$$|\tilde{u}(z,\omega)| \ll \left|\tilde{A}_s(z,\omega)\right| \iff z_a \ll 8z_0\,. \tag{4.75}$$

This means that, if the periodic perturbations are arranged with a period of $z_a \ll 8z_0$, they will show no resonance effects and this perturbation can also be treated as a continuous perturbation. Because $z_0 \propto \tau^2$ according to (4.61), the period of the perturbation has to be made ever smaller for shorter pulses. To develop a feeling for the order of magnitude of the soliton period, we consider some examples below.

In modelocked lasers with significant soliton pulse shaping, the periodic perturbation per cavity roundtrip can lead to so-called Kelly sidebands in the modelocked spectrum [92Kel4]. These are narrow peaks superimposed on the soliton pulse spectrum and are more common for ultrafast fiber lasers [94Den].

Example: Soliton Period in Optical Communications

Let us consider the propagation of pulses with different durations inside a glass fiber at a center frequency of 1.55 μm. We assume that the beam radius w inside the fiber is 4 μm. According to Table 2.4, the negative dispersion at 1.55 μm is $D_\lambda = 17\text{ps/km nm}$. According to (2.85), we thus have

$$|k_n''| = \frac{2\pi c}{\omega^2}D_\lambda = \frac{\lambda^2}{2\pi c}D_\lambda\,,$$

whence

$$|k_n''| = \frac{(1.55\times10^{-6}\text{m})^2}{2\pi\times3\times10^8\text{m/s}}17\frac{10^{-12}\text{s}}{10^3\text{m}\times10^{-9}\text{m}} = 2.17\times10^{-26}\frac{\text{s}^2}{\text{m}} \approx 22\frac{\text{ps}^2}{\text{km}}\,. \tag{4.76}$$

Table 4.3 Soliton propagation inside a fiber with $k_n'' \approx 22$ ps²/km

$\tau_p (ps)$	z_0 (4.77) (km)	F_p (4.78) (J/cm²)	E_p (4.79) (pJ)	f_{rep} (GHz)	$P_{av} = E_p f_{rep}$ (mW)
10	2.33	7.5×10^{-6}	3.8	10	38
50	58	1.5×10^{-6}	0.75	10	7.5
100	233	7.5×10^{-7}	0.38	10	3.8

For the soliton period, (4.61) implies

$$z_0 = \frac{\pi}{2} \frac{\tau^2}{|k_n''|} \overset{(4.72)}{=} 23.3\tau_p^2 \left[\text{ps}^2\right] \text{ m} . \tag{4.77}$$

We further assume that all pulses are solitons. Then, by (4.57), the required pulse fluence F_p is

$$\tau_p = 1.7627 \times \frac{2|k_n''|}{kn_2 F_p} \Longrightarrow F_p = \frac{1.7627}{\tau_p} \frac{2|k_n''|}{kn_2} = 7.5 \times 10^{-5} \frac{1}{\tau_p\left[\text{ps}\right]} \frac{\text{J}}{\text{cm}^2} , \tag{4.78}$$

where we have taken $n_2 = 2.5 \times 10^{-16}$cm²/W (see Table 4.1). The pulse energy E_p is then given by

$$E_p = \pi w^2 F_p \overset{w=4\,\mu\text{m}}{=} 5.03 \times 10^{-7} F_p \left[\frac{\text{J}}{\text{cm}^2}\right] \text{J} \tag{4.79}$$

For a given pulse repetition rate f_{rep}, the average power P_{av} is determined by $P_{av} = E_p f_{rep}$. The results for different pulse durations are summarized in Table 4.3, with $|k_n''|$ given by (4.76).

One can see that the soliton period becomes relatively short for a pulse duration of 10 ps. To increase the soliton period, the second order dispersion has to be reduced. Through additional Ge doping, the refractive index in the fiber core can be increased, and this strongly influences the dispersion. In addition to that, the dispersion can be varied considerably with appropriate waveguide geometries.

4.4 Self-Steepening

4.4.1 Higher-Order Nonlinear Effects

For ultrashort optical pulses with a width of $\tau_p \leq 100$ fs, it becomes typically necessary to consider higher-order dispersion and nonlinear effects. Therefore, (4.46) should be modified. The spectral width $\Delta\omega$ of such pulses becomes comparable to the carrier frequency ω_0 and several approximations made in the derivation of (4.46) become questionable.

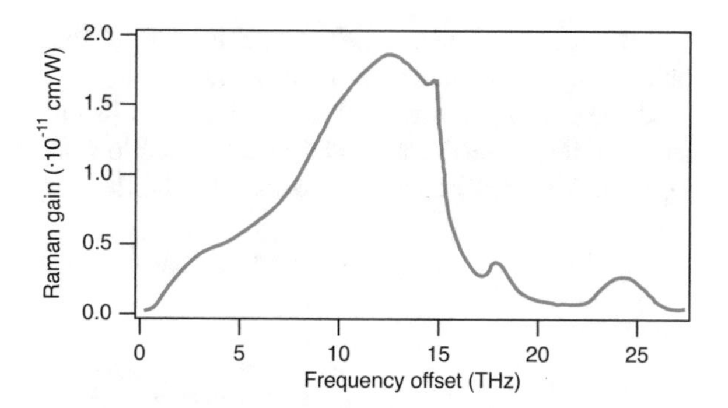

Fig. 4.27 Raman gain spectrum for fused silica for a pump wavelength $\lambda_p = 526$ nm and as a function of the frequency shift from the pump frequency. After [73Sto]

Furthermore, the spectrum of such short pulses is wide enough ($\geq$1 THz) for gain from stimulated Raman scattering (Raman gain) to be able to amplify their low-frequency components by transferring energy from the high-frequency components of the same pulse (Fig. 4.27). This effect is sometimes called intrapulse Raman scattering. As a result, the pulse spectrum shifts to the red side as the pulse propagates inside the fiber, a phenomenon referred to as the self-frequency shift [86Mit]. The physical origin of this effect is related to the retarded nature of the nonlinear response [86Gor]. The inclusion of the delayed nonlinear response by following a perturbative approach [87Kod2] leads to three additional terms in (4.46). The resulting equation is

$$\frac{\partial A(z,t')}{\partial z} - \frac{i}{2}k_n'' \frac{\partial^2 A(z,t')}{\partial t'^2} - \frac{1}{6}k_n''' \frac{\partial^3 A(z,t')}{\partial t'^3} \\ = -i\gamma \left[|A(z,t')|^2 A(z,t') - \frac{i}{\omega_0}\frac{\partial}{\partial t'}\left[|A(z,t')|^2 A(z,t')\right] \right. \\ \left. - T_R A(z,t') \frac{\partial |A(z,t')|^2}{\partial t'} \right], \tag{4.80}$$

where $\gamma = n_2\omega_0/cA_{\text{eff}}$ is the nonlinearity coefficient and T_R is related to the slope of the Raman gain. Equation (4.80) is sometimes called the generalized nonlinear Schrödinger equation.

We can identify the origin of the three new higher-order terms in (4.80). The term proportional to k_n''' results from including the cubic term in the expansion of the propagation constant. The term proportional to γ/ω_0 results from including the first derivative of the slowly varying part P_{nl} of the nonlinear polarization. It is responsible for self-steepening and shock formation at a pulse edge [67DeM, 73Gri, 81Tzo, 83And, 84Yan, 86Gol, 86Man, 87Kod, 87Ohk]. The last term proportional to γT_R has its origin in the delayed Raman response and is responsible for the self-frequency shift.

Self-steepening is an important higher-order nonlinear effect and is governed by the second term on the right-hand side of (4.80). It results from the intensity dependence of the group velocity [67DeM, 67Jon, 67Ost, 73Gri] and leads to an asymmetry in SPM-broadened spectra [67Shi, 69Gus, 70Alf, 70Cub, 71She]. In this section, we will discuss the effects of self-steepening on the shape and spectrum of ultrashort pulses propagating in single-mode fibers.

If the contribution of molecular vibrations (Raman response) to the nonlinear susceptibility is neglected, (4.80) can be used to study self-steepening after setting $T_{\mathrm{R}} = 0$. We can define the normalized amplitude U by

$$A(z, t_1) = \sqrt{P_{\mathrm{peak}}}\, U(z, t_1) \ ,$$

where

$$t_1 = \frac{t'}{\tau} = \frac{t - z/v_{\mathrm{g}}}{\tau} \ ,$$

and P_{peak} is the peak power of the incident pulse. Equation (4.80) then takes the form

$$\begin{aligned} -\mathrm{i}\frac{\partial U(z, t_1)}{\partial z} &= \frac{\operatorname{sgn} k_n''}{2L_{\mathrm{D}}} \frac{\partial^2 U(z, t_1)}{\partial t_1^2} - \mathrm{i}\frac{\operatorname{sgn} k_n'''}{6L_{\mathrm{D}}'} \frac{\partial^3 U(z, t_1)}{\partial t_1^3} \\ &\quad - \frac{1}{L_{\mathrm{nl}}}\left[|U(z, t_1)|^2 U(z, t_1) - \mathrm{i}s\frac{\partial}{\partial t_1}\left(|U(z, t_1)|^2 U(z, t_1)\right)\right] , \end{aligned} \tag{4.81}$$

where L_{D} and L_{D}' are the two length scales introduced in Sect. 2.5.10, viz.,

$$L_{\mathrm{D}} = \frac{\tau^2}{|k_n''|} \ , \quad L_{\mathrm{D}}' = \frac{\tau^3}{|k_n'''|} \ ,$$

and L_{nl} is the nonlinear length given by

$$L_{\mathrm{nl}} = \frac{1}{\gamma P_{\mathrm{peak}}} \ .$$

The self-steepening effects are governed by the last term in (4.81), where the parameter s is defined by

$$s = \frac{1}{\omega_0 \tau} = \frac{T}{2\pi\tau} \ ,$$

and T is the optical period given by $2\pi/\omega_0$. Since $T \approx 2$ fs in the visible region, s is a small parameter and exceeds 0.01 for $\tau \leq 30$ fs.

4.4.2 Optical Shock Front

It is instructive to first consider the case without dispersion by setting $k_n'' = k_n''' = 0$. In this specific case, (4.81) can be solved analytically [67DeM,83And]. By defining a normalized distance $Z = z/L_{\mathrm{nl}}$, (4.81) becomes

$$\frac{\partial U(Z, t_1)}{\partial Z} + s\frac{\partial}{\partial t_1}\left[|U(Z, t_1)|^2 U(Z, t_1)\right] = -\mathrm{i}\,|U(Z, t_1)|^2 U(Z, t_1) \ . \tag{4.82}$$

Using $U = \sqrt{I} \exp(-i\phi)$ in (4.82) and separating the real and imaginary parts, we obtain

$$\frac{\partial I(Z, t_1)}{\partial Z} + 3sI(Z, t_1) \frac{\partial I(Z, t_1)}{\partial t_1} = 0 , \tag{4.83}$$

$$\frac{\partial \phi(Z, t_1)}{\partial Z} + sI(Z, t_1) \frac{\partial \phi(Z, t_1)}{\partial t_1} = I(Z, t_1) . \tag{4.84}$$

Since the intensity (4.83) is decoupled from the phase (4.84), it can be solved using the method of characteristics. The general solution is [67DeM,81Tzo]

$$I(Z, t_1) = f(t_1 - 3sI(Z, t_1) Z) , \tag{4.85}$$

where we have used the initial condition that

$$I(0, t_1) = f(t_1) ,$$

and $f(t_1)$ describes the initial pulse shape at $z = 0$. Equation (4.85) shows that each point t_1 moves along a straight line from its initial value, and the slope of the line is intensity dependent. This feature leads to pulse distortion.

As an example, consider the case of a Gaussian pulse for which

$$I(0, t_1) = f(t_1) = \exp\left(-t_1^2\right) . \tag{4.86}$$

From (4.85) and (4.86), the pulse shape after a distance Z is obtained using

$$I(Z, t_1) = \exp\left[-(t_1 - 3sI(Z, t_1) Z)^2\right] . \tag{4.87}$$

Figure 4.28 shows the calculated pulse shapes at $z = 3$ mm and 6 mm for $s = 0.01$. As the pulse propagates inside the fiber, it becomes asymmetric with its peak shifted towards the trailing edge. As a result, the trailing edge becomes steeper and steeper with increasing z. Physically, the group velocity of the pulse is intensity dependent, in such a way that the peak moves at a lower speed than the wings.

Self-steepening of the trailing edge of the pulse eventually creates an optical shock analogous to the development of an acoustical shock on the leading edge of a sound wave [67DeM]. This happens when the intensity drop of the trailing edge of the pulse becomes instantaneous. The critical distance corresponding to the shock formation can be obtained from (4.87) by requiring that $\partial I / \partial t_1$ be infinite at the shock location. It is then given by [83And]

$$z_s = \left(\frac{e}{2}\right)^{1/2} \frac{L_{nl}}{3s} = 0.39 \frac{L_{nl}}{s} \approx \frac{1.2\tau}{\gamma P_{peak} T} .$$

For picosecond pulses with $\tau = 1$ ps and $P_{peak} \sim 1$ W, the shock distance is $z_s \sim$ 100 km. However, for femtosecond pulses with $\tau \leq 100$ fs and $P_{peak} \geq 1$ kW, z_s typically becomes less than 1 m. As a result, significant self-steepening of a pulse can occur over a fiber a few centimeters long.

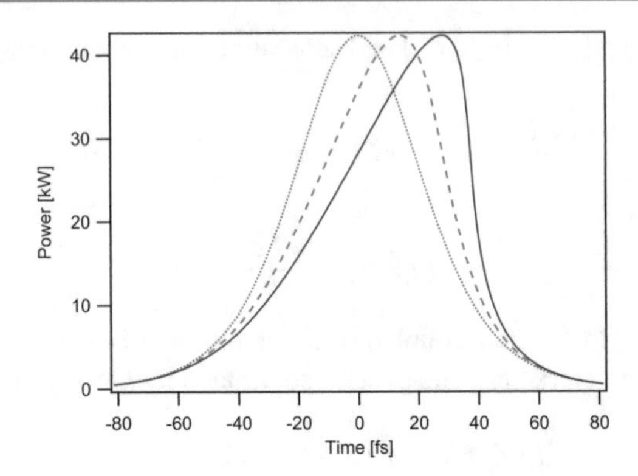

Fig. 4.28 Self-steepening of a Gaussian pulse for the case of no dispersion. The *dotted curve* shows the input pulse shape at $z = 0$, the *dashed curve* shows its deformation with propagation for $z = 3$ mm, and the *solid curve* for $z = 6$ mm, calculated for 50 fs pulses at a center wavelength of 800 nm and propagating in a fiber with a core diameter of 1.7 μm, $n_2 = 2.5 \times 10^{-20}$ m^2/W

The optical shock corresponding to an infinitely sharp trailing edge never occurs in practice because of dispersion. As the pulse edge becomes steeper, the dispersive terms in (4.81) become increasingly more important and cannot be ignored. The effect of GDD on self-steepening is discussed later in this section. The shock distance is also affected by the fiber loss α. In the case without dispersion, fiber loss delays the formation of optical shock. If $\alpha z_s > 1$, the shock does not develop at all [83And].

Self-steepening also affects SPM-induced spectral broadening. For the case without dispersion, the phase $\phi(Z, t_1)$ can be obtained by solving (4.84) analytically. The spectrum is obtained by taking the Fourier transform of $U(Z, t_1)$.

Figure 4.29 shows the calculated spectrum at $z = 6$ mm for $s = 0.01$. The most notable feature is the asymmetry, i.e., the peak amplitude is larger for red-shifted peaks compared with that of blue-shifted peaks. The other notable feature is that SPM-induced spectral broadening is larger on the blue side than on the red side. Both of these features can be understood qualitatively from the self-steepening-induced changes in the pulse shape. First, the spectrum is asymmetric because the pulse shape is asymmetric. Second, the steeper trailing edge of the pulse implies greater spectral broadening on the blue side since SPM generates blue components near the trailing edge. Self-steepening stretches the blue portion in frequency. The amplitude of the high-frequency peaks decreases because the same energy is distributed over a wider spectral range. This asymmetry can be observed in fiber grating pulse compression (e.g., Fig. 4.12).

4.4.3 Effect of GDD on Optical Shock

The spectral features seen in Fig. 4.29 are significantly affected by GDD, which cannot be ignored when ultrashort optical pulses propagate inside silica fibers. The pulse evolution in this case is studied by solving (4.81) numerically.

Figure 4.30 shows the pulse shapes and spectra at $z = 6$ mm and 12 mm for the case of an unchirped Gaussian pulse propagating with normal dispersion. The self-

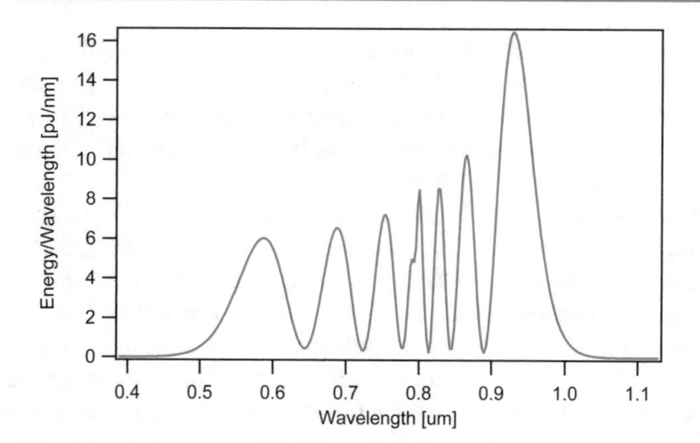

Fig. 4.29 Calculated spectrum at $z = 6$ mm for $s = 0.01$ without dispersion. Self-steepening is responsible for the asymmetry in the SPM-broadened spectrum, calculated for 50 fs pulses at a center wavelength of 800 nm and propagating in a fiber with a core diameter of 1.7 μm, $n_2 = 2.5 \times 10^{-20}$ m^2/W

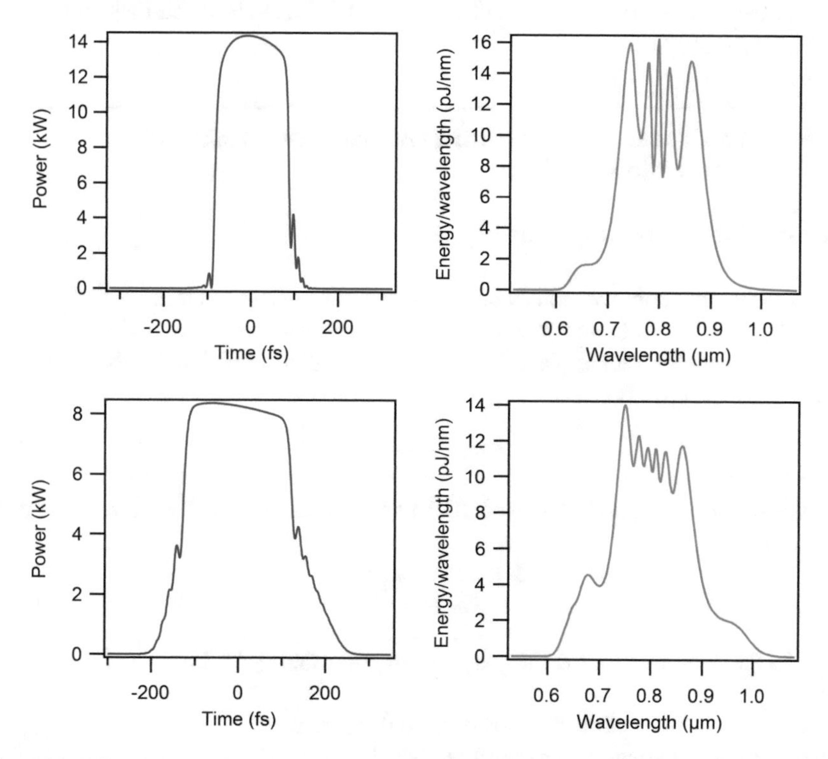

Fig. 4.30 Pulse shapes and spectra at $z = 6$ mm (*upper row*) and 12 mm (*lower row*) for a pulse propagating in the normal dispersion regime of a fiber calculated for 50 fs pulses at a center wavelength of 800 nm propagating in a fiber with a core diameter of 1.7 μm, $n_2 = 2.5 \times 10^{-20}$ m^2/W, and a GDD of 3.5 fs^2/100 μm

steepening parameter is $s = 0.01$ in order to facilitate comparison with the case with no dispersion. In the absence of GDD, the pulse shape and spectrum shown in the upper row of Fig. 4.30 reduce to those shown in Figs. 4.28 and 4.29 for the case $z = 6$ mm. A direct comparison shows that both the shape and the spectrum are significantly affected by the GDD.

The lower row shows the pulse shape and spectrum at $z = 12$ mm. The qualitative changes induced by the GDD are self-evident. For this value of z, the propagation distance z exceeds the shock distance z_s. It is the GDD that dissipates the shock by broadening the steepened trailing edge, a feature clearly seen in the asymmetric pulse shapes of Fig. 4.30. Although the pulse spectra do not exhibit deep oscillations (seen in Fig. 4.29 for the case without dispersion), the longer tail on the blue side is a manifestation of self-steepening. With a further increase in the propagation distance z, the pulse continues to broaden while the spectrum remains nearly unchanged.

Equation (4.81) assumes a frequency-independent nonlinear susceptibility and is valid only if the response time T_{nl} is much shorter than the pulse width τ. The effects of a finite response time are most dramatic in the context of solitons, where they can lead to new phenomena such as soliton decay [87Gru, 87Hod] and self-frequency shift [86Gor, 86Mit].

4.5 Nonlinear Propagation in a Saturable Absorber or Saturable Amplifier

4.5.1 Saturable Amplifier

Let us consider a homogeneously broadened solid-state laser material, which we can describe by a 2-level atom (or an ideal 4-level system). The rate equation for such a laser system yields a saturable gain. The nonlinear gain coefficient for the amplitude is given by

$$g = \frac{g_0}{1 + I/I_{sat}} , \tag{4.88}$$

where g_0 is the small-signal gain coefficient and I_{sat} is the saturation intensity given by [86Sie]

$$I_{sat} = \frac{h\nu}{\sigma \tau_L} , \tag{4.89}$$

with $h\nu$ the photon energy, σ the gain cross-section, and τ_L the lifetime of the excited atom.

The gain cross-section is defined through its relation with the amplitude gain coefficient per unit length $g(z)$ [86Sie] given by

$$g(z) \equiv \frac{N}{V}\sigma , \tag{4.90}$$

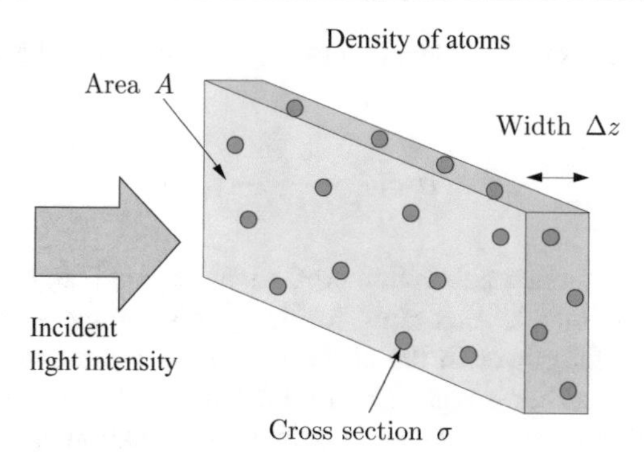

Fig. 4.31 An ion-doped solid-state laser

Table 4.4 The gain cross-section σ and the lifetime of the laser level τ_L for different laser materials. A more complete selection of solid-state laser materials is given in Table 5.1

Laser material	σ (cm^2)	τ_L
Semiconductor laser	$\approx 10^{-14}$	$\approx$1 ns
Dye laser: rhodamine 6G	$\approx 10^{-16}$	$\approx$5 ns
Ti:sapphire @ 780 nm	3.8×10^{-19}	2.5 μs
Cr:LiSAF @ 830 nm	4×10^{-20}	67 μs
Nd:YAG @ 1.064 μm	6.5×10^{-19}	250 μs
Yb:YAG @ 1.03 μm	2×10^{-20}	950 μs
Nd:glass @ 1.054 μm	4×10^{-20}	350 μs

where N/V is the number of inverted atoms within a unit volume V, and $g(z)$ is usually expressed in units cm^{-1}. The units of σ are therefore cm^2. The gain cross-section is a material parameter and is independent of the final laser configuration (Table 4.4).

4.5.2 Saturable Absorber

All atoms that are not inverted contribute in the form of a saturable absorber. A saturable absorber is a material which has decreasing light absorption with increasing light intensity. Most materials show some saturable absorption, but often only at very high optical intensities. For pulse generation with lasers, we need saturable absorbers which show this effect at intensities that are typical inside laser cavities, and as we will see later in much more detail in Chap. 7, semiconductor saturable absorbers are ideally suited for this. The key parameters for a saturable absorber are its wavelength range (where it absorbs), its dynamic response (how fast it recovers), and its saturation intensity and fluence (at what intensity or pulse energy it saturates).

For the simplest case of a 2-level atom, by analogy with (4.88), the amplitude absorption coefficient is given by

$$\alpha = \frac{\alpha_0}{1 + I/I_{\text{sat}}} , \tag{4.91}$$

where α_0 is the unsaturated absorption coefficient or small-signal absorption coefficient (for the amplitude). According to (4.91), a saturable absorber has a high absorption if the light power in the absorber is low, and a low absorption if the light power in the absorber is high. This means that the absorption is saturated, i.e., reduced, by a high intensity. The loss coefficient for small intensities is then given by

$$\alpha_0 = \frac{N_0}{V}\sigma , \tag{4.92}$$

where N_0/V is the doped ion density.

Because the stimulated transition probability from laser level $2 \to 1$ is the same as the transition probability from laser level $1 \to 2$, the absorption cross-section σ_{A} is equal to the gain cross-section σ_{L}:

$$\sigma_{\text{A}} = \sigma_{\text{L}} = \sigma . \tag{4.93}$$

Therefore, the saturated intensity for an absorber is also the same as for an amplifier, when both rely on the same material.

Example: Semiconductor Saturable Absorber

A semiconductor saturable absorber will be discussed in more detail in Sect. 7.4. Here we only give a short introduction. A semiconductor can absorb light if the photon energy is sufficient to excite carriers from the valence band to the conduction band. Under conditions of strong excitation, the absorption is saturated because possible initial states of the pump transition are depleted, while the final states are partially occupied (Fig. 7.10). Within typically 60–300 fs after the excitation, the carriers in each band thermalize, and this already leads to a partial recovery of the absorption. On a longer time scale (i.e., typically between a few picoseconds and a few nanoseconds) they will be removed by recombination and trapping. The presence of two different time scales can be rather useful for pulse generation and will be discussed later.

In a semiconductor the absorption from the valence band to the conduction band is reduced if free charges already occupy certain states. Through Pauli's exclusion principle the number of possible states in the conduction band is reduced and the absorption is thereby bleached. As long as we can neglect these saturation effects, we have

$$\alpha_0(E) = \frac{D(E)}{V}\sigma , \tag{4.94}$$

where $D(E)/V$ is the density of states and where we assume we have no saturation effects in the absorption and V is the volume. $D(E)$ determines the total number of quantum mechanical states N for electrons and holes for a certain energy:

$$D(E) = \frac{\mathrm{d}N}{\mathrm{d}E}\mathrm{d}E\,.$$

For a semiconductor (with a direct band gap), the density of states near the band gap depends on the quantum confinement. For the bulk (3D), we have

$$D_3(\Delta E) = \frac{V}{2\pi^2}\left(\frac{2m^*}{\hbar^2}\right)^{3/2}\sqrt{\Delta E} = \frac{8\pi V\sqrt{2(m^*)^3}}{h^3}\sqrt{\Delta E} \propto \sqrt{\Delta E}\,. \tag{4.95}$$

For a quantum well (2D), we have

$$D_2(\Delta E) = L^2\frac{m^*}{\pi\hbar^2}\,, \tag{4.96}$$

constant as a function of ΔE, and for an ideal quantum wire (1D),

$$D_1(\Delta E) = L\frac{\sqrt{2m^*}}{\pi\hbar}\frac{1}{\sqrt{\Delta E}} \propto \frac{1}{\sqrt{\Delta E}}\,, \tag{4.97}$$

where $V = L^3$ is the volume, ΔE is the energy difference to the band gap, and m^* is the effective mass of the charges, i.e., electrons or holes. A full derivation will be given in Sect. 7.4.6.

In semiconductors, we normally also have to take into account saturation effects in the absorption. Using perturbation theory, we can determine the transition probability W from an initial state in the valence band to a final state in the conduction band:

$$W = \sum_i \frac{2\pi}{\hbar}\frac{1}{V}\int |\langle f|H_{ww}|i\rangle|^2 D_{\mathrm{T}}(E_f)\,\delta(E_f - E_i - \hbar\omega)\,\mathrm{d}E_f\,,$$

where $D_{\mathrm{T}}(E_f)$ is given by

$$D_{\mathrm{T}}(E_f) = D(E_f - E_i)\left\{\left[1 - f_{\mathrm{C}}(E_f)\right]f_{\mathrm{V}}(E_i) - f_{\mathrm{C}}(E_f)\left[1 - f_{\mathrm{V}}(E_i)\right]\right\},$$

where the first term $\left[1 - f_{\mathrm{C}}(E_f)\right]f_{\mathrm{V}}(E_i)$ is the probability of absorption and the second term $f_{\mathrm{C}}(E_f)\left[1 - f_{\mathrm{V}}(E_i)\right]$ the probability of emission. The function $f(E)$ stands for Fermi–Dirac statistics. Therefore $f_{\mathrm{C}}(E_f)$ is the probability of finding an electron with the energy E_f in the conduction band, and $f_{\mathrm{V}}(E_i)$ is the probability of finding an electron with the energy E_i in the valence band. The absorption coefficient α then follows from

$$\alpha = \frac{W\hbar\omega}{|\bar{S}|} = \frac{\text{absorbed power}}{\text{incident power}}\,.$$

In the simplest case one differentiates between two types of saturable absorbers: the slow and the fast. For the slow saturable absorber, the saturated absorption recovers slowly compared to the pulse duration. For the fast saturable absorber, one assumes that it recovers much more quickly compared to the pulse duration, and the saturable absorption coefficient is given by (4.91), which is directly proportional to the pulse intensity well above the saturation intensity. This will be discussed in more detail in Chap. 7.

4.5.3 Nonlinear Pulse Propagation in a Saturable Absorber or Amplifier

According to (4.91), the absorption is intensity dependent, so pulse propagation in a saturable absorber becomes nonlinear. Let us first take a look at the nonlinear transmission through a saturable absorber with a continuous-wave (cw) beam.

(a) Nonlinear Transmission of a Continuous-Wave Beam

We consider the nonlinear transmission of a continuous-wave laser beam through a fast saturable absorber according to (4.91). The propagation direction is determined by the z-axis. The intensity along the propagation direction changes as follows, according to (4.91):

$$\frac{\mathrm{d}I(z)}{\mathrm{d}z} = -2\alpha(I)I(z) = -\frac{2\alpha_0}{1 + I(z)/I_{\mathrm{sat}}} I(z) \ . \tag{4.98}$$

This differential equation can be solved by separation of the variables:

$$\frac{1}{I}\left(1 + \frac{I}{I_{\mathrm{sat}}}\right)\mathrm{d}I = -2\alpha_0 \mathrm{d}z \ . \tag{4.99}$$

Integration then yields

$$\int_{I_{\mathrm{in}}}^{I_{\mathrm{out}}} \left(\frac{1}{I} + \frac{1}{I_{\mathrm{sat}}}\right)\mathrm{d}I = -2\alpha_0 \int_0^d \mathrm{d}z \ , \tag{4.100}$$

where the thickness of the absorber is given by d. Carrying out the integration in (4.100) results in

$$\ln \frac{I_{\mathrm{out}}}{I_{\mathrm{in}}} + \frac{1}{I_{\mathrm{sat}}}(I_{\mathrm{out}} - I_{\mathrm{in}}) = -2\alpha_0 d \ , \tag{4.101}$$

and with

$$T \equiv \frac{I_{\mathrm{out}}}{I_{\mathrm{in}}} \ , \tag{4.102}$$

where T is the transmission, it follows that

$$\ln T + \frac{I_{\mathrm{in}}}{I_{\mathrm{sat}}}(T - 1) = -2\alpha_0 d \ . \tag{4.103}$$

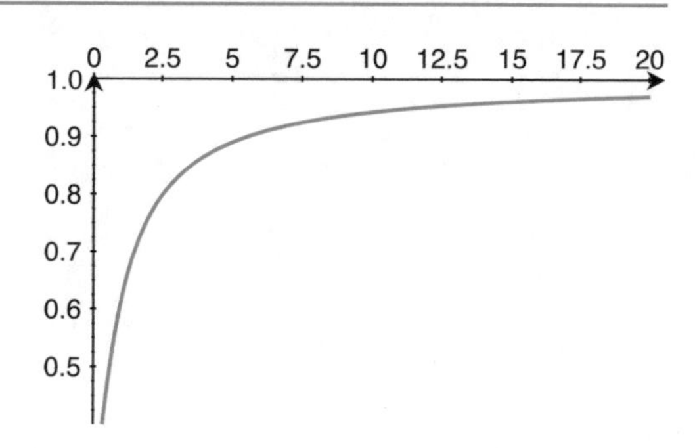

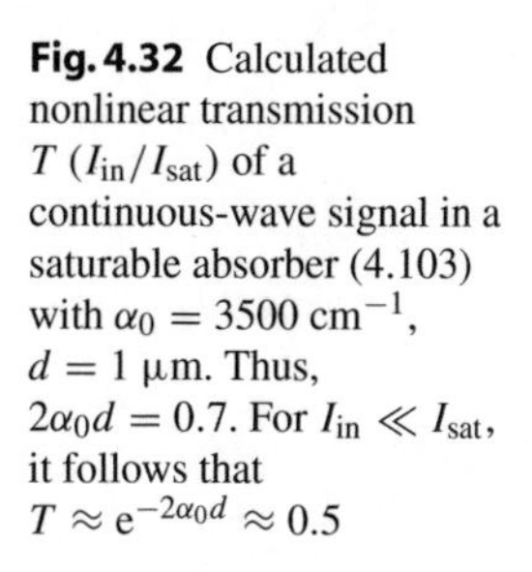
Fig. 4.32 Calculated nonlinear transmission $T\left(I_{\text{in}}/I_{\text{sat}}\right)$ of a continuous-wave signal in a saturable absorber (4.103) with $\alpha_0 = 3500\ \text{cm}^{-1}$, $d = 1\ \mu\text{m}$. Thus, $2\alpha_0 d = 0.7$. For $I_{\text{in}} \ll I_{\text{sat}}$, it follows that $T \approx \mathrm{e}^{-2\alpha_0 d} \approx 0.5$

This equation has to be solved numerically for the transmission $T\left(I_{\text{in}}/I_{\text{sat}}\right)$. For $I_{\text{in}} \ll I_{\text{sat}}$ and using (4.103),

$$I_{\text{in}} \ll I_{\text{sat}} \implies T\left(I_{\text{in}}/I_{\text{sat}} \approx 0\right) = \mathrm{e}^{-2\alpha_0 d} . \tag{4.104}$$

An example is illustrated in Fig. 4.32. It is important to note, that only far above the saturated intensity, i.e., $I_{\text{in}}/I_{\text{sat}}$ greater than approximately 5 to 10, is the transmission strongly saturated with $T \approx 1$.

(b) Nonlinear Pulse Propagation

If absorption or amplification were linear effects, i.e., if α or g were not intensity dependent, then the pulse shape would not change during propagation, as long as the absorber or amplifier bandwidth were much broader than the spectral width of the pulse. But because we consider nonlinear absorption or amplification, the pulse shape will change. For example, in a saturable amplifier, the pulse shape changes because the leading edge of a pulse will saturate the amplifier, whereas the trailing edge of the pulse becomes less amplified. As will be discussed later, this saturation effect can be used to significantly shorten the pulse because it allows the high-intensity part of the pulse to experience less loss than the low-intensity tails. We can usually neglect the change in the pulse shape if $I \ll I_{\text{sat}}$.

In addition to a change in the pulse envelope, the spectrum will change, too. Through the Kramers–Kronig relation, every intensity-dependent absorption or amplification will also generate an intensity-dependent refractive index. Through a pulse, the refractive index therefore also becomes time dependent, so SPM is generated according to Sect. 4.1. We must therefore also consider spectral changes.

A more detailed treatment of nonlinear pulse propagation in a saturable absorber or gain will be given in Chap. 7.

Laser Rate Equations, Steady-State Solutions, Relaxation Oscillations, and Transfer Functions

5

5.1 What Do We Need to Know About Lasers?

5.1.1 Diode-Pumped Solid-State Laser

Here we will assume the reader to have a basic knowledge of lasers. Good introductions can be found in many textbooks, such as [86Sie,91Sal2,98Sve]. Today most ultrafast lasers are diode-pumped solid-state lasers. A comprehensive overview of the relevant lasers can be found in the introductory chapter of [07Kel] and are summarized in Table 5.1. Diode-pumped solid-state lasers are typically pumped by high-power semiconductor diode arrays or bars (see Fig. 5.1). Since the individual diode lasers that form the array or bar are not coupled, the spatial coherence of the emitted beam is significantly reduced. An output beam with such low spatial coherence can still be modeled with Gaussian beam optics using a modified wavelength (discussed in the introduction to [07Kel]). Even though the spatial coherence of these diode arrays is reduced, it is still much brighter than a normal flashlamp.

Moreover, such arrays have much narrower spectral emission than a typical flashlamp and can therefore be tuned directly into an absorption line of the solid-state laser. This results in more efficient pump absorption and reduced heating effects, especially when the pump energy is as close as possible to the laser energy (see Fig. 5.2). Therefore, one important advantage of diode-pumped solid-state lasers is that they can convert fairly low-cost, low-beam-quality optical pump power from high-power diode laser arrays or bars into a near-diffraction-limited output beam with good efficiency.

We typically distinguish between longitudinal and transverse pumping, as shown in Fig. 5.3. Generally, in the low-power regime, we obtain a much better pump efficiency with longitudinal pumping, because we only pump the laser mode volume in the laser crystal. We normally distinguish between different types of lasers such as 3-level and 4-level lasers (Fig. 5.4). The 4-level laser is ideal, but with good laser

© The Author(s), under exclusive licence to Springer Nature Switzerland AG 2021
U. Keller, *Ultrafast Lasers*, Graduate Texts in Physics,
https://doi.org/10.1007/978-3-030-82532-4_5

Table 5.1 Relevant solid-state laser materials for short and ultrashort pulse generation. Center lasing wavelength λ_0. Gain cross-section σ_L. Upper state lifetime τ_L. FWHM gain bandwidth $\Delta\lambda_g$

Laser material	Reference	λ_0 [nm]	σ_L [10^{-20} cm^2]	τ_L [μs]	$\Delta\lambda_g$ [nm]
Transition metal doped solid-state lasers (Cr^{2+}, Cr^{3+}, Cr^{4+}, Ti^{3+})					
Ti^{3+}:sapphire	[86Mou,96Koe]	790	41	3.2	≈ 230
Cr^{3+}:LiSAF	[89Pay,91Sch1,96Koe]	850	5	67	180
Cr^{3+}:LiCAF	[88Pay]	758	1.23	175	115
Cr^{3+}:LiSGAF	[92Smi]	835	3.3	88	190
Cr^{3+}:LiSCAF	[92Cha]	≈800		80	≈155
Cr^{3+}:alexandrite	[79Wal,89Sam,96Koe]	750	0.76	260	≈100
Cr^{4+}:forsterite	[88Pet,93Car,96Koe]	1240	14.4	2.7	170
Cr^{4+}:YAG	[88Ang,95Kuc]	1378	33	4.1	≈250
Cr^{2+}:ZnSe	[97Pag,03Dru]	2500	90	7	600
Cr^{2+}:ZnS	[97Pag,03Dru]	2350	140	4.5	500
Rare earth doped solid-state lasers (Nd^{3+}, Yb^{3+}, Er^{3+})					
Nd:YAG	[64Geu,97Zay]	1064	33	230	0.6
Nd:YLF	[69Har,97Zay,96Koe]	1047 1314	18	480 480	1.2
Nd:YVO_4	[87Fan,88Ris,98For] [97Zay]	914 1064 1342	4 300 60	 100 100	 0.8
Nd:$YAlO_3$		930	4.1	160	2.5
Nd:LSB (10 at.%)	[91Kut,94Mey]	1064	13	118	4
Nd:$GdVO_4$	[94Jen]	1064	76	90	
Nd:(Gd,Y)VO_4	[03Qin]	1064			
Nd:phosphate glass (LG-760)	[99Sch]	1054	4.5	323	24.3
Nd:silicate glass (LG-680)	[99Sch]	1059.7	2.54	361	35.9
Nd:fluorophosphate glass (LG-812)	[78Sch]	1054	2.6	495	26.1
Dual glass gain laser Nd:fluorophosphate & silicate		1064			
Yb:YAG	[93Fan]	1030	2.0	950	6.3
		1050			
Yb:YVO_4	[04Kis,04Kra]	1037	1.25	250	
Yb:$LuVO_4$	[05Liu,06Riv]	1022–1055	0.2–1.0	256	
Yb:CAlGO	[05Pet]	1016–1050	0.75		
Yb:KGW	[97Kul1,97Kul2,03Kul]	1026	2.8	≈250	≈25
Yb:KYW	[97Kul1,97Kul2,02Puj, 00Dem]	1025	3.0	≈250	≈25

Table 5.1 (continued)

Laser material	Reference	λ_0 [nm]	σ_L [10^{-20} cm^2]	τ_L [µs]	$\Delta\lambda_g$ [nm]
Yb:KLuW					
Yb:NGW		1025	2	240–400	
Yb:NaYW	[07Gar]	≈1030	3	309	
Yb:NLM	[05Man]	1000	3.5	285	
Yb:GdCOB	[99Mou]	1045	0.36	2600	44
Yb:BOYS	[02Dru1]	1062	0.2	1100	60
Yb:SIS	[03Dru]	1040	0.44	820	73
Yb:CaF_2	[04Pet]				
Yb:YSO	[95Jac]	1000–1010			
Yb:LSO	[95Jac]				
Yb:GSO	[06Xue]	1088	0.42	1110	70
Yb:Lu_2O_3	[04Gri]	1030	0.42	1110	70
Yb:BCBF	[96Sch]	1034	1.3	1170	24
Yb:LSB	[05Rom]	1016–1050	0.75		
Yb:Y AB	[00Wan]	1040	0.8	680	25
Yb:S-FAP	[94Del]	1047	7.3	1260	4
Yb:FAB	[94Del]	1043	5.9	1100	4.1
Yb:LSO	[95Jac]				
Yb:YSO	[04Jac,95Jac]	1000–1010			
Yb:GSO	[06Xue]	1088	0.42	1110	70
Yb:phosphate glass		1025–1060	0.049	1300	62
Yb:silicate glass		1025–1060	0.093	1100	77
Yb:fluoride phosphate		1025–1060	0.158	1300	81
Er,Yb:glass	[91Lap,96Kig,99Lap]	1535	0.8	7900	55

design, we can also obtain good performance with 3-level lasers. Important examples of 4-level lasers are Nd-doped lasers (e.g., Nd:YAG, Nd:vanadate, etc.) and the Ti:sapphire laser. Important examples of 3-level lasers are Yb-doped lasers (e.g., Yb:YAG, etc.).

For an ideal four-level system with a homogeneously broadened laser medium, and with the assumption of a linear laser resonator that favors a standing wave inside the cavity and thus explains the factor of two for the intracavity intensity in (5.1), we can describe the saturated gain coefficient g in the laser by

$$g = \frac{g_0}{1 + 2I/I_{sat}}, \qquad (5.1)$$

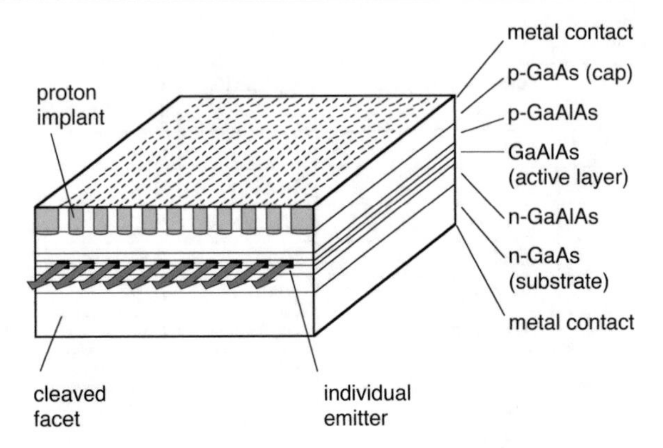

Fig. 5.1 Semiconductor diode laser array. Separate semiconductor lasers are stacked in one array in the horizontal direction (shown here), and for further power scaling they can even be stacked in the vertical direction. The brightness (i.e., power per unit area) has been continuously improved since the 1980s. The individual diode lasers are not coupled, so the spatial coherence of this source is significantly reduced

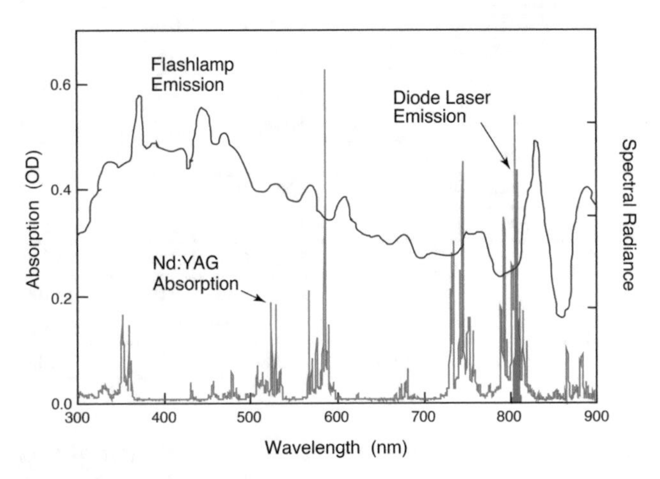

Fig. 5.2 Efficient pumping with diode lasers by pumping directly into an absorption line of the solid-state laser, e.g., Nd:YAG

where g_0 is the small-signal gain coefficient and I_{sat} the saturation intensity of a four-level laser material, given by

$$I_{sat} = \frac{h\nu}{\sigma \tau_L}, \tag{5.2}$$

with $h\nu$ the photon energy, σ the gain cross-section, and τ_L the spontaneous lifetime of the excited laser level. Note here that we continue with the discussion of an ideal four-level laser material. See Sect. 5.1.5 for the required modification for a three-level laser material.

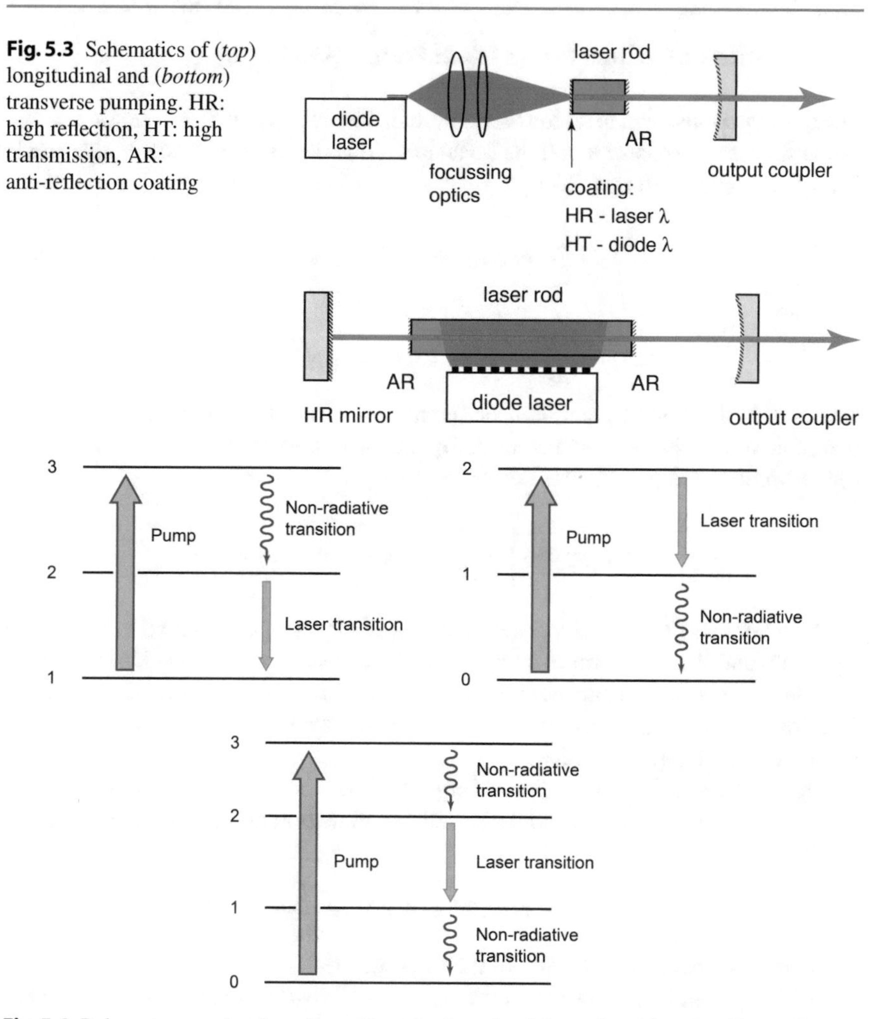

Fig. 5.3 Schematics of (*top*) longitudinal and (*bottom*) transverse pumping. HR: high reflection, HT: high transmission, AR: anti-reflection coating

Fig. 5.4 Relevant energy levels and transitions for three-level (*upper*) and four-level lasers (*lower*)

With increasing intensity I inside the laser resonator, g will decrease (i.e., saturate) until a steady-state intensity I_s is reached with $g\,(t \to \infty) = l$, where l summarizes the total resonator losses (including the resonator output coupling). The steady-state intensity I_s then follows from (5.1) with $g = l$

$$I_s = \frac{I_{sat}}{2}\left(\frac{g_0}{l} - 1\right) . \tag{5.3}$$

Note that we typically use amplitude gain and loss coefficients if not stated otherwise.

5.1.2 Rate Equations for an Ideal Four-Level Laser

The rate equations of an ideal four-level system are given for both the photon number inside the laser resonator n and the inversion N inside the laser mode (single-mode operation is assumed) by [86Sie]:

$$\frac{\mathrm{d}n}{\mathrm{d}t} = K\,(n+1)\,N - \gamma_{\mathrm{c}} n\,, \tag{5.4}$$

$$\frac{\mathrm{d}N}{\mathrm{d}t} = R_{\mathrm{p}} - KnN - \gamma_{\mathrm{L}} N\,, \tag{5.5}$$

where K is the coupling constant or spontaneous transition probability, R_{p} is the pumping rate, γ_{L} is the spontaneous decay rate or atomic decay rate, T_{R} is the cavity roundtrip time and γ_{c}, which is defined by

$$I(t) = I\,(0)\exp\left[-\frac{2l}{T_{\mathrm{R}}}\,(t-t_0)\right] \Longrightarrow \gamma_{\mathrm{c}} = \frac{2l}{T_{\mathrm{R}}}\,, \tag{5.6}$$

is the cavity loss rate or cavity decay rate. The KnN terms correspond to stimulated emission, and the KN term to spontaneous emission. Note that the KnN term for stimulated emission appears both in the rate equation for photons and the inversion. This means that the decrease in inversion by stimulated emission translates directly to an increase in photon number.

The stimulated transition probability W^{stim} of an inverted atom is given by $W_1^{\mathrm{stim}} \equiv W_{12} = W_{21} = Kn$ and the stimulated transition probability for N inverted atoms is given by

$$W^{\mathrm{stim}} = NW_1^{\mathrm{stim}} = KnN = \frac{I}{h\nu}\sigma N\,, \tag{5.7}$$

where σ is the gain cross-section of the laser material.

The important laser parameters for ultrafast solid-state lasers are summarized in Table 5.1.

5.1.3 Steady-State Solutions (Four-Level Laser)

The steady-state values above the threshold condition of the laser satisfy

$$\frac{\mathrm{d}n_{\mathrm{s}}}{\mathrm{d}t} = 0\,, \quad \frac{\mathrm{d}N_{\mathrm{s}}}{\mathrm{d}t} = 0\,. \tag{5.8}$$

With $n = n_{\mathrm{s}}$ and $N = N_{\mathrm{s}}$, the rate equations (5.4) and (5.5) reduce to

$$K\,(n_{\mathrm{s}}+1)\,N_{\mathrm{s}} - \gamma_{\mathrm{c}} n_{\mathrm{s}} = 0\,, \tag{5.9}$$

$$R_{\mathrm{p}} - Kn_{\mathrm{s}}N_{\mathrm{s}} - \gamma_{\mathrm{L}} N_{\mathrm{s}} = 0\,. \tag{5.10}$$

Solving (5.9) for n_s yields

$$n_s = \frac{N_s}{\frac{\gamma_c}{K} - N_s} \equiv \frac{N_s}{N_{th} - N_s} , \tag{5.11}$$

where N_{th} is the threshold inversion given by

$$N_{th} \equiv \frac{\gamma_c}{K} . \tag{5.12}$$

Solving (5.11) for N_s yields

$$N_s = N_{th} \frac{n_s}{n_s + 1} , \tag{5.13}$$

and with (5.10) it follows that

$$N_s = \frac{R_p \tau_L}{1 + K \tau_L n_s} , \tag{5.14}$$

where we have used the fact that $\tau_L \equiv 1/\gamma_L$ is the spontaneous or upper state lifetime. This equation can be solved for n_s again:

$$n_s = \frac{R_p \tau_L - N_s}{K \tau_L N_s} . \tag{5.15}$$

To simplify the subsequent calculations, we introduce the following parameters:

- The threshold pumping rate $R_{p,th}$ is defined by

$$N_s \equiv N_{th} \frac{R_p}{R_{p,th}} . \tag{5.16}$$

- The normalized pumping rate is defined by

$$r \equiv \frac{R_p}{R_{p,th}} . \tag{5.17}$$

As long as the photon number is small, i.e., $K \tau_L n_s \ll 1$, (5.14) yields

$$N_s \approx R_p \tau_L , \quad \text{for } n_s \ll 1 \text{ and } r < 1 . \tag{5.18}$$

At the laser threshold, it then follows, using $R_p = R_{p,th}$ and $N_s = N_{th}$, that

$$R_{p,th} \approx \frac{N_{th}}{\tau_L} . \tag{5.19}$$

Above threshold n_s becomes larger by many orders of magnitude, whence $n_s \gg 1$. Therefore it follows from (5.13) that

$$N_s \approx N_{th} \ , \quad \text{for } n_s \gg 1 \text{ and } r > 1 \ . \tag{5.20}$$

This means that above threshold (i.e., $r > 1$) the inversion is clamped to N_{th} even when the laser is pumped with higher pump power. Therefore, above threshold, with (5.15), $N_s \approx N_{th}$ (5.20), and $R_{p,th} \approx N_{th}/\tau_L$ (5.19), it follows that

$$n_s \approx \frac{R_p \tau_L - N_{th}}{K \tau_L N_{th}} = \frac{1}{K N_{th}} \left(R_p - \frac{N_{th}}{\tau_L} \right) = \frac{1}{K N_{th}} \left(R_p - R_{p,th} \right) \ . \tag{5.21}$$

Equation (5.21) can be further transformed using (5.12), (5.17), and (5.19) to

$$n_s \approx \frac{R_{p,th}}{K N_{th}} (r-1) \overset{(5.19)}{=} \frac{N_{th}}{K \tau_L N_{th}} (r-1) \overset{(5.12)}{=} \frac{\tau_c N_{th}}{\tau_L} (r-1) \ , \tag{5.22}$$

where we have used the fact that

$$\tau_c \equiv \frac{1}{\gamma_c} \tag{5.23}$$

is the photon cavity lifetime or simply the cavity lifetime.

In summary, the steady-state solutions satisfy the following conditions as shown in Fig. 5.5:

$$\boxed{\left.\begin{array}{l} n_s = \dfrac{r}{1-r} \\ N_s \approx r N_{th} \end{array}\right\} \text{ below threshold, i.e., } r < 1} \ , \tag{5.24}$$

$$\boxed{\left.\begin{array}{l} n_s \approx \dfrac{\tau_c}{\tau_L} N_{th} (r-1) = \dfrac{\gamma_L}{K} (r-1) \\ N_s \approx N_{th} = \dfrac{\gamma_c}{K} \end{array}\right\} \text{ above threshold, i.e., } r > 1} \ . \tag{5.25}$$

5.1.4 Gain Saturation (Four-Level Laser)

The roundtrip amplitude gain coefficient g in a linear cavity is given by

$$2g = 2g(z)\, 2L_g = 2L_g \frac{N_s}{V} \sigma \ , \tag{5.26}$$

where N_s/V is the inversion density for a given pump volume V inside the laser mode, σ is the gain cross-section and L_g the length of the gain medium. Here we

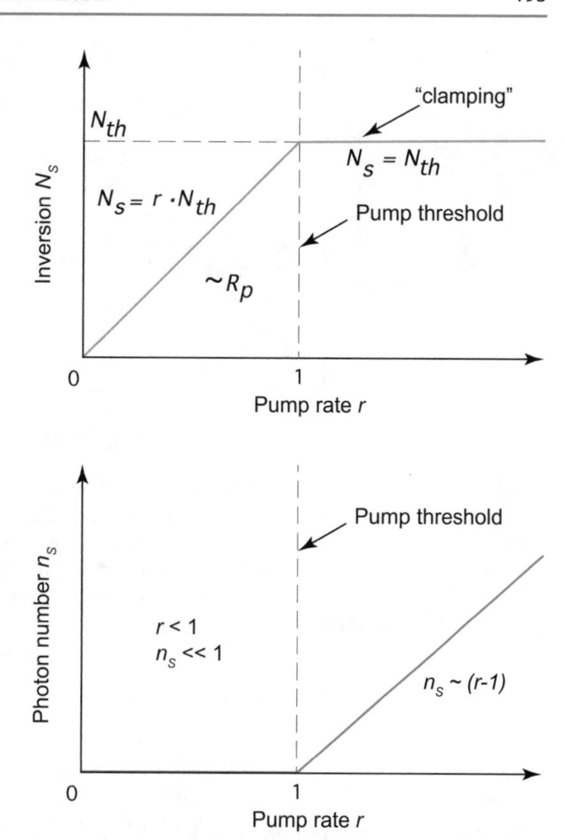

Fig. 5.5 Steady-state solution as a function of the pump rate r for the inversion N_s (*top*) and for the intracavity photon number n_s (*bottom*). At the lasing threshold, r becomes 1

assume a linear cavity, where we have a double pass through the gain medium, i.e., instead of L_g for a single pass, we have $2L_g$ for a double pass. Note that in the literature the gain coefficient g is often given for the intensity and not the amplitude, and this results in a difference by a factor of 2.

Using the solution for N_s in (5.14), we obtain

$$2g = 2L_g \frac{N_s}{V}\sigma = \frac{2L_g R_p \tau_L \sigma}{V(1 + K\tau_L n_s)} . \tag{5.27}$$

We can then introduce the saturation intensity I_{sat} to obtain, with (5.27),

$$\boxed{g = \frac{g_0}{1 + I/I_{sat}} \quad \text{or} \quad g = \frac{g_0}{1 + 2I/I_{sat}}} , \tag{5.28}$$

where g_0 is the small-signal gain coefficient and the second term is valid for a standing wave cavity with an additional factor of 2 for the intracavity intensity. Comparing (5.27) with (5.28) results in

$$g_0 = L_g \frac{R_p}{V} \tau_L \sigma \tag{5.29}$$

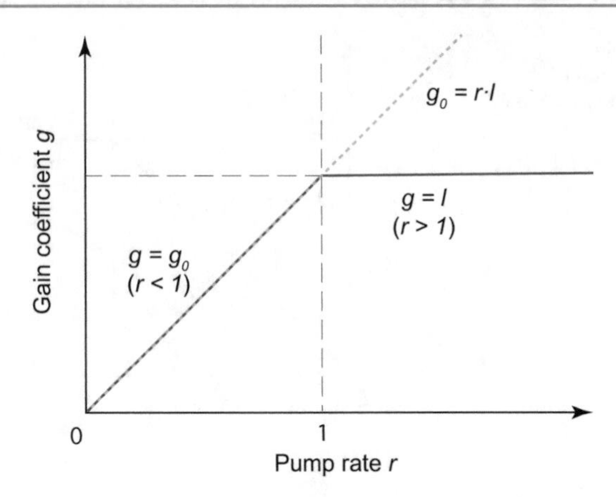

Fig. 5.6 Gain coefficient g as a function of the pump rate r. g_0 is the small-signal gain and l the total cavity loss coefficient. Above threshold the gain is saturated to a constant value for which the gain is equal to the loss. At this point the inversion is also clamped to a constant value even if we continue to pump harder (see Fig. 5.5). Furthermore, in this regime each additional absorbed pump photon is converted into stimulated emission and the output power increases linearly with increased pump power, i.e., with larger r

and

$$\frac{I}{I_{\text{sat}}} = K\tau_{\text{L}} n_{\text{s}} \; . \tag{5.30}$$

The small-signal gain coefficient in (5.29) only depends on the laser material and the pump rate, i.e., the harder the laser material is pumped, the higher the small-signal gain. Below the laser threshold, the saturated gain is equal to the small-signal gain because the intracavity intensity is dominated by spontaneous emission and therefore negligible [see (5.28)]. As a result at threshold we have the condition

$$g_{0,\text{th}} = g_{\text{th}} = L_{\text{g}} \frac{R_{\text{p,th}}}{V} \tau_{\text{L}} \sigma = l \; , \tag{5.31}$$

and above threshold it follows from (5.29) and (5.31) that

$$g_0 = \frac{R_{\text{p}}}{R_{\text{p,th}}} g_{0,\text{th}} = \frac{R_{\text{p}}}{R_{\text{p,th}}} L_{\text{g}} \frac{R_{\text{p,th}}}{V} \tau_{\text{L}} \sigma = rl \; , \tag{5.32}$$

as shown in Fig. 5.6.

5.1.5 Three-Level Laser

A growing number of applications in science and industry are currently pushing the development of ultrafast laser technologies that enable high average powers.

Yb-doped solid-state lasers currently achieve higher pulse energies and average powers than any other diode-pumped solid-state lasers. In contrast to Nd-doped lasers, Yb-doped solid-state lasers are three-level lasers, which are discussed in this section.

First we consider a two-level system in thermal equilibrium. If γ_{12} and γ_{21} are the transfer rates from level 1 to level 2 and from level 2 to level 1, respectively, the rate equation can be written as

$$\dot{N}_2 = \gamma_{12} N_1 - \gamma_{21} N_2 \,, \qquad N = N_1 + N_2 \,, \tag{5.33}$$

where $\dot{N}_2 = \mathrm{d}N_2/\mathrm{d}t$ and N_i gives the density of atoms in level i. In steady state, the first equation can be reduced to $N_2/N_1 = \gamma_{12}/\gamma_{21}$. Since we are in thermal equilibrium this means that

$$\frac{N_2}{N_1} = \frac{\gamma_{12}}{\gamma_{21}} = \mathrm{e}^{-h\nu/k_\mathrm{B}T} \,. \tag{5.34}$$

Therefore, in thermodynamic equilibrium, we can neglect γ_{12} for a typical laser.

The next step is to pump the system. We assume a three-level system with the following properties (see Fig. 5.4, left panel of three-level lasers): level 1 is the ground level of the laser transition and at the same time the lower pump level, while level 2 is the upper laser level, and level 3 is the upper pump level. It is assumed that the transition from level 3 to level 2 is fast, so that the approximation

$$N_1 + N_2 \approx N = N_1 + N_2 + N_3 \tag{5.35}$$

is justified. To pump the system, we choose a (not realistic for the moment) mechanism that transfers R_p atoms per second and per unit volume from level 1 to level 3. This leads to the following rate equation:

$$\dot{N}_2 = \gamma_{12} N_1 - \gamma_{21} N_2 + R_\mathrm{p} \,, \qquad N = N_1 + N_2 \,, \tag{5.36}$$

where N_i is now the total density of atoms in the pumped volume in level i. Looking at the steady state, we have $\dot{N}_2 = 0$, viz.,

$$0 = \gamma_{12} N_1 - \gamma_{21} N_2 + R_\mathrm{p} = \gamma_{12} N_1 + \gamma_{12} N_2 - \gamma_{12} N_2 - \gamma_{21} N_2 + R_\mathrm{p} \,,$$

and with $N = N_1 + N_2$ we then obtain

$$0 = \gamma_{12} N - (\gamma_{12} + \gamma_{21}) N_2 + R_\mathrm{p} \,,$$

and therefore

$$N_2 = \frac{\gamma_{12}}{\gamma_{12} + \gamma_{21}} N + \frac{R_\mathrm{p}}{\gamma_{12} + \gamma_{21}} \approx R_\mathrm{p} \tau_\mathrm{L} \,, \tag{5.37}$$

where τ_L is the upper state lifetime of the laser material, and we have to include reabsorption from the ground state, i.e.,

$$\frac{1}{\tau_L} = \gamma_{12} + \gamma_{21} \, . \tag{5.38}$$

Therefore an inversion can be achieved in this three-level laser system by chosing R_p appropriately high.

Let us look more closely at the pumping mechanism. It is clear that the pumping efficiency decreases when the population density N_1 of the lower pump level is reduced. The total absorbed pump power is

$$P_{p,abs} = P_{p,0}\left(1 - e^{-\sigma_p N_1 L_g}\right) , \tag{5.39}$$

where σ_p is the absorption cross-section at the pump wavelength and L_g is the length of the gain medium. Note that $P_{p,abs}$ is only a function of the average area density in level 1 and independent of the distribution of the absorbing atoms over the length of the crystal. We see that a good model for the pumping mechanism is

$$V_p R_p = \frac{P_{p,0}}{h\nu_p}\left(1 - e^{-\sigma_p N_1 L_g}\right) , \tag{5.40}$$

where V_p is the pump volume and $h\nu_p$ is the pump photon energy. This may be approximated for small absorption (not useful for a laser) or for high absorption. Moreover, typically, this approximation cannot be used if the crystal length is much longer than one absorption length. Longer crystals lead to big reabsorption losses in the weakly pumped region of the crystal. To give some numbers: an absorption efficiency of 99% in thermal equilibrium is reduced to about 90% when half of the atoms are in the upper laser level.

In summary, the equations for a pumped three-level system are

$$\dot{N}_2 = \gamma_{12} N_1 - \gamma_{21} N_2 + \frac{P_{p,0}}{V_p h\nu_p}\left(1 - e^{-\sigma_p N_1 L_g}\right) , \qquad N = N_1 + N_2 \, , \tag{5.41}$$

where V_p is the pump volume.

When building a cavity around the crystal, we introduce two new terms into the atomic rate equation: decay of atoms in level 2 due to stimulated emission into the laser mode and uptransfer of atoms in the lower laser level because of reabsorption of photons in the laser mode. Additionally, a rate equation for the number of photons in the cavity n is needed. We then have

$$\begin{aligned} \dot{N}_2 &= \gamma_{12} N_1 - \gamma_{21} N_2 + \frac{P_{p,0}}{V_L h\nu_p}\eta'(N_1) - \frac{m\sigma_{em} N_2 L_g}{V_L T_R} n + \frac{m\sigma_{abs} N_1 L_g}{V_L T_R} n \, , \\ N &= N_1 + N_2 \, , \qquad \dot{n} = \frac{m\sigma_{em} N_2 L_g}{T_R} n - \frac{m\sigma_{abs} N_1 L_g}{T_R} n - \gamma_c n \, , \end{aligned} \tag{5.42}$$

where $\eta'(N_1)$ is the previously described pump absorption efficiency (5.41)

$$\eta'(N_1) \equiv \left(1 - \mathrm{e}^{-\sigma_\mathrm{p} N_1 L_\mathrm{g}}\right) , \tag{5.43}$$

σ_em and σ_abs are the stimulated emission cross-section and reabsorption cross-section, respectively, $\sigma_\mathrm{tot} = \sigma_\mathrm{em} + \sigma_\mathrm{abs}$, T_R is the cavity roundtrip time, L_g is the length of the laser crystal, V_L is the volume of the laser mode inside the crystal, m is the number of passes through the gain medium per cavity roundtrip, and $\gamma_\mathrm{c} = 2l/T_\mathrm{R}$ is the photon loss rate due to output coupling, etc. N_1 and N_2 now denote the population densities in the volume of the laser mode inside the crystal (instead of the pumped volume).

5.2 Relaxation Oscillations in a Four-Level Laser

5.2.1 Linearized Rate Equations

For now we continue our discussion with an ideal four-level laser with the rate equations given by (5.4) and (5.5). If the laser is briefly disturbed from its steady state, then the photon number and the inversion move away from equilibrium, and it takes a certain time to reach steady state again. To better understand this behavior, we will derive the linearized rate equations from (5.4) and (5.5). Here we follow very closely the discussion in [86Sie].

We consider small deviations from the photon number n and the population number or inversion N of the corresponding steady-state values n_s and N_s. Initially, we thus perturb each value by a small amount from its steady-state value and derive linearized rate equations for the fluctuations in the photon number $\delta n(t)$ and the inversion $\delta N(t)$ which are assumed to be much smaller than their steady-state values:

$$\begin{aligned} n(t) &= n_\mathrm{s} + \delta n(t) , \quad \delta n(t) \ll n_\mathrm{s} , \\ N(t) &= N_\mathrm{s} + \delta N(t) , \quad \delta N(t) \ll N_\mathrm{s} . \end{aligned} \tag{5.44}$$

We consider small perturbations while the laser is running above the threshold condition. We can therefore neglect the spontaneous emission in (5.4):

$$\begin{aligned} \frac{\mathrm{d}\delta n(t)}{\mathrm{d}t} &= K\left[N_\mathrm{s} + \delta N(t)\right]\left[n_\mathrm{s} + \delta n(t)\right] - \gamma_\mathrm{c}\left[n_\mathrm{s} + \delta n(t)\right] \\ &= \underbrace{KN_\mathrm{s}n_\mathrm{s}}_{=\gamma_\mathrm{c} n_\mathrm{s}} + \underbrace{KN_\mathrm{s}\delta n(t)}_{=\gamma_\mathrm{c}\delta n(t)} + Kn_\mathrm{s}\delta N(t) + \underbrace{K\delta N(t)\delta n(t)}_{\approx 0} - \gamma_\mathrm{c} n_\mathrm{s} - \gamma_\mathrm{c}\delta n(t) \\ &= Kn_\mathrm{s}\delta N(t) = \gamma_\mathrm{L}(r-1)\,\delta N(t) , \end{aligned}$$

whence

$$\boxed{\frac{\mathrm{d}\delta n(t)}{\mathrm{d}t} = \gamma_\mathrm{L}(r-1)\,\delta N(t)} . \tag{5.45}$$

The rate equation (5.5) implies

$$\begin{aligned}\frac{\mathrm{d}\delta N(t)}{\mathrm{d}t} &= R_\mathrm{p} - K\left[N_\mathrm{s} + \delta N(t)\right]\left[n_\mathrm{s} + \delta n(t)\right] - \gamma_\mathrm{L}\left[N_\mathrm{s} + \delta N(t)\right] \\ &= R_\mathrm{p} - \underbrace{K N_\mathrm{s} n_\mathrm{s}}_{=\gamma_\mathrm{L}(r-1)N_\mathrm{s}} - \underbrace{K N_\mathrm{s}\delta n(t)}_{=\gamma_\mathrm{c}\delta n(t)} - \underbrace{K n_\mathrm{s}\delta N(t)}_{=\gamma_\mathrm{L}(r-1)\delta N(t)} - \underbrace{K\delta N(t)\delta n(t)}_{\approx 0} \\ &\quad -\gamma_\mathrm{L} N_\mathrm{s} - \gamma_\mathrm{L}\delta N(t) \\ &= R_\mathrm{p} - \gamma_\mathrm{L} r N_\mathrm{s} - \gamma_\mathrm{L} r\delta N(t) - \gamma_\mathrm{c}\delta n(t)\ .\end{aligned}$$

With $R_\mathrm{p} = r\,R_\mathrm{p,th}$ from (5.17) and $R_\mathrm{p,th} = N_\mathrm{th}/\tau_\mathrm{L} = \gamma_\mathrm{L} N_\mathrm{s}$ from (5.19), we have

$$\boxed{\frac{\mathrm{d}\delta N(t)}{\mathrm{d}t} = -\gamma_\mathrm{L} r\delta N(t) - \gamma_\mathrm{c}\delta n(t)}\ . \tag{5.46}$$

The two differential equations (5.45) and (5.46) are the linearized rate equations [86Sie].

5.2.2 Ansatz for Solution After Perturbation

If the laser is disturbed briefly, we can assume that the perturbed photon numbers and the perturbed inversion will approach their steady-state values again. We therefore assume an exponential ansatz for $\delta n(t)$ and $\delta N(t)$, i.e.,

$$\delta n(t) \propto \mathrm{e}^{st}\ , \quad \delta N(t) \propto \mathrm{e}^{st}\ , \tag{5.47}$$

where $\mathrm{Re}(s) < 0$ is required for a stable solution. Inserting this ansatz into the linearized rate equations (5.45) and (5.46) yields

$$\begin{aligned} s\delta n(t) - \gamma_\mathrm{L}\,(r-1)\,\delta N(t) &= 0\ , \\ s\delta N(t) + \gamma_\mathrm{L} r\delta N(t) + \gamma_\mathrm{c}\delta n(t) &= 0\ . \end{aligned} \tag{5.48}$$

This homogeneous system of equations has a non-trivial solution, if the following condition is fulfilled:

$$\det\begin{pmatrix} s & -\gamma_\mathrm{L}\,(r-1) \\ \gamma_\mathrm{c} & s + \gamma_\mathrm{L} r \end{pmatrix} = 0\ . \tag{5.49}$$

This results in a quadratic equation for s :

$$s^2 + \gamma_\mathrm{L} r s + \gamma_\mathrm{L}\gamma_\mathrm{c}\,(r-1) = 0\ ,$$

with the solutions

$$s = s_{1,2} = -\frac{r\gamma_\mathrm{L}}{2} \pm \sqrt{\left(\frac{r\gamma_\mathrm{L}}{2}\right)^2 - \gamma_\mathrm{L}\gamma_\mathrm{c}\,(r-1)}\ . \tag{5.50}$$

There are two different regimes for a stable solution with $\mathrm{Re}(s) < 0$, depending on the values under the square root.

5.2.3 Over-Critically Damped Lasers

If s is real, there are no relaxation oscillations. This case is commonly referred to as the over-critically damped case. Typical examples for this case are gas lasers. According to (5.50), this corresponds to the case

$$\left(\frac{r\gamma_L}{2}\right)^2 > \gamma_L\gamma_c(r-1) ,$$

and therefore

$$\frac{r\gamma_L}{4} > \gamma_c\frac{r-1}{r} . \tag{5.51}$$

Here we have assumed $r > 1$, which corresponds to the above laser threshold regime. Note that $r\gamma_L$ is the stimulated decay rate of the excited atom.

For the borderline case where the stimulated decay rate is much larger than the cavity decay rate, i.e., $r\gamma_L \gg \gamma_c$, the relation (5.50) and the stability condition $s < 0$ imply

$$\begin{aligned} s_{1,2} &= -\frac{r\gamma_L}{2} \pm \frac{r\gamma_L}{2}\sqrt{1-\frac{4\gamma_L\gamma_c(r-1)}{r^2\gamma_L^2}} = -\frac{r\gamma_L}{2}\left(1 \mp \sqrt{1-4\frac{\gamma_c}{r\gamma_L}\frac{r-1}{r}}\right) \\ &\overset{r\gamma_L\gg\gamma_c}{\approx} -\frac{r\gamma_L}{2}\left[1 \mp \left(1-2\frac{\gamma_c}{r\gamma_L}\frac{r-1}{r}\right)\right] , \end{aligned}$$

where we have made the approximation $\sqrt{1-x} \approx 1 - x/2$ for $x \ll 1$, whence

$$s_1 = -r\gamma_L + \gamma_c\frac{r-1}{r} , \qquad s_2 = -\gamma_c\frac{r-1}{r} .$$

For $r\gamma_L \gg \gamma_c$, there are thus two different decay rates for the perturbations:

$$\left.\begin{aligned} s_1 &= -r\gamma_L \\ s_2 &= -\gamma_c\frac{r-1}{r} \end{aligned}\right\} \quad \text{for } r\gamma_L \gg \gamma_c . \tag{5.52}$$

The time-dependent reaction of the laser to perturbations from outside has two over-critically damped relaxation constants: one to restore the steady-state inversion, which is proportional to the stimulated decay rate of the excited atoms, i.e., $s_1 = -r\gamma_L$, and one that is proportional to the photon decay rate inside the laser resonator, i.e., $s_2 = -\gamma_c$. Since only real solutions exist, no oscillations develop and the system is over-critically damped. After a perturbation, the laser will go back to steady state, either within a cavity-photon lifetime or within the stimulated lifetime of the laser level.

If the pump power of such lasers is suddenly switched on, the output power will slowly increase and pass into steady state without the laser giving way to strong oscillations or producing large power spikes.

Example: He–Ne Laser
For a He–Ne laser we can assume $\lambda = 632.8\,\text{nm}$, $\tau_L \approx 100\,\text{ns}$, $2l \approx 0.02$ (with a 2% output coupler), and $T_R = 2\,\text{ns}$, so that according to (5.6) with $\tau_c = 1/\gamma_c$, it follows that

$$\tau_c = T_R/2l \approx 100\,\text{ns} \implies \tau_L \approx \tau_c \ ,$$

where l is the amplitude-cavity loss coefficient within a cavity roundtrip, and T_R is the cavity roundtrip time. If the He–Ne laser is pumped sufficiently strongly above the threshold, the condition (5.51) is satisfied. In this case, after a perturbation, the He–Ne laser will fall back into steady state within approximately 100 ns and without any of the relaxation oscillations which are typically observed for solid-state lasers, to be discussed next.

5.2.4 Under-Critically Damped Lasers

If s becomes imaginary, there are significant relaxation oscillations. This situation is commonly referred to as the under-critically damped case. Typical examples for this case are solid-state lasers. According to (5.50), this corresponds to the case

$$\left(\frac{\gamma_L r}{2}\right)^2 < \gamma_L \gamma_c (r-1) \ ,$$

and therefore

$$\frac{r\gamma_L}{4} < \gamma_c \frac{r-1}{r} \ . \tag{5.53}$$

In this regime, (5.50) can be rewritten in the form

$$s = s_{1,2} = -\frac{r\gamma_L}{2} \pm \mathrm{i}\sqrt{\gamma_L \gamma_c (r-1) - \left(\frac{r\gamma_L}{2}\right)^2} \ . \tag{5.54}$$

Thus, the exponential ansatz for the solution has to be expanded to

$$n(t) = n_s + n_1 \mathrm{e}^{-\gamma_{\text{relax}} t} \cos(\omega_{\text{relax}} t) \ , \tag{5.55}$$

which corresponds to a damped oscillation that is commonly referred to as a relaxation oscillation. According to (5.54) and (5.55), the real part of s describes the attenuation rate γ_{relax} of the relaxation oscillations as

$$\boxed{\gamma_{\text{relax}} = \frac{r\gamma_L}{2}} \ , \tag{5.56}$$

and the imaginary part of s determines the relaxation oscillation frequency f_{relax} as

$$f_{\text{relax}} = \frac{1}{2\pi}\sqrt{\gamma_L \gamma_c (r-1) - \left(\frac{r\gamma_L}{2}\right)^2} \overset{\gamma_c \gg r\gamma_L}{\approx} \frac{1}{2\pi}\sqrt{\gamma_L \gamma_c (r-1)} \ . \tag{5.57}$$

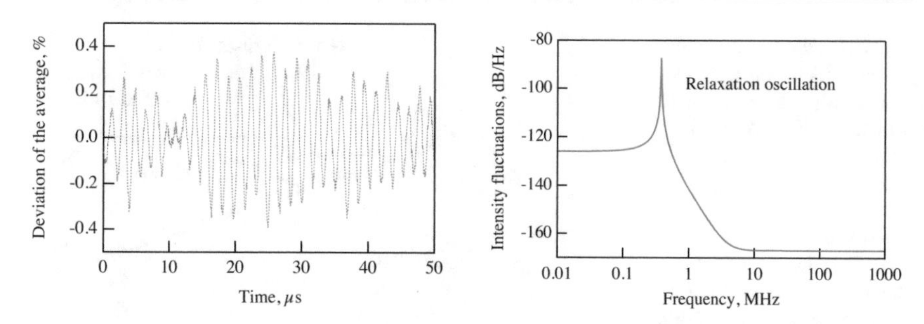

Fig. 5.7 Relaxation oscillations in the time and spectral domain of the intensity fluctuations

For the approximation $\gamma_c \gg r\gamma_L$, we can further reduce (5.57):

$$\boxed{f_{\text{relax}} \approx \frac{1}{2\pi}\sqrt{\frac{r-1}{\tau_L}\frac{1}{\tau_c}} = \frac{1}{2\pi}\sqrt{\frac{1}{\tau_{\text{stim}}}\frac{1}{\tau_c}}\,, \quad \text{for } \gamma_c \gg r\gamma_L}\,, \tag{5.58}$$

where $\tau_{\text{stim}} = \tau_L/(r-1)$. Thus, the relaxation oscillation frequency is determined by the geometric mean of the decay rate of the inversion, i.e., $1/\tau_{\text{stim}}$, and of the photons inside the cavity, i.e., $1/\tau_c$.

Example: Nd:YAG Laser

For a Nd:YAG laser we can assume $\lambda = 1.06\,\mu\text{m}$, $\tau_L \approx 250\,\mu\text{s}$, $2l \approx 0.02$ (with a 2% output coupler), and $T_R = 10\,\text{ns}$. Thus according to (5.6) with $\tau_c = 1/\gamma_c$, it follows that

$$\tau_c = T_R/2l = 500\,\text{ns}\,, \quad \text{for a 100 MHz cavity}\,.$$

Then,

$$\tau_L \gg \tau_c \implies \gamma_L \ll \gamma_c\,.$$

Under normal operating conditions (i.e., for r not too large, meaning for the Nd:YAG laser $r < 1000$), the condition in (5.53) is clearly satisfied.

The relaxation oscillation can be easily measured with a microwave spectrum analyzer (Fig. 5.7). With a photodiode and a microwave spectrum analyzer we can spectrally resolve the intensity fluctuations of a laser. Even small relaxation oscillations can be detected with the large signal-to-noise ratio of good microwave spectrum analyzers. Figure 5.8 shows typical relaxation oscillations of a diode-pumped cw Nd:YLF laser [94Wei]. From (5.58), one can see that the relaxation frequency increases well above thresold (i.e., $r \gg 1$) with the square root of the pumping rate and therefore the pump power (see also Fig. 5.10).

One obtains the strength of the relative relaxation fluctuations $\Delta P_{\text{relax}}/P$ by integration of the relaxation oscillations over their time duration. Equation (5.55) then yields

$$\frac{\Delta P_{\text{relax}}}{P} \propto \int_0^\infty e^{-\gamma_{\text{relax}}t}\,dt \propto \frac{1}{\gamma_{\text{relax}}} = \frac{2}{r\gamma_L}\,, \tag{5.59}$$

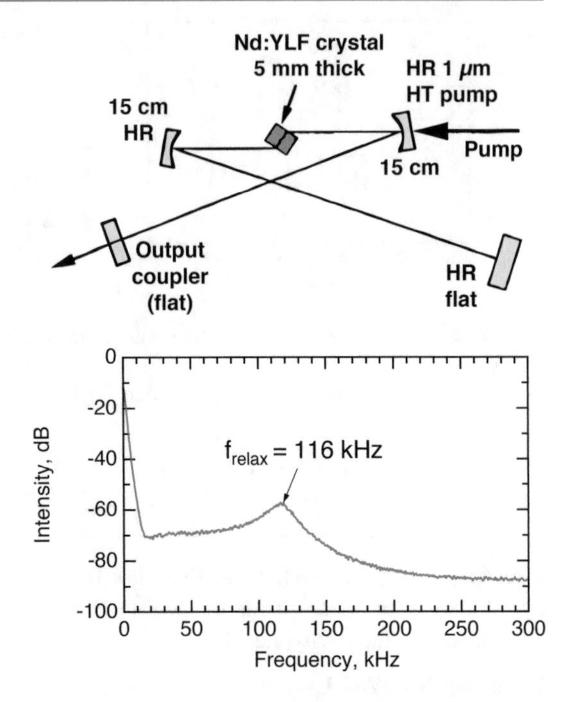

Fig. 5.8 *Upper*: Schematic of the Nd:YLF laser cavity. HR: highly reflective mirror with a specified radius of curvature. HT pump: high transmission for pump wavelength. *Lower*: Microwave spectrum analyzer measurement for the frequency-dependent intensity fluctuations of this cw Nd:YLF laser (resolution bandwidth 3 kHz). The maximum is reached at the relaxation frequency. After [94Wei]

where γ_{relax} is given by (5.56). This means that $\Delta P_{\text{relax}}/P$ decreases with increasing pump power and increases with increasing lifetime of the laser level. For example, a Nd:YAG laser with $\tau_L \approx 250\,\mu\text{s}$ has weaker relaxation oscillations than a Nd:YLF laser with $\tau_L \approx 450\,\mu\text{s}$. One can also determine the Q-factor of the relaxation oscillations:

$$Q_{\text{relax}} \approx \frac{\omega_{\text{relax}}}{\gamma_{\text{relax}}} \approx \frac{2\sqrt{\gamma_L \gamma_c (r-1)}}{r\gamma_L} = \sqrt{\frac{4(r-1)\gamma_c}{r^2 \gamma_L}} \approx 10\text{–}100\,. \tag{5.60}$$

5.2.5 Examples of Relaxation Oscillations Using Different Laser Materials

Consider a linear laser cavity with the following parameters: cavity length $l = 1.5\,\text{m}$, pump power $P_{\text{pump}} = 5\,\text{W}$, pump threshold $P_{\text{thresh}} = 0.5\,\text{W}$ at 10% output coupling. Here we investigate the following gain media: Nd:YLF with $\tau_L = 480\,\mu\text{s}$, Nd:$YVO_4$ with $\tau_L = 50\,\mu\text{s}$, and Ti:sapphire with $\tau_L = 3.2\,\mu\text{s}$, where τ_L is the upper state lifetime.

(a) Verify for all three cases whether the corresponding laser oscillators are over- or under-critically damped.

(b) Calculate the frequency and damping rate of the relaxation oscillations.
(c) What pump powers would be necessary to completely suppress the relaxation oscillations?

The solution is given as follows.

(a) According to Sects. 5.2.3 and 5.2.4, in order to determine whether the three proposed laser oscillators are over- or under-critically damped, we need to evaluate the sign of the argument D of the square root in the expression for s, given by (5.50), viz.,

$$s = s_{1,2} = -\frac{r\gamma_L}{2} \pm \sqrt{\left(\frac{r\gamma_L}{2}\right)^2 - \gamma_L\gamma_c(r-1)} = -\frac{r\gamma_L}{2} \pm \sqrt{D} .$$

If $D \geq 0$, the oscillator is over-critically damped and no relaxation oscillations occur. If $D < 0$, we are in the under-critical (oscillating) case. Now, let us evaluate D for the given parameters. With $r = P_{pump}/P_{thresh} = 10$, $\gamma_L = 1/\tau_L$, $\gamma_c = 10^7\ s^{-1}$, we obtain the following table:

	Nd:YLF	Nd:YVO_4	Ti:sapphire
τ_L [μs]	480	50	3.2
γ_L [μs^{-1}]	2.08×10^{-3}	0.02	0.3125
D [$10^{12} s^{-2}$]	−0.19	−1.79	−25.68

Thus, in a laser cavity with the given parameters, all three laser materials are under-critically damped (D is always less than 0).

(b) From (5.56) and (5.57), $\gamma_{relax} = r\gamma_L/2$ and

$$\omega_{relax} = \sqrt{\gamma_L\gamma_c(r-1) - \left(\frac{r\gamma_L}{2}\right)^2} = \sqrt{|D|} .$$

We thus obtain the following values:

	Nd:YLF	Nd:YVO_4	Ti:sapphire
γ_{relax} [μs^{-1}]	0.01	0.1	1.6
ω_{relax} [MHz]	0.43	1.34	5.07

This shows that the relaxation oscillation frequency is higher with a shorter upper state lifetime. This also explains why, for example, semiconductor lasers with an upper state lifetime in the nanosecond regime have much higher relaxation frequencies (i.e., typically in the gigahertz regime).

(c) The relaxation oscillations are suppressed when $D \geq 0$. We thus look for the solution to $D = 0$:

$$\left(\frac{r\gamma_L}{2}\right)^2 = \gamma_L\gamma_c(r-1) \implies r^2\frac{\gamma_L^2}{4} - \gamma_L\gamma_c r + \gamma_L\gamma_c = 0 ,$$

which gives us the solutions

$$r_{1,2} = \frac{2\gamma_c}{\gamma_L} \pm 2\sqrt{\left(\frac{\gamma_c}{\gamma_L}\right)^2 - \left(\frac{\gamma_c}{\gamma_L}\right)} = \frac{2\gamma_c}{\gamma_L}\left[1 \pm \sqrt{1 - \left(\frac{\gamma_L}{\gamma_c}\right)}\right].$$

Mathematically, there are two valid solutions for r. The solution with the negative sign results in a pump power P_{pump} that is only slightly larger than the pump threshold, which is not relevant for practical applications. As a result, the solution with positive sign is physically more meaningful in a regime with $r \gg 1$, so in the following we will consider this case. This gives us the following results:

	Nd:YLF	Nd:YVO_4	Ti:sapphire
r	1.92×10^4	1999	127
P_{pump}	9.6 kW	999.5 W	63.5 W

This result shows that the relaxation oscillations for the Ti:sapphire laser are suppressed at much lower pump powers compared to Nd:YLF and Nd:YVO_4. This means that from this selection of lasers the Ti:sapphire laser would be the best choice for a low-noise laser oscillator.

By adding a "white noise term" to the linearized rate equation, it is possible to analyze the reaction of the laser to external perturbations. With the white noise we add a constant noise for all frequencies, i.e.,

$$\tilde{r}_{var}(\omega) = \text{const.}$$

With the inverse Fourier transform we obtain $r_{var}(t)$. We therefore add the noise term $r_{var}(t)$ to the linearized rate equations (5.45) and (5.46) as follows:

$$\frac{d\delta N(t)}{dt} = -\gamma_L r \delta N(t) - \gamma_c \delta n(t) + r_{var}(t) , \tag{5.61}$$

$$\frac{d\delta n(t)}{dt} = \gamma_L(r-1)\delta N(t) .$$

The relaxation oscillations are then given by the maximum of $\delta\tilde{n}(\omega)$, which is the Fourier transform of $\delta n(t)$. We solve the second equation for $\delta N(t)$ and insert in the first:

$$\frac{d^2\delta n(t)}{dt^2}\frac{1}{\gamma_L(r-1)} = -\frac{d\delta n}{dt}\frac{r}{r-1} - \gamma_c\delta n(t) + r_{var}(t) ,$$

$$\frac{d^2\delta n(t)}{dt^2} + \frac{d\delta n}{dt}\gamma_L r = -\gamma_L\gamma_c(r-1)\delta n(t) + \gamma_L(r-1)r_{var}(t) .$$

Now, we perform the Fourier transform, using the property $F\{\mathrm{d}/\mathrm{d}t\} \to \mathrm{i}\omega$ and $F\{\mathrm{d}^2/\mathrm{d}t^2\} \to -\omega^2$:

$$[-\omega^2 + \mathrm{i}\omega\gamma_\mathrm{L} r + \gamma_\mathrm{L}\gamma_\mathrm{c}(r-1)]\delta\tilde{n}(\omega) = \gamma_\mathrm{L}(r-1)\tilde{r}_\mathrm{var}(\omega) \ ,$$

$$\delta\tilde{n}(\omega) = \frac{\gamma_\mathrm{L}(r-1)\tilde{r}_\mathrm{var}(\omega)}{-\omega^2 + \mathrm{i}\omega\gamma_\mathrm{L} r + \gamma_\mathrm{L}\gamma_\mathrm{c}(r-1)} \ .$$

With $\tilde{r}_\mathrm{var}(\omega) = \text{const.}$, only the denominator depends on ω. Thus, in order to find the maximum of the expression above, we only need to find the minimum for the denominator. More precisely, with $\chi(\omega) = -\omega^2 + \mathrm{i}\omega\gamma_\mathrm{L} r + \gamma_\mathrm{L}\gamma_\mathrm{c}(r-1)$, we need to find the value of ω for which $|\chi(\omega)|^2 = \min$, that means $\frac{\mathrm{d}}{\mathrm{d}\omega}\left(|\chi(\omega)|^2\right) = 0$:

$$\begin{aligned}|\chi(\omega)|^2 &= [-\omega^2 + \mathrm{i}\omega\gamma_\mathrm{L} r + \gamma_\mathrm{L}\gamma_\mathrm{c}(r-1)][-\omega^2 - \mathrm{i}\omega\gamma_\mathrm{L} r + \gamma_\mathrm{L}\gamma_\mathrm{c}(r-1)] \\ &= \omega^4 + \omega^2[\gamma_\mathrm{L}^2 r^2 - 2\gamma_\mathrm{L}\gamma_\mathrm{c}(r-1)] + \gamma_\mathrm{L}^2\gamma_\mathrm{c}^2(r-1)^2 \ .\end{aligned}$$

Calculating the derivative and setting it equal to zero, we obtain

$$4\omega^3 + 2\omega[\gamma_\mathrm{L}^2 r^2 - 2\gamma_\mathrm{L}\gamma_\mathrm{c}(r-1)] = 0 \ ,$$

$$2\omega^2 = -\gamma_\mathrm{L}^2 r^2 + 2\gamma_\mathrm{L}\gamma_\mathrm{c}(r-1) \ ,$$

which yields the solution

$$\omega_\mathrm{m} = \pm\sqrt{\gamma_\mathrm{L}\left(\gamma_\mathrm{c}(r-1) - \frac{1}{2}\gamma_\mathrm{L} r^2\right)} \ .$$

Note that ω_m refers to the frequency of the noise peak for the situation when a constant noise term $r_\mathrm{var} = 1$ has been added to the linearized rate equation (5.46) that describes the inversion.

For the following table and in Fig. 5.9, we use $r = 10$ and $\gamma_\mathrm{c} = 10^7\mathrm{s}^{-1}$.

	Nd:YLF	Nd:YVO$_4$	Ti:sapphire
$\tau_\mathrm{L}[\mu\mathrm{s}]$	480	50	3.2
$\gamma_\mathrm{L}[\mu\mathrm{s}^{-1}]$	2.08×10^{-3}	0.02	0.3125
ω_m [MHz]	0.43	1.33	4.82
$\lvert\tilde{n}_1(\omega_\mathrm{m})\rvert$, $r_\mathrm{var} \equiv 1$	2.08×10^{-6}	6.73×10^{-7}	1.77×10^{-7}

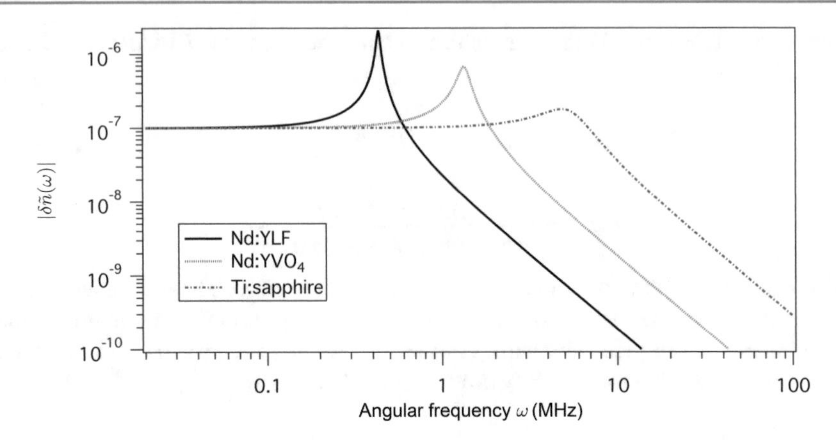

Fig. 5.9 Plot of $|\delta\tilde{n}(\omega)|$ as a function of ω for an added white noise in the inversion with $\tilde{r}_{\mathrm{var}}(\omega) = 1$ [see modified rate equation (5.61)]. The laser materials with larger τ_{L} are the most susceptible to external perturbations (i.e., show the more resonant behavior around the relaxation oscillator frequency ω_{m}). As a result, out of these three materials, Ti:sapphire would be the best choice for a low-noise laser oscillator

5.2.6 Measurement of the Small-Signal Gain

In this section, we discuss a simple, in-situ technique to calculate the small-signal gain of typical solid-state continuous-wave (cw) lasers such as Nd:YAG or Nd:YLF by measuring the frequency of the relaxation oscillation noise peak [94Wei]. The laser's small-signal gain can be directly calculated from the frequency of the relaxation oscillation with knowledge of the upper state lifetime, the cavity roundtrip time, and total losses, whose values can typically be measured. When the laser is pumped many times above threshold, the internal losses do not need to be known accurately because the total losses are dominated by the output coupler. This technique needs to be compared to the traditional method of changing output couplers to determine the intracavity losses. For lasers with a relatively high small-signal gain, evaluating the small-signal gain via the pump threshold is highly inaccurate. In particular, diode-pumped solid-state lasers have a very high small-signal gain of several 100% per roundtrip, and total cavity losses of only a few percent.

The value of a laser's small-signal gain is a fundamental parameter that is often important for laser design and operation. For example, knowing the value of the small-signal gain allows one to calculate the optimum output coupler in cw lasers, and also to predict the modelocked laser pulse duration and the necessary hold-off to prevent lasing in Q-switched lasers. It is also an important design parameter for influencing the dynamics of passively modelocked solid-state lasers.

In a classic four-level laser, it is well established that one can determine both the laser's small-signal gain and the internal loss by measuring the pump power at threshold and the laser's slope efficiency as a function of pump power, with two or more different output couplers. However, for certain lasers and cavity designs, it is impossible to accurately measure the threshold and slope efficiency, e.g., in a lamp-pumped Nd:YAG laser where thermal lensing allows only a limited setting of the pump powers, or in a monolithic cavity, such as a non-planar ring oscillator [85Kan], where the output coupler is directly coated onto the laser crystal and cannot be changed or varied. The laser gain can also be measured by probing the gain element with another laser at the same wavelength. However, this is also experimentally more complicated.

For a typical solid-state laser such as Nd:YAG and Nd:YLF, the upper-state lifetime τ_L (hundreds of microseconds) is much longer than the cavity decay time τ_c (tens to hundreds of nanoseconds). Therefore, the approximation $\tau_L \gg \tau_c$ and $r\gamma_L \ll \gamma_c$ is fulfilled if the laser is not pumped too far above threshold. Equation (5.57) then implies

$$f_{\mathrm{relax}} \approx \frac{1}{2\pi}\sqrt{\gamma_L\gamma_c(r-1)} = \frac{1}{2\pi}\sqrt{\frac{r-1}{\tau_L\tau_c}} = \frac{1}{2\pi}\sqrt{\frac{\frac{g_0}{l}-1}{\tau_L\frac{T_R}{2l}}} = \frac{1}{2\pi}\sqrt{\frac{2g_0-2l}{\tau_L T_R}}\,, \tag{5.62}$$

where, according to (5.32), we have used

$$r = \frac{g_0}{l}\,,$$

and where g_0 is the small-signal gain coefficient. Using $\tau_c = 1/\gamma_c$ with (5.6), it follows that

$$\tau_c = \frac{T_R}{2l}\,, \tag{5.63}$$

where γ_c is the photon cavity loss rate or decay rate as defined by the rate equation (5.4). For $g_0 \gg l$, the relaxation frequency becomes independent of the output coupler or l, and only depends on the pump power, i.e., $f_{\mathrm{relax}} \propto \sqrt{g_0} \propto \sqrt{P_{\mathrm{pump}}}$ (see Fig. 5.10). We can then solve (5.62) for the small-signal gain coefficient:

$$2g_0 \approx 4\pi^2\tau_L T_R f_{\mathrm{relax}}^2 + 2l\,. \tag{5.64}$$

The small-signal gain for the intensity is then given by

$$G_0 = \mathrm{e}^{2g_0} \approx \exp\left(4\pi^2\tau_L T_R f_{\mathrm{relax}}^2 + 2l\right)\,. \tag{5.65}$$

It should also be noted that, when r becomes very large, the second term in (5.57) can no longer be ignored. When this term dominates, f_{relax} begins to decrease with increasing r. We can calculate this inflection point by setting the derivative of (5.57) with respect to r to zero and solving for r, which gives the value $r_{\mathrm{infl}} = 2\tau_L/\tau_c$. If we put this value back into (5.57) and solve for the inflection frequency, we get $f_{\mathrm{infl}} \approx 1/2\pi\tau_c$, with the assumption that $\tau_L \gg \tau_c$. Under normal operating conditions for most lasers, these values are typically never reached. Additionally, under extreme pumping conditions, other nonlinear effects can occur to shift f_{relax}.

To check the accuracy of this theory, the output power and f_{relax} were measured as a function of pump power for a Ti:sapphire laser pumped Nd:YLF laser with

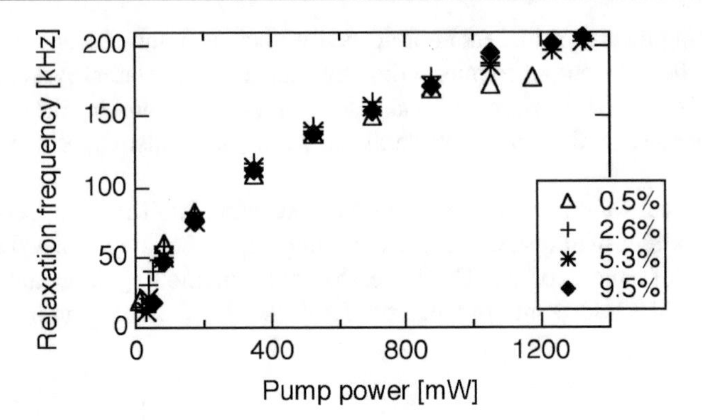

Fig. 5.10 Continuous wave (cw) Nd:YLF laser. Measured relaxation frequencies as a function of the pump power P_{pump} with different output couplers [94Wei]

several different values of output couplers [94Wei]. Figure 5.8 shows a schematic of the laser cavity. Output couplers with measured values of power transmission T_{out} of 0.5%, 2.6%, 5.3%, and 9.5% were used. The longitudinal mode noise peak was measured with a microwave spectrum analyzer while slightly tapping on one of the laser cavity mirrors to enhance the mode beating noise. We obtained 250 MHz, giving a roundtrip time of $T_R = 4\,\text{ns}$. The upper-state lifetime of the Nd:YLF of $\tau_L = 450\,\mu\text{s}$ was measured (by chopping the pump beam and observing the fluorescence decay time), in good agreement with published values [69Har]. To measure f_{relax}, the laser intensity was monitored with a silicon PIN photodiode whose output went into a low-noise, 1-MHz bandwidth amplifier and then into a microwave spectrum analyzer. The output coupler alignment was adjusted to give maximum output power (always with TEM_{00} spatial mode) and maximum f_{relax}.

Figure 5.10 shows the measured relaxation oscillation frequencies versus pump power for the different output couplers. Figure 5.8 shows the typical spectrum analyzer data used to determine the noise peak of f_{relax}. This value is used to calculate the small-signal gain coefficient according to (5.62) using $2l \approx T_{out} + 2 \cdot 0.8\%$ (i.e., with an 0.8% internal amplitude loss). Figure 5.11 shows the calculated small-signal gain versus pump power for the different output couplers. As expected, we see that the small-signal gain coefficient is relatively linear with pump power and independent of the output coupler. Note that, for the higher output couplers or when we are far above threshold, we can ignore the internal loss value with very small change in the calculated small-signal gain.

We can also check the validity of dropping the second term of (5.57). For the 0.5% output coupler (note that the output coupler transmission is given for power values), the additional 0.8% estimated internal amplitude loss coefficient, a roundtrip time of $T_R = 4\,\text{ns}$, and an upper-state lifetime of the Nd:YLF of $\tau_L = 450\,\mu\text{s}$, we calculate $r_{infl} = 2\tau_L/\tau_c = 2925$ and the corresponding $f_{infl} \approx 1/2\pi\tau_c = 517\,\text{kHz}$, with τ_c given in (5.63). The highest value for r here is approximately 230. This results in a 2% difference in the exact and approximate solution in (5.57). For output

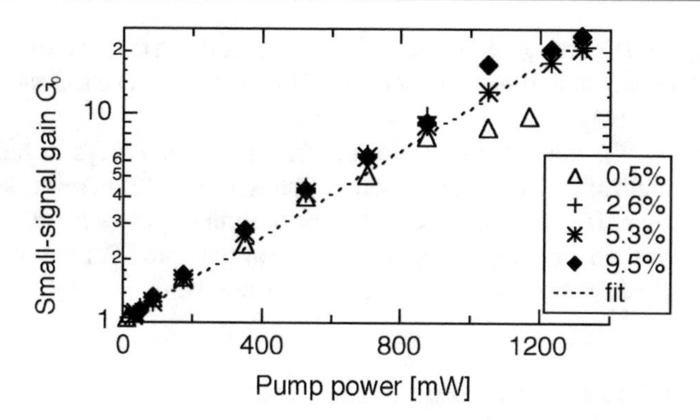

Fig. 5.11 Continuous wave (cw) Nd:YLF-laser. Calculated small-signal gain G_0 from the measured relaxation oscillation frequencies (Fig. 5.10) on a log scale. As expected, the small-signal gain increases exponentially with the pump power [94Wei]

couplers with higher values, r_{infl} and f_{infl} are much higher and this error in (5.57) can be neglected.

In conclusion, we have discussed a useful and simple method to measure the small-signal gain for solid-state lasers by measuring the relaxation oscillation frequency. This technique gives a much more accurate value of the small-signal gain compared with the conventional technique of changing output couplers and measuring slope efficiency. The relaxation oscillation frequency method allows for in-situ measurement of the small-signal gain under any specific operating conditions by measuring the value of f_{relax} for the laser and then applying (5.62). This technique also takes advantage of the sensitivity, dynamic range, and frequency accuracy of typical spectrum analyzers to measure the frequency of the relatively small relaxation oscillation noise peak. This approach can be particularly useful for characterizing a laser when changing output couplers is not easy or impossible, such as for a monolithic laser.

5.3 Transfer Function Analysis

For a transfer function analysis we apply a small modulation with variable frequency to the pump power and monitor the corresponding modulation of the average output power of the laser. The relation between pump and output modulation defines the frequency-dependent transfer function. From this function we can determine the frequency and the damping of the relaxation oscillations. Here we follow very closely the discussion in [04Sch].

As we will show later Q-switching instabilities in passively modelocked lasers can be interpreted as the occurrence of undamped relaxation oscillations (see Sect. 9.7). We show experimentally that the damping of the relaxation oscillations approaches zero as the pump power is decreased towards the Q-switched modelocking threshold. By comparing the damping with theoretically expected values we can check our understanding of the dynamics in a stable modelocking regime above the Q-switched modelocking threshold. We will compare the dynamics with and without the SESAM in the cavity in order to demonstrate its influence on the relaxation oscillations. In this section we will do the analysis for the cw laser, i.e., without a saturable absorber inside the laser cavity.

Furthermore, the noise properties of ultrafast lasers and ways to reduce noise with active stabilization will be discussed in Chaps. 11 and 12. In this case it is very useful to characterize how the modulation of the pump power affects the output power, the arrival time of the pulses, and the optical spectrum. Therefore also in this case transfer-function measurements and analysis will be very useful.

5.3.1 Rate Equations for Power and Gain

We first describe the behavior of a three-level laser system and then explain the simplifications for a four-level system. The rate equations of a three-level laser system (Sect. 5.1.5) can be rewritten as (assuming power gain and loss coefficient)

$$\frac{\mathrm{d}P}{\mathrm{d}t} = \frac{g-l}{T_\mathrm{R}} P \,, \tag{5.66}$$

$$\frac{\mathrm{d}g}{\mathrm{d}t} = \frac{g_0 - g}{\tau_\mathrm{L}} - \frac{P}{E_\mathrm{sat,L}} g + \frac{\eta(g) P_\mathrm{p}}{E_\mathrm{sat,L}} \,, \tag{5.67}$$

where P is the average intracavity laser power

$$P = \frac{h\nu_\mathrm{L} n}{T_\mathrm{R}} \,, \tag{5.68}$$

with T_R the cavity roundtrip time and $h\nu_\mathrm{L}$ the energy difference between the laser levels. g is the time-dependent roundtrip power gain coefficient

$$g = m L_\mathrm{g} (\sigma_\mathrm{tot} N_2 - \sigma_\mathrm{abs} N) \,, \quad \sigma_\mathrm{tot} \equiv \sigma_\mathrm{em} + \sigma_\mathrm{abs} \,, \tag{5.69}$$

while g_0 is the small-signal gain coefficient for the unpumped crystal (< 0 for three-level system), i.e., for $P = 0$. l is the linear loss coefficent per roundtrip. Note that the quantities P and g are described on the time scale of several roundtrip times T_R. τ_L is the upper state lifetime of the laser material, where we have to include the reabsorption from the ground state:

$$\frac{1}{\tau_\mathrm{L}} = \gamma_{12} + \gamma_{21} \,. \tag{5.70}$$

The saturation energy is given by

$$E_\mathrm{sat,L} = A_\mathrm{L} \frac{h\nu_\mathrm{L}}{m\sigma_\mathrm{tot}} = A_\mathrm{L} \frac{h\nu_\mathrm{L}}{m\left(\sigma_\mathrm{em} + \sigma_\mathrm{abs}\right)} \,, \tag{5.71}$$

where A_L is the effective laser mode area, and m the number of passes through the gain medium per cavity roundtrip (i.e., $m = 2$ for a linear standing wave cavity). Furthermore, g_0 and the pump efficiency $\eta(g)$ (including absorption efficiency and quantum defects) are given by

$$g_0 = mNL_g\left(\frac{\gamma_{12}}{\gamma_{12}+\gamma_{21}}\sigma_{tot} - \sigma_{abs}\right), \quad \eta(g) = \frac{\nu_L}{\nu_p}\eta'\left(\frac{\sigma_{em}}{\sigma_{tot}}N - \frac{E_{sat,L}}{V_L h\nu_L}g\right), \tag{5.72}$$

where η' is given by (5.43) for $N_1 + N_2 \approx N$ (5.35).

For the steady state solution, we find

$$g_s = l\,, \quad P_s = \frac{g_0 - l}{l}\frac{E_{sat,L}}{\tau_L} + \frac{\eta(l)}{l}P_p\,. \tag{5.73}$$

From this, we obtain the slope efficiency η_{SE} and the pumping threshold:

$$\eta_{SE} \equiv \frac{\partial P_{out}}{\partial P_p} = \frac{T_{oc}}{l}\eta(l)\,, \quad P_{p,th} = (l - g_0)\frac{1}{\eta(l)}\frac{E_{sat,L}}{\tau_L}\,, \tag{5.74}$$

where $\eta(l) = \eta(g = l)$ according to (5.72). In the case of a four-level laser, there is no reabsorption loss in the crystal. Therefore we have $g_0 = 0$ and $\sigma_{tot} \equiv \sigma_{em}$. Additionally, the absorption efficiency is typically only weakly dependent on the gain and it may be approximated by 1.

Derivation of (5.66) and (5.67)
By (5.42), we have

$$\dot{N}_2 = \gamma_{12}N_1 - \gamma_{21}N_2 + \frac{P_{p,0}}{V_L h\nu_p}\eta'(N_1) - \frac{m\sigma_{em}N_2L_g}{V_L T_R}n + \frac{m\sigma_{abs}N_1L_g}{V_L T_R}n\,,$$

$$N = N_1 + N_2\,, \qquad \dot{n} = \frac{m\sigma_{em}N_2L_g}{T_R}n - \frac{m\sigma_{abs}N_1L_g}{T_R}n - \gamma_c n\,, \quad \gamma_c = \frac{l}{T_R}\,.$$

Now, $P = h\nu_L n/T_R$, so

$$\begin{aligned}\dot{P} &= P(\sigma_{em}N_2 - \sigma_{abs}N_1)\frac{mL_g}{T_R} - \frac{l}{T_R}P \\ &= \frac{mL_g(\sigma_{tot}N_2 - \sigma_{abs}N) - l}{T_R}P \\ &= \frac{g - l}{T_R}P\,,\end{aligned}$$

which is (5.66). Here we have used

$$\sigma_{em}N_2 - \sigma_{abs}N_1 = \sigma_{em}N_2 - \sigma_{abs}(N - N_2) = \sigma_{tot}N_2 - \sigma_{abs}N$$

and

$$g \equiv mL_g(\sigma_{tot}N_2 - \sigma_{abs}N) \ .$$

The latter implies

$$\begin{aligned}
\dot{g} &= mL_g\sigma_{tot}\dot{N}_2 \\
&= mL_g\sigma_{tot}\Big\{\gamma_{12}(N-N_2) - \gamma_{21}N_2 + \frac{P_{p,0}}{V_L h\nu_p}\eta'(N-N_2) \\
&\qquad - mL_g\frac{n}{V_L T_R}\big[\sigma_{em}N_2 - \sigma_{abs}(N-N_2)\big]\Big\} \\
&= mL_g\sigma_{tot}\Big[-\underbrace{N_2(\gamma_{12}+\gamma_{21})}_{=N_2/\tau_L} + \gamma_{12}N + \frac{P_{p,0}}{V_L h\nu_p}\eta' - \underbrace{mL_g\frac{n}{V_L T_R}(N_2\sigma_{tot} - N\sigma_{abs})}_{=gn/V_L T_R}\Big] \\
&= mL_g\sigma_{tot}\left[\left(\gamma_{12}N - \frac{N_2}{\tau_L}\right) + \frac{P_{p,0}}{V_L h\nu_p}\eta'\right] - mL_g\sigma_{tot}\frac{n}{V_L T_R}g \ .
\end{aligned}$$

We now use $P = h\nu_L n/T_R$ to eliminate n, whence

$$\dot{g} = mL_g\sigma_{tot}\left[\left(\gamma_{12}N - \frac{N_2}{\tau_L}\right) + \frac{P_{p,0}}{V_L h\nu_p}\eta'\right] - mL_g\sigma_{tot}\frac{P}{h\nu_L V_L}g \ .$$

By (5.71),

$$E_{sat,L} = A_L\frac{h\nu_L}{m\sigma_{tot}} = \frac{V_L}{L_g}\frac{h\nu_L}{m\sigma_{tot}} \ ,$$

and using $g = mL_g(\sigma_{tot}N_2 - \sigma_{abs}N)$, we have

$$\begin{aligned}
\dot{g} &= mL_g\sigma_{tot}\left[\left(\gamma_{12}N - \frac{N_2}{\tau_L}\right) + \frac{P_{p,0}}{V_L h\nu_p}\eta'\right] - \frac{P}{E_{sat,L}}g \\
&= -\frac{g + mL_g\sigma_{abs}N}{\tau_L} + mL_g\sigma_{tot}\gamma_{12}N + mL_g\sigma_{tot}\frac{P_{p,0}}{V_L h\nu_p}\eta' - \frac{P}{E_{sat,L}}g \\
&= -\frac{g}{\tau_L} + mL_g\left(\gamma_{12}\sigma_{tot}N - \frac{1}{\tau_L}\sigma_{abs}N\right) + mL_g\sigma_{tot}\frac{P_{p,0}}{V_L h\nu_p}\eta' - \frac{P}{E_{sat,L}}g \\
&= \frac{g_0 - g}{\tau_L} + mL_g\sigma_{tot}\frac{P_{p,0}}{V_L h\nu_p}\eta' - \frac{P}{E_{sat,L}}g \ ,
\end{aligned}$$

where

$$g_0 \equiv mL_gN\left(\frac{\gamma_{12}}{\gamma_{12}+\gamma_{21}}\sigma_{tot} - \sigma_{abs}\right) \ .$$

Hence,

$$\dot{g} = \frac{g_0 - g}{\tau_L} + P_{p,0}\eta' \frac{mL_g\sigma_{tot}}{V_L h\nu_p} - \frac{P}{E_{sat,L}} g$$
$$= \frac{g_0 - g}{\tau_L} + \frac{P_{p,0}\eta}{E_{sat,L}} - \frac{P}{E_{sat,L}} g \,,$$

where we have used

$$\frac{mL_g\sigma_{tot}}{V_L h\nu_p} = \frac{\nu_L}{\nu_P} \frac{1}{E_{sat,L}}$$

and defined

$$\eta \equiv \frac{\nu_L}{\nu_P}\eta' \,.$$

We thus arrive at the macroscopic differential equations (5.66) and (5.67):

$$\dot{P} = \frac{g - l}{T_R} P \,,$$

$$\dot{g} = \frac{g_0 - g}{\tau_L} + \frac{P_p\eta}{E_{sat,L}} - \frac{P}{E_{sat,L}} g \,.$$

5.3.2 Relaxation Oscillations

In order to find the transfer function, we linearize the differential equations (5.66) and (5.67) around the steady state and also allow the pump power to vary. We define δP, δg, and δP_p by

$$\delta P \equiv P(t) - P_s \,, \quad \delta g \equiv g(t) - g_s \,, \quad \delta P_p \equiv P(t) - P_{ps} \,, \tag{5.75}$$

where P_s, g_s, and P_{ps} are the steady-state values. For the linearization around the steady state, we then have

$$\alpha \equiv \frac{\partial \dot{P}}{\partial P} = \frac{g - l}{T_R} = 0 \quad \text{at steady state} \,, \tag{5.76}$$

where $\dot{P} = \mathrm{d}P/\mathrm{d}t$. Then

$$\beta \equiv \frac{\partial \dot{P}}{\partial g} = \frac{P}{T_R} \,, \tag{5.77}$$

$$-\gamma \equiv \frac{\partial \dot{g}}{\partial P} = -\frac{g}{E_{sat,L}} \,, \tag{5.78}$$

where $\dot{g} = \mathrm{d}g/\mathrm{d}t$,

$$-\varepsilon \equiv \frac{\partial \dot{g}}{\partial g} = -\frac{1}{\tau_\mathrm{L}} + \frac{P_\mathrm{p}}{E_\mathrm{sat,L}} \frac{\partial \eta}{\partial g}(g) - \frac{P}{E_\mathrm{sat,L}} , \tag{5.79}$$

and

$$\zeta \equiv \frac{\partial \dot{g}}{\partial P_\mathrm{p}} = \frac{\eta(g)}{E_\mathrm{sat,L}} . \tag{5.80}$$

For small deviations from the steady-state and small variations of the pump power, we therefore find

$$\frac{\mathrm{d}}{\mathrm{d}t}\begin{pmatrix} \delta P \\ \delta g \end{pmatrix} = \begin{pmatrix} \alpha & \beta \\ -\gamma & -\varepsilon \end{pmatrix}\begin{pmatrix} \delta P \\ \delta g \end{pmatrix} + \zeta \begin{pmatrix} 0 \\ \delta P_\mathrm{p} \end{pmatrix} . \tag{5.81}$$

We can then determine the eigenvalues λ :

$$(\alpha - \lambda)(-\varepsilon - \lambda) - (-\gamma\beta) = 0 ,$$

giving

$$\lambda^2 + (\varepsilon - \alpha)\lambda + \beta\gamma - \alpha\varepsilon = 0 .$$

This has solutions

$$\lambda_{1,2} = -\frac{1}{2}(\varepsilon - \alpha) \pm \sqrt{\alpha\varepsilon - \beta\gamma + \frac{1}{4}(\varepsilon - \alpha)^2} . \tag{5.82}$$

For typical solid-state lasers, the radicand is negative. We find for the frequency of the relaxation oscillations (imaginary part of the eigenvalues):

$$\boxed{\omega_\mathrm{ro} = \sqrt{\beta\gamma - \alpha\varepsilon - \frac{1}{4}(\varepsilon - \alpha)^2}} . \tag{5.83}$$

The damping of the oscillations is given by the real part of the eigenvalues and inversely proportional to the damping time of the relaxation oscillations τ_ro :

$$\boxed{\tau_\mathrm{ro} = \frac{2}{\varepsilon - \alpha}} . \tag{5.84}$$

5.3.3 Transfer Function

In the stable regime, we define the transfer function $\chi(\omega)$ as the ratio of the complex amplitudes of the oscillation of output power and pump power, when a harmonic modulation is applied to the pump power according to

$$\delta P_{\mathrm{p}} = \widehat{\delta P_{\mathrm{p}}} \sin(\omega_{\mathrm{m}} t) , \tag{5.85}$$

where $\widehat{\delta P_{\mathrm{p}}}$ is the peak pump power modulation, and ω_{m} is the modulation frequency of the pump power.

By solving the differential equations, we find the transfer function:

$$\boxed{\chi(\omega_{\mathrm{m}}) \equiv \frac{\delta P_{\mathrm{out}}}{\delta P_{\mathrm{p}}} = \frac{T_{\mathrm{oc}} \beta \zeta}{-\omega_{\mathrm{m}}^2 + (\varepsilon - \alpha)\mathrm{i}\omega_{\mathrm{m}} + \beta\gamma - \alpha\varepsilon}} , \tag{5.86}$$

where T_{oc} is the output coupler power transmission, and the output power P_{out} is given by $P_{\mathrm{out}} = T_{\mathrm{oc}} P$.

(a) Derivation of the Transfer Function

From (5.81) and (5.85),

$$\begin{cases} \mathrm{i}\omega_{\mathrm{m}} \widehat{\delta P} = \alpha \widehat{\delta P} + \beta \widehat{\delta g} \implies \widehat{\delta g} = \dfrac{\mathrm{i}\omega_{\mathrm{m}} - \alpha}{\beta} \widehat{\delta P} , \\ \mathrm{i}\omega_{\mathrm{m}} \widehat{\delta g} = -\gamma \widehat{\delta P} - \varepsilon \widehat{\delta g} + \zeta \widehat{\delta P_{\mathrm{p}}} . \end{cases}$$

Hence,

$$\mathrm{i}\omega_{\mathrm{m}} \widehat{\delta g} = -\gamma \frac{\beta}{\mathrm{i}\omega_{\mathrm{m}} - \alpha} \widehat{\delta g} - \varepsilon \widehat{\delta g} + \zeta \widehat{\delta P_{\mathrm{p}}} ,$$

and

$$\left(\mathrm{i}\omega_{\mathrm{m}} + \frac{\beta\gamma}{\mathrm{i}\omega_{\mathrm{m}} - \alpha} + \varepsilon \right) \widehat{\delta g} = \zeta \widehat{\delta P_{\mathrm{p}}} .$$

Then, from above,

$$\begin{aligned} \widehat{\delta P} &= \frac{\beta}{\mathrm{i}\omega_{\mathrm{m}} - \alpha} \widehat{\delta g} = \frac{\beta}{\mathrm{i}\omega_{\mathrm{m}} - \alpha} \frac{\zeta}{\mathrm{i}\omega_{\mathrm{m}} + \dfrac{\beta\gamma}{\mathrm{i}\omega_{\mathrm{m}} - \alpha} + \varepsilon} \widehat{\delta P_{\mathrm{p}}} \\ &= \frac{\beta\zeta}{\mathrm{i}\omega_{\mathrm{m}}(\mathrm{i}\omega_{\mathrm{m}} - \alpha) + \beta\gamma + \varepsilon(\mathrm{i}\omega_{\mathrm{m}} - \alpha)} \widehat{\delta P_{\mathrm{p}}} . \end{aligned}$$

Finally,

$$\chi(\omega_{\mathrm{m}}) = \frac{\widehat{\delta P_{\mathrm{out}}}}{\widehat{\delta P_{\mathrm{p}}}} = T_{\mathrm{oc}} \frac{\widehat{\delta P}}{\widehat{\delta P_{\mathrm{p}}}} = \frac{T_{\mathrm{oc}} \beta \zeta}{-\omega_{\mathrm{m}}^2 + (\varepsilon - \alpha)\mathrm{i}\omega_{\mathrm{m}} + \beta\gamma - \alpha\varepsilon} ,$$

which is (5.86).

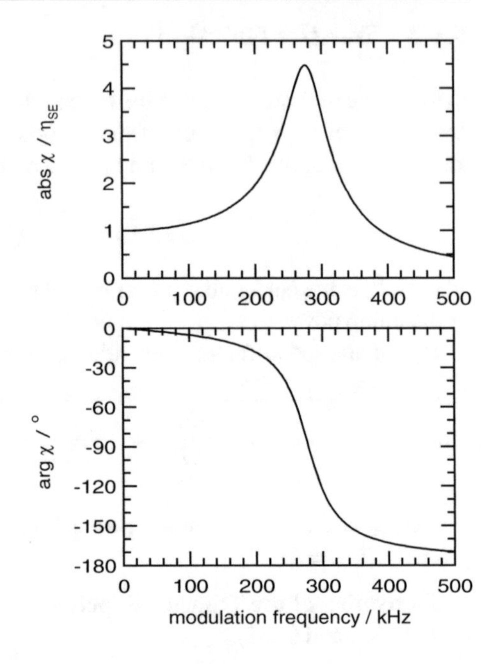

Fig. 5.12 Typical shape of a transfer function according to (5.90). The transfer function is completely characterized by the resonance frequency f_{res} (5.91) (frequency where the modulus of the transfer function has a peak), the FWHM peak width of approximately $2\omega_{1/2}$ (5.98), and the peak height $\widehat{\chi}$ (5.89). η_{SE} is the slope efficiency (5.74)

(b) Properties of the Transfer Function

Analyzing the modulus of the transfer function (Fig. 5.12), we identify the following characteristic parameters. The resonance frequency (frequency of the peak) is given by

$$f_{\text{res}} = \frac{1}{2\pi}\left[\beta\gamma - \alpha\varepsilon - \frac{1}{2}(\varepsilon-\alpha)^2\right]^{1/2} . \tag{5.87}$$

The parameter

$$\begin{aligned} \xi &\equiv f_{1/2}^2 - f_{\text{res}}^2 \\ &= \frac{\sqrt{3}}{4\pi^2}(\varepsilon-\alpha)\left[\beta\gamma - \alpha\varepsilon - \frac{1}{4}(\varepsilon-\alpha)^2\right]^{1/2} \\ &= \frac{\sqrt{3}}{\pi}\frac{f_{\text{ro}}}{\tau_{\text{ro}}} \end{aligned} \tag{5.88}$$

characterizes the width of the resonance peak. Here $f_{1/2}$ is the (higher) frequency, where the modulus of the transfer function has fallen to half its peak value. We use this type of width parameter (instead of the full width at half maximum, FWHM, for example) because the parameter ξ is directly related to the damping time of the relaxation oscillations.

The peak height of the modulus of the transfer function is

$$\widehat{\chi} = \frac{1}{2\pi} \frac{T_{\text{oc}}\beta\zeta}{(\varepsilon - \alpha)\left[\beta\gamma - \alpha\varepsilon - \frac{1}{4}(\varepsilon - \alpha)^2\right]^{1/2}}$$

$$= \frac{1}{4\pi} T_{\text{oc}}\beta\zeta \frac{\tau_{\text{ro}}}{f_{\text{ro}}} . \tag{5.89}$$

We can rewrite the modulus and phase of the transfer function, which we now interpret as a function of the frequency f instead of the angular frequency:

$$\boxed{\begin{aligned} |\chi(f)| &= \frac{\widehat{\chi}\xi}{\left[3\left(f^4 - 2f_{\text{res}}^2 f^2 + \frac{1}{3}\xi^2 + f_{\text{res}}^4\right)\right]^{1/2}} \\ \arg\chi(f) &= -\arctan \frac{f\left\{2\left[\left(f_{\text{res}}^4 + \frac{1}{3}\xi^2\right)^{1/2} - f_{\text{res}}^2\right]\right\}^{1/2}}{\left(f_{\text{res}}^4 + \frac{1}{3}\xi^2\right)^{1/2} - f^2} \end{aligned}} \tag{5.90}$$

For $f = 0$, i.e., when the pump power is not modulated, the transfer function must equal the slope efficiency. A typical transfer function is shown in Fig. 5.12.

We typically have $\varepsilon^2 \ll \beta\gamma$. In this case, the resonance frequency f_{res} can be approximated as

$$\boxed{f_{\text{res}} \approx \frac{1}{2\pi}\sqrt{\beta\gamma} = \frac{1}{2\pi}\left(\frac{Pg}{T_{\text{R}} E_{\text{sat,L}}}\right)^2} . \tag{5.91}$$

It follows for the frequency of the relaxation oscillations that

$$\boxed{f_{\text{ro}} \approx f_{\text{res}}} , \tag{5.92}$$

and we also have for the damping time of the relaxation oscillations

$$\boxed{\tau_{\text{ro}} = \frac{\sqrt{3}}{\pi}\frac{f_{\text{ro}}}{\xi} \approx \frac{\sqrt{3}}{\pi}\frac{f_{\text{res}}}{\xi}} , \tag{5.93}$$

where ξ is given by (5.88).

Derivation of (5.90)

Reparametrization (i.e., *a*, *b*, and *c*) of the Transfer Function

From (5.86) we can define the new parameters a, b, and c as follows:

$$\chi(\omega) = \frac{T_{\mathrm{oc}}\beta\zeta}{-\omega^2 + (\varepsilon - \alpha)\mathrm{i}\omega + \beta\gamma - \alpha\varepsilon} \equiv \frac{a}{-\omega^2 + b\mathrm{i}\omega + c}\,, \tag{5.94}$$

whence

$$|\chi(\omega)| = \frac{a}{\sqrt{(-\omega^2 + c)^2 + b^2\omega^2}}\,, \quad \arg\chi(\omega) = -\arctan\frac{b\omega}{c - \omega^2}\,. \tag{5.95}$$

Resonance Frequency

From (5.95),

$$\begin{aligned}
\frac{\partial}{\partial\omega}|\chi(\omega)| = 0 &\iff \frac{\partial}{\partial\omega}\left[(-\omega^2 + c)^2 + b^2\omega^2\right] = 0 \\
&\iff 4\omega^2 - 4c + 2b^2 = 0 \\
&\iff \omega = \sqrt{c - \frac{1}{2}b^2}\,.
\end{aligned}$$

Hence,

$$\omega_{\mathrm{res}} = \sqrt{c - \frac{1}{2}b^2}\,. \tag{5.96}$$

Peak Height
From (5.95) and (5.96),

$$\widehat{\chi} = |\chi(\omega_{\mathrm{res}})| = \frac{a}{\sqrt{(-c + b^2/2 + c)^2 + b^2(c - b^2/2)}} = \frac{a}{\sqrt{b^4/4 + b^2(c - b^4/2)}}\,,$$

whence

$$\widehat{\chi} = \frac{a}{|b|}\frac{1}{\sqrt{c - b^2/4}}\,. \tag{5.97}$$

Peak Width
From (5.95) and (5.97),

$$\begin{aligned}
|\chi(\omega)| = \frac{1}{2}\widehat{\chi} &\iff \frac{a}{\sqrt{(-\omega^2 + c)^2 + b^2\omega^2}} = \frac{a}{2|b|}\frac{1}{\sqrt{c - b^2/4}} \\
&\iff (-\omega^2 + c)^2 + b^2\omega^2 = 4b^2\left(c - \frac{b^2}{4}\right) \\
&\iff \omega^4 - 2c\omega^2 + c^2 + b^2\omega^2 - 4b^2c + b^4 = 0\,.
\end{aligned}$$

Defining $s \equiv \omega^2$, this becomes

$$s^2 + (-2c + b^2)s + c^2 - 4b^2c + b^4 = 0 ,$$

with solutions

$$\begin{aligned} s &= \frac{1}{2}\Big[2c - b^2 \pm \sqrt{(b^2 - 2c)^2 - 4(c^2 - 4b^2c + b^4)}\Big] \\ &= c - \frac{b^2}{2} \pm \sqrt{\frac{1}{4}(b^2 - 2c)^2 - (c^2 - 4b^2c + b^4)} . \end{aligned}$$

Here, we select the positive root, whence finally,

$$\omega_{1/2}^2 = c - \frac{b^2}{2} + \sqrt{\frac{1}{4}(b^2 - 2c)^2 - (c^2 - 4b^2c + b^4)} . \tag{5.98}$$

Now,

$$\begin{aligned} 4\pi^2\xi &\equiv \omega_{1/2}^2 - \omega_{\text{res}}^2 \\ &= \sqrt{\frac{1}{4}(b^2 - 2c)^2 - (c^2 - 4b^2c + b^4)} \\ &= \sqrt{\frac{1}{4}(b^4 - 4cb^2 + c^2) - (c^2 - 4b^2c + b^4)} \\ &= \sqrt{-\frac{3}{4}b^4 + 3b^2c} . \end{aligned}$$

Hence,

$$4\pi^2\xi = \sqrt{3}|b|\sqrt{c - \frac{b^2}{4}} . \tag{5.99}$$

Rewriting the transfer function in (5.95) with the new parameters, we obtain

$$\begin{aligned} |\chi(\omega)| &= \frac{a}{\sqrt{(-\omega^2 + c)^2 + b^2\omega^2}} = \frac{a}{\sqrt{\omega^4 - 2c\omega^2 + c^2 + b^2\omega^2}} \\ &= \frac{a}{\sqrt{\omega^4 + (-2c + b^2)\omega^2 + c^2}} = \frac{a}{\sqrt{\omega^4 - 2\omega_{\text{res}}^2\omega^2 + c^2}} , \end{aligned}$$

where we have used $-2c + b^2 = -2\omega_{\text{res}}^2$ from (5.96). Since

$$\widehat{\chi} = \frac{a}{|b|}\frac{1}{\sqrt{c - b^2/4}}$$

from (5.97), we find

$$|\chi(\omega)| = \frac{\widehat{\chi}|b|\sqrt{c-b^2/4}}{\sqrt{\omega^4 - 2\omega_{\rm res}^2\omega^2 + c^2}} = \frac{4\pi^2\widehat{\chi}\xi}{\sqrt{3}\sqrt{\omega^4 - 2\omega_{\rm res}^2\omega^2 + c^2}} . \tag{5.100}$$

Using

$$\omega_{\rm res}^4 + \frac{1}{3}(4\pi^2\xi)^2 = c^2 - b^2c + \frac{1}{4}b^4 + b^2\left(c - \frac{1}{4}b^2\right) = c^2 ,$$

we have

$$|\chi(\omega)| = \frac{4\pi^2\widehat{\chi}\xi}{\sqrt{3}\sqrt{\omega^4 - 2\omega_{\rm res}^2\omega^2 + \omega_{\rm res}^4 + \frac{1}{3}(4\pi^2)^2\xi^2}} .$$

Rewriting this as a function of f instead of $\omega = 2\pi f$, we obtain

$$|\chi(f)| = \frac{4\pi^2\widehat{\chi}\xi}{\sqrt{3}\sqrt{(2\pi)^4 f^4 - 2(2\pi)^4 f_{\rm res}^2 f^2 + (2\pi)^4 f_{\rm res}^4 + \frac{1}{3}(2\pi)^4\xi^2}} ,$$

and cancelling the factor of $4\pi^2$ in the numerator and denominator leads to (5.90). Similarly, for arg $\chi(f)$.

5.3.4 Transfer Function Measurement

To measure the transfer function of the laser, we sinusoidally modulate the current of the pump laser diode and measure the amplitude and phase of the resulting modulation of pump and output power for a range of modulation frequencies. For example, when a Ti:sapphire laser is used as a pump laser, then the pump power cannot easily be modulated electrically. Instead, an acousto-optic modulator can be inserted into the pump beam.

The setup is shown in Fig. 5.13. It consists of the pump laser and means to modulate its power. The amplitude and frequency of the modulation are controlled by a signal generator. The amplitude has to be chosen small enough to stay within the validity range of the linearized rate equations. Pump and output power are measured with silicon photodiodes and a transimpedance amplifier. A lock-in amplifier measures the amplitude and phase of the modulation on the signal. The signal source can be manually switched between pump and laser output. The resulting function was scaled not by an absolute calibration of the amplified signal to the powers, but by requiring that the low-frequency limit must be identical to the slope efficiency $\eta_{\rm SE}$ (5.74).

1.064 μm Nd:YVO$_4$ and 1.321 μm Nd:YLF Lasers (Four-Level Lasers)

Both the Nd:YVO$_4$ and Nd:YLF laser cavities are shown in Fig. 5.14, which are chosen to also work for passive modelocking using an intracavity saturable absorber

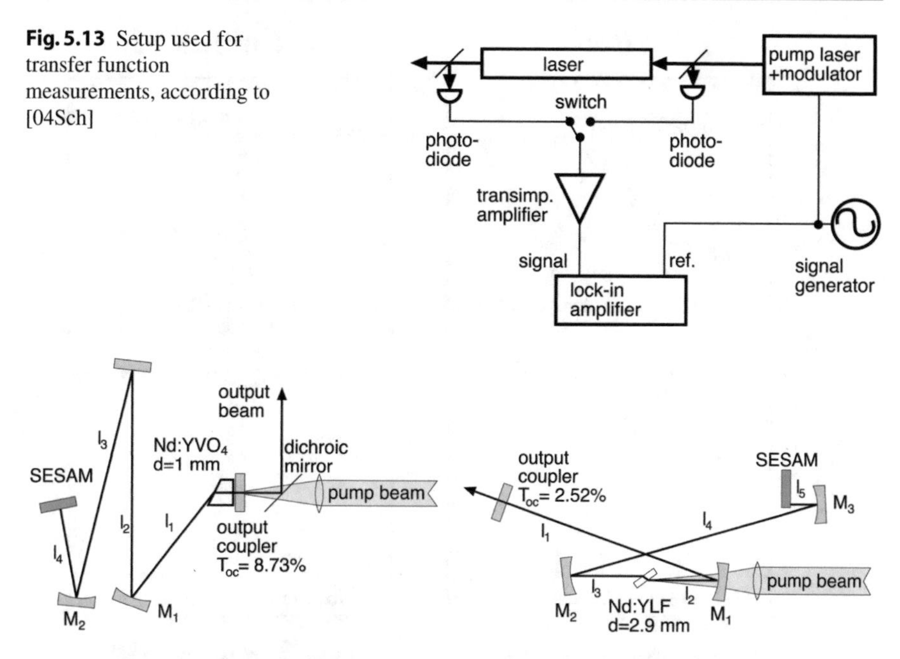

Fig. 5.13 Setup used for transfer function measurements, according to [04Sch]

Fig. 5.14 Nd:YVO_4 and Nd:YLF laser cavities used for transfer function analysis. For cw operation the SESAM is replaced by a high reflector. The cavity design is discussed in more detail in [04Sch]

(i.e., a SESAM, to be discussed in more detail in Chap. 7). For cw operation the SESAM is replaced by a highly reflective mirror. The Nd:YVO_4 laser is diode pumped at 808 nm through the output-coupling mirror. A dichroic mirror is used to separate the output laser beam from the pump beam. The crystal has a Brewster interface to avoid reflection losses. The Nd:YLF laser cavity has a Brewster-angled Nd:YLF crystal which is pumped with a Ti:sapphire laser at 797 nm through one of the folding mirrors.

The transfer functions of the modelocked and the cw laser (where the SESAM is replaced with a highly reflecting mirror) were measured at different intracavity power levels. Shown are the transfer functions at the lowest and the highest measured intracavity power (Figs. 5.15 and 5.16). The modulus of the measured transfer functions was fitted with the function given in (5.90), using f_{res}, $\widehat{\chi}$, and ξ as fit parameters. The measured data as well as the fitted functions were afterwards scaled to meet $|\chi(0)| = \eta_{\mathrm{SE}}$, the slope efficiency (5.74). The phase data allow for additional confirmation of the obtained values of f_{res} and ξ. Note that $\widehat{\chi}$ has no influence on the phase. The lines in the phase plots show the functions given in (5.90) using the parameters from the above-mentioned fitting procedure, rather than doing another fit.

The dynamic behavior of both lasers is as expected from the theory. Compared with the cw laser, the passively modelocked lasers exhibit a weaker damping of the relaxation oscillations, because the SESAM introduces a lower loss for higher pulse energies. If we reduce the pump power, we approach a critical point where the

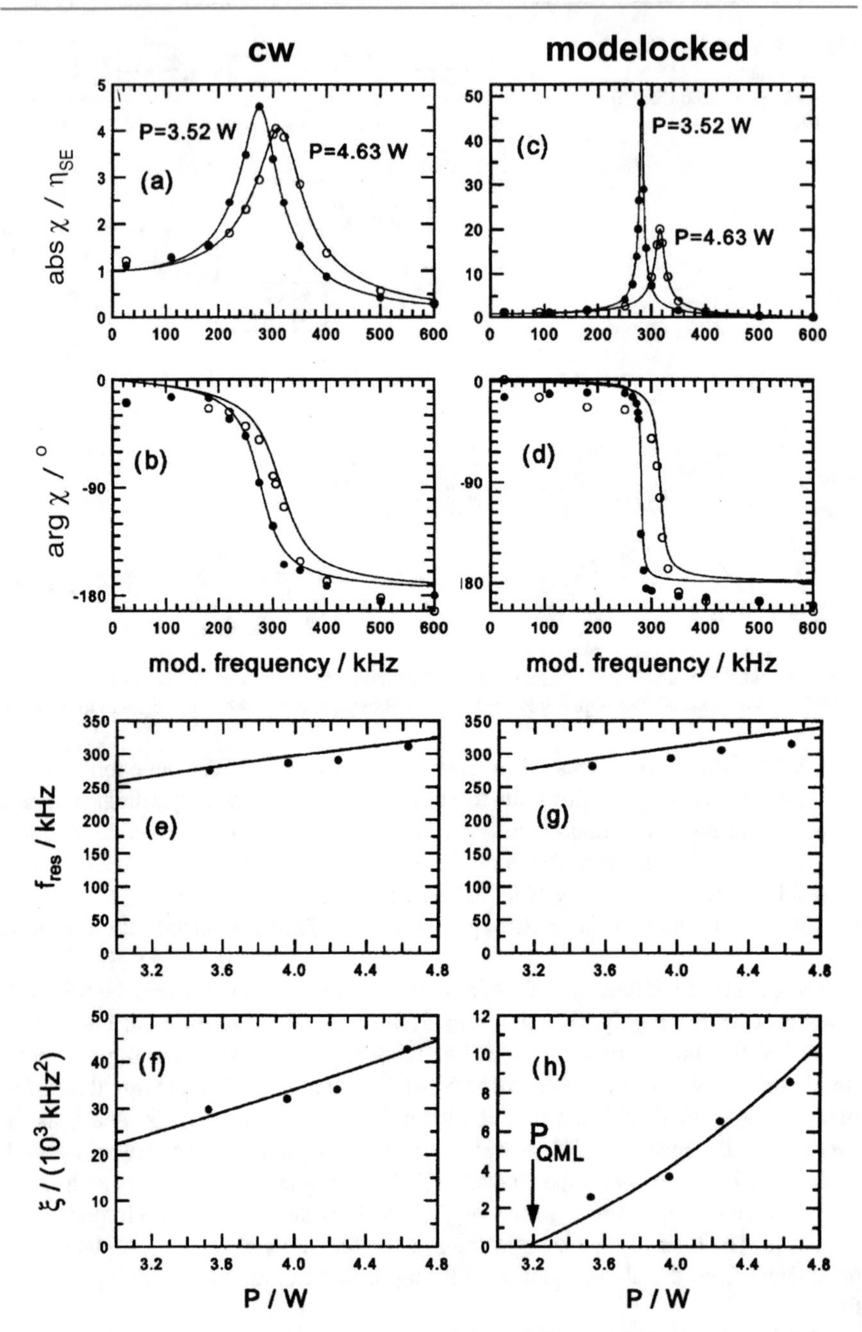

Fig. 5.15 $Nd:YVO_4$ laser: **a**, **b** Transfer functions of the cw and **c**, **d** the SESAM modelocked $Nd:YVO_4$ laser at the highest and lowest measured intracavity power. For clarity, the other curves are not shown here. **e–h** Parameters as functions of the intracavity power for the cw and the modelocked laser with f_{res} (5.91) the resonance frequency i.e., approximately the relaxation oscillation frequency $f_{ro} \approx f_{res}$ (5.92) and $\xi \equiv f_{1/2}^2 - f_{res}^2$ (5.88) characterizing the width of the resonance peak. η_{SE} is the slope efficiency (5.74). The Q-switched modelocking (QML) threshold power will be discussed in Chap. 9. According to [04Sch]

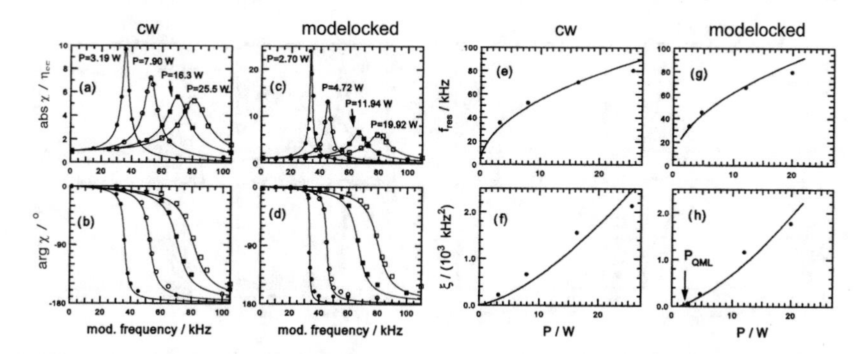

Fig. 5.16 Nd:YLF laser with a pulse repetition rate of 119 MHz and, following the same figure caption as Fig. 5.15, with several different intracavity power levels for cw operation (i.e., 3.19 W, 7.90 W, 16.3 W, 25.5 W) and for modelocking (i.e., 2.70 W, 4.72 W, 11.94 W, 19.92 W). At higher power the relaxation oscillations are more strongly damped for both cw and modelocked operation and the additional saturable absorber reduces the damping of the relaxation oscillations compared to cw operation. According to [04Sch]

damping becomes zero. Below this point, fluctuations of the pulse energy grow exponentially and lead the laser into the Q-switched modelocked regime. By comparing the measured transfer functions between the Nd:YLF and the Nd:YVO_4, we can see a lower influence of the SESAM on the pulse-energy dynamics for the Nd:YLF laser. In this case the properties of the transfer functions, the measured low QML threshold, and the long pulse length all suggest that the modulation depth of the SESAM is unusually small.

1.534 μm Er:Yb:Glass Laser (Three-Level Laser)

Er:Yb:glass is a three-level gain material that uses ytterbium as a sensitizer to increase the pump efficiency and erbium to generate gain in the 1.5-μm region. For the transfer function measurements a similar laser cavity is used as for the Nd:YLF laser (Fig. 5.14) and is described in more detail in [04Sch]. The repetition rate of the laser is 61 MHz, and the pulse length with SESAM modelocking is 3.0 ps. The laser generates up to 55 mW of output. The QML threshold is at 24-mW output power. The relaxation oscillations were too weak to appear in the RF spectrum, but they are expected to be at around 25 kHz for maximum output power.

The Er:Yb laser can be described by a quasi-three-level system (erbium ions) that is pumped through energy transfer from ytterbium ions. Note that this Yb→Er energy transfer is fairly efficient, because the transfer rate is seven times higher than the spontaneous decay rate of the Yb level alone, which is assumed to be 1 ms. We therefore need to have an additional rate equation which describes this Yb→Er transfer. It can be shown [04Sch] that the transfer function of the Er:Yb laser is the product of a three-level-laser transfer function $\chi_{\mathrm{Er}}(\omega)$ and the transfer function of the ytterbium system $\chi_{\mathrm{Yb}\rightarrow\mathrm{Er}}(\omega)$. The function $\chi_{\mathrm{Yb}\rightarrow\mathrm{Er}}(\omega)$ describes a low-pass filter with a cut-off frequency f_{co} and can be described by

$$\chi_{\mathrm{Yb}\rightarrow\mathrm{Er}}(f) = \frac{A}{1 + \mathrm{i}f/f_{\mathrm{co}}}, \tag{5.101}$$

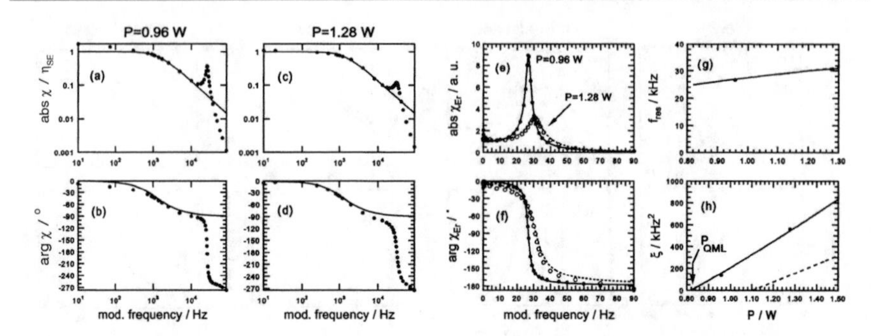

Fig. 5.17 Er:Yb:glass laser with a pulse repetition rate of 61 MHz. Measured transfer functions $\chi(f) = \chi_{\mathrm{Er}}(f)\chi_{\mathrm{Yb}\to\mathrm{Er}}(f)$ for the SESAM modelocked laser for two different intracavity powers (i.e., 0.96 W and 1.28 W). **(g)-(h)** f_{res} (5.91) the resonance frequency, i.e., approximately the relaxation oscillation frequency $f_{\mathrm{ro}} \approx f_{\mathrm{res}}$ (5.92) and $\xi \equiv f_{1/2}^2 - f_{\mathrm{res}}^2$ (5.88) characterizing the width of the resonance peak. η_{SE} is the slope efficiency (5.74). The Q-switched modelocking (QML) threshold power will be discussed in Chap. 9. According to [04Sch]

where A is a constant. Note that A is not important because the transfer function is normalized by $|\chi(0)| = \eta_{\mathrm{SE}}$, the slope efficiency (5.74).

The measured transfer functions are shown in Fig. 5.17 at two different intracavity powers of the modelocked laser. Transfer functions of the cw laser have not been measured because the output power of the laser was not stable enough, apparently due to spectral hole burning and uncontrolled hopping between the axial modes. The passively modelocked laser is much more stable. For the total transfer function $\chi(f) = \chi_{\mathrm{Er}}(f)\chi_{\mathrm{Yb}\to\mathrm{Er}}(f)$ the low-frequency part was fitted with $\chi_{\mathrm{Yb}\to\mathrm{Er}}(f)$ (5.101) to match the transfer functions at low frequencies. Good agreement was obtained for both power levels with $f_{\mathrm{co}} = 1.33$ kHz. With the fitted low-pass filter function, the transfer function $\chi_{\mathrm{Er}}(f)$ for the laser transition can be determined.

The solid curves in Fig. 5.17g–h show $f_{\mathrm{res}}(P)$ and $\xi(p)$ as a function of intracavity power P, taking into account additional nonlinear loss in the SESAM. The dashed curves (not distinguishable from the solid curve in Fig. 5.17g) are plotted with the additional nonlinear loss reduced to the amount expected from two-photon absorption in the SESAM.

The Yb→Er energy transfer strongly reduces the relaxation oscillations in this laser (Fig. 5.17) because the cut-off frequency $f_{\mathrm{co}} = 1.33$ kHz is significantly lower than the relaxation frequency above 20 kHz. This results in lower noise performance for frequency comb generation, as will be discussed in Chap. 12.

6 Active Modelocking

For the analytical treatment in this chapter, we mainly rely on the theoretical framework of Hermann A. Haus [75Hau2, 84Hau] and Anthony E. Siegman [70Kui1, 70Kui2, 86Sie]. Their theories date back to the 1970s. The master equation of Haus describes the temporal evolution of the pulse envelope inside the laser resonator in the slowly varying pulse envelope approximation. Analytical solutions have the benefit of providing a deeper physical understanding. In this case, however, additional approximations such as parabolic gain spectrum, linearization of the loss modulation, linearized dispersion operators, and so on, are required for the solution of the Haus master equation. Linearization assumes that the order of the optical elements (such as modulator, gain crystal, output coupler, prism pair for dispersion compensation) within the laser resonator does not play a role. In addition, we have to assume that at steady state each of these optical elements induces a small change in the pulse envelope per cavity roundtrip. These approximations are typically fulfilled for continuous wave (cw) modelocked lasers. Finally, numerical simulations and experiments can confirm the validity of these approximations. This chapter will provide a first introduction to the Haus master equation. We have ourselves used his theoretical framework to explain new operating regimes in active and passive modelocking that have not been explored before, such as soliton modelocking.

6.1 Modelocking

6.1.1 Basic Principle of Modelocking

The boundary condition of the optical resonator defines the axial modes of the laser [86Sie]. Normally, for a homogeneously broadened laser, the laser only oscillates in one axial mode for stable continuous wave (cw) operation. As we will show in this

© The Author(s), under exclusive licence to Springer Nature Switzerland AG 2021

U. Keller, *Ultrafast Lasers*, Graduate Texts in Physics,
https://doi.org/10.1007/978-3-030-82532-4_6

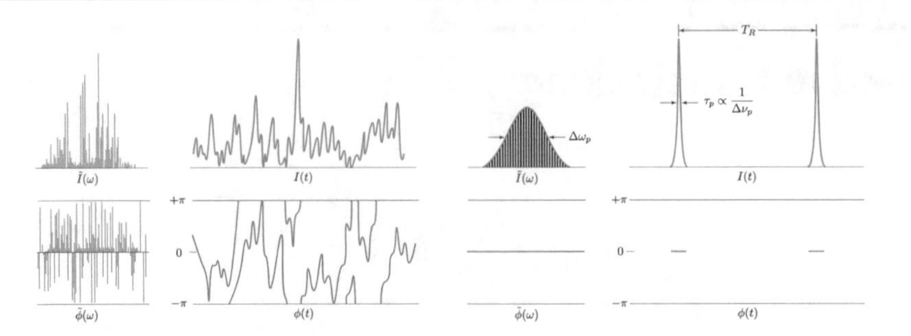

Fig. 6.1 Numerical simulations for different optical spectra with intensity $\tilde{I}(\omega)$ and phase $\tilde{\phi}(\omega)$. *Left*: Multi-mode laser without modelocking: only noise in intensity and phase. *Right*: Multi-mode laser with modelocking (e.g., constant phase and well-defined intensity spectrum): short pulse in the time domain. Note that only two pulses in the pulse train (6.2) are shown here

chapter an intracavity modulator can cause the laser to oscillate in more than one axial mode, and the intensity and phase of the different axial modes then determine the pulse duration.

For example if several axial modes oscillate inside a laser, but have an arbitrary amplitude and phase distribution, then the Fourier transform to the time domain will only cause amplitude and phase fluctuations (see Fig. 6.1 left). In contrast, we can consider the case where the axial modes with a constant axial mode spacing of $\Delta\omega_{\mathrm{ax}}$ have a defined amplitude $\tilde{E}(\omega)$ and phase distribution $\tilde{\phi}(\omega)$ given by

$$\tilde{E}_{\mathrm{train}}(\omega) = \tilde{E}(\omega)\mathrm{e}^{\mathrm{i}\tilde{\phi}(\omega)} \sum_{m=-\infty}^{\infty} \delta\left(\omega - m\Delta\omega_{\mathrm{ax}}\right) , \tag{6.1}$$

where m is an integer number as shown in Fig. 6.1 (right), with $\tilde{\phi}(\omega) = 0$. In this case the optical spectrum of the pulse train $\tilde{I}_{\mathrm{train}}(\omega) = |\tilde{E}_{\mathrm{train}}(\omega)|^2$ is a frequency comb spectrum. Note the definition of the δ-function is given in Appendix C, the δ-comb (i.e., the Dirac comb) in Appendix D, and the Fourier transform in Appendix A. In principle, it is assumed that the reader is familiar with these basic mathematical concepts.

With the inverse Fourier transform to the time domain, we then obtain a train of short pulses

$$I_{\mathrm{train}}(t) = I_{\mathrm{p}}(t) * \sum_{m=-\infty}^{\infty} \delta\left(t - mT_{\mathrm{R}}\right) = \sum_{m=-\infty}^{\infty} I_{\mathrm{p}}\left(t - mT_{\mathrm{R}}\right) , \tag{6.2}$$

where $I_{\mathrm{p}}(t)$ is thus the individual pulse with the time-dependent intensity which is repeated inside the pulse train after a time T_{R}. Note that according to Appendix A, we have used the fact that the product of two terms in the frequency domain becomes equal to the convolution in the time domain. Then, according to Appendix D, the Dirac comb becomes a Dirac comb in the frequency domain once again. The individual pulses $I_{\mathrm{p}}(t)$ have a pulse length τ_{p} which is determined by the width of the optical spectrum $\tilde{I}(\nu)$ given by $\Delta\nu_p$ according to Table 2.1, and a pulse repetition rate

$$f_{rep} = \Delta\nu_{ax} = \frac{\Delta\omega_{ax}}{2\pi} = 1/T_R \; .$$

Note first that the individual pulse alone has a continuous optical spectrum $\tilde{I}_p(\omega)$. A frequency comb spectrum $\tilde{I}_{train}(\omega)$ given in (6.1) with $\tilde{I}_{train}(\omega) = |\tilde{E}_{train}(\omega)|^2$ and as shown in Fig. 6.1 (right) is only obtained when averaged over many pulses in the pulse train.

Figure 6.1 indicates with basic Fourier transform properties that, for ultrashort pulse generation, the laser should oscillate in as many axial modes as possible, and the respective axial modes should be phase-locked. Different techniques and methods exist to establish this condition inside a laser. All of these techniques are called "modelocking", because the modes that form a frequency comb are phase-locked and have a well-defined stationary phase distribution function $\tilde{\phi}(\omega)$. We then distinguish between active and passive modelocking. In this chapter, we will discuss active modelocking.

Note second that the frequency comb of a modelocked laser is equidistant as shown in (6.1). In reality there is an additional degree of freedom of the modelocked frequency comb shown in Fig. 6.2 which only became important much later when with passively modelocked lasers the pulses have become so short that only a few optical cycles remain underneath the pulse envelope (see Sect. 9.10.1). My group in collaboration with Telle [99Tel] realized that there is an additional degree of freedom of a modelocked frequency comb which needs to be equidistant, but can also have an offset. In the time domain, this frequency offset corresponds to a shift in the position of the maximum electric field (the carrier) underneath the pulse envelope. We therefore referred to this frequency offset as the carrier envelope offset (CEO) frequency f_{CEO}. This has been used to stabilize the electric field below the pulse envelope inside a modelocked laser. The offset of the electric field relative to the pulse envelope has been referred to as the carrier envelope offset phase, or more briefly, the carrier envelope phase (CEP) (see Fig. 6.2) [99Tel].

The stabilization of the CEO phase has become important in the regime of few-cycle pulse durations, where the slowly-varying pulse envelope approximation also starts to fail. We will see in the following discussions about active and passive modelocking that modelocking only sets a steady state condition on the pulse envelope, which means that at steady state the pulse envelope will not change after one cavity roundtrip. There is no locking mechanism on the electric field underneath the pulse envelope, and therefore the electric field will experience arbitrary CEO phase fluctuations from pulse to pulse [03Hel]. Each pulse from a modelocked laser will therefore have a different CEO phase. This normally has no impact because the slowly-varying envelope approximation is valid as long as the pulse duration is much longer than one optical cycle. For example, at a center wavelength of 800 nm, an optical cycle is only 2.7 fs and therefore pulse durations well above 30 fs will have negligibly small electric field amplitude changes when the CEO phase is changing. Thus nonlinear experiments will not be affected by arbitrary CEO phase fluctua-

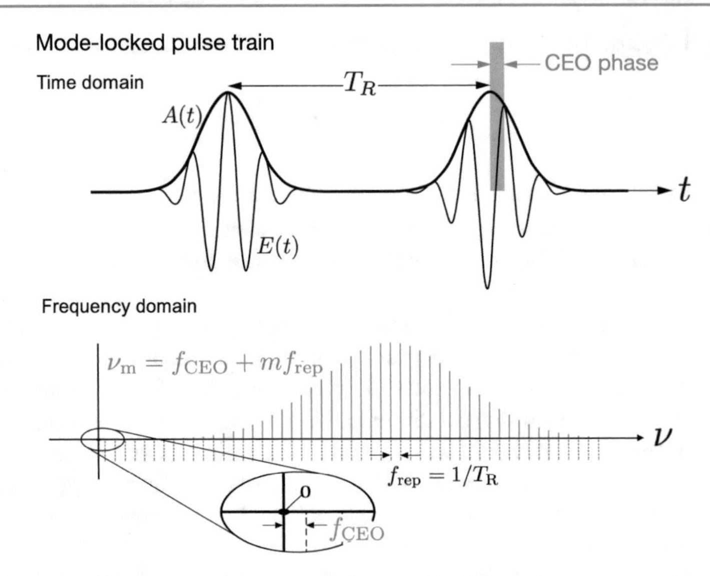

Fig. 6.2 Modelocked pulse train with the carrier envelope offset (CEO) phase for an ultrashort pulse with only a few optical cycles [99Tel]. The modelocked pulse train spectrum has two degrees of freedom: the frequency comb spacing (i.e., the pulse repetition rate $f_{rep} = 1/T_R$) and the frequency comb offset f_{CEO}. This will be discussed in more detail in Chap. 12. $A(t)$ is the pulse envelope, $E(t)$ is the electric field

tions. However, this changes for pulse durations in few optical cycle regime. This has been extremely important for highly nonlinear interactions such as high harmonic generation, attosecond pulse generation and strong laser field experiments.

6.1.2 Modelocked Frequency Comb and Axial Cavity Modes

Let us now first derive the conditions for the axial cavity modes for a single-frequency laser. The stability condition in a laser resonator requires the electromagnetic wave to have the same complex amplitude after a roundtrip in the cavity. This condition for self-sustained oscillation can be mathematically formulated by

$$e^{g-l} e^{i\phi} = 1 , \tag{6.3}$$

where g is the roundtrip amplitude gain coefficient, l the roundtrip amplitude loss coefficient, and ϕ the roundtrip phase inside the laser resonator. At steady state with gain equals loss (i.e., $g = l$), we have for the roundtrip phase the steady-state condition

$$\phi = 2\pi m , \quad m = 0, 1, 2, \ldots , \tag{6.4}$$

For a linear Fabry–Pérot cavity of length L, the phase that the light accumulates during one roundtrip is given by

$$\phi(\lambda_m) = 2k_m n(\lambda_m) L = \frac{4\pi}{\lambda_m} n(\lambda_m) L , \tag{6.5}$$

where λ_m is the wavelength and k_m the vacuum wave number of the axial or longitudinal modes. Note that there is an additional factor of 2 in (6.5) because the effective cavity roundtrip length is $2L$ for a linear Fabry–Pérot cavity of length L. With (6.5) we can then express the axial wavelength in the form

$$\boxed{\lambda_m = \frac{2n(\lambda_m) L}{m} , \quad m = 0, 1, 2, \ldots} . \tag{6.6}$$

According to (6.4), the distance between two axial modes is given by

$$\Delta\phi = \phi(\lambda_{m+1}) - \phi(\lambda_m) = 2\pi . \tag{6.7}$$

For a small axial mode spacing and together with a first-order Taylor approximation of this phase difference, we find

$$\Delta\phi \approx \left.\frac{\mathrm{d}\phi}{\mathrm{d}\omega}\right|_{\omega=\omega_m} \Delta\omega_{\mathrm{ax}} = 2\pi . \tag{6.8}$$

Using (6.5), the first derivative of the phase is given by

$$\Delta\phi \approx \left.\frac{\mathrm{d}\phi}{\mathrm{d}\omega}\right|_{\omega=\omega_m} \Delta\omega_{\mathrm{ax}} = 2\frac{\mathrm{d}k}{\mathrm{d}\omega} L \Delta\omega_{\mathrm{ax}} = \frac{2L}{\upsilon_{\mathrm{g}}} \Delta\omega_{\mathrm{ax}} = 2\pi , \tag{6.9}$$

where υ_{g} is the group velocity (Chap. 2). Equation (6.9) then yields the axial mode spacing in the form

$$\boxed{\Delta\omega_{\mathrm{ax}} = 2\pi\frac{\upsilon_{\mathrm{g}}}{2L} , \quad \Delta\nu_{\mathrm{ax}} = \frac{\upsilon_{\mathrm{g}}}{2L} , \quad |\Delta\lambda_{\mathrm{ax}}| = \frac{\lambda^2}{c}\frac{\upsilon_{\mathrm{g}}}{2L}} . \tag{6.10}$$

We have to distinguish between continuous wave (cw) and modelocked operation. For cw operation, the axial mode spacing can vary for different wavelengths because of group velocity dispersion. For modelocked lasers, the group velocity at the center wavelength $\upsilon_{\mathrm{g}} = \upsilon_{\mathrm{g}}(\lambda_0)$ will determine the axial mode spacing, and we obtain an equidistant mode spacing over the full pulse spectrum as in (6.1) and (6.2) and Fig. 6.2. The axial mode spacing at the center wavelength λ_0 will then determine the pulse repetition rate

$$f_{\mathrm{rep}} = \Delta\nu_{\mathrm{ax}} = \frac{\upsilon_{\mathrm{g}}(\lambda_0)}{2L} . \tag{6.11}$$

This corresponds to the frequency at which a pulse with the group velocity $\upsilon_g(\lambda_0)$ would oscillate inside a Fabry–Pérot cavity.

The pulse train emitted by a modelocked laser can thus be envisaged in terms of a pulse that propagates back and forth inside the cavity, and each time the pulse hits an outcoupling end-mirror of the cavity, a small fraction gets transmitted through this output coupler. This results in a pulse train, and the delay between subsequent pulses in the pulse train is determined by the roundtrip time T_R of the pulse inside the cavity as given in (6.2).

6.1.3 Difference Between Q-Switching and Modelocking

Ideally, a single Q-switched pulse (Chap. 8) is a single-frequency pulse, while a single modelocked pulse always consists of a continuous spectrum that extends over at least two axial modes of the laser. Let us now discuss in more detail the fundamental differences between Q-switching and modelocking (see Fig. 6.3):

- Q-switching. To generate a stable Q-switched pulse, one forces the laser to oscillate in only one axial mode by either choosing the laser resonator to be sufficiently short or by favoring a single axial mode through injection seeding, for example. If the Q-switched laser oscillates in several axial modes, then these modes are not phase-locked, and usually increase the intensity noise in the pulse. Thus, for Q-switching, the pulse repetition rate is always lower than the resonator roundtrip frequency, and the pulse duration is always longer than the resonator roundtrip time. The pulse repetition rate and pulse duration are given either by an active loss modulator or by a saturable absorber (see Fig. 8.2). If the laser is pumped continuously, we also refer to it as cw Q-switching.

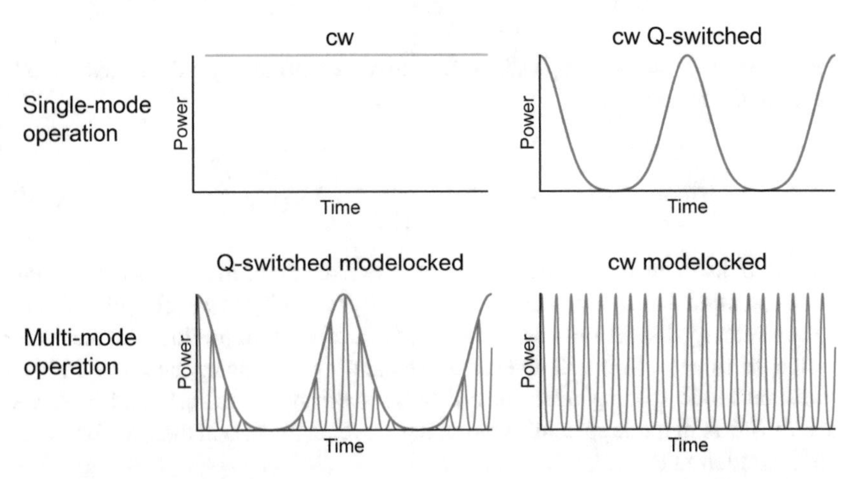

Fig. 6.3 The different laser states with single or multiple axial laser cavity modes. Cw modelocked or cw Q-switched means that the laser is continuously pumped during Q-switching and modelocking (cw = continuous wave) [96Kel]

- Modelocking. To generate an ultrashort pulse, the spectrum has to be extended over at least two axial modes and the frequency comb is phaselocked. Thus, the pulse repetition rate is given by the axial mode distance [see (6.10)], i.e., by the resonator roundtrip time. Modelocked pulses are always shorter than the resonator roundtrip time (Fig. 6.3).
- Q-switched modelocking. Similar to passive Q-switching, the stability of the modelocked axial modes is influenced by a saturable absorber. Under certain circumstances, the laser is not only modelocked by the saturable absorber, but also Q-switched, i.e., short pulses with a pulse repetition rate given by the resonator roundtrip frequency are additionally modulated by Q-switched macropulses (Fig. 6.3).

In this textbook, we will discuss in more detail the conditions under which these different modes of operation occur.

6.2 Basic Principles of Active Modelocking

Figure 6.4 shows the principle of an actively modelocked solid-state laser with a time-constant gain saturation. Inside the laser resonator is a gain medium, which amplifies the laser light, and an optical modulator (see Sect. 6.3), which varies the loss in the resonator periodically over time. If the modulator is switched off, the laser emits cw light with constant power through one of the partially reflective end-mirrors, i.e., the output coupler, of the resonator. For fundamental modelocking where only one pulse is circulating inside the laser resonator, the loss modulator has to be placed at the end of the standing-wave linear cavity. Thus, the pulse inside the laser resonator only passes through the loss modulator once per cavity roundtrip. Whenever the pulse hits the output coupler, a small fraction is coupled out and we obtain a stable pulse train with a pulse repetition period given by the cavity roundtrip time T_R of the pulse inside the laser resonator.

For active modelocking, an external signal is applied to an optical loss modulator, typically using the acousto-optic or electro-optic effect. Such an electronically driven loss modulation produces a sinusoidal loss modulation with a period given by the cavity roundtrip time T_R. The gain at steady state in a typical solid-state laser is only saturated by the average power and exhibits no dynamic gain saturation between the pulses. In this case net gain is only obtained around the minimum of the loss modulation and therefore only supports pulses that are significantly shorter than the cavity roundtrip time (Fig. 6.4). This means that a modelocked laser produces an equidistant pulse train, with a period defined by the roundtrip time of a pulse inside the laser cavity T_R and a pulse duration τ_p. In the frequency domain, this results in a phaselocked frequency comb with a constant mode spacing that is equal to the pulse repetition rate $f_{rep} = 1/T_R$ (Fig. 6.2). The spectral width of the envelope of this frequency comb is inversely proportional to the pulse duration [see (6.1), (6.2), and Table 2.1].

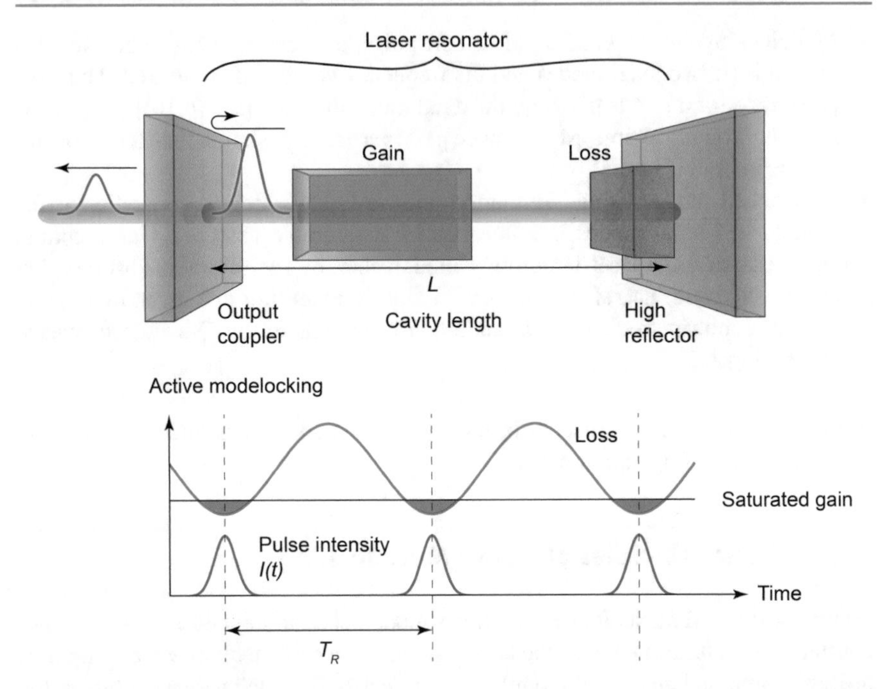

Fig. 6.4 Basic principle of an actively modelocked laser. We assume fundamental modelocking with only one pulse per cavity roundtrip. This means that the pulse travels only one time per cavity roundtrip through the loss modulator. T_R: cavity roundtrip time

We will now discuss the fundamental principle of active modelocking in more detail. To discuss the effect of the loss modulation, we refer to Fig. 6.5. With the loss modulation switched on at exactly the roundtrip frequency of the optical pulse inside the laser resonator, it is energetically favorable for the laser to move into a pulsed mode of operation, with the pulses going through the loss modulator when the lowest loss is generated. From the time domain in Fig. 6.5, it becomes clear that the pulse experiences less loss per roundtrip in the modulator with shorter pulses, which will continue to shorten the pulse further. However, this is balanced by the pulse broadening effect from the gain, as shown in the frequency domain of Fig. 6.5. Every optical gain material possesses a finite gain bandwidth, which can be approximately described with a parabolic frequency dependence of the gain. If the pulse becomes shorter, its spectral width increases. Therefore, on average, the pulse experiences less gain per roundtrip in the gain medium due to the finite gain bandwidth. The amplification process introduces spectral narrowing because the frequencies away from the gain maximum experience less amplification. Therefore, the finite gain bandwidth will balance the pulse shortening process of the modulator and, at steady state, a balanced and stable pulse duration is generated. At steady state, every pass through the loss modulator will shorten the pulse further and every pass through the gain will then broaden the pulse by the same amount again, so that a stable pulse duration is maintained after one cavity roundtrip.

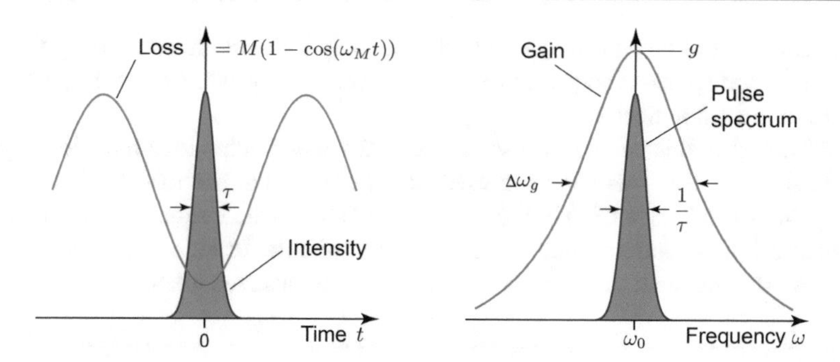

Fig. 6.5 *Left*: Time domain. Loss modulation at the position of the modulator. *Right*: Frequency domain. Spectral dependence of the gain profile of the laser medium. A steady-state solution is obtained by the balance of pulse shortening in the modulator and pulse broadening in the gain

In the spectral domain the principle of pulse generation can be explained as follows (see Sect. 6.4.3). For active modelocking, all axial modes in a laser are forced to oscillate in phase using a periodic loss modulator. Through the periodic loss modulation, optical energy is transferred into the neighboring axial modes as long as the loss modulator is modulated with the roundtrip frequency of the laser resonator. This corresponds to phaselocked injection seeding because the energy in the modulation sidebands is phaselocked to the original mode. In this way, the gain is increased for the adjacent axial modes and the laser can oscillate over a much broader coherent bandwidth.

For picosecond pulse generation in diode-pumped solid-state lasers, we do not normally have to introduce dispersion compensation inside the laser cavity to obtain the short pulse duration. Active modelocking is then very well described by the balance between loss modulation and amplification (see Sect. 6.4). However, for shorter pulse generation in the visible and near-infrared regime, negative dispersion has to be introduced with prism pairs, GTIs, or chirped mirrors (see Chap. 3) to compensate for the positive dispersion introduced by the gain material and the loss modulator. Note that, for longer wavelengths, the material dispersion can become negative, as demonstrated for example in [21Bar]. In this case the pulse peak intensity inside the laser can become so high that additional nonlinear effects have to be taken into account. This is further emphasized by the fact that the laser mode area inside the laser gain material is generally optimized for a low laser threshold and therefore for a small mode area to obtain high inversion for a given limited pump power. Thus nonlinear effects in the gain material very often have to be taken into account.

The most important nonlinear effect is self-phase modulation (SPM) due to the nonlinear refractive index (see Chap. 4). The effect of SPM on active modelocking will be discussed in Sect. 6.5. SPM will drive the modelocking process to instability if it becomes too large. It is therefore more advantageous to balance SPM with negative GDD to obtain stable soliton pulse formation (Sect. 6.6). This will typically lead to shorter pulses than with loss modulation alone and is referred to as soliton modelocking.

Note that we generally discuss a homogeneously broadened laser which only lases in one axial mode for cw operation. However, in a standing wave cavity we always experience spatial hole burning due to the standing wave in the gain medium. This leads to inhomogeneous spectral broadening, which can be beneficial for modelocking, as discussed in Sect. 6.7.

At the end of this chapter, we shall briefly discuss synchronous modelocking to obtain short pulses. This has been used for dye lasers but has lost its significance today with the advance of SESAM modelocked solid-state lasers, which is a much simpler technique and has made ultrafast lasers suitable for industrial applications. We now continue with the discussion of the different loss modulators.

6.3 Optical Loss Modulators

There are two types of periodic loss modulators, the amplitude modulator and the phase modulator, which we shall treat separately.

6.3.1 Acousto-Optic Modulator (AOM)

For active modelocking, we typically use an acousto-optic modulator (AOM) to introduce a sinusoidal amplitude loss modulation (see Fig. 6.4). Most of the acousto-optic loss modulators use an acoustic standing wave, which generates an intensity-dependent strain in a transparent material and therefore a periodic modulation in the refractive index through the acousto-optic effect. This time-dependent periodic index modulation acts as a diffraction grating for the incoming light with a periodic loss modulation for the non-diffracted beam. The strength of the loss modulation can be controlled with the intensity of the acoustic wave.

The acousto-optic modulator as shown in Fig. 6.6 typically consists of a "transducer" and an acousto-optic substrate. The transducer generates a longitudinal acoustic wave of frequency ω_a in the substrate, which propagates with the sound velocity υ_a. The acoustic wave is reflected again at the end of the substrate (the glass/air boundary is like a 100% reflector for the acoustic wave), and thus generates a standing wave. A standing wave can be described as a superposition of two plane waves propagating in opposite directions. These two counter-propagating acoustic plane waves then generate a standing wave which is modulated by $2\omega_a$. The acoustic wave deforms the material, slightly changes the binding energy of the valence electrons, and hence introduces a small change in the refractive index that is approximately proportional to the acoustic wave intensity. Through the acousto-optic effect (see, e.g., [84Yar, 91Sal2]), the standing wave generates a periodic phase grating in space and time with a frequency $2\omega_a$. This spatially modulated refractive index change acts like a diffraction grating for the incoming laser beam at a center radial frequency of ω_0 (Fig. 6.6).

According to Fig. 6.6 a typical AOM is operated under the Bragg condition, where the incoming optical beam has a Bragg angle θ_B. The amplitude loss of the incoming optical beam $E_0 e^{i\omega_0 t}$ can then also be considered as a reflection at each of the two counter-propagating acoustic waves with an optical frequency shift due to the

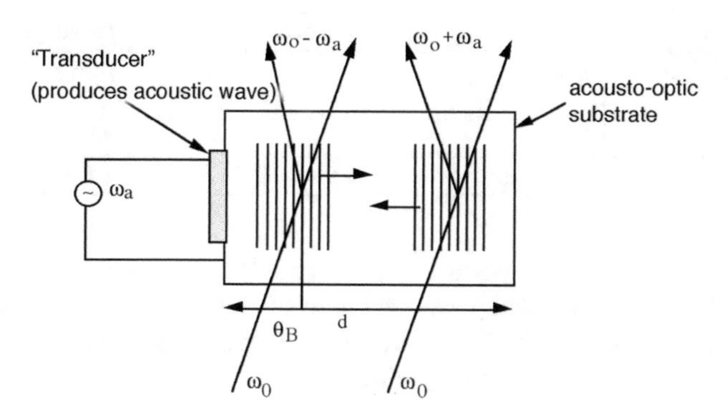

Fig. 6.6 Acousto-optic loss modulator

Doppler shift at the moving acoustic wave front:

$$\text{amplitude loss} \propto \left[E_0 e^{i(\omega_0+\omega_a)t} + E_0 e^{i(\omega_0-\omega_a)t} \right] . \tag{6.12}$$

The intensity loss is then given by

$$\text{intensity loss} \propto \left| E_0 e^{i(\omega_0+\omega_a)t} + E_0 e^{i(\omega_0-\omega_a)t} \right|^2 . \tag{6.13}$$

This yields

$$\begin{aligned} \left| E_0 e^{i(\omega_0+\omega_a)t} + E_0 e^{i(\omega_0-\omega_a)t} \right|^2 &= \left[E_0 e^{i(\omega_0+\omega_a)t} + E_0 e^{i(\omega_0-\omega_a)t} \right] \\ &\quad \times \left[E_0 e^{-i(\omega_0+\omega_a)t} + E_0 e^{-i(\omega_0-\omega_a)t} \right] \\ &= E_0^2 + E_0^2 e^{2i\omega_a t} + E_0^2 e^{-2i\omega_a t} + E_0^2 \\ &= 2E_0^2 (1 + \cos 2\omega_a t) . \end{aligned} \tag{6.14}$$

This means that the counterpropagating acoustic waves introduce a periodic loss modulation at twice the acoustic frequency.

We distinguish between two types of acousto-optic loss modulators. In one case the substrate is not resonant with a matched transducer–substrate interface, and one typically needs relatively high power (i.e., typically 10 W of high-frequency electrical input power) for the transducer to reach a sufficient loss modulation. In the other case the substrate is an acoustic resonator (i.e., similar to an optical Fabry–Pérot resonator, but now with an acoustic wave), and one only needs about a 1 W high-frequency electrical input power to obtain a sufficient loss modulation. The trade-off is that we need active temperature control of the modulator substrate to keep the acoustic resonance frequency at a specific value. In both cases we need to adapt the loss modulation to the roundtrip frequency of the optical pulse inside the laser resonator.

Table 6.1 Acousto-optic parameters for different substrates [90Kel1,87Kin], and see Fig. 3.11 in [73Aul]

	Sapphire	$LiNbO_3$	Fused quartz
$\alpha \propto \omega_a^2$ at 250 MHz [dB/μs]	0.015	0.019	0.56
υ_a [km/s]	11.1	7.33	5.96
M [s^3/kg]	0.22×10^{-15}	6.99×10^{-15}	1.5×10^{-15}

In most cases, the resonant acoustic substrate is used for active modelocking of lasers because of the lower electrical power requirements for the loss modulation.

The maximum amplitude of the induced phase grating is given by the maximum change in the refractive index Δn_0 [90Kel1]:

$$\Delta n_0 \approx \frac{1}{\sqrt{2}} \sqrt{\frac{M}{\alpha d} \frac{P_a}{A}} , \tag{6.15}$$

where M is the acoustic figure of merit, α the acoustic loss coefficient, d the thickness of the substrate, A the transducer area, and P_a the absorbed electric power, under the assumption that no loss occurs in the transducer and $\alpha d \ll 1$. The loss coefficient increases quadratically with the acoustic carrier frequency, i.e., $\alpha \propto \omega_a^2$. This means that it becomes more difficult to manufacture a high-frequency acousto-optic loss modulator. Examples for different substrate materials are given in Table 6.1.

For the Bragg condition, an optical light beam experiences the maximum diffraction efficiency. Energy and momentum conservation determine the Bragg condition:

$$\omega_{out} = \omega_{in} \pm \omega_a , \tag{6.16}$$

$$k_{in} = K_a + k_{out} , \tag{6.17}$$

where k_{in} and k_{out} are the vectors of the incoming and outgoing optical propagation constants, respectively, and K_a is the vector of the acoustic propagation constant (e.g., according to Fig. 6.6). K_a is given by the period of the acoustic wave Λ_a:

$$K_a = \frac{2\pi}{\Lambda_a} = \frac{\omega_a}{\upsilon_a} , \tag{6.18}$$

where υ_a is the acoustic phase velocity. According to (6.17) and the approximation $\omega_{out} \approx \omega_{in} = \omega_0$, which follows when $\omega_a \ll \omega_0$, the angle of incidence θ_B for the Bragg condition is given for the first diffraction order by

$$\sin \theta_B = \frac{K_a}{2k_{in}} . \tag{6.19}$$

When the Bragg condition holds, the diffraction efficiency and thus the maximum loss modulation is given by (see, e.g., [84Yar])

$$\boxed{\eta_0 = \sin^2 \theta_m , \quad \theta_m \equiv \frac{\pi}{\lambda_0} \Delta n_0 L \cos \theta_B} , \tag{6.20}$$

where λ_0 is the vacuum light wavelength and L the transducer length (i.e., the beam width of the acoustic wave).

In most cases, fused quartz is used as the substrate material because of its excellent optical quality with minimal loss inside the laser resonator (Table 6.1). $LiNbO_3$ is not used very often, because the material is more easily damaged by self-focusing. Sapphire exhibits low acoustic loss at high frequencies, but it is relatively expensive to make.

Note that, similarly to the acousto-optic loss modulator, one can generate a phase modulation using an electro-optic modulator [84Yar]. A laser can be modelocked using either an amplitude or a phase modulation.

6.4 Active Modelocking Without SPM and GDD

For a better physical understanding we provide a theoretical discussion about the influence of the different intracavity elements on the pulse envelope taking into account the gain, constant loss (such as from the output coupler), loss modulation, group delay dispersion (GDD), and self-phase modulation (SPM). The goal is to understand how each element influences the modelocking process. For this reason we first discuss in this section the steady-state solution for modelocking when we neglect SPM and GDD. In this section we will use two different theoretical models, first based on Kuizenga and Siegman (i.e., the Gaussian pulse analysis) and then continue with the Haus master equation. Later on we will continue with the master equation approach because it is a very powerful theoretical tool to explore the different regimes such as in active modelocking with GDD or SPM and in passive modelocking. Therefore this initial discussion serves also as an introduction to the master equation formalism.

Neglecting GDD typically gives a good approximation for diode-pumped solid-state lasers in the picosecond pulse-width regime, because the intracavity dispersive pulse broadening is mostly introduced by a reasonably short laser gain material. The theoretical background on the way dispersion (and GDD) affects pulse propagation is given in Chap. 2. With the longer gain rods in flashlamp-pumped lasers, GDD can become more significant even in the picosecond regime. Especially diode-pumped solid-state lasers typically have a much smaller laser mode area inside the gain material, and therefore SPM may already have some influence for picosecond pulses [95Bra2, 95Kar3]. This will be discussed in Sect. 6.5. The combination of GDD and SPM in the femtosecond domain will be discussed in Sect. 6.6.

6.4.1 Gaussian Pulse Analysis

Here we follow the derivation by Kuizenga and Siegman in [86Sie, Chap. 27] and [70Kui1], where they assumed that the steady state solution for active modelocking without SPM and GDD is given by a Gaussian pulse shape. This modelocking theory

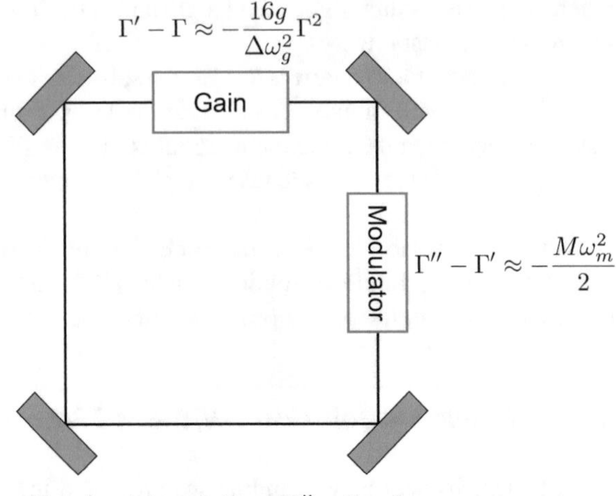

Fig. 6.7 Schematic diagram of active modelocking for a Gaussian pulse showing the change in the complex Gaussian parameter after gain and modulator with a parabolic approximation. At steady-state after one cavity roundtrip the total change is zero

is also referred to as the Kuizenga–Siegman theory. This ansatz is later confirmed by Haus (see Sect. 6.4.2). The ansatz for a Gaussian pulse is given by (2.35)

$$E(t) = \exp\left(-\Gamma t^2 + \mathrm{i}\omega_0 t\right) , \tag{6.21}$$

where Γ is the complex Gaussian parameter (2.36) (see Chap. 2) (Fig. 6.7).

Because active modelocking without SPM describes a linear system, i.e., does not depend on the optical intensity, the problem can also be solved in the spectral domain. The Fourier transform of (6.21) is then given by (Appendix A)

$$\tilde{E}(\omega) = \exp\left[-\frac{(\omega-\omega_0)^2}{4\Gamma}\right] , \tag{6.22}$$

where we have dropped the constant factors because we are only interested in the change in pulse duration. For the pulse propagation through the gain material (see also Appendix F), we have

$$\tilde{E}'(\omega) = \exp\left[g(\omega)\right]\tilde{E}(\omega) = \exp\left[\frac{g}{1+4(\omega-\omega_0)^2/\Delta\omega_g^2}\right]\tilde{E}(\omega) , \tag{6.23}$$

where $g(\omega)$ is the frequency-dependent laser gain coefficient, g is the amplitude gain coefficient per resonator roundtrip, ω_0 is the center frequency of the laser gain bandwidth, and $\Delta\omega_g$ is the FWHM gain bandwidth. In the following we assume that the gain maximum determines the center frequency of the actively modelocked pulse as shown in Fig. 6.5. The frequency-dependent laser gain coefficient $g(\omega)$ describes a Lorentzian gain profile. The laser gain bandwidth $\Delta\omega_g$ is assumed to be significantly larger than the pulse spectrum, i.e., $4(\omega-\omega_0)^2/\Delta\omega_g^2 \ll 1$, so that with $(1+x)^{-1} \approx 1 - x$ in the first-order Taylor approximation, we find

$$\frac{g}{1+4(\omega-\omega_0)^2/\Delta\omega_g^2} \approx g - \frac{4g}{\Delta\omega_g^2}(\omega-\omega_0)^2 . \tag{6.24}$$

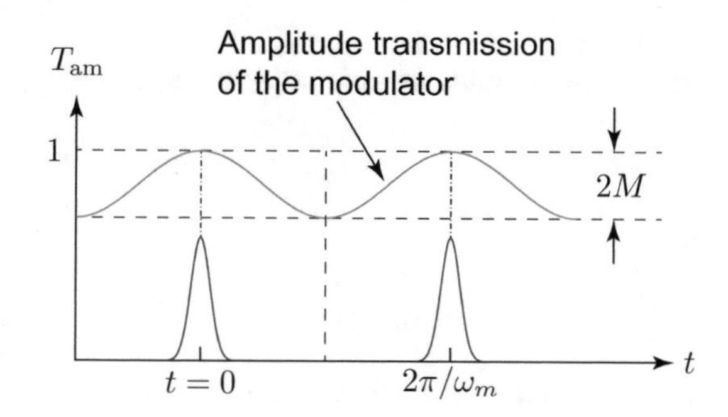

Fig. 6.8 Pulse generation through active modelocking with the amplitude transmission of the loss modulator $m_{AM}(t)$

In the following, we neglect the frequency-independent terms for the time being. Thus,

$$\tilde{E}'(\omega) \approx \exp\left[-\frac{4g}{\Delta\omega_g^2}(\omega-\omega_0)^2\right]\tilde{E}(\omega) \equiv \exp\left[-\frac{(\omega-\omega_0)^2}{4\Gamma'}\right], \tag{6.25}$$

where Γ' is given by

$$\frac{1}{\Gamma'} = \frac{1}{\Gamma} + \frac{16g}{\Delta\omega_g^2}. \tag{6.26}$$

Usually, the gain coefficient per resonator roundtrip g is very small, and (6.26) can be further simplified to

$$\Gamma' - \Gamma = -\frac{16g}{\Delta\omega_g^2}\Gamma'\Gamma \approx -\frac{16g}{\Delta\omega_g^2}\Gamma^2. \tag{6.27}$$

Pulse propagation through a loss modulator is given by

$$E''(t) = m(t)E'(t). \tag{6.28}$$

In the case of an amplitude loss modulator, $m(t) = m_{AM}(t)$:

$$m_{AM}(t) = \exp[-l(t)] = \exp\Big[-M(1-\cos\omega_m t)\Big], \tag{6.29}$$

according to (6.14).

In the case of a frequency modulator, $m(t) = m_{FM}(t)$ is given by

$$m_{FM}(t) = \exp(-iM\cos\omega_m t), \tag{6.30}$$

where ω_m is the modulation frequency and M is the modulation strength (Fig. 6.8).

The case of the phase modulator will be solved at the end of this section. Usually, a phase modulator does not generate bandwidth-limited pulses, because the phase modulation produces a chirp. In contrast an amplitude modulator can generate bandwidth-limited pulses, and is therefore used more often.

(a) Amplitude Loss Modulator
For amplitude modulation as in (6.29), we use the parabolic approximation

$$\cos x \approx 1 - x^2/2$$

to obtain

$$m_{\mathrm{AM}}(t) \approx \exp\left(-\frac{M\omega_{\mathrm{m}}^2}{2}t^2\right) , \quad |t|^2 \ll T_{\mathrm{m}} , \tag{6.31}$$

where T_{m} is the modulation period. With this, (6.28) simplifies to

$$E''(t) = m_{\mathrm{AM}}(t)E'(t) \approx \exp\left(-\frac{M\omega_{\mathrm{m}}^2}{2}t^2\right) E'(t) \equiv \exp\left(-\Gamma''t^2\right) . \tag{6.32}$$

Thus, the Gaussian pulse shape is maintained after the amplitude modulator. From (6.32), it then follows that

$$\Gamma'' - \Gamma' \approx \frac{M\omega_{\mathrm{m}}^2}{2} . \tag{6.33}$$

The steady-state condition requires that

$$\Gamma'' - \Gamma = 0 . \tag{6.34}$$

Equations (6.27), (6.33), and (6.34) imply

$$\Gamma'' - \Gamma = 0 = -\frac{16g}{\Delta\omega_{\mathrm{g}}^2}\Gamma_{\mathrm{s}}^2 + \frac{M\omega_{\mathrm{m}}^2}{2} , \tag{6.35}$$

where Γ_{s} is the steady-state Gaussian parameter. Equation (6.35) determines the steady-state solution:

$$\Gamma_{\mathrm{s}} = \sqrt{\frac{M}{g}\frac{\omega_{\mathrm{m}}\Delta\omega_{\mathrm{g}}}{4\sqrt{2}}} , \tag{6.36}$$

which then determines the steady-state FWHM pulse duration $\tau_{\mathrm{p,s}}$ of an actively modelocked laser:

$$\boxed{\begin{aligned} \tau_{\mathrm{p,s}} &= \sqrt{\frac{2\ln 2}{\mathrm{Re}\,(\Gamma_{\mathrm{s}})}} = \sqrt{\frac{2\sqrt{2}\ln 2}{\pi^2}}\sqrt[4]{\frac{g}{M}}\sqrt{\frac{1}{f_{\mathrm{m}}\Delta f_{\mathrm{g}}}} \\ &= 0.4457\sqrt[4]{\frac{g}{M}}\sqrt{\frac{1}{f_{\mathrm{m}}\Delta f_{\mathrm{g}}}} \end{aligned}} \tag{6.37}$$

where Δf_{g} is the FWHM gain bandwidth, f_{m} is the loss modulation frequency, g is the saturated roundtrip amplitude gain coefficient, and M is the modulation strength. The theory and the result of (6.37) were published for the first time by

Kuizenga and Siegman [70Kui1,70Kui2], and the theory is also referred to as the Kuizenga–Siegman modelocking theory.

Example: Nd:YAG Laser
In this case,

$$\lambda_g = 1.064\,\mu\text{m}\ ,\quad \Delta\lambda_g = 0.45\,\text{nm} \implies \Delta f_g = \Delta\lambda_g c/\lambda_g^2 = 1.19\times 10^{11}\,\text{Hz}\ ,$$
$$f_m = 100\,\text{MHz}\ ,\quad g = 0.05\ ,\quad \text{i.e., } 2g \approx T_{\text{out}} \approx 10\%\ ,\quad M = 0.2\ ,$$

where T_{out} is the intensity transmission of the laser output coupler. Equation (6.37) yields a steady-state pulse length of $\tau_{p,s} = 91\,\text{ps}$.

Example: Nd:YLF Laser
In this case,

$$\lambda_g = 1.047\,\mu\text{m}\ ,\quad \Delta\lambda_g = 1.3\,\text{nm} \implies \Delta f_g = \Delta\lambda_g c/\lambda_g^2 = 3.56\times 10^{11}\,\text{Hz}\ ,$$
$$f_m = 100\,\text{MHz}\ ,\quad g = 0.05\ ,\quad \text{i.e., } 2g \approx T_{\text{out}} \approx 10\%\ ,\quad M = 0.2\ .$$

Equation (6.37) yields a steady-state pulse length of $\tau_{p,s} = 53\,\text{ps}$.

These are typical pulse durations which have been achieved with flashlamp-pumped Nd:YAG and Nd:YLF lasers. Additional compression techniques (see Chaps. 2 and 3) have been used to achieve pulse durations in the region of only a few picoseconds.

(b) Phase Modulator
We can again make a parabolic approximation $\cos x \approx 1 - x^2/2$. Equation (6.30) then implies

$$m_{\text{FM}}(t) \approx \exp\left[\mp \text{i}M\left(1 - \frac{\omega_m^2}{2}t^2\right)\right]\ . \tag{6.38}$$

For more about the signs here, the interested reader is referred to [70Kui1]. The frequency-independent part of this phase-shift can be accounted for with a minor variation in the total effective length of the laser cavity so that it can be neglected here. The frequency-dependent part, on the other hand, leads to a change in the imaginary part of the Gaussian parameter, i.e., it imposes a chirp on the pulse. The net change in the Gaussian pulse parameter after passing through a phase modulator can then be written

$$\Gamma'' - \Gamma' \approx \pm \text{i}\frac{M\omega_m^2}{2}\ . \tag{6.39}$$

The total change $\Gamma'' - \Gamma$ in the Gaussian parameter during one complete roundtrip is given by

$$\Gamma'' - \Gamma = 0 = -\frac{16g}{\Delta\omega_g^2}\Gamma_s^2 \pm \text{i}\frac{M\omega_m^2}{2}\ . \tag{6.40}$$

Thus, the steady-state solution is

$$\Gamma_s = \sqrt{\pm \text{i}}\sqrt{\frac{M}{g}\frac{\omega_m\Delta\omega_g}{4\sqrt{2}}}\ . \tag{6.41}$$

The real part of this steady-state Gaussian parameter is

$$\mathrm{Re}\,(\Gamma_s) = \frac{1}{\sqrt{2}}\sqrt{\frac{M}{g}\frac{\omega_m \Delta\omega_g}{4\sqrt{2}}}\,, \tag{6.42}$$

whence the FWHM pulse width in the FM modelocked situation is given by

$$\boxed{\begin{aligned}\tau_{p,s,FM} &= \sqrt{\frac{2\ln 2}{\mathrm{Re}\,(\Gamma_s)}} = \sqrt[4]{2}\sqrt{\frac{2\sqrt{2}\ln 2}{\pi^2}}\sqrt[4]{\frac{g}{M}}\sqrt{\frac{1}{f_m \Delta f_g}} \\ &= 0.53\sqrt[4]{\frac{g}{M}}\sqrt{\frac{1}{f_m \Delta f_g}}\end{aligned}} \tag{6.43}$$

This is slightly longer than for amplitude loss modulation, and because the Gaussian parameter is complex, we will also obtain chirped pulses. This means that amplitude loss modulation is generally preferable.

6.4.2 Derivation and Solution of the Haus Master Equation

In the previous section we used a Gaussian pulse as ansatz for the solution. We will now show that this ansatz is correct based on the Haus master equation approach [75Hau2, 84Hau]. For this we use linearized operators to derive a differential equation in the time domain. In the following, the changes ΔA in the pulse envelope that occur during a resonator roundtrip due to gain, loss, and amplitude modulation will be determined. These linearized operators are also summarized in Appendix F.

We assume that the changes in the pulse envelope per resonator roundtrip are small, i.e., $\Delta A < 20\%$. In the following, it makes sense to introduce a second time scale T. The time scale t describes the detailed time dependence of the pulse envelope A, and the longer time scale T describes the temporal changes over the course of several resonator roundtrip times T_R. Hence, $t \ll T_R$ and $T \gg T_R$.

Regarding the gain (see Appendix F for a detailed derivation), the change in the envelope after passing through the gain material is

$$\boxed{\Delta A_1 = g\left(1 + \frac{1}{\Omega_g^2}\frac{\partial^2}{\partial t^2}\right)A(T,t)}\,. \tag{6.44}$$

Here it is useful to introduce the notation Ω_g (half-width at half-maximum, HWHM) to get rid of the factor of 2:

$$\Omega_g \equiv \frac{\Delta\omega_g}{2}\,, \tag{6.45}$$

where $\Delta\omega_g$ is the full-width at half-maximum (FWHM) of the gain profile, and g is the spatially integrated saturated gain coefficient of the laser material of length L_g, i.e., $g = g(z)L_g$, with $g(z)$ the gain coefficient.

For the amplitude loss modulator, using (6.14), it follows that

$$A_{\text{out}}(t) = \exp\big[-M(1-\cos\omega_m t)\big]A_{\text{in}}(t) . \tag{6.46}$$

For small modulations, we expand the exponential function $e^x \approx 1 + x$ for $x \ll 1$:

$$A_{\text{out}}(t) \approx \big[1 - M(1-\cos\omega_m t)\big]A_{\text{in}}(t) ,$$

whence

$$\Delta A_2 = A_{\text{out}}(t) - A_{\text{in}}(t) \approx -M(1-\cos\omega_m t)\,A_{\text{in}}(t) .$$

This implies

$$\boxed{\Delta A_2 \approx -M(1-\cos\omega_m t)\,A(T,t)} . \tag{6.47}$$

The time-independent constant loss is found from

$$A_{\text{out}}(t) = e^{-l}A_{\text{in}}(t) , \tag{6.48}$$

where l is the time-independent amplitude loss coefficient in the laser cavity, which takes into account the output coupler loss and any other scattering losses inside the cavity. As before, for small losses, the exponential function can be expanded $e^x \approx 1 + x$ for $x \ll 1$:

$$A_{\text{out}}(t) = e^{-l}A_{\text{in}}(t) \approx (1-l)A_{\text{in}}(t) ,$$

whence

$$\Delta A_3 = A_{\text{out}}(t) - A_{\text{in}}(t) \approx -lA_{\text{in}}(T,t) .$$

This implies

$$\boxed{\Delta A_3 \approx -lA(T,t)} . \tag{6.49}$$

Master Equation

Haus' master equation formalism [75Hau2, 84Hau] is based on linearized differential operators that describe the temporal evolution of a pulse envelope inside the laser cavity. At steady-state, we then obtain the differential equation

$$T_R\frac{\partial A(T,t)}{\partial T} = \sum_i \Delta A_i = 0 , \tag{6.50}$$

where ΔA_i are the changes in the pulse envelope due to different elements in the cavity, such as gain, loss modulator, dispersion, and so on. Equation (6.50) basically means that, at steady state, after one laser cavity roundtrip, the pulse envelope cannot

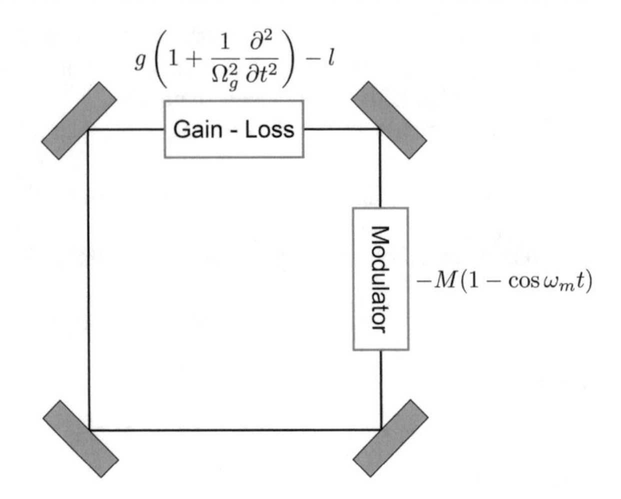

Fig. 6.9 Schematic diagram for all linearized operators acting on the pulse envelope for active modelocking without SPM and GDD. We separate the time-independent loss l from the modulator loss

change and all the small changes due to the different elements in the cavity have to sum up to zero. Each element is modeled as a linearized operator as discussed above (Fig. 6.9).

Equations (6.44), (6.47), (6.49), and (6.50) then imply

$$\boxed{T_{\mathrm{R}}\frac{\partial A(T,t)}{\partial T}=\left[g\left(1+\frac{1}{\Omega_{\mathrm{g}}^2}\frac{\partial^2}{\partial t^2}\right)-l-M\left(1-\cos\omega_{\mathrm{m}}t\right)\right]A(T,t)=0}\,. \tag{6.51}$$

This differential equation is analogous to Schrödinger's equation. Equation (6.51) implies

$$\left[\frac{g}{\Omega_{\mathrm{g}}^2}\frac{\partial^2}{\partial t^2}-M\left(1-\cos\omega_{\mathrm{m}}t\right)\right]A=(l-g+\lambda)\,A\;, \tag{6.52}$$

where λ are the eigenvalues (see the explanation below), and this corresponds to a Schrödinger equation with

$$\left[-\frac{\hbar^2}{2m}\frac{\partial^2}{\partial x^2}+V(x)\right]\Psi=E\Psi\;. \tag{6.53}$$

Comparing (6.52) and (6.53), it becomes apparent that the loss modulation $-M[1-\cos(\omega_{\mathrm{m}}t)]$ acts on the amplitude of the pulse just as a periodic potential acts on the wave function of a particle. We know from quantum mechanics that a periodic potential, e.g., the potential of a crystal grating, does not possess bound states. This has been shown for the example of Bloch wave functions in a solid. But if $V(x)$ is deep enough, i.e., if M is large enough, then it would take a long time for a localized wave packet to tunnel out of a valley. For a deep, cosine-shaped potential, we can make a parabolic approximation as before in (6.31):

$$M\left(1-\cos\omega_{\mathrm{m}}t\right)\approx M\frac{\omega_{\mathrm{m}}^2t^2}{2}\;. \tag{6.54}$$

Thus, the solution is determined by the harmonic oscillator [75Hau2]

$$A_n(t) = \sqrt{\frac{W_n}{2^n\sqrt{\pi}n!\tau}} H_n\left(\frac{t}{\tau}\right) e^{-t^2/2\tau^2} , \tag{6.55}$$

where H_n is the Hermite polynomial of degree n, with $H_0 = 1$, W_n a constant, and

$$\tau = \sqrt[4]{\frac{D_g}{M_s}} , \tag{6.56}$$

with D_g the gain dispersion parameter defined by (Appendix F)

$$D_g \equiv \frac{g}{\Omega_g^2} , \tag{6.57}$$

and M_s the curvature of the loss modulation defined by

$$M_s \equiv \frac{M\omega_m^2}{2} . \tag{6.58}$$

The corresponding eigenvalues are given by

$$\lambda_n = g - l - M\omega_m^2\tau^2\left(n + \frac{1}{2}\right) , \tag{6.59}$$

with the solution ansatz

$$A_n(T, t) = A_n(t)e^{\lambda_n T/T_R} . \tag{6.60}$$

The solution with $\mathrm{Re}(\lambda) = 0$ is a stable pulse, if all other solutions A_n have eigenvalues with $\mathrm{Re}\,(\lambda_n) < 0$.

For $\lambda_0 = 0$, it follows that $\lambda_n < 0$ for $n \geq 1$, by (6.59). This means that only the fundamental mode can oscillate in a stable mode of operation and all other modes are unstable. The stable solution for $n = 0$ yields a Gaussian pulse according to (6.55) with $H_0 = 1$. Then, using (6.56)–(6.58), the FWHM pulse width is given by

$$\boxed{\tau_p = 1.665 \cdot \tau = 1.665\sqrt[4]{\frac{2g}{M}}\sqrt{\frac{1}{\omega_m\Omega_g}} = 0.446\sqrt[4]{\frac{g}{M}}\sqrt{\frac{1}{f_m\Delta f_g}}} , \tag{6.61}$$

and the Gaussian pulse shape for the pulse envelope by

$$\boxed{A(t) = A_0\exp\left(-\frac{t^2}{2\tau^2}\right)} . \tag{6.62}$$

Equation (6.61) corresponds to the result (6.37) derived above under the assumption that the solution is a Gaussian pulse, which we have confirmed here. In (6.61), we considered that $\omega_m = 2\pi f_m$ and that $\Omega_g = 2\pi \Delta f_g/2$, with $\Delta\omega_g = 2\pi \Delta f_g$ the FWHM spectral width of the gain spectrum.

The stable solution follows from (6.59) with $\lambda_0 = 0$:

$$0 = g - l - \frac{1}{2} M \omega_m^2 \tau^2 ,$$

whence

$$\boxed{g = l + \frac{1}{2} M \omega_m^2 \tau^2} . \tag{6.63}$$

This is the gain–loss balance equation which is modified in comparison to cw operation. For cw operation at steady state we have "gain equals loss," i.e., $g = l$. For active modelocking we have an additional loss with the loss modulator which appears as an additional term in the new gain–loss balance equation. The gain has been increased by the modulator loss to allow for several axial modes to reach threshold and to support lasing in many axial modes simultaneously, even though the gain material is homogeneously broadened (see Fig. 6.10). This will be discussed again descriptively in the next section.

We can transform (6.63) using (6.56) and (6.58):

$$\boxed{\frac{1}{2} M \omega_m^2 \tau^2 \overset{(6.58)}{=} M_s \tau^2 \overset{(6.56)}{=} \frac{D_g}{\tau^2}} , \tag{6.64}$$

and therefore obtain

$$\boxed{g = l + \frac{D_g}{\tau^2}} . \tag{6.65}$$

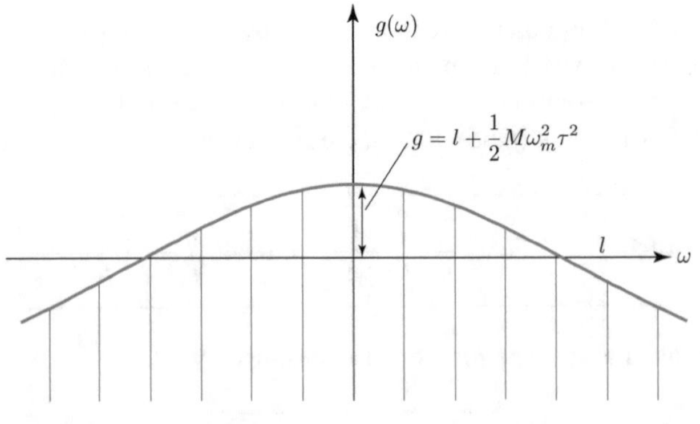

Fig. 6.10 Gain–loss balance equation for active modelocking without SPM and GDD (6.63). Homogeneous gain coefficient at steady state with a loss modulator. Gain is increased which supports many axial modes for modelocking. For cw operation at steady state, we have "gain equals loss," i.e., the simplest gain–loss balance equation $g = l$

Comparing (6.63) and (6.65) leads to the simple description we made in the introduction (see Fig. 6.5). At steady state, an optimum pulse duration is obtained, which is a balance between the pulse shortening of the loss modulator and the pulse broadening of the amplification in the gain material (6.64). Note that (6.65) also indirectly describes the gain–loss balance, but is not the preferred way because it does not actually contain the modulator loss and the gain dispersion D_g is actually loss-free. But of course also this equation is correct and equivalent to (6.63) because of the identity (6.64).

For the steady-state solution, the deviation from the condition $g = l$ is very small, because typically

$$\frac{g-l}{g} = \frac{1}{2}\frac{M}{g}\omega_m^2\tau^2 \ll 1 , \tag{6.66}$$

since $f_m\tau_p \ll 1$. For example, for a typical flashlamp-pumped Nd:YAG laser, we have $f_m = 100\,\text{MHz}$ and $\tau_p = 100\,\text{ps}$. Because this change in the gain is very small, the pulse energy E_p is still determined by the homogeneous gain and the steady state condition $g = l$:

$$\frac{g_0}{1+\dfrac{E_p}{P_{sat}T_R}} = l , \tag{6.67}$$

where $P_{sat} = I_{sat}A_L$ is the saturation power, with I_{sat} the saturation intensity of the gain material and A_L the laser mode area inside the gain medium. $P = E_p/T_R$ is the intracavity laser power. Note that, for a standing wave cavity, E_p should be replaced by $2E_p$, i.e., according to (5.28). We therefore still obtain a stable solution even though many axial modes have a higher gain than l.

6.4.3 Explanation of Active Modelocking in the Spectral Domain

Equation (6.47) implies

$$\Delta A(t) \approx -M\left(1-\cos\omega_m t\right)A(t) = m(t)A(t) . \tag{6.68}$$

The Fourier transform of a product is a convolution, so

$$\Delta\tilde{A}(\omega) = \tilde{m}(\omega) * \tilde{A}(\omega) , \tag{6.69}$$

with

$$\begin{aligned}\tilde{m}(\omega) &= F\left\{-M + M\cos\omega_m t\right\} = -2\pi M\delta(\omega) + \frac{M}{2}\left[F\left\{e^{i\omega_m t}\right\} + F\left\{e^{-i\omega_m t}\right\}\right] \\ &= -2\pi M\delta(\omega) + 2\pi\frac{M}{2}\left[\delta\left(\omega-\omega_m\right) + \delta\left(\omega+\omega_m\right)\right] .\end{aligned} \tag{6.70}$$

Assuming $\tilde{A}(\omega)$ in (6.69) corresponds to only one axial mode, energy is taken away from this axial mode by the modulator and is given to the modulation sidebands [see (6.70) and Fig. 6.11]. However, through the convolution in (6.69), this happens for every axial mode in the pulse spectrum.

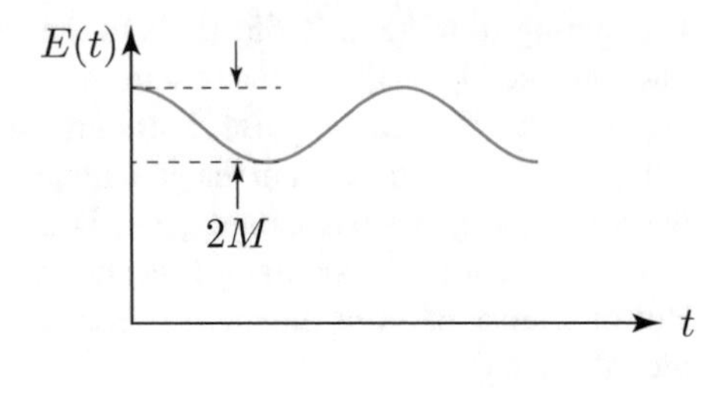

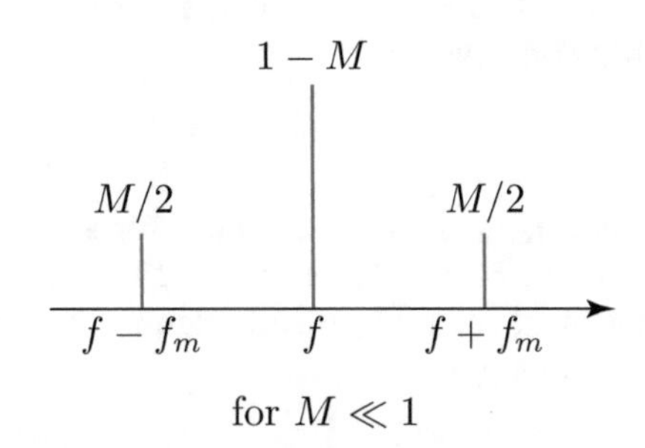

Fig. 6.11 With the modulator, energy is transferred into the modulation sidebands, which again are axial modes, if the modulation frequency is equal to the resonator roundtrip frequency. In this way, the other axial modes can have a larger gain and reach the laser threshold, as shown in Fig. 6.10

6.5 Active Modelocking with SPM, but Without GDD

Having understood active modelocking without SPM and GDD, we now want to address the question of how SPM affects active modelocking. We will use the master equation formalism to continue the discussion in this section. We will see that SPM will support shorter pulse durations but will also introduce a chirp in the Gaussian pulse shape, and with too much SPM modelocking will become unstable.

6.5.1 Master Equation

In contrast to the derivation of the master equation in Sect. 6.4.2, we will now also consider self-phase modulation (SPM) (see Appendix F or Chap. 4):

$$\boxed{\Delta A_4 \approx -\mathrm{i}\delta |A|^2 A(T,t)} \,. \tag{6.71}$$

Applying the approach of the Haus master equation again, we have

$$T_{\mathrm{R}} \frac{\partial A(T,t)}{\partial T} = \sum_i \Delta A_i = 0 \,,$$

and (6.44), (6.47), (6.49), and (6.71) yield (see also Fig. 6.12)

$$\boxed{\left[g - l + D_{\mathrm{g}} \frac{\partial^2}{\partial t^2} - M\left(1 - \cos \omega_{\mathrm{m}} t\right) - \mathrm{i}\delta \left|A(T,t)\right|^2 + \mathrm{i}\psi\right] A(T,t) = 0} \,. \tag{6.72}$$

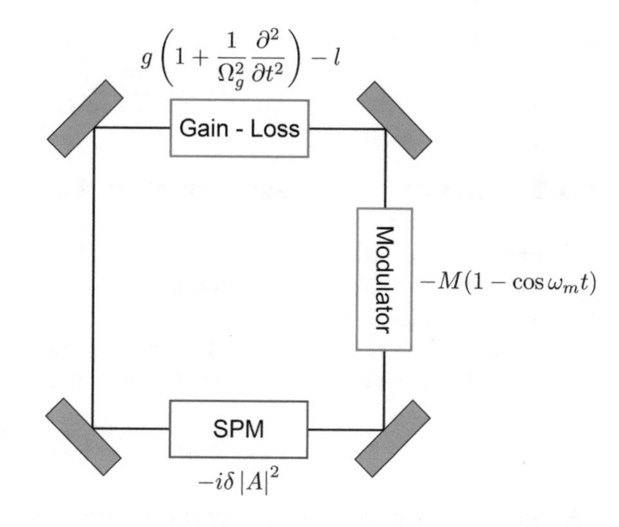

Fig. 6.12 Schematic diagram for all linearized operators acting on the pulse envelope for active modelocking with SPM, but still without GDD. By "gain–loss," we refer to the time-independent loss l in the laser cavity, such as the output coupler

In (6.72), we have included an additional phase shift ψ to find a self-consistent solution. When the laser is in pulsed operation, only the pulse envelope has to be unchanged after one resonator roundtrip, but not the optical phase. Up to now, we have always assumed this constant phase shift to be $\psi = 0$, i.e., we have not taken advantage of this additional degree of freedom. The phase shift is an additional degree of freedom for the pulse, which is given by

$$E(z,t) = A(z,t)\mathrm{e}^{\mathrm{i}[\omega_0 t - k(\omega_0)z + \psi]} .$$

Thus, it becomes clear that the full spectrum, including all axial modes, experiences a constant phase shift.

Note that the arrival time of the laser pulse is not changed by the additional phase shift ψ, because the envelope $A(z,t)$ determines the group velocity and thus the arrival time of the pulse. The phase shift ψ only affects the phase velocity and therefore only the electric field is shifted within the pulse envelope (see Fig. 6.2). As already discussed, this phase shift changes the CEO phase of the pulse and no locking of the CEO phase occurs during the modelocking process. We have included the phase shift ψ, which can be set to an arbitrary value in the master equation to find a self-consistent solution.

The additional phase shift results in the following change in the pulse envelope:

$$A_{\text{out}}(t) = \mathrm{e}^{\mathrm{i}\psi} A_{\text{in}}(t) , \tag{6.73}$$

whence

$$A_{\text{out}}(t) = \mathrm{e}^{\mathrm{i}\psi} A_{\text{in}}(t) \approx (1 + \mathrm{i}\psi)\, A_{\text{in}}(t) .$$

Then

$$\Delta A_5 = A_{\text{out}}(t) - A_{\text{in}}(t) \approx \mathrm{i}\psi A_{\text{in}}(T,t)$$

and we conclude that

$$\boxed{\Delta A_5 \approx \mathrm{i}\psi A(T,t)}\,. \tag{6.74}$$

6.5.2 Solution of the Master Equation: A Chirped Gaussian Pulse

It is important to realize that, in the positive GDD regime, too much SPM can introduce instabilities in the pulse formation in the time domain. However, SPM produces an additional spectral broadening which counteracts the spectral narrowing of the amplification, and this can lead to shorter pulses than without SPM.

The steady-state solution of the master equation (6.72) satisfies the condition

$$T_{\mathrm{R}}\frac{\partial}{\partial T}A(T,t) = 0\,.$$

We make a solution ansatz for a chirped Gaussian pulse:

$$\boxed{A(t) = A_0 \exp\left[-\frac{1}{2}\frac{t^2}{\tau^2}(1-\mathrm{i}\beta)\right]}\,, \tag{6.75}$$

where β is the chirp parameter. This ansatz is motivated by the fact that the solution of the master equation without SPM is a Gaussian pulse. SPM without GDD cannot support a soliton, and it is therefore realistic to assume that the pulse shape is still dominated by the modulator. We further assume that SPM causes a chirp, because SPM introduces a complex term into the master equation.

To validate that this ansatz is indeed a solution to the differential equation (6.72), we want to insert the ansatz of a chirped Gaussian pulse (6.75) into the master equation. To simplify this calculation, we first determine how the individual terms of (6.72) act on the chirped Gaussian pulse:

$$\begin{aligned}\frac{\mathrm{d}^2}{\mathrm{d}t^2}A(t) &= \frac{\mathrm{d}}{\mathrm{d}t}\left[A(t)\left\{-\frac{t}{\tau^2}(1-\mathrm{i}\beta)\right\}\right]\\ &= \frac{\mathrm{d}}{\mathrm{d}t}A(t)\left\{-\frac{t}{\tau^2}(1-\mathrm{i}\beta)\right\} + A(t)\frac{\mathrm{d}}{\mathrm{d}t}\left\{-\frac{t}{\tau^2}(1-\mathrm{i}\beta)\right\}\\ &= A(t)\left\{\frac{t^2}{\tau^4}(1-\mathrm{i}\beta)^2 - \frac{1-\mathrm{i}\beta}{\tau^2}\right\}\,.\end{aligned} \tag{6.76}$$

We again make a parabolic approximation for the modulator (6.54):

$$-M(1-\cos\omega_{\mathrm{m}}t) \approx -\frac{M\omega_{\mathrm{m}}^2}{2}t^2\,. \tag{6.77}$$

Equation (6.75) implies

$$|A|^2 = |A_0|^2\exp\left(-\frac{t^2}{\tau^2}\right) \approx |A_0|^2\left(1-\frac{t^2}{\tau^2}\right)\,, \tag{6.78}$$

where the parabolic approximation is introduced as well. The amplitude is given by the peak intensity I_p, which is determined by the peak power P_peak and the mode area πw_0^2 in the laser crystal:

$$|A_0|^2 = I_\mathrm{p} = \frac{P_\mathrm{peak}}{\pi w_0^2} \,, \tag{6.79}$$

where we have assumed that the pulse envelope is normalized accordingly. After inserting (6.76)–(6.79) into the master equation (6.72), we obtain the steady-state solution:

$$g - l - \frac{M\omega_\mathrm{m}^2}{2} t^2 + D_\mathrm{g} \frac{(1-\mathrm{i}\beta)^2}{\tau^4} t^2 - D_\mathrm{g} \frac{1-\mathrm{i}\beta}{\tau^2} - \mathrm{i}\delta \frac{P_\mathrm{peak}}{\pi w_0^2} + \mathrm{i}\delta \frac{P_\mathrm{peak}}{\pi w_0^2} \frac{t^2}{\tau^2} + \mathrm{i}\psi = 0 \,. \tag{6.80}$$

Comparing coefficients of the terms in t^2, this implies

$$\boxed{-M\frac{\omega_\mathrm{m}^2}{2} + D_\mathrm{g} \frac{(1-\mathrm{i}\beta)^2}{\tau^4} + \mathrm{i}\delta \frac{P_\mathrm{peak}}{\pi w_0^2} \frac{1}{\tau^2} = 0} \,, \tag{6.81}$$

while the coefficients of the constant terms yield

$$\boxed{g - l - D_\mathrm{g} \frac{1-\mathrm{i}\beta}{\tau^2} - \mathrm{i}\delta \frac{P_\mathrm{peak}}{\pi w_0^2} + \mathrm{i}\psi = 0} \,. \tag{6.82}$$

Both the real and the imaginary parts of (6.81) and (6.82) have to be zero. This results in four equations for the four unknowns τ, β, ψ, and $g - l$.

(a) Special Case with $\beta = 0$ and $\delta = 0$ (without SPM)

In this case, it should be possible to retrieve the previous result. Equation (6.82) implies

$$g - l - \frac{D_\mathrm{g}}{\tau^2} + \mathrm{i}\psi = 0 \,, \tag{6.83}$$

where the real and the imaginary parts must both be zero, i.e.,

$$\psi = 0 \,, \tag{6.84}$$

and (6.81) implies

$$-\frac{M\omega_\mathrm{m}^2}{2} + \frac{D_\mathrm{g}}{\tau^4} = 0 \,. \tag{6.85}$$

The pulse duration is therefore given by

$$\tau^4 = \frac{2D_\mathrm{g}}{M\omega_\mathrm{m}^2} \,, \tag{6.86}$$

which implies

$$\tau = \sqrt[4]{\frac{2D_{\mathrm{g}}}{M}}\sqrt{\frac{1}{\omega_{\mathrm{m}}}} . \tag{6.87}$$

With D_{g} given by (see Appendix F)

$$D_{\mathrm{g}} = \frac{g}{\Omega_{\mathrm{g}}^2} ,$$

it follows that

$$\boxed{\tau = \sqrt[4]{\frac{2g}{M}}\sqrt{\frac{1}{\omega_{\mathrm{m}}\Omega_{\mathrm{g}}}} \overset{(2.103)}{\Longrightarrow} \tau_{\mathrm{p,FWHM}} = 1.665\tau} . \tag{6.88}$$

The FWHM pulse duration is given, according to the discussion in Chap. 2 (see Table 2.1) for a Gaussian pulse shape, with a factor of $2\sqrt{\ln 2} = 1.665$. As expected, (6.88) corresponds to the previous result (6.37) and (6.61). For the special case without SPM, (6.83) yields

$$\boxed{g - l - \frac{D_{\mathrm{g}}}{\tau^2} = 0} , \tag{6.89}$$

which is also identical to (6.63). Equation (6.86) then implies

$$\frac{1}{\tau^2} = \tau^2 \frac{1}{\tau^4} = \tau^2 \frac{M\omega_{\mathrm{m}}^2}{2D_{\mathrm{g}}} . \tag{6.90}$$

Using (6.89), we then have the gain–loss balance equation

$$\boxed{g - l - \frac{M\omega_{\mathrm{m}}^2}{2}\tau^2 = 0} , \tag{6.91}$$

Equations (6.89) and (6.91) correspond to the previous results (6.63) and (6.65).

(b) General Case

The real part of (6.81) yields

$$-\frac{M\omega_{\mathrm{m}}^2}{2} + D_{\mathrm{g}}\frac{1-\beta^2}{\tau^4} = 0 , \tag{6.92}$$

and its imaginary part yields

$$-2\beta D_{\mathrm{g}}\frac{1}{\tau^4} + \phi_{\mathrm{nl}}\frac{1}{\tau^2} = 0 , \tag{6.93}$$

where the nonlinear phase shift per resonator roundtrip ϕ_{nl} is given by (4.4)

$$\boxed{\phi_{\mathrm{nl}} = \delta \frac{P_{\mathrm{peak}}}{\pi w_0^2} = 2 \frac{2\pi}{\lambda_0} L_{\mathrm{K}} n_2 \frac{P_{\mathrm{peak}}}{\pi w_0^2}} . \tag{6.94}$$

The second equality follows from (4.5), which describes the self-phase modulation (SPM) coefficient δ upon propagation through a Kerr material of length L_{K}. The factor of 2 originates from the fact that, per resonator roundtrip, we pass twice through this material. Equation (6.93) can be solved for β:

$$\boxed{\beta = \frac{\tau^2 \phi_{\mathrm{nl}}}{2 D_{\mathrm{g}}}} . \tag{6.95}$$

Equation (6.95) shows that the chirp of the pulse increases with increasing SPM.

Equation (6.95) can be inserted into (6.92), and then solved for the pulse duration. Hence,

$$D_{\mathrm{g}} \frac{1}{\tau^4} = \frac{M \omega_{\mathrm{m}}^2}{2} + \frac{\phi_{\mathrm{nl}}^2}{4 D_{\mathrm{g}}} ,$$

and it follows that

$$\boxed{\tau^4 = \frac{D_{\mathrm{g}}}{\dfrac{M \omega_{\mathrm{m}}^2}{2} + \dfrac{\phi_{\mathrm{nl}}^2}{4 D_{\mathrm{g}}}}} . \tag{6.96}$$

Equation (6.96) tells us that, as expected, the pulse width is further shortened by SPM [95Bra2]. The pulses are additionally chirped (6.95), so it may be possible to further compress them outside the resonator.

The real part of (6.82) yields

$$g - l - \frac{D_{\mathrm{g}}}{\tau^2} = 0 . \tag{6.97}$$

Equation (6.96) gives

$$\frac{1}{\tau^2} = \tau^2 \frac{1}{\tau^4} = \tau^2 \frac{1}{D_{\mathrm{g}}} \left(\frac{M \omega_{\mathrm{m}}^2}{2} + \frac{\phi_{\mathrm{nl}}^2}{4 D_{\mathrm{g}}} \right) , \tag{6.98}$$

and thus, using (6.97), we obtain the gain–loss balance equation

$$\boxed{g - l - \frac{M \omega_{\mathrm{m}}^2}{2} \tau^2 - \frac{\phi_{\mathrm{nl}}^2}{4 D_{\mathrm{g}}} \tau^2 = 0} , \tag{6.99}$$

with an additional SPM term in comparison to (6.63) and Fig. 6.10, which supports the shorter pulse durations. Similarly to the situation with the modulator (see Fig. 6.10), the gain is also increased by SPM. The formal similarity between the influence of SPM and the amplitude modulation is not surprising since the nonlinear SPM interaction also increases the pulse spectrum.

The imaginary part of (6.82) yields

$$D_g \frac{\beta}{\tau^2} - \phi_{nl} + \psi = 0 \,. \tag{6.100}$$

Using (6.95) for the chirp β, we find

$$\psi = \phi_{nl} - D_g \frac{\beta}{\tau^2} = \phi_{nl} - \frac{D_g}{\tau^2} \frac{\tau^2 \phi_{nl}}{2D_g} = \frac{\phi_{nl}}{2} \,. \tag{6.101}$$

(c) Energy Balance and Gain–Loss Balance

The gain–loss balance equation (6.99) can also be derived from the energy balance. At steady state, the pulse envelope does not change after one resonator roundtrip and therefore the pulse energy remains constant:

$$\boxed{\frac{\partial}{\partial T} \int_0^{T_R} |A(T,t)|^2 \, dt = 0} \,. \tag{6.102}$$

This energy balance equation can be simplified to

$$\int_0^{T_R} \left(\frac{\partial A}{\partial T} A^* + A \frac{\partial A^*}{\partial T} \right) dt = \int_0^{T_R} 2\mathrm{Re} \left(\frac{\partial A}{\partial T} A^* \right) dt = 0 \,. \tag{6.103}$$

Taking the real part of $\partial A / \partial T$ directly from the master equation, (6.72) with the parabolic approximation of the loss modulation (6.54) results in

$$\int_0^{T_R} \left[(g - l) A A^* + D_g \frac{\partial^2 A}{\partial t^2} A^* - \frac{M}{2} \omega_m^2 t^2 A A^* \right] dt = 0 \,, \tag{6.104}$$

which yields

$$(g - l) \int_0^{T_R} |A|^2 dt + D_g \int_0^{T_R} \frac{\partial^2 A}{\partial t^2} A^* dt - \frac{M}{2} \omega_m^2 \int_0^{T_R} t^2 |A|^2 dt = 0 \,. \tag{6.105}$$

We can now proceed to integrate (6.105).

With the partial integration

$$\int_0^{T_R} u v' dt = u v \Big|_0^{T_R} - \int_0^{T_R} u' v \, dt \,,$$

and using $u \equiv \partial A / \partial t$ and $v \equiv A^*$, it follows that

$$\int_0^{T_R} \frac{\partial^2 A}{\partial t^2} A^* dt = - \int_0^{T_R} \frac{\partial A}{\partial t} \frac{\partial A^*}{\partial t} dt = - \int_0^{T_R} \left| \frac{\partial A}{\partial t} \right|^2 dt \,. \tag{6.106}$$

Due to the fact that we are dealing with pulses which are shorter than the resonator roundtrip time T_R, the first term in the partial integration is zero, because $A^*(T_R) = 0$ and $A^*(0) = 0$ on the integration boundary. Moreover, using (6.75), we have

$$\int_0^{T_R} |A|^2 \mathrm{d}t = \int_{-\infty}^{+\infty} A_0^2 \exp\left(-\frac{t^2}{\tau^2}\right) \mathrm{d}t = \sqrt{\pi}\tau A_0^2 . \tag{6.107}$$

Because we always integrate short pulses, the integration from 0 to T_R is approximately equal to the integration from $-\infty$ to $+\infty$. Furthermore,

$$\begin{aligned}\int_0^{T_R} \left|\frac{\partial A}{\partial t}\right|^2 \mathrm{d}t &= \int_0^{T_R} \frac{t^2}{\tau^4} A_0^2 (1+\beta^2) \exp\left(-\frac{t^2}{\tau^2}\right) \mathrm{d}t \\ &= \frac{A_0^2}{\tau^4}(1+\beta^2) \int_0^{T_R} t^2 \exp\left(-\frac{t^2}{\tau^2}\right) \mathrm{d}t ,\end{aligned} \tag{6.108}$$

where

$$\int_{-\infty}^{+\infty} t^2 \exp\left(-\frac{t^2}{\tau^2}\right) \mathrm{d}t = \sqrt{\pi}\frac{1}{2}\tau^3 . \tag{6.109}$$

Inserting (6.106) to (6.109) into (6.105), we obtain

$$(g-l)\sqrt{\pi}\tau - D_g \frac{1}{\tau^4}(1+\beta^2)\sqrt{\pi}\frac{1}{2}\tau^3 - \frac{M\omega_{\mathrm{m}}^2}{2}\sqrt{\pi}\frac{1}{2}\tau^3 = 0 , \tag{6.110}$$

which can be simplified to

$$g - l - D_g \frac{1}{2\tau^2}(1+\beta^2) - \frac{M\omega_{\mathrm{m}}^2}{4}\tau^2 = 0 . \tag{6.111}$$

With (6.95), this results in

$$g - l - D_g \frac{1}{2\tau^2} - \frac{\phi_{\mathrm{nl}}^2}{8D_g}\tau^2 - \frac{M\omega_{\mathrm{m}}^2}{4}\tau^2 = 0 . \tag{6.112}$$

Using (6.96), we have

$$\frac{1}{\tau^2} = \tau^2 \frac{1}{\tau^4} = \frac{\tau^2}{D_g}\left(\frac{M\omega_{\mathrm{m}}^2}{2} + \frac{\phi_{\mathrm{nl}}^2}{4D_g}\right) , \tag{6.113}$$

and thus

$$g - l - \frac{M\omega_{\mathrm{m}}^2}{4}\tau^2 - \frac{\phi_{\mathrm{nl}}^2}{8D_g}\tau^2 - \frac{\phi_{\mathrm{nl}}^2}{8D_g}\tau^2 - \frac{M\omega_{\mathrm{m}}^2}{4}\tau^2 = 0 , \tag{6.114}$$

and further

$$g - l - \frac{M\omega_{\mathrm{m}}^2}{2}\tau^2 - \frac{\phi_{\mathrm{nl}}^2}{4D_g}\tau^2 = 0 ,$$

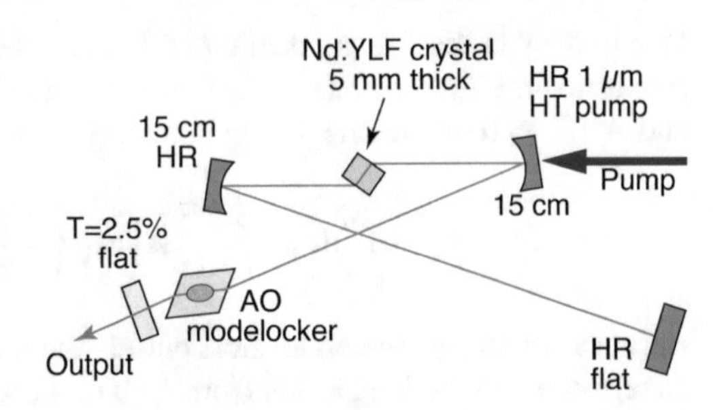

Fig. 6.13 Experimental setup of an actively modelocked Nd:YLF laser according to [95Bra2]. Mode radius inside the Nd:YLF crystal 127 μm × 87 μm, crystal length $L_K = 5\,\text{mm}$, output coupler $T = 2.5\%$, AOM loss ≈ 20%

which is equivalent to (6.99). Note that this equation has been derived without parabolic approximations from the Gaussian pulse (where the modulator was still approximated parabolically, although this would not be necessary for this derivation) and yields the same result as (6.99).

(d) Stability of the Solution

Up to now we have not considered the stability of the solutions. Equation (6.96) would predict that, with more and more SPM, the pulse can be further shortened. However, this is not the case for an arbitrary increase because instabilities start to occur. Numerical simulations [86Hau] have shown that the maximum pulse shortening is expected to be by a factor of about 2. This is significantly different to SPM with negative GDD, where the pulse shortening can be much greater [95Kar2].

6.5.3 Example: Nd:YLF Laser

The laser gain material is a π-oriented Nd:YLF crystal, lasing at 1047 nm. The Nd:YLF crystal plate with a thickness of 5 mm is inserted at the Brewster angle (see Fig. 6.13). The Nd doping is 1.5%. The pump source for the Nd:YLF laser is a cw Ti:sapphire laser at a wavelength of 793 nm, resulting in an absorption depth of 2.3 mm. The measured spot radius of the pump beam in the tangential plane is 20 μm.

For the Nd:YLF laser, we have the following data:

$$\lambda_g = 1.047\,\mu\text{m}\,, \quad \Delta\lambda_g = 1.3\,\text{nm}\,, \quad \Delta f_g = \Delta\lambda_g c/\lambda_g^2 = 3.56 \times 10^{11}\,\text{Hz}\,,$$

whence

$$\Omega_g = 2\pi\frac{\Delta f_g}{2} = 1.12 \times 10^{12}\,\text{Hz}\,, \quad f_{\text{rep}} = f_m = 250\,\text{MHz}\,, \quad g = 0.0126\,,$$

so that $2g \approx T_{\text{out}} \approx 2.52\%$, and $M = 0.2$. We also have

$$D_g = \frac{g}{\Omega_g^2} = \frac{0.0126}{\left(1.12 \times 10^{12}\,\text{Hz}\right)^2} = 1.01 \times 10^{-26}\,\text{s}^2\,,$$

$$\frac{M\omega_m^2}{2} = \frac{0.2\,(2\pi \times 250\,\text{MHz})^2}{2} = 2.47 \times 10^{17}\,\text{s}^{-2}\,,$$

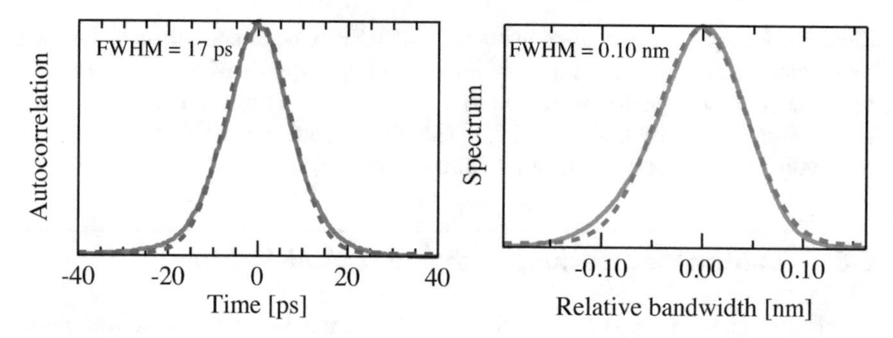

Fig. 6.14 Actively modelocked Nd:YLF laser at a wavelength of 1.047 μm with an average output power of 622 mW for a pump power of 1.5 W [95Bra2]. *Left*: Non-collinear intensity autocorrelation with a Gaussian fit (*dashed line*). *Right*: The laser spectrum is measured using a scanning Fabry–Pérot interferometer. Optical spectrum with a Gaussian fit (*dashed line*). The time–bandwidth product is about 0.46

$$n_2 = 1.72 \times 10^{-16}\frac{\text{cm}^2}{\text{W}}, \quad L_\text{K} = 5\text{ mm},$$

according to Table 4.1, and

$$w_\text{t} = 127\,\mu\text{m}, \quad w_\text{s} = 87\,\mu\text{m}, \quad P_\text{out} = 622\,\text{mW},$$

whence

$$P_\text{intra} = P_\text{out}/T_\text{out} = 24.7\,\text{W},$$

where w_t and w_s are the tangential and sagittal beam waist, respectively, and P_intra is the intracavity average power.

Using $\tau_\text{p} = 17\,\text{ps}$ (see Fig. 6.14a), we have the peak intensity

$$I_\text{p} = \frac{P_\text{intra}}{f_\text{rep}\tau_\text{p}\pi w_\text{t} w_\text{s}} = 1.67 \times 10^7\frac{\text{W}}{\text{cm}^2},$$

$$\phi_\text{nl} = 2\frac{2\pi}{\lambda_\text{g}}n_2 I_\text{p} L_\text{K} = 1.72 \times 10^{-4}$$

and

$$\frac{\phi_\text{nl}^2}{4D_\text{g}} = \frac{\left(1.72 \times 10^{-4}\right)^2}{4 \times 1.005 \times 10^{-26}\text{s}^2} = 7.4 \times 10^{17}\text{s}^{-2}.$$

Equation (6.96) implies

$$\tau_\text{p,FWHM} = 1.665 \times \sqrt[4]{\frac{D_\text{g}}{\frac{M\omega_\text{m}^2}{2} + \frac{\phi_\text{nl}^2}{4D_\text{g}}}} = 1.665 \times \sqrt[4]{\frac{1.005 \times 10^{-26}\,\text{s}^2}{2.467 \times 10^{17}\text{s}^{-2} + 7.4 \times 10^{17}\text{s}^{-2}}}$$

$$= 18\,\text{ps}.$$

Thus, the theory is in very good agreement with the experimental results. Without SPM, one would only obtain a pulse duration of $\tau_{p,FWHM} \approx 24\,\mathrm{ps}$, which follows from the previous equation with $\psi = \phi_{nl}/2 = 0$. Thus, in this example, SPM leads to pulse shortening by a factor of 1.3. The chirp could then be compensated for externally to further shorten the final pulse duration.

6.6 Soliton Modelocking with Active Modelocking

In this section we discuss the solution for active modelocking with soliton pulse shaping, i.e., with both GDD and SPM. The interplay between GDD and SPM allows for soliton formation which can become the dominant pulse shaping process. Initially, it was surprising that we could obtain stable soliton pulses with relatively large SPM and much shorter pulses than other theories predicted. The theoretical explanation using soliton perturbation theory as discussed in this section with an intracavity loss modulator and in Sect. 9.5 with a slow saturable absorber explained the underlying physics and was referred to as soliton modelocking [95Kar2,95Jun2,96Kar]. In this case the loss modulator or the slow saturable absorber only started and stabilized the modelocking process.

Silberberg and Haus [86Hau] investigated the impact of GDD and SPM on active and passive modelocking. They found that the two processes together may lead to pulse shortening by as much as a factor of 2.5 until instabilities arise. However, this is not true. Here, we give a simplified explanation as to why, for the case of large enough negative GDD and SPM, pulse shortening can be much greater due to soliton pulse shaping. In this case, we have actually observed pulse shortening by as much as a factor of 30 [94Kop2].

Soliton formation is discussed in more detail in Sect. 4.3 and will be used in the following discussion. As we will show here the intracavity loss modulator only starts and stabilizes the modelocking process and much shorter pulse durations can be achieved. This modelocking technique is referred to as soliton modelocking and was initially discovered with passive modelocking using slow saturable absorbers. The final pulse duration satisfies the soliton condition and is no longer a balance between the pulse broadening in the gain and the pulse shortening in the loss modulator. With constant pulse energy the soliton pulse duration scales directly in proportion to the GDD (4.57). Therefore with an intracavity prism pair (Fig. 3.9) we can tune the pulse duration in a soliton modelocked laser, keeping the pulse energy constant and keeping the pulse transform-limited as shown in Fig. 6.15. The pulses become shorter with smaller GDD, which is ultimately limited by the onset of instabilities as shown with soliton perturbation theory.

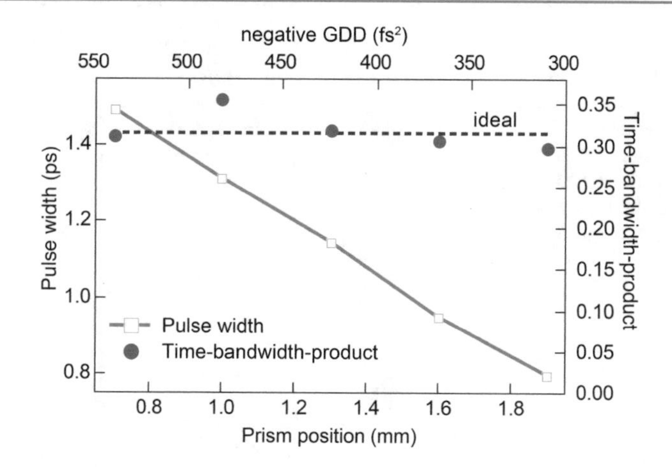

Fig. 6.15 Pulse duration as a function of intracavity GDD for soliton modelocking, shown here with a Nd:glass laser [94Kop2]. An intracavity prism pair for dispersion compensation allows for continuous tuning of the transform-limited soliton pulse duration, which is directly proportional to the group delay dispersion (GDD) according to (4.57)

6.6.1 Derivation and Solution of the Master Equation with SPM and GDD

For soliton modelocking the final pulse formation is dominated by soliton formation when positive SPM is combined with negative GDD. With a sinusoidal loss modulation from an AOM as shown in Figs. 6.4 and 6.5 with much shorter pulse durations than the modulation period, the AOM modulation has a weaker pulse-shaping effect. For femtosecond pulse durations τ_p we typically have $\tau_p \ll T_R$, where T_R is the cavity roundtrip time. Therefore the sinusoidal loss modulation at a frequency $\omega_m = 2\pi/T_R$ has hardly any pulse-shaping action on this pulse. In this case, the intracavity acousto-optic loss modulator (AOM) only starts and stabilizes the modelocking process. Therefore, depending on the gain bandwidth, much shorter pulses can be generated with active modelocking [95Kar2].

If we also want to consider SPM and GDD, we have to include the following terms in the master equation (6.51) (see Fig. 6.16):

- SPM (Appendix F):

$$\Delta A_4 = -\mathrm{i}\delta |A|^2 A(T,t) . \tag{6.115}$$

- GDD (Appendix F):

$$\Delta A_5 = \mathrm{i}D \frac{\partial^2}{\partial t^2} A(T,t) . \tag{6.116}$$

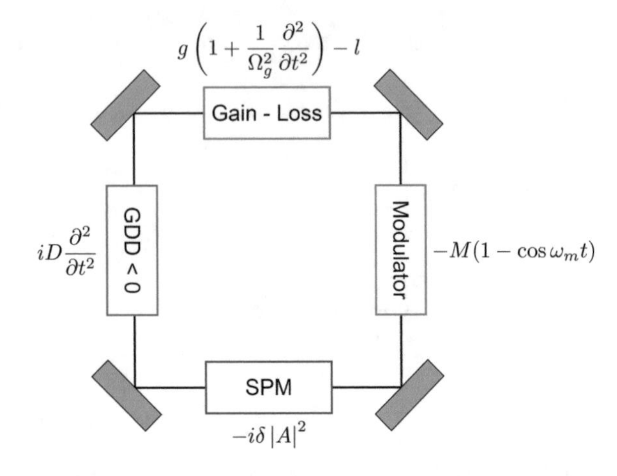

Fig. 6.16 Schematic diagram for all linearized operators acting on the pulse envelope for active modelocking with positive SPM and negative GDD. Time-independent loss l is separated from modulator loss

Using the master equation (6.50), we then have a more complicated differential equation:

$$\boxed{\begin{aligned} T_{\mathrm{R}}\frac{\partial}{\partial T}A(T,t) = &\left[\mathrm{i}D\frac{\partial^2}{\partial t^2} - \mathrm{i}\delta\,|A(T,t)|^2\right]A(T,t) \\ &+\left[g - l + D_{\mathrm{g}}\frac{\partial^2}{\partial t^2} - M\,(1-\cos\omega_{\mathrm{m}}t)\right]A(T,t) = 0 \end{aligned}} \tag{6.117}$$

The steady-state solution satisfies the condition $T_{\mathrm{R}}\partial A(T,t)/\partial T = 0$.

The first part of this differential equation corresponds to the nonlinear Schrödinger equation (4.46), whose solution for GDD < 0 and $n_2 > 0$ is given by the first-order soliton (4.47). Laser materials which have a sufficiently large gain bandwidth Ω_{g} to generate ultrashort pulses also have $D_{\mathrm{g}} = g/\Omega_{\mathrm{g}}^2 \ll 1$. Furthermore, when the pulses become extremely short, the cosine-shaped loss modulation has no significant effect on the pulse. In this case the first part of this differential equation can become dominant. Therefore the second part of this differential equation can simply be considered as a perturbation of the soliton. Mathematically, the solution of the master equation can then be found using the soliton-perturbation calculation [95Kar2]. In contrast to before for active modelocking, a Gaussian-shaped pulse has to be replaced by a hyperbolic secant-shaped pulse, the soliton of the nonlinear Schrödinger equation. The other effects, like gain and loss modulation, only have to compensate for the inevitable loss, and keep the soliton-like pulse stable.

The following computer simulations as shown in Fig. 6.17 were conducted for a parameter set of an Nd:YAG laser which shows the temporal development of the laser pulse inside the resonator over 1000 and 10 000 resonator roundtrips. We start with a 68 ps Gaussian pulse. After several 1000 roundtrips, a steady-state 24 ps soliton-like pulse forms. We can clearly see that, after only about 100 resonator roundtrips, a soliton-like pulse has already formed out of the long Gaussian pulse. The rest of the initial pulse energy, which could not be transferred into the soliton-like pulse, forms the continuum. For the soliton, the dispersive and nonlinear effects are

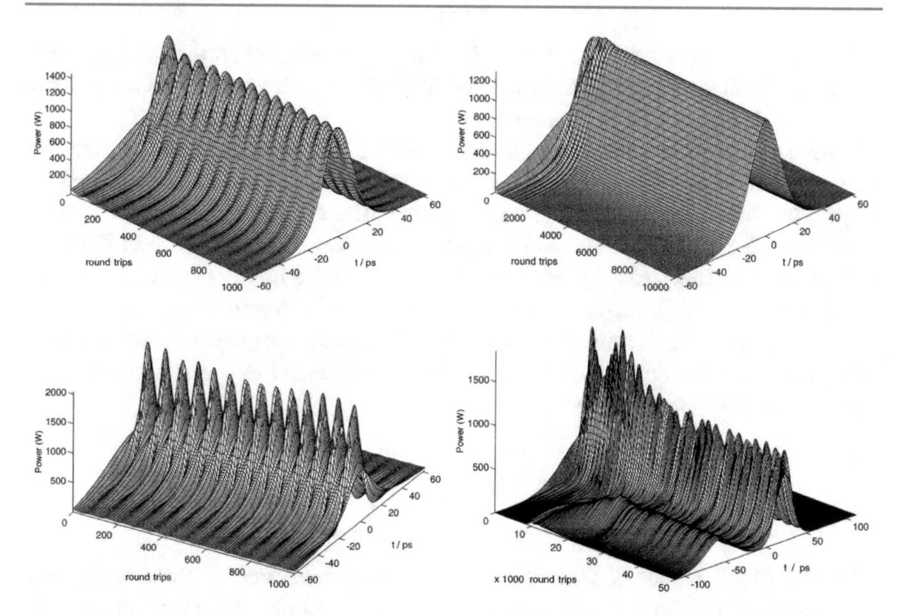

Fig. 6.17 Temporal evolution of the intensity of a pulse in an actively modelocked Nd:YAG laser with self-phase modulation and negative dispersion for 1000 and 10 000 resonator roundtrips. The initial pulse is Gaussian with a duration of 68 ps. *Upper*: The laser works in the stable region with a sufficiently high negative dispersion that the continuum decays and the laser reaches steady state after about 4000 roundtrips. *Lower*: The dispersion has been reduced to produce even shorter pulses, but the laser is becoming unstable. After [95Kar2]

balanced so that a constant pulse duration is maintained. Since the continuum has a much smaller intensity than the soliton pulse, the intensity-dependent self-phase modulation (SPM) is much weaker than the dispersion. Consequently, the continuum is temporally broadened since the self-phase modulation is insufficient to balance the dispersive pulse broadening. However, if the continuum is sufficiently broadened, it experiences the increased loss due to the modulator. The continuum therefore sees higher loss per roundtrip than the soliton, and therefore decreases over time.

In the soliton pulse frame of reference, the soliton changes its phase periodically due to the Kerr effect [see (4.47) and (4.50) in Sect. 4.3], but the continuum does not. This results in periodic interferences between soliton and continuum, which can be seen in Fig. 6.17. With the decay of the continuum, the amplitude of the interferences decreases until, in the end, only the soliton is still visible.

To obtain shorter soliton pulses, we could choose to reduce the amount of dispersion, because for a constant pulse energy this leads to a shorter soliton pulse (Fig. 6.15). The result of the computer simulation is depicted in the lower panels in Fig. 6.17. As can already be seen from the first 1000 roundtrips, an even shorter pulse forms again very quickly. However, the dispersion is too small to broaden the constantly regenerated continuum fast enough to be absorbed in the loss modulator. Therefore, the continuum does not decay with time and leads to instabilities. After about 10 000 roundtrips, the continuum has grown so far that it spontaneously breaks the symmetry of the total pulse shape, which has been maintained until then. Subsequently, a longer satellite pulse forms, which propagates in front of the soliton.

This phenomenon can also be observed in the experiment, when the laser becomes unstable. So there is a limit to how short the soliton pulses can become and still remain stable against the continuum.

The continuum is a much longer continuum pulse which contains all the lost energy generated by the perturbation in the loss modulator, output coupler, and gain dispersion inside the laser resonator. Therefore, the continuum pulse has a much lower intensity and is no longer held together through the balance between SPM and negative GDD. This strongly broadened continuum pulse finally experiences higher loss through the AOM (acousto-optic modulator) and is therefore suppressed.

The steady-state solution ansatz for (6.117) in this case is given by a soliton pulse and a much weaker continuum pulse which can be treated as a perturbation to the soliton [see (4.47) in Sect. 4.3]:

$$\boxed{A(T,t) = \left(A_0 \operatorname{sech}\frac{t}{\tau}\right)\exp\left(\mathrm{i}\phi_0\frac{T}{T_\mathrm{R}}\right) + \text{continuum}}\,, \tag{6.118}$$

where ϕ_0 (4.52) is half of the nonlinear phase shift generated by SPM during one resonator roundtrip [see (4.56) in Sect. 4.3], A_0 is the amplitude of the soliton pulse envelope, and T_R the cavity roundtrip time.

The soliton perturbation theory can be used to show that, with sufficient negative dispersion, this continuum decays exponentially, and the solution of (6.117) is given by a soliton [95Kar2]. Sufficient negative dispersion means that, for the given pulse energy, a soliton can be formed. The pulse length of an ideal soliton is given by [see (4.57) in Sect. 4.3]

$$\tau = \frac{4\,|D|}{F_\mathrm{p,L}\delta}\,, \tag{6.119}$$

where D is the dispersion parameter [see (4.39) in Sect. 4.3], δ is the SPM coefficient [see (4.5) in Sect. 4.3], and $F_\mathrm{p,L}$ is the pulse fluence (i.e., energy density) inside the laser material, assuming SPM is generated in the laser gain crystal. According to Table 2.3 the group delay dispersion (GDD) is given by the second order dispersion with $\phi = k_\mathrm{n} L_\mathrm{d}$:

$$\mathrm{GDD} = \frac{\mathrm{d}^2\phi}{\mathrm{d}\omega^2} = \frac{\mathrm{d}^2 k_\mathrm{n}}{\mathrm{d}\omega^2} L_\mathrm{d} = 2D\,.$$

Equation (6.119) and Fig. 6.15 suggest that, with ever smaller GDD, the pulse length can be shortened indefinitely. However, there is a limit to this, and the soliton perturbation theory can give an analytical solution which specifies how far the dispersion can be reduced before instabilities occur. The minimum normalized dispersion $D_\mathrm{n,min}$ is given by [95Kar2]

$$\boxed{D_\mathrm{n,min} = \frac{|D|}{D_\mathrm{g}} = \frac{2}{9}\sqrt[3]{\frac{(9\phi_0/2)^2}{D_\mathrm{g} M_\mathrm{s}}}}\,, \tag{6.120}$$

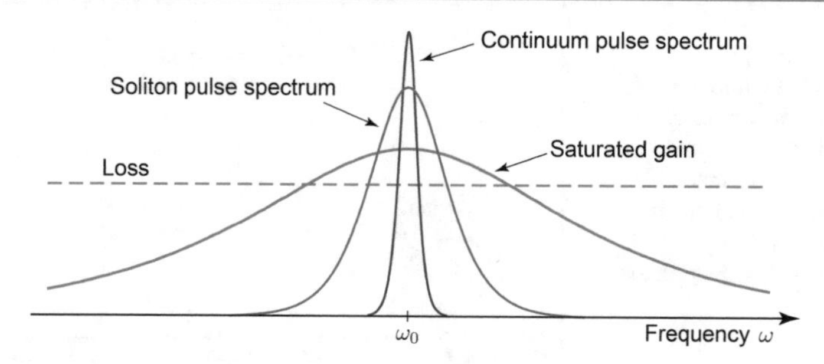

Fig. 6.18 The gain is saturated by the soliton. However, since the continuum is spectrally narrower, it will see a higher gain and will increase. Additional loss is only introduced through the loss modulator, which prevents the soliton from becoming unstable

with M_s given in (6.58), and therefore the minimum FWHM pulse duration with (4.57) is

$$\boxed{\tau_{\min} = 1.763 \sqrt[6]{\frac{2D_g^2}{9\phi_0 M_s}}} \,. \tag{6.121}$$

In comparison, the FWHM pulse length τ_p without GDD and SPM is given by (6.61) as

$$\tau_p = 1.665 \sqrt[4]{\frac{D_g}{M_s}} \,. \tag{6.122}$$

The pulse-shortening factor $R_{\max}$ is then given by (6.121) and (6.122) as

$$R_{\max} = \frac{\tau_p}{\tau_{\min}} = \frac{1.665}{1.763} \sqrt[12]{\frac{(9\phi_0/2)^2}{D_g M_s}} = \frac{1.66}{1.76} \sqrt[4]{\frac{9}{2} D_{n,\min}} \,. \tag{6.123}$$

In contrast to active modelocking with negligible SPM and negative GDD, the pulse envelope is now determined by a soliton, i.e., we obtain an unchirped sech^2 pulse shape instead of a chirped Gaussian pulse shape.

A physically descriptive explanation of this result can be summarized as follows. The solution is a soliton from the first part of the master equation (6.117). This soliton is perturbed by the second part of the master equation, i.e., gain, loss, gain dispersion, and the modulator. The ansatz then is a soliton and a perturbation pulse, which contains the energy that does not fit into the soliton pulse. Because the second part of the equation gives a much weaker contribution, this perturbation pulse is very weak. The relatively weak perturbation pulse is very quickly broadened by GDD, because the intensity of this perturbation pulse is not sufficient to maintain a constant pulse duration through the balance with SPM. The perturbation pulse experiences dispersive pulse broadening, becomes increasingly longer, and is therefore referred to as 'the continuum'. The continuum has a narrow spectrum and therefore experiences a larger amplification than the broadband soliton (see Fig. 6.18). What prevents the continuum from increasing further is that the longer continuum pulse experiences

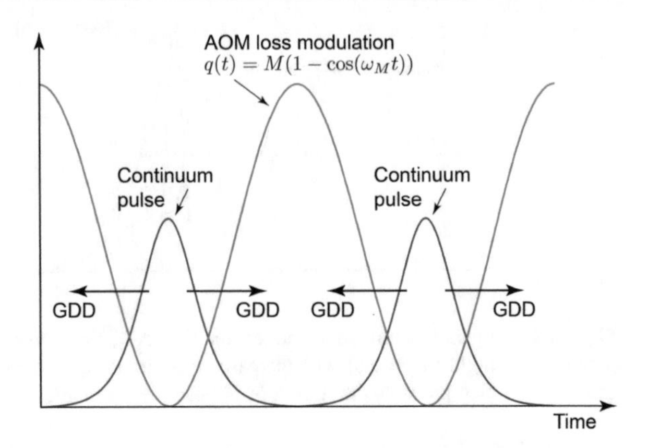

Fig. 6.19 Through dispersion but without SPM, the continuum is broadened very quickly. The continuum, which becomes longer and longer with GDD alone, is finally suppressed by the acousto-optic loss modulator

more loss by the loss modulator (Fig. 6.19). Therefore, soliton modelocking alone without a loss modulator (or a saturable absorber, as described in Chap. 9 for passive modelocking) would be unstable.

Therefore the soliton is stable as long as the soliton loss l_s is smaller than the continuum loss l_c. Both can be determined from soliton perturbation theory. The soliton loss is given by [see [95Kar2], (3.2)]

$$\boxed{l_s = \frac{D_g}{3\tau^2} + \frac{\pi^2}{24} M\omega_m^2 \tau^2} \,. \tag{6.124}$$

The gain–loss balance is then given by $g = l + l_s$, where l is all the rest of the cavity loss (e.g., output coupler). The first part of the soliton loss is due to the finite gain bandwidth which effectively broadens the soliton pulse, and the second part is the loss in the loss modulator. This second part is very similar to the gain–loss balance equation for active modelocking without SPM and GDD (6.63):

$$g = l + \frac{1}{2} M\omega_m^2 \tau^2 \,.$$

However, for very short soliton pulses the modulator loss has hardly any influence on the soliton pulse and can be considered negligible. This is in contrast to the much longer continuum pulse which experiences strong loss in the modulator.

6.6.2 Soliton Modelocked Nd:glass Laser Stabilized with an Intracavity AOM

Active modelocking of Nd:glass lasers usually results in pulse trains with a pulse length of 5 to 10 ps [86Yan, 88Bas, 92Hug]. We have generated pulses as short as 310 fs from actively modelocked Nd:phosphate and silicate lasers [94Kop2] (Fig. 6.20). The pulses have become so short that the loss modulation had to be synchronized to the arrival time of the modelocked pulses to keep the short pulse

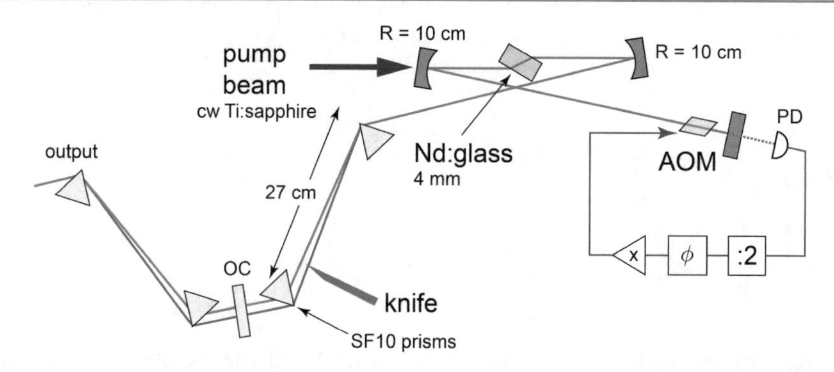

Fig. 6.20 Standard cavity design for femtosecond pulse generation with a solid-state laser. Here an Nd:glass plate, 4 mm thick, is inserted at the Brewster angle. The Nd:glass laser, operates with SPM in the gain material and negative GDD introduced with the prism pairs. The acousto-optic modelocker (AOM) is synchronized to the pulse repetition rate of the laser resonator with a photodiode (PD). Output coupler OC, radius of curvature *R* [94Kop2]

centered at the minimum loss modulation. As shown in Fig. 6.20 the drive signal for the AOM is generated from the pulse train itself, and this is referred to as regenerative active modelocking [92Kaf]. This is necessary in this short pulse duration regime because the modulator is no longer the dominant pulse-forming mechanism and any pulse jitter with respect to the modulation minimum introduces additional perturbation to the soliton modelocking.

For the Nd:phosphate glass laser, net gain reshaping was introduced by a knife edge inserted between the two prisms as shown in Figs. 6.20 and 6.21. Because of the angular dispersion between the prisms, this knife edge acts as an intracavity filter and flattens the gain profile (Fig. 6.21). This reduces the effect of spectral narrowing during the amplification in the gain medium, and soliton modelocking can shorten the pulses further. As shown in Fig. 6.22 soliton pulse shapes are obtained.

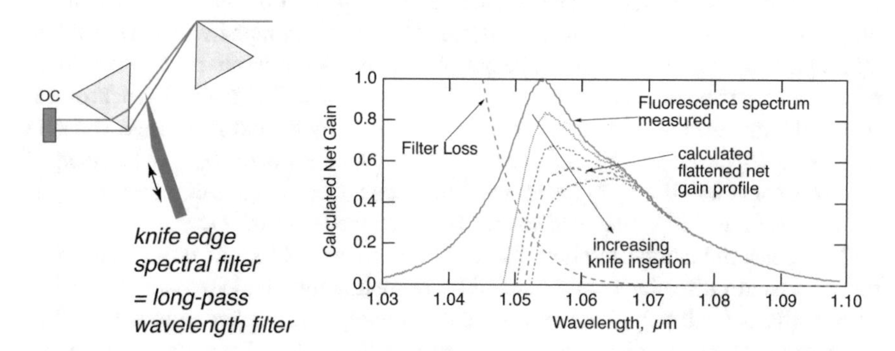

Fig. 6.21 *Left*: Gain reshaping by inserting a knife edge into the laser beam between the two prisms. Because of the angular dispersion the knife edge as shown here introduces more loss for the short wavelength part of the spectrum. *Right*: The measured fluorescence signal (*solid line*) of a quasi-homogeneously broadened Nd:phosphate glass is proportional to the saturated gain. Inserting the knife edge introduces a tunable long-pass wavelength filter which flattens the net gain profile over a bandwidth of approximately 10 nm [94Kop2]

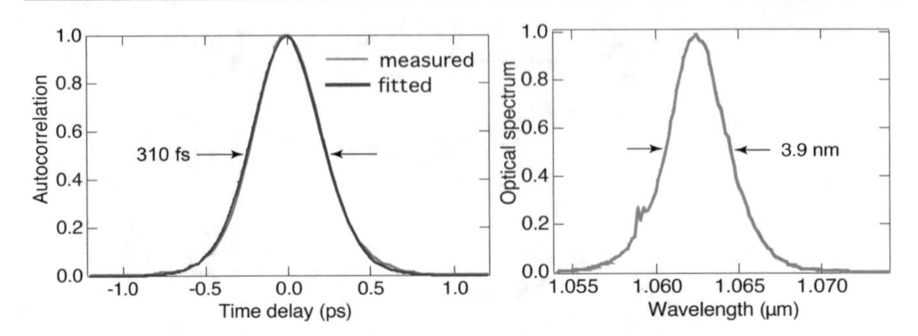

Fig. 6.22 Soliton modelocked Nd:phosphate glass laser started and stabilized with an acousto-optic modulator with some gain reshaping using a knife edge (see Fig. 6.21). Non-collinear measured autocorrelation trace fitted to an ideal sech^2 pulse shape of an ideal soliton. The time–bandwidth product is 0.32, close to the value 0.315 for a transform-limited soliton pulse (see Table 2.1). Absorbed power is 930 mW, output power 70 mW at 1% output coupling. Total cavity losses are 2%. The Nd:phosphate glass laser is quasi-homogeneously broadened (see Sect. 7.8) [94Kop2]

After flattening out the net gain profile (Fig. 6.21), we can estimate from calculations that the equivalent linewidth is increased by a factor of 10 to 20. Based on (6.61), the modelocked pulse duration would then be reduced from 14 ps (Nd:phosphate with $\Omega_g \approx 2\pi \times 2.1 \times 10^{-12}\,\mathrm{s}^{-1}$, $M = 0.01$, and $g = 0.01$) by a factor of 3 to 4.5, since it scales with the square root of the effective linewidth. This, however, does not explain the pulse duration of 310 fs (Fig. 6.22).

The formation of a stable pulse in an Nd:glass laser with a high gain bandwidth needs several hundred thousand resonator roundtrips, which is longer than for a Nd:YAG laser (Fig. 6.17). The computer simulations in Fig. 6.17 were conducted for a parameter set of an Nd:YAG laser with an approximately 20 times smaller gain bandwidth.

Soliton modelocking was observed with an Nd:phosphate glass laser as shown in Figs. 6.20 to 6.22 [94Kop2]. The calculated laser mode in the Nd:glass crystal has elliptical semi-major axes 30 µm × 50 µm. The laser is pumped with a Ti:sapphire laser at a wavelength of 800 nm. The negative dispersion is adjusted with two SF10 prisms with an apex-to-apex distance of 27 cm (see Chap. 3). To avoid synchronization problems, the acousto-optic modulator is driven by the output pulses. The gain dispersion D_g (Appendix F), and thus the normalized dispersion D_n, can be changed using a wavelength filter. By inserting a knife edge, the influence of gain filtering is reduced and the gain is flattened so that D_g is reduced (Fig. 6.21).

The measured autocorrelation is better fitted to an ideal sech^2 pulse shape than to a Gaussian pulse shape as predicted by normal active modelocking (Fig. 6.22). This confirms the theory that in this case the actively modelocked pulses with SPM and sufficient negative dispersion are given by soliton pulses. Furthermore, the measured pulse shortening agrees well with the prediction [see Fig. 6.23 and (6.123)] using the following laser parameters: $M = 0.01$, $f_{rep} = f_m = 240\,\mathrm{MHz}$, $g = 0.01$ ($2l_{tot} = 2\%$, whence $2g = 2l \approx 0.02$), $\Omega_g = 2\pi \times 4.0\,\mathrm{THz}$, $\phi_0 = 0.01116$, and $P_{out} = 70\,\mathrm{mW}$, with a mode radius in the gain material of 30 µm and 50 µm. Without GDD and SPM, one would predict a modelocked pulse duration of 10 ps from

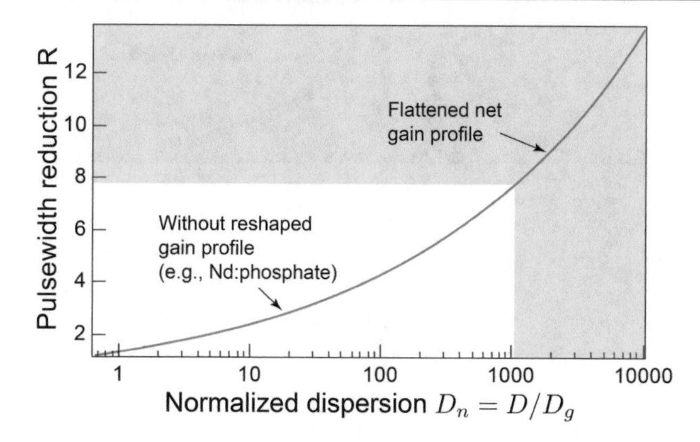

Fig. 6.23 Maximum pulse shortening $R_{\max}$ (6.123) as a function of the normalized dispersion $D_{\mathrm{n}} = |D|/D_{\mathrm{g}}$ (6.120). After [95Kar2]

(6.122). With the filter, the gain is flattened. We estimate a 5 to 10 times greater gain bandwidth. Therefore, we predict, again without GDD and SPM, a pulse length of about 4.5 ps ($5\Omega_{\mathrm{g}}$) to 3.2 ps ($10\Omega_{\mathrm{g}}$) using (6.122). With SPM and GDD, we would predict a minimum pulse length of 520 fs ($5\Omega_{\mathrm{g}}$) to 330 fs ($10\Omega_{\mathrm{g}}$) from (6.121). The normalized dispersion is approximately 2.5×10^3 (6.120) for 10-fold broadening, which corresponds to pulse shortening by a factor of about 10 (see Fig. 6.23), in good agreement with the final measurement of 310 fs (see Fig. 6.22).

Note that we obtained similar pulse durations with an inhomogeneously broadened Nd:silicate glass laser without any gain reshaping with the knife edge [94Kop2]. The theory here, however, assumes homogeneously broadened lasers. The beneficial effects of inhomogeneously broadened lasers for modelocking will be discussed next.

6.7 Modelocking with Homogeneously Versus Inhomogeneously Broadened Gain

So far we have always assumed a homogenously broadened laser gain. However, this is not the case for a standing wave laser cavity which is typically used for ultrafast solid-state lasers. Independently of the gain material, the standing wave intensity introduces effective inhomogeneous gain broadening. In this section we address the question of how this affects modelocking and the final pulse duration. This discussion also applies to passive modelocking. The good news is that generally inhomogenous gain broadening helps modelocking and supports shorter pulses. However, there are trade-offs which will ultimately depend on the specific laser parameters.

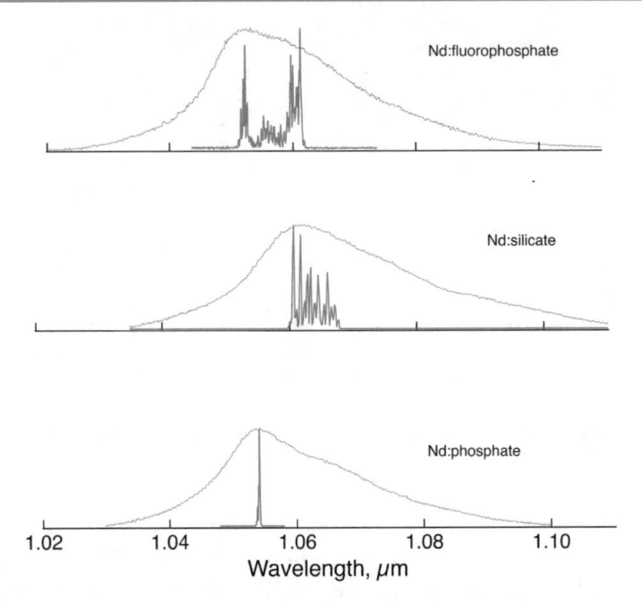

Fig. 6.24 Measured fluorescence spectra (*dotted curves*) showing the gain bandwidth of the laser material. Measured cw lasing output spectra (*solid curves*) for three different Nd:glass laser materials, reflecting the inhomogeneous broadening of the fluorophosphate and the silicate and the homogeneous broadening of the phosphate. From [95Kop1, Fig. 6.1]

6.7.1 Homogeneously Versus Inhomogeneously Broadened Gain: CW Lasing

For a homogeneously broadened ion-doped solid-state laser, the dopant ions that are responsible for the lasing transition are embedded in the same host crystal environment, such that all of them have the same emission linewidth. Typically these ions deform the crystal structure, which generates local electric fields that can shift the energy levels. Thus many solid-state lasers are phonon-broadened because the lattice vibrations also introduce some fluctuations into these local electric fields and therefore broaden the emission linewidth. The laser with such a homogeneously broadened gain material then lases only in one axial mode well above threshold. The bandwidth of this axial mode is typically much narrower than the phonon-broadened gain spectrum and determined by the Q-factor of the laser cavity. For example, the gain broadening of the Ti:sapphire, Nd:YAG, and Yb:YAG lasers is homogeneous.

Some Nd:glass lasers, such as the Nd:phosphate laser (Fig. 6.24), have been optimized for quasi-homogeneous behaviour. Other glass materials such as silicate and fluorophosphate, however, are inhomogeneously broadened where spectral hole burning is taking place and the laser lases in more than one axial mode well above threshold (Fig. 6.24). There will be significant mode-beating noise for inhomogeneously broadened lasers, so homogeneous broadening is clearly an advantage for stable cw lasing.

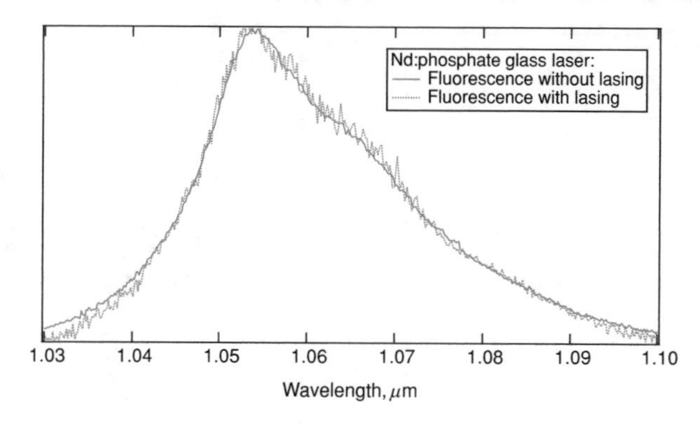

Fig. 6.25 Measured normalized fluorescence spectra show the gain bandwidth of a homogeneously broadened laser material with lasing (*dashed line*) and without lasing (*solid line*) when the laser cavity is blocked at the same pump power. There is much more noise in lasing conditions because the inversion is clamped to a much lower level, well above threshold. We can clearly see the same wavelength-dependent gain spectra for homogeneous broadening

Homogeneous broadening only results in single axial mode operation because the wavelength-dependent gain spectrum is saturated without changing its shape (Fig. 6.25). This is also illustrated in Fig. 6.26, where we measured the fluorescence spectra in lasing conditions while tuning the Nd:phosphate laser. We measured the signal from the side of the crystal perpendicular to the laser mode which also shows the scattered lasing signal as it was tuned with an intracavity element. This scattered signal is due to impurities and defects in the laser crystal. The lasing signal is much stronger in comparison to the fluorescence spectra.

6.7.2 Modelocking Results Better for Inhomogeneous Gain Broadening

Now that we understand how modelocking works in the spectral domain (see Sect. 6.4.3 and Fig. 6.11), we expect to more easily modelock an inhomogeneously broadened laser material. This follows from Fig. 6.24, which shows that the cw lasing spectrum covers many different axial modes with inhomogeneous gain broadening, so that the amplitude modulation only needs to lock the phases of these modes and thus transfer less energy into additional modes via sideband generation. Indeed, this has been observed with a modelocked Nd:glass laser, where the gain material was changed from Nd:phosphate to Nd:silicate (see Fig. 6.20) [94Kop2]. For the homogeneously broadened Nd:phosphate laser material, we needed the additional gain flattening with the knife edge (Fig. 6.21) to obtain pulses as short as 310 fs. However, using the inhomogeneously broadened Nd:silicate laser, no intracavity knife-edge filter was required, and pulses as short as 330 fs have been generated.

Therefore we can draw the following conclusion: although the modelocking theory derived up to now is only valid for homogeneously broadened gain materials, it should be noted that even shorter pulses can be generated in inhomogeneously broad-

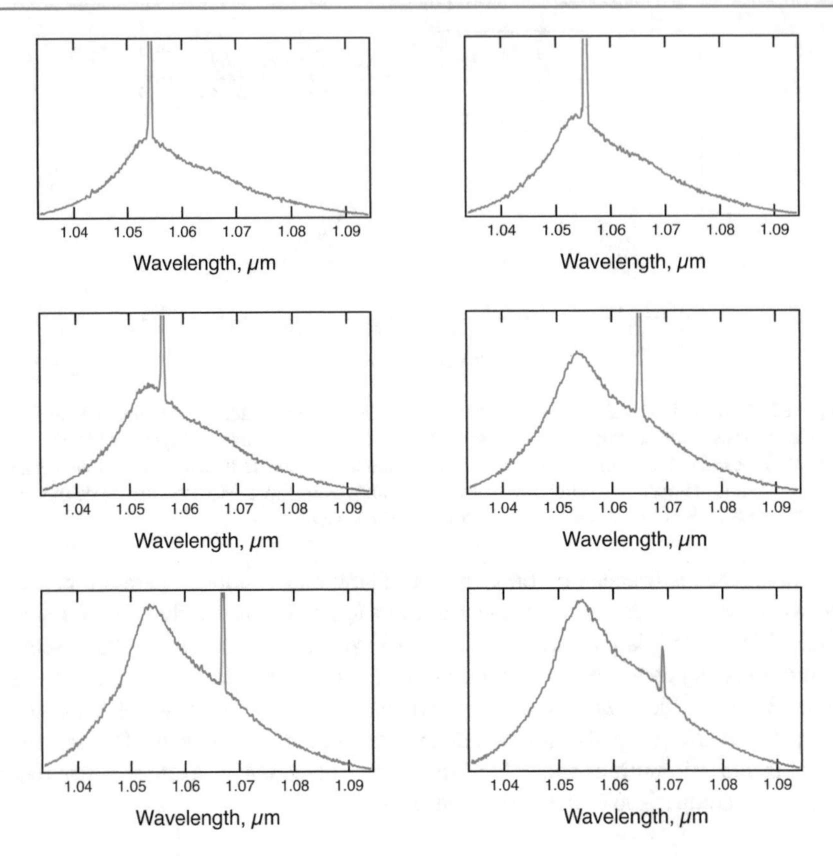

Fig. 6.26 Measured scattered spectra while tuning the homogeneously broadened Nd:phosphate laser. We clearly do not see any spectral hole burning during the tuning of the lasing wavelength

ened lasers. This is because, even in cw operation, the laser emits in a much broader spectrum (6.23) and the loss modulator does not need to transfer as much energy into the sidebands. This is equivalent to a stronger loss modulation or a smaller gain dispersion, which would result in shorter pulses, according to (6.37).

6.7.3 Modelocking with Spatial Hole Burning

In a standing wave laser cavity, even a homogeneously broadened laser becomes inhomogeneously broadened due to spatial hole burning. With spatial hole burning [95Bra2,95Kar3], the laser already oscillates in cw operation with a much broader spectrum. Considerably shorter pulses are thereby generated. For the modelocking theory, one can assume a smaller gain dispersion D_g by replacing the gain bandwidth with the bandwidth of the cw spectrum. We then obtain a pulse shortening based on (6.37). A detailed description is given in [95Bra2] and [95Kar3].

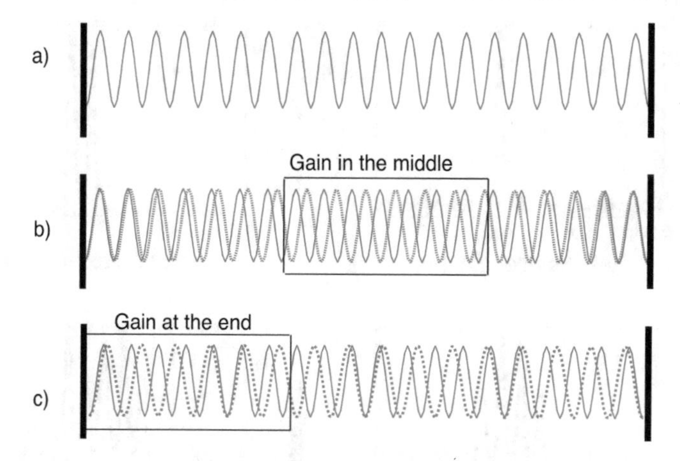

Fig. 6.27 Schematic description of enhanced spatial hole burning. **a** Standing wave pattern of one axial mode inside a linear laser cavity. **b** An adjacent axial mode fills in the undepleted gain regions of a gain-in-the-middle (GM) cavity. **c** The axial mode that can fill in the undepleted gain regions of a gain-in-the-end (GE) cavity has a much larger frequency difference than the first mode

We have systematically investigated the difference between both actively and passively modelocked lasers with gain-at-the-end (GE) and gain-in-the-middle (GM) for the example of Nd:YLF lasers (Fig. 6.27) [95Bra2]. The GE laser generates pulse widths approximately three times shorter than a comparable GM cavity. This is due to enhanced spatial hole burning, which effectively flattens the saturated gain and allows for a larger lasing bandwidth compared to a GM cavity (Figs. 6.28 and 6.29).

We first investigated enhanced spatial hole burning by measuring the cw mode spectrum, and observed that the mode spacing in GE cavities depends primarily on the crystal length, which was also confirmed with an Nd:LSB crystal where the pump absorption length was significantly shorter than the crystal length. In modelocked operation, pulse widths of 4 ps for passive modelocking and 5 ps for active modelocking have been demonstrated with GE cavities, compared to 11 ps for passive and 17 ps for active modelocking with GM cavities.

Additionally, the time–bandwidth product for the GE cavity is approximately twice the ideal product for a sech^2 pulse shape, while the GM cavity has nearly ideal time–bandwidth limited pulses. The results for the GM cavity compare well with existing theories, taking into account the added effect of a pump-power dependent gain bandwidth, which increases the bandwidth of Nd:YLF from 360 GHz to more than 500 GHz. A rigorous theoretical treatment of the effects due to spatial hole burning is presented in [95Kar3].

We can give a simplified schematic picture as to why a laser cavity with gain-at-the-end (GE) experiences 'enhanced' spatial hole burning compared to laser cavities with gain-in-the-middle (GM). A laser with an ideal homogeneously-broadened gain medium in a uni-directional ring laser using for example an intracavity Faraday isolator will only lase in one axial mode, and experiences no spatial hole burning (Fig. 6.29).

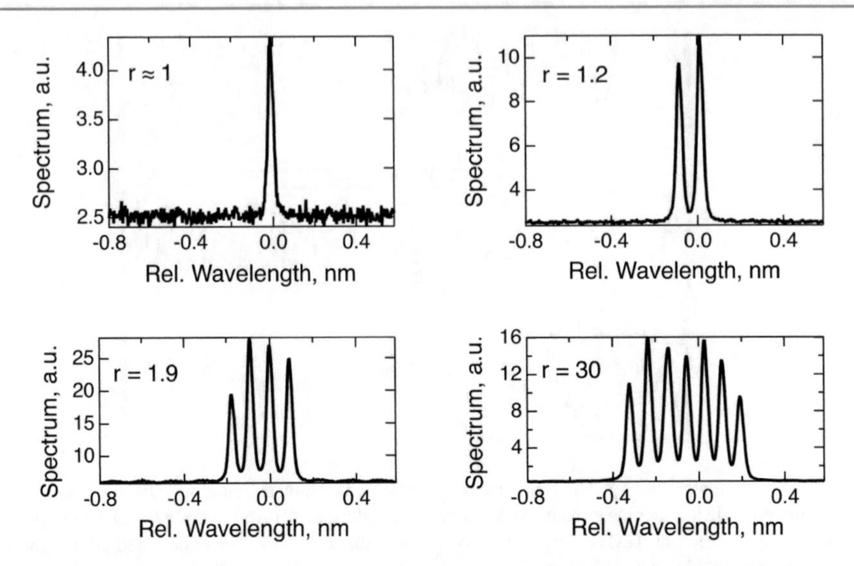

Fig. 6.28 Enhanced spatial hole burning in a GE Nd:YLF laser cavity which increases inhomogeneous gain broadening. As the pump power is increased, we increase the pump parameter r, defined by how many times the laser is pumped above threshold, and increase the number of axial modes in cw lasing [95Bra2]

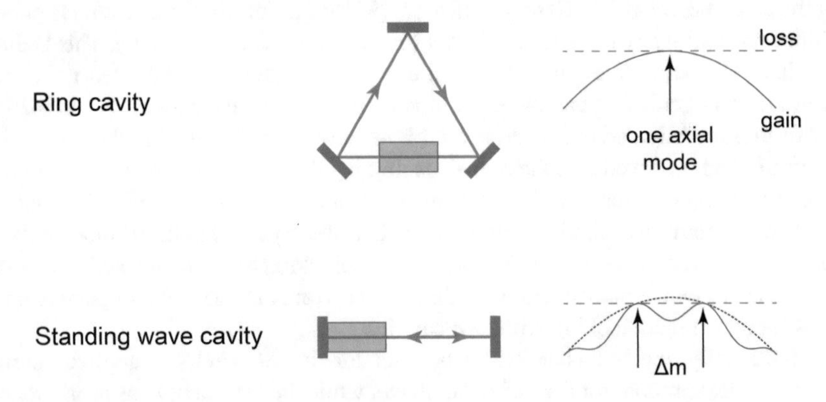

Fig. 6.29 In a ring cavity, the laser is ideally homogeneously broadened, and the first axial mode to lase saturates the entire gain spectrum equally to "gain equals loss." In a standing-wave cavity, the spatial hole burning causes an effective periodic modulation of the gain because additional axial modes can fill in the undepleted gain regions. As the gain medium is moved closer to the end of the cavity (see Fig. 6.26), the spacing Δm between lasing modes increases [95Bra2]

For a standing wave cavity, the first mode to lase sets up a standing wave in the laser gain (Fig. 6.27). A second axial mode will then see a space-dependent gain, with lower gain at the peaks of the standing wave due to gain saturation, and higher gain at nodes of the standing wave, where the gain is unsaturated. The boundary condition at the end of the standing-wave cavity requires a common node for all axial modes. Then, with a gain medium in the middle of the laser cavity and one mode running, the intensity modulation of the next adjacent axial mode will be exactly 180° out of phase with the first mode in the middle of the cavity, and experiences the largest spatial overlap with the remaining undepleted gain (Fig. 6.27b). With gain-in-the middle, then, adjacent axial modes will typically lase and deplete the available gain.

However, the frequency separation between modes that fill the available spatial holes increases as the gain region is moved from the middle of the cavity to the end of the cavity (Fig. 6.27c). It is a well-known effect that the position of the gain affects the standing-wave pattern and this has been used to increase the performance of single-frequency microchip lasers [90Zay1]. With the gain medium at the end mirror, the next axial mode to lase has therefore a much larger frequency difference compared to gain-in-the-middle. In addition, the more closely the gain medium is placed to the cavity end, the more modes are required to fully deplete the gain due to the common boundary condition, i.e., the common node, at the end of the cavity.

For this GE case, we have experimentally and theoretically demonstrated that pumping the laser far above threshold results in a much broader overall cw-lasing bandwidth with a large frequency separation between the lasing modes which ultimately allows for shorter pulse generation, as shown in Fig. 6.28, for which the laser spectrum is measured using a scanning Fabry–Pérot interferometer. The Nd:YLF laser with the GE cavity has a broad cw spectrum consisting typically of 6–8 longitudinal modes, each spaced by approximately 23 GHz (see Fig. 6.28). The number of lasing longitudinal modes increases with increasing pump power. In contrast the GM-cavity laser always shows a narrow spectrum which cannot be resolved by the same scanning Fabry–Pérot interferometer.

6.8 Synchronous Modelocking

Synchronous pumping provides another possibility for active modelocking, where the gain is modulated synchronously with the pulse roundtrip time. This technique has been used for lasers with shorter upper state lifetime than the cavity roundtrip time, i.e., $\tau_L < T_R$. Examples are dye lasers and semiconductor lasers. For this synchronous modelocking technique, another modelocked pump laser is used, which in the early days of dye lasers was a modelocked argon or Nd:YAG laser. Since 1992, it has also been possible to pump with shorter pulses (10–100 fs), because a modelocked Ti:sapphire laser has a sufficiently high output power to synchronously pump another laser.

The basic principle is shown in Fig. 6.30. The pump laser of length L delivers a pulse train with a pulse separation given by $2L/\upsilon_g$, where υ_g is the group velocity of the pulses inside the modelocked laser. This periodic pulse train generates a gain

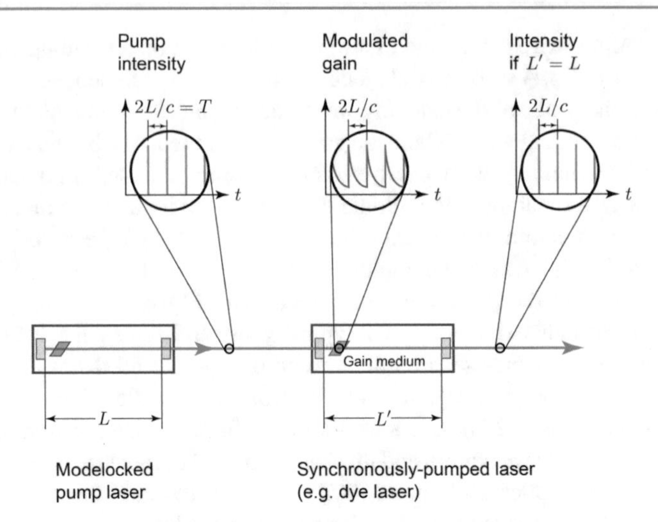

Fig. 6.30 Schematic description of synchronous modelocking [88Kai, p. 8, Fig. 2.4]. A modelocked pump laser delivers a short pulse train with a period $2L/\upsilon_g$. This pump pulse train generates a periodic modulation of the gain in the dye medium. If the roundtrip of the pulses in the dye laser is synchronized to this modulation (i.e., $L' = L$), short dye-laser pulses are generated. Here it was assumed that $\upsilon_g = c$

modulation with the same period $2L/\upsilon_g$. The gain modulation $g(t)$ is characterized by a steep rise, which is essentially given by the pump pulse duration, and a slower decay, which is determined by the lifetime of the excited laser level τ_L. Essential for this technique of synchronous pumping is the choice of the resonator length L' of the dye laser. If the optical cavity lengths $L' = L$ are chosen equal with an active electronic feedback loop, then this yields a synchronization of the resonator roundtrip time with the gain modulation. This in turn leads to pulses which always arrive in the dye medium in the proximity of the maximum gain. In this way, by stimulated emission, the gain is depleted very fast, resulting in a significant modulation of the gain $g(t)$.

Synchronously pumped dye lasers have a multitude of advantages. They generate stable pulses with a duration of about 2 to 3 ps, even when they have been pumped with > 100 ps pump pulses, and they are completely tunable across the gain bandwidth of the dye material. Typical commercially available dyes can emit in the region of about 400 nm to 1.3 μm, and they can be pumped with modelocked argon ion lasers as well as with modelocked Nd:YAG lasers. Today they are no longer used because solid-state lasers can produce broadly tunable pulses in the near-infrared and, because of their large output power, the accessible spectrum can be extended with nonlinear conversion such as with optical parametric amplifiers and oscillators.

Another advantage of synchronously pumped dye lasers is that they can be synchronized temporally to other synchronously pumped dye lasers, as long as they are optically pumped by the same pump laser [77Jai]. The only disadvantage is that, with conventional pump lasers and pump pulse durations of 50 to 150 ps, the pulse duration cannot be reduced under 1 ps. This disadvantage can be eliminated by a combination of passive modelocking and synchronous pumping. The ability to synchronize is preserved, but the tunability of the laser is typically lost again.

Another advantage is that for two synchronously pumped dye lasers the pulse repetition rate can be slightly detuned while still obtaining stable pulse generation. This allowed for equivalent time sampling measurements with two pulse trains with a well defined difference in pulse repetition rate. See the more detailed discussions in Sects. 9.9 and 11.6.4.

Today synchronous modelocking is not so important anymore, because high-power ultrafast solid-state lasers and nonlinear frequency conversion can cover a spectrum from THz to hard X-rays, and pulse durations reaching even the attosecond regime. In addition, all these laser sources can be stabilized with respect to each other to very high accuracy (see Chaps. 11 and 12). Furthermore, new concepts for dual-comb modelocking allow for the generation of two pulse trains with an adjustable pulse repetition rate difference from a single laser cavity (Sect. 9.9).

6.9 Selected Results of Active Modelocked Solid-State Lasers

Table 6.2 summarizes some active modelocking results with different solid-state materials. Active modelocking is no longer a very active field of research because SESAM modelocking generates shorter pulses and better stability and is much easier to implement. Today most industrial ultrafast diode-pumped solid-state lasers are based on SESAM modelocking (see Chaps. 7 and 9 for more detail). Note that we have not discussed actively modelocked semiconductor or integrated waveguide lasers. These lasers are beyond the scope of this book.

Note that the analytical solutions for the different theoretical models discussed in this chapter for active modelocking are also summarized together with passive modelocking in Sect. 9.6.

Table 6.2 Summary of a selection of active modelocked solid-state lasers using different solid-state laser materials. Modelocking ML, acousto-optic modulator AOM, electro-optic phase modulator EOM, frequency modulation FM, center lasing wavelength λ_0, measured pulse duration at FWHM τ_p, average output power $P_{av,out}$, pulse repetition rate f_{rep}

Laser material	ML technique	λ_0	τ_p	$P_{av,out}$	f_{rep}	Remarks	Reference
Ti:sapphire ($Ti^{3+}:Al_2O_3$)	Active AOM	814 nm	150 fs	600 mW	80.5 MHz		[91Cur],
		780 nm	1.3 ps	2 W			[92Kaf]
Cr^{2+}:ZnSe	Active AOM	2.5 μm	4.4 ps	82 mW	81 MHz		[00Car]
Nd:YAG ($Nd^{3+}:Y_3Al_5O_{12}$)	Active AOM	1.064 μm	25 ps		100 MHz	Lamp-pumped: pulse shortening due to intracavity etalon	[86Ros]
	Active FM	1.064 μm	12 ps	65 mW	350 MHz		[89Mak1]
	Active EOM	1.32 μm	8 ps	240 mW	1 GHz		[91Zho]
	Active AOM	1.32 μm	53 ps	1.5 W	200 MHz	Lamp-pumped and harmonic modelocked	[88Kel]
Nd:YLF ($Nd^{3+}:LiYF_4$)	Active AOM	1.053 μm	37 ps	6.5 W	100 MHz	Lamp-pumped	[87Bad]
			18 ps	12 mW	230 MHz		[89Mak2]
		1.047 μm	9 ps	150 mW	500 MHz		[90Kel1]
			7 ps	135 mW	2 GHz		[90Wei]
			6.2 ps	20 mW	1 GHz	Ti:sapphire laser pumped	[90Wal]
	Active EOM	1.053 μm	13 ps	350 mW	5.4 GHz	Ti:sapphire laser pumped	[91Sch2]
			4 ps	400 mW	237.5 MHz		[92Wei]
			4.5 ps	400 mW	2.85 GHz		[92Wei]
		1.3 μm	8 ps	240 mW	1 GHz		[91Zho]
	Active MQW	1.047 μm	200 ps	27 mW	100 MHz	Semiconductor multiple quantum well (MQW) modulator	[95Bro3]
	Active piezoelectric diffraction modulator	1.047 μm	7 ps	160 mW	160 MHz		[90Juh]

(continued)

Table 6.2 (continued)

Laser material	ML technique	λ_0	τ_p	$P_{av,out}$	f_{rep}	Remarks	Reference
Nd:BEL ($Nd:La_2Be_2O_5$)	Active AOM	1070 μm	7.5 ps	230 mW	250 MHz		[91Li]
	Active FM	1070 μm	2.9 ps	30 mW	238 MHz	Harmonic modelocking	[91God]
			3.9 ps	30 mW	20 GHz	Harmonic modelocking	[91God]
Nd:YAP ($Nd:Y_3AlO_3$)	Active AOM	1.08 μm	10 ps	220 mW			[96Guy]
Nd:KGW ($Nd:Y_3AlO_3$)	Active FM	1.067 μm	12 ps		200 MHz		[95Flo]
Nd:glass (Nd phosphate)	Active AOM	1.054 μm	7 ps	20 mW		Ar-ion laser pumped	[86Yan]
		1.054 μm	≈ 10 ps	0.3 mW			[88Bas]
		1.054 μm	9 ps	30 mW	240 MHz		[92Hug]
		1.063 μm	310 fs	70 mW	240 MHz	Ti:sapphire laser pumped, regeneratively actively modelocked	[94Kop2]
	Active FM	1.054 μm	9 ps	14 mW	235 MHz		[91Hug]
Er:Yb:glass	Active AOM	≈ 1.53 μm	90 ps	7 mW	100 MHz		[94Cer]
	Active FM	1.53 μm	9.6 ps	3 mW	2.5 GHz	3rd harmonic modelocking	[94Lon]
		≈ 1.5 μm	9.6–30 ps	3 mW	2.5 and 5 GHz	3rd harmonic modelocking	[95Lap]
		1.534 μm	48 ps	1 mW	5 GHz	Dual wavelength with 165 GHz separation	[98Lon]

7 Saturable Absorbers for Solid-State Lasers

7.1 Introduction

In 1992, a new class of practical ultrafast picosecond and femtosecond lasers became possible through the invention of the semiconductor saturable absorber mirror (SESAM) device [92Kel2,96Kel]. This invention solved a difficult 25-year-old technical challenge to demonstrate stable operation of passively modelocked diode-pumped solid-state lasers and allowed for ultrafast lasers to move from complicated laboratory devices to reliable turnkey systems in industrial applications. These lasers allow for many new and relevant applications, for example, in precision material processing, processing of materials ranging from very hard and difficult to work with materials (ceramics, carbides, etc.) to mainstream semiconductor devices (wafer scribing and dicing), to novel mixed materials and thin-film technologies (new organic light-emitting diode materials for electronic displays used in mobile devices and TVs, new types of photovoltaic devices to accelerate the adoption of fossil-fuel free energy), medical and biomedical applications such as multiphoton imaging, precise, low-damage tissue ablation used in many types of corrective eye surgery in ophthalmology, precision diagnostics and measurements in general and frequency comb applications more specifically.

Passive modelocking will be discussed in more detail in Chap. 9. A passive modelocked laser produces a constant train of nearly ideal light pulses with picosecond or femtosecond durations, without requiring an external modulator signal as demonstrated with active modelocking (see Chap. 6). Instead of an active modulator, a saturable absorber generates a self-amplitude modulation when a short pulse passes through. Passive modelocking was first demonstrated in 1966 using a dye saturable absorber [66DeM], only six years after the first laser was demonstrated. However, these early techniques suffered from a fundamental problem: self-Q-switched modelocking behavior, where a large overlying Q-switched pulse modulated the train of much shorter modelocked pulses. This meant that the laser would turn itself off at

© The Author(s), under exclusive licence to Springer Nature Switzerland AG 2021

U. Keller, *Ultrafast Lasers*, Graduate Texts in Physics,
https://doi.org/10.1007/978-3-030-82532-4_7

regular intervals. The underlying modelocked pulses could only be used in special circumstances and very limited applications.

Early theory suggested that continuous wave (cw) passive modelocking of solid-state lasers, without this self-Q-switching, would be difficult or even impossible. For example, the theoretical model by Haus in 1976 predicted that 'stable modelocking is unachievable' and 'steady-state modelocking is prevented by relaxation oscillations' [76Hau, p. 174].

We solved this long-standing challenge with the invention of the SESAM. Since then, we have developed the theoretical underpinnings of the performance of SESAMs in solid-state lasers, worked out design guidelines for application to practical laser systems, and taken this knowhow to demonstrate unprecedented laser performance [10Kel]: shortest pulses (below 6 femtoseconds [99Sut,00Mat]), highest average power and pulse energy direct from a modelocked laser oscillator (350 watts [19Sal] and 80 microjoules [14Sar]), and highest pulse repetition rate (160 gigahertz [02Kra2]).

Fundamental work on the reliability of these SESAM devices was also critical for commercial applications. We performed a systematic study of lifetime and damage of SESAMs with an understanding of the underlying processes. New SESAM design guidelines with improved performance enabled new world-record power and pulse energy performance of ultrafast laser oscillators. The simplicity of SESAM modelocked lasers, combined with diode-laser-pumped schemes developed during the 1990s, has resulted in many new, practical, commercially available ultrafast laser systems. These laser systems are being used extensively in many relevant applications, where expensive, power-hungry, maintenance-intensive lasers are being replaced. Today the SESAM technology has established itself as a key enabling approach for ultrafast lasers, and most if not all commercially available ultrafast lasers use SESAM technology in industrial-compatible laser systems. The SESAM approach has also been adopted across other technology platforms such as fiber lasers and optically-pumped semiconductor lasers.

In this chapter we develop a theoretical framework for the important saturable absorber parameters for slow and fast saturable absorbers independent of the saturable absorber material. We will derive model functions with important saturable absorber parameters that we will use later for passive pulse generation. These parameters need to be adapted to the different laser gain materials and cavity designs to obtain stable pulse generation. Semiconductor saturable absorbers are ideally suited for ultrafast solid-state lasers and therefore we shall focus on this class of saturable absorbers in this chapter. First we give a short introduction to the required semiconductor physics background and then review the specific SESAM device design. At the end we will give a very brief outlook for other interesting saturable absorber materials.

7.2 Slow and Fast Saturable Absorbers

7.2.1 Saturable Absorber Parameters and Rate Equation

A saturable absorber is a material which has decreasing light absorption with increasing light intensity and pulse fluence. The pulse fluence is the pulse energy per laser mode area. Most materials show some saturable absorption, but often only at very high pulse fluences. We typically need saturable absorbers which show this effect at intensities and pulse fluences typical in solid-state laser cavities and do not introduce Q-switching instabilities, which will be discussed in more detail in Sect. 9.7. For example saturable absorber dye cells introduced Q-switching instabilities with solid-state lasers. We need lower saturation fluence to prevent such instabilities, and semiconductor saturable absorbers are ideally suited for this because of their high gain cross-section (see Sect. 4.5). The absorber cross-section σ_A for semiconductor materials is about two orders of magnitude greater than for dyes (see Table 4.4), and therefore the material saturation fluence $F_{sat,A}$ is much lower. According to (4.90) we obtain

$$F_{sat,A} = \tau_A I_{sat,A} = \frac{h\nu}{\sigma_A}, \tag{7.1}$$

where τ_A is the absorber recovery time, $I_{sat,A}$ the absorber saturation intensity, and $h\nu$ the photon energy. Note that sometimes we drop the subscript A when it is obvious in the context of the discussion. In addition, semiconductor materials can be designed to absorb over a broad range of wavelengths (from the visible to the mid-infrared).

We typically operate a diode-pumped solid-state laser in a standing wave cavity, where we place the saturable absorber at one end mirror of the laser cavity. Epitaxial growth techniques allow for wafer-scale integration of the semiconductor saturable absorber into or on top of a highly reflective mirror structure. We then obtain one path through the saturable absorber per cavity roundtrip and only one modelocked pulse per cavity roundtrip. We can control the absorption recovery dynamics with different epitaxial growth parameters (see Sect. 7.4) and we can change the fundamental material properties when we integrate the saturable absorber with a mirror device structure (see Sect. 7.4). This means that fundamental material properties such as τ_A and $F_{sat,A}$ (7.1) can be substantially modified. Such designed saturable absorber properties of SESAMs have made it possible to both passively Q-switch and modelock many lasers.

Some important saturable absorber parameters are summarized in Table 7.1. These parameters can be measured as discussed in more detail in Sect. 7.7 and are shown here with an example in Fig. 7.1. The basic parameters for a saturable absorber device are its wavelength range (where it absorbs), its dynamic response (how fast it recovers), and its saturation intensity and fluence (at what intensity or pulse fluence it starts to saturate). In our notation, we assume that the saturable absorber is integrated with a mirror structure, so we are interested in the nonlinear reflectivity change.

We distinguish between the time- and fluence-dependent nonlinear reflectivity $R(F_{p,A}, t)$ and the time-integrated (i.e., pulse-averaged) nonlinear reflectivity $R(F_{p,A})$, where $F_{p,A}$ is the incident pulse fluence on the saturable absorber. If the

Table 7.1 Saturable absorber parameters with their defining equations and units without taking into account inverse saturable absorption (ISA). For additional ISA see Sect. 7.3.5. Note that sometimes we drop the subscript A when it is obvious in the context of the discussion

Quantity	Symbol	Defining equation or measurement	Unit
Saturation fluence	$F_{\text{sat,A}}$	Measurement $R(F_{\text{p,A}})$ or $T(F_{\text{p,A}})$	J/cm^2
Recovery time	τ_{A}	Measurement $R(t)$ or $T(t)$	s
Incident beam area	A_{A}	Measurement	cm^2
Saturation energy	$E_{\text{sat,A}}$	$E_{\text{sat,A}} = A_{\text{A}} F_{\text{sat,A}}$	J
Saturation intensity	$I_{\text{sat,A}}$	$I_{\text{sat,A}} = F_{\text{sat,A}}/\tau_{\text{A}}$	W/cm^2
Modulation depth	ΔR or ΔT	Measurement $R(F_{\text{p,A}})$ or $T(F_{\text{p,A}})$	
Nonsaturable loss	ΔR_{ns} or ΔT_{ns}	Measurement $R(F_{\text{p,A}})$ or $T(F_{\text{p,A}})$	
Incident pulse energy	E_{p}	Measurement	J
Incident pulse fluence	$F_{\text{p,A}}$	$F_{\text{p,A}} = E_{\text{p}}/A_{\text{A}}$	J/cm^2
Incident intensity	$I_{\text{A}}(t)$	$F_{\text{p,A}} = \int I_{\text{A}}(t)\text{d}t$	W/cm^2

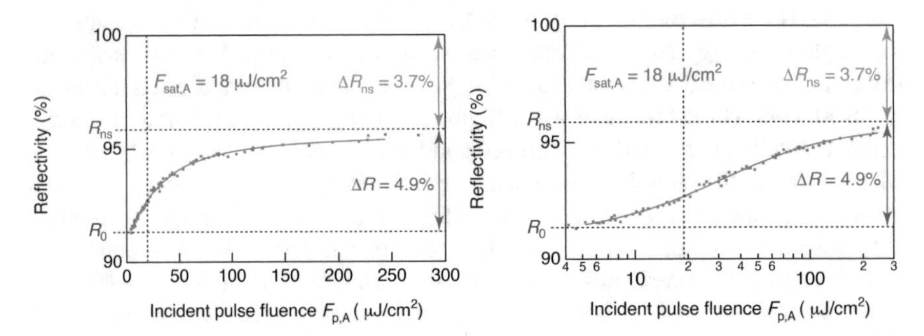

Fig. 7.1 Example of the pulse-averaged nonlinear reflectivity $R(F_{\text{p,A}})$ of a SESAM on a linear scale (*left*) and a logarithmic scale (*right*) for a given pulse duration. The pulse fluence $F_{\text{p,A}}$ is the incident pulse energy per unit surface area on the SESAM. A saturable absorber is integrated with an otherwise highly-reflecting mirror. The reflectivity is thereby increased by the saturation of the absorber. The nonsaturable loss $\Delta R_{\text{ns}} = 1 - R_{\text{ns}}$ should be kept as small as possible

saturable absorber is used in transmission, we simply characterize the absorber by nonlinear transmission measurements, i.e., $T(F_{\text{p,A}}, t)$ and $T(F_{\text{p,A}})$. All the relevant saturable absorber parameters can be determined experimentally—in principle—without any need to understand the microscopic properties of the nonlinearities. Examples for both the saturation fluence $F_{\text{sat,A}}$ and the absorber recovery time τ_{A} are shown in Figs. 7.1 and 7.2. The relevant parameters of the saturable absorber depend not only on the material properties but also on the specific device structure in which the absorber is integrated, which gives significantly more design freedom (see Sect. 7.5). A better understanding of the underlying physics helps to better optimize these devices.

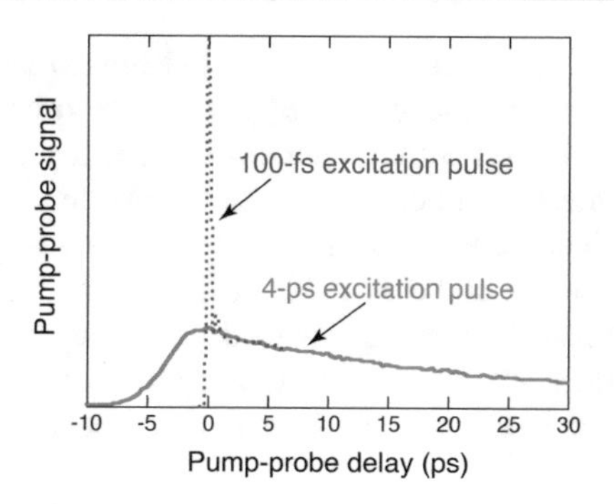

Fig. 7.2 Measured dynamic response of a SESAM which typically shows a bitemporal recovery for femtosecond pulses. The nonlinear reflectivity $R(t)$ is measured with a pump–probe technique and represents the impulse response of the saturable absorber mirror. Typically, this also depends on the incoming pulse fluence, i.e., $R(t) = R(F_{p,A}, t)$

Standard pump–probe measurement techniques determine the impulse response of the saturable absorber mirror $R(F_{p,A}, t)$ and therefore also the absorber recovery time τ_A (Fig. 7.2). In the picosecond regime, we typically only have to consider one picosecond recovery time, because much faster femtosecond nonlinearities in the saturable absorber give negligible modulation depth. This is shown for a semiconductor saturable absorber in Fig. 7.2, where the impulse response was measured for two different excitation pulse durations using a semiconductor saturable absorber. For excitation with a 4 ps pulse, the pump–probe trace clearly shows no significant modulation depth with a fast time constant. In the 100 fs pulse regime, we normally have to consider more than one absorber recovery time. In this case, the slow component normally helps to start the initial pulse formation process. The modulation depth of the fast component can help to support shorter pulse durations at steady state. Further improvements of the saturable absorber normally require some better understanding of the underlying physics of the nonlinearities, which can be very interesting and rather complex (see Sect. 7.4 for semiconductors).

The pulse-averaged nonlinear reflectivity $R(F_{p,A})$ is shown in Fig. 7.1. For illustration, the same curve is plotted on a linear fluence scale (Fig. 7.1 left) and a logarithmic scale (Fig. 7.1 right). This curve is completely described by three parameters: (i) the low reflectivity R_0 for pulses with 'zero' pulse fluence, (ii) the reflectivity R_{ns} for 'infinitely' high pulse fluences when all saturable absorption is bleached, and (iii) the saturation fluence $F_{sat,A}$, described in detail later. Here we neglect higher order nonlinearities such as two-photon absorption which will be discussed later in this chapter. The modulation depth ΔR and the nonsaturable losses ΔR_{ns} in reflectivity are defined by

$$\Delta R = R_{ns} - R_0 \tag{7.2}$$

and

$$\Delta R_{ns} = 1 - R_{ns} \ . \tag{7.3}$$

The definitions above imply that R_0 and R_{ns} are not experimentally accessible, but extrapolated values from the measured data using a proper model function. The saturation fluence $F_{sat,A}$ is the fluence required to begin to saturate the absorption. For an infinitely thin absorber, the reflectivity for a pulse with fluence $F_p = F_{sat,A}$ is increased by $1/e$ (37%) of ΔR with respect to R_0. Normally, it takes a pulse fluence of about 5 times the saturation fluence for a fully saturated absorber.

The measured nonlinear reflectivity shown in Fig. 7.1 is used to fit a model function to obtain the key parameters of the saturable absorber. The saturation of a saturable absorber without any nonsaturable loss can be described by the following simplified rate equation which takes into account one recovery time τ_A of the saturable absorber [89Agr] (a more detailed justification for the approximations in this equation is given in Sect. 7.2.2):

$$\boxed{\frac{dq(t)}{dt} = -\frac{q(t) - q_0}{\tau_A} - \frac{q(t)P(t)}{E_{sat,A}}}, \tag{7.4}$$

where $P(t)$ is the time-dependent power of the incoming pulse in the time reference frame of the moving pulse and with the following saturable absorber notation:

- $q(t)$ time-dependent amplitude absorption coefficient, spatially averaged over the thickness of the saturable absorber. Note that $q(t)$ only describes the saturable part of the loss coefficient and does not include any nonsaturable losses. For example, Fig. 7.1 shows a saturable absorber which also has a non-saturable loss ΔR_{ns}.
- q_0 unsaturated amplitude absorption coefficient, which is also the maximum loss coefficient of the saturable absorber.

Thus $q(t)$ is the partially saturated absorber coefficient at any instant within the pulse. For a saturable absorber mirror, we have

$$I_{out}(t) = R(t)I_{in}(t) = e^{-2q(t)}I_{in}(t) , \tag{7.5}$$

where $R(t) = R(F_{p,A}, t)$ is the time-dependent nonlinear reflectivity, $q(t)$ is the amplitude loss coefficient at time t given by (7.4), $I_{in}(t)$ the time-dependent input intensity, and $I_{out}(t)$ the time-dependent output intensity in the thin slab approximation (see Sect. 7.3.1).

In most cases we consider a saturable absorber which is directly integrated with a mirror. The reflectivity of this nonlinear mirror is thereby increased by the saturation of the absorber, as is shown in Fig. 7.1 for a typical case. We then have

$$\Delta R = 1 - e^{-2q_0} \approx 2q_0 , \quad q_0 \ll 1 . \tag{7.6}$$

The factor of two was introduced because ΔR describes the intensity reflectivity of the SESAM according to (7.5).

As long as the pulse duration τ_p for the incoming pulse on a saturable absorber mirror is much shorter than the recovery time of the absorber τ_A (Fig. 7.2), we can make the slow saturable absorber approximation (see next section for more details). In this case the following model function for the pulse-averaged nonlinear reflectivity is well suited to describe the nonlinear reflectivity of a thin saturable absorber mirror with one incoming beam of short pulses with a pulse duration τ_p and a pulse fluence of F_p assuming $\tau_p \ll \tau_A$ [04Hai]:

$$R(F_p) = R_{ns} \frac{\ln\left[1 + \frac{R_0}{R_{ns}}\left(e^S - 1\right)\right]}{S}, \quad S = \frac{F_p}{F_{sat,A}} . \tag{7.7}$$

Note that this is the pulse-averaged reflectivity, i.e., the final time-integrated reflectivity a single pulse experiences after being reflected from the saturable absorber mirror. In addition, we assume a constant incoming pulse fluence over the beam profile and no inverse saturable absorption (ISA) such as two-photon absorption. Corrections for a Gaussian beam profile will be discussed in Sect. 7.3.4 and for inverse saturable absorption (ISA), such as two-photon absorption (TPA), in Sect. 7.3.5. All model functions for the nonlinear reflectivity will be summarized in Table 7.4.

A fit with this model function and a correction for the incoming Gaussian beam profile is shown in Fig. 7.1 for a SESAM. This demonstrates excellent agreement even though rather strong approximations are made (to be discussed later in Sects. 7.2.2 and 7.3.1). The fit parameters are ΔR, $F_{sat,A}$, and R_{ns}. The latter is the reflectivity for high pulse energies and determines the nonsaturable loss $\Delta R_{ns} = 1 - R_{ns}$ as in (7.3). ΔR is the maximum change of nonlinear reflectivity given by (7.2), also referred to as the maximum modulation depth of the SESAM device.

For absorbers with ΔR smaller than about 10%, we can simplify (7.7) to

$$R(F_p) = R_{ns} - \frac{\Delta R}{S}\left(1 - e^{-S}\right) \tag{7.8}$$

(see derivation in Sect. 7.3.3). The experimental methods for SESAM characterization as shown for example in Figs. 7.1 and 7.2 will be discussed in Sect. 7.3.5. Note that a time-dependent nonlinear reflectivity can also be measured with a pump–probe configuration, but in this case a different model function for the nonlinear reflectivity is required, as discussed also in Sect. 7.3.2.

7.2.2 Justification for the Simplified Saturable Absorber Rate Equation

The simplified rate equation for the saturable absorber (7.4) is based on the original work on pulse propagation in laser amplifiers which goes back to the 1960s [63Fra, 64Wit]. Here we refer to a more recent paper [89Agr] which we have found very helpful for semiconductor saturable absorbers. A number of approximations are made for the gain in an optical amplifier. To begin with, we assume (i) a two-level system, which means that we can also use the rate equation for saturable absorbers, and secondly, for semiconductor materials, we assume (ii) that we can neglect intraband relaxation processes (see Sect. 7.4.2). We can thus use a simple carrier-density rate equation of the form

$$\frac{\partial N}{\partial t} = D\nabla^2 N + \frac{J}{eV} - \frac{N}{\tau_c} - \frac{a(N - N_0)}{\hbar\omega}|E|^2 ,$$

where N is the carrier density (for electrons and holes), D is the diffusion coefficient, J is the injection current, which is directly related to the pump rate in an optically pumped amplifier, e is the elementary charge, V is the volume of the active region, τ_c is the spontaneous carrier lifetime, $\hbar\omega$ is the photon energy, a is the gain coefficient, N_0 is the carrier density required for transparency, and E is the electric field amplitude. Here we have assumed (iii) that we can neglect any parasitic loss inside the optical amplifier (and therefore the nonsaturable loss in the saturable absorber).

For our applications, we can neglect the diffusion of the carriers out of the laser mode area (see discussion below), and for the saturable absorber, we can neglect the injection current, i.e., we can set $J = 0$. We are then left with the spontaneous decay rate, i.e., $-N/\tau_c$, and the stimulated emission if the carrier density is above the transparency level. We then make the additional approximation (iv) of a uniform carrier density along the transverse dimension and a spatially averaged gain coefficient in the thin slab approximation, which will be discussed in more detail in Sect. 7.3.1. The rate equation then reduces to

$$\frac{\partial N}{\partial t} = -\frac{N}{\tau_c} - \frac{g(N)}{\hbar\omega}|A|^2 ,$$

where the gain is defined by

$$g(N) = \sigma(N - N_0) ,$$

and we have additionally assumed (v), the slowly-varying envelope approximation (see Sect. 2.5.2) with A as the pulse envelope. If we assume that $|A|^2$ is normalized for the pulse power $P(t)$, for which we moved into the reference frame of the moving laser pulse, and E_{sat} is the saturation energy (defined earlier), we then obtain the final rate equation for the gain coefficient in the form

$$\frac{\partial g}{\partial t} = \frac{g_0 - g}{\tau_c} - \frac{gP(t)}{E_{sat}} ,$$

where we introduced the small-signal gain g_0 for a pumping rate to obtain transparency in the optical amplifier. We have also assumed (vi) that we may neglect any pulse broadening and group velocity dispersion of the pulse propagation through the optical amplifier. Because we are working in the limit of a thin slab amplifier, this last approximation is also justified. The thin slab approximation is discussed in more detail in Sect. 7.3.1.

Because we used a two-level system, we can also apply this rate equation directly for a saturable absorber (7.4) for which q_0 is the low-energy, unsaturated absorption coefficient and τ_A the absorber relaxation time after excitation (neglecting the much faster coherent intraband relaxation processes).

For modelocking applications, we will show that the second approximation neglecting coherent intraband relaxation is justified because these really fast processes are typically averaged out in the final self-amplitude modulation due to the fact that the pulse durations are typically longer (see, e.g., Fig. 7.2). Moreover, in the femtosecond domain, we normally work with soliton formation, which is responsible for the fast pulse formation (see the discussion of soliton modelocking in Sect. 9.5). The two-level approximation is normally also justified for semiconductor energy bands, once we neglect these ultrafast coherent relaxation processes [89Agr].

7.2.3 Slow Saturable Absorber

In this case, defined by $\tau_p \ll \tau_A$, one can neglect the recovery of the absorber within the pulse duration, and the differential equation (7.4) reduces to

$$\frac{dq(t)}{dt} \approx -\frac{q(t)P(t)}{E_{sat,A}} . \tag{7.9}$$

This differential equation can be solved relatively easily for $q(t)$:

$$\int_{q_0}^{q(t)} \frac{dq}{q} = -\int_0^t \frac{P(t')}{E_{sat,A}} dt' = -\int_0^t \frac{E_p}{E_{sat,A}} f(t') dt' = -\frac{E_p}{E_{sat,A}} \int_0^t f(t') dt' , \tag{7.10}$$

where $P(t)$ and $E_p(t)$ are the time-dependent power and energy during a pulse:

$$E_p(t) = E_p f(t) , \quad \int_0^{T_R} f(t) dt = 1 , \tag{7.11}$$

with E_p the pulse energy and T_R the resonator roundtrip time. We also have

$$\int_{q_0}^{q(t)} \frac{dq}{q} = \ln \frac{q(t)}{q_0} . \tag{7.12}$$

Therefore, the time-dependent loss coefficient $q(t)$ experienced by a pulse at time t during a passage through the absorber becomes

$$q(t) = q_0 \exp\left[-\frac{E_p}{E_{sat,A}} \int_0^t f(t') dt'\right] . \tag{7.13}$$

One can further show that the total loss q_p which the laser pulse experiences on its passage through this saturable absorber is given by its time-integrated (i.e., pulse-averaged) value as follows (see below for the derivation) [99Hon1]:

$$\boxed{q_p(E_p) = \int_0^{T_R} q(t) f(t) dt = q_0 \frac{F_{sat,A} A_A}{E_p} \left[1 - \exp\left(-\frac{E_p}{F_{sat,A} A_A}\right)\right]} , \tag{7.14}$$

where E_{p} is the incoming pulse energy, $F_{\mathrm{sat,A}}$ is the saturation fluence of the absorber, and A_{A} is the incident beam area on the saturable absorber. For the borderline cases $E_{\mathrm{p}} \to 0$ and $E_{\mathrm{p}} \to \infty$, we have

$$\lim_{E_{\mathrm{p}}\to 0} q_{\mathrm{p}}\left(E_{\mathrm{p}}\right) = \lim_{E_{\mathrm{p}}\to 0} q_0 \frac{E_{\mathrm{sat,A}}}{E_{\mathrm{p}}}\left[1-\left(1-\frac{E_{\mathrm{p}}}{E_{\mathrm{sat,A}}}\right)\right] = q_0$$

and

$$\lim_{E_{\mathrm{p}}\to\infty} q_{\mathrm{p}}\left(E_{\mathrm{p}}\right) = 0 \ ,$$

which means that no nonsaturable absorptions are considered in $q(t)$, as already mentioned above.

The relationship between $q_{\mathrm{p}}\left(E_{\mathrm{p}}\right)$ and $R\left(E_{\mathrm{p}}\right)$ then follows directly from (7.5):

$$R\left(E_{\mathrm{p}}\right) = \mathrm{e}^{-2q_{\mathrm{p}}(E_{\mathrm{p}})} \approx 1 - 2q_{\mathrm{p}}\left(E_{\mathrm{p}}\right) \ , \quad q_0 \ll 1 \ , \tag{7.15}$$

and a pulse energy dependent modulation depth $\Delta R\left(E_{\mathrm{p}}\right)$ is given by

$$\Delta R\left(E_{\mathrm{p}}\right) = 2q_{\mathrm{p}}\left(E_{\mathrm{p}}\right) \ , \quad q_0 \ll 1 \ . \tag{7.16}$$

For a fully saturated absorber we then obtain $\Delta R\left(E_{\mathrm{p}} \to \infty\right) \equiv \Delta R = 2q_0$.

Using (7.14) for a strongly saturated absorber with approximately $E_{\mathrm{p}} \geq 5E_{\mathrm{sat,A}}$, we thus have

$$\boxed{q_{\mathrm{p}}\left(E_{\mathrm{p}} \geq 5E_{\mathrm{sat,A}}\right) \approx q_0 \frac{E_{\mathrm{sat,A}}}{E_{\mathrm{p}}} \implies R\left(E_{\mathrm{p}} \geq 5E_{\mathrm{sat,A}}\right) \approx 1 - \Delta R \frac{E_{\mathrm{sat,A}}}{E_{\mathrm{p}}}} \ . \tag{7.17}$$

Note that the total amplitude loss coefficient $q_p(E_p)$ for a slow saturable absorber (i.e., $\tau_p \ll \tau_A$) does not depend on the pulse shape. This loss occurs because a part of the pulse has to be absorbed to saturate the absorber. The absorbed energy E_{abs} is then

$$E_{\mathrm{abs}} = 2q_{\mathrm{p}}\left(E_{\mathrm{p}}\right) E_{\mathrm{p}} \ . \tag{7.18}$$

where the factor of two comes from the squaring of the field to obtain the intensity.

Derivation of (7.14)

For a saturable absorber, we have according to (7.5)

$$I_{\mathrm{out}}(t) = R(t) I_{\mathrm{in}}(t) = \mathrm{e}^{-2q(t)} I_{\mathrm{in}}(t) \ ,$$

where $R(t)$ is the time-dependent nonlinear reflectivity and $q(t)$ is the amplitude loss coefficient at time t given by (7.13). Integrating this equation yields

$$F_{\mathrm{out}} = \int I_{\mathrm{out}}(t)\mathrm{d}t = \int R(t) I_{\mathrm{in}}(t)\mathrm{d}t = \int \mathrm{e}^{-2q(t)} I_{\mathrm{in}}(t)\mathrm{d}t \ .$$

For $q(t) \ll 1$, we have

$$F_{\text{out}} = \int \left[1 - 2q(t)\right] I_{\text{in}}(t)\text{d}t = \int I_{\text{in}}(t)\text{d}t - 2\int q(t) I_{\text{in}}(t)\text{d}t = F_{\text{in}} - 2\int q(t) I_{\text{in}}(t)\text{d}t \;,$$

whence the total pulse-averaged reflectivity R_{tot} is

$$R_{\text{tot}} = \frac{\int I_{\text{out}}(t)\text{d}t}{\int I_{\text{in}}(t)\text{d}t} = \frac{F_{\text{out}}}{F_{\text{in}}} = 1 - \frac{2}{F_{\text{in}}}\int q(t) I_{\text{in}}(t)\text{d}t \;.$$

Using

$$R_{\text{tot}} = \text{e}^{-2q_{\text{p}}} \approx 1 - 2q_{\text{p}} \;,$$

it follows that q_{p} in (7.14) is given by

$$q_{\text{p}} = \frac{1}{F_{\text{in}}}\int q(t) I_{\text{in}}(t)\text{d}t = \int q(t) f(t)\text{d}t \;,$$

where

$$f(t) \equiv \frac{I_{\text{in}}(t)}{F_{\text{in}}} = \frac{P_{\text{in}}(t)}{E_{\text{p,in}}} \;, \qquad \int f(t)\text{d}t = \frac{1}{F_{\text{in}}}\int I_{\text{in}}(t)\text{d}t = 1 \;.$$

Thus, we can now calculate $q_{\text{p}}\left(E_{\text{p}}\right)$, where E_{p} is identical to the incoming pulse energy $E_{\text{p,in}}$ on the absorber:

$$\begin{aligned} q_{\text{p}}\left(E_{\text{p}}\right) &= \int_0^{T_{\text{R}}} q(t) f(t)\text{d}t = \int_0^{T_{\text{R}}} q_0 \exp\left[-\frac{E_{\text{p}}}{E_{\text{sat,A}}}\int_0^t f\left(t'\right)\text{d}t'\right] f(t)\text{d}t \\ &= q_0 \frac{E_{\text{sat,A}}}{E_{\text{p}}}\int_0^{T_{\text{R}}} \frac{E_{\text{p}}}{E_{\text{sat,A}}} f(t) \exp\left[-\frac{E_{\text{p}}}{E_{\text{sat,A}}}\int_0^t f\left(t'\right)\text{d}t'\right]\text{d}t \;. \end{aligned}$$

Using

$$\frac{\text{d}}{\text{d}t}\exp\left[-\frac{E_{\text{p}}}{E_{\text{sat,A}}}\int_0^t f\left(t'\right)\text{d}t'\right] = -\frac{E_{\text{p}}}{E_{\text{sat,A}}} f(t) \exp\left[-\frac{E_{\text{p}}}{E_{\text{sat,A}}}\int_0^t f\left(t'\right)\text{d}t'\right] ,$$

it follows that

$$q_{\text{p}}\left(E_{\text{p}}\right) = q_0\left(-\frac{E_{\text{sat,A}}}{E_{\text{p}}}\right)\int_0^{T_{\text{R}}} \frac{\text{d}}{\text{d}t}\exp\left[-\frac{E_{\text{p}}}{E_{\text{sat,A}}}\int_0^t f\left(t'\right)\text{d}t'\right]\text{d}t \;.$$

The integration is then straightforward:

$$\begin{aligned} q_{\text{p}}\left(E_{\text{p}}\right) &= -q_0 \frac{E_{\text{sat,A}}}{E_{\text{p}}} \exp\left[-\frac{E_{\text{p}}}{E_{\text{sat,A}}}\int_0^t f\left(t'\right)\text{d}t'\right]\Bigg|_0^{T_{\text{R}}} \\ &= -q_0 \frac{E_{\text{sat,A}}}{E_{\text{p}}}\left\{\exp\left[-\frac{E_{\text{p}}}{E_{\text{sat,A}}}\int_0^{T_{\text{R}}} f\left(t'\right)\text{d}t'\right] - \exp\left[-\frac{E_{\text{p}}}{E_{\text{sat,A}}}\int_0^0 f\left(t'\right)\text{d}t'\right]\right\} \\ &= -q_0 \frac{E_{\text{sat,A}}}{E_{\text{p}}}\left(\text{e}^{-E_{\text{p}}/E_{\text{sat,A}}} - 1\right) = q_0 \frac{E_{\text{sat,A}}}{E_{\text{p}}}\left(1 - \text{e}^{-E_{\text{p}}/E_{\text{sat,A}}}\right) . \end{aligned}$$

Since $E_{\text{sat,A}} = F_{\text{sat,A}} A_{\text{A}}$, (7.14) follows.

Simpler Derivation of (7.14)

The loss coefficient q_f after the pulse is given by

$$q_f = q_0 \exp\left(-\frac{E_\mathrm{p}}{E_\mathrm{sat,A}}\right) .$$

The absorbed power P_abs is

$$P_\mathrm{abs} = 2q(t)P(t) .$$

The absorbed energy E_abs is

$$\begin{aligned} E_\mathrm{abs} &= \int P_\mathrm{abs}(t)\mathrm{d}t \overset{(7.9)}{=} -2E_\mathrm{sat,A}\int \mathrm{d}q = 2E_\mathrm{sat,A}\left(q_0 - q_f\right) \\ &= 2E_\mathrm{sat,A}q_0\left[1-\exp\left(-\frac{E_\mathrm{p}}{E_\mathrm{sat,A}}\right)\right] . \end{aligned}$$

Hence,

$$q_\mathrm{p}\left(E_\mathrm{p}\right) = \frac{1}{2}\frac{E_\mathrm{abs}}{E_\mathrm{p}} = \frac{E_\mathrm{sat,A}}{E_\mathrm{p}}q_0\left[1-\exp\left(-\frac{E_\mathrm{p}}{E_\mathrm{sat,A}}\right)\right] .$$

7.2.4 Fast Saturable Absorber

In a fast saturable absorber, with $\tau_\mathrm{p} \gg \tau_\mathrm{A}$, the absorption will adjust itself 'immediately' to the value for a given incoming power P. Therefore, in the differential equation (7.4), we can set $\mathrm{d}q/\mathrm{d}t = 0$. This yields

$$0 = -\frac{q(t)-q_0}{\tau_\mathrm{A}} - \frac{q(t)P(t)}{E_\mathrm{sat,A}} . \tag{7.19}$$

We can now solve this equation very easily for $q(t)$:

$$q(t) = \frac{q_0}{1+\dfrac{P(t)\tau_\mathrm{A}}{E_\mathrm{sat,A}}} . \tag{7.20}$$

Using

$$P_\mathrm{sat,A} = \frac{E_\mathrm{sat,A}}{\tau_\mathrm{A}} \implies \frac{P(t)}{P_\mathrm{sat,A}} = \frac{I_\mathrm{A}(t)}{I_\mathrm{sat,A}} , \tag{7.21}$$

it follows that

$$\boxed{q(t) = \frac{q_0}{1+\dfrac{I_\mathrm{A}(t)}{I_\mathrm{sat,A}}}} . \tag{7.22}$$

In cw operation, a fast saturable absorber is saturated by the intensity I_A:

$$q_{cw} = \frac{q_0}{1 + \frac{I_A}{I_{sat,A}}} .$$

Assuming the absorber is only weakly saturated, i.e., $I_A(t) \ll I_{sat,A}$, we can deduce from (7.22) with the approximation that $(1+x)^{-1} \approx 1 - x$, for $x \ll 1$:

$$\boxed{q(t) = \frac{q_0}{1 + \frac{I_A(t)}{I_{sat,A}}} \approx q_0 \left[1 - \frac{I_A(t)}{I_{sat,A}}\right] = q_0 - \gamma_A I_A(t)} , \tag{7.23}$$

with

$$\boxed{\gamma_A \equiv \frac{q_0}{I_{sat,A}}} . \tag{7.24}$$

Although (7.24) only explains the fast saturable absorber for weak saturation, we have to describe the fast saturable absorber using this 'ideal' saturation behaviour if we hope later to derive an analytical solution for passive modelocking with a fast saturable absorber.

For a soliton pulse with a $\mathrm{sech}^2(t/\tau)$ pulse shape, the total loss for a fully saturated fast absorber becomes (see derivation below)

$$\boxed{q_p\left(E_{p,in}\right) = q_0\left(1 - \frac{1}{3}\frac{E_{p,in}}{\tau A_A I_{sat,A}}\right) = q_0\left(1 - \frac{1}{3}\frac{F_{p,A}}{F_{sat,A}}\frac{\tau_A}{\tau}\right)} , \tag{7.25}$$

where we have used $F_{p,A} = E_{p,in}/A_A$ and $I_{sat,A} = F_{sat,A}/\tau_A$.

For a 'fully saturated ideally fast' absorber as defined in Fig. 7.3, this yields a total loss of

$$\boxed{q_p \approx \frac{q_0}{3}} . \tag{7.26}$$

This will be derived below. In contrast to the case of the slow saturable absorber [see (7.8)], $q_p\left(E_p\right)$ now depends on the pulse shape.

Note that a 'fully saturated ideally fast' absorber is in principle a contradiction in terms. In principle, the linear approximation is only valid for a weakly saturated absorber (see also Fig. 7.3). This approximation is nevertheless also used in the fully saturated regime to obtain an analytic solution from the master equation, as will be discussed in Sect. 9.4. However, the analytic solution still describes the experimental results reasonably well, so we accept this somewhat unattractive approach.

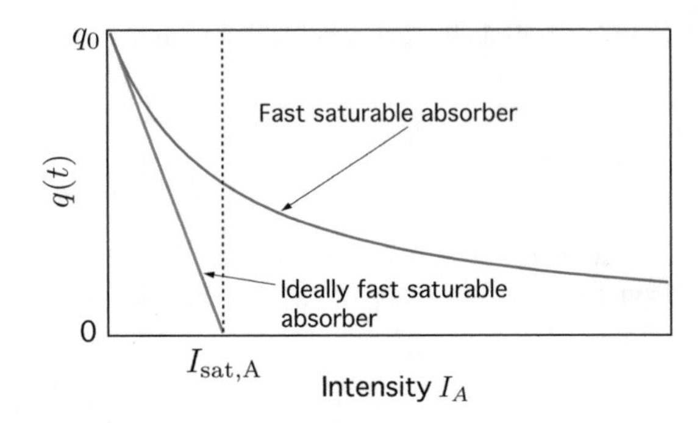

Fig. 7.3 A fast saturable absorber. An 'ideal fast' saturable absorber is described by $q_0 - \gamma I_A(t)$, where γ is given by $\gamma = q_0/I_{\text{sat,A}}$

Derivation of (7.25) and (7.26)

According to (7.23),

$$q(t) = q_0 - \gamma_A I_A(t) \ , \quad I_A(t) = I_{\text{p,A}} \operatorname{sech}^2(t/\tau) \ .$$

Thus, analogously to the derivation of (7.14), the total loss is given by

$$q_p\left(E_p\right) = \frac{1}{F_{\text{p,A}}} \int_{-T_R/2}^{T_R/2} q(t) I_{\text{in,A}}(t)\mathrm{d}t \ ,$$

where $I_{\text{in,A}} \equiv I_A$. The integration limits are set symmetrically around zero, because the soliton pulse is also symmetric with respect to $t = 0$. When $\tau_p \ll T_R$, we can consider the integration limits as $T_R \to \infty$, whence

$$q_p\left(F_{\text{p,A}}\right) = \frac{1}{F_{\text{p,A}}} \int q(t) I_A(t)\mathrm{d}t = \frac{1}{F_{\text{p,A}}} \int \left[q_0 - \gamma_A I_{\text{p,A}} \operatorname{sech}^2\left(\frac{t}{\tau}\right)\right] I_{\text{p,A}} \operatorname{sech}^2\left(\frac{t}{\tau}\right) \mathrm{d}t \ .$$

The quantity $F_{\text{p,A}}$ is found from

$$F_{\text{p,A}} = \int I_A(t)\mathrm{d}t = I_{\text{p,A}} \int \operatorname{sech}^2\left(\frac{t}{\tau}\right) \mathrm{d}t = I_{\text{p,A}}\, \tau \tanh\left(\frac{t}{\tau}\right)\Bigg|_{-\infty}^{+\infty} = 2\tau I_{\text{p,A}} \ .$$

The first summand in $q_p\left(F_{\text{p,A}}\right)$ is

$$\int q_0 I_{\text{p,A}} \operatorname{sech}^2\left(\frac{t}{\tau}\right) \mathrm{d}t = q_0 I_{\text{p,A}}\, \tau \tanh\left(\frac{t}{\tau}\right)\Bigg|_{-\infty}^{+\infty} = 2q_0 I_{\text{p,A}} \tau \ ,$$

and the second is

$$\int \operatorname{sech}^4\left(\frac{t}{\tau}\right) \mathrm{d}t = \left[\frac{1}{3}\tau \tanh\left(\frac{t}{\tau}\right) \operatorname{sech}^2\left(\frac{t}{\tau}\right) + \frac{2}{3}\tau \tanh\left(\frac{t}{\tau}\right)\right]_{-\infty}^{+\infty} = \frac{4}{3}\tau \ .$$

Note the following points:

- The first term in the square brackets provides no contribution, because $\mathrm{sech}^2(t/\tau)$ vanishes at the integration limits, i.e., at $\pm\infty$.
- To evaluate the integral, we use the recursion formula

$$\int \mathrm{sech}^{-n}(ax)\mathrm{d}x = -\frac{1}{a\,(n+1)}\sinh(ax)\,\mathrm{sech}^{-(n+1)}(ax) + \frac{n+2}{n+1}\int \mathrm{sech}^{-(n+2)}(ax)\mathrm{d}x\,,$$

for $n = -4$.

Finally, we obtain

$$q_\mathrm{p} = \frac{1}{2\tau I_\mathrm{p,A}}\left(2q_0 I_\mathrm{p,A}\tau - \frac{4}{3}\gamma_\mathrm{A} I_\mathrm{p,A}^2\tau\right) = q_0 - \frac{2}{3}\gamma_\mathrm{A} I_\mathrm{p,A}\,.$$

Using $\gamma_\mathrm{A} = q_0/I_\mathrm{sat,A}$ from (7.24) and the relation $E_\mathrm{p,in} = 2A_\mathrm{A}\tau I_\mathrm{p,A}$, it follows that

$$q_\mathrm{p}\left(E_\mathrm{p,in}\right) = q_0\left(1 - \frac{2}{3}\frac{I_\mathrm{p,A}}{I_\mathrm{sat,A}}\right) = q_0\left(1 - \frac{1}{3}\frac{E_\mathrm{p,in}}{\tau A_\mathrm{A} I_\mathrm{sat,A}}\right)\,,$$

which is equivalent to (7.25).

Now, we further assume that the absorber is fully saturated at the peak intensity I_p, i.e., $q(t=0) = 0$. Then using (7.24), it follows for an ideally fast saturable absorber (as defined in Fig. 7.3) that

$$q(t=0) = 0 = q_0 - \gamma_\mathrm{A} I_\mathrm{p,A} \implies \gamma_\mathrm{A} I_\mathrm{p,A} = q_0\,.$$

Thus,

$$q_\mathrm{p} = q_0 - \frac{2}{3}q_0 = \frac{1}{3}q_0\,.$$

7.2.5 Summary of Relevant Equations

Tables 7.2 and 7.3 summarise the important equations for slow and fast saturable absorbers, respectively.

Table 7.2 Summary of important equations for slow saturable absorbers. The loss does not depend on the specific pulse shape

$$\frac{dq(t)}{dt} = -\frac{q(t)-q_0}{\tau_A} - \frac{q(t)I(t)}{F_{sat,A}} \tag{7.4}$$

$$\tau_p \ll \tau_A \implies \frac{dq(t)}{dt} \approx -\frac{q(t)P(t)}{E_{sat,A}} \tag{7.9}$$

$$q(t) = q_0 \exp\left[-\frac{E_p}{E_{sat,A}}\int_0^t f\left(t'\right)dt'\right] \tag{7.13}$$

$$E_{abs} = 2q_p\left(E_p\right)E_p \tag{7.18}$$

$$q_p\left(E_p\right) = \frac{q_0}{S}\left(1-e^{-S}\right), \quad S \equiv \frac{F_p}{F_{sat,A}} = \frac{E_p}{A_A F_{sat,A}} \tag{7.14}$$

$$F_p \geq 5F_{sat,A}\ (S \gg 1) \implies q_p \approx \frac{q_0}{S}, \quad R(F_p) \approx 1 - \frac{\Delta R}{S} \tag{7.17}$$

Table 7.3 Summary of important equations for fast saturable absorbers. Loss depends on pulse shape

$$\frac{dq(t)}{dt} = -\frac{q(t)-q_0}{\tau_A} - \frac{q(t)I(t)}{F_{sat,A}} \tag{7.4}$$

$$\tau_p \gg \tau_A \implies 0 = -\frac{q(t)-q_0}{\tau_A} - \frac{q(t)P(t)}{E_{sat,A}} \tag{7.19}$$

$$q(t) = \frac{q_0}{1 + I_A(t)/I_{sat,A}} \tag{7.22}$$

$$I_{sat,A} = F_{sat,A}/\tau_A \tag{7.21}$$

Weakly saturated fast absorber

$$I_A(t) \ll I_{sat,A} \implies q(t) \approx q_0 - \gamma_A I_A(t) \tag{7.23}$$

$$\gamma_A \equiv q_0/I_{sat,A} \tag{7.24}$$

Total loss depends on pulse shape, and for the $\text{sech}^2(t/\tau)$ pulse shape, we obtain

$$q_p\left(F_p\right) = q_0\left(1 - \frac{1}{3}\frac{F_p}{F_{sat,A}}\frac{\tau_A}{\tau}\right) \tag{7.25}$$

Ideal fast saturable absorber approximation, fully saturated, $\text{sech}^2(t/\tau)$ pulse shape

$$q_p \approx q_0/3 \tag{7.26}$$

7.3 Nonlinear Reflectivity Model Functions

All the nonlinear reflectivity model functions that we continue to use in this textbook are summarized in Table 7.4 in Sect. 7.3.6. Their detailed derivation is discussed next.

7.3.1 Approximations

We make the following approximations:

1. We do not take into account standing wave patterns or Fabry–Pérot effects in the SESAM. The reflectivity is calculated as transmission through twice an absorber of length $d/2$ (see Fig. 7.4). The main difference between the results of a complete transfer matrix calculation and the simple traveling wave approximation is in the scaling of the effective saturation fluence and modulation depth. Fortunately, these deviations scale linearly with the field enhancement factor for the modulation depth and inverse linearly for the saturation fluence. Consequently, from the applied model function, not only can we measure macroscopic SESAM parameters, but we can also extract microscopic values for material comparison or SESAM design. We then obtain a simple differential equation

$$\frac{\mathrm{d}I(z,t)}{\mathrm{d}z} = -2\alpha(z,t)I(z,t) , \tag{7.27}$$

where $\alpha(z,t)$ is the amplitude absorption coefficient within the thin slab layer.
2. We then make the approximation of a thin slab with a spatially integrated amplitude absorption coefficient $q(t)$ given by

$$q(t) = \int_0^d \alpha(z,t)\mathrm{d}z . \tag{7.28}$$

Here $q(t)$ is the time-dependent or partially saturated absorption coefficient at any instant within the pulse. The time is in the reference frame of the moving pulse.
3. We consider a slow saturable absorber so that according to (7.9) in Sect. 7.2.3 we then obtain

$$\frac{\mathrm{d}\alpha(z,t)}{\mathrm{d}t} = -\frac{\alpha(z,t)I(t)}{F_{\mathrm{sat,A}}} . \tag{7.29}$$

With these approximations, the coupled differential equations (7.27) and (7.29) become easier to solve.

Equation (7.27) can be simplified by moving the intensity I to the left-hand side and integrating with respect to z over the thin slab:

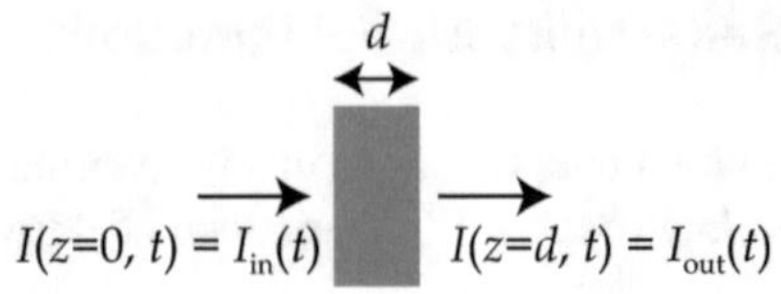

d

$I(z{=}0, t) = I_{in}(t)$ $I(z{=}d, t) = I_{out}(t)$

Fig. 7.4 Thin slab with thickness d used to model the saturable absorber with an absorption amplitude coefficient $\alpha(z, t)$

$$\frac{1}{I(z,t)}\mathrm{d}I(z,t) = -2\alpha(z,t)\mathrm{d}z \implies \int_{I_{in}}^{I_{out}} \frac{1}{I}\mathrm{d}I = \ln\frac{I_{out}(t)}{I_{in}(t)} = -2\int_0^d \alpha(z,t)\mathrm{d}z = -2q(t)$$
$$\implies \frac{I_{out}(t)}{I_{in}(t)} = \exp\left[-2q(t)\right]$$
$$\implies I_{out}(t) = \mathrm{e}^{-2q(t)} I_{in}(t) , \tag{7.30}$$

in agreement with (7.5). This is exactly the result for a thin slab with spatially integrated amplitude absorption coefficient $q(t)$ according to (7.5). We then substitute (7.29) into (7.27) and integrate over the slab thickness using (7.28):

$$\frac{\mathrm{d}I(z,t)}{\mathrm{d}z} = -2\alpha(z,t)I(z,t) \overset{(7.29)}{=} 2F_{sat,A}\frac{\mathrm{d}\alpha(z,t)}{\mathrm{d}t}$$
$$\implies \int \mathrm{d}I = 2F_{sat,A}\frac{\mathrm{d}}{\mathrm{d}t}\int \alpha(z,t)\mathrm{d}z$$
$$\implies I_{out}(t) - I_{in}(t) \overset{(7.28)}{=} 2F_{sat,A}\frac{\mathrm{d}}{\mathrm{d}t}\left[q(t)\right] ,$$

whence

$$\frac{\mathrm{d}q(t)}{\mathrm{d}t} = \frac{I_{out}(t) - I_{in}(t)}{2F_{sat,A}} . \tag{7.31}$$

The two coupled equations (7.30) and (7.31) will be the working equations we shall be using from this point on. The spatially averaged absorption coefficient means that we can consider the nonlinear transmission of an arbitrarily thin slab in Fig. 7.4 (see also [86Sie, Chap. 10]).

7.3.2 Time-Dependent Reflectivity

The time-dependent nonlinear reflectivity $R(t)$ (or transmission) within the approximation discussed above is given by (7.30), resp. (7.5):

$$\boxed{R(t) = \frac{I_{out}(t)}{I_{in}(t)} = \mathrm{e}^{-2q(t)}} , \tag{7.32}$$

and $q(t)$ is given by (7.31). Before we go on, it is useful to introduce the following notation:

$$F_{in}(t) \equiv \int_{-\infty}^{t} I_{in}(t')\mathrm{d}t' , \quad F_{out}(t) \equiv \int_{-\infty}^{t} I_{out}(t')\mathrm{d}t' , \tag{7.33}$$

which are the accumulated pulse fluences until time t within the input and output pulses, respectively. Here we assume we only have one pulse in the pulse train. Then we introduce with (7.32) the following parameter

$$R_0 \equiv e^{-2q_0} , \tag{7.34}$$

because for $t \to -\infty$ we have $q(t \to -\infty) = q_0$, i.e., before the leading edge of the pulse we have $R(-\infty) = R_0$. This is the unsaturated low-fluence reflectivity of the saturable absorber mirror introduced in Fig. 7.1. In addition we introduce

$$S_{\text{in}}(t) \equiv \frac{F_{\text{in}}(t)}{F_{\text{sat,A}}} , \quad S_{\text{out}}(t) \equiv \frac{F_{\text{out}}(t)}{F_{\text{sat,A}}} . \tag{7.35}$$

For the derivation of $R(t)$ we insert (7.30) into (7.31) and solve the equation by separation of variables, i.e., moving $q(t)$ to the left-hand side of the equality and integrating both sides with respect to t, which gives

$$\frac{dq(t)}{dt} = \frac{1}{2F_{\text{sat,A}}}\left[e^{-2q(t)} I_{\text{in}}(t) - I_{\text{in}}(t)\right] = \frac{I_{\text{in}}(t)}{2F_{\text{sat,A}}}\left[e^{-2q(t)} - 1\right] .$$

This equation can be integrated in the form

$$\int_{q_0}^{q(t)} \frac{2}{e^{-2q} - 1} dq = \frac{1}{F_{\text{sat,A}}} \int_{-\infty}^{t} I_{\text{in}}(t) dt = \frac{F_{\text{in}}(t)}{F_{\text{sat,A}}} ,$$

where we used the fact that $q(t \to -\infty) = q_0$. The integral on the left-hand side then gives

$$\int_{q_0}^{q(t)} \frac{2}{e^{-2q} - 1} dq = -\left[\ln(e^x - 1)\right]_{x=2q_0}^{x=2q(t)} = \ln \frac{e^{2q_0} - 1}{e^{2q(t)} - 1} .$$

Hence,

$$F_{\text{in}}(t) = F_{\text{sat,A}} \ln \frac{e^{2q_0} - 1}{e^{2q(t)} - 1} ,$$

and using (7.32) and (7.34) we obtain,

$$F_{\text{in}}(t) = F_{\text{sat,A}} \ln \frac{1 - 1/R_0}{1 - 1/R(t)} . \tag{7.36}$$

Equation (7.36) can be solved for $R(t)$ as follows:

$$\begin{aligned} \exp \frac{F_{\text{in}}(t)}{F_{\text{sat,A}}} = e^{S_{\text{in}}(t)} = \frac{1 - 1/R_0}{1 - 1/R(t)} &\Longrightarrow \frac{1}{R(t)} = 1 - \left(1 - \frac{1}{R_0}\right) e^{-S_{\text{in}}(t)} \\ &\Longrightarrow \frac{R_0}{R(t)} = R_0 - (R_0 - 1) e^{-S_{\text{in}}(t)} \\ &\Longrightarrow R(t) = \frac{R_0}{R_0 - (R_0 - 1) e^{-S_{\text{in}}(t)}} . \end{aligned}$$

So, finally, we obtain

$$R(t) = \frac{1}{1 + \left(\frac{1}{R_0} - 1\right) e^{-F_{\text{in}}(t)/F_{\text{sat,A}}}} . \tag{7.37}$$

We can modify (7.37) by taking into account also a nonsaturable loss as follows:

$$\boxed{R(F_\text{p}, t) = \frac{R_\text{ns}}{1 + \left(\frac{R_\text{ns}}{R_0} - 1\right) e^{-F_{\text{in}}(t)/F_{\text{sat,A}}}}} , \tag{7.38}$$

where $F_{\text{in}}(t)$ is defined by (7.33) and $F_\text{p} = F_{\text{in}}(\infty) = \int_{-\infty}^{\infty} I_{\text{in}}(t')\text{d}t'$. This corresponds to the reflection a weak probe pulse experiences at the time t when a strong pump pulse with an input pulse fluence $F_{\text{in}}(t)$ has saturated the absorber to level $q(t)$. So (7.38) convoluted with the probe pulse is what we would measure for a pump–probe measurement which will be discussed in more detail in Sect.7.7.3. In such measurements it is assumed that the probe pulse does not affect any changes in the reflectivity, which means that the incoming probe fluence is much smaller than $F_{\text{sat,A}}$. However, the model function in (7.7) is not based on such a measurement.

7.3.3 Pulse-Averaged Reflectivity

We can measure the nonlinear reflectivity with the change in reflection of one beam, which also saturates the absorber mirror. In this case, the quantity of interest is the energy or fluence loss defined by the pulse-integrated or pulse-averaged reflectivity

$$\boxed{R(F_\text{p}) = \frac{F_{\text{out}}(\infty)}{F_{\text{in}}(\infty)}} . \tag{7.39}$$

The quantities $F_{\text{in}}(t)$ and $F_{\text{out}}(t)$ are given in (7.33) and for $t \to \infty$ we obtain $F_{\text{in}}(\infty)$ and $F_{\text{out}}(\infty)$.

Derivation of $F_{\text{out}}(t)$

Here we follow the same derivation as for $F_{\text{in}}(t)$ in (7.36) [86Sie, Chap. 10]. From (7.30), it follows that

$$I_{\text{in}}(t) = I_{\text{out}}(t)\text{e}^{2q(t)} .$$

Inserting this into (7.31), we obtain

$$\frac{\text{d}q(t)}{\text{d}t} = \frac{I_{\text{out}}(t) - I_{\text{in}}(t)}{2F_{\text{sat,A}}} \Longrightarrow \frac{\text{d}q(t)}{\text{d}t} = \frac{1}{2F_{\text{sat,A}}} I_{\text{out}}(t)\left[1 - \text{e}^{2q(t)}\right] .$$

We then separate the integration as follows:

$$\int_{q_0}^{q(t)} \frac{2}{1-\mathrm{e}^{2q}}\mathrm{d}q = -\ln(\mathrm{e}^{-x}-1)\Big|_{x=2q_0}^{x=2q(t)} = \ln\frac{\mathrm{e}^{-2q_0}-1}{\mathrm{e}^{-2q(t)}-1},$$

and

$$\frac{1}{F_{\mathrm{sat,A}}}\int_{-\infty}^{t} I_{\mathrm{out}}(t)\mathrm{d}t = \frac{F_{\mathrm{out}}(t)}{F_{\mathrm{sat,A}}},$$

where we used the fact that $q(t \to -\infty) = q_0$. We obtain the solution

$$\frac{F_{\mathrm{out}}(t)}{F_{\mathrm{sat,A}}} = \ln\frac{R_0-1}{R(t)-1}, \quad \text{using} \quad \mathrm{R_0} = \mathrm{e}^{-2q_0}, \quad \mathrm{R(t)} = \mathrm{e}^{-2q(t)}. \tag{7.40}$$

Derivation of $R(F_p)$

Using (7.39), for $F_{\mathrm{out}}(t)$ (7.40), and for $F_{\mathrm{in}}(t)$ (7.36), we obtain with $R(t \to \infty) \equiv R_{\mathrm{f}}$

$$R(F_{\mathrm{p}}) = \frac{F_{\mathrm{out}}(\infty)}{F_{\mathrm{in}}(\infty)} = \frac{\ln\dfrac{R_0-1}{R_{\mathrm{f}}-1}}{\ln\dfrac{1-1/R_0}{1-1/R_{\mathrm{f}}}}. \tag{7.41}$$

Using $R(t)$ given in (7.38), $R_{\mathrm{f}} = R(t \to \infty)$, $F_{\mathrm{p}} = F_{\mathrm{in}}(t \to \infty)$, and the saturation parameter $S = S_{\mathrm{in}}(t \to \infty)$, we obtain

$$R_{\mathrm{f}} = \frac{R_0}{R_0-(R_0-1)\mathrm{e}^{-S}}, \quad S = \frac{F_{\mathrm{p}}}{F_{\mathrm{sat,A}}}. \tag{7.42}$$

Since this implies

$$R_{\mathrm{f}}-1 = \frac{R_0}{R_0-(R_0-1)\mathrm{e}^{-S}} - 1 = \frac{(R_0-1)\mathrm{e}^{-S}}{R_0-(R_0-1)\mathrm{e}^{-S}},$$

we have

$$\frac{R_0-1}{R_{\mathrm{f}}-1} = \frac{R_0-(R_0-1)\mathrm{e}^{-S}}{\mathrm{e}^{-S}} = R_0\mathrm{e}^{S}-(R_0-1) = 1+R_0(\mathrm{e}^{S}-1).$$

Regarding the denominator of (7.41), we have with (7.42)

$$R_0\left(1-\frac{1}{R_{\mathrm{f}}}\right) = (R_0-1)\mathrm{e}^{-S}, \quad R_0\left(1-\frac{1}{R_0}\right) = R_0-1,$$

whence

$$\frac{1-1/R_0}{1-1/R_{\mathrm{f}}} = \mathrm{e}^{S}.$$

Putting everything together in (7.41), we thus obtain

$$R(F_\mathrm{p}) = \frac{\ln\left[1 + R_0(\mathrm{e}^S - 1)\right]}{S} . \tag{7.43}$$

Adding Nonsaturable Losses

A real absorber device will always show some nonsaturable losses, which means that the device will not reach 100% reflectivity, even for arbitrarily high fluences. The origins of these losses are residual transmission losses through the Bragg mirror, scattering losses from rough interfaces, nonsaturable defect absorption, free-carrier absorption, Auger recombination, and many more. In most cases, they are homogeneously distributed over the absorber layer or transmission losses. These losses can be accounted for by including a scaling factor R_ns in (7.43) [04Hai]:

$$R(F_\mathrm{p}) = R_\mathrm{ns} \frac{\ln\left[1 + \dfrac{R_0}{R_\mathrm{ns}}(\mathrm{e}^S - 1)\right]}{S} .$$

This is identical to (7.7). Numerical simulations have confirmed that this is valid. For an absorber with very high nonsaturable losses ($> 10\%$), the characteristic curve is still described by (7.7). For $R_\mathrm{ns} = 90\%$ (99%), we observe only a 5% (0.5%) increase in F_sat due to losses.

Derivation of the Model Function Given in (7.8)

Assuming that the modulation depth is small, i.e., $2q_0 \approx \Delta R$, and (7.7), we can directly write

$$R(F_\mathrm{p}) \approx 1 - 2q_\mathrm{p}(F_\mathrm{p}) , \tag{7.44}$$

where $q_\mathrm{p}(F_\mathrm{p})$ is the total loss a pulse experiences in a slow saturable absorber. According to (7.14) with $S = F_\mathrm{p}/F_\mathrm{sat,A}$ we have

$$q_\mathrm{p}(F_\mathrm{p}) = q_0 \frac{1}{S}\left(1 - \mathrm{e}^{-S}\right) .$$

We then get

$$R(F_\mathrm{p}) \approx 1 - 2q_\mathrm{p}(F_\mathrm{p}) = 1 - \frac{2q_0}{S}\left(1 - \mathrm{e}^{-S}\right) .$$

With $\Delta R \approx 2q_0$ and $R_\mathrm{ns} \leq 1$, we obtain

$$R(F_\mathrm{p}) \approx R_\mathrm{ns} - \frac{\Delta R}{S}\left(1 - \mathrm{e}^{-S}\right) .$$

In this case, the nonsaturable loss can easily be taken into account, replacing 1 by R_ns because the saturable absorber is only saturated to $R_\mathrm{ns} < 1$. This is then equivalent to (7.8).

7.3.4 Correction for Gaussian Beam Profile

So far all calculations have been done with a constant fluence, i.e., a flat top beam profile with constant fluence F_p and radius w. However, all the measurements in the lab normally use a Gaussian beam profile. This requires a correction for the model functions $R(F_p)$ [04Hai]. For a given pulse energy E_p, it then follows that

$$F_p = \frac{E_p}{\pi w^2} , \tag{7.45}$$

where w is the $(1/e^2)$ beam radius. In most applications the laser beam has a Gaussian profile, viz.,

$$F_p^{\text{Gauss}}(r) = F_{\text{peak}} \exp\left(-\frac{2r^2}{w^2}\right) = 2F_p \exp\left(-\frac{2r^2}{w^2}\right) , \tag{7.46}$$

where F_{peak} is the peak fluence. The pulse energy E_p is given by

$$E_p = \int F_p^{\text{Gauss}}(r) 2\pi r \, \mathrm{d}r = \frac{1}{2} F_{\text{peak}} \cdot \pi w^2 .$$

Therefore,

$$F_p = \frac{E_p}{\pi w^2} = \frac{1}{2} F_{\text{peak}} .$$

The peak fluence in the Gaussian beam is $2F_p$ so saturation already occurs at lower fluences. In the wings of the pulse, the fluence and the saturation are much weaker. To calculate the reflectivity for a pulse, we have to integrate over the spatial energy distribution with $E_{\text{in}} = E_p$:

$$R^{\text{Gauss}} = \frac{E_{\text{out}}}{E_{\text{in}}} = \frac{1}{E_p} \int_0^\infty 2\pi r \, R\left(F_p^{\text{Gauss}}(r)\right) F_p^{\text{Gauss}}(r) \mathrm{d}r ,$$

where R is the model function given in (7.7). This is the flat-top beam profile with a constant fluence. We later emphasize this fact by replacing R with $R^{\text{flat top}}$. Substituting

$$z = 2F_p \mathrm{e}^{-2r^2/w^2} , \quad \mathrm{d}z = -\frac{8r}{w^2} F_p \mathrm{e}^{-2r^2/w^2} \mathrm{d}r ,$$

we have

$$R^{\text{Gauss}} = \frac{1}{E_p} \int_0^{2F_p} \mathrm{d}z \frac{\pi w^2}{2} R(z) = \frac{1}{2F_p} \int_0^{2F_p} \mathrm{d}z \, R(z) = \int_0^1 \mathrm{d}z' R(2F_p z') .$$

Therefore,

$$\boxed{R^{\text{Gauss}}(F_p) = \int_0^1 \mathrm{d}z \, R^{\text{flat top}}(2F_p z)} , \tag{7.47}$$

where $R^{\text{flat top}}$ is the model function for a flat-top beam profile given by (7.7). For good accuracy, it is important to fit the measured nonlinear reflectivity to $R^{\text{Gauss}}(F_{\text{p}})$ given in (7.47). This is a valid transformation for any function $R^{\text{Gauss}}(F_{\text{p}})$ and $R^{\text{flat top}}$ as long as lateral diffusion can be neglected.

7.3.5 Inverse Saturable Absorption (ISA)

Over the years, we have observed a deviation from the model function [04Hai]. At higher fluences, the reflectivity decreases with increasing fluence according to a second order process like two-photon absorption (TPA), leading to a rollover in the reflectivity curve (Fig. 7.5). The maximum reflectivity is already reached at a certain fluence before the saturable absorber can be fully saturated, and this increases the effective nonsaturable loss and lowers the damage threshold, but has the benefit of less Q-switching instability (to be discussed in Chap. 9). In the femtosecond regime, the most significant part is due to two-photon absorption (TPA) in the SESAM structure, and not limited to the absorber layer alone. However, thermal effects, free-carrier absorption, and other sources of induced absorption also contribute.

From a practical point of view, we take this rollover into account by multiplying the model function given in (7.7) by $\exp(-F_{\text{p}}/F_2)$:

$$\boxed{R(F_{\text{p}})\Big|_{\text{with ISA}} = R(F_{\text{p}})\exp\left(-\frac{F_{\text{p}}}{F_2}\right)}, \tag{7.48}$$

which yields excellent agreement with experiments. Here, F_2 is the fluence when the SESAM reflectivity has dropped to 37% (1/e) due to induced absorption. Figure 7.5 shows the calculated reflectivity $R(F_{\text{p}})$ based on (7.7), multiplied by the ISA factor in (7.48). When $F_2 \to \infty$, (7.48) tends to 1, describing the limit of no ISA or TPA.

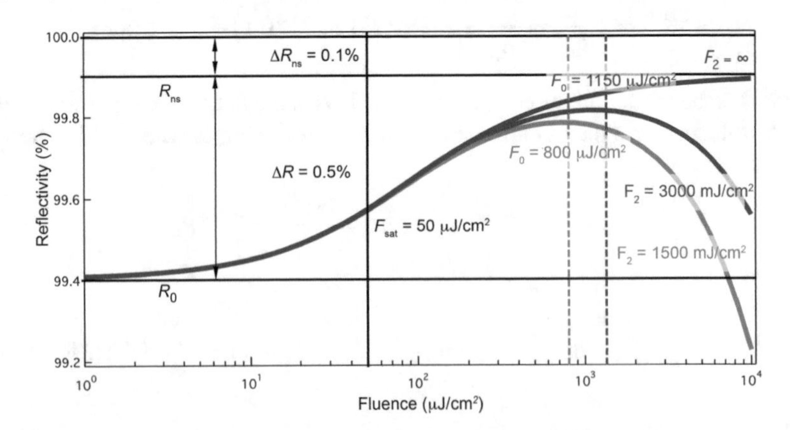

Fig. 7.5 Nonlinear reflectivity with inverse saturable absorption (ISA). Influence of different levels of ISA with different F_2 values (7.50), but keeping the other parameters constant. F_0 is the pulse fluence on the saturable absorber for maximum reflectivity at the rollover (7.52)

For the derivation of (7.48), we assume only some weak TPA. This is justified because of the small thickness of the SESAM. Moreover, the entire nonlinear absorption of a few percent has little effect on the field distribution, which can therefore simply be calculated without TPA. For higher nonlinear absorption, we would have to take into account the fact that the absorption caused by TPA depends on the optical field distribution, which is itself modified by the absorption, so that the exact calculation becomes complicated. For low nonlinear absorption and for a given incident intensity I_{in}, we can then calculate the absorbed intensity (by TPA) I_{abs} for low absorption by simply integrating over the structure as follows:

$$\begin{aligned}
\frac{\mathrm{d}I}{\mathrm{d}z} = -\beta I^2 &\Longrightarrow \mathrm{d}I = -\beta(z)I(z)^2\mathrm{d}z \\
&\Longrightarrow I_{\mathrm{out}} - I_{\mathrm{in}} = -\int \beta(z)I(z)^2\mathrm{d}z \equiv -I_{\mathrm{abs}} \\
&\Longrightarrow I_{\mathrm{abs}} = \int \beta(z)I(z)^2\mathrm{d}z \ ,
\end{aligned}$$

where z is the depth in the structure, $I(z)$ the local intensity, and $\beta(z)$ the local TPA coefficient that depends on the material in the structure. The local intensity $I(z)$ is calculated from the normalized electric field $|E_{\mathrm{n}}(z)|^2$ [normalized so that the incident beam satisfies $|E_{\mathrm{n}}(z=0)|^2 = 1$]:

$$I(z) = \frac{c\varepsilon_0}{2}n(z)|E(z)|^2 = n(z)|E_{\mathrm{n}}(z)|^2 I_{\mathrm{in}} \ .$$

Here, c is the vacuum speed of light, ε_0 is the dielectric constant, and $n(z)$ is the local refractive index. We then have the relative power loss q_{TPA} caused by TPA:

$$q_{\mathrm{TPA}} = \frac{I_{\mathrm{abs}}}{I_{\mathrm{in}}} \ , \tag{7.49}$$

given by the ratio of the absorbed to the incoming intensity. This is based on the argument that the absorbed power must be missing in the reflected beam, as long as no significant power is transmitted through the substrate. We actually work with pulses of fluence F_{p} and FWHM pulse duration τ_{p}, so we have to integrate the TPA effect over the temporal shape:

$$\begin{aligned}
q_{\mathrm{TPA}} &= \frac{\int I_{\mathrm{abs}}(t)\mathrm{d}t}{\int I_{\mathrm{in}}(t)\mathrm{d}t} = \frac{\iint \beta(z)I^2(z)\mathrm{d}z\mathrm{d}t}{F_{\mathrm{p}}} \\
&= \frac{1}{F_{\mathrm{p}}}\iint \beta(z)\left[n(z)|E_{\mathrm{n}}(z)|^2 I_{\mathrm{in}}(t)\right]^2\mathrm{d}z\mathrm{d}t \\
&= \frac{1}{F_{\mathrm{p}}}\int \beta(z)n^2(z)|E_{\mathrm{n}}(z)|^4\mathrm{d}z \int I_{\mathrm{in}}^2(t)\mathrm{d}t \ .
\end{aligned}$$

For sech^2-shaped pulses, we have

$$\int I_{\text{in}}^2(t)\text{d}t \approx 0.585\frac{F_{\text{p}}^2}{\tau_{\text{p}}} .$$

This means that the averaged TPA effect is about 0.585 times the effect for a top-hat pulse with the same fluence and duration. Thus, we obtain

$$q_{\text{TPA}} = 0.585\frac{F_{\text{p}}^2}{\tau_{\text{p}}}\int \beta(z)n^2(z)|E_{\text{n}}(z)|^4\text{d}z . \tag{7.50}$$

Using (7.8), we then simply add the loss of the TPA into the final equation taking into account both nonsaturable losses and the inverse saturable absorber

$$R(F_{\text{p}}) \approx R_{\text{ns}} - \frac{\Delta R}{S}(1-\text{e}^{-S}) - \frac{F_{\text{p}}}{F_2} . \tag{7.51}$$

With F_2 defined by (7.49), (7.50) and $q_{\text{TPA}} \equiv F_{\text{p}}/F_2$, which then implies [05Gra2]

$$F_2 = \frac{\tau_{\text{p}}}{0.585\int \beta(z)n^2(z)|E_{\text{n}}(z)|^4\text{d}z} . \tag{7.52}$$

For $\exp(-q_{\text{TPA}})$, we receive (7.48).

For weak TPA, the rollover in the nonlinear reflectivity $R(F_{\text{p}})$ takes place at higher fluences. When we can assume that this rollover happens in a regime $F_{\text{p}} \gg F_{\text{sat,A}}$, we can very easily determine the fluence F_0 at the maximum saturation level:

$$\left.\frac{\text{d}R(F_{\text{p}})}{\text{d}F_{\text{p}}}\right|_{F_{\text{p}}=F_0} = 0 . \tag{7.53}$$

Using (7.51) and making the approximation of weak TPA, viz.,

$$\left[1-\exp\left(-\frac{F_{\text{p}}}{F_{\text{sat,A}}}\right)\right] \approx 1 , \quad F_{\text{p}} \gg F_{\text{sat,A}} ,$$

we then obtain an equation for F_0 as follows:

$$\begin{aligned}\left.\frac{\text{d}}{\text{d}F_{\text{p}}}R(F_{\text{p}})\right|_{F_{\text{p}}=F_0} &\approx \left.\frac{\text{d}}{\text{d}F_{\text{p}}}\left(R_{\text{ns}} - \frac{F_{\text{p}}}{F_2} - \frac{\Delta R}{F_{\text{p}}}F_{\text{sat,A}}\right)\right|_{F_{\text{p}}=F_0} \\ &\approx \left.\left(-\frac{1}{F_2} + \frac{\Delta R F_{\text{sat,A}}}{F_{\text{p}}^2}\right)\right|_{F_{\text{p}}=F_0} = 0 ,\end{aligned}$$

whence

$$\frac{1}{F_2} = \frac{\Delta R F_{\text{sat,A}}}{F_0^2} ,$$

and finally,

$$\boxed{F_0 \approx \sqrt{F_2 F_{\mathrm{sat,A}} \Delta R}} \,. \tag{7.54}$$

7.3.6 Summary of Relevant Model Functions for Nonlinear Reflectivity

Table 7.4 summarises the important model functions that describe the nonlinear reflectivity of the saturable absorber mirror using the slow saturable absorber approximation. The measurement techniques are described in Sect. 7.7. The fit parameters for measured nonlinear reflectivity using these model functions deliver the important parameters to describe passive Q-switching and passive modelocking. Some of the fit parameters are summarized in Table 7.1, and we have shown that we need the additional parameters F_2 and F_0 for inverse saturable absorption (see Sect. 7.3.5) when the pulses become shorter and for example two-photon absorption becomes significant.

Note that these model functions can be used for any saturable absorber material and mirror design in the slow saturable absorber approximation. When we use semiconductor saturable absorbers, we call these nonlinear mirrors SESAMs, emphasizing the fact that the relevant saturable absorber is a semiconductor material. We have made different approximations, but our experience has been that a much greater variety of SESAM designs can be fitted extremely well with these model functions and the fitting parameters describe the macroscopic device rather than the intrinsic material parameters.

We can measure the macroscopic SESAM parameters using the model functions summarized in Table 7.4 without fully understanding the underlying saturable absorber dynamics, which can become rather complex. However, for parameter optimization, it is useful to better understand the ultrafast relaxation dynamics in the saturable absorbers. An introduction to semiconductor saturable absorbers and their relaxation dynamics is given in Sect. 7.4. In addition, the parameters can also be substantially modified by the SESAM device design (see Sect. 7.5), i.e., the positions inside the SESAM where the absorber layers are placed and the way the electric field enhancement inside the SESAM can be modified with different top reflectors. These two topics will be discussed in more detail in the next two sections. Note that this can be easily applied to any other saturable absorber material for which we have a similarly outstanding fabrication control of the multi-layer nonlinear mirror stack as with semiconductor epitaxial growth. For semiconductor materials we have superior epitaxial growth with highly controllable methods for systematically assembling dissimilar materials into artificial nonlinear mirror structures with atomic-scale precision and adjustable recovery times ranging from femtoseconds to nanoseconds.

Table 7.4 Summary of important model functions for the nonlinear reflectivity with the slow absorber approximation. See the discussion in Sect. 7.2 for more definitions of parameters, i.e., Table 7.1 and Fig. 7.1

$$R(F_p) = R_{ns}\frac{\ln\left[1+\frac{R_0}{R_{ns}}(e^S-1)\right]}{S}, \quad S = \frac{F_p}{F_{sat,A}} \tag{7.7}$$

Pulse-averaged nonlinear reflectivity model function of a SESAM with spatially averaged thin absorbers, in the slow absorber approximation without ISA, and Gaussian beam profile

$$R(F_p) \approx R_{ns} - 2q_p(F_p) = R_{ns} - \frac{\Delta R}{S}(1-e^{-S}) \tag{7.8}$$

Simplification of (7.7) for $\Delta R \leq 10\%$, where q_p is the total loss a pulse experiences in a slow saturable absorber (7.14)

$$R(F_p)\Big|_{\text{with ISA}} = R(F_p)\exp\left(-\frac{F_p}{F_2}\right) \tag{7.48}$$

Correction for inverse saturable absorption (ISA), where F_2 is given in (7.52) for two-photon absorption (TPA). For $\Delta R \leq 10\%$,

$$R(F_p)\Big|_{\text{with ISA}} \approx R(F_p) - \frac{F_p}{F_2} = R_{ns} - \frac{\Delta R}{S}(1-e^{-S}) - \frac{F_p}{F_2} \tag{7.51}$$

$$R^{\text{Gauss}}(F_p) = \int_0^1 dz\, R^{\text{flat top}}(2F_p z) \tag{7.47}$$

Correction for a Gaussian beam profile with pulse fluence $F_p = E_p/\pi w^2$. $R^{\text{flat top}}$ is the model function for a flat-top beam profile given by (7.7) or (7.48).

$$F_0 \approx \sqrt{F_2 F_{sat,A}\Delta R} \tag{7.54}$$

Maximum reflectivity fluence with ISA, when $\Delta R \leq 10\%$ and $F_2 \gg F_{sat,A}$

$$R(F_p, t) = \frac{R_{ns}}{1+\left(\frac{R_{ns}}{R_0}-1\right)e^{-F_{in}(t)/F_{sat,A}}} \tag{7.38}$$

Nonlinear reflectivity model function of a SESAM measured by pump–probe spectroscopy. $F_{in}(t)$ is defined by (7.33): $F_{in}(t) \equiv \int_{-\infty}^{t} I_{in}(t')dt'$ with $F_p = \int_{-\infty}^{\infty} I_{in}(t')dt'$

7.4 Semiconductor Saturable Absorbers

7.4.1 Semiconductor Saturable Absorber Materials

Semiconductor saturable absorber mirrors (SESAM devices) [92Kel2,96Kel] are well-established for passive modelocking or Q-switching of many kinds of solid-state lasers [99Kel,04Kel,07Kel]. Since both linear and nonlinear optical properties of these devices can be engineered over a broad range, the device performance can be readily optimized for a wide variety of laser designs and operating regimes. The main device parameters such as operation wavelength λ, modulation depth ΔR, saturation fluence F_{sat}, and absorber recovery time τ_A can be custom designed over a wide range for either stable cw modelocking [99Hon1] or pure Q-switching [99Spu1], or a combination of both [99Kel].

Semiconductor materials offer a great deal of flexibility in choosing the laser absorption and emission wavelength. The SESAM and the optically pumped vertical-external-cavity surface-emitting laser (VECSEL) (see Sect. 9.8.1 for more details) have very similar layer structures. Without an external cavity mirror, the VCSEL is normally electrically pumped. The operation wavelength can range from around 400 nm in the UV using GaN-based materials to around 2.5 μm in the mid-infrared using GaInAsSb-based materials. Standard high-performance semiconductor material systems which can be grown today cover the infrared wavelength range from 800 nm to 1.5 μm. More recently, high-quality GaSb-based SESAMs have been grown at 2 μm [21Hei] and 2.4 μm [21Bar]. Semiconductor compounds used for different wavelength regimes are AlGaAs (800 to 870 nm), InGaAs (870 to about 1150 nm), GaInNAs (1.1–1.5 μm), InGaAsP (1.5 μm range), and GaInAsSb (2–3 μm). A broader wavelength range for a given material composition may only be obtained at the expense of increased defect concentrations, owing to increased lattice mismatch with a given substrate material. On the other hand, this can be optimized for saturable absorber applications.

The InGaAs/GaAs/AlGaAs semiconductor material system, with properties summarised in Fig. 7.6, offers the materials that are best suited to the 800 nm–1.1 μm wavelength range because of the near-perfect lattice match between GaAs and

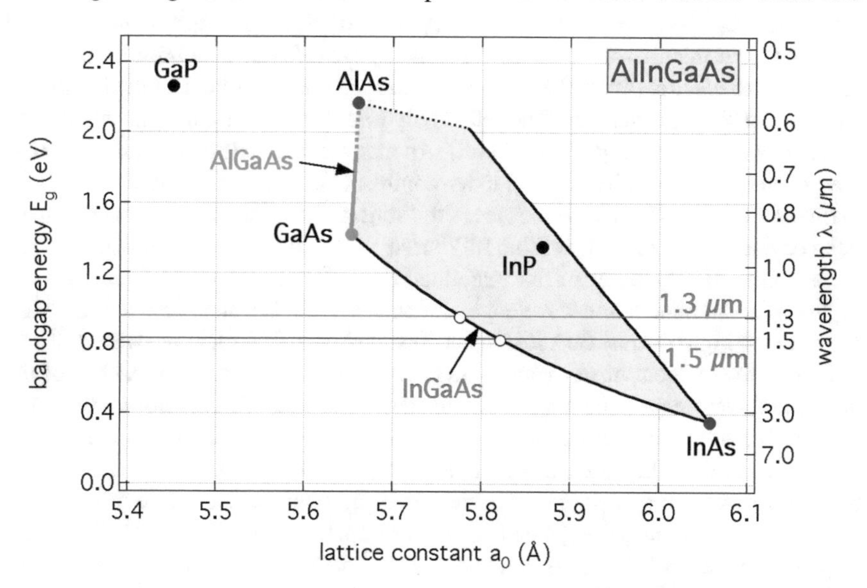

Fig. 7.6 Bandgap versus lattice constant for the GaAs–AlAs–InAs system [06Kel]

AlGaAs. With the inclusion of phosphorus, as Fig. 7.6 indicates, the bandgap opens up into the visible, allowing the fabrication of red diode lasers. A red VECSEL has been reported with this approach [05Has]. InGaAs layers can be grown on GaAs substrates with low compressive strain, and this enhances the quantum well gain, and shifts the bandgap to the 1 μm region. This system has been used successfully for diode lasers and SESAMs for many years. However, at an operating wavelength of about 1 μm, InGaAs saturable absorber layers that exceeded the critical thickness have surface striations that introduce too much scattering loss to be used inside a laser [94Kel]. Low-temperature MBE growth resulted in strain-relaxed structures, with surfaces that were optically flat, but with greatly increased defect densities. For SESAM applications, this is actually advantageous, and has been exploited to optimize the fast dynamic response of the SESAM. For high-power SESAMs, however, it has also become important to do some strain compensation even at an operating wavelength of about 1 μm [16Alf].

InGaAs saturable absorbers have been grown on AlAs/GaAs Bragg mirrors, even at an operating wavelength of 1.3 μm [96Flu2,97Flu] and 1.55 μm [98Flu, 01Har1,02Kra1,03Spu,03Zel,04Zel]. However, these highly strained layers with high indium content exceed the critical thickness, and show significant nonsaturable loss due to strain and defect formation. Optimized low-temperature MBE growth, however, allowed improved InGaAs SESAMs to support stable modelocking in diode-pumped solid-state lasers [02Kra1,03Spu,03Zel,04Zel].

SESAMs and VECSELs at wavelengths of 1.3 and 1.55 μm based on the GaInAsP/InP material system (Fig. 7.7) suffer from low refractive index contrast and poor temperature characteristics. Due to the low refractive index contrast, many InP/GaInAsP mirror pairs are required to form distributed Bragg reflectors (DBRs). This demands very precise control of the growth to achieve DBRs with uniform and accurate layer thickness. The presence of many DBR layers also introduces a high resistance to electrically pumped devices and increases the effective cavity length, resulting in slower dynamics. The reflectivity of a thinner DBR may be augmented by using a layer of gold [03Hoo,17Gui]. An alternative to lattice-matched DBRs is wafer fusion, where high refractive index contrast GaAs/AlAs DBRs are fused to the GaInAsP active layers. The properties of the various material compositions that have been developed to make monolithic InP-based VCSELs are summarised in [01Sag].

A family of GaAs-based nitride materials (Fig. 7.8) have emerged for laser devices in the telecommunication wavelength range between 1.3 and 1.55 μm because they can use high contrast GaAs/AlGaAs DBR mirrors [02Har,02Rie,04Bal,08Ero]. Adding a few percent of nitrogen to InGaAs has two advantages: a redshift of the absorption wavelength and a reduction of the lattice mismatch to GaAs. The drawback is that nitrogen incorporation decreases the crystalline quality, which is a big challenge for the fabrication of active devices [04Buy,08Ero]. However, SESAMs are passive devices relying on fast defect-induced nonradiative carrier recombination to allow for short pulse generation. For example, an optically pumped GaInNAs VECSEL at around 1.3 μm has been passively modelocked with a GaInNAs SESAM [06Rut]. GaInNAs SESAMs centered at 1.3 μm have been demonstrated for solid-state laser modelocking at high repetition rates. The first GaInNAs SESAM was

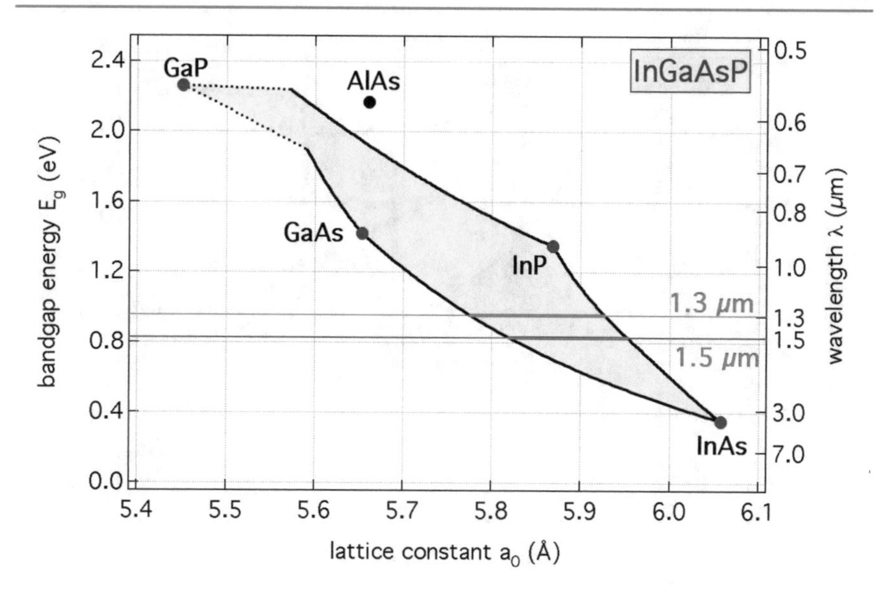

Fig. 7.7 Bandgap versus lattice constant for the InP–InAs–GaAs system [06Kel]

reported to modelock a quasi-cw pumped Nd:YLF and Nd:YALO laser at 1.3 μm [02Sun] and self-starting stable passive cw modelocking of a solid-state laser with a 1.3 μm GaInNAs SESAM was demonstrated soon afterwards [04Liv]. A detailed study of the absorber properties and the modelocking behavior revealed that GaInNAs SESAMs can provide both low saturation fluences and low nonsaturable loss [04Liv,05Gra3,04Sch2]. These improved GaInNAs-SESAM properties supported stable cw modelocking at repetition rates of 5 and 10 GHz and successfully suppressed the increased tendency for Q-switching instabilities at higher pulse repetition rates [05Spu2] (Fig. 7.8).

A GaInNAs material emitting around 1.5 μm is even more challenging because of the many nonradiative defects introduced by the increased nitrogen concentration to match the desired wavelength. In 2003, GaInNAs SESAMs at 1.5 μm were shown to modelock Er-doped fiber lasers [03Okh]. Early attempts to modelock a 1.5 μm solid-state lasers were not possible due to the high nonsaturable loss [03Har]. One approach to overcome this limitation was to introduce antimony into GaInNAs. This allows for both an increase in the emitting wavelength at lower nitrogen concentrations and a better surface morphology of the quantum well. For the first time, successful modelocking of solid-state lasers at 1.54 μm was finally demonstrated using a better GaInNAs SESAM [05Rut].

A major challenge of 2–3 μm SESAM development is that it relies on the GaSb optoelectronic platform, which is much less mature than the conventional InGaAs platform used for near-IR SESAMs. Diode-pumped transition metal (Cr^{2+}) doped ZnS/ZnSe solid-state lasers have shown wide wavelength tunability covering the entire 2–3 μm range. The performance of 2.4 μm SESAMs has been lacking [15Sor], but more recently watt-level and sub-100 fs modelocked 2.4 μm Cr:ZnS laser oscillators have become possible with novel GaSb-based SESAMs [21Bar]. Note that,

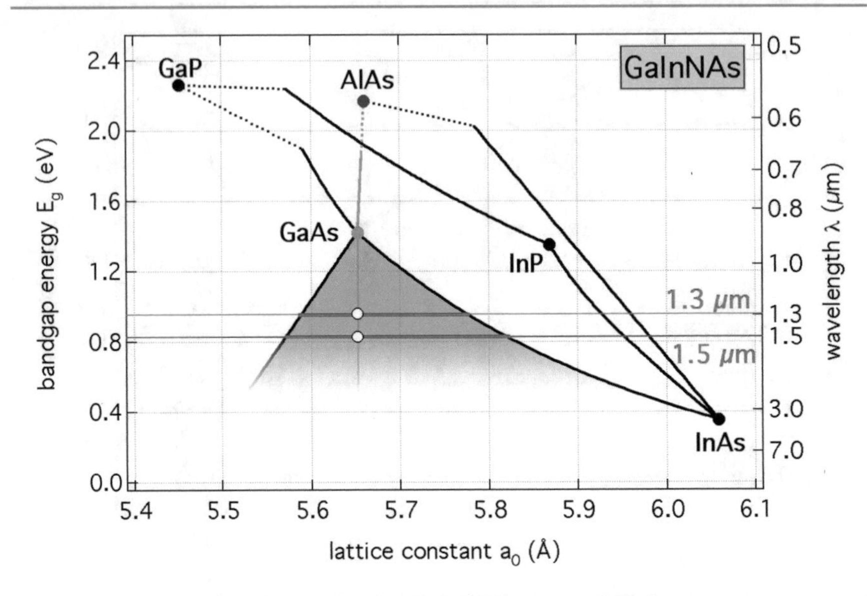

Fig. 7.8 Bandgap versus lattice constant for the GaInNAs system [06Kel]

the measured recovery time of the 2.4 μm GaSb-based SESAM is much faster than the InGaAs-based near-IR SESAM, mainly due to the stronger Auger recombination at lower bandgap.

7.4.2 Introduction to Semiconductor Relaxation Dynamics

Ultrafast semiconductor spectroscopy techniques [99Sha] have provided a very good understanding of semiconductor relaxation dynamics. In ultrafast semiconductor dynamics, it is often convenient to distinguish between excitonic excitations, i.e., Coulomb-bound electron–hole pairs at the band edge [63Ell], and unbound electron–hole pairs in the continuum of the spectrum. Laser pulses with a temporal width well below 100 fs have a spectral bandwidth which is much broader than the spectral width of the exciton resonance and the exciton binding energy in most III–V semiconductors. In addition, low-temperature (LT) MBE growth will smear out the excitonic absorption features. Therefore, saturable absorber applications with either sub-100 fs pulses or with special materials such as LT-grown materials mostly involve continuum excitations. For this reason, we shall focus on ultrafast continuum nonlinearities and dynamics. Exciton dynamics will be discussed only for the special case of a saturable absorber based on the quantum confined Stark effect. For a comprehensive, in-depth review of ultrafast semiconductor spectroscopy, the interested reader is referred to [99Sha].

The semiconductor electronic structure gives rise to strong interactions among optical excitations on ultrafast time scales, and very complex dynamics. Despite the complexity of the dynamics, different time regimes can be distinguished in the evolution of optical excitations in semiconductors. These different time regimes are

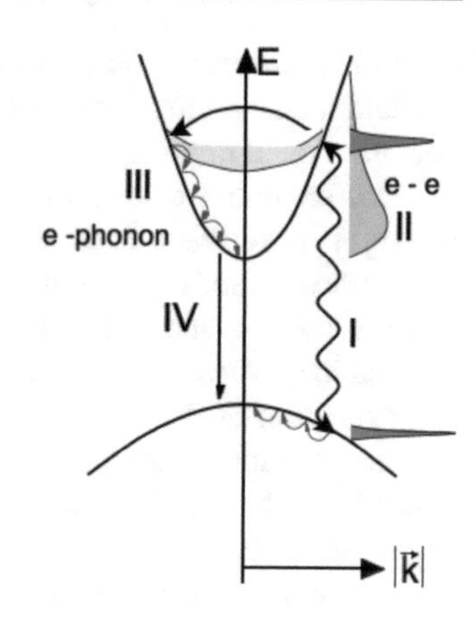

Fig. 7.9 Schematic dispersion diagram $E(k)$ of a 2-band bulk semiconductor, with $E = \hbar\omega$ and the electron momentum $\hbar k$ for an electron wave function. This shows the time regimes I–IV after optical excitation (see text for more details). Electron–electron (e–e) scattering, electron–phonon (e–phonon) scattering

illustrated schematically in Fig. 7.9. The figure shows the energy dispersion diagram $E(k) = \hbar\omega(k)$ of a 2-band bulk semiconductor, which is typical for a III–V semiconductor material, with a higher conduction band and a lower valence band separated by an energy gap. For a low-energy excitation within the bands, we can assume a parabolic band approximation, where the curvature of the bands defines the effective mass of the electrons and holes.

Optical excitation with an ultrafast laser pulse prepares the semiconductor in the coherent regime (time regime I in Fig. 7.9). In this regime, a well-defined phase relation exists between the optical excitations and the electric field of the laser pulse and among the optical excitations themselves. The coherence among the excitations in the semiconductor gives rise to a macroscopic polarization (dipole moment density). Since the macroscopic polarization enters as a source term in Maxwell's equations, it leads to an electric field which is experimentally accessible. The magnitude and decay of the polarization provide information about the properties of the semiconductor in the coherent regime. The irreversible decay of the polarization is due to scattering processes, i.e., electron–electron and electron–phonon scattering, and is usually described by the so-called dephasing or transverse relaxation time. For a mathematical definition of this time constant, the reader is referred to [63Ell, 75All, 99Sha].

After the loss of coherence, ultrafast spectroscopy of semiconductors is solely concerned with the dynamics of the population, i.e., electron and hole distributions. In this incoherent regime, the time regimes II–IV in Fig. 7.9 can be distinguished, as described here. The initial electron and hole distributions are non-thermal in most cases, i.e., they cannot be described by Fermi–Dirac statistics with a well-defined temperature [85Oud, 86Kno]. The excited electrons in the conduction band are excited over an energy spectrum defined by the conservation of energy and momentum, the excitation pulse bandwidth, and the dispersion of the conduction and valence band. The photon momentum $\hbar k_{\text{photon}}$ is much smaller than the electron momentum $\hbar k$ in the semiconductor crystal, because $k_{\text{photon}} = 2\pi/\lambda$ with wavelength $\lambda \approx 1\ \mu\text{m}$ in the infrared, while for the electron $k \leq 2\pi/a$, with lattice constant $a \approx 1$ Å. This results in a 'vertical' transition of the electron with absorption of a photon from the valence band to the conduction band, i.e., no change in the electron momentum with photon absorption, as shown in Fig. 7.9. Depending on the excitation energy, the average electron energy in the conduction band can be well above the band minimum.

With an initial energy distribution well above the band minimum, scattering among charge carriers is mainly responsible for the redistribution of energy within the carrier distributions and for the formation of thermal distributions. This thermalization is shown as time regime II in Fig. 7.9, for the example of a thermalizing electron distribution, where thermalization occurs through scattering among the electrons. For excitation of the continuum, thermalization usually occurs on a time scale of 100 fs under most experimental conditions. The exact thermalization time strongly depends on the carrier density, the excess photon energy with respect to the band edge, and the type of carrier [99Sha, 87Sch].

In general, the carriers have a temperature different from the lattice temperature after e–e thermalization has been completed. It is assumed in Fig. 7.9 that the carriers have a higher temperature than the lattice. For this case, Fig. 7.9 shows schematically the cooling of carriers by the emission of phonons, i.e., energy transfer to the lattice. Cooling defines the time regime III. Typical time constants are in the picosecond and tens of picosecond range. Depending on the excitation parameters, e–e and e–phonon scattering takes place at the same time, and this tends to slow down the initial fast e–e relaxation process.

Finally, the optically excited semiconductor returns to thermodynamic equilibrium by the recombination of electron–hole pairs. Recombination is shown as time regime IV in Fig. 7.9. In a perfect semiconductor crystal, recombination proceeds via the emission of photons or Auger processes at high carrier densities. In a good quality semiconductor, i.e., with a low level of defect states, these recombination processes take place on time scales of hundreds of picoseconds and longer. These slow recombination processes as well as the relatively slow carrier cooling is described in more detail in [99Sha].

Another ultrafast process is encountered if high densities of deep level traps are incorporated in a semiconductor. The trapping of carriers into deep levels can proceed on sub-picosecond time scales (not shown in Fig. 7.9) [99Kel]. Since carrier trapping is important in many saturable absorber applications, it is discussed in more detail below.

We note that the different time regimes overlap temporally. For example, a scattering process may destroy the coherence and contribute to thermalization. Nevertheless, it is very useful to distinguish between the different time regimes because they provide a convenient means for describing the complex semiconductor dynamics. The schematic picture of the different time regimes also demonstrates that two or more time constants are usually required to describe the temporal response of a semiconductor absorber. For example, we recall that thermalization typically takes place on the 100 fs time scale, while carrier trapping proceeds on time scales from a few hundred femtoseconds to a few tens of picoseconds. This results in the measured self-amplitude modulation (SAM) of a semiconductor saturable absorber, as shown in Fig. 7.10.

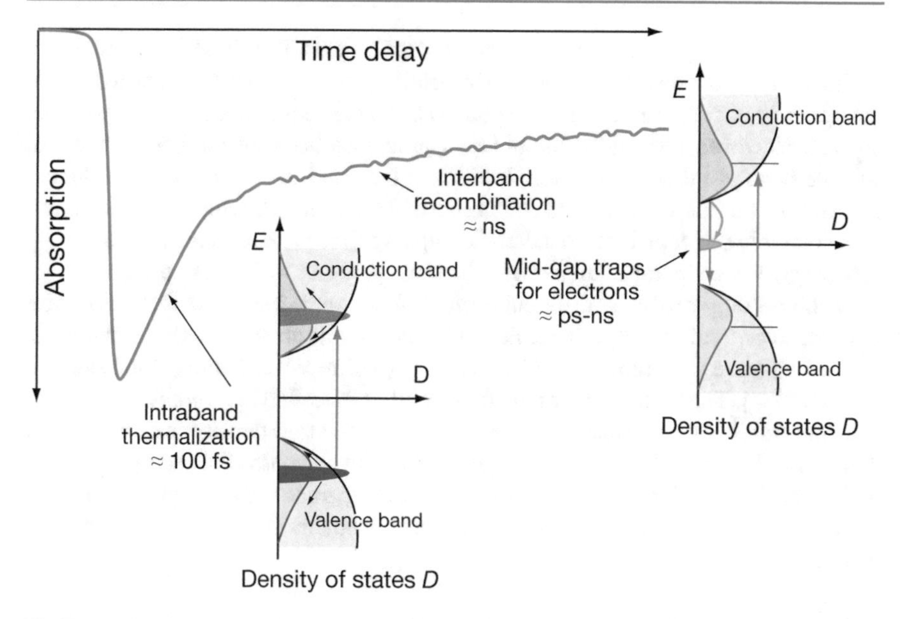

Fig. 7.10 Absorption saturation due to bandfilling. Typical change of absorption as a function of time after a short laser pulse has saturated the optical transition from valence band to conduction band

We can obtain a fast self-amplitude modulation (SAM) in a SESAM based on absorption bleaching with bandfilling as shown in Fig. 7.10. A semiconductor can absorb light if the photon energy is sufficient to excite carriers from the valence band to the conduction band. Under conditions of strong excitation, the absorption is saturated because possible initial states of the pump transition are depleted while the final states are partially occupied. This is shown schematically in Fig. 7.10 with the dark red and blue shaded area over an energy range given by the pulse spectral width. Initially, these electron (red) and hole (blue) distributions are not thermalized and lead to an initial fast reduction in absorption which is referred to as bandfilling and caused by the Pauli exclusion principle. Within typically 60–300 fs after the excitation, the carriers in each band thermalize to a Fermi–Dirac distribution with a higher temperature than the lattice. This hot carrier distribution will thermalize to the lattice temperature via electron–phonon interaction. This is referred to as intraband thermalization. It already leads to a partial recovery of the absorption which depends on how far the pulse energy is above the bandgap. On a longer time scale, typically between a few ps and a few ns depending on defect engineering, they will be removed by recombination and trapping.

7.4.3 Fast Saturable Absorbers with Carrier Trapping Engineering

Processes that remove electrons from the conduction band of a semiconductor lead to a faster recovery of the saturable absorber and therefore a faster saturable absorber. Semiconductor saturable absorber applications in ultrashort pulse generation often require picosecond absorber recovery times (see Chap. 9). The simplest way to obtain such short absorber recovery times with bandfilling nonlinearities would be to remove the optically excited carriers from the bands as fast as required after they have been created. However, intrinsic recombination processes between the conduction and valence band are usually too slow to deplete the band states of a semiconductor. Initially, we therefore introduced defect states in the band gap which give rise to fast carrier trapping to deplete the bands. The trapping time is determined by the density and the type of such traps. Higher trap densities give rise to faster trapping.

In this section we discuss different methods for obtaining useful defects for carrier trapping to give sufficiently fast semiconductor saturable absorbers. This includes (a) low-temperature (LT) molecular beam epitaxy [88Smi,94Liu], and (b) ion implantation [89Zie]. These are standard methods for the controlled incorporation of defect and trap states. In ion-implanted semiconductors, the trap density and the type of defect are determined by the implantation dose. Less controlled incorporation of defects occurs close to surfaces and this tends to degrade the damage threshold of such SESAMs because the non-zero field enhancement at the surface lowers the damage threshold.

(a) Low-Temperature (LT) MBE Growth

The growth temperature controls the defect density in LT-grown semiconductors, where higher defect densities are incorporated at lower temperatures [94Liu]. Semiconductor saturable absorbers can be produced either by molecular beam epitaxy (MBE) or by metal-organic chemical vapor deposition (MOVPE). MBE gives us the additional flexibility to grow semiconductors at lower temperatures, down to about 200 °C, while MOVPE usually requires growth temperatures of about 600 °C to break up the incident molecules on the wafer surface during growth. Lower growth temperatures lead to microscopic structural defects that act as stable traps for excited carriers and thus reduce the recovery time, which is beneficial for use in ultrafast lasers.

GaAs is the best understood LT-grown III–V semiconductor because of its semi-insulating property, as required for GaAs integrated transistors. A detailed review of the properties of LT GaAs motivated by this application can be found in [92Wit,93Wit]. More details about carrier trapping in LT-grown semiconductors for saturable absorber applications at high-excitation levels are given in [96Sie]. The high excitation level is important for this application because we typically want to operate the absorber close to the fully saturated regime with an excitation fluence of about 5 times the saturation fluence (see Sect. 7.2). LT-grown semiconductor saturable absorbers were first used for passive modelocking of diode-pumped solid-state lasers in 1991 [92Kel2], and a more systematic study of growth-temperature-dependent recovery times on the modelocking dynamics was presented in 1993 [93Kel]. There was clearly a trade-off between efficient self-starting of the modelocking process and the onset of Q-switching instabilities with different recovery times of the LT-grown InGaAs saturable absorber. LT-grown InGaAs saturable absorbers had the additional benefit that the strain of InGaAs on a GaAs/AlGaAs DBR was relaxed during LT growth. Therefore, an optically flat surface was even achieved for thicker InGaAs absorber layers. More details on modelocking will be given in Chap. 9.

Figure 7.11a illustrates the basic trap mechanism in LT-grown GaAs [94Liu, 95Liu]. LT growth of GaAs is performed at temperatures of 200–400 °C, as compared to about 600 °C in standard MBE. During LT growth of GaAs, excess arsenic is incorporated in the form of arsenic antisites (As on Ga lattice site: As_{Ga}) at densities as high as 10^{20} cm^{-3}. In undoped LT GaAs, more than 90% of the antisites are neutral, while the rest are singly ionized due to the presence of Ga vacancies (V_{Ga}) which are the native acceptors in the material (see Fig. 7.11a). The ionized arsenic antisites (As_{Ga}^{+}) have been identified as electron traps.

Figure 7.11b shows what happens with post-growth annealing at higher temperatures (typically 600 °C and higher) after LT-growth. Such annealing converts the arsenic antisite point defects into arsenic clusters, so-called As precipitates [90Mel]. The carrier trapping times in as-grown LT GaAs can be in the sub-picosecond regime and show the expected decrease with decreasing growth temperature [91Gup, 92Gup]. Sub-picosecond recovery times of nonlinear transmission or reflectivity changes are also found in annealed LT-grown GaAs, indicating that arsenic precipitates efficiently deplete the band states [97McI, 99Hai2].

Besides an ultrafast carrier trapping and absorber recovery time, other important saturable absorber parameters are the modulation depth and the nonsaturable losses, which remain even at the highest pump energy fluences. Optimized materials combine an ultrafast recovery time with high modulation and low nonsaturable losses.

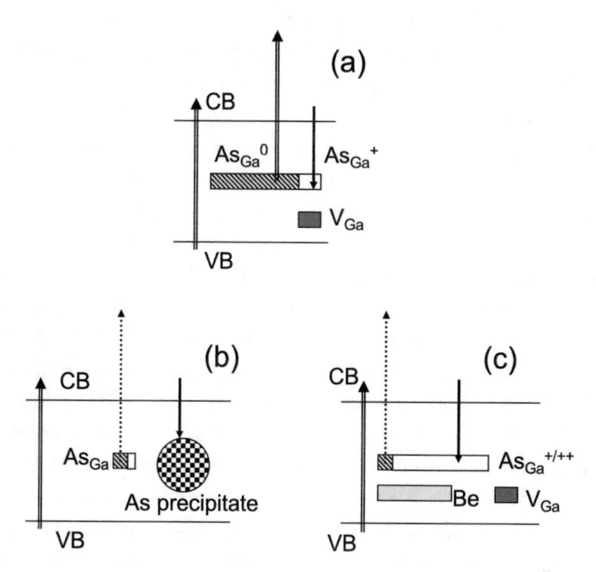

Fig. 7.11 Defect engineering in LT GaAs. Electronic structure of undoped as-grown (**a**), undoped annealed (**b**), and beryllium-doped (i.e., Be is incorporated as an interstitial acceptor for p-doping in GaAs) as-grown LT-grown GaAs (**c**). The *double arrows* mark strong optical absorption transitions. Weak optical absorption transitions are indicated by *dotted arrows*. Trapping processes are shown by *downward arrows*. Conduction band CB, valence band VB, arsenic antisites As_{Ga} (i.e., As on Ga lattice site), neutral As antisites As_{Ga}^{0}, ionized As antisites As_{Ga}^{+}, Ga vacancies V_{Ga}

This material optimization issue has been addressed. The nonlinearity of continuum transitions has to be investigated in different modifications of GaAs (Fig. 7.11). The preparation of the semiconductor layers ensured that the modulation depth and the nonsaturable losses were determined by nonlinear absorption changes.

It has been shown that standard as-grown LT GaAs with an ultrafast carrier trapping time suffers from a small absorption modulation and high nonsaturable absorption losses [99Hai2]. Note that high nonsaturable absorption decreases the modulation depth and causes high nonsaturable losses when the semiconductor absorber is integrated within a mirror structure. The high nonsaturable absorption mainly results from strong defect absorption from neutral As antisites (As_{Ga}^0) to the conduction band (As_{Ga}^0–CB transition in Fig. 7.11a), whose saturation fluence has been shown to be extremely high, i.e., 1.7 mJ/cm^2 [99Hai2]. Therefore, as-grown undoped LT GaAs with femtosecond recovery times and high As_{Ga}^0 density suffers from high nonsaturable absorption losses. The goals of material optimization are (i) to reduce the nonsaturable absorption by reducing the density of neutral As antisites and (ii) to maintain a fast trapping and absorber recovery time.

We have demonstrated two different ways to reach these goals. Annealing of LT GaAs significantly reduces the density of neutral As antisites and nonsaturable absorption (Fig. 7.11b). The simultaneous reduction of the density of useful ionized As antisite electron traps does not substantially increase the absorber recovery time due to the presence of the As precipitates. With decreasing MBE growth temperature, the recovery time decreases in as-grown LT GaAs due to the increasing density of ionized As antisites (As_{Ga}^+), which act as electron traps. However, sub-picosecond recovery times in undoped as-grown LT GaAs are only obtained at the expense of low absorption modulation and high nonsaturable loss. For a given recovery time, annealed LT GaAs has a much larger modulation depth and much lower nonsaturable loss. Since annealing reduces the density of As_{Ga}^+ electron traps, a fast recovery time is maintained by the As precipitates formed upon annealing.

Figure 7.11 provides insight into the microscopic origin of the weak optical nonlinearity in undoped as-grown LT-GaAs with a fast recovery time. It is found that the linear above-bandgap absorption significantly increases with increasing defect density in as-grown LT-GaAs. We recall that, in undoped as-grown LT-GaAs, more than 90% of the arsenic antisites are neutral (As_{Ga}^0) while the rest is ionized (As_{Ga}^+) [94Liu]. For below-bandgap energies, the As_{Ga}^0–CB transition has a much higher absorption cross-section than the transition from the valence band (VB) to the ionized As_{Ga}^+ [88Sil]. Consequently, we attribute the excess linear absorption in undoped as-grown GaAs to the As_{Ga}^0–CB transition. Our quantitative data allowed us to determine the absorption cross-section σ of this transition at 830 nm: $\sigma = 1.4 \times 10^{-16}$ cm^2. From σ, the saturation fluence $F_{sat} = \hbar\omega/\sigma$ of the As_{Ga}^0–CB transition can be determined quantitatively: $F_{sat} = 1.7$ mJ/cm^2. At this fluence, the absorption of the As_{Ga}^0–CB transitions decreases significantly. In contrast, the saturation fluence of the interband transition in GaAs is typically below 50 μJ/cm^2. The comparison shows that the As_{Ga}^0–CB transition is hardly decreased by fluences which almost fully saturate the interband transition. Therefore, the As_{Ga}^0–CB absorption fully contributes to the nonsaturable loss.

Surprisingly, we found that the absorption modulation decreases with increasing As_{Ga}^0 density. This decrease cannot be due to nonsaturable trap absorption through the As_{Ga}^0–CB transition because this contributes equally to both the linear absorption and the nonsaturable absorption. We concluded that there must be another mechanism which gives rise to additional nonsaturable absorption. A possible reason could be free-carrier absorption due to carriers high in the CB. These carriers can be generated by the As_{Ga}^0–CB transition in as-grown undoped LT GaAs. A quantitative analysis showed that this additional nonsaturable loss can make up about 40% of the total nonsaturable absorption in as-grown undoped LT GaAs.

Figure 7.11c therefore shows an alternative method for the optimization of LT GaAs for ultrafast all-optical switching applications which takes advantage of Be doping. Beryllium is incorporated as an interstitial acceptor for p-doping in GaAs, which means that the Be-atom has fewer valence electrons. Doping with Be acceptors reduces the density of neutral As antisites (As_{Ga}^0) and increases the density of ionized As antisites (As_{Ga}^+) [97Spe]. The latter effect ensures ultrafast recovery times. Therefore, Be-doped LT GaAs combines ultrafast recovery times with high modulation depth and low nonsaturable losses [99Hai1].

In conclusion, both annealed LT GaAs and Be-doped LT GaAs combine ultrafast recovery times with high modulation depth and low nonsaturable losses. These materials are well suited for saturable absorber devices in laser physics and for all-optical switching applications [99Sie]. At this point the trapping dynamics have been carefully studied within GaAs bulk material.

For SESAMs at an operating wavelength around 1 μm, InGaAs quantum wells are grown at low temperatures (i.e., LT InGaAs) and typically a large part of the AlAs cladding around them as well. There is less knowledge about the stability, trapping dynamics, and carrier mobility with regards to point defects in superlattice structures. More recently there have been some theoretical predictions about GaAs/AlAs superlattices in [18Jia] which showed that antisite defects (i.e., Ga_{As}, Al_{As}, or As_{Ga}) are energetically more favorable than vacancy and interstitial defects. Furthermore, in [97Gab] it was shown that in AlAs the defects are rather fixed at their positions and cannot be easily moved by annealing. We used this effect with MIXSEL structures, which are similar to optically pumped VECSELs but with an additional saturable absorber layer integrated into the same epitaxial growth (see Sect. 9.8.2 for more details). Femtosecond pulses from a MIXSEL were initially only achieved with a single low-temperature grown InGaAs quantum well embedded in LT AlAs spacer layers to obtain a faster absorber saturation compared to the gain [13Man]. The defects in the LT AlAs barrier layers turned out to be very robust even with long annealing times.

(b) Ion Implantation

Picosecond and sub-picosecond carrier trapping times have also been found in semiconductors implanted with various ion species [89Joh, 91Lam, 95Gan, 95Kro, 96Jag, 99Led]. A decrease in the trapping time with increasing ion dose was observed at lower doses [89Joh, 91Lam, 96Jag]. At higher ion doses, the trapping time can increase with the dose [99Led]. The correlation of trapping times with structural properties of ion-implanted semiconductors has given more insight into this unexpected dose dependence of the trapping time [99Tan]. This work indicates that, not only the defect density, but also the type of defect depends on the ion dose. Both the density and the type of defect affect carrier trapping, leading to longer trapping times if less effective traps are generated at higher ion doses.

Studies of the modulation depth ΔR, the nonsaturable losses ΔR_{ns}, and the recovery time τ_A in ion-implanted GaAs have shown that ΔR decreases and ΔR_{ns} increases with decreasing recovery time and higher defect concentration [99Led]. Nevertheless, if the ion species, ion dose, and annealing conditions are properly chosen, combinations of ΔR, ΔR_{ns}, and τ_A can be obtained which are appropriate for saturable absorber applications. Ion-implanted GaAs is therefore an alternative to annealed or Be-doped LT GaAs as a material for saturable absorber devices.

7.4.4 Fast Saturable Absorbers with Quantum-Confined Stark Effect

A quantum well absorber with intrinsically fast self-modulation can be realised using the nonlinear response of a semiconductor excited in the spectral transparency region at an energy less than that of the exciton resonance. It is well known from atomic physics that virtual emission and reabsorption of non-resonant photons leads to a light-induced shift of the atomic energy levels. This light shift, or optical Stark shift, was first observed in a solid system by Mysyrowicz et al. [86Mys] using pump and probe spectroscopy of GaAs/AlGaAs multiquantum well structures at low temperature. These authors reported a blueshift of the heavy-hole and light-hole exciton resonances induced by irradiation with a sub-picosecond pulse tuned to the transparency region of the wells. No carriers are injected; the excitonic shift arises from the coupling of the exciton to virtual biexciton states, and persists only for the duration of the non-resonant pump pulse.

The quantum well therefore exhibits fast saturable absorption on the low energy wing of the exciton resonance, with enhanced transmission that recovers on the timescale of the pulse duration. To exploit the optical Stark effect as a fast self-absorption modulation mechanism, it is necessary to work at small detuning within the low energy wing of the exciton resonance, where the modulation depth is greatest. This has the advantage that, at small detuning, the shift is larger. However, simultaneously there is resonant excitation of the quantum well, creating a population of real carriers. A SESAM based on the optical Stark effect was used to generate stable sub-500 fs pulses from a modelocked optically pumped semiconducting disk laser [05Hoo].

7.4.5 Saturable Absorber Optimization with Quantum Confinement

Bandfilling nonlinearities as described in Sect. 7.4.2 depend on the density of states for the photoexcited electrons. The density of states can be adjusted with the confinement level, going from bulk materials (3D) to quantum wells (2D) to quantum wires (1D) and quantum dots (0D) (see Fig. 7.12 for 3D and Table 7.5). In Sect. 7.4.6, we derive the density of states.

As we will see later, very often we need to optimize the semiconductor saturable absorber for low saturation fluence. Therefore, generally, quantum well semiconductor saturable absorbers are preferred for higher excitation levels well above the bandgap. Quantum wells with a typical thickness of around 10 nm can also be placed at different positions within the standing wave intensity profile inside a SESAM structure which can be used to further adjust the saturation fluence. An infinitely thin quantum well at the node of a standing wave will introduce no absorption because at the node the electromagnetic field is zero, and this will therefore effectively produce an infinitely large saturation fluence. Thus the saturation fluence within a saturable absorber mirror can be adjusted with such spatially dependent field enhancements (see Sect. 7.5). Quantum wires are less important for SESAM applications because we work with rather large laser mode sizes on the SESAM. However, quantum dots are very interesting because they give additional degrees of freedom for the design, as discussed below.

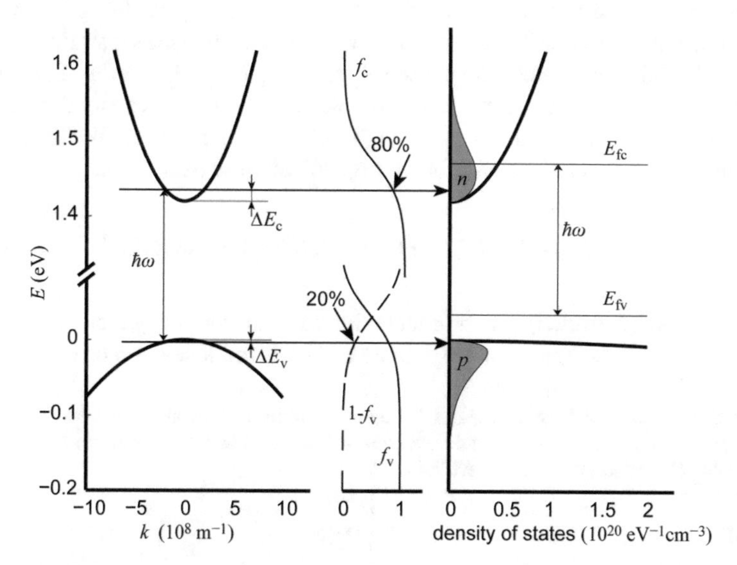

Fig. 7.12 Bulk GaAs is transparent 10 nm above the bandgap. *On the left*, the energy dispersion is shown using the effective mass approximation. *In the middle*, the Fermi functions are given for the conduction and valence band at transparency. *On the right*, the density of states is shown. Note that, at transparency, the Fermi energy in the conduction band E_{fc} and valence band E_{fv} are separated by exactly the photon energy. After [08Maa1]

(a) Bulk Saturable Absorber

In order to optimize absorber properties like saturation fluence and modulation depth, a better understanding of the saturation is needed. We will see that the product of the modulation depth with the saturation fluence is proportional to the transparency density N_0 and therefore strongly related to the density of states. The transparency density N_0 depends on the material composition, wavelength, temperature, and confinement. For bulk GaAs at 300 K, the saturation requirements are depicted in Fig. 7.12.

On the left in Fig. 7.12, the energy dispersion for bulk GaAs is shown using the effective mass approximation. Since the curvature of the valence band is less, the hole effective mass is greater than the electron effective mass (approximately 7 times). The arrow labeled with $\hbar\omega$ indicates the optical transition connecting identical k points in the two bands. As explained before, the photon momentum $\hbar k_{\text{photon}}$ is much smaller than the electron momentum $\hbar k$ in the semiconductor crystal, because $k_{\text{photon}} = 2\pi/\lambda$ with wavelength $\lambda \approx 1\ \mu\text{m}$ in the infrared, while for the electron $k \leq 2\pi/a$, with lattice constant $a \approx 1$ Å. This results in a 'vertical' transition of the electron with absorption of a photon from the valence band to the conduction band, i.e., no change in the electron momentum with photon absorption.

In thermal equilibrium, the Fermi level is inside the bandgap. After pumping the semiconductor, the thermal equilibrium is disturbed, but the conduction band electrons and valence band holes are in thermal equilibrium among themselves. A quasi-equilibrium is formed on a timescale of 100 fs–10 ps. The semiconductor is transparent if the probability of stimulated emission equals the probability of absorption. One can show that this condition is fulfilled when the separation between the quasi-Fermi levels equals the photon energy, i.e., $E_{\text{fc}} - E_{\text{fv}} = \hbar\omega$. The number of electrons should equal the number of holes, i.e., $n = p$. The carrier densities are computed with the density of states $D(E)$ (Table 7.5) and the Fermi functions f_{c} and f_{v} for the conduction and valence band, respectively (see Sect. 7.4.6):

$$n = \int_{E_{\text{c}}}^{\infty} D(E) f_{\text{c}}(E) \mathrm{d}E\ , \quad p = \int_{-\infty}^{0} D(E)\big[1 - f_{\text{v}}(E)\big] \mathrm{d}E\ , \qquad (7.55)$$

with E_{c} the minimum energy of the conduction band and assuming that the maximum energy of the valence band can be set to zero. The Fermi levels can be computed by

Table 7.5 Summary of density of states for an electron in a semiconductor with the effective mass approximation and parabolic band energies. m_{e}^* is the effective mass of the electron. $E = \hbar^2 k^2 / 2\pi m_{\text{e}}^*$. Derivation is given in Sect. 7.4.6

Bulk 3D	$D_3(E) = \frac{V}{2\pi^2}\left(\frac{2m_{\text{e}}^*}{\hbar^2}\right)^{3/2}\sqrt{E}\mathrm{d}E$
Quantum well 2D	$D_2(E) = L^2 \frac{m_{\text{e}}^*}{\pi\hbar^2}\mathrm{d}E$
Quantum wire 1D	$D_1(E) = L\frac{\sqrt{2m_{\text{e}}^*}}{\pi\hbar}\frac{1}{\sqrt{E}}\mathrm{d}E$

numerically solving (7.55) with $E_{fc} - E_{fv} = \hbar\omega$ and the density of states given in Table 7.5. As can be seen from the derivation in Sect. 7.4.6, (7.55) is also valid in 2D and 1D.

As an example, we consider bulk GaAs for a wavelength of 863 nm. We take into account the following parameters for bulk GaAs: bandgap 1.42 eV (873 nm), electron effective mass $0.07m_e$ (m_e is the free electron mass), hole effective mass $0.5m_e$, and photon energy 1.44 eV (863 nm). We then obtain a transparency density of $N_0 = 2.2 \times 10^{18}\ \text{cm}^{-3}$, a Fermi energy $E_{fc} = 1.47$ eV in the conduction band, and $E_{fv} = 0.034$ eV in the valence band. It turns out that the conduction band Fermi energy is inside the conduction band, whereas the valence band Fermi energy is still inside the bandgap. Using the computed Fermi energies, the probability of finding an electron in the conduction band is 80% and the probability of finding a hole in the valence band is 20%. This means that the conduction band is more highly saturated than the valence band, as expected, because of the lower density of states. The fluence needed to obtain transparency can be computed by multiplying the transparency density by the photon energy and the thickness of the layer. Assuming we have a piece of GaAs with a thickness $d = 20$ nm, the transparency fluence F_t needed to generate these carriers is

$$F_t = dN_0\hbar\omega \approx 1\ \mu\text{J/cm}^2\ . \tag{7.56}$$

We can also express the pulse fluence needed for transparency in terms of the macroscopic SESAM parameters. The absorbed pulse energy per unit area for a pulse fluence F_p is equal to $F_p[1 - R(F_p)]$. For a fully saturated SESAM with $F_p \gg F_{sat}$ we obtain from Table 7.4 and (7.8) and with $R_{ns} = 1$,

$$F_t = \lim_{F\to\infty} F[1 - R(F)] \approx F_{sat}\Delta R\ . \tag{7.57}$$

Therefore (7.56) and (7.57) result in

$$F_{sat}\Delta R \approx dN_0\hbar\omega\ , \tag{7.58}$$

which relates the SESAM parameters (which we can measure) to an intrinsic material property, the transparency density N_0. For example, a SESAM with 20 nm bulk GaAs saturable absorber layer and a modulation depth of 1% would have a saturation fluence of 100 μJ/cm^2 according to (7.58).

(b) Quantum Well (QW) Saturable Absorber

In quantum well (QW) absorbers, (7.58) can be written as

$$F_{sat}\Delta R \approx N_0\hbar\omega\ , \tag{7.59}$$

with N_0 the two-dimensional transparency density expressed in cm^{-2}. We typically use the QW absorbers with bandfilling nonlinearities, as before in the bulk material.

Excitonic effects have generally not been used because low-temperature MBE growth and high excitation levels tend to wash these effects out. For a reduced saturation fluence, we can benefit from the density of states. One option is to operate closer to the bandgap. Wavelength-dependent measurements on QW-SESAMs have shown that the modulation depth ΔR is constant, whereas the saturation fluence F_{sat} decreases for photon energies closer to the bandgap [05Gra3]. Another option is the choice of a material with a broad absorption edge like GaInNAs [05Rut], which has a measured transparency fluence (7.57) of $F_{sat}\Delta R = 20$ nJ/cm^2. One further step is to use quantum dots to reduce the density of states even more.

(c) Quantum Dot (QD) Saturable Absorber

QD saturable absorbers help to optimize various parameters: the absorption wavelength can be adjusted with different QD sizes, and the modulation depth can be easily regulated by changing the QD density or number of QD layers. Furthermore, the typical inhomogeneous linewidth broadening leads to a broader absorption range. For an operating wavelength in the near-infrared around 1 μm, self-assembled InAs quantum dots (QDs) can be used [08Maa2]. These are grown using Stranski–Krastanov growth, which critically depends on many MBE growth parameters such as substrate temperature, arsenic pressure, growth rate in number of monolayers per second (ML/sec), and indium monolayer coverage (ML coverage). The indium coverage is controlled by the opening time of the indium source shutter: a longer shutter time results in higher ML coverage, which in turn results in a higher dot density [95Sol]. For a systematic study, we grew the QDs at temperatures of 380°C, 400°C, and 430°C, while all other layers were grown at 600°C. The MBE growth temperature is typically measured using band-edge absorption. Generally, the growth parameters are much more critical for QDs than for QWs.

QD-SESAMs have an additional degree of design freedom compared with bulk and QW saturable absorbers. For QDs the product corresponding to (7.58) and (7.59) can be written as

$$F_{sat}\Delta R \approx \gamma N \hbar\omega \,, \tag{7.60}$$

where N is the dot density and γ the average carriers needed per dot to obtain transparency. For example, $\gamma = 1$ at 960 nm and a dot density of $N = 10^{11}$ cm^{-2} results in $F_{sat}\Delta R = 20$ nJ/cm^2, which is comparable to GaInNAs QWs. The modulation depth can be tuned with the ML coverage, i.e., the QD density, while the saturation fluence remains approximately constant (Fig. 7.13). This is in agreement with (7.60) and in contrast to QWs (7.59) and bulk (7.58) where the product $F_{sat}\Delta R$ is a material constant, i.e., the transparency density N_0. In Fig. 7.13, the measurement data (dots) is fitted (solid lines) with the model function as summarized in Table 7.4, taking into account the Gaussian beam profile (see Sect. 7.3.4) and inverse saturable absorption (7.48). Note that the measurements were made with a femtosecond Ti:sapphire laser, whereas using picosecond pulses, we would have less induced absorption and therefore the rollover would occur at a higher fluence.

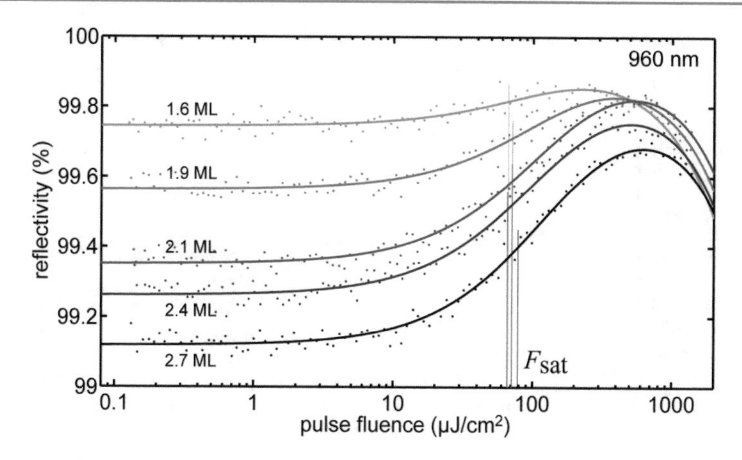

Fig. 7.13 Nonlinear reflectivity measurements of QD-SESAMs at a center wavelength of 960 nm with 140 fs pulses. The *upper curve* has the lowest monolayer (ML) coverage (dot density) and therefore the smallest modulation depth. By increasing the dot density, the modulation depth also increases, while the saturation fluence remains approximately constant. A saturation fluence of around 65–70 $\mu J/cm^2$ was measured [08Maa2]

Due to its inherent inhomogeneous size distribution, the dots have broader uniform spectral properties than quantum wells. Nevertheless, the modulation depth increases slightly when operated at shorter wavelengths, due to excited state absorption. The pump–probe measurements show that a higher dot density results in faster recombination most likely because of defect recombination. Post-growth annealing reduces the saturation fluence, but it tends to make the saturable absorbers slower. The small modulation depth can be increased by growing more than one QD absorber layer, if necessary.

For GaAs absorbers, we obtained fast recovery with low-temperature MBE growth (below 350°C) to create point defects to trap the carriers (see Sect. 7.4.3). Low-temperature growth is not required for QDs to have fast recovery times. Rafailov et al. have measured recovery times of approximately 1 ps [04Raf]. As shown in Fig. 7.14, we measured a strong bitemporal response with a large fast component which quickly drops to 50% of its original value and is followed by a slow component in the picosecond regime depending on the growth parameters. The slow component is determined by carrier recombination and carrier escape. The sample grown with 1.6 ML coverage has a slow component of approximately 500 ps which is the expected recombination time for InAs QDs [00Pai]. The measurement shows that, by increasing the ML coverage, the recombination becomes faster, which can be explained by a higher defect density and thus a faster recombination [95Ser], or a higher interdot transfer probability because of the smaller distances between the dots [05Maz]. The fast processes are due to transitions in the dots. There are several processes that can explain the fast relaxation mechanism. Auger processes are much more likely than phonon relaxation [91Ben], and recovery times of 0.5 ps have been measured [01Ber]. Another very fast process is thermal hole activation [02Quo], where the hole is removed and stimulated recombination drops (a SESAM

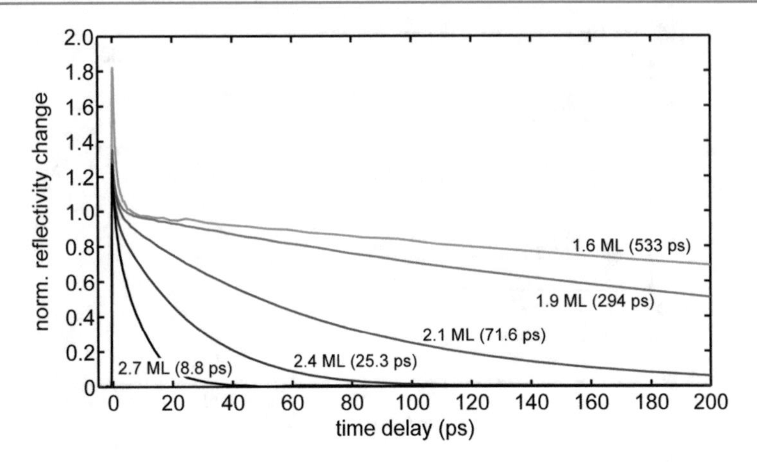

Fig. 7.14 Pump–probe measurements of the five samples with different ML coverage, measured at a center wavelength of 960 nm. The strong bitemporal recovery is fitted to two exponential decay times with slow and fast components. The measurements are normalized so that the amplitude of the slow exponential decay is unity. Ideal for passive modelocking in the soliton modelocking regime are recovery times <30 ps [08Maa2]

is saturated when the stimulated emission is equal to the absorption). We measured a fast relaxation time constant between 0.7 and 1.2 ps.

(d) Novel Concepts

Typical QD SESAMs operating in the 1 μm range are based on self-assembled InAs QD layers grown at temperatures below 400°C. This leads to an increased number of growth-related defects, especially arsenic anti-site point defects. However, the optical quality of InAs or InGaAs QDs improves with growth temperatures above 430°C, and this is measurable by increased photoluminescence intensities. The question is: how can we still obtain a fast recovery for such QDs?

Semiconductor MBE growth technology offers more than we have discussed so far. Instead of non-radiative recombination centers, the radiative recombination is enhanced by a p-type δ-doping, which generates a high hole density in QDs by modulation doping [20Fin]. This allows for faster recombination and can reduce the slow component of the recovery dynamics, because no hole transport is needed anymore, and the hole density can be kept constantly high, even though the electron density becomes low on longer time scales. The effect of p-doping should not be confused with the positive effect of p-doping of low-temperature grown GaAs absorbers. There, the p-doping improves the performance due to the reduction in the density of neutral anti-site defects (As^0_{Ga}) in comparison to charged ones (As^+_{Ga}) (see Fig. 7.11c) and has nothing to do with the enhancement of radiative recombination.

With a new concept of p-type δ-doping in close proximity of the QD layer and subsequent post-growth annealing (Fig. 7.15), we obtained similar fast recovery times, as before, using low-temperature grown QW-SESAMs, but with a very high temperature insensitivity suitable for monolithic integration with longer annealing times

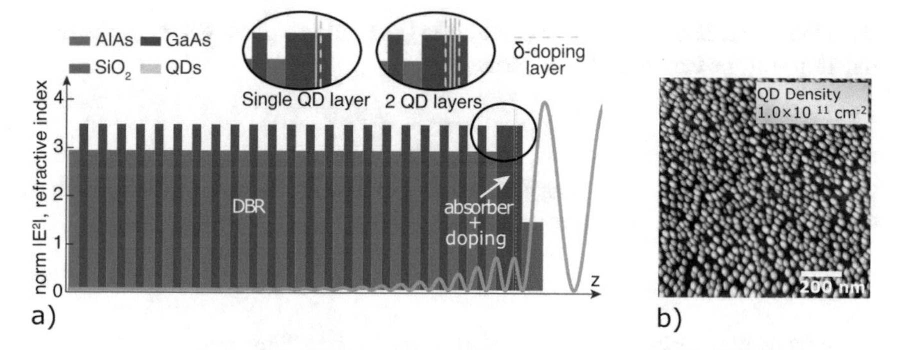

Fig. 7.15 **a** Schematic layer stack of a complete QD-SESAM structure. Refractive index profile (colored as shown) and electric field intensity pattern at a wavelength of 1030 nm (*grey*). One or two QD layers and δ-doping were incorporated. **b** AFM scan ($1 \times 1\ \mu m^2$) of $In_xGa_{1-x}As$ QDs grown at 480 °C with an indium content of 60% and a nominal deposited thickness of 1.4 nm [20Fin]

[20Fin]. The intention is to get a high percentage of holes from the δ-doping layer to the QDs, but to avoid an additional non-radiative recombination effect caused by the dopant impurity itself.

This clearly demonstrates that the fast decay rates in saturable absorbers based on non-radiative recombination centers can be successfully replaced by an enhanced radiative recombination in high optical quality QD absorbers using a spatially separated p-type high-density modulation doping. We expect more novel concepts to allow for even better SESAMs in the future, because we can benefit from a large commercial effort in semiconductor photonics and their investment in epitaxial growth.

7.4.6 Derivation of the Density of States

The density of states is very important for the design of the saturable absorber parameters (see Table 7.5). It depends on the quantum confinement. We consider a crystal cube with volume $V = L^3$, where L is the side of the cube. The cube is macroscopic and contains all the carriers in the saturable absorber. However, for the derivation of the density of states, we still need to be able to count the states, and this can be done by assuming infinitely high potential barriers on the boundaries of the crystal. The electrons are then described by standing wave functions $\psi(x)$ with the boundary condition that they vanish on the faces of the cube, i.e.,

$$\psi(\mathbf{r} = 0) = 0\,, \quad \psi(x + L, y, z) = 0\,, \quad \psi(x, y + L, z) = 0\,, \quad \psi(x, y, z + L) = 0\,. \tag{7.61}$$

From quantum mechanics, for an electron in an infinitely deep quantum well with thickness L, we then have the electron wave function given by $\psi_m(x) \propto \sin k_m x$, and with the boundary condition of (7.61), we find that k_m is quantized, i.e.,

$$k_m = m\frac{\pi}{L}\,, \quad m = 0, 1, 2, \ldots\,. \tag{7.62}$$

Here the k-values can only be positive. It is better to use plane waves as the basic functions for the electrons, so we write

$$\sin k_m x \sim \mathrm{e}^{\mathrm{i}k_m x} - \mathrm{e}^{-\mathrm{i}k_m x} ,$$

where k_m can have positive or negative values. For the plane waves, we then introduce the periodic boundary conditions

$$\psi(x+L, y+L, z+L) = \psi(x, y, z) . \tag{7.63}$$

The quantization of k_m is then given as follows: in the x direction,

$$\psi(x+L) = \psi(x) \implies \mathrm{e}^{\mathrm{i}k(x+L)} = \mathrm{e}^{\mathrm{i}kx} \implies k_m L = m2\pi ,$$

whence

$$k_m = m\frac{2\pi}{L} , \quad m = 0, \pm 1, \dots . \tag{7.64}$$

The same occurs in the y and z directions. We can therefore count the number of states in the full k-space very easily. The distance between the different states in the k-space is given by $\Delta k = 2\pi/L$, according to (7.64). We thus obtain for the one-dimensional density of states $D_1(\mathbf{k})$, with one state within Δk:

$$D_1(\mathbf{k}) = 2\frac{1}{\Delta k} = \frac{L}{\pi} , \tag{7.65}$$

where the factor of 2 comes from the spin, i.e., the Pauli exclusion principle allows for 2 electrons per electron wave function $\psi_m(x) \propto \exp(k_m x)$, one with spin up and one with spin down. Therefore, for the higher dimensions, we obtain

$$D_2(\mathbf{k}) = 2\frac{1}{(\Delta k)^2} = \frac{L^2}{2\pi^2} , \quad D_3(\mathbf{k}) = 2\frac{1}{(\Delta k)^3} = \frac{L^3}{4\pi^3} . \tag{7.66}$$

For the density of states for the k-space, i.e., $k = |\mathbf{k}|$, we then obtain (see Fig. 7.16):

$$\begin{aligned} D_1(k) &= D_1(|\mathbf{k}|) = 2\mathrm{d}k\, D_1(\mathbf{k}) = \frac{2L}{\pi}\mathrm{d}k , \\ D_2(k) &= D_2(|\mathbf{k}|) = 2\pi k\mathrm{d}k\, D_2(\mathbf{k}) = \frac{L^2}{\pi}k\mathrm{d}k , \\ D_3(k) &= D_3(|\mathbf{k}|) = 4\pi^2 k^2\mathrm{d}k\, D_3(\mathbf{k}) = \frac{L^3}{\pi}k^2\mathrm{d}k . \end{aligned} \tag{7.67}$$

With the effective mass approximation for the electron in a parabolic energy band (Figs. 7.9 and 7.12) inside an infinitely deep quantum well, the energy is simply given by the kinetic energy

$$E = \frac{\hbar^2}{2m_\mathrm{e}^*}k^2 , \tag{7.68}$$

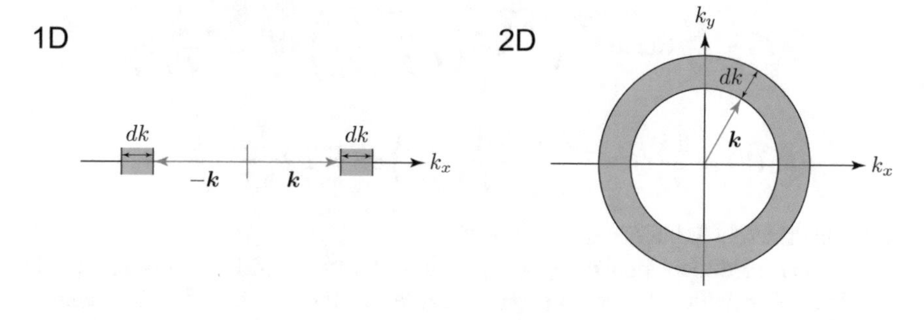

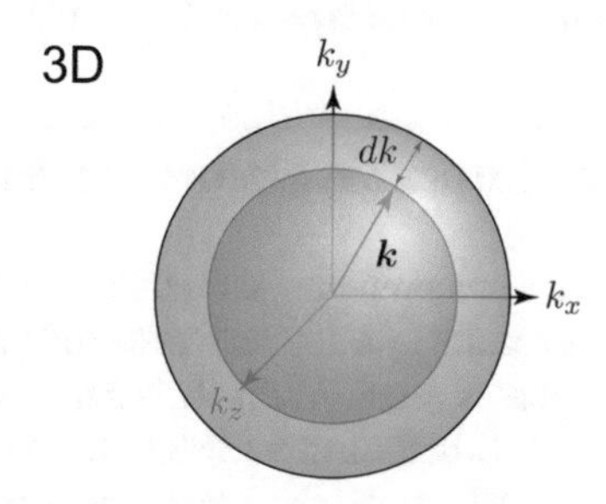

Fig. 7.16 k-space area with $k = |\mathbf{k}|$. In the 3D case, this is a spherical shell of thickness dk and radius k, where we assume that L is large and $\Delta\mathbf{k} \to \mathbf{dk}$ becomes very small. The 3D k-space area is then given by the spherical surface with radius k times the thickness dk of the shell (i.e., $4\pi^2 k^2 \mathrm{d}k$) as used in (7.67). Analogously for 2D, the k-space area is given by the circumference of a circle with radius k times dk (i.e., $2\pi k \mathrm{d}k$), and for 1D, 2 times dk at a constant distance k in the positive and negative k-directions

where m_e^* is the effective mass of the electron. Using (7.67), we can then derive the density of states in E-space. Equation (7.68) implies

$$\frac{\mathrm{d}E}{\mathrm{d}k} = \frac{\hbar^2}{m_\mathrm{e}^*}k \,, \quad k = \sqrt{\frac{2m_\mathrm{e}^*}{\hbar^2}E} \,, \tag{7.69}$$

whence

$$\frac{\mathrm{d}E}{\mathrm{d}k} = \sqrt{\frac{2}{m_\mathrm{e}^*}}\hbar\sqrt{E} \,,$$

and finally,

$$\mathrm{d}k = \sqrt{\frac{m_\mathrm{e}^*}{2}}\frac{1}{\hbar\sqrt{E}}\mathrm{d}E \,. \tag{7.70}$$

We then use (7.67) with (7.69) and (7.70) to obtain

$$D_3(E) = \frac{L^3}{\pi}\left(\frac{2m_\mathrm{e}^*}{\hbar^2}E\right)\left(\sqrt{\frac{m_\mathrm{e}^*}{2}}\frac{1}{\hbar\sqrt{E}}\right)\mathrm{d}E = \frac{L^3}{2\pi^2}\left(\frac{2m_\mathrm{e}^*}{\hbar^2}\right)^{3/2}\sqrt{E}\mathrm{d}E \,,$$

$$D_2(E) = \frac{L^2}{\pi} k \mathrm{d}k = \frac{L^2}{\pi} \sqrt{\frac{2m_\mathrm{e}^* E}{\hbar^2}} \left(\sqrt{\frac{m_\mathrm{e}^*}{2}} \frac{1}{\hbar\sqrt{E}} \right) \mathrm{d}E = L^2 \frac{m_\mathrm{e}^*}{\pi \hbar^2} \mathrm{d}E \,,$$

$$D_1(E) = \frac{2L}{\pi} \mathrm{d}k = \frac{2L}{\pi} \left(\sqrt{\frac{m_\mathrm{e}^*}{2}} \frac{1}{\hbar\sqrt{E}} \right) \mathrm{d}E = L \frac{\sqrt{2m_\mathrm{e}^*}}{\pi \hbar} \frac{1}{\sqrt{E}} \mathrm{d}E \,,$$

as summarized in Table 7.5 using $L^3 = V$.

For the zero-dimensional case, i.e., quantum dots, we consider the discrete levels of a single atom with only two states per energy level. However, in reality the quantum dots in semiconductors have a finite size (see Fig. 7.15).

7.5 Semiconductor Saturable Absorber Mirror (SESAM)

We typically integrate the semiconductor saturable absorber with a highly reflective mirror structure, and this results in a device whose reflectivity increases as the incident optical intensity increases (Sect. 7.2.1). This general class of devices are called semiconductor saturable absorber mirrors (SESAMs). Since both linear and nonlinear optical properties of SESAMs can be engineered over a wide range, the device performance can be readily optimized for a wide variety of laser designs and operating regimes. The main device parameters, such as center operation wavelength λ_0, modulation depth ΔR, saturation fluence $F_{\mathrm{sat,A}}$, nonsaturable losses ΔR_{ns}, absorber recovery time τ_A, and the rollover parameter F_2 can be custom designed over a wide range for either stable cw modelocking or pure Q-switching, or a combination of both (see Chaps. 8 and 9). These device parameters can be measured with high accuracy (Sect. 7.7).

One important parameter of a SESAM device is its saturation fluence, which has typical values in the range of 1 to 100 μJ/cm^2 depending on the SESAM design. Lower saturation fluence is particularly relevant for many modelocked solid-state lasers to reduce Q-switching instabilities. For example, at higher gigahertz pulse repetition rates, it becomes harder to saturate the SESAM device, as the intracavity pulse energy becomes increasingly lower. Semiconductor saturable absorbers are ideally suited because of the large absorber cross-section (in the range of 10^{-14} cm^2) and correspondingly small saturation fluence (in the range of 10 to 100 μJ/cm^2). The saturation fluence can be further reduced with different materials, i.e., GaInNAs absorbers can show decreased saturation fluence [97Kon,04Liv,05Rut], and quantization levels, i.e., quantum wells and quantum dots. In addition, the SESAM structure can be described as a multilayer design which determines the exact position of the saturable absorber layers with respect to the standing wave pattern of the incident light. The peak of the field intensity at the absorber layer then determines the saturation fluence (see Sect. 7.5.2) [96Kel,05Spu3].

The SESAM device benefits from the nanotechnology of semiconductor epitaxial material deposition, allowing control of layers of material down to sub-nanometer accuracy with defect and doping engineering. This allows for very precise optical devices with practically complete control of the key design parameters, such as the magnitude and phase of the optical absorption and reflection, plus adjustment of key parameters such as the saturation fluence and its temporal response. Once the device design has been optimized, semiconductor wafer fabrication can be used to produce these devices economically, and ultimately in large volumes.

7.5.1 SESAM Design: A Historical Perspective

The evolution of the concept of the SESAM device began with coupled cavity modelocking (see Chap. 9). To understand and appreciate this evolution, it helps to be aware of the state of knowledge in the early 1990s. It is worth remembering that the current status of our understanding is usually taken for granted. If we look back to around 1990, the current approach was often assumed impossible, and we had no idea how to solve long-term ongoing problems. Relevant and important examples are [10Kel]:

- Diode-pumped solid-state lasers could not be passively modelocked without Q-switching instabilities. Theory even further established the commonly accepted knowledge that this could not work.
- Novel concepts based on coupled cavity modelocking were assumed to finally resolve these problems, but added a lot of other problematic issues.
- Novel Ti:sapphire lasers generated pulses without any known explanation or any visible saturable absorber inside, and contradicted the established modelocking theories.
- Stable passive modelocking was obtained with long, open net gain windows after the pulse, and it was not clear why noise would not grow in these windows and ultimately destabilize the pulse generation process.

All these problems ultimately resulted in an explosion of activity which made modelocking of solid-state lasers a hot topic of research for more than two decades.

We worked on resonant passive modelocking (RPM), which is a coupled cavity modelocking technique for which the nonlinearity inside the coupled cavity is based on a semiconductor saturable absorber (Fig. 7.17). For RPM, we used an early SESAM design in a coupled cavity. However, this SESAM had a far too large modulation depth and loss, due to its 75 quantum wells, to be introduced inside the main cavity, and in any case, we were all somewhat stuck on the assumption that a simple intracavity saturable absorber would not work for a diode-pumped solid-state laser. Nevertheless, the more detailed understanding of how RPM worked helped us to design the first working intracavity saturable absorber device which we called the antiresonant Fabry–Pérot saturable absorber (A-FPSA); in retrospect, not a good choice for an acronym [10Kel].

The RPM operation can be understood on the basis of the intensity-dependent coupled cavity reflectivity as shown schematically in Fig. 7.17. From inside the main cavity, this nonlinear coupled cavity's reflectivity R_{nl} looks like that of a Fabry–Pérot cavity (Fig. 7.18). For an amplitude nonlinearity (i.e., the case for RPM) the laser

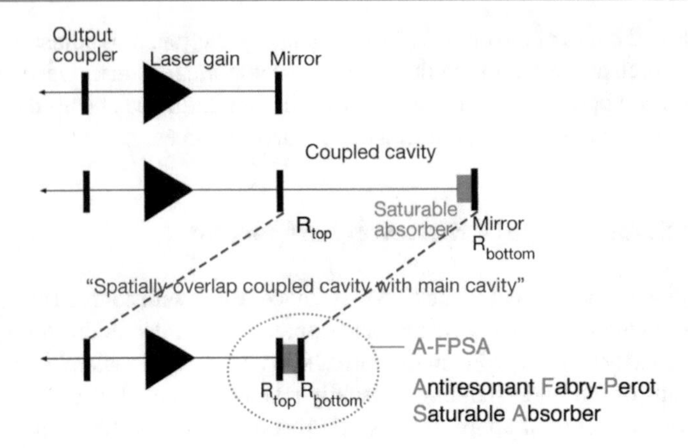

Fig. 7.17 Moving from coupled cavity modelocking with a saturable absorber, i.e., resonant passive modelocking (RPM), towards SESAM modelocking with the first SESAM design: an antiresonant Fabry–Pérot saturable absorber (A-FPSA) [92Kel2]

experiences an intensity-dependent reflectivity change at the center frequency ν_0 of the pulses at the maximum reflectivity which corresponds to the antiresonance regime of the Fabry–Pérot reflector. The coupled cavity in RPM can therefore be treated as an intracavity antiresonant Fabry–Pérot saturable absorber (A-FPSA) [92Kel2].

In fact, modelocking with an A-FPSA is equivalent to a 'monolithic' RPM, as shown schematically in Fig. 7.17, because when we spatially overlap the coupled cavity with the main cavity, we obtain the design guidelines for an A-FPSA. The Fabry–Pérot thickness is designed for antiresonance for the center frequency ν_0 of the modelocked laser pulses according to Fig. 7.18. Based on the RPM Nd:YLF laser parameters [92Kel1] we had a coupling mirror R_{top} between the two coupled cavities with only 2% transmission. If we then choose the equivalent top mirror with 98% reflectivity (Fig. 7.22), we obtain the first SESAM device that passively modelocked a diode-pumped solid-state laser. This first SESAM design was based on 50 InGaAs/GaAs quantum wells grown at low temperatures by molecular beam epitaxy (MBE) (red in Fig. 7.22) with a highly reflecting bottom mirror and a 98% top mirror. This worked right away and resulted in stable self-starting passive modelocking of a Nd:YLF laser generating 3.3 ps pulses [92Kel2]. Although this formal relation with RPM exists, it is simpler and equally correct to view the entire A-FPSA as an effective intracavity saturable absorber. We therefore used the acronym A-FPSA for the first SESAM device.

The common accepted guideline for 'creating a new acronym' in science is that something new has to be discovered or invented to justify a new name. The SESAM was a new intracavity saturable absorber device that solved the long-standing Q-switching problem for solid-state lasers. Therefore, it was appropriate to create such an acronym, and today this has also been widely accepted in the scientific community. The SESAM is a relatively good and very appropriate acronym which describes the most important characteristics of this intracavity saturable absorber device that is based on semiconductor nonlinearities integrated with a highly reflecting mirror. The

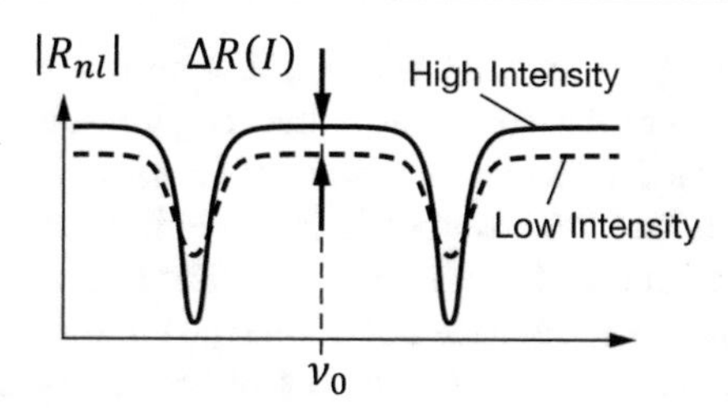

Fig. 7.18 The coherent superposition of the pulse in the main cavity with the pulse from the nonlinear coupled cavity shown in the middle panel of Fig. 7.17 can be described by modeling the laser coupling mirror R_{top} and the external cavity as a nonlinear intensity dependent Fabry–Pérot with a reflectivity R_{nl}. The pulses have a center frequency ν_0 for which the Fabry–Pérot reflectivity is maximal, which is the antiresonance of a Fabry–Pérot. At higher intensity the saturable absorber inside the Fabry–Pérot cavity becomes saturated with less loss. This results in a higher finesse Fabry–Pérot reflectivity and a saturable absorber modulation $\Delta R(I)$ at the pulse center frequency ν_0 at the antiresonance of the Fabry–Pérot [91Hau2,92Kel1]

SESAM is a multilayer structure optimized for a certain amplitude and phase reflectivity response, and therefore has a large number of different specific layer structures. The A-FPSA is only one option out of many SESAM designs and we therefore refer to this type of SESAM as the antiresonant high-finesse SESAM (Fig. 7.22), which will become clear from the following discussions.

The semiconductor material has been so successful because the absorber cross-section is around 10^{-14} cm^2, whereas the diode-pumped solid-state laser gain cross-section is typically around 10^{-19} to 10^{-22} cm^2. This leads to a significantly lower saturation energy of the absorber, which is highly beneficial for the stable mode-locking of such lasers (see Chap. 9). Thus the saturation fluence of a SESAM can be designed to be in the range of 10 μJ/cm^2 without pushing it too much into resonance (see Sect. 7.5.2).

7.5.2 Resonant Versus Antiresonant SESAM

The difference between a resonant and an antiresonant SESAM is the field intensity enhancement in the absorber layers inside the SESAM structure. This is illustrated in Fig. 7.19 which shows two different SESAM designs to explain the basic difference between a resonant and an antiresonant SESAM. The distributed AlAs/GaAs Bragg reflector (DBR) serves as a highly reflective bottom mirror in the SESAM (see Sects. 3.6.1 and 7.5.4 for more details). The DBR contains quarter-wave layers of a low-index material (AlAs, $n_{\text{AlAs}} = 2.96$ at 960 nm) and a high-index material (GaAs, $n_{\text{GaAs}} = 3.54$ at 960 nm) indicated by the different grey-shaded sections with different heights proportional to the refractive index. By increasing the number of pairs, the reflection bandwidth remains constant, but the reflectivity increases. In the near-infrared regime we typically have to use 30 mirror pairs of AlAs and GaAs layers which results in about 0.03% mirror reflection loss for the resonant SESAM and even less for the antiresonant SESAM. This mirror loss adds to the total nonsaturable losses which need to be much lower than the output coupler for an efficient laser. On top of this AlAs/GaAs DBR mirror is a spacer layer with a single InGaAs quantum well saturable absorber (i.e., the solid black line inside the last GaAs spacer layer). The thickness of the last GaAs spacer layer after the DBR mirror is chosen such that the absorber is in the anti-node of the standing wave pattern of the electric field as shown in Fig. 7.19. The thickness of the spacer layer determines the field intensity enhancement at the quantum well absorber.

We can make the analogy with a Fabry–Pérot formed by the last GaAs spacer layer with the two reflections of the bottom DBR and the GaAs/air interface top reflections. In this case by making the roundtrip phase change in this Fabry–Pérot equal to $\pi(2m-1)$, the SESAM is antiresonant, and by making the roundtrip phase change equal to $2\pi m$, the SESAM is resonant, where m is an integer.

We can define a field intensity enhancement factor ξ for the SESAM by [05Spu3]

$$\xi = |\mathscr{E}(z)|^2 , \tag{7.71}$$

where z is the position of the absorber. For the examples shown in Fig. 7.19, in the antiresonant SESAM, the enhancement is $\xi \approx 0.32$, while in the resonant case $\xi = 4$. The roundtrip phase includes the phase shift from the lower DBR and the air interface (see a more detailed discussion in Sect. 7.5.4).

The SESAM modulation depth is proportional to the enhancement factor, i.e., $\Delta R \propto \xi$, and the SESAM saturation fluence is inversely proportional to it, i.e., $F_{\text{sat}} \propto 1/\xi$. The product $F_{\text{sat}}\Delta R$ is thus a constant which is simply a material parameter for bulk and quantum wells [see (7.58) and (7.59) in Sect. 7.4.5], but can be changed for quantum dots with the dot density N (7.60). Note that a quantum well absorber at the node of the standing wave results in $\xi \approx 0$ with practically no effective absorption in the SESAM device. Shifting the quantum well layer within the standing wave pattern allows for an additional degree of freedom for adjusting the saturation fluence. Note at this point we have only shown two SESAM designs without any additional dielectric top coating. Such examples will be discussed later in this chapter. Note also that there are several trade-offs when we move the saturable absorber towards a more resonant design. This can be explained as follows.

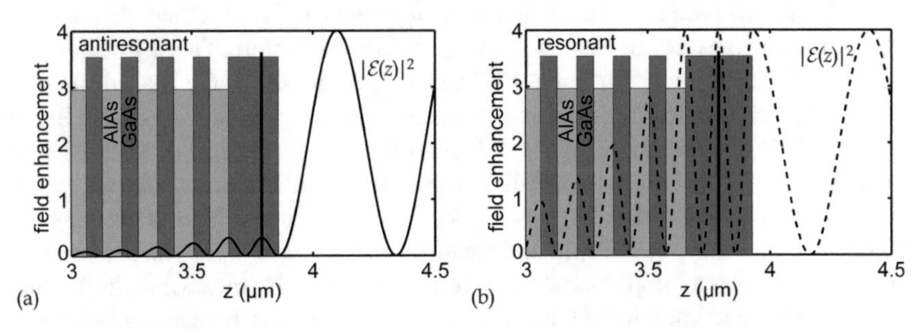

Fig. 7.19 Refractive index pattern and field intensity enhancement $|\mathscr{E}(z)|^2$ (7.71) of **a** an antiresonant SESAM with a roundtrip phase $\pi(2m-1)$ in the last layer with the single quantum well saturable absorber (i.e., thin dark line in last layer), and **b** a resonant SESAM with a roundtrip phase $2\pi m$ [08Maa1,05Spu3]

In Fig. 7.20, the field intensity enhancement ξ and the GDD are shown as a function of the wavelength for the two SESAM designs shown in Fig. 7.19. The resonant structure is extremely wavelength sensitive, having the advantage that one can tune the modulation depth with the wavelength. On the other hand, the structure is more difficult to grow. A 1% error in the layer thickness during growth will shift the resonance by 10 nm. The gray line indicates the mirror reflectivity, which has a bandwidth of over 80 nm, while the dashed line is the resonant SESAM and the solid line is the antiresonant SESAM.

There are other trade-offs for the resonant SESAM design. The larger field intensity enhancement inside the semiconductor layers results in a lower damage fluence and a stronger two-photon absorption with a smaller rollover parameter F_2. Note that a large field intensity enhancement at the SESAM surface for the resonant design makes the damage issues even worse. According to the model functions as summarized in Table 7.4 and (7.54), the rollover occurs at a much lower fluence and this effectively increases the nonsaturable losses for shorter pulses as shown in Fig. 7.21. This counteracts further pulse shortening in SESAM modelocking because the maximum reflectivity at the rollover shifts to lower fluence F_0 according to $F_0 \approx \sqrt{F_2 F_{\text{sat}} \Delta R}$ (7.54). Another benefit of the rollover is that it can help to stabilize against Q-switching instabilities (to be discussed in Sect. 9.7).

These two examples clearly show the benefit of the antiresonant SESAM design, which is well suited for modelocking of ultrashort pulses. As we will show later, ideally we want a constant GDD over the full spectral bandwidth of the modelocked pulses in the laser cavity. Furthermore, a rollover in the nonlinear reflectivity at lower pulse fluences tends to limit further pulse shortening but can also be a benefit for suppressing Q-switching instabilities in SESAM modelocking. The multilayer mirror design however allows for more sophisticated SESAMs with properties in between the resonant and antiresonant design discussed so far [05Spu3].

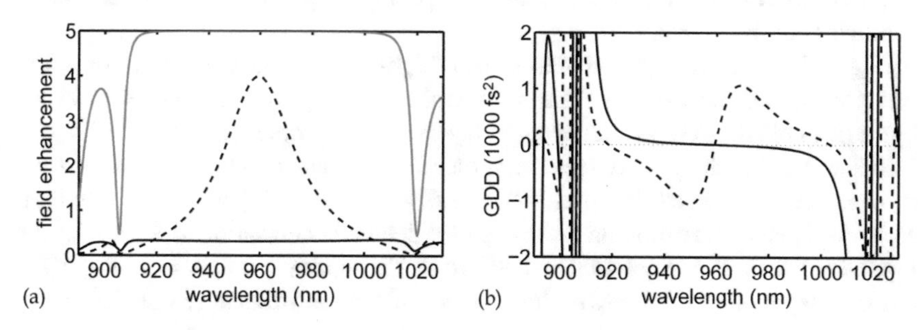

Fig. 7.20 Wavelength dependence of SESAM properties. **a** Field intensity enhancement as a function of wavelength (7.71). **b** GDD as a function of wavelength. The *gray curve* indicates the DBR reflectivity, the *solid black curve* is the antiresonant SESAM, and the *dashed black curve* is the resonant SESAM [08Maa1,05Spu3]

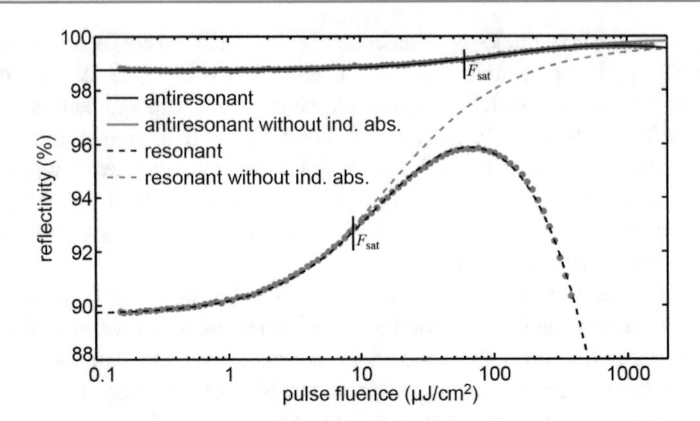

Fig. 7.21 Example of nonlinear reflectivity for an antiresonant (*solid curves*) and a resonant (*dashed curves*) SESAM having the same saturable absorber. *Dots* are measured values. *Black curves* show the least-square fit of the model functions given in Table 7.4 taking into account the Gaussian beam profile of the incoming beam and the induced saturable absorption. *Gray curves* show the fit function without induced absorption (ind. abs.), thus the reflectivity obtained by longer picosecond pulses without any significant two-photon absorption and rollover [08Maa1]

7.5.3 Antiresonant High-Finesse SESAM

The antiresonant high-finesse SESAM is based on a Fabry–Pérot structure which is typically formed by a lower semiconductor DBR and a dielectric top mirror, with a saturable absorber and possibly transparent spacer layer in between (Fig. 7.22). The thicknesses of the total absorber and spacer layers are adjusted so that the Fabry–Pérot is operated at antiresonance. Operation at antiresonance results in a device that is broadband and has minimal group delay dispersion (Sect. 7.5.2). The bandwidth of the SESAM is limited by either the free spectral range of the Fabry–Pérot or the bandwidth of the mirrors.

Figure 7.22 shows a typical antiresonant high-finesse SESAM design for a laser wavelength of about 1.05 μm. The bottom mirror is a highly reflecting DBR formed by 16 pairs of AlAs/GaAs quarter-wave layers with a complex reflectivity of $R_b e^{i\phi_b}$. In this case, the phase shift seen from the absorber layer to the bottom mirror is $\phi_b = \pi$ (see detailed explanation in Sect. 7.5.4 and Fig. 7.25a) with a reflectivity of $R_b \approx 98\%$, and to the top mirror $\phi_t = 0$ (see detailed explanation in Sect. 7.5.4 and Fig. 7.25b) with $R_t \approx 96\%$ [95Bro1,95Bro4]. The multiple quantum well (MQW) absorber layer has a thickness d chosen to satisfy the antiresonance condition:

$$\phi_{rt} = \phi_b + \phi_t + 2k\bar{n}d = (2m + 1)\pi \ , \tag{7.72}$$

where ϕ_{rt} is the roundtrip phase inside the Fabry–Pérot, $\bar{n}$ is the average refractive index of the absorber-spacer layer (i.e., MQW saturable absorber area in Fig. 7.22), $k = 2\pi/\lambda$ is the wave vector, λ is the wavelength in vacuum, and m is an integer. From (7.72), it follows that

$$d = m\frac{\lambda}{2\bar{n}} \ . \tag{7.73}$$

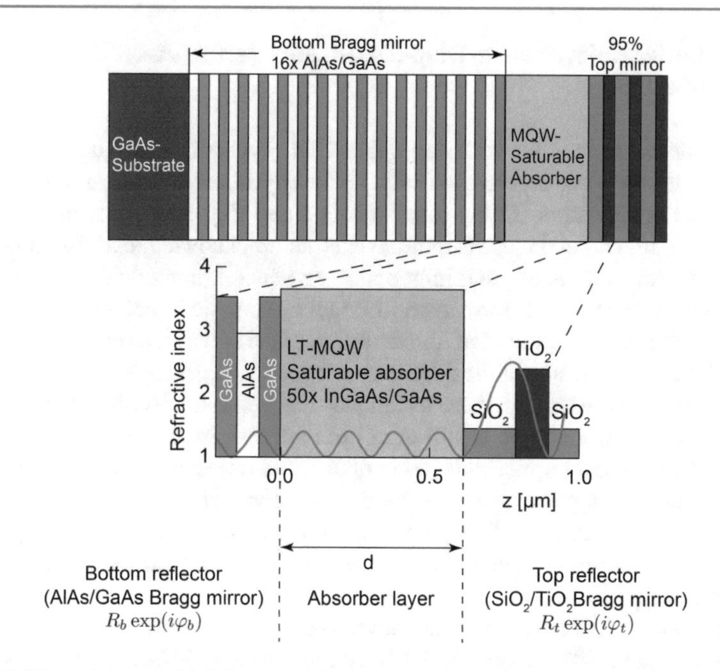

Fig. 7.22 The antiresonant high-finesse SESAM design. A specific design for a $\lambda \approx 1.05\ \mu$m center wavelength laser. The bottom mirror is a highly reflecting distributed Bragg mirror (DBR). The enlarged section also shows the calculated standing-wave intensity pattern of an incident electromagnetic wave centered at 1.05 μm. The Fabry–Pérot is formed by the lower AlAs/GaAs Bragg reflector, the absorber layer of thickness $d = 4 \times \lambda_{\bar{n}}/2$, and a top SiO_2/TiO_2 Bragg reflector, where $\bar{n}$ is the average refractive index of the absorber layer and $\lambda_{\bar{n}} = \lambda/\bar{n}$

From the calculated intensity distribution in Fig. 7.22, we see that $m = 4$.

The saturable absorber layer inside the antiresonant high-finesse SESAM (see Fig. 7.22) is typically extended over several periods of the standing wave pattern of the incident electromagnetic wave. This results in about a factor of 2 increase in the saturation fluence and intensity compared to the material value measured without standing wave effects. We typically measure a saturation fluence of around 60 $\mu J/cm^2$ [95Bro1] for an AR-coated device. With a top reflector, the effective saturation fluence is increased as given by (13) and (14) of [95Bro1]. For a relatively high top reflector above 95% at antiresonance, the effective saturation fluence is typically increased by about 2 orders of magnitude.

For a center wavelength around 800 nm, we typically use an AlGaAs/AlAs Bragg mirror with a small enough Ga content to introduce no significant absorption. These mirrors have less reflection bandwidth than the GaAs/AlAs Bragg mirrors because of the lower refractive index difference. However, we have demonstrated pulses as short as 19 fs from a Ti:sapphire laser [95Jun1] with such a device. In this case, the bandwidth of the modelocked pulse extended slightly beyond the bandwidth of the lower AlGaAs/AlAs mirror, because the much broader SiO_2/TiO_2 Bragg mirror on top reduces the bandwidth-limiting effects of the lower mirror. Reducing the top mirror reflectivity increases the minimum attainable pulse width due to the lower mirror bandwidth.

7.5.4 Reflectivity, Phase, Dispersion, and Penetration Depth of a DBR

A high reflectivity coating with negligible GDD typically consists of a stack of dielectric quarter-wave layers, also called a Bragg mirror, Bragg reflector, or distributed Bragg reflectors (DBR) (see Sect. 3.7 and Fig. 3.28). For the design of such a highly reflecting dielectric multilayer structure, knowledge of the reflectivity, phase, and dispersion is of great importance. In general, numerical methods based on the transmission matrix formalism [85Mac] or analytical methods based on the coupled-mode equations are used for the design of the structure. The latter, however, only yields accurate results if the difference between the refractive indices of the layers is not too great. Although these techniques are well established and documented in many textbooks, the topic is still of current interest, motivated by a wide range of applications in laser optics, integrated optics, and optoelectronics. In addition, in many textbooks, analytic expressions for the reflectivity and the penetration depth at the Bragg wavelength typically do not take into account a higher index step before the DBR. This causes a significant error, as we will show next, and is relevant for SESAM design.

From the design of an antiresonant or resonant SESAM, we need to better understand the complex reflectivity at the DBR. Here we follow closely [95Bro1]. Starting with the coupled-mode equations, we can derive analytical expressions for the reflectivity and the penetration depth of a dielectric Bragg mirror. In contrast to previous coupled-mode calculations, we include the interfaces with the adjacent media and show that they can have a large effect on the penetration depth and phase. The simple analytical formulas that are thereby derived are very accurate, as comparisons with numerical calculations show.

(a) Complex DBR Reflectivity and Transmission in a Periodic Dielectric Medium

We use the coupled-mode equations to find a matrix which connects the forward $u^+(z)$ and backward $u^-(z)$ travelling field amplitudes at any point in the Bragg medium (see Fig. 7.23). We then derive the boundary conditions at the interface to a medium with constant refractive index and use them to obtain an expression for the reflectivity of the entire structure. These boundaries have generally been neglected in previous analytical treatments based on coupled-mode theory. A comparison with numerical calculations, however, shows that, without inclusion of the boundaries, a large error can occur, especially in the phase of the reflectivity.

Figure 7.23 shows the DBR structure and defines the parameters used in the following derivation. We also define an average refractive index $\bar{n}$ by

$$\Lambda\bar{n} = d_{\mathrm{H}}n_{\mathrm{H}} + d_{\mathrm{L}}n_{\mathrm{L}} \,. \tag{7.74}$$

We consider a wave $u(z)$ with the complex propagation constant β as follows:

$$u(z) = u_0 \exp(\mathrm{i}\beta z) \,, \quad \beta = \bar{n}k + \mathrm{i}\alpha \,, \tag{7.75}$$

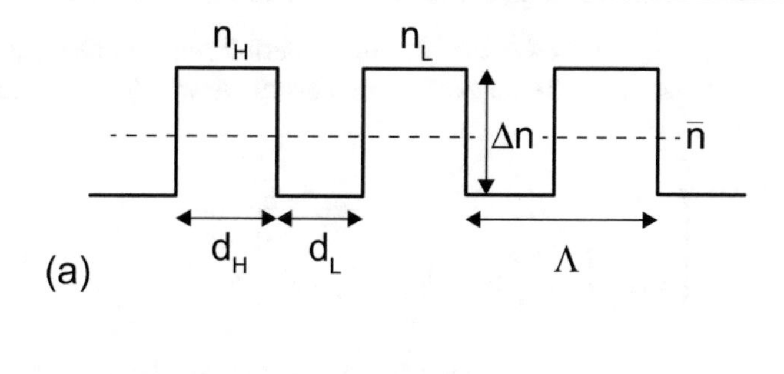

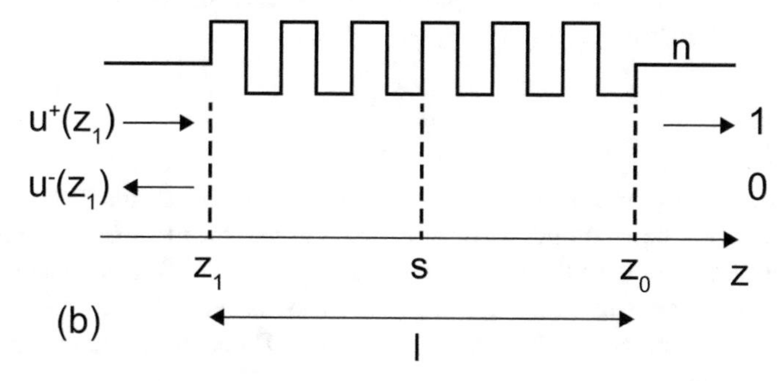

Fig. 7.23 Schematic view of a DBR. **a** A medium with a periodically varying refractive index by alternating two media with high and low indices, n_{H} and n_{L}, respectively, with the periodicity Λ. **b** Embedded DBR with the forward $u^{+}(z)$ and backward $u^{-}(z)$ travelling field amplitudes at any point in the Bragg medium, assuming only a wave travelling in the positive z-direction at the right boundary at z_0

where α is the amplitude absorption coefficient. A periodic perturbation of the medium generates sidebands with the wave numbers $\beta \pm 2\pi/\Lambda$, which are in resonance with counter-propagating waves if

$$|\beta| = \frac{\pi}{\Lambda} = \bar{n}k_{\mathrm{B}} , \tag{7.76}$$

which is the Bragg condition with the Bragg wave number k_{B}. Using (7.74), we can also obtain the Bragg wavelength λ_{B} with $k_{\mathrm{B}} = 2\pi/\lambda_{\mathrm{B}}$:

$$d_{\mathrm{H}}n_{\mathrm{H}} = \frac{\lambda_{\mathrm{B}}}{4} , \quad d_{\mathrm{L}}n_{\mathrm{L}} = \frac{\lambda_{\mathrm{B}}}{4} . \tag{7.77}$$

We can then define a matrix M, that connects the amplitudes travelling in positive and negative directions at two different points inside an unmodulated medium (see, e.g., basic microwave theory and optical filter textbooks):

$$\begin{pmatrix} u^{+}(z_1) \\ u^{-}(z_1) \end{pmatrix} = M(z_1, z_0) \begin{pmatrix} u^{+}(z_0) \\ u^{-}(z_0) \end{pmatrix} , \quad M(\Delta z) = \begin{pmatrix} \mathrm{e}^{\mathrm{i}\beta\Delta z} & 0 \\ 0 & \mathrm{e}^{-\mathrm{i}\beta\Delta z} \end{pmatrix} , \tag{7.78}$$

where $\Delta z = z_1 - z_0$. To calculate M in a periodically modulated medium (Fig. 7.23a), we start with the coupled-mode equations, which are normally written in the form

$$\frac{\mathrm{d}}{\mathrm{d}z}\begin{pmatrix} u^+ \\ u^- \end{pmatrix} = \begin{pmatrix} \mathrm{i}\beta & \kappa \exp\left[\mathrm{i}\frac{2\pi}{\Lambda}(z-s)\right] \\ \kappa \exp\left[-\mathrm{i}\frac{2\pi}{\Lambda}(z-s)\right] & -\mathrm{i}\beta \end{pmatrix}\begin{pmatrix} u^+ \\ u^- \end{pmatrix}, \tag{7.79}$$

with the coupling coefficent κ, which is, for the case of the rectangular-shaped structure shown in Fig. 7.23a, given by

$$\kappa = \frac{2\Delta n}{\lambda}. \tag{7.80}$$

The parameter s is introduced to obtain the correct phase of the reflectivity, i.e., the phase relation between the forward $u^+(z)$ and backward $u^-(z)$ travelling field amplitudes at any point in the Bragg medium. Equation (7.79) can be solved in the well-known manner by splitting off the fast phase factors $\exp[\pm\mathrm{i}(2\pi/\Lambda)(z-s)]$ and looking for eigenvectors of the remaining system. Then, for the matrix defined in (7.78), we obtain

$$M(z_1, z_0) = \frac{1}{\Gamma}\begin{pmatrix} \exp\left(\mathrm{i}\frac{\pi}{\Lambda}\Delta z\right)\left[\Gamma\cosh(\Gamma\Delta z) + \mathrm{i}\delta\sinh(\Gamma\Delta z)\right] & \exp\left[\mathrm{i}\frac{\pi}{\Lambda}(z_1+z_0+2s)\right]\kappa\sinh(\Gamma\Delta z) \\ \exp\left[-\mathrm{i}\frac{\pi}{\Lambda}(z_1+z_0+2s)\right]\kappa\sinh(\Gamma\Delta z) & \exp\left(-\mathrm{i}\frac{\pi}{\Lambda}\Delta z\right)\left[\Gamma\cosh(\Gamma\Delta z) - \mathrm{i}\delta\sinh(\Gamma\Delta z)\right] \end{pmatrix}, \tag{7.81}$$

where we have introduced the detuning parameter δ and the propagation constant in the Bragg medium Γ, given by

$$\delta \equiv \beta\frac{\pi}{\Lambda}, \quad \Gamma \equiv +\sqrt{\kappa^2 - \delta^2}. \tag{7.82}$$

Using (7.81) we can obtain the amplitude reflectivity r of a DBR with length $l = z_0 - z_1 = -\Delta z$, as shown in Fig. 7.23:

$$\begin{pmatrix} u^+(z_1) \\ u^-(z_1) \end{pmatrix} = M(z_1, z_0)\begin{pmatrix} 1 \\ 0 \end{pmatrix}, \quad r = \frac{u^-(z_1)}{u^+(z_1)}, \tag{7.83}$$

whence

$$r = \exp\left[-\mathrm{i}\frac{2\pi}{\Lambda}(z_1 - s)\right]\frac{-\kappa\sinh(\Gamma l)}{\Gamma\cosh(\Gamma l) - \mathrm{i}\delta\sinh(\Gamma l)}. \tag{7.84}$$

In the literature, the phase factor in front of the expression for r is often omitted. It is $+1$ if $z_1 - s = m\Lambda$ for any integer m, i.e., the Bragg reflector thus starts with a layer with higher refractive index of the Bragg mirror layer pair, and -1 if $z_1 - s = (m + 1/2)\Lambda$, i.e., the Bragg reflector starts with a layer with lower refractive index of the Bragg mirror layer pair. Equation (7.84) is the usually quoted formula for the reflectivity of a Bragg mirror as shown in Fig. 7.23.

(b) Boundary Conditions at the End of a Periodic Medium

However, (7.84) is not correct (within the approximations of the coupled-mode theory) for a DBR as shown in Figs. 7.24 and 7.25, where we have different boundary conditions at the end of a periodic medium. In (7.84), we have not taken into account the transmission matrices at the end of the periodic medium, i.e., at $z = z_0$ in Fig. 7.24. Neglecting this matrix may lead to a large error in the phase and the penetration depth into the mirror.

The calculation of the reflectivity at an interface between two adjacent dielectric media with transmission matrices is based on the requirement of continuity of the wave amplitudes and their derivatives. In the same manner, we can calculate the transmission matrix of the amplitudes for the case of a Bragg reflector adjacent to an unperturbed medium with refractive index n_1 (Fig. 7.24a). The derivatives of the waves in the unperturbed medium are, from (7.78),

$$\frac{\mathrm{d}}{\mathrm{d}z}\begin{pmatrix} u_1^+ \\ u_1^- \end{pmatrix}_{z_0} = \begin{pmatrix} \mathrm{i}\beta_1 & 0 \\ 0 & -\mathrm{i}\beta_1 \end{pmatrix}\begin{pmatrix} u_1^+ \\ u_1^- \end{pmatrix}_{z_0}, \tag{7.85}$$

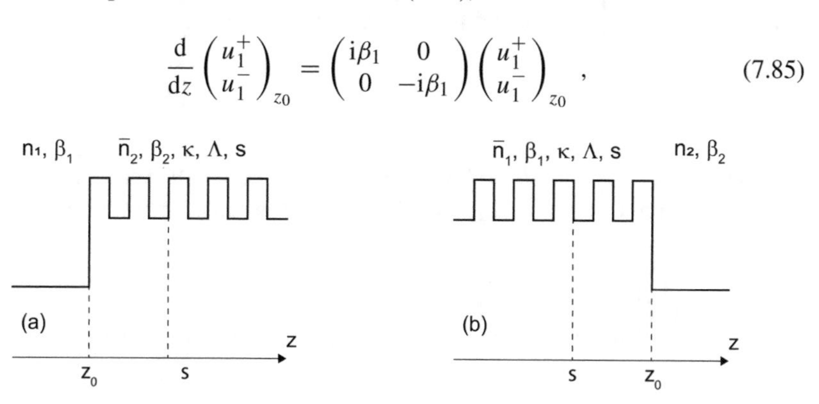

Fig. 7.24 Boundary condition at the end of a periodic medium: Schematic view of a Bragg medium adjacent to a medium with constant refractive index

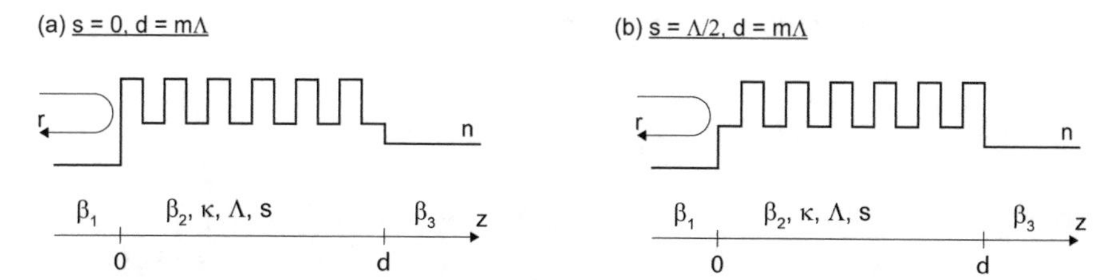

Fig. 7.25 DBR with different boundary conditions than in Fig. 7.23. Correct solutions for the complex reflectivity r and transmission t are given in (7.92) for case **a** with $s = 0$, i.e., $z_1 - s = m\Lambda$ for any integer m in Fig. 7.23b when the Bragg reflector starts with a layer with the higher refractive index of the Bragg mirror layer pairs, and (7.93) for case **b** with $s = \Lambda/2$, i.e., $z_1 - s = (m + 1/2)\Lambda$, when the Bragg reflector starts with a layer with lower refractive index. For $d \to \infty$ the DBR intensity reflectivity becomes 100% and the amplitude reflectivity is $r = -1$ (**a**) and $r = +1$ (**b**), according to (7.92) and (7.93)

and in the Bragg medium (Fig. 7.24b), according to (7.79),

$$\frac{\mathrm{d}}{\mathrm{d}z}\begin{pmatrix} u_2^+ \\ u_2^- \end{pmatrix}_{z_0} = \begin{pmatrix} \mathrm{i}\beta_2 & \kappa \exp\left[\mathrm{i}\frac{2\pi}{\Lambda}(z_0 - s)\right] \\ \kappa \exp\left[-\mathrm{i}\frac{2\pi}{\Lambda}(z_0 - s)\right] & -\mathrm{i}\beta_2 \end{pmatrix} \begin{pmatrix} u_2^+ \\ u_2^- \end{pmatrix}_{z_0} . \tag{7.86}$$

Continuity of the amplitudes at $z = z_0$ therefore requires

$$u_1^+ + u_1^- = u_2^+ + u_2^- , \tag{7.87}$$

and the continuity of the derivatives leads to

$$\begin{aligned} \mathrm{i}\beta_1(u_1^+ - u_1^-) = &\left[\mathrm{i}\beta_2 + \kappa \exp\left(-\mathrm{i}\frac{2\pi}{\Lambda}(z_0 - s)\right)\right] u_2^+ \\ &+ \left[-\mathrm{i}\beta_2 + \kappa \exp\left(\mathrm{i}\frac{2\pi}{\Lambda}(z_0 - s)\right)\right] u_2^- . \end{aligned} \tag{7.88}$$

We can solve (7.87) and (7.88) for u_1^+ and u_1^-, and obtain the transmission matrix for the transition from the Bragg medium to the unperturbed medium (Fig. 7.24a):

$$\begin{pmatrix} u_1^+ \\ u_1^- \end{pmatrix} = \frac{1}{2\beta_1} \begin{pmatrix} \beta_1 + \beta_2 - \mathrm{i}\kappa \exp\left[-\mathrm{i}\frac{2\pi}{\Lambda}(z_0 - s)\right] & \beta_1 - \beta_2 - \mathrm{i}\kappa \exp\left[\mathrm{i}\frac{2\pi}{\Lambda}(z_0 - s)\right] \\ \beta_1 - \beta_2 + \mathrm{i}\kappa \exp\left[-\mathrm{i}\frac{2\pi}{\Lambda}(z_0 - s)\right] & \beta_1 + \beta_2 + \mathrm{i}\kappa \exp\left[\mathrm{i}\frac{2\pi}{\Lambda}(z_0 - s)\right] \end{pmatrix} \begin{pmatrix} u_2^+ \\ u_2^- \end{pmatrix} . \tag{7.89}$$

For $\kappa = 0$, (7.89) is identical to the well-known Fresnel matrix for reflection at dielectric interfaces. In the same way, for the transmission matrix from the unperturbed to the Bragg medium (Fig. 7.24b), we obtain

$$\begin{aligned} \begin{pmatrix} u_1^+ \\ u_1^- \end{pmatrix} = &\left\{2\left[\beta_1 - \kappa \sin\left(\frac{2\pi}{\Lambda}(z_0 - s)\right)\right]\right\}^{-1} \\ &\begin{pmatrix} \beta_1 + \beta_2 + \mathrm{i}\kappa \exp\left[\mathrm{i}\frac{2\pi}{\Lambda}(z_0 - s)\right] & \beta_1 - \beta_2 + \mathrm{i}\kappa \exp\left[\mathrm{i}\frac{2\pi}{\Lambda}(z_0 - s)\right] \\ \beta_1 - \beta_2 - \mathrm{i}\kappa \exp\left[-\mathrm{i}\frac{2\pi}{\Lambda}(z_0 - s)\right] & \beta_1 + \beta_2 - \mathrm{i}\kappa \exp\left[-\mathrm{i}\frac{2\pi}{\Lambda}(z_0 - s)\right] \end{pmatrix} \begin{pmatrix} u_2^+ \\ u_2^- \end{pmatrix} . \end{aligned} \tag{7.90}$$

With the matrices as defined in (7.81), (7.89), and (7.90), we are in principle able to calculate the reflectivity of a Bragg mirror, taking into account the influence of the two interfaces at the boundaries of the mirror.

We consider a Bragg mirror of length d, as shown in Fig. 7.25. The Bragg mirror starts at $z = 0$, where we can calculate the reflectivity by multiplying the matrices (7.81), (7.89), and (7.90). We consider two important cases, where the mirror consists of an integer number of periods (i.e., $d = m\Lambda$) (i.e., we always consider a number of Bragg mirror layer pairs of high and low refractive index) and starts with a layer either with higher index (i.e., $s = 0$, Fig. 7.25a) or with a lower index (i.e., $s = \Lambda/2$, Fig. 7.25b). After multiplying the matrices and using the abbreviations C and S defined by

$$C \equiv \cosh(\Gamma d) , \quad S \equiv \sinh(\Gamma d) , \tag{7.91}$$

we obtain, for the amplitude reflectivity and transmission, in the first case ($s = 0$, Fig. 7.25a)

$$\boxed{\begin{aligned} r &= -\frac{\Gamma C\beta_2(\beta_3-\beta_1)+\kappa\beta_2 S(\beta_3+\beta_1+2\mathrm{i}\kappa)+\mathrm{i}\delta S\left[\mathrm{i}\kappa(\beta_3+\beta_1)-\beta_2^2+\beta_1\beta_3-\kappa^2\right]}{\Gamma C\beta_2(\beta_3+\beta_1)+\kappa\beta_2 S(\beta_3-\beta_1+2\mathrm{i}\kappa)+\mathrm{i}\delta S\left[\mathrm{i}\kappa(\beta_3-\beta_1)-\beta_2^2-\beta_1\beta_3-\kappa^2\right]} \\ t &= \frac{2(-1)^m\beta_1\beta_2\Gamma}{\Gamma C\beta_2(\beta_3+\beta_1)+\kappa\beta_2 S(\beta_3-\beta_1+2\mathrm{i}\kappa)+\mathrm{i}\delta S\left[\mathrm{i}\kappa(\beta_3-\beta_1)-\beta_2^2-\beta_1\beta_3-\kappa^2\right]} \end{aligned}} \tag{7.92}$$

and in the second case ($s = \Lambda/2$, Fig. 7.25b)

$$\boxed{\begin{aligned} r &= \frac{\Gamma C\beta_2(\beta_1-\beta_3)+\kappa\beta_2 S(\beta_1+\beta_3-2\mathrm{i}\kappa)+\mathrm{i}\delta S\left[\mathrm{i}\kappa(\beta_1+\beta_3)+\beta_2^2-\beta_1\beta_3+\kappa^2\right]}{\Gamma C\beta_2(\beta_1+\beta_3)+\kappa\beta_2 S(\beta_1-\beta_3+2\mathrm{i}\kappa)+\mathrm{i}\delta S\left[\mathrm{i}\kappa(\beta_1-\beta_3)-\beta_2^2-\beta_1\beta_3-\kappa^2\right]} \\ t &= \frac{2(-1)^m\beta_1\beta_2\Gamma}{\Gamma C\beta_2(\beta_1+\beta_3)+\kappa\beta_2 S(\beta_1-\beta_3+2\mathrm{i}\kappa)+\mathrm{i}\delta S\left[\mathrm{i}\kappa(\beta_1-\beta_3)-\beta_2^2-\beta_1\beta_3-\kappa^2\right]} \end{aligned}} \tag{7.93}$$

Equations (7.92) and (7.93) are much more complicated than the usually quoted formula (7.84) for the reflectivity of a Bragg mirror, but they yield the correct phase and penetration depth as we will show below. We can rewrite (7.92) for a DBR structure with

$$r = \frac{-\kappa S(1+\mathrm{i}\kappa/\beta)+\mathrm{i}\delta S(\mathrm{i}\kappa/\beta-\kappa^2/2\beta^2)}{\Gamma C-\mathrm{i}\delta S(1+\kappa^2/2\beta^2)+\kappa S(\mathrm{i}\kappa)/\beta} , \quad \text{assuming } \beta_1=\beta_2=\beta_3=\beta . \tag{7.94}$$

It is interesting to note that (7.84) is not correct even in this case since the transmission matrices (7.89) and (7.90) are not unity for a nonzero coupling coefficient κ. As expected, however, (7.94) is identical to (7.84) if $\kappa\beta$ goes to zero.

Figure 7.26 compares the analytical and numerical solutions, calculated using the transfer matrix formalism. We use two different DBR structures (Fig. 7.26 top row): on the left, the 'classical' DBR consisting of air–(GaAs/AlAs), 30 pairs with a low field intensity enhancement inside the DBR for the Bragg wavelength (dashed line), which corresponds to the case of Fig. 7.25a with $s = 0$; on the right, the 'resonant' DBR air–(AlAs/GaAs), 30 pairs with a relatively high field intensity enhancement inside the DBR for the Bragg wavelength (dashed line), which corresponds to the case of Fig. 7.25b with $s = \Lambda/2$. We use these names because of the different field enhancements inside the DBR. It may be surprising to see that the small difference as shown in Figs. 7.25 and 7.26 can make such a large difference in the field intensity enhancement. Note that we always assume that the full Bragg mirror consists of a certain number of high and low refractive index pairs (i.e., $d = m\Lambda$ as shown in Fig. 7.25).

We have the cases as shown in Fig. 7.25 and the numbers n_H and n_L of layer pairs is sufficiently large to ensure that the total reflectivity is close to 100% and the field intensity enhancement is 4 outside the DBR, as expected from a perfect standing wave formed by two counter-propagating plane waves with equal amplitude. The calculated reflectivity and phase in Fig. 7.26 middle panels, for the 'classical' DBR, and bottom panels, for the 'resonant' DBR, then clearly show that the typically quoted solution in (7.84) shows a larger error than (7.92) and (7.93), which takes into account the additional interface from the semiconductor DBR to air. Thus, with (7.92) or (7.93), near the Bragg wavelength (i.e., in the high reflectivity part of the DBR), we obtain a very accurate result, whereas the commonly used formula (7.84) leads to a large error in the phase. Figure 7.27 gives another example for a classical and resonant TiO_2/SiO_2 DBR for a Bragg wavelength at 1050 nm. Note that the error in the derivative of the phase, which is directly related to the group delay or the penetration depth, is significant.

Considering the field intensity in the 'standard' DBR, we see in Fig. 7.26 (top left) that the ξ factor is approximately 0.34, and the GDD (from middle panel) and ξ are nearly flat over the center wavelength range (as can be seen better in Fig. 7.20). This device does not exhibit any resonant-like behavior in either the field intensity or the GDD. Note that we can make the following comments on this DBR structure: the last quarter-wave layer is a high-index layer, and accordingly the field intensity at the air interface ends on or very near a node, i.e., a field null.

Now consider the 'resonant' DBR (Fig. 7.26 top right and bottom panel). This results in several key feature changes to the field intensity in the structure and the device's GDD behavior. First, the ξ factor now equals four, i.e., it is enhanced with respect to the 'standard' DBR (Fig. 7.26 top left) by a factor of more than 11. Secondly, the device exhibits some peaking of the group delay, resulting in significant GDD in the device away from its design-operating point (as can be seen better in Fig. 7.20). This peaking of the group delay is 'resonant-like' behavior. However, the field intensity in the device is never larger than the external intensity. Note also that the air–device interface characteristically ends on or near a peak (anti-node) of the field intensity.

Note that we chose to add a quarter-wave layer of the low-index material of the DBR. However, we are free to add a quarter-wave layer of any index material, and still achieve similar results, i.e., an increase in the ξ factor to nearly 4, and an enhanced GDD response.

It has often been stated that the coupled-mode theory only leads to exact results for a medium with small periodic variations of the refractive index. This is in principle true, but we have shown that the coupled-mode equations lead to very accurate results when the transmission matrices at the boundaries of the Bragg medium are taken into account in the correct way (Figs. 7.26 and 7.27). At these boundaries, the difference between the refractive indices can be arbitrarily large, corresponding to a situation which is often found in reality for SESAMs [95Bro1].

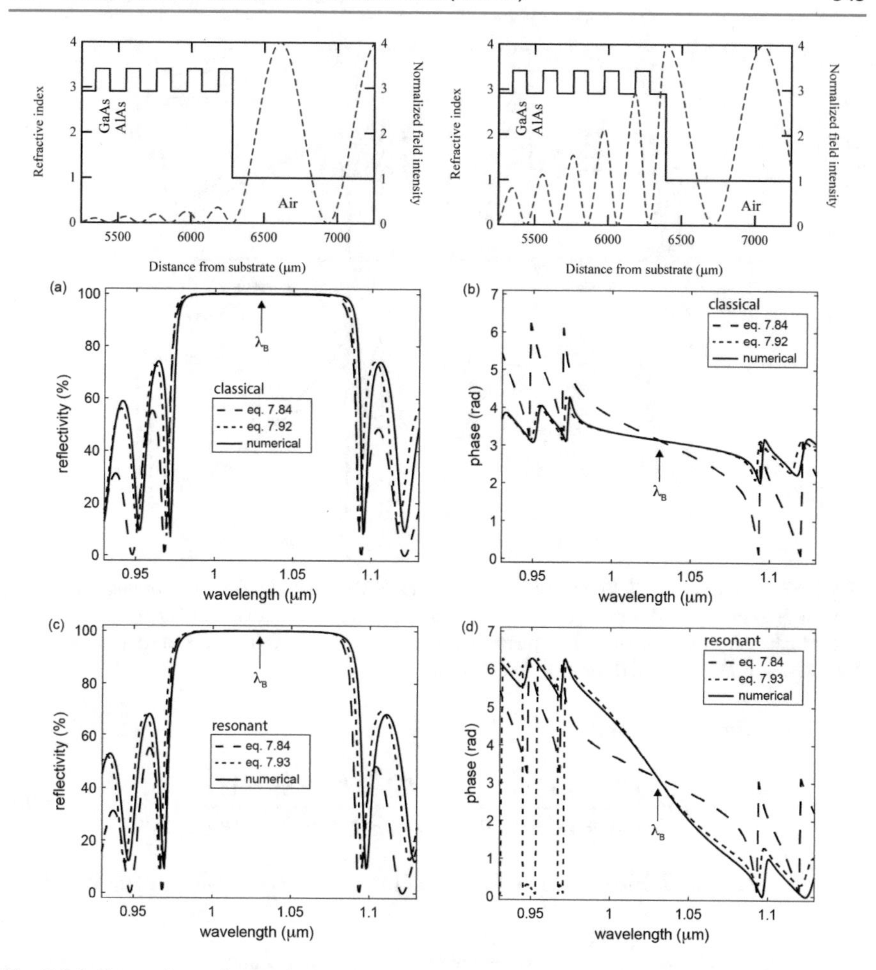

Fig. 7.26 Comparison of numerical and analytical solutions for the reflectivity and phase for two DBRs as shown in the *top row*: *left* 'classical' DBR, *right* 'resonant' DBR. We assume the DBR consists of 30 pairs of alternating quarter-wave layers of AlAs as low index material ($n = 2.91$ at 1030 nm, quarter-wave layer thickness 88.5 nm) and GaAs as high index material ($n = 3.41$ at 1030 nm, quarter-wave layer thickness 75.5 nm). The *middle panel* shows results for the 'classical' DBR and the *bottom panel* for a 'resonant' DBR

(c) Group Delay and Penetration Depth L_{eff} into an Infinitely Thick DBR

As the next step, we derive an analytical expression for the group delay and the penetration depth of an infinitely thick, i.e., $d \to \infty$, Bragg mirror. The reflectivity can be calculated in the same way, but the matrix in (7.90) is now the unit matrix and $\cosh(\Gamma d) \approx \sinh(\Gamma d)$. We obtain

$$r = \frac{(\beta_1 - \beta_2 + \mathrm{i}\kappa)(\Gamma - \mathrm{i}\delta) - (\beta_1 + \beta_2 + \mathrm{i}\kappa)\kappa}{(\beta_1 + \beta_2 - \mathrm{i}\kappa)(\Gamma - \mathrm{i}\delta) - (\beta_1 - \beta_2 - \mathrm{i}\kappa)\kappa}, \tag{7.95}$$

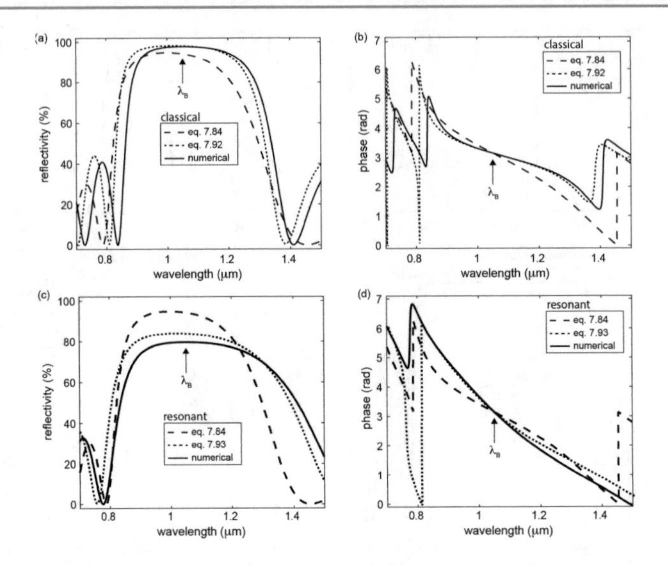

Fig. 7.27 Similar to Fig. 7.26 result for a 4-pair TiO_2/SiO_2 DBR at a Bragg wavelength 1050 nm, with SiO_2 as the low-index material ($n = 1.45$, quarter-wave layer thickness 181 nm) and with TiO_2 as the high-index material ($n = 2.4$, quarter-wave layer thickness 109 nm). **a** and **b** show results for the 'classical' DBR, **c** and **d** for a 'resonant' DBR

for $s = 0$ (Fig. 7.29a), and

$$r = \frac{(\beta_1 - \beta_2 - \mathrm{i}\kappa)(\Gamma - \mathrm{i}\delta) - (\beta_1 + \beta_2 - \mathrm{i}\kappa)\kappa}{(\beta_1 + \beta_2 + \mathrm{i}\kappa)(\Gamma - \mathrm{i}\delta) - (\beta_1 - \beta_2 + \mathrm{i}\kappa)\kappa} , \tag{7.96}$$

for $s = \Lambda/2$ (Fig. 7.29b). The penetration depth L_{eff} into the mirror is defined by

$$\left.\frac{\mathrm{d}\Phi}{\mathrm{d}k}\right|_{k=k_{\mathrm{B}}} = \left.\frac{\mathrm{d}}{\mathrm{d}k}\arg(r)\right|_{k=k_{\mathrm{B}}} = 2\bar{n}_2 L_{\mathrm{eff}} , \tag{7.97}$$

where Φ is the phase of the reflectivity, $r(k) = |r| \exp[\mathrm{i}\Phi(k)]$, and $\bar{n}_2$ is the average refractive index of the mirror as defined in (7.74). It has the following meaning: near the Bragg wave number k_{B}, the phase can be written as the linear approximation

$$\Phi(k) \approx \begin{cases} \pi + 2\bar{n}_2 L_{\mathrm{eff}}(k - k_{\mathrm{B}}) , & s = 0 , \\ 2\bar{n}_2 L_{\mathrm{eff}}(k - k_{\mathrm{B}}) , & s = \Lambda/2 , \end{cases} \tag{7.98}$$

which is equivalent to the situation with a constant phase of the reflectivity, but with the interface shifted by a distance L_{eff} into the mirror. Indeed, the free spectral range $\Delta\lambda$ and optical length L_{opt} of a Fabry–Pérot consisting of a layer with thickness d and refractive index n sandwiched between two Bragg mirrors are given by

$$\Delta\lambda = \frac{\lambda^2}{2L_{\mathrm{opt}}} , \qquad L_{\mathrm{opt}} = nd + \bar{n}_{\mathrm{t}} L_{\mathrm{eff}}^{\mathrm{t}} + \bar{n}_{\mathrm{b}} L_{\mathrm{eff}}^{\mathrm{b}} . \tag{7.99}$$

Here $L_{\text{eff}}^{\text{t,b}}$ denote the penetration depths of the top and bottom mirrors, respectively. We carry out the derivative in (7.97) using $\kappa = 2\Delta n/\lambda$, (7.95), and (7.96), and end up with the surprisingly simple expressions

$$\boxed{L_{\text{eff}} = \frac{n_1\lambda_{\text{B}}}{4\bar{n}_2\Delta n}}\,, \tag{7.100}$$

for $s = 0$ (Fig. 7.29a), and

$$\boxed{L_{\text{eff}} = \frac{\bar{n}_2\lambda_{\text{B}}}{4n_1\Delta n} + \frac{\lambda_{\text{B}}\Delta n}{2\pi^2 n_1\bar{n}_2}}\,, \tag{7.101}$$

for $s = \Lambda/2$ (Fig. 7.29b). The second term in (7.101) is in general negligible because it is much smaller than the Bragg wavelength λ_{B}. We can compare these results to the usually quoted formula

$$L_{\text{eff}} = \frac{\lambda_{\text{B}}}{4\Delta n}\,. \tag{7.102}$$

Obviously, the correct penetration depth can be obtained simply by multiplying L_{eff} from (7.102) by either $n_1/\bar{n}_2$ or by $\bar{n}_2/n_1$ depending on s. It is somewhat surprising that the penetration depth depends so strongly on whether the Bragg mirror starts with a layer with higher or lower refractive index (see Table 7.6), but it becomes more understandable when we look at the z-dependent field intensity enhancement inside the DBR and the reflectivity phase, as shown in Figs. 7.26 and 7.27.

The expressions for the penetration depth (7.100) and (7.101) are suprisingly simple and very useful for the design of small-cavity optoelectronic devices relying on Fabry–Pérot effects.

Table 7.6 Calculated penetration depth into TiO_2/SiO_2 and GaAs/AlAs DBRs as shown in Figs. 7.26 and 7.27 using different equations. An acceptable agreement with the numerical solution is only obtained with the correct analytical equations (7.100) or (7.101)

DBR structure	n_{H}	n_{L}	λ_{B} [μm]	(7.102) [μm]	(7.100) or (7.101) [μm]	Numerical (7.97) [μm]
Air–TiO_2/SiO_2 $s = 0$ classical	2.4	1.45	1.05	0.276	0.153	0.149
Air–TiO_2/SiO_2 $s = \Lambda/2$ resonant	2.4	1.45	1.05	0.276	0.527	0.497
Air–GaAs/AlAs $s = 0$ classical	3.41	2.91	1.03	0.515	0.164	0.164
Air–GaAs/AlAs $s = \Lambda/2$ resonant	3.41	2.91	1.03	0.515	1.625	1.627

7.5.5 Antiresonant Low-Finesse SESAM

The top reflector of the antiresonant high-finesse SESAM (Fig. 7.22) is an adjustable parameter that determines the intensity entering the semiconductor saturable absorber and therefore the effective macroscopic saturation fluence of the device. We have since demonstrated a more general category of SESAM designs, in one limit, for example, by replacing the top mirror with an AR coating [95Bro2]. Using the incident laser mode area as an adjustable parameter, we can adapt the incident pulse fluence to the saturation fluence of the device. However, to reduce the nonsaturable insertion loss of the device, we typically have to reduce the thickness of the saturable absorber layer.

A special intermediate design, which we refer to as the antiresonant low-finesse SESAM (Fig. 7.28) [95Jun1,95Hon], is achieved with no additional topcoating, resulting in a top reflector formed by the Fresnel reflection at the semiconductor–air interface, which is typically around 30%. Figure 7.28 shows some specific designs for a wavelength around 1.05 μm and 860 nm.

The antiresonant low-finesse SESAM on the left of Fig. 7.28 can be explained similarly to the high-finesse SESAM (Fig. 7.22). The bottom mirror is a DBR formed by 25 pairs of AlAs/GaAs quarter-wave layers with a complex reflectivity of $R_b e^{i\varphi_b}$ with $R_b \approx 100\%$ and $\varphi_b = \pi$, according to Fig. 7.25a with $s = 0$ and $d = 25\Lambda$. The top reflector is formed by the semiconductor–air interface, and therefore $\varphi_t = 0$ for the Fresnel reflection from high-index to air. The thickness d of the spacer and absorber layers are adjusted for antiresonance (7.72). With the roundtrip phase (7.72), we obtain a thickness of $\lambda_B/2\bar{n}$ for $m = 1$ for antiresonance, with $\bar{n}$ the average refractive index of the last three layers, i.e., AlAs, LT InGaAs, and GaAs, and λ_B the Bragg wavelength of the DBR. The residual reflection from the different spacer and absorber layers is negligible in comparison to the accumulated reflection from the lower multilayer DBR and the semiconductor–air interface. This is also confirmed by the numerical calculation of the standing wave intensity pattern (red line in Fig. 7.28). Because there is no special surface passivation layer, it is advantageous for a higher damage threshold to have a node of the standing wave intensity pattern at the SESAM–air surface [96Kop2].

Note that the SESAMs in Fig. 7.28 can also be viewed as a classical DBR (see Fig. 7.26 top left) with a thin absorber layer integrated into the antiresonant DBR structure with the lower field intensity enhancement (see also Fig. 7.19). Therefore the low-finesse SESAM device for a center wavelength around 860 nm as shown on the right in Fig. 7.28 was referred to as the saturable Bragg reflector (SBR) [95Tsu]. This SESAM is in principle the same as shown in Fig. 7.28 on the left, as discussed above, and also similar to the AR-coated saturable absorber Bragg reflector device published in 1995 [95Bro2]. The additional AR coating also acted as a passivation layer for the broadband operation inside the Ti:sapphire laser, because the node of the standing wave profile (Fig. 7.28) is only obtained for the Bragg wavelength.

However, the SBR acronym is not appropriate because it is not a new device, the acronym is far too restrictive, and it has been used before for a different device. A Bragg mirror is by definition a quarter-wave-layer stack. But we can easily change

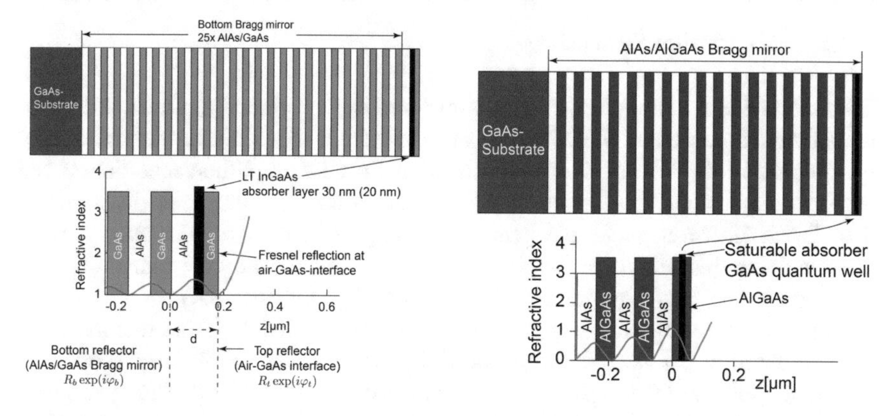

Fig. 7.28 Antiresonant low-finesse SESAM. *Left*: A specific design for an approximately 1.05 μm center wavelength laser. In contrast to the high-finesse SESAM in Fig. 7.22, the Fabry–Pérot is formed here by the lower AlAs/GaAs Bragg reflector, the absorber-spacer layer of thickness $\bar{n}d = \lambda/2$, and the Fresnel reflection from the semiconductor–air interface. Again, the thickness d is adjusted for antiresonance (7.72). *Right*: Another specific design for an approximately 860 nm center wavelength laser where a single InGaAs quantum well layer is embedded inside the last high index layer of the DBR

the thickness of the last layer as shown for example in Fig. 7.19 without changing its function. Furthermore, in principle the Bragg reflector can be replaced by a metal reflector to obtain a broad-bandwidth SESAM design (see Sect. 7.5.7). An earlier version of a nonlinear or saturable Bragg reflector design was introduced by Kim et al. in 1989 [89Kim], where the narrow bandgap DBR material actually provided saturable absorption due to band filling. This results in a distributed absorption over many DBR layers, which saturates the overall DBR reflectivity. However, this device would introduce too much loss inside a diode-pumped solid-state laser. Therefore, only one or a few thin absorbing layers are required inside the quarter-wave layers of the Bragg reflector. The effective saturation fluence of the device can then be varied by changing the position of the buried absorber layers within the mirror structure, taking into account the fact that a very thin absorber layer at the node of a standing wave does not introduce any absorption. It is generally much easier to grow a high quality highly reflecting DBR and then grow the additional absorber layers on top to achieve many different SESAM parameters.

One limitation of the low-finesse SESAM devices is based on the bandwidth of the lower Bragg mirror, and potentially higher insertion loss than in the high-finesse SESAM. For example, pulses as short as 19 fs have been generated with the high-finesse SESAM, compared to 34 fs with the low-finesse SESAM using the same lower DBR [95Jun1]. Replacing the lower Bragg mirror with a broadband silver mirror [96Flu1] resulted in pulses as short as 6.5 fs [97Jun3,98Sut] and 5.8 fs [99Sut] with a KLM-assisted Ti:sapphire.

7.5.6 Dispersive SESAM

The SESAM is a multilayer structure optimized for a certain amplitude and phase reflectivity response and therefore has many different specific layer structures as a possible solution. We can thus envision to obtain both amplitude modulation and negative dispersion. Ideally, we could consider a chirped mirror as discussed in Sect. 3.6.3. However, so far the many layers required for such a design prevented us from doing this with a semiconductor MBE. Instead, an early dispersive SESAM was based on a Gires–Tournois interferometer (GTI) (see Sect. 3.4).

We can obtain negative dispersion with a GTI-type SESAM, and this has been used to obtain a compact femtosecond diode-pumped Cr:LiSAF laser [96Kop2]. The total thickness of the spacer layers for the resonant SESAM is chosen so that a Gires–Tournois-like Fabry–Pérot is formed by the top semiconductor–air interface and the DBR as the two dominant reflecting surfaces. This Fabry–Pérot structure has a resonance which is close to the laser wavelength $\lambda = 840$ nm. The resulting group delay around this resonance (solid line in Fig. 7.29 top panel) indicates that a

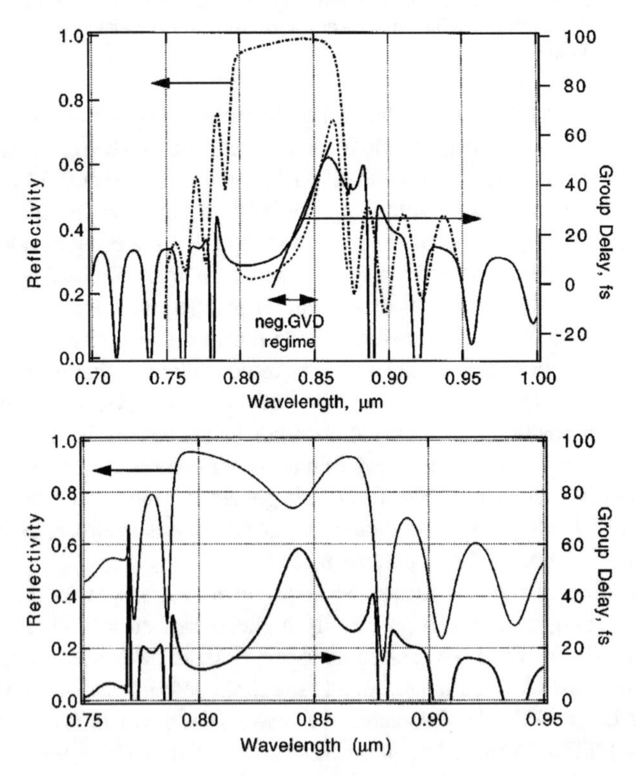

Fig. 7.29 Dispersive SESAM based on the GTI-effect [96Kop2]. *Top*: Measured reflectivity and group delay of the resonant SESAM, calculated group delay (*solid line*), where a positive slope corresponds to a negative GDD. *Bottom*: Calculated reflectivity curve and group delay for a design with an absorber layer positioned at the field maximum of the standing wave inside the Gires–Tournois structure, which would not work [96Kop2]

negative GDD of about $\partial T_g/\partial\omega = -400\ \text{fs}^2$ is obtained at the laser wavelength. This is also confirmed by the measured group delay (dashed line in Fig. 7.29 top panel), which is in good agreement with the design data.

However, the position of the saturable absorber layer inside the structure is an important design parameter. We would prefer to place the saturable absorber layer at a high field location to reduce the effective saturation fluence of the device. However, this would cause a pronounced absorption dip at the resonance wavelength (Fig. 7.29 bottom panel) and the extra loss would push the lasing wavelength away from the negative GDD regime towards one of the two reflection maxima. Therefore, we put the absorber layer close to the field minimum. Different wavelengths have different field node positions, resulting from the lower DBR phase (see Fig. 3.28). As we approach the resonance wavelength, the intensity maximum inside the dispersive SESAM rises, but at the same time the field node moves closer to the center of the absorbing layer and this reduces the absorption caused by the absorber layer. The resulting reflectivity curve as a function of wavelength no longer shows a reflectivity dip (Fig. 7.29 bottom panel) and the laser operates in the negative GDD regime. Because of the negative dispersion from such a dispersive SESAM, transform-limited pulses as short as 160 fs have been generated without any additional prism pairs for dispersion compensation [96Kop2].

7.5.7 Ultrabroadband SESAMs

Different SESAM designs have been extensively used for passive pulse generation in solid-state lasers. However, the $Al_xGa_{1-x}As/AlAs$ Bragg mirror in an antiresonant low-finesse SESAM design limits the pulse generation with Ti:sapphire lasers to 34 fs [95Bro2]. In this case, a single GaAs quantum well layer was directly imbedded into an $Al_xGa_{1-x}As/AlAs$ Bragg mirror. An antiresonant high-finesse SESAM design was able to reduce the $Al_xGa_{1-x}As/AlAs$ Bragg mirror bandwidth limitation and supported 19 fs [95Jun1]. In the latter case, the SESAM consists of a lower $Al_xGa_{1-x}As/AlAs$ Bragg mirror, a GaAs saturable absorber layer, and a SiO_2/TiO_2 top reflector with a reflectivity of 96%. This reflector has a larger reflectivity bandwidth than the lower DBR because the reflectivity bandwidth of Bragg mirrors increases rapidly with the ratio of the refractive indices of the materials forming the mirror. So far, the shortest pulses have only been obtained when the lower Al_xGa_{1-x} As/AlAs Bragg mirror is replaced by a silver mirror [96Flu1,97Jun1,99Sut,00Zha]. Since the semiconductor saturable absorber cannot be grown directly on a silver mirror by MBE, post-growth processing was applied to fabricate an antiresonant low-finesse SESAM. Such devices then supported sub-6 fs pulses [99Sut].

To overcome the refractive index limitation of Al_xGa_{1-x} As/AlAs DBRs, we have also investigated fluoride heteroepitaxy to increase the refractive index difference (Fig. 7.30). For broadband SESAMs based on DBR, a material pair is needed which (a) differ greatly in their refractive indices, (b) can be epitaxially grown in stacks, and (c) allow for growth of a saturable absorber material on top of the Bragg mirror or embedded in the mirror. For the best choice of a material pair fulfilling these

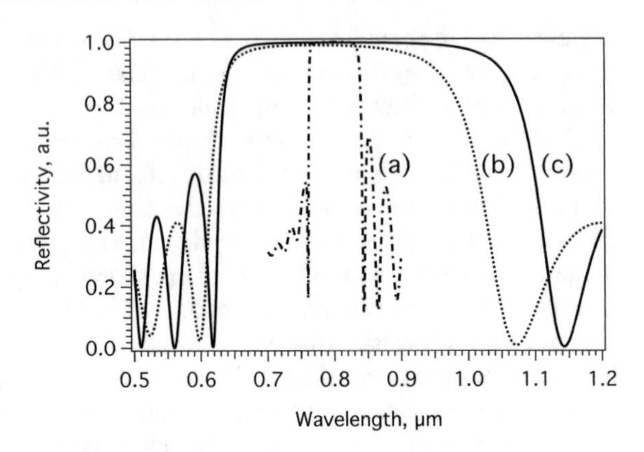

Fig. 7.30 Calculated reflectivity spectra for different DBRs. **a** 25 pairs $Al_{0.17}Ga_{0.83}As$/AlAs on GaAs, **b** 4 pairs ZnTe/BaF_2 on ZnTe, and **c** 4 pairs $Al_{0.8}Ga_{0.2}As/CaF_2$ on GaAs. After [00Sch]

requirements, we have to look at optoelectronic properties such as refractive index, bandgap energy, and material properties like lattice constant, stability in air, and intermixing.

Figure 7.31 presents the refractive indices versus lattice constants of commonly used semiconductors and insulators. Material pairs only made of III–V semiconductors will provide small Bragg mirror reflection bandwidth since their refractive indices do not differ very much. Besides the $Al_xGa_{1-x}As$/AlAs Bragg mirrors mentioned above, ZnX (X = S, Se, Te) mixed crystals with additional Mg or Cd incorporated have been used in DBRs as well [98Taw]. However, both III–V and II–VI Bragg reflectors have small spectral bandwidth and require many mirror pairs to obtain reflectivities above 99%. An increase in refractive index ratio can be provided with ZnTe/BaF_2 Bragg mirrors. These materials are nearly lattice matched and can theoretically support a high reflection bandwidth of up to 310 nm (Fig. 7.30). However, Fig. 7.31 clearly shows that the highest difference in refractive indices is given by a combination of $Al_xGa_{1-x}As$ with a group IIa fluoride, e.g., CaF_2.

For the wavelength range of interest (0.5–2 µm), fluorides are nonabsorbing and have no significant dispersion. Most of them can be grown epitaxially by high temperature evaporation. Some of the fluorides exhibit more (LiF) or less (BaF_2) hygroscopic behavior. Initially, $Al_xGa_{1-x}As/(Ca,Sr)F_2$ Bragg reflectors were demonstrated with a 90% reflectance around a center wavelength of 870 nm for only three periods of layer pairs [90Fon]. However, they suffered from serious crack formation after the epitaxy due to the thermal stress between the two materials, because there is a large difference between their thermal expansion coefficients. A crack-free GaAs/fluoride Bragg mirror was demonstrated for the first time using CaF_2–BaF_2–CaF_2 fluoride $\lambda/4$ layers with only thin CaF_2 parts [96Shi]. Since the difference between the refractive indices of CaF_2 (1.43) and BaF_2 (1.47) (Fig. 7.31) is very small, the reflection at the CaF_2–BaF_2 interfaces can be neglected. BaF_2 can be grown epitaxially on CaF_2 and has a smaller stiffness constant, which was expected to help to overcome the crack problem caused by thermal mismatch strain. In addition, a fluoride stack can be further improved with low-temperature (LT) growth, which increases the maximum crack-free CaF_2 thickness. This allowed for the growth of four pair crack-free $Al_xGa_{1-x}As/CaF_2$ Bragg reflectors with above 98% reflectivity

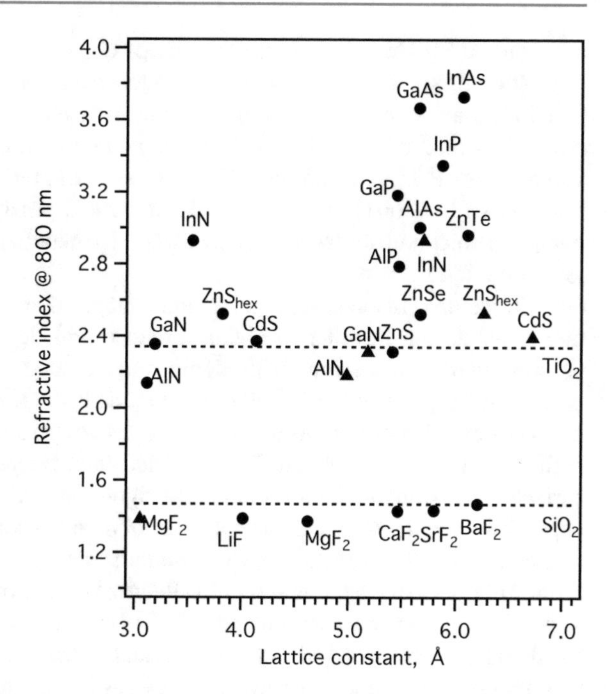

Fig. 7.31 Refractive indices (n or n_o with •, n_e with ▲) verses lattice constants of commonly used semiconductors and insulators. For birefringent crystals we either plot the ordinary (n_o) or the extraordinary (n_e) refractive index. The values for TiO_2 and SiO_2 are given by *dotted lines*, since these materials are amorphous in broadband Bragg mirrors. After [00Sch]

at a center wavelength of 880 nm and a bandwidth of about 400 nm [98Shi]. Later, the interface roughness could be further improved using (111) substrate orientation compared to (100) oriented growth [99Sch2]. Here we use the Miller indices (111) resp. (100) for the definition of the crystal surface plane. This was all necessary for the first successful demonstration of a GaAs saturable absorber epitaxially grown on CaF_2 [00Sch].

This resulted in the first ultrabroadband monolithically grown AlGaAs/CaF_2 SESAM covering nearly the entire gain spectrum of a Ti:sapphire laser [02Sch1]. A large high reflectivity bandwidth of more than 300 nm was provided by this SESAM using only six material layers. This fluoride SESAM had a modulation depth of 2.2%, a fast recovery time constant of less than 150 fs, and a slow recovery time constant of 1.2 ps. Using this SESAM inside a Ti:sapphire laser produced self-starting pulses as short as 9.5 fs. Unfortunately, the fluoride SESAM center wavelength shifted to longer wavelength, which ultimately limited this pulse duration. In principle, even shorter pulses can be supported with such a fluoride SESAM.

7.5.8 SESAM Optimization with Standing Wave Field Enhancement

One important limit of a SESAM device is its saturation fluence, which has typical values in the range of 10–100 $\mu J/cm^2$. Lower saturation fluence is particularly relevant for fundamentally modelocked solid-state lasers with gigahertz pulse repetition rates (Sect. 9.10.3). It becomes harder to saturate the SESAM device in such a laser, as the intra-cavity pulse energy becomes increasingly lower, requiring laser mode sizes on the SESAM device on the order of only a few microns (i.e., close to the diffraction limit). Additionally, the very short cavity length required for such high repetition rates does not leave much room for cavity design with arbitrary small mode sizes on the SESAM device. Furthermore, the threshold for cw modelocking without Q-switching instabilities (QML threshold) is decreased with a low saturation fluence [99Hon1]. It is also worth noting that passively modelocked optically pumped semiconductor lasers (Sect. 9.8) can also benefit from a SESAM device with low saturation fluence.

Here we investigate specific structural changes that can be made to the SESAM multilayer design in order to obtain a device with lower saturation fluence, and describe two novel designs with significantly reduced saturation fluence without pushing it fully into a high-finesse resonant device. Of course, improved absorber materials can also be combined with improved design structures. Figure 7.32 shows both the refractive index and the calculated field intensity enhancement near the surface of the structure on the left. On the right-hand side, we show the wavelength-dependent field intensity enhancement factor ξ and group delay dispersion (GDD).

The classical SESAM design with the antiresonant low-finesse SESAM (Fig. 7.32, top and Fig. 7.19) has a larger saturation fluence because of the antiresonant structure. A possible design for a wavelength of 1314 nm consists of a 30-pair AlAs/GaAs standard DBR grown on a GaAs substrate, and ending on a GaAs quarter-wave layer, then an approximately half-wave layer consisting of the following: 91 nm GaAs, 10 nm GaInNAs, absorber layer and another 91 nm of GaAs. This SESAM device exhibits a low enhancement factor of $\xi = 0.34$ and it shows flat, non-resonant-like behavior. Note that the field intensity at the device–air interface ends on a node.

Moving to a full high-finesse resonant structure would lower the saturation fluence. Such devices have the trade-off that their optical properties (F_{sat}, ΔR, losses, and GDD) become increasingly narrowband as ξ increases, and are increasingly sensitive to ambient changes and growth tolerances due to the sharp resonance. For example, the GDD would no longer show a flat response over the center of the mirror reflectivity, but would show a peaked, resonant response (see Sects. 7.5.2 and 7.5.4). However, for devices which require only a limited operation bandwidth of a few nanometers, some enhancement of the ξ factor can be useful, limited by the above trade-offs. Two SESAM devices were introduced to achieve an enhancement of the intra-device fields, without introducing a top mirror element. These were referred to as the low-field-enhancement resonant-like SESAM device (LOFERS) and enhanced SESAM device (E-SESAM) (Fig. 7.32) [05Spu3].

Consider the LOFERS device shown in Fig. 7.32 (middle panels). We start with a 30-pair DBR, but extend the top quarter-wave layer of GaAs by approximately an additional quarter-wave layer. The absorber is placed near the surface, and a very thin GaAs cap layer is added to protect the absorber layer, e.g., an 80.8 nm GaAs spacer layer, 10 nm GaInNAs absorber layer, and 5 nm GaAs cap. The field in the absorber is nearly equal to the outside field: $\xi = 3.90$. Relative to the classical low-finesse SESAM device ($\xi = 0.34$), this is a field enhancement by a factor of approximately 11, implying that the saturation fluence should be 11 times lower and the modulation depth 11 times greater. We can see that the stronger dependence of ξ and GDD on λ

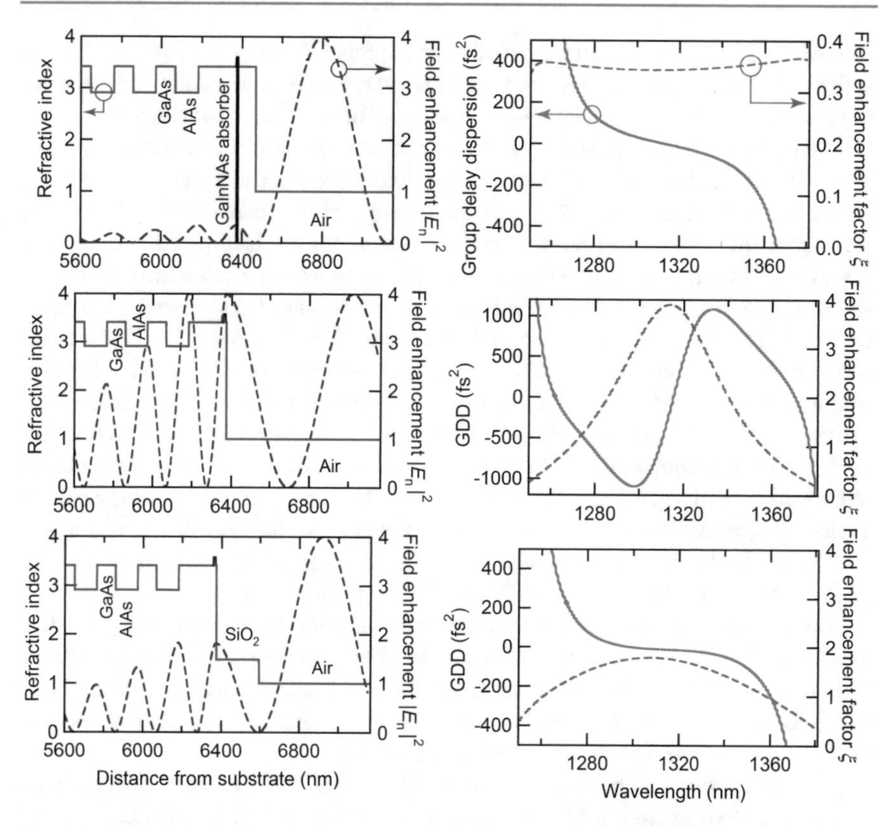

Fig. 7.32 SESAM designs with different field intensity enhancement: antiresonant low-finesse SESAM (*top*), LOFERS (*middle*), and E-SESAM (*bottom*). After [05Spu3]

indicates that the operation bandwidth and consequently the growth tolerances of a LOFERS are reduced. Moreover, the GDD values of up to over 1000 fs^2 can lead to non-transform limited pulses. In spite of these limitations, the LOFERS can be a very useful device for a laser whose emission (tuning range or bandwidth) is limited to a few nanometers, as within this range the variation of the main absorber parameters is acceptably low.

When we consider the E-SESAM (Fig. 7.32 bottom), we find an intermediate solution between the LOFERS and the antiresonant low-finesse SESAM. In this case, a LOFERS-type structure is completed with a single low-index (i.e., lower than the low-index DBR material) dielectric quarter-wave layer. In our example here, we deposit a quarter-wave layer of SiO_2 at the surface. This yields a relatively high enhancement factor of $\xi = 1.77$, slightly less than half of the LOFERS, but still more than five times enhanced compared to an antiresonant low-finesse SESAM. The GDD of this structure is now nearly wavelength independent and close to zero in the regime of interest. The field intensity has a node on or near the surface of the device. In addition, the wavelength dependence of ξ is greatly reduced. Therefore, this device can be a better choice for tunable lasers or lasers with larger operating bandwidth. Moreover, the growth tolerances are relaxed compared to the LOFERS, at the expense of a post-processing step. However, the enhancement layer can serve as passivation, and it can be used to compensate for possible growth errors.

The LOFERS and E-SESAM concepts can be further extended to a broader range of device parameters. As a general design guideline, we can make the following formulations. First, one can select and design an appropriate classical SESAM device (i.e., the antiresonant low-finesse SESAM), which has a ξ factor of approximately 0.34. If this does not have the desired modulation depth or saturation fluence, we can add an approximate quarter-wave layer of any optical material to obtain a LOFERS structure. If we choose to add the high index material (GaAs), we can add this to the structure during its epitaxial growth, without removing it from the growth chamber, resulting in a field enhancement of up to 11 times the classical SESAM device. Note that we can either add an approximate quarter-wave of GaAs to the classical SESAM device (corresponding to a final layer of approximately three-quarter wave thickness of GaAs) or similarly we can 'shorten' the last GaAs layer of the classical SESAM device by approximately a quarter wave. In either case, the absorber is positioned near the peak of the field, either near the air interface of the last layer, or at the next peak at approximately a half-wave inward from the GaAs–air interface.

As an alternative the LOFERS structure can be fabricated after epitaxial growth of the classical SESAM structure, by deposition of any appropriate dielectric material layer of approximately quarter-wave thickness. The lower the index of refraction of the dielectric, the lower the ξ factor. In all cases, the standing-wave field intensity in the LOFERS is near a peak at the device–air interface.

For applications which require a generally flatter GDD response and do not require ξ factors as high as from a LOFERS, one can choose an E-SESAM structure. The E-SESAM would nominally consist of the LOFERS described above (i.e., DBR plus quarter-wave or three-quarter-wave high index layer, with absorber near a field peak) plus an approximate quarter-wave layer of a material with typically lower refractive index, which we call the enhancement layer. Note that if this enhancement layer was GaAs, we would be back to a structure nearly identical to the classical SESAM device, with a ξ factor of about 0.34 and a flat GDD close to zero. Choosing the refractive index of the enhancement layer to be lower than the refractive index of both DBR materials, this ξ factor can be adjusted to any value between 0.34 to greater than 2. The E-SESAM generally shows a flatter GDD response and relaxed growth tolerances compared to the LOFERS structure.

Even more generally (and requiring typically full matrix-formalism calculations), the layer thicknesses can be adjusted so that we can obtain structures with ξ, F_{sat}, and GDD values in an even wider range (i.e., the layer thickness must not be restricted to quarter-wave multiples). The appropriate modulation depth can then be achieved by adjusting the thickness of the absorber layer within the physical limits. Further, it is possible to decrease the modulation depth by moving the absorber layer away from the peak of the field intensity. However, this is accompanied by an increase in the saturation fluence. Note that, although we have described typical structures based on GaAs and AlAs, other material combinations would certainly be possible to achieve similar structures and similar performance values.

7.6 SESAM Damage

7.6.1 SESAM Damage Measurements

We define damage of a SESAM under test as an irreversible change in the structure resulting in a dramatic drop in the measured reflectivity as shown in Fig. 7.33. We define the damage fluence threshold F_d by the minimum fluence where this irreversible reflectivity drop occurs in 1 s, i.e., time-to-damage in Fig. 7.33 is 1 s. In order to measure the damage fluence and the time-to-damage of a sample, we use the same setup as the one described in Sect. 7.7.1 for nonlinear reflectivity measurements with one incident beam. We set the fluence to a constant value and track the reflectivity of the sample versus time (Fig. 7.33). We used different SESAM designs with different topcoatings for a systematic study of SESAM damage (Fig. 7.34 and Table 7.7).

We measured lifetime curves of the different SESAMs according to Fig. 7.33 for which the damage fluence could be measured with the available maximum fluence. The results are presented in Fig. 7.36. For all SESAMs, the lifetime curves seem to follow an exponential behavior. The fit for the lifetime curves was performed using a single exponential function:

$$t(F) = t_0 + e^{-(F-F_3)/F_4} . \tag{7.103}$$

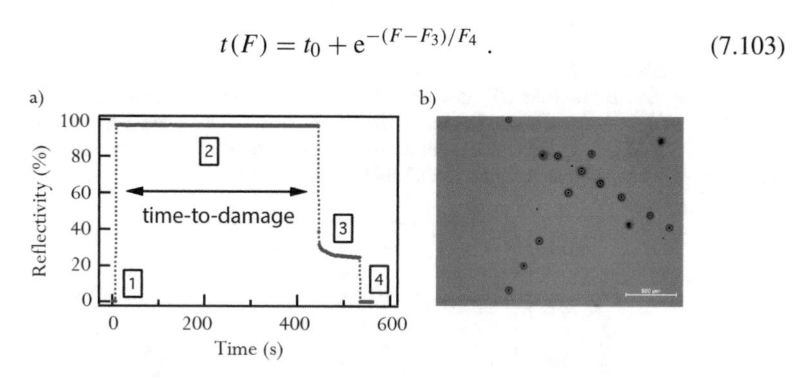

Fig. 7.33 **a** Typical time-to-damage measurement. In this example, we measured a time-to-damage of 445 s at a fluence of 27 mJ/cm^2, corresponding to a fluence 65% lower than the damage threshold (i.e., time-to-damage of 1 s). **b** Image of a tested SESAM showing multiple damage spots resulting from the performed characterization. After [12Sar4]

Table 7.7 Nonlinear reflectivity measurement results for antiresonant low-finesse SESAMs as shown in Fig. 7.33 with no topcoating (NTC), with a semiconductor topcoating (SCTC), and with a dielectric topcoating (DTC). DTC2 and DTC3 refer to two resp. three pairs of SiO_2/Si_3N_4 in the dielectric topcoating

	F_{sat} (μJ/cm^2)	ΔR (%)	ΔR_{ns} (%)	F_2 (mJ/cm^2)
NTC	72	2.05	< 0.1	3200
SCTC	279	0.52	< 0.1	5500
DTC2	168	0.71	< 0.1	31 700
DTC3	247	0.43	< 0.1	346 000

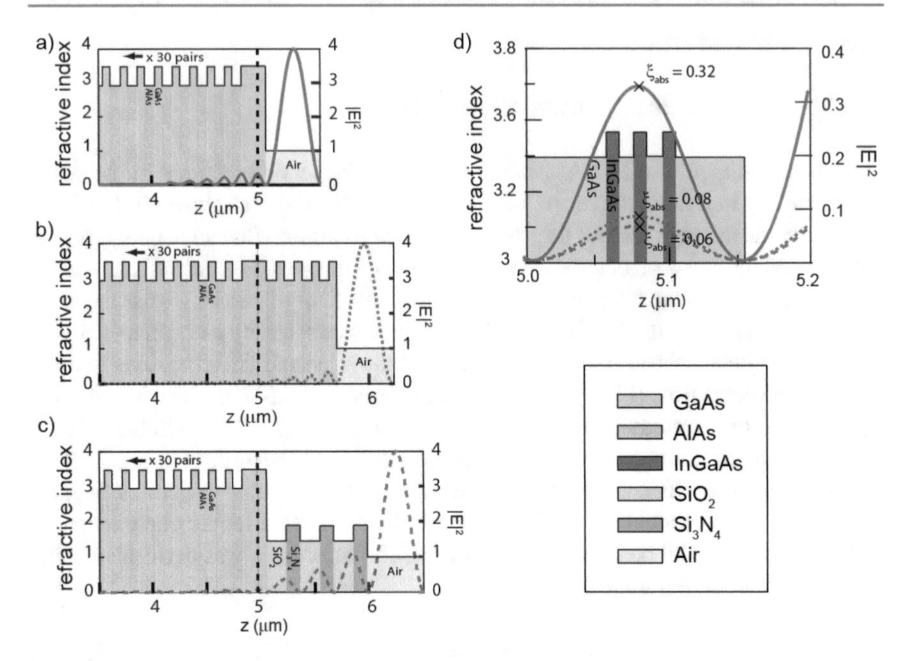

Fig. 7.34 Different SESAM structures used for the damage study. **a** Antiresonant low-finesse SESAM design with 3 InGaAs QW saturable absorber layers and no topcoating (NTC). **b** The same SESAM as in **a**, but with a 4-quarter-wave-pair GaAs/AlAs semiconductor topcoating (SCTC). **c** The same SESAM as in **a**, but with a 3-quarter-wave pair SiO_2/Si_3N_4 dielectric topcoating (DTC3). **d** Field intensity enhancement in the absorbers ξ_{abs} [the same as ξ in (7.71)]. After [12Sar4]

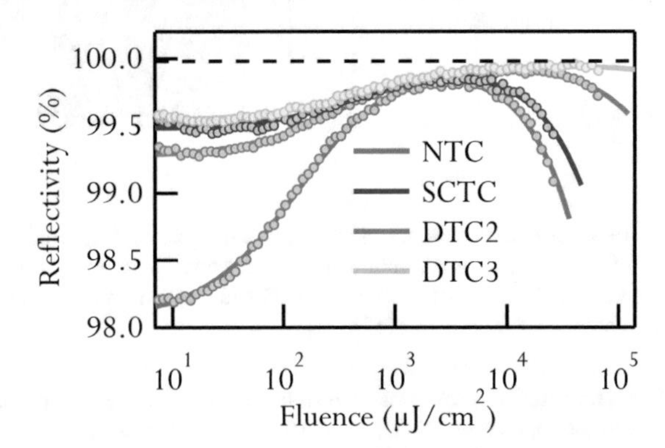

Fig. 7.35 Nonlinear reflectivity measurement of the different SESAMs as shown in Fig. 7.34, illustrating the effect of the different applied topcoatings. NTC no topcoating, SCTC semiconductor topcoating, DTC2 and DTC3 refer to two resp. three pairs SiO_2/Si_3N_4 dielectric topcoating. Fit to the model function (7.7) with ISA (7.48) and Gaussian beam profile [(7.47) in Sect. 7.3.4] results in the SESAM parameters shown in Table 7.7. After [12Sar4]

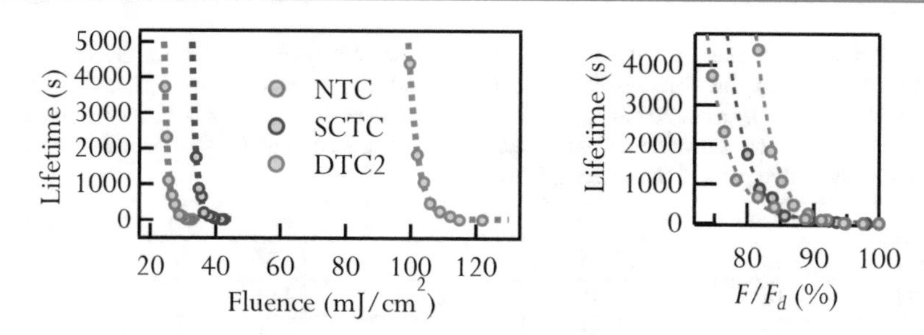

Fig. 7.36 Lifetime curves of different representative SESAMs defined in Fig. 7.34. *Left*: Lifetime curve with fluence axis in mJ/cm^2, where one can see the large shift to higher fluences for the sample with a dielectric topcoating (DTC2). *Right*: Same lifetime (7.103) curves, but with the fluence axis normalized to the damage threshold of the sample. After [12Sar4]

The fit parameters are presented in Table 7.8. The fit parameter t_0 is a small offset of the exponential in the y-axis needed to correctly fit the measurements. The damage fluence F_d is defined by (7.103) for $t(F_d) = 1$ s. For every SESAM, the measurements were made for fluence levels ranging from the instantaneous damage fluence to around 80% of this value, leading to maximum time-to-damage values of 1–2 h.

In Fig. 7.36 (left), we can see that the lifetime of the sample with a dielectric topcoating (DTC2) is shifted to higher fluences (increased F_3), confirming the advantage of this type of topcoating for increasing the damage threshold of SESAMs. In addition, the slope of the exponential is steeper for the topcoated samples, and in particular the dielectric topcoated one shows best performance (see Fig. 7.36 right).

We can use these measurements to estimate the lifetime of SESAMs at a given lower fluence. SESAMs in multi-100-W average power laser oscillators have operated SESAMs with saturation parameters $S = F_p/F_{sat} \approx 20$–100, which is higher than the typical value of $S \approx 5$. It is therefore interesting to evaluate their lifetime at this high saturation level. The longest time-to-damage measurement was performed using the SESAM NTC in Fig. 7.33, but with one QW instead of 3 QWs as absorber. The lifetime of this sample is presented on a logarithmic scale as a function of the saturation parameter in Fig. 7.37 and extended to very low saturation parameters. We can see that, at saturation parameters of $10 < S < 50$, the suggested lifetimes are on the order of 700 000–150 000 hr. Testing a sample with a dielectric topcoating like the ones presented in Figs. 7.34 and 7.36 (DTC2, for example) should result in even longer lifetimes at much lower fluences. Nevertheless, considering that the longest time-to-damage point was taken at a fluence over an order of magnitude greater than the evaluated points, our measurements seem to give a good approximation of the lifetime of SESAMs at more standard fluences. However, we can see here that it is crucial to take points at long lifetimes (> 10 h) for a correct extension of the lifetime curves.

In order to better understand the influence of the absorber thickness on the lifetime, we compared the damage behavior of a SESAM NTC (Fig. 7.34a) with a single QW absorber, one with 3 QWs, and one with a DBR mirror (without absorber section) in order to evaluate whether the damage thresholds are dependent on the absorber

Table 7.8 Fit parameters for the lifetime curves presented in Fig. 7.36 using the fit function described in (7.103). The damage fluence F_d is defined for $t(F_d) = 1$ s from (7.103). This means that it takes 1 s for measurable damage at F_d. For antiresonant low-finesse SESAMs as shown in Fig. 7.33 with no topcoating (NTC), with a semiconductor topcoating (SCTC), and with a dielectric topcoating with two pairs of SiO_2/Si_3N_4 (DTC2).

	t_0 (s)	F_4 (mJ/cm^2)	F_3 (mJ/cm^2)	F_d (mJ/cm^2)	$S_d = F_d/F_{sat}$
NTC	48	1.15	14	32.6	450
SCTC	14	1.31	24	44.1	158
DTC2	24	2.97	74	122	726

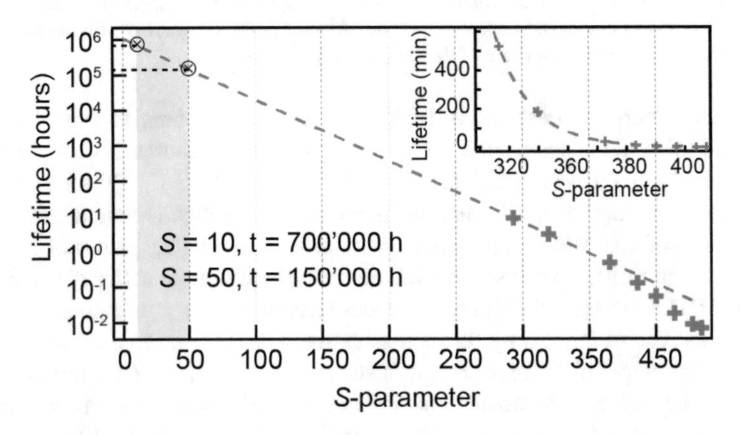

Fig. 7.37 Lifetime of a SESAM NTC (with no topcoating) in Fig. 7.33, but with one QW instead of 3 QWs as absorber, extended to still relatively high saturation parameters of 10 and 50. S is the saturation level of the SESAM, i.e., $S = F_p/F_{sat}$. More typical SESAM saturation is at $S \approx 5$. After [12Sar4]

geometry. All the samples have no topcoating and are in an antiresonant configuration (Fig. 7.34a). The results are presented in Fig. 7.38 and Table 7.9.

We can see that the damage behavior of a DBR is similar to that of the characterized SESAMs. Damage occurs deep in the rollover regime, and this rollover occurs at comparable fluence levels as for the SESAMs. This seems to indicate that this mechanism is related to the inverse saturable absorption (ISA) process (Sect. 7.3.5), which is mainly caused by two-photon absorption (TPA) in the GaAs layers of the structure, i.e., the material with largest TPA coefficient. Note also that the absorbers seem to have a tendency to slightly reduce the lifetime. One possible reason is that the SESAMs have additional GaAs spacers in the absorber section where the electric field is strong, and therefore represent an important contribution to the ISA. This would also explain the 50% higher damage fluence for the DBR (Table 7.9).

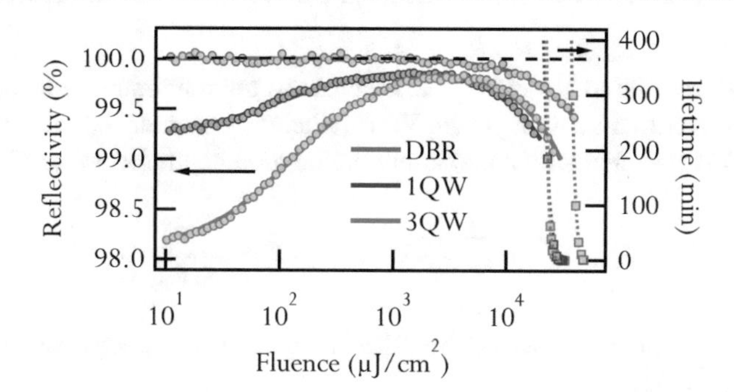

Fig. 7.38 Nonlinear reflectivity and lifetime measurements for a DBR and for antiresonant low-finesse SESAMs with no topcoating (NTC) (Fig. 7.34a) with 1 QW and 3 QWs as absorber. After [12Sar4]

Table 7.9 Nonlinear reflectivity fit parameters and damage thresholds of the samples with different absorber sections presented in Fig. 7.38

	F_{sat} (μJ/cm^2)	F_2 (mJ/cm^2)	F_d (mJ/cm^2)	$S_d = F_d/F_{sat}$
DBR	–	7400	48	–
1QW	57	2600	32.1	560
3QW	72	3200	32.6	453

7.6.2 SESAM Damage Theory

For all SESAMs under test, we observed that damage occurs at fluences deep in the rollover regime, suggesting a damage mechanism related to the absorbed energy due to inverse saturable absorption (ISA) (Sect. 7.3.5). It is interesting to note that, in the case of the sample with 3 QWs and a two-pair dielectric topcoating (DTC2), F_2 is greatly increased and the damage curve is also shifted to higher values (Fig. 7.36). The energy absorbed by the sample due to ISA can be evaluated if we accept to make a number of approximations.

For any given fluence, one can calculate the absorbed fluence F_{abs} as a function of the incoming fluence F_p on the SESAM:

$$F_{abs}(F_p) = F_p\left[1 - R(F_p)\right] . \tag{7.104}$$

We approximate $R(F_p)$ using the expression for a flat-top beam including the extra rollover factor (see Table 7.4). This expression can be numerically corrected for a Gaussian beam (Sect. 7.3.4). However, using the approximate expression for a flat-top beam leads to a small error in our calculation and allows us to obtain an approximate analytical expression. We simplify the model function $R(F_p)$ (7.7) and use (7.8) with ISA, i.e., (7.51), and for $S \gg 1$, i.e., $F_p \gg F_{sat,A}$, we obtain

$$R(F_p) = R_{ns} - \frac{\Delta R}{S}(1 - e^{-S}) - \frac{F_p}{F_2} \approx R_{ns} - \frac{\Delta R}{S} - \frac{F_p}{F_2} , \tag{7.105}$$

since $1 - e^{-S} \approx 1$ for $S = F_p / F_{sat,A} \gg 1$.

Inside a laser, the SESAM is operated close to the maximum reflectivity of the SESAM for maximum output power. We can therefore assume that, inside the laser, the SESAM is operated with a maximum modulation depth following (7.54):

$$F_0 \approx \sqrt{F_2 F_{sat,A} \Delta R} \Longrightarrow \Delta R \approx \frac{F_0^2}{F_2 F_{sat,A}} . \quad (7.106)$$

For an incoming pulse fluence on the SESAM F_p close to F_0, we then obtain, using (7.105) and (7.106):

$$R(F_p \approx F_0) \approx R_{ns} - 2\frac{F_p}{F_2} . \quad (7.107)$$

Therefore, using (7.104) and (7.107), we obtain

$$F_{abs}(F_p \approx F_0) = F_p \left(1 - R_{ns} + 2\frac{F_p}{F_2}\right) \overset{R_{ns} \approx 1}{\longrightarrow} 2\frac{F_p^2}{F_2} . \quad (7.108)$$

If we assume that damage occurs at a fixed absorbed fluence, we have a condition for the incident damage fluence F_d at which damage occurs [12Sar4]:

$$F_{abs} \approx 2\frac{F_d^2}{F_2} = \text{constant} . \quad (7.109)$$

Therefore F_d scales with $\sqrt{F_2}$ (assuming $R_{ns} \approx 1$):

$$\boxed{F_d \propto \sqrt{F_2}} . \quad (7.110)$$

This implies that for a high damage fluence a high F_2 is required. This means negligible ISA as shown in Fig. 7.5 with no rollover in the nonlinear reflectivity with $F_2 \to \infty$. According to (7.106) this also implies a large F_0, i.e., a SESAM with a high-fluence rollover.

This can be achieved with an optimized SESAM device design, i.e., a larger F_2 with a dielectric topcoating compared to no topcoating (see Fig. 7.35). In addition the two-photon absorption (TPA) coefficient β_{TPA} depends on the material and scales as E_g^{-3}, where E_g is the bandgap energy in a semiconductor between 1.4 and 3.7 eV [85Str2]. The photon energy is less than the direct bandgap E_g but greater than $E_g/2$ so that TPA can occur. The rollover parameter scales as $F_2 \propto 1/\beta_{TPA} \propto E_g^3$, so a large bandgap material will reduce TPA and shift the rollover fluence to higher values.

7.6.3 SESAM Design for High Average Power Thin-Disk Lasers

The current state-of-the-art for high-power/high-energy ultrafast laser oscillators is based on SESAM modelocked Yb:YAG thin-disk lasers generating multi-100-W average output power and pulse energies close to 100 μJ [19Sal]. The intracavity average power is well above 1 kW. High damage-fluence SESAMs for high average power modelocked lasers require low nonsaturable losses (i.e., $R_{ns} \approx 1$), high F_2 (7.110), and therefore high F_0. The SESAM damage studies in Sect. 7.6.1 showed that a SESAM with a dielectric topcoating have a high F_0 which is the preferred approach. In comparison to semiconductor topcoatings, the dielectric materials have less two-photon absorption (TPA) and less defect-related loss.

A thicker topcoating requires a thicker saturable absorber section within the SESAM design to obtain a sufficiently large modulation depth ΔR. At a center wavelength of around 1 μm, we typically use InGaAs quantum wells (QWs) on GaAs/AlGaAs DBRs. This increases the strain as the number of InGaAs QWs grows, degrading the SESAM performance (Fig. 7.39a). Therefore strain compensation needs to be applied [16Alf]. In addition, power-scaling is achieved with a larger cavity mode size on the SESAM and a thin-disk gain crystal, which increases the requirements for the SESAM to have a flat surface [16Die].

With shorter pulses the inverse saturable absorption is dominated by TPA. The SESAMs that will support significantly shorter pulses, i.e., less than 500 fs, at high average power will require a more demanding set of parameters. In particular, high average power operation needs SESAMs with higher saturation fluence, high-damage threshold, a rollover shifted towards higher fluence, and lower non-saturable losses. However, there is a trade-off. Shorter pulses need fast recovery times, i.e., a few picoseconds, low TPA, and a large ΔR (> 1%), limited by the onset of Q-switching instabilities or multi-pulsing instabilities. These stability requirements for SESAM modelocking will be discussed in Chap. 9. It becomes a major challenge to achieve such SESAM parameters simultaneously, because the combination of

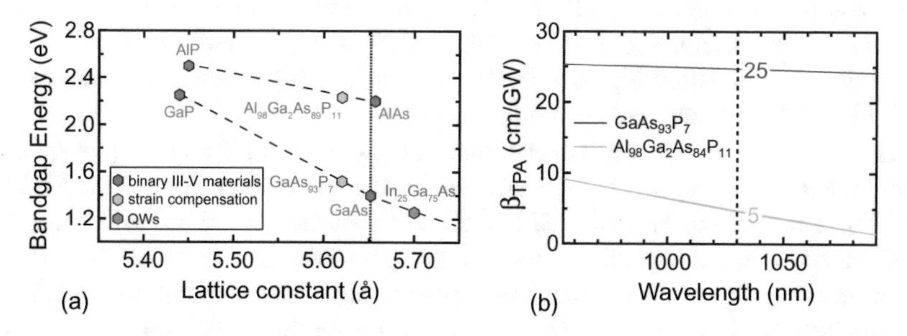

Fig. 7.39 Strain compensation for InGaAs SESAMs. **a** Bandgap energy E_g and crystal lattice constant of semiconductor materials of interest. *Dotted vertical line*: perfect matching to the lattice constant of the GaAs substrate. In *light blue*, compounds which can be used to compensate the tensile strain introduced by the InGaAs quantum wells (*green*). **b** TPA coefficient β_{TPA} for low-bandgap GaAsP and large-bandgap AlGaAsP strain compensation materials calculated as in [85Str2]. Low values for β_{TPA} are preferred to reduce the rollover fluence. The numerical inset values refer to a center wavelength of 1030 nm (*dashed line*). After [16Alf]

higher rollover, increased modulation depth, and faster recovery times in antiresonant SESAM designs requires a top reflector with higher reflectivity and an increased number of absorber layers. We therefore have to address the following issues:

(a) Strain compensation with lower TPA: For strain-compensated SESAMs with reduced TPA, we developed large-bandgap quaternary AlGaAsP materials (e.g., $E_g > 2.2$ eV for $Al_{0.98}Ga_{0.02}As_{1-x}P_x$) with low TPA coefficient β_{TPA} (Fig. 7.39b). Strain compensation provided by large-bandgap AlGaAsP materials with negligible TPA coefficients allows us to significantly increase the number of QWs from the typical value of 3 to 8, without increasing the nonsaturable losses. In accordance with our guidelines, this increase in the number of absorber layers, together with the higher reflectivity of the dielectric topcoating, gives a higher F_{sat}, a higher reflectivity rollover, and a higher damage threshold. In addition, the excellent crystal quality of this SESAM results in a higher heat conductivity and makes it suitable for post-processing with better heat sinks, as described in more detail in [16Die].
(b) Fast SESAMs with low nonsaturable losses: Normally we observe a trade-off between the fast recovery time of the saturable absorption and the nonsaturable losses. These nonsaturable losses can be reduced through annealing, which produces As precipitates that still form good electron traps but have no strong transition cross-section for absorption into the conduction band (Sect. 7.4.3). But once again, there is a trade-off because this annealing tends to make the saturable absorbers slower, ultimately requiring a compromise [99Hai2,99Sie]. We discovered that low-temperature grown InGaAs QW embedded in LT AlAs barriers can generate electron traps with even faster recombination, which also remain fast with longer annealing times [13Man].

With such overall optimization steps, much better high-average-power InGaAs SESAM parameters for a center wavelength of around 1 μm could be achieved, such as an F_{sat} of 334 μJ/cm^2, a ΔR of 1.3%, a sufficiently fast recovery time of 17 ps, extremely low nonsaturable loss of 0.14%, an F_0 of 8.2 mJ/cm^2, an F_2 of 9 J/cm^2, and a damage threshold higher than 100 mJ/cm^2. A more detailed description of such SESAM optimization is given in [16Alf].

The thermal lens of a SESAM becomes an issue for high average power lasers (see Sect. 9.10.2), and this is even affected by the gas-lens in front of the SESAM [18Die]. The thermal lens of a SESAM can be further optimized and controlled by silicate bonding a sapphire window to the front face of a SESAM [21Lan]. The magnitude and sign of the thermal lensing can be adjusted by tuning the thickness of the sapphire. Such sapphire SESAMs in a high-power thin-disk laser have achieved 230 W average output power, which is a 70 W improvement compared to a state-of-the-art standard SESAM in the same cavity. This improvement can be attributed to the inverse thermal lensing of the sapphire SESAM which paves the way for high-power optics with adjustable thermal properties. Thermal management of SESAM structures is a key aspect of the power scaling of ultrafast oscillators.

7.7 SESAM Characterization

Precise knowledge of the time-dependent nonlinear optical reflectivity is required to optimize SESAMs for self-starting passive modelocking. In this section, we discuss two methods for wide dynamic range nonlinear reflectivity measurements and describe the pump–probe setup we use to measure the temporal response of the SESAMs.

7.7.1 One-Beam Measurement of Nonlinear Reflectivity

We have demonstrated a first measurement system for optical characterization as shown in Fig. 7.40 [04Hai]. The laser is a pulsed laser source, typically a modelocked laser with sufficient energy to saturate the SESAM under test. Ideally, we want to characterize the SESAM under identical conditions to those inside a laser cavity that is passively modelocked by the same SESAM. One approach is based on the one-beam measurement, where we use the output beam of a sufficiently high-power modelocked laser source to measure the nonlinear SESAM reflectivity of this one beam as function of the incident pulse fluence (Fig. 7.40).

The pulsed laser source should therefore operate at the same center wavelength and pulse duration. The pulse fluence incident on the SESAM should ideally be at least 10 times F_{sat} for a precise parameter extraction, but depending on the induced absorption (e.g., a large F_2), fluences of up to 50 times F_{sat} can be required for a precise characterization. In this case we use the model functions (7.7) given in Sect. 7.2 and summarized in Table 7.4, with a correction factor for ISA (7.48) which results in

$$R^{\text{flat-top}}(F_{\mathrm{p}}) = R_{\mathrm{ns}} \frac{\ln\left[1 + \dfrac{R_0}{R_{\mathrm{ns}}}(\mathrm{e}^S - 1)\right]}{S} \exp\left(-\frac{F_{\mathrm{p}}}{F_2}\right) , \quad S = \frac{F_{\mathrm{p}}}{F_{\mathrm{sat,A}}} . \tag{7.111}$$

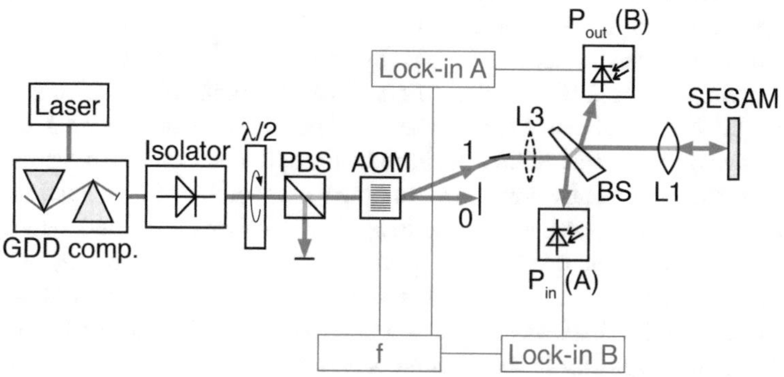

Fig. 7.40 Nonlinear reflectivity measurement setup [04Hai]. An AOM is used for lock-in detection. Beam splitter BS, acousto-optic modulator AOM, polarizing beam splitter PBS, laser diode LD. A CCD camera is used to measure the incident spot size. Figure from [08Maa1]

This is the model function for a flat-top beam profile with a constant incident pulse fluence F_p. For a Gaussian beam profile we can numerically calculate the correct model function using (7.111) according to (7.47):

$$R^{\text{Gauss}}(F_p) = \int_0^1 dz\, R^{\text{flat-top}}(2F_p z) \ . \tag{7.112}$$

In a modelocked laser, the intracavity fluence on the SESAM is typically 3–10 times F_{sat}. So we need to achieve a higher fluence than inside the cavity for fewer errors in the fitting procedure with the model functions. This is usually obtained either by strong focusing onto the SESAM or by employing a lower repetition rate of the evaluation laser. At high repetition rates, some SESAMs do not fully recover between successive laser pulses, and an evaluation at the precise repetition rate may even be needed. The pulse duration of the laser should be shorter than the recovery time of the SESAM under test, because the fitted model function (7.111) as derived in Sect. 7.2 uses the slow saturable absorber approximation. The angle of incidence onto the SESAM is perpendicular, so some light is reflected back into the laser source. An isolator prevents such back-reflections from entering the laser source, which would lead to modelocking instabilities.

The core of the nonlinear reflectivity measurement setup shown in Fig. 7.40 is a beam splitter BS (an uncoated small-angle wedged glass plate) in front of the SESAM. The reflection on the front side of the BS (A) is proportional to P_{in} and the reflection on the back side of the BS (B) is proportional to RP_{in}. The reflectivity of the SESAM is $R = B/A$ and can be computed as a function of the incident pulse fluence, which is changed by a variable attenuator through a combination of a half-wave plate, a polarizer, and an acousto-optic modulator (AOM). The detectors must be able to measure voltages over at least four orders of magnitude with an accuracy of better than 0.1%. A lock-in detection with two separate lock-in amplifiers has to be used to reduce the influence of stray light and noise. The AOM is used for modulation of the incident beam required for the lock-in scheme. However, it is very challenging to achieve sufficient accuracy, because the required performance is close to the linearity limit of the lock-in amplifiers.

The most challenging part is the requirement for a wide-dynamic range attenuator which can be realized in many ways, e.g., by graded neutral density filters, acousto-optical modulators (AOMs), or polarization splitters in combination with polarization rotation optics. Neutral density filter wheels, however, may cause thermal lensing at high average power levels. AOMs can introduce amplitude noise at stronger attenuation levels and are typically limited to two orders of magnitude of attenuation. In our setup as shown in Fig. 7.40, we choose to use the polarization method, which relies on two optical elements: one element rotates the linear polarization state, while the second element selects the desired polarization. Polarization rotation can be achieved using, for example, a half-wave plate or a polarizing beam splitter (PBS). Because half-wave plates typically have a limited bandwidth, we prefer to use PBSs which have a bandwidth above 100 nm. The dynamic range is determined by the extinction ratio of the first PBS.

An improved version of a nonlinear reflectivity measurement is illustrated in Fig. 7.41. The isolator consists of two polarizing beam splitters PBS 2 and PBS 3 and a Faraday rotator. To obtain good isolation, we used Glan Laser PBSs, which have a 250 nm–2.3 μm wavelength range and an extinction ratio of 1:100 000. The lens L1 focuses the incident beam onto the SESAM and typical beam radii are between 5 and 20 μm. We employ a photodetector (PD) with a large detection area (typically $7 \times 7\ \text{mm}^2$) to measure a collimated beam with large beam radius.

The measurement part consists of a non-polarizing beam splitter cube (BS), a lens, a chopper wheel, and a photodetector (PD). Instead of detecting A and B simultaneously by two different detectors (see Fig. 7.40), the signals are separated in time and measured with the same detector system. The separation in time is achieved by a chopper wheel which simultaneously chops both arms and is put close to the 50:50 beam splitter. The signal is amplified and measured with an analog-to-digital (AD) converter and recorded with a computer. The chopper frequency is typically in the range of 100 Hz, and a low-cost 14 bit AD converter is sufficient to measure photovoltages with 0.01% accuracy (when the photocurrent amplifier is set to obtain a full scale for the reference signal, 14 bits results in 0.006% resolution, and averaging over more points can even increase this value).

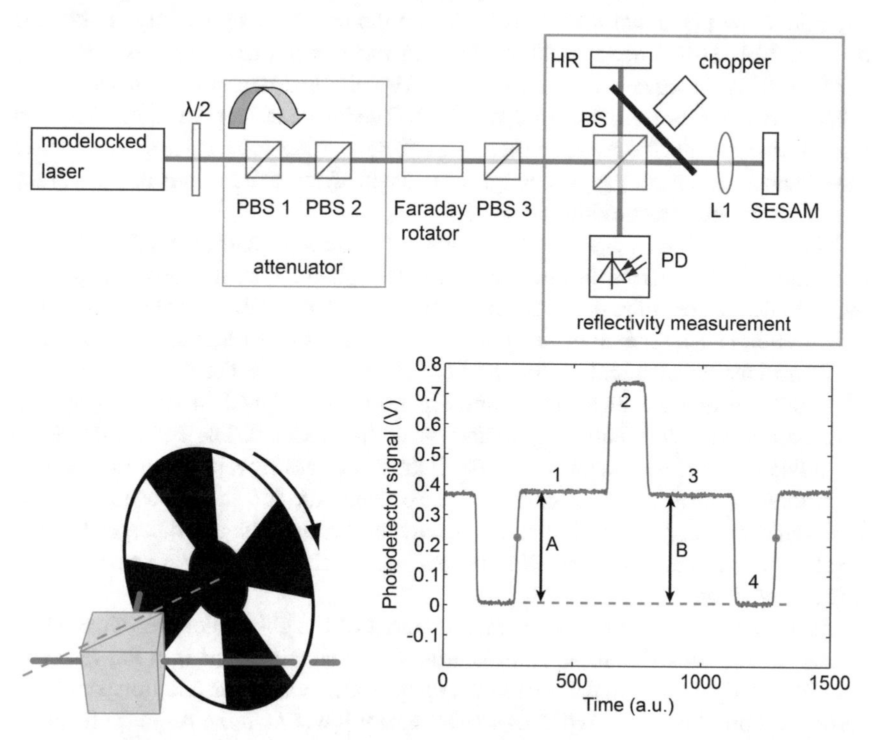

Fig. 7.41 Improved nonlinear reflectivity measurement setup [08Maa3]. The output of the mode-locked laser source propagates through a variable attenuator to set the pulse energy, then through the isolator to eliminate back reflections, and finally enters the measurement part. A mechanical chopper greatly improves sensitivity. Beam splitter BS, polarizing beam splitter PBS, highly reflecting mirror HR, photodetector PD

In our measurement system, we lifted the chopper wheel so that its axis was a few centimeters above the beam heights (Fig. 7.41 bottom left). During one chopper wheel cycle, four different states occur (Fig. 7.41 bottom right): (1) only reference beam measured, (2) both beams measured, (3) only sample beam measured, and (4) both beams blocked. The signal in phase 4 corresponds to a background signal from photodiode dark current and environmental background light, which is then discriminated from the measurement signal in the other phases. In [04Hai], a lock-in detection was required to reject the background signal. The PD signal was amplified by a computer-controlled variable pre-amplifier to use the full range of the AD converter. The absolute gain and the offset have no influence on the measurement accuracy, and are only necessary to provide a linear response. Since the reflectivity R is encoded in only one optical/electrical signal, the constraints on the amplifier have become negligible. This is in contrast to the method of Haiml et al., in which the same gain and no offset have to be achieved by both amplifiers [04Hai].

The computer algorithm first detects the rising edges, the red dots in Fig. 7.41 (bottom right), and then takes the mean value of the data points on the flat levels, the red lines in Fig. 7.41 (bottom right). As both beams are blocked in phase 4, we can precisely measure the offset of the photodiode. Levels A and B are obtained by subtracting the signal level in state 1, and the nonlinear reflectivity is obtained as $R = B/A$. This is done for 500 periods in succession (taking approximately 5 s per fluence) and averaged to minimize detector noise and laser noise. This averaged reflectivity has a standard deviation of 0.01%. The incident fluence can be computed from the level A and the pre-amplifier gain setting. An accuracy of 5% for the fluence measurement is typically good enough, as this will afterwards result in an inaccuracy of 5% for the fitted saturation fluence F_{sat}.

The beam splitter as well as all the other optical components are slightly tilted to eliminate reflections from the interfaces hitting the photodiode. We observed up to 2% measurement error when parasitic reflections or scattered light reached the detector, so preventing such reflections from hitting the detector is particularly important.

To achieve an identical optical-to-electrical response of the detector from the sample beam and the reference beam, the setup is designed in such a way that both beams have an identical spot size on the photodiode. The beam that enters the reflectivity measurement part is collimated and has a waist (1 mm radius) on the reference mirror. The sample beam is focused onto the sample using lens L1. Both returning beams have the same q parameter (beam width and divergence). The overlap of both backward traveling beams can be checked using a camera at the detector position.

An obvious sanity check is the measurement of a highly reflective mirror (HR) instead of the SESAM in the sample arm, which should result in a flat response over the full dynamic range. Because the sample arm includes an additional lens, the response from this arm is typically a few percent lower than the response from the reference arm. We introduce a calibration factor such that the reflectivity $R = CB/A$ is constant. In case of systematic errors, C can be a function of the fluence F.

For a measurement, the SESAM has to be positioned at the waist at zero degree incidence. First, the SESAM is aligned perpendicularly to the beam without the lens

L1 by using an aperture. Then the lens is inserted and the beam is again aligned with the aperture. The final alignment concerns the fine tuning of the SESAM position, which needs to be exactly in the focus of the laser beam. The SESAM is moved along the propagation direction and the beam waist is where the nonlinearity is the largest and thus the measured saturation fluence is the smallest.

One has to be careful when replacing the SESAM under test, because slight misalignments can result in measurement errors. We therefore use two alignment beams from a laser pointer to memorize the sample position and tilting angle: the first beam is reflected by the sample and is directly aligned with an aperture; the second beam propagates through two lenses (focal length 5 cm), with the sample in-between. A linear translation of the sample causes a deviation in the angle of the collimated output beam that is aligned to a second aperture. The first beam is sensitive to angular errors and has a reproducibility better than 30 mrad. The second beam is mainly sensitive to position errors and has a reproducibility of 10 μm which is well below the Rayleigh length (140 μm). We obtain a 0.02% reproducibility of the reflectivity measurement.

Figure 7.42 shows the measurement using the setup shown in Fig. 7.41. The SESAM operates at 1030 nm and at a high saturation fluence, which enables the realization of passively modelocked thin-disk lasers exceeding 10 μJ pulse energy [08Mar]. The SESAM has InGaAs quantum wells which absorb the laser light. At a sufficient fluence the carriers start filling the bands and the wells become transparent. The laser source is a modelocked Yb:Lu_2O_3 thin-disk laser, running at a lower repetition rate of 65 MHz [07Mar2]. The pulses are 570 fs long, which will cause induced saturable absorption at higher fluences. We used a lens with a focal length of 25 mm to obtain a 10.3 μm beam waist (radius) on the sample, which allowed measurements up to 6800 μJ/cm^2. The HR flatness was improved by choosing the calibration function $C(F)$ as a second order polynomial in log F. The HR measured with

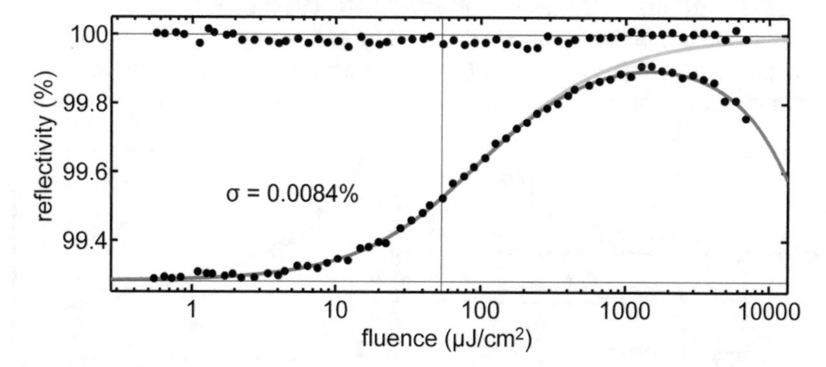

Fig. 7.42 Measurement at a center wavelength of about 1 μm [08Maa1] of a highly reflective mirror (HR) with calibration. The reflectivity is close to 100% with a flatness of 0.055%. SESAM measurement with small modulation depth, measured over four orders of magnitude. The *grey curves* show the fit model function according to (7.111) and (7.112) with and without the additional induced absorption factor (7.48), taking into account the correction for a Gaussian beam profile (7.112)

this correction factor has a flatness of 0.055% (Fig. 7.42). From the fit we obtained $F_{sat} = 54.7 \pm 1.4$ μJ/cm^2, $\Delta R = 0.722 \pm 0.005\%$, $F_2 = 3.18 \pm 0.13$ J/cm^2, and very small nonsaturable losses $\Delta R_{ns} = -0.003 \pm 0.005\%$. The systematic error in the measurement of the nonsaturable losses ΔR_{ns} depends on the accuracy of the HR reference measurement, so we can only conclude that $\Delta R_{ns} < 0.1\%$.

Moving this high precision nonlinear reflectivity measurement technique from the near-infrared to the mid-infrared regime is more challenging. Similar precision was, however, achieved for a wavelength range of 1.9–3 μm [21Hei]. This allowed for full characterization of 2 μm [21Hei] and 2.4 μm [21Bar] SESAMs using Sb materials.

7.7.2 Pump–Probe Measurement of Recovery Time

The recovery dynamics of the SESAM are important for pulse formation and need to be characterized precisely. In order to reach the shortest pulse duration, sufficiently fast recovery has to be achieved. The recovery is measured with a time-resolved differential reflection setup (also referred to as a pump–probe setup), as shown in Fig. 7.43a. The pulsed laser beam is first split into a strong pump beam and a weak probe beam. A typical intensity ratio should be many orders of magnitude to make sure that the probe pulses do not affect the measurement. For SESAMs the probe pulse fluence needs to be much smaller than the saturation fluence $F_{sat,A}$ to not saturate the reflectivity and therefore not influence the measurement.

The two beams should overlap on the SESAM under test. We can achieve this by using a beam profiler and adjusting the two lenses in front of the sample. The temporal overlap is adjusted with a computer controlled translation stage. The pump beam hits the sample first, under a small angle, and after a variable delay, the probe beam reflects off the sample and is measured with a photodiode.

Both pump and probe beams are chopped by acousto-optic modulators (AOMs). By chopping the pump we can directly measure with the probe beam the modulation this pump beam induces on the sample. Ideally, the pump beam is chopped at more

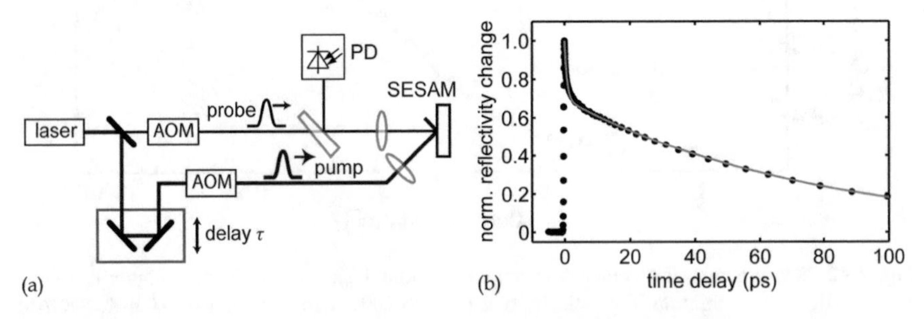

Fig. 7.43 The pump–probe setup used to measure the recovery time of the SESAM [08Maa1]. **a** By modulating the pump at high frequencies (i.e., ideally > 1 MHz), we directly measure the reflectivity change of the SESAM $\Delta R_{pp}(\tau)$. The probe beam is also modulated but at much lower frequency to suppress stray light from the pump beam. Lock-in detection at the difference frequency makes the setup very sensitive. **b** A typical SESAM response. Acousto-optic modulator AOM

than 1 MHz because the laser intensity noise typically becomes negligible and the measurement can be shot-noise limited. Chapter 11 explains this in more detail, and Fig. 11.20 shows the principle. In addition, we chop the probe beam at much lower frequency to suppress the scattered pump light that accidentally hits the detector. The probe beam is detected with a lock-in amplifier at the difference frequency. Note that the chopping of the probe beam will also translate the noise sidebands up around the chopping frequency. Therefore the probe chop frequency should be much smaller than the high-frequency pump modulation.

A typical pump–probe signal from a SESAM is shown in Fig. 7.43b. The laser source used for the measurement is a modelocked wavelength-tunable Ti:sapphire laser with an 80 MHz pulse repetition rate. The 140 fs pulses (measured directly after the laser) are nearly transform-limited, with a time–bandwidth product of 0.34. The pump spot has a diameter of 20 μm and the probe is slightly smaller, with a diameter of 15 μm to make sure we only probe the saturated region. This results in fluences of 5.1 and 0.16 μ J/cm^2 for pump and probe, respectively.

The normalized pump–probe response can be fitted well with a bitemporal response function with two time constants

$$\Delta R_{\mathrm{pp}}(t) = A\mathrm{e}^{-t/\tau_1} + (1 - A)\mathrm{e}^{-t/\tau_2} , \tag{7.113}$$

where t is the time delay between pump and probe pulse, A is the amplitude of the component with time constant τ_1 and $1 - A$ is the amplitude of the component with time constant τ_2. In the example above, $\tau_2 = 0.77$ ps, $\tau_1 = 74$ ps, and $A = 70\%$. A more detailed discussion of the carrier dynamics is given in Sect. 7.4.2.

7.7.3 Pump–Probe Measurement of Nonlinear Reflectivity

The nonlinear reflectivity of the SESAM can also be measured with the pump–probe setup (Fig. 7.43) as function of the incident pump pulse fluence. Ideally, both pump and probe beams should have the same polarization to match the SESAM modelocking conditions inside a laser. The nonlinear reflectivity is then given at zero delay between the pump and the probe pulse, i.e., $\Delta R_{\mathrm{pp}}(F_{\mathrm{p}}, t \approx 0)$ (7.113) for a given pump fluence on the SESAM. In this case we need a different model function to fit the nonlinear reflectivity $\Delta R_{\mathrm{pp}}(F_{\mathrm{p}}, t \approx 0)$.

A low-fluence probe pulse at time t experiences a SESAM reflection $R(F_{\mathrm{p}}, t)$ when a high-fluence pump pulse with an incident fluence F_{p} saturates the SESAM. This pump-induced probe reflectivity is given by (7.38):

$$R(F_{\mathrm{p}}, t) = \frac{R_{\mathrm{ns}}}{1 + \left(\dfrac{R_{\mathrm{ns}}}{R_0} - 1\right)\mathrm{e}^{-F_{\mathrm{in}}(t)/F_{\mathrm{sat,A}}}} \exp\left(-\frac{F_{\mathrm{in}}(t)}{F_2}\right) , \tag{7.114}$$

taking into account the inverse saturable absorber factor (7.48), and $F_{\mathrm{in}}(t)$ is defined by (7.33):

$$F_{\text{in}}(t) \equiv \int_{-\infty}^{t} I_{\text{in}}(t')\mathrm{d}t' \,, \quad F_{\text{p}} = \int_{-\infty}^{\infty} I_{\text{in}}(t')\mathrm{d}t' \,, \tag{7.115}$$

assuming that the incident instantaneous intensity $I_{\text{in}}(t)$ describes only one pump pulse (i.e., not a pulse train). Assuming a probe pulse with a probe fluence F_{probe} with an instantaneous intensity $I_{\text{probe}}(t)$, we obtain the probe reflectivity by a convolution of $I_{\text{probe}}(t)$ with $R(F_{\text{p}}, t)$ given in (7.114):

$$\Delta R_{\text{pp}}(F_{\text{p}}, t) = \frac{1}{F_{\text{probe}}} \int R(F_{\text{p}}, t') I_{\text{probe}}(t' - t)\mathrm{d}t' \,. \tag{7.116}$$

The one-beam and the pump–probe measurements result in the same saturable absorber parameters as long the correct model functions are used to fit the measured nonlinear reflectivity and the pump and the probe pulse completely overlap in space and time [13Fle]. This can be achieved with collinear pump and probe pulses with the same spot size and taking the reflectivity at zero time delay. This was shown experimentally in [13Fle]. The nonlinear reflectivity needs to be fitted to the correct model functions which are different for the one-beam and the pump–probe experiment. The model function for the one-beam measurement is given by (7.111) and for the pump–probe measurement by (7.114) and (7.116) for time zero [i.e., $\Delta R_{\text{pp}}(F_{\text{p}}, t = 0)$]. In both cases these model functions need to be numerically modified by the Gaussian beam profile with (7.112).

The requirement of perfect overlap in space and time of the pump and probe pulses adds a potential risk for errors in the fitted saturable absorber parameters using the pump–probe measurement. However novel dual-comb measurement techniques [20Wil2,21Nus] make this approach more attractive. In this case the nonlinear pump–probe reflectivity can be measured very fast within less than a millisecond and without any mechanical delay lines using equivalent time sampling (see Sect. 11.6.4) and dual-comb modelocked lasers (see Sect. 9.9). Such measurements enable fast and robust determination of all the nonlinear reflectivity and recovery time parameters of the devices from a single setup, and show good agreement with conventional nonlinear reflectivity measurements. In addition, we showed that the inverse saturable absorption is much less pronounced in the cross-polarized configuration compared to the co-polarized configuration. Furthermore, there is a dependence on the relative angle between the pump polarization and the rotation angle of the SESAM [21Nus].

7.8 Novel Saturable Absorber Materials

There is an ever ongoing effort to develop novel saturable absorber materials. It would go beyond the scope of this book to summarize all these efforts. In this chapter we defined a number of macroscopic parameters that are important for passive pulse generation. Their definition and measurement system are independent of the absorber materials and can be applied to any new material development (Sect. 7.7). So far, SESAMs have been able to provide the required operating parameters which have not been obtained with other saturable absorber materials to date. In most cases, there is an issue with nonsaturable losses, high saturation fluence, and/or low damage threshold. Throughout this chapter, we have given many examples for different SESAM parameters so this can serve as a reference in the future.

It is important to use similar measurement systems and fit model functions to obtain results that can be compared. There is a lot of interest in carbon nanotube (CNT) and graphene saturable absorbers [12Sun, 11Bao]. Saturable absorption in CNTs has shown ultrafast recovery times, i.e., of picosecond order, which has resulted in some of the first Nd:glass and Er/Yb:glass lasers with passive mode-locking [05Sch3]. The fast response time of such polymer-embedded carbon nanotubes supported pulse durations as short as 68 fs at 1570 nm from the Er/Yb:glass laser [05Sch3].

We used a single-wall carbon nanotube saturable absorber integrated into a DBR and measured a saturation fluence of 57 $\mu J/cm^2$ and a fast recovery time of 600 fs [07Fon]. However, for a measured modulation depth of 0.5%, there was also a non-saturable loss of 1.1% due to scattering, which is too much for ultrafast solid-state lasers. Ultrafast fiber lasers typically have a much larger output coupling rate which is less sensitive to larger nonsaturable losses.

We also integrated a single graphene layer onto a DBR structure with different field enhancements [13Zau] and performed a detailed saturable absorber characterization as described in this chapter. However, the result was disappointing because the damage threshold was never above the saturation fluence of the device. Not being able to fully saturate the nonlinear mirror means that we have to operate the saturable absorber with more loss, which is acceptable in fiber lasers where we have much higher gain (and larger output couplers) than for diode-pumped bulk solid-state lasers. Most likely, the deposition onto the DBR was the main problem in this case. Another full characterization [11Bao] reveals very high saturation fluences and relatively high non-saturable losses, even with a single monolayer of material. More recent measurements revealed an unconventional double-bended saturation [17Win].

There is much more materials research ongoing [12Nov]. The SESAM has the advantage that it benefits from a large commercial material effort for semiconductor-based photonics.

8 Q-Switching

One of the first successful methods for the generation of laser pulses was already proposed shortly after the invention of the laser in 1960 and is called Q-switching [61Hel]. Our emphasis in this chapter is on short picosecond pulse generation with Q-switching. The pulse duration in Q-switching scales with the photon cavity lifetime, which becomes shorter with a shorter cavity length. The ultimate laser for scaling the cavity length down is demonstrated with the microchip laser [89Zay].

We first discuss the fundamental principle of active Q-switching, then the theory, closely following the account in [86Sie]. We then discuss passive Q-switching, the theoretical background, and in more detail, passively Q-switched microchip lasers, following closely [99Spu1].

The shortest Q-switched pulses have been obtained with a passively Q-switched Nd:YVO_4 microchip laser, using very short laser cavities and a semiconductor saturable absorber mirror (SESAM). With a Nd:YVO_4-crystal thickness of only 185 μm, pulses as short as 37 ps have been generated with 53 nJ pulse energy, which corresponds to 1.4 kW peak power and 8.5 mW average power [99Spu1]. Reducing the gain crystal length even further to 110 μm, pulses shorter than 22 ps have been generated but at a lower pulse energy of only 0.9 nJ because the crystal has become shorter than the pump absorption length [12But]. These are pulse durations that have become comparable to modelocked lasers but with very compact laser cavities and with sufficient peak power for supercontinuum generation in a nonlinear fiber (i.e., a very bright white light source).

Concerning the notation, there is one main difference in this chapter compared to most other chapters: all loss and gain coefficients are given for the intensity (or power) and not for the amplitude, and are therefore a factor of 2 larger! The notation is summarised here:

© The Author(s), under exclusive licence to Springer Nature Switzerland AG 2021

U. Keller, *Ultrafast Lasers*, Graduate Texts in Physics,
https://doi.org/10.1007/978-3-030-82532-4_8

l total nonsaturable intensity loss coefficient per resonator roundtrip, i.e., without the saturable absorber, but including output coupler loss and any additional parasitic loss, and also the nonsaturable loss of the saturable absorber

q saturable intensity loss coefficient of the saturable absorber per cavity roundtrip with no nonsaturable loss (i.e., $R_{ns} = 1$)

q_0 unbleached intensity loss coefficient of the saturable absorber per cavity roundtrip, i.e., maximum q at low intensity

g saturated intensity gain coefficient per resonator roundtrip

g_0 small-signal gain for a homogeneous gain material, where, in the steady-state, $g = g_0/(1 + 2I/I_{sat})$ (with a factor of 2 for a linear standing-wave resonator)

8.1 Active Q-Switching

8.1.1 Fundamental Principle of Active Q-Switching

The method of active Q-switching relies on the following considerations (see Fig. 8.1). We assume a pump pulse with a length of approximately 100 μs (e.g., from a flashlamp) with a sufficiently high energy to generate a large inversion while the laser resonator is blocked and no stimulated emission can take place (see Fig. 8.1a and b). As soon as we remove the intracavity block (see Fig. 8.1c), we reduce the high loss and stimulated emission starts to build up very fast because the large inversion is far above the steady-state level of the now low-loss cavity. As shown in Fig. 8.1d, the intracavity power continues to increase and the inversion becomes more and more depleted until we reach the low-loss cavity steady-state inversion for which the saturated gain is equal to the total cavity loss. At this point the laser power has reached its maximum level, which is far above the steady-state value. Thus, the inversion becomes even more depleted and we have less gain than loss. In this situation, the intracavity power decays with the photon cavity decay rate. At the end there is no inversion left for lasing and we have to wait for the next pump pulse to repeat the full process again. This process generates a short, energetic laser pulse (Fig. 8.2), whose duration is typically in the microsecond to nanosecond region. With microchip lasers, even sub-100-picosecond pulses can be obtained (see examples in Table 8.1).

The optical switch, which 'suddenly' reduces the losses, switches the Q-factor of the resonator from a low to a high value. Thus, the name Q-switching. One can choose between different optical switches, for example rotating mirrors or prisms, electrically controlled Kerr and Pockels cells, and acousto-optic switches.

A Q-switched pulse is typically asymmetric, with a fast rise time and a slow decay time (Fig. 7.2). The initial intensity will increase with the small-signal gain g_0, which is directly proportional to the generated inversion. Without further pumping, the decay time only depends on the photon cavity lifetime τ_c inside the resonator

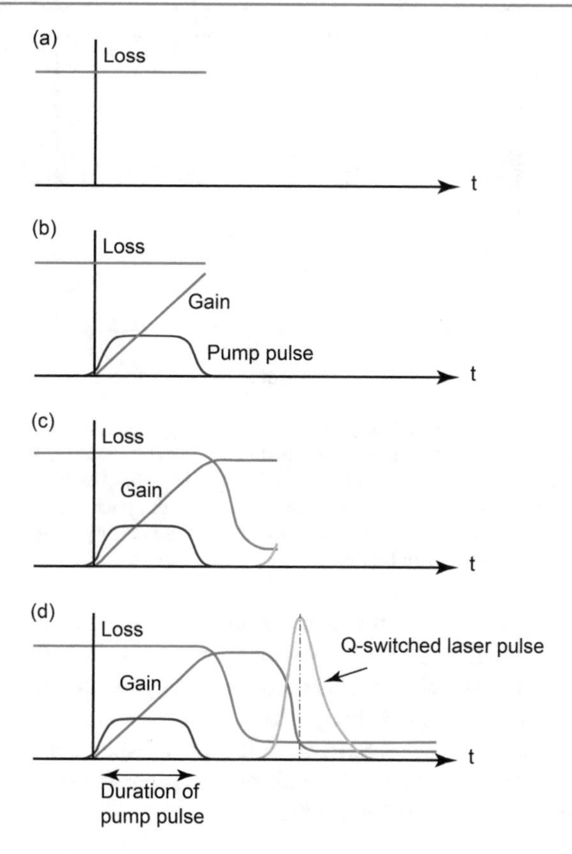

Fig. 8.1 Principle of active Q-switching. Here we assume that, at a certain time, both high loss and pump power are switched off. For example the high loss can be a beam block inside the laser cavity

and decays with a rate of $\gamma_c = 1/\tau_c = l/T_R$, according to (5.6), but now with l as the intensity instead of the amplitude loss coefficient), and T_R is the resonator roundtrip time.

For power scaling of Q-switched pulses, one tries to generate as large an initial inversion as possible in as short a time as possible. Spontaneous emission limits the possible maximum inversion because it depletes the inversion during the pumping process. Therefore, laser materials with long upper state lifetimes are ideal for intense Q-switched pulses. With $g_0 = rl$ (Fig. 5.6), we obtain a larger small-signal gain for $r \gg 1$, where r is the pump parameter, defined by the ratio of pump power P to the threshold power P_{th} in the low-loss cavity, i.e., $r = P/P_{th}$. Figure 8.2 clearly shows a faster rise time in the pulse for a larger r, which corresponds to a larger initial inversion, a larger initial small-signal gain, and a shorter cavity photon lifetime.

A larger cavity photon decay rate $\gamma_c = l/T_R$ shortens the trailing edge of the pulse, but elongates the rising edge when a larger ouput coupler is used. Therefore, there is an optimum γ_c for which the pulse duration becomes minimal. Clearly, one

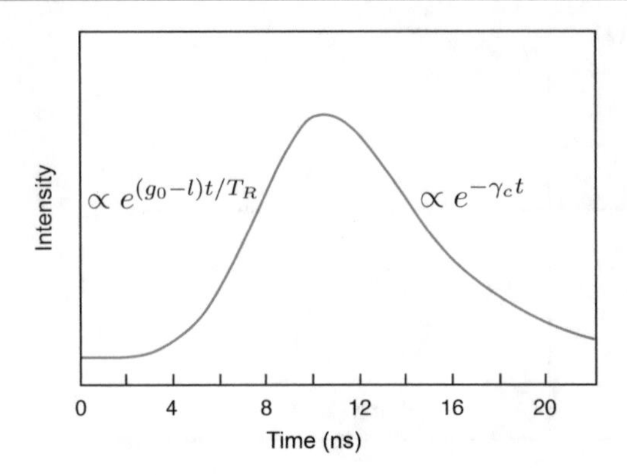

Fig. 8.2 Asymmetric Q-switched pulse. g_0 is the small-signal gain coefficient after the pump pulse with high intracavity loss such that there is no stimulated emission (i.e., intracavity photon number $n \approx 0$), l is the total intracavity intensity loss coefficient for the high-Q cavity (i.e., high intracavity loss switched off), T_R is the cavity roundtrip time, and γ_c is the photon cavity decay rate given by $\gamma_c = 1/\tau_c = l/T_R$, according to (5.6), but now with l as the intensity instead of amplitude loss coefficient, and τ_c the photon cavity lifetime

obtains a larger γ_c through shorter laser resonators, i.e., when T_R is smaller, and higher output coupling, i.e., when l is larger, because $\gamma_c = l/T_R$. The switching speed of the optical Q-switch will ultimately limit the pulse duration. Only electro-optic switches with a typical switching speed of ≤ 10 ns and acousto-optic switches with a typical switching speed ≤ 50 ns are sufficiently fast. In contrast, it is possible to realize faster switches with saturable absorbers, because the loss modulation is generated by the Q-switched pulse itself.

8.1.2 Acousto-Optic Q-Switched Diode-Pumped Solid-State Laser

Figure 8.3 shows a typical cavity layout for a Q-switched diode-pumped solid-state laser. The laser crystal is pumped longitudinally, so ideally only one TEM_{00} mode is excited. Short pulses are generated with a high small-signal gain, i.e., with a preferably small pump beam radius compared with the absorption length in the laser crystal. Additionally, a relatively high output coupling and a short resonator length are chosen. To minimize the optical switching speed of the acousto-optic modulator, the laser beam diameter inside the modulator has to be small. If the Q-switch is too slow, we can obtain multiple pulses, as discussed in more detail in [86Sie].

For example, with an acousto-optic switch and a diode-pumped Nd:YLF laser as in Fig. 8.3, pulse durations down to 700 ps with a pulse repetition rate $f_{rep} = 1$ kHz, a peak power $P_{peak} = 15$ kW, an average power $P_{av} = 10.5$ mW, and a pulse energy $E_p = 10.5\ \mu$J have been generated. With an Nd:Vanadate (Nd:YVO_4) laser, pulse durations down to 600 ps with $f_{rep} = 1$ kHz, $P_{peak} = 5$ kW, $P_{av} = 3$ mW, and $E_p = 3\ \mu$J have been produced [93Pla].

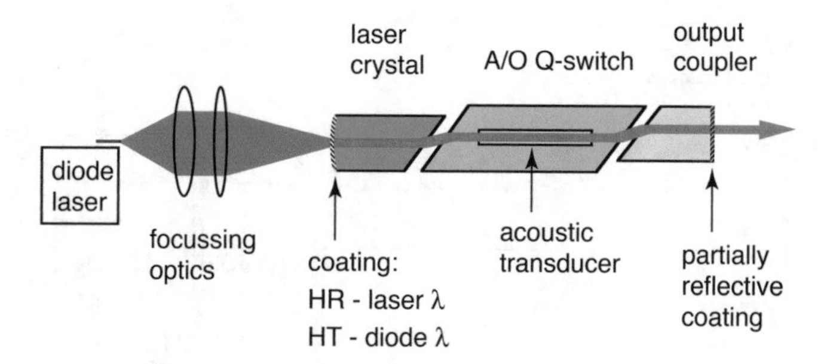

Fig. 8.3 Resonator design of a Q-switched diode-pumped Nd:YLF or Nd:YVO_4 laser [93Pla]. HR is high reflector, HT is high transmission, A/O is acousto-optic

8.1.3 Pulsed Single-Frequency Laser

Q-switched lasers only produce stable pulses if the laser oscillates in a single axial mode. High power single-frequency pulses can be achieved, for example, with injection seeding. This is done with a weak continuous wave (cw) laser beam coupled into the main laser resonator. In this way, one axial mode, which corresponds to the wavelength of the seed laser, obtains higher gain and will therefore saturate the laser gain in single-mode operation.

Single-axial-mode operation can also be achieved using other methods, e.g., using so-called microchip lasers [89Zay], where the laser resonator is so short that the axial mode spacing is larger than the gain bandwidth of the laser. Another approach is based on a unidirectional ring laser without a standing wave and therefore no spatial hole burning. In this case, only a single axial mode can saturate a homogeneously broadened gain medium and we can obtain stable Q-switching.

8.1.4 Actively Q-Switched Microchip Laser

Very short pulses can be achieved with Q-switched microchip lasers due to their short resonator lengths [95Zay, 91Zay]. Active Q-switching can be achieved for example by inserting a tunable etalon into the resonator [91Zay]. The etalon transmission curve is then quickly detuned compared to the resonator mode frequency. Due to the large resonator mode frequency distance in microchip lasers, the laser can then be Q-switched. In this way, pulse durations of 6 ns have been achieved.

Much shorter pulses have been obtained in a configuration with an electro-optic crystal inside the resonator of a microchip laser, as shown in Fig. 8.4 [92Zay, 95Zay]. Pulses with an energy of 12 μJ, pulse durations of 115 ps, and repetition rates of up to 1 kHz have been generated in this way using a Nd:YVO_4 microchip laser [95Zay]. These are the shortest pulses generated by active Q-switching with a solid-state laser. Here, the electro-optic crystal was used as a tunable etalon by analogy with the configuration using a piezo-electrically modulated mirror.

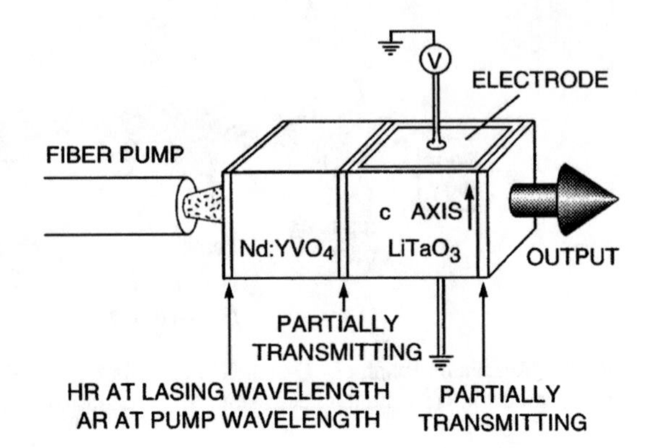

Fig. 8.4 Actively Q-switched microchip laser using an electro-optic modulator [95Zay]

8.2 Theory for Active Q-Switching

8.2.1 Rate Equations

We consider the rate equations for an ideal four-level laser system (Chap. 5). The photon number $n(t)$ inside the resonator satisfies the following rate equation due to stimulated emission and resonator losses:

$$\frac{\mathrm{d}n}{\mathrm{d}t} = KNn - \gamma_\mathrm{c} n \,. \tag{8.1}$$

Here we can neglect the spontaneous emission above the laser threshold. The stimulated transition probability W^stim of an inverted atom is given through the coupling constant K by $W^\mathrm{stim} \equiv W_{12} = W_{21} = Kn$. The population inversion $N(t)$ satisfies the following rate equation due to the pumping rate R_p, the stimulated emission, and the spontaneous emission $\gamma_\mathrm{L} N$, where $\gamma_\mathrm{L} = 1/\tau_\mathrm{L}$ and τ_L is the upper state lifetime of the laser material (see examples in Table 5.1):

$$\frac{\mathrm{d}N}{\mathrm{d}t} = R_\mathrm{p} - \gamma_\mathrm{L} N - KnN \,, \tag{8.2}$$

where the pumping rate is given by

$$R_\mathrm{p} = \frac{P_\mathrm{abs}}{h\nu_\mathrm{pump}} \,, \tag{8.3}$$

with P_abs the absorbed pump power and $h\nu_\mathrm{pump}$ the pump photon energy.

8.2.2 Inversion Build-Up Phase

As shown in Fig. 8.5, the losses inside the resonator during the inversion build-up phase are so high that the threshold condition is not reached. We therefore have

$$\text{Approximation in the build-up phase: } n(t) \approx 0 \,, \quad R_\mathrm{p} = \text{const.} \tag{8.4}$$

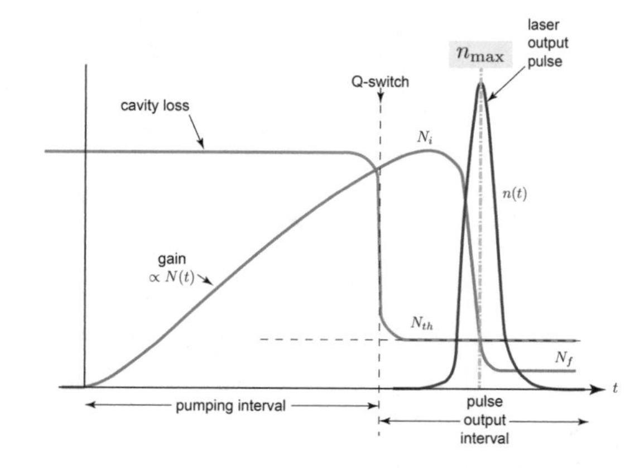

Fig. 8.5 Basic principle of active Q-switching with the approximation that the loss is instantaneously reduced at $t = 0$. In contrast to Fig. 8.1, we use continuous pumping

Equation (8.2) then implies

$$\frac{\mathrm{d}N}{\mathrm{d}t} \approx R_\mathrm{p} - \gamma_\mathrm{L} N = R_\mathrm{p} - \frac{N}{\tau_\mathrm{L}} . \tag{8.5}$$

The solution to this simple differential equation is

$$\boxed{N(t) = R_\mathrm{p}\tau_\mathrm{L}\left[1 - \exp\left(-t/\tau_\mathrm{L}\right)\right] = N_\mathrm{max}\left[1 - \exp\left(-t/\tau_\mathrm{L}\right)\right]} , \tag{8.6}$$

where N_max is the maximum reachable inversion (Fig. 8.6):

$$N_\mathrm{max} = R_\mathrm{p}\tau_\mathrm{L} . \tag{8.7}$$

Due to the presence of the spontaneous emission rate $-\gamma_\mathrm{L} N$ in (8.5), the inversion does not increase to arbitrarily high values with a constant pumping rate. Indeed, at a constant pumping rate, the initially inverted atoms will spontaneously decay with a rate given by $1/\tau_\mathrm{L}$. Therefore, the pumping efficiency rapidly decreases for longer pumping intervals. In addition, lasers with a short upper state lifetime τ_L are not good laser materials for Q-switching. For example, gas lasers, dye lasers, and semiconductor lasers have τ_L in the range of 1–10 ns. Therefore, the usable pump duration is too short to generate a sufficiently large inversion for short, energetic Q-switched pulses. In such a case, an extremely powerful short-pulsed pump laser would be required to nevertheless generate a sufficiently large inversion. In contrast, diode-pumped solid-state lasers have τ_L in the range of 100 μs to more than 1 ms.

Active Q-switching allows for a simple adjustment of the pulse repetition rate by changing the modulation frequency, typically in the range of up to 100 kHz. For low repetition rates, the pulse energy E_p is constant, which follows from (8.6) and Fig. 8.6:

$$E_\mathrm{p} = \text{const.} \iff T_\mathrm{rep} \gtrsim 3\tau_\mathrm{L} , \text{ or } f_\mathrm{rep} = \frac{1}{T_\mathrm{rep}} \lesssim \frac{1}{3\tau_\mathrm{L}} . \tag{8.8}$$

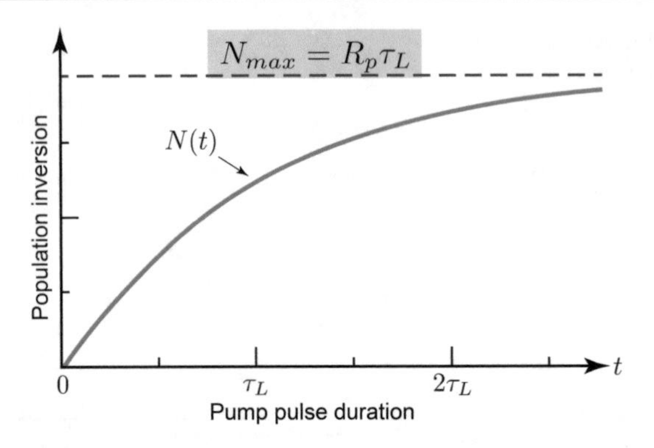

Fig. 8.6 Inversion as a function of pump duration according to (8.6) and [86Sie]. The maximum reachable inversion is given by $N_{\max} = R_p\tau_L$, where τ_L is the upper state lifetime of the laser material (see examples in Table 5.1). Typically, one can assume that a pump duration of approximately $3\tau_L$ is needed to generate the maximum inversion [compare with (8.8)]

If there is not sufficient time left between the pulses to bring the inversion back to its original maximum value, then the pulse energy will be reduced. Additionally, the highest pulse energies can be generated with lasers with a long upper state lifetime of the laser level. One can partially avoid such limitations if for example a short high-energy pulse is used for pumping.

8.2.3 Pulse Build-Up Phase: Leading Edge of the Pulse

According to Fig. 8.5, we make a first approximation that, at $t = 0$, the losses are instantaneously reduced. The initial conditions for the rate equations (8.1) and (8.2) are

$$N(t = 0) = N_i \ , \tag{8.9}$$

$$n(t = 0) = n_i \approx 1 \ , \tag{8.10}$$

where n_i is on the order of 1, which corresponds to the original fluctuation values of photons inside the resonator. We assume that there is only one axial mode here.

We further assume that the inversion is not strongly reduced during the pulse build-up phase, because the photon number is still very small:

$$\text{Approximation in the build-up phase: } N(t) \approx N_i \approx \text{const.} \tag{8.11}$$

Using an initial inversion ratio $r = N_i/N_{th}$, this implies the rate equation (8.1) for the pulse build-up phase:

$$\frac{dn}{dt} \approx K\,(N_i - N_{th})\,n = KN_{th}(r-1)n = \frac{r-1}{\tau_c}n \ , \tag{8.12}$$

where $N_{\text{th}} = \gamma_{\text{c}}/K$ is the threshold inversion for $t > 0$. The solution of this simplified differential equation is then

$$n(t) \approx n_{\text{i}} \exp\left(\frac{r-1}{\tau_{\text{c}}} t\right) \stackrel{\tau_{\text{c}}=T_{\text{R}}/l,\ g_0=rl}{=} n_{\text{i}} \exp\left[(g_0 - l)\frac{t}{T_{\text{R}}}\right] . \tag{8.13}$$

This is also shown in Fig. 8.2 for the leading edge of the Q-switched pulse.

8.2.4 Dynamics During the Pulse Duration

The Q-switched pulse is usually so intense and short that the pumping rate and the spontaneous decay rate of the inversion can be neglected during that time:

$$\text{Approximation during pulse:} \quad R_{\text{p}} - \gamma_{\text{L}} N \ll KnN . \tag{8.14}$$

Therefore, the rate equations (8.1) and (8.2) simplify to

$$\frac{\text{d}n}{\text{d}t} = K(N - N_{\text{th}})\,n , \quad \frac{\text{d}N}{\text{d}t} \approx -KnN . \tag{8.15}$$

Dividing one of these by the other, we obtain

$$\frac{\text{d}n}{\text{d}N} \approx \frac{K(N - N_{\text{th}})\,n}{-KnN} = \frac{N_{\text{th}} - N}{N} ,$$

whence

$$\text{d}n \approx \frac{N_{\text{th}} - N}{N}\text{d}N . \tag{8.16}$$

With the initial conditions (8.9) and (8.10), viz.,

$$N(t=0) = N_{\text{i}} = rN_{\text{th}} , \quad n(t=0) = n_{\text{i}} \approx 1 , \tag{8.17}$$

integration of (8.16) yields

$$\int_{n_{\text{i}}}^{n(t)} \text{d}n \approx \int_{N_{\text{i}}=rN_{\text{th}}}^{N(t)} \frac{N_{\text{th}} - N}{N}\text{d}N ,$$

and hence,

$$n(t) \approx N_{\text{i}} - N(t) - \frac{N_{\text{i}}}{r} \ln \frac{N_{\text{i}}}{N(t)} , \quad N_{\text{i}} = rN_{\text{th}} , \tag{8.18}$$

where we neglected n_{i} because $n_{\text{i}} \ll n(t)$. The pulse reaches the maximum intensity, when the inversion is depleted to the steady state N_{s} by increased stimulated emission, because at this point the steady-state condition 'gain equals loss' is satisfied. As soon as the loss becomes greater than the gain, the intensity inside the resonator can no longer increase:

$$n(t) = n_{\max} \text{ for } g = l \iff N(t) = N_{\text{th}} \,. \tag{8.19}$$

Thus, (8.18) implies

$$\boxed{n_{\max} \approx \frac{r - 1 - \ln r}{r} N_{\text{i}} \,, \quad N_{\text{i}} = r N_{\text{th}}} \,. \tag{8.20}$$

The output peak power $P_{\text{p,out}}$ of a Q-switched laser is then given by

$$\boxed{P_{\text{p,out}} = \frac{n_{\max} h\nu}{\tau_{\text{c}}}} \,, \tag{8.21}$$

where $h\nu$ is the photon energy and $n_{\max}$ is determined by (8.20).

When $N(t)$ has reached the final inversion N_{f}, then according to Fig. 8.5, the photon number has to be approximately zero, i.e.,

$$n(t_{\text{f}})|_{t_{\text{f}}:\, N(t_{\text{f}}) = N_{\text{f}}} \approx 0 \,. \tag{8.22}$$

Using (8.18), we obtain N_{f} as a function of N_{i} and r :

$$N_{\text{i}} - N_{\text{f}} - \frac{N_{\text{i}}}{r} \ln \frac{N_{\text{i}}}{N_{\text{f}}} \approx 0 \,. \tag{8.23}$$

The pulse energy E_{p} can now be estimated as

$$\boxed{E_{\text{p,out}} \approx E_{\text{p}} \approx (N_{\text{i}} - N_{\text{f}})\, h\nu} \,, \tag{8.24}$$

where $h\nu$ is the photon energy and $N_{\text{f}} = N_{\text{f}}\,(r, N_{\text{i}} = r N_{\text{th}})$ is determined using (8.23). If the resonator losses are dominated by the output coupling, all photons inside the resonator will eventually be coupled out after a certain time, and the pulse energy in (8.24) is also the output pulse energy $E_{\text{p,out}}$ (Fig. 8.7).

One can define the pumping efficiency η by

$$\eta \equiv \frac{\text{Q-switched pulse energy}}{\text{stored energy}} = \frac{(N_{\text{i}} - N_{\text{f}})\, h\nu}{N_{\text{i}} h\nu} = \frac{N_{\text{i}} - N_{\text{f}}}{N_{\text{i}}} \,, \tag{8.25}$$

whence

$$\frac{N_{\text{f}}}{N_{\text{i}}} = 1 - \eta \,.$$

Thus, (8.23) can be transformed as follows:

$$1 - \frac{N_{\text{f}}}{N_{\text{i}}} - \frac{1}{r} \ln \frac{N_{\text{i}}}{N_{\text{f}}} \approx 0 \,,$$

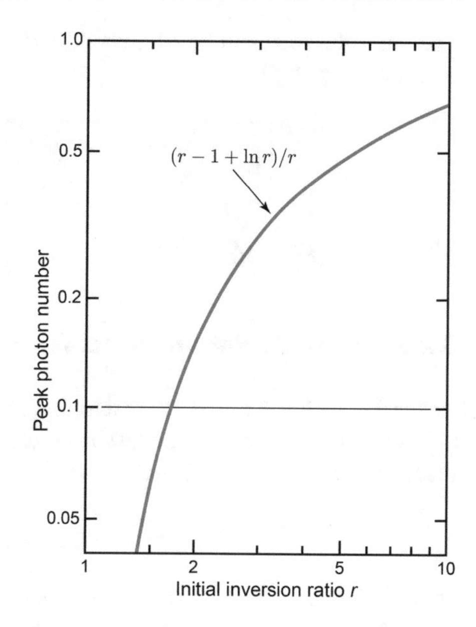

Fig. 8.7 Normalized maximum photon number $n_{\max}/N_i$ as a function of the initial inversion ratio $r = N_i/N_{th}$, where N_i is the initial inversion right after the high intracavity loss has been switched off and N_{th} is the threshold inversion of the high-Q cavity. High peak intensities in Q-switched pulses are obtained for $n_{\max} \approx N_i$, and according to (8.20), can only be reached if $r \gg 1$

and we obtain

$$\boxed{\eta - \frac{1}{r}\ln\left(\frac{1}{1-\eta}\right) \approx 0} \,. \tag{8.26}$$

Equation (8.26) tells us that the pumping efficiency η only depends on the initial pump parameter, i.e., $\eta = \eta(r)$ (Fig. 8.8). Through $\eta(r)$, the pulse energy is also given by (8.24) as

$$\boxed{E_{p,out} = E_p \approx \eta(r) N_i h\nu} \,. \tag{8.27}$$

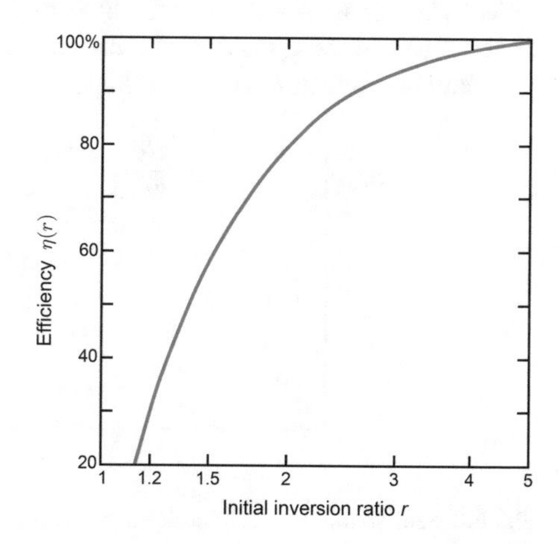

Fig. 8.8 Pumping efficiency $\eta(r)$ according to (8.26). Only for large r is the initial inversion N_i depleted by the Q-switched pulse. For small r, only a small part of the inversion is depleted, so the pump energy is not used efficiently

The pulse duration can be estimated by the ratio of the pulse energy (8.27) to the peak intensity (8.21):

$$\boxed{\tau_{\mathrm{p}} \approx \frac{E_{\mathrm{p,out}}}{P_{\mathrm{p,out}}} \approx \frac{\eta(r)N_{\mathrm{i}}}{n_{\max}}\tau_{\mathrm{c}} \approx \frac{r\eta(r)}{r-1-\ln r}\tau_{\mathrm{c}}}\,, \tag{8.28}$$

where $\eta(r)$ is given by (8.26).

8.2.5 Pulse Depletion Phase: Trailing Edge of the Pulse

According to Fig. 8.5 and the discussion in the last section, it becomes clear that, after the pulse maximum, the loss is larger than the gain. Thus, the photons inside the resonator decay with the resonator decay rate:

$$n(t) = n_{\max} \exp\left(-t/\tau_{\mathrm{c}}\right) \; . \tag{8.29}$$

Here, we assume that $g \approx 0$, which means, according to Fig. 8.8, that $r \geq 3$. Otherwise, the decay will be slower and the pulse length longer. This is also shown in Fig. 8.2 for the trailing edge of the Q-switched pulse.

8.3 Passive Q-Switching

8.3.1 Fundamental Principle of Passive Q-Switching

In passive Q-switching, a saturable absorber is used as the Q-factor modulator, i.e., the absorber has a high absorption for low intensities and a lower absorption for high intensities (see Chap. 7). Figure 8.9 shows the intensity-dependent nonlinear reflectivity for an absorber embedded inside a mirror—a saturable reflector. We assume that, in cw operation, the intensity is below the saturation intensity I_{sat} of the saturable reflector. Thus, the slope $\mathrm{d}R/\mathrm{d}I$ around $I \approx 0$ (Fig. 8.9), where R is

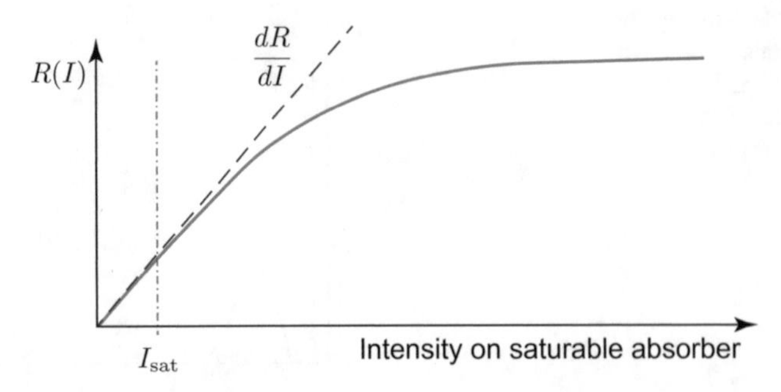

Fig. 8.9 Saturable reflector. Intensity dependent reflectivity $R(I)$ of a saturable reflector mirror [95Kar4]

the reflectivity, determines how much small intensity fluctuations change the losses inside the resonator. Depending on how much the losses are reduced per resonator roundtrip, the intensity fluctuations will show a damped behaviour and go back to steady state with damped relaxation oscillations (Chap. 5).

Let us now remind ourselves briefly what the relaxation oscillations mean. Relaxation oscillations are a periodic increase in the intensity inside the resonator, which are typically weakly damped and coupled with a reduction of the population inversion through increased stimulated emission. An intracavity intensity above the steady-state value leads to a reduction of the population inversion below the threshold value, which then prevents a further increase in the intensity. This is the first half-period of the relaxation oscillation. In the second half-period, the pumping process again supplies the missing population inversion. But in the meantime, due to the missing inversion, the intensity inside the resonator has fallen, so the pump increases the inversion slightly above the threshold value, and the whole process can start all over again. As we have seen, these relaxation oscillations are normally damped and subside over time. We will show in this chapter that a saturable absorber inside the resonator can also lead to a destabilization of these relaxation oscillations, which can in turn produce stable Q-switched pulses. The qualitative interplay of gain and loss saturation due to the generated Q-switched pulses is shown in Fig. 8.10.

The condition under which the relaxation oscillations become destabilized and potentially stable Q-switching can become possible can be summarized as follows, in a similar way to [76Hau, 95Kar4]:

$$\boxed{\left|\frac{\mathrm{d}R}{\mathrm{d}I}\right| I > \frac{T_\mathrm{R}}{\tau_\mathrm{stim}} \approx r\frac{T_\mathrm{R}}{\tau_\mathrm{L}}}\,, \tag{8.30}$$

where T_R is the resonator roundtrip time, τ_stim is the stimulated lifetime of the laser level, τ_L is the spontaneous lifetime of the laser level, and r is the pump parameter, i.e., how many times above the threshold is being pumped. This condition can be understood as follows: on the left of the inequality in (8.30), we have the reduction of the loss per resonator roundtrip due to saturation of the absorber. On the right, we see how much the gain can saturate itself per resonator roundtrip until steady state 'gain equals loss' can be reached again. If the gain cannot compensate the reduction

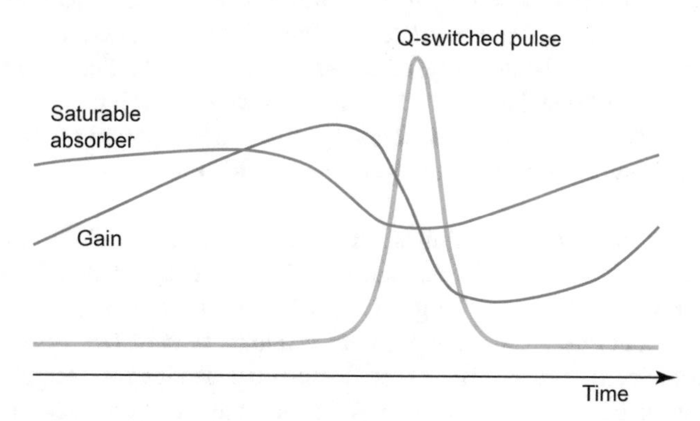

Fig. 8.10 Dynamics of a passively Q-switched laser

of the loss, then the gain will become larger than the loss, and thus the intensity inside the resonator will increase very fast, according to Fig. 8.10. This ultimately leads to passive Q-switching. The result (8.30) implies that the condition for passive Q-switching is much easier to fulfill for lasers with a long lifetime of the excited laser level τ_L, i.e., for example, for solid-state lasers. Before we continue with a more detailed analysis, we present a few examples of passively Q-switched lasers.

8.3.2 Passively Q-Switched Microchip Laser

In comparison, significantly higher pulse energies have been achieved with Cr^{4+}:YAG saturable absorbers: using Nd:YAG resulted in 11 μJ pulse energy and 337 ps pulse duration [94Zay], and using liquid-nitrogen cooled Yb:YAG resulted in 12.1 mJ pulse energy and 3.3 ns pulse duration [18Guo].

Adjustable saturable absorber parameters can be obtained with a semiconductor saturable absorber. In addition, the semiconductor saturable absorber can be integrated into a mirror structure, referred to as a SESAM [92Kel2,96Kel] and therefore does not substantially increase the cavity length. Thus, it is not surprising that the shortest pulse durations of only 37 ps [99Spu1] (Fig. 8.11) and 22 ps [12But] are produced with SESAM Q-switched microchip lasers. The SESAM Q-switched microchip laser was first demonstrated with Nd:LSB, generating 180 ps [96Bra1] (Fig. 8.11) and higher pulse energies with SESAM Q-switching have been obtained with Yb:YAG generating 530 ps with 1.1 μJ pulse energy [01Spu2] and with Er:Yb:glass generating 840 ps with 11.2 μJ pulse energy [01Har1].

Table 8.1 summarizes more results for passively Q-switched microchip lasers.

8.3.3 Passively Q-Switched Monolithic Ring Laser

A monolithic ring laser (Fig. 8.12a) [85Kan] has been passively Q-switched with evanescent wave coupling into a semiconductor saturable absorber (Fig. 8.12b) [95Bra1]. A Nd:YAG MISER (monolithic isolated single-mode end-pumped ring laser) [85Kan] or later also called NPRO (nonplanar ring oscillator) used in this experiment had an axial mode spacing of 6.0 GHz, a lasing wavelength of 1064 nm, and a geometrical size of approximately 0.2 cm × 1 cm × 1 cm. The NPRO resonator is formed inside the laser crystal by three total internal reflection surfaces (B, C, and D in Fig. 8.12a) and a surface A which is coated for high transmission of the pump laser at 809 nm and 2% output coupling at 1064 nm. Applying a magnetic field to the crystal leads to Faraday rotation in YAG, which forces the laser to operate unidirectionally.

By bringing an uncoated nonlinear semiconductor reflector close to one of the total internal reflection points, i.e., less than 1 μm, light is coupled through the evanescent wave into the sample (Fig. 8.12b). The uncoated nonlinear semiconductor reflector is a saturable absorber layer with a thickness d of 0.6 μm grown on a GaAs/AlAs dielectric mirror. The saturable absorber bandgap is designed for 1.06 μm and consists of 50 pairs of InGaAs/GaAs multiple quantum wells MBE-grown

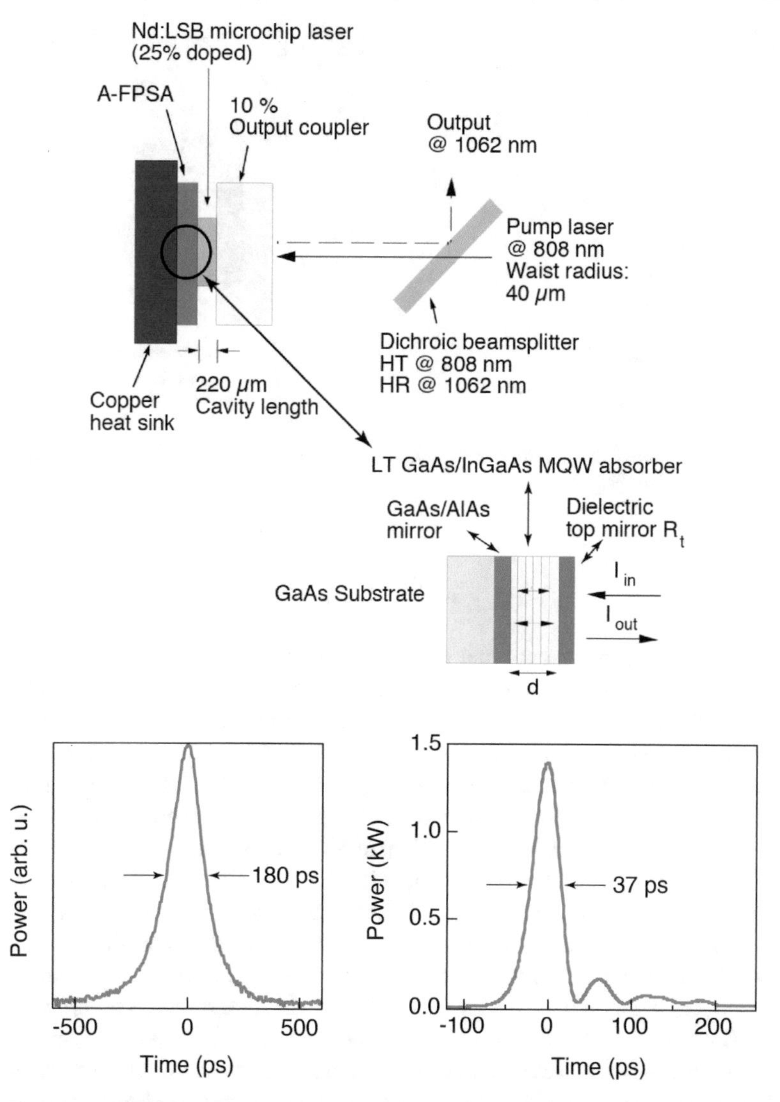

Fig. 8.11 Single-frequency pulse at a wavelength of 1.06 μm from the first SESAM Q-switched microchip laser (here the SESAM design was based on an A-FPSA, an antiresonant Fabry–Pérot saturable absorber). The microchip laser is an Nd:LSB laser, which can be more highly doped than an Nd:YAG laser. Pulses as short as 180 ps have been generated [96Bra1]. In a very similar setup, it was possible to generate pulses as short as 37 ps with an Nd:vanadate laser [99Spu1]. More details are shown in Table 8.1

at low temperature (between 300 and 400°C), which results in an absorber recovery time of about 40 ps. A saturation fluence of the AR-coated sample was measured to be $F_{\mathrm{sat,A}} \approx 50\,\mu\mathrm{J/cm}^2$ using 1.4 ps pulses from a modelocked Ti:sapphire laser. However, because the carrier lifetime is much shorter than the Q-switched pulse duration of approximately 100 ns, only the saturation power

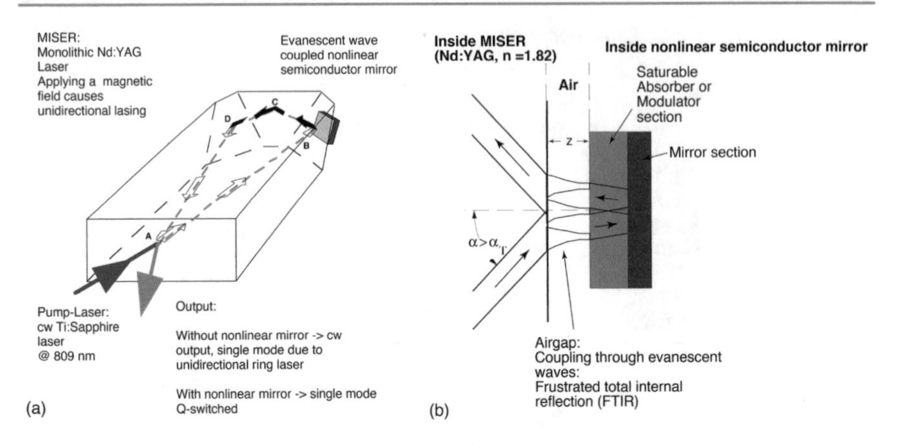

Fig. 8.12 Monolithic Nd:YAG ring laser [85Kan] Q-switched with a saturable absorber, which acts as a nonlinear mirror [95Bra1]

$$P_{\mathrm{sat,A}} = \frac{F_{\mathrm{sat,A}}}{\tau_{\mathrm{A}}} A = 375\,\mathrm{W}$$

and the evanescent wave coupling will determine the onset of passive Q-switching (discussed later). The laser mode area A inside the MISER is on the order of cm^2.

The NPRO was pumped with a cw Ti:sapphire laser at 809 nm, polarized parallel to the crystal plane to minimize reflection losses at the front surface. The pump radius was measured to be $33\,\mu\mathrm{m}$. A diode-pumped cw MISER has been demonstrated before and is available as a commercial product. The threshold for Q-switching was observed at a cw pump power of 790 mW and an average intracavity power of 5.6 W with the semiconductor saturable absorber placed at a distance $z \approx 0.3\,\mu\mathrm{m}$ estimated by the additional intracavity loss. The roundtrip small-signal gain G_0 was 170%, which was determined with the measured relaxation oscillation of the cw pumped MISER with a cavity roundtrip time $T_{\mathrm{R}} = 167\,\mathrm{ps}$, a laser upper state lifetime $\tau_{\mathrm{L}} = 230\,\mu\mathrm{s}$, and a total cw cavity loss of 3.1%. This determines a pump parameter $r = \ln G_0/l_0 \approx 9.1$, and therefore the threshold condition for passive Q-switching, where l_0 is the total intracavity loss coefficient when Q-switched, given by 3.1% plus 2.7% additional loss introduced by the semiconductor saturable absorber. As we will see shortly, the threshold for passive Q-switching is determined by (see Sect. 8.4)

$$-T_{\mathrm{L}} P \left.\frac{\mathrm{d}q}{\mathrm{d}P}\right|_{\mathrm{cw}} = \frac{T_{\mathrm{L}} q_0}{\chi} \frac{r-1}{\left(1+\dfrac{r-1}{\chi}\right)^2} \approx \frac{T_{\mathrm{L}} q_0}{\chi}(r-1) > r\,, \tag{8.31}$$

where q and q_0 are the saturated and unsaturated absorber loss coefficients, respectively, P is the intracavity power, $T_{\mathrm{L}} = \tau_{\mathrm{L}}/T_{\mathrm{R}} \approx 1.4 \times 10^6$ is the laser upper state lifetime normalized to the roundtrip time, and $\chi = P_{\mathrm{sat}}^{\mathrm{eff}}/P_{\mathrm{L}}$, with P_{L} the saturation power of the gain medium. Here we have $P_{\mathrm{L}} = Ah\nu/\sigma\tau_{\mathrm{L}} \approx 0.63\,\mathrm{W}$ and therefore $\chi = 1.3 \times 10^4$. The approximation made in (8.31) is valid because $\chi \gg r$.

Table 8.1 Q-switched microchip lasers using different techniques. First microchip laser introduced by Zayhowski and Mooradian [89Zay]. 'Best' refers to pulse duration, highest average output power, highest pulse repetition rate, etc. The result for which 'best' applies is in *bold letters*. The lasers are assumed to be diode pumped, if not otherwise stated. SESAM: passive Q-switching using semiconductor saturable absorber mirrors (SESAMs). EOM: electro-optic phase modulator. Center lasing wavelength λ_0, measured pulse duration τ_p, pulse energy E_p, pulse repetition rate f_{rep}

Laser material	ML technique	λ_0	τ_p	E_p	f_{rep}	Remarks	Reference
Nd:YAG	Active EOM	1.064 μm	6 ns			First Q-switched microchip laser. 7 W peak, 3.5 mW	[91Zay2]
			270 ps	6.8 μJ	5 kHz	Pulse width increases with increased f_{rep} for 5–500 kHz: at 500 kHz, 13.3 ns	[92Zay]
	Cr^{4+}:YAG		337 ps	11 μJ	6 kHz		[94Zay]
			148 ps		4 kHz	Optimization for maximum energy [04Bur]	[03Zay]
$Nd:YVO_4$	Active EOM	1.064 μm	115 ps	**12 μJ**	1 kHz		[95Zay]
	SESAM	1.064 μm	56 ps	62 nJ	85 kHz	1.1 kW peak, 5.3 mW average	[97Bra]
			68 ps	0.37 μJ	160 kHz	5.4 kW peak, 58 mW average	[97Bra]
			143 ps	48 nJ	160 kHz	Output coupling (OC) SESAM	[01Spu1]
			37 ps	**53 nJ**	160 kHz	1.4 kW peak, 8.5 mW average	[99Spu1]
			117 ps	0.16 μJ	510 kHz	1.4 kW peak, 82 mW average	[99Spu1]
			181 ps	0.28 μJ	440 kHz	1.6 kW peak, 120 mW average	[99Spu1]
			2.64 ns	45 nJ	**7.8 MHz**	17 W peak, 350 mW average	[99Spu1]
			50 ps	1 μJ	40 kHz	20 kW peak	[07Nod]
			22 ps	**0.9 nJ**		110μm thick gain crystal	[12But]
			200 ps	130–400 nJ	<2 MHz		[09Ste]
		1.34 μm	230 ps		53 kHz	450 W peak	[97Flu]
Nd:LSB	SESAM	1.062 μm	180 ps	0.1 μJ	110 kHz		[96Bra1]
	Cr^{4+}:YAG	1.063 μm	650 ps	3.56 μJ			[05Voi]
Yb:YAG	SESAM	1.03 μm	530 ps	1.1 μ J	12 kHz	1.9 kW peak	[01Spu2]
	Cr^{4+}:YAG	1.03 μm	3.3 ns	**12.1 mJ**		3.7 MW peak, **liquid-nitrogen cooled**	[18Guo]
Er:Yb:glass	SESAM	1.535 μm	1.2 ns	45 nJ	47 kHz	2.1 mW average	[98Flu]
		≈1.5 μm	840 ps	**11.2 μJ**	1.4 kHz	16 mW average, 10.6 kW peak	[01Har1]
	p-i-n modulator	1.553 μm	56 ns	470 nJ	10 kHz	Previously p-i-n modulator for ML [95Bro3]	[01Han]
Cr:YAG	SESAM	≈1.5 μm	42 ns		1 MHz	Diode-pumped $Nd:YVO_4$ pumped, Q-switched ML	[01Sch]

The experimentally determined threshold for Q-switching results in the condition $1.9 > 1$, so it fulfills the theoretical condition for Q-switching.

At a pump power of 3.1 W and an average output power of 545 mW, single-frequency Q-switched pulses as short as 95 ns are obtained with a repetition rate of 750 kHz and a pulse energy of 0.73 μJ [95Bra1]. The intracavity peak power at the reflecting surface B is then 360 W, which is still well below the effective saturation power $P_{\text{sat}}^{\text{eff}}$ of 8.1 kW. Above 3.1 W pump power, it became more difficult to maintain unidirectional lasing, and no stable Q-switching was observed with counter-propagating laser beams. With a distance variation of 0.6 μm, the pulse width changes from about 100 ns to 350 ns, and the repetition rate from 700 kHz to 1.2 MHz. This experiment also demonstrates that an evanescently coupled saturable absorber could also be used to Q-switch or potentially modelock a waveguide laser.

8.4 Theory for Passive Q-Switching

In this section, we discuss a model for passively Q-switched microchip lasers and derive simple equations for the pulse width, repetition rate, and pulse energy. The validity of the model has been checked experimentally by systematically varying the relevant device parameters. We used the model to derive practical design guidelines for realizing operating parameters that can be varied over wide ranges by adapting the parameters of the semiconductor saturable absorber mirror (SESAM) and choosing the appropriate gain medium. Applying these design guidelines, we obtained Q-switched pulses as short as 37 ps, which are the shortest pulses ever generated in a Q-switched solid-state laser. We closely follow [99Spu1].

The history of Q-switching goes back to 1961, when Hellwarth [61Hel] predicted that a laser can emit short pulses if the losses of an optical resonator are rapidly switched from a high to a low value. The experimental proof was provided a year later [62McC,62Col]. Theories were then established, first for rapid Q-switching (instantly switched intracavity loss, with two rate equations for the photon density and the inversion) [63Vuy,63Wag], and then for the case of passive Q-switching [66Eri,67Eri] by introducing a third rate equation for the saturable absorber. Motivated by their work with Q-switched microchip lasers [91Zay], J. J. Zayhowski and P. L. Kelley analyzed the rate equations for rapid Q-switching [91Zay]. In 1995, J. J. Degnan extended the work of Erickson and Szabo for passive Q-switching [95Deg], optimizing the output coupler for maximum pulse energy using the Lagrange multiplier technique.

Here we present simple derivations for the relevant quantities in a passively Q-switched microchip laser, namely pulse width, pulse energy, peak power, and repetition rate. We will use the rate equations as derived and discussed in Chap. 5.

8.4.1 Rate Equations

First we establish a simple model for the passively Q-switched laser in Fig. 8.13 using rate equations as discussed in Chap. 5. Starting with the laser rate equations,

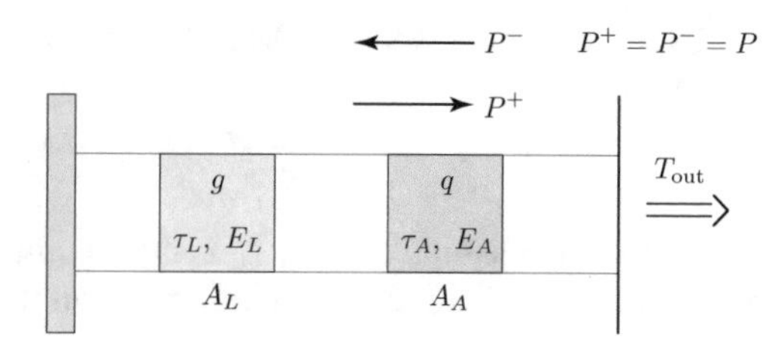

Fig. 8.13 Simple model of a passively Q-switched laser

we can add an additional equation for the saturable absorber as shown schematically in Fig. 8.13, discussed in Sect. 7.2.2 using (7.4):

$$\frac{\mathrm{d}n}{\mathrm{d}t} = \left(K_\mathrm{L}N_\mathrm{L} - K_\mathrm{A}N_\mathrm{A} - \frac{1}{\tau_\mathrm{c}}\right)n\,, \tag{8.32}$$

$$\frac{\mathrm{d}N_\mathrm{L}}{\mathrm{d}t} = -\frac{N_\mathrm{L}}{\tau_\mathrm{L}} - K_\mathrm{L}nN_\mathrm{L} + R_\mathrm{p}\,, \tag{8.33}$$

$$\frac{\mathrm{d}N_\mathrm{A}}{\mathrm{d}t} = -\frac{N_\mathrm{A} - N_\mathrm{A0}}{\tau_\mathrm{A}} - K_\mathrm{A}nN_\mathrm{A}\,, \tag{8.34}$$

with the following notation:

- n photon number
- N_L number of inverted laser atoms—the same as in (8.1) and (8.2), but also used when there is a saturable absorber inside the laser resonator
- K_L coupling constant for induced emission in the amplifier—the same as in (8.1) and (8.2), but also used when there is a saturable absorber inside the laser resonator
- K_A coupling constant for the absorption of the saturable absorber
- R_p pumping rate $R_\mathrm{p} = P_\mathrm{abs}/h\nu_\mathrm{pump}$, with P_abs the absorbed pump power and $h\nu_\mathrm{pump}$ the pump photon energy
- N_A number of absorbing atoms inside the resonator, i.e., number of absorber atoms that are not yet bleached
- N_A0 total number of absorber atoms inside the resonator, so that $-(N_\mathrm{A} - N_\mathrm{A0})$ in (8.34) is the number of bleached absorber atoms, which temporarily reduce the absorption
- τ_A relaxation time of the saturable absorber, which describes the decay of the absorber atoms into the ground state, from where they can absorb again

Equations (8.32) and (8.33) follow directly from (8.1) and (8.2), where we added the absorption due to the absorber atoms in (8.32) in the form $-K_\mathrm{A}nN_\mathrm{A}$. We neglect the spontaneous emission above threshold. An additional rate equation (8.34) is added, which describes the dynamics of the absorber. The first term in (8.32) describes the relaxation of the absorber to thermal equilibrium after absorption bleaching. The second term describes the reduction in the number of absorbing atoms with the absorption of a photon.

The intracavity power of the laser P is related to the photon number in the resonator as follows:

$$n = \frac{P}{h\nu} T_{\mathrm{R}} \overset{T_{\mathrm{R}}=2L/c}{=} \frac{2L}{ch\nu} P \,, \tag{8.35}$$

where $h\nu$ is the photon energy, T_R is the roundtrip time, L the resonator length, and c the velocity of light. Furthermore, we recall that

$$g = L_{\mathrm{g}} \frac{N_{\mathrm{L}}}{V} \sigma_{\mathrm{L}} \overset{V=A_{\mathrm{L}} L_{\mathrm{g}}}{=} \frac{N_{\mathrm{L}}}{A_{\mathrm{L}}} \sigma_{\mathrm{L}} \,, \tag{8.36}$$

where A_{L} is the average mode area in the laser material, σ_{L} is the gain cross-section, L_{g} is the length of the amplifier, and N_{L}/V is the inversion density inside the amplifier. Furthermore, the probability for stimulated emission W^{stim} is given by

$$W^{\mathrm{stim}} = K_{\mathrm{L}} n = \frac{I}{h\nu} \sigma_{\mathrm{L}} = \frac{P}{A_{\mathrm{L}} h\nu} \sigma_{\mathrm{L}} \,, \tag{8.37}$$

where K_{L} is the coupling constant. Equation (8.35) also implies

$$K_{\mathrm{L}} = \frac{\sigma_{\mathrm{L}}}{A_{\mathrm{L}} T_{\mathrm{R}}} \,, \tag{8.38}$$

and analogously for the saturable absorber,

$$K_{\mathrm{A}} = \frac{\sigma_{\mathrm{A}}}{A_{\mathrm{A}} T_{\mathrm{R}}} \,, \tag{8.39}$$

with A_{A} the average mode area in the absorber and σ_{A} the absorber cross-section. Thus, with $T = t/T_{\mathrm{R}}$, the rate equations (8.32)–(8.34) imply the following new rate equations for the power P, gain coefficient g, and absorber loss coefficient q:

$$T_{\mathrm{R}} \frac{\mathrm{d}P(t)}{\mathrm{d}t} = \left[g(t) - l(t) - q(t)\right] P(t) \,, \tag{8.40}$$

$$\frac{\mathrm{d}g(t)}{\mathrm{d}t} = -\frac{g(t) - g_0}{\tau_{\mathrm{L}}} - \frac{g(t)P(t)}{E_{\mathrm{sat,L}}} \,, \tag{8.41}$$

$$\frac{\mathrm{d}q(t)}{\mathrm{d}t} = -\frac{q(t) - q_0}{\tau_{\mathrm{A}}} - \frac{q(t)P(t)}{E_{\mathrm{sat,A}}} \,, \tag{8.42}$$

with g_0 given by (5.29), and where the absorber loss coefficient q is given by analogy with the gain coefficient by

$$q = \frac{N_{\mathrm{A}}}{A_{\mathrm{A}}} \sigma_{\mathrm{A}} \,.$$

This corresponds to the rate equations for a four-level laser as shown in (5.66) and (5.67). In addition, (8.42) is equivalent to (7.4). The saturation energies of the laser material $E_{\mathrm{sat,L}}$ and the absorber $E_{\mathrm{sat,A}}$ are given in a linear standing wave cavity by (5.71) with $m = 2$:

$$E_{\text{sat,L}} = \frac{h\nu}{2\sigma_{\text{L}}} A_{\text{L}} \,, \quad E_{\text{sat,A}} = \frac{h\nu}{2\sigma_{\text{A}}} A_{\text{A}} \,. \tag{8.43}$$

Here the rate equations for a four-level laser system are stated in terms of experimentally relevant parameters such as the laser power $P(t)$, the intensity gain coefficient per cavity roundtrip $g(t)$, and the intensity saturable loss coefficient per cavity roundtrip $q(t)$. We assume that all changes per cavity roundtrip are small, i.e., $g, q, l \ll 1$, and we neglect spontaneous emission into the laser mode above the lasing threshold.

In this simple model, we have neglected the effects of inhomogeneous amplification due to spatial hole burning (see Sect. 6.7), which does not change anything for the general behaviour as far as it is of interest in this section. In addition, we again want to point out that the simple rate equations (8.40)–(8.42) are only valid if the gain resp. the losses per roundtrip in the resonator are small. Otherwise, we are dealing with a spatially and temporally varying system. If the laser passes into a pronounced gigantic solution it may be that the above approximation for small changes per roundtrip will no longer be valid. This then has to be examined separately from case to case. In any case, the threshold for the appearance of the oscillation should be very well described by (8.40)–(8.42) if the resonator losses per roundtrip do not exceed 20, as confirmed by numerical simulations.

We can integrate the rate equations for a cw pumped four-level gain material numerically. For integration, we used the GEAR method [75Chu], since the set of differential equations is rather stiff, i.e., the dynamics of the system occurs on time scales which differ by many orders of magnitude.

During the transit time of the pulse, the absorber relaxation and the pumping of the gain can be neglected in (8.40) to (8.42). Then, with these approximations the peak power and the pulse width can be determined from the following simplified set of rate equations for passive Q-switching :

$$T_{\text{R}} \frac{\mathrm{d}P}{\mathrm{d}t} = (g - l - q)P \,, \tag{8.44}$$

$$\frac{\mathrm{d}g}{\mathrm{d}t} = -\frac{gP}{E_{\text{sat,L}}} \,, \tag{8.45}$$

$$\frac{\mathrm{d}q}{\mathrm{d}t} = -\frac{qP}{E_{\text{sat,A}}} \,. \tag{8.46}$$

The same approximations have also been made by [95Deg, 65Sza]. Division of (8.46) by (8.45) gives a relationship between the saturable absorption $q(t)$ and the gain $g(t)$, which can be integrated to express $q(t)$ as a function of $g(t)$. We have

$$\frac{\mathrm{d}q}{\mathrm{d}t}\frac{\mathrm{d}t}{\mathrm{d}g} = \frac{qP}{E_{\text{sat,A}}}\frac{E_{\text{sat,L}}}{gP} \,,$$

which implies

$$\frac{\mathrm{d}q}{\mathrm{d}g} = \beta \frac{q}{g} \,, \quad \beta \equiv \frac{E_{\text{sat,L}}}{E_{\text{sat,A}}} \,.$$

From this

$$\frac{1}{q}\mathrm{d}q = \beta\frac{1}{g}\mathrm{d}g \ ,$$

and integration yields

$$\ln q - \ln q_0 = \beta(\ln g - \ln g_{\mathrm{i}}) \ ,$$

whence

$$\ln\frac{q}{q_0} = \beta\ln\frac{g}{g_{\mathrm{i}}} = \ln\left(\frac{g}{g_{\mathrm{i}}}\right)^{\beta} \ ,$$

and finally,

$$q = q_0\left(\frac{g}{g_{\mathrm{i}}}\right)^{\beta} \ , \tag{8.47}$$

where the initial conditions are the initial gain g_{i} and an absorber in the ground state. β is the ratio of the saturation energy of the gain and the saturation energy of the absorber:

$$\beta \equiv \frac{E_{\mathrm{sat,L}}}{E_{\mathrm{sat,A}}} \ . \tag{8.48}$$

Analogously, we can divide (8.44) by (8.45), and make use of (8.47). Subsequent integration yields the (instantaneous) power as a function of the gain during the Q-switch cycle [95Deg]. We have

$$T_{\mathrm{R}}\frac{\mathrm{d}P}{\mathrm{d}t}\frac{\mathrm{d}t}{\mathrm{d}g} = -(g - q - l)P\frac{E_{\mathrm{sat,L}}}{gP} \ ,$$

which implies

$$\mathrm{d}P = -\frac{E_{\mathrm{sat,L}}}{T_{\mathrm{R}}}\frac{g - q - l}{g}\mathrm{d}g \ .$$

Equation (8.47) then implies

$$\mathrm{d}P = -\frac{E_{\mathrm{sat,L}}}{T_{\mathrm{R}}}\frac{g - q_0(g/g_{\mathrm{i}})^{\beta} - l}{g}\mathrm{d}g \ .$$

To integrate the right-hand side, it is better to simplify

$$\frac{g - q_0(g/g_{\mathrm{i}})^{\beta} - l}{g} = -\frac{q_0 g^{\beta-1}}{g_{\mathrm{i}}^{\beta}} - \frac{l}{g} + 1$$

which has the integral

$$-l\ln\frac{g}{g_{\mathrm{i}}} - \frac{q_0}{\beta}\left(\frac{g^{\beta}}{g_{\mathrm{i}}^{\beta}} - \frac{g_{\mathrm{i}}^{\beta}}{g_{\mathrm{i}}^{\beta}}\right) + (g - g_{\mathrm{i}}) = -l\ln\frac{g}{g_{\mathrm{i}}} - \frac{q_0}{\beta}\left[\left(\frac{g}{g_{\mathrm{i}}}\right)^{\beta} - 1\right] + (g - g_{\mathrm{i}}) \ .$$

Therefore,

$$P = \frac{l E_{\text{sat,L}}}{T_{\text{R}}} \ln \frac{g}{g_{\text{i}}} - \frac{E_{\text{sat,L}}}{T_{\text{R}}}(g - g_{\text{i}}) + \frac{E_{\text{sat,L}}}{T_{\text{R}}} \frac{q_0}{\beta} \left[\left(\frac{g}{g_{\text{i}}} \right)^{\beta} - 1 \right] .$$

Then inserting (8.48), we obtain

$$P(g) = -\frac{E_{\text{sat,L}}}{T_{\text{R}}}(g - g_{\text{i}}) + \frac{l E_{\text{sat,L}}}{T_{\text{R}}} \ln \frac{g}{g_{\text{i}}} + \frac{E_{\text{sat,A}}}{T_{\text{R}}} q_0 \left[\left(\frac{g}{g_{\text{i}}} \right)^{\beta} - 1 \right] . \tag{8.49}$$

This expression is equivalent to [95Deg, Eq. (6)]. If we neglect the last term in (8.49) which is due to the saturable absorber, (8.48) reduces to [86Sie, Eq. (17)], where the case of a rapidly Q-switched laser is analyzed. Integrating (8.49), one can get an expression for the pulse energy, which is consistent with (8.60). With the help of the peak power, also from (8.49), the pulse energy, and an assumed pulse, (8.44) can be derived.

8.4.2 Model for SESAM Q-Switched Microchip Laser

We consider a microchip laser cavity consisting of a crystal (or glass) of length L_{g} to provide the time dependent roundtrip intensity gain coefficient $g(t)$, a saturable absorber giving a saturable loss coefficient $q(t)$, i.e., unbleached value q_0 and bleached value 0. We assume that all changes per cavity roundtrip are small and we neglect the spontaneous emission into the laser mode above the lasing threshold, as discussed in Sect. 8.4.1.

For a SESAM saturable absorber, the saturation energy $E_{\text{sat,A}}$ of the absorber is assumed to be small compared to the saturation energy of the gain medium $E_{\text{sat,L}}$, viz.,

$$E_{\text{sat,A}} \ll E_{\text{sat,L}} . \tag{8.50}$$

This is not satisfied for a Nd:YAG microchip laser, using Cr^{4+}YAG as a saturable absorber, where $E_{\text{sat,A}}$ and $E_{\text{sat,L}}$ are of the same order of magnitude. We typically have $\beta \gg 1$, i.e., in the 1 μm experiments discussed in Sect. 8.5.3, $\beta \approx 1000$. We have an output coupler with transmission T_{out} and output coupling coefficient l_{out}, defined by $T_{\text{out}} = 1 - \exp(-l_{\text{out}}) \approx l_{\text{out}}$, and a parasitic loss coefficient l_{p} (which is usually dominated by the nonsaturable loss of the SESAM). l_{out} and l_{p} determine the total nonsaturable loss coefficient per roundtrip $l = l_{\text{out}} + l_{\text{p}}$. As the gain and loss per roundtrip are usually not more than a few percent, we can represent the gain factor $G = \exp(g) \approx 1 + g$, for $g \ll 1$ and for losses $L = 1 - \exp(-l) \approx l$ for $l \ll 1$.

We neglect spatial hole burning (SHB), which alters the saturation of the gain and causes a spatial variation of the inversion that can lead to multiple longitudinal mode output [90Zay1,95Bra2]. The transverse beam profiles of the pump and laser modes are approximated by top-hat functions of equal area A, although deviations from this will have to be discussed in Sect. 8.5.

To obtain an expression for the gain reduction of a Q-switched pulse, we make use of (8.49) and note that the power is approximately zero after the pulse. Note that it is not exactly zero because, for this case, the power would remain zero forever according to (8.40). A few seed photons due to spontaneous emission are always present to initiate the Q-switching process. But this number is orders of magnitude smaller than the typical photon number present in a Q-switched pulse. This yields an expression for the final gain coefficient after the pulse and the gain reduction of the pulse, viz.,

$$\Delta g = g_i - g_f , \tag{8.51}$$

and initially the absorber is in an unbleached state:

$$g_i = l + q_0 . \tag{8.52}$$

Equation (8.49) thus implies

$$0 \approx \Delta g + l \ln \frac{l + q_0 - \Delta g}{l + q_0} , \tag{8.53}$$

where we have neglected the last term in (8.49), assuming $\Delta g \ll g_i$ and hence

$$\left(\frac{g_f}{g_i}\right)^\beta = \left(\frac{g_i - \Delta g}{g_i}\right)^\beta = \left(1 - \frac{\Delta g}{g_i}\right)^\beta \approx 1 .$$

From (8.49) with $g = g_f$, we have

$$P(g = g_f) = 0 = g_i - g_f + l \ln \frac{g_f}{g_i} + \frac{E_{sat,A}}{E_{sat,L}} q_0 \left[\left(\frac{g_f}{g_i}\right)^\beta - 1\right] , \tag{8.54}$$

and therefore

$$0 \approx g_i - g_f + l \ln \frac{g_f}{g_i} .$$

This then results in (8.53) using (8.51) and (8.52).

Numerical solutions of (8.53) for Δg as a function of l are depicted in Fig. 8.14 for different values of q_0. One can see that the approximation $\Delta g \approx 2q_0$ holds for values of q_0 up to $q_0 < 20\%$.

8.4.3 Pulse Energy

Figure 8.15 shows numerical solutions for the output pulse energy E_p normalized to the saturation energy of the gain medium $E_L \equiv E_{sat,L}$ as a function of the total nonsaturable losses l using (8.53) for Δg and for fixed initial gain

$$g_i = l_{out} + l_p + q_0 , \tag{8.55}$$

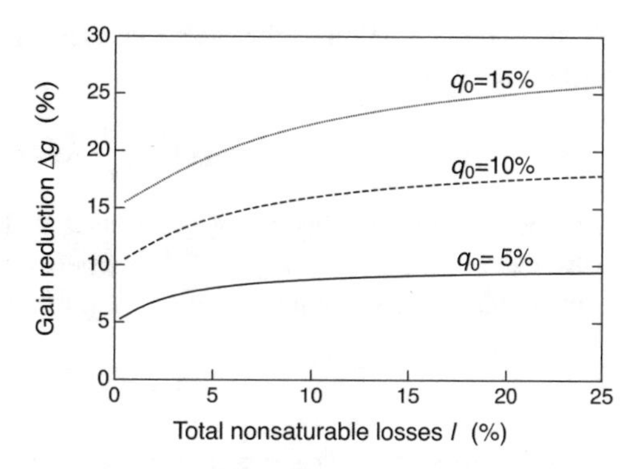

Fig. 8.14 Dependence of the gain reduction Δg on the total nonsaturable losses obtained from the rate equations. For $l \geq q_0$, the approximation $\Delta g \approx 2q_0$ holds with at least 20% accuracy

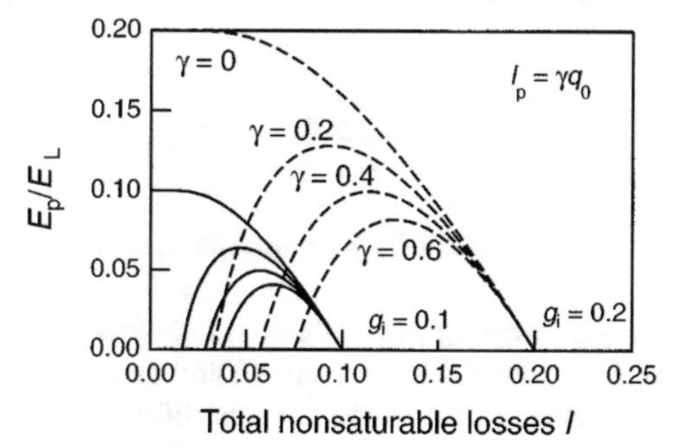

Fig. 8.15 Output pulse energy E_p normalized to the saturation energy of the gain medium $E_L \equiv E_{sat,L}$ versus the total nonsaturable losses, for parasitic losses $l_p = \gamma q_0$ and two different fixed initial gains, obtained from the rate equations. *Dashed curves* correspond to $g_i = 0.2$ and *solid curves* to $g_i = 0.1$. The *solid curves* have the same values for γ and on the same order as the *dashed curves*. As soon as there are non-vanishing parasitic losses, the maximum pulse energy is achieved for l close to $l \approx q_0 = g_i/2$ [99Spu1]

and for parasitic losses l_p given by

$$l_p = \gamma q_0 , \tag{8.56}$$

for various proportionality factors γ. The maximum pulse energy is obtained for values of l close to $g_i/2$.

It is instructive to derive an equation for the pulse energy by referring to the reduction of the gain during the evolution of a pulse (without using rate equations). The stored energy in the pumped gain material is proportional to the excitation density N_2, the photon energy $h\nu_L$ at the lasing wavelength, and the pumped volume AL_g:

$$E_{stored} = AL_g N_2 h\nu_L . \tag{8.57}$$

In the case of Q-switched microchip lasers using a SESAM, the length L_g of the gain medium is nearly equal to the cavity length (see Sect. 8.5). In a standing wave cavity, the intensity gain coefficient per roundtrip is $g = 2\sigma_L N_2 L_g$, where σ_L is the emission cross-section of the laser material. This leads to the following expression for the stored energy:

$$E_{\text{stored}} = \frac{h\nu_{\text{L}}}{2\sigma_{\text{L}}} A g = E_{\text{sat,L}} g \,, \tag{8.58}$$

with the saturation energy $E_{\text{sat,L}}$ of the laser medium given by (8.43). If a Q-switched pulse reduces the gain by Δg in (8.50), where g_{i} and g_{f} are the values of the intensity gain coefficient just before and after the pulse, respectively, it releases the energy

$$E_{\text{released}} = E_{\text{sat,L}} \Delta g \,, \tag{8.59}$$

according to (8.58). The output energy is obtained by multiplying by the output coupling efficiency:

$$E_{\text{p}} = E_{\text{sat,L}} \Delta g \frac{l_{\text{out}}}{l_{\text{out}} + l_{\text{p}}} \leq E_{\text{sat,L}}\, g_{\text{i}} \,. \tag{8.60}$$

As Δg cannot exceed g_{i}, the quantity $E_{\text{L}} g_{\text{i}}$ represents an upper limit for the attainable pulse energy.

Δg, the only unknown quantity in (8.60), could be obtained by solving the rate equations (see Sects. 8.4.1 and 8.4.2), but simple and useful equations can also be derived as follows. The pulse cycle consists of four different phases (see Fig. 8.16):

- Phase 1. The absorber is in its unbleached state. A pulse can start to develop as soon as the pump has lifted the gain to the unsaturated value of the losses:

$$g_{\text{i}} = l + q_0 \,.$$

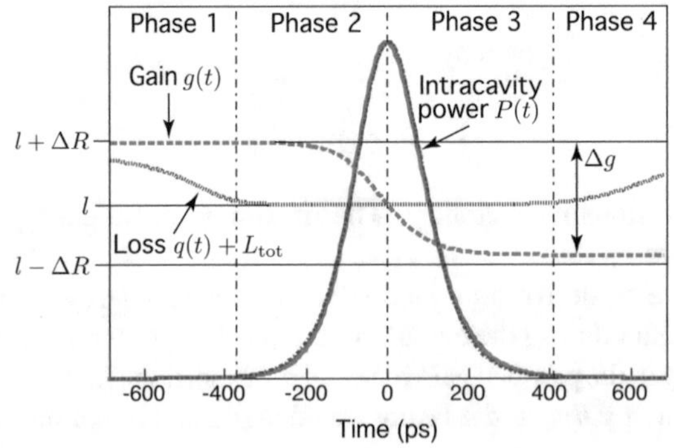

Fig. 8.16 Evolution of power, loss, and gain on the time scale of the pulse width, obtained from a numerical integration of the rate equations (see Sect. 8.4.1) [99Spu1]. As soon as the gain exceeds the loss, the power grows. The peak of the Q-switched pulse is reached when the gain equals the total losses. The parameters used here are taken for a 200 μm thick Nd:YVO_4 crystal and are as follows: $T_{\text{R}} = 2.61\,\text{ps}$, $l = 14\%$, $\tau_{\text{L}} = 50\,\mu\text{s}$, $F_{\text{sat,L}} = 37.3\,\text{mJ/cm}^2$, $\tau_{\text{A}} = 200\,\text{ps}$, $F_{\text{sat,A}} = 40\,\mu\text{J/cm}^2$, $r = 3$, $q_0 = 5\%$, and $A = (35\,\mu\text{m})^2\pi$

Now the intracavity power P grows slowly, starting from spontaneous emission noise, until the intensity is sufficient to bleach the absorber. This point is reached long before the power has reached the maximum value, because the saturation energy of the SESAM is typically chosen to be at least one order of magnitude lower than the pulse energy. In addition, the absorber recovery time is normally at least as long as the pulse duration, so that the absorber has little influence on the further evolution of the pulse.

- Phase 2. As soon as the SESAM is fully bleached ($q = 0$), the power grows much more quickly, because the net gain is now $g_i - l - q \approx q_0$, until the gain starts to be depleted. The pulse maximum is reached when the net gain is zero, i.e., $g = l$.
- Phase 3. With further depletion of the gain, the net gain becomes negative and the intracavity power decays. Nevertheless, the pulse still extracts significant energy in this phase.
- Phase 4. After the pulse, the absorber recovers and the gain has to be pumped to the threshold level again, before phase 1 of the subsequent pulse cycle can start. This takes much longer than the absorber recovery time τ_A. Therefore, the absorber is fully recovered by the beginning of phase 1. The time between two pulses is also much longer than the pulse width τ_p (see Fig. 8.17). Typical values in the experiments (see Sect. 8.5) are of the order of $\tau_A = 200\,\text{ps}$, $\tau_p = 100\,\text{ps}$, and $1/f_{rep} = 10\,\mu\text{s}$.

The gain reduction in the third phase depends on q_0 and l. The exact dependence can be calculated from the rate equations, as was done in Sect. 8.4.1, and the results are shown in Fig. 8.14. In most practical cases, the typical regime of operation is $q_0 \approx l$, where the pulse energy is optimized and also the pulses are nearly symmetric (see Sect. 8.4.4). For $l \geq q_0$, we have (see Fig. 8.14)

$$\Delta g \approx 2q_0 \ .$$

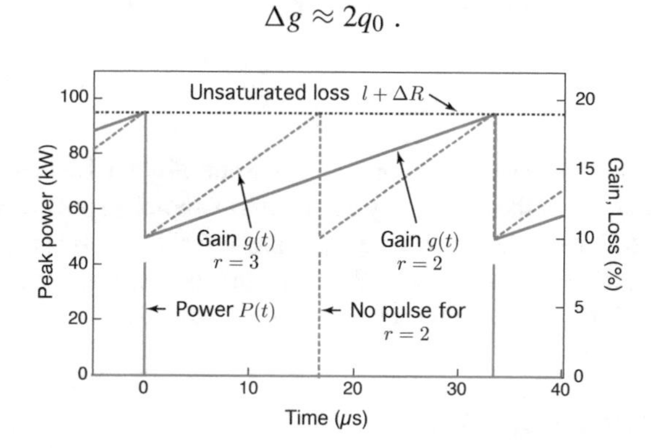

Fig. 8.17 Evolution of power, loss, and gain on the time scale of the repetition period for pump parameters $r = 2$ and $r = 3$, obtained by numerical integration of the rate equations (see Sect. 8.4.1) [99Spu1]. As soon as the gain reaches the unsaturated loss level, a pulse is emitted and the gain is reduced by Δg. The next pulse is emitted when the pump has replaced the extracted energy. The repetition rate is proportional to $r - 1$. Besides r, the simulation parameters are the same as in Fig. 8.16

Even for $q_0 = l$, this expression holds with about 20% accuracy. Inserting in (8.60), we obtain a simple expression for the pulse energy:

$$E_p \approx \frac{h\nu_L}{2\sigma_L} A 2q_0 \frac{l_{out}}{l_{out} + l_p}, \quad q_0 \leq l. \tag{8.61}$$

Δg and thus the pulse energy can be increased by increasing q_0 and l (see Fig. 8.14). However, the available gain limits the value of $g_i = l + q_0$. In addition, we have to take into account the fact that the parasitic losses introduced by the SESAM are typically proportional to the modulation depth. In Fig. 8.15, we have plotted the normalized output pulse energy versus l for different proportionality factors and initial gains (see Sect. 8.4.1). As long as there are parasitic losses, optimized pulse energy is achieved for values of l close to $l \approx q_0$, and (8.61) can be applied.

8.4.4 Pulse Duration and Pulse Shape

As the saturation energy of the absorber is typically chosen to be very small compared to the pulse energy, phase 1 has only little influence on the pulse width, i.e., on the FWHM. For a rough estimate, we assume the gain during phase 2 to stay undepleted at the value g_i, whereas the absorber is fully bleached ($q = 0$). This results in a net gain of $g_i - l - q \approx q_0$. From (8.50), the net gain during phase 3 becomes $g_f - l - q = q_0 - \Delta g \approx -q_0$, as the absorber is still fully saturated ($q = 0$). Therefore, the growth rate and the decay rate of the Q-switched pulse are both q_0/T_R, where T_R is the cavity roundtrip time. This would result in an estimated FWHM pulse duration of $2(\ln 2)T_R/q_0$ assuming a Gaussian pulse shape. The exact result for the pulse width, taking into account the decrease in growth and decay of the pulse during saturation of the gain, contains another factor of about two, as was noted earlier by Zayhowski [91Zay]:

$$\tau_p \approx \frac{3.52 \times T_R}{q_0}. \tag{8.62}$$

Figure 8.11 depicts a measured Q-switched pulse and Fig. 8.16 shows the result of a numerical simulation for the intracavity power of a Q-switched pulse together with an ideal sech^2 fit. Both experiment and theory confirm that the pulses are symmetric in the regime of interest ($l \geq q_0$), and that the sech^2 function appears to fit the simulated pulse very well. For $l < q_0$, the pulses become asymmetric.

8.4.5 Pulse Repetition Rate

To obtain the repetition rate f_{rep}, we divide the average output power by the output pulse energy. The average power is given by $P_{av} = \eta_s(P_p - P_{p,th})$, where P_p is the pump power, $P_{p,th}$ the threshold pump power, and η_s the slope efficiency, which is the product of the quantum efficiency, the pump efficiency, and the output coupling efficiency. This results in

$$f_{\text{rep}} = \frac{\eta_s(P_p - P_{p,\text{th}})}{E_p} \propto r - 1 \ . \tag{8.63}$$

Here, $r = P_p/P_{p,\text{th}}$ is the pump parameter. In the numerically obtained curves in Fig. 8.17, the $r - 1$ dependence of the repetition rate is clearly visible: the repetition rate for $r = 3$ is twice as large as for $r = 2$.

In order to write (8.63) in another useful form, we connect the pump power and the threshold pump power to the small-signal gain coefficient g_0 and the loss coefficient:

$$P_p = \frac{h\nu_p A}{2\sigma_L \tau_L \eta_p} g_0 \ , \qquad P_{p,\text{th}} = \frac{h\nu_p A}{2\sigma_L \tau_L \eta_p} (l + q_0) \ . \tag{8.64}$$

Here, $h\nu_p$ is the pump photon energy, η_p is the pumping efficiency, i.e., how many pump photons are required on average to excite one ion, and τ_L is the upper state lifetime of the gain medium. Equation (8.64) can be derived by expressing the small-signal gain in terms of the small-signal excitation density N_{20}, viz., $g_0 = 2\sigma_L N_{20} L_g$, and using the fact that the decay rate due to spontaneous emission, $AL_g N_{20}/\tau_L$, has to be compensated by the pump rate $\eta_p P_p / h\nu_p$. Inserting (8.57) and (8.64) into the relation (8.63) and using $\Delta g \approx 2q_0$, we obtain

$$f_{\text{rep}} = \frac{g_0 - (l + q_0)}{\Delta g \tau_L} \approx \frac{g_0 - (l + q_0)}{2q_0 \tau_L} \ . \tag{8.65}$$

Equation (8.65) might be more appropriate for everyday use than (8.63), as the quantities can be estimated more easily. Note, however, that g_0 is proportional to τ_L, so a change in τ_L has little influence on f_{rep} far above threshold ($g_0 \gg l + q_0$). In addition, far above threshold, we can neglect the term $l + q_0$ in (8.65) and it reduces to [97Bra, Eq. (2)]:

$$f_{\text{rep}} \approx \frac{g_0}{2\Delta R \tau_L}$$

8.4.6 Remarks on Three-Level Lasers

So far, we have only considered four-level lasers. The rate equations for a three-level laser are given in Sect. 5.1.5. With a few changes, we can therefore generalize the results of this chapter to three-level lasers. Here we have ground state absorption in an unpumped crystal, and the gain coefficient per roundtrip for a linear cavity is

$$g = 2\left(\sigma_L N_2 - \sigma_L^{\text{abs}} N_1\right) L_g \ , \tag{8.66}$$

where σ_L^{abs} is the absorption cross-section of the three-level gain medium at the laser wavelength and N_1 is the density of ions in the ground state. Assuming that all the ions are either in level one or two for most of the time, we can write: $N_1 = N_D - N_2$, where N_D is the dopant density. We obtain

$$g = 2\left(\sigma_{\mathrm{L}} + \sigma_{\mathrm{L}}^{\mathrm{abs}}\right) N_2 L_{\mathrm{g}} - l_{\mathrm{g}} \ , \tag{8.67}$$

where $l_{\mathrm{g}} = 2\sigma_{\mathrm{L}}^{\mathrm{abs}} N_{\mathrm{D}} L_{\mathrm{g}}$ is the loss coefficient per roundtrip caused by ground state absorption. Using (8.52), (8.58), and (8.67), the expression for the stored energy in a three-level laser medium is

$$E_{\mathrm{stored}} = \frac{h\nu_{\mathrm{L}}}{2\left(\sigma_{\mathrm{L}} + \sigma_{\mathrm{L}}^{\mathrm{abs}}\right)} A \left(g + l_{\mathrm{g}}\right) = E_{\mathrm{sat,L}} \left(g + l_{\mathrm{g}}\right) \ , \tag{8.68}$$

with the modified expression for the saturation energy of the gain medium

$$E_{\mathrm{sat,L}} = \frac{h\nu}{2\left(\sigma_{\mathrm{L}} + \sigma_{\mathrm{L}}^{\mathrm{abs}}\right)} A \ . \tag{8.69}$$

With these modifications, we can still use expressions (8.60) to (8.65) [except (8.61) and (8.64)], i.e., the equations for the gain reduction Δg, pulse energies E_{released} and E_{p}, the pulse duration τ_{p}, and the repetition rate f_{rep}.

We see that the main effect of the ground state absorption is to reduce the saturation energy $E_{\mathrm{sat,L}}$ and thus the pulse energy E_{p}, although we note that the typical tendency of three-level lasers to operate at longer wavelengths (where the reabsorption is weaker) could in practice even increase the pulse energy, particularly in the case of high dopant concentrations.

8.5 Passively Q-Switched Microchip Lasers Using SESAMs

8.5.1 SESAM Design

The use of a semiconductor saturable absorber as a passive Q-switch in a microchip laser (rather than, e.g., a doped crystal such as Cr:YAG) is advantageous for several reasons. First, the SESAM [96Kel, 92Kel2] is used as an end mirror (Figs. 8.11, 8.20) and has an effective penetration depth of less than a few microns (see Sect. 7.5.4). Thus, we can add a saturable absorber to the microchip laser with only a negligible increase in the cavity length. The cavity length is given by the gain crystal length which is typically > 200 μm, ideally as short as the absorption length of the gain material. Therefore, we maintain a shorter cavity length and obtain shorter Q-switched pulse widths (8.62) in comparison with other approaches that require larger bulk modulation elements [96Zay]. Secondly, the bandgap of the absorber layer can be adapted to the laser wavelength. It is therefore possible to adapt the system of the absorber to other laser materials at different lasing wavelengths, in contrast to bulk absorbers like Cr^{4+}YAG, which are limited to specific spectral regions. In addition, there is enough design freedom to independently adjust the absorber modulation and saturation intensity ($I_{\mathrm{sat,A}}$), energy ($E_{\mathrm{sat,A}}$), and fluence ($F_{\mathrm{sat,A}}$):

$$I_{\mathrm{sat,A}} = \frac{E_{\mathrm{sat,A}}}{A\tau_{\mathrm{A}}} = \frac{F_{\mathrm{sat,A}}}{\tau_{\mathrm{A}}} \tag{8.70}$$

even if the mode size on the absorber (A) is fixed. τ_A is the recovery time of the absorber

In microchip lasers the laser mode size, which is nearly constant over the whole length of the flat–flat cavity, cannot be changed over a wide range, because the transverse modes are determined mainly by the thermal lens, thermal expansion, and gain guiding [90Zay2,94Lon2], and the mode sizes in the laser medium and the absorber are always equal in these short cavities. The pulse width, the repetition rate, and the pulse energy are mainly determined by the saturable loss coefficient q_0 (see Sect. 8.4), a parameter which can be varied over a wide range, still keeping $E_{sat,A}$ and $I_{sat,A}$ small. Thus, both the pulse width and the repetition rate can be reliably varied over several orders of magnitude.

The SESAM consists of a semiconductor saturable absorber layer grown on a highly reflecting bottom mirror. For microchip lasers, the SESAM normally has a high reflection coating for the pump wavelength at the top, which avoids heating of the SESAM by absorption of residual pump light and allows for double pass of the pump light through the gain medium to increase pump absorption efficiency. The same coating has a pre-designed reflectivity for the laser wavelength to control the amount of saturable absorption and the saturation intensity of the device. The modulation depth of a SESAM is defined as the maximum reflectivity change of the absorber

$$\Delta R = 1 - \exp(-q_0) \ . \tag{8.71}$$

For $q_0 \ll 1$, (8.71) becomes $\Delta R \approx q_0$. A more general introduction for SESAMs is given in Chap. 7.

1μm SESAM for Q-switched Microchip Lasers

For the 1 μm SESAM Q-switched microchip lasers as shown in Table 8.1 with Nd:YVO_4, SiO_2/HfO_2 dielectric topcoatings with a reflectivity of 25% or 50% were used. Both MOCVD and MBE have been used with a growth temperature of around 480°C for the saturable absorber section. Both designs consist of an $In_{0.25}Ga_{0.75}As$/GaAs multi-quantum-well (MQW) absorber, grown on top of a high reflecting AlAs/GaAs Bragg mirror (25 pairs). The antiresonant SESAM design and the resulting standing wave pattern in the absorber are shown in Fig. 8.18. Because InGaAs with a bandgap of 1 μm is not lattice matched to GaAs (Fig. 7.39), there is a critical thickness for the absorber layer, above which cross hatches due to strain significantly degrade the surface quality [93Kel]. This critical thickness is generally larger for low-temperature grown MBE layers [92Wit], but still with a smaller modulation depth and higher nonsaturable losses [95Bro1]. Additional strain compensation barrier layers are therefore advantageous [16Alf], as discussed in more detail in Sect. 7.6.3.

Figure 8.19 shows the measured change in reflectivity versus incident pulse fluence and the temporal pulse response of one of the SESAMs used in the experiments of Sect. 6.5.3. The same MOCVD grown absorber with 18 quantum wells was coated with two different top reflectors for the lasing wavelength. The SESAM coated with a 25% reflector has a modulation depth $\Delta R = 10.3\%$ and a saturation fluence of $F_{sat,A} = 36\,\mu J/cm^2$ (see Fig. 8.19 top left). For the corresponding SESAM with a

Fig. 8.18 SESAM design for passive Q-switching at a center wavelength of 1 μm [99Spu1]: structure of a SESAM with a dielectric top reflector and the standing wave intensity pattern in this structure. The effective penetration depth is only of the order of the 1 μm center wavelength. Quantum well QW

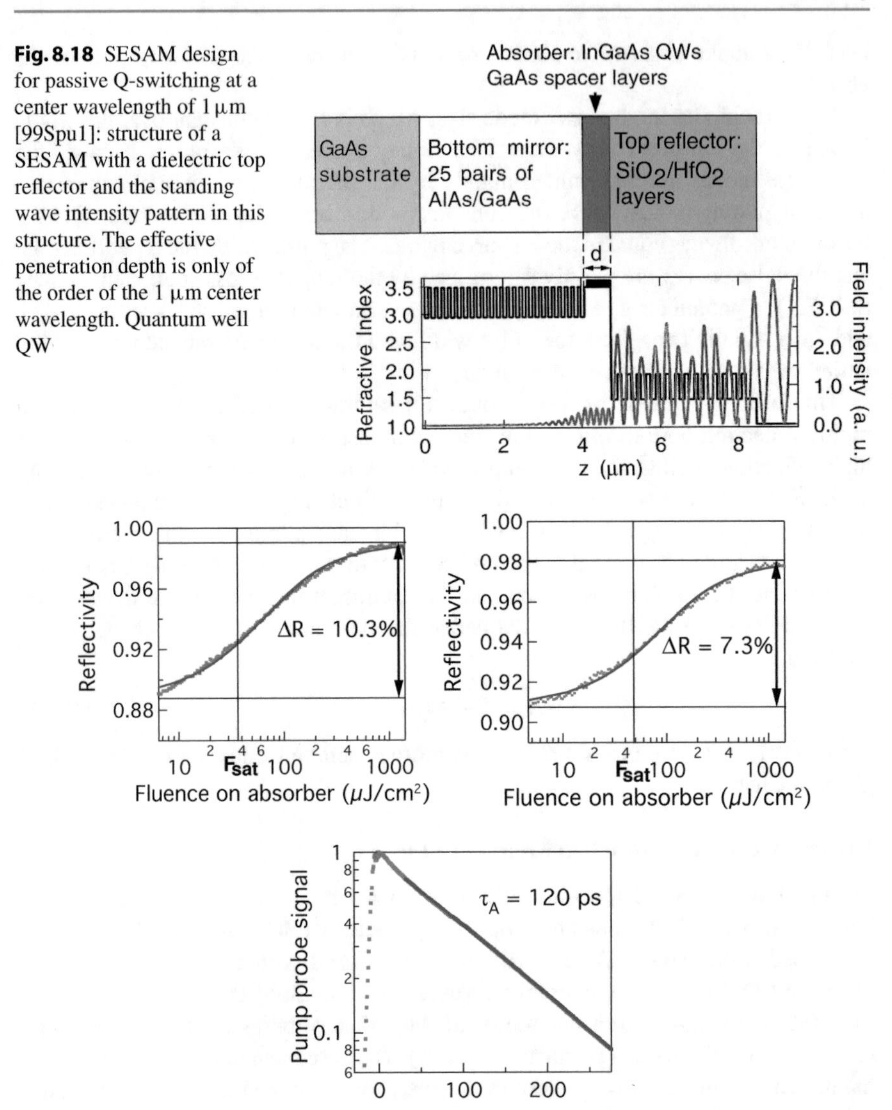

Fig. 8.19 Measured change in reflectivity versus incident pulse fluence (*dots*) and theoretical fit (*solid curve*), according to (7.111) and (7.112), for a SESAM consisting of 18 InGaAs quantum wells and a 25% ($\Delta R = 10.3\%$) and a 50% ($\Delta R = 7.3\%$) top reflector, *top left* and *top right*, respectively. *Bottom*: Measured temporal impulse response (*dots*) together with an exponential fit (*solid line*) [99Spu1]

50% top reflector, we measured $\Delta R = 7.3\%$ and $F_{\text{sat,A}} = 47\,\mu\text{J/cm}^2$ (Fig. 8.19 top right). The absorber lifetime was determined to be $\tau_A = 120\,\text{ps}$ (Fig. 8.19 bottom), independently of the top reflector, but depending on the excitation density.

1.5 μm SESAM

The SESAMs for the eyesafe microchip laser were grown by MOCVD. They consist of a bulk $In_{0.58}Ga_{0.42}As_{0.9}P_{0.1}$ absorber and an InP cap layer, grown on a bottom mirror, formed of 40 pairs of $In_{0.65}Ga_{0.35}As_{0.73}P_{0.27}$/InP layers. The dielectric SiO_2/HfO_2 top coating had a designed reflectivity of 50% or 70% at a laser wavelength of 1.53 μm and a high reflectivity at the pump wavelength of 975 nm.

8.5.2 Laser Setup

We discuss two typical setups for a passively Q-switched microchip laser. The laser, similar to the ones presented in Fig. 8.20, consists of an a-cut 200 μm thin 3-at.% Nd-doped, AR-coated plane parallel Nd:YVO_4 gain element sandwiched between the SESAM and the output coupler or between the output coupling SESAM and a high reflector, respectively.

There are a number of additional complicating issues for an output coupling SESAM. The device has to be grown on a non-absorbing substrate, which at a laser wavelength of about 1 μm can be an undoped GaAs wafer, polished on both sides. The rear side of the wafer has to be anti-reflection (AR) coated in order to avoid residual back reflections and thus additional etalon effects, which could degrade the nonlinear response. The residual losses in the GaAs wafer can reach a few percent. Note that these are extracavity losses and do not affect the internal laser dynamics. As an alternative, the substrate could be etched away. This would be especially relevant in wavelength regimes where there are no transparent substrates available or if the nonlinear losses in the substrate are high.

The experimental setup for the Nd:YVO_4 experiments at 1 μm is shown in Fig. 8.20. The microchip crystal is sandwiched between an output coupler and a SESAM. The top reflector of the SESAM is a high reflector for the pump wavelength and a partial reflector for the lasing wavelength. The crystal is pumped by a 2 watt, 200 μm stripe-size broad-area emitter diode through a dichroic beamsplitter that transmits the pump light and reflects the output beam. The diode pump laser is focused down to radii of 20 μm and 50 μm inside the microchip crystal for the fast

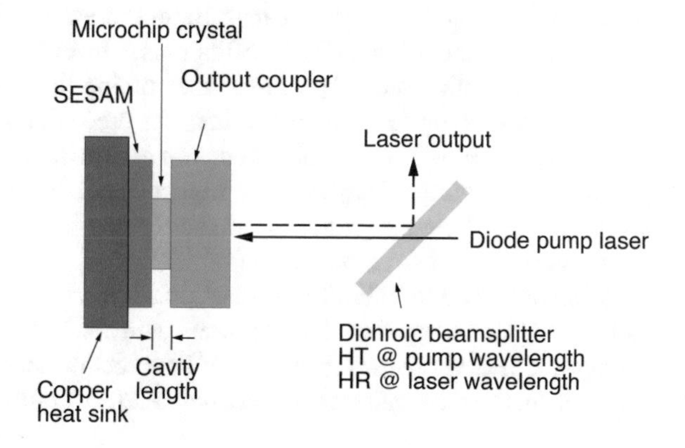

Fig. 8.20 Cavity setup of the microchip laser containing the high reflecting SESAM. Highly reflective HR, highly transmissive HT

and the slow axes, respectively. The pump wavelength is 808 nm for $Nd:YVO_4$ or Nd:LSB crystals of different thicknesses.

We also discuss a Yb:YAG microchip laser which is a three-level laser. There we used a 30 μm stripe-size broad-area emitter (JDS Uniphase AG) focused to radii of 27 μm × 56 μm in the 20% doped, 200 μm thick Yb:YAG crystal [01Spu2].

The setup for the 1.5 μm eyesafe microchip laser is similar to the one shown in Fig. 8.20. We used a 1 mm thick Er:Yb:glass disc, pumped by a Ti:sapphire laser, although diode pumped operation has also been demonstrated [98Flu,01Har1].

The main material properties of microchip lasers with several gain media are summarized in Table 8.2.

8.5.3 SESAM Q-Switching Results

Before going into details, we refer to Table 8.1, which is intended to give a selective overview of examples of the performance of different Q-switched microchip lasers. With regards to SESAM Q-switching, 37 ps pulses have been achieved with an MBE-grown SESAM (growth temperature 480°C) with 35 InGaAs quantum wells with a thickness of 8.7 nm and a 50% top reflector, and a 185 μm thick uncoated $Nd:YVO_4$ crystal. With a 10% output coupler and an incident pump power of 460 mW, single-frequency Q-switched pulses of 37 ps FWHM (see Fig. 8.11) and a peak power of 1.4 kW have been generated [99Spu1]. The oscillation after the pulse is due to detector ringing, identified by comparing the trace with the impulse response of a much shorter modelocked pulse (see, e.g., Fig. 10.5). The average output power was 8.5 mW, the repetition rate 160 kHz, and the pulse energy 53 nJ. In this case, the modulation depth of the saturable absorber was approximately 13%. Equation (8.62) can be used to calculate a pulse width of 64 ps with the given parameters. The deviation from the measured 37 ps can be explained by the Fabry–Pérot effect of the air gap between SESAM and crystal, which can enhance the coupling into the SESAM and therefore the modulation depth. Shorter pulses were achieved when the cavity length was further reduced to only 110 μm below the absorption length of the $Nd:YVO_4$ crystal [12But].

The pulse widths were measured with a 50 GHz sampling oscilloscope and a 45 GHz p-i-n photodetector (see Chap. 10 for more details). The time resolution was better than 15 ps, confirmed with 150 fs pulses from a modelocked Nd:glass laser. Because of the microsecond pulse-to-pulse timing jitter, the oscilloscope had to be triggered on each pulse itself (rather than the preceding pulse) in order to preserve the time resolution. This was obtained with a 10 m optical delay line for the signal, providing the required delay after the trigger signal. Note that passively Q-switched lasers typically have a significantly higher timing jitter than passively modelocked lasers (see Chap. 11 for more details).

As summarized in Table 8.1 shortest pulses have been achieved with $Nd:YVO_4$ and higher pulse energies and peak powers with Yb:YAG and Er:Yb:glass lasers. For example, a SESAM Q-switched Yb:YAG microchip laser can generate higher pulse energies, due to the smaller cross-sections. A 20% Yb-doped YAG crystal was used

Table 8.2 Material properties of $Nd:YVO_4$, Nd:LSB, Nd:YAG, Yb:YAG, and Er:Yb:glass (QX/Er; Kigre, Inc.). Data are taken from [96Kig, 94Bei, 94Mey, 92Koe] and Casix, crystals and materials catalog (Casix, Inc., Fuzhou, Fujian, China). The parameter m is defined in [91Tai] as the number of axial modes within the gain bandwidth [see (8.74)]

Parameter	$Nd:YVO_4$	Nd:LSB	Nd:YAG	Yb:YAG	Er:Yb:glass
Emission wavelength [nm]	1064	1062	1064	1030	1535
Doping D	< 4% [Casix]	≤100%	~1.1% (typically) 0.4–1.1% [Casix]	typically 20% (a wide range is possible)	Er 7.3×10^{19} cm^{-3} Yb 1.8×10^{21} cm^{-3}
Absorption length	90 μm ($D = 3\%$)	110 μm ($D = 25\%$)	1.2 mm ($D = 1.1\%$) [94Bei]	650 μm	650 μm
Absorption bandwidth [nm]	4 (@808 nm)	3.0	0.8 ($D = 1.1\%$) [94Bei]	18 @941nm [92Koe]	0.8
Gain bandwidth [nm]	1.0	4.0	0.7 [94Bei]	≈10	37
Upper-state lifetime [μs]	50 ($D = 3\%$)	87 ($D = 25\%$)	230 [92Koe]	950	7900
Gain cross-section [10^{-19} cm^2]	15.6 [92Koe]	1.3 ($D = 25\%$)	3.3	0.21 [92Koe]	0.08
Absorption cross-section [10^{-19} cm^2]	2.7	0.71	0.7 [94Mey]	0.056	0.17 @ 977nm [96Kig]
Saturation fluence F_L [mJ/cm^2]	37.3	720	333	≈3900	≈8090
Thermal conductivity [20°C, W/m K]	5.2	2.8 ($D = 10\%$)	14 [92Koe]	14 [92Koe]	0.7 (@25°C)
dn/dT [10^{-6}/K]	8.5 (@20°C)	4.4 ($D = 10\%$)	7.3	7.3	−2.1 [96Kig]
Refractive index at emission wavelength given above	1.76	1.82	1.82 [92Koe] 0.82 @1μm [92Koe]	1.82 [92Koe]	1.52
m	0.31	1.42	2.5 [94Bei]	19	26.3

in order to get a relatively short absorption length. The crystal thickness was 200 μm to achieve single-longitudinal-mode operation, and the SESAM had relatively low modulation depth to remain below the damage threshold. With this laser, pulses were generated with 530 ps duration, 1.9 kW peak power, and 1.1 μJ energy with a pulse repetition rate of 12 kHz at a center wavelength around 1 μm [01Spu2].

In the following, we will first present measurements on lasers with $Nd:YVO_4$ as gain material at a wavelength of 1.064 μm. Then, at the end of this chapter, we will describe experiments done on an eyesafe Er:Yb:glass microchip laser. The parameters of these gain materials (i.e., cross-sections, thermal properties, emission bandwidth, etc., see Tables 5.1 and 8.2) and the operating parameters are very different, but our model describes all these lasers with good accuracy.

1 μm SESAM Q-switched $Nd:YVO_4$ Microchip Laser: Parameter Study

Figures 8.21, 8.22, 8.23, and 8.24 present data from SESAM Q-switched microchip lasers with a 435 μm thick $Nd:YVO_4$ crystal [99Spu1]. In this parameter study, the pump power, output coupler, pump mode size, SESAM, and cavity length are varied. We then compare the measured Q-switched pulse performance with our theoretical model. The crystal was AR-coated on both sides to avoid etalon effects between crystal and output coupler or crystal and SESAM. This was necessary to be sure about the exact value of the output coupling and the coupling into the SESAM, because changing the absorber could result in varying air gaps of different thicknesses (and therefore different etalon effects) between the crystal and neighboring elements. For this study we have two 1 μm SESAMs as described in Sect. 8.5 and characterization results shown in Fig. 8.19.

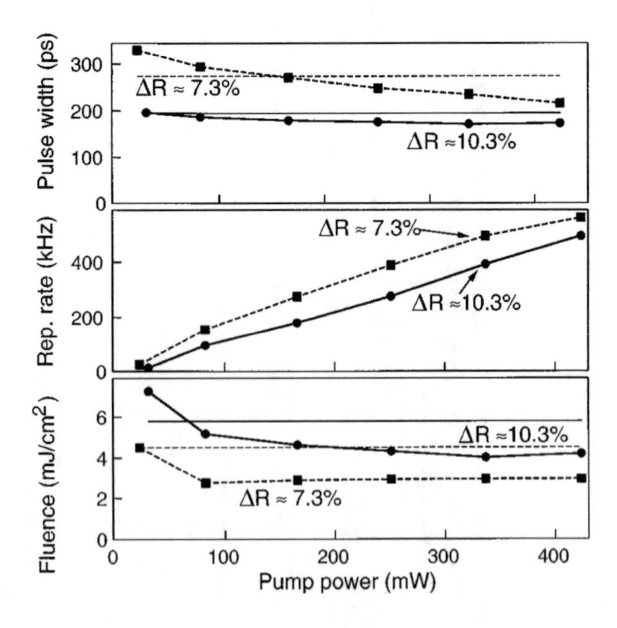

Fig. 8.21 SESAM Q-switched $Nd:YVO_4$ microchip laser [99Spu1]. Pulse width, repetition rate, and output pulse fluence as a function of the pump power for two different SESAM modulation depths ΔR. *Thin horizontal lines* represent the corresponding theoretical values. The experiments were carried out with a 435 μm thick 3% doped $Nd:YVO_4$ crystal, AR-coated on both sides, an output coupler $T_{out} = 9\%$, parasitic losses $l_p = 3\%$ for the cavity with the higher $\Delta R = 10.3\%$ and $l_p = 2\%$ for the lower $\Delta R = 7.3\%$

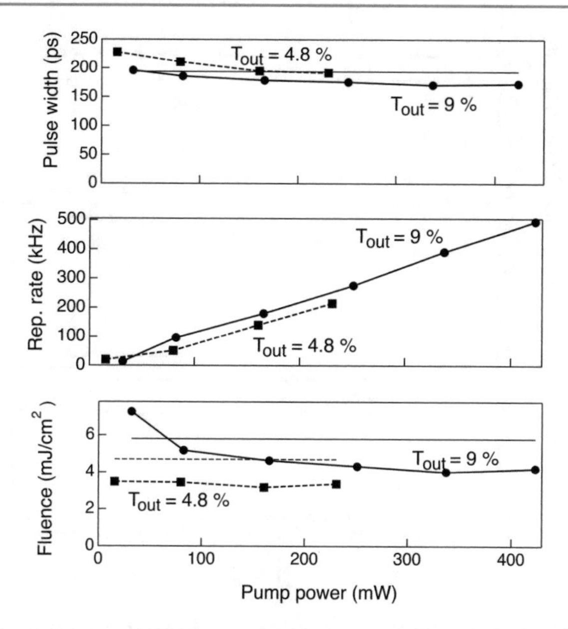

Fig. 8.22 SESAM Q-switched Nd:YVO_4 microchip laser [99Spu1]. Pulse width, repetition rate, and output pulse fluence as a function of the pump power for two different output couplers T_{out}. *Thin horizontal lines* represent the corresponding theoretical values. The experiments were carried out with a 435 μm thick 3% doped Nd:YVO_4 crystal, AR-coated on both sides, $\Delta R = 10.3\%$ and parasitic loss $l_p = 3\%$

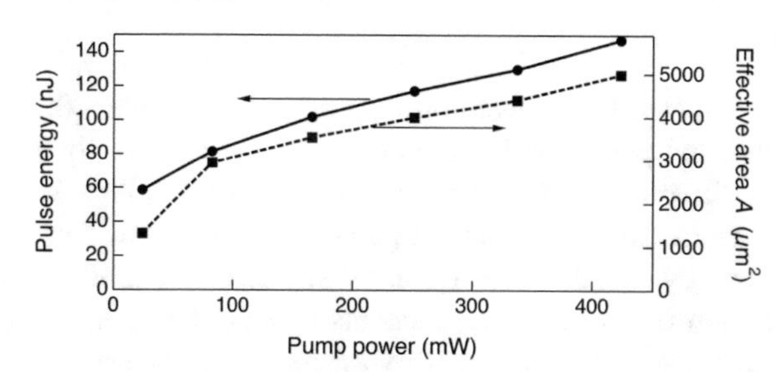

Fig. 8.23 SESAM Q-switched Nd:YVO_4 microchip laser [99Spu1]. Pulse energy and effective laser area as a function of the pump power. The measurement was carried out with a 435 μm thick 3% doped Nd:YVO_4 crystal, AR-coated on both sides, $\Delta R = 7.3\%$, and $T_{out} = 9\%$. The pulse energy is strongly correlated with the effective area, resulting in a constant fluence $F = E_P/A$ (see Fig. 8.21)

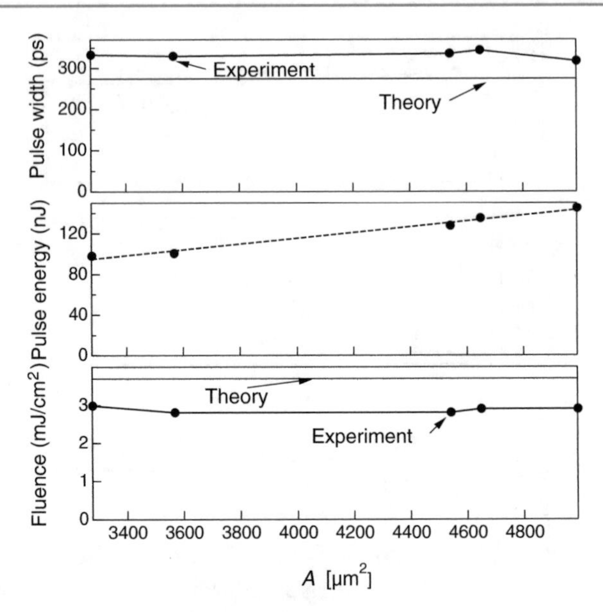

Fig. 8.24 SESAM Q-switched Nd:YVO$_4$ microchip laser [99Spu1]. Pulse width, pulse energy, and output pulse fluence as a function of the effective area, which was controlled by varying the pump spot size. *Thin horizontal lines* represent the corresponding theoretical values. The experiments were carried out with a 435 µm thick 3% doped Nd:YVO$_4$ crystal, AR-coated on both sides, $\Delta R = 7.3\%$, $T_{\mathrm{out}} = 4.2\%$, and $l_{\mathrm{p}} = 2\%$. The pump power was held constant at 160 mW. The *dashed line* is a linear fit to the pulse energy, which appears to be directly proportional to the effective area, without any additive constant

The average output power, repetition rate, emission wavelength, effective laser area in the cavity, and pulse width were measured simultaneously as a function of pump power for different SESAMs and different output couplers. The pulse energy was determined from the average output power and the repetition rate, the fluence from the pulse energy and the effective area. The effective mode area in the laser was calculated from beam size measurements outside the laser, monitored with a beam profiler. The emission wavelength was measured because it determines the effective emission cross-section: one cannot always be sure to have oscillation at the maximum emission cross-section, because the mode spacing is on the order of the gain bandwidth, due to the short cavity length. In the experiments depicted in Figs. 8.21, 8.22, and 8.23, the laser operated near the maximum of the gain curve, resulting in a saturation energy of $F_{\mathrm{sat,L}} = 37.3\,\mathrm{mJ/cm^2}$.

Figure 8.21 shows the pulse width, the repetition rate, and the fluence versus pump power for two SESAMs with different modulation depths ΔR. The same MOCVD-grown absorber with 18 InGaAs quantum wells was used with different topcoatings of 25% and 50%, resulting in modulation depths of 10.3% and 7.3%, respectively. The squared dots with dashed lines correspond to the SESAM with $\Delta R = 7.3\%$ and the circle dots with solid lines to the one with $\Delta R = 10.3\%$. In both cases, the output coupling T_{out} was 9%. The absorber lifetime was $\tau_{\mathrm{A}} = 120\,\mathrm{ps}$ (independently of the dielectric topcoating). The saturation fluences of the absorbers $F_{\mathrm{sat,A}} = E_{\mathrm{sat,A}}/A$

were measured to be 36 and 47 μJ/cm^2, respectively. This means that, for typical mode areas of about 4000 μm^2, the absorber is already cw-saturated at 12 W and 16 W, respectively (8.70), i.e., at power levels far below the typical peak powers of the pulses of several hundreds of watts (phase 1, Sect. 8.4.3). The parasitic losses l_p (calculated from the slope efficiencies) were 3% for the setup with the larger modulation depth and 2% for the one with $\Delta R = 7.3\%$.

The pulse width in Fig. 8.21 appears to be nearly independent of the pump power, as expected, and fits to within 20% of the theoretical value [thin straight lines, (8.62)]. With the exception of the first point, close to the threshold, the fluence is nearly independent of the pump power. The general deviation of about 30% from the theoretical curve (8.61) observed in all measurements is attributed to the transverse effects and spatial hole burning that were neglected in Sect. 8.4. Both of them lower the saturation energy of the gain medium. For one and the same absorber, the repetition rate can be varied over more than one order of magnitude by just varying the pump power without significantly changing the pulse width and fluence.

Figure 8.22 shows data similar to those in Fig. 8.21, but for different output couplers. The same SESAM as in Fig. 8.21 with $\Delta R = 10.3\%$ was used with output couplers of 4.8% and 9%, respectively. The parasitic losses l_p were 3%. The pulse width is nearly independent of the output coupler transmission and pump power and fits very well to the theoretical curve. For the larger T_{out}, the output coupling efficiency, and hence also the fluence, is higher. Again, the deviation of the measured fluence from the theory is systematically about 30% and is ascribed to the same reasons as for Fig. 8.21.

Figures 8.21 and 8.22 show the fluence rather than the pulse energy versus pump power, because the effective area of the laser mode changes significantly with pump power due to varying thermal conditions. Figure 8.23 shows the pulse energy and the effective area as a function of the pump power for laser parameters $\Delta R = 7.3\%$ and $T_{out} = 9\%$. The pulse energy changes by more than a factor of three due to an increase in the mode area A, while the fluence $F = E_{out}/A$ stays constant. In contrast to [90Zay2], we have an increasing effective laser area with increasing pump power, which might be due to different thermal properties. The cavity (Fig. 8.20) has a highly asymmetric and especially non-transverse heat flow. The SESAM has a much better heat conductivity than the output coupler and the surrounding air. Therefore, the main heat sink will be the SESAM, mounted on a copper block. The only poorer heat conductor on the SESAM side is the dielectric coating. For example, we observed different pump power dependencies on the effective area for different coating thicknesses and materials.

In Fig. 8.24, we varied the mode area A while keeping all laser parameters constant ($T_{out} = 4.2\%$, $\Delta R = 7.3\%$). A was changed simply by moving the whole microchip laser out of the pump focus. The pump power was held constant at 160 mW. Thus, the pumping intensity and therefore the pump parameter r, the inversion, and the average output power decreased with increasing A. As expected, this did not affect the pulse fluence and duration, while the pulse energy increased with the mode area and the repetition rate was reduced accordingly. The dashed line in Fig. 8.24 is a linear fit (without additive constant) for the pulse energy versus A, confirming the A dependence of (8.61).

To visualize (8.62) the cavity roundtrip time T_R was varied (Fig. 8.25). This was done in a slightly different, non-monolithic cavity setup, shown in the inset of Fig. 8.25. A curved output coupler ($T_{out} = 1\%$, radius of curvature ROC = 10 cm) was used to ensure cavity stability even for longer cavity lengths, without having to change the crystal thickness and therefore many cavity parameters for every data point. The crystal, 200 μm thick uncoated 3% doped Nd:YVO_4, was bonded to a SESAM consisting of 9 quantum wells, 8.7 nm thick and with no dielectric topcoating. T_R was changed by varying the length of the air gap between the output coupler and the crystal. The pulse width turned out to be directly proportional to the cavity length (Fig. 8.25). Changing the cavity length might slightly change the location of the spot on the SESAM. The bad data points in Fig. 8.25 can thus be explained by inhomogeneities of the absorber. The modulation depth of the absorber can be determined with the help of the slope of the linear fit to the data points and (8.62). This results in $\Delta R = 4.5\%$, which is in excellent agreement with the modulation depth determined with the standard one-beam nonlinear reflectivity measurement introduced in Sect. 7.7.1, giving $\Delta R = 4.4\%$.

1.5 μm SESAM Q-switched Er:Yb:Glass Microchip Laser: Parameter Study

In order to test the applicability of our model to lasers operating in a very different parameter range, a number of experiments with SESAM Q-switched Er:Yb:glass microchip lasers were done for a center wavelength of around 1.5 μm. The quasi-three-level laser medium Er:Yb:glass [65Sni] has a broad emission spectrum, enhancing the tendency for multiple longitudinal mode oscillation, and small emission cross-sections, resulting in higher pump thresholds, lower small-signal gain, and larger pulse fluences (even getting close to the damage threshold of the SESAM),

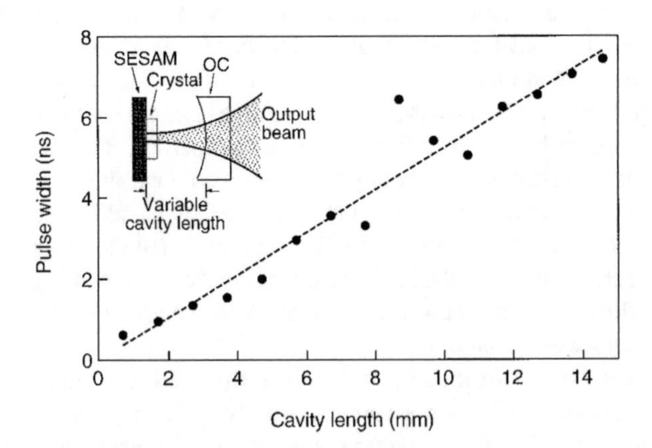

Fig. 8.25 SESAM Q-switched Nd:YVO_4 laser [99Spu1]. Pulse width as a function of cavity length, measured with a cavity as shown in the *inset*. A 200 μm thick 3% doped Nd:YVO_4 crystal was bonded to an uncoated SESAM with an effective modulation depth $\Delta R = 4.4\%$. Resonator stability was ensured by a curved 1% output coupler (ROC = 10 cm), also for large air gaps between crystal and output coupler. With the linear fit (*dashed line*) and (8.62), the modulation depth was calculated to be $\Delta R = 4.5\%$

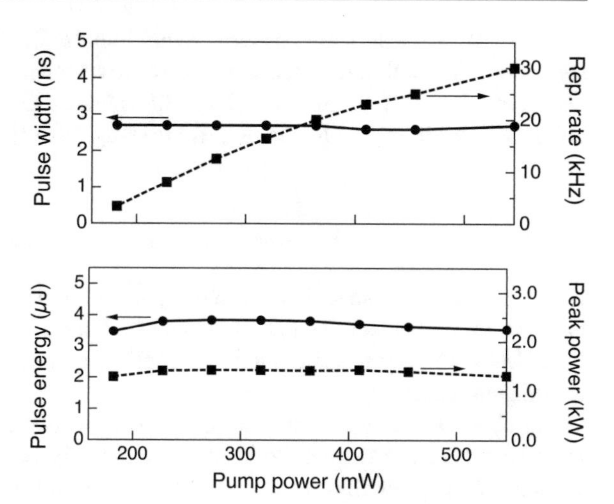

Fig. 8.26 SESAM Q-switched Er:Yb:glass microchip laser [99Spu1]. Pulse width, repetition rate, pulse energy, and peak power as a function of the pump power of an eyesafe 1.5 μm microchip laser using a 1 mm thick Er:Yb:glass (Kigre Inc. QX/Er) plate as gain medium, and $\Delta R \approx 1.5\%$, $T_{\text{out}} = 5\%$. The parasitic losses l_p were estimated to be 1.5%. The pulse energy, pulse width, and peak power appear to be independent of the pump power

compared to Nd-doped crystals. The thermal characteristics differ significantly from those of $Nd:YVO_4$ (see Table 8.2). In particular, the heat conductivity is relatively poor, and the refractive index falls with rising temperature, resulting in a negative thermal lens induced by the pump. The cavity mode is therefore stabilized mainly by gain guiding and thermal expansion.

Figure 8.26 shows the pulse width, repetition rate, pulse energy, and peak power for the 1.5 μm Er:Yb:glass microchip laser. A 1 mm thick Er:Yb:glass disc (Kigre Inc. QX/Er) was used, pumped by a Ti:sapphire laser with a beam radius of 50 μm.

From the measured pulse width of 2.7 ns, the effective modulation depth ΔR (which was influenced by Fabry–Pérot effects at the interfaces) was estimated to be 1.2%. The parasitic losses were 1.5%, and the output coupling 5%. As expected, the pulse width shows no pump power dependence (Fig. 8.26). In contrast to the results for the $Nd:YVO_4$ lasers, the pulse energy was also independent of the pump power. This observation is compatible with the assumption of constant pulse fluence (as predicted by the model) and constant mode area. The latter was not measured, but can be expected to be smaller than the pump mode area and not to increase significantly with the pump power, since for a three-level gain medium it is limited by reabsorption in the wings of the transverse mode profile. The measured pulse energy of 4 μJ is compatible with the assumption that A is one third of the pump area and the laser parameters given above. Thus, the experiments are also in reasonable agreement with the theory for the Er:Yb:glass laser. Again, the repetition rate can be varied over a wide range without changing the pulse width and energy, just by changing the pump power.

8.5.4 Design Guidelines

As we have seen, the model presented in Sect. 8.4 describes the performance of SESAM Q-switched microchip lasers with reasonable accuracy, we now apply the theoretical results to derive practical guidelines for the design of such lasers.

First, we address some general requirements. To achieve stable passive Q-switching, the modulation depth of the absorber must be large enough. More precisely, the decay rate of intensity fluctuations must exceed the relaxation rate of the gain. The condition for Q-switching is [76Hau,95Kar4]

$$-\frac{1}{T_{\mathrm{R}}} I \frac{\mathrm{d}q}{\mathrm{d}I}\bigg|_{\text{cw steady state}} > \frac{r}{\tau_{\mathrm{L}}}\bigg|_{\text{cw steady state}} . \tag{8.72}$$

Here, τ_{L} is the upper state lifetime of the gain medium and I the intracavity intensity. Our experience is that Q-switching is easily achieved in microchip lasers, except for SESAMs with very small modulation depth, e.g., smaller than about 3% for Nd-doped gain materials.

An obvious but important criterion for various design problems is that the gain must be sufficient to reach the laser threshold. The steady-state small-signal gain $g_0 = 2\sigma_{\mathrm{L}} N_{20} L_{\mathrm{g}}$ [or (8.66) for three-level lasers] has to exceed the initial gain g_{i} given by $l + q_0$ to start laser oscillation and Q-switching. However, for a given g_0, g_0 obviously presents an upper limit for g_{i}, and thus also for the gain reduction Δg.

The cavity roundtrip time is related to the length of the gain medium by $T_{\mathrm{R}} = 2L_{\mathrm{g}}/\upsilon_{\mathrm{g}}$, where υ_{g} is group velocity in the gain material. L_{g} is usually chosen to be on the order of the absorption length L_{abs}, which depends on the doping level. For L_{g} much larger than L_{abs}, one increases the pulse width without any other benefit, while a smaller L_{g} would compromise the laser efficiency.

The requirement of generating sufficient gain ($g_0 > g_{\mathrm{i}}$) finally leads to the condition

$$\frac{\sigma_{\mathrm{L}} N_{\mathrm{D}} c}{2n} > \frac{l + \Delta R}{T_{\mathrm{R}}} . \tag{8.73}$$

This permits a pump parameter of $r = 2$ for full inversion, which means that the small-signal excitation density N_{20} equals the dopant density N_{D} (we assume that sufficient pump power is available). This gain criterion will be an ultimate limit for the following design guidelines.

The beam quality of the pump laser is not usually a critical parameter for a microchip laser due to the short cavity length. In addition, a circular output beam profile is easily achieved, even with an elliptical pump spot.

In the following sections we discuss how different design objectives can be reached.

Single-Longitudinal-Mode Operation

Q-switched operation can occur in single or in multiple longitudinal cavity modes. The latter case is less attractive because it leads to mode beating noise and increased timing jitter. The microchip approach is advantageous for single-mode operation because of the large mode spacing which is caused by the short cavity length. A figure of merit for laser materials with respect to single-mode operation is the parameter m defined in [91Tai] as the number of axial modes within the gain bandwidth, assuming a microchip cavity length L_g equal to the absorption length $L_{abs} = 1/\alpha$ of the laser material:

$$m = \frac{\Delta f_g}{c/2nL_g} = \frac{2n\Delta f_g}{\alpha c} , \tag{8.74}$$

where Δf_g is the FWHM gain bandwidth and α the intensity absorption coefficient.

As can be seen from Table 8.2, Nd:YVO_4 has a particularly low (thus favorable) value of the m parameter, 4.6 times smaller than for Nd:LSB, 6.3 times smaller than for Nd:YAG, 63 times smaller than for Yb:YAG, and 85 times smaller than for Er:Yb:glass. Generally, as host materials, glasses have larger m values due to their broader emission spectra. Note, however, that the use of a suitable intracavity filter can ensure single-mode operation even for materials with high m parameter [86Sie]. For example, this was achieved in an Er:Yb:glass laser by incorporating a piece of $LiNbO_3$ which, in conjunction with the output coupler, formed a Fabry–Pérot structure [98Flu].

Short Pulses

We have seen that, for $\Delta R \approx q_0$ as given by (8.71), (8.62) implies that the pulse duration is given

$$\tau_p \approx \frac{3.52 T_R}{q_0} . \tag{8.75}$$

For small τ_p, we obviously need a large modulation depth and a short roundtrip time. Here we are ultimately limited by the gain criterion (8.73). The roundtrip time is related to the absorption length L_{abs} and thus limited for a given gain medium by the maximum possible doping level. L_g can be reduced below L_{abs}, but this compromises the efficiency of the laser and is not useful anyway if the reduced gain would require a smaller value of ΔR. The use of a high gain material (with large σ_L and large possible doping level) is favorable for short pulse generation. The gain bandwidth is not a limiting factor, even for high gain materials. In particular, Nd:YVO_4, with its small possible absorption length (down to 90 μm [95Cas]) and large σ_L, is well suited for this purpose. In addition, the narrow emission bandwidth results in a small m parameter, so single-mode operation is achieved even at high output powers, and is only limited by spatial hole burning [90Zay1,95Bra2].

The SESAM of maximum modulation depth ΔR with good optical quality was about 13% (at 1 μm laser wavelength), limited by cross hatches due to strain in the absorber (see Sect. 8.5.1). The limitations of the crystal thickness imposed by the absorption length, gain criterion, and polishing technique lead to L_g on the order of 100 μm (for Nd:YVO_4).

High Pulse Energy

For $\Delta R \approx q_0$ as in (8.71), the pulse energy is given by (8.61):

$$E_p = \frac{h\nu_L}{2\sigma_L} A 2\Delta R \frac{l_{out}}{l_{out} + l_p} . \tag{8.76}$$

In order to achieve high pulse energies, we need a large modulation depth, laser mode area, and output coupling efficiency, as well as a small emission cross-section. A large laser mode area can be achieved with a large pump spot, but is limited by transverse mode instabilities and a high threshold power. The main way to achieve high pulse energies is to choose a gain material with a small gain cross-section, resulting in a high saturation energy. Therefore, in the 1 μm region Yb:YAG and in the 1.5 μm region, Er:Yb:glass with its small laser cross-sections yields much higher pulse energies than Nd-doped crystals (see Table 8.2).

On the other hand, the limiting factor can be the damage fluence of the SESAM, which is on the order of 30 mJ/cm^2 for a low-finesse SESAM and pulse durations of around 30 ps (see Sect. 8.5.1). The damage threshold can be higher for high-reflecting dielectric topcoatings. It scales inversely with the field intensity enhancement ξ (7.71) [95Bro1, Eq. (14)], which is the ratio of the peak intensity inside the SESAM to the intensity outside. The damage fluence also scales with $\tau_p^{1/2}$ for pulse durations above about 30 ps [76Bet].

As stated in Sect. 8.4.3 and shown in Fig. 8.15, maximum pulse energy is obtained for values of l close to $q_0 \approx \Delta R$ if the nonsaturable losses of the SESAM are accounted for. Then (8.61) reduces to $E_p \approx 2E_{sat,L} l_{out}$. But g_0 must still exceed g_i.

Generally, another way to increase the pulse energy by using a smaller gain cross-section would be to apply an intracavity filter which enforces operation at a wavelength away from the gain maximum, or even at a different transition. This was done in [97Bra], forcing a Nd:YVO_4 laser to oscillate on the 1.34 μm line (where σ_L is lower than at 1 μm), but still benefiting from the good thermal and crystal properties.

High Peak Power

For some applications, it is interesting to have a high peak power P_{peak}, given for $\Delta R \approx q_0$ [see (8.71)] by

$$P_{peak} = S_p \frac{E_p}{\tau_p} \propto \frac{(\Delta R)^2 A}{T_R \sigma_L} \eta_{OC} . \tag{8.77}$$

Here, $S_p$is the pulse shape factor ($S_p = 0.88$ for sech^2 pulses), and $\eta_{OC} = l_{out}/(l_{out} + l_p)$ the output coupling efficiency. A large modulation depth is obviously important. Equation (8.75) also suggests using a small roundtrip time T_R, but this is beneficial only as long as enough gain can be generated to allow for a large modulation depth, which is contained quadratically in (8.75). In Table 8.1, we see that the achieved peak powers do not differ very much, even if different gain media and crystal thicknesses are used.

8.5.5 Summary of the Q-Switched Microchip Laser

Note that we have only discussed SESAM Q-switching results from microchip lasers as these provide the shortest pulse durations due to their short cavity length. These results also demonstrate that passively Q-switched microchip lasers can bridge the gap in terms of pulse durations between modelocking and Q-switching. Thus, the shorter the laser cavity, the shorter the pulses that can be generated. Microchip lasers [89Zay] are single axial frequency lasers using a miniature, monolithic, flat–flat, solid-state cavity whose mode spacing is greater than the medium-gain bandwidth. They rely on gain-guiding, temperature effects, and/or other nonlinear optical effects to define the transverse dimension of the lasing mode. The microchip lasers are longitudinally pumped with a diode laser. Table 8.1 summarizes the results obtained with actively and passively Q-switched microchip lasers. The shortest pulses of only 37 ps or even 22 ps were obtained with $Nd:YVO_4$ passively Q-switched with a SESAM attached to the microchip laser with a crystal length of 184 μm [99Spu1] and 110 μm [12But], respectively. In the second case, the pulse energy was limited

Table 8.3 Summary of design guidelines

	Equation	Result	Reference
Short pulse	$\tau_p \approx 3.5 T_R/\Delta R$	$\tau_p = 37\,\text{ps}$, 53 nJ	[99Spu1]
• short cavity T_R	(8.62) and (8.71)	$f_{rep} = 160\,\text{kHz}$	
• large modulation depth ΔR		$\Delta R = 13\%$	
• large emission cross-section σ_L			
• laser material $Nd:YVO_4$		$\tau_p = 22\,\text{ps}$, 0.9 nJ	[12But]
High pulse energy	$E_p \approx \frac{h\nu_L}{\sigma_L} A \Delta R \eta_{out}$		
• small emission cross-section σ_L	(8.61) and (8.71)		
• large modulation depth ΔR			
• large mode area A			
• laser material Yb:YAG		$E_p = 1.1\,\mu\text{J}$	[01Spu2]
• laser material Er:Yb:glass		$E_p = 11.2\,\mu\text{J}$	[01Har1]
Repetition rate	$f_{rep} \approx g_0/2\Delta R \tau_L$	$f_{rep} = 300\,\text{Hz}$–7.8 MHz	[01Spu2] for 7.8 MHz result [98Flu] for 300 Hz result
• adjustable by pump power	$g_0 \gg l + q_0$ (8.65) and (8.71)		

because the crystal length was below the absorption length. Using different laser crystal thicknesses ranging from 185 to 440 μm, the pulse duration could be changed from 37 ps to 2.6 ns and the pulse repetition rate from 160 kHz to 7.8 MHz. Such a laser can therefore be easily adapted to different application requirements.

Active Q-switched microchip lasers generated pulses as short as 115 ps [95Zay]. Generally, the pulse energies in actively Q-switched microchip lasers tend to be higher, e.g., 12 μJ in [95Zay], although a SESAM Q-switched Er:Yb:glass microchip laser resulted in 11.2 μJ pulse energy [01Har1]. More results are summarized in Table 8.1, and the design guidelines are summarized in Table 8.3.

9 Passive Modelocking

9.1 Introduction and Basic Principle

9.1.1 Basic Principle

Modelocking is a technique to generate ultrashort pulses from lasers. Typically, an intracavity loss modulator (i.e., a loss modulator inside a laser cavity) is used to collect the laser light in short pulses around the minimum of the loss modulation with a period given by the cavity roundtrip time $T_R = 2L/\upsilon_g$, where L is the laser cavity length and υ_g the group velocity, i.e., the propagation velocity of the peak of the pulse intensity. The factor of two is introduced for a linear cavity. We distinguish between active and passive modelocking (Fig. 9.1).

For active modelocking (Chap. 6), an external signal is applied to an optical loss modulator, typically using the acousto-optic or electro-optic effect. Such an electronically driven loss modulation produces a sinusoidal loss modulation with a period given by the cavity roundtrip time T_R. The saturated gain in a typical solid-state laser at steady state then only supports net gain around the minimum of the loss modulation and therefore only provides net gain for pulses that are significantly shorter than the cavity roundtrip time.

For passive modelocking, a saturable absorber is used to obtain a self-amplitude modulation (SAM) of the light inside the laser cavity. Saturable absorbers are discussed in more detail in Chap. 7. Just as a reminder, such an absorber introduces some loss to the intracavity laser radiation, which is relatively large for low intensities, but significantly smaller for a short pulse with high intensity. Thus, a short pulse then produces a loss modulation because the high intensity at the peak of the pulse saturates the absorber more strongly than its low-intensity wings. This results in a loss modulation with a fast initial loss saturation, i.e., reduction of the loss, determined by the pulse duration and typically a somewhat slower recovery which depends on the detailed mechanism of the absorption and recovery processes in the saturable absorber. In effect, the circulating pulse saturates the gain in a typical solid-state

© The Author(s), under exclusive licence to Springer Nature Switzerland AG 2021

U. Keller, *Ultrafast Lasers*, Graduate Texts in Physics,
https://doi.org/10.1007/978-3-030-82532-4_9

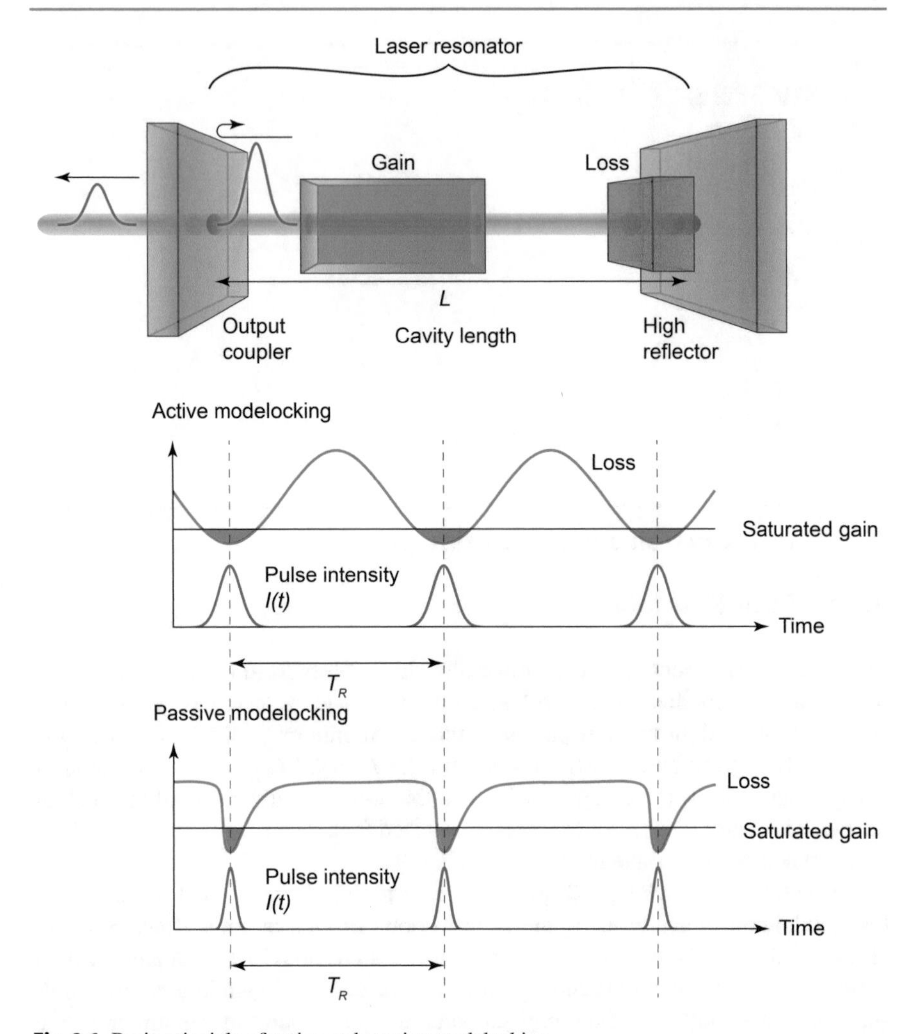

Fig. 9.1 Basic principle of active and passive modelocking

laser to a level which is just sufficient to compensate the loss for the pulse itself, while any other circulating low-intensity light experiences more loss than gain, and thus dies out during the following cavity roundtrips.

The obvious remaining question is: how does passive modelocking start? Ideally, from normal noise fluctuations in the laser. One noise spike is strong enough to significantly reduce its loss in the saturable absorber and will thus be more strongly amplified during the following cavity roundtrips, so that the stronger noise spike continues to further reduce its loss and continues its growth until it reaches a steady state where a stable pulse train will be formed. Self-starting of the modelocking process sets certain conditions on the saturable absorber, which will be discussed in this chapter as well.

Generally, we can obtain much shorter pulses with passive modelocking than with active modelocking, because the recovery time of the saturable absorber can be very fast, resulting in a fast loss modulation. We learned in Chap. 6 on active modelocking that the final pulse duration is given by the balance between the pulse shortening in the loss modulation and the pulse broadening in the gain. The balance is determined by the 'curvatures' of the loss modulation and the gain spectrum. Therefore, from Fig. 9.1 we would expect shorter pulses from passive modelocking, because of the 'stronger curvature' of the SAM compared to the sinusoidal loss modulation obtained with active modelocking. Modelocked pulses are much shorter than the cavity roundtrip time and so, in the case, can produce an ideal fast loss modulation which is inversely proportional to the pulse envelope. Very often, however, the absorber has a slower recovery time that has to be taken into account. We will therefore distinguish between fast and slow saturable absorber modelocking. In comparison, any electronically driven loss modulation is significantly slower due to its sinusoidal loss modulation.

9.1.2 Starting Passive Modelocking

The laser can reduce its loss if it passes into modelocked operation because the loss introduced by the saturated absorber is smaller in pulsed operation. Typically, the saturable absorber is designed to fit the laser parameters in such a way that, in pulsed operation, the saturation $E_p/E_{sat,A}$ in the absorber is on the order of 5, where E_p is the pulse energy inside the laser cavity and $E_{sat,A}$ is the saturation energy of the absorber (see Chap. 7). This can be achieved by adjusting the mode area of the laser beam on the absorber or by adjusting the saturation energy of the absorber to given powers within the laser resonator.

Modelocking in a cw-pumped laser starts through normal intensity fluctuations. One fluctuation is slightly bigger than the others and can therefore reduce the loss in the saturable absorber. Thereby, this intensity fluctuation in particular is amplified, so the absorber can be saturated even more strongly, etc. In this way, after many roundtrips, a modelocked pulse is slowly formed. At the beginning these pulses are very long, such that the initial reduction of the loss through the saturable absorber is comparable to the one of a cw beam. A cw beam saturates the absorber much less than a pulsed beam, because the absorber recovery time τ_A is much shorter than the cavity roundtrip time T_R. Assuming that the saturable absorber is fully saturated by the final pulses, i.e., $E_p \approx 5E_{sat,A}$, the saturable absorber will be much less saturated in cw operation, i.e., $I \ll I_{sat,A}$, if $T_R \gg \tau_A$, with $I_{sat,A}$ given by

$$I_{sat,A} = \frac{F_{sat,A}}{\tau_A} = \frac{E_{sat,A}}{A\tau_A} . \tag{9.1}$$

Here $F_{sat,A}$ is the saturation fluence, $E_{sat,A}$ if the saturation energy of the saturable absorber, and A is the mode area on the saturable absorber.

In a cw modelocked laser, we obtain one pulse per cavity roundtrip time T_R. Initially, the noise pulse is much longer than the final modelocked pulse duration and on the order of T_R. Moreover, the initial pulse energy of the noise pulse is lower. The average power P_{av} for a given pulse energy and pulse repetition rate or frequency f_{rep} is given by

$$P_{av} = E_p f_{rep} = \frac{E_p}{T_R} . \tag{9.2}$$

Therefore the intensity is given by $I = F_p/T_R$, with $F_p = E_{p/A}$ the pulse fluence with a certain mode area that is changing during propagation inside and outside a laser cavity. We can typically assume a Gaussian beam profile in a cw modelocked laser.

For self-starting passive modelocking, we need to have a sufficiently large change in the loss due to small intensity fluctuations. As we will see for ultrafast solid-state lasers, the average power will not change much during the cw modelocking build-up phase, because the loss modulation $\Delta R \approx 2q_0$ is small, whence the loss of a fully saturated saturable absorber is also small, i.e., for an ideal fast saturable absorber it is roughly $q_0/3$ and for a slow saturable absorber roughly q_0/S, where S is the saturation parameter $S = F_p/F_{sat,A}$. More details are given in Chap. 7 and summarized in Tables 7.2 and 7.3. As long as these saturated absorber losses are much smaller than the output coupler, the average output power will not change much with or without cw modelocking.

In Sect. 4.5.3, we determined the nonlinear transmission through a saturable absorber for a cw beam (see Fig. 4.32). We obtain optimal self-starting conditions for

$$\frac{dT}{dI} \text{ maximal } \Longleftrightarrow I \ll I_{sat} . \tag{9.3}$$

With $T_R \geq 5\tau_A$ and $F_p = 5F_{sat,A}$, the condition $I = F_p/T_R \leq F_{sat,A}/\tau_A = I_{sat,A}$ of (9.3) is easily satisfied, not taking into account the fact that the initial noise pulse is only a small fluctuation on top of a cw signal, i.e., $F_{p,noise} \ll F_p$. In any case, we would like to make sure that the saturable absorber is not already saturated by the cw beam, i.e., $I \ll I_{sat,A}$. During the modelocking build-up time, the noise pulse $F_{p,noise}(t)$ will grow in pulse energy and become shorter. Therefore, the saturable absorber will ultimately become fully saturated if all parameters are designed correctly.

For stable self-starting modelocking, the build-up time for passive modelocking should be less than around 1 ms for good stability. We can measure the modelocking build-up time (Fig. 9.2) when we use an intracavity mechanical chopper to study the self-starting dynamics and monitor both the average output power and the second-harmonic generation (SHG) signal. The delay of the second-harmonic signal determines the modelocking build-up time, which is typically much longer than the cw power build-up time. This was measured for an Nd:YLF laser [93Kel] and it was shown that the modelocking build-up time becomes shorter when the recovery time of the saturable absorber τ_A becomes larger within a range of 3 to 30 ps and for a cavity roundtrip time of $T_R = 4.5$ ns which satisfies (9.3). The modelocking build-up time changed from less than 100 μs (for $\tau_A = 30$ ps) to up to 10 ms. During the

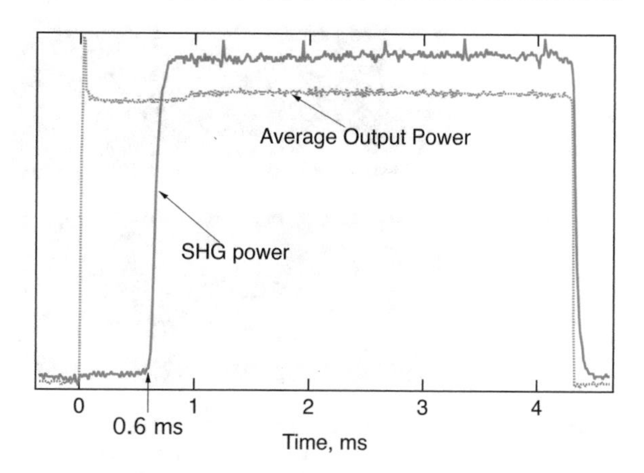

Fig. 9.2 Measured modelocking build-up time from a passively modelocked Nd:YLF laser. We defined the modelocking build-up time by the delay of the second-harmonic generation (SHG), here shown with 0.6 ms. From [94Kel]

build-up phase, the saturation parameter S is reduced by the shorter recovery time of the saturable absorber for the long noise pulse $F_{\mathrm{p,noise}}(t)$:

$$S_{\text{noise}} \sim \int \frac{F_{\mathrm{p,noise}}(t)}{F_{\mathrm{sat,A}}} \mathrm{e}^{-t/\tau_{\mathrm{A}}} \mathrm{d}t \ . \tag{9.4}$$

Therefore the longer recovery time τ_{A} of the saturable absorber helps for stable self-starting modelocking with a recommended upper limit of $T_{\mathrm{R}} \geq 5\tau_{\mathrm{A}}$.

It was interesting to observe that the inhomogeneous gain broadening of the laser also affected the modelocking build-up time [94Kel]. From the discussion in Sect. 6.7, we should expect a shorter modelocking build-up time for an inhomogeneously broadened gain. With the position of the gain crystal in an Nd:YLF standing wave cavity, we can adjust spatial hole burning and therefore the level of inhomogeneous gain broadening (see Sect. 6.7.3). For a gain-at-the-end (GE) cavity, we have cw lasing with a broader spectrum (see Fig. 6.28). Indeed, for a GE cavity, we observed a shorter modelocking build-up time and shorter pulse durations (but not transform-limited pulses) in comparison to the gain-in-the-middle (GM) cavity [94Kel]. No negative dispersion compensation was used in this case.

9.1.3 Historical Development

Just six years after the first laser was demonstrated in 1960, De Maria and coworkers produced the first passively modelocked pulses with an intracavity saturable absorber, in the picosecond regime with a flashlamp-pumped Nd:glass laser [66DeM]. However, this result had an underlying problem: looking more closely, one realizes that they did not measure a constant pulse train, but a strongly amplitude modulated pulse train (Fig. 9.3), where the picosecond pulses lie beneath much longer 'macro pulses', and these macro pulses are repeated at a repetition rate in the kHz regime. This mode of operation is called Q-switched modelocking. This remained a problem for passively modelocked solid-state lasers for more than 20 years.

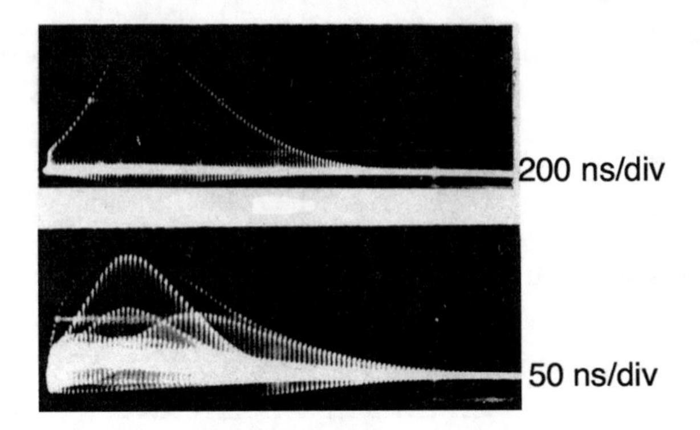

Fig. 9.3 Passively modelocked Nd:glass laser using a dye saturable absorber inside the laser resonator. Picosecond pulses were achieved beneath much longer pulses, shown here with a sweep speed of 50 ns/div. Today referred to as Q-switched modelocking. From [66DeM]

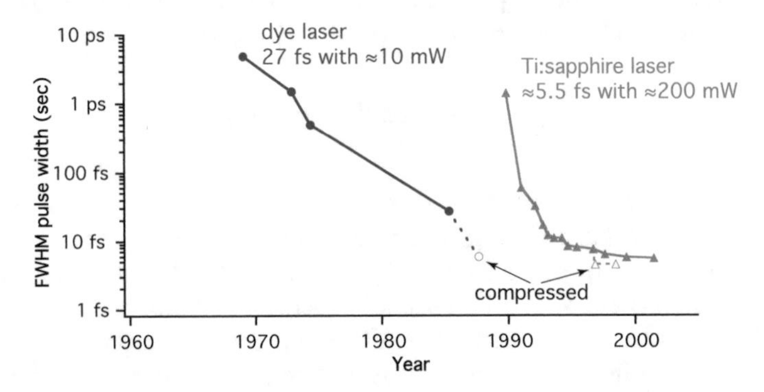

Fig. 9.4 Historical development of ultrashort laser pulse generation with the shortest pulses at a typical average output power at that time. For many years, this field was dominated by dye lasers. This situation changed with the Ti:sapphire laser, which was commercialized in the late 1980s

For many years, the frontier of ultrafast lasers was determined by passively modelocked dye lasers (Fig. 9.4). The ultrafast dye lasers supported sub-picosecond pulses for the first time and ultimately generated pulses as short as 27 fs at a center wavelength of 620 nm and an average output power of around 10 mW [86Val]. With additional optical amplification and external compression techniques, these pulses were then further reduced to 6 fs [87For]. A slow dye saturable absorber was used inside the dye laser resonator. Here, through saturation of the amplifier and absorber, a short net gain window is opened, which can support an ultrashort pulse (Fig. 9.5a). This will be discussed in more detail in Sect. 9.3. Passive modelocking with dynamic gain saturation has also been used in semiconductor diode lasers [85Smi] and in color center lasers [89Isl], which have comparable gain cross-section to the dye lasers (Table 9.1). In these cases, amplifier and absorber recover relatively slowly, typically within a few nanoseconds. Such ultrafast dye lasers allowed for the exploration of new sub-picosecond dynamics in solid-state physics and chemistry for the first time. However, the effort involved in these experiments was much greater than for ultrafast solid-state lasers today. Due to the fact that

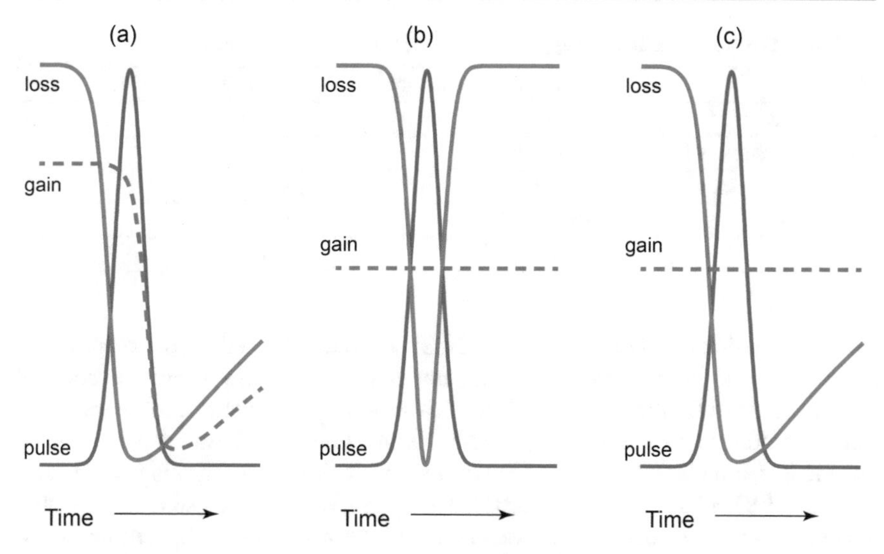

Fig. 9.5 Summary of different modelocking techniques [99Kel]. **a** Passive modelocking with a slow saturable absorber and dynamic gain saturation [72New,75Hau3], **b** passive modelocking with a fast saturable absorber [75Hau1], and **c** passive modelocking with a slow saturable absorber without dynamic gain saturation in the picosecond regime [01Pas] and in the femtosecond regime with soliton modelocking [95Kar1,95Jun2,96Kar]

the chemical composition of the dye liquid already changes after a short time, the nonlinear optical properties degrade very fast. Therefore, such ultrafast dye lasers were mostly suitable for scientific applications.

In the 1980s, there was a renaissance in diode pumped solid-state lasers because high-power diode array pump lasers became commercially available. This greatly increased the interest in ultrafast solid-state laser technology. However, for a long time solid-state lasers were not passively modelocked because the available saturable absorbers were either destroyed by the high intracavity intensity, generated too much loss, brought the laser to "self-Q-switching", or caused other instabilities. It was generally assumed by the experts in the field that flashlamp- and diode-pumped solid-state lasers could not be passively modelocked, so active modelocking was used. The high small-signal gain and small laser mode area in the gain material of diode-pumped solid-state lasers reduced the pulse duration for active modelocking into the sub-10 ps regime for Nd-doped YAG, YLF, and vanadate lasers for the first time (see Table 6.2).

Novel coupled cavity modelocking techniques that were used to passively modelock color center lasers also stimulated new efforts with diode-pumped solid-state laser. The soliton laser [84Mol] was one of the early examples of coupled cavity modelocking, in which soliton pulse compression inside a nonlinear coupled cavity with a negative-GDD fiber for a color center laser resulted in passive modelocking. This technique was extended to a coupled cavity with positive GDD [89Kea], which was later referred to as additive pulse modelocking (APM) [89Mar, 89Ipp]. There were theoretical predictions that any nonlinear coupled cavity should

Table 9.1 Gain cross-section and upper state lifetime for different laser materials

Material	Gain cross-section [cm^2]	Upper state lifetime
Semiconductor	$\approx 10^{-14}$	$\approx$1 ns
Dye (Rhodamine 6G)	1.4×10^{-16}	$\approx$10 ns
Color center (NaCl:OH$^-$)	0.9×10^{-16}	$\approx$150 ns [86Pin]
Ti:sapphire	4×10^{-19}	3 μs
Nd:YLF	1.8×10^{-19}	480 μs

passively modelock a laser [88Blo1,88Blo2]. This stimulated the work on resonant passive modelocking (RPM), where an amplitude nonlinearity inside the coupled cavity was used for the first time [90Kel2,91Hau2]. For example, a lossy semiconductor saturable absorber mirror (SESAM) with a high modulation depth was used for the first time inside a coupled cavity to modelock both a Ti:sapphire [90Kel2] and an Nd:YLF laser [91Kel2,92Kel1]. That SESAM introduced so much loss that it could never have been used inside the main laser resonator—however, this work ultimately stimulated the correct SESAM design to passively modelock solid-state lasers without Q-switching instabilities. Before this happened, however, Kerr lens modelocking (KLM) was discovered with the Ti:sapphire laser.

Kerr lens modelocking (KLM) was a breakthrough that became possible because of the discovery of a new solid-state laser material, the Ti:sapphire laser, which had broad gain bandwidth supporting sub-10 fs pulses [82Mou,86Mou]. To passively modelock the Ti:sapphire laser, new methods had to be developed, because in this case the saturation of the gain could not help with pulse formation. Ti:sapphire is a material with an emission cross-section which is about 1000 times smaller than that of laser dyes (Table 9.1). We would not expect a laser consisting of a Ti:sapphire crystal as the gain medium and a dye jet as the saturable absorber medium to operate like an ultrafast dye laser, since there is no dynamic gain saturation, as shown in Fig. 9.5a. The Ti:sapphire gain is only saturated by the average intensity (Fig. 9.5b, c). In spite of this, Sarukura et al. [91Sar] passively modelocked a Ti:sapphire with an intracavity HITCI absorber dye and produced stable sub-100 fs pulses. With no significant gain saturation in the Ti:sapphire laser, the roughly 1.2 ns recovery time of the HITCI absorber dye is not fast enough to explain the short pulse duration. This result was presented at the Ultrafast Phenomena VII meeting in 1990, just before the CLEO conference where Sibbett's "magic modelocking" result was presented as a postdeadline result. Sarukura's result did not have the same impact because there was still a saturable absorber in the laser, whereas in Sibbett's result there was no saturable absorber visible and it was thus initially referred to as "magic modelocking".

Sibbett's group with their "magic modelocking" result actually discovered a new fast saturable absorber (according to Fig. 9.5b) by accident in the Ti:sapphire laser [91Spe], but this fast saturable absorber was not a separate device and they initially explained their results in terms of coupled cavity modelocking [91Spe]. Additional experiments then explained their experimental result with a fast saturable absorber model [91Kel], now generally referred to as Kerr lens modelocking (KLM).

In addition, Sarukura's result could also be explained by KLM, and the slow dye saturable absorber was only responsible for the self-starting process [92Kel1]. With KLM, new world record pulse durations were achieved, and for the first time the long-standing world record of 6 fs [87For] was surpassed with a KLM modelocked Ti:sapphire laser using double chirped mirrors (DCMs) for dispersion compensation [99Sut]. Pulses as short as 5.8 fs were generated with an average output power of about 300 mW [99Sut,00Sut]. Improved DCMs then resulted in pulse durations of around 5 fs [01Ell]. KLM will be discussed in detail in Sect. 9.4 and pulse generation in the few-cycle regime in Sect. 9.10.1.

The problem with KLM, however, is that modelocking is typically not self-starting and requires critical cavity designs with very small position tolerances (i.e., less than 100 μm) of certain cavity components. Therefore, the normal intensity fluctuations in the laser have to be enhanced with movable mirrors or other loss modulators. Unfortunately, KLM is also very weak for picosecond pulses, which are of importance for industrial applications. Therefore, the search went on for other applicable saturable absorbers for solid-state lasers.

Initially, it was assumed that a fast saturable absorber is needed for stable pulse generation for solid-state lasers without any significant dynamic gain saturation (Fig. 9.5b) [92Kel1]. Through the fast saturable absorber, a short net gain window is opened, which is approximately as short as the pulse.

In 1992, the first type of SESAM (semiconductor saturable absorber mirror [92Kel2,96Kel,99Kel]) device was invented, and stable self-starting modelocking of a diode pumped solid-state laser was demonstrated for the first time. The SESAM is a novel family of optical devices (see Chap. 7) that allow for very simple, self-starting passive modelocking of ultrafast solid-state and semiconductor lasers and Q-switching of microchip lasers. Previously, Q-switching instabilities had prevented stable passive modelocking of solid-state lasers for more than 25 years. All important parameters of the semicondcutor saturable absorber can be suitably adjusted through material and device structure (Chap. 7). Thus, passively modelocked solid-state and fiber lasers using intracavity SESAMs became a very attractive alternative to KLM, and are today widely used in the above 10 fs regime in both research and commercial environments. SESAM modelocked solid-state laser oscillators have increased pulse energies by four orders of magnitude, up to more than 10 μJ for the first time in 2007 [07Mar,08Sud], and increased pulse repetition rates by more than two orders of magnitude, to 160 GHz [02Kra2]. Scaling the average power and pulse energy will be discussed in more detail in Sect. 9.10.2. Suppression of Q-switching instabilities for passive modelocking becomes more difficult for gigahertz pulse repetition rates. Examples of how to overcome this challenge are discussed in more detail in Sect. 9.10.3.

Extending this technique to passively modelocked optically pumped semiconductor (OPS) lasers based on vertical external cavity surface emitting lasers (VECSELs), we have pushed the performance of these types of lasers into the multi-watt level using new concepts and devices—again an improvement of more than two orders of magnitude [06Kel,15Til]. Wafer-scale integration will make them extremely attractive for high-volume applications [07Maa]. Such ultrafast lasers will be discussed in more detail in Sect. 9.8.

The use of SESAMs resulted in the discovery of a different form of modelocking process called soliton modelocking (Fig. 9.5c) [95Kar1,95Jun2,96Kar] (Sect. 9.5). For a long time, it was believed that the use of a fast saturable absorber is necessary to modelock a solid-state laser, because modelocking relies on a short net gain window that only supports the pulse and discriminates against the noise that might grow outside the pulse [92Kel1,98Kar]. However, SESAM modelocked solid-state lasers resulted in much shorter pulses than the recovery time of the SESAM (compare Fig. 9.5c) [92Kel2,96Kel]. The soliton-like pulse shaping leads to stable pulsing even in the presence of a considerable open net gain window following the pulse. The latter is no longer predominantly shaped by the saturable absorber, but the absorber is still essential for pulse stability. Note that this regime of operation is significantly different from what was discussed before in the context of the dye lasers (Fig. 9.5a), where the interplay between loss and gain saturation always leads to a short net gain window in time. The same is very much true for the fast saturable absorber (Fig. 9.5b), where the soliton-like pulse shaping leads to pulses about a factor of two shorter than without soliton-like pulse shaping [91Hau]. The open net gain windows that form in these cases are only about one pulse width long. For soliton modelocking, on the other hand, the net gain window can remain open for 10 to 30 times the pulse duration, depending on the specific laser parameters [98Kar]. Therefore, soliton modelocked Ti:sapphire lasers stabilized with slow SESAMs were used for the generation of pulses as short as 13 fs—no KLM with its critical cavity alignment was necessary for this result [96Kar].

Experiments with SESAM modelocked solid-state lasers in the picosecond regime also showed that even without soliton effects the pulse duration can be at least 20 times shorter than the absorber recovery time. This finding may seem quite surprising because on the trailing edge of the pulse there is no shaping action of the absorber (Fig. 9.5c). There is even net gain, because the loss caused by the absorber is very small for the trailing edge, assuming a fully saturated absorber. Thus one might expect this net gain to either prevent the pulse from getting so short or destabilize it by amplifying its trailing wing more and more. Indeed, there are a number of publications (e.g., [98Kal,98Akh]) in which it is assumed that stable pulse generation is not possible under such circumstances. To understand why stable pulses with a duration far below the absorber recovery time are possible, we have to consider that the action of the absorber steadily delays the pulse [01Pas]: it mainly attenuates its leading wing, thus shifting the pulse center backwards in each cavity roundtrip. Note that this shift does not apply to any structures well behind the pulse maximum, because the absorber is then already fully saturated. In effect, the pulse is constantly moving backward and can swallow any noise growing behind it. This noise has only a limited time in which it can experience gain before it merges with the pulse itself. The same mechanism can also prevent the trailing edge of the pulse from growing. This will be discussed in more detail in Sect. 9.5.4.

Most cases can be treated with the 'master equation' formalism of Hermann Haus. Therefore, we will follow this formalism again here. However, there are limitations, as already discussed in Chap. 6:

1. Haus' master-equation formalism also includes the slowly-varying-envelope approximation. As soon as the pulse length is much shorter than 10 fs, this approximation is no longer valid. Additionally, especially in this regime, the changes per element (amplifier, dispersion compensation, output coupling, etc.) in the laser resonator are relatively large. In this case, the order of the optical elements also becomes important. Therefore, only numerical methods can describe this case [92Kra2].
2. Modelocking with a slow saturable absorber without dynamic gain saturation is the typical case for SESAM-modelocked solid-state lasers. This case can also no longer be treated analytically without soliton formation [01Pas], and will be discussed in more detail in Sect. 9.5.4.

9.2 Coupled Cavity Modelocking

Coupled cavity modelocking was very important for a time before the discovery of KLM and SESAM modelocking. Today these techniques are no longer used, simply because the other techniques are so much better. However, for educational reasons I would still like to give a short summary. We can distinguish between three different types of coupled cavity modelocking, as shown schematically in Fig. 9.6.

The soliton laser [84Mol] is an early example of ultrashort pulse generation by coupled cavity modelocking, in which the nonlinear external cavity medium supports soliton pulse formation in a net negative GDD regime. The pulse inside the coupled cavity becomes shorter via soliton formation and the superposition of the returning shorter pulse from the nonlinear coupled cavity and the pulse from the main cavity then results in more pulse shortening, until a steady state condition is reached. Active cavity length stabilization of the coupled cavity was required for stable pulse generation.

Following early soliton laser experiments by Mollenauer in which a color center laser was coupled to a nonlinear fiber cavity in the soliton regime, Blow and Wood [88Blo2] demonstrated theoretically with numerical models that any nonlinear coupled cavity can produce modelocking. Confirming Blow and Wood's theoretical prediction, enhanced modelocking of a synchronously pumped color center laser was demonstrated [88Blo1,89Kea], using a fiber with positive GDD inside a coupled cavity (Fig. 9.6b). Similarly, Fujimoto et al. modelocked a cw pumped Ti:sapphire laser [89Goo]. Later, Ippen and Haus [89Mar,89Ipp] introduced a simple analytical model, called additive pulse modelocking (APM), which explains the pulse shortening process through interference of the pulse in the main cavity with the self-phase-modulated reinjected longer pulse from the coupled cavity containing a Kerr nonlinearity. The nonlinear phase allows for constructive interference at the peak of the pulse and destructive interference in the wings, effectively shortening

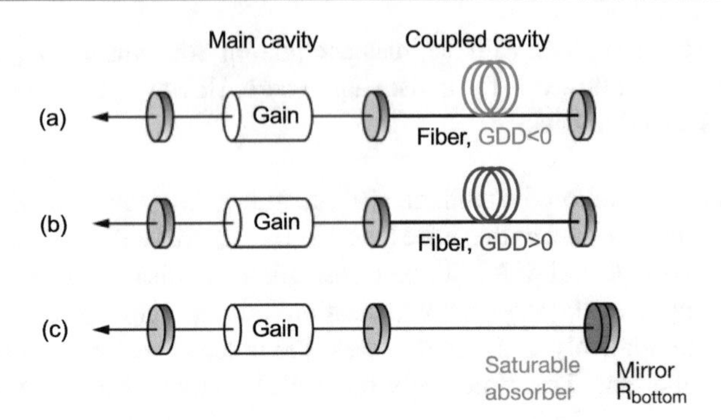

Fig. 9.6 Coupled cavity modelocking based on a main cavity with the laser gain and a nonlinear coupled cavity with **a** a fiber with negative GDD, i.e., a soliton laser [84Mol], **b** a fiber with positive GDD, i.e., additive pulse modelocking (APM) [89Kea,89Ipp], and **c** a real saturable absorber such as a semiconductor mirror, i.e., resonant passive modelocking (RPM) [90Kel2]

the pulse in the main cavity. This is similar to a nonlinear Michelson interferometer [86Oue,89Mor], where one branch contains a Kerr medium and with which a CO_2 laser was passively modelocked. Under certain conditions, APM of a cw pumped laser can be self-starting [90Ipp], and many different APM lasers have been demonstrated (see review given with [07Kel, Table 2.1.2]). Active cavity length stabilization of the coupled cavity was required for APM lasers.

Furthermore, Blow and Wood's theoretical prediction [88Blo2] and the early APM laser results inspired other coupled-cavity modelocking techniques using a self-amplitude modulation in a semiconductor saturable absorber with a self-amplitude modulation (Fig. 9.6c) [90Kel2], an active modelocker [89Fre], and a moving mirror [90Fre] in the coupled cavity.

However, the use of a real saturable absorber with a self-amplitude modulation in the coupled cavity (Fig. 9.6c) removed the requirement for active cavity length stabilization. We modelocked both a Ti:sapphire [90Kel2] and an Nd:YLF laser [91Kel2,92Kel1] using a semiconductor saturable absorber mirror in the coupled cavity. Because the nonlinearity in the coupled cavity is based on a fast intensity-dependent reflectivity change of the semiconductor reflector, a resonant nonlinearity, we refer to it as resonant passive modelocking (RPM). This also emphasizes the fact that, for both the soliton laser and for APM, only a nonlinear phase shift was used. The widespread success of APM was strongly enhanced by the easy availability of fibers. In contrast to RPM, APM requires a feedback system to control the cavity length and to maintain stable modelocking. However, even with an active feedback loop in place, a cw pumped APM laser has a tendency to drop out of modelocked operation, induced by sudden mechanical vibrations, and this is difficult to track with a simple feedback circuit. In addition, self-starting modelocking was more challenging.

For an all-solid-state ultrafast laser technology, semiconductor saturable absorbers have the advantage that they are compact, fast, and can cover bandgaps from the visible to the mid-infrared (see Chap. 7). It was initially a surprising result that for RPM no cavity length stabilization of the coupled cavity was required to produce stable modelocked pulses [90Kel2]. But it could be shown experimentally that the laser

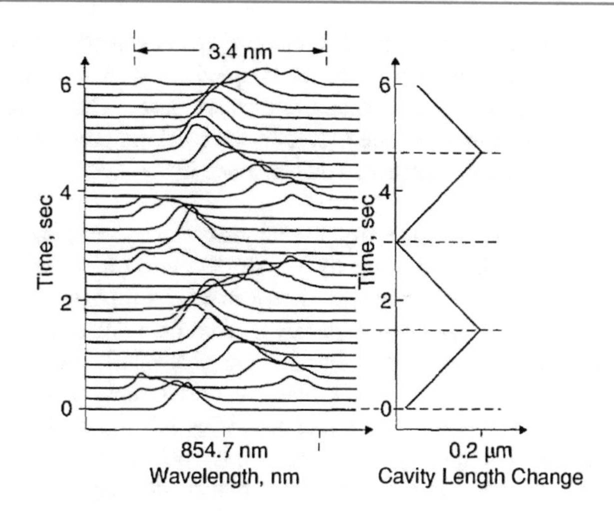

Fig. 9.7 RPM is self-stabilized by optical frequency adjustments [90Kel2]. Measured optical spectrum when the coupled cavity length was changed for a Ti:sapphire laser

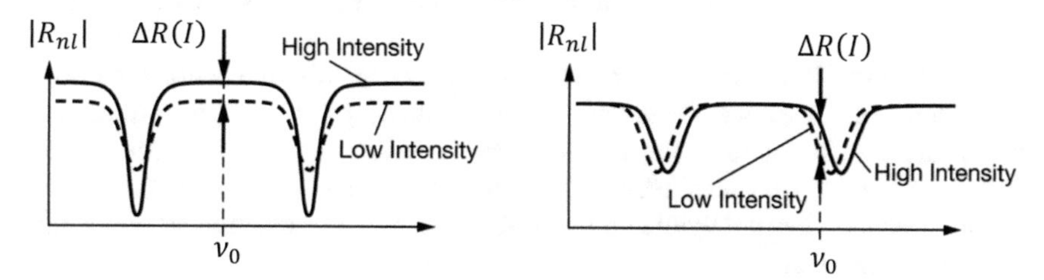

Fig. 9.8 Coupled cavity reflectivity as seen from the main cavity for RPM (*left*) and APM (*right*). The Fabry–Pérot is operated at maximum reflectivity for RPM and at a critical lower reflectivity point for APM which is not stable on its own [92Kel1]

operates in stable modelocked pulse trains by self-adjusting its optical frequency in response to phase shifts due to external cavity length variations (Fig. 9.7) [90Kel2]. This is, in fact, a simple property of a linear Fabry–Pérot cavity, where the resonance wavelength shifts with the Fabry–Pérot cavity length. When we look from the main cavity towards the nonlinear coupled cavity, we observe a Fabry–Pérot-like reflectivity as shown in Fig. 9.8a. Therefore, RPM is self-stabilized by optical frequency adjustments, which means that the coherent superposition of the pulse in the main cavity and the pulse in the coupled cavity is maintained by small optical self-frequency shifts compensating for any relative cavity length fluctuations.

The coherent superposition of the pulse in the main cavity with the pulse from the nonlinear coupled cavity can be described by modeling the laser coupling mirror and the external cavity as a nonlinear intensity-dependent mirror. The reflectivity $R_{\rm nl}$ of this mirror is then referred to as the nonlinear coupled-cavity reflectivity (Fig. 9.8 left). The amplitude coupled-cavity reflectivity is then given by the Fabry–Pérot reflectivity. With the additional condition that the resonance frequency has to be an axial mode of the coupled cavity, the coupled-cavity reflectivity has a periodicity of

$c/2\delta L$, where δL is the cavity length detuning. A more detailed derivation is given in [92Kel1]. A saturated absorber inside the coupled cavity then reduces the loss at high intensity and the nonlinear reflectivity thus increases. Therefore, this can be described as a fast or slow saturable absorber modelocking process with a normal intracavity saturable absorber mirror, which resulted in the invention of the SESAM device (see Sect. 7.5.1).

In the case of a phase nonlinearity, the intensity-dependent phase shift can be treated as an intensity-dependent cavity length detuning $\delta L(I)$, which shifts the frequency for maximum reflectivity as shown in Fig. 9.8 (right). Assuming that the laser is self-adjusting its center frequency to maximum reflectivity, no intensity-dependent reflectivity is introduced by the nonlinear phase shift. Therefore, in order to obtain an intensity-dependent reflectivity, the laser must be held at a center frequency ν_0 that does not correspond to maximum gain (Fig. 9.8 right). An APM laser is operated at this center frequency. From Fig. 9.8 (right), it also becomes clear that only on one slope is the reflectivity higher at higher intensity and lower at lower intensity. Therefore, interferometric absolute cavity length stabilization is required. This explains why the APM laser has a tendency to drop out of modelocked operation induced by sudden mechanical vibrations that cannot be tracked by a simple feedback circuit.

Note that in RPM the nonlinear phase shift associated with the nonlinear absorption change in the semiconductor mirror (determined by the Kramers–Kronig relations) has no effect due to the self-frequency adjustments to maximum coupled cavity reflectivity. However, even without the self-frequency adjustments, the nonlinear phase shift has negligible effect on RPM as long as the nonlinear phase shift is much smaller than π.

The performance of an RPM Nd:YLF laser was improved using a low-temperature MBE-grown multiple quantum well semiconductor mirror [92Kel1], resulting in better surface quality and a faster saturable absorber recovery time. The RPM technique enabled the discovery of SESAM modelocking (see Sect. 7.5), for which the nonlinear coupled cavity was replaced by a single intracavity saturable absorber (Fig. 7.17). RPM allowed us to adjust the SESAM parameters for a regime of stable modelocking, without any Q-switching instabilities. Initially, it was surprising how little loss modulation was required to obtain stable passive modelocking with SESAMs. In this chapter we discuss the underlying laser physics theory which was partially newly developed after the first demonstration of SESAM modelocking in 1992 [92Kel1].

9.3 Passive Modelocking with a Slow Saturable Absorber and Dynamic Gain Saturation

For historical reasons we start with the theoretical treatment of passive modelocked dye lasers. In this case we have to take into account dynamic gain saturation which we can neglect later for typical solid-state lasers. More recently this modelocking technique has become more important also for diode pumped semiconductor lasers. We discuss in more detail colliding pulse modelocking (CPM) which allowed for the generation of world-record sub-30 fs pulse durations for the first time [86Val].

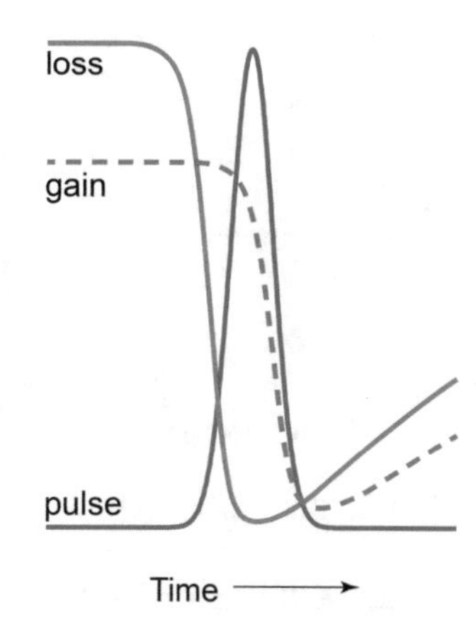

Fig. 9.9 Modelocking with a slow saturable absorber. Pulse shaping through critical balance of saturable absorption and gain. Absorber and gain recover only slowly after the pulse [72New, 75Hau3]

This laser enabled many breakthroughs in fundamental science. The detailed design of such a CPM laser is actually very interesting and can motivate novel ideas for ultrafast solid-state lasers. Note that the final analytical solutions are summarized in Sect. 9.6.

9.3.1 Modelocking Conditions

Application: Dye Laser and Semiconductor Laser
Figure 9.9 illustrates three criteria for stable modelocking with a slow saturable absorber and dynamic gain saturation:

1. At the beginning the loss is bigger than the gain, i.e.,

$$l_0 > g_0 \,, \tag{9.5}$$

   where l_0 and g_0 are the unsaturated loss and gain coefficients after one laser resonator roundtrip (assuming both recover completely from one pulse to the next).
2. The absorber has to saturate faster than the gain, i.e.,

$$E_{\mathrm{sat,A}} < E_{\mathrm{sat,L}} \,. \tag{9.6}$$

   Because $E_{\mathrm{sat,A}} = A_{\mathrm{A}}\hbar\omega/\sigma_{\mathrm{A}}$ (7.1) and accordingly $E_{\mathrm{sat,L}} = A_{\mathrm{L}}\hbar\omega/\sigma_{\mathrm{L}}$, we obtain the condition

$$E_{\mathrm{sat,A}} < E_{\mathrm{sat,L}} \iff \frac{A_{\mathrm{A}}}{\sigma_{\mathrm{A}}} < \frac{A_{\mathrm{L}}}{\sigma_{\mathrm{L}}} \,, \tag{9.7}$$

   where σ_{A} and σ_{L} are the absorber and laser gain cross-sections, and A_{A} and A_{L} the effective cavity mode area in the saturable absorber and in the laser gain.
3. The saturable absorber has to recover faster than the saturated gain:

$$\tau_{\mathrm{A}} < \tau_{\mathrm{L}} \,. \tag{9.8}$$

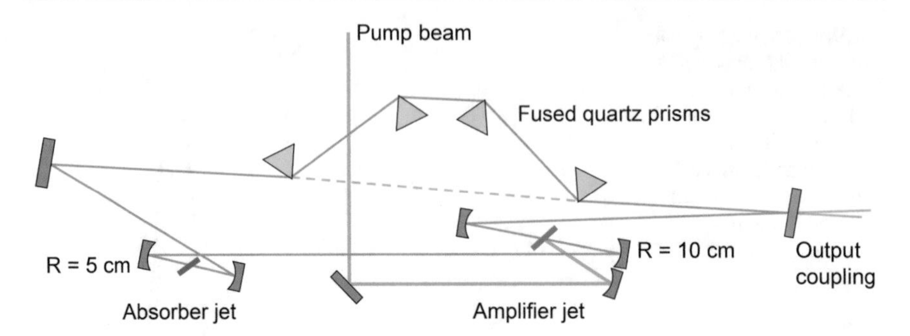

Fig. 9.10 The colliding pulse modelocking (CPM) dye laser [86Val]

9.3.2 Example: Colliding Pulse Modelocking (CPM)

The first picosecond pulses generated with dye lasers and dye absorbers were surprising and initially not properly understood, because the recovery time of the saturable dye absorber was in the nanosecond regime, until it was suggested that gain saturation might help with the pulse generation process (Fig. 9.9) [72New]. The breakthrough for the generation of stable femtosecond pulses with pulse durations below 100 fs was achieved in 1980 with the introduction of the 'colliding pulse modelocked' (CPM) dye laser [81For]. This type of laser, which is passively modelocked by using an absorber dye, is shown schematically in Fig. 9.10, along with a lab photo in Fig. 9.11. Both the gain and the absorber are based on dyes which have similar gain and absorber cross-sections. However, in a ring laser, a stronger saturation of the absorber can be achieved by two pulses propagating in opposite directions and then colliding in the saturable absorber. Two pulses saturate the absorber more strongly than one pulse alone, and the condition in (9.6) can be fulfilled. For the first time, it was shown that four pulse formation processes are involved in the generation of femtosecond pulses: (i) saturation of the absorption, (ii) saturation of the gain, (iii) self-phase-modulation (SPM) of the laser pulses, and (iv) group delay dispersion (GDD) through the components of the laser resonator.

Through suitable compensation of SPM and GDD, a compression of the laser pulses is possible through soliton effects, as is known from pulse propagation in optical glass fibers (Sect. 4.3). For the first pulse generation of less than 100 fs, 'magic mirrors', i.e., I was told that mirrors with a slightly shifted center wavelength (see Sect. 3.6) were used for GDD compensation in [81For]. With the introduction of prism pairs (see Sect. 3.2 and [84For]), it became possible to continuously vary and optimize the negative dispersion. Exploiting all of these effects, it was possible to generate pulses of only 27 fs duration directly from a CPM dye laser [86Val].

A considerable disadvantage of the CPM laser is that the emission is limited to wavelengths between 605 and 635 nm, although both dyes have a much broader tunability. The critical balance between the saturation of the absorber and the gain (Fig. 9.9) needed to obtain a short laser pulse is only satisfied in a narrow wavelength region, which strongly confines the tunability of the CPM dye laser. For the short

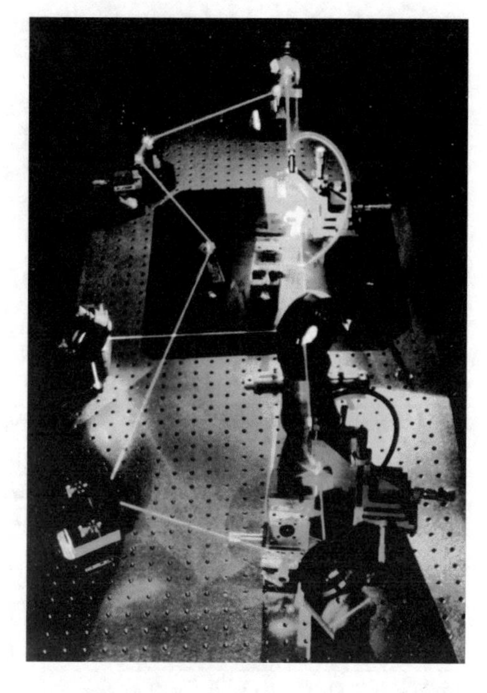

Fig. 9.11 CPM dye laser with Rhodamine 6G as the gain dye and DODCI as the saturable absorber dye. This laser was built by U. Keller as a summer student at AT&T Bell Laboratory, Murray Hill, New Jersey in 1985

pulse duration regime of less than 30 fs, this laser is no longer tunable at all, and has a fixed emission at a wavelength of 620 nm (Table 9.3).

The CPM dye laser is a modelocked ring laser with the following elements (Figs. 9.10 and 9.11):

- An amplifier jet (thickness around 350 μm), which contains the laser dye R6G (Rhodamine 6G) and was initially pumped via a pump mirror by a cw argon ion laser with a power of about 5 W at a wavelength of 514 nm.
- The thin absorber jet (about 20 μm thick), which contains the saturable absorber dye DODCI (3,3-diethylocadicarbocyanin) and is placed at the focal point of another mirror configuration. The ring configuration favors the formation and propagation of two pulses in opposite directions, meeting in the absorber. This means an intensity enhancement in the absorber causes more effective bleaching compared to a single pulse in the resonator. In addition, a smaller focus in the absorber jet favors faster saturation as well.

The two pulses that collide inside the saturable absorber jet and propagate in opposite directions are only amplified by the same amount if the distance between absorber and amplifier is about a quarter of the resonator length. In this way, the pulses meeting in the absorber see the same gain in the amplifier jet, because the time before a new pulse in the amplifier always corresponds to half the roundtrip time, and the amplifier can recover to the same gain value for both pulses. The two laser pulses are coupled out of the ring resonator via an output coupler (more typical transmission 2.5–3.5%).

Table 9.2 Emission cross-section σ_L and recovery time of the saturation τ_L of R6G, and the absorption parameters σ_A, τ_A, $\tilde{\sigma}_A$, and $\tilde{\tau}_A$ of DODCI and its photo-isomer, respectively, at $\lambda = 620$ nm

Rhodamine 6G (R6G)	$\sigma_L = 1.36 \times 10^{-16}$ cm^2	$\tau_L = 4$ ns
DODCI	$\sigma_A = 0.52 \times 10^{-16}$ cm^2	$\tau_A = 2.2$ ns
Photo-isomer	$\tilde{\sigma}_A = 1.08 \times 10^{-16}$ cm^2	$\tilde{\tau}_A \approx 1$ ns

Table 9.3 Typical and best performance parameters for a CPM Rhodamine-6G laser with a DODCI-dye absorber. Valdmanis achieved the world-record result when both dye liquids were freshly mixed, because the dye photodegrades in about 1–3 weeks [86Val]. Typical values are given based on personal experience in the lab, and the best 27 fs pulses were around 30 to 35 fs long with an output coupler T_{out} of 3.5% (i.e., typical best)

CPM Rhodamine 6G-laser	Typical	Best [86Val]
FWHM pulse duration τ_p	100 fs	27 fs
Emission wavelength λ_L	620 nm	620 nm
Average output power P_{av}	10 mW	20 mW for 30 to 35 fs pulses
Resonator length L	≈ 3 m	≈ 3 m
Repetition rate f_{rep}	100 MHz	100 MHz
Pulse peak output power $P_{peak} = E_p/\tau_p$	1 kW	6.7 kW for 30 fs and 20 mW
Output pulse energy E_p with $P_{av} = E_p f_{rep}$	0.1 nJ	0.2 nJ for typical best and $T_{out} = 3.5\%$
Pump power	5 W at 514 nm	4 W at 514 nm
Amplifier jet thickness (at focus) and pressure		80 μm, 300 μm nozzle, 40 psi
Absorber jet thickness (at focus) and pressure		30–50 μm, 120 μm nozzle, 15 psi
Prisms for GDD compensation		Quartz glass

The typical and the world-record (i.e., best or typical best) laser performance are summarized in Table 9.3.

The CPM laser mainly differs in two points from previous passively modelocked dye laser constructions, which could neither reach the sub-100 femtosecond regime nor exhibit the pulse stability of the CPM laser: (i) the ring resonator, and (ii) the thin absorber jet and the transient interference grating, generated by the pulses propagating in opposite directions, which increases the saturation

9.3.3 Pulse Formation through Saturable Absorption and Gain

Both the absorber and the amplifier dye show strong nonlinear behavior at high intensities in the dye jets (about 80 GW/cm^2). For a two-level model of the dye molecule and for a slow saturable absorber (i.e., pulse duration τ_p much shorter than the recovery time of the dye) the saturation (or bleaching) of the absorption $q(t)$ is described by (7.13):

$$\boxed{q(t) = \sigma_A N_A d_A e^{-E_p(t)/E_{sat,A}}} \,, \tag{9.9}$$

where σ_A is the absorption cross-section of the dye, N_A is the number of absorbing molecules per cm^3, d_A is the thickness of the absorbing layer, and

$$E_p(t) = E_p \int_{-\infty}^{t} f(t')dt'$$

is the time-dependent energy of the intracavity laser pulse according to (7.13). It then follows that

$$E_p(t) \propto \int_{-\infty}^{t} |A(t')|^2 \, dt' \,, \tag{9.10}$$

where $A(t')$ is the pulse envelope, and the pulse envelope is normalized such that $|A(t)|^2$ is the instantaneous intracavity power.

Similarly, the population inversion in the amplifier dye, generated by the pump laser, is depleted by the laser pulse. The saturation of the gain $g(t)$ is described by analogy with the slow saturable absorber model (7.13) because the recovery time of the gain is, according to Table 9.2, in the nanosecond regime and therefore much longer than the final pulse duration. Hence,

$$\boxed{g(t) = \sigma_L N_L d_L e^{-E_p(t)/E_{sat,L}}} \,, \tag{9.11}$$

where σ_L is the emission cross-section of the amplifier dye (Table 9.2), N_L is the population inversion per cm^3, and d_L is the thickness of the gain medium.

Parabolic Gain Approximation and *s*-Parameter

The total net gain g_T obtained by the combined effect of amplifier and absorber dye is, according to (9.9) and (9.11), given by

$$g_T(t) = g(t) - q(t) - l_R = g_0 e^{-E_p(t)/E_{sat,L}} - q_A e^{-E_p(t)/E_{sat,A}} - l_R \,, \tag{9.12}$$

where $g_0 = \sigma_L N_L d_L$, $q_A = \sigma_A N_A d_A$, and l_R is the loss in the resonator without absorber, i.e., $l_0 = q_A + l_R$, in (9.5). We will now consider the s-parameter introduced by New [72New]. This parameter describes the ratio of the saturation energies from the absorber and amplifier:

$$s \equiv \frac{E_{sat,L}}{E_{sat,A}} \,. \tag{9.13}$$

Then, the total net gain $g_T(x)$ can be expanded in powers of x up to the second order, i.e., $e^x \approx 1 + x + x^2/2$, in (9.12):

$$g_T(E_p) = g_0 \left[1 - \frac{E_p}{E_{sat,L}} + \frac{1}{2}\left(\frac{E_p}{E_{sat,L}}\right)^2\right] - q_A \left[-\frac{E_p}{E_{sat,A}} + \frac{1}{2}\left(\frac{E_p}{E_{sat,A}}\right)^2\right] - l_R$$

and therefore

$$g_{\mathrm{T}}(E) = g_0 - q_{\mathrm{A}} - l_{\mathrm{R}} + \left(q_{\mathrm{A}} - \frac{g_0}{s}\right) E + \frac{1}{2}\left(\frac{g_0}{s^2} - q_{\mathrm{A}}\right) E^2 . \tag{9.14}$$

with $E \equiv E_{\mathrm{p}}(t)/E_{\mathrm{sat,A}}$. The energy dependence for the total net gain $g_{\mathrm{T}}(E)$ is parabolic. To obtain pulse formation on both the leading and the trailing pulse edge, g_{T} has to be less than zero at both the beginning, i.e., $E = 0$, and the end, i.e., $E = E_{\mathrm{p}}/E_{\mathrm{sat,A}}$, of the pulse, where E_{p} is the total energy of the pulse (Fig. 9.9). Therefore, l_0 and g_0 cannot be chosen arbitrarily, but have to satisfy the conditions $g_{\mathrm{T}}(0) \leq 0$ and $g_{\mathrm{T}}(E_{\mathrm{p}}/E_{\mathrm{sat,A}}) \leq 0$.

The pulse duration becomes shorter, the bigger the gain $g_{\mathrm{T}}^{\max}$ in the maximum of the gain parabola. Let us therefore take a look at the influence of the s-parameter on $g_{\mathrm{T}}^{\max}$. With

$$g_{\mathrm{T}}^{\max} = g_{\mathrm{T}}\left(E_{\max}\right) , \qquad \left.\frac{\partial g_{\mathrm{T}}}{\partial E}\right|_{E=E_{\max}} = 0 ,$$

and for $l_0 \approx q_{\mathrm{A}} \approx g_0$, but still $l_0 > g_0$ from (9.5), it follows from (9.14) that

$$\left.\frac{\partial g_{\mathrm{T}}}{\partial E}\right|_{E=E_{\max}} = l_0\left(1 - \frac{1}{s}\right) - l_0\left(1 - \frac{1}{s^2}\right) E_{\max} = 0 ,$$

whence

$$E_{\max} = \frac{1 - 1/s}{1 - 1/s^2} .$$

This in turn implies

$$g_{\mathrm{T}}^{\max} \equiv g_{\mathrm{T}}\left(E_{\max}\right) = l_0\left(1 - \frac{1}{s}\right) E_{\max} - \frac{l_0}{2}\left(1 - \frac{1}{s^2}\right) E_{\max}^2 ,$$

and finally,

$$g_{\mathrm{T}}^{\max} = \frac{l_0}{2}\frac{(1 - 1/s)^2}{1 - 1/s^2} . \tag{9.15}$$

This function is plotted for constant l_0 in Fig. 9.12. One can see that the maximum gain increases with an increasing s-parameter and asymptotically approaches $l_0/2$.

The goal of every version of a passively modelocked dye laser is therefore to optimize the s-parameter. Values in the range

$$3 < s < 6 \tag{9.16}$$

can reasonably be reached. Equations (9.7) and (9.13) then imply

$$3 < \frac{\sigma_{\mathrm{A}} A_{\mathrm{L}}}{\sigma_{\mathrm{L}} A_{\mathrm{A}}} < 6 . \tag{9.17}$$

For an even bigger s-parameter, $g_{\mathrm{T}}^{\max}$ is only weakly dependent on s (see Fig. 9.12).

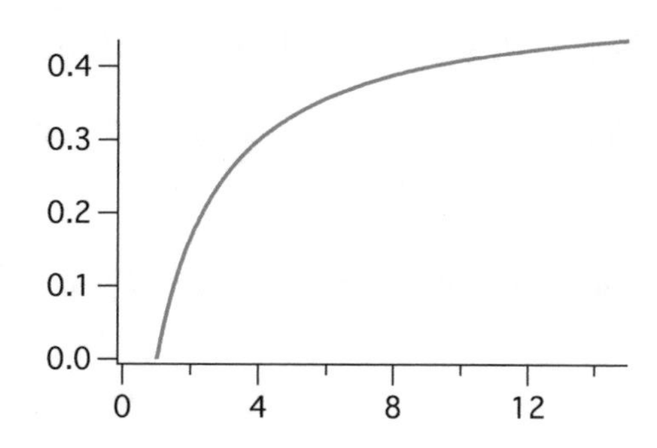

Fig. 9.12 $g_T^{max}(s)$ for constant loss (here $l_0 = 1$) according to (9.15)

In the CPM dye laser, the s-parameter is determined by the following parameters:

(i) the cross-sections σ_L, σ_A of the amplifier and absorber,
(ii) the beam mode areas A_L and A_A of the laser in the gain and absorber media, and
(iii) the interaction of the pulses of the ring laser propagating in opposite directions in the absorber.

The s-parameter increases if the DODCI photo-isomer is taken into account (see Table 9.2). However, the quantum yield for the formation of the photo-isomer in the absorber dye is not known. The s-parameter can be increased if the beam mode area A_A in the absorber is chosen smaller than A_L in the amplifier. Therefore, in the CPM dye laser, mirrors with a radius of curvature of $R = 5$ cm are typically chosen for the absorber, whereas $R = 10$ cm is a typical value for the amplifier (see Fig. 9.10). The s-parameter is thereby approximately increased by a factor of 4.

In the bidirectional CPM ring laser, the pulses propagating in opposite directions meet in the absorber and form a standing wave. Through the intensity enhancement in the peak of the standing wave, the absorber is bleached more strongly. This results in an increase of the s-parameter by a factor of about 2–3. Because the pulses do not meet in the amplifier, no intensity enhancement occurs. However, the described interference only has an effect if the transient grating expands over the full absorber thickness. This is only the case if the absorber jet is thin compared to the spatial extent of the femtosecond pulses. Therefore, for a pulse duration of $\tau_p = 100$ fs, we need an absorber thickness d_A satisfying

$$d_A < \frac{c}{n}\tau_p \approx 20 \ \mu\text{m} , \tag{9.18}$$

where $n = 1.43$ is the refractive index of the solvent (ethylene glycol) and c is the velocity of light in vacuum. The use of thin nozzles for the absorber jet is one of the central points in the construction of CPM dye lasers. Through the counter-propagating pulses, the thin nozzle, and the different radii of curvature of the mirrors in the absorber and the amplifier section, the s-parameter in such a CPM dye laser can be optimized to a value about 12 times higher than in previous constructions.

9.3.4 Master Equation and Solution

We can use Haus' master equation formalism [75Hau3] as introduced for active modelocking (see Sect. 6.4.2) to treat the problem analytically.

Gain
The gain is given by (see Appendix F and (6.51))

$$\Delta A_1 = g\left(1 + \frac{1}{\Omega_g^2}\frac{\partial^2}{\partial t^2}\right)A(T,t) .$$

Because the gain is also time-dependent (9.11), we have to modify this linearized operator as follows:

$$\Delta A_1 = \left[g(t) + \frac{g_0}{\Omega_g^2}\frac{\partial^2}{\partial t^2}\right]A(T,t) , \tag{9.19}$$

where g_0 is the maximum unsaturated gain and we later assume that $g_0 \approx l_0$, i.e., gain and absorption recover completely between the pulses and the self-amplitude modulation is assumed to be small. $g(t)$ in (9.19) is given by (9.10) and (9.11):

$$g(t) = g_0 e^{-E_p(t)/E_{sat,L}} = g_0 \exp\left[-\frac{\sigma_L}{A_L h\nu}\int_{-\infty}^{t}\left|A(t')\right|^2 dt'\right] . \tag{9.20}$$

Note that we have neglected the proportionality factor in (9.10) and $E_{sat,L} = A_L F_{sat,L} = A_L h\nu/\sigma_L$ is given in analogy to (7.1).

Self-Amplitude Modulation (SAM)
We have

$$A_{out}(t) = \exp\left[-q(t)\right]A_{in}(t) .$$

Once again, we use the parabolic approximation for small loss (see Appendix F):

$$\Delta A_2 = -q(t)A(T,t) , \tag{9.21}$$

where $q(t)$ is given by (9.9) and (9.10) as

$$q(t) = q_A \exp\left[-\frac{\sigma_A}{A_A h\nu}\int_{-\infty}^{t}\left|A(t')\right|^2 dt'\right] , \tag{9.22}$$

and q_A is the unsaturated loss of the absorber jet. Note that we have neglected the proportionality factor in (9.10) by analogy with (9.20).

Time Shift of the Pulse
Through the saturable absorber, the pulse can be slightly shifted in time, which changes the resonator roundtrip time. For passive modelocking, this does not play a role, and we can consider this effect with a time shift of the pulse t_D:

$$\Delta A_3 = t_D \frac{\partial}{\partial t} A(T,t) \, . \tag{9.23}$$

For a first estimate, we neglect SPM and GDD. Since

$$T_R \frac{\partial A(T,t)}{\partial T} = \sum_i \Delta A_i \, , \tag{9.24}$$

the master equation reads

$$\boxed{T_R \frac{\partial A(T,t)}{\partial T} = \left[g(t) - q(t) + \frac{g_0}{\Omega_g^2} \frac{\partial^2}{\partial t^2} + t_D \frac{\partial}{\partial t} \right] A(T,t)} \, . \tag{9.25}$$

The steady-state solution has to satisfy the condition $T_R \partial A(T,t)/\partial T = 0$.

Motivated by experimental results, we make the following solution ansatz:

$$A(t) = A_0 \operatorname{sech}\left(\frac{t}{\tau}\right) \, . \tag{9.26}$$

This solution ansatz would be more strongly justified if we also considered SPM and negative GDD. A possible time shift of the pulse (9.23) has to be taken into account for a self-consistent solution.

From Appendix E, we have

$$\frac{\mathrm{d}}{\mathrm{d}x} \operatorname{sech}(x) = -\tanh(x) \operatorname{sech}(x) \, , \tag{9.27}$$

$$\frac{\mathrm{d}^2}{\mathrm{d}x^2} \operatorname{sech}(x) = \operatorname{sech}(x) \left[1 - 2 \operatorname{sech}^2(x) \right] \, . \tag{9.28}$$

Then, from (9.26),

$$\tau \frac{\partial}{\partial t} A(t) = -\tanh\left(\frac{t}{\tau}\right) A(t) \, , \tag{9.29}$$

$$\tau^2 \frac{\partial^2}{\partial t^2} A(t) = A(t) \left[1 - 2 \operatorname{sech}^2\left(\frac{t}{\tau}\right) \right] = A(t) \left\{ 1 - 2 \left[1 - \tanh^2\left(\frac{t}{\tau}\right) \right] \right\} \, . \tag{9.30}$$

We expand $g(t)$ (9.20) and $q(t)$ (9.22) up to the second order in $\int_{-\infty}^{t} \left| A(t') \right|^2 \mathrm{d}t'$, i.e., using the expansion $\mathrm{e}^x \approx 1 + x + x^2/2$:

$$g(t) \approx g_0 \left[1 - \frac{\sigma_L}{A_L h\nu} \int_{-\infty}^{t} \left| A(t') \right|^2 \mathrm{d}t' + \frac{\sigma_L^2}{2 (A_L h\nu)^2} \left(\int_{-\infty}^{t} \left| A(t') \right|^2 \mathrm{d}t' \right)^2 \right] \, ,$$

$$q(t) \approx q_A \left[1 - \frac{\sigma_A}{A_A h\nu} \int_{-\infty}^{t} \left| A(t') \right|^2 \mathrm{d}t' + \frac{\sigma_A^2}{2 (A_A h\nu)^2} \left(\int_{-\infty}^{t} \left| A(t') \right|^2 \mathrm{d}t' \right)^2 \right] \, .$$

From Appendix E, we also have

$$\int \mathrm{sech}^2(ax)\mathrm{d}x = \frac{\tanh(ax)}{a} .$$

This yields

$$\int_{-\infty}^{t} |A(t')|^2 \, \mathrm{d}t' = \tau A_0^2 \tanh\left(\frac{t}{\tau}\right)\Bigg|_{-\infty}^{t} = \tau A_0^2 \left[\tanh\left(\frac{t}{\tau}\right) + 1\right] ,$$

because

$$\lim_{x\to\pm\infty} \tanh(x) = \lim_{x\to\pm\infty} \frac{\mathrm{e}^{-x}\left(\mathrm{e}^{2x}-1\right)}{\mathrm{e}^{-x}\left(\mathrm{e}^{2x}+1\right)} = \pm 1 .$$

Additionally, the pulse energy E_p (9.10) is given by

$$E_\mathrm{p} \propto \int_{-\infty}^{+\infty} |A(t')|^2 \, \mathrm{d}t' = \tau A_0^2 \tanh\left(\frac{t}{\tau}\right)\Bigg|_{-\infty}^{+\infty} = 2\tau A_0^2 ,$$

whence

$$\int_{-\infty}^{t} |A(t')|^2 \, \mathrm{d}t' = \tau A_0^2 \left[\tanh\left(\frac{t}{\tau}\right) + 1\right] = \frac{E_\mathrm{p}}{2} \left[\tanh\left(\frac{t}{\tau}\right) + 1\right] . \tag{9.31}$$

Bringing together all the terms in the master equation (9.25):

$$\begin{aligned} 0 = g_0 &- \frac{\sigma_\mathrm{L} g_0}{h\nu} \frac{E_\mathrm{p}}{2A_\mathrm{L}} \left[\tanh\left(\frac{t}{\tau}\right) + 1\right] + \frac{\sigma_\mathrm{L}^2 g_0 E_\mathrm{p}^2}{8\,(h\nu)^2 A_\mathrm{L}^2} \left[\tanh\left(\frac{t}{\tau}\right) + 1\right]^2 \\ &- q_\mathrm{A} + \frac{1}{A_\mathrm{A}} \frac{\sigma_\mathrm{A} q_\mathrm{A}}{h\nu} \frac{E_\mathrm{p}}{2} \left[\tanh\left(\frac{t}{\tau}\right) + 1\right] - \frac{\sigma_\mathrm{A}^2 q_\mathrm{A} E_\mathrm{p}^2}{8\,(h\nu)^2 A_\mathrm{A}^2} \left[\tanh\left(\frac{t}{\tau}\right) + 1\right]^2 \\ &+ \frac{g_0}{\Omega_\mathrm{g}^2 \tau^2} \left\{1 - 2\left[1 - \tanh^2\left(\frac{t}{\tau}\right)\right]\right\} - \frac{t_\mathrm{D}}{\tau} \tanh\left(\frac{t}{\tau}\right) . \end{aligned} \tag{9.32}$$

Since (9.32) must be valid for all t, the coefficients of different powers of $\tanh^2(t/\tau)$ must all vanish. The terms proportional to $\tanh^2(t/\tau)$ yield

$$\frac{\sigma_\mathrm{L}^2 g_0 E_\mathrm{p}^2}{8\,(h\nu)^2 A_\mathrm{L}^2} - \frac{\sigma_\mathrm{A}^2 q_\mathrm{A} E_\mathrm{p}^2}{8\,(h\nu)^2 A_\mathrm{A}^2} + \frac{2g_0}{\Omega_\mathrm{g}^2 \tau^2} = 0 . \tag{9.33}$$

This can then be solved for the pulse duration τ, and using (9.7) and $\Omega_\mathrm{g} = \Delta\omega_\mathrm{g}/2$, we obtain

$$\tau = \frac{4}{\pi} \frac{1}{\Delta\nu_\mathrm{g}} \frac{E_\mathrm{sat,L}}{E_\mathrm{p}} \frac{1}{\sqrt{\dfrac{E_\mathrm{sat,L}^2}{E_\mathrm{sat,A}^2} \dfrac{q_\mathrm{A}}{g_0} - 1}} . \tag{9.34}$$

Since $E_{\mathrm{sat,L}} \gg E_{\mathrm{sat,A}}$, the discussion in Sect. 9.3.3, i.e., $s = E_{\mathrm{sat,L}}/E_{\mathrm{sat,A}} \approx 10$–$20$, and $q_{\mathrm{A}} \approx g_0$, the relation (9.34) can be further simplified to

$$\boxed{\tau \approx \frac{4}{\pi} \frac{1}{\Delta\nu_{\mathrm{g}}} \frac{E_{\mathrm{sat,A}}}{E_{\mathrm{p}}}} , \tag{9.35}$$

where $\Delta\nu_{\mathrm{g}}$ is the FWHM bandwidth of the gain medium. Under the condition that the saturable absorber is well saturated, i.e., $E_{\mathrm{p}} > E_{\mathrm{sat,A}}$, it follows that

$$\tau \le \frac{4}{\pi} \frac{1}{\Delta\nu_{\mathrm{g}}} . \tag{9.36}$$

Due to the relatively large gain bandwidth of Rhodamine 6G, i.e., $\Delta\nu_{\mathrm{g}} \approx 4 \times 10^{13}$ Hz, which implies $\Delta\lambda_{\mathrm{g}} \approx 50$ nm, the modelocked pulses become very short, i.e., according to (9.36), $\tau \le 32$ fs, and thus $\tau_{\mathrm{p}} = 1.76\tau \le 56$ fs. The shortest pulses demonstrated were 27 fs [86Val], but more typically pulse durations as short as about 60 fs are observed, which is in good agreement with the theoretical prediction. The lower limit is determined by dispersion compensation.

Martinez [85Mar] has expanded this relatively simple theory and also taken into account SPM and GDD. SPM and negative GDD further shorten the pulses compared to the case when only self-amplitude modulation (SAM) is considered. However, Martinez only predicted an additional pulse-shortening by less than a factor of 2. Later, we will show that, with sufficient negative GDD, shorter pulses can in principle be generated.

9.4 Passive Modelocking with a Fast Saturable Absorber

Here we develop the basic understanding of passive modelocking with a fast saturable absorber which ideally introduces a loss modulation directly proportional to the pulse intensity. In comparison to active modelocking with a sinusoidal loss modulation we obtain a much larger curvature with a fast self-amplitude modulation (SAM) (Figs. 9.1 and 9.13). We therefore expect also much shorter pulses using the gain–loss balance argument discussed for active modelocking. We will see that this is indeed the case and this analogy with active modelocking is correct.

Once again we want to understand how the different cavity elements affect the final pulse duration and we first discuss the solution for the case without SPM and GDD. We then add SPM and GDD with and without soliton pulse shaping. The final analytical solutions are summarized in Sect. 9.6.

Note in this section, we will normalize the pulse envelope as follows:

$$P(t) = |A(t)|^2 .$$

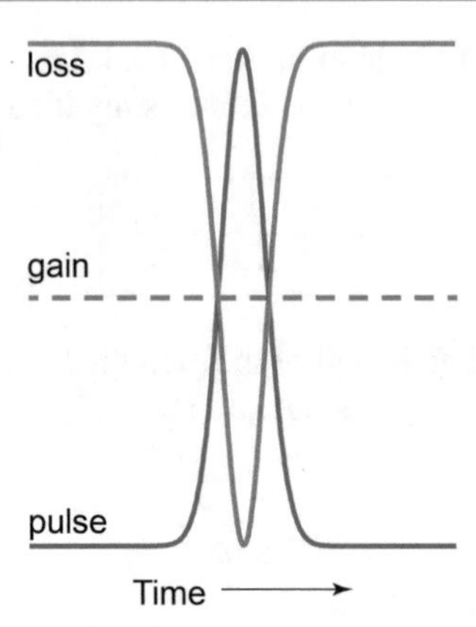

Fig. 9.13 Modelocking with a fast saturable absorber and without any dynamic gain saturation. The gain is only saturated by the average intracavity intensity in typical solid-state lasers (*green dashed line*). Pulse generation is mainly supported by the saturable absorber. The total intracavity loss coefficient (*red line*) is given by $l(t) = l_0 - \bar{\gamma}_A |A(t)|^2$ (9.43) for an ideal fast saturable absorber and is therefore proportional to the instantaneous pulse intensity [75Hau1]

Thus, the SPM coefficient δ (4.5) and the absorber coefficient γ_A (7.28) are replaced by

$$\delta \to \bar{\delta} \equiv \frac{\delta}{A_L}, \qquad \gamma_A \to \bar{\gamma}_A \equiv \frac{\gamma_A}{A_A}. \tag{9.37}$$

We assume here that the dominant SPM is produced in the laser material.

9.4.1 Definition of an Ideally Fast Saturable Absorber

Application: Solid-State Laser

In a solid-state laser, the gain cross-section is typically much smaller than in dye lasers, with values around 10^{-19}cm^2 compared to 10^{-16} cm^2. This means that $E_{\text{sat,L}}$ will be much higher [see (9.7)] and as a result the dynamic gain saturation will be negligible [see (9.11)]. Therefore, we have to assume a constant gain that is saturated by the average intensity inside the gain medium (see Figs. 9.5b and 9.13). In the model of modelocking with a fast saturable absorber, one assumes that the saturable absorber recovers 'ideally' fast (see Sect. 7.2.4 and Fig. 7.3), so that the saturation follows the pulse profile (7.23):

$$q(t) = q_0 - \bar{\gamma}_A |A(t)|^2. \tag{9.38}$$

As mentioned, we assume

$$g(t) = g, \tag{9.39}$$

where g is the cw saturated gain coefficient.

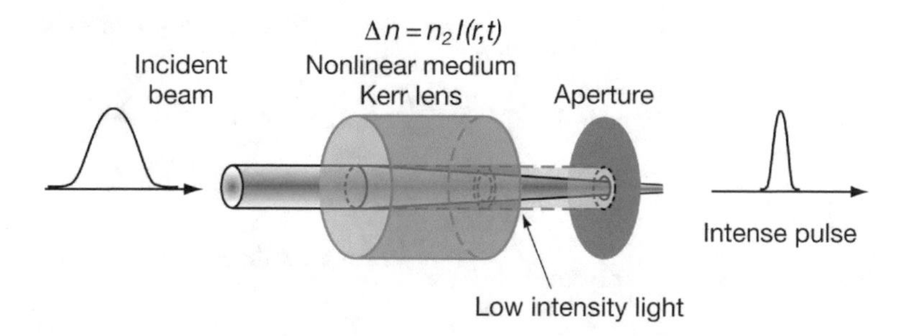

Fig. 9.14 Kerr lens modelocking (KLM) [91Spe, 91Kel, 91Neg, 91Sal1]. KLM is obtained due to the Kerr lens (4.13) at an intracavity focus in the gain medium or in another material, where the refractive index increases with increasing intensity $\Delta n = n_2 I(r,t)$, with n_2 the nonlinear refractive index and $I(r,t)$ the radial- and time-dependent intensity of a short-pulsed laser beam. In combination with a hard aperture inside the cavity, the cavity is designed so that the Kerr lens reduces the laser mode area for high intensities at the aperture and therefore forms an effective fast saturable absorber. In most cases, however, soft-aperture KLM is used where the reduced mode area in the gain medium briefly improves the overlap with the (strongly focused) pump beam, and therefore the effective gain. A significant change in mode size is only achieved by operating the laser cavity near one of the stability limits of the cavity

9.4.2 Example: Kerr Lens Modelocking (KLM)

An example of such a fast saturable absorber is shown in Fig. 9.14 and a Kerr lens modelocked laser system based on a fast saturable absorber is shown in Fig. 9.15. An artificial saturable absorber is formed by the Kerr effect, viz.,

$$\Delta n(r,t) = n_2 \frac{|A(r,t)|^2}{A_\mathrm{L}} , \tag{9.40}$$

and an additional aperture (Fig. 9.14). This saturable absorber is said to be 'artificial', because the Kerr effect alone produces no absorption. Like the linear polarization, the Kerr effect is a reactive process, and thus very fast. Estimates yield a time constant on the order of 1 fs. With this time constant, this artificial saturable absorber fulfills the condition of a fast saturable absorber.

9.4.3 Master Equation

For the case of an 'ideally' fast saturable absorber (see Sect. 7.2.4 and Fig. 7.3, and Sect. 9.4.1) with (9.38), an analytical solution can be found [75Hau1, 91Hau]. According to Table F.4 in Appendix F, we obtain the following terms in the master equation (Fig. 9.15):

- Gain and loss:

$$\Delta A_1 = g \left(1 + \frac{1}{\Omega_\mathrm{g}^2} \frac{\partial^2}{\partial t^2} \right) A - l_0 A . \tag{9.41}$$

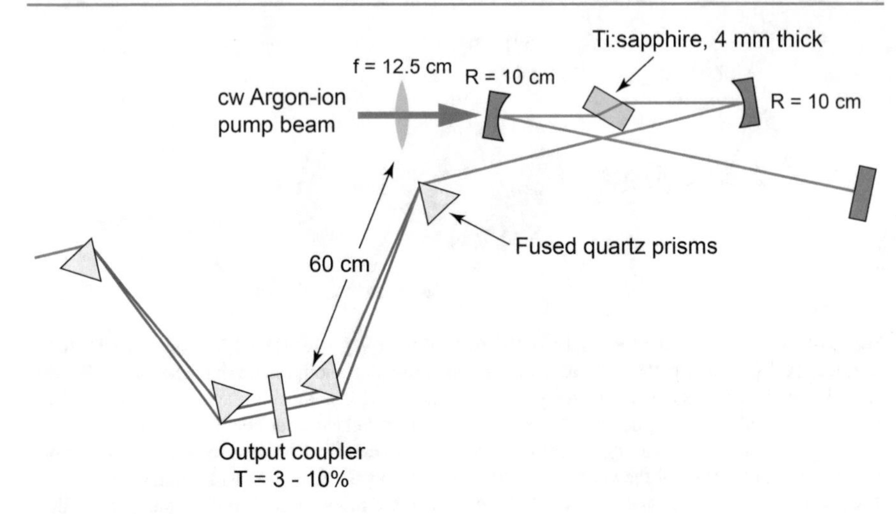

Fig. 9.15 KLM Ti:sapphire laser with prism pairs for dispersion compensation and for pulse generation of roughly 10 fs pulses. Laser mode radii in the crystal are approximately 35 μm × 20 μm [93Asa]

Fig. 9.16 Schematic diagram for the linearized operators acting on the pulse envelope for modelocking with an 'ideally' fast saturable absorber

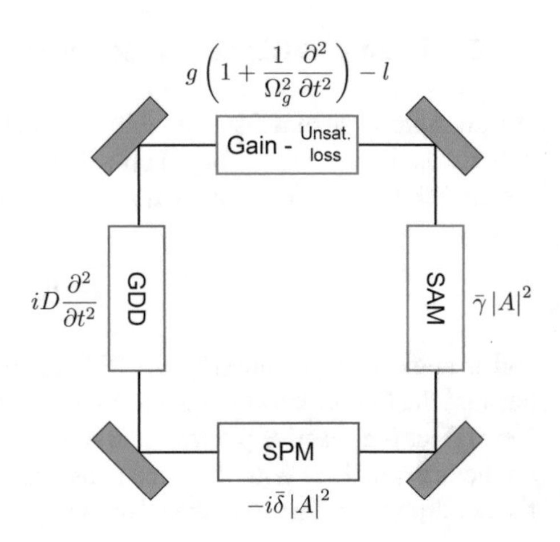

- Ideally fast saturable absorber:

$$\Delta A_3 = \bar{\gamma}_A |A|^2 A \, , \tag{9.42}$$

where, according to (9.38), the total loss is given by

$$\boxed{l(t) = l_0 - \bar{\gamma}_A \, |A(t)|^2} \, . \tag{9.43}$$

Apart from q_0, l_0 also contains all non-saturable resonator loss, e.g., through the output coupler, non-saturable absorber loss, etc.

- SPM:

$$\Delta A_2 = -i\bar{\delta} \, |A|^2 A \, . \tag{9.44}$$

- GDD:

$$\Delta A_4 = \mathrm{i}D\frac{\partial^2}{\partial t^2}A\,. \tag{9.45}$$

- TOD:

$$\Delta A_5 = -D'''\frac{\partial^3}{\partial t^3}A\,. \tag{9.46}$$

- Phase shift:

$$\Delta A_6 = -\mathrm{i}\psi A\,. \tag{9.47}$$

- Master equation:

$$T_{\mathrm{R}}\frac{\partial A(T,t)}{\partial T} = \sum_i \Delta A_i = 0\,. \tag{9.48}$$

9.4.4 Solution without SPM and GDD

From (9.41)–(9.48), it follows that, without GDD and SPM,

$$\boxed{T_{\mathrm{R}}\frac{\partial A}{\partial T} = g\left(1+\frac{1}{\Omega_{\mathrm{g}}^2}\frac{\partial^2}{\partial t^2}\right)A - l_0 A + \bar{\gamma}_{\mathrm{A}}|A|^2 A = 0}\,. \tag{9.49}$$

For the steady-state solution, we have

$$\frac{\partial A}{\partial T} = 0\,.$$

We make the ansatz

$$\boxed{A(t) = A_0\,\mathrm{sech}\left(\frac{t}{\tau}\right)}\,. \tag{9.50}$$

Equation (9.28) implies

$$\left\{g + D_{\mathrm{g}}\frac{1}{\tau^2}\left[1-2\,\mathrm{sech}^2\left(\frac{t}{\tau}\right)\right] - l_0 + \bar{\gamma}_{\mathrm{A}}A_0^2\,\mathrm{sech}^2\left(\frac{t}{\tau}\right)\right\}A = 0\,, \tag{9.51}$$

with $D_{\mathrm{g}} \equiv g/\Omega_{\mathrm{g}}^2$, which yields two equations. One comes from the constant terms:

$$g + D_{\mathrm{g}}\frac{1}{\tau^2} - l_0 = 0\,,$$

whence

$$\boxed{l_0 = g + \frac{D_{\mathrm{g}}}{\tau^2}}\,. \tag{9.52}$$

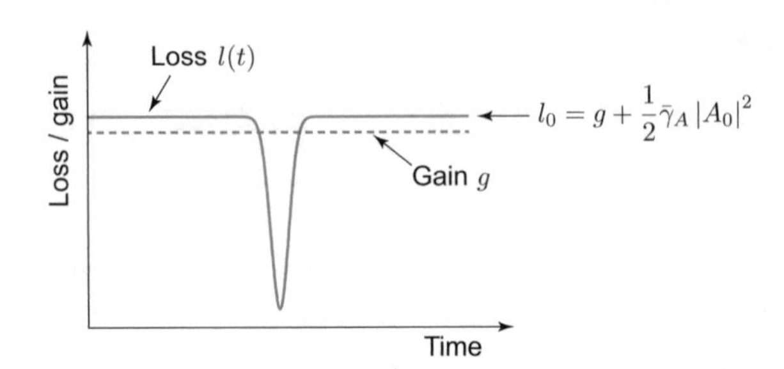

Fig. 9.17 Gain–loss balance. According to (9.54), a necessary stability condition is that $l_0 - g > 0$, where $l(t)$ is given by (9.43)

This is the same result as for active modelocking (6.65) which we will discuss in more detail below. As discussed before, with active modelocking this is not the most physically meaningful gain–loss balance equation because the saturable loss is not explicitly included. It therefore makes more sense to use the identity

$$\bar{\gamma}_A |A_0|^2 = 2D_g/\tau^2 , \tag{9.53}$$

which follows from the terms proportional to $\operatorname{sech}^2(t/\tau)$ in (9.51). We can therefore rewrite (9.52) for a better gain–loss balance equation:

$$\boxed{l_0 = g + \frac{1}{2}\bar{\gamma}_A |A_0|^2} . \tag{9.54}$$

This gain–loss balance equation clearly shows that, without saturation of the absorber, the loss is higher than the saturated gain. Therefore a net gain window is only formed within the short time window of the optical pulse as shown in Fig. 9.17. Equation (9.54) is one necessary stability criterion, but as we will discuss later, not sufficient to generate stable modelocked pulses. In the frequency domain, this gain–loss balance equation is similar to active modelocking as shown in Fig. 6.10, where the loss modulator is replaced by the self-amplitude modulation to increase the gain, such that more axial modes reach lasing threshold and can be modelocked.

The saturated gain level lies only slightly below the unsaturated loss level [see Fig. 9.17 and (9.54)], i.e.,

$$g < l_0 . \tag{9.55}$$

In the limit of very long pulses $\tau \to \infty$, we obtain the cw solution:

$$g = \frac{l_0}{1 + \dfrac{1}{\Omega_g^2 \tau^2}} \implies g = l_0 \text{ for } \tau \to \infty . \tag{9.56}$$

For shorter pulses, the difference between the unsaturated loss l_0 and the saturated gain g becomes larger.

For the pulse length, we have with (9.53)

$$\tau^2 = \frac{2g}{\Omega_g^2 \bar{\gamma}_A |A_0|^2} . \tag{9.57}$$

With the pulse energy E_p given by (4.54),

$$E_p = \int P(t)\mathrm{d}t = \int |A|^2 \mathrm{d}t = A_0^2 \int \mathrm{sech}^2\left(\frac{t}{\tau}\right)\mathrm{d}t = 2A_0^2\tau\ , \tag{9.58}$$

it follows from (9.57) that

$$\boxed{\tau = \frac{4g}{\Omega_g^2 \bar{\gamma}_A E_p} = \frac{4D_g}{\bar{\gamma}_A E_p}}\ . \tag{9.59}$$

9.4.5 Comparison with Active Modelocking without SPM and GDD

First let us compare the final pulse durations obtained with active and passive modelocking. There is one fundamental difference regarding the pulse shape, which is Gaussian for active modelocking (6.62) and a soliton pulse shape for passive modelocking. This results in different numerical factors for the FWHM pulse duration τ_p, i.e., $\tau_p = 1.665\tau$ for a Gaussian pulse (6.61) and $\tau_p = 1.763\tau$ for a soliton (4.48):

$$\tau_{p,\mathrm{Gauss}} = 2\sqrt{\ln 2}\tau \approx 1.665\tau \iff \tau_{p,\mathrm{sech}} = 1.763\tau\ .$$

The pulse shapes can be compared as follows:

$$A(t) = A_0 \exp\left(-\frac{t^2}{2\tau^2}\right) \iff A(t) = A_0\ \mathrm{sech}\left(\frac{t}{\tau}\right)\ .$$

The pulse durations in both cases resulted in

$$\tau^4 = \frac{D_g}{M_s} = \frac{2D_g}{M\omega_m^2} \iff \tau^2 = \frac{2D_g}{\bar{\gamma}_A |A_0|^2}\ ,$$

where M_s is the curvature of the loss modulation (6.58) for the pulse duration (6.56) with active modelocking using a sinusoidal loss modulation with a modulation frequency ω_m. Here we have used (9.53) for the final pulse duration in passive modelocking for this comparison.

Motivated by this comparison we can define a curvature for the fast saturable absorber modulation M_{SAM}:

$$\boxed{\frac{1}{M_s} = \frac{2}{M\omega_m^2} \iff \frac{1}{M_{\mathrm{SAM}}} \equiv \frac{2\tau^2}{\bar{\gamma}_A |A_0|^2}} \tag{9.60}$$

This definition makes sense when we look at Fig. 9.13 and can visualize that the curvature of the loss modulation becomes stronger with shorter pulses and with larger modulation amplitude given by $\bar{\gamma}_{\mathrm{A}}\,|A_0|^2$. We can then make the following parameter comparison:

$$\boxed{M\omega_{\mathrm{m}}^2 \iff \frac{\bar{\gamma}_{\mathrm{A}}\,|A_0|^2}{\tau^2}}\,. \tag{9.61}$$

This analogy means that modelocking with a saturable absorber is similar to modelocking with a loss modulator, only that in the first case the loss modulation is induced by the laser pulse itself. This is therefore also referred to as self-amplitude modulation (SAM). Equation (9.60) thus shows that the effective SAM becomes stronger with shorter pulse durations. This is consistent with the 'curvature' argument in active modelocking: the greater the loss curvature, the shorter the pulses, because the final pulse duration is a balance between the curvature of the gain and the curvature of the loss modulation.

This parameter analogy (9.61) is also consistent with regard to the gain–loss balance equations. With the analogy of (9.61) we can show that the gain–loss balance is consistent for both passive modelocking (9.54), viz.,

$$l_0 = g + \frac{1}{2}\bar{\gamma}_{\mathrm{A}}\,|A_0|^2\,,$$

and active modelocking (6.63), viz.,

$$g = l + \frac{1}{2}M\omega_{\mathrm{m}}^2\tau^2\,.$$

Furthermore, equation (9.52), viz.,

$$l_0 = g + \frac{D_{\mathrm{g}}}{\tau^2}\,,$$

is the same for active modelocking (6.65) and passive modelocking with an ideal fast saturable absorber (without SPM and GDD). This means that for both cases at steady state, an optimum pulse duration is obtained, which is a balance between the pulse shortening of the loss modulation and the pulse broadening of the amplification in the gain material.

9.4.6 Solution with SPM and GDD for Soliton Formation

From (9.41)–(9.48), it follows that, with GDD and SPM,

$$\boxed{T_{\mathrm{R}}\frac{\partial A}{\partial T} = g\left(1 + \frac{1}{\Omega_{\mathrm{g}}^2}\frac{\partial^2}{\partial t^2}\right)A - l_0 A + \bar{\gamma}_{\mathrm{A}}|A|^2 A + \mathrm{i}D\frac{\partial^2}{\partial t^2}A - \mathrm{i}\delta|A|^2 A - \mathrm{i}\psi A = 0}\,. \tag{9.62}$$

Note that we include an additional degree of freedom with a phase shift ψ similar to (6.72) in active modelocking. The steady-state solution of this master equation can be satisfied with the following ansatz:

$$\boxed{A(t) = A_0 \left[\operatorname{sech}\left(\frac{t}{\tau}\right)\right]^{1+\mathrm{i}\beta}} . \tag{9.63}$$

This ansatz can also be transformed to

$$A(t) = A_0 \left[\operatorname{sech}\left(\frac{t}{\tau}\right)\right]^{1+\mathrm{i}\beta} = A_0 \operatorname{sech}\left(\frac{t}{\tau}\right) \exp\left\{\mathrm{i}\beta \ln\left[\operatorname{sech}\left(\frac{t}{\tau}\right)\right]\right\} . \tag{9.64}$$

The last equality is proved by noting that

$$\exp\left\{\mathrm{i}\beta \ln\left[\operatorname{sech}\left(\frac{t}{\tau}\right)\right]\right\} = \exp\left\{\ln\left[\operatorname{sech}\left(\frac{t}{\tau}\right)^{\mathrm{i}\beta}\right]\right\} = \left[\operatorname{sech}\left(\frac{t}{\tau}\right)\right]^{\mathrm{i}\beta} .$$

Equation (9.64) shows that, with this ansatz, the chirp is not linear. The fact that this ansatz satisfies the master equation was first published by Martinez et al. [85Mar].

We will now calculate the first and second derivatives of the above ansatz (9.63). It follows from (9.27) that

$$\frac{\mathrm{d}}{\mathrm{d}t}A(t) = A_0\,(1+\mathrm{i}\beta)\left[\operatorname{sech}\left(\frac{t}{\tau}\right)\right]^{\mathrm{i}\beta}\left[-\frac{1}{\tau}\tanh\left(\frac{t}{\tau}\right)\operatorname{sech}\left(\frac{t}{\tau}\right)\right] , \tag{9.65}$$

and from (9.63),

$$\tau\frac{\mathrm{d}}{\mathrm{d}t}A(t) = -\,(1+\mathrm{i}\beta)\tanh\left(\frac{t}{\tau}\right)A(t) . \tag{9.66}$$

From this it follows that

$$\begin{aligned}
\tau\frac{\mathrm{d}^2}{\mathrm{d}t^2}A(t) &= -\,(1+\mathrm{i}\beta)\frac{\mathrm{d}}{\mathrm{d}t}\left[\tanh\left(\frac{t}{\tau}\right)A(t)\right] \\
&= -\,(1+\mathrm{i}\beta)\left[A(t)\frac{\mathrm{d}}{\mathrm{d}t}\tanh\left(\frac{t}{\tau}\right) + \tanh\left(\frac{t}{\tau}\right)\frac{\mathrm{d}}{\mathrm{d}t}A(t)\right] .
\end{aligned}$$

Moreover (see Appendix E),

$$\frac{\mathrm{d}}{\mathrm{d}x}\tanh(x) = \operatorname{sech}^2(x) ,$$

equation (9.66) implies

$$\tau^2\frac{\mathrm{d}^2}{\mathrm{d}t^2}A(t) = -\,(1+\mathrm{i}\beta)\left\{A(t)\operatorname{sech}^2\left(\frac{t}{\tau}\right) + \tanh\left(\frac{t}{\tau}\right)\left[-\,(1+\mathrm{i}\beta)\tanh\left(\frac{t}{\tau}\right)A(t)\right]\right\} . \tag{9.67}$$

Then using (see Appendix E)

$$\tanh^2(x) = 1 - \mathrm{sech}^2(x) \;,$$

equation (9.67) implies

$$\begin{aligned}
\tau^2 \frac{\mathrm{d}^2}{\mathrm{d}t^2} A(t) &= -(1+\mathrm{i}\beta)\, A(t) \left\{ \mathrm{sech}^2\left(\frac{t}{\tau}\right) - (1+\mathrm{i}\beta)\left[1 - \mathrm{sech}^2\left(\frac{t}{\tau}\right)\right]\right\} \\
&= -(1+\mathrm{i}\beta)\, A(t) \left[(2+\mathrm{i}\beta)\,\mathrm{sech}^2\left(\frac{t}{\tau}\right) - (1+\mathrm{i}\beta)\right] \\
&= A(t)\left[-\left(2+3\mathrm{i}\beta-\beta^2\right)\mathrm{sech}^2\left(\frac{t}{\tau}\right) + (1+\mathrm{i}\beta)^2\right] .
\end{aligned} \tag{9.68}$$

Inserting our ansatz into the master equation (9.62) thus yields

$$\begin{aligned}
0 = g &+ \frac{D_\mathrm{g}}{\tau^2}\left[-\left(2+3\mathrm{i}\beta-\beta^2\right)\mathrm{sech}^2\left(\frac{t}{\tau}\right) + (1+\mathrm{i}\beta)^2\right] - l_0 \\
&+ \bar{\gamma}_\mathrm{A} A_0^2\,\mathrm{sech}^2\left(\frac{t}{\tau}\right) + \mathrm{i}\frac{D}{\tau^2}\left[-\left(2+3\mathrm{i}\beta-\beta^2\right)\mathrm{sech}^2\left(\frac{t}{\tau}\right) + (1+\mathrm{i}\beta)^2\right] \\
&- \mathrm{i}\bar{\delta} A_0^2\,\mathrm{sech}^2\left(\frac{t}{\tau}\right) - \mathrm{i}\psi \;.
\end{aligned} \tag{9.69}$$

By comparing coefficients, the constant terms in (9.69) yield

$$-\,\mathrm{i}\psi + g - l_0 + \frac{(1+\mathrm{i}\beta)^2}{\tau^2}\left(D_\mathrm{g} + \mathrm{i}D\right) = 0 \;, \tag{9.70}$$

while the $\mathrm{sech}^2(t/\tau)$ terms in (9.69) yield

$$\frac{1}{\tau^2}\left(D_\mathrm{g} + \mathrm{i}D\right)\left(2+3\mathrm{i}\beta-\beta^2\right) = \left(\bar{\gamma}_\mathrm{A} - \mathrm{i}\bar{\delta}\right) A_0^2 \;. \tag{9.71}$$

The two complex equations (9.70) and (9.71) result in four real equations for the four (real) unknowns ψ, β, τ, and A_0^2.

The imaginary part of (9.70) only determines the phase shift ψ between two consecutive pulses. However, the real part of (9.70) describes the balance between gain and loss in the laser:

$$\boxed{g - l_0 + \frac{1-\beta^2}{\tau^2} D_\mathrm{g} - \frac{2\beta D}{\tau^2} = 0} \;. \tag{9.72}$$

This equation may be confusing at first glance, because it contains the dispersion, which is actually loss-free, but it does not include the gain through saturated absorption [similar to (9.52)]. Note that the dispersion changes the chirp and the pulse duration, and therefore indirectly also the gain. Generally, it makes sense physically to set up the gain–loss balance as a function of the gain and the saturable absorber, as shown later in (9.98).

The real part of (9.71) results in

$$\frac{1}{\tau^2}\left[\frac{g}{\Omega_g^2}\left(2-\beta^2\right)-3\beta D\right]=\bar{\gamma}_A A_0^2\,, \tag{9.73}$$

and thus

$$\frac{1}{\tau^2}=\frac{\bar{\gamma}_A A_0^2}{\left[\frac{g}{\Omega_g^2}\left(2-\beta^2\right)-3\beta D\right]}\,. \tag{9.74}$$

The right-hand side of (9.73) describes the pulse shortening per roundtrip, and the terms on the left-hand side describe the pulse broadening due to the gain filter D_g and the dispersion D. Note that, in the case of a stronger chirp $\beta^2 > 2$, the gain filter also shortens the pulse.

The imaginary part of (9.71) yields

$$3\frac{1}{\tau^2}D_g\beta+\frac{1}{\tau^2}D\left(2-\beta^2\right)=-\bar{\delta}A_0^2\,,$$

and thus a quadratic equation for β:

$$\beta^2-3\frac{D_g}{D}\beta-2-\frac{\bar{\delta}}{D}A_0^2\tau^2=0\,.$$

The solution of this quadratic equation is given by

$$\boxed{\beta=\frac{1}{2}\left[3\frac{D_g}{D}\pm\sqrt{9\left(\frac{D_g}{D}\right)^2+8+4\frac{\bar{\delta}A_0^2\tau^2}{D}}\right]}\,. \tag{9.75}$$

The signs are chosen such that the solution makes sense physically, e.g., the sign in the solution for β is chosen such that the normalized pulse duration is positive [see (9.81) below].

Special Case of an Unchirped Soliton, i.e., $\beta = 0$

From (9.72), it follows that, for $\beta = 0$,

$$\boxed{g-l_0+\frac{g}{\Omega_g^2}\frac{1}{\tau^2}=0}\,, \tag{9.76}$$

which is identical to (9.52). Further, for $\beta = 0$, (9.73) implies

$$\frac{1}{\tau^2}\frac{2g}{\Omega_g^2} = \bar{\gamma}_A A_0^2 ,$$

and using (9.58),

$$\frac{1}{\tau}\frac{4g}{\Omega_g^2} = \bar{\gamma}_A E_p .$$

Hence,

$$\boxed{\tau = \frac{4g}{\Omega_g^2 \bar{\gamma}_A E_p} = \frac{4D_g}{\bar{\gamma}_A E_p}} . \tag{9.77}$$

As expected, when $\beta = 0$, the result reduces to (9.59). However, the soliton condition must also be fulfilled. According to the solution ansatz of (9.63), the solution also corresponds to an unchirped soliton pulse when $\beta = 0$. According to (4.57), we have for the soliton

$$\tau = \frac{4|D|}{\bar{\delta} E_p} . \tag{9.78}$$

If we compare this with the unchirped solution for a fast saturable absorber (9.77), we must have

$$\boxed{\tau = \frac{4|D|}{\bar{\delta} E_p} = \frac{4D_g}{\bar{\gamma}_A E_p}} , \tag{9.79}$$

and thus for an unchirped soliton,

$$\boxed{\frac{|D|}{\bar{\delta}} = \frac{D_g}{\bar{\gamma}_A}} . \tag{9.80}$$

This means that for an unchirped soliton the ratio of dispersion and SPM is perfectly balanced with the ratio of gain dispersion and self-amplitude modulation (SAM). This is also an additional motivation for introducing the parameter D_g as the gain dispersion.

Discussion of the General Solution, i.e., $\beta \neq 0$

For the discussion of the results, it is useful to introduce the following parameters:

- The normalized dispersion D_n defined by

$$\boxed{D_n \equiv \frac{D}{g/\Omega_g^2} = \frac{D}{D_g}} . \tag{9.81}$$

- The normalized pulse length τ_n defined by

$$\boxed{\tau_n \equiv \frac{E_p}{2D_g}\tau} . \tag{9.82}$$

With this, (9.71) becomes

$$\frac{1}{\tau_n}(1+iD_n)\left(2+3i\beta-\beta^2\right) = \bar{\gamma}_A - i\bar{\delta} . \tag{9.83}$$

Hence, for the real part,

$$\frac{1}{\tau_n}\left(2-\beta^2\right) - \frac{D_n}{\tau_n}3\beta = \bar{\gamma}_A , \tag{9.84}$$

and for the imaginary part,

$$\frac{D_n}{\tau_n}\left(2-\beta^2\right) + \frac{1}{\tau_n}3\beta = -\bar{\delta} . \tag{9.85}$$

Dividing (9.84) by (9.85), we obtain

$$\begin{aligned}
&\frac{\left(2-\beta^2\right) - D_n 3\beta}{D_n\left(2-\beta^2\right)+3\beta} = -\frac{\bar{\gamma}_A}{\bar{\delta}} , \\
&\bar{\delta}\left(2-\beta^2\right) - \bar{\delta}D_n 3\beta = -\bar{\gamma}_A D_n\left(2-\beta^2\right) - \bar{\gamma}_A 3\beta , \\
&\left(\bar{\delta}+\bar{\gamma}_A D_n\right)\left(2-\beta^2\right) = 3\beta\left(\bar{\delta}D_n - \bar{\gamma}_A\right) ,
\end{aligned}$$

and therefore define a new parameter χ as follows:

$$\boxed{\frac{3\beta}{2-\beta^2} = \frac{\bar{\delta}+\bar{\gamma}_A D_n}{\bar{\delta}D_n - \bar{\gamma}_A} \equiv \frac{1}{\chi}} . \tag{9.86}$$

For the special case $\beta = 0$, this implies

$$\frac{\bar{\delta}+\bar{\gamma}_A D_n}{\bar{\delta}D_n - \bar{\gamma}_A} = 0 ,$$

and thus

$$\bar{\delta}+\bar{\gamma}_A D_n = \bar{\delta}+\bar{\gamma}_A\frac{D}{D_g} = 0 .$$

So finally, for $\beta = 0$,

$$-\frac{D}{\bar{\delta}} = \frac{D_g}{\bar{\gamma}_A} \Longrightarrow -D_n = \frac{\bar{\delta}}{\bar{\gamma}_A} , \tag{9.87}$$

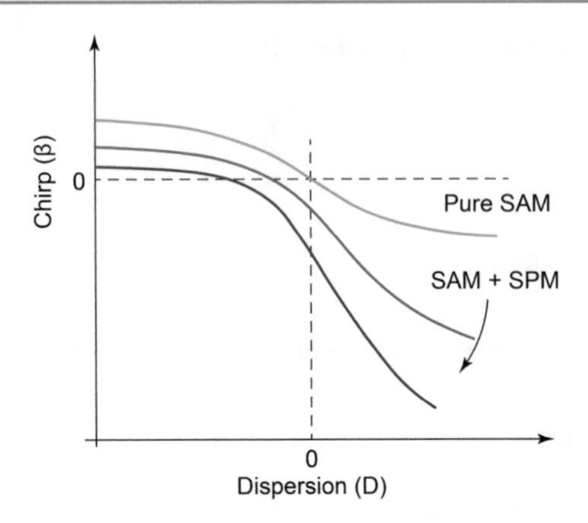

Fig. 9.18 Chirp as a function of the dispersion (see [91Hau] and [94Ipp, Fig. 11]). Self-phase modulation (SPM) increases the chirp in the positive GDD regime. Self-amplitude modulation (SAM) has no big influence on the chirp in the negative GDD regime, but in the positive GDD regime the chirp is reduced by a stronger SAM [91Hau, Fig. 3]. An unchirped pulse, i.e., $\beta = 0$, is a soliton (9.79)

which is identical to the solution of (9.80).

We can solve (9.86) for the chirp parameter:

$$3\beta\chi - 2 + \beta^2 = 0\,.$$

Hence,

$$\boxed{\beta = -\frac{3}{2}\chi \pm \sqrt{\left(\frac{3}{2}\chi\right)^2 + 2}}\,. \tag{9.88}$$

Equation (9.84) implies

$$\tau_{\mathrm{n}} = \frac{2 - \beta^2 - 3\beta D_{\mathrm{n}}}{\bar{\gamma}_{\mathrm{A}}}\,, \tag{9.89}$$

which is identical to (9.74). The sign in the solution for β is chosen such that the normalized pulse duration is positive. The conditions are summarized in Table 1 in the paper of Haus et al. [91Hau].

Thus, using (9.89), we have

$$\tau_{\mathrm{n}} = \frac{-2D_{\mathrm{n}} + D_{\mathrm{n}}\beta^2 - 3\beta}{\bar{\delta}}\,. \tag{9.90}$$

Equations (9.85) and (9.89) imply

$$\boxed{\tau_{\mathrm{n}} = \frac{2 - \beta^2 - 3\beta D_{\mathrm{n}}}{\bar{\gamma}_{\mathrm{A}}} = \frac{-2D_{\mathrm{n}} + D_{\mathrm{n}}\beta^2 - 3\beta}{\bar{\delta}}}\,. \tag{9.91}$$

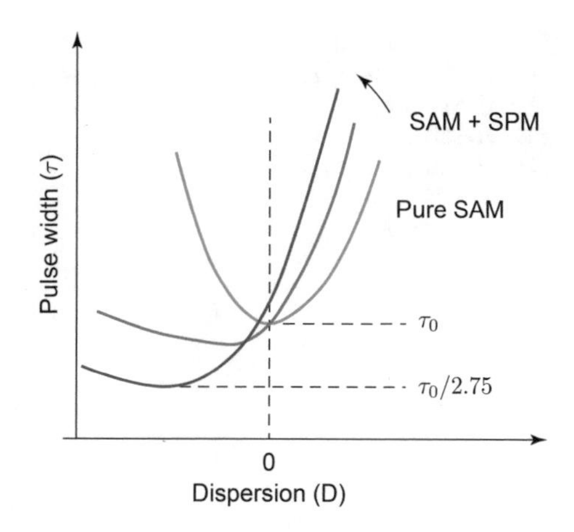

Fig. 9.19 Pulse duration as a function of group delay dispersion (GDD). Self-amplitude modulation (SAM). Self-phase modulation (SPM). See [94Ipp, Fig. 11] and [91Hau]

Positive GDD regime ($D_n > 1$). Equations (9.86) and (9.88) show that the pulse is strongly chirped [91Hau, Figs. 2 and 3]. Usually, a filter broadens a pulse (i.e., there is gain dispersion), but if this pulse is strongly chirped, a filter can further shorten it.

Influence of the SAM strength on the pulse duration. In the negative GDD regime, the pulse duration does not depend very strongly on SAM, but in the positive GDD regime, one can obtain much shorter pulses with a strong SAM.

Influence of the SPM strength on the pulse duration. For both GDD regimes, the strength of SPM has a considerable influence on the pulse duration. In the negative GDD regime, SPM shortens the pulse duration, but in the positive GDD regime the pulse duration is longer (see Fig. 9.19).

Stability of the Solution

The solution given by (9.91) only makes sense physically if the solution is stable. For example, for active modelocking we showed that a Gaussian pulse is the only stable solution to the master equation (Chap. 6). This is generally difficult to show. A simple criterion is given by the condition that the loss for cw perturbations l_{cw} is greater than the temporally averaged loss for the pulses, i.e.,

$$l_{\text{cw}} > \frac{1}{T_{\text{R}}} \int_0^{T_{\text{R}}} l(t) \text{d}t \,. \tag{9.92}$$

Using (9.56), viz.,

$$g = \frac{1}{T_{\text{R}}} \int_0^{T_{\text{R}}} g(t) \text{d}t = \frac{1}{T_{\text{R}}} \int_0^{T_{\text{R}}} l(t) \text{d}t \,,$$

we have

$$l_{\text{cw}} > g \implies g - l_{\text{cw}} < 0 \,. \tag{9.93}$$

Then from (9.72) with $l_{\text{cw}} \approx l_0$,

$$g - l_0 = -\left(1 - \beta^2\right) \frac{g}{\Omega_{\text{g}}^2 \tau^2} + \frac{2\beta D}{\tau^2} < 0 \,, \tag{9.94}$$

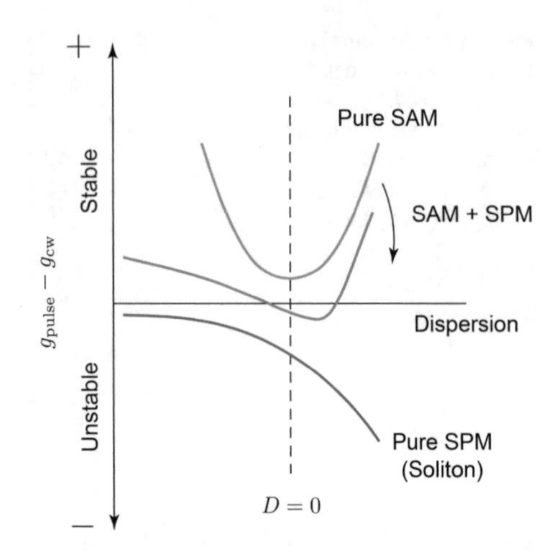

Fig. 9.20 Stability of modelocking according to (9.95) (see [94Ipp, Fig. 11] and [91Hau]). Self-amplitude modulation (SAM). Self-phase modulation (SPM)

and thus

$$\boxed{S \equiv \left(1-\beta^2\right) - 2\beta D_{\mathrm{n}} > 0}\,, \tag{9.95}$$

where S is the stability parameter (Fig. 9.20).

These results are discussed further in Haus et al. [91Hau]. Compare also with Sects. 6.6 and 9.5.

Balance between Gain and Loss (An Alternative Approach)

As already mentioned, (9.72) may seem confusing at first glance, because the dispersion should not directly enter the balance between loss and gain. However, the dispersion changes the chirp and the pulse duration, and therefore indirectly also the gain. Generally, it makes sense physically to set up the gain–loss balance as a function of the gain and the saturable absorber.

We therefore discuss an alternative approach and derive the equation for the energy balance of the pulse starting from the master equation (9.62) (by analogy with Sect. 6.5.2c):

$$T_{\mathrm{R}} \frac{\partial}{\partial T} \int_{-\infty}^{+\infty} |A(T,t)|^2 \,\mathrm{d}t = 0\,. \tag{9.96}$$

This equation simplifies to

$$\int \left(\frac{\partial A}{\partial T} A^* + A \frac{\partial A^*}{\partial T} \right) \mathrm{d}t = 2\mathrm{Re} \int \frac{\partial A}{\partial T} A^* \mathrm{d}t = 0\,.$$

Using the master equation (9.49), we have

$$\begin{aligned}
2\mathrm{Re} \int T_{\mathrm{R}} \frac{\partial A}{\partial T} A^* \mathrm{d}t &= 2 \int \left(g A A^* + \frac{g}{\Omega_{\mathrm{g}}^2} A^* \frac{\partial^2}{\partial t^2} A - l_0 A A^* + \bar{\gamma}_{\mathrm{A}} |A|^2 A A^* \right) \mathrm{d}t \\
&= 2(g - l_0) \int |A|^2 \mathrm{d}t + \frac{2g}{\Omega_{\mathrm{g}}^2} \int A^* \frac{\partial^2}{\partial t^2} A \mathrm{d}t + 2\bar{\gamma}_{\mathrm{A}} \int |A|^4 \mathrm{d}t\,.
\end{aligned}$$

By integration by parts of the second integral,

$$T_{\mathrm{R}}\frac{\partial}{\partial T}\int_{-\infty}^{+\infty}|A(T,t)|^2\,\mathrm{d}t = 2\,(g-l_0)\int|A|^2\mathrm{d}t - \frac{2g}{\Omega_{\mathrm{g}}^2}\int\left|\frac{\partial A}{\partial t}\right|^2\mathrm{d}t + 2\bar{\gamma}_{\mathrm{A}}\int|A|^4\mathrm{d}t\,. \quad (9.97)$$

Using the solution ansatz (9.63), we obtain

$$\int|A|^2\mathrm{d}t = A_0^2\int\mathrm{sech}^2\left(\frac{t}{\tau}\right)\mathrm{d}t = A_0^2\tau\tanh\left(\frac{t}{\tau}\right)\Bigg|_{-\infty}^{+\infty} = 2A_0^2\tau\,.$$

Equation (9.66) then yields

$$\int\left|\frac{\partial A}{\partial t}\right|^2\mathrm{d}t = \int\left|-\frac{1}{\tau}\,(1+\mathrm{i}\beta)\tanh\left(\frac{t}{\tau}\right)A(t)\right|^2\mathrm{d}t$$
$$= \frac{A_0^2}{\tau^2}\left(1+\beta^2\right)\int\tanh^2\left(\frac{t}{\tau}\right)\mathrm{sech}^2\left(\frac{t}{\tau}\right)\mathrm{d}t\,.$$

Using $\tanh^2(x) = 1-\mathrm{sech}^2(x)$, we have

$$\int\left|\frac{\partial A}{\partial t}\right|^2\mathrm{d}t = \frac{A_0^2}{\tau^2}\left(1+\beta^2\right)\left[\int\mathrm{sech}^2\left(\frac{t}{\tau}\right)\mathrm{d}t - \int\mathrm{sech}^4\left(\frac{t}{\tau}\right)\mathrm{d}t\right].$$

Since

$$\int\mathrm{sech}^2\left(\frac{t}{\tau}\right)\mathrm{d}t = 2\tau$$

and

$$\int\mathrm{sech}^4\left(\frac{t}{\tau}\right)\mathrm{d}t = \int\cosh^{-4}\left(\frac{t}{\tau}\right)\mathrm{d}t$$
$$= \frac{\tau}{3}\sinh\left(\frac{t}{\tau}\right)\mathrm{sech}^3\left(\frac{t}{\tau}\right)\Bigg|_{-\infty}^{+\infty} + \frac{2}{3}\int\mathrm{sech}^2\left(\frac{t}{\tau}\right)\mathrm{d}t = \frac{4\tau}{3}\,,$$

it follows that

$$\int\left|\frac{\partial A}{\partial t}\right|^2\mathrm{d}t = \frac{A_0^2}{\tau^2}\left(1+\beta^2\right)\left[\int\mathrm{sech}^2\left(\frac{t}{\tau}\right)\mathrm{d}t - \int\mathrm{sech}^4\left(\frac{t}{\tau}\right)\mathrm{d}t\right] = \frac{2A_0^2\tau}{3\tau^2}\left(1+\beta^2\right).$$

Equation (9.97) can then be written, using

$$\int|A|^4\mathrm{d}t = \frac{4\tau}{3}A_0^4\,,$$

as follows:

$$T_{\mathrm{R}}\frac{\partial}{\partial T}\int_{-\infty}^{+\infty}|A(T,t)|^2\,\mathrm{d}t=\left[2\left(g-l_0\right)-\frac{2}{3}\frac{g}{\Omega_{\mathrm{g}}^2\tau^2}\left(1+\beta^2\right)+\frac{4}{3}\bar{\gamma}_{\mathrm{A}}A_0^2\right]2A_0^2\tau\ .$$

At steady state, this gives

$$T_{\mathrm{R}}\frac{\partial}{\partial T}\int_{-\infty}^{+\infty}|A(T,t)|^2\,\mathrm{d}t=0\ ,$$

and hence finally,

$$\boxed{l_0=g-\frac{1}{3}\frac{g}{\Omega_{\mathrm{g}}^2\tau^2}\left(1+\beta^2\right)+\frac{2}{3}\bar{\gamma}_{\mathrm{A}}A_0^2}\ . \tag{9.98}$$

Inserting (9.73) into (9.98), we again obtain (9.72). For the special case $\beta=0$ and $\delta=0$, (9.98) and (9.53) imply (9.52). Equation (9.98) implies that the pulses are stabilized by the saturable absorber, i.e., $l-g>0$, and that chirp and filter, i.e., gain dispersion, have a destabilizing effect. The gain is saturated by the pulse energy E_{p}:

$$g=\frac{g_0}{1+E_{\mathrm{p}}/E_{\mathrm{sat,L}}}\ . \tag{9.99}$$

9.4.7 Problem for Self-Starting Modelocking

A KLM Ti:sapphire laser does not usually start by itself, and additional methods have to be employed to guarantee reliable starting of passive modelocking. Typical methods are acousto-optic loss modulators, additional shaking of resonator elements to induce increased intensity fluctuations, and therefore a stronger noise pulse that can more strongly saturate the saturable absorber, and initiate the modelocking process. The reason why modelocking with a fast saturable absorber does not easily start by itself is shown schematically in Fig. 9.21.

Ippen introduced a phenomenological modelocking driving force (Fig. 9.21). For an ideally fast saturable absorber, the self-amplitude modulation scales with peak intensity (9.38), which is very weak for long pulses and increases inversely proportional to the pulse duration. Therefore the modelocking driving force is small for long noise pulses and self-starting of modelocking becomes more challenging.

For a slow saturable absorber the self-amplitude modulation does not depend on the pulse shape and pulse duration, but only on the pulse energy (7.14). Therefore the modelocking driving force is constant for all pulse durations. Typically, reliable self-starting modelocking is obtained.

Active modelocking does not depend on a self-amplitude modulation. The sinusoidal loss modulation with a period given by the cavity roundtrip time (Fig. 6.4)

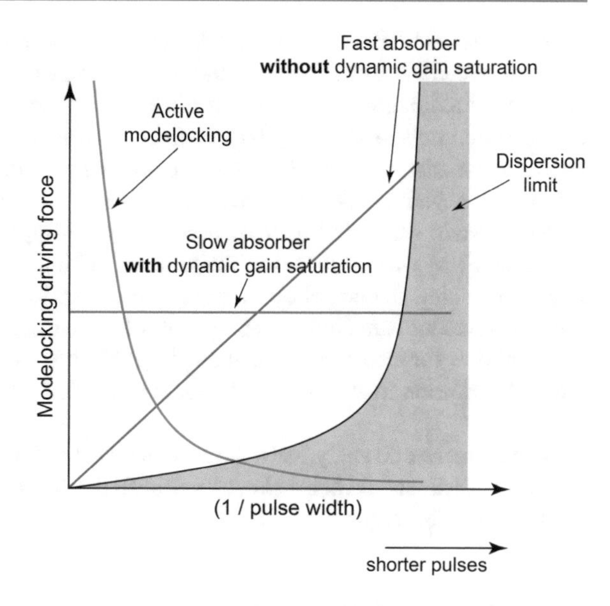

Fig. 9.21 Modelocking driving force. After Ippen [94Ipp, Fig. 10]

results in a large loss for very long pulses and becomes negligible for very short pulses. As the pulses become shorter, dispersive broadening becomes a strong limitation for further pulse shortening.

9.5 Passive Modelocking with Slow Saturable Absorber and Without Dynamic Gain Saturation

Here we address the question of how a slow saturable absorber can support much shorter pulses than the recovery time of the saturable absorber without the help of dynamic gain saturation. This is the case for typical modelocked solid-state lasers with a constant gain saturation by the average intensity inside the laser gain medium. In this case there is a net gain window that remains open for much longer than the final pulse duration. How is this possible? What is the pulse shaping mechanism for the trailing edge of the pulse? Why does the noise not grow after the pulse that destabilizes the shorter pulse?

As we will learn, soliton modelocking is the solution which occurs with a balance between GDD and SPM. Solitons are described in Sect. 4.3. Typical material properties result in positive SPM, i.e., a nonlinear refractive index $n_2 > 0$, and require negative GDD for soliton formation. How to obtain negative GDD is explained in Chap. 3. There are good reasons why soliton formation with positive GDD can be interesting. However, this requires negative SPM (i.e., $n_2 < 0$). We will show that cascaded quadratic nonlinearities can provide an interesting solution. Soliton modelocking is very successful in comparison to fast saturable absorber modelocking because the requirements on the intracavity saturable absorber are fully relaxed and self-starting of modelocking is typically not a problem.

What was new about soliton modelocking? Previously it was shown that SPM and negative GDD further shorten the pulses compared to the case when only self-amplitude modulation (SAM) is considered. For example, Martinez only predicted an additional pulse-shortening by less than a factor of 2 for dye lasers [85Mar]. In addition, for fast saturable absorber modelocking it was predicted that SPM and SAM with negative GDD can only reduce the pulse duration by about a factor of 2.75 at most, as shown in Fig. 9.19, according to [94Ipp] and [91Hau]. Furthermore, too much SPM was predicted to make modelocking unstable even before the full pulse shortening can be achieved, as shown in Fig. 9.20. The following discussion will show that we can obtain an analytical solution using soliton perturbation theory, which allows for stable pulses that are 10 to 30 times shorter than the SAM action alone. In addition much more SPM can be accepted before modelocking becomes unstable.

We summarize all analytical solutions for all the different modelocking techniques in Sect. 9.6. This should help to keep track of all the different solutions for the different modelocking techniques.

9.5.1 Soliton Modelocking

In soliton modelocking, the pulse shaping is done solely by soliton formation, i.e., the balance of group delay dispersion (GDD) and self-phase modulation (SPM) at steady state, with no additional requirements on the cavity stability regime, such as KLM, for example. In contrast to KLM, we use only the time-dependent part of the Kerr effect at the peak intensity and not the radially dependent part. The radially dependent part of the Kerr effect is responsible for KLM because it forms the nonlinear lens that reduces the beam diameter at an intracavity aperture inside the gain medium (Fig. 9.14). Thus, this nonlinear lens forms an effective fast saturable absorber because the intensity-dependent beam diameter reduction at an aperture introduces less loss at high intensity and more loss at low intensity. However, such a radially dependent effective saturable absorber couples the modelocking mechanism with the cavity mode. In contrast, soliton modelocking does not depend on the transverse Kerr effect, and so has the advantage that the modelocking mechanism is decoupled from the cavity design, and no critical cavity stability regime is required—it basically works over the full cavity stability range.

Soliton modelocking was discovered because of the unexpected short pulse durations of 130 fs from a SESAM-modelocked Nd:glass laser in 1993 [93Kel]. The initial explanation at that time of a novel Kerr-shift modelocking technique was too strongly focused on the intracavity knife-edge gain filter (see Figs. 6.20 and 6.21), and did not fully reveal the new pulse-shortening opportunities with stronger SPM and slow saturable absorbers. Additional experiments with a SESAM-modelocked diode pumped Cr:LiSAF laser [94Kop1] also resulted in much shorter pulses than expected from prior theories (see Sect. 9.3.4) [85Mar].

At this point in time, many experts in the field believed that there might be a hidden fast saturable absorber inside the SESAM helping the modelocking pro-

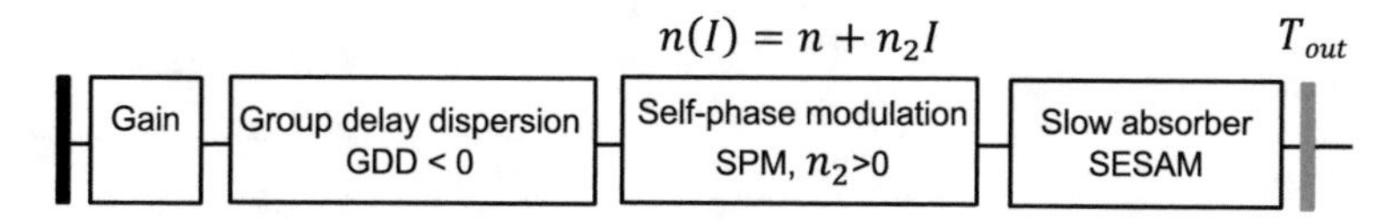

Fig. 9.22 The different effects acting on the pulse envelope inside the resonator

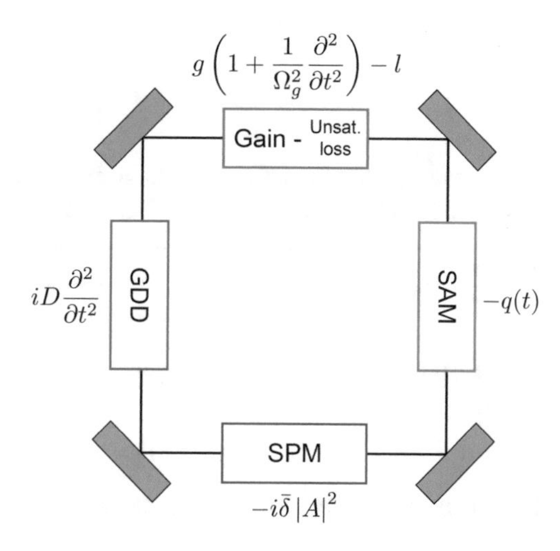

Fig. 9.23 Schematic diagram for all actions on the pulse envelope for soliton modelocking with positive SPM and negative GDD. In contrast to Fig. 6.16, we replace the modulator by a slow saturable absorber $q(t)$

cess. This motivated us to perform a more systematic study using both active modelocked and SESAM-modelocked Nd:glass lasers (see Sect. 6.6) [94Kop2], and a more detailed characterization of the SESAM recovery times [94Kel]. Even with active modelocking, much shorter pulses were obtained which were not in agreement with theoretical predictions.

To explain the results we used soliton perturbation calculations that had been developed for optical communication. This theory was fundamentally different from prior theoretical predictions [85Mar] because intracavity SPM allowed for much more pulse shortening before the onset of SPM-induced instabilities. We show next that this strongly relaxes the requirements on the saturable absorber that only starts and stabilizes soliton modelocking, and that a slow saturable absorber is sufficient for ultrashort pulse durations that can be more than 10 times shorter than the recovery time of the saturable absorber. This was then also confirmed with a specially designed slow SESAM in a soliton modelocked Ti:sapphire laser (see Sect. 9.5.2).

In soliton modelocking, an additional loss mechanism, such as a saturable absorber [95Kar1], or an acousto-optic modulator [95Kar2], is necessary to start the modelocking process and to stabilize the soliton pulse-forming process. Figure 9.22 shows all the elements inside a laser resonator required for stable soliton modelocking. We typically use the gain material for SPM with a positive nonlinear refractive index $n_2 > 0$. For soliton formation, we therefore need negative GDD, which is typically introduced with prism pairs or dispersive mirrors (see Chap. 3). Similarly to Sect. 6.6, we can use the master equation with the linearized operators, as shown in Fig. 9.23.

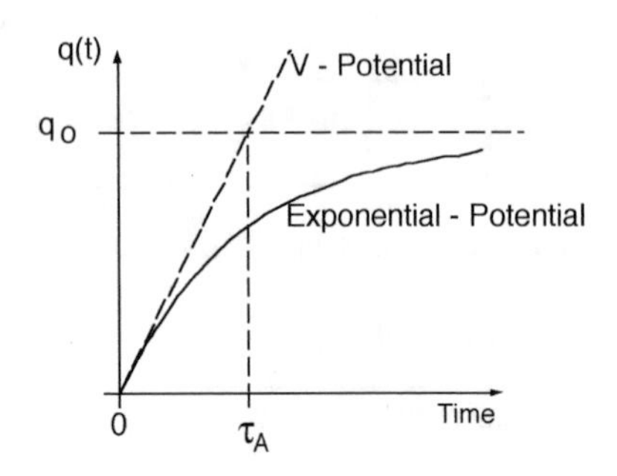

Fig. 9.24 Slow saturable absorber with an exponential decay

In analogy with Sect. 6.6, we can derive the master equation for a slow saturable absorber with SPM and negative GDD [see (6.117)]:

$$\boxed{T_R \frac{\partial}{\partial T} A(T,t) = \left[\mathrm{i}D \frac{\partial^2}{\partial t^2} - \mathrm{i}\delta \left| A(T,t) \right|^2 \right] A(T,t) + \left[g - l + D_g \frac{\partial^2}{\partial t^2} - q(t) \right] A(T,t) = 0}\,, \tag{9.100}$$

where $q(t)$ describes the time-dependent loss of the slow saturable absorber [see (7.13)]:

$$A_{\mathrm{out}}(T,t) = \mathrm{e}^{-q(t)} A_{\mathrm{in}}(T,t)\,. \tag{9.101}$$

For small loss modulation, we also have

$$A_{\mathrm{out}}(T,t) = [1 - q(t)]\, A_{\mathrm{in}}(T,t) \implies \Delta A(T,t) = -q(t) A(T,t)\,. \tag{9.102}$$

The steady-state solution satisfies the condition $T_R \partial A(T,t)/\partial T = 0$.

The first part of this differential equation corresponds to the nonlinear Schrödinger equation (Sect. 4.3), whose solution for GDD < 0 and $n_2 > 0$, given by the first-order soliton (4.48). For laser materials with a sufficiently large gain bandwidth, i.e., $D_g \ll 1$, and for small loss modulation, the first part of this differential equation can become dominant. Thus, we can treat the solution using the soliton perturbation calculation. For an analytical solution the exponential potential is approximated by the V-potential (see Fig. 9.24). By analogy with Sect. 6.6, the solution ansatz is given by a soliton (4.47) and a perturbation pulse, which is called the continuum:

$$\boxed{A(T,t) = A_0 \operatorname{sech}\left(\frac{t}{\tau}\right) \exp\left(\mathrm{i}\phi_0 \frac{T}{T_R}\right) + \text{continuum}}\,, \tag{9.103}$$

where the pulse length of an ideal soliton is given by (4.57):

$$\tau = \frac{4|D|}{\delta F_{p,L}}\,, \tag{9.104}$$

with D the dispersion parameter (4.39), δ the SPM coefficient (4.5), and $F_{p,L}$ the pulse fluence in the laser material. The FWHM pulse duration τ_p is then given by (4.48) as $\tau_p = 1.763\tau$.

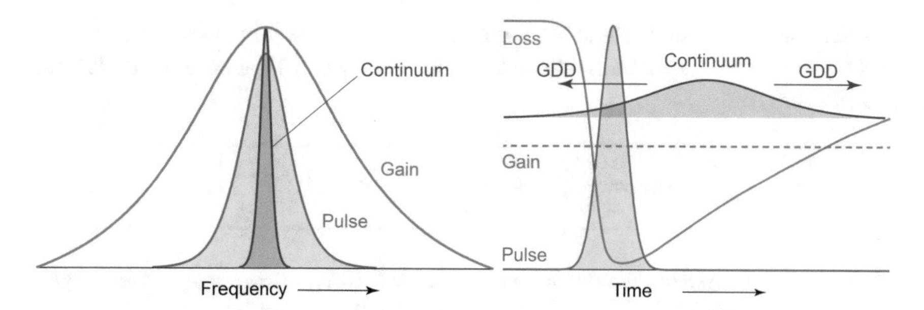

Fig. 9.25 Soliton modelocking. Soliton and continuum in the spectral domain and time domain. The continuum experiences more loss through the slow saturable absorber, and is therefore unable to destabilize the soliton. The saturable absorber only starts and stabilizes the modelocking, whereas the soliton formation determines the final pulse duration

Using the soliton perturbation calculation, it can be shown that the continuum with the exception of SPM has to satisfy the same differential equation (9.100) [95Kar1,96Kar,95Kar2]. The intensity of the continuum is not sufficient to be able to profit from SPM. Therefore, the continuum is very quickly broadened by negative GDD (see Fig. 9.25). Without the slow saturable absorber, the soliton would not be stable. The continuum is spectrally less broad, and therefore experiences a higher gain than the soliton (Fig. 9.25), so the continuum would destabilize the soliton. The slow saturable absorber now has to be fast enough to ensure that the increasing continuum, which is more and more broadened by negative GVD, experiences enhanced loss by the absorber, and so can never reach the threshold condition (Fig. 9.25). Hence, the slow saturable absorber stabilizes the soliton.

The solution is a pulse which is a soliton given by (4.47)

$$\boxed{A_\mathrm{s}(T,t) = \sqrt{\frac{F_\mathrm{p,L}}{2\tau}}\,\mathrm{sech}\left(\frac{t}{\tau}\right)\mathrm{e}^{\mathrm{i}\phi_0 T/T_\mathrm{R}}}\,, \tag{9.105}$$

where $F_\mathrm{p,L}$ is the pulse fluence in the laser gain crystal, and τ determines the pulse length of the soliton (4.57):

$$\tau_\mathrm{FWHM} = 1.763\tau = 1.763\frac{4|D|}{\delta F_\mathrm{p,L}}\,, \tag{9.106}$$

with ϕ_0 the soliton phase per resonator roundtrip (4.52) given by

$$\phi_0 = \frac{|D|}{\tau^2} = \frac{\delta F_\mathrm{p,L}}{4\tau}\,, \tag{9.107}$$

assuming that SPM is only significant inside the laser gain crystal, and $F_\mathrm{p,L}$ is the pulse fluence inside the gain.

On its passage through the slow saturable absorber, the soliton pulse experiences a loss l_s given by the gain bandwidth and the total loss q_p (7.14) (according to [95Jun2] and [95Kar2]):

$$\boxed{l_s = \frac{D_g}{3\tau^2} + q_p = \frac{D_g}{3\tau^2} + q_0 \frac{F_{sat,A}}{F_{p,A}} \left(1 - e^{-F_{p,A}/F_{sat,A}}\right)} , \tag{9.108}$$

where $F_{p,A}$ is the pulse fluence incident on the SESAM, and $D_g = g/\Omega_g^2$ is the gain dispersion (Appendix F). The gain–loss balance is then given by

$$g = l + l_s , \tag{9.109}$$

where l_s is the additional loss experienced by the soliton and l is the rest of the cavity losses. The first part of l_s describes the loss caused by the finite gain bandwidth. The second part of l_s is determined by the slow saturable absorber loss q_p (7.14).

The soliton is stable as long as the loss for the soliton is lower than the loss for the continuum. The loss for the continuum follows directly from the soliton perturbation calculation [95Jun2,96Kar]. This stability criteria will be discussed in more detail with the following example of a soliton modelocked Ti:sapphire laser.

The soliton loss l_s (9.108) is consistent with the fast saturable absorber result, for which we obtained the following gain–loss balance (9.98):

$$l_0 = g - \frac{1}{3} \frac{g}{\Omega_g^2 \tau^2} \left(1 + \beta^2\right) + \frac{2}{3} \bar{\gamma}_A A_0^2 .$$

When we compare this equation with the soliton loss l_s (9.108), the fast saturable absorber loss (i.e., $2\bar{\gamma}_A A_0^2/3$) is replaced by the slow saturable absorber loss q_p (7.14). In addition, according to (9.103) and (9.105), with soliton modelocking we obtain transform-limited soliton pulses, i.e., $\beta = 0$.

In soliton modelocking the net gain window can remain open for significantly more than 10 times longer than the ultrashort pulse, depending on the specific laser parameters, e.g., 30 times for the example we will discuss in the next section [95Jun2, 96Kar]. This strongly relaxed the requirements on the saturable absorber and we can obtain ultrashort pulses even in the 10 fs regime with SESAMs that have much longer recovery times. For example, we generated pulses as short as 34 fs in a Ti:sapphire laser [95Bro2], and 90 fs in an Nd:glass laser without gain reshaping [95Kop1]. A top-coated SESAM with a dielectric mirror reduced the bandwidth limitation of the lower AlGaAs/GaAs DBR and allowed for even shorter 19 fs pulses from a soliton modelocked Ti:sapphire laser [95Jun1]. When the lower semiconductor DBR was replaced by a metal mirror, a soliton modelocked Ti:sapphire laser even generated pulses as short as 16 fs [96Flu1].

With the same modulation depth, one can obtain almost the same minimal pulse duration as with a fast saturable absorber, as long as the absorber recovery time is roughly less than ten times longer than the final pulse width. In addition, high dynamic range autocorrelation measurements showed excellent pulse pedestal suppression

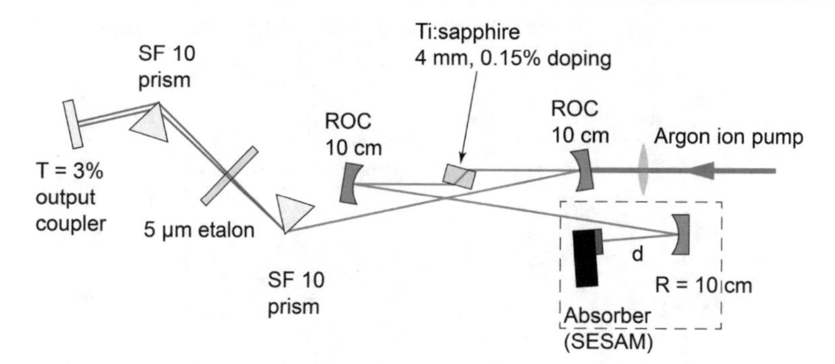

Fig. 9.26 Laser resonator setup of a Ti:sapphire laser. In 1995, we still used the acronym A-FPSA for the SESAM

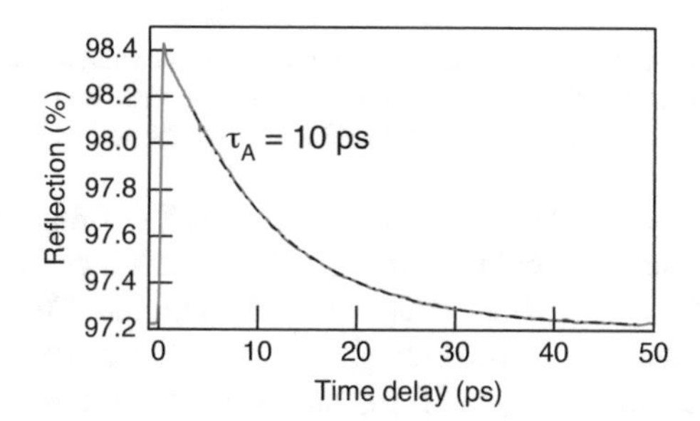

Fig. 9.27 Impulse response of a slow GaAs saturable absorber integrated into an antiresonant SESAM. This was measured with 100 fs long pulses at a center wavelength of 800 nm. The GaAs was grown at low temperatures with MBE on a AlGaAs/AlAs Bragg mirror, which strongly reduces the lifetime of the electrons in the conduction band, and thus the recovery time of the saturable absorber [95Jun2]

over more than seven orders of magnitude in 130 fs pulses of an Nd:glass laser [95Kop1] and very similar to or even better than KLM pulses in the 10 fs pulse width regime [97Jun2]. Even better performance is predicted theoretically if the saturable absorber also shows a negative refractive index change coupled with the absorption change, as is the case for semiconductor materials [96Kar].

9.5.2 Example: Soliton Modelocked Ti:sapphire Laser

The theory of soliton modelocking was also confirmed experimentally with a Ti:sapphire laser [95Jun2]. The goal here was not to generate the shortest pulses, but to determine the pulse length and the stability limit for known parameters. To do this, we used a Ti:sapphire laser (Fig. 9.26) with a 5 μm thick Fabry–Pérot (etalon), which had a bandwidth of 98 nm.

The SESAM was a 330 nm thick GaAs-semiconductor material which had been grown at low temperatures, and with a recovery time of 10 ps (Fig. 9.27). This GaAs saturable absorber was grown on an AlAs/Al(x)Ga($1-x$)As Bragg mirror

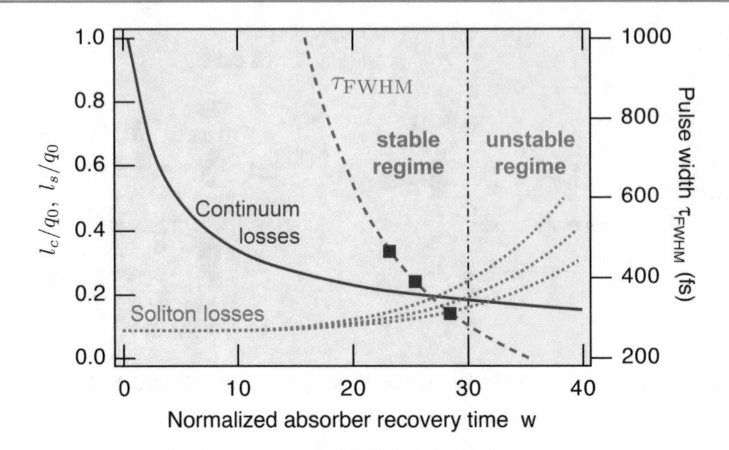

Fig. 9.28 Soliton modelocking stability region. The solution of the soliton is shown with the *dashed line* according to (9.105). Experimental results are shown by *squares*. Soliton modelocking is stable as long as the loss of the continuum is greater than the loss for the soliton [95Jun2]

($x = 0.2$). The loss was reduced with a 80% SiO_2/TiO_2 Bragg mirror which was coated onto the semiconductor structure. The GaAs absorber was within a Fabry–Pérot structure which satisfied an antiresonance condition. The impulse response (Fig. 9.27) was measured with 100 fs long pulses at a center wavelength of 800 nm. We obtained a time constant $\tau_A = 10$ ps, and the maximum modulation depth ΔR was 1.2%, giving $q_0 = 0.006$ ($\Delta R \approx 2q_0$). The non-saturable loss was 1.6%. The saturation fluence $F_{sat,A}$ was 0.6 mJ/cm^2.

Figure 9.28 shows the stability region for soliton modelocking as a function of the normalized absorber recovery time w, defined by

$$w \equiv \frac{\tau_A}{\tau}\sqrt{\frac{q_0}{\phi_0}} = \tau_A\sqrt{\frac{q_0}{|D|}}\,, \tag{9.110}$$

where we have used (4.50) for ϕ_0. Here $q_0 = 0.006$ is the maximum amplitude modulation depth. Experimentally, one can change the dispersion $|D|$ continuously inside the Ti:sapphire laser, and thus with (9.110), the normalized absorber recovery time w. The soliton pulse is stable as long as the loss for the continuum l_c is greater than the loss for the soliton l_s, as shown in Fig. 9.28.

In this example we used an additional intracavity filter with a 5 μm etalon (see Fig. 9.26). This filter modifies the effective gain dispersion as follows:

$$D_g = \frac{g}{\Omega_g^2} + \frac{1}{\Omega_f^2} \approx \frac{1}{\Omega_f^2}\,, \tag{9.111}$$

where Ω_f is the HWHM bandwidth of the filter. The gain–loss balance is then given by $g = l + l_s$, where l_s is the additional loss experienced by the soliton (9.108), for which the loss caused by the finite gain bandwidth is now dominated by the additional filter. The second part of (9.108) is determined by the slow saturable absorber loss q_p (7.14).

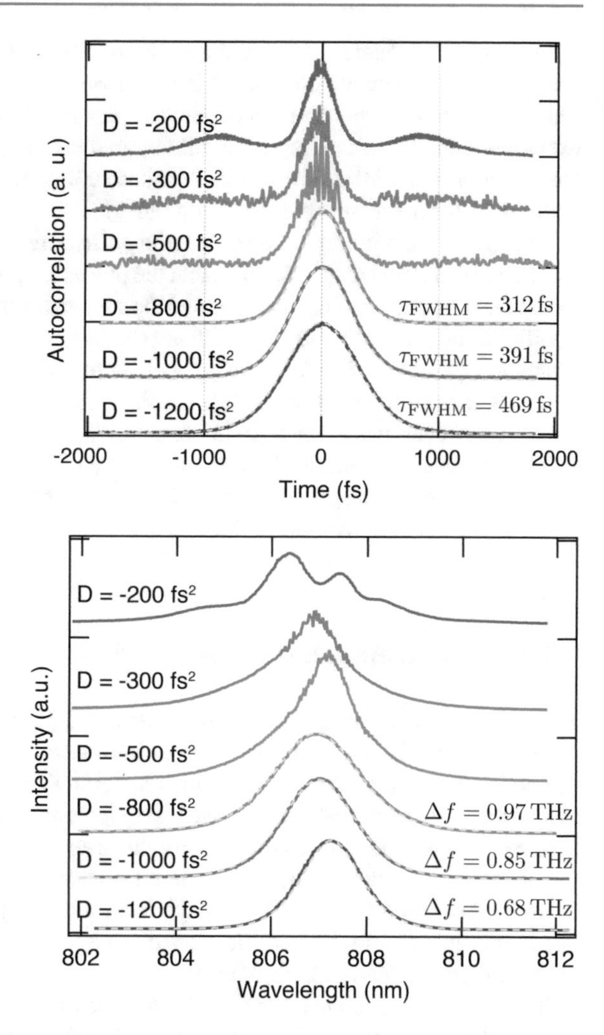

Fig. 9.29 Autocorrelation and spectrum for different dispersion parameters [95Jun2]

The pulse fluence in the Ti:sapphire laser is so high that the saturable absorber is completely saturated. According to (7.17), $q_\mathrm{p}\left(F_\mathrm{p,A} \geq 5F_\mathrm{sat,A}\right) \approx q_0 F_\mathrm{sat,A}/F_\mathrm{p,A}$ and we obtain from (9.108) [95Jun2]

$$l_\mathrm{s} \approx q_0 \frac{F_\mathrm{sat,A}}{F_\mathrm{p,A}} + q_0 \left(\frac{w}{w_0}\right)^4 , \quad \text{with } w_0^2 = \frac{4\sqrt{3}\Omega_\mathrm{f}\tau_\mathrm{A}^2 q_0^{3/2}}{\delta F_\mathrm{p,L}} , \tag{9.112}$$

using for τ (4.57) and for w (9.110). This means that the soliton loss increases very fast with w^4 (see Fig. 9.28). This is not surprising since the dispersion decreases according to (9.110) with increasing w, and thus the pulses get shorter (9.106). When the pulses become shorter, their spectral width also becomes broader, and the solitons therefore experience more loss in the filter.

Figure 9.28 shows the loss for the continuum, which follows directly from the soliton perturbation calculation [95Jun2,96Kar]. In contrast to the soliton, the loss of the continuum decreases with increasing w. For higher w values, the dispersion decreases, and hence the continuum is not broadened so quickly by the dispersion. The slow saturable absorber generates less loss for shorter continuum pulses than for longer continuum pulses, still assuming that the continuum pulse is longer than the recovery time of the slow saturable absorber (i.e., the saturable absorber is only really slow for the much shorter soliton pulse).

The dashed line in Fig. 9.28 depicts the pulse length of an ideal soliton (9.105), and the experimental results confirm that the pulses are given by solitons. Figure 9.29 shows the measured autocorrelations and the spectra for pulses in the proximity of the stability limit. According to Fig. 9.28 and (9.110), the stability limit is at a negative dispersion of $D = -750\ \mathrm{fs}^2$, which is also confirmed by Fig. 9.29.According to Fig. 9.29, for a sufficiently small negative dispersion, a higher-order soliton develops. Usually, however, this is not a desirable solution.

The pulses exhibit an ideal soliton shape even if the autocorrelation is measured over many orders of magnitude (Fig. 9.30). This means that one obtains ideal solitons, although the net gain window in this case is open more than 30 times longer than the pulse duration.

9.5.3 Soliton Modelocking with GDD > 0 and $n_2 < 0$

For soliton modelocking, negative group delay dispersion (GDD) is required when $n_2 > 0$ (Fig. 9.22). We therefore have to introduce additional intracavity negative dispersion to compensate for the positive material dispersion of bulk gain materials inside the laser cavity at near-infrared wavelengths. However, with a negative n_2 (i.e., $n_2 < 0$) we can potentially simplify the laser cavity as shown schematically in Fig. 9.31. How can we obtain negative n_2? Typical materials all provide a positive n_2 when the laser wavelength is far enough from the bandgap to avoid additional loss due to multiphoton absorption. Here we discuss a solution with cascaded quadratic nonlinearities (CQNs) [93Sch,96Ste].

There are some good reasons for soliton modelocking with negative n_2. For example, negative GDD becomes increasingly challenging when scaling to high laser repetition rates with more restricted cavity design space to prevent Q-switching instabilities and when scaling to high average power with damage in GTI-type dispersive mirrors [15Sar]. The field enhancement inside GTIs lowers the damage threshold and we begin to see overheated laser spots on such mirrors with an infrared viewer in multi-100-W average power thin-disk lasers. Furthermore, the Kerr effect yields a positive nonlinear refractive index n_2 when the laser wavelength is far enough from the bandgap to avoid multiphoton absorption. Using wide bandgap materials to satisfy this constraint means only moderate values of n_2 are typically accessible, since n_2 scales inversely with the fourth power of the bandgap [10Chr]. Thus, achieving sufficient nonlinearity for soliton modelocking becomes increasingly difficult for lower energy lasers, e.g., at gigahertz pulse repetition rates, because the average power P is given by $P = E_\mathrm{p} f_\mathrm{rep}$, where E_p is the pulse energy and f_rep is the pulse repetition rate.

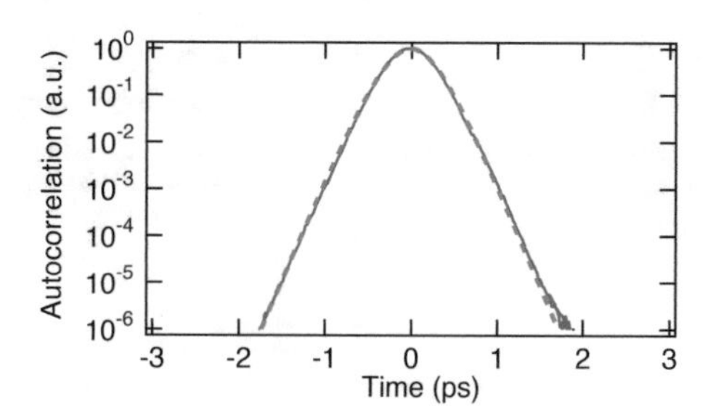

Fig. 9.30 'High dynamic range' autocorrelation [95Jun2]

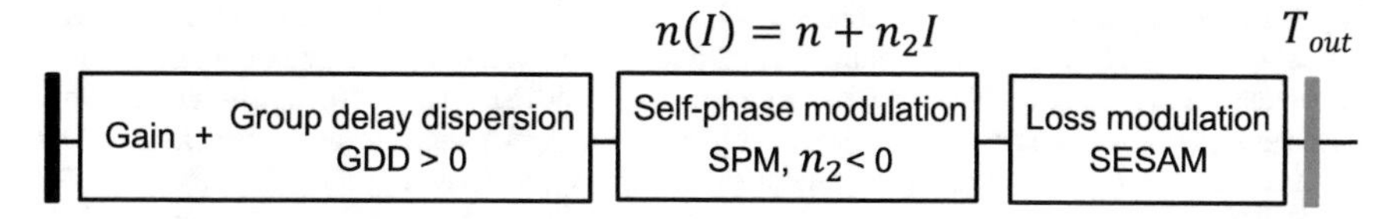

Fig. 9.31 Soliton modelocking with negative n_2 and positive GDD, exploiting the normal material dispersion of the gain medium inside the laser cavity. Additional negative GDD is potentially no longer required. Negative n_2 can be obtained with cascaded quadratic nonlinearities (CQNs)

An alternative to Kerr nonlinearity is offered by cascaded quadratic nonlinearities (CQNs) [93Sch, 96Ste]. By tuning a second-harmonic generation (SHG) medium far from phase-matching, a large and negative effective nonlinear refractive index can be achieved, supporting a Kerr-like soliton with positive GDD and negative n_2. In contrast to phase-matched SHG (see Fig. 9.32 left), a significant phase mismatch is introduced with CQN, leading to a periodic exchange of energy between the fundamental and the second harmonic signal (Fig. 9.32 right). As soon as the two signals are out of phase the SHG interferes destructively and reduces the conversion efficiency.

Therefore with such an approach, we can achieve soliton modelocking without additional negative GDD compensation and we can simply use the positive GDD in the gain material for soliton formation (see Figs. 9.22 and 9.31). In crystals that lack a center of inversion, we can obtain, in addition to the third order nonlinearity, a cascaded $\chi^{(2)}$ second order nonlinearity that contributes to the nonlinear refraction, whence this effect is referred to as a cascaded quadratic nonlinearity (CQN).

The CQN nonlinearities arise from phase-mismatched SHG, and can be characterized by the effective second-order nonlinear coefficient d_{eff} and the phase mismatch $\Delta k = k_2 - 2k_1$, where $k_j = n_j(\omega_j)\omega_j/c$ is the carrier wave vector of wave j. The characteristic property of cascading interaction is a large phase mismatch Δk, in which a $\chi^{(3)}$-like SPM can be obtained instead of efficient SHG. For large $|\Delta k|$, the effective $\chi^{(3)}$ susceptibility associated with this cascading process yields a contribution to a nonlinear n_2 given by [96Ste]

$$n_2^{\text{casc}} = -\frac{1}{\Delta k} \frac{4\pi d_{\text{eff}}^2}{n_1^2 n_2 \lambda_1 \varepsilon_0 c} \propto -\frac{1}{\Delta k} , \tag{9.113}$$

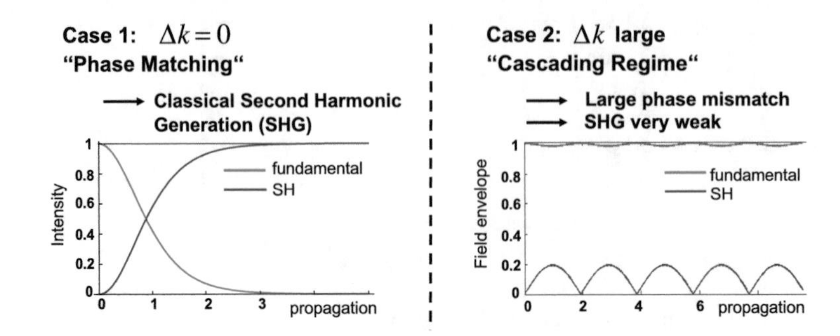

Fig. 9.32 Principle of cascaded quadratic nonlinearities (CQN). Case 1 *on the left* shows phase matching for classical second-harmonic generation (SHG), where the energy is transferred from the fundamental harmonic (FH) wave to the second harmonic (SH) wave with high efficiency. Case 2 *on the right* shows the cascading regime with large phase mismatch Δk and weak SHG, where we observe a weak modulation of the fundamental and a weak periodic energy transfer to the SH signal. This introduces an effective n_2 which can be positive or negative depending on the phase mismatch Δk, given by n_2^{casc} (9.113)

where subscript 1 denotes the first harmonic (FH) and subscript 2 denotes the second harmonic (SH). An exception to this notation is the nonlinear refractive index: we use the superscripts on n_2^{NL} and n_2^{casc} to avoid confusion with the linear refractive index n_j of wave $j = 2$, the SH. Depending on the sign of the phase mismatch Δk, we can obtain a positive or a negative n_2^{casc}.

CQNs offer numerous potential advantages for modelocking due to the large, adjustable, and negative (self-defocusing) nonlinear refractive indices that can be obtained. Such cascaded $\chi^{(2)}$ nonlinearities have previously been exploited for various applications, including pulse compression [99Liu2,02Ash], Raman-like frequency shifting [04Ild], and supercontinuum generation [07Lan, 11Phi, 12Zho]. CQNs have also been used for modelocking, such as a Kerr lens modelocked (KLM) Nd:YAG [95Cer,98Zav], Cr:forsterite [99Qia], Nd:GdVO$_4$ [05Hol], Yb:KYW [14Mei], Nd:YVO$_4$ at 1.06 μm [11Sch] and at 1.34 μm [11Ili], a SESAM modelocked Yb:YAG [05Agn2], soliton modelocked Yb:CALGO [14Phi], and Yb:YAG [18Sal] lasers.

For example, in a soliton-modelocked 210-W Yb:YAG thin-disk laser [18Sal] the damage in GTI mirrors could be addressed using the CQN to cancel the positive intracavity SPM from the gain medium and the air in the long high-power MHz cavity. A simple nonlinear crystal plate was added into the laser cavity. This nonlinear crystal was an AR-coated LBO crystal plate with a thickness of 5 mm which was tilted away from perfect phase matching to obtain negative n_2 with cascaded quadratic nonlinearities (CQN) according to (9.113) and Fig. 9.32. This can result in a negative CQN SPM coefficient

$$\gamma_{\mathrm{CQN}} = k n_2^{\mathrm{casc}} L \,, \tag{9.114}$$

where n_2^{casc} is given by (9.113) and L is the length of the CQN interaction in a nonlinear crystal. The SHG conversion efficiency can be described by [08Boy]

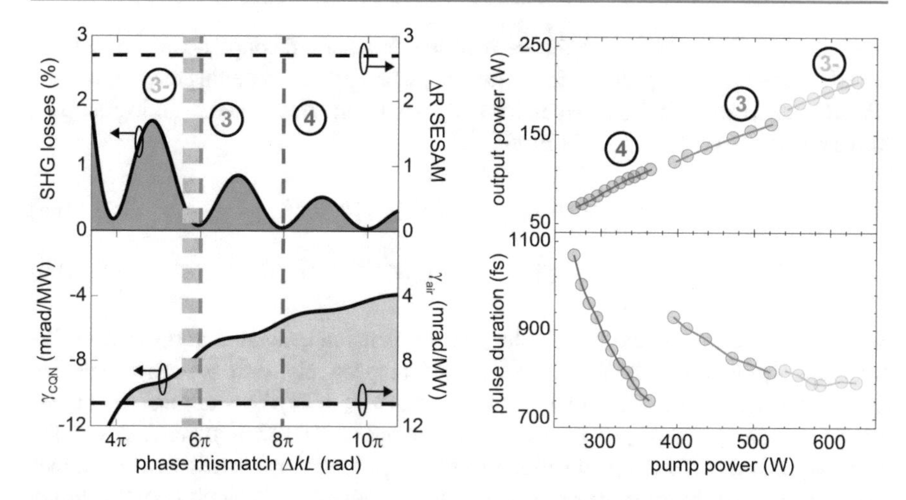

Fig. 9.33 *Left*: Round-trip SHG losses and CQN SPM coefficient γ_{CQN} as a function of phase mismatching of an intracavity LBO plate (i.e., ΔkL). By operating the nonlinear crystal plate in a SHG minimum, fewer than 0.1% losses can be obtained while cancelling most of the SPM from air. *Right*: Laser performance with output power and pulse duration as a function of the pump power. Different colors and numbers refer to different phase mismatch values ΔkL. For section 3, up to 80% of SPM in air could be cancelled [18Sal]

$$\eta_{SHG} \propto \mathrm{sinc}^2\left(\frac{\Delta kL}{2\pi}\right), \tag{9.115}$$

which is an additional loss inside the laser cavity. Note that this simple equation breaks down if the intensity is high enough. When the intensity is high and the efficiency is high close to phase-matched interaction Δk, the phase-matching curve narrows and is no longer exactly a sinc^2 function [84Eck]. For intracavity cascaded devices we are often in this situation. This is a minor effect for the current discussion because it mainly affects the effective intracavity losses and the sinc^2 function above still gives a good first approximation.

Figure 9.33 shows the SHG losses η_{SHG} (9.115) and the CQN SPM coefficient γ_{CQN} (9.114) as a function of the phase mismatch ΔkL in the tilted nonlinear crystal plate with a length L. There is a straightforward tunability of Δk by adapting the tilt of the nonlinear crystal plate during laser operation. The SHG conversion efficiency η_{SHG} is modulated by the sinc^2 function and lower losses can be obtained when the nonlinear crystal is angle-tuned to these low-loss values. The laser performance is shown with average output power and pulse durations for the different phase mismatch adjustments. An intracavity SESAM was used to stabilize soliton modelocking with negative n_2.

Theory Behind (9.113) and Trade-Off Between SPM and Loss

Here we only provide a short discussion about the theory and the trade-off between SPM and loss. The interested reader is referred to the more detailed theoretical discussion given in [14Phi]. The theory is based on standard coupled wave equations for SHG, assuming co-propagating waves, neglecting self-steepening effects for simplicity, but allowing for both dispersion and diffraction effects. The resulting coupled wave equations are given by [08Boy]

$$\left(\partial_z + \hat{D}_1\right) A_1 = -\mathrm{i}\kappa_1 A_2 A_1^* \mathrm{e}^{-\mathrm{i}\Delta k z} , \tag{9.116a}$$

$$\left(\partial_z + \hat{D}_2\right) A_2 = -\mathrm{i}\kappa_2 A_1^2 \mathrm{e}^{\mathrm{i}\Delta k z} , \tag{9.116b}$$

where $\Delta k \equiv k_2(\omega_2) - 2k_1(\omega_1)$ is the phase mismatch evaluated at the carrier frequencies ω_1 and ω_2 of the FH and SH pulses, respectively. Here we have neglected $\chi^{(3)}$ effects for compactness. In these equations, $\partial_z = \partial/\partial z$, the coupling coefficients are defined as $\kappa_j = \omega_1 d_{\mathrm{eff}}/n_j c$, and $\hat{D}_j$ are the operators describing diffraction, dispersion, and other linear optical effects of the co-propagating electric field envelopes A_1 and A_2. For the relevant case where group velocity mismatch (GVM), group delay dispersion (GDD), and Poynting vector walk-off (PVW) effects are dominant, the $\hat{D}_j$ are given by

$$\hat{D}_j = \rho_j \frac{\partial}{\partial x} + \delta_j \frac{\partial}{\partial t} - \mathrm{i}\frac{\beta_j}{2}\frac{\partial^2}{\partial t^2} , \tag{9.117}$$

where ρ_j is the PVW coefficient (for transverse coordinate x), $\delta_j = \upsilon_j^{-1} - \upsilon_{\mathrm{ref}}^{-1}$ for group velocities υ_j and arbitrary reference velocity υ_{ref}, and $\beta_j = (\partial^2 k_j/\partial\omega^2)|_{\omega=\omega_j}$ is the GDD coefficient for wave j. These operators give rise to a frequency dependence of the phase mismatch.

For CQN, it is advantageous to use perturbative frequency-domain analysis, assuming small nonlinear changes to the fundamental harmonic (FH) beam [98Zav, 00Ime]. Typically, GDD and higher orders of dispersion can be neglected when calculating Δk. Furthermore, we can neglect the contributions to Δk from diffraction and dispersion by making a first-order expansion of Δk in its spatial and optical frequency variables. Under these approximations, we can define an accumulated phase mismatch for each spatial and temporal frequency.

As mentioned above, the characteristic property of cascading interactions is a large phase mismatch Δk. This property enables a perturbative analysis of the interaction based on multiple-scale analysis, as introduced in [02Con]. In [02Con], the combined effects of cascading and diffraction were considered to first order in this perturbation, yielding effective lensing effects. A similar procedure [14Phi] can be performed with the operators $\hat{D}_j$ introduced in (9.116) for linear optical propagation, capturing both spatial and temporal effects. In this more general situation, the approximation is based on the length scale $L_{\mathrm{coh}} = \pi/|\Delta k|$ being sufficiently short compared with the other characteristic lengths of the interaction. These characteristic lengths include group velocity walk-off, Poynting vector walk-off, pulse and beam spreading due to dispersion and diffraction, respectively, and the nonlinear length characterizing the rate of coupling between the fields [13Phi].

To include losses and the non-instantaneous character of the nonlinearity, we go to second order in the perturbation and account for boundary conditions of the crystal and laser cavity. The influence of GVM effects can be minimized by keeping $L_{\mathrm{GVM}} \gg L_{\mathrm{coh}}$.

For the cascading interaction itself, there is, at a given pulse intensity, a trade-off between crystal length and phase mismatch in terms of the following:

1. Accumulating sufficient negative SPM via cascading, obtained via small Δk and/or large L, and characterized by (9.113)].
2. Minimizing nonlinear losses due to SHG, which are characterized, per pass through the crystal, by (9.118), and are *independent* of length L.
3. Adding unwanted GDD by using a longer crystal. This GDD scales with L, as does the SPM.

A useful relation for the nonlinear losses in terms of the cascading-induced SPM which emphasizes the above trade-off is given by [14Phi]

$$a_{\mathrm{NL}} = \frac{4}{3\Delta k L}\phi_{\mathrm{NL}} \,, \tag{9.118}$$

where ϕ_{NL} is the accumulated nonlinear phase from the cascading process, given by

$$\phi_{\mathrm{NL}} = \int_0^L \frac{2\pi}{\lambda_1} n_2^{\mathrm{casc}} I_1(z) \mathrm{d}z \,, \tag{9.119}$$

for FH average intensity $I_1(z)$ and effective nonlinear refractive index n_2^{casc} given by (9.113). This loss corresponds to the total loss of a pulse which is Gaussian in space, sech-shape in time. The phase ϕ_{NL} corresponds to the effective SPM in time for soliton formation, which includes a factor of 1/2 to account for averaging in space. This factor does not show up in (9.119) because we assume that $I_1(z)$ is the average rather than the peak intensity.

Regarding point 3 above, bulk temporal solitons are supported via the positive GDD and negative SPM present in the SHG crystal, so that in the absence of any other cavity dispersion or nonlinearity, increasing the crystal length at a given pulse intensity does not alter the soliton formation, at least to the extent that lensing and diffraction effects remain negligible. However, when there is GDD and SPM from other intracavity elements (in particular, the gain crystal), the SHG crystal must 'work harder' to enable soliton formation than it would otherwise, for example, by being operated closer to phase-matching (and hence incurring larger nonlinear losses).

Regarding point 2 above, the tolerable nonlinear losses are coupled to the SESAM parameters, in particular its modulation depth, saturation fluence, and the rollover characteristics of its nonlinear reflectivity. By analogy with conventional SESAM modelocking, multi-pulsing can occur when operating in the rollover regime.

Following this optimization strategy we can, according to point 1 above, obtain a large self-phase modulation (SPM), by using a longer CQN crystal and operating further from phase-matching with a large Δk. According to point 2 and (9.118), the losses are reduced because the magnitude of the SHG at the beginning of the nonlinear crystal is reduced. This enabled stable SESAM-assisted soliton modelocking of a diode pumped Yb:CALGO laser with negative n_2^{casc} at a repetition rate of 540 MHz, producing 100 fs pulses with 760 mW of average output power [14Phi].

Reduced Loss with Apodized Quasi-Phase-Matched CQN Devices

There are trade-offs. An inherent property of using CQNs for modelocking is the need for SHG, since it requires an interplay between the fundamental or first harmonic (FH) and the second harmonic (SH). Therefore, a key feature of a CQN interaction using ultrashort pulses is the presence of a weak SH pulse propagating with a stronger FH pulse. Energy is lost when exciting the SH pulse (9.118). Since these losses are intensity-dependent, they can destabilize the modelocking.

These SHG losses can be reduced, for a given amount of self-phase modulation (SPM), by using a longer CQN crystal and operating further from phase-matching, as was done in [14Phi]. However, this design procedure has several drawbacks, since it introduces additional dispersion, linear losses, and optical delay to the pulses and requires a long crystal, which prohibits ultracompact devices. For low loss optical cavities utilizing CQNs, we are thus presented with a fundamental trade-off between SPM and nonlinear losses (9.118).

This trade-off can be solved by slowly switching on the SHG, as for impedance matching in double chirped mirrors. In optics this is typically referred to as apodization. Apodization is a technical term used when we change mathematical filter functions to smooth out the switching on and off of artifacts. For example, in optical imaging the diffraction pattern on hard round apertures results in significant diffraction with side lobes due to the Airy disk, which degrade images. Apodization techniques are used to suppress these side lobes, for example.

Slowly switching on SHG requires a chirp of $\Delta k(z)$ along the propagation direction z which slowly varies the nonlinear conversion efficiency η_{SHG} (9.115). This can be achieved with quasi-phase matching (QPM) in periodically poled nonlinear crystals, such as for example, in a PPLN (i.e., periodically poled $\mathrm{LiNbO_3}$) crystal. In QPM the crystal has a periodic poling with an alternating positive and negative effective nonlinear coefficient $\pm d_{\mathrm{eff}}$ with a period Λ. The nonlinear conversion efficiency η_{SHG} can be increased with a longer interaction length even with a nonzero Δk because the destructive interference of a newly generated SH wave after an interaction length of $L_{\mathrm{c}} = \pi/|\Delta k|$ is prevented by changing the sign of d_{eff}. Therefore, the period Λ of the QPM is given by

$$\Lambda = 2L_{\mathrm{c}} = \frac{2\pi}{|\Delta k|} \,. \tag{9.120}$$

The phase mismatch Δk can be varied via a chirped QPM period $\Lambda(z)$:

$$\Delta k(z) = k_2 - 2k_1 - \frac{2\pi}{\Lambda(z)} = k_2 - 2k_1 - K_{\mathrm{g}}(z) \,, \tag{9.121}$$

where $K_{\mathrm{g}} = 2\pi/\Lambda$ is the wave vector of the periodic structure.

If the QPM period $\Lambda(z)$ varies slowly enough, the fields can be considered a local eigenmode associated with the position-dependent value of the QPM grating, and the fields can adiabatically follow the eigenmode as it evolves with $\Delta k(z)$. The loss associated with the rapid switching on of SHG (9.118) can be greatly reduced. This capability has enabled adiabatic frequency conversion, in which $\Delta k(z)$ is swept slowly through phase-matching in order to achieve efficient frequency conversion [09Suc]. For example, it was also shown that both high efficiency and broad bandwidth can be achieved at the same time when the phase mismatch is adiabatically varied along the propagation. Furthermore, best performance was obtained with nonlinear chirp apodization, where the poling period is varied smoothly, monotonically, and rapidly at the edges of the QPM device. This is referred to as the apodized QPM crystal [13Phi]. Note that the accumulated CQN SPM is not affected by the sign change of the periodic poling because $n_2^{\text{casc}} \propto d_{\text{eff}}^2$ (9.113).

In Fig. 9.34 the QPM period $\Lambda(z)$ varies with respect to longitudinal position (propagation direction z): the regions near the edges (input and output sides) have in comparison a long period for a large Δk (i.e., small η_{SHG} and small n_2^{casc} with small $K_{\text{g}} = 2\pi/\Lambda$), while the region in the middle has a short period for a smaller Δk (i.e., larger η_{SHG} and larger n_2^{casc} with a larger $K_{\text{g}} = 2\pi/\Lambda$). The QPM periods in Regions 1 and 3 are chirped. This 'apodized' longitudinal QPM profile is designed to adiabatically excite ('switch on') and later de-excite ('switch off') the SH wave in a CQN-like interaction, allowing a large SPM in the middle region where the SH is strong, while minimizing the energy lost from the fundamental pulse. Using an additional apodized QPM section at the beginning and the end of the crystal [15Phi] (Fig. 9.34), we therefore transiently excite large CQNs within the crystal, utilize their advantageous properties for pulse formation and stabilization, and then convert the energy back to the resonating FH laser pulse before the end of the crystal in order to suppress losses.

Furthermore, a two-dimensionally patterned fanout QPM grating (Fig. 9.34) [15Phi] gives some additional adjustment of the effective negative SPM. Translating this crystal perpendicular to the beam propagation along x changes the QPM period $\Lambda(x, z)$ and therefore the effective negative n_2. This device therefore combines the longitudinal apodizing concept for low loss and the transverse fanout structure to enable continuous tuning of the nonlinearity. The transverse fanout structure has been used previously in an OPO [98Pow]. With this apodized 2D-QPM device, we thus have access to all the advantageous properties of CQNs, and overcome their fundamental drawback to enable a multitude of new applications.

With this new approach, using the 2D-fanout QPM device for low-loss tunable CQN generation, we have been able to scale gigahertz femtosecond lasers into the 10 GHz regime in a simple straight cavity, as shown in Fig. 9.35, solving also the Q-switching problem (see Sect. 9.7.5) [17May]. Such lasers cannot accept much loss, the SESAM modulation is small, and therefore the apodized QPM device was absolutely required to keep the losses low. In this case, we used the large negative n_2^{casc}, not only for soliton modelocking, but also for generating a defocusing Kerr lens to stabilize against Q-switching instabilities and prevent damage in the intracavity optics. Pulses as short as 108 fs have been generated at 10.4 GHz with 812 mW of average output power [20Kru].

This approach therefore offers a flexible and compact route to managing nonlinearities inside laser cavities, while suppressing the losses that could otherwise prevent

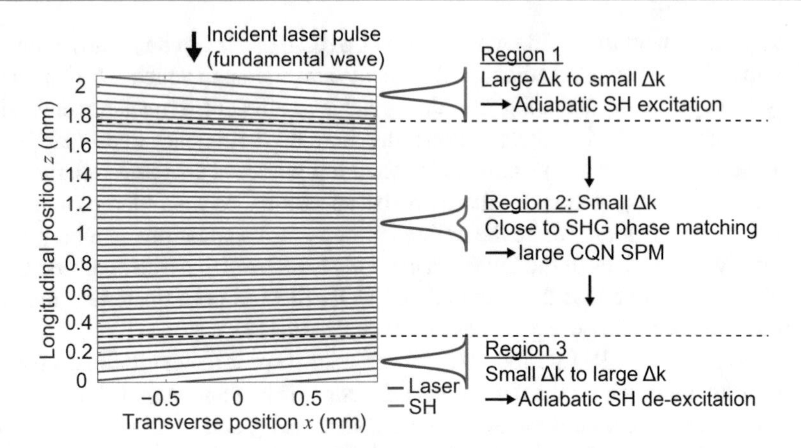

Fig. 9.34 Concept of an apodized 2D-fanout QPM device for low-loss tunable CQN generation. The QPM period $\Lambda(x, z)$ (9.120) is adjusted along the pulse propagation direction z to slowly switch on and off SHG at the beginning and the end of the QPM device (i.e., regions 1 and 3). Such an apodized QPM device solves the trade-off problem, lowers the loss and in region 2 enables a large negative n_2^{casc} closer to SHG phase matching (i.e., small Δk) and therefore a large CQN SPM (9.114). In addition, the QPM period $\Lambda(x, z)$ is varied perpendicularly to z along the transverse position x in a fanout structure to make negative SPM adjustable with translation of the device along x. For clarity, the transverse profile of every 15th domain is shown, i.e., longitudinal position (mm) domains 1, 16, 31, …, and the thickness of the lines is not to scale (each domain is actually about 3.5 μm long). QPM, quasi-phase matching [15Phi]

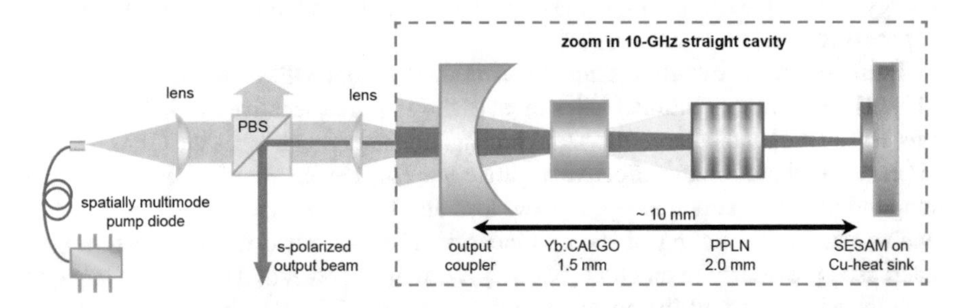

Fig. 9.35 Soliton modelocking with negative n_2. Illustration of the 10 GHz cavity containing a curved output coupler mirror, Yb:CALGO gain crystal, PPLN, and SESAM. A polarizing beam splitter (PBS) is used to couple the linearly polarized light out of the cavity. The PPLN is a 2D-fanout QPM device (Fig. 9.34) with which the effective negative n_2 (i.e., n_2^{casc}) can be continuously tuned by translating the crystal perpendicular to the laser beam. PPLN, periodically poled lithium niobate for quasi-phase matching (QPM) [17May, 20Kru]

or deteriorate modelocked operation, and is particularly interesting for highly compact bulk, fiber, and waveguide lasers with gigahertz repetition rates and operating wavelengths from the near- to mid-infrared spectral regions.

9.5.4 What Happens Without Soliton Formation

We address the question of how short the generated pulses can be without soliton effects and find a limit which depends on the absorber recovery time. It turns out that, even without soliton effects, the pulse duration can be at least 20 times shorter than the absorber recovery time [01Pas]. We also want to discuss the nature of the instability occurring for too long a recovery time and describe the optimum conditions for short pulses. In the following, we closely follow the discussion in [01Pas].

Analytical results for the pulse durations of passively modelocked lasers without soliton pulse shaping have been derived [75Hau3,85Mar,98Akh], but only in the case of weak absorber saturation. In most experimental cases, however, the absorber is fully saturated with a saturation parameter $S > 3$, i.e., at least at three times its saturation fluence, where the assumption of weak saturation is not valid. In this situation, we can still simulate the pulse formation numerically, using a model which repeatedly propagates the pulse through the laser cavity, taking into account the effects of gain, linear cavity loss, and saturable absorption [01Pas]. The gain is assumed to have a Gaussian spectral shape (not affected, e.g., by spatial hole burning), and it saturates according to the average power (Fig. 9.5c). Dynamic gain saturation during a pulse (Fig. 9.5a) is neglected, because this is typically a very weak effect in solid-state lasers. We ignore effects of Kerr nonlinearity and dispersion in the cavity, as well as phase changes on the absorber. Typically, we are interested in the steady-state situation which may be reached after a large number of cavity roundtrips. To find this steady state, the program applies more and more cavity roundtrips until a number of pulse parameters (energy, duration, spectral width, and center wavelength) no longer change significantly during a roundtrip.

Using this model, we found that a useful guideline is to use the equation [01Pas]

$$\tau_{\mathrm{p}} \approx \frac{1.07}{\Delta v_{\mathrm{g}}} \sqrt{\frac{2g}{\Delta R}} \tag{9.122}$$

as an estimate for the resulting steady-state pulse duration. Here, Δv_{g} is the FWHM gain bandwidth (assuming a Gaussian-shaped gain spectrum) and g is the amplitude gain coefficient per cavity roundtrip. In the steady state, g has to balance the overall cavity losses, which consist of the output coupler transmission, other smaller parasitic losses, and the saturable absorber loss, given by (7.17), provided that the saturation parameter S is at least about 5. We found that (9.122) reasonably matches the results from numerical simulations if the absorber is operated at roughly 3–5 times the saturation fluence (Fig. 9.36). For significantly weaker or stronger absorber saturation, the pulse duration gets somewhat longer. The absorber recovery time was assumed to be 100 ps, i.e., much longer than the obtained pulse durations, and surprisingly the exact value of the recovery time has little influence on the pulse duration.

Note that (9.122) has the same form as an equation derived in [75Hau1] for a weakly saturated fast absorber, except that the constant factor (adapted to our notation) has been changed from 0.66 to 1.07 (see also Sect. 9.4). For comparison, we did similar simulations for a fast saturable absorber and found that under optimum

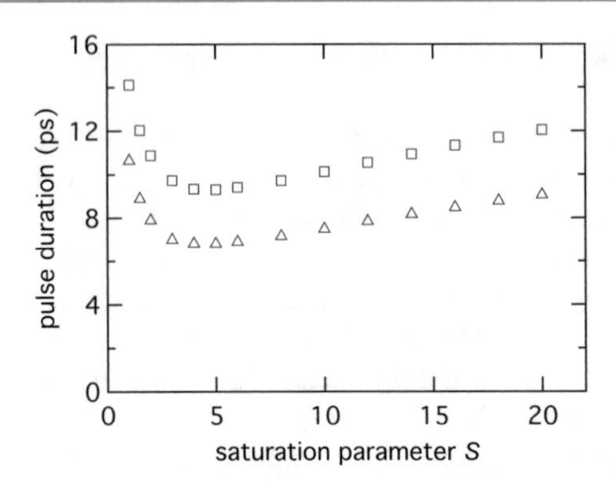

Fig. 9.36 Pulse durations from numerical simulations of a laser with 1 nm gain bandwidth, 5% linear cavity losses, and a SESAM modulation depth of 1% (*rectangles*) and 2% (*triangles*), where the saturation parameter of the SESAM is varied along the *horizontal axis*. Values around 3 to 5 give the shortest pulses, and the resulting durations agree with (9.122) [01Pas]

saturation conditions the obtained pulse duration is only about 15% shorter (for the same modulation depth), compared to the slow absorber. Taking into account the fact that the cavity losses caused by the fast absorber (of equal modulation depth) are somewhat greater (see Tables 7.2 and 7.3), one could make the modulation depth of the slow absorber somewhat larger, which would further reduce the difference in achievable pulse durations. The shortest pulses with a slow absorber are obtained, according to (9.122), by making the nonsaturable cavity loss (output coupler transmission and parasitic linear losses) smaller than the saturable loss. This, however, compromises the power efficiency of the laser.

Numerical simulations like those discussed above show that the resulting pulse duration can be much shorter than the recovery time of the absorber, even if soliton effects are absent. This finding may seem quite surprising because, on the trailing edge of the pulse, there is no shaping action of the absorber. There is even net gain (Fig. 9.5c), because the loss caused by the absorber is very small for the trailing edge (always assuming a fully saturated absorber). Thus one might expect this net gain either to prevent the pulse from getting so short or to destabilize it by amplifying its trailing wing more and more. Indeed, there are a number of publications (e.g., [97Kal,98Akh]) where it was assumed that stable pulse generation is not possible under such circumstances. However, we were able to show that this is not necessarily the case, so that previously often used stability criteria are not correct [01Pas].

To understand why stable pulses with a duration far below the absorber recovery time are possible, we have to consider that the action of the absorber steadily delays the pulse (see Fig. 9.37): it mostly attenuates its leading wing, thus shifting the pulse center backwards in each cavity roundtrip. Note that this shift does not apply to any structures well behind the pulse maximum, because the absorber is then already fully saturated. In effect, the pulse is constantly moving backward and can swallow up any noise growing behind itself. This noise has only a limited time in which it can experience gain before it merges with the pulse itself. The same mechanism can also prevent the trailing edge of the pulse from growing. Such a stabilizing mechanism has been discussed in the context of soliton modelocking (see Sect. 9.5.1), but has

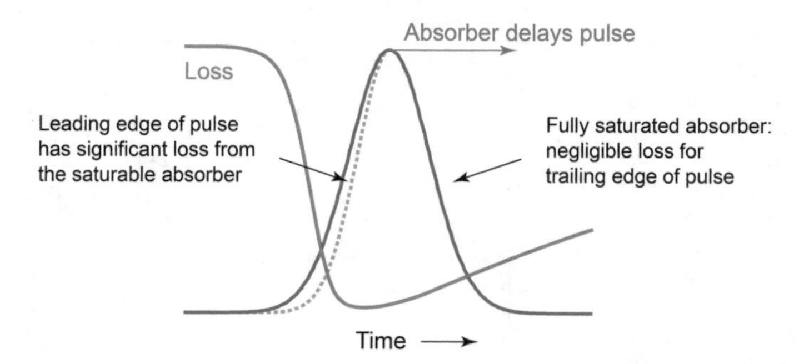

Fig. 9.37 The fully saturated slow absorber shifts the pulse backwards in time. This stabilizes slow saturable absorber modelocking with an open net-gain window (Fig. 9.5c) that is longer than more than 20 times the final pulse duration without soliton formation

not been applied to the situation of passively modelocked lasers without soliton effects. A similar temporal shift and its stabilizing influence on the pulses have been discussed for actively modelocked lasers [99Kar].

To get a quantitative picture, we first estimate the net gain behind the pulse. We always assume a strongly saturated absorber ($S > 3$) and that the absorber recovery time τ_A is significantly longer than the pulse duration τ_p. The average loss for the pulse caused by the slow absorber is given by (7.17):

$$2q_p \approx \frac{\Delta R}{S} . \tag{9.123}$$

The pulse-to-pulse dynamic saturable absorber loss (Fig. 9.37) has a modulation depth of ΔR followed by a loss close to zero after passage of the pulse. Therefore, there is a net gain directly after the pulse, which follows an exponential function, and which we can approximate by a linear function between the time directly after the pulse and the point where the net gain becomes zero. The slope of this function is $\Delta R/\tau_A$. Thus the zero net gain is obtained at a time

$$\frac{\Delta R/S}{\Delta R/\tau_A} = \frac{\tau_A}{S} \tag{9.124}$$

after the pulse.

For the temporal shift caused by the absorber in each cavity roundtrip, i.e., for one reflection at the absorber, we cannot get an analytical solution. This shift is proportional to the modulation depth and to the pulse duration, while it has a more complicated dependence on the saturation parameter. Numerical simulations show that the temporal shift is approximately given by [01Pas]

$$\Delta t \approx 0.12 \Delta R \tau_p , \quad \text{for } S \approx 3 , \tag{9.125}$$

if the pulse has a sech^2 shape and the saturation parameter S is about 3 (a typical value). The shift is smaller for weaker or stronger saturation. It is also slightly weaker

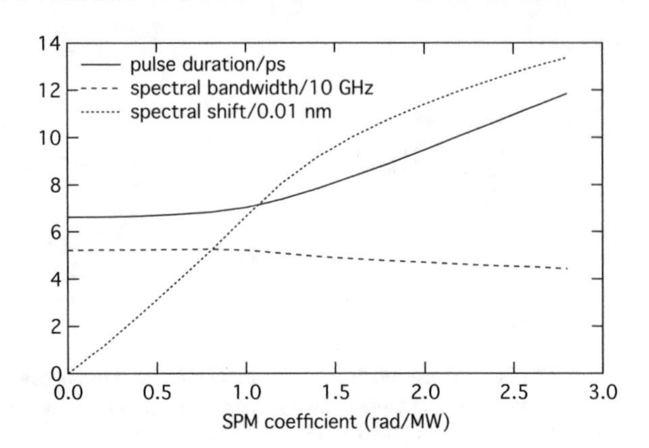

Fig. 9.38 Effect of a variable degree of SPM on the obtained pulse parameters from an Nd:YVO_4 laser without dispersion (see the text for details). For higher SPM coefficients, stable pulses are not obtained. Surprisingly, there is no SPM-induced pulse shortening for slow-saturable absorber modelocking with GDD = 0 [01Pas]

for Gaussian pulses. This means that, for a modulation depth of 2%, we obtain a temporal shift of the pulse of about $0.0024\tau_p$. So it will take many roundtrips to fully shift the pulse over the net gain window given by (9.124). Given that the modelocking build-up time is typically in the range greater than 1 μs (Sects. 9.1.2 and 9.5.1), it takes also many roundtrips to destabilize the trailing edge of the pulse.

It is known for passively modelocked lasers with fast saturable absorbers that some additional influence of SPM can somewhat decrease the pulse duration (see Sect. 9.4) [86Hau], although the pulses become unstable when there is too much SPM. The decrease in the pulse duration is usually explained by the fact that SPM tends to spectrally broaden the pulse. One might expect the influence of SPM to be similar for lasers modelocked with slow saturable absorbers. However, our numerical simulations show that this is not the case (Fig. 9.38) [01Pas].

For Fig. 9.38, we used a slow-absorber modelocked Nd:YVO_4 laser with zero GDD and the following properties: standing-wave cavity with 100 MHz repetition rate, gain bandwidth 1 nm, no spatial hole burning, output coupler transmission 5%, operation roughly 10 times above threshold with about 0.14 W average output power, SESAM with 2% modulation depth, 50 ps recovery time, saturation parameter $S = 4$, no phase changes in the absorber, no dispersion in the cavity, and a variable amount of SPM. The simulations (Fig. 9.38) show that the pulse duration without SPM is about 6.6 ps, and a moderate amount of SPM makes the pulses somewhat longer (not shorter!), while the pulse bandwidth is decreased and the time–bandwidth product is somewhat increased.

The SPM-induced pulse broadening can be explained as follows: SPM (with positive n_2, as is usually the case) decreases the instantaneous frequency in the leading wing and increases the frequency in the trailing wing. The absorber always attenuates the leading wing, thus removing the lower frequency components, with the effect that the center frequency increases and the pulse bandwidth decreases. This also explains the longer pulses as shown in Fig. 9.38. It becomes apparent once again that the temporal asymmetry caused by a slow saturable absorber (but not by a fast absorber) plays a crucial role in the modelocking process.

In conclusion, the effect of SPM on passively modelocked picosecond lasers can be important for fast saturable absorbers and with soliton modelocking. For a slow absorber with negligible GDD, it is never beneficial as it always tends to make the pulses longer and may also destabilize them, particularly for absorbers with small modulation depth. Thus one should always keep the effect of SPM small, particularly by using short laser crystals. A rule of thumb is that the nonlinear phase shift for the peak should be at most a few mrad per 1% of modulation depth. It is clear that SPM could hardly be made weak enough in the sub-picosecond domain—for this reason (and not only due to the effect of dispersion), soliton modelocking is usually required in the sub-picosecond domain, because the nonlinear phase changes can be much larger there (see Sect. 9.5.1).

9.6 Summary of Analytical Solutions for all Modelocking Techniques

Table 9.4 summarizes the main theory results for all the different modelocking techniques that we have discussed in this textbook. We can provide analytical solutions using linearized operators that initially were introduced by H. A. Haus. This table allows for a more direct comparison between the different analytical results which also makes the analogy more visible. For example, we made such a comparison between active modelocking and passive modelocking with an ideal fast saturable absorber without SPM and GDD in Sect. 9.4.5.

For the linearized second order dispersion operator we introduced a dispersion parameter D (4.39) which is half of the intracavity roundtrip GDD. According to Table 2.3, the group delay dispersion (GDD) is given by the second order dispersion with $\phi = k_n L_d$:

$$\mathrm{GDD} = \frac{\mathrm{d}^2\phi}{\mathrm{d}\omega^2} = \frac{\mathrm{d}^2 k_n}{\mathrm{d}\omega^2} L_d = 2D \ .$$

With an ideal fast saturable absorber the shortest pulses can be generated. However, such an ideal fast saturable absorber is in reality not easily obtained and has some fundamental issues for self-starting modelocking (Sect. 9.4.7). The best results have been achieved with KLM. Note that KLM has a lot of trade-offs as discussed in Sect. 9.4.2, and has limited commercial applications beyond the scientific laser market so far. SESAM-assisted soliton modelocking can be considered the best approach for femtosecond pulse generation from ultrafast solid-state lasers and has been successfully commercialized for the industrial laser market. The requirements on the saturable absorber parameters are strongly relaxed because a slow saturable absorber only starts and stabilizes the modelocking.

We have discussed soliton modelocking using both an active loss modulator and a slow saturable absorber for starting and stabilizing the modelocking process. The theoretical treatment is very similar and is based on soliton perturbation theory, where a transform-limited soliton pulse is the final solution when the soliton loss is smaller than the loss for the perturbative continuum pulse. This was not the first time that soliton pulse formation was considered in passive modelocking. However, previously additional SPM with negative dispersion predicted only an additional pulse shortening by a factor of two before SPM starts to destabilize modelocking [85Mar, 91Hau]. In contrast, with soliton modelocking the pulse shortening can be as large as 10 to 30 times and much more SPM can be used for stable pulse generation.

Table 9.4 Summary of active and passive modelocking results using an analytical model with linearized operators that act on the pulse envelope for the different intracavity elements inside the laser cavity. The steady-state solution is given with the requirement that the pulse envelope does not change after one cavity roundtrip. The linearized operators with their more basic parameters are summarized in Appendix F and Table F.4

	Equation	Important parameters
Active modelocking without SPM and GDD (Sect. 6.4)		
Pulse envelope (6.62)	$A(t) = A_0 \exp\left(-\frac{t^2}{2\tau^2}\right)$	$D_\mathrm{g} \equiv \frac{g}{\Omega_\mathrm{g}^2}$ (6.57), gain dispersion
Pulse duration, FWHM (6.56)	$\tau_\mathrm{p} = 1.665\tau,\ \tau = \sqrt[4]{\frac{D_\mathrm{g}}{M_\mathrm{s}}}$	$M_\mathrm{s} \equiv \frac{M\omega_\mathrm{m}^2}{2}$ (6.58) curvature of loss modulation
Gain–loss balance (6.63) (gain equals loss)	$g = l + \frac{1}{2} M\omega_\mathrm{m}^2 \tau^2$	
Master equation (6.51)	$\left[g\left(1 + \frac{1}{\Omega_\mathrm{g}^2}\frac{\partial^2}{\partial t^2}\right) - l - M\left(1 - \cos\omega_\mathrm{m} t\right)\right] A(T,t) = 0$	
Gain dispersion and loss balance (6.64)	$\frac{1}{2} M\omega_\mathrm{m}^2 \tau^2 \overset{(6.58)}{=} M_\mathrm{s}\tau^2 \overset{(6.56)}{=} \frac{D_\mathrm{g}}{\tau^2}$	

Table 9.4 (continued)

	Equation	Important parameters
Active modelocking with SPM but no GDD (Sect. 6.5)		
Pulse envelope (6.75)	$A(t) = A_0 \exp\left[-\frac{1}{2}\frac{t^2}{\tau^2}(1 - \mathrm{i}\beta)\right]$	$\phi_{\mathrm{nl}} = \delta \frac{P_{\mathrm{peak}}}{\pi w_0^2} = 2\frac{2\pi}{\lambda_0} L_{\mathrm{K}} n_2 \frac{P_{\mathrm{peak}}}{\pi w_0^2}$ (6.94) nonlinear phase shift per resonator roundtrip
Pulse duration, FWHM (6.96)	$\tau_{\mathrm{p}} = 1.665\tau,\ \tau^4 = \frac{D_{\mathrm{g}}}{\frac{M\omega_{\mathrm{m}}^2}{2} + \frac{\phi_{\mathrm{nl}}^2}{4D_{\mathrm{g}}}}$	$\beta = \frac{\tau^2 \phi_{\mathrm{nl}}}{2D_{\mathrm{g}}}$ (6.95) chirp parameter for Gaussian pulse
Gain–loss balance (6.99) (gain equals loss)	$g - l - \frac{M\omega_{\mathrm{m}}^2}{2}\tau^2 - \frac{\phi_{\mathrm{nl}}^2}{4D_{\mathrm{g}}}\tau^2 = 0$	
Master equation (6.72)	$\left[g - l + D_{\mathrm{g}}\frac{\partial^2}{\partial t^2} - M(1 - \cos\omega_{\mathrm{m}} t) - \mathrm{i}\delta\,\lvert A(T,t)\rvert^2 + \mathrm{i}\psi\right] A(T,t) = 0$	

Table 9.4 (continued)

	Equation	Important parameters
Active modelocking with SPM and GDD (soliton modelocking) (Sect. 6.6)		
Pulse envelope (6.118)	$A(T,t)=\left(A_0\,\mathrm{sech}\,\frac{t}{\tau}\right)\exp\left(\mathrm{i}\phi_0\frac{T}{T_\mathrm{R}}\right)+\text{continuum}$	
Pulse duration, FWHM (6.121)	$\tau_\mathrm{min}=1.763\sqrt[6]{\frac{2D_\mathrm{g}^2}{9\phi_0 M_\mathrm{s}}}$	$\phi_0=\frac{\lvert D\rvert}{\tau^2}=\frac{\delta F_\mathrm{p,L}}{4\tau}$ (9.107) Soliton phase per resonator roundtrip
Gain–loss balance (6.124) (gain equals loss)	$g=l+l_\mathrm{s}\,,\; l_\mathrm{s}=\frac{D_\mathrm{g}}{3\tau^2}+\frac{\pi}{24}M\omega_\mathrm{m}^2\tau^2$	
Master equation (6.117)	$T_\mathrm{R}\frac{\partial}{\partial T}A(T,t)=\left[\mathrm{i}D\frac{\partial^2}{\partial t^2}-\mathrm{i}\delta\,\lvert A(T,t)\rvert^2\right]A(T,t)+\left[g-l+D_\mathrm{g}\frac{\partial^2}{\partial t^2}-M\left(1-\cos\omega_\mathrm{m}t\right)\right]A(T,t)=0$	

Table 9.4 (continued)

	Equation	Important parameters
Passive modelocking with dynamic gain saturation (Sect. 9.3)		
Pulse envelope (9.26)	$A(t)=A_0\,\mathrm{sech}\left(\frac{t}{\tau}\right)$	
Pulse duration, FWHM (9.34) and (9.35)	$\tau=\frac{4}{\pi}\frac{1}{\Delta\nu_g}\frac{E_{\mathrm{sat,L}}}{E_p}\frac{1}{\sqrt{\frac{E_{\mathrm{sat,L}}^2}{E_{\mathrm{sat,A}}^2}\frac{q_A}{g_0}-1}}\ \xrightarrow[E_{\mathrm{sat,L}}\gg E_{\mathrm{sat,A}}]{q_A\approx g_0}\ \tau\approx\frac{4}{\pi}\frac{1}{\Delta\nu_g}\frac{E_{\mathrm{sat,A}}}{E_p}$	
Loss modulation (9.22)	$q(t)=q_A\exp\left[-\frac{\sigma_A}{A_A h\nu}\int_{-\infty}^{t}\left\|A(t')\right\|^2\mathrm{d}t'\right]$	
Dynamics gain saturation (9.20)	$g(t)=g_0\mathrm{e}^{-E_p(t)/E_{\mathrm{sat,L}}}=g_0\exp\left[-\frac{\sigma_L}{A_L h\nu}\int_{-\infty}^{t}\left\|A(t')\right\|^2\mathrm{d}t'\right]$	$E_p(t)\propto\int_{-\infty}^{t}\left\|A(t')\right\|^2\mathrm{d}t'$ (9.10)
Net gain window (9.14)	$g_T(E)=g_0-q_A-l_R+\left(q_A-\frac{g_0}{s}\right)E+\frac{1}{2}\left(\frac{g_0}{s^2}-q_A\right)E^2$ $g_T^{\max}=\frac{l_0}{2}\frac{(1-1/s)^2}{1-1/s^2}$ (9.15)	$E\equiv\frac{E_p(t)}{E_{\mathrm{sat,A}}}$ $s\equiv\frac{E_{\mathrm{sat,L}}}{E_{\mathrm{sat,A}}}$ (9.13)
Master equation (9.25)	$\left[g(t)-q(t)+\frac{g_0}{\Omega_g^2}\frac{\partial^2}{\partial t^2}+t_D\frac{\partial}{\partial t}\right]A(T,t)=0$	

Table 9.4 (continued)

	Equation	Important parameters
Passive modelocking with ideal fast saturable absorber, no SPM and no GDD (Sects. 9.4 and 9.4.4)		
Pulse envelope (9.50)	$A(t) = A_0 \operatorname{sech}\left(\frac{t}{\tau}\right)$	$P(t) = \|A(t)\|^2$
Pulse duration, FWHM (9.59)	$\tau_p = 1.763\tau\ ,\ \ \tau = \frac{4g}{\Omega_g^2 \bar{\gamma}_A E_p} = \frac{4D_g}{\bar{\gamma}_A E_p}$	$\delta \rightarrow \bar{\delta} \equiv \frac{\delta}{A_L}\ ,\ \ \gamma_A \rightarrow \bar{\gamma}_A \equiv \frac{\gamma_A}{A_A}$ (9.37) A_A and A_L are the effective mode areas in absorber and laser gain
Gain–loss balance (9.54) (gain equals loss)	$l_0 = g + \frac{1}{2}\bar{\gamma}_A \|A_0\|^2$	$\frac{1}{M_s} = \frac{2}{M\omega_m^2} \iff \frac{1}{M_{SAM}} \equiv \frac{2\tau^2}{\bar{\gamma}_A \|A_0\|^2}$ (9.60) Analogy with active modelocking M_{SAM} = curvature of self-amplitude modulation (SAM)
Master equation (9.49)	$g\left(1 + \frac{1}{\Omega_g^2}\frac{\partial^2}{\partial t^2}\right)A - l_0 A + \bar{\gamma}_A \|A\|^2 A = 0$	
Loss dynamics (9.43)	$l(t) = l_0 - \bar{\gamma}_A \|A(t)\|^2$	

Table 9.4 (continued)

	Equation	Important parameters
Passive modelocking with ideal fast saturable absorber with SPM and GDD (Sects. 9.4 and 9.4.6)		
Pulse envelope (9.63)	$A(t) = A_0 \left[\operatorname{sech}\left(\frac{t}{\tau}\right) \right]^{1+i\beta}, \quad P(t) = \lvert A(t) \rvert^2$	
Pulse duration, FWHM (9.91)	$\tau_p = 1.763\tau$, $\tau_n = \frac{2-\beta^2-3\beta D_n}{\bar{\gamma}_A} = \frac{-2D_n + D_n\beta^2 - 3\beta}{\bar{\delta}}$	$\delta \to \bar{\delta} \equiv \frac{\delta}{A_L}, \quad \gamma_A \to \bar{\gamma}_A \equiv \frac{\gamma_A}{A_A}$ (9.37) A_A and A_L are the effective mode areas in absorber and laser gain
Gain–loss balance (9.98) (gain equals loss)	$l_0 = g - \frac{1}{3}\frac{g}{\Omega_g^2 \tau^2}\left(1+\beta^2\right) + \frac{2}{3}\bar{\gamma}_A A_0^2$	$\beta = -\frac{3}{2}\chi \pm \sqrt{\left(\frac{3}{2}\chi\right)^2 + 2}$ (9.88) chirp parameter for soliton pulse
Master equation (9.62)	$g\left(1 + \frac{1}{\Omega_g^2}\frac{\partial^2}{\partial t^2}\right)A - l_0 A + \bar{\gamma}_A \lvert A \rvert^2 A +$ $iD\frac{\partial^2}{\partial t^2}A - i\bar{\delta}\lvert A \rvert^2 A - i\psi A = 0$	$D \equiv \frac{1}{2}k_n'' L_d$ (4.39) dispersion parameter
Loss dynamics (9.43)	$l(t) = l_0 - \bar{\gamma}_A \lvert A(t) \rvert^2$	
Balance for $\beta = 0$ (9.80)	$\frac{\lvert D \rvert}{\bar{\delta}} = \frac{D_g}{\bar{\gamma}_A}$	$D_n \equiv \frac{D}{g/\Omega_g^2} = \frac{D}{D_g}$ (9.81)
Stability criteria (9.95)	$S \equiv \left(1-\beta^2\right) - 2\beta D_n > 0$	$\frac{3\beta}{2-\beta^2} = \frac{\bar{\delta} + \bar{\gamma}_A D_n}{\bar{\delta} D_n - \bar{\gamma}_A} \equiv \frac{1}{\chi}$ (9.86)

Table 9.4 (continued)

	Equation	Important parameters
Soliton modelocking: passive modelocking with slow saturable absorber, constant gain, with SPM and GDD (Sect. 9.5)		
Pulse envelope (9.105)	$A_s(T,t) = \sqrt{\frac{F_{p,L}}{2\tau}}\operatorname{sech}\left(\frac{t}{\tau}\right)e^{i\phi_0 T/T_R}$	$\phi_0 = \frac{\vert D\vert}{\tau^2} = \frac{\delta F_{p,L}}{4\tau}$ (4.52) Soliton phase per resonator roundtrip
Pulse duration, FWHM (9.106) and (4.57)	$\tau_{FWHM} = 1.76\tau = 1.76\frac{4\vert D\vert}{\delta F_{p,L}}$	$D \equiv \frac{1}{2}k_n'' L_d$ (4.39) dispersion parameter
Gain–loss balance (9.108) (gain equals loss)	$g = l + l_s\,,\quad l_s = \frac{D_g}{3\tau^2} + q_p = \frac{D_g}{3\tau^2} + q_0\frac{F_{sat,A}}{F_{p,A}}\left(1 - e^{-F_{p,A}/F_{sat,A}}\right)$	
Master equation (9.100)	$\left[iD\frac{\partial^2}{\partial t^2} - i\delta\,\vert A(T,t)\vert^2\right]A(T,t) + \left[g - l + D_g\frac{\partial^2}{\partial t^2} - q(t)\right]A(T,t) = 0$	
Passive modelocking with slow saturable absorber, constant gain, with SPM and GDD, but no soliton formation (Sect. 9.5.4)		
Pulse envelope	$A(t) = A_0\exp\left(-\frac{t^2}{2\tau^2}\right)$	
Pulse duration, FWHM (9.122)	$\tau_p = 1.665\tau\,,\quad \tau_p \approx \frac{1.07}{\Delta v_g}\sqrt{\frac{2g}{\Delta R}}$	

Active modelocking has been replaced by SESAM modelocking for most solid-state lasers because better performance can be achieved with a fully passive approach. SESAM damage is not an issue even for ultrafast laser oscillators in the multi-100 W average output power regime. Special SESAM design criteria in this high average power regime are given in Sect. 7.6 and the performance frontier of such lasers is summarized in Sect. 9.10.2. Soliton modelocking becomes more challenging for very high pulse repetition rates with diode pumped solid-state lasers, which will be discussed in more detail in Sect. 9.10.3.

The success of ultrafast solid-state lasers means that they have replaced passively modelocked dye lasers (Sect. 9.3). Therefore, the passive modelocking technique based on the more critical balance between dynamic gain and loss saturation has become less important (Fig. 9.5a) . Most solid-state lasers only experience a constant gain saturation determined by the average intensity inside the laser crystal (Fig. 9.5). However, optically pumped semiconductor lasers have become more important and have been successfully commercialized, replacing the more expensive Ti:sapphire laser for many industrial applications. Passive modelocking of optically pumped semiconductor lasers will be discussed in Sect. 9.8. For this case dynamic gain saturation is relevant and the basic laser physics knowhow could be transferred from the dye lasers.

A fundamental issue for passive modelocked solid-state lasers is Q-switching instabilities. This will be discussed in the next section. This issue can be addressed with the right saturable absorber parameters and the appropriate cavity designs. This results in reliable and stable passive modelocking.

9.7 Q-Switching Instabilities of Passively Modelocked Solid-State Lasers

9.7.1 Q-Switching Instabilities: A More Serious Issue for Solid-State Lasers

Solid-state lasers can be passively modelocked by incorporation of a saturable absorber, which, however, at the same time tends to introduce Q-switching instabilities that can drive the laser into the regime of stable Q-switched modelocking (in the following referred to as QML) or simply make stable modelocking impossible [99Hon1]. In the QML regime of operation, the laser output consists of modelocked pulses underneath a Q-switched envelope (Fig. 9.39). The modelocked pulse repetition rates in diode pumped solid-state laser oscillators are on the order of 10 MHz to 10 GHz, determined by the laser cavity length, while typical Q-switching modulations have frequencies in the kHz region, approximately given by the relaxation oscillation frequency (see Chap. 5).

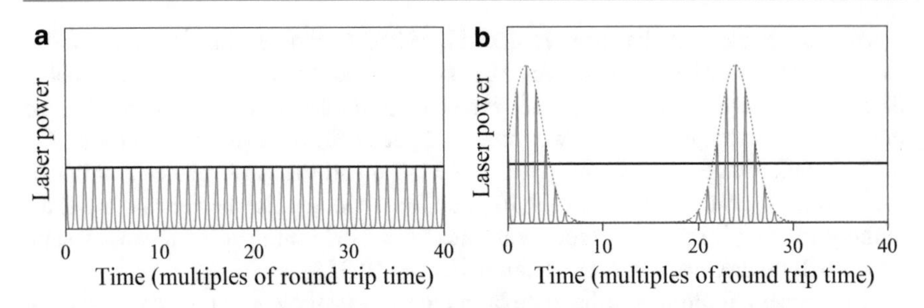

Fig. 9.39 Instantaneous and average laser power versus time for **a** a stable cw modelocked laser and for **b** a modelocked laser exhibiting large Q-switching instabilities. The average laser power (*bold*) is the same for both lasers [99Hon1]

Stable QML is possible, but most often noise is substantially increased due to Q-switching instabilities. Such Q-switching instabilities are unwanted for most applications, where constant pulse energy and a high repetition rate are required. However, stable QML can be a benefit for some applications, such as nonlinear frequency conversion, precise fabrication of microstructures, or medical surgery. In this case the QML regime may be attractive due to the significantly increased pulse energy, which is still concentrated in ultrashort modelocked pulses. In any case, it is important to understand the solid-state laser dynamics with respect to QML and Q-switching instabilities in order to suppress or exploit this QML mode of operation.

For a long time, Q-switching instabilities have been observed when solid-state lasers with small gain cross-sections (i.e., $< 10^{-19}$ cm^2) were passively modelocked with an intracavity saturable absorber. Early examples are ruby lasers [65Moc], using dye and color filter glass saturable absorbers, and Nd:glass lasers [66DeM], using dye saturable absorbers. Q-switching instabilities could not be suppressed in solid-state lasers, such as Nd:YAG lasers, because of the limited parameter regime of available saturable absorbers in the past. In the attempt to passively modelock an Nd:YAG laser, the theoretical model by Haus in 1976 predicted that "stable modelocking is unachievable" and "steady-state modelocking is prevented by relaxation oscillations" [76Hau, p. 174].

The first intracavity saturable absorber that generated stable self-starting cw modelocked pulses from a solid-state laser with small gain cross-sections and long upper state lifetimes, such as Nd:YAG and Nd:YLF lasers, was based on semiconductor nonlinearities [92Kel2]. Since then, many different lasers have been SESAM-modelocked with excellent stability. Semiconductor saturable absorbers have the advantage that the relevant absorber parameters can be varied over several orders of magnitude (see Chap. 7). Therefore, semiconductor saturable absorbers allowed for a systematic investigation of the Q-switching stability limits in passively modelocked solid-state lasers.

We can directly observe the onset of Q-switching instabilities when we observe the output of the passively modelocked laser with a photodetector and a microwave spectrum analyzer (Fig. 9.40). This gives a quantitative measure for the onset of self-Q-switching instabilities. In the microwave frequency domain, the detected intensity of the modelocked laser produces harmonics at multiples of the pulse repetition rate, while noise and relaxation oscillations produce noise sidebands which peak at the relaxation oscillation frequency with a certain attenuation with respect to the laser harmonic (see Chap. 5). With increasing pump power, a SESAM-modelocked solid-state laser typically goes through three stages, which are demonstrated with an Nd:YLF laser and shown in Fig. 9.40 [94Kel]:

- First, at the lowest pump power of 0.6 W (Fig. 9.40), the SESAM produces strong self-Q-switching instabilities with dominant noise sidebands around the enhanced axial mode beating noise peak, which later becomes the first-harmonic signal at the pulse repetition rate. No stable pulse formation is observed in this case on any other diagnostic system, such as an autocorrelation or a sampling scope. Note that the SESAM introduces axial mode beating noise because some noise spikes can start to saturate the SESAM and therefore more than one axial mode can temporarily reach the lasing threshold.
- Second, in an intermediate regime with stable QML at a pump power of 1.4 W (Fig. 9.40), self-Q-switching becomes stable, typically producing μs-long macro pulses forming an amplitude envelope over the modelocked pulse train. The self-Q-switching produces very strong modulation sidebands (typically around -1 dBc) at around the relaxation oscillation frequency. Note that the theory for the relaxation oscillation is based on a small signal perturbation which is no longer the case with stable QML. However, even in this strong signal regime, the relaxation oscillation frequency gives a good first approximation for the repetition rate of the Q-switched macro pulses. Short modelocked pulses are observed on an autocorrelator (see Chap. 10 for more details about pulse characterization techniques).
- Third, well above the Q-switched modelocking (QML) threshold at a pump power of 2 W (Fig. 9.40), the laser operates in stable cw modelocking with the relaxation oscillation-noise sidebands typically > 40 dB below the first harmonic of the modelocked pulses. The strength of the relaxation oscillation noise sidebands with respect to the first harmonic gives a quantitative measure of how well self-Q-switching instabilities are suppressed and how stably the laser is cw modelocked.

As shown above, the microwave spectrum analyzer measurements for the QML threshold allowed for the confirmation of the first analytical treatment of Q-switching instabilities in solid-state lasers modelocked by saturable absorbers. First results were published in 1995 [95Kar4] and in more detail in 1999 [99Hon1] for both positive and negative GDD. We have worked out two very simple conditions which determine a QML threshold. For picosecond solid-state lasers with stable cw modelocking, the following condition applies [99Hon1, Eq. (13)]:

$$\boxed{E_{\mathrm{p}}^2 > E_{\mathrm{sat,L}} E_{\mathrm{sat,A}} \Delta R} \,. \tag{9.126}$$

For this equation we used (7.11) (see Table 7.4) assuming a small modulation depth, less than 10%. We further assumed a fully saturated absorber, so that $\exp(-S) \approx 0$, with $S = E_{\mathrm{p}}/E_{\mathrm{sat,A}}$. Finally, in addition we assumed that $\tau_{\mathrm{L}} \gg T_{\mathrm{R}}$ because for

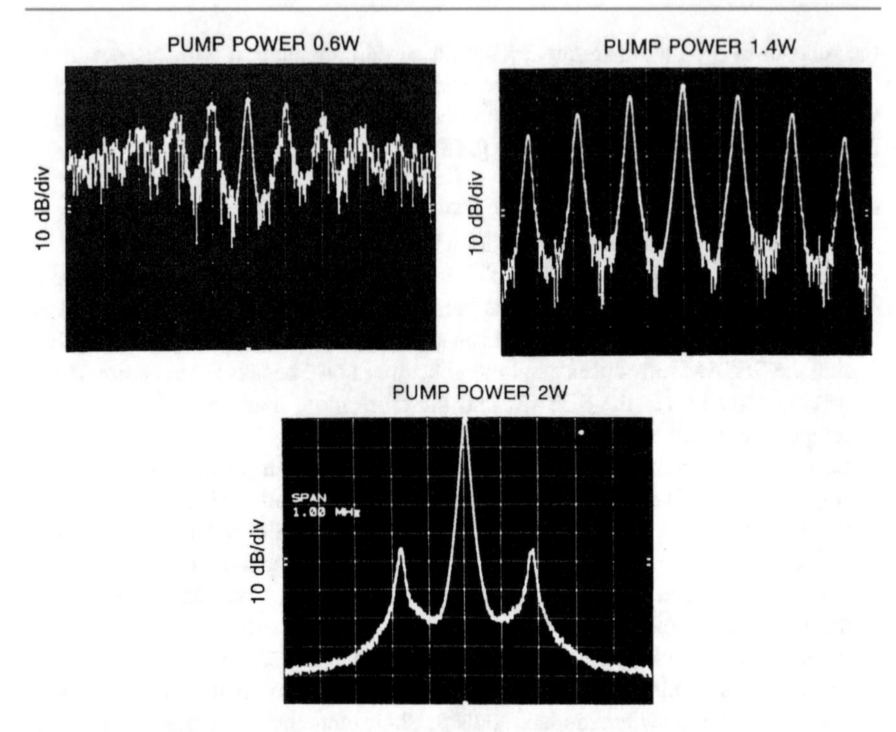

Fig. 9.40 SESAM modelocked Nd:YLF laser. Microwave spectrum analyzer at a center frequency of 219.4 MHz, a frequency span of 1 MHz, and a resolution bandwidth of 10 kHz, showing the transition from stable cw modelocking to Q-switching instabilities [94Kel]

typical solid-state lasers the upper-state lifetime τ_L is much longer than the cavity roundtrip time T_R.

For soliton modelocking using the same approximations as above but now in the femtosecond regime, we have a more relaxed condition given by [99Hon1, Eq. (27)]:

$$\boxed{2gE_{\mathrm{sat,L}}K^2E_{\mathrm{p}}^3 + E_{\mathrm{p}}^2 > E_{\mathrm{sat,L}}E_{\mathrm{sat,A}}\Delta R}\,, \tag{9.127}$$

where g is the saturated amplitude gain coefficient per cavity roundtrip, E_{p} is the pulse energy inside the laser resonator, and ΔR is the modulation depth of the saturable absorber. For the femtosecond laser, the QML threshold is typically 5 times lower, because the soliton effect together with the gain filter also stabilizes pulse generation against Q-switching instabilities. In (9.127), we have

$$K \equiv \frac{0.315}{1.763}\frac{\pi n_2 L_{\mathrm{K}}}{k_n'' L_{\mathrm{d}} A_{\mathrm{eff,L}}\lambda_0 \Delta\nu_{\mathrm{g}}}\,, \tag{9.128}$$

where L_{K} is the length of the Kerr medium for one cavity roundtrip inside the laser resonator, $\Delta\nu_{\mathrm{g}}$ is the FWHM gain bandwidth of the laser, λ_0 is the center wavelength of the pulse, $k_n''L_{\mathrm{d}}$ is the total second-order dispersion per resonator roundtrip, and $A_{\mathrm{eff,L}}$ is the spatially average effective laser beam cross-section in the laser material.

The basic idea for the additional stabilization in the femtosecond region is the following. If the energy of a pulse increases by relaxation oscillations, the spectrum of the pulse is broadened by SPM (or also SAM). A broader spectrum, however, will experience a smaller average gain due to the finite bandwidth of the laser material. This effect has a much smaller influence on picosecond lasers, because SPM is much weaker. These two conditions will be derived in detail in Sects. 9.7.2 and 9.7.3.

However, for some high repetition rate lasers, the QML threshold was found to be significantly lower than expected. It was shown that, for some Er:Yb:glass lasers, this could be explained with modified saturation characteristics of the SESAMs, taking into account inverse saturable absorption (see Sect. 7.3.5) [04Sch,05Gra2].

Inverse saturable absorption (ISA) causes a rollover of the nonlinear reflectivity for higher pulse fluences (Fig. 7.5). This rollover lowers the QML threshold [85Har, 99Tho,05Gra2]. With precise nonlinear reflectivity measurements as explained in Sect. 7.7, with many different SESAMs it was found that, for femtosecond pulses, the observed strong rollover can be explained by TPA [05Gra2]. For picosecond pulses, however, the rollover is much stronger than expected from TPA alone. Thus, an additional mechanism must be at work here, such as free-carrier absorption, Auger recombination, hot-carrier generation, or lattice heating.

Considering the rollover caused by ISA, the threshold condition against Q-switching instabilities is modified [05Gra2, Eq. (21)]:

$$\boxed{E_\mathrm{p}^2 > \frac{E_\mathrm{sat,A}\Delta R}{\dfrac{1}{E_\mathrm{sat,L}} + \dfrac{1}{A_\mathrm{A} F_2}}}\,, \tag{9.129}$$

where A_A is the mode area on the SESAM, and F_2 is the inverse slope of the ISA effect (Sect. 7.3.5). For $F_2 \to \infty$, i.e., without the rollover, we obtain once again the simpler (9.126). This condition will be derived in detail in Sect. 9.7.4. Note also that we used here the approximations of $\tau_\mathrm{L} \gg T_\mathrm{R}$, small modulation depth of less than 10%, and a fully saturated absorber so that $\exp(-S) \approx 0$, with $S = E_\mathrm{p}/E_\mathrm{sat,A}$.

It is useful to reformulate the two QML threshold conditions without (9.126) and with (9.129) ISA as follows:

$$\boxed{E_\mathrm{p} > E_\mathrm{sat,L}\frac{\Delta R}{S}}\,, \quad \text{without ISA}\,, \tag{9.130}$$

$$\boxed{E_\mathrm{p} > E_\mathrm{sat,L}\frac{\Delta R}{S}\left(1 - \frac{S^2}{S_0^2}\right)}\,, \quad \text{with ISA}\,, \tag{9.131}$$

using the saturation parameter

$$S = \frac{E_\mathrm{p}}{E_\mathrm{sat,A}}\,, \quad S_0 \equiv \sqrt{\frac{F_2 \Delta R}{F_\mathrm{sat,A}}}\,. \tag{9.132}$$

9.7.2 Q-Switching Instabilities Without ISA: Derivation and Discussion of (9.126)

To derive a stability criterion against QML, we start from the rate equations for the intracavity power, gain, and saturable absorption. We obtain this criterion by performing a linearized stability analysis analogous to the derivation of the relaxation oscillations (Chap. 5): stability means that the relaxation oscillations are damped. The rate equations for the modelocked laser can be written as [see (8.40)–(8.42) and [99Hon1]], assuming amplitude gain and loss coefficients,

$$\frac{\mathrm{d}P}{\mathrm{d}t} = 2\frac{g - l - q_{\mathrm{p}}(E_{\mathrm{p}})}{T_{\mathrm{R}}} P , \tag{9.133}$$

$$\frac{\mathrm{d}g}{\mathrm{d}t} = -\frac{g - g_0}{\tau_{\mathrm{L}}} - \frac{P}{E_{\mathrm{sat,L}}} g , \tag{9.134}$$

$$\frac{\mathrm{d}q}{\mathrm{d}t} = -\frac{q - q_0}{\tau_{\mathrm{A}}} - \frac{P}{E_{\mathrm{sat,A}}} q , \tag{9.135}$$

where P is the average intracavity laser power and T_{R} the cavity roundtrip time. $E_{\mathrm{p}} = P T_{\mathrm{R}}$ is the energy of a modelocked pulse in the cavity. Here we assume a four-level laser. The modification in the rate equations for a three-level laser is given by (5.67). We always assume fundamental modelocking, i.e., the presence of a single pulse in the cavity. g and q are the time-dependent roundtrip amplitude gain and saturable absorption coefficient, respectively, while g_0 and q_0 denote the corresponding quantities in equilibrium with no intracavity power, i.e., for $P = 0$. l is the linear loss per roundtrip. Note that the quantities P, E_{p}, and g are described on the time scale of several roundtrip times T_{R}, whereas (9.135) for $q(t)$ must be solved for the time elapsed during a pulse (Chap. 7). τ_{L} and τ_{A} are the upper state lifetime of the laser medium and the absorber recovery time, respectively.

$E_{\mathrm{sat,L}}$ is the saturation energy of the gain, which is defined as the product of saturation fluence $F_{\mathrm{sat,L}} = h\nu/m\sigma_{\mathrm{L}}$ and the effective laser mode area inside the gain medium $A_{\mathrm{eff,L}}$. We define the effective mode area as $A_{\mathrm{eff,L}} = \pi w^2$, where w is the $1/\mathrm{e}^2$ Gaussian beam radius with respect to intensity. The factor m in the definition of $F_{\mathrm{sat,L}}$ is the number of passes through the gain element per cavity roundtrip. Due to this factor, the gain saturation depends on the geometry of the laser cavity. For a ring cavity, m is equal to 1, while for a simple standing wave cavity, m is 2. For a cavity with multiple passes through the gain medium, m can be greater than 2. For simplicity, we assume the gain material to be homogeneously broadened. Note, however, that inhomogeneous line broadening also influences the saturation behavior of the gain. The measured emission cross-section σ_{L} at a certain wavelength usually represents an average for all ions. However, the laser ions with the largest cross-sections participate more strongly in the Q-switching cycle, so the gain saturation is stronger than that assumed by using the averaged cross-section. As we will see later, this should lead to a weaker tendency for QML compared to the predictions of our model. We also neglect spatial hole burning effects, assuming, e.g., that the gain medium is located in the middle of a standing wave laser cavity (see Sect. 6.7).

$E_{\mathrm{sat,A}}$ denotes the absorber saturation energy and is defined by the product of absorber saturation fluence $F_{\mathrm{sat,A}}$ and effective laser mode area on the saturable absorber, i.e., $A_{\mathrm{eff,A}} F_{\mathrm{sat,A}}$. This can be measured and corresponds to the pulse fluence which is necessary to bleach the saturable absorption to $1/\mathrm{e}$ of its maximum amount q_0 (see Chap. 7 for more details).

The quantity $q_{\mathrm{p}}(E_{\mathrm{p}})$ in (9.133) represents the roundtrip amplitude loss introduced by the saturable absorber for the given intracavity pulse energy. We make two assumptions to determine q_{p}. First, we assume a slow absorber, i.e., the duration τ_{p} of the modelocked pulses must be shorter than the absorber recovery time τ_{A}, although we found that the results remain valid even for $\tau_{\mathrm{A}} \approx \tau_{\mathrm{p}}$. Second, the absorber recovery time τ_{A} must be much shorter than the cavity roundtrip time T_{R}. This can easily be achieved for semiconductor absorbers since τ_{A} can be custom-designed within a wide range. Then for (9.135) we obtain the pulse energy amplitude loss per round trip [see (7.14) or the summary given in Table 7.2]:

$$q_{\mathrm{p}}(E_{\mathrm{p}}) = q_0 \frac{F_{\mathrm{sat,A}} A_{\mathrm{eff,A}}}{E_{\mathrm{p}}} \left[1 - \exp\left(- \frac{E_{\mathrm{p}}}{F_{\mathrm{sat,A}} A_{\mathrm{eff,A}}} \right) \right] . \tag{9.136}$$

Using this result, we can describe the modelocked laser by the following two coupled rate equations:

$$\dot{P} \equiv \frac{\mathrm{d}P}{\mathrm{d}t} = 2 \frac{g - l - q_{\mathrm{p}}(E_{\mathrm{p}})}{T_{\mathrm{R}}} P , \tag{9.137}$$

$$\dot{g} \equiv \frac{\mathrm{d}g}{\mathrm{d}t} = \frac{g_0 - g}{\tau_{\mathrm{L}}} - \frac{P}{E_{\mathrm{sat,L}}} g . \tag{9.138}$$

By linearizing these equations for small deviations δP and δg from the steady state values P and g, by analogy with the examination of relaxation oscillations in Chap. 5, we obtain the stability criterion against Q-switching instabilities:

$$-2E_{\mathrm{p}} \frac{\partial q_{\mathrm{p}}(E_{\mathrm{p}})}{\partial E_{\mathrm{p}}} < \frac{T_{\mathrm{R}}}{\tau_{\mathrm{L}}} + \frac{E_{\mathrm{p}}}{E_{\mathrm{sat,L}}} . \tag{9.139}$$

Derivation of (9.139)
Linearizing the system (9.137) and (9.138) around P and g by inserting $P + \delta P$ and $g + \delta g$, we get

$$\begin{pmatrix} \delta \dot{P} \\ \delta \dot{g} \end{pmatrix} = \begin{pmatrix} \alpha & \beta \\ -\gamma & -\varepsilon \end{pmatrix} \begin{pmatrix} \delta P \\ \delta g \end{pmatrix} , \tag{9.140}$$

where

$$\alpha \equiv \frac{\partial \dot{P}}{\partial P} \overset{(9.137)}{=} 2 \frac{g - l - q_{\mathrm{p}}(E_{\mathrm{p}})}{T_{\mathrm{R}}} + 2 \frac{P}{T_{\mathrm{R}}} \left[- \frac{\partial q_{\mathrm{p}}(E_{\mathrm{p}})}{\partial E_{\mathrm{p}}} \frac{\partial E_{\mathrm{p}}}{\partial P} \right] , \tag{9.141}$$

and the first term on the right-hand side of (9.141) is zero at steady state. Using $E_p = PT_R$, we have $\partial E_p/\partial P = T_R$, and therefore

$$\alpha = -2P\frac{\partial q_p(E_p)}{\partial E_p} , \tag{9.142}$$

$$\beta \equiv \frac{\partial \dot{P}}{\partial g} = 2\frac{P}{T_R} , \tag{9.143}$$

$$-\gamma \equiv \frac{\partial \dot{g}}{\partial P} = -\frac{g}{E_{sat,L}} , \tag{9.144}$$

$$-\varepsilon \equiv \frac{\partial \dot{g}}{\partial g} = -\frac{1}{\tau_L} - \frac{P}{E_{sat,L}} . \tag{9.145}$$

The eigenvalues are then given by

$$(\alpha - \lambda)(-\varepsilon - \lambda) - (-\gamma\beta) = 0 ,$$

and hence,

$$\lambda^2 + (\varepsilon - \alpha)\lambda + \beta\gamma - \alpha\varepsilon = 0 .$$

The two eigenvalues are

$$\lambda_{1,2} = -\frac{1}{2}(\varepsilon - \alpha) \pm \sqrt{-\beta\gamma + \alpha\varepsilon + \frac{1}{4}(\varepsilon - \alpha)^2} . \tag{9.146}$$

For typical solid-state lasers, the radicand is negative. For the frequency of the relaxation oscillations (the imaginary part of the eigenvalues), we find

$$\omega_{relax} = \sqrt{\beta\gamma - \alpha\varepsilon - \frac{1}{4}(\varepsilon - \alpha)^2} , \tag{9.147}$$

while the damping of the oscillations is given by the real part of λ, viz.,

$$\mathrm{Re}(\lambda) = -\frac{1}{2}(\varepsilon - \alpha) .$$

For stable modelocking, the relaxation oscillations have to be damped, which means that $\mathrm{Re}(\lambda) < 0$, and therefore

$$\varepsilon - \alpha > 0 . \tag{9.148}$$

This reads

$$\frac{1}{\tau_L} + \frac{P}{E_{sat,L}} + 2P\frac{\partial q_p(E_p)}{\partial E_p} > 0 ,$$

whence

$$-2P\frac{\partial q_{\mathrm{p}}(E_{\mathrm{p}})}{\partial E_{\mathrm{p}}} < \frac{1}{\tau_{\mathrm{L}}} + \frac{P}{E_{\mathrm{sat,L}}} . \tag{9.149}$$

Using $P = E_{\mathrm{p}}/T_{\mathrm{R}}$, we have

$$-2E_{\mathrm{p}}\frac{\partial q_{\mathrm{p}}(E_{\mathrm{p}})}{\partial E_{\mathrm{p}}} < \frac{T_{\mathrm{R}}}{\tau_{\mathrm{L}}} + \frac{E_{\mathrm{p}}}{E_{\mathrm{sat,L}}} . \tag{9.150}$$

The physical understanding of the stability criterion against Q-switching instabilities given by (9.139) goes back to the criterion $\varepsilon - \alpha > 0$ in the derivation, which means in principle that δq_{p} should be smaller than δg.

If the pulse energy rises slightly due to relaxation oscillations, this pulse energy fluctuation first grows exponentially due to the stronger bleaching of the absorber. On the other hand, the increased pulse energy starts to saturate the gain. The laser is stable against QML if the gain saturation is sufficiently strong to stop the exponential rise. So far, the theory neglects gain filtering and nonlinear effects that act on the pulse bandwidth. This point, however, becomes important for soliton-like pulses in the femtosecond regime. We will discuss the consequences of this in Sect. 9.7.3.

The nonlinear reflectivity $R(E_{\mathrm{p}})$ of the SESAM is related to the pulse energy loss per roundtrip $q_{\mathrm{p}}(E_{\mathrm{p}})$, as discussed in Chap. 7 and summarized in Table 7.4. Remember that we did not include any nonsaturable losses in $q(t)$ or $q_{\mathrm{p}}(E_{\mathrm{p}})$. We make the approximation of a small modulation depth, less than 10%. Then, from (7.8) (see Table 7.4), it follows that

$$R(F_{\mathrm{p}}) \approx R_{\mathrm{ns}} - 2q_{\mathrm{p}}(F_{\mathrm{p}}) = R_{\mathrm{ns}} - \frac{\Delta R}{S}(1 - \mathrm{e}^{-S}) ,$$

where S is the saturation parameter, i.e., $S = F_{\mathrm{p}}/F_{\mathrm{sat,A}}$ and $F_{\mathrm{p}} = E_{\mathrm{p}}/A_{\mathrm{eff,A}}$. The inequality (9.150) can then be rewritten with the nonlinear reflectivity as

$$\boxed{E_{\mathrm{p}}\frac{\mathrm{d}R(E_{\mathrm{p}})}{\mathrm{d}E_{\mathrm{p}}} < \frac{T_{\mathrm{R}}}{\tau_{\mathrm{L}}} + \frac{E_{\mathrm{p}}}{E_{\mathrm{sat,L}}}} . \tag{9.151}$$

We can further simplify (9.151) assuming a fully saturated absorber, so that $\exp(-S) \approx 0$. We then obtain, using (7.8) (see Table 7.4) mentioned again just above,

$$R(E_{\mathrm{p}}) \approx R_{\mathrm{ns}} - \Delta R/S .$$

In addition, for most solid-state lasers with long upper state lifetimes τ_{L}, we have $\tau_{\mathrm{L}} \gg T_{\mathrm{R}}$. With these two approximations, we have

$$\begin{aligned} E_{\mathrm{p}}\frac{\mathrm{d}R(E_{\mathrm{p}})}{\mathrm{d}E_{\mathrm{p}}} &= E_{\mathrm{p}}\frac{\mathrm{d}}{\mathrm{d}E_{\mathrm{p}}}\left(1 - \frac{A_{\mathrm{eff,A}}\Delta R}{E_{\mathrm{p}}}F_{\mathrm{sat,A}}\right) \\ &= E_{\mathrm{p}}\,\Delta R\,E_{\mathrm{sat,A}}/E_{\mathrm{p}}^2 , \end{aligned}$$

with $E_{\text{sat,A}} = A_{\text{eff,A}} F_{\text{sat,A}}$. Using $\tau_L \gg T_R$ in (9.151),

$$\frac{E_p \, \Delta R \, E_{\text{sat,A}}}{E_p^2} < \frac{E_p}{E_{\text{sat,L}}} ,$$

whence finally,

$$E_p^2 > E_{\text{sat,L}} E_{\text{sat,A}} \Delta R ,$$

as given in (9.126).

We then define the critical intracavity pulse energy $E_{p,c}$ for a QML threshold by

$$E_{p,c} \equiv \sqrt{E_{\text{sat,L}} E_{\text{sat,A}} \Delta R} = \sqrt{F_{\text{sat,L}} A_{\text{eff,L}} F_{\text{sat,A}} A_{\text{eff,A}} \Delta R} . \tag{9.152}$$

This is the minimum intracavity pulse energy to obtain stable cw modelocking, i.e., for $E_p > E_{p,c}$, we obtain stable cw modelocking and for $E_p < E_{p,c}$, we obtain QML instabilities.

Note that by adopting both approximations, i.e., (7.7) and neglecting the lifetime-dependent term in (9.151) and (9.126), we obtain a slightly stricter stability criterion: a laser fulfilling the stability condition with these approximations will always fulfill the exact condition. For good stability of a modelocked laser against unwanted fluctuations of pulse energy, operation close to the stability limit is not recommended.

9.7.3 Q-Switching Instabilities with Soliton Modelocking but Without ISA: Derivation and Discussion of (9.127)

When we checked experimental results from various passively modelocked solid-state lasers, we found an excellent agreement for picosecond lasers. However, we discovered that Nd:glass [97Aus] and Yb:glass [98Hon] lasers can show stable cw modelocking in a regime where they should actually be Q-switch modelocked according to the criterion in (9.126). We explain this in the following by the interplay of soliton effects and gain filtering. Additionally, there could be effects caused by the saturable absorber, namely stronger saturation of the fast component [98Kar] for fluctuations towards higher pulse energies (increasing the tendency for QML), or spectral filtering effects in the bleached SESAM. However, at least in the experiments discussed in [99Hon1], we believe that the previously mentioned interplay of soliton effects and gain filtering dominates.

The basic idea is as follows: if the energy of an ultrashort pulse rises slightly due to relaxation oscillations, SPM and/or SAM broadens the pulse spectrum. A broader spectrum, however, reduces the effective gain due to the finite gain bandwidth, which provides some negative feedback and thus decreases the critical pulse energy which is necessary for stable cw modelocking.

For a quantitative description of this idea we modify the rate equations (9.133)–(9.135) by introducing the effective gain for a modelocked laser running with a spectral distribution $I(\nu)$, viz.,

$$g_{\mathrm{eff}} \equiv \frac{\int g(\nu) I(\nu) \mathrm{d}\nu}{\int I(\nu) \mathrm{d}\nu} , \tag{9.153}$$

where $g(\nu)$ describes the frequency-dependent gain spectrum of the laser material. If we assume that the pulse spectrum adapts quickly to changes in E_{p}, i.e., within a time which is much shorter than the period of the relaxation oscillations, we can write g_{eff} as a function of E_{p}, not depending on the history of the pulse. To derive an analytical expression for the effective gain, we make an approximation around the peak of the gain spectrum with a Gaussian-shaped gain and pulse spectra, assuming that the pulse spectrum is centered at the peak of the gain spectrum. We then find for the effective gain

$$g_{\mathrm{eff}}(E_{\mathrm{p}}) = \frac{g}{\sqrt{1 + \left[\frac{\Delta\nu(E_{\mathrm{p}})}{\Delta\nu_{\mathrm{g}}}\right]^2}} , \tag{9.154}$$

where $\Delta\nu(E_{\mathrm{p}})$ and $\Delta\nu_g$ are the FWHM (full width at half maximum) of the pulse and gain spectrum, respectively. This results in the following modified set of rate equations from (9.137) and (9.138) with $P = E_{\mathrm{p}}/T_{\mathrm{R}}$:

$$T_{\mathrm{R}} \frac{\mathrm{d}E_{\mathrm{p}}}{\mathrm{d}t} = 2\Big[g_{\mathrm{eff}}(E_{\mathrm{p}}) - l - q_{\mathrm{p}}(E_{\mathrm{p}})\Big] E_{\mathrm{p}} , \tag{9.155}$$

$$\frac{\mathrm{d}g}{\mathrm{d}t} = -\frac{g_{\mathrm{eff}}(E_{\mathrm{p}}) - g_0}{\tau_{\mathrm{L}}} - \frac{E_{\mathrm{p}}}{E_{\mathrm{sat,L}} T_{\mathrm{R}}} g_{\mathrm{eff}}(E_{\mathrm{p}}) . \tag{9.156}$$

Linearizing these equations by inserting $E_{\mathrm{p}} \to E_{\mathrm{p}} + \delta E_{\mathrm{p}}$ and $g \to g + \delta g$, as in Sect. 9.7.2 similarly to (9.140), we can write

$$\begin{pmatrix} \delta \dot{E}_{\mathrm{p}} \\ \delta \dot{g} \end{pmatrix} = \begin{pmatrix} \alpha & \beta \\ -\gamma & -\varepsilon \end{pmatrix} \begin{pmatrix} \delta E_{\mathrm{p}} \\ \delta g \end{pmatrix} . \tag{9.157}$$

Equations (9.155) and (9.156) lead to

$$\alpha \equiv \frac{\mathrm{d}\dot{E}_{\mathrm{p}}}{\mathrm{d}E_{\mathrm{p}}} \overset{(9.155)}{=} 2\frac{g_{\mathrm{eff}} - l - q_{\mathrm{p}}(E_{\mathrm{p}})}{T_{\mathrm{R}}} + 2\frac{E_{\mathrm{p}}}{T_{\mathrm{R}}} \left(\frac{\mathrm{d}g_{\mathrm{eff}}}{\mathrm{d}E_{\mathrm{p}}} - \frac{\mathrm{d}q_{\mathrm{p}}}{\mathrm{d}E_{\mathrm{p}}}\right) , \tag{9.158}$$

and the first term on the right-hand side of (9.158) is zero at steady state, so we have

$$\alpha = 2\frac{E_{\mathrm{p}}}{T_{\mathrm{R}}} \left(\frac{\mathrm{d}g_{\mathrm{eff}}}{\mathrm{d}E_{\mathrm{p}}} - \frac{\mathrm{d}q_{\mathrm{p}}}{\mathrm{d}E_{\mathrm{p}}}\right) , \tag{9.159}$$

$$\beta \equiv \frac{\partial \dot{E}_{\mathrm{p}}}{\partial g} = 2\frac{2}{T_{\mathrm{R}}} \frac{\partial g_{\mathrm{eff}}}{\partial g} E_{\mathrm{p}} = 2\frac{g_{\mathrm{eff}}(E_{\mathrm{p}})}{g T_{\mathrm{R}}} E_{\mathrm{p}} , \tag{9.160}$$

$$-\gamma \equiv \frac{\partial \dot{g}}{\partial E_{\mathrm{p}}} = -\frac{g_{\mathrm{eff}}(E_{\mathrm{p}})}{E_{\mathrm{sat,L}} T_{\mathrm{R}}} + \frac{E_{\mathrm{p}}}{E_{\mathrm{sat,L}} T_{\mathrm{R}}} \frac{\mathrm{d}g_{\mathrm{eff}}}{\mathrm{d}E_{\mathrm{p}}} , \tag{9.161}$$

$$-\varepsilon \equiv \frac{\partial \dot{g}}{\partial g} = \frac{g_{\text{eff}}(E_\text{p})}{g\tau_\text{L}} + \frac{E_\text{p}}{E_{\text{sat,L}}T_\text{R}}\frac{g_{\text{eff}}(E_\text{p})}{g} . \tag{9.162}$$

With the stability requirement (9.148), viz., $\varepsilon - \alpha > 0$, we then obtain

$$2E_\text{p}\left[\frac{\text{d}g_{\text{eff}}(E_\text{p})}{\text{d}E_\text{p}} - \frac{\text{d}q_\text{p}(E_\text{p})}{\text{d}E_\text{p}}\right] < \frac{T_\text{R}}{\tau_\text{L}}\frac{g_{\text{eff}}}{g} + \frac{E_\text{p}}{E_{\text{sat,L}}}\frac{g_{\text{eff}}}{g} \approx \frac{E_\text{p}}{E_{\text{sat,L}}}\frac{g_{\text{eff}}}{g} . \tag{9.163}$$

Making the approximation that the upper state lifetime τ_L is much longer than the cavity roundtrip time T_R for most solid-state lasers, we can neglect the first part on the right-hand side of (9.163). Following the derivation in Sect. 9.7.2 we make the additional approximations as follows: assume a small modulation depth, less than 10%. Then, from (7.8) (see Table 7.4), it follows that $R(E_\text{p}) \approx R_\text{ns} - 2q_\text{p}(E_\text{p})$:

$$R(F_\text{p}) \approx R_\text{ns} - 2q_\text{p}(F_\text{p}) = R_\text{ns} - \frac{\Delta R}{S}(1 - \text{e}^{-S}) ,$$

where $S = E_\text{p}/E_{\text{sat,A}}$. We can further simply (9.163) assuming a fully saturated absorber, so that $\exp(-S) \approx 0$. We then obtain with using (7.8) (see Table 7.4) and repeated again just above, $R(E_\text{p}) \approx R_\text{ns} - \Delta R/S$. Therefore, we have

$$2q_\text{p}(E_\text{p}) \approx \frac{\Delta R}{S} = \frac{\Delta R}{E_\text{p}}E_{\text{sat,A}}$$

and hence,

$$2\frac{\text{d}q_\text{p}}{\text{d}E_\text{p}} = -\Delta R\frac{E_{\text{sat,A}}}{E_\text{p}^2} . \tag{9.164}$$

We can then simplify (9.163) to

$$\left(\frac{g_{\text{eff}}}{g} - 2E_{\text{sat,L}}\frac{\text{d}g_{\text{eff}}}{\text{d}E_\text{p}}\right)E_\text{p}^2 > E_{\text{sat,L}}E_{\text{sat,A}}\Delta R . \tag{9.165}$$

Equation (9.165) is a generalization of condition (9.126). The only modification is the bracket term on the left, which contains gain filtering and spectral broadening effects in the laser. From (9.165), we see that $g_{\text{eff}}/g < 1$ and $\text{d}g_{\text{eff}}/\text{d}E_\text{p} < 0$. The term g_{eff}/g alone would shift the stability limit to higher pulse energies and therefore increase the QML threshold. However, the term $-\text{d}g_{\text{eff}}/\text{d}E_\text{p}$ is always greater than zero and significantly increases the magnitude of the bracket term on the left side of (9.165). It thus substantially decreases the critical pulse energy, e.g., by a factor of 4, see [99Hon1, Sect. 3]. Most picosecond lasers are well described by the simpler (9.126), because the pulse spectra are narrow and the nonlinearities are negligible due to the low peak intensities. Indeed, we observed no spectral broadening of the pulse bandwidth with increasing pulse energy in our Nd:YLF experiments.

It is convenient to rewrite (9.165) by introducing the new parameter

$$f \equiv \frac{\Delta v(E_\mathrm{p})}{\Delta v_\mathrm{g}} ,$$

which is the ratio of the pulse bandwidth and the gain bandwidth. f depends on the pulse energy and on the pulse duration. With

$$\frac{\mathrm{d}g_\mathrm{eff}}{\mathrm{d}E_\mathrm{p}} = \frac{\partial g_\mathrm{eff}}{\partial f}\frac{\mathrm{d}f}{\mathrm{d}E_\mathrm{p}} = -gf\left(1+f^2\right)^{-3/2}\frac{\mathrm{d}f}{\mathrm{d}E_\mathrm{p}} ,$$

we can rewrite (9.165) as

$$\left[\left(1+f^2\right)^{-1/2} + 2E_\mathrm{sat,L}gf\left(1+f^2\right)^{-3/2}\frac{\mathrm{d}f}{\mathrm{d}E_\mathrm{p}}\right]E_\mathrm{p}^2 > E_\mathrm{sat,L}E_\mathrm{sat,A}\Delta R . \quad (9.166)$$

We can now calculate f and $\mathrm{d}f/\mathrm{d}E_\mathrm{p}$ from the soliton equation (4.57), assuming that the dominant SPM is generated inside the laser gain material (i.e., the length of the Kerr medium L_K is the cavity roundtrip length of the laser gain crystal):

$$\tau_\mathrm{p} = 1.7627 \times \frac{4\,|D|}{\delta F_\mathrm{p,L}} = 1.7627 \times \frac{2D\lambda_0 A_\mathrm{eff,L}}{\pi n_2 L_\mathrm{K}}\frac{1}{E_\mathrm{p}} , \quad (9.167)$$

where D is the dispersion parameter (4.39) and $\delta = kn_2L_\mathrm{K}$. The quantity $2D \equiv k_n'' L_\mathrm{d}$ denotes the amount of negative intracavity GDD per roundtrip, which compensates for the chirp introduced by self-phase modulation in the soliton pulse. We then use the constant time–bandwidth product for soliton pulses $\Delta\nu\,\tau_\mathrm{p} = 0.315$ (Table 2.1):

$$f \equiv \frac{\Delta v}{\Delta v_\mathrm{g}} = \frac{\pi n_2 L_\mathrm{K}}{2DA_\mathrm{eff,L}\lambda_0\Delta v_\mathrm{g}}\frac{0.315}{1.763}E_\mathrm{p} \quad (9.168)$$

and

$$\frac{\mathrm{d}f}{\mathrm{d}E_\mathrm{p}} = \frac{\pi n_2 L_\mathrm{K}}{2DA_\mathrm{eff,L}\lambda_0\Delta v_\mathrm{g}}\frac{0.315}{1.763} \equiv K , \quad (9.169)$$

and therefore

$$f = KE_\mathrm{p} .$$

Here, τ_p is the FWHM soliton duration, λ_0 is the center wavelength of the modelocked pulse, n_2 is the nonlinear refractive index, and L_K is the propagation length in the Kerr medium per roundtrip. The gain medium itself usually acts as the Kerr medium, so in a standing wave cavity L_K is twice the length of the laser crystal L_g. For $f^2 = K^2E_\mathrm{p}^2 \ll 1$, which is very clearly satisfied if the spectrum of the modelocked pulse fills less than about 40% of the available gain bandwidth, we can rewrite (9.166) with $\left(1+f^2\right)^{-1/2} \approx 1$ and $\left(1+f^2\right)^{-3/2} \approx 1$ to get the more compact form

$$2gE_\mathrm{sat,L}K^2E_\mathrm{p}^3 + E_\mathrm{p}^2 > E_\mathrm{sat,L}E_\mathrm{sat,A}\Delta R , \quad (9.170)$$

where g is the saturated amplitude gain coefficient.

9.7.4 Q-Switching Instabilities with ISA: Discussion of (9.129)

Theoretical results for the QML threshold [99Hon1] have generally been found to be in good agreement with experimental values. However, for gigahertz picosecond Er:Yb:glass lasers at a center wavelength of 1.55 μm [03Zel,08Oeh], the QML threshold was found to be significantly lower than expected due to the inverse saturable absorption (ISA) saturation characteristics of the SESAMs [04Sch] (see also Sect. 7.3.5 for an introduction to ISA). ISA causes a nonlinear rollover in the SESAM reflectivity (see Fig. 7.5).

For picosecond pulse durations, however, ISA due to two-photon absorption would be too weak to be significant for practical values of the pulse fluence. Therefore, it was surprising that a significant rollover was observed even in this regime [04Sch,05Gra2], while many earlier experiments on other SESAM modelocked lasers did not provide evidence for this effect. Note that the need to suppress Q-switching instabilities has enforced the use of SESAMs with reduced modulation depth (maximum nonlinear reflectivity change) to below 1%, particularly in the context of lasers with multi-GHz repetition rates [(9.126) for low pulse energies] [04Pas]. Precise nonlinear reflectivity measurements (Sect. 7.7.1) of several SESAMs in a wide range of pulse durations between 190 fs and 20 ps revealed that the observed strong rollover for femtosecond pulses is consistent with theoretical expectations based on two-photon absorption. For picosecond pulses the measured rollover is still much stronger than expected from two-photon absorption alone. This means that an additional mechanism must be at work here, such as free-carrier absorption, Auger recombination, hot-carrier generation, or lattice heating.

This rollover in ISA can be detrimental in some cases (in particular for the generation of very short pulses), but very helpful in other cases. For example, the unexpectedly strong rollover for picosecond pulses explains why it was possible to achieve stable operation of SESAM-modelocked picosecond Er:Yb:glass lasers with pulse repetition rates up to 50 GHz [04Zel] and even 100 GHz [08Oeh], which according to (9.126) should not have been stable against Q-switching instabilities.

An inverse saturable absorption reduces the tendency for Q-switching instabilities. This is not only due to the reduced effective modulation depth, but also (and more importantly) due to the reduced slope of the nonlinear reflectivity curve in the rollover regime. In this section, we show quantitatively how an induced nonlinear absorption in a SESAM lowers the threshold power for stable modelocking without Q-switching instabilities.

With ISA, we have a modified nonlinear reflectivity (7.48). For a small SESAM modulation of less than 10%, this results in a modified loss coefficient according to (7.51) and (7.8) (see Table 7.4):

$$R(E_{\mathrm{p}}) \approx R_{\mathrm{ns}} - 2q_{\mathrm{p}}(E_{\mathrm{p}}) = R_{\mathrm{ns}} - \Delta R\left[1 - \exp\left(-\frac{E_{\mathrm{p}}}{E_{\mathrm{sat,A}}}\right)\right]\frac{E_{\mathrm{sat,A}}}{E_{\mathrm{p}}} - \frac{E_{\mathrm{p}}}{A_{\mathrm{A}}F_2}\,. \tag{9.171}$$

Now we consider the stability of a modelocked laser against QML with the saturation energy $E_{\mathrm{sat,L}}$ and the upper-state lifetime τ_{L} of the gain medium. The dynamical variables are the intracavity average power $P = E_p/T_R$ (with the cavity roundtrip time T_R) and the gain coefficient g. We use the stability criterion from (9.139) from Sect. 9.7.2, viz.,

$$\frac{T_R}{\tau_L} + \frac{E_p}{E_{sat,L}} > -2E_p \frac{\partial q_p(E_p)}{\partial E_p} \, .$$

Here, the right-hand side describes the destabilizing effect that the saturable absorber typically favors increased pulse energies by reducing its loss ($\partial q_p/\partial E_p < 0$). On the left-hand side, the second term usually dominates for operation well above threshold, and it describes the stabilizing effect of gain saturation. Stability against Q-switching is achieved when the stabilizing effect of gain saturation is greater than the destabilizing effect of the absorber.

Using our absorber model with induced nonlinear losses (9.171), we have for high saturation so that $\exp(-S) \approx 0$, with $S = E_p/E_{sat,A}$,

$$-2\frac{\partial q_p}{\partial E_p} = +\Delta R \frac{E_{sat,A}}{E_p^2} - \frac{1}{A_A F_2} \, .$$

This leads to the stability condition

$$\frac{T_R}{\tau_L} + \frac{E_p}{E_{sat,L}} > E_p \left(\Delta R \frac{E_{sat,A}}{E_p^2} - \frac{1}{A_A F_2} \right) . \tag{9.172}$$

Once again, we can neglect the first term in (9.172): for typical solid-state lasers we have $\tau_L \gg T_R$ because the upper-state lifetime in most solid-state lasers is much longer than the cavity roundtrip time. We then obtain a modified stability criterion with (9.129) [05Gra2]:

$$E_p^2 > \frac{E_{sat,A} \Delta R}{\dfrac{1}{E_{sat,L}} + \dfrac{1}{A_A F_2}} \, ,$$

which is equivalent to (9.129).

For $F_2 \to \infty$, i.e., without induced nonlinear losses, we retrieve the simpler equation (9.126) [99Hon1]. With finite F_2, the threshold for stable modelocking is reduced, and for small enough values of F_2 the saturation energy $E_{sat,L}$ of the gain medium is no longer important. This is because $|\partial q_p/\partial E_p|$ becomes zero at the rollover, so that the destabilizing effect of the absorber is effectively removed by the inverse saturable absorption, and gain saturation is no longer needed to stabilize the laser. In this regime, minimization of the mode size in the gain medium, which leads to a minimized value of $E_{sat,L}$, is no longer necessary for stable operation. Indeed, it has been found experimentally that two different 10 GHz Er:Yb:glass lasers, which used the same SESAM but very different mode areas in the gain media, exhibited quite similar QML thresholds.

When designing lasers, it is often convenient to fix a certain value of $S = E_p/E_{sat,A}$ for the desired output power, and this value of S can later be achieved

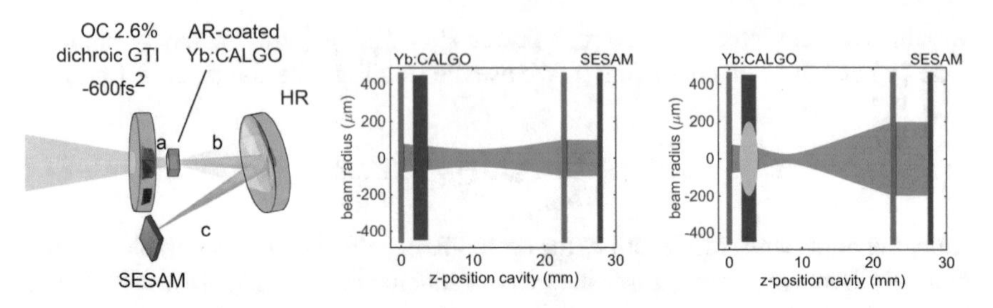

Fig. 9.41 All-optical Q-switching limited with proper cavity design. *Left*: 5-GHz Yb:CALGO laser cavity. Output coupler OC and highly reflecting mirror HR. *Center* and *right*: Unfolded cavity with the Kerr lens (4.13) in the gain crystal which increases the mode size on the SESAM for higher intensity. Shown here for low intensity (*center*) and high intensity (*right*) [15Kle]

by adjusting the mode size on the absorber [see (9.131) and (9.132)]. With strong enough (but not excessive) saturation on the SESAM, Q-switching instabilities can be fully eliminated without minimization of the mode size in the gain medium. The discovered stronger rollover makes it possible to reach this goal even in the picosecond regime, where TPA alone would normally be much too weak. This has enabled very stable multi-GHz 1.55-μm Er:Yb:glass lasers with very low noise properties [05Sch2], used to achieve new world record optical communication [11Hil].

9.7.5 Special Cavity Designs to Prevent Q-Switching Instabilities

Passively modelocked diode pumped solid-state lasers with pulse repetition rates in the gigahertz regime suffer from an increased tendency for Q-switching instabilities because, for a given average output power, the pulse energy E_{p} becomes smaller. Therefore Q-switching instabilities become harder to suppress, according to (9.126). SESAM with lower saturation fluence $F_{\mathrm{sat,A}}$ and ISA with a rollover can help. However, the problem becomes increasingly difficult with higher pulse repetition rates because the laser mode area must normally be more strongly focused on the SESAM.

We presented a passive stabilization mechanism, an all-optical Q-switching limiter, to reduce the impact of Q-switching instabilities and increase the potential output power of SESAM modelocked lasers in the gigahertz regime [15Kle]. With a suitable cavity design, the Kerr lens (4.13) inside the gain material can increase the mode size on the SESAM with higher intensities and hence lower the saturation and loss modulation in the SESAM. This gives negative feedback for Q-switching instabilities, and hence a negative saturable absorber effect which limits the onset of Q-switching instabilities. No critical cavity alignment is required, because this Q-switching limiter acts well within the cavity stability regime. Using a suitable cavity design, a high-power diode pumped Yb:CALGO solid-state laser (see Fig. 9.41 left) generated sub-100 fs pulses with an average output power of 4.1 W at a pulse repetition rate of 5 GHz [15Kle].

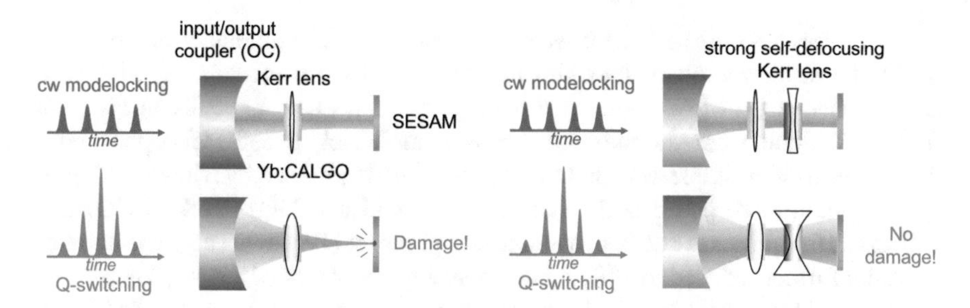

Fig. 9.42 Linear straight cavity Yb:CALGO laser. *Left*: Kerr lens (4.13) with $n_2 > 0$ focuses the laser cavity mode on the SESAM with higher intensity. *Right*: Additional CQN material with $n_2 < 0$, which defocuses the laser cavity mode on SESAM with higher intensity. This suppresses Q-switching instabilities and damage. Cascading of quadratic nonlinearities CQN

9.7.6 Negative n_2 to Prevent Q-Switching Instabilities

For a linear cavity, we can never obtain an all-optical Q-switching limiter as shown in Fig. 9.41. A straight-cavity design with the SESAM as one of the end mirrors (Fig. 9.42 left) greatly simplifies the cavity design, but self-focusing of the beam and hence damage will occur in the cavity when operating in the conventional soliton modelocking regime. Using an additional material inside the straight cavity with a negative n_2 can solve the Q-switching problem (Fig. 9.42 right) [17May].

Instead of relying on the intrinsic (weak) nonlinearity of the gain medium and negative dispersion compensating elements (conventional soliton modelocking regime, Fig. 9.42 left), a material with a relatively large negative nonlinear refractive index n_2 is added to stabilize against Q-switching instabilities with an increasing defocusing lens effect with increasing peak intensity (Fig. 9.42 right). A compelling approach to obtain such a tunable self-defocusing nonlinearity is via the cascading of quadratic nonlinearities (CQN) (see Sect. 9.5.3). For example, a two-dimensionally patterned apodized quasi-phase-matching (QPM) CQN device, as shown in Fig. 9.34, can provide an effective nonlinear index $n_{2,\mathrm{eff}}$ ranging from -1×10^{-18} m^{2}W^{-1} to -7×10^{-18} m^{2}W^{-1}. This can be achieved in a periodically poled lithium niobate (PPLN) device. These values are approximately two orders of magnitude higher than the intrinsic third-order material nonlinearities of conventional wide bandgap materials. For example, the nonlinear index of the gain material Yb:CALGO amounts only to $n_{2,\mathrm{intrinsic}}^{\mathrm{CALGO}} \approx +8 \times 10^{-20}$ m^{2}W^{-1}.

We demonstrated this Q-switching stabilization scheme as shown in Fig. 9.42 with a diode pumped 10 GHz Yb:CALGO laser as shown in Fig. 9.35. A 1.5 mm-long Yb:CALGO gain crystal was used, pumped by an internally wavelength-stabilized spatially-multimode diode array with an $M^2 \approx 36$ capable of providing up to 60 W at 980 nm. The pump beam is focused through a 12 mm-radius pump-transparent mirror that acts as a 2.8% output coupler for the 1050 nm laser center wavelength. In order to minimize the thermal load inside the laser cavity, the vertically polarized pump light, which would only be weakly absorbed in the gain crystal, is removed using a polarizing beam splitter. At the same time, the beam splitter allows the vertically polarized laser beam to exit. In an intracavity 2D-fanout-QPM CQN device (Fig. 9.34), we create a large, low-loss self-defocusing nonlinearity, which simultaneously provides SESAM-assisted soliton modelocking in the normal dispersion regime, i.e., GDD>0, and suppresses Q-switching-induced damage as shown schematically in Fig. 9.42 (right). We demonstrated femtosecond passive modelocking at 10 GHz pulse repetition rates from a simple straight laser cavity, directly pumped by a low-cost highly spatially-multimode pump diode. The 10.6-GHz Yb:CaGdAlO$_4$ (Yb:CALGO) laser delivers 166 fs pulses at 1.2 W of average output power [17May]. The pulse duration was later further reduced to 108 fs with an average power of 812 mW [20Kru].

This enables a new class of femtosecond modelocked diode pumped solid-state lasers with repetition rates at 10 GHz and beyond. This solved the long-standing Q-switching problem for multi-GHz SESAM modelocked diode-pumped solid-state lasers in a simple straight cavity.

9.8 Passively Modelocked Diode-Pumped Semiconductor Lasers

9.8.1 Optically Pumped VECSELs

A diode-pumped semiconductor laser bridges the gap between semiconductor lasers and solid-state lasers. Significant power scaling from diode-pumped semiconductor lasers with more than 0.5 W average output power in a fundamental mode has been obtained with the optically pumped vertical-external-cavity surface-emitting laser (VECSEL) in 1997 [97Kuz]. The trade-off for optically pumped semiconductor lasers is increased complexity, because electrical pumping is given up for power scaling. However, with regard to diode-pumped ion-doped solid-state lasers, this optically pumped semiconductor laser has clear benefits: the semiconductor gain medium allows for a flexible choice of emission wavelength via bandgap engineering and offers a wealth of possibilities from the semiconductor processing world. Almost arbitrary optical layer structures can be integrated vertically with the gain section; in particular multi-purpose mirrors, combining functions such as dispersion control, dual-wavelength reflection for pump and laser wavelength, antireflection layers, and even semiconductor saturable absorbers. One additional advantage of diode-pumped VECSELs is that they can convert fairly low-cost, low-beam-quality optical pump power from high-power diode laser bars into a near-diffraction-limited output beam with good efficiency. The combination of the mature optical pumping technology used extensively for diode-pumped solid-state lasers with efficient heat removal of solid-state thin-disk lasers has resulted in VECSEL performances that surpass anything possible to date with conventional semiconductor lasers [17Gui].

The advantage of the near-IR 1-μm wavelength region is based on the mature GaAs, AlAs, and InGaAs material systems. The AlAs/GaAs distributed Bragg reflector (DBR) has a small lattice mismatch, enabling low-strain monolithic fabrication, and has a high refractive index contrast, allowing for a high reflectivity with a small number of layer pairs. This small number of layer pairs results in a thin DBR structure, and the good thermal conductivity of GaAs and AlAs allows very good heat removal,

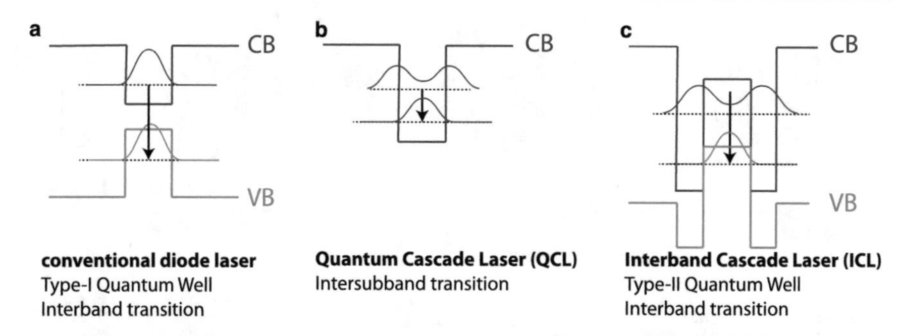

Fig. 9.43 The most common semiconductor lasers for the 2 μm to mid-IR spectral region can be grouped into three different gain types. (**a**) Conventional diode lasers with interband transitions from the conduction band (CB) to the valence band (VB). The bandgap of standard III–V semiconductor materials sets a long wavelength limit to 3.7 μm, but the increased Auger loss currently limits ultrafast type-I VECSELs to 2 μm. (**b**) Quantum cascade lasers (QCL) are based on intersubband transitions within the CB, and are therefore well suited for the mid-IR and THz regime. (**c**) Interband cascade lasers (ICL) with type-II quantum wells are not limited by the III–V bandgap energy and there is no significant Auger loss. CB: conduction band, VB: valence band

often the key limitation for power scaling. The optical gain is usually provided by InGaAs/GaAs type-I quantum wells (QWs) (Fig. 9.43) or InAs/GaAs quantum dots (QDs). The emission wavelength is then limited to about 920 nm to 1180 nm, given by the increasing strain introduced by the high In-content InGaAs QWs. The compressive strain caused by the InGaAs QWs can be compensated by tensile-strained GaAsP layers. These semiconductor lasers are optically pumped into the barrier layers, and for InGaAs QW gain structure, low-cost near-800 nm diode laser arrays are used.

For longer emission wavelengths around 1.5 μm, one can also use InGaAs QWs with diluted concentrations of nitrogen or InAs/GaAs quantum dots (QDs) for emission wavelengths up to 1.55 μm [17Gui]. Other than that, one has to switch to the InP-based material system to reach telecom wavelengths, but it should be noted that it suffers from a low-index-contrast DBR.

Industrial semiconductor lasers for 2 μm to mid-IR sensing applications are based on quantum cascade lasers (QCLs) or interband cascade lasers (ICLs) (Fig. 9.43). The interband cascade laser (ICL) combines advantages of conventional diode lasers and QCLs. Passive modelocking of QCLs has been not possible so far due to the very short picosecond upper lasing lifetime of intersubband transitions (Fig. 9.43b). In contrast to QCLs, the upper state lifetime in ICLs is over three orders of magnitude longer in the nanosecond regime, as they are based on an interband transition from conduction band (CB) to valence band (VB) (Fig. 9.43c). In contrast to conventional type-I quantum wells (Fig. 9.43a) ICLs are based on a type-II quantum-well band structure, which enables mid-IR transitions with III–V semiconductor materials. The trade-off of type II to type-I quantum-well lasers, however, is lower gain, due to the reduced spatial overlap of the electron and hole wave functions. The key benefit of type-II gain quantum wells, in contrast to type-I in the mid-IR, is negligible Auger loss.

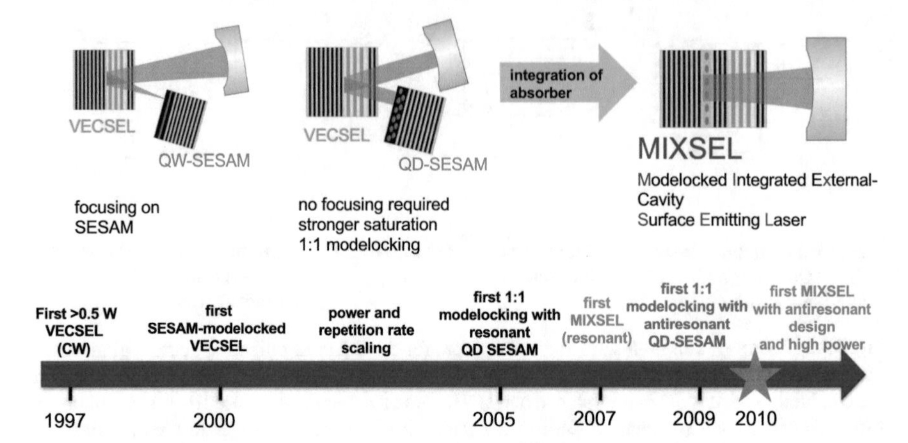

Fig. 9.44 History of diode-pumped high-average-power semiconductor surface emitting lasers, also referred to as semiconductor disk lasers (SDLs) by analogy with the heat removal concept of thin disk Yb-doped solid-state lasers. Mentioned here are key milestone results. For SESAM-modelocked VECSELs, we typically use the VECSEL chip structure as a folding mirror inside a standing wave laser cavity, with one end defined by the SESAM and the other end by the output coupler. The MIXSEL allows for a simple linear straight cavity, with the MIXSEL chip and the output coupler as the cavity end mirrors. QW quantum well, QD quantum dot

Consequently, it is not surprising that the best modelocking and cw performance is achieved in the 1-μm wavelength region [17Gui, 15Til]. To date the highest average output power from a diode-pumped VECSEL is 106 W at an emission wavelength of 1028 nm using InGaAs type-I QW gain [12Hei].

9.8.2 Optically-Pumped MIXSELs and SESAM-Modelocked VECSELs

The next step towards even more compact and less expensive ultrafast lasers in the gigahertz regime can be achieved by modelocking such optically pumped semiconductor lasers. The first SESAM-modelocked VECSEL was demonstrated in 2000 (see Fig. 9.44) [00Hoo]. Passive modelocking results were summarized in review articles in 2006 [06Kel] and in 2015 [15Til]. Passive modelocking is based on dynamic gain saturation (see Fig. 9.5a) in a similar way to dye lasers and in contrast to diode-pumped solid-state lasers, where the gain is saturated only by the average power (Fig. 9.5c). Thus the SESAM parameters have to be designed more carefully for stable modelocking, typically requiring low saturation fluences (as discussed in Sect. 7.5). This is the main reason why in early experiments the mode size on the SESAM was always significantly smaller than in the gain, to saturate the absorber faster than the gain (see Fig. 9.44).

Electrically-pumped modelocked VECSELs have also been demonstrated, but at significantly lower output power [14Zau]. Passive modelocking has also been demonstrated with carbon nanotubes [13Seg] and graphene [13Zau] saturable absorbers. However, damage and scattering losses have remained significant issues so far. Even without a saturable absorber, modelocking has been reported [11Che], referred to as 'self-modelocking' or 'SESAM-free modelocking'. The mechanism for this self-modelocking is presumed to be a Kerr effect in the VECSEL gain material, but there is an ongoing and unresolved discussion in the community [16Gaa, 16Sha]. This approach is not expected to be robust enough for industrial applications.

Initially, we focused on average power scaling of SESAM-modelocked VECSELs in the near-IR regime (Fig. 9.44), achieving historic milestone results for a picosecond VECSEL with more than 2 W average output power in 2005 [05Asc1] and a femtosecond VECSEL with more than 1 W in 2011 [11Hof]. The complexity of such lasers was further reduced with a wafer-scale integrated modelocked VECSEL, where both the gain and the absorber are integrated into the same wafer, referred to as the MIXSEL (Modelocked Integrated eXternal-cavity Surface Emitting Laser) in 2007 [07Maa]. However, significant output power was only achieved with a diode-pumped antiresonant MIXSEL in 2010 (Fig. 9.44), generating more than 6 W average power [10Rud].

In an antiresonant MIXSEL structure, we have a similar field enhancement in the absorber and in the gain, as shown in Fig. 9.45 for a more recent MIXSEL chip structure. Antiresonance reduces problematic issues such as strong frequency dependent dispersion (see Sect. 7.5.2), and very low tolerances on semiconductor epitaxial growth errors [09Bel]. For a MIXSEL (Fig. 9.45), we add one more distributed Bragg reflector (DBR) between the active gain layers and the saturable absorber layers which reflects the residual pump light. If sufficient pump absorption is obtained, one can consider leaving this pump DBR out. The integrated absorber parameters have to be adjusted so that they still fulfill the modelocking stability criterion to saturate the absorber faster than the gain. This is more challenging for an antiresonant MIXSEL. Stable modelocking with the same mode size in the gain and the absorber in a SESAM-modelocked VECSEL (see Fig. 9.44 with 1:1 modelocking) was therefore an important milestone result before a successful antiresonant MIXSEL integration was possible [04Lor].

There has been always a trade-off between shorter pulse duration and average output power [17Alf]. To date the shortest pulses obtained with a reasonably high average power of 100 mW from any ultrafast diode-pumped semiconductor laser is currently around 100 fs [16Wal]. For 270 mW average output power, the pulse duration was increased to 216 fs [18Alf]. The pulse duration from a diode-pumped MIXSEL was initially limited to the picosecond regime by the slow recovery time of the quantum dot (QD) saturable absorber. With a fast AlAs/InGaAs/AlAs saturable absorber (see Sect. 7.4.3a) integrated into the MIXSEL, we obtained pulses as short as 184 fs with 115 mW average output power [17Alf] and finally 144 fs with 30 mW [19Alf] with a more broadband MIXSEL chip design, as shown in Fig. 9.45. The laser system is a highly nonlinear system that tends to break up into multiple pulses which are hard to detect and can result in false results [18Wal]. We will discuss ways to properly characterize pulse durations and noise in general in Chaps. 10, 11, and 12.

The highest pulse repetition rate demonstrated for a SESAM-modelocked VECSEL is 50 GHz [06Lor], and 100 GHz for a MIXSEL [14Man]. Even higher repetition

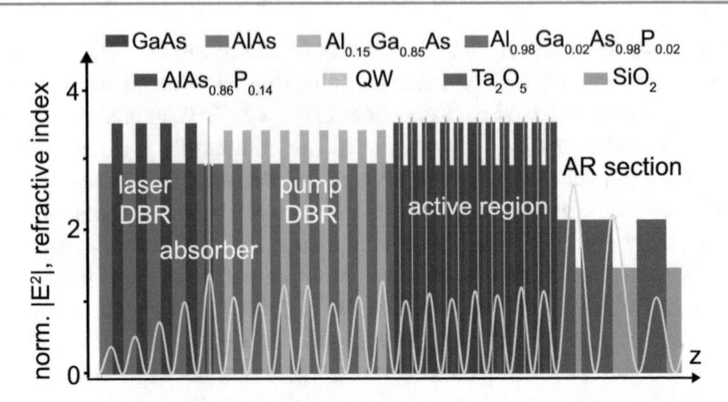

Fig. 9.45 InGaAs type-I QW MIXSEL chip structure showing different refractive indices in color and the standing wave intensity profile of the laser mode (*white line*). The lower, highly reflecting GaAs/AlAs DBR is shown only with the last few top layers. The simple straight linear laser cavity is formed with this MIXSEL chip and an output coupler generating pulses as short as 144 fs with an average output power of 30 mW [19Alf]. Note that the InGaAs QW gain layers are close, but not exactly at the antinodes [19Alf]

rates can be achieved with harmonic modelocking [10Qua, 12Wil], but this increases the noise and reduces stability. We therefore always focus on fundamental modelocking, i.e., one pulse per cavity roundtrip. For low pulse repetition rates below 1 GHz, these semiconductor lasers have a tendency to show multi-pulse instabilities, but with a multi-gain-pass cavity geometry, pulse repetition rates around 100 MHz have been achieved [12Zau].

Interesting cavity designs, such as the colliding pulse modelocked VECSEL [16Lau, 17Lin2], the dual-comb modelocked MIXSEL [17Lin1], or the DBR-free MECSEL [16Kah], show the innovative potential of these lasers.

For the long-wavelength regime (i.e., above 2 μm) the best power result for a cw VECSEL has been achieved using 10 type-I InGaSb gain quantum wells. Such a laser generated up to 20.7 W of cw power at 2.02 μm [15Hol]. The optical gain offered by the GaSb material system at 2 μm is sufficient to support the losses introduced by an intracavity semiconductor saturable absorber mirror (SESAM). This allowed for first SESAM-VECSEL modelocking at 1.96 μm [10Har] even in the femtosecond regime, with 384 fs pulses at a pulse repetition rate of 890 MHz and 25 mW of average output power [11Har]. These results are the longest wavelength at which a semiconductor surface emitting laser could be modelocked to date. GaSb-based type-I quantum wells have been demonstrated up to 3.7 μm [12Kri], but with substantially increased Auger losses, preventing modelocking. However, the significantly increasing Auger recombination loss in long-wavelength (i.e., above 2 μm) type-I quantum wells can still be used to obtain fast saturable absorbers [21Bar]. Therefore, a combination of type-I saturable absorber and type-II gain quantum wells inside a MIXSEL chip structure looks very promising for this long-wavelength regime.

9.9 Dual-Comb Modelocking

A dual-comb modelocked laser, in the most general sense, means that two modelocked pulse trains can be generated from a single laser cavity. Depending on the application, these two pulse trains are mainly distinguished by their center wavelength and/or their pulse repetition rate. With the revolution in frequency metrology with optical frequency combs (see Chap. 12), the term 'dual comb' has become used just for two combs with the same spectrum, but with a slightly different pulse repetition rate. We will therefore distinguish between 'dual-comb' and 'dual-modelocked' laser, with the latter being the more general expression.

The mathematical description of a modelocked pulse train is well known from many older textbooks (e.g., [86Sie]) and without any noise can be described in the time and frequency domain by a δ-comb, because a Fourier transform of a δ-comb results once again in a δ-comb (D.6) (see more details in Appendix D):

$$F\left\{\sum \delta\left(t-mT\right)\right\} \sim \sum \delta\left(\nu - m f_{\mathrm{rep}}\right) , \tag{9.173}$$

where m is an integer, $T = 1/f_{\mathrm{rep}}$ is the pulse repetition period (i.e., the cavity roundtrip time for fundamental modelocking with only one pulse inside the cavity) with f_{rep} the pulse repetition frequency. In the time-domain, the intensity of a modelocked pulse train without any noise is then given by

$$I_{\mathrm{train}}(t) = I_{\mathrm{p}}(t) * \sum_{m=-\infty}^{\infty} \delta\left(t-mT\right) , \tag{9.174}$$

where $I_{\mathrm{p}}(t) = |A_{\mathrm{p}}(t)|^2$ is the time-dependent intensity of a single pulse, and $A_{\mathrm{p}}(t)$ is the single pulse envelope. For the electric field in the time and frequency domain, we then obtain

$$E_{\mathrm{train}}(t) = E_{\mathrm{p}}(t) * \sum_{m=-\infty}^{\infty} \delta\left(t-mT\right) , \tag{9.175}$$

$$\tilde{E}_{\mathrm{train}}(\nu) = \tilde{E}_{\mathrm{p}}(\nu) \sum_{m=-\infty}^{\infty} \delta\left(\nu - m f_{\mathrm{rep}} - f_{\mathrm{CEO}}\right) , \tag{9.176}$$

where we have taken into account the fact that the convolution operator $*$ goes over into a multiplication with the Fourier transform (Appendix A). Here, $E_{\mathrm{p}}(t)$ and $\tilde{E}_{\mathrm{p}}(\nu)$ denote the electric field of the single pulse in the time and frequency domains, respectively, and the modelocked pulse train spectrum is described by $\nu_m = f_{\mathrm{CEO}} + m f_{\mathrm{rep}}$ (see Fig. 6.2). The modelocked spectrum then has two degrees of freedom with the frequency comb spacing f_{rep} and the frequency comb offset f_{CEO} [99Tel], as discussed in more detail in Chap. 12. How noise actually affects the modelocked pulse train is discussed in Chaps. 11 and 12.

Early work on dual-modelocked generation goes back to the 1970s with synchronous modelocked dye lasers using a modelocked argon-ion pump laser

(Fig. 6.30). Synchronous pumping allowed for the generation of two synchronized pulse trains with a freely tunable center wavelength, which at that time was mainly motivated for nondegenerate picosecond pump–probe spectroscopy [77Jai]. When the two wavelength-duplexed pulse trains shared the same gain (i.e., a dye jet) using space duplexing with angle dispersion of an intracavity prism (Fig. 9.46a), it was found that the competition between the two wavelengths in the gain started to limit stability and maximum wavelength separation [77Jai]. A better solution for dual-wavelength tunability was based on two fully spatially separated synchronously pumped laser cavities.

Synchronous modelocking also allows for the generation of two pulse trains with a small difference in the pulse repetition rate, which was used for equivalent time sampling [87Elz] (see the more detailed discussion in Sect. 11.6.4). This work was then extended to KLM Ti:sapphire lasers initially with limited wavelength tunability [93Eva] and later with improved wavelength tunability [95Leit]. Again it was confirmed that, for stable and tunable dual-wavelength modelocked generation, the pump laser has to pump the two combs independently [95Leit]. The pulse repetition rates in this case were still optimized to be the same with some small gain sharing.

Progress in active stabilization of pulse repetition rates of modelocked lasers in the 1980s (see Chap. 11) [86Rod,87Cot] resulted in many different dual- and multi-comb generations using separate laser cavities at that time limited to the relative timing jitter stabilization of the pulse envelope, i.e., the intensity of the pulse trains (9.174). The frequency metrology revolution after 1999 (see Chap. 12 for more detail) [99Tel, 00Apo,00Jon] allowed for fully stabilized frequency combs for which both the f_{CEO} and the f_{rep} could be stabilized, i.e., the electric field of the pulse trains (9.175) and (9.176). Equivalent time sampling concepts from the 1980s [87Elz,85Wei,88Wei] were then extended into frequency metrology applications. However, at this point, dual-comb applications relied on two separate modelocked lasers with at least four stabilization loops for f_{CEO} and f_{rep} in each laser, and this substantially increased cost and complexity.

A more compact solution was invented in 2015 with polarization duplexing of two passively modelocked pulse trains from a single laser cavity (Fig. 9.46b) [15Lin]. This allowed for dual-comb spectroscopy in water vapor in 2017 [17Lin1] and acetylene in 2019 [19Nur] from a free-running dual-comb modelocked laser. This means no stabilization was required for sufficient accuracy, which is important for industrial applications. The 2017 dual-comb spectroscopy demonstration from a free-running laser [17Lin1] resulted in a paradigm shift for dual-comb applications. After 2015 dual-comb modelocking from a single laser cavity became a hot topic, using different multiplexing techniques and applied to many different lasers (see the more recent review articles, e.g., [20For,20Lia]).

The free running dual-comb stability with an adjustable pulse repetition rate difference is expected to be best if all laser cavity elements are shared by the two modelocked combs, but with the beams spatially separated in the gain and absorber. For the first demonstration of polarization duplexing [15Lin], an intracavity birefringent crystal was used to separate the two combs on the gain and the saturable absorber (i.e., the MIXSEL chip) (Fig. 9.46b). The single MIXSEL chip in the dual-

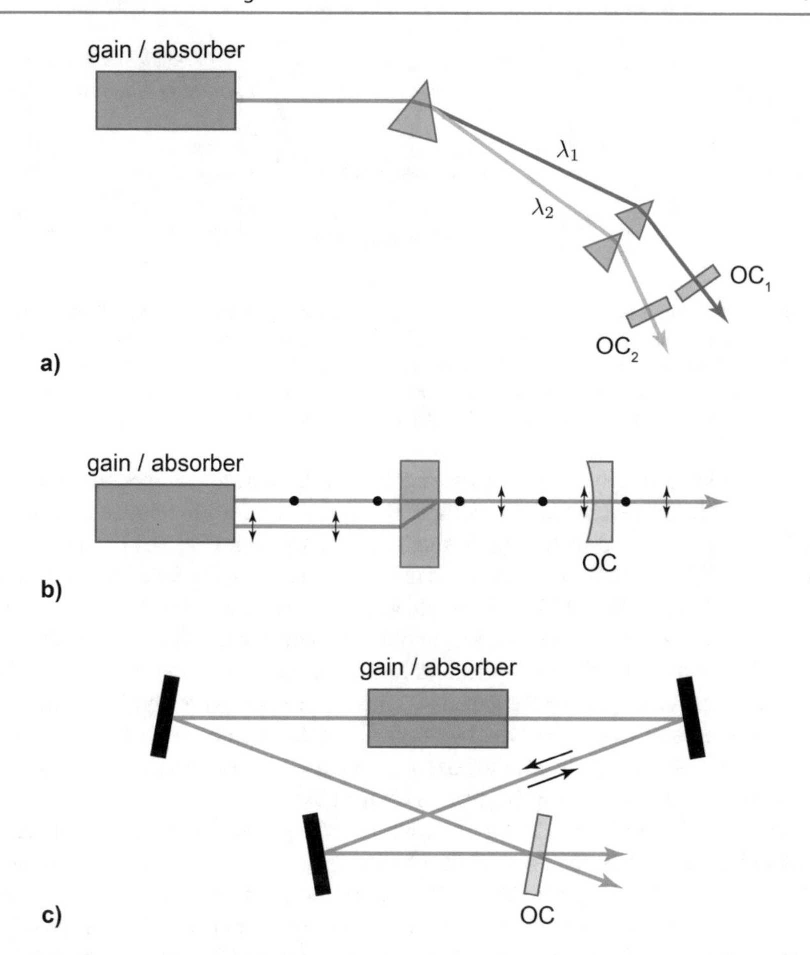

Fig. 9.46 Different methods resulting in dual-comb modelocking from a single cavity based on a partial combination of wavelength, polarization, circulation-direction, and space duplexing. **a** Wavelength/space duplexing of synchronous modelocked dye [77Jai] and KLM Ti:sapphire [93Eva, 95Leit] lasers. **b** Polarization duplexing of MIXSEL [15Lin] and solid-state (i.e., Yb:CaF_2) [20Wil2] lasers. **c** Ring laser for circular direction duplexing for Ti:sapphire [16Ide] and fiber [16Meh] lasers. OC: output coupler

comb MIXSEL provided both the gain and the saturable absorber (see Sect. 9.8), which made this cavity design extremely simple, and the optical path length difference in one or two birefringent crystals for different Δf_{rep} can be adjusted over a wide range [16Lin]. For a diode-pumped solid-state laser, the cavity design had to be modified with a beam separation both in the gain and the SESAM for stable modelocking with a fixed $\Delta f_{rep} \neq 0$ [20Wil2].

Interesting from a laser physics point of view is the fact that the SESAM decouples the noise stabilization in the two combs because of the small time shift of the pulse after the full SESAM saturation (Fig. 9.47a) [16Lin]. This time shift is decoupled for the spatially separated beams on the SESAM, and some amount of partial

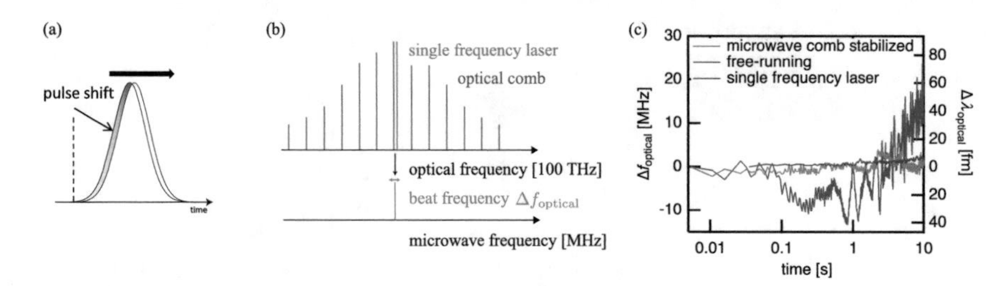

Fig. 9.47 Dual-comb SESAM modelocking. Spatially separated beams on the SESAM decouple optical frequency noise stabilization. **a** Time shift of a pulse after full SESAM saturation is responsible for the noise decoupling [16Lin]. **b** Frequency beat signal $\Delta f_{\text{optical}}$ of a modelocked optical frequency comb with stable single frequency laser oscillator. **c** Measured optical frequency comb stability for free-running and microwave stabilized dual-comb MIXSEL [17Lin1]

spatial overlap can fully lock the pulse repetition rates before modelocking becomes unstable with too much spatial overlap. The stability of a modelocked frequency comb can be measured with a stable single frequency laser (Fig. 9.47b). For a free-running MIXSEL, the beat frequency is only sufficiently stable for a short time (i.e., less than 100 ms in Fig. 9.47c). A simple microwave comb stabilization for a dual-comb MIXSEL maintains a much longer optical comb stability because the SESAM decouples the noise for the two optical frequency combs and the microwave stabilization stabilizes the difference frequencies. Thus, after microwave comb stabilization, the only degree of freedom in the two optical combs is an equal optical frequency drift which is very unlikely with the decoupled noise. A more detailed discussion of optical frequency comb noise is given in Chap. 12.

In the dual-comb KLM Ti:sapphire ring laser (Fig. 9.46c) [16Ide] the difference in the pulse repetition rate Δf_{rep} for the clockwise and counter-clockwise pulse train is less obvious. The origin of the difference in the pulse repetition rates between the two pulse trains was explained by the non-uniform intensity distribution of the pump laser in the Ti:sapphire crystal, a slight offset of the Ti:sapphire crystal from the midpoint, and a bi-directional Kerr effect due to the first two effects. Control of the difference in pulse repetition rate was improved in bidirectionally modelocked fiber lasers [16Meh, 19Fel].

9.10 Performance Frontiers in Ultrafast Lasers

9.10.1 Pulse Generation in the Few-Cycle Regime

The development of tunable broadband solid-state lasers began in the 1960s. For ultrafast solid-state lasers, a broad bandwidth is one requirement. In general, broadband tunable solid-state lasers can be achieved with a strong electron–phonon coupling of the lasing ion to the host lattice. Table 5.1 gives an overview of many different solid-state lasers and their parameters. In 1963, the first tunable "vibronic" transition metal laser, a nickel-doped magnesium fluoride laser, was reported [63Joh]. Many different lasers followed, but they only worked at cryogenic temperatures, which

Table 9.5 The race for few-cycle pulse generation from a Ti:sapphire laser in the 1990s. τ_p is the FWHM pulse duration, P_{av} the average output power, and f_{rep} the pulse repetition rate

τ_p	P_{av}	f_{rep}	Remarks	Reference
140 fs			KLM started with dye saturable absorber (not understood, assumed to have a CPM Ti:sapphire laser)	[91Sar]
60 fs	300 mW		First demonstration of KLM (but KLM not understood)	[91Spe]
73 fs	240 mW	134 MHz	First experimental evidence for KLM, self-starting due to RPM	[91Kel]
47 fs	110 mW			[92Riz1]
39 fs	1.5 W	82 MHz		[92Kaf]
32 fs	320 mW	≈100 MHz	LaFN28-glass prisms, 2 cm Ti:sapphire crystal thickness	[92Hua1]
33 fs			Schott F2 prisms, 8 mm Ti:sapphire crystal thickness	[92Kra1]
22 fs	950 mW	100 MHz	Schott LaK31 prisms, 2 cm Ti:sapphire crystal thickness	[92Lem]
17 fs	500 mW	80 MHz	Schott LaKL21 prisms, 9 mm Ti:sapphire crystal thickness	[92Hua2]
12.3 fs			Fused silica prisms, 4 mm Ti:sapphire crystal thickness	[93Cur]
11 fs	500 mW		Fused silica prisms, 4.5 mm Ti:sapphire crystal thickness	[93Asa]
11 fs	300 mW	100 MHz	Chirped mirrors, no prisms	[94Sti]
8.5 fs	≈1 mW		Metal mirrors and fused silica prisms	[94Zho]
8.2 fs	100 mW	80 MHz	Chirped mirrors only	[95Sti]
7.5 fs	150 mW		Chirped mirrors, ring cavity	[96Xu2]
6.5 fs	200 mW	86 MHz	Fused silica prisms and double-chirped mirrors, KLM is self-starting with SESAM	[97Jun3]
5.8 fs	300 mW	85 MHz	Fused silica prisms and double-chirped mirrors, KLM is self-starting with SESAM, pulse duration measured with SPIDER [00Mat]	[99Sut,00Mat]
≈5 fs	200 mW	90 MHz	CaF_2 prisms, double chirped mirrors, pulse duration measured with fit to IAC (not very accurate)	[99Mor1, 99Mor2]
≈5 fs	120 mW	65 MHz	Similar to [99Mor1] but with double-Z cavity with second focus in a glass plate for additional SPM	[01Ell]

was a serious drawback for practical applications. In addition, the discovery of the Nd:YAG laser in 1964 [64Geu], and the popularity of dye lasers [66Sor,66Sch] and color center lasers [87Mol], diverted research away from tunable transition metal lasers. Renewed research efforts in tunable solid-state lasers produced the first demonstration of the alexandrite laser ($Cr:BeAl_2O_4$) in 1979 [79Wal] and of the Ti:sapphire laser ($Ti:Al_2O_3$) by Moulton in 1982 [82Mou].

The room-temperature Ti:sapphire laser material has an exceptionally wide tuning range of over 400 nm (680–1100 nm), a relatively large gain cross-section (peak value around 4×10^{-19} cm^2, about half of Nd:YAG), little excited-state absorption of the laser radiation, and a high optical quality. These properties were the basis for the success of the Ti:sapphire laser, leading to its crucial role in the race for few-cycle pulse generation from a passively modelocked laser directly without any external pulse compression, as shown in Table 9.1. The shortest pulses are generated with fast saturable absorber modelocking using a self-focusing effect with the intensity-dependent Kerr lens (4.13), thus referred to as Kerr lens modelocking (KLM) (see Sect. 9.4).

In 1990, two important experimental results were presented at two international conferences on Ultrafast Phenomena and CLEO, reporting sub-100 fs pulse generation of Ti:sapphire lasers, which caught the attention of the ultrafast community. In one experiment, a Ti:sapphire laser was passively modelocked with an intracavity saturable absorber dye jet [91Sar] using a 'CPM approach' that should not have worked. In the other experiment, 'self-modelocking' of a Ti:sapphire laser was observed by simply misaligning an otherwise empty laser cavity [91Spe]. Any missing saturable absorber resulted in the temporary description of magic modelocking, because this result was really a huge surprise. Both results were later explained by KLM [91Kel,92Kel3].

(a) Moving from Magic Modelocking to KLM

The 'self-modelocked' Ti:sapphire laser appeared to be the simplest femtosecond laser ever invented because no saturable absorber seemed to be present [91Spe]. This also resulted in the name of 'magic modelocking'. At the same time, this magically modelocked laser had the potential for far higher output power, and tunability, even though the laser intermittently self-modelocked and then fell back into cw operation. Initially, this result was explained by a nonlinear coupling between fundamental and higher-order modes inside the laser cavity, because the laser had to be slightly misaligned up to the level of the appearance of higher order modes before the laser could spontaneously start to modelock, while falling back into a clean fundamental laser mode [91Spe]. This starting was somewhat mysterious, resulting from a certain misalignment of the laser cavity and a certain amount of mechanical perturbation.

Using a weakly coupled nonlinear quantum well cavity, i.e., weak RPM, we achieved a clear separation of the modelocking and starting processes [91Kel]. This experiment showed for the first time that a resonant nonlinearity could self-start the modelocking, but not participate in the final pulse formation, because with an intracavity aperture, we could switch between picosecond and femtosecond pulse generation simply by partially closing the intracavity mechanical aperture. The nonlinear cavity then provided reliable self-starting modelocking without any cavity misalignment. Furthermore, after this self-focusing femtosecond modelocking has been initiated, the coupled semiconductor absorber cavity can be blocked and the modelocking sustains in the main cavity until the cavity is interrupted or larger mechanical vibrations occur. If the main cavity is then opened with the semiconductor absorber blocked, only cw behavior is obtained. This demonstrated that no misalignment of the laser cavity is necessary and that self-focusing is important in modelocking. Furthermore, it was suggested that an intensity-dependent Kerr lens inside the Ti:sapphire laser might reduce the loss inside the laser, when the pulse duration self-adjusts from picosecond to femtosecond. This work resulted in a more detailed theoretical prediction by Piché and co-workers [91Sal1], who received the [91Kel] paper for peer review. Their theoretical model confirmed that a nonlinear phase effect, such as Kerr-induced self-focusing, can lead to lower losses at high intensity as shown in Fig 9.14. The name Kerr lens modelocking was introduced by

Negus et al. [91Neg], who specifically redesigned a Ti:sapphire cavity to take full advantage of the self-focusing effect and demonstrated with an intracavity slit (similar to our intracavity aperture) fast saturable-absorber-like modelocking. Everything happened extremely fast and it was a really exciting time.

It became clear that, for certain resonator parameters, an asymmetric Kerr-lens nonlinearity can compensate for the linear misalignment introduced in the self-modelocked Ti:sapphire laser. We can model the self-focusing effect inside the laser cavity (Fig. 9.14) using an ABCD matrix approach. The intensity-dependent refractive index in the Ti:sapphire rod can be modeled with a Gaussian duct, using a parabolic approximation for the lateral refractive index change, and assuming a constant average beam size in the laser crystal. The RPM Ti:sapphire laser cavity as described in [92Kel3] is astigmatically compensated, the output coupler of the main cavity is approximately 3.5%, and the coupled cavity length is equal to the main cavity length. At typical intracavity peak powers of around 1 MW (average output power roughly 330 mW at about 5 W pump power, repetition rate 134 MHz, pulse duration 70 fs), the reduction of the beam radius in the tangential plane is close to 10%. This reduction only happens in the tangential plane, which makes a slit more appropriate than a round aperture. The slit acts as an intensity-dependent loss when the mode is reduced with increased intensity. Therefore, combining an intracavity slit close to the mirror output coupler with the self-focusing effect inside the Ti:sapphire laser rod produces a fast saturable absorber (Fig. 9.14). The nonlinear phase shift $\Delta\phi = kln_2$, where k is the propagation constant, l is the crystal length, and n_2 is the refractive index change due to self-phase modulation, gives some first indication of the importance of self-focusing. In this case, we obtain $\Delta\phi \approx 0.7$ radians in steady state.

A similar effect can be obtained without a real aperture or slit inside the cavity but relying on an effective soft-aperture in the gain. The Ti:sapphire laser is typically longitudinally pumped and therefore has a spatially limited gain beam profile inside the laser crystal. When the laser cavity is misaligned so that the laser cavity mode is not fully overlapping with the gain profile, an additional Kerr lens can pull the laser mode back into full alignment and increase the gain. This is also referred to as soft-aperture KLM [93Pic].

(b) Moving from CPM to KLM

Dyes have been used for many years as gain and saturable absorber media for femtosecond lasers. A careful balance of gain and absorber saturation forms a short net gain window (Fig. 9.5a) in a passively modelocked CPM laser (see Sect. 9.3). Ti:sapphire is a material with an emission cross-section which is about 1000 times smaller than that of laser dyes. We would not expect a laser consisting of a Ti:sapphire crystal as the gain medium and a dye jet as the saturable absorber medium to operate like a CPM laser [92Kel3], since the saturation energy of the gain medium is 1000 times higher than the saturation energy of the absorber, whence the absorber would be incapable of shaping the pulse, as shown in Fig. 9.5a. The gain is only saturated by the average power and no dynamic pulse-to-pulse gain saturation is relevant, as shown in Fig. 9.5c. In spite of this, Sarukura et al. [91Sar] passively modelocked a Ti:sapphire with an intracavity saturable absorber dye and produced stable sub-100 fs pulses. With no significant dynamic gain saturation in the Ti:sapphire laser, the roughly 1.2 ns recovery time of the saturable absorber dye is not fast enough to explain the short pulse duration (see Sect. 9.5).

As was explained in [92Kel3], at low concentrations, the dye jet only starts KLM. This can be shown in a number of ways. Sarukura et al. have investigated the starting characteristics as a function of the dye concentration and found a direct correspondence between the starting time and the dye concentration. Although the self-starting tuning range was originally reported to be 750–820 nm, we found that it is possible to extend this range using different dyes, or even a mixture of dyes with different absorption bandwidths. We confirmed that, with only HITCI in the jet, the self-starting range is 750–820 nm, as indicated by the HITCI absorption bandwidth. However, the laser will sustain modelocking up to 870 nm if carefully tuned, without introducing any perturbations. This is analogous to the weak RPM starting mechanism discussed above in Sect. 9.10.1a, since the HITCI is effectively being 'removed' by the operation at long wavelengths. Therefore, the HITCI is only self-starting the laser, and not participating significantly in the pulse shaping or the modelocking in steady state. When IR140 is added, a dye with more long-wavelength absorption than the HITCI dye, the self-starting range is extended to 870 nm. Moreover, by analogy with the weak RPM starting process, we find that if the dye concentration is too high, complex unstable operation is observed, suggesting that the dye modelocking is interfering with the KLM process.

(c) Fast Saturable Absorbers Based on Kerr Nonlinearity

It is important to emphasize that the modelocking process based on self-focusing with a Kerr lens is not self-starting: another starting mechanism has to be applied. However, once started the stability of these sub-100 fs pulses is excellent. Different starting mechanisms have been used, such as higher-order transverse mode beatings [91Spe], weak dye saturable absorber [91Sar], weak RPM [91Kel], moving mirror, acousto-optic modelocker [91Cur], and synchronous pumping [91Spi]. Generally, self-starting can be improved when operating the laser cavity at the stability limit, where an additional small Kerr lens introduces a larger cavity mode size change. This results in a coupling of the saturable absorber action with the cavity stability regime, which makes this approach less user friendly. Very small misalignments of the laser cavity components can strongly reduce the ability to start KLM.

Furthermore, a fast saturable-absorber-like modelocking is achieved with a nonlinear polarization rotation. An intracavity birefringent Kerr material combined with a polarizer results in higher transmission at higher intensity. Pulses as short as 60 fs were produced with a Nd:glass fiber laser [93Obe]. However, being based on the Kerr nonlinearity, this modelocking technique is also not self-starting: a SESAM was used for self-starting.

Nonlinear effects such as Kerr nonlinearities in laser cavities were discussed as early as 1968 [68Lau,74Smi]. In addition, Kerr-induced polarization rotation for passive modelocking was proposed and demonstrated as early as 1972 [72Dah], followed by other applications [77Sal]. With the newly available femtosecond solid-state lasers, however, the intracavity peak intensities have become sufficiently high to seriously exploit these relatively weak non-resonant nonlinearities. This is one reason why KLM was discovered by accident: those who discovered it were unaware of these early publications. The Kerr nonlinearity has the advantage of being broadband and fast, but the disadvantage of not producing self-starting modelocking.

(d) Higher Order Dispersion Compensation for Few-Cycle Pulse Generation
Within a very short time the pulse duration from KLM Ti:sapphire lasers was pushed into the few-cycle regime (see Table 9.5 and Fig. 9.4 in the introduction to this chapter). In comparison to the progress towards shorter pulse durations with dye lasers as shown in Fig. 9.4, the progress with the Ti:sapphire laser was amazingly fast, and a race between several competing groups took place. By minimizing higher-order intracavity dispersion in Ti:sapphire lasers, several groups produced pulses shorter than 30 fs (see Table 9.5).

However, limited control of higher-order intracavity dispersion is the main obstacle in the generation of extremely short pulses in the few-cycle regime. For optimum pulse formation in a KLM laser, a small but constant intracavity dispersion is required over the full spectral content of the pulse. In a typical sub-10 fs Ti:sapphire laser, the uncompensated roundtrip GDD caused by the laser crystal alone amounts to more than 100 fs delay between the long and short wavelength components covered by the gain material. To compensate for this GDD, a component with the opposite sign of dispersion has to be inserted into the cavity. Between these sections where the GDD has the opposite sign, the pulse width will be periodically stretched and recompressed from the sub-10 fs to the 100 fs range, for which the linearized master equation starts to fail to predict the final pulse duration. Moreover, the position of the different components starts to become relevant. This all leads to rather complex pulse durations and spectra which cannot be fitted by a simple sech^2 pulse shape.

Different schemes have been proposed for providing dispersion to compensate for material effects and self-phase modulation inside a laser cavity (see Chap. 3). The prism compressor was initially the most often employed scheme and introduces adjustable dispersion. However, generally speaking, it is only possible to compensate dispersion to second order with a prism pair. For optimum performance of a laser, the prism material has to be carefully chosen and the total amount of material inside the cavity should be kept to a minimum. This explains the initial progress towards shorter pulses (see remarks in Table 9.5), moving to different prism materials. Pulse durations of about 8.5 fs have been demonstrated using only prism compensation and metal mirrors [94Zho]. For shorter pulse operation, higher-order dispersion has to be better compensated for.

Chirped mirrors solved two problems (see Sect. 3.6): first, they give a much broader reflection bandwidth than standard Bragg mirrors, and second, they can be designed for a specific second and third order GDD. Such chirped mirrors introduce some small residual group delay oscillations with periods of several tens of nanometers. The magnitude of these residual dispersion oscillations increases with the bandwidth of the mirrors and for the longer wavelength in the pulse spectrum. The combined action of SPM and dispersion provides a means of energy transfer within the pulse spectrum inside a modelocked laser. Effectively, this concentrates energy at the minima of the group delay oscillations as indicated by the strongly modulated spectra typical of sub-10 fs laser pulses; the pulse shape is strongly affected by the dispersion oscillations and significantly deviates from the simple picture predicted from the soliton-like pulse shaping. Moreover, group delay oscillations can ultimately limit the intracavity bandwidth and the shortest achievable pulse width. Different chirped mirrors with shifted dispersion characteristics can be combined to reduce the total intracavity group delay oscillations. Access to such a selection of different chirped mirrors determined the race towards the end.

Using both prisms and chirped mirrors for dispersion compensation has the advantage that the chirped mirrors only have to compensate for higher-order dispersion. This results in a broader bandwidth of the chirped coatings. Some of the shortest pulses have been generated with a combination of chirped mirrors and prism dispersion compensation. Ultrabroadband SESAM even supported self-starting modelocking.

In the few-cycle regime, the pulse duration measurement becomes very challenging and unfortunately the less carefully measured pulses tend to result in shorter pulse durations. This ultimately saturated the race for new world-record results with Ti:sapphire lasers, as shown in Table 9.5.

9.10.2 High Average Power

(a) Motivation for Multi-100-W-Average-Power Ultrafast Lasers

High pulse repetition rates and high pulse energy from diode-pumped solid-state lasers have become very interesting tools for both scientific research and industrial applications (e.g., materials processing). With both high pulse repetition rate f_{rep} and high pulse energy E_{p} these ultrafast laser systems need to handle high average power because $P_{\text{av}} = E_{\text{p}} f_{\text{rep}}$. With Yb-doped solid-state and fiber lasers, average power well above 1 kW was obtained both at cw and with sub-picosecond pulse durations. Such multi-100-W-average-power ultrafast laser systems have become an indispensable tool for many applications, as briefly discussed in the following.

For example, for scientific applications, high harmonic generation (HHG) [88Fer, 94Lew] was discovered thanks to cutting-edge laser technology and has enabled table-top vacuum ultraviolet (VUV) to soft X-ray sources, spanning a wavelength range from 100 nm to 1 nm with femtosecond to attosecond pulses. Conversion efficiency in HHG is typically low (i.e., $\ll 10^{-4}$), resulting in pulse energies in the nanojoule range. This means that the pulse energy generated in the VUV to soft X-ray range with HHG is similar to the pulse energy directly generated from traditional laser oscillators in the visible and near-infrared, but at a much lower pulse repetition rate. This is at least 10 000 times lower when comparing 1 kHz amplifiers with 100 MHz oscillators. Typical HHG experiments using attosecond pulses suffer from either substantially increased acquisition time and/or decreased signal-to-noise ratio because of these low pulse repetition rates. In addition, the signal-to-noise ratio is determined by the average photon flux, and the minimal pulse energy has very often to be increased to obtain any measurable signal above the noise floor. The higher pulse energies reach a level where for example, space-charge effects (i.e., electromagnetic forces between the generated charged particles within the interaction volume) start to blur the dynamics to be investigated. This is especially important for solid-state physics experiments using for example, time-resolved angle-resolved photoemission spectroscopy (ARPES) (see, e.g., [06Kor, 15Loc]) and attosecond transient absorption (see, e.g., [13Sch, 16Luc]). Therefore for such applications attosecond pulses with higher pulse repetition rates (i.e., above 100 kHz), higher average photon flux for better signal-to-noise, and still moderate pulse energies (i.e., around 1 nJ) are greatly beneficial.

For industrial applications, for example, material processing is an important market that traditionally uses nanosecond laser pulse durations, but has been moving towards picosecond and femtosecond pulses due to higher precision and reduced heat effects (i.e., thermal damage) [97Liu, 97Lin]. Such ultrafast lasers can create features with micrometer precision (microprocessing) [16Dre]. They offer a key advantage in such processing because ablation of the target area occurs faster than the transfer of heat to the surrounding material. Therefore, this is typically referred to as cold ablation, for which the heat-affected zone surrounding the removed material is minimized. This results in more precise and sharper features. Note there are many more applications that benefit from multi-100-W-average-power ultrafast laser systems, but the goal was to give just a few examples here.

(b) The Success Story of Yb-Doped Laser Amplifiers

The average output power of ultrafast laser systems increased dramatically due to the development of diode-pumped Yb-doped lasers, because they can be pumped with a very low quantum defect (i.e., the pump energy minus the lasing energy is low) and have very little excited-state absorption and upconversion [91Lac]. Such gain materials are ideally suited for high-average-power operation as they allow for efficient and cost-effective diode pumping at a high doping concentration, while maintaining a good material quality. For efficient heat removal, the laser material geometries have been optimized for a high surface-to-volume ratio. This has enabled new world-record results using fiber [18Mul], thin disk [20Die], and slab [10Rus] geometries supporting more than 1-kW average power [see Fig. 9.48 (right) and Table 9.6]. Such amplifier systems are usually multi-stage chirped pulse amplification (CPA) [85Str] systems in which a low-power oscillator is pulse-picked, temporally stretched, directed through a chain of laser amplifiers, and then compressed (Fig. 9.48 and Table 9.6).

A crucial aspect of these multi-100-W-average-power laser systems is the need for effective heat removal without introducing thermo-optic distortions. For higher pulse energies alone, the pulse repetition rate is typically reduced starting from low-energy laser oscillators in the 100 MHz regime and going to high-energy amplifier systems in the 1-kHz regime (or even lower for even higher pulse energies) due to average power limitations of the amplifier.

In Fig. 9.48 on the left the typical operation parameters for laser oscillators and amplifiers are shown. The diagonal lines correspond to constant average power $P_{\mathrm{av}} = E_{\mathrm{p}} f_{\mathrm{rep}}$, where E_{p} is the pulse energy and f_{rep} is the pulse repetition rate. Typically, in the past the output energy of a modelocked laser oscillator was amplified with an external amplifier without significantly increasing the average output power. So in

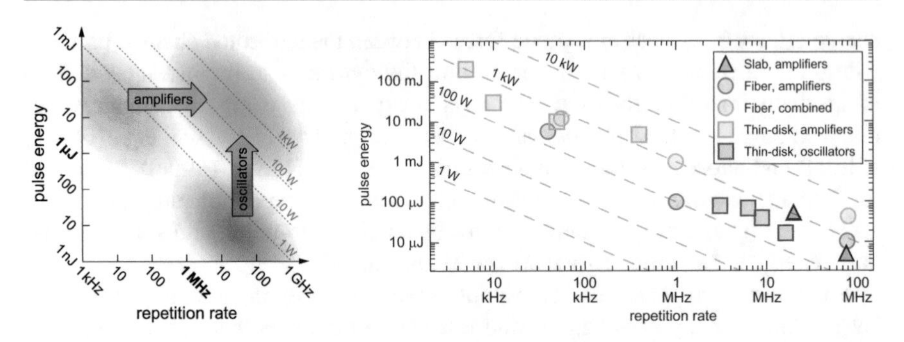

Fig. 9.48 High-energy ultrafast lasers can be separated into two different classes: oscillators and amplifiers. Typical operating regimes are shown according to their pulse energy E_p and pulse repetition rate f_{rep}. *Diagonal lines* correspond to constant average power $P_{av} = E_p f_{rep}$. *Left*: Before 2000, these two classes of lasers typically operated in two well separated operating regimes in terms of repetition rate and pulse energy [08Sud]. *Right*: World record results as of 2020 and Table 9.6. See text for further explanation

terms of the performance graph showing E_p as a function of f_{rep}, the amplification moves the pulse energy along the dashed diagonal lines of constant average power.

Historic milestone results for average-power scaling of ultrafast laser amplifier systems had the following pulse performance with regenerative amplifiers: femtosecond pulses with 116 µJ at 40 kHz using a thin-disk Yb:YAG [07Lar] and 1.7 µJ at 250 kHz using Ti:sapphire [92Nor]. The pulse repetition rates of regenerative amplifiers were limited by their high-voltage Pockels cell and the onset of chaotic instabilities. Similar power scaling limitations were found with a cavity-dumped Yb:KYW thin-disk laser, which generated high energy femtosecond pulses with 3 µJ at 1 MHz [07Pal]. Further improvements with large mode area Pockels cells resulted in new world-record performance with thin-disk Yb:YAG generating 200 mJ pulse energy at 5 kHz pulse repetition rate [17Nub]. However, scaling towards higher pulse repetition rates is more difficult in comparison to other techniques (see Table 9.6).

Fiber CPA systems are more promising for high-power megahertz pulse repetition rates. For example, historic milestone results with pulse energies as high as 1.8 µJ at 73 MHz [05Ros] and 100 µJ at 900 kHz [07Ros] have been demonstrated. More recent record results are summarized in Table 9.6 reaching 1 mJ in pulse energy at around 1 MHz pulse repetition rate [16Mul].

Coherent beam combination enables even higher energy and higher average power, but this even further increases the complexity of the overall system. In 2018, a record result in terms of average output power has been achieved by combining the output of multiple fiber amplifiers, generating 3.5 kW of average power [18Mul]. In 2019, 12 individual CPA fiber amplifiers were combined using subsequent spatiotemporal coherent combination of 96 (!) amplified pulse replicas to a single pulse to support a new world-record pulse energy of 23 mJ at an average power of 674 W. Each of the 12 fiber CPAs amplified a short pulse train with 8 pulses which results in $8 \times 12 = 96$ pulse replicas. At lower pulse energies, a compressed pulse duration of 235 fs was demonstrated [19Sta]. In 2021, 16 individual CPA fiber amplifiers were combined to generate an average power of 1 kW with 120 fs externally compressed pulses with a pulse energy of 10 mJ at a pulse repetition rate of 100 kHz [21Sta]. In the future the complexity may be reduced with multi-core fibers [20Ste].

(c) The Success Story of Ultrafast Yb-Doped Thin-Disk Laser (TDL) Oscillators
Thin-disk technology [94Gie] has enabled an alternative and potentially simpler approach: directly generating sub-picosecond pulses with multi-100-W average power from a single modelocked thin-disk laser oscillator (Fig. 9.49). The thin-disk laser, which was introduced in 1994 by Giesen et al. [94Gie], solves the thermal problems occurring in conventional high power rod or slab lasers. In a thin-disk laser head, the gain medium has the shape of a very thin disk, which is coated on one side with a highly reflective coating and on the other with an antireflective coating for both pump and laser wavelength. The reflective side is mounted directly onto a water-cooled heat sink and the disk is used in reflection. The thickness of the disk is in the range of about 100 μm, and therefore very small compared to the diameters of the pump and the laser beam, which are in the range of about 1 cm. In this geometry, the disk can be cooled very effectively and the heat flow is mainly one-dimensional towards the heat sink, i.e., collinear to the direction of the laser beam. The thermal gradients in the transverse direction are weak, and thermal lensing is strongly reduced compared to a rod or slab design. The pump absorption in a single pass through the thin disk is weak, but highly efficient operation can be achieved by arranging multiple passes through the disk (typically 16 to 32 passes, resulting in over 90% absorption of the pump light in the crystal [07Gie]).

In 2000, the first SESAM modelocked thin-disk laser was demonstrated [00Aus]. In contrast to the Ti:sapphire laser, this laser is directly diode-pumped and has a much higher electrical-to-optical efficiency. In subsequent work the pulse energy from ultrafast lasers was scaled from typically around 1 nJ to 80 μJ using thin-disk laser oscillators, which is nearly 5 orders of magnitude. With regard to average power, we increased from typically 100 mW to more than 400 W (Fig. 9.49 and Table 9.7).

Compared to ultrafast Ti:sapphire laser oscillators, Yb-doped laser materials can be highly doped for more efficient heat removal in a thin-disk geometry which increases the average output power by at least three orders of magnitude. The trade-off is that Yb-doped gain materials with high average power performance have had a much narrower gain bandwidth compared to Ti:sapphire, so far. Therefore, as of 2020, they have not delivered pulse energies well above 1 μJ with pulse durations below 100 fs [15Sar] (Table 9.8 and Fig. 9.50 updated from [15Sar] in 2020).

Few-cycle pulses with high average power and high pulse energy are desirable for fundamental science applications. There have been two approaches to reach this goal: one is based on optical parametric chirped pulse amplification (OPCPA) and the other on external pulse compression. We can benefit from the picosecond pulses from Yb-doped laser systems to pump an OPCPA. However, until recently, sufficient peak power was only achieved at repetition rates of 1 kHz or less, and only very recently at 100 kHz. For example, a milestone result was a mid-IR OPCPA system operating at 100 kHz repetition rate and delivering 16.5 fs pulses (i.e., 2.2 cycles) centered at a wavelength of 2.2 μm with an average power of 25 W [20Pup]. The corresponding

Table 9.6 Historic milestone results with world-record performance at the time of publication of ultrafast laser systems with pulse duration shorter than 2 ps and in terms of high average power with high pulse energy. These results have been partially used in Fig. 9.48 (right). Average power P_{av}, pulse repetition rate f_{rep}, FWHM pulse duration τ_p, pulse energy E_p, pulse peak power P_{peak}, chirp pulse amplification CPA

Technology	P_{av}	f_{rep}	τ_p	E_p	P_{peak}	Reference
Innoslab amplifier	400 W	75.8 MHz	682 fs	5 μJ	6.8 MW	[09Rus]
	1.1 kW	20 MHz	615 fs	55 μJ	80 MW	[10Rus]
Fibre CPA	830 W	78 MHz	640 fs	11 μJ	12 MW	[10Eid]
	230 W	40 kHz	200 fs	5.7 mJ	22 GW	[14Kle]
	100 W	1 MHz	270 fs	100 μJ	370 MW	[15Zha]
Fibre CPA, combined	700 W	56 kHz	262 fs	12 mJ	35 GW	[16Kie]
	1 kW	0.99 MHz	260 fs	1 mJ	3.3 GW	[16Mul]
	3.5 kW	89 MHz	430 fs	44 μJ	–	[18Mul]
	674 W	–	–	23 mJ	–	[19Sta]
	1 kW	100 kHz	120 fs	10 mJ	–	[21Sta]
	388 W	500 kHz	6.9 fs	776 μJ	–	[21Mul]
Thin-disk regenerative amplifier	300 W	10 kHz	1.6 ps	30 mJ	1.6 GW	[13Tei]
	540 W	50 kHz	1.14 ps	10 mJ	8.3 GW	[16Sch2]
	1.014 kW	5 kHz	1.08 ps	200 mJ	–	[17Nub]
Thin-disk amplifier, multi-pass	1.9 kW	400 kHz	1.3 ps	4.8 mJ	3.25 GW	[20Die]
Thin-disk oscillator	275 W	16.3 MHz	583 fs	16.9 μJ	25.6 MW	[12Sar]
	242 W	3.03 MHz	1.07 ps	80 μJ	66 MW	[14Sar]
	350 W	8.88 MHz	0.94 ps	39 μJ	37 MW	[19Sal]
	430 W	6.29 MHz	769 fs	68 μJ	78 MW	[20Sal]

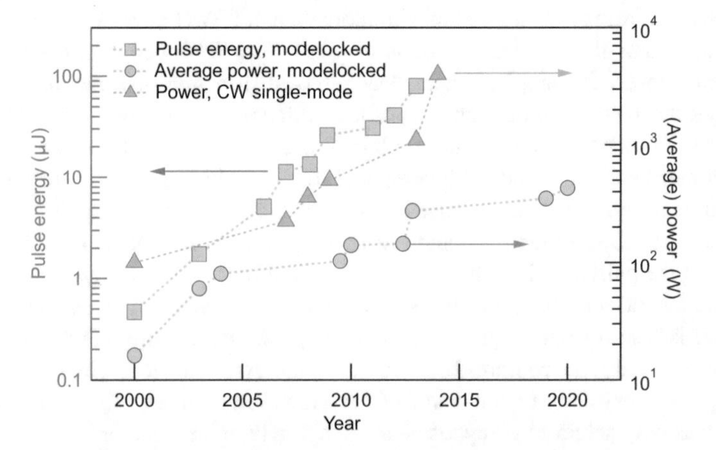

Fig. 9.49 Historical evolution of average power and pulse energy from modelocked thin-disk laser (TDL) oscillators since their first demonstration in the year 2000, together with the evolution of the maximum power in cw operation with a single-transverse fundamental mode from a single-disk resonator. Table 9.7 summarizes these results with references

Table 9.7 Historical evolution results and references from thin-disk laser (TDL) oscillators, as shown in Fig. 9.49. Note that only the key milestone results in terms of average power P_{av} or pulse energy E_p are listed in the table. Results without references are based on personal communications

	Cw, single-mode TDL oscillators		Modelocked TDL oscillators, average power		
Year	P_{av} [W]	Reference	P_{av} [W]	E_p [μJ]	Reference
2000	100	–	16.2	0.47	[00Aus]
2003	–	–	60	1.75	[03Inn]
2004	–	–	80	–	[04Bru]
2006	–	–	–	5.12	[06Mar]
2007	225	[07Gie]	–	11.25	[08Mar]
2008	360	[08Men]	–	–	–
2008	–	–	–	13.4	[08Neu]
2008	–	–	–	25.9	[08Neu2]
2009	500	–	103	–	–
2010	–	–	141	–	–
2011	–	–	–	30.7	[11Bau]
2012	–	–	145	41	[12Bau]
2013	1100	[13Pen]	–	80	[14Sar]
2014	4000	[15Kuh]	–	–	–
2019	–	–	350	–	[19Sal]
2020	–	–	430	–	[20Sal]

peak power was 14 GW, which is a record value for high repetition rate mid-IR systems.

External pulse compression at multi-100-W average power has become a significant research focus, moving away from waveguide to bulk materials because of the much higher input pulse energies (see Sect. 4.1.4). For example, the demonstration of an efficient (i.e., above 90%) multi-pass cell compression schemes with minimal impact on the spatial and temporal beam quality [17Wei, 17Wei2] or an argon-filled Herriott cell [19Rus] are expected to make Yb-doped lasers even more important for the ultrafast scientific community. For example, an argon-filled Herriott-type multi-pass cell pulse compressor generated 31 fs, 1 mJ pulses with an average power of 1 kW using a coherently combined fiber CPA system with an overall compression efficiency of 96% [20Gre]. This clearly shows a feasible path forward towards few-cycle pulse duration approaching 1-kW average power based on Yb-doped lasers typically operating at a center wavelength of around 1 μm. Research for new laser materials, however, must continue if more progress towards less complex laser systems is to be made.

(d) The Laser Physics behind High Average Power Ultrafast TDL Oscillators

For successful energy and power scaling of femtosecond laser oscillators, we have to maintain a high power fundamental transverse mode of operation with a broadband gain material and stable femtosecond pulse formation at a high intracavity power level. Modelocked thin-disk laser oscillators have faced three main chal-

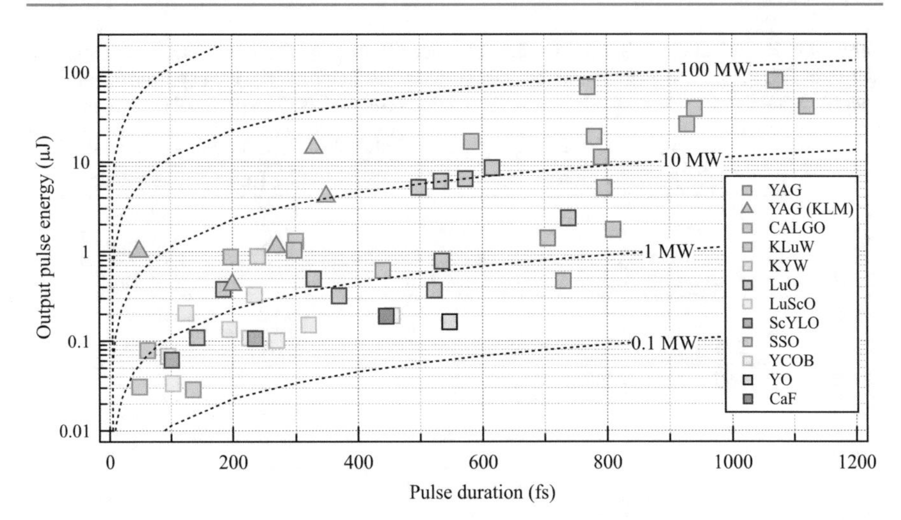

Fig. 9.50 Output pulse energy versus pulse duration of modelocked thin-disk laser oscillators with different Yb-doped gain materials, as demonstrated until 2020. This illustrates the ongoing challenge of reducing the pulse duration with above 10 μJ to the sub-100 fs regime. *Dashed curves* represent constant peak power. An overview of the different results shown in this graph is given in Table 9.8

lenges for average power scaling: managing the strong intracavity optical nonlinearities, achieving shorter pulses given the limited choice of high-power-compatible gain materials, and damage of intracavity components, such as SESAM and GTI mirrors. In Fig. 9.50, we clearly see a trade-off between shorter pulses and average output power. Scaling this performance needs a good understanding of the underlying laser physics, which is summarized in the following sections. Understanding the fundamental issues has helped address these challenges.

Ultrafast TDL Oscillator Physics: Nonlinearity Management

While the average power of passively modelocked thin-disk laser oscillators continuously increased, the pulse energy initially appeared to be limited to values below 2 μJ. Beyond this level, pulse instabilities and break-up into multiple pulses were observed. In 2006, the Kerr nonlinearity of the air inside the thin-disk laser cavity was identified as the origin of these instabilities. Covering the Yb:YAG thin-disk laser with an enclosure that was then flooded with helium, which has negligible nonlinearity, eliminated this effect. A pulse energy of 5 μJ at a repetition rate of 12 MHz was then achieved [06Mar], limited at that time by the power available from the pump diodes.

Therefore, one long-standing issue in modelocked thin-disk lasers has been the optical nonlinearity of air. The intracavity pulses are so intense that the weak intensity-dependent refractive index of air can be enough to destabilize the normal ultrashort-pulse formation process. Consequently, the highest powers to date have been achieved by operating in a near-vacuum environment, with the corresponding experimental complexity.

A new solution to this problem was demonstrated by adding a specially designed nonlinear crystal with a negative optical nonlinearity inside the laser resonator, which

Table 9.8 Selected results as shown in Fig. 9.50. FWHM pulse duration τ_p, average output power P_{av}, pulse energy E_p, peak power P_{peak}, pulse repetition rate f_{rep}

Material	τ_p [fs]	P_{av} [W]	E_p [μJ]	P_{peak} [MW]	f_{rep} [MHz]	Reference
Yb:CaF_2	445	6.6	0.19	0.38	34.7	[14Dan]
Yb:CAF	285	17.8	1.78	5.50	10	[16Dan]
Yb:CALGO	30	0.15	1.2×10^{-3}	0.04	124	[18Mod]
Yb:KLuW	440	21.3	0.61	1.23	34.7	[08Pal]
Yb:KYW	240	22	0.88	3.23	25	[02Bru]
Yb:Lu_2O_3	523	24	0.37	0.62	65	[07Mar2]
Yb:Lu_2O_3	370	21	0.32	0.75	65	[07Mar2]
Yb:Lu_2O_3	738	141	2.35	2.80	60	[10Bae]
Yb:Lu_2O_3	535	63	0.78	1.28	81	[09Bae2]
Yb:Lu_2O_3	329	40	0.49	1.32	81	[09Bae2]
Yb:Lu_2O_3	142	7	0.11	0.68	64	[12Sar3]
Yb:Lu_2O_3	616	82	8.60	12.30	9.5	[17Gra]
Yb:Lu_2O_3	534	90	6.10	10.10	14.8	[17Gra]
Yb:LuO_3/Yb:ScO_3	103	1.4	0.03	0.29	41.7	[14Sch]
Yb:LuO_3/Yb:ScO_3	124	8.6	0.21	1.44	41.7	[14Sch]
Yb:$LuScO_3$	227	7.2	0.11	0.42	66.5	[09Bae]
Yb:$LuScO_3$	321	10.1	0.15	0.42	66.5	[09Bae]
Yb:$LuScO_3$	235	23	0.33	1.23	70	[11Sar]
Yb:$LuScO_3$	19	9.5	0.14	0.61	70	[11Sar]
Yb:$LuScO_3$	96	5.1	0.07	0.63	77.5	[12Sar2]
Yb:SSO	298	27.8	1.03	3.04	27	[12Wen]
Yb:Y_2O_3	547	7.4	0.16	0.26	45	[12Tok]
Yb:YAG	1120	145	41.43	32.55	3.5	[12Bau]
Yb:YAG	705	80	1.40	1.75	57	[04Bru]
Yb:YAG	928	76	24.6	24.85	2.9	[08Neu2]
Yb:YAG	1360	50	13.40	8.67	3.8	[08Neu]
Yb:YAG	796	63	5.12	5.66	12.3	[06Mar]
Yb:YAG	810	60	1.75	1.90	34.3	[03Inn]
Yb:YAG	791	45	11.25	1.25	4	[08Mar]
Yb:YAG	730	16.2	0.47	0.56	34.6	[00Aus]
Yb:YAG	583	275	16.87	25.47	16.3	[12Sar]
Yb:YAG	1070	242	80.67	66.00	3.03	[14Sar]
Yb:YAG	940	350	39.00	36.50	8.9	[19Sal]
Yb:YAG	780	210	19.10	21.50	11	[18Sal]
Yb:YAG	769	430	68.40	78.20	6.29	[20Sal]
Yb:YAG	200	17	0.43	1.87	40	[11Pro]
Yb:YAG	270	45	1.13	3.66	40	[11Pro]
Yb:YAG	330	270	14.36	38.29	18.8	[14Bro]
Yb:YAG	140	155	10.00	62.00	15.6	[16Bro]
Yb:YAG	323	21	1.18	3.20	17.8	[17Sal]
Yb:YAG	570	28	1.57	2.42	17.8	[17Sal]
Yb:YAG	426	66	7.10	14.70	9.3	[19Gra]
Yb:YAG	586	87	9.80	14.70	8.9	[19Gra]
Yb:YCOB	270	2	0.10	0.33	19.7	[10Hec]
Yb:YCOB	455	4.7	0.19	0.37	24.4	[10Hec]

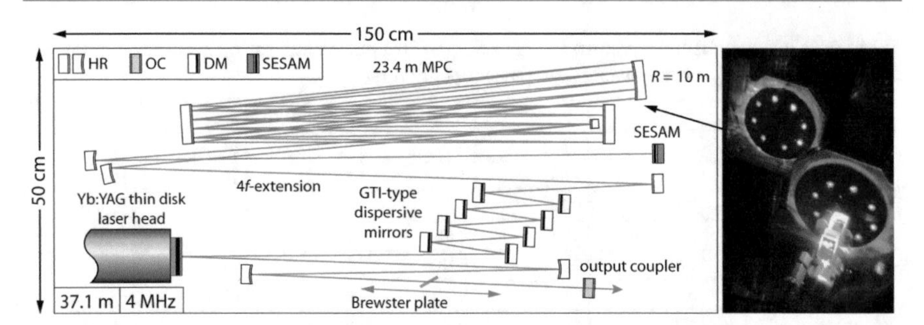

Fig. 9.51 SESAM-assisted soliton-modelocked, diode-pumped Yb:YAG thin-disk laser generating 11.3 μJ pulse energy at a center wavelength of 1.03 μm. Cavity setup at a pulse repetition rate of 4 MHz using a Herriott-type multiple-pass cavity (MPC). The image *on the right* was taken with an IR-sensitive camera. Highly reflective mirror HR, output coupler (10%) OC, dispersive GTI-mirror DM [08Mar, Fig. 3]

can generate an equal and opposite nonlinearity to that of air (see Sect. 9.5.3). Stable soliton modelocking can then be obtained with positive GDD. This practical approach enabled a SESAM-modelocked Yb:YAG thin-disk laser with 210-W average power and 780 fs pulse duration [18Sal].

Ultrafast TDL Oscillator Physics: Scaling to a Few Megahertz Pulse Repetition Rate

A higher pulse energy can be achieved by significantly reducing the laser repetition rate. The repetition rate can be reduced by lengthening the beam path inside the laser cavity. A simple way to achieve this is to insert a Herriott-type multiple-pass cavity (MPC) (Fig. 9.51) [64Her]. Such a cavity consists of at least two mirrors which reflect an incoming laser beam many times between them before it exits the cavity with the same nominal beam parameter. Folding up this MPC reduces the size further. With this approach a long beam path can be realized inside a compact thin-disk laser cavity [15Sar]. For the first time the pulse energy was increased beyond 10 μJ using such a Herriott-type MPC for a 4 MHz pulse repetition rate in 2008 (Fig. 9.51) [08Mar]. This was then even further increased to 80 μJ with 3 MHz in 2014 (see Table 9.6) [14Sar].

Ultrafast TDL Oscillator Physics: Scaling to Shorter Pulses

In another set of experiments, shorter pulses than those typically obtained from SESAM-modelocked Yb:YAG thin-disk lasers were pursued. Extensive research has been devoted towards new gain materials for this purpose (see Table 5.1 for a more detailed overview of different solid-state laser materials). Materials such as Yb:Lu_2O_3 or Yb:CALGO appear promising for approaching the average power performance of Yb:YAG while supporting shorter pulses because of their broader emission cross-sections. Nonetheless, a trade-off remains between high-power-compatible and short-pulse-compatible gain materials (Fig. 9.50 and Table 9.8).

An alternative approach to shorter pulses, while still using the well established Yb:YAG material, is to modify the modelocking mechanism (Fig. 9.52). For example, pulses in the experimental setup short as 330 fs with up to 270 W average power were obtained in a Kerr-lens modelocked (KLM) oscillator [14Bro], demonstrating the capacity of KLM to generate shorter pulses because of its fast nonlinear response.

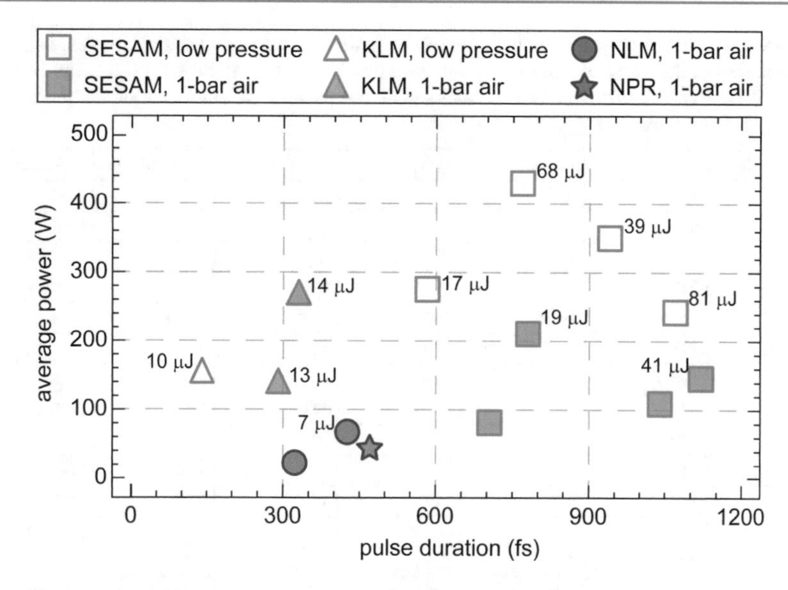

Fig. 9.52 Overview of high-average-power and high-energy thin-disk laser oscillators modelocked using different techniques: SESAM (soliton modelocking), KLM (Kerr lens modelocking), NLM (nonlinear mirror modelocking), and NPR (nonlinear polarization rotation). Results and references are summarized in Table 9.9

Ideally, a modelocking mechanism would simultaneously offer the advantages of the SESAM (robust operation with stability of the laser cavity decoupled from the pulse formation mechanism) and of KLM (which offers a fast saturable loss mechanism). The frequency-doubling nonlinear mirror modelocking (NLM) technique, first proposed in 1988 by K.A. Stankov [88Sta], may offer both of these advantages. In earlier work, this technique was successfully used with bulk crystal lasers, although the laser peak power achievable was limited. In the first proof-of-principle thin-disk demonstration, a moderate average power of 21 W was obtained, with pulses as short as 323 fs [17Sal], which was then improved with a more reliable SESAM-assisted starting of the NLM generating 66 W with pulses as short as 426 fs (Fig. 9.52 and Table 9.9) [19Gra].

Ultrafast TDL Oscillator Physics: Preventing SESAM and GTI Mirror Damage Power scaling also leads to kilowatt-level average powers on the intracavity optics. This can be troublesome for the SESAM (Sect. 7.5) and for dispersion-compensating GTI mirrors. Dispersive GTI mirrors have a more complicated thin-film structure than standard highly reflective Bragg mirrors, which leads to higher field enhancements and therefore to stronger thermal effects and lower laser-induced damage thresholds (Chap. 3). As shown in Sect. 7.5, SESAM damage can be more easily resolved. Even simple Bragg mirrors need to be optimized for higher damage threshold with different coating materials. For example, in the experimental setup shown in Fig. 9.51, we observed stronger heating at the laser beam spots on the mirror of the Herriott multi-pass cavity when viewed with an infrared camera. This ultimately leads to damage.

Table 9.9 World record results at the time of publication for ultrafast laser oscillators, i.e., pulse duration shorter than 2 ps, in terms of high average power with high pulse energy. Results as shown in Fig. 9.52. FWHM pulse duration τ_p, average output power P_{av}, pulse energy E_p, peak power P_{peak}, pulse repetition rate f_{rep}

ML technique	P_{av} [W]	E_p [μJ]	τ_p [fs]	f_{rep} [MHz]	P_{peak} [MW]	Reference
SESAM	430	68.4	769	6.29	78.2	[20Sal]
SESAM	350	39.0	940	8.9	36.5	[19Sal]
SESAM	275	16.9	583	16.3	25.4	[12Sar]
SESAM	242	80.7	1070	3.0	66.0	[14Sar]
SESAM	210	19.10	780	11.0	21.50	[18Sal]
SESAM	145	41.4	1120	3.5	32.6	[12Bau]
SESAM	108	30.7	1040	3.5	26.0	[11Bau]
SESAM	80	1.4	705	57.0	1.8	[04Bru]
KLM	270	14.4	330	18.8	38.2	[14Bro]
KLM	140	13.2	290	10.6	40.0	[19Poe]
KLM	155	9.9	140	15.6	62.0	[16Bro]
NPR	44	–	400–500 fs	–	–	[15Bor]
NLM	20.8	1.2	323	17.8	3.2	[17Sal]
NLM + SESAM	66	7.1	426	9.3	14.7	[19Gra]

Ultrafast TDL Oscillator Physics: Average Power Scaling Towards 1 kW
Reaching new milestones in terms of average output power is a core goal of this line of research. Therefore, it is beneficial for the modelocked laser to require minimum intracavity power and group delay dispersion for a given output power (while still satisfying the various requirements for stable pulse formation).

These considerations favor a cavity design in which the laser is reflected multiple times on the laser gain material disk, to provide higher gain and enable higher output coupling rates. Simultaneously reducing the required dispersion can be accomplished by the SPM cancellation method described above using cascade second harmonic generation [18Sal], or by operating in a low-pressure environment. The latter approach turns out to yield reduced thermal lensing effects at the same time. This was initially rather surprising. But it works for the following reason. In high-power operation, the pumped region of the thin disk can heat up to around 100°C. This causes the air near the disk to be heated up, which in turn causes a gas-lensing effect because of the thermo-optic coefficient of air [18Die]. Under normal operation, this gas lens contributes a significant fraction of the overall thermal lens of the disk, but is removed by operation in vacuum.

Investigation of this effect showed that SESAM modelocking in a low-pressure environment with a multipass gain cavity represents a particularly promising approach for power scaling. This yielded new world record output power for a thin-disk laser oscillator. The laser produced 350 W average power with 940 fs pulses, and a pulse energy of 40 μJ using 3 passes on the thin gain disk in 2019 [19Sal]. A 5-pass cavity then resulted in 430 W average power with 770 fs pulses, and a pulse energy of 68 μJ [20Sal]. Most recently, it has been shown that the thermal lens of a SESAM can be further optimized and controlled by silicate bonding a sapphire window to the front face of a SESAM [21Lan]. This paves the way for high-power optics with adjustable thermal properties (see Sect. 7.6.1).

9.10.3 Gigahertz Pulse Repetition Rates

(a) Motivation and Gigahertz Fiber Lasers

Many applications can benefit from compact inexpensive ultrafast sources with sufficient average power in the gigahertz pulse repetition rate regime:

- optical information technology, e.g., optical telecommunication, optical interconnects, optical clock distribution, all-optical-switching, quantum cryptography,
- ultrafast measurements, e.g., high-speed electro-optic sampling,
- high-precision measurements, e.g., frequency metrology, optical clocks,
- biomedical applications, e.g., multi-photon microscopy.

Modelocked fiber lasers have a number of apparent advantages [13Fer]. They potentially benefit from existing telecom components with lower cost and higher performance due to the demands of the existing market for telecom components. However, we see a number of disadvantages and pitfalls for gigahertz pulse repetition rates. Foremost, it is difficult to achieve stable, gigahertz repetition rates directly from the fiber oscillator. For a pulse repetition rate $f_{\mathrm{rep}} = 1$ GHz, we only have a cavity length $L = 15$ cm in air, assuming a standing-wave linear cavity with a cavity roundtrip time $T_{\mathrm{R}} = 2L/c$, where c is the light velocity in air. A short fiber length (i.e., less than 10 cm) requires high doping and strong pumping for sufficient gain. This has allowed for fundamental SESAM-modelocking up to 3 GHz with Yb:fiber lasers [12Che] and even up to close to 20 GHz with reduced pulse quality [11Mar].

Typically, harmonic modelocking is used with gigahertz fiber lasers, which requires multiple pulses circulating in the cavity at the same time. This is a source of chaotic noise and instabilities, and also requires additional control elements in the cavity, including potentially an active modulator with a high-frequency electronic driver [08Qui,07Yos].

Therefore, ultrafast fiber lasers are fundamentally not well-suited for short laser cavities, as they require a certain length of gain to achieve efficient lasing. Furthermore, they typically require elements which are not integrated into a fiber format to achieve the control functions for transform-limited pulse generation. That is, one must couple into and out of the fiber for various components, reducing the apparent simplicity and increasing the cost of the system. Additionally, these interfaces increase cavity loss in the laser, resulting in an increase in the background noise floor and potential damage at high peak powers. This all results typically in final ultrafast fiber lasers with additional stabilization that makes them larger in size and lower in output power than a diode-pumped solid-state laser for gigahertz repetition rates. Typical performance parameters have been summarized in Table 9.10.

Table 9.10 Examples of gigahertz modelocked fiber lasers which are not harmonic modelocked. Center wavelength λ_0, pulse repetition rate f_{rep}, FWHM pulse duration τ_p, average output power P_{av}, pulse energy E_p, peak power P_{peak}

Gain	λ_0 [nm]	f_{rep} [GHz]	τ_p [fs]	P_{av} [mW]	E_p [pJ]	P_{peak} [W]	Reference
Yb:fiber	1026	3	206	53	17.6	75	[12Che]
Er:fiber	1560	4.2	680	0.63	0.15	0.19	[11Mar]
Er:fiber	1565	5.2	680	0.17	0.03	0.04	[05Yam]
Er:fiber	1565	9.7	865	1.58	0.16	0.17	[12Mar]
Er:fiber	1560	9.6	940	2.5	0.26	0.24	[11Mar]
Er:fiber	1560	19.5	790	6.3	0.32	0.36	[11Mar]

(b) SESAM and Laser Cavity Designs to Suppress Self-Q-Switching Instabilities
Passively modelocked diode-pumped solid-state lasers are more challenging to scale into the gigahertz pulse repetition regime because of increased Q-switching instability issues. Until the late 1990s, the repetition rate of passively modelocked solid-state lasers was limited to a few gigahertz. Q-switching instabilities impaired performance at the highest pulse repetition rates (Sect. 9.7) [99Hon1]. During the early 2000s, the consequent exploitation of the flexibility of SESAMs with low saturation fluences (see Sect. 7.5.8) [05Spu3] allowed the development of passively modelocked solid-state lasers with pulse repetition rates up to approximately 160 GHz. Typical cavity designs are shown in Figs. 9.53 and 9.54, and early results are summarized in Table 9.11 for picosecond and Table 9.12 for femtosecond pulse generation. Very good picosecond pulse quality, comparatively high average output powers of several watts, and even wavelength tunability with Er:Yb:glass lasers in the ITU-specified C-band from approximately 1525 to 1565 nm has been demonstrated [04Pas].

The guidelines to suppress Q-switching instabilities according to (9.126), (9.130), and (9.132) are given by

$$E_p^2 > E_{sat,L} E_{sat,A} \Delta R \iff E_p > E_{sat,L} \frac{\Delta R}{S} \,, \quad S \equiv \frac{E_p}{E_{sat,A}} \,, \tag{9.177}$$

where S is the saturation parameter of the SESAM. According to (9.177), this means that the saturation of the SESAM should be large (i.e., S large) with an upper limit due to multipulse instabilities [97Aus, 18Wal]. Note that, for these modelocking stability requirements, we have assumed a fully saturated absorber (i.e., $S > 3$) and a SESAM modulation depth ΔR less than 10%.

The onset of multipulse instabilities can be explained as follows [97Aus]. The reflectivity of the SESAM increases with increasing pulse fluence and eventually goes into saturation. Given a pulse fluence many times the saturation fluence $F_{sat,A}$, the reflectivity is strongly saturated and therefore similar to the reflectivity for a pulse of half the fluence. Therefore, the SESAM provides reduced discrimination between single and double pulsing at increased incident pulse fluence. In addition, the limited gain bandwidth of the laser preferentially supports double pulsing over single pulses because the two longer pulses with soliton modelocking have a narrower spectrum with more gain than a single shorter intracavity soliton. Therefore, the laser generates two intracavity pulses over a single pulse for a sufficiently broad modelocked spectrum with a given gain bandwidth and a given saturation level of the saturable absorber.

What is special about gigahertz pulse repetition rates? For a given average output power P_{av} with high pulse repetition rates f_{rep}, the pulse energy E_p becomes relatively low because $P_{av} = E_p f_{rep}$. This means that (9.177) is a more difficult stability criterion. A large S can in this case still be obtained using SESAMs with a low saturation fluence $F_{sat,A}$ relative to the incoming pulse fluence on the saturable absorber $F_{p,A}$, i.e., $S = F_{p,A}/F_{sat,A} > 3$. This also explains why the criteria (9.130), i.e., the second one in (9.177), with the saturation parameter of the saturable absorber, is a more useful form of the QML threshold condition for gigahertz pulse repetition rates.

For gigahertz pulse repetition rates, it can be even more useful to reformulate the QML threshold criteria from (9.177) as follows [99Kra1]:

$$I_L > I_{L,crit} = f_{rep} F_{sat,L} \frac{\Delta R}{S} , \tag{9.178}$$

where I_L is the intensity inside the gain medium, $I_{L,crit}$ the minimum intensity inside the gain medium for stable modelocking, and $F_{sat,L}$ is the saturation fluence of the gain material. According to the rate equations for a three-level laser discussed in Sect. 5.3.1, and using (5.71), we obtain

$$F_{sat,L} = \frac{h\nu}{m(\sigma_L + \sigma_A)} , \tag{9.179}$$

where σ_L and σ_A are the emission and absorption cross-sections, respectively, at the laser wavelength. In a four-level laser, like Nd:YVO_4, we have $\sigma_A = 0$. Here, m stands for the number of passes through the gain medium within one round trip, e.g., $m = 2$ for simple standing-wave cavities.

Furthermore, it can also be useful to reformulate (9.177) as follows:

$$P_{out,crit} \geq f_{rep} T_{out} \sqrt{F_{sat,L} F_{sat,A} A_L A_A \Delta R} , \tag{9.180}$$

where T_{out} is the power output coupling rate, P_{out} is the average output power, and A_L and A_A are the mode areas in the gain and saturable absorber.

The SESAM and cavity design requirements are discussed in [02Kra2]. With regard to the SESAM parameters, we can summarize the requirements for gigahertz solid-state lasers as follows:

1. Modulation depth ΔR: For cavity losses on the order of 0.5–1%, a modulation depth on the order of 0.2–0.4% is typically sufficient. This is usually achieved with a quantum-well absorber layer of 5–10 nm thickness. A larger modulation depth ΔR may be desirable for generating shorter pulses, but it increases the QML tendency (9.178). In addition, a larger ΔR will tend to increase the nonsaturable losses in typical SESAMs, decreasing the intracavity power and thus increasing the QML tendency with the lower I_L in (9.178).

2. Saturation fluence $F_{\mathrm{sat,A}}$: The small pulse energies in multi-gigahertz lasers (i.e., $E_{\mathrm{p}} = P_{\mathrm{av}}/f_{\mathrm{rep}}$) and $S > 3$ typically requires fairly small mode areas A_{A} which are sometimes difficult to achieve due to geometrical restrictions. Therefore, SESAM designs with low saturation fluence in the range 1–10 μJ/cm^2 are desirable.
3. Cavity mode size: With (9.180), it helps to keep the mode sizes small in both the saturable absorber and the gain medium. This can be relaxed if $F_{\mathrm{sat,A}}$ and $F_{\mathrm{sat,L}}$ are small. Note that S cannot be too large, otherwise multipulse instabilities start to occur [97Aus, 18Wal]. In addition, small mode sizes are limited by damage in both the gain and saturable absorber.
4. Recovery time τ_{A}: We require $\tau_{\mathrm{A}} \ll T_{\mathrm{R}}$, where $T_{\mathrm{R}} = 1/f_{\mathrm{rep}}$ is the roundtrip time. Note that for gigahertz lasers the roundtrip time can be rather short e.g., 20 ps for 50 GHz. The recovery time becomes an issue when $\tau_{\mathrm{A}} > T_{\mathrm{R}}$, because the SESAM recovery during one roundtrip is then incomplete. This effectively lowers ΔR. In this case, τ_{A} needs to be reduced and this also helps to generate shorter pulses with limited soliton pulse formation.
5. Reflectivity rollover with ISA: As shown with the modified QML criteria (9.129), operating the SESAM on the maximum reflectivity at the rollover will help to suppress Q-switching. However, this tends to increase the losses and reduce the output power because of the low output coupling rates in typical gigahertz diode-pumped solid-state lasers.
6. Soliton modelocking can substantially reduce the QML threshold: This results from the modified QML criteria (9.127). However, low peak intensities in gigahertz lasers typically result in weak soliton pulse-shaping effects. The Kerr effect yields only moderate positive values for the nonlinear refractive index n_2 when the laser wavelength is far enough from the bandgap to avoid multiphoton absorption [10Chr]. Thus, achieving sufficient nonlinearity for soliton modelocking becomes increasingly difficult for lower energy lasers with lower peak intensities. The effective nonlinear refractive index can be substantially increased using cascaded quadratic nonlinearities (CQNs) with apodized QPM structures to keep the intra-cavity losses low enough for gigahertz solid-state lasers, as shown in Sect. 9.5.3 [17May, 20Kru]. The effective negative n_2 from CQN interactions even helps to suppress Q-switching instabilities in a straight linear two-mirror laser cavity (Sect. 9.7.6).

(c) Picosecond Pulse Generation at Gigahertz Repetition Rates
Initially, gigahertz pulse generation was limited to the picosecond regime using gain materials with relatively large gain cross-sections for solid-state lasers (see Table 5.1) and therefore a small saturation energy $E_{\mathrm{sat,L}}$, which more strongly suppressed Q-switching instabilities [see (9.177)]. The results are summarized in Table 9.11. Note that these lasers have different gain cross-sections, according to Table 5.1, summarized again here for Nd:YVO$_4$ with $\sigma_{\mathrm{L}} = 300 \times 10^{-20}$ cm^2, Nd:GdVO$_4$ with $\sigma_{\mathrm{L}} = 76 \times 10^{-20}$ cm^2, Nd:YLF with $\sigma_{\mathrm{L}} = 18 \times 10^{-20}$ cm^2, and Er:Yb:glass with $\sigma_{\mathrm{L}} = 0.8 \times 10^{-20}$ cm^2.

Here Er:Yb:glass is an exception with a very small σ_{L}. However, the Yb to Er pump energy transfer results in an effective low-pass filter of the transfer function and

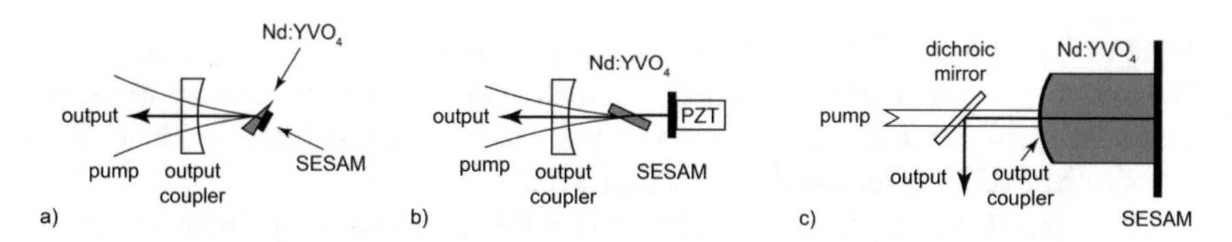

Fig. 9.53 Typical cavity designs for picosecond SESAM-modelocked $Nd:YVO_4$ lasers with pulse repetition rates up to about 160 GHz. Stable modelocking in a simple straight linear cavity has become possible using SESAMs with low saturation fluence and taking advantage of the relatively large gain cross-section (i.e., small saturation energy $E_{sat,L}$) of $Nd:YVO_4$ for solid-state lasers. Note that a pulse repetition rate of 5 GHz requires a cavity length of about 3 cm in air

greatly reduces relaxation oscillations, as discussed in Sect. 5.3. The lower relaxation oscillation peak in Er:Yb:glass lasers, as shown in Fig. 5.17, greatly suppresses self-Q-switching instabilities. This also results in extremely low-noise modelocking performance, which will be discussed in more detail in Chap. 12. This also explains a remarkable low-noise result with a monolithic 1-GHz femtosecond Er:Yb:glass laser [16Sho].

Gigahertz $Nd:YVO_4$ Lasers at 1.06 μm

In 1999, for the first time, pulse repetition rates above 10 GHz from passively modelocked ion-doped solid-state lasers were generated with $Nd:YVO_4$ lasers at a center wavelength around 1.06 μm in a straight linear cavity [99Kra1]. This was obtained with the cavity design shown in Fig. 9.53a, where according to (9.178) the intensity I_L in the laser medium is increased by moving the gain crystal close to the SESAM at the beam waist of the laser cavity mode [or according to (9.180), A_L was reduced] and also taking advantage of the stronger spatial hole burning with a 'gain-at-the-end' cavity (Sect. 6.7.3). This supports a pulse duration of 8.3 ps with diode-pumping, even though a SESAM modulation depth of only $\Delta R \approx 0.4\%$ was used. With a relatively large SESAM saturation fluence at that time (i.e., $F_{sat,A} = 60\ \mu J/cm^2$), a small output coupler of 0.4% was needed to obtain a sufficiently large saturation parameter S for the available pump power and average output power. This cavity design (Fig. 9.53a) supported pulse repetition rates from 2.7 GHz to 13 GHz by simply changing the cavity length. Note that (9.177)–(9.180) did explain qualitatively but not quantitatively the measured QML threshold intensity because there was significant spatial hole burning helping SESAM modelocking and lowering the QML threshold. The reader is referred to the more detailed discussion of modelocking with spatial hole burning in Sect. 6.7.3.

A higher average output power at 10 GHz was obtained with a different cavity (Fig. 9.53b), where the cavity length of about 15 mm can be more easily tuned with an additional piezoelectric translator (PZT) substantially changing the repetition rate from 9 to 11 GHz. This result is explained in more detail in [02Kra2]. The average output power was optimized and relatively high with 2.1 W using a much larger output coupler of 1.8% and very low nonsaturable losses from the SESAM of less than 0.1%. At that time this low nonsaturable loss was achieved with a top-coated SESAM, at the expense of a higher saturation fluence of $F_{sat,A} = 100\ \mu J/cm^2$. According to (9.180) both the higher output coupler and the higher $F_{sat,A}$ then requires a lower modulation depth of the SESAM (i.e., $\Delta R \approx 0.24\%$) to prevent self-Q-switching instabilities. The low modulation depth to suppress Q-switching instabilities resulted in longer modelocked pulse durations of around 14 ps, which were significantly longer than for other SESAM modelocked Nd:YVO_4 lasers.

The large cavity length adjustment was useful to synchronously pump an optical parametric oscillator (OPO). Even with picosecond pulses the peak power was sufficient for efficient nonlinear frequency conversion. For example, a synchronously pumped OPO was demonstrated, producing picosecond pulses broadly tunable around 1.55 μm with up to 350 mW average output power [02Lec,04Lec]. Such all-solid-state synch-pumped OPOs can reach the S-, C-, and L-bands in telecommunications. With an additional Yb-doped fiber amplifier for the OPO pump laser (i.e., a SESAM modelocked Nd:YVO_4 laser using the cavity design shown in Fig. 9.53c [00Kra2]), the repetition rate of the OPO was even pushed up to 80 GHz [05Lec2]. In this case the OPO cavity length was adjusted for synchronization.

The highest repetition rate of nearly 160 GHz [02Kra2] has been achieved with a quasi-monolithic laser cavity (Fig. 9.53c), consisting of a roughly 0.44 mm long Nd:YVO_4 crystal with the output coupler mirror coating evaporated on a curved end and a SESAM directly attached to the other (flat) end. This laser generated 2.7 ps pulses with a 157 GHz repetition rate and 45 mW average power. The pulses are remarkably short, despite the low modulation depth of the SESAM; the latter is due to spatial hole burning which increases the effective gain bandwidth (Sect. 6.7). With a 2.7 ps pulse duration and a pulse spacing of only 6.4 ps, subsequent pulses begin to overlap to some extent. With this picosecond laser, we reached the fundamental limit for the repetition rate which is given by the gain bandwidth, because a higher repetition rate would require shorter pulses. With this quasi-monolithic laser cavity (Fig. 9.53c) different pulse repetition rates were achieved with a different Nd:YVO_4 crystal length ranging from 2.3 mm to 440 μm to cover a pulse repetition rate from 29 GHz [99Kra2] to 160 GHz [02Kra2], as summarized in Table 9.11.

Gigahertz Er:Yb:glass Lasers at 1.5 μm

In the telecom wavelength ranges (around 1.3 and 1.55 μm), where only a few solid-state gain media are available, it was very difficult to demonstrate multi-GHz pulse repetition rates [97Col,95Lap]. The QML threshold for a 10 GHz Er:Yb:glass laser as shown in Fig. 9.54 had been found to be considerably lower (i.e., about a factor of 20) than initially predicted [02Kra1]. This laser generated 15 mW of average output power and the QML threshold was at 6 mW. The relaxation oscillations were too weak to appear in the microwave spectrum analyzer and were expected to be at around 300 kHz for maximum output power. Stable modelocking was possible as a combination of the following reasons: (i) improved SESAM designs with low saturation fluences (see Chap. 7), (ii) a deeper understanding of the Q-switching instabilities with inverse saturable absorption (Sect. 9.7.4) with the modified QML threshold criteria (9.129), and (iii) the low-pass filter of the Yb→Er transfer function (5.101) as shown and discussed in Fig. 5.17 for a 61 MHz modelocked Er:Yb:glass laser. For the 10 GHz laser a cut-off frequency of $f_{co} = 1.2$ kHz was measured [04Sch]. For the short 10 GHz laser cavity the relaxation oscillations become higher with an estimated $f_{res} \approx 400$ kHz which are more strongly suppressed by the low-pass filter of the Yb→Er energy transfer.

Table 9.11 Picosecond gigahertz modelocked diode-pumped solid-state lasers. ML modelocking technique. Center wavelength λ_0, pulse repetition rate f_{rep}, FWHM pulse duration τ_p, average output power P_{av}

Gain	ML	λ_0 [μm]	τ_p [ps]	P_{av} [mW]	f_{rep} [GHz]	Reference
Nd:YLF	SESAM	1.34	21	127	1.4	[06Zel]
$Nd:YVO_4$	SESAM	1.064	13.7	**2100**	10	[02Kra2]
			8.3	198	12.6	[99Kra1]
			6.8	81	29	[99Kra2]
			5.5, 5.3, 4.8	60, 80, 30	39, 49, 59	[00Kra1]
			2.7	288	40	[05Lec1]
			2.7	65	77	[00Kra2]
			2.7	45	**157**	[02Kra2]
		1.34	7, 7.3	45, 40	5, 10	[05Spu2]
$Nd:GdVO_4$	SESAM	1.064	18.9	3460	0.37–3.4	[04Kon]
			4.4	≈60	2.5–2.7	[05Agn]
			12	500	9.66	[04Kra]
Er:Yb:glass	SESAM	1.534	3.8	12	10.5	[02Kra1]
			5	50	10	[10Oeh]
		1.534	≈1.8	> 20	**8.8–13.3**	[04Ern]
		full C-band	1.9	25	25	[03Spu]
		1.534	4.3	18	40	[03Zel]
		1.533	2	7.5	50	[04Zel]
		1.536	3	10.7	77	[07Zel]
		1.535	1.6	35	100	[08Oeh]
		≈1.53	1–4	≤35	**10–100**	[10Oeh]
		1.55	2	≤15	10	[16Res]

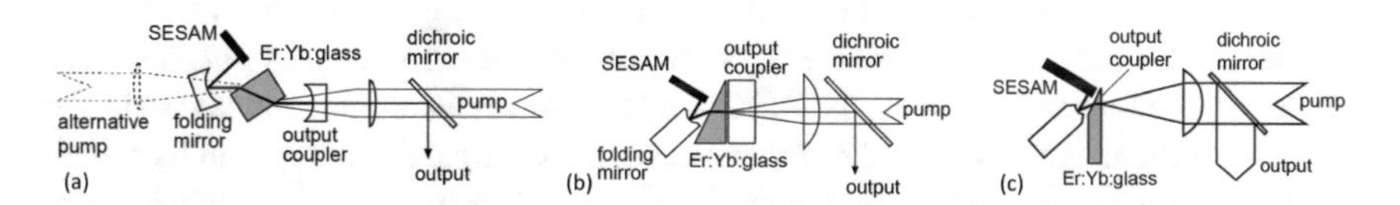

Fig. 9.54 Typical cavity designs for picosecond SESAM-modelocked diode-pumped Er:Yb:glass lasers with pulse repetition rates: **a** 10 to 50 GHz, **b** 77 GHz, **c** 100 GHz [10Oeh]

Figure 9.54 shows the different cavity designs used for SESAM-modelocked Er:Yb:glass lasers covering a pulse repetition rate from 10 to 100 GHz. Full C-band tuning (i.e., a range of roughly 40 nm) at 40 GHz [03Spu] was achieved with an additional intracavity etalon filter, and the pulse repetition rate was tuned between 8.8 and 13.3 GHz by simply changing the cavity length [04Ern]. Table 9.11 summarizes the different results with picosecond pulse generation.

In contrast to $Nd:YVO_4$ lasers, no linear two-mirror cavity was possible, and an additional folding mirror was required to obtain a smaller spot size on the SESAM and still maintain a small spot size in the gain, i.e., both A_L and A_A needed to be small, but $A_A < A_L$ to prevent Q-switching instabilities (9.180). However, with the correctly adapted cavity designs a remarkable low-noise performance was obtained with SESAM modelocked Er:Yb:glass lasers with even up to 100 GHz pulse repetition rates [08Oeh, 10Oeh].

After our initial demonstrations of the low-noise performance with different performance parameters, as shown in Table 9.11, gigahertz SESAM-modelocked diode-pumped Er:Yb:glass lasers were commercialized (and referred to as ERGO lasers), generating picosecond pulses (i.e., shorter than 2 ps) at 10, 12.5, and 25 GHz pulse repetition rates for the telecom market with very low timing jitter. Based on suppressed relaxation oscillations by the low-pass filter effect for the Yb to Er pump energy transfer (see Sect. 5.3 and Fig. 5.17), extremely low noise performance can be achieved. For example, the timing jitter of 10 GHz SESAM modelocked Er:Yb:glass lasers can be very close to the quantum noise limit as shown in Fig. 11.16 [05Sch2].

Further progress with SESAMs at 1.55 μm resulted in a low saturation fluence of 9 $\mu J/cm^2$ using a quantum dot saturable absorber inside the SESAM (i.e., QD-SESAM) [16Res]. A quantum well (QW) and a QD-SESAM were compared at a pulse repetition rate of 10 GHz with the following parameters [16Res]: QD-SESAM with $\Delta R \approx 0.24\%$, $F_{\mathrm{sat,A}} = 9\ \mu\mathrm{J/cm}^2$; QW-SESAM with $\Delta R \approx 0.5\%$, $F_{\mathrm{sat,A}} = 15\ \mu\mathrm{J/cm}^2$ with a measurable nonsaturable loss of 0.1%. The lasing threshold with the QD-SESAM was lower, which means that the nonsaturable losses were even lower than 0.1%.

Typically, highly strained InGaAs SESAMs have been used for the reported results. Antimonide SESAMs on InP-based DBRs can provide an attractive alternative. For example, a 10 GHz Er:Yb:glass laser has been also successfully modelocked with an AlGaAsSb SESAM on an InP substrate at a center wavelength of 1535 nm generating 4.7 ps pulses [06Gra].

(d) New Challenge: Transverse Mode Instabilities for a Short Cavity Length

This was initially a surprise [10Oeh] because previous work had been forgotten! It then became clear from the literature that laser cavities in general exhibit frequency degeneracies of transverse cavity modes for certain resonator lengths [99Zha2]. By modeling the laser cavity with the ABCD matrix formalism, one can calculate the position of the degeneracy points within the stability range. For a laser cavity which is radially symmetric with respect to the intracavity laser mode, it is straightforward to calculate the resonance frequencies of Hermite–Gaussian laser modes using

$$\nu_{m,n,q} = \frac{c}{2L_{\mathrm{eff}}}\left[q + (m+n+1)\frac{\cos^{-1}\left(\pm\sqrt{(A_1+D_1)/2}\right)}{\pi}\right], \tag{9.181}$$

where c is the speed of light, L is the optical resonator length, q is the longitudinal mode number, m and n are the transverse mode indices, and

$$\begin{pmatrix} A_1 & C_1 \\ B_1 & D_1 \end{pmatrix}$$

is a single-pass ABCD cavity matrix, as described in [86Sie]. Degeneracies arise when different cavity modes have the same frequency such that

$$\nu_{m+\Delta m,n+\Delta n,q+\Delta q} = \nu_{m,n,q} \ . \tag{9.182}$$

Cavity elements inside a laser resonator usually have a certain range with respect to their position along the optical axis, for which the resonator will be stable. For multi-GHz lasers the stability range of the SESAM used as an end mirror is on the order of a few 100 μm. To obtain stable modelocking, the SESAM usually has to be moved once within this range in order to find an optimum mode size on the device leading to the best saturation of the absorber. The stability range is usually limited at one end by the growing mode size in the gain, continuously raising the laser threshold. At the other end, it is limited by the mode size on the SESAM approaching zero, which results in an unstable resonator (and potential damage of the SESAM). Within this range, a rather smoothly varying laser output power is expected.

However, we observed multiple pronounced power drops while moving the SESAM within the stability range (Fig. 9.55). When measuring the dependence between output power and end-mirror position in cw-operation (in order to avoid Q-switched modelocking instabilities) by scanning a high-reflecting mirror through the stability range with a spatial resolution of about 100 nm, we observed that the output power drops significantly at certain distinct points.

The larger power drops actually have a well-known origin. Laser cavities generally exhibit frequency degeneracies of transverse cavity modes for certain resonator lengths (9.182) within their stability range. For lasers designed to operate in the fundamental mode (TEM_{00}) this is particularly important, because at such degeneracy points, higher order spatial modes can couple resonantly to the fundamental mode and become predominant in the cavity. For a passively modelocked laser, this can lead to instabilities, as the SESAM is no longer sufficiently saturated by these modes, and Q-switching can occur. When higher order spatial modes occur, they introduce an efficient power-loss channel, as some modes are quenched through effects of spatial hole burning in the inversion of the gain medium, leading to output power drops or even a complete laser shut-off. This is particularly distracting in a cavity with a more limited longitudinal stability range (with respect to the SESAM position), as for short cavity lengths.

For stable operation of high repetition rate modelocked lasers, it is essential to accurately examine the cavity's stability range, as certain cavity lengths and their vicinity have to be avoided [10Oeh]. To detect the beam quality deteriorations of a 100 GHz laser cavity, the stability range was scanned in cw operation with a resolution of about 200 nm and both the output power and the output beam profile of the laser were recorded. Figure 9.55 shows multiple large power drops within the stability range, together with the associated beam profiles. In every power dip, the beam quality is clearly degraded, while at normal power levels the output beam is nearly diffraction limited. The positions of the individual degeneracies are reproducible over multiple scans for a particular fixed pump power. Such frequency degeneracies of cavity modes have to be avoided for stable modelocking.

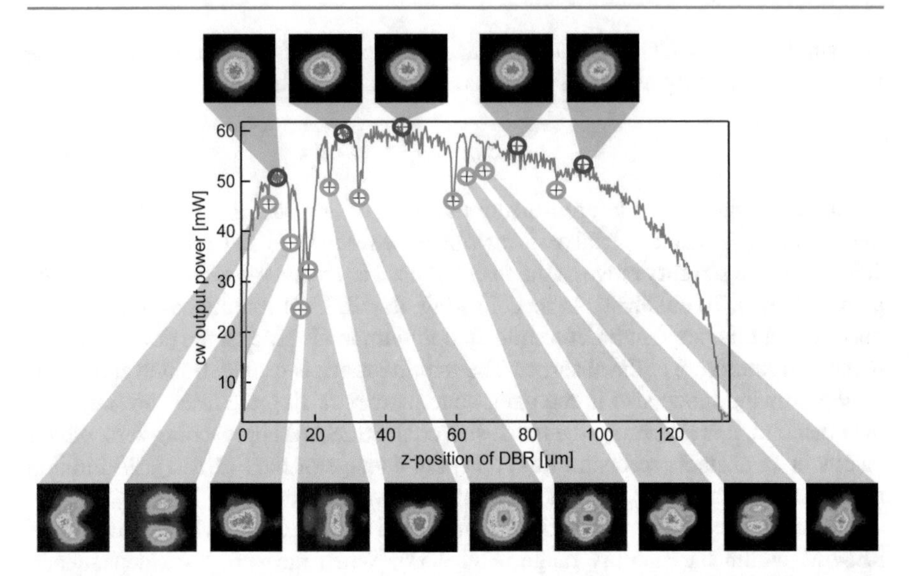

Fig. 9.55 Stability range (with respect to the end-mirror position) of a 100 GHz laser cavity (see Fig. 9.54) scanned in cw operation with a spatial resolution of 200 nm. The observed power drops can by clearly associated with strong beam quality degradations of the laser output. At normal power levels, the output beam is diffraction limited [10Oeh]

(e) Femtosecond Pulse Generation at Gigahertz Repetition Rates

The frequency comb revolution in 1999 (see Chap. 12 for more details) motivated research efforts in femtosecond gigahertz lasers. Possible diode-pumped solid-state laser materials (see Table 5.1) that have the necessary gain bandwidth to support femtosecond pulses, such as Yb-doped crystals and glasses, show that the gain cross-sections of these lasers are rather small compared to Nd:YVO_4, and Q-switching instabilities are therefore expected to become more severe according to (9.177)–(9.180). The gigahertz results with femtosecond pulse durations are summarized in Table 9.13. In contrast to the early picosecond gigahertz laser results (Table 9.11), the SESAM design with low saturation fluence has made significant progress, and stable passive modelocking was obtained with Yb-doped solid-state lasers with a gain cross-section at 1064 nm as low as $\sigma_L = 1 \times 10^{-20}$ cm^2 (see Table 5.1), i.e., 300 times lower than for an Nd:YVO_4 laser.

Ti:sapphire lasers have a larger gain cross-section $\sigma_L = 41 \times 10^{-20}$ cm^2 (Table 5.1) and typically small gain spot sizes can be easily obtained with perfect Gaussian beam pumping from a 'green' laser (i.e., argon-ion laser, SHG of a diode-pumped 1-μm solid-state or VECSEL laser). Thus, best performance with regard to short pulse durations can be obtained with KLM Ti:sapphire lasers (Table 9.12 and Fig. 9.56). For example, a 10 GHz four-mirror ring cavity with a 30 mm roundtrip cavity length is pumped with 8.5 W from a green laser and generates 1 W of average output power with 42 fs pulses [08Bar]. Such a laser has been used for stabilized frequency comb generation [11Hei]. A similar 5-GHz four-mirror ring cavity achieved 24 fs pulses [07Bar2], and a 2-GHz six-mirror ring cavity achieved 23 fs with external dispersion compensation in chirped mirror pairs [99Bar]. Ultrabroad bandwidth

Table 9.12 Femtosecond gigahertz KLM Ti:sapphire lasers. Center wavelength λ_0, pulse repetition rate f_{rep}, FWHM pulse duration τ_p, average output power P_{av}, pulse energy E_p, peak power P_{peak}

λ_0 [nm]	f_{rep} [GHz]	τ_p [fs]	P_{av} [mW]	E_p [nJ]	P_{peak} [kW]	Reference
788	10	42	1060	0.1	3.5	[08Bar]
798	5	24	1150	0.23	8.5	[07Bar]
782	2	23	300	0.15	5.7	[99Bar]
782	1	6	≈960	0.96	136	[06For]

to support pulses as short as 6 fs has been generated with a KLM four-mirror ring laser with intracavity chirped mirrors and a repetition rate between 550 MHz and 1.35 GHz [06For].

With regard to diode-pumped solid-state lasers, the results are summarized in Table 9.13 and Fig. 9.56. The best results for average output power and pulse duration can be achieved with Yb-doped solid-state lasers, as shown in Fig. 9.56. More than 1 W of average output power was first obtained with Yb:KGW in 2010, but the pulse duration was limited to typically more than 280 fs using SESAM modelocking [10Pek]. In comparison, Yb:CALGO provides an exceptionally broad and smooth absorption and emission cross-section which is ideally suited for sub-100 fs pulse generation. Additionally beneficial is the relatively low group delay dispersion (GDD $\approx +108$ fs^2/mm, measured with a white light interferometer). The high thermal conductivity and low quantum defect of Yb:CALGO reduce undesired thermal lensing. In comparison to the previously more widely used Yb:KGW/KYW, the tendency for Q-switching instabilities is slightly increased due to the higher gain saturation fluence. Thus more careful cavity and SESAM designs are required.

Usually, only high-brightness pump lasers are used for high repetition rate lasers, since a tight focus is required for a small A_L in the gain to avoid Q-switching instabilities (9.180). However, single-mode pump diodes are usually limited in power. Instead of a cumbersome combination of multiple low-power pump diodes [10Li] or using highly sensitive tapered amplifiers used in [10Pek, 11Pek, 12Pek], it is more attractive to use commercially available transverse-multimode fiber-coupled laser diodes [13Kle]. This simple and robust pumping scheme offers high power and reasonably good high brightness, and it reduces the overall complexity and cost. A high brightness can potentially also be obtained with more progress in close to single-mode high-power pump diodes.

To avoid self-Q-switching instabilities, we typically still have to use a folded cavity with small mode-sizes in both the gain and the SESAM (i.e., a z-shaped cavity as shown in Fig. 9.57a) to avoid self-Q-switching instability (9.180). For the initial demonstration in 2014 [14Kle2], the SESAM was based on an antiresonant design with a single InGaAs quantum well absorber, generating a modulation depth of 1.4%, a saturation fluence of 11.3 μJ/cm^2, and nonsaturable losses below 0.3%. The output coupler was 1.3% and the average output power 2.95 W. GTI cavity mirrors were used for negative GDD and the SESAM was used to start and stabilize soliton modelocking. In this case, we typically obtain close to transform-limited soliton-shaped output pulses with a pulse duration of 59 fs. With a slightly longer

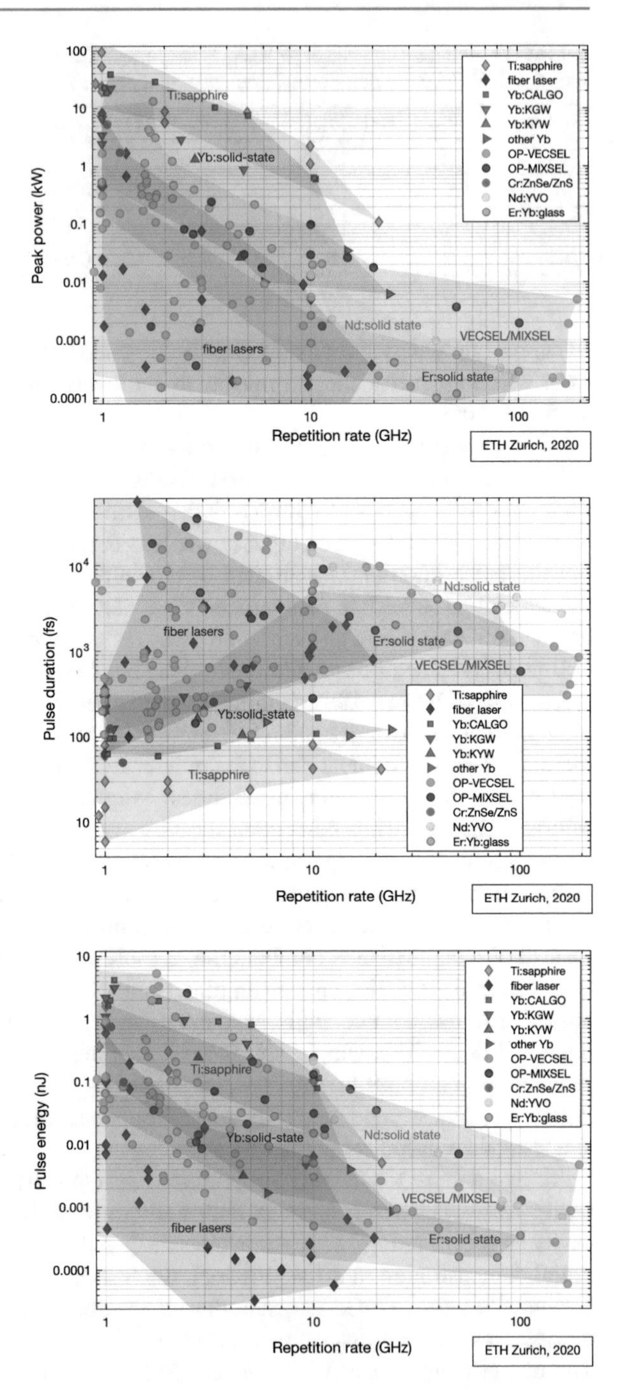

Fig. 9.56 Overview graphs as a function of gigahertz pulse repetition rates from diode-pumped femtosecond solid-state lasers with regard to peak power, pulse duration, and pulse energy, as of September 2020

Table 9.13 Diode-pumped femtosecond solid-sate lasers with gigahertz pulse repetition rates. ML modelocking technique. Center wavelength λ_0, pulse repetition rate f_{rep}, FWHM pulse duration τ_p, average output power P_{av}. The $+n_2$ in the CALGO result means that a CQN device has been used (Figs. 9.34 and 9.35) to stabilize against Q-switching instabilities (Sect. 9.7.6). Bold numbers are current milestone results with regard to pulse duration, average power, and pulse repetition rate

Gain	ML	λ_0 [μm]	τ_p [fs]	P_{av} [mW]	f_{rep} [GHz]	Reference
Cr:LiSAF	SESAM	0.857	146	3	1	[01Kem]
		0.865	55	110	1	[10Li]
Cr:YAG	KLM	1.54	115	150	2.64	[01Tom]
		1.52	68	138	2.33	[03Tom]
	SESAM	1.52	200	82	0.9, 1.8, 2.7	[97Col]
		1.52	75	280	1	[98Mel]
Yb:KYW	KLM	1.047	200	115	1	[09Was]
		1.046	146	14.6	4.6	[12End]
	SESAM	1.045	162	680	2.8	[10Yam]
Yb:KGW	SESAM	1.041	281	1100	1	[10Pek]
		1.042	290	2200	1	[11Pek]
		1.043	396	1900	4.8	[12Pek]
		1.042	278	770	1	[12Sch2]
		1.046	125	3430	1.1	[13Kle]
Yb:LuO	KLM	1.076	161	10	6	[13End]
Yb:YO	KLM	1.080	152	60	15	[15End]
		1.080	140	–	**23.8**	[19Kim]
Yb:CALGO	SESAM	1.06	**59**	**2950**	1.8	[14Kle2]
		1.06	**64**	**1700**	1.03	[14Kle3]
		1.054	96	4100	5.1	[15Kle]
	$+n_2 < 0$	1.0534	166	1200	10.6	[17May]
	$+n_2 < 0$	1.041	108	812	10.4	[20Kru]
	$+n_2 < 0$	1.042	**149**	**1180**	10.45	
		1.051	172	1440	10.52	
Cr:ZnSe/ZnS		2.4	125	800	1.06	[17Vas]
		2.4	50	120	1.22	[17Vas]
Er:Yb:glass		1.55	> 107	65	1	[16Sho]

cavity and a 2% output coupler, average power as high as 3.5 W was achieved with slightly longer pulse durations of 64 fs.

In the femtosecond domain with SESAM-assisted soliton modelocking, a higher pulse repetition rate of 5 GHz was initially only achieved with a special cavity design that acted like a ‘Q-switch limiter,’ as discussed in Sect. 9.7.5 and shown in Fig. 9.41. Only with additional large negative n_2 using cascaded quadratic nonlinearities (CQNs) (Sect. 9.5.3) has a straight linear two-mirror cavity for SESAM modelocking become possible with Yb:CALGO, with pulse repetition rates up to 10 GHz. The highest pulse repetition rate of 23.8 GHz was obtained with a KLM

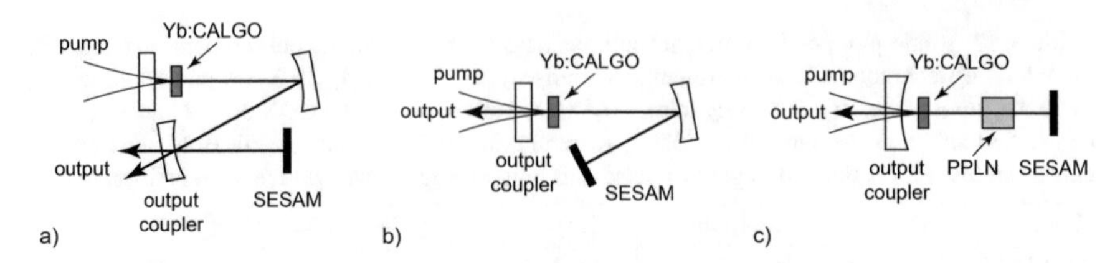

Fig. 9.57 Typical cavity designs for femtosecond SESAM-modelocked diode-pumped Yb-doped solid-state lasers with gigahertz pulse repetition rates: **a** z-shaped cavity for an ultrafast Yb:CALGO laser with 1.8 GHz [14Kle2] and 1 GHz [14Kle3], **b** 5-GHz three-mirror Yb:CALGO cavity with an all-optical Q-switch limiter as shown in Fig. 9.42 [15Kle], **c** 1-GHz two-mirror Yb:CALGO cavity using self-defocusing effect in an apodized CQN PPLN crystal (Fig. 9.34) providing large negative n_2 with low losses [17May,20Kru]

Yb:YO [19Kim], which is a rather impressive result given that the KLM couples the cavity design with the saturable absorber modulation.

A fully monolithic cavity design with record-low noise has also been demonstrated with a 1-GHz SESAM-modelocked Er:Yb:glass laser [16Sho]. In contrast to Fig. 9.53a for Nd:YVO_4 lasers, a folded three-mirror cavity was still required to suppress Q-switching instabilities. The monolithic cavity spacer layer was not the gain material because, in a three-level laser, significant re-absorption occurs in an unpumped section. An 8.5 cm long CaF_2 spacer layer had excellent transparency at the laser wavelength of 1.55 μm and zero GDD at 1545 nm. A GTI coating on the curved mirror on the CaF_2 spacer layer then only needed to compensate for the third order dispersion. Both a short Er:Yb:glass crystal and a SESAM were attached to the CaF_2 spacer layer. The SESAM had a modulation depth of about 0.5%. The modelocked output spectrum had a 24 nm bandwidth with a Fourier limit of about 107 fs. However, no autocorrelation measurement was provided in [16Sho].

Pulse Duration Measurements

10

10.1 Electronic Measurements of the Pulse Duration

10.1.1 Cables and Connectors

Typically 50 Ω BNC with RG-58 cables are used in a laboratory. These cables have significant signal attenuation for microwave signals above around 2 GHz. High-performance radio frequency (RF) and microwave cables and connectors up to about 60 GHz are commercially available. However, at these frequencies the loss is no longer negligible and one should always keep the cable length as short as possible. Table 10.1 gives a summary of microwave connectors: the most important are BNC-, SMA-, K-, and V-connectors.

Fiber-optic cables and fiber connectors (ST-, FC-, etc., connectors) are also often used for transmission of light, and offer the potential for much higher bandwidth than electrical cables, e.g., allowing for transmission of $\geq$100 fs pulses over short distances. Ultimately, dispersion and nonlinear effects start to limit unperturbed transmission (see Chaps. 3 and 4). Many standards exist for different fiber connectors, which can often be traced back to systems developed by large companies such as AT&T, NTT, etc., from the early days of fiber-optic development.

10.1.2 Fast Photodiode

A circuit corresponding to a photodiode is shown in Fig. 10.1. A photodiode is an ideal current source with an additional series resistor R_s and a capacitor C_p connected in parallel.

© The Author(s), under exclusive licence to Springer Nature Switzerland AG 2021

U. Keller, *Ultrafast Lasers*, Graduate Texts in Physics,
https://doi.org/10.1007/978-3-030-82532-4_10

Table 10.1 Electric high-frequency connectors

Connector type	Frequency region	Compatibility
Bayonet Neill-Concelman Connector (BNC)	DC–2 GHz	
Sub-Miniature C (SMC)	DC–7 GHz	
Amphenol Precision Connector (APC)	DC–18 GHz	
Type N (Neill) 50 Ω	DC–18 GHz	
Sub-Miniature A (SMA)	DC–24 GHz	3.5 mm, 2.92 mm, Wiltron K
3.5 mm	DC–34 GHz	SMA, 2.92 mm, Wiltron K
2.92 mm or Wiltron K	DC–40 GHz	SMA, 3.5 mm
2.4 mm	DC–50 GHz	1.85 mm, Wiltron V
1.85 mm or Wiltron V	DC–65 GHz	2.4 mm

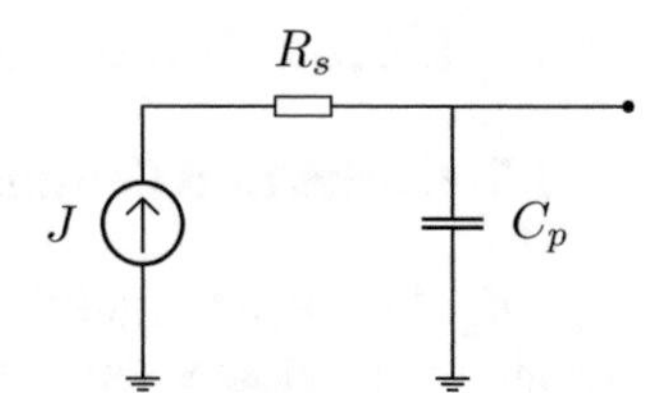

Fig. 10.1 Equivalent circuit of an ideal photodiode. Ideal current source with a series resistor R_s and a parallel capacitor C_p

The electric current J generated in a photodiode is given by the incident optical power P_{opt}:

$$J = \frac{q\eta P_{opt}}{h\nu} , \tag{10.1}$$

where q is the electron charge, η is the quantum efficiency of the photodetector, ν is the optical frequency, and $h\nu$ the photon energy of the incident light. Usually, the series resistor R_s is relatively small (around 1–2 Ω) and can be neglected in the ideal case.

The photodiode is connected to a measurement device, e.g., an oscilloscope, a sampling-scope, a high-frequency spectrum analyzer, to measure the generated voltage (Fig. 10.2). In RF and microwave technology, transmission lines with a characteristic impedance of 50 Ω and high-frequency devices with 50 Ω input impedance (Fig. 10.2) are standardly used. This also means that a photodiode circuit very often uses a 50 Ω input impedance (that is, an additional parallel resistor of 50 Ω in Fig. 10.1). With matched 50 Ω cable and input impedances, small to negligible signal reflections of the propagating signal are generated. In this case, the length of the cable between photodiode and measurement device does not play a significant role, except for loss and bandwidth limitation.

Kirchhoff's law at a nodal point requires that

$$\sum_i J_i = 0 . \tag{10.2}$$

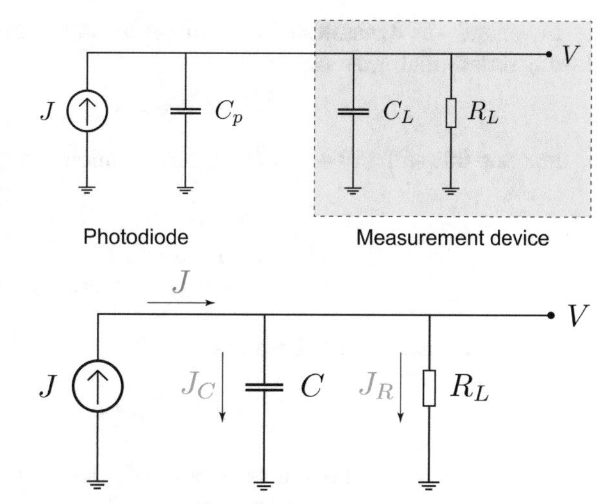

Fig. 10.2 Equivalent circuit of a photodiode connected to a microwave measurement device, i.e., $R_L = 50\ \Omega$. *Upper*: The measurement device has an input resistor R_L and an input capacity C_L, i.e., for the photodiode this is a load resistor and a load capacitor. *Lower*: Simplified equivalent circuit with $C = C_L + C_p$. Note that the photodiode circuit very often also uses a 50 Ω input impedance (not shown here) to minimize reflection

With this, it follows from Fig. 10.2 (lower) that

$$-J + C\frac{\mathrm{d}V}{\mathrm{d}t} + \frac{V}{R_L} = 0\,. \tag{10.3}$$

Here, we have used the following relation for a capacitor:

$$V = \frac{1}{C}Q\,,$$

where V is the voltage across the capacitor, C is the capacity, and Q is the charge on the capacitor. The electric current is given by

$$J = \frac{\mathrm{d}Q}{\mathrm{d}t}$$

and the current J_C through the capacitor is

$$J_C = C\frac{\mathrm{d}V}{\mathrm{d}t}\,.$$

The current J_R which passes through the resistor is given by Ohm's law $V = R_L J_R$, where R_L is the load resistor of the photodiode.

The differential equation (10.3) can then be transformed to

$$\frac{\mathrm{d}V}{\mathrm{d}t} + \frac{V}{R_L C} = \frac{J}{C}\,. \tag{10.4}$$

The differential equation (10.4) describes the dynamics of a first-order linear system.

In general, the dynamics of a first-order linear system are described by the following differential equation:

$$\tau \dot{y} + y = x \, . \tag{10.5}$$

Comparing this with (10.4) results in these relations:

$$\begin{aligned} y(t) &= V(t) \, , \\ x(t) &= R_{\mathrm{L}} J(t) \, , \\ \tau &= R_{\mathrm{L}} C = \text{const.} \end{aligned} \tag{10.6}$$

In the Fourier space, (10.5) becomes

$$\tau \mathrm{i}\omega \tilde{y} + \tilde{y} = \tilde{x} \, , \tag{10.7}$$

where $\tilde{y}$ and $\tilde{x}$ are the Fourier transforms of y and x, respectively. Referring back to the introduction to linear system theory in Sect. 2.3.2, the solution of (10.7) then yields the transfer function of the linear system:

$$\tilde{h}(\omega) = \frac{\tilde{y}}{\tilde{x}} \, . \tag{10.8}$$

A fast photodiode is characterized by its impulse response. If we assume an infinitely short optical pulse hitting a fast photodiode, we have

$$x(t) = \delta(t) \implies \tilde{x}(\omega) = 1 \, . \tag{10.9}$$

In this case the output signal will equal the transfer function in the spectral domain $\tilde{y}(\omega) = \tilde{h}(\omega)$. Equations (10.7) and (10.8) then simplify to a complex Lorentz function:

$$\tilde{h}(\omega) = \frac{1}{1 + \mathrm{i}\omega\tau} \, . \tag{10.10}$$

The spectral power density is given by

$$\tilde{H}(\omega) = \left| \tilde{h}(\omega) \right|^2 = \frac{1}{1 + (\omega\tau)^2} \, , \tag{10.11}$$

and the 3 dB bandwidth by

$$\boxed{\tilde{H}(\omega_{3\,\mathrm{dB}}) = \frac{1}{2} \max\left\{ \tilde{H}(\omega) \right\} = \frac{1}{2} \implies \omega_{3\,\mathrm{dB}} = \frac{1}{\tau}} \, . \tag{10.12}$$

The transfer function $\tilde{h}(\omega)$ (10.10) implies an impulse response $y(t) = h(t)$:

$$h(t) = \frac{1}{\tau} \mathrm{e}^{-t/\tau} \, , \tag{10.13}$$

where τ is the e^{-1} width for the impulse response (not the FWHM). The validity of (10.13) can be verified by applying a Fourier transform, which will give (10.10).

With these results for the photodiode, the impulse response $h(t)$ given by (10.6), (10.9), and (10.13) with $x(t) = R_{\mathrm{L}}\delta(t)$ is

$$h(t) = \frac{1}{C}\mathrm{e}^{-t/R_{\mathrm{L}}C}\,, \tag{10.14}$$

where the e^{-1} time constant τ and the 3 dB bandwidth of the spectral power density are given by

$$\boxed{\tau = R_{\mathrm{L}}C\,, \quad f_{3\,\mathrm{dB}} = \frac{\omega_{3\,\mathrm{dB}}}{2\pi} = \frac{1}{2\pi R_{\mathrm{L}}C}}\,. \tag{10.15}$$

For an arbitrary optical pulse $P_{\mathrm{opt}}(t)$, an instantaneous pulse of current $J(t)$ is generated as in (10.1), and this produces a response voltage $V(t)$ given by

$$V(t) = h(t) * J(t) = \int h\left(t'\right) J\left(t - t'\right) \mathrm{d}t' = \int \frac{1}{C}\mathrm{e}^{-t'/R_{\mathrm{L}}C} J\left(t - t'\right) \mathrm{d}t'\,. \tag{10.16}$$

Very often in RF and microwave technology, the time resolution in switching is considered. In this case, one does not examine the impulse response of a linear system (e.g., a photodiode), but the step response, which in a first-order linear system is given by

$$V(t) \propto 1 - \mathrm{e}^{-t/\tau}\,. \tag{10.17}$$

The rise time of (10.17) or the decay time of (10.14) is determined by τ. Most of the time, however, the rise time (or analogously, the decay time) is described by the time span Δt during which the signal increases from 10 to 90% of the maximum value. With $\tau = 1/2\pi f_{3\,\mathrm{dB}}$ from (10.15) and using (10.17), it follows that

$$\boxed{\Delta t\,[\mathrm{ps}] = \ln 9 \times \tau\,[\mathrm{ps}] = 2.2 \times \tau\,[\mathrm{ps}] = 2.2 \times \frac{1}{2\pi f_{3\,\mathrm{dB}}} \approx \frac{350\ \mathrm{GHz}}{f_{3\,\mathrm{dB}}[\mathrm{GHz}]}}\,. \tag{10.18}$$

For example, a 60 GHz photodetector has a step response (i.e., 10–90% rise time) of 5.8 ps, obtained from (10.18) with $f_{3\,\mathrm{dB}} = 60$ GHz.

Most photodiodes are limited by their $R_{\mathrm{L}}C$ time constant, where in a high-frequency measurement system the load resistor is $R_{\mathrm{L}} = 50\ \Omega$. Thus, it is possible to reduce the $R_{\mathrm{L}}C$ time constant by reducing the capacity C of the capacitor, where C is given approximately by a coplanar capacitor (see Fig. 10.3):

$$C = \frac{\varepsilon\varepsilon_0 A}{d}\,, \tag{10.19}$$

and $\varepsilon_0 = 8.854 \times 10^{-14}$ F/cm, ε is the dielectric constant, e.g., for GaAs, $\varepsilon = 13$, A is the area of the capacitor, and d is the distance between the conducting layers,

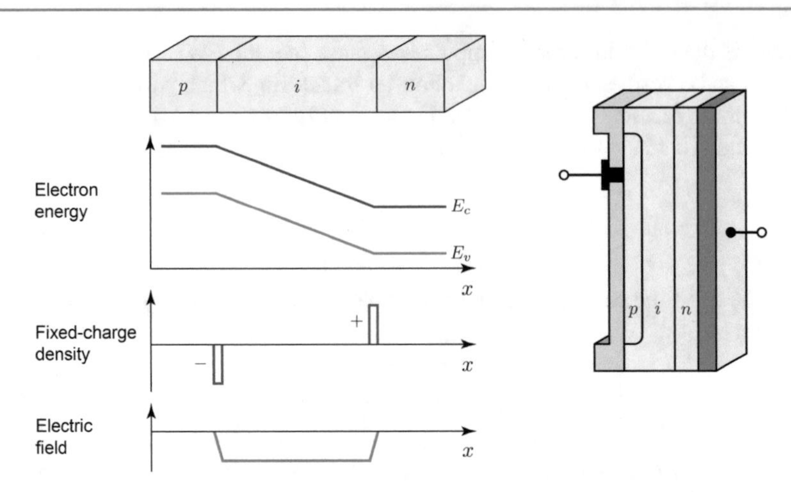

Fig. 10.3 Typical configuration of a semiconductor p-i-n photodiode

i.e., the p- and n-doped layers in Fig. 10.3. Thus, one can reduce the capacity with a smaller area and a thicker insulating layer. From a certain point onward, charge transport phenomena become important. The charge carriers are generated by the photons in the insulating layer and need a certain time, i.e., a transit time, to reach the conducting layers.

This transit time is limited by slow diffusion (in areas without electric fields) and drift (in areas with electric fields). Diffusion can be neglected as long as an electric field acts on the free charge carriers. This is the case if charge carriers are only generated within the capacitor area and the incident intensity is small enough to avoid saturation effects. In GaAs, the saturation velocity υ_s, i.e., the steady-state drift velocity, is $\upsilon_s \approx 10^7$ cm/s. A transit time τ_{trans} of 1 ps therefore needs a relatively thin insulating layer of $d = \upsilon_s \tau_{\mathrm{trans}} \approx 1$ μm.

A 60 GHz photodiode in a 50 Ω system then has a maximum allowable capacitance of $C \approx 50$ fF according to (10.15). To reach this small capacitance, a GaAs photodiode with $d = 1$ μm needs to have a circular area with a radius approximately 12 μm, from (10.19). Therefore, it is understandable that in most cases the optical signal has to be tightly focused onto the small area of the photodetector to reach the necessary time resolution.

10.1.3 Estimating the Time Resolution and Measurement Bandwidth

Although in most cases the impulse responses are not given by Gaussian functions, e.g. as for the photodiode in (10.14), the assumption of a Gaussian impulse response is usually a good approximation.

In the special case of Gaussian impulse responses of the form

$$h(t) = \exp\left(-t^2/\tau^2\right) , \tag{10.20}$$

one can show relatively easily that the total time resolution of several linear systems, such as photodiode + amplifier + oscilloscope, is

$$\boxed{\tau_{\text{tot}} = \sqrt{\tau_1^2 + \tau_2^2 + \tau_3^2 + \cdots}} \,. \tag{10.21}$$

This is proven as follows. Equation (10.21) follows directly from the fact that the Fourier transform of $h(t)$ in (10.20) is again a Gaussian function $\tilde{h}(\omega)$ and that the transfer functions can be multiplied in the Fourier space:

$$\begin{aligned}\tilde{h}_{\text{tot}}(\omega) &= \exp\left(-\frac{\omega^2\tau_1^2}{4}\right)\exp\left(-\frac{\omega^2\tau_2^2}{4}\right)\cdots \\ &= \exp\left[-\frac{\omega^2\left(\tau_1^2+\tau_2^2+\cdots\right)}{4}\right] = \exp\left(-\frac{\omega^2\tau_{\text{tot}}^2}{4}\right) .\end{aligned}$$

Assuming we have a Gaussian impulse response,

$$h(t) = \exp\left(-4\ln 2\frac{t^2}{\tau_{\text{FWHM}}^2}\right) . \tag{10.22}$$

The Fourier transform of a Gaussian function is then given by (see Appendix A):

$$\exp\left(-\Gamma t^2\right) \overset{\text{Fourier transform}}{\longrightarrow} \exp\left(-\frac{\omega^2}{4\Gamma}\right) . \tag{10.23}$$

Hence, the Fourier transform of (10.22) is given by

$$\tilde{h}(\omega) = \exp\left(-\frac{\omega^2\tau_{\text{FWHM}}^2}{16\ln 2}\right) . \tag{10.24}$$

The FWHM bandwidth of the spectral power density then follows directly from (10.24):

$$\tilde{H}(\omega) = \left|\tilde{h}(\omega)\right|^2 = \exp\left(-\frac{\omega^2\tau_{\text{FWHM}}^2}{8\ln 2}\right) \equiv \exp\left(-4\ln 2\frac{\omega^2}{\Delta\omega_{\text{FWHM}}^2}\right) . \tag{10.25}$$

The 3 dB bandwidth is then

$$\omega_{3\,\text{dB}} = \frac{\Delta\omega_{\text{FWHM}}}{2} , \tag{10.26}$$

or

$$f_{3\,\text{dB}} = \frac{\omega_{3\,\text{dB}}}{2\pi} = \frac{\Delta\omega_{\text{FWHM}}}{4\pi} . \tag{10.27}$$

Equation (10.25) then implies

$$\frac{\tau_{\mathrm{FWHM}}^2}{8\ln 2} = \frac{4\ln 2}{16\pi^2 f_{3\,\mathrm{dB}}^2}$$

and thus

$$\tau_{\mathrm{FWHM}} f_{3\,\mathrm{dB}} = \frac{\sqrt{2}\ln 2}{\pi} = 0.312\,, \tag{10.28}$$

or

$$\boxed{\tau_{\mathrm{FWHM}}\left[\mathrm{ps}\right] \approx \frac{312\ \mathrm{GHz}}{f_{3\,\mathrm{dB}}[\mathrm{GHz}]}}\,. \tag{10.29}$$

If we compare (10.29) with (10.18), it follows that for a given bandwidth $f_{3\,\mathrm{dB}}$ the FWHM pulse width τ_{FWHM} of a Gaussian impulse response is almost the same as the 10–90% rise time Δt of a step response of an ideal first-order linear system (e.g., a photodiode). To simplify matters for the estimation of a high-frequency measurement system, one often describes the photodiode by a Gaussian impulse response, where the FWHM Gaussian pulse width τ_{FWHM} is given by (10.29).

Examples
In the approximation of a Gaussian impulse response, a 1 GHz oscilloscope has an estimated time resolution of $\tau_1 = 312$ ps by (10.29). Use of an additional 500 MHz photodetector, i.e., $\tau_2 = 624$ ps by (10.29), results in the approximation of a Gaussian impulse response in a combined time resolution of

$$\sqrt{(312\ \mathrm{ps})^2 + (624\ \mathrm{ps})^2} \approx 700\ \mathrm{ps}\,,$$

according to (10.21).

Fast commercially available photodiodes can have a bandwidth of about 60 GHz and a minimal time resolution of around 5 ps by (10.29). However, this time resolution typically requires the use of a special high-speed sampling oscilloscope.

10.1.4 Sampling Oscilloscope

A sampling technique can be used to reach a higher time resolution electronically (see Fig. 10.4). This technique measures the time response of a repetitive signal with a picosecond time resolution. The sampling scope has two input signals: the signal to be measured $V(t)$ and the trigger signal. The trigger signal usually corresponds to a signal synchronized to $V(t)$ with the same period. Commercially available sampling heads have a measurement bandwidth of around 50 GHz, which results in a minimal time resolution of around 6 ps by (10.29). Combining a 50 GHz sampling unit with a 60 GHz photodiode, one can reach a time resolution of about 8 ps, according to (10.21) and (10.29). An example is shown in Fig. 10.5.

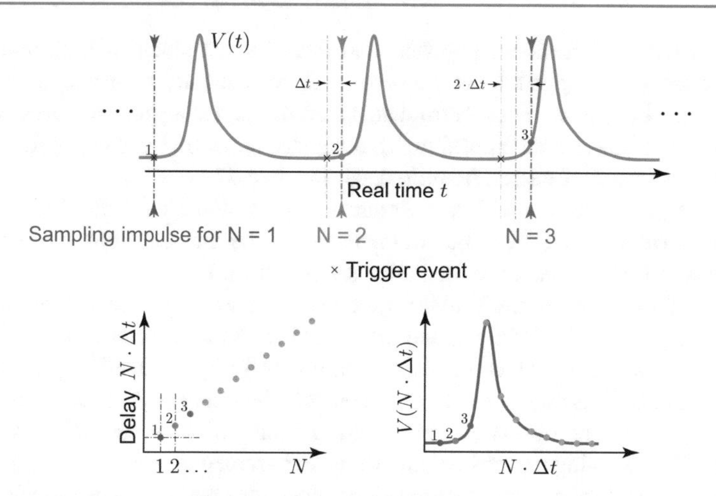

Fig. 10.4 Sampling technique. The measurement event is typically synchronized with a multiple of the pulse repetition rate, but with a small offset frequency added. This means that for every pulse the voltage is measured at a slightly different temporal position within the voltage pulse signal and therefore slowly maps the high-speed voltage signal at a rate given by the offset frequency

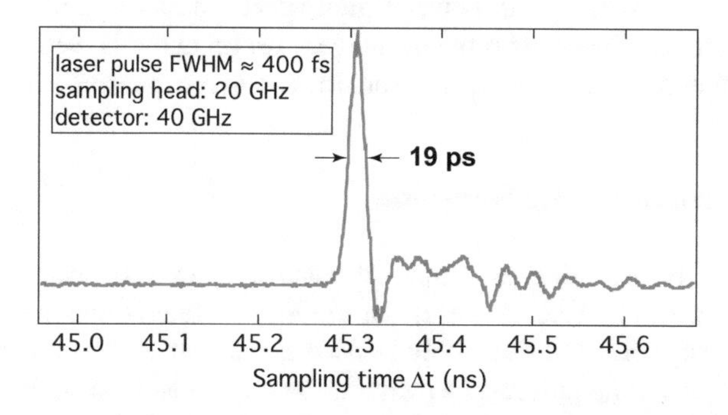

Fig. 10.5 Measured impulse response of a fast 40 GHz photodiode with a 20 GHz sampling oscilloscope. The optical pulse is 400 fs long, and has been generated with a modelocked Nd:glass laser. The measured pulse width is about 19 ps. Using the approximation of a Gaussian-shaped impulse response, as in (10.22) and (10.29), one obtains an estimated measurable pulsewidth of $\sqrt{(0.4\ \text{ps})^2 + (7.8\ \text{ps})^2 + (15.6\ \text{ps})^2} = 17.5$ ps, which approximately corresponds to the measurement value. The voltage oscillations after the pulse are due to higher order nonlinear effects from the electronic detection system

10.1.5 Microwave Spectrum and Signal Analyzers

Spectrum and signal analyzers measure the response in the frequency domain. A traditional spectrum analyzer measures the power spectrum of an input signal with a certain resolution bandwidth (RBW) over a certain frequency range (the frequency span) with an optimized sweep speed. The time for the sweep over the entire frequency span is the sweep time. The RBW defines the frequency resolution and the noise floor and is the main limitation of the sweep speed. Normally, the user specifies the RBW and the frequency span and then the instrument will automatically optimize the sweep speed. This, however, can be manually adjusted for different applications. In contrast to a spectrum analyzer, a signal analyzer measures both the magnitude and phase of an input signal. Such instruments are typically used for noise characterization, as discussed in more detail in Chap. 11.

Using a photodetector with a spectrum analyzer then means that the power of the photocurrent is measured and not the power of the incident light. (Note that the photocurrent is proportional to the incident light power.)

The signal-to-noise ratios in microwave spectrum analyzers are so high that they are also very useful in finding stable modelocking. As long as a laser is not really modelocked, we can enhance the mode beating noise by mechanically tapping on cavity elements. This noise can often be seen with the microwave spectrum analyzer. Once a signal can be observed, even if it is not stable at the beginning, it can be optimized by changing the various experimental parameters. For example, the mode beating noise will increase to a strong signal once the laser becomes modelocked.

The noise floor can be reduced with a small RBW, although this typically leads to long measurement times with low frequency sweep speed. Generally speaking, it is good to have a spectrum analyzer which has a low noise floor over a wide frequency range with a fast sweep speed. Novel digital spectrum and signal analyzers offer more options today. These instruments are very useful and it is well worth reading their application notes to benefit fully from these rather expensive instruments.

10.1.6 Equivalent-Time Sampling

In equivalent-time sampling, an ultrashort pulse samples a repetitive signal waveform, as shown in Fig. 10.6. A picture of the input signal is thereby accumulated over many wave cycles and it can only be used to measure signals that are periodic. Assuming the sampling pulse has a repetition rate of f_0 and the signal waveform a repetition rate of exactly Nf_0, i.e., an integer multiple N of the probe repetition rate, the probe pulse interacts with the signal waveform every N th period at a fixed point within its cycle. Over many repetitions, the pulses sample the signal waveform at the same time within the cycle, producing an equal sampling signal for each pulse. The resulting average intensity of the sampling signal is proportional to the signal under test using a measurement bandwidth that is much less than the pulse repetition rate.

To detect the entire signal waveform, the signal frequency is increased by a small amount Δf. The probe pulses are then slowly delayed with respect to the signal, sampling successively delayed points, so that the average sampling intensity changes in proportion to the signal waveform, but repeating at a rate Δf. A sampling detector averages the output intensity with a low-pass filter over a measurement time much longer than the sampling pulse period, eliminating the individual pulses and measuring only the change in intensity due to the signal under test. This is shown schematically in Fig. 10.6 for $N = 1$.

What is the difference between equivalent and real-time sampling? A real-time oscilloscope captures the entire waveform between two successive trigger events.

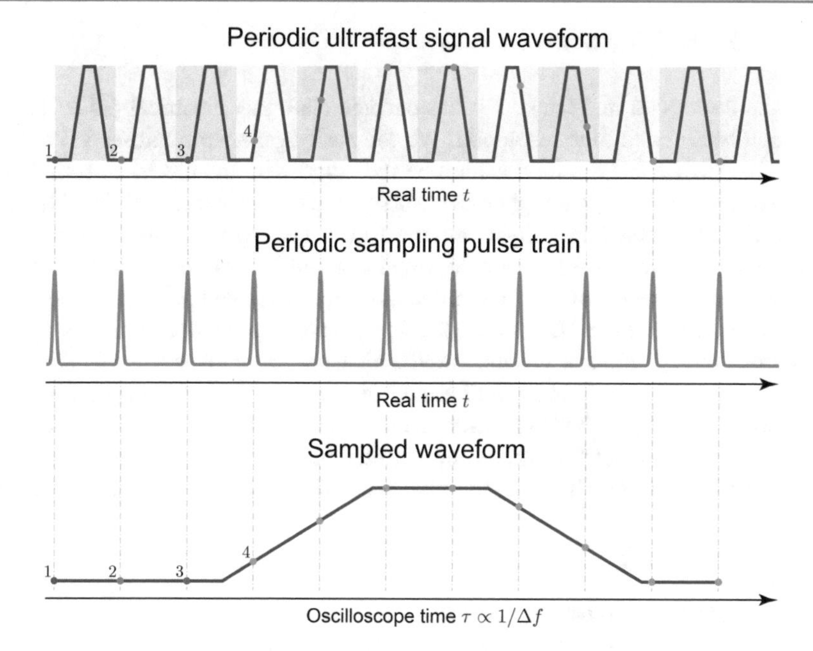

Fig. 10.6 Equivalent time sampling [88Wei]. Short sampling pulses are synchronized to a high-frequency generator. The principle of measurement is analogous to strobe light, where fast processes are slowed down so that they can be measured with slow electronics

For example, a digital storage oscilloscope is based on real-time sampling. Thus a repetitive signal waveform is not required in real-time sampling.

As another example, equivalent-time sampling was used for electro-optic sampling in the 1980s [88Wei], for which a high-frequency voltage signal can be measured with the electro-optic effect using picosecond modelocked pulses in the near-infrared as the sampling pulses. This was also the first time that a modelocked pulse train was stabilized to an external microwave signal [86Rod]. The residual noise in the stabilized pulse repetition rate results in some fluctuations in the arrival time of the pulses, which is referred to as the timing jitter.

The time resolution is given by the sampling pulse and timing jitter. Electronic sampling pulses generated with a 'nonlinear transmission line' [91Rod] can reach pulse lengths in the region of 1 ps. This would correspond to a measurement bandwidth of about 312 GHz, by (10.29), if we neglect synchronization issues. With synchronized laser pulses, the pulse lengths can be in the region of 10 fs, which would correspond to a measurement bandwidth of around 31 THz by (10.29) (also neglecting synchronization issues). However, the timing jitter can only be neglected if the electronic signal to be measured has been generated by the laser pulse. Otherwise, one has to synchronize the repetition rate of the laser pulse with the signal to be measured, and this has to be done very carefully [86Rod,89Rod]. In most cases the time resolution is limited by the synchronization of signal and sampling pulse.

10.2 Optical Autocorrelation

From the discussion in Sect. 10.1 it has become clear that electrical pulse characterization has limited time resolution. We can easily generate femtosecond optical pulses, but their pulse duration cannot be measured with the electrical techniques discussed before. In addition, photodetectors measure the intensity of the pulse and without interferometric techniques we have no access to the phase of the pulse. Because in principle if we measure the amplitude and phase of an optical spectrum, a simple Fourier transform will determine the pulse shape and pulse duration.

In this section we first focus on optical pulse characterization based on optical autocorrelation techniques. The measured optical spectrum gives some feedback on the assumptions for the deconvolution and the final pulse duration. We therefore normally always expect both a measurement of the optical spectrum and the optical autocorrelation. An interferometric autocorrelation gives some additional information with regard to the chirp of the pulse and will be discussed in this section as well.

10.2.1 Pulse Spectrum

The optical spectrum is typically measured by using angular dispersion resulting from the interaction of light with a diffraction grating or a prism. Such measurement instruments are called optical spectrum analyzers (OSA). These measurement instruments differ in their intensity measurement. If only one detector is used and the direction of the spectrally resolved parts is varied by rotating the diffraction grating or the prism, the detector always measures only one specific spectral component at a certain time. Such measurement devices are also called monochromators. On the other hand, if a detector array or a CCD camera is used, the spectrally dispersed light can be measured simultaneously without having to move certain elements mechanically. Such a measurement instrument is called an optical multichannel analyzer (OMA). An OMA usually has a lower wavelength resolution than a monochromator.

10.2.2 Intensity Autocorrelation

Experimental Setup

As we have seen in the previous section, the electronic time resolution of commercial systems is currently limited to between 1 and 10 ps. If one wants to measure optical pulses with a shorter durations, one has to use another method. Optical intensity autocorrelation using second harmonic generation (SHG) [67Web, 67Arm] is well suited for this task (Fig. 10.7). The incident beam with an optical pulse train is split into two beams with equal intensity using a beam splitter. One of these beams is delayed against the other by changing the optical path length in one arm. These two beams are then focused into a nonlinear crystal and superimposed in space and time. The nonlinear interaction in the crystal generates a frequency-doubled signal that

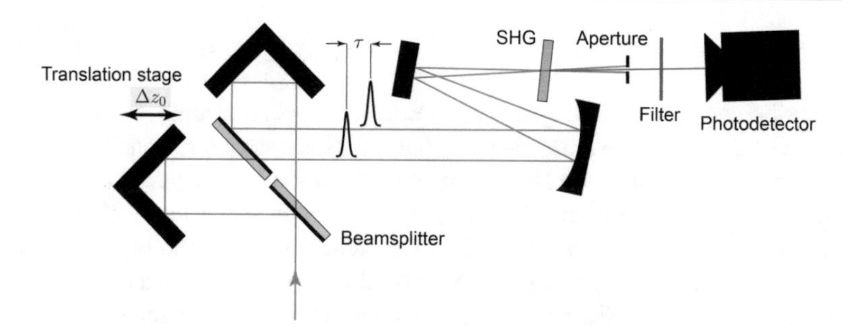

Fig. 10.7 Optical intensity autocorrelation using second harmonic generation (SHG) in a non-collinear configuration. This is a background-free intensity autocorrelation with no signal when the two pulses do not overlap. The aperture can filter out the spatially separated non-collinear beams. The filter transmits the frequency-doubled signal $I_{2\omega}(\tau)$, where τ is the delay between the two pulses and given by $2\Delta z_0/c$, with c the speed of light in vacuum (i.e., a good approximation for the group velocity in air)

depends on the delay between the two pulses, reaching a minimum at no temporal overlap and a maximum for perfect temporal overlap at a zero delay between the pulses. Therefore, by slowly varying the path length difference, one can measure the intensity of the frequency-doubled light $I_{2\omega}(\tau)$, where τ is the delay between the two pulses, using a photomultiplier and a normal low-speed oscilloscope. This information can then be used to determine the temporal intensity profile of the input pulse.

The measured SHG intensity $I_{2\omega}(\tau)$ corresponds mathematically to an autocorrelation function. As we will discuss below, we lose any pulse phase information with this measurement. For an error free detection, the SHG efficiency is required to be constant over the full pulse spectrum, and this can become problematic in the few-cycle regime. This is below 10 fs for a center wavelength of about 800 nm.

Second Harmonic Generation (SHG) in a Nonlinear Crystal

When an electromagnetic field is applied to bound valence electrons of a medium, these electrons are forced into an oscillation. At lower electric field strengths, the binding potential of these electrons can be approximated by a harmonic potential which results in a polarization P of the material, proportional to the applied field E:

$$P = \varepsilon_0 \chi E \, , \tag{10.30}$$

where ε_0 is the vacuum permittivity and χ is the electric susceptibility. Thus, for this linear interaction, the induced polarization oscillations emits electromagnetic radiation at the same frequency as the incident electromagnetic wave. The superposition of these two waves in the forward direction then results in a modified phase velocity, as described by the linear refractive index.

For higher intensities, as typically obtained with pulsed laser radiation, the harmonic binding potential is not a good approximation and higher order terms have to be taken into account. This results in a nonlinear polarization for the incident field E which can be described by a Taylor series:

$$P_i = P_i^L + P_i^{NL} = \varepsilon_0 \chi_{ij}^{(1)} E_j + \varepsilon_0 \chi_{ijk}^2 E_i E_k + \cdots . \quad (10.31)$$

The coefficient $\chi_{ijk}^{(2)}$ of the quadratic term is a 3-dimensional tensor. Solving Maxwell's equations with the ansatz of (10.31) for the polarization one finds that the new wave equation now also allows for second harmonic generation (SHG) of an incident wave or the addition/subtraction of the frequencies of two incident waves. For efficient SHG we also need to take into account phase matching such that all frequency doubled waves generated at each oscillating electron will add up constructively which means that the phase velocity at the fundamental frequency and at the second harmonic frequency is the same. This can be obtained in birefringent crystals with, for example, $n_o(\omega_0) = n_e(2\omega_0)$, where o stands for the ordinary and e for the extraordinary beam and polarization. There are many excellent textbooks that give an introduction to nonlinear optics and birefringent crystals.

For now, we limit the discussion to the second-order nonlinearity and the quadratic part in the Taylor series (10.31). We do this because for the optical autocorrelation measurement we typically only use second harmonic generation (SHG). In this case, we can write the nonlinear polarization in the form

$$P_{NL} = 2dE^2 , \quad (10.32)$$

where a specific crystal symmetry in a non-centrosymmetric crystal is used to simplify (10.32) with a material parameter d as the relevant second-order nonlinear coefficient. In nonlinear optics, we have to make sure we use the real part of the electric field, because for two complex numbers z_1 and z_2, we generally have

$$\mathrm{Re}(z_1 z_2) \neq \mathrm{Re} z_1 \mathrm{Re} z_2 . \quad (10.33)$$

This was already discussed in Chap. 1 for the derivation of the electromagnetic energy density. The nonlinear polarization then has two contributions: one for optical rectification and the other for second harmonic generation (SHG), given by $P_{SHG} = dE^2(t)$. The superposition of the two time-delayed pulses then gives an autocorrelator amplitude signal

$$E_{2\omega}(t,\tau) \propto \left[E(t) + E(t-\tau)\right]^2 . \quad (10.34)$$

Calculation of the Background-Free Optical Intensity Autocorrelation

Non-collinear autocorrelation has the advantage that the frequency-doubled signal is spatially separated from the fundamental signal and that therefore the fundamental signals can be easily removed from the nonlinear signal. Therefore, one can neglect frequency-doubling of the individual beams $E_1(t)$ and $E_2(t-\tau)$ and obtain a background free autocorrelation. For the second harmonic signal, according to (10.34), we then obtain

$$E_{2\omega}(t,\tau) \propto E_1^2(t) + E_2^2(t-\tau) + 2E_1(t)E_2(t-\tau) , \quad (10.35)$$

and we only need to take into account the last term because of the non-collinear SHG in Fig.10.7. This is referred to as a background-free intensity autocorrelation.

Therefore, according to (10.35) and Fig. 10.7 for a background-free intensity autocorrelation the measured intensity of the frequency-doubled signal $I_{2\omega}(\tau)$ depends on the temporal overlap of the two pulses:

$$I_{2\omega}(\tau) \propto \int_{-\infty}^{+\infty} |E_1(t)E_2(t-\tau)|^2\,\mathrm{d}t \overset{E_1=E_2=E}{\propto} \int_{-\infty}^{+\infty} |E(t)E(t-\tau)|^2\,\mathrm{d}t\,. \tag{10.36}$$

Here we assumed that we measure $I_{2\omega}(\tau)$ with a slow detector and a slow oscilloscope, which means that we have to integrate the signal intensity $|E_{2\omega}(t,\tau)|^2$ over time. When the two pulses do not overlap, we have no signal.

The FWHM pulse length of the autocorrelation is given by τ_A, and under the assumption of a constant phase and a certain pulse shape it is possible to determine the FWHM pulse length of the original optical pulse τ_p. This is summarized in Table 10.2 for different pulse shapes. By fitting the measured autocorrelation $I_{2\omega}(\tau)$ and the measured spectrum to a known function, the pulse shape can be estimated. In addition, the theories for pulse generation will predict certain pulse shapes. However, this estimation is not unambiguous (more on that in Sect. 10.2.3 and when we discuss the FROG and SPIDER techniques), because the autocorrelation (10.36) loses all information about the phase and the chirp of the pulse. This can be seen from (10.36) where the phase factors drop out. Thus (10.36) is equivalent to

$$\boxed{I_{2\omega}(\tau) \propto \int_{-\infty}^{+\infty} I(t)I(t-\tau)\mathrm{d}t}\,. \tag{10.37}$$

It can be shown that, with an autocorrelation, we always obtain a symmetric function $I_{2\omega}(\tau) = I_{2\omega}(-\tau)$. If the measurement shows a non-symmetric autocorrelation, then this is due to a bad alignment of the autocorrelator. Furthermore, for an accurately measured autocorrelation, we have to take the following points into consideration:

1. For a fixed time delay τ, the detector averages over several laser pulses. Therefore, to avoid ambiguities, the laser should have no large fluctuations in pulse shape and phase.
2. The e.m. field used in the equation is ideally the one obtained directly from the ultrafast laser system. However, dispersion and nonlinearities in the beam guiding optics will change this. Especially for short pulses, this can lead to severe pulse deformation. To minimize unwanted dispersion effects, it is recommended to use reflective optics and thin beam splitters. To minimize nonlinearities, it is better to use free space propagation.
3. Dispersion in the nonlinear crystal for SHG is another serious issue. To minimize group delay dispersion and therefore temporal pulse walk-off between the fundamental and second harmonic signal, the nonlinear crystal thickness has to be reasonably small.

Table 10.2 FWHM pulse length of the original optical pulse τ_p as a function of the FWHM pulse length of the autocorrelation τ_A for different pulse shapes

$I(t), x \equiv t/T$	τ_p/τ_A	τ_p/T
Rectangle $I(t) = \begin{cases} 1, & \lvert t\rvert \le T/2, \\ 0, & \lvert t\rvert > T/2, \end{cases}$	1	1
Parabola $I(t) = \begin{cases} 1 - x^2, & \lvert t\rvert \le T/2, \\ 0, & \lvert t\rvert > T/2, \end{cases}$	0.8716	$\sqrt{2}$
Diffraction function $I(t) = \dfrac{\sin^2 x}{x^2}$	0.7511	2.7811
Gaussian $I(t) = \mathrm{e}^{-x^2}$	0.7071	$2\sqrt{\ln 2}$
Triangle $I(t) = \begin{cases} 1 - \lvert x\rvert, & \lvert t\rvert \le T, \\ 0, & \lvert t\rvert > T, \end{cases}$	0.7071	$2\sqrt{\ln 2}$
Hyperbolic secant $I(t) = \operatorname{sech}^2(x)$	0.6482	1.7627
Lorentzian $I(t) = \dfrac{1}{1+x^2}$	0.5000	2
One-sided exponential $I(t) = \begin{cases} \mathrm{e}^{-x}, & t \ge 0, \\ 0, & t < 0, \end{cases}$	0.5000	ln 2
Symmetric two-sided exponential $I(t) = \mathrm{e}^{-2\lvert x\rvert}$	0.4130	ln 2

4. Bandwidth limitation in the nonlinear conversion efficiency has to be reduced. The conversion efficiency in the nonlinear crystal should ideally be constant over the full spectral range of the pulse. To check this, one can for example compare the spectral intensity of the second harmonic $I_{2\omega_0}(\nu)$ with the incident fundamental wave $I_{\omega_0}(\nu)$. When the spectral bandwidth limitations are negligible, we need to have

$$I_{2\omega_0}(\omega) = \left|F\left\{E_{2\omega_0}(t)\right\}\right|^2 \propto \left|F\left\{E_{\omega_0}^2(t)\right\}\right|^2 = \left|\tilde{E}_{\omega_0}(\omega) * \tilde{E}_{\omega_0}(\omega)\right|^2 , \quad (10.38)$$

where $F\{E\}$ is the Fourier transform of E.

Time Calibration of the Optical Autocorrelation

As shown in Figs. 10.7 and 10.8, we can control the temporal overlap of the two pulses by a spatial translation Δz_0. For a fast refreshing time of the autocorrelation measurement, we typically use a periodic oscillation in this delay such that the full intensity autocorrelation is measured several times per second. This can be obtained, for example, with a loudspeaker and a retro-reflector. We simply have to make sure that the autocorrelation pulse duration is measured well within the linear delay response of this loud speaker motion (which is sinusoidal).

We need a time calibration for the measured $I_{\omega_0}(\nu)$ on the oscilloscope. This can be achieved as follows. If we change the offset of the periodic delay by an additional mechanical delay Δz_0, we can observe on the oscilloscope how the pulse is shifted in time by $\Delta\tau$. As long as this delay is caused by free space propagation and we can

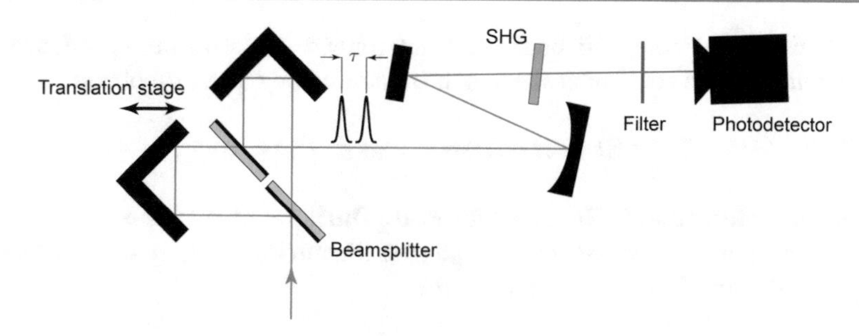

Fig. 10.8 Setup of an interferometric autocorrelator (IAC). The filter transmits the frequency-doubled signal

neglect any dispersion, we can determine the 'true' time delay Δt_{true} caused by the additional spatial translation Δz_0

$$\Delta t_{\text{true}} = \frac{2\Delta z_0}{c}, \tag{10.39}$$

where c is the speed of light. Then the time calibration of the optical autocorrelation is given by

$$\Delta t_{\text{true}} \Longleftrightarrow \Delta\tau . \tag{10.40}$$

10.2.3 Interferometric Autocorrelation (IAC)

Experimental Setup

Figure 10.8 shows the experimental setup for the measurement of the interferometric autocorrelation (IAC). In contrast to the background-free intensity autocorrelation, the two delayed pulses propagate collinearly and interfere in the nonlinear crystal. The setup can be explained as follows.

The laser pulse enters a Michelson interferometer, where it is split into two beams. These beams can be delayed against each other by changing the path length in one arm, before they are spatially recombined again. The two collinear and spatially overlapping beams are focused into a nonlinear crystal (suitable materials for the near-infrared regime are, e.g., KDP or BBO). Through the nonlinear interaction between the two pulses, we can measure a second harmonic signal that depends on the time delay between the two pulses. For example, when the fundamental pulse is in the red spectral domain, one will now also find a time dependent blue signal. A photodetector with a high-frequency-pass filter, which absorbs the fundamental spectrum with the center frequency ω_0, then measures the intensity of the second harmonic signal. This measured intensity as a function of the delay time between the fundamental collinear pulses is called the interferometric autocorrelation, because the two collinear pulses have the same polarization and will therefore interfere with each other. If we consider only the fundamental interference signal assuming a Michelson interferometer with equal intensities in the interferometer arms, we obtain

$$I_{\text{int}} = 2I + 2I \cos \Delta\phi \ , \quad \Delta\phi = k\Delta z = \omega_0 \tau \ .$$

Nonlinear Interaction Is Required for Pulse Duration Measurement
Without the nonlinear crystal and the high-pass filter in Fig. 10.8, we would measure an interferogram of the fundamental pulses:

$$I(\tau) = \frac{\int |E(t) + E(t-\tau)|^2 \, \mathrm{d}t}{2\int |E(t)|^2 \, \mathrm{d}t} = 1 + \frac{\int \left[E^*(t)E(t-\tau) + E(t)E^*(t-\tau)\right] \mathrm{d}t}{2\int |E(t)|^2 \, \mathrm{d}t} \ . \tag{10.41}$$

According to the convolution theorem for the Fourier transform, we then obtain

$$F\{I(\tau)\} = F\{1\} + \frac{\tilde{E}^*(-\omega)\tilde{E}(-\omega) + \tilde{E}(\omega)\tilde{E}^*(\omega)}{2\int |E(t)|^2 \, \mathrm{d}t} = F\{1\} + \frac{I(-\omega) + I(\omega)}{2\int |E(t)|^2 \, \mathrm{d}t} \ . \tag{10.42}$$

We have no dependence on the delay between the two arms and the interferogram (10.42) does not contain more information than the intensity spectrum. A measurement of the pulse length is not possible. This method is used, for example, in Fourier spectrography to measure the spectral power density with very high sensitivity.

Calculation of the Interferometric Autocorrelation (IAC)
The intensity of the second harmonic signal is determined from (10.34)

$$I_{2\omega}(\tau) = \int \left| \left[E(t) + E(t-\tau)\right]^2 \right|^2 \mathrm{d}t \ . \tag{10.43}$$

The fundamental E field of the pulse can be described by $E(t) = A(t)\mathrm{e}^{\mathrm{i}\omega_0 t}$, where $A(t)$ is the complex pulse envelope. The pulse envelope is real if the pulse is transform limited. We then obtain

$$\boxed{I_{2\omega}(\tau) = 2I_0 + 4I_1(\tau) + 4\mathrm{Re}\left\{\left[I_2(\tau) + I_2^*(-\tau)\right]\mathrm{e}^{\mathrm{i}\omega_0\tau}\right\} + 2\mathrm{Re}\left(I_3\mathrm{e}^{\mathrm{i}2\omega_0\tau}\right)} \ , \tag{10.44}$$

where

$$\boxed{\begin{aligned} I_0 &\equiv \textstyle\int |A(t)|^4 \, \mathrm{d}t \\ I_1(\tau) &\equiv \textstyle\int |A(t)A(t-\tau)|^2 \, \mathrm{d}t \\ I_2(\tau) &\equiv \textstyle\int |A(t)|^2 \, A(t)A^*(t-\tau)\mathrm{d}t \\ I_3(\tau) &\equiv \textstyle\int A^2(t)\left[A^2(t-\tau)\right]^* \mathrm{d}t \end{aligned}} \tag{10.45}$$

In the interferometric autocorrelation, we can distinguish between four different terms which also have a real physical meaning: I_0 contains the background part of the signal, $I_1(\tau)$ corresponds to the common intensity autocorrelation [see (10.36)], and finally the two remaining integrals describe the interference terms. From a mathematical point of view these different terms are all integrals of the basic form

$$K(\tau) = \int g(t)h(t-\tau)\mathrm{d}t \,, \tag{10.46}$$

with complex functions $g(t)$ and $h(t)$. $K(\tau)$ is the cross-correlation of the functions $g(t)$ and $h(t)$, or when $g(t) = h(t)$, $K(\tau)$ is the autocorrelation.

Because $A(t)$ is the pulse envelope we have a special condition:

$$A(t) \overset{t\to\pm\infty}{\longrightarrow} 0 \,. \tag{10.47}$$

Therefore, the integral terms I_1, I_2, and I_3 will vanish for $\tau \to \infty$ in the cases considered here. Furthermore, we notice that in the case of equal path length (i.e., $\tau = 0$) I_0 is equal to $I_1(0)$. We then define the normalized interferometric autocorrelation function as follows:

$$\boxed{I_{\mathrm{IAC}}(\tau) \equiv \frac{I_{2\omega}(\tau)}{I_{2\omega}(\infty)} = \frac{\int \left| [E(t)+E(t-\tau)]^2 \right|^2 \mathrm{d}t}{2\int |E(t)|^4\,\mathrm{d}t}} \,, \tag{10.48}$$

or

$$\boxed{I_{\mathrm{IAC}}(\tau) = 1 + \frac{1}{I_1(0)}\left\{2I_1(\tau) + 2\mathrm{Re}\left\{\left[I_2(\tau) + I_2^*(-\tau)\right]\mathrm{e}^{\mathrm{i}\omega_0\tau}\right\} + \mathrm{Re}\left(I_3\mathrm{e}^{\mathrm{i}2\omega_0\tau}\right)\right\}} \,. \tag{10.49}$$

The upper and lower envelopes of the IAC (see Fig. 10.9) are obtained by replacing $\omega_0\tau$ by 2π or π, respectively, in (10.49). However, we want to point out that, especially in the case of chirped pulses, this is often only a good approximation for sufficiently small τ.

Without explicitly calculating the IAC, one can already draw the following conclusions:

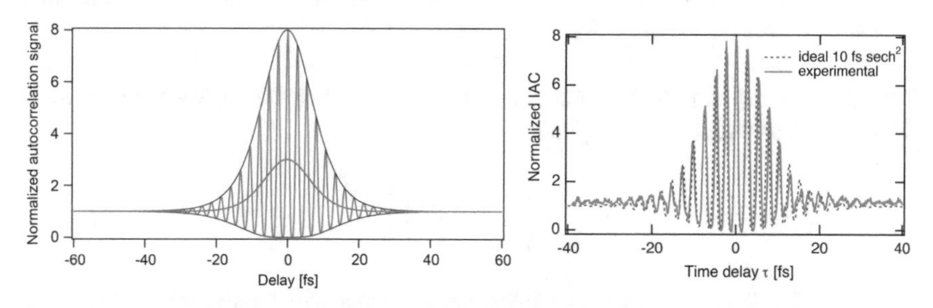

Fig. 10.9 *Left*: Calculated IAC for an unchirped 10 fs sech^2 pulse at a center wavelength of 800 nm. In addition to the IAC, we also show the intensity autocorrelation (*green*). *Right*: Measured IAC from a passively modelocked Ti:sapphire laser [96Flu1]

1. $I_{\mathrm{IAC}}(\tau)$ is a symmetric function, i.e.,

$$I_{\mathrm{IAC}}(\tau) = I_{\mathrm{IAC}}(-\tau) . \tag{10.50}$$

Thus, a possible asymmetry of the pulse will not show itself in an asymmetry of the IAC.
2. Because the integral terms I_1, I_2, and I_3 will vanish for $\tau \to \infty$, $I_{\mathrm{IAC}}(\tau)$ will tend to 1 for large values of $|\tau|$. For $\tau = 0$, all the integrals I_k in (10.45) are equal:

$$\begin{aligned} I_0 &\equiv \int |A(t)|^4 \,\mathrm{d}t , \\ I_1(\tau = 0) &\equiv \int \left|A^2(t)\right|^2 \mathrm{d}t = \int |A(t)|^4 \,\mathrm{d}t = I_0 , \\ I_2(\tau = 0) &\equiv \int |A(t)|^2 A(t)A^*(t)\mathrm{d}t = \int |A(t)|^4 \,\mathrm{d}t = I_0 , \\ I_3(\tau = 0) &\equiv \int A^2(t) \left[A^2(t)\right]^* \mathrm{d}t = \int |A(t)|^4 \,\mathrm{d}t = I_0 . \end{aligned} \tag{10.51}$$

Therefore, the right-hand side of (10.49) takes on the maximum value 8. Thus, one obtains a ratio of maximum to constant background of 8:1. This proportion is clearly shown in Fig. 10.9, and summarized as follows:

$$\boxed{I_{\mathrm{IAC}}\,(\tau \to \pm\infty) = 1 , \quad I_{\mathrm{IAC}}(\tau)\big|_{\max} = I_{\mathrm{IAC}}(0) = 8 , \quad I_{\mathrm{IAC}}(\tau)\big|_{\min} = 0} . \tag{10.52}$$

3. In the slowly varying envelope approximation, the field strength will only change slightly in the time duration of an oscillation period $2\pi/\omega_0$, both pulses will interfere destructively at the time π/ω_0, and the integral will take on the minimal value 0. After about the same time, a maximum will follow again due to constructive interference.
4. For a transform-limited pulse with a well defined spectrum given by the Fourier transform of a definite pulse shape with zero phase, we have a coherence time given by the pulse duration. We will therefore observe the interference over the full pulse duration, as shown in Fig. 10.9.
5. For a chirped pulse, the pulse envelope is no longer a real function. The spectral bandwidth becomes larger than for a transform-limited pulse with the same pulse duration. Therefore, the coherence time over which interferences are fully resolved becomes shorter than the pulse duration.

The case of chirped pulses is shown in more detail in Fig. 10.10. We take a look at the example of a chirped sech pulse:

$$A(t) = \left[\operatorname{sech}\left(\frac{t}{\tau_\mathrm{p}}\right)\right]^{1+\mathrm{i}\beta} , \tag{10.53}$$

where β describes the chirp. The interference terms will reduce within the pulse duration τ_p, because the coherence time is given by

$$\tau_{\mathrm{coh}} = \frac{1}{\Delta\nu} . \tag{10.54}$$

Figure 10.10 then shows differently chirped sech^2 pulses. All of these pulses have the same power spectrum and their intensity profile is described by (10.53). However, because β varies between these pulses, so does the spectral phase in the frequency domain and the variation of the instantaneous frequency in the time domain. The spectral domain representation of the different sech^2 pulses is slightly asymmetric. This is because the optical power spectrum and the spectral phases are plotted as a function of the wavelength. This is how the spectra would be measured with a typical optical spectrometer in the laboratory. On a frequency scale, these graphs would be symmetric.

In the time domain graphs, i.e., the middle column in Fig. 10.10, it can be seen how the pulses broaden with increasing GDD for a fixed power spectrum. The temporal phase shown there is the phase of the envelope, i.e., the phase without the fast varying $\omega_0 t$ term. Its derivative thus describes how much the instantaneous frequency deviates from the carrier frequency.

The curvature of the temporal phase decreases for increasing GDD. This can be understood from the fact that the more the pulse is broadened in time, the longer the time interval over which its bandwidth spreads. The change in the instantaneous frequency per unit time therefore has to decrease for larger amounts of GDD.

The interferometric autocorrelations (IAC) are shown in the right-hand column of Fig. 10.10. For reference, the interferometric autocorrelations are shown together with their upper and lower envelope functions (blue). The green curves show the interferometric autocorrelations when the fast-oscillating parts (with frequencies ω_0 and $2\omega_0$) are neglected. In this case the autocorrelations reduce to the intensity autocorrelations, but with an added background (i.e., the non-background-free intensity autocorrelation) because the fundamental beams and their SHG arrive collinearly. The intensity autocorrelation contribution spreads out with increasing GDD, reflecting the broadening of the pulses in the time domain. The visibility of the interference is only given within the coherence time of the pulses. Once the pulses in the time domain become longer with the chirp but without changing the spectral width $\Delta\nu$, they are also no longer transform-limited. Therefore the interferometric autocorrelation is reduced to the intensity autocorrelation in its wings at larger GDD values, because the interference modulations in the autocorrelation signal are constrained along the delay axis by the coherence length of the corresponding spectrum. These examples demonstrate that an IAC provides more information about the pulse than a simple intensity autocorrelation.

According to (10.37), the non-collinear intensity autocorrelation $I_{2\omega}(\tau)$ also occurs as a term in the IAC as given by (10.49) with $I_1(\tau)$. This opens up the possibility of obtaining the interference-free intensity autocorrelation with the measurement setup of the IAC. For example, if one simply operates the periodic delay in the Michelson interferometer sufficiently fast, the photodetector can no longer resolve the faster oscillating part with $\omega_0\tau$ and $2\omega_0\tau$, and therefore averages out the minima and maxima of the interference of the IAC. Using (10.49), we then obtain the non-background-free intensity autocorrelation $I_{2\omega}^{\text{collinear}}(\tau)$:

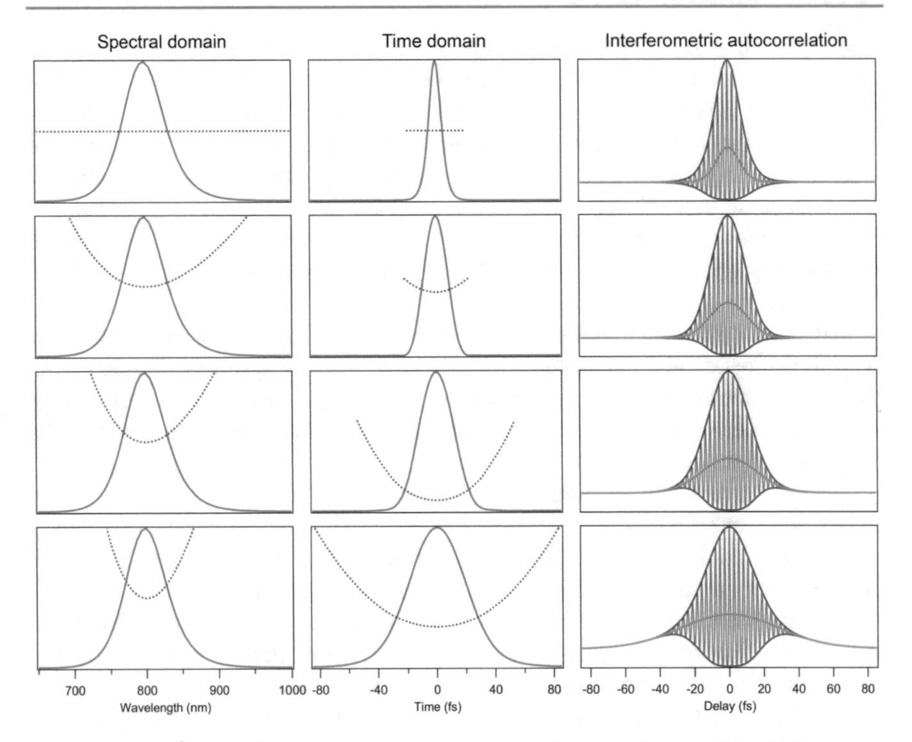

Fig. 10.10 Sech^2 pulses in the spectral domain and the time domain and their interferometric autocorrelation for different amounts of GDD. *Top row*: Transform-limited 10 fs sech^2 pulses. In the *rows below*, 50, 100, and 200 fs^2 of pure GDD have been added to this pulse, respectively. The *blue dotted lines* represent the phase in the respective domains. Scaling of the axes is maintained for all figures of the same type. Pulse intensity profiles and interferometric autocorrelations are normalized to their peak value

$$I_{2\omega}^{\text{collinear}}(\tau) = 1 + \frac{2I_1(\tau)}{I_1(0)} = 1 + \frac{2\int I(t)I(t-\tau)\mathrm{d}t}{\int I^2(t)\mathrm{d}t}, \tag{10.55}$$

which is shown in Fig. 10.10. In this case, the ratio of maximum to constant background is 3:1, which follows directly from (10.55):

$$\boxed{\frac{I_{2\omega}^{\text{collinear}}(\tau = 0)}{I_{2\omega}^{\text{collinear}}(\tau \to \infty)} = \frac{3}{1}}. \tag{10.56}$$

The non-background-free intensity autocorrelation only obtains a contrast ratio of 3:1. In contrast, one of the big advantages of the background-free autocorrelation, i.e., the non-collinear autocorrelation, is the high dynamic range over which the pulse duration can be measured (see Sect. 10.2.4). The disadvantages of the intensity autocorrelation are that the information about the phase is lost and that there is a relative insensitivity against the pulse shape, in contrast to the IAC, where the fields enter with the fourth power.

Note that, in principle, the fast oscillations in the IAC measurement can be removed after the measurement with some simple mathematical filter: from the Fourier transform of the IAC, one eliminates the oscillation parts at ω_0 and $2\omega_0$. After the inverse Fourier transform, one then obtains the intensity autocorrelation.

Analytical Examples: Gaussian Pulse and Soliton Pulse

In the case of a chirped Gaussian pulse, we can calculate the IAC function analytically. We assume a pulse envelope given by

$$A(t) = \exp\left[-\frac{1}{2}\left(\frac{t}{\tau_p}\right)^{2(1+i\beta)}\right] , \tag{10.57}$$

and then determine the IAC:

$$\begin{aligned} I_{\text{IAC}}(\tau) = 1 + &\left\{2 + \exp\left[-\frac{\beta^2}{2}\left(\frac{\tau}{\tau_p}\right)^2\right]\cos(2\omega_0\tau)\right\}\exp\left[-\frac{1}{2}\left(\frac{\tau}{\tau_p}\right)^2\right] \\ &+4\exp\left[-\frac{3+\beta^2}{8}\left(\frac{\tau}{\tau_p}\right)^2\right]\cos\left[\frac{\beta}{4}\left(\frac{\tau}{\tau_p}\right)^2\right]\cos(\omega_0\tau) . \end{aligned} \tag{10.58}$$

If the terms oscillating with ω_0 and $2\omega_0$ are omitted, the intensity autocorrelation is obtained. Note that this means that all terms that include the chirp parameter β will drop out. As can be seen here, or also directly from (10.43), for the above definition of the chirp, the intensity autocorrelation of the frequency-modulated pulse does not change. Thus, without the IAC, one would have no possibility of determining the chirp. While the FWHM of the IAC becomes smaller for such a chirped pulse, the pulse width can still be determined directly from the intensity autocorrelation, for the above reason.

With a little more effort (contour integration), the IAC of an unchirped sech pulse can also be calculated analytically. We assume a soliton pulse envelope

$$A(t) = \operatorname{sech}\left(t/\tau_p\right) ,$$

and then calculate the IAC:

$$\begin{aligned} I_{\text{IAC}}(\tau) = 1 + [2+\cos(2\omega_0\tau)]\,&\frac{3\left[\left(\tau/\tau_p\right)\cosh\left(\tau/\tau_p\right) - \sinh\left(\tau/\tau_p\right)\right]}{\sinh^3\left(\tau/\tau_p\right)} \\ +&\frac{3\left[\sinh\left(2\tau/\tau_p\right) - \left(2\tau/\tau_p\right)\right]}{\sinh^3\left(\tau/\tau_p\right)}\cos(\omega_0\tau) . \end{aligned}$$

We thus obtain the FWHM pulse duration by evaluating the IAC signal $I_{\text{IAC}}(\tau) = 4$ assuming the normalized maximum is at 8:

$$\tau_{\text{IAC}} = 1.897\tau_{p,\text{sech}} \implies \frac{\tau_{p,\text{sech}}}{\tau_{\text{IAC}}} = 0.527 . \tag{10.59}$$

10.2.4 High-Dynamic-Range Autocorrelation

Noncollinear autocorrelation measurements (Fig. 10.7) result in a background-free autocorrelation signal. This is in contrast to the interferometric autocorrelation (Sect. 10.2.3), for which the autocorrelation signal has a constant signal even when the two pulses do not overlap. Therefore, the noncollinear autocorrelation can be used to characterize the pulse shape in the wings more carefully.

For a high-dynamic-range autocorrelation measurement, we use a slow delay scan Δz_0 in Fig. 10.7 with a mechanical chopper wheel in one of the fundamental beams before the SHG crystal. This allows for lock-in detection to measure the much weaker SHG signal in the pulse wings. This results in a high dynamic range which can easily go beyond a signal-to-noise ratio of 10^6 as shown in Fig. 10.11. These measurements were made with a mechanical chopper wheel in one of the fundamental beam paths. The lock-in detector uses a homodyne detection scheme with a narrow filter around the chopping frequency which reduces the noise in the measured signal.

An even better signal-to-noise can be achieved at a higher megahertz chopping frequency because the noise of the modelocked laser is lower at that frequency offset (see Chap. 11 for more details). These higher modulation frequencies can be achieved with an acousto-optic modulator, for example. In addition, we can reject any scattered

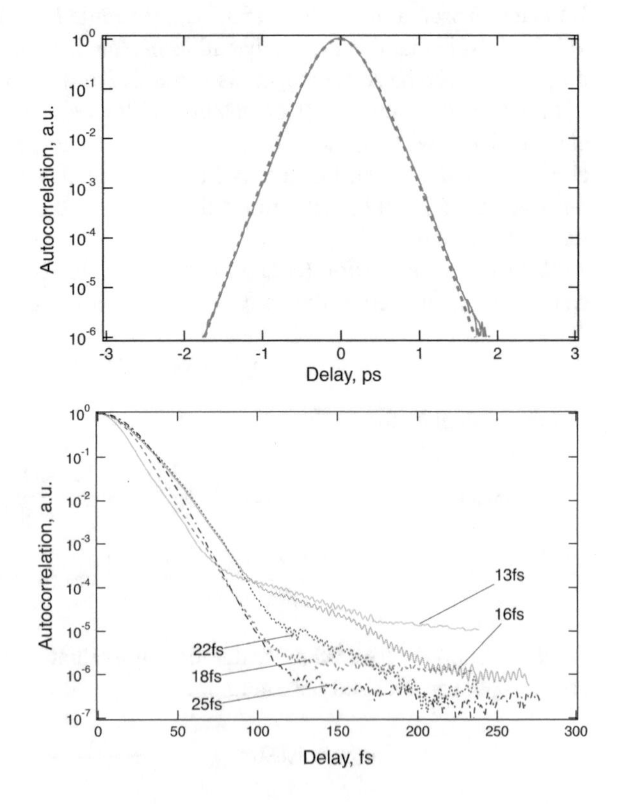

Fig. 10.11 Measured high-dynamic-range autocorrelation for soliton modelocked pulses with different pulse durations. *Upper*: High quality transform-limited pulse [95Jun2]. *Lower*: Degradation of pulse quality as pulses become shorter, as shown here for FWHM pulse durations ranging between 25 and 13 fs. This was observed for both Kerr lens and soliton modelocking, as discussed in more detail in [97Jun2]

light in the lock-in detection when the two fundamental beams are modulated at two different frequencies and the SHG signal is detected at the difference frequency.

Numerical simulations for passive modelocking have shown that higher-order dispersion, too much self-phase modulation, and bandwidth limitations tend to degrade the quality of the pulses and generate pulse pedestals, i.e., the short transform-limited pulse sits on a longer and much weaker pulse which is referred to as a pulse pedestal (see, e.g., [97Jun2]). Very often, only high-dynamic-range autocorrelation measurements can reveal these pulse pedestals, as shown in Fig. 10.11 (lower). For example, the measured high-dynamic-range autocorrelation shows no significant deviations from an ideal soliton pulse shape over 6 orders of magnitude even though the net gain time window in soliton modelocking is open for longer than 35 times the pulse duration of 300 fs (Fig. 10.11 upper). The additional pulse pedestals for even shorter pulses, as shown in Fig. 10.11 (lower), have been observed for both a fast and a slow saturable absorber. For a slow saturable absorber, soliton formation was the dominant pulse shaping mechanism and the slow saturable absorber only started and stabilized the soliton modelocking [95Jun2,97Jun2] (see Sect. 9.5).

10.2.5 Temporal Smearing in Noncollinear Autocorrelation

For noncollinear autocorrelation, we have a temporal smearing because of the nonzero crossing angle θ_0 between the fundamental beams in the nonlinear crystal. When the two fundamental pulses fully overlap in time in the middle of the beam, we can see from Fig. 10.12 that one of the pulses precedes the other on the right, and the other precedes the one on the left [96Taf]. This results in a range of delays between the two pulses across the beam waist.

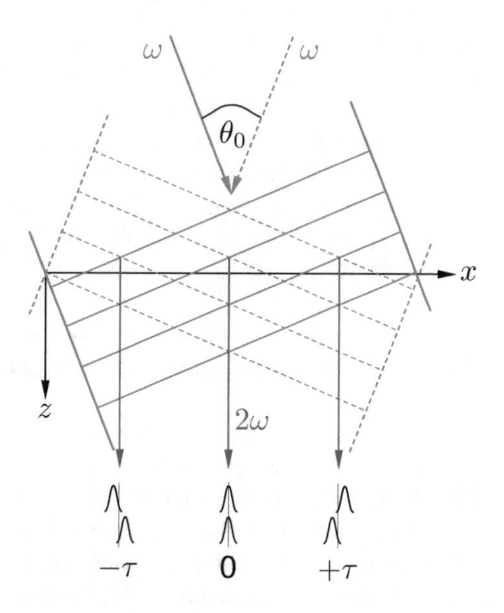

Fig. 10.12 Geometrical smearing of the delay between the two fundamental pulses due to the crossing angle θ_0 within the SHG crystal

We can calculate this geometrical smearing of the time delay between the two fundamental pulses using Fig. 10.12. The propagation direction of the SHG signal is along z. The two fundamental beams are focused within the nonlinear crystal with a beam waist w_0. For pulses around 10 fs at about 800 nm, the nonlinear crystal has been chosen very short to limit GDD effects. Therefore, we can use the plane wave approximation close to the beam waist for Gaussian beams. From Fig. 10.12, there must then be a temporal smearing δt given by

$$\delta t = \frac{2}{\upsilon_\mathrm{p}} w_0 \tan\frac{\theta_0}{2} \approx \theta_0 \frac{w_0}{\upsilon_\mathrm{p}} , \qquad (10.60)$$

where υ_p is the phase velocity inside the nonlinear crystal for the fundamental beam. To keep the time smearing small, we need to keep the crossing angle θ_0 small, and therefore the approximation for small θ_0 in (10.60) is justified. Assuming Gaussian intensity pulses, the product of two Gaussian pulses results again in an SHG Gaussian pulse, and the time smearing of (10.60) results in a longer measured pulse duration τ_m given by

$$\tau_\mathrm{m}^2 = \tau_\mathrm{p}^2 + \delta t^2 . \qquad (10.61)$$

This geometrical time smearing only becomes important for short pulse durations, as can be seen from a few numerical examples. For a crossing angle $\theta_0 = 2°$, a beam waist $w_0 = 30\ \mu$m, and a refractive index of the nonlinear crystal of 1.5, we obtain a time smearing of 5.2 fs, according to (10.60). For a pulse duration of 10 fs, this results in a measured pulse duration of 11.3 fs, according to (10.61). However, this becomes even more severe for few-cycle pulse measurements with pulse durations well below 10 fs in the near infrared. For example, at a center wavelength of 800 nm, the optical cycle is 2.7 fs. Thus for a 5 fs pulse with the same measurement parameters discussed above, we obtain a measured pulse duration of 7.2 fs, which is obviously significantly longer than 5 fs. Reducing the beam waist to 10 μm, we then obtain 5.3 fs, which is much closer to the reality.

Note that tight focus and small crossing angle are conflicting requirements. The smaller the focus (the tighter the focusing), the larger the crossing angle will be if the two incoming beams need to be spatially separated on the focusing optics. Therefore, non-collinear geometries (with spatially integrating detectors) are not suitable for measuring few-fs pulses.

10.3 Frequency-Resolved Optical Gating (FROG)

A fundamental problem with autocorrelation is that, from the measured autocorrelation trace, one cannot unambiguously extract the underlying pulse shape. There are trials to iteratively reconstruct the pulse shape by using additional information such as a simultaneously recorded spectrum of the pulse. This eliminates the arbitrariness of the choice of a certain pulse shape for the determination of the pulse duration. However, the basic difficulty is that information about any temporal asymmetry of the pulse is then lost, and is contained neither in the always symmetric autocorrelation trace, nor in the spectrum. This can lead to considerable uncertainties in the estimation of important pulse parameters, such as the pulse duration. There is great interest in finding ways to avoid such uncertainties during the measurement.

Techniques that allow for a complete reconstruction of the pulse shape have to be able to determine the phase and the amplitude of the electric field of the pulse simultaneously. In principle, this can be achieved in both the temporal or spectral domain. These techniques are called complete characterization techniques. Examples of such techniques are frequency-resolved optical gating (FROG) [93Kan, 97Tre] and spectral phase interferometry for direct electric-field reconstruction (SPIDER) [98Iac, 99Iac]. In the following, these two methods will be introduced in more detail. Based on our experience, SPIDER is easier and more reliable for very short pulses in the sub-10 fs region because no corrections are required for a frequency-dependent nonlinear conversion efficiency [99Gal, 00Gal2].

10.3.1 Basic Principle

The basic concept behind the FROG technique is to measure the spectrally resolved autocorrelation for each time delay between the two pulses. This means that the autocorrelation signal is not directly measured with a simple intensity detector, but is previously decomposed into its Fourier components in a spectrometer. Then, a complete optical spectrum of the nonlinear signal is recorded for every delay. There exist many different versions of FROG. All have in common that they measure a two-dimensional data field with autocorrelation delay and frequency as the parameters. What can be mathematically proven is that the 2D phase retrieval problem is solvable: if one has a 2D intensity map ('modulus squared') with finite support, it is possible to reconstruct the underlying complex data set. However, the step from this 2D complex data set to the (1D) complex electric field of the pulse has not been proven to yield an unambiguous result. For certain FROG techniques, an ambiguity exists regarding the time direction, but this can be eliminated relatively easy by measuring the pulses with and without an additional known dispersive element in the beam. For all practical purposes, this uncertainty in the determination of the pulse shape therefore only plays a secondary role. In principle, there are other ambiguities, such as the phase between the main pulse and a disconnected satellite pulse. However, in practice this is usually not so relevant.

In the following, we will describe second harmonic generation FROG (SHG-FROG) in detail, while the other versions which rely for example on the Kerr effect or nonlinear polarization rotation, will be discussed very briefly. In principle, all approaches are very similar but have different benefits and trade-offs. We will follow very closely the discussion in the paper by Kane [99Kan].

The pulse entering the SHG-FROG setup shown in Fig. 10.13 can be written as

$$E(t) = \mathrm{Re}\left\{\sqrt{I(t)}\exp\left[\mathrm{i}\omega_0 t - \mathrm{i}\varphi(t)\right]\right\} ,$$

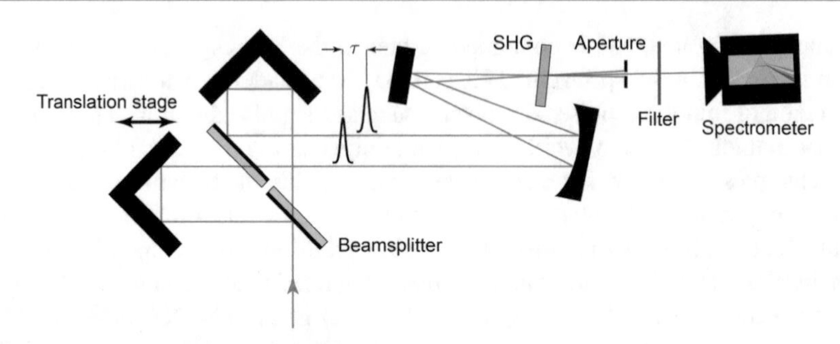

Fig. 10.13 The second harmonic generation FROG (SHG-FROG) setup. The experimental apparatus used in SHG-FROG is nearly identical to a background-free intensity autocorrelation. The only difference is that the spectrally integrating detector has been replaced by a spectrometer

where $I(t)$ and $\varphi(t)$ are the time dependent intensity and phase, and ω_0 is the center frequency of the optical spectrum, typically referred to as the carrier frequency. This pulse is then split into two identical pulses using a beam splitter. These identical pulses are focused into a nonlinear material and overlapped in space and time. The SHG of each individual beam is blocked with the aperture, and a high-frequency-pass filter removes any fundamental signal. We then measure a nonlinear signal

$$E_{\mathrm{sig}}(t,\tau) = E(t)\Gamma\left[E(t-\tau)\right] \ ,$$

where $E(t)$ is referred to as the probe and Γ is the gate function regulating the passage of the probe signal through the setup. This regulation of the probe signal depends on the nonlinear interaction and therefore converts the pulse into the gate. For SHG-FROG (see Fig. 10.13), $\Gamma\left[E(t-\tau)\right] = E(t-\tau)$, because an SHG crystal is used and the second harmonic is collected. Thus the nonlinear signal corresponds to the crossterm of (10.34).

In principle, any kind of nonlinearity can be used for FROG traces. Figure 10.14 summarizes several experimental configurations that have been used with FROG. For all these various configurations, the generated field is given by [00Tre]

$$E_{\mathrm{sig}}(t,\tau) \propto \begin{cases} E(t)E(t-\tau) & \text{for second harmonic generation (SHG)}\,, \\ E(t)\left|E(t-\tau)\right|^2 & \text{for polarization gating (PG)}\,, \\ E(t)^2E^*(t-\tau) & \text{for self-diffraction (SD)}\,, \\ E(t)^2E(t-\tau) & \text{for third harmonic generation (THG)}\,. \end{cases} \tag{10.62}$$

Transient grating (TG) FROG is mathematically equivalent to SD FROG. Different nonlinear interactions are used for the different FROG configurations (Fig. 10.14), but they all result in a signal that is spectrally resolved.

For PG FROG, the two polarizers in Fig. 10.14 are oriented at 0 and 90°, respectively, such that an intense 45°-polarized gate beam induces birefringence in the nonlinear crystal and hence a polarization rotation of the 0°-polarized beam, which can then leak through the second 90° polarizer. The output pulse that is measured by the spectrometer is representative of the input pulse, only with orthogonal polarization.

For SD FROG, the two fundamental beams interfere and yield a sinusoidal intensity pattern which induces a refractive index grating inside the medium. Then each beam diffracts off the grating. The pulses at the output indicate the signal pulses, here in the $2\mathbf{k}_1 - \mathbf{k}_2$ and $2\mathbf{k}_2 - \mathbf{k}_1$ directions.

For TG FROG, the geometry is more complex because the fundamental beam needs to be split into three beams, and these must fully overlap in space and time to generate a transient grating in a third order nonlinearity material which is probed by the third beam. The diffracted four-wave-mixing output signal must then be spectrally resolved as a function of the time delay. TG FROG is mathematically equivalent to SD FROG.

For SHG FROG, we use the same configuration as in noncollinear autocorrelation. The only difference is that we spectrally resolve the SHG signal as a function of the time delay between the two fundamental pulses. Note that in the setup of Fig. 10.14

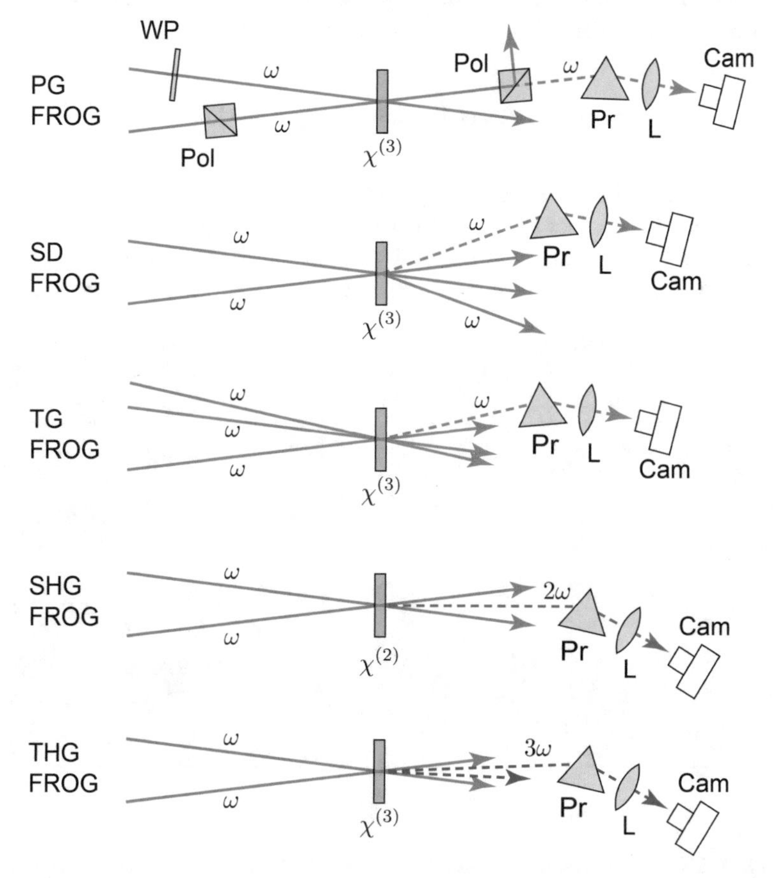

Fig. 10.14 Summary of different FROG configurations. Polarization gating (PG), self-diffraction (SD), transient grating (TG), second harmonic generation (SHG), and third harmonic generation (THG). Waveplate WP, polarizer Pol, prism Pr, lens L, Camera Cam. The Pr-L-Cam configuration gives a spectrally resolved measurement

a prism is used to select the crossterm of the SHG emission, rather than an aperture in combination with a high-frequency pass filter. For THG FROG, we use the third harmonic generation instead of the second. Assuming two different incoming fundamental frequencies ω_1 and ω_2, we can then obtain the third harmonic with either $2\omega_1 + \omega_2$ or $2\omega_2 + \omega_1$, which results in two different beam directions, as shown in Fig. 10.14.

The signal is spectrally resolved in a spectrometer. Thus, we have to perform a Fourier transform into the frequency domain, and obtain

$$\tilde{E}(\omega, \tau) = F\left\{E_{\text{sig}}(t, \tau)\right\} = \int E_{\text{sig}}(t, \tau) \exp(-\mathrm{i}\omega t)\mathrm{d}t \ .$$

A detector, such as a CCD array, produces the FROG trace by recording the spectral intensity of the signal at each time delay. Thus, these data can be represented as the magnitude squared of the Fourier transform of $E_{\text{sig}}(t, \tau)$:

$$I_{\text{FROG}}(\omega, \tau) = \left| \int E_{\text{sig}}(t, \tau) \exp(-\mathrm{i}\omega t)\mathrm{d}t \right|^2 \ . \qquad (10.63)$$

Examples of SHG FROG traces $I_{\text{SHG FROG}}(\omega, \tau)$ according to (10.62) and (10.63) are shown in Fig. 10.15 for a Gaussian and a sech^2 pulse shape. In addition, we show how the FROG traces change with an additional 100 fs^2 of GDD, which significantly broadens the pulse in the time domain, but does not affect the spectral width.

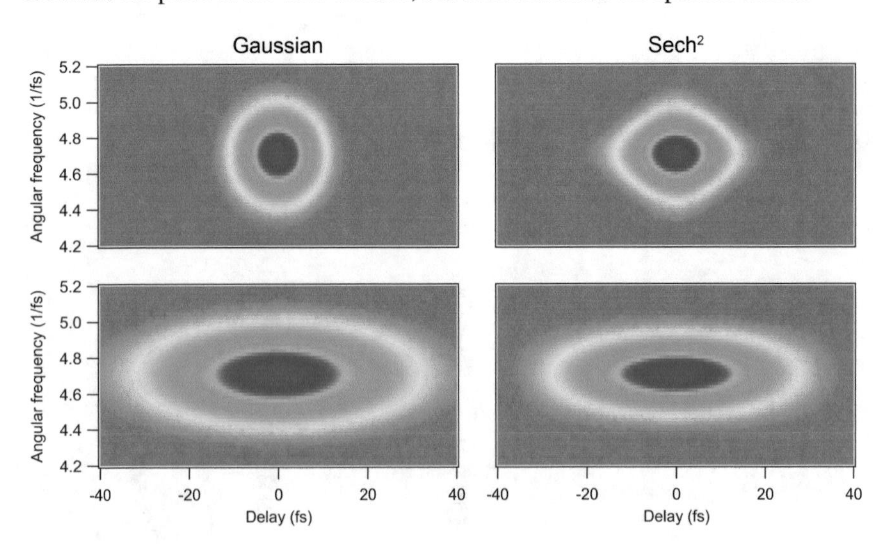

Fig. 10.15 SHG FROG traces calculated for Gaussian and sech^2 pulses. The *first row* shows transform-limited 10 fs pulses with the corresponding shapes. In the *second row*, 100 fs^2 of GDD has been applied to these pulses

Since $I_{\mathrm{FROG}}(\omega, \tau)$ is real, it has no direct phase information. The goal of the FROG inversion algorithm is therefore to determine the phase by solving the equation

$$\sqrt{I_{\mathrm{FROG}}(\omega, \tau)}\phi(\omega, \tau) = \int E_{\mathrm{sig}}(t, \tau)\exp(-\mathrm{i}\omega t)\mathrm{d}t \,, \qquad (10.64)$$

for $\phi(\omega, \tau)$ which is a complex function of unit magnitude.

Thus, for all FROG algorithms, (10.62) and (10.64) define the two constraints that must be satisfied, shown schematically in Fig. 10.16 [97Tre] . Equation (10.62) is called the 'physical constraint' and is used in the FROG algorithms both to obtain the next guess for $E(t)$ and to construct the new signal field. It is applied in the time domain. Equation (10.64) is the 'intensity constraint', which is applied in the frequency domain. The goal of the algorithm is to minimize the difference between the measured FROG trace and the FROG trace calculated from the current pulse $E(t)$ [see (10.65)].

Figure 10.17 shows the general form of FROG trace inversion algorithms. An initial guess is provided for $E(t)$ to get the algorithm started. $E_{\mathrm{calc}}(t, \tau)$, the new $E_{\mathrm{sig}}(t, \tau)$, is calculated and Fourier transformed into $\sqrt{I_{\mathrm{calc}}(\omega, \tau)}\phi_{\mathrm{calc}}(\omega, \tau)$. $\sqrt{I_{\mathrm{calc}}(\omega, \tau)}$ is replaced by the square root $\sqrt{I_{\mathrm{FROG}}(\omega, \tau)}$ of the measured FROG trace. The next guess for $E(t)$ is then calculated from $\sqrt{I_{\mathrm{FROG}}(\omega, \tau)}\phi_{\mathrm{calc}}(\omega, \tau)$ after an inverse Fourier transform back to the time domain. One reconstructs the electric field of the pulse, which is N amplitudes and N phases = $2N$ variables. Note that reconstructing the complex 2D signal field is a problem with N^2 variables, but they are not truly independent.

Once the new estimate for $E(t)$ is obtained, a new spectrogram is constructed. The process is repeated (see Fig. 10.17) until the FROG trace error reaches an acceptable minimum

$$\varepsilon_{\mathrm{FROG}} \equiv \left\{ \frac{1}{N^2} \sum_{i=1}^{N} \sum_{j=1}^{N} \left[I_{\mathrm{calc}}\left(\omega_i, \tau_j\right) - I_{\mathrm{FROG}}\left(\omega_i, \tau_j\right) \right]^2 \right\}^{1/2} , \qquad (10.65)$$

where $\varepsilon_{\mathrm{FROG}}$ represents the rms error per element of the spectrogram, $I_{\mathrm{calc}}\left(\omega_i, \tau_j\right)$ is the current iteration of the spectrogram, $I_{\mathrm{FROG}}\left(\omega_i, \tau_j\right)$ is the measured spectrogram, and ω_i and τ_j are the i th frequency and j th delay in the frequency and delay vectors, respectively.

The various FROG algorithms differ in the way that $E(t)$ is calculated from $\sqrt{I_{\mathrm{FROG}}(\omega, \tau)}\phi_{\mathrm{calc}}(\omega, \tau)$ [00Tre]. The most widely used FROG inversion algorithm is the generalized projection algorithm depicted in Fig. 10.16. There are several reasons for this. First, it virtually guarantees that the error always decreases for each iteration. Second, the generalized projection algorithm is very robust [97Tre]. Third, it is reasonably fast, being much faster than brute force minimization. Finally, it converges well even in the presence of noise.

Like previous algorithms, generalized projection works by alternating between two sets, S_1 and S_2 (see Fig. 10.16) [97Tre]. For FROG trace inversion, S_1 is the

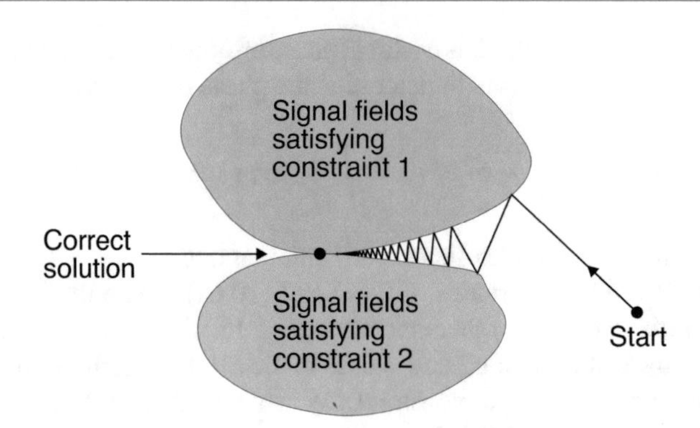

Fig. 10.16 Schematic of the generalized projection algorithm

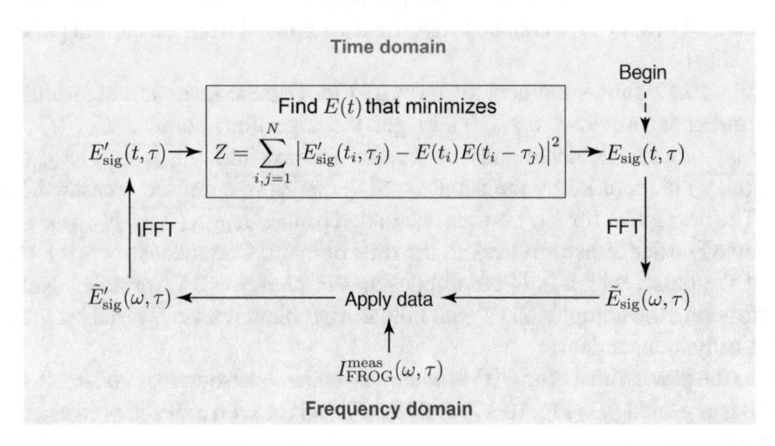

Fig. 10.17 The generalized projection algorithm. The error metric Z is given for the example of SHG FROG. Fast Fourier transform FFT, inverse fast Fourier transform IFFT

set of all $E_{\mathrm{sig}}(t, \tau)$ satisfying the nonlinear material response, and S_2 is the set of all complex functions with magnitude $I_{\mathrm{FROG}}(\omega, \tau)$. Unlike previous algorithms, generalized projection finds the next guess by ensuring that the distance between S_1 and S_2 is minimized for each iteration of the algorithm. DeLong et al. developed a generalized projection algorithm for FROG by using a minimization algorithm to find the $E(t)$ that minimizes the Euclidean distance from the signal field that satisfies the intensity constraint $E'_{\mathrm{sig}}(t, \tau) = \sqrt{I_{\mathrm{FROG}}(\omega, \tau)}\phi_{\mathrm{calc}}(\omega, \tau)$. That is, one minimizes

$$Z = \sum_{i,j=1}^{N} \left| E'_{\mathrm{sig}}(t_i, \tau_j) - E(t_i)\Gamma\left[E\left(t_i - \tau_j\right)\right] \right|^2$$

with respect to $E(t)$ to obtain the next estimate of $E(t)$ [97Tre]. Details of the FROG technique can be found in a review article by Trebino [97Tre], and in his book *Frequency-Resolved Optical Gating: The Measurement of Ultrashort Laser Pulses* [00Tre].

10.3.2 Second Harmonic Generation FROG (SHG-FROG)

In principle, any nonlinear process can be used to make a FROG trace [00Tre]. SHG-FROG has the main advantage that it is very sensitive and works for pulse energies as small as about 1 pJ for multipulse averaging, which is only slightly less sensitive than a simple SHG autocorrelator [00Tre]. In addition, SHG-FROG achieves the best signal-to-noise ratios because its signal beam is centered around a different frequency and scattered light from the fundamental beam is easily filtered out. Therefore, this technique is typically used to characterize unamplified pulses from modelocked laser oscillators. The main disadvantages are that the SHG-FROG traces are unintuitive and symmetrical with respect to delay. Thus, the direction of time is not determined and the actual pulse may in fact be the time-reversed version of the retrieved pulse.

The most important issue with SHG-FROG is that the SHG efficiency needs to be constant over the full pulse spectrum. Thus, the SHG crystal needs to have sufficient bandwidth, and for practical reasons this means that the nonlinear crystal must be sufficiently thin, since the bandwidth is inversely proportional to the crystal thickness [00Tre, eq. 3.51, p. 55]:

$$\delta\lambda_{\mathrm{FWHM}} = \frac{0.44\lambda_0/L}{\left| n'(\lambda_0) - \dfrac{1}{2}n'(\lambda_0/2) \right|} ,$$

where λ_0 is the center wavelength of the fundamental pulse, L is the nonlinear crystal thickness, and $n'(\lambda) = \mathrm{d}n/\mathrm{d}\lambda$. Here we have used the fact that the nonlinear conversion efficiency scales with $L^2 \mathrm{sinc}^2 (L\Delta k/2)$, and Δk is the phase mismatch given by

$$\Delta k(\lambda) = 2k_1 - k_2 = 2\left[2\pi \frac{n(\lambda)}{\lambda}\right] - 2\pi \frac{n(\lambda/2)}{\lambda/2} = \frac{4\pi}{\lambda}\left[n(\lambda) - n(\lambda/2)\right] .$$

We have perfect phase-matching at the center wavelength, i.e., $\Delta k(\lambda_0) = 0$. The phase-matching bandwidth is then found by noting that the sinc^2 curve will decrease by a factor of 2 when $\Delta k L/2 = \pm 1.39$. In addition, the factor $1/\lambda$ can be expanded to first order in the wavelength, which leads to the expression for $\delta\lambda_{\mathrm{FWHM}}$.

In the sub-10-femtosecond regime, the pulse shape becomes more complex both in the time domain and in the spectral domain, and simple autocorrelation and IAC measurements are not sufficient for pulse characterization. Therefore, FROG or SPIDER techniques are typically applied. However, with FROG, we do in principle need constant SHG efficiency and this becomes practically impossible in the sub-10-femtosecond regime. Theoretical corrections have been proposed. See the example and discussion at the end of this section.

Figure 10.18 shows the calculated SHG FROG traces according to (10.62) and (10.63). The input pulses are shown on the left in both the spectral and time domain, assuming a sech^2 pulse shape and a constant power spectrum. First a transform-limited 10 fs pulse is shown with a constant phase over the full spectrum. Then a

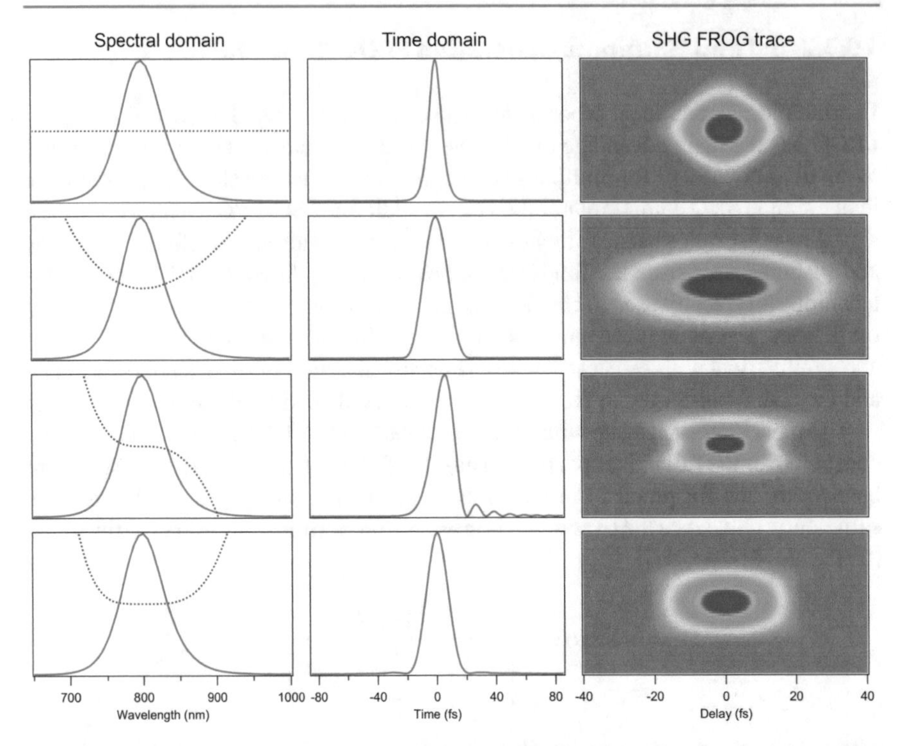

Fig. 10.18 Calculated SHG FROG traces for given input pulses with a sech^2 pulse shape shown *on the left*. In the spectral domain, different orders of dispersion (from top to bottom: none, GDD, TOD, FOD) are added to the pulse. See text for explanation

positive 100 fs^2 GDD is added to the pulse, and this results in a quadratic spectral phase as a function of frequency, and broadens the pulse in the time domain. A third example is given for adding a positive third order dispersion (TOD) of 1000 fs^3, and this results in a cubic spectral phase as a function of frequency. As can be seen in the time domain, the TOD no longer acts symmetrically on the pulse. There is a real pulse tail only on one side of the pulse. Note that the SHG FROG shows a symmetric change by definition. For the last example, we added a fourth order dispersion (FOD) of 1000 fs^4 to the 10 fs pulse. The pulse wings become symmetric again.

In reality, few-cycle pulse shapes become much more complex, and the standard noncollinear autocorrelations are no longer sufficient because they require a priori knowledge of the pulse shape. The temporal parameters have usually been obtained by fitting an analytical pulse shape with constant phase to an autocorrelation measurement. The particular fitting function is motivated by theoretical models of the pulse formation process. For passively modelocked lasers, for example, which allow for parabolic approximation of these formation mechanisms, one expects a sech^2 temporal and spectral pulse shape (see Chap. 9). For lasers obeying such a model, the a priori assumption of a theoretically predicted pulse shape is well motivated and leads to good estimates of the pulse duration, as long as the measured spectrum also agrees with the theoretical prediction. Passively modelocked solid-state lasers in the

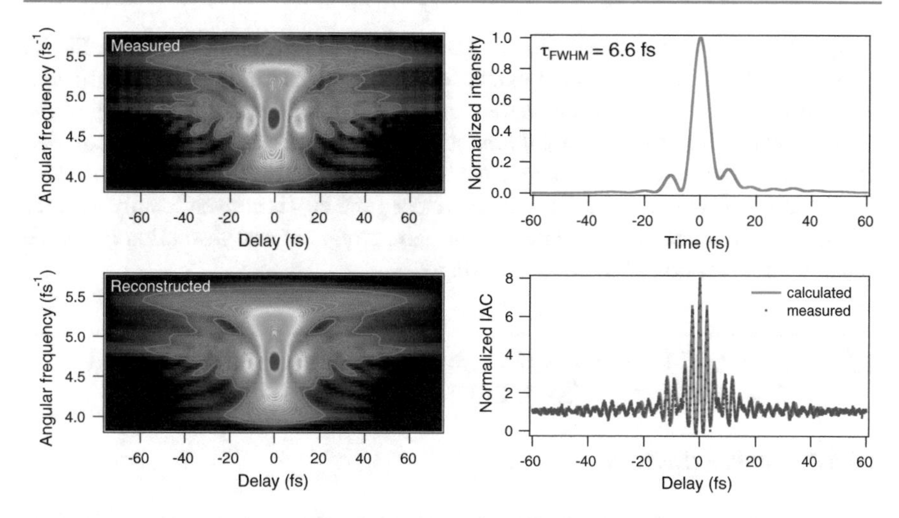

Fig. 10.19 Measured and reconstructed SHG FROG trace from a real pulse generated with a passively modelocked Ti:sapphire laser [00Gal2]. The reconstructed amplitude and phase are used to calculate the pulse intensity and IAC. An IAC has also been measured

near-infrared with pulse durations well above 10 fs normally generate pulses close to the ideal sech^2 shape. Therefore, autocorrelation is still a good standard diagnostic for such laser sources.

In the sub-10 fs near-infrared regime (see Sect. 9.10.1), however, theoretical models have to consider higher order effects, and thus become more complicated. In this case, no simple analytical pulse shapes can be expected. Experimentally, this situation is clearly indicated by more complex pulse spectra, deviating from the ideal sech^2 shapes and more complex FROG traces, as shown in Fig. 10.19. Additionally, even after dispersion compensation, broadband pulses usually exhibit an uncompensated chirp. Chirped mirrors, for example, lead to a sinusoidal modulation of the spectral phase that cannot be completely compensated for by conventional techniques (see Sect. 3.6). Although close to the transform limit, such pulses are distorted by the uncompensated phase structure. Figure 10.19 shows the reconstructed pulse using SHG FROG for a pulse generated with a passively modelocked Ti:sapphire laser [00Gal2]. With the reconstructed pulse amplitude and phase, the normalized intensity is calculated and a FWHM pulse duration is determined to be 6.6 fs. In addition, an IAC trace is calculated with the reconstructed pulse from the FROG measurement, and this shows good agreement with a measured IAC trace. The measured power spectrum (not shown in Fig. 10.19) could support a transform-limited pulse of 5.3 fs, which is much shorter than the measured 6.6 fs pulse. In addition, the pulse shows strong wings which can ultimately degrade the time resolution in ultrafast pump–probe measurements. This experimental result shows a very typical complexity for a few-cycle pulse duration.

In the sub-10 fs range, two serious problems arise with the FROG technique. The first is bandwidth limitation of the optics and the detection system involved, in particular the bandwidth limitation of the SHG process. This problem is reduced by using extremely thin nonlinear optical crystals. In contrast to autocorrelation, FROG allows one to correct to a certain extent for bandwidth limitations. Still, particular care has to be given to the accurate determination of the spectral calibration of the setup. A second, more fundamental limitation is the reduction of temporal resolution caused by the finite beam-crossing angle in the nonlinear crystal (see Sect. 10.2.5). Unlike sub-10 fs autocorrelators, a noncollinear geometry is conventionally used for FROG measurements. In the noncollinear geometry, temporal resolution has to be traded for suppression of interference fringes.

10.4 Spectral Phase Interferometry for Direct Electric Field Reconstruction (SPIDER)

10.4.1 Basic Principle

Spectral phase interferometry for direct electric-field reconstruction (SPIDER) [98Iac,99Iac] is based on another approach than FROG. The measurement of the spectral phase of a pulse with SPIDER is accomplished by self-referencing spectral interferometry. To access the spectral phase, it is necessary to produce two time-delayed replicas of the pulse with a spectral shear $\delta\omega$ between their carrier frequencies, i.e., $E(\omega)$ and $E(\omega + \delta\omega)$ as shown in Fig. 10.20. The spectral shear between the central frequencies of the two replicas is generated by upconversion of the two replicas $E(\omega)$ with a strongly chirped pulse using sum-frequency generation (SFG) in a nonlinear optical crystal. The detailed explanation of the experimental realization

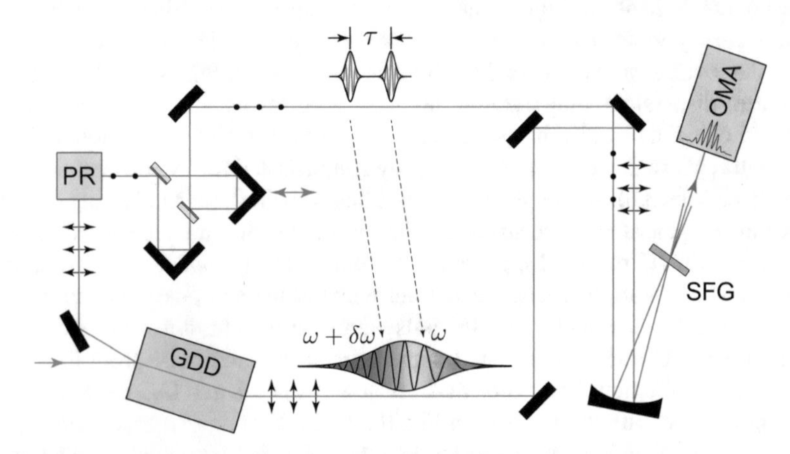

Fig. 10.20 The broadband near-infrared SPIDER setup [99Gal]. All reflections are on silver-coated mirrors and the *filled circles* and *arrows* on the beam path indicate the polarization direction of the beam. GDD is obtained with a SF10 glass block. Periscope for polarization rotation PR, sum frequency generation SFG, type II upconversion BBO crystal, optical multichannel analyzer OMA typically used as the spectrometer

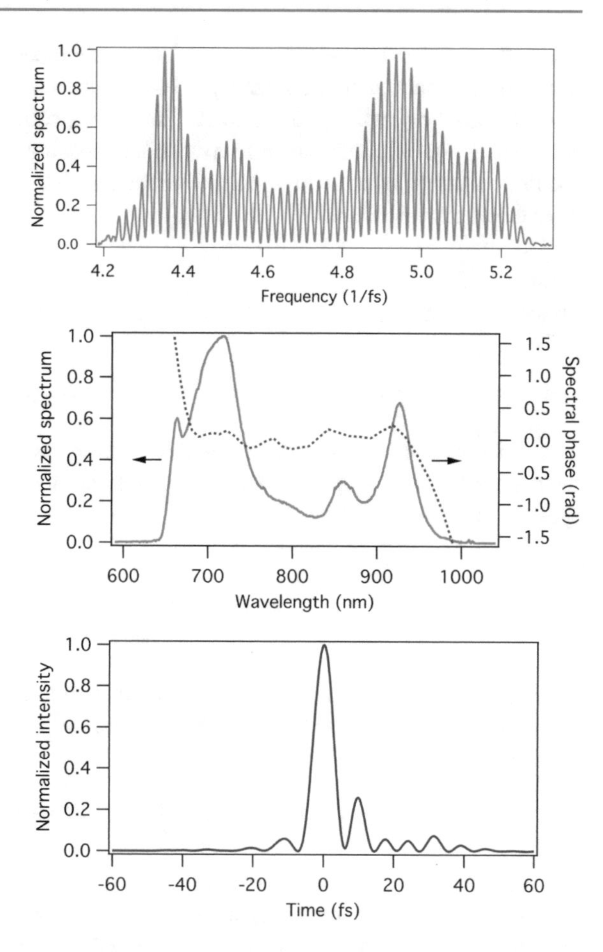

Fig. 10.21 This is the 5.9 fs pulse from a SESAM-assisted KLM Ti:sapphire laser [99Gal]. The *top graph* shows the SPIDER interferogram, the *middle* one shows the measured spectrum and the retrieved phase from the SPIDER interferogram, and the *bottom graph* shows the calculated temporal pulse profile using the measured spectrum and retrieved phase

is given in Sect. 10.4.2. Figure 10.21 then shows such a measured interferogram for a 5.9 fs pulse. We first discuss how the phase can be determined from this SPIDER interferogram.

The SPIDER interferogram $I_{\text{SPIDER}}(\omega)$ (see Fig. 10.21 for an example) contains the phase information of the pulse and reads

$$I_{\text{SPIDER}}(\omega) = |E(\omega)|^2 + |E(\omega+\delta\omega)|^2 \\ +2\,|E(\omega)E(\omega+\delta\omega)|\cos\left[\varphi(\omega+\delta\omega)-\varphi(\omega)+\omega\tau\right], \tag{10.66}$$

where $E(\omega)$ is the electric field of the pulse and τ is the delay between the replicas.

The phase information can be extracted from the Takeda algorithm [82Tak] and was originally developed for spatial interferograms. The Fourier transform to the time domain of the SPIDER interferogram consists of a peak at zero time and two side peaks located near τ and $-\tau$. The two side peaks contain equivalent phase information and are equal [98Iac]. One of the peaks is isolated by applying a suitable filter function, and an inverse Fourier transform back to the frequency domain yields a complex function $c(\omega)$ whose complex phase gives access to the pulse spectral phase:

$\arg c(\omega) = \varphi(\omega + \delta\omega) - \varphi(\omega) + \omega\tau$. A separate measurement of the linear phase term $\omega\tau$ by spectral interferometry of the short unsheared pulse replicas is subtracted from the previous expression to yield the spectral phase difference $\Delta\varphi(\omega) = \varphi(\omega + \delta\omega) - \varphi(\omega)$. From an arbitrarily chosen starting frequency ω_0, one obtains the spectral phase $\varphi(\omega)$ at evenly spaced frequencies $\omega_i = \omega_0 + i\delta\omega$ by the following concatenation procedure:

$$\begin{aligned}
&\varphi(\omega_0) = \varphi_0 \ , \\
&\varphi(\omega_1) = \Delta\varphi(\omega_0) + \varphi(\omega_0) = \varphi(\omega_0 + \delta\omega) - \varphi(\omega_0) + \varphi(\omega_0) = \varphi(\omega_0 + 1 \cdot \delta\omega) \ , \\
&\vdots \\
&\varphi(\omega_{i+1}) = \Delta\varphi(\omega_i) + \varphi(\omega_i) = \Delta\varphi(\omega_i) + \Delta\varphi(\omega_{i-1}) + \ldots + \Delta\varphi(\omega_0) + \varphi_0 \ .
\end{aligned} \tag{10.67}$$

The constant φ_0 remains undetermined but is only an offset to the spectral phase. It does not affect the temporal pulse shape and we may thus set it equal to zero. The spectral phase is written as

$$\varphi(\omega_{i+1}) = \sum_{k=0}^{i} \Delta\varphi(\omega_k) \ . \tag{10.68}$$

If $\delta\omega$ is small relative to features in the spectral phase, $\Delta\varphi(\omega)$ corresponds in first order to the first derivative of the spectral phase, and we can approximate the spectral phase by

$$\varphi(\omega) \approx \frac{1}{\delta\omega} \int_{\omega_0}^{\omega} \Delta\varphi(\omega')\mathrm{d}\omega' \ . \tag{10.69}$$

The integral expression for the spectral phase has the advantage that all the measured sampling points of the interferogram can be used, rather than just using a subset with sampling according to the spectral shear, as with the concatenation method.

10.4.2 Experimental Realization

The optical setup is shown in Fig. 10.20. The two replicas of the incident pulse are generated in a Michelson interferometer. The spectral shear between the central frequencies of the two replicas is generated by upconversion of the two replicas with a strongly chirped pulse using sum-frequency generation (SFG) in a nonlinear optical crystal. The strongly chirped pulse is generated by propagating another copy of the original pulse through a thick glass block. The spectral shear arises because the time delay between the replicas assures that each replica is upconverted with a different portion of the chirped pulse containing a different instantaneous frequency.

Three design parameters determine the range of pulse durations that can be measured by a given SPIDER apparatus: the delay τ, the spectral shear $\delta\omega$, and the group-delay dispersion $\mathrm{GDD}_{\mathrm{up}}$ used to generate the strongly linearly chirped upconverter pulse. These three parameters are related by

$$\delta\omega = \frac{\tau}{\mathrm{GDD_{up}}} \,. \tag{10.70}$$

The delay τ, which determines the positions of the two side peaks of the Fourier transform of the interferogram, is chosen in such a way as to ensure that the side peaks are well separated from the center peak. On the other hand, the fringe spacing of the interferogram is proportional to $2\pi/\tau$, so τ must be small enough for the spectrometer to be able to fully resolve the fringes.

The stretching factor $\mathrm{GDD_{up}}$ is then chosen such that the spectral shear $\delta\omega$, which determines the sampling interval of the reconstructed spectral phase, is small enough to ensure correct reconstruction of the electric field in the time domain according to the Whittaker–Shannon sampling theorem [00Dor]. The constrained relationship for τ and $\delta\omega$ expressed by (10.70) means that, with a particular SPIDER setup, only pulses with a limited range of pulse durations can be measured.

The linear phase term $\omega\tau$, which will later be subtracted from the phase term from the SPIDER signal, is obtained by deriving it from the interferogram created by the second harmonics of the two pulse replicas. This procedure has the advantage that this signal spectrally overlaps with the SPIDER interferogram, whence the same wavelength calibration can be used and most of the calibration dependence cancels out. It is not necessary to measure the linear phase term for every single pulse as it is a characteristic of the setup.

An additional independent measurement of the pulse spectrum provides all the necessary information to determine the electric field in the time domain by a Fourier transform of

$$E(\omega) = \sqrt{I(\omega)}\mathrm{e}^{\mathrm{i}\varphi(\omega)} \,, \tag{10.71}$$

where $I(\omega)$ is the measured spectrum and $\varphi(\omega)$ the reconstructed spectral phase. The electric field in the time domain is thus determined unambiguously except for the carrier–envelope offset phase (CEO) [99Tel].

The setup shown in (Fig. 10.20) can measure pulses with a pulse duration in the range of 5 to about 40 fs in the near infrared. The group delay dispersion $\mathrm{GDD_{up}}$ is generated by propagation through 6.5 cm of SF10 glass. Depending on the bandwidth of the pulse to be measured, the delay τ is chosen between 150 and 250 fs. The broad bandwidth associated with sub-10 fs pulses requires ultra-broadband optics throughout our setup, and in particular the up-conversion bandwidth of the nonlinear crystal must be sufficiently large. For the up-conversion, we therefore chose a 0.05 mm thick BBO crystal in a type-II phase-matching configuration [99Gal]. The SPIDER interferogram is measured at high spectral resolution, in our case better than 0.5 nm, and, using the procedure described above, the spectral phase can be reliably reconstructed.

However, it is important to note that SPIDER does not need to have a constant nonlinear conversion efficiency, as is required for example in FROG. We only extract the phase encoded in the interferogram (Fig. 10.21, top trace) but not the amplitude. As long as the interferences are fully resolved, we obtain this information. This is a major advantage of the SPIDER measurements in the ultrashort pulse width regime. This technique only becomes more difficult if the spectrum shows very

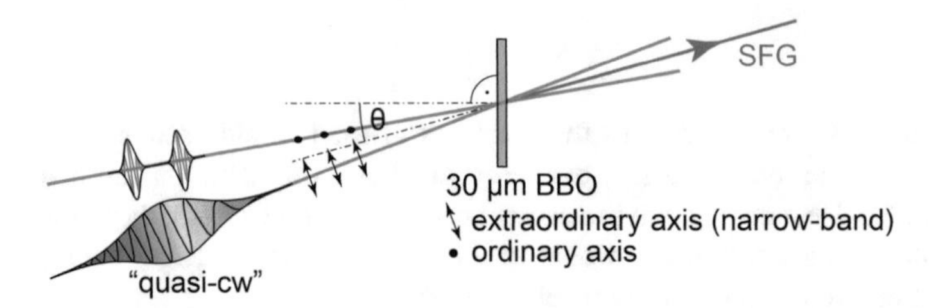

Fig. 10.22 Noncollinear beam crossing angle in SFG in SPIDER for pulses with a center wavelength around 800 nm

strong intensity modulation, as is typically observed in ultra-broadband continuum generation.

In addition, SPIDER does not have the temporal smearing limitation (Sect. 10.2.5) because of the noncollinear beam crossing angle in the SFG crystal (Figs. 10.20 and 10.22). For SPIDER measurements, type-II phase matching allows for particularly broad conversion bandwidth along one polarization direction in sum-frequency mixing in a nonlinear crystal such as BBO. The two short replicas of the pulse to be characterized are therefore polarized along this axis (ordinary axis for a negatively uniaxial crystal in Fig. 10.22). Since only a narrowband slice of the strongly linearly chirped upconversion pulse is needed, bandwidth is a minor concern for this pulse, which is polarized along the extraordinary axis for type-II interaction in a negatively uniaxial crystal. A noncollinear interaction between the short pulses and the upconversion pulse can be used to avoid the need for an additional recombination beamsplitter. The temporal smearing due to the noncollinear geometry and the group-velocity mismatch between the ordinary and extraordinary wave will lead to an increased temporal overlap of the short pulse replicas with the upconversion pulse. As long as the upconversion pulse can still be considered quasi-monochromatic over this broadened time window, this has no negative effects on the SPIDER measurement. In contrast to autocorrelation or crosscorrelation based techniques this does not negatively affect the time resolution of the measurement. While the noncollinear crossing angle is not a critical parameter for SPIDER, it should still be chosen moderately small to maintain a good overlap of the interacting beams.

Figure 10.23 shows a comparison with an IAC measurement for a 5.9 fs pulse. The top trace shows the calculated IAC trace (solid line, red) using the measured pulse information, i.e., the spectral phase from the SPIDER measurement and the spectral amplitude from the measured power spectrum. We obtain a much better agreement with the measured IAC in the pulse wings than when we simply fit the central part of the IAC with a transform-limited sech^2 4.5 fs pulse shape. There is significant disagreement with the IAC measurement in the latter case. A more correct pulse characterization with SPIDER results in a significantly longer FWHM pulse duration of 5.9 fs versus 4.5 fs. This was one of the challenges in the run for a new world record pulse duration from a passively modelocked Ti:sapphire laser, as described in Sect. 9.10.1. The more carefully characterized pulses normally resulted in longer pulses, so it was less likely that a new world record would result.

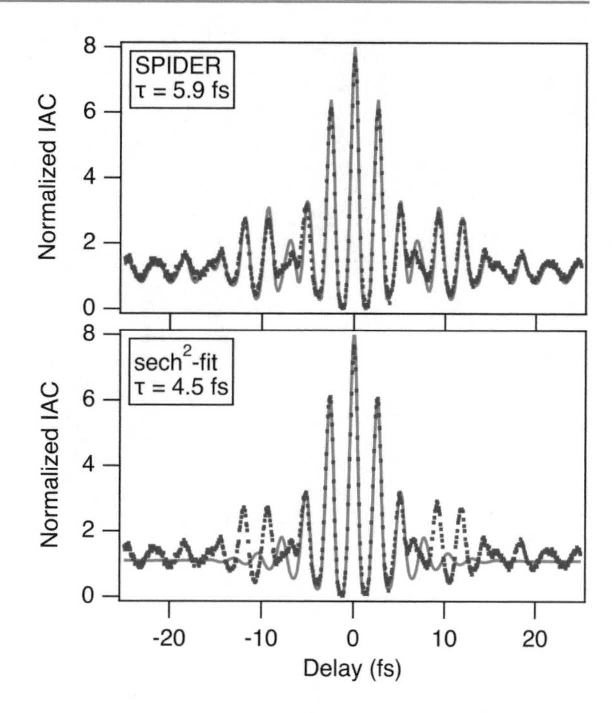

Fig. 10.23 Measured IAC (*dots*, *blue*) and calculated IAC from SPIDER measurements (*top*) and from a transform-limited $sech^2$ pulse fit (*bottom*)

In general, it is best practice to use a combination of different pulse characterization techniques for cross-checking. Both SPIDER and FROG have different trade-offs, as discussed in more detail in [00Gal2]. Commercial systems have become available for all the techniques described so far. The different instruments are normally optimized for a certain pulse duration range and it is advised to study the manual when using such systems.

11 Intensity Noise and Timing Jitter of Modelocked Lasers

11.1 Introduction

In this chapter we first introduce intensity noise and timing jitter and discuss how we measure these fluctuations as a function of (noise) frequency. Normally, for modelocked lasers, in the low-frequency regime the noise is dominated by mechanical perturbation of the laser resonator and the high-frequency part is dominated by nonlinear effects and quantum noise. Typically, a laser has very specific noise peaks superimposed on an approximately $1/f$-noise which normally rolls off with increasing noise frequencies f. Such peaks might be due to mechanical resonances and relaxation oscillations, for example (see Figs. 5.7 and 5.8). In solid-state lasers the noise typically rolls off into the detector limited shot noise well below 10 MHz. The fully characterized frequency-dependent noise is important for ultrafast measurements for which an optimal signal-to-noise ratio is needed. This will also be discussed towards the end of this chapter.

Note that we distinguish between noise frequencies f and optical frequencies ν. For pedagogical reasons we distinguish between these two frequencies with different symbols. However, we will use the same symbol for the angular frequency $\omega = 2\pi f$ or $\omega = 2\pi\nu$. Calling f the noise frequency is also in principle optional and very often we simply refer to f as the frequency and for modelocked lasers as the offset frequency (for reasons explained in this chapter). A clear distinction between these frequencies will become even more important in Chap. 12 when we discuss optical frequency noise.

Given that this is a textbook on ultrafast lasers, I would like to make a personal comment. Unfortunately, in physics departments at many universities, a sound education in the theory of random signals and noise has been replaced by other hot topics in physics. This is unfortunate because it has a huge impact in experimental physics and if it is not understood correctly, it can give rise to misleading physical interpretations of experimental data. I personally was able to benefit from this, after

© The Author(s), under exclusive licence to Springer Nature Switzerland AG 2021

U. Keller, *Ultrafast Lasers*, Graduate Texts in Physics,
https://doi.org/10.1007/978-3-030-82532-4_11

my basic education in physics at ETH Zurich, when working for my Ph.D. under the supervision of a professor in electrical engineering at Stanford university (Prof. D. M. Bloom), where I was introduced to the noise characterization of modelocked lasers. This chapter is largely based on that experience.

11.1.1 Definition of Intensity Noise and Timing Jitter

In this chapter we discuss the noise of a modelocked pulse train (Fig. 11.1). As discussed in Chap. 6 for active modelocking and in Chap. 9 for passive modelocking, the pulse repetition rate for fundamental modelocking is given by the cavity roundtrip time. Thus any fluctuations in the optical path length introduce noise in the arrival time of the pulses, which we refer to as timing jitter $\Delta T(t)$, as shown in Fig. 11.1. Absolute timing jitter is defined by the deviation of the temporal pulse positions (defined by the times with maximum optical power, or better as some 'center of gravity') from the perfect timing of a hypothetical ideal clock with period T. Timing jitter can also be specified relative to some other oscillator, such as another laser or an electronic oscillator. In such cases, the term residual jitter is sometimes used. The intensity noise is determined by the intensity fluctuations $\Delta I(t)$ of the individual pulses within the pulse train (Fig. 11.1). In a modelocked laser, the intensity noise can be understood as the noise of the average power, which is therefore related to the noise of the pulse energy and repetition rate. The intensity noise is typically normalized to the average value, and in this case, the term relative intensity noise (RIN) is also used.

There are a number of effects that couple different kinds of noise in a modelocked laser. The optical path length can for example change with mechanical fluctuations of the cavity components and with nonlinear effects from both the pump and the laser beam. Many effects couple the intensity noise with the timing jitter. Normally, this noise is much greater than any fundamental quantum noise [07Pas]. The various coupling mechanisms are rather complex because many different effects can couple different kinds of noise with each other in a modelocked laser. Examples will be discussed in Sect. 11.4.

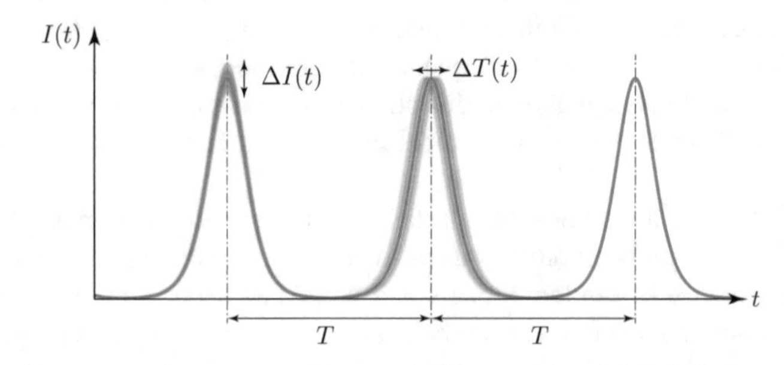

Fig. 11.1 Modelocked pulse train with intensity $I(t)$ and with a constant pulse period T. Intensity noise $\Delta I(t)$ and timing jitter $\Delta T(t)$ are shown schematically as a small deviation from a perfectly stable pulse train

In this chapter we will not discuss any fluctuations of the carrier envelope offset (CEO) phase as shown in Fig. 6.2 and the phase noise of all the relevant cavity modes. This will be part of the frequency comb discussion in Chap. 12. Neither will we discuss here the other parameters of the pulses that can also exhibit noise, in particular the pulse duration, center frequency, chirp, etc.

Taking the first harmonic of the periodic pulse train, we have a sinusoidal signal with a certain amplitude and phase which can be measured and expressed with a voltage signal $V(t)$ of the form

$$V(t) = V_0\left[1 + \alpha(t)\right]\exp\left[\mathrm{i}\left(\omega t + \varphi(t)\right)\right], \tag{11.1}$$

where $\alpha(t)$ is the normalized amplitude fluctuation and $\varphi(t)$ is the phase fluctuation, as shown in Fig. 11.2. Typically, we assume that these fluctuations are small, i.e., $\alpha(t) \ll 1$ and $\varphi(t) \ll 1$, even though this is not always the case. In Fig. 11.2, we also show the phasor representation of the amplitude and phase noise, which is motivated by the fact that we often work with complex amplitudes, i.e., phasors. There are many textbooks on phase noise and frequency stability in oscillators within the RF and microwave engineering fields, e.g., [09Rub]. The noise characterization of laser oscillators has greatly benefited from that knowledge, and many old microwave engineering concepts have been directly applied to lasers.

The timing jitter is sometimes also referred to as the phase noise, which follows by comparing Figs. 11.1 and 11.2. The phase noise can be translated into timing noise when we relate a phase change 2π to one period T. For ultrafast measurements, we are interested in the time resolution, which is determined by both the pulse duration

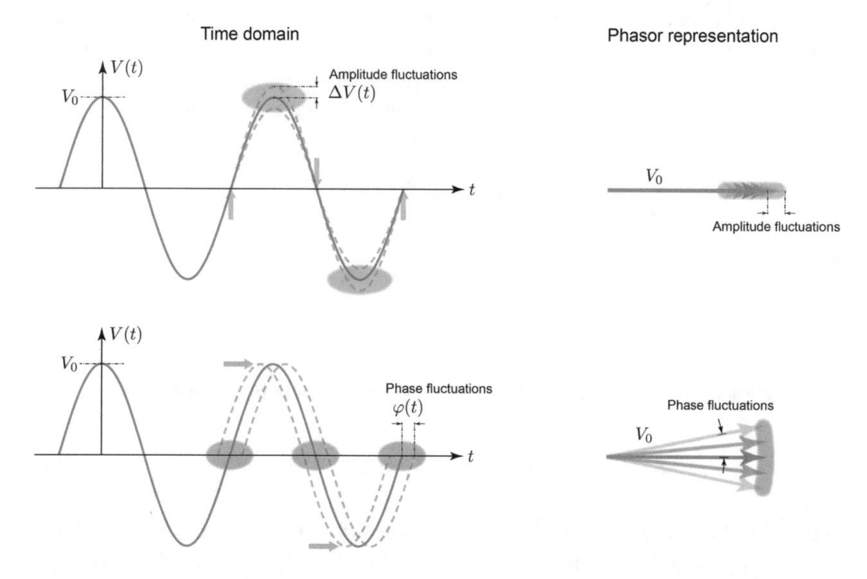

Fig. 11.2 First harmonic signal from the pulse train shown as a voltage signal $V(t)$ in the time domain and in a phasor representation. The amplitude and phase noise are shown schematically with $\Delta V(t) = V_0\alpha(t)$ and $\varphi(t)$, where $\alpha(t)$ is the normalized amplitude fluctuation [09Rub]

and the timing jitter during the measurement time, and therefore we will mostly keep to the timing jitter notation in the following discussion.

In contrast to noise in continuous-wave (cw) lasers, pulse parameters like energy, temporal position, duration, etc., are not defined on a continuous time scale, but only on a discrete temporal grid. Therefore, these noise spectra can be specified only up to half the pulse repetition rate, corresponding to the Nyquist frequency. However, as we will show in this chapter this is not normally a problem, as the noise usually rolls off into the measurement system noise floor well below 10 MHz, i.e., typically below half of the pulse repetition rate.

11.1.2 Basic Mathematical Principles for Noise

For a measured quantity X, e.g., intensity or arrival time of the pulse, we have a certain probability distribution $p(X)$ for the noise fluctuations. For a simple approximation, we very often assume a Gaussian probability distribution for the timing jitter. For a noise quantity X, we can define a Gaussian probability distribution

$$p(X) = \frac{1}{\sigma\sqrt{2\pi}} \exp\left[-\frac{1}{2}\left(\frac{X - X_0}{\sigma}\right)^2\right] , \tag{11.2}$$

where $p(X)$ is the probability density function (PDF), X_0 is the mean or expectation of the distribution, and σ the standard deviation, which is not the FWHM. The PDF is normalized to unity. For example, when the quantity X is described by the timing jitter ΔT, then $X_0 = 0$. For the Gaussian PDF given in (11.2), the X values within less than one standard deviation from the mean account for 68.3% of the set, whereas those within two standard deviations from the mean account for 95.4%, and those within three standard deviations account for 99.7%. The mean value X_0 is given by

$$X_0 \equiv \langle X \rangle = \int X p(X) \mathrm{d}X , \tag{11.3}$$

and the root mean square (rms) value by

$$X_{\mathrm{rms}} = \sqrt{\langle X^2 \rangle} . \tag{11.4}$$

Thus the root mean square deviation from the mean value is given by

$$\Delta X_{\mathrm{rms}} = \sqrt{\langle (X - \langle X \rangle)^2 \rangle} = \sqrt{\langle (X - X_0)^2 \rangle} . \tag{11.5}$$

The statistical average as defined by (11.3) is mathematically sound, but very often it cannot be measured because we do not normally know the specific PDFs. However, if the quantities are stationary, there is a well defined mean value and if the quantities are ergodic, the time-domain averages are equivalent to the statistical expectations. This is summarized in Table 11.1. This means that the average of a random process

Table 11.1 Relevant statistical and time-domain averages which are equivalent for processes that are stationary and ergodic. $p(X)$ is the probability density function (PDF) and $X(t)$ is typically assumed to be a real function, otherwise the autocorrelation function $R_X(\tau)$ is defined by $\langle X^*(t)X(t+\tau)\rangle$ or $\langle X(t)X^*(t-\tau)\rangle$

Parameter	Statistical expectation	Time-domain average
Mean value (or average)	$\langle X\rangle = \int Xp(X)\mathrm{d}X$	$\langle X(t)\rangle = \overline{X(t)} = \lim\limits_{T_o\to\infty}\frac{1}{T_o}\int_{-T_o/2}^{T_o/2} X(t)\mathrm{d}t$
Average power	$\langle \|X\|^2\rangle$	$\langle \|X(t)\|^2\rangle = \overline{\|X(t)\|^2} = \lim\limits_{T_o\to\infty}\frac{1}{T_o}\int_{-T_o/2}^{T_o/2} \|X(t)\|^2\mathrm{d}t$
Variance	$\sigma_X^2 = \langle \|X-\langle X\rangle\|^2\rangle$	$\langle \|X-\langle X\rangle\|^2\rangle = \overline{\|X(t)-\overline{X(t)}\|^2}$ $= \lim\limits_{T_o\to\infty}\frac{1}{T_o}\int_{-T_o/2}^{T_o/2} \|X(t)-\overline{X(t)}\|^2\mathrm{d}t$
Autocorrelation	$R_X(\tau) \equiv \langle X(t)X(t+\tau)\rangle$	$R_X(\tau) = \overline{X(t)X(t+\tau)}$ $= \lim\limits_{T_o\to\infty}\frac{1}{T_o}\int_{-T_o/2}^{T_o/2} X(t)X(t+\tau)\mathrm{d}t$

parameter over time and the average over the statistical ensemble are the same. A stationary stochastic process means that the probability distribution does not change with time, and this in turn implies that the mean and variance do not change over time.

The noise can be measured in the time or in the frequency domain. As we will see in the next section, it is normally easier to measure the noise in the frequency domain. We continue to use the definition of the Fourier transform as introduced in Sect. 2.2, and summarized in Table 11.2 with important mathematical definitions and theorems that we will continue to use.

Typically, the noise of modelocked lasers is ergodic, but not necessarily stationary. For example, the timing jitter of a free-running passively modelocked laser can have a long-term drift for which the mean value $\overline{X(t)} = \overline{\Delta T(t)}$ and the variance as defined in Table 11.1 does not exist. When we keep the observation time T_o large but still limited, we can define a finite-time interval time average for different starting times t_1 by

$$\langle X(t)\rangle_{t_1,T_o} \equiv \overline{X(t)}_{t_1} \equiv \frac{1}{T_o}\int_{t_1}^{t_1+T_o} X(t)\mathrm{d}t\ . \tag{11.6}$$

This new mean value $\langle X(t)\rangle_{t_1,T_o}$ can become a function of t_1 and T_o. For example, for a free-running passively modelocked laser, the arrival time of the pulse has a long-term drift which can be described similarly to a random walk, because there is no restoring force for the pulse repetition rate. The longer the measurement, the greater this drift. The lack of stationarity produces divergences in the spectral densities $S_X(f)$ at low Fourier frequencies. Normally, this problem can be addressed by limiting the averaging time T_o and hence also the lower frequency limit for the spectral noise. From an experimental point of view, this is less of a problem for

Table 11.2 Frequency-domain description and the connection to the time domain. The frequency that we measure with the microwave spectrum analyzer is f in units of Hz. The Fourier transform has a factor of $1/2\pi$ with respect to ω in units of rad/s. Parseval's theorem tells us that the total energy of a signal can be calculated by summing power-per-sample over time or spectral power over frequency. We use the two-sided power spectral density (PSD). Note that the PSD is called the power spectral density, or spectral density of the power, or spectral power density

Fourier transform $F\{X(t)\}$	$\tilde{X}(\omega) \equiv F\{X(t)\} = \int X(t)\mathrm{e}^{-\mathrm{i}\omega t}\,\mathrm{d}t$
Inverse Fourier transform $F^{-1}\{\tilde{X}(\omega)\}$	$X(t) \equiv F^{-1}\{\tilde{X}(\omega)\} = \frac{1}{2\pi}\int \tilde{X}(\omega)\mathrm{e}^{\mathrm{i}\omega t}\,\mathrm{d}\omega$ $= \int \tilde{X}(f)\mathrm{e}^{\mathrm{i}2\pi f t}\,\mathrm{d}f$
Power spectral density (PSD) $S_X(\omega)$	$S_X(\omega) = \lvert\tilde{X}(\omega)\rvert^2$ $S_X(\omega) = F\{R_X(\tau)\} = \int R_X(\tau)\mathrm{e}^{-\mathrm{i}\omega\tau}\,\mathrm{d}\tau$
Parseval's theorem	$\int \lvert X(t)\rvert^2\mathrm{d}t = \frac{1}{2\pi}\int \lvert\tilde{X}(\omega)\rvert^2\mathrm{d}\omega = \int \lvert\tilde{X}(f)\rvert^2\mathrm{d}f$
Variance for laser noise with $\langle X\rangle = 0$	$\sigma_X^2 \equiv \langle\lvert X\rvert^2\rangle = R_X(0) = \frac{1}{2\pi}\int S_X(\omega)\mathrm{d}\omega = \int S_X(f)\mathrm{d}f$

actively modelocked lasers and for timing-stabilized passively modelocked lasers. A mathematical solution that can take into account long-term drifts was introduced with the Allan variance, which will be discussed in more detail in Chap. 12.

It is important to note that the simple definition of the power spectral density (or spectral density of the power) $S_X(\omega) \equiv |\tilde{X}(\omega)|^2$ as shown in Table 11.2 is not really defined for functions with an infinite temporal extent. For example, in the case $X(t) = \overline{X} = \text{constant}$, we have $\tilde{X}(\omega) = 2\pi\overline{X}\delta(\omega)$ and $S_X(\omega) \equiv |\tilde{X}(\omega)|^2 \propto |\delta(\omega)|^2$, which is not defined. This problem can be resolved with the following definition of the two-sided power spectral density:

$$S_X(\omega) = \frac{1}{2\pi}\lim_{T_o\to\infty}\left|\frac{1}{T_o}\int_{-T_o/2}^{T_o/2} X(t)\mathrm{e}^{-\mathrm{i}\omega t}\,\mathrm{d}t\right|^2 . \tag{11.7}$$

This definition is not very convenient to calculate and it is easier to calculate the power spectral density with the Fourier transform of the autocorrelation function $R_X(\tau)$, as shown in Table 11.2, i.e., the Wiener–Khinchin theorem:

$$S_X(\omega) = F\{R_X(\tau)\} = |\tilde{X}(\omega)|^2 . \tag{11.8}$$

Symmetries play an important role in Fourier transforms. The autocorrelation function is a symmetrical, or even, function with $R_X(-\tau) = R_X(\tau)$ because $R_X(\tau) = \langle X(t)X(t+\tau)\rangle = \langle X(t)X(t-\tau)\rangle$ in Table 11.2. Therefore, $F\{R_X(\tau)\}$ is also an even function, i.e., $S_X(\omega) = S_X(-\omega)$. It is a general property of Fourier transforms that if a function is even, its transform is even, and if odd, its transform is odd. For even functions we therefore have the following Fourier transform pair of the two-sided PSD $S_X(\omega)$ and the autocorrelation $R_X(\tau)$, where we used the fact that for even functions the cosine transform is the same as the Fourier transform:

$$\begin{aligned} S_X(\omega) &= F\{R_X(\tau)\} = \int R_X(\tau)\mathrm{e}^{-\mathrm{i}\omega\tau}\mathrm{d}\tau = 2\int_0^\infty R_X(\tau)\cos(\omega\tau)\mathrm{d}\tau \\ R_X(\tau) &= F^{-1}\{S_X(\omega)\} = \frac{1}{2\pi}\int S_X(\omega)\mathrm{e}^{\mathrm{i}\omega\tau}\mathrm{d}\omega = \frac{1}{\pi}\int_0^\infty S_X^{\text{two-sided}}(\omega)\cos(\omega\tau)\mathrm{d}\omega \end{aligned} \tag{11.9}$$

Note that the cosine transform, as defined above, takes no account of the even function to the left of the origin, because the integration limits are set from 0 to ∞, which also explains the factor of 2.

We will continue to use the two-sided power spectral density to discuss intensity and timing jitter noise in the context of modelocked lasers, where we characterize the noise around the laser harmonics, and the zero noise frequency is set at the harmonic peak. The noise frequency is then the offset frequency to the harmonic peak. This gives a direct relation to the output of a spectrum analyzer. We will see many examples later in this chapter.

There are different ways to describe noise fluctuations such as peak-to-peak, variance, or root mean square (rms) values. However, such specifications are always meaningless if no information is given about the time duration over which the noise is accumulated. In addition, the detection bandwidth is also important. For example, if a sinusoidal signal at a specific frequency is superimposed on white noise, i.e., frequency-independent noise, then the signal-to-noise ratio will be reduced when we increase the detection bandwidth, because the signal remains constant, whereas the noise increases linearly with the bandwidth. *Therefore, the variance, peak-to-peak, or rms value is meaningless if it is not specified for a specific frequency interval* $[f_1, f_2]$, *where the upper limit* f_2 *is given by the detection bandwidth and the lower limit* f_1 *by the reciprocal of the measurement time*. This will be discussed in more detail in the following sections.

11.2 Measurement Techniques for Intensity Noise and Timing Jitter

11.2.1 General Remarks on Intensity Noise

The intensity noise of a laser is usually measured with a photodiode. The three principal sources of noise are laser shot noise, Johnson (i.e., thermal) noise, and excess laser noise. Shot noise originates from the discrete nature and random arrival of photons from the laser, and therefore, the discrete production of charge carriers in the photodiode. Johnson noise is the thermal noise based on the random thermal motion of the electrons in the conductor. Finally, excess laser noise is the one we want to characterize here. Because there is no statistical correlation between these noise contributions in the photodiode, they need to be added on a mean-square or

Table 11.3 Sources of noise in the photocurrent J measured with a photodiode. B is the detection bandwidth, q is the electron (hole) charge, J_{av} is the average photocurrent, k_B is the Boltzmann constant, T is the absolute temperature, and the photodiode is connected to a load resistor R_L

Source of noise	Spectral noise density	Units and comments
Shot noise	$\overline{J_{SN}^2}/B = 2qJ_{av}$	A^2/Hz
Johnson (or thermal) noise	$\overline{J_{JN}^2}/B = 4k_BT/R_L$	A^2/Hz
Excess laser noise	$\overline{J_{LN}^2}/B$	A^2/Hz, depends on the specific laser

power basis for the total noise:

$$\overline{J_N^2} = \overline{J_{SN}^2} + \overline{J_{JN}^2} + \overline{J_{LN}^2} \ , \tag{11.10}$$

where the different noise terms are summarized in Table 11.3.

Table 11.3 shows that both the shot noise and the thermal noise are white noise, i.e., frequency-independent noise. For a photocurrent connected to a load resistor R_L, we obtain a total noise power without the excess laser noise given by

$$P_{noise} = \left(\overline{J_{SN}^2} + \overline{J_{JN}^2}\right) BR_L = 2qJ_{av}BR_L + 4k_BTB \ , \tag{11.11}$$

where B is the detection bandwidth and J_{av} is the average photocurrent. We will see later that we can neglect excess laser noise with high frequency modulation techniques (see Sect. 11.5). Ideally, we would like to make some shot-noise limited detection, which means that, according to (11.11), we want to have a sufficiently high average photocurrent to compete against thermal noise in the detection electronics. However, the photocurrent per unit detection area is limited by the onset of detector saturation, and large-area devices tend to have a reduced detection bandwidth which limits measurements at higher harmonics of modelocked lasers.

In the visible spectral region, medium-sized p-i-n silicon photodiodes are widely used because of their demonstrated ability to handle fairly large power levels with high surface uniformity. For longer wavelengths, e.g., 1.5 μm, InGaAs detectors are commonly used. These offer a much better compromise between bandwidth and photocurrent than, e.g., germanium-based devices. Velocity-matched devices [96Lin] even allow bandwidths of tens of gigahertz combined with photocurrents of tens of milliamperes. For other spectral regions, the detector quality can become an issue for noise characterization. For example, there is a growing interest in applications of high-speed photodetectors operating in the long wavelength range from around 2 μm to the mid-infrared. In this regime quantum cascade detectors based on intersubband transitions can be a solution. For example, an impulse response of 13 ps was measured with a 3 dB bandwidth of more than 20 GHz for a center wavelength of 4.3 μm [21Hil].

For electronic processing, one often has to separate the DC part of the photocurrent from the noise part. The calibration may then require a separate measurement, for instance, involving a mechanical chopper or acousto-optic modulator, producing a well-defined power modulation. However, this is not an issue for modelocked lasers because we already have a strongly modulated signal with well defined harmonics.

11.2.2 General Remarks on Timing Jitter

The measurement of frequency noise, timing jitter (Fig. 11.1), or phase noise (Fig. 11.2) always requires some kind of reference. Different frequency noise measurement schemes can rely on different kinds of references and can be grouped as follows. We use a modelocked laser for the noise measurements and refer to it as the device under test (DUT):

1. Techniques based on passive devices such as filters or resonators, which can provide a direct conversion from the frequency fluctuations of the DUT to power fluctuations. Sometimes referred to as a direct or homodyne detection technique.
2. Interferometric techniques, where the instantaneous phase of the DUT is measured via interference with a reference signal which itself is derived from the output of the DUT, typically with some time delay or temporal averaging. This is the basis of self-heterodyning methods. This can be very misleading, as discussed in Sect. 11.2.5.
3. Heterodyne techniques with a reference signal generated by an auxiliary oscillator of superior frequency stability. Alternatively, one may employ two similar optical oscillators. Assuming uncorrelated noise processes with the same power density for both lasers, the measured noise power densities are twice those of the single DUT.

In the simplest case (1), frequency fluctuations are converted into power fluctuations with frequency-dependent transmission of a discriminator in a restricted frequency range. This discriminator can be electrical or optical. For an optical discriminator, for example, the slope of a Fabry–Pérot interferometer (FPI) or the absorption of certain atoms or molecules near the resonance peak can be used. If the oscillator or the filter is tuned such that the mean frequency of the oscillator is on the slope, preferably near the inflection point, the power transmitted by the filter varies, to first order, linearly with the frequency of the signal as $P(\nu - \nu_S) \approx P(\nu_S) + D(\nu - \nu_S)$, where D is the sensitivity or discriminator slope at ν_S.

11.2.3 Measurement Based on Microwave Spectrum Analyzers

A frequently used technique for noise measurements is based on the direct spectral analysis of a photodiode signal with a microwave spectrum analyzer (Fig. 11.3). A photodetector converts the intensity fluctuations into fluctuations of the photocurrent which can be characterized with an electronic spectrum analyzer. Modern spectrum analyzers show a quantity directly related to the spectral density of the fluctuations of

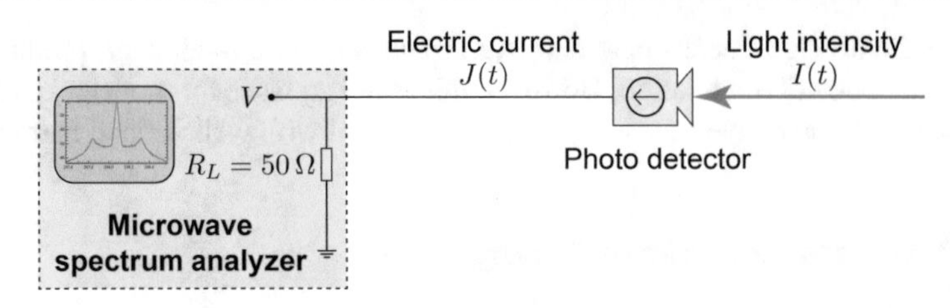

Fig. 11.3 Noise measurement based on spectral power densities. A spectrum analyzer measures the electric power in a $50\,\Omega$ load resistor R_{L} generated by the photocurrent $J(t) \propto I(t)$. The frequency dependence of the measured electric power, i.e., the power spectral density, determines the intensity noise and the timing jitter

the signal. We can use power spectrum techniques to measure laser intensity noise and timing jitter, which are well-known techniques in the characterization of microwave synthesizers and other microwave instruments. These techniques were first used to characterize the noise of cw modelocked lasers in the early 1980s [84Klu, 86vdL, 86Rod].

We typically start with the measurement of fluctuating time-dependent quantities such as the laser output intensity $I(t)$, or the temporal positions of laser pulses $\Delta T(t)$. When we measure the noise with a microwave spectrum analyzer, as shown in Fig. 11.3, we use a spectral description in the form of power spectral densities $S_I(f)$ or $S_{\Delta T}(f)$, as defined in Table 11.2 as a function of $\omega = 2\pi f$, where f is the noise frequency in units of Hz. It is important to realize that these are the spectral power densities of the photocurrent J, which is again proportional to the optical intensity.

Note that such power spectral density techniques are much more reliable than any conclusions from a stable autocorrelation measurement. An autocorrelator (Chap. 10) usually gives inadequate information about amplitude noise, because the bandwidth of the receiver (the photomultiplier and its circuit) is limited. No information can be gained regarding the higher-frequency background intensity noise and the timing jitter of the laser.

As we will show in the next few sections, a modelocked pulse train displays sharp peaks of the power density at integer multiples of the repetition rate (i.e., it has the form of a Delta-comb, Appendix D) and noise sidebands around them, which contain information on intensity noise and timing jitter. A simplified analysis, using the approximation of small timing excursions (compared to the pulse period) and ignoring various correlations between different fluctuating quantities, leads to the simple result that the intensity noise causes equal contributions to the sidebands of all orders (see Fig. 11.6), while the contributions of the timing phase noise scale with the square of the sideband order (see Fig. 11.9). Thus, we can in principle separate the contributions of intensity noise and timing jitter from each other when we compare the noise sidebands from the first harmonic with the ones at a sufficiently high harmonic, where the timing jitter dominates the sidebands.

The simplicity of this method, both in terms of experimental setup and data analysis, makes it very attractive. Note that the phase reference is the local oscillator of the spectrum analyzer, which also exhibits phase noise. It turns out that the phase noise of state-of-the-art spectrum analyzers can easily exceed the timing phase noise of certain types of modelocked lasers. For example, this has been demonstrated for passively modelocked Er:Yb:glass lasers [05Pas, 05Sch2]. Other technical issues can arise from the difficulty to detect high enough harmonics for multi-gigahertz lasers and from possible AM/PM conversion, i.e., conversion from intensity to phase noise, in the photodetector. Finally, note that noise data obtained with electronic spectrum analyzers require certain corrections related to envelope detection, logarithmic averaging, and the effective noise bandwidth of the RF filter, i.e., on the order of 1–2 dB relative to the measured data. Other errors can result from inadequate detector modes or signal power levels. However, nowadays there are commercial instruments available that offer a completely integrated solution for this kind of measurement.

11.2.4 Measurement Based on Electronic Reference Signals

Another class of timing jitter measurement techniques is based on electronic phase detectors. Typically, a microwave mixer is used as phase detector. Its inputs are the photodiode signal and a sinusoidal reference signal from a stable electronic oscillator. The relative phase of these must be adjusted to be $\pi/2$ on average. As long as the phase deviations from this condition are small enough (well below 1 rad), the average voltage output of the mixer is proportional to the phase deviation. For larger phase differences, of course, the method is not suitable. Therefore, it is applicable to cases where the phase excursion remains small for all times, as is typically the case, e.g., for the residual jitter in actively modelocked lasers, for timing-stabilized passively modelocked lasers and for synchronously pumped lasers but not for free-running passively modelocked lasers.

Technical difficulties arise from parasitic effects such as mixer offsets, which lead to incomplete suppression of amplitude noise, and excess noise of electronic components. Such problems can be addressed with chopper schemes [89Rod] and/or with correlation methods [00Rub], allowing the suppression to a high degree of excess noise from electronic components like mixers and amplifiers [03Iva]. Other methods involve the conversion of the microwave reference into an optical amplitude or phase modulation [01Juo]. Another issue is the stability of the electronic oscillator. For high sensitivity, one may employ an ultralow-noise sapphire-loaded cavity oscillator [98Iva], which however is rather expensive, or replace this with a second modelocked laser, which should either be superior in terms of noise, or have noise properties similar to those of the first laser. With such a scheme and the combination of two phase detectors operating on different harmonics of the photodiode signal, timing stabilization and measurement have become possible with record-low relative rms timing jitter [02She].

11.2.5 Measurement Based on Optical Cross-Correlations

Optical cross-correlations are sometimes used to investigate the timing jitter of pulsed lasers. Here, the inputs are two modelocked lasers carrying femtosecond pulse trains, and the optical phase detector is a nonlinear crystal [02She2], or a combination of two nonlinear crystals [03Sch2] for better intensity noise suppression. Note that such a phase detector can work only for timing errors in a very narrow range, on the order of the pulse duration. This is therefore not a useful technique for free-running lasers. However, two timing-stabilized femtosecond lasers can then be synchronized with even lower residual timing jitter, using such a balanced optical cross-correlation technique [03Sch2], which can go beyond the limits set by shot noise in all photodiode-based methods.

However, when only one modelocked laser is used, typically very little information about the timing jitter can be gained with an optical cross-correlation. This can be explained with the following experiment for which we use a Michelson interferometer to split the incoming pulse train into two parts and to delay one part with respect to the other with a different path length in one of the two interferometer arms. When these two parts are recombined at the exit of the interferometer, we can measure an optical cross-correlation proportional to the autocorrelation function $R_{\mathrm{I}}(\tau) = \langle I(t)I(t+\tau)\rangle$ (Table 11.1), where we have assumed for simplicity that we split the pulse train into two equal parts $I(t)$ with a time delay τ. The cross-correlation then becomes an autocorrelation in a nonlinear crystal and is maximum when the two pulses in the pulse train overlap, and is therefore related to the timing jitter $\Delta T(\tau)$. The power spectral density $S_{\mathrm{I}}(f)$ is then the Fourier transform of this autocorrelation function $S_{\mathrm{I}}(f) = F\{R_{\mathrm{I}}(\tau)\} = \int R_{\mathrm{I}}(\tau)\mathrm{e}^{-\mathrm{i}2\pi f\tau}\mathrm{d}\tau$, according to Table 11.2. This means that the maximum time delay $\tau_{\max}$ determines the lower frequency limit $f_{\mathrm{l}} = 1/\tau_{\max}$ from which the noise can be integrated.

Let us assume a pulse repetition period $T = 10$ ns for a pulse repetition rate of 100 MHz, which is a typical value for a standard ultrafast solid-state laser. When we want to perform a cross-correlation of two consecutive pulses, we introduce a path length delay of 10 ns in one arm of the interferometer. The cross-correlation of two consecutive pulses then only measures the timing jitter over $\tau_{\max} = 10$ ns and therefore the spectral noise above $1/10$ ns $= 100$ MHz. This is the case even when we average the cross-correlation over two consecutive pulses over a long time. As we will show later, the timing jitter noise spectrum of a modelocked solid-state laser typically rolls off into the shot noise floor at a frequency lower than 100 kHz. To measure this relevant low-frequency noise in the Michelson interferometer, we would need to increase $\tau_{\max}$ to 1 s and even to 10 s (for a lower frequency limit f_{l} equal to 1 Hz or even 0.1 Hz), requiring an optical delay line of about a 10^8 to 10^9 m [89Kel]. We simply cannot keep such a large interferometer stable for such a measurement. Thus, for a meaningful timing jitter measurement for megahertz or even gigahertz solid-state lasers using this kind of optical cross-correlation technique, we need to have two independent lasers.

This situation changes somewhat when there is significant noise at much higher noise frequencies. This is, for example, the case for semiconductor diode lasers which have multi-gigahertz relaxation oscillation frequencies [88Pet2]. To measure the frequency noise above, for example, above 1 MHz, in order to fully resolve the relaxation frequency noise, we then need to introduce a path length difference of less than 30 cm, which is still challenging with regard to mechanical stability requirements, but potentially can be managed if special care is taken. Nevertheless, there is still significant noise below 1 MHz which still cannot be measured with this technique.

11.2.6 Measurement with an Indirect Phase Comparison Method

The noise performance of modelocked solid-state lasers has become so good that a microwave reference oscillator actually has more noise. We can therefore obtain much more sensitive timing jitter results when the relative timing jitter between two modelocked lasers is measured [05Pas]. The setup for comparing the timing of two lasers is shown in Fig. 11.4. The outputs of the two lasers are detected with fast photodetectors and mixed with the signal from a tunable electronic oscillator f_{osc} to obtain mixing products at much lower frequencies of, e.g., 200 kHz. These are then simultaneously recorded with a two-channel digital storage oscilloscope (or a sampling card in a PC). Both mixing products are actually affected by the phase noise of the electronic reference oscillator, which may be stronger than the phase noise arising from the timing noise of the lasers. However, with suitable Fourier transform techniques, one can extract the relative phase of the two recorded data traces, and this relative phase reflects the timing difference of the lasers, with no influence from the phase noise of the electronic oscillator.

It has been shown that the system phase noise resulting from thermal and digitizing noise is almost white, so that particularly at low noise frequencies the sensitivity of this method is much higher than for other methods which are affected by phase noise from a reference oscillator. Note that such a time-domain method allows for noise characterization at very low frequencies, limited only by the available sampling memory. For low noise frequencies, another time-domain method has been discussed in [99Tsu2].

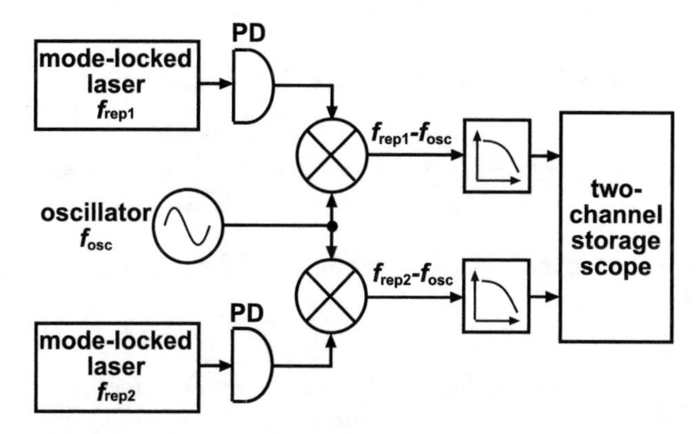

Fig. 11.4 Relative timing jitter measurement in the time domain. The photodiode signals are down-converted to low frequencies, low-pass filtered, and digitally recorded. PD is the photodiode. According to [05Pas, 07Pas]

11.3 Noise Measurements with Power Spectral Densities

11.3.1 Ideal Laser Without Amplitude Fluctuations and Timing Jitter

A modelocked pulse train with no noise can be described with a simple δ-comb by

$$P(t) = \sum_{n=-\infty}^{\infty} P_{\mathrm{p}}(t - nT) = \left[\sum \delta(t - nT)\right] * P_{\mathrm{p}}(t) , \tag{11.12}$$

where $P(t)$ is the time-dependent power, T is the pulse repetition period, and $P_{\mathrm{p}}(t)$ is the time-dependent power of an individual pulse with $P_{\mathrm{p}}(t) \propto |A(t)|^2$, with $A(t)$ the pulse envelope, as introduced in Chap. 2. For the last part of (11.12), we left out the summation limits, assuming the sum is over integers n from $-\infty$ to ∞. As usual, $*$ is the convolution operator, discussed in Appendix D on the Delta-comb.

For microwave spectrum analyzer noise characterization, we can neglect the pulse shape $P_{\mathrm{p}}(t)$ because the pulse duration is well below 1 ps, which is itself well below the response time of the photodetector and spectrum analyzer. We can therefore represent the pulses with a δ-function (see Appendix C). We also typically refer to intensity noise, which is directly proportional to the power. We can thus simplify (11.12) as follows:

$$I(t) = I_0 T \sum \delta(t - nT) , \tag{11.13}$$

where the time-averaged intensity is given by

$$\overline{I(t)} = I_0 T \frac{1}{T} \int_{-T/2}^{T/2} \sum \delta(t - nT) \mathrm{d}t = I_0 . \tag{11.14}$$

To determine the power spectral density $S_I(\omega)$ (Table 11.2), we need to first determine the Fourier transform of $I(t)$, which follows from (D.6) in Appendix D:

$$\tilde{I}(\omega) = 2\pi I_0 \sum \delta(\omega - n\omega_T) , \quad \omega_T = 2\pi/T . \tag{11.15}$$

The power spectral density $S_I(\omega)$ is then given by (Table 11.2)

$$S_I(\omega) = |\tilde{I}(\omega)|^2 = (2\pi)^2 I_0^2 \sum_{n=-\infty}^{\infty} \delta(\omega - n\omega_T) \sum_{m=-\infty}^{\infty} \delta(\omega - m\omega_T) . \tag{11.16}$$

Using the definition of the δ-function, that $\delta(\omega) = 0$ for $\omega \neq 0$ (see Appendix C), it follows that

$$\sum_{m=-\infty}^{\infty} \delta(\omega - m\omega_T) f(\omega) = \sum_{m=-\infty}^{\infty} \delta(\omega - m\omega_T) f(m\omega_T) , \tag{11.17}$$

whence

$$\sum_{m=-\infty}^{\infty} \delta(\omega - m\omega_T) \sum_{n=-\infty}^{\infty} \delta(\omega - n\omega_T) = \sum_{m=-\infty}^{\infty} \left[\delta(\omega - m\omega_T) \sum_{n=-\infty}^{\infty} \delta(m\omega_T - n\omega_T) \right]$$
$$= \sum_{m=-\infty}^{\infty} \left[\sum_{n=-\infty}^{\infty} \delta(\omega - m\omega_T)\delta(m\omega_T - n\omega_T) \right] . \tag{11.18}$$

The function $\delta(m\omega_T - n\omega_T)$ is only nonzero if $n = m$, so the sum reduces to

$$\sum_{n=-\infty}^{\infty} \delta(\omega - m\omega_T)\delta(m\omega_T - n\omega_T) = \delta(\omega - m\omega_T) . \tag{11.19}$$

We then obtain

$$\boxed{S_I(\omega) = 4\pi^2 I_0^2 \sum_{m=-\infty}^{\infty} \delta(\omega - m\omega_T) , \quad \omega_T = 2\pi/T} , \tag{11.20}$$

as shown in Fig. 11.5. Dealing with laser and electrical signals we typically determine the power spectral density (PSD) $S_P(\omega)$ in units of watts/hertz. Note that concerning the PSD $S_I(\omega)$ we still speak of power spectral densities, even though the units for $S_I(\omega)$ are watts/(cm^2 · hertz). As we will show in the next section, the relative intensity noise is usually expressed in units of dBc per Hz.

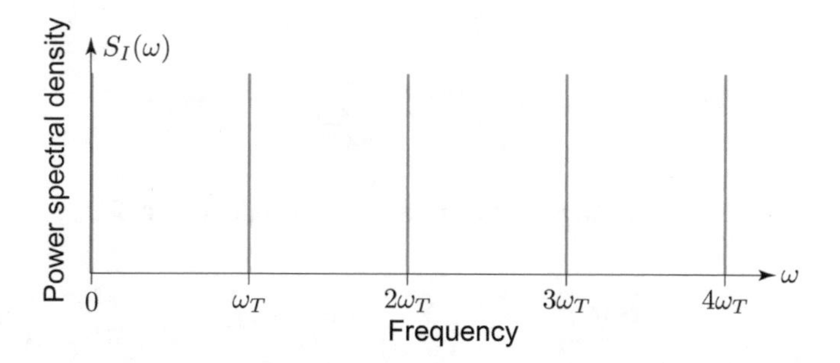

Fig. 11.5 Without amplitude and timing fluctuations, the power spectral density (PSD) for the modelocked laser $S_I(\omega)$ is described by the δ pulse comb of (11.20). In reality, the measurement device has a limited measurement bandwidth, and the δ pulses are replaced by the filter function of the measurement device, e.g., a microwave spectrum analyzer. $\omega_T = 2\pi f_T$, where f_T is the pulse repetition rate of the laser

11.3.2 Pulse Train with Intensity Noise

Now, we will also take into account the normalized intensity noise of the laser $N(t)$:

$$I(t) = I_0 T\left[1 + N(t)\right] \sum_{n=-\infty}^{+\infty} \delta(t - nT) = I_0(t) + I_{\mathrm{N}}(t) , \tag{11.21}$$

where $I_0(t)$ describes the intensity without intensity noise and $I_{\mathrm{N}}(t)$ describes the intensity with intensity noise. Equation (11.21) implies

$$I_0(t) = I_0 T \sum_{n=-\infty}^{+\infty} \delta(t - nT) \tag{11.22}$$

and

$$I_{\mathrm{N}}(t) = I_0 T N(t) \sum_{n=-\infty}^{+\infty} \delta(t - nT) = I_0(t) N(t) . \tag{11.23}$$

Therefore $N(t)$ is the normalized intensity noise with a mean value $\langle N(t) \rangle = 0$. Using (11.16), we have

$$S_I(\omega) = \left|\tilde{I}(\omega)\right|^2 = \left|\tilde{I}_0(\omega) + \tilde{I}_{\mathrm{N}}(\omega)\right|^2 , \tag{11.24}$$

where $\tilde{I}_0(\omega)$ is given by (11.15). The Fourier transform of $I_{\mathrm{N}}(t)$, i.e., given by (11.23), follows with

$$\tilde{I}_{\mathrm{N}}(\omega) = F\left\{I_{\mathrm{N}}(t)\right\} = 2\pi I_0 \tilde{N}(\omega) * \sum_{n=-\infty}^{+\infty} \delta(\omega - n\omega_T) , \tag{11.25}$$

where $\tilde{N}(\omega) = F\{N(t)\}$. Then, by the results in Appendix D, we obtain

$$\tilde{I}_{\mathrm{N}}(\omega) = 2\pi I_0 \sum_{n=-\infty}^{+\infty} \tilde{N}(\omega - n\omega_T) . \tag{11.26}$$

Using (11.16) and (11.24), it follows that

$$\begin{aligned} S_I(\omega) &= \left|\tilde{I}_0(\omega)\right|^2 + \left|\tilde{I}_{\mathrm{N}}(\omega)\right|^2 + 2\tilde{I}_0(\omega)\tilde{I}_{\mathrm{N}}(\omega) \approx S_{I_0}(\omega) + S_{I_{\mathrm{N}}}(\omega) , \\ S_{I_0}(\omega) &= \left|\tilde{I}_0(\omega)\right|^2 , \\ S_{I_{\mathrm{N}}}(\omega) &= \left|\tilde{I}_{\mathrm{N}}(\omega)\right|^2 , \end{aligned} \tag{11.27}$$

where S_{I_0} is the power spectral density of the laser pulse without intensity noise, and $S_{I_{\mathrm{N}}}$ is the power spectral density of the intensity noise.

One can easily see that

$$\tilde{I}_0(\omega)\tilde{I}_{\mathrm{N}}(\omega) = \tilde{I}_0(0)\tilde{I}_{\mathrm{N}}(0) \ll \tilde{I}_0(0)\ . \tag{11.28}$$

Regarding the power spectral density of the normalized intensity fluctuations $S_{\mathrm{N}}(\omega)$, (11.26) now implies

$$S_{I_{\mathrm{N}}}(\omega) = \left|\tilde{I}_{\mathrm{N}}(\omega)\right|^2 = 4\pi^2 I_0^2 \sum_{n=-\infty}^{+\infty} \tilde{N}(\omega - n\omega_T) \sum_{m=-\infty}^{+\infty} \tilde{N}(\omega - m\omega_T)\ . \tag{11.29}$$

Because the bandwidth of the fluctuation spectra Δf_{N}, i.e., $\Delta f_{\mathrm{N}} < 1$ MHz, does not usually overlap with the neighboring harmonics of the pulse repetition rate, i.e., $f_T \gg 1$ MHz, we have

$$\tilde{N}(\omega - n\omega_T)\tilde{N}(\omega - m\omega_T) = 0\ , \quad n \neq m\ , \tag{11.30}$$

from which it follows that

$$S_{I_{\mathrm{N}}}(\omega) = 4\pi^2 I_0^2 \sum_{n=-\infty}^{+\infty} \left[\tilde{N}(\omega - n\omega_T)\right]^2 . \tag{11.31}$$

Finally, using (11.20) and (11.27) we obtain

$$\boxed{S_I(\omega) = 4\pi^2 I_0^2 \sum_{n=-\infty}^{+\infty} \left\{\delta(\omega - n\omega_T) + \left[\tilde{N}(\omega - n\omega_T)\right]^2\right\}}\ , \tag{11.32}$$

as shown in Figs. 11.6 and 11.7.

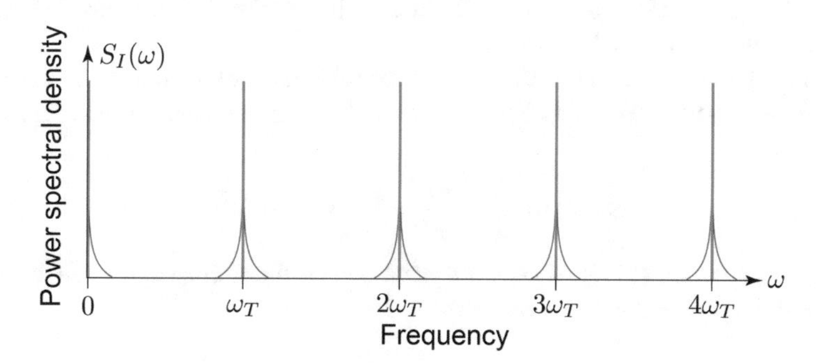

Fig. 11.6 The PSD $S_I(\omega)$ according to (11.32) with intensity noise. In contrast to the ideal laser (Fig. 11.5), additional intensity noise generates constant sidebands proportional to $\left[\tilde{N}(\omega - n\omega_T)\right]^2$ in all δ-pulses in the power spectral density $S_I(\omega)$

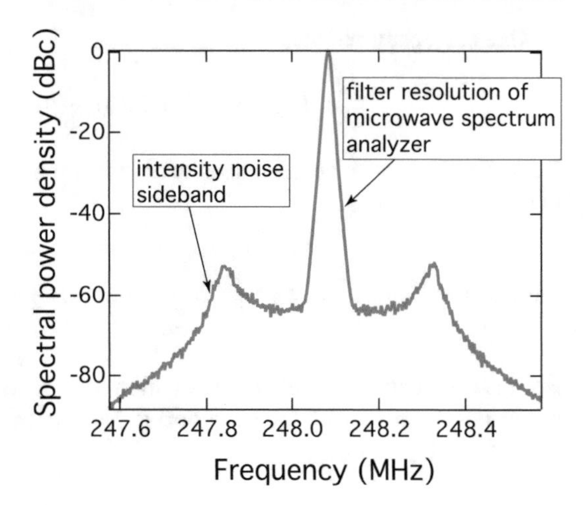

Fig. 11.7 Power spectral density of the intensity noise $S_I(\omega)$ of a pulsed Nd:YLF laser (pumped by a Ti:sapphire laser) with a pulse repetition rate of 248.08 MHz, measured using a microwave spectrum analyzer at a fixed filter bandwidth (resolution bandwidth). The intensity noise sidebands show strong relaxation oscillations at an offset frequency of around 240 kHz

The unit dB (=decibel) is defined by

$$\text{dB} \equiv 10 \log \frac{P_2}{P_1} , \tag{11.33}$$

where P_1 and P_2 are two power values, and dBm is defined by

$$\text{dBm} \equiv 10 \log \frac{P}{1\ \text{mW}} . \tag{11.34}$$

Similarly to dBm we an define the unit dBc

$$\text{dBc} \equiv 10 \log \frac{P}{P_\text{c}} , \tag{11.35}$$

where P_c is the carrier power and dBc means the number of dB below the carrier power, as shown in Figs. 11.7 and 11.8.

From (11.31) we can obtain the power spectral density of the normalized (or relative) intensity noise $N(t) = I_\text{N}(t)/I_0(t)$ [see (11.23)] around the first laser harmonic as follows:

$$S_\text{N}(\omega) = \frac{S_{I_\text{N}}(\omega)}{S_{I_0}(\omega_T)} = |\tilde{N}(\omega - \omega_T)|^2 . \tag{11.36}$$

The rms fluctuation (11.4) or standard deviation of the normalized intensity noise $N(t)$ with $\langle N(t) \rangle = 0$ is given by Table 11.2:

$$\boxed{\sigma_\text{N} = \sqrt{\langle N^2(t) \rangle} = \sqrt{R_\text{N}(0)} = \sqrt{\frac{1}{\pi} \int_0^\infty S_\text{N}(\omega) \text{d}\omega}} . \tag{11.37}$$

Note that, because we use two-sided power spectral densities, we have to multiply the integral over only positive frequencies by a factor of 2.

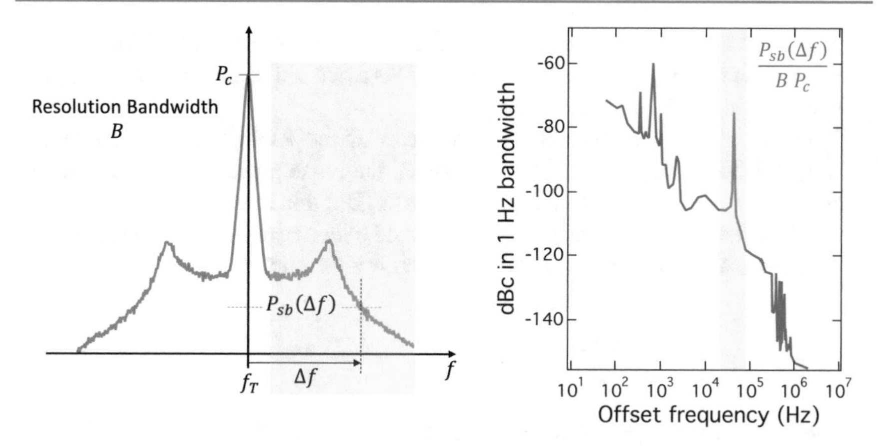

Fig. 11.8 Measurement of intensity noise sidebands. Double-logarithmic plot of the power spectral density of the normalized intensity noise $S_N(\Delta f) = P_{sb}(\Delta f)/B P_c$ for the example of a flashlamp-pumped cw Nd:YLF laser (Quantronix) [90Kel3] (*right*). The *left panel* shows the actual display on the microwave spectrum analyzer for a given resolution bandwidth B and a frequency span given by the *yellow shaded area*. Note that $S_N(\Delta f)$ is also referred to as the relative intensity noise RIN(Δf) according to (11.41), and normally the offset frequency from the laser harmonic Δf is simply replaced by the noise frequency f, because the noise sidebands are equal for all harmonics (see Fig. 11.6)

During a measurement, only the intensity noise within a certain spectral region $[\omega_1, \omega_2] = 2\pi\,[f_1, f_2]$ is acquired. The lower bound f_1 is given by the measurement duration (i.e., inversely proportional to the integration time of a measurement) and the upper bound f_2 is determined by the measurement bandwidth (e.g., the measurement bandwidth of the photodiode). Therefore, the rms normalized intensity noise in this spectral region is given by (see Table 11.2 and note the factor of 2 because we use a two-sided PSD)

$$\sigma_N\,[\omega_1, \omega_2] = \sqrt{\frac{1}{\pi}\int_{\omega_1}^{\omega_2} S_N(\omega)\mathrm{d}\omega}\ . \tag{11.38}$$

The variance of the normalized intensity noise can therefore be determined from the measurement, as shown in Fig. 11.8, in the following way:

$$\sigma_N\,[f_1, f_2] = \sqrt{\frac{P_{sb}\,[f_1, f_2]}{P_c}}\ , \tag{11.39}$$

where $P_{sb}\,[f_1, f_2]$ is the power in the integrated intensity noise sidebands for the frequency span $[f_1, f_2]$, and P_c is the carrier peak power of the filter function (corresponding to the δ-peak) around the first harmonic of the pulse repetition rate $f_T = \omega_T/2\pi$, with f the offset frequency or noise frequency, i.e., effectively setting the first harmonic frequency $f_T = 0$:

$$\frac{P_{sb}\,[f_1, f_2]}{P_c} = 2\int_{nf_T+f_1}^{nf_T+f_2} \frac{P_{sb}(f)/P_c}{B}\mathrm{d}f\ . \tag{11.40}$$

Here B is the measurement bandwidth, i.e., the resolution bandwidth of the spectrum analyzer. The factor of 2 arises because $P_{\mathrm{sb}}(f)$ is a two-sided power spectral density, as shown in Figs. 11.7 and 11.8.

Very often intensity noise is given in terms of the RIN (i.e., relative intensity noise), which is defined by $\langle P_{\mathrm{n}}^2 \rangle / P_0^2$, with P_{n} the noise power and P_0 the average optical power and in units of dB, i.e., 10 log RIN. But this is equivalent to $\langle P_{\mathrm{sb}}^2 \rangle / P_c^2$, where the sideband noise power P_{sb} is integrated over a certain frequency span. RIN is sometimes also expanded to RIN(f), which is equivalent to

$$\mathrm{RIN} = \frac{\langle P_{\mathrm{n}}^2 \rangle}{P_0^2} = \frac{\langle P_{\mathrm{sb}}^2 \rangle}{P_c^2} \Longrightarrow \mathrm{RIN}(f) = \frac{\langle P_{\mathrm{sb}}^2 \rangle (f)}{P_c^2} . \tag{11.41}$$

The shot noise floor depends on the average photocurrent and follows from Table 11.3 with

$$\mathrm{RIN}_{\mathrm{SN}} = \frac{2q J_{\mathrm{av}} B}{J_{\mathrm{av}}^2} = \frac{2qB}{J_{\mathrm{av}}} = 3.2 \times 10^{-16} \frac{\mathrm{mA}}{J_{\mathrm{av}}} B . \tag{11.42}$$

This then results in an average photocurrent $J_{\mathrm{av}} = 1$ mA in a shot noise floor of -155 dBc in 1 Hz bandwidth.

To access significant noise in typical solid-state lasers we need to reduce the frequency span around the laser harmonics to much less than 1 MHz. Figure 11.7 shows a typical measurement of intensity noise sidebands around the first laser harmonic of the pulse repetition rate with a swept-tuned microwave spectrum analyzer [04Agi]. With an analog swept-tuned spectrum analyzer the time-dependent input signal is spectrally filtered with a mixer and a voltage-controlled local oscillator (LO). This local oscillator has a certain filter function with an adjustable resolution bandwidth (RBW) and the voltage-controlled center frequency is swept through a full frequency span. With a digital approach the time-domain signal is digitized to perform a fast Fourier transform to display the input signal in the frequency domain, which has the advantage of obtaining both the spectrally resolved amplitude and phase signal from the input signal. However, the analog approach normally gives a higher sensitivity and a larger dynamic range.

The analog swept-tuned spectrum analyzer has several important control settings such as the RBW of the filter function (see Fig. 11.7), the frequency span, the center frequency, and the video bandwidth (VBW), etc. The VBW is the bandwidth of an additional noise filter on the final output signal which acts as a video filter in a spectrum analyzer. The effect is most noticeable when noise sidebands are measured, because with a smaller VBW, the peak-to-peak variations of the noise are reduced. This is equivalent to averaging or smoothing a noisy trace. There is a trade-off between how quickly the display can update the full frequency span during a measurement with a given RBW and VBW. This sweep time (ST) is given by [04Agi]

$$\mathrm{ST} = k \frac{(\mathrm{Span})}{(\mathrm{RBW})(\mathrm{VBW})} , \tag{11.43}$$

where k is a constant, and (Span) is the frequency span. The spectrum analyzer normally sets the sweep time automatically to account for both the RBW and VBW settings. Note that the VBW does not affect the average noise level, whereas the RBW increases the noise power approximately proportionally to the RBW for small RBW. However, decreasing the VBW too much increases the sweep time, and this can become a problem for nonstationary noise processes (see Sect. 11.5.1).

Normally, a full frequency span from 1 Hz to 1 MHz is done with several measurements using different frequency spans each with an optimal selection of RBW and VBW. The lowest usable noise frequency within a certain frequency span is on the order of three times the RBW to make sure that the noise in the sideband is not affected by the filter function (see Fig. 11.7). The noise power in the sideband $P_{\mathrm{sb}}(f)$ is then normalized to the carrier power P_{c} at the laser harmonic and by the RBW, which then results in units of dBc/Hz (i.e., dBc, the number of dB below the carrier) [see (11.39) and (11.40)]. This is therefore a normalized or relative noise measurement which allows for comparable measurements using different photodetector receivers. In modelocked lasers we benefit from the strong carrier signal P_{c} at the laser harmonics. This results in calibrated relative noise signals with regard to the average optical signal (i.e., $P_{\mathrm{sb}}/P_{\mathrm{c}}$), independently of the different photodetectors.

Only the noise floor of the measurement can be affected between different measurement parameters. For example, the shot noise depends on the specific photocurrent (11.42), whereas the excess laser noise in dBc is independent of the photocurrent. In any case, for all measurements we always need to make sure that the measurement system noise floor is well below the noise sideband signals, and ideally this is only limited by shot noise. The shot noise floor is lower with increasing photocurrent according to (11.42). We then finally combine all the noise sideband measurements together in a double logarithmic plot, as shown in Fig. 11.8.

There are fully integrated noise measurement systems commercially available. One of the early systems was developed by Hewlett Packard with the HP3047A, which automatically corrects for all electronic issues and is a very powerful instrument. It has both a microwave spectrum analyzer and a computer running the measurement software integrated into one instrument supporting noise analysis within a frequency span of [0.02 Hz, 40 MHz]. It also includes an interface box (HP35601A), which results in an absolute system noise floor of -169 dBm/Hz. The photodetector circuit has to be designed to a shot noise that is larger than this noise floor to obtain a high-dynamic range noise measurement (see, for example, [01Sco]). The interface box also includes a 5 MHz to 1.6 GHz mixer for timing jitter measurements.

11.3.3 Pulse Train with Timing Jitter

In this case we neglect the intensity fluctuations and only consider the fluctuations in the arrival time of the pulses given by $\Delta T(t)$, i.e., the timing jitter. Therefore, the light intensity is given by

$$I(t) = I_0 T \sum_{n=-\infty}^{+\infty} \delta(t - nT - \Delta T(t)) \ . \tag{11.44}$$

In the following, we will use the identity (see Appendix D)

$$\sum_{n=-\infty}^{+\infty} \delta(t - nT) = \frac{1}{T} \sum_{n=-\infty}^{+\infty} \mathrm{e}^{\mathrm{i}n\omega_T t} , \quad \omega_T = \frac{2\pi}{T} .$$

Using this, (11.44) can be transformed to

$$I(t) = I_0 \sum_{n=-\infty}^{+\infty} \mathrm{e}^{\mathrm{i}n\omega_T [t-\Delta T(t)]} = I_0 \sum_{n=-\infty}^{+\infty} \mathrm{e}^{\mathrm{i}n\omega_T t} \mathrm{e}^{-\mathrm{i}n\omega_T \Delta T(t)} . \tag{11.45}$$

For small jitter $n\omega_T \Delta T(t) \ll 1$, the exponential function of the timing jitter can be linearized:

$$I(t) \approx I_0 \sum_{n=-\infty}^{+\infty} \mathrm{e}^{\mathrm{i}n\omega_T t} \left[1 - \mathrm{i}n\omega_T \Delta T(t)\right] . \tag{11.46}$$

This implies

$$I(t) \approx I_0 \sum_{n=-\infty}^{+\infty} \mathrm{e}^{\mathrm{i}n\omega_T t} - I_0 \sum_{n=-\infty}^{+\infty} \mathrm{e}^{\mathrm{i}n\omega_T t} \mathrm{i}n\omega_T \Delta T(t) \equiv I_0(t) + I_{\Delta T}(t) , \tag{11.47}$$

where $I_{\Delta T}$ is the intensity of the timing jitter. The photocurrent generated by timing jitter is proportional to $I_{\Delta T}(t)$. The Fourier transform of $I_{\Delta T}(t)$ then follows directly from the shift theorem for the Fourier transform:

$$\Delta\tilde{T}(\omega) = F\left\{\Delta T(t)\right\} , \quad \Delta\tilde{T}(\omega - \omega_0) = F\left\{\mathrm{e}^{\mathrm{i}\omega_0 t} \Delta T(t)\right\} .$$

Equation (11.47) then implies

$$\tilde{I}_{\Delta T}(\omega) = F\left\{I_{\Delta T}(t)\right\} = -I_0 \sum_{n=-\infty}^{+\infty} \mathrm{i}n\omega_T \Delta\tilde{T}(\omega - n\omega_T) , \tag{11.48}$$

whence the power spectral density is

$$\begin{aligned} S_{I_{\Delta T}}(\omega) &= \tilde{I}_{\Delta T}(\omega) \tilde{I}^*_{\Delta T}(\omega) \\ &= (-\mathrm{i})(+\mathrm{i}) I_0^2 \sum_{n=-\infty}^{+\infty} n^2 \omega_T^2 \left[\Delta\tilde{T}(\omega - n\omega_T)\right]^2 , \end{aligned} \tag{11.49}$$

where we have once again assumed, as before with the intensity noise (11.30), that the individual jitter sidebands do not overlap with the neighboring sidebands:

$$\Delta\tilde{T}(\omega - n\omega_T) \Delta\tilde{T}(\omega - m\omega_T) = 0 , \quad n \neq m . \tag{11.50}$$

Thus, as before with the intensity noise [see (11.29)–(11.31)], the following multiplication simplifies to

$$\sum_{n=-\infty}^{+\infty} n\omega_T \Delta\tilde{T}(\omega - n\omega_T) \sum_{m=-\infty}^{+\infty} m\omega_T \Delta\tilde{T}(\omega - m\omega_T) = \sum_{n=-\infty}^{+\infty} n^2\omega_T^2 \left[\Delta\tilde{T}(\omega - n\omega_T)\right]^2 .$$

Equation (11.49) with (11.20) then yields

$$\boxed{S_\mathrm{I}(\omega) = 4\pi^2 I_0^2 \sum_{n=-\infty}^{+\infty} \left\{\delta(\omega - n\omega_T) + \frac{1}{4\pi^2} n^2\omega_T^2 \left[\Delta\tilde{T}(\omega - n\omega_T)\right]^2\right\}} . \quad (11.51)$$

By analogy with the intensity noise, the rms timing jitter ΔT_rms or standard deviation $\sigma_{\Delta T}$ is given by (see Table 11.2, with $X = \Delta T$)

$$\boxed{\Delta T_\mathrm{rms} = \sigma_{\Delta T} = \sqrt{\left\langle \Delta T^2(t) \right\rangle} = \sqrt{R_{\Delta T}(0)} = \sqrt{\frac{1}{\pi}\int_0^\infty S_{\Delta T}(\omega)\mathrm{d}\omega}} , \quad (11.52)$$

with $S_{\Delta T}(\omega) = |\Delta\tilde{T}(\omega)|^2$ according to Table 11.2. Therefore, with (11.49) and from the measured noise sidebands, we obtain

$$\boxed{\Delta T_\mathrm{rms}[f_1, f_2] = \sigma_{\Delta T}[f_1, f_2] = \frac{1}{f_T}\sqrt{2\int_{f_1}^{f_2} L_1(f)\mathrm{d}f} = \frac{1}{nf_T}\sqrt{2\int_{f_1}^{f_2} L_n(f)\mathrm{d}f}} , \quad (11.53)$$

where $L_1(f)$ is defined as the measured normalized timing jitter sideband, i.e., normalized power spectral density, around the first harmonic of the pulse repetition rate $f_T = \omega_T/2\pi$, with f the offset frequency or noise frequency, and $L_n(f)$ is the normalized power spectral density around the n th harmonic of the pulse repetition frequency (Fig. 11.9):

$$L_1(f) = \frac{P_\mathrm{sb}^{(1)}(f)}{P_\mathrm{c}^{(1)}} , \quad L_n(f) = \frac{P_\mathrm{sb}^{(n)}(f)}{P_\mathrm{c}^{(n)}} . \quad (11.54)$$

Using (11.53) and Fig. 11.10, we obtain numerically

$$\Delta T_\mathrm{rms}[130\ \mathrm{Hz}, 20\ \mathrm{kHz}] = \sigma_{\Delta T}[130\ \mathrm{Hz}, 20\ \mathrm{kHz}] \approx \sigma_{\Delta T}[130\ \mathrm{Hz}, \infty] = 9\ \mathrm{ps} .$$

Assuming the fluctuations satisfy a Gaussian statistic, we have

$$\Delta T(t) = \frac{1}{\sqrt{2\pi}\sigma_{\Delta T}} \exp\left(-\frac{t^2}{2\sigma_{\Delta T}^2}\right) . \quad (11.55)$$

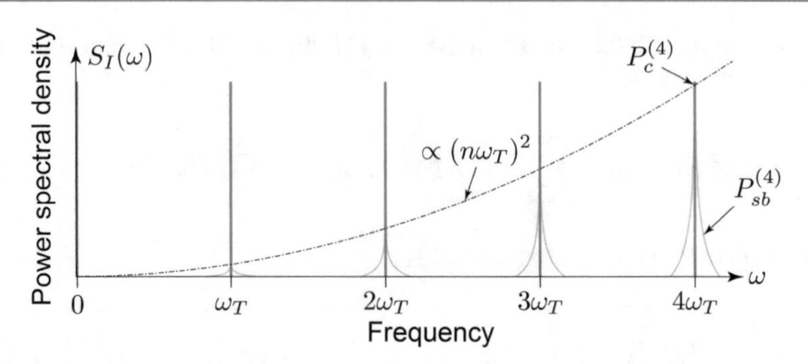

Fig. 11.9 The PSD $S_I(\omega)$ according to (11.51) with timing jitter. In contrast to the ideal laser (Fig. 11.2), additional timing jitter generates sidebands to all δ-pulses, and these increase quadratically with higher harmonics $n\omega_T$. For the lower harmonics, the intensity noise usually dominates

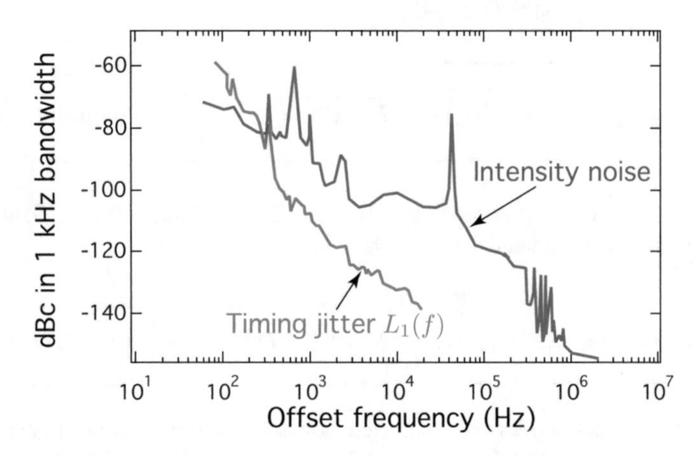

Fig. 11.10 Relative intensity noise and timing jitter of a passively modelocked NaCl color center laser with 275 fs pulse duration, pumped by a cw Nd:YLF laser [89Isl]. The intensity noise is determined by the pump laser. Comparison clearly shows that, around the first harmonic, the fluctuation sidebands are dominated by intensity noise. The timing jitter was measured at the 40th harmonic, where the sidebands $L_{40}(f)$ are dominated by the timing jitter but are still in the linearized region of the fluctuations, i.e., $L_n(f) \propto n^2$

The rms timing fluctuations correspond to a FWHM timing fluctuation $\tau_{\Delta T}$ given by

$$\Delta T(t) = \exp\left[-4\ln 2\left(\frac{t}{\tau_{\Delta T}}\right)^2\right] . \tag{11.56}$$

Comparing (11.55) and (11.56) yields

$$\frac{t^2}{2\sigma_{\Delta T}^2} = 4\ln 2\frac{t^2}{\tau_{\Delta T}^2} ,$$

$$\boxed{\tau_{\Delta T} = \sqrt{8\ln 2}\sigma_{\Delta T} = 2.355\sigma_{\Delta T}} . \tag{11.57}$$

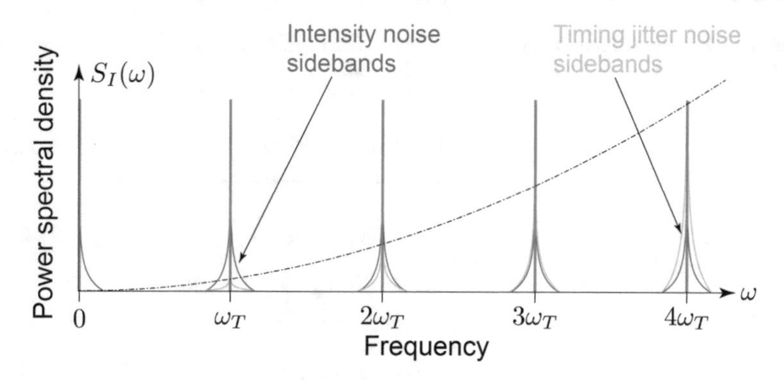

Fig. 11.11 The PSD $S_L(\omega) = S_I(\omega)$ according to (11.59) with intensity noise and timing jitter. At lower harmonics, the intensity noise is usually much stronger, and it therefore dominates the sidebands

The previously measured variance of the timing jitter of 9 ps then results by (11.57) in a FWHM timing jitter for a Gaussian distribution of approximately 21 ps with the frequency range [130 Hz, 20 kHz].

11.3.4 Summary of Intensity Noise and Timing Jitter

In summary, by (11.21) and (11.44), we have

$$\boxed{I(t) = I_0 T\left[1 + N(t)\right] \sum_{n=-\infty}^{+\infty} \delta\left(t - nT - \Delta T(t)\right)}, \tag{11.58}$$

and, under the assumption that the intensity and timing fluctuations are not correlated, the power spectral density of the pulsed laser is given, using (11.32) and (11.51), by

$$\boxed{S_I(\omega) \approx 4\pi^2 I_0^2 \sum_{n=-\infty}^{+\infty} \left\{\delta(\omega - n\omega_T) + \left[\tilde{N}(\omega - n\omega_T)\right]^2 + \frac{1}{4\pi^2} n^2 \omega_T^2 \left[\Delta\tilde{T}(\omega - n\omega_T)\right]^2\right\}}, \tag{11.59}$$

as shown schematically in Fig. 11.11 (Tables 11.4 and 11.5).

11.4 Noise Characteristics of Modelocked Lasers

11.4.1 Some Basic Remarks on Noise, Stabilization, and Coupling Mechanisms

There are some fundamental differences between the characteristics of intensity noise and timing jitter. Gain saturation provides a restoring force for the output power, which normally pulls the output power towards the steady-state solution,

Table 11.4 Summary of intensity noise with the notation in Tables 11.1 and 11.2, and $X(t) = N(t)$, the relative intensity noise. Here we use two-sided PSDs

Parameter	Intensity	Equation
Time domain	$I(t) = I_0(t) + I_N(t) = I_0 T\left[1 + N(t)\right] \sum_{n=-\infty}^{+\infty} \delta(t - nT)$	(11.21)
	$I_N(t) = I_0(t)N(t)$	(11.23)
Mean value	$\langle N(t)\rangle = 0$	
Power spectral density (PSD)	$S_I(\omega) \approx S_{I_0}(\omega) + S_{I_N}(\omega)$	(11.27)
	$S_{I_0}(\omega) = \lvert\tilde{I}_0(\omega)\rvert^2 = 4\pi^2 I_0^2 \sum_{m=-\infty}^{\infty} \delta(\omega - m\omega_T),\ \ \omega_T = 2\pi/T$	(11.20)
	$S_{I_N}(\omega) = \lvert\tilde{I}_N(\omega)\rvert^2 = 4\pi^2 I_0^2 \sum_{m=-\infty}^{\infty} \left[\tilde{N}(\omega - n\omega_T)\right]^2$	(11.31)
Normalized (relative) PSD around one harmonic	$S_N(\omega) = \dfrac{S_{I_N}(\omega)}{S_{I_0}(\omega_T)} = \left\lvert\tilde{N}(\omega - \omega_T)\right\rvert^2$	(11.36)
Variance (with zero mean value)	$\sigma_N^2 = \left\langle N^2(t)\right\rangle = R_N(0) = \dfrac{1}{\pi}\displaystyle\int_0^{\infty} S_N(\omega)\mathrm{d}\omega$	(11.37)
rms noise in $[f_1, f_2]$ (Fig. 11.8)	$\sigma_N\left[\omega_1, \omega_2\right] = \sqrt{\dfrac{1}{\pi}\displaystyle\int_{\omega_1}^{\omega_2} S_N(\omega)\mathrm{d}\omega}$	(11.38)
	$\sigma_N\left[f_1, f_2\right] = \sqrt{\dfrac{P_{sb}\left[f_1, f_2\right]}{P_c}}$	(11.39)
	$\dfrac{P_{sb}\left[f_1, f_2\right]}{P_c} = 2\displaystyle\int_{nf_T+f_1}^{nf_T+f_2} \dfrac{P_{sb}(f)/P_c}{B}\mathrm{d}f$	(11.40)

Table 11.5 Summary of timing jitter noise $X(t) = \Delta T(t)$ with the notation in Tables 11.1 and 11.2. Here we use two-sided PSDs

Parameter	Timing jitter	Equation
Time domain	$I(t) = I_0 T \sum_{n=-\infty}^{+\infty} \delta(t - nT - \Delta T(t))$	(11.44)
Mean value	$\langle \Delta T(t) \rangle = 0$	
Power spectral density (PSD) of $I(t)$	$S_I(\omega) = 4\pi^2 I_0^2 \sum_{n=-\infty}^{+\infty} \left\{ \delta(\omega - n\omega_T) + \frac{1}{4\pi^2} n^2 \omega_T^2 \left[\Delta \tilde{T}(\omega - n\omega_T) \right]^2 \right\}$	(11.51)
Variance (with zero mean value)	$\alpha_{\Delta T}^2 = \langle \Delta T^2(t) \rangle = R_{\Delta T}(0) = \frac{1}{\pi} \int_0^{\infty} S_{\Delta T}(\omega) \mathrm{d}\omega$	(11.52)
rms noise in $[f_1, f_2]$ (Fig. 11.10)	$\Delta T_{\mathrm{rms}}[f_1, f_2] = \sigma_{\Delta T}[f_1, f_2] = \frac{1}{f_T} \sqrt{2 \int_{f_1}^{f_2} L_1(f) \mathrm{d}f} = \frac{1}{n f_T} \sqrt{2 \int_{f_1}^{f_2} L_n(f) \mathrm{d}f}$	(11.53)
	$L_1(f) = \frac{P_{\mathrm{sb}}^{(1)}(f)}{P_{\mathrm{c}}^{(1)}}, \quad L_n(f) = \frac{P_{\mathrm{sb}}^{(n)}(f)}{P_{\mathrm{c}}^{(n)}}$	(11.54)

even under the influence of noise. This has been discussed in Chap. 5 with regard to the relaxation oscillations and in Sect. 9.7 with regard to Q-switching instabilities for passively modelocked lasers. An actively modelocked laser has a timing reference given by the intracavity modulator. Therefore, there is also a restoring force for the residual timing jitter in this case. On the other hand, the timing jitter has no restoring force in a passively modelocked laser, so the pulse repetition frequency undergoes a random walk, which can lead to significant drifts.

Noise can be stabilized with an active feedback loop. For example, the timing jitter can be stabilized to an external reference oscillator with an electronic phase-lock loop, as proposed by D. Cotter in 1985 [87Cot] and demonstrated in 1986 by D. M. Bloom and coworkers [86Rod]. The electronic feedback typically controls the phase of the optical loss modulation for active modelocking or the laser cavity length for passive modelocking. However, in principle, any parameter that affects the pulse repetition rate can be used for such electronic feedback control.

For an electronic feedback control of any noise, we need a control parameter to generate a measurable error signal that can be used to stabilize a specific laser noise (see Sect. 12.6.1 for more details). Examples for control parameters are cavity length, pump power, intracavity wedges for dispersion, tilting of cavity mirrors or prisms, etc. By transfer function analysis (e.g., see Sect. 5.3 with the pump power modulation as the control parameter), we can determine the usefulness and the frequency bandwidth of such control parameters. In this case a small sinusoidal modulation of a possible control parameter as a function of frequency can be used to measure the effect on

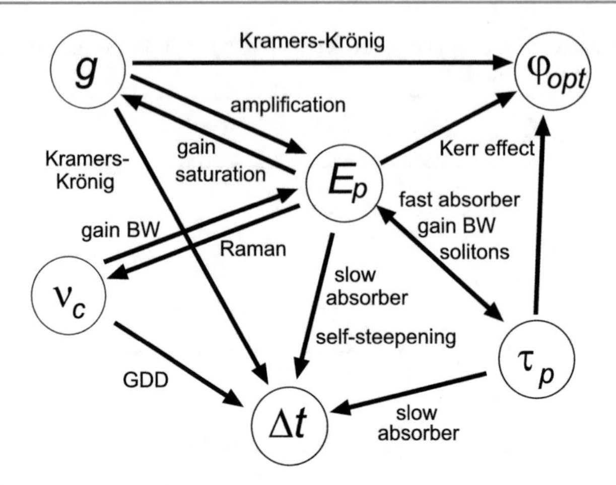

Fig. 11.12 Illustration of noise coupling effects with different interactions of various pulse parameters, according to [07Pas]. Gain coefficient g, pulse energy E_{p}, pulse duration τ_{p}, optical phase φ_{opt}, shift of temporal position of pulse Δt, bandwidth BW, the optical carrier frequency, i.e., the optical center frequency of the pulse ν_{c}. Note that we distinguish between the optical frequency ν and the Fourier frequency (or the 'noise frequency') f of the noise spectrum, whereas ω is either $2\pi\nu$ or $2\pi f$

the quantity that we want to stabilize. This has been done for several SESAM-modelocked solid-state lasers [04Sch]. However, more detailed discussion of the required electronic feedback loop would go beyond the scope of this textbook.

In the following sections, we will present the results of several noise measurements which demonstrate very low noise performance for modelocked solid-state lasers in general. In addition, the noise can be reduced to the quantum limit with additional stabilization.

The underlying noise dynamics can be rather complex because there are significant noise coupling mechanisms which transfer noise of one quantity into noise of some other quantity. In addition, there are nonlinear dynamics and coupling effects. In the following, a few examples of such noise coupling effects are discussed and shown schematically in Fig. 11.12 [07Pas].

As shown in Chap. 5, the evolution of pulse energy and gain is closely tied to the effects of laser amplification and gain saturation: any deviation of the gain from the net loss in the cavity leads to changes in the pulse energy, while larger pulse energies tend to reduce the gain. For example, for solid-state lasers the pulse energy is typically far below the gain saturation energy and the resulting dynamics leads to relaxation oscillations which appear as visible peaks in power spectral density measurements. Note, however, that this dynamics can be coupled to other variables. For example, the gain can also depend on the spectral width of the pulses.

The pulse energy can then couple to the pulse duration and other variables. For example, a higher pulse energy can lead to a lower loss on a saturable absorber (see Sect. 9.7) and a shorter pulse duration with constant pulse energy can lead to a lower loss on a fast saturable absorber, but not on a slow absorber. In lasers with a soliton circulating in the cavity [93Hau], pulse duration and pulse energy are coupled to each other by the interplay of dispersion and nonlinearity. The optical phase can be affected by intensity fluctuations (and thus by fluctuations in the pulse energy or duration) via the Kerr nonlinearity, and also by gain fluctuations due to refractive index changes, as described by the Kramers–Krönig relations.

The timing (temporal position) of pulses is not necessarily only affected by the cavity length. There is a time delay from the slow saturable absorber which depends on the pulse energy and duration (Chap. 7 and Sect. 9.7). The timing can also be modified via the self-steepening effect (Chap. 4) (which is important for very short pulses), by center frequency fluctuations if there is dispersion, and by gain fluctuations via the Kramers–Krönig effects. The center frequency can affect the pulse energy via the frequency-dependent laser gain and/or cavity losses.

11.4.2 Ultrafast Dye Lasers

The power spectral density for the noise was first characterized for synchronous modelocking (Fig. 6.30) with a synchronously pumped rhodamine 6G dye laser in 1984 in the group of D. von der Linde [84Klu]. This is one reason why this technique is sometimes also referred to as the 'von der Linde' technique for noise characterization. However, this technique was previously extensively used for the noise characterization of microwave oscillators.

The pump laser for the synchronously pumped dye laser was an acousto-optically modelocked argon ion laser. The detailed power spectral density measurements revealed that temporal jitter of the actively modelocked pump pulses was directly transferred to the synchronously pumped dye laser [84Klu]. Interestingly, the authors also mentioned that they observed that 'the argon laser pulses can be shifted with respect to the RF drive voltage of the modulation by a few hundred picoseconds' and added that 'no detailed investigation was made' (see [84Klu, p. 276]). This opened the door for a valid laser stabilization patent application by Cotter in 1984, which was granted with a US patent in 1987 [87Cot].

The fact that with synchronously pumped lasers the timing jitter is determined by the modelocked pump laser has also been used to pump two different dye lasers for the generation of two synchronized pulse trains at two independently tunable wavelengths [77Jai]. Jain et al. also reported that pumping two different lasers was advantageous because, when the two wavelengths shared a common gain medium, the minimal wavelength separation was limited by competition instabilities.

Synchronously pumped dye lasers became less important with the invention of the colliding pulse modelocked (CPM) dye laser in 1981 [81For] (see Sect. 9.3.2), which is pumped by a cw argon-ion laser. An optimized CPM rhodamine 6G laser with reliable dispersion compensation [85Val, 86Val] became the workhorse of femtosecond optics and spectroscopy before the Ti:sapphire laser became commercially available around 1989 (see Sect. 9.10.1). The CPM dye laser was first characterized by the 'von der Linde' technique in 1986 [86vdL]. This was also the first time that the noise of a cw modelocked laser was investigated and compared to that of a synchronously pumped dye laser. The actively modelocked argon-ion laser had a 0.15% relative timing jitter $\Delta T/T$ and a 19 ps timing jitter for 0.4 ms, i.e., with a lower integration frequency $f_1 = 2.5$ kHz (see Sect. 11.3). The intensity noise was given with regard to the relative rms pulse energy noise of 0.16% within 0.8 μs, i.e., $f_1 = 5$ kHz, which was then increased to 0.3% for the synchronously pumped dye laser. In contrast, the CPM laser generally showed better long-term power stability. However, in the first preliminary investigation by von der Linde [86vdL], significant amplitude noise peaks at megahertz frequencies were observed with the CPM dye laser, but the origin of this modulation was not understood.

The timing jitter and intensity noise of a CPM dye laser was then investigated in more detail in 1990 [90Nus,91Har]. The measured timing jitter was 1.8 ps rms in a noise frequency bandwidth of $f_1 = 2$ Hz to $f_2 = 1$ kHz, showing superior performance compared to synchronously pumped lasers, taking into account the fact that noise in such lasers substantially increases with a lower f_1 value. The spurious noise peaks in the spectral noise sidebands was observed again (see Fig. 11.13 right) [86vdL]. A more detailed intensity noise characterization showed that these modulations coupled from the cw argon-ion pump laser (Fig. 11.13, left) into the CPM dye laser (Fig. 11.13, right). The frequencies of these substantial noise modulations could be changed with the CPM cavity length and resulted from harmonic mixing of the cavity mode frequencies of pump and dye lasers. The noise peaks could be eliminated by operating the argon-ion pump laser in a single longitudinal mode using an intracavity etalon (Fig. 11.13, left, squares) [90Nus]. Similar pump-laser-induced noise was observed in a passively modelocked color-center laser pumped by a cw Nd:YAG laser [90Kel3]. Such noise peaks can substantially reduce the signal-to-noise ratio in ultrafast measurements and should therefore be avoided (see Sect. 11.6).

Note that, with semiconductor laser diodes, the mode partition noise as observed in the cw argon-ion laser (Fig. 11.13a) has been studied extensively [88Pet2]. It was shown that the intensity noise is substantially increased when the relative portion of the modal power fluctuates in a multi-mode laser with essentially constant

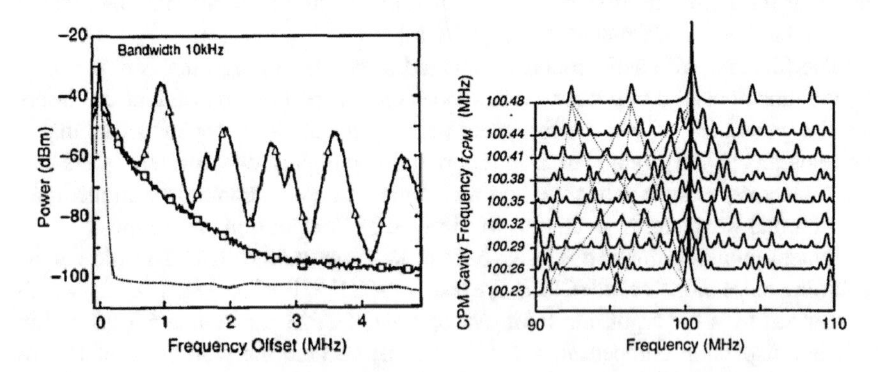

Fig. 11.13 Power spectral density measured with a microwave spectrum analyzer. Intensity noise peaks from a cw argon-ion laser (*left*) used to pump a CPM dye (*right*). The argon-ion noise peaks can be removed with an intracavity etalon. The spurious noise peaks in the CPM dye laser noise power spectrum shift and change into even more peaks when the cavity length is varied. These peaks disappear when an etalon is used inside the argon-ion pump laser. More details are given in [90Nus]

output power. Maximum noise is observed for two-mode emission. However, with an increasing number of lasing modes, the spectral noise density decreased again. Therefore, it is generally preferred to have a single lasing pump mode or then go to the other extreme of very many longitudinal lasing cavity modes. Going to single-mode operation was also the best solution for the 'green problem' [86Bae] for high average power intracavity doubled diode-pumped solid-state lasers.

11.4.3 Flashlamp-Pumped Solid-State Lasers

Early noise measurements were performed on flashlamp-pumped actively mode-locked Nd:YAG lasers [86Rod, 89Rod]. The timing jitter was initially reduced by more than 20 dB in a spectral range of 50–250 Hz [86Rod], by stabilizing the phase of the acousto-optic modelocker driver with a stable reference oscillator. With this phase adjustment, the timing of the modelocked pulse train could be shifted. This result was further improved with better electronic feedback loops, leading to a timing jitter reduction from 20.6 to 0.3 ps rms in [0.25 Hz, 25 kHz] [86Rod].

Figure 11.8 shows an amplitude noise sideband of a flashlamp-pumped actively modelocked Nd:YLF laser which rolls off into the system noise floor at around 1 MHz. Clearly visible is the strong relaxation oscillation peak between 10 and 100 kHz. At lower frequencies, we see discrete spectral lines peak at multiples of 60 Hz (resp. 50 Hz in Europe) and other noise peaks which usually come from the power supply and flashlamp current drivers. In comparison, the timing jitter noise sideband rolls off even faster with increasing frequency and becomes negligible at an upper noise frequency of 100 kHz (Fig. 11.10). These noise sidebands are typical for solid-state lasers with relaxation oscillations well below 100 kHz and negligible noise power for frequencies above several megahertz. This is in contrast to semiconductor diode lasers, which exhibit multi-gigahertz relaxation oscillations (see, e.g., [88Pet2]).

These early spectrally resolved noise measurements showed that both the intensity noise and timing jitter sidebands exhibit very similar spectral noise features, which were introduced by the pump laser in a passively modelocked color center laser (see Fig. 11.14) [90Fin, 90Kel3]. This was later confirmed many times with many different modelocked lasers. Therefore, any low-noise optimization efforts need to start with noise reduction on the pump laser. Furthermore, Fig. 11.14 (right) shows a strong increase in timing jitter noise in the low-frequency part of the noise spectrum. This shows the benefits of an actively modelocked laser which has a restoring force for the timing of the pulse to the external modulation frequency. In contrast to synchronously pumped lasers, the APM laser showed much more low-frequency timing noise in comparison to the modelocked pump laser, because this modelocking technique critically depends on the relative cavity length stabilization of the nonlinear coupled cavity (see Sect. 9.2). This strong increase in the low-frequency timing jitter noise was, for example, not observed for resonant passive modelocking [92Kel1].

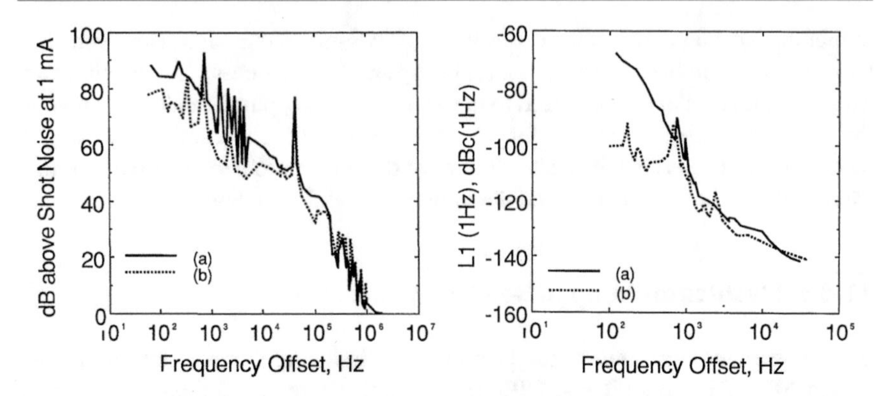

Fig. 11.14 Intensity noise (*left*) and timing jitter (*right*) characterization of an additive pulse modelocked (APM) color-center laser (**a**) pumped by an actively modelocked flashlamp-pumped Nd:YLF laser (**b**) using noise power spectral measurements with a microwave spectrum analyzer [90Kel3]. Measurements were made as described in Sect. 11.3

11.4.4 Diode-Pumped SESAM-Modelocked Solid-State Lasers

Diode-pumping can significantly improve noise performance in comparison to flashlamp-pumped lasers. Special measures are required to reduce the noise of high-power diode laser arrays. Typically, the electrical noise of the laser diode driver is the main source of noise and is transferred throughout the system. However, a narrow absorption linewidth of the solid-state laser can cause some significant excess noise when the optical frequency of the diode pump laser is not stabilized. Fully stabilized diode-pumped SESAM-modelocked solid-state lasers became commercially available around 1995. For low-noise performance, any Q-switching instabilities of passively modelocked solid-state lasers (see Sect. 9.7) have to be avoided, and the laser has to be operated well above the Q-switching modelocking threshold.

Pushing the noise performance to the quantum noise limit requires more effort. To better understand, to reduce noise in free-running lasers, and finally to optimize the electronic feedback stabilization, it is very useful to characterize how the modulation of the pump power affects the output power. Such transfer-function measurements have been made for several cw and SESAM-modelocked solid-state lasers, such as Nd:YVO_4, Nd:YLF, and Er:Yb:glass [04Sch]. The theoretical treatment goes back to the linearized differential equations for small variations around the steady state, as discussed in Chap. 5 and Sect. 9.7. Relaxation oscillations then show up as resonance peaks with a phase shift of $-180°$ in the measured output power of the laser using lock-in detection.

Figure 11.15 left and center panels show the result for a cw and SESAM-modelocked Nd:YVO_4 laser. The dynamic behavior of this laser is as expected from theory. Compared with the cw laser (Fig. 11.15 left), the passively modelocked laser (Fig. 11.15 center) exhibits a weaker damping of the relaxation oscillations, because the SESAM introduces lower loss for higher pulse energies. If we reduce the pump power, we approach a critical point where the damping becomes zero. Below this

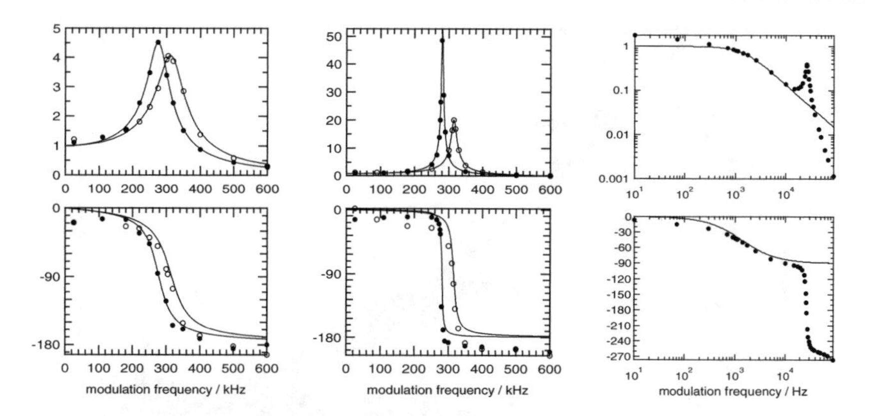

Fig. 11.15 Transfer-function measurements of the output power as a function of pump power modulation frequency. Both the amplitude (*top*) and phase (*bottom*) response are shown as measured with a lock-in detector. *Left*: Cw Nd:YVO_4 laser, *Centre*: SESAM modelocked Nd:YVO_4 laser. Different pump power levels in the low (*solid dots*) and the high (*open dots*) power regime. According to [04Sch, Fig. 4]. *Right*: Transfer-function measurement of a SESAM modelocked Er:Yb:glass laser. According to [04Sch, Fig. 8]

point, fluctuations of the pulse energy grow exponentially and lead the laser into the Q-switched modelocked regime.

Figure 11.15 (right panels) shows the result for the SESAM-modelocked Er:Yb:glass laser, which is significantly different because of the energy transfer from the Yb energy reservoir to the Er ion. This introduces a low-pass filter for perturbations of the pump power. Apart from this, the Q-switching dynamics for the Er:Yb:glass laser are similar to those of a directly pumped Er:glass laser. In addition, from the measured transfer functions, we could see that the relaxation oscillations were damped more strongly than initially predicted. This was explained quantitatively by a nonlinear loss in the SESAM that increases linearly with the incident pulse energy. This effect has indeed been verified directly by saturation measurements on the SESAM. See a more detailed discussion in Sect. 5.3.

High-frequency optical communication applications greatly benefit from low-noise modelocked sources at multi-gigahertz pulse repetition rates [11Hil]. With improved SESAM designs taking into account the roll-over for additional stabilization for Q-switching instabilities, a low-noise SESAM-modelocked diode-pumped Er:Yb:glass laser was demonstrated with pulse repetition rates as high as 10.7 GHz at a center wavelength of 1534 nm, with a pulse duration of 3.8 ps and an average output power of 15 mW [02Kra1]. Tuning over the entire C-band for telecom applications was also demonstrated at 2 GHz. This laser became commercially available, i.e., ERGO-XG, Time-Bandwidth Products, and was used to demonstrate what was at that time the world record 26 Tbit/s optical data transmission rate [11Hil]. The high data rate was achieved with wavelength-division multiplexing (WDM) using 336 subcarriers with a carrier spacing of 12.5 GHz from the SESAM-modelocked Er:Yb:glass laser. The modelocked spectrum was broadened in a highly nonlinear fiber and then equalized in power to generate this broad comb spectrum.

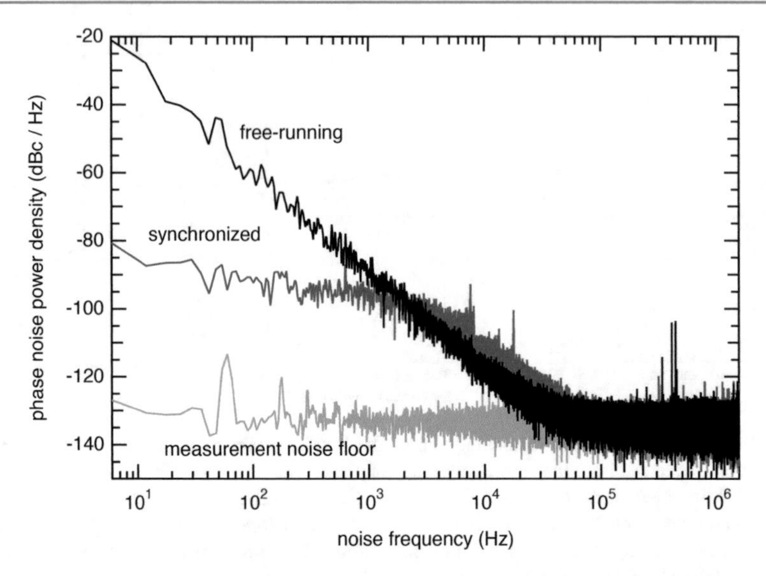

Fig. 11.16 SESAM-modelocked diode-pumped Er:Yb:glass laser with a pulse repetition rate of 10 GHz. Power spectral densities for the measured relative timing jitter of two free-running lasers and two timing-synchronized lasers is shown with the measurement noise floor. The relative timing jitter of the two free-running lasers is 190 fs rms [100 Hz, 1.56 MHz], and for the synchronized lasers is 26 fs rms [6 Hz, 1.56 MHz] [05Sch2]

The noise performance of these lasers has become much better than any microwave oscillators. Therefore two of these 10 GHz SESAM-modelocked Er:Yb:glass lasers have been used to measure the relative timing jitter, as shown in Fig. 11.16, using the technique described in Fig. 11.4 (Sect. 11.2.6). For such a short laser cavity, the relaxation oscillations are expected to be higher. Indeed transfer-function measurements for a 10 GHz SESAM modelocked Er:Yb:glass laser (see [04Sch, Fig. 11]) did not resolve a relaxation oscillation peak, in contrast to the 61 MHz laser shown in Fig. 11.15 (right). This can be explained as follows. The relaxation frequency of the laser scales proportionally to $\sqrt{1/\tau_c}$ [see (5.58)], where $\tau_c = T_R/2l$ is the cavity photon lifetime (see Sect. 5.1), T_R is the cavity roundtrip time, and l is the total amplitude cavity loss coefficient. Therefore, a shorter laser cavity with a smaller T_R generates a higher relaxation oscillation frequency. In addition, the additional low-pass filter from the energy transfer between the Yb and the Er ions tends to dampen these relaxation oscillations, because the coupling with the noise from the pump laser is reduced (Fig. 11.15c). This all helps to reduce the overall noise performance of these lasers.

The extraordinary low-noise performance of modelocked solid-state lasers keeps making further progress, with ever-improving levels of performance. This makes characterizing the intensity noise and timing jitter of modern lasers a challenging task [01Sco]. For example, a fully monolithic SESAM-modelocked Yb:Er:glass laser with a 1 GHz pulse repetition rate showed extremely low output power fluctuations of less than 0.03%, measured over 140 minutes without any further stabilization [16Sho]. Even sub-fs timing jitter has been demonstrated with an additional balanced optical cross-correlator (Sect. 11.2.5) which synchronized a Ti:sapphire laser with a Cr:forsterite laser with a residual timing jitter of 300 as in a noise frequency bandwidth of [10 mHz, 2.3 MHz] [03Sch2].

11.4.5 Argon-Ion Versus Diode-Pumped Ti:Sapphire Lasers

The importance of pump noise for the final laser performance was also confirmed with Ti:sapphire lasers. Initially, they were pumped with cw argon-ion lasers, which were widely available in all laser labs from the previously used CPM dye lasers. Much progress was made with regard to noise and overall efficiency when these argon-ion pump lasers were replaced by high-power frequency-doubled diode-pumped Nd:YVO_4 lasers. The excess noise of these pump lasers, also referred to as the 'green problem' [86Bae], was resolved with single-mode operation.

A comprehensive and direct comparison of how much noise is introduced in a passively modelocked Ti:sapphire laser using either a 5 W argon-ion laser (Coherent I-310) or a 5 W single-mode frequency-doubled diode-pumped solid-state laser (Coherent Verdi-V5) is presented in [01Sco]:

- Both pump lasers have similar low-frequency performance (less than about 50 Hz), but from 100 Hz to 1 MHz, there is up to a 30 dB difference in the amplitude noise. The diode-pumped solid-state laser shows large noise peaks at 100 kHz and its harmonics originating from the switching power supply. These can be removed when powered from a battery source. The measured spectral power densities for the intensity noise in a bandwidth of 1 Hz to 40 MHz results in 0.21% rms for the argon-ion laser and 0.011% rms for the green solid-state laser (see [01Sco, Figs. 9 and 10]). The intensity noise of the two pump lasers rolls off into the shot noise level at about 1 MHz.
- Pump noise is directly transferred to the intensity noise of a Kerr-lens modelocked (KLM) Ti:sapphire laser. The correlation with the intensity noise of the pump lasers (see [01Sco, Fig. 10]) is very strong and the attenuation of the pump noise transfer above 300 kHz due to the 3.2 μs fluorescent lifetime of the Ti:sapphire gain is also evident. The integrated intensity noise in [1 Hz, 40 MHz] resulted in 0.5% rms for the argon-ion laser pumped KLM Ti:sapphire laser and in 0.015% rms for the diode-pumped KLM Ti:sapphire laser. Again the intensity noise of the KLM Ti:sapphire laser rolls off into the shot noise at about 1 MHz.
- Timing jitter is also directly affected by the pump intensity noise (see [01Sco, Figs. 15 and 16]), but is further increased in the low-frequency regime by environmental disturbances such as mount vibrations, air currents, etc. The noise is limited beyond 7 MHz by the shot noise and in the regime of 300 kHz to 7 MHz by the noise of the reference microwave oscillator. For an integration bandwidth of [1 Hz, 40 MHz], the argon-ion pumped laser had a timing jitter of 1.04 ps rms, and the diode-pumped laser 0.74 ps. Again a strong correlation is observed between the intensity noise of pump lasers with the timing jitter of KLM Ti:sapphire lasers, which confirms the strong coupling between different noise quantities in these lasers, as discussed above (see Sect. 11.4.1).

These results demonstrate the much better noise performance of diode-pumped solid-state lasers in comparison to argon-ion lasers. The high power single mode operation of the green pump laser is achieved with a unidirectional ring cavity of a diode-pumped Nd:YVO_4 laser with intracavity frequency doubling. Measurements showed that single-mode operation has much lower noise than multi-mode operation (see [14Sut, Fig. 1a]). The noise reduction for single-mode versus multi-mode operation can be as large as 10 to 20 dB in the noise frequency regime from 1 to 100 kHz, which couples into the Ti:sapphire laser. The pump noise transfer has a low-pass filter with a frequency cutoff around 300 kHz, which is inversely proportional to the 3.2 μs upper state lifetime of the Ti:sapphire laser.

More recently, intracavity frequency doubled optically pumped semiconductor disk lasers (see Sect. 9.8) have reached an average green output power of more than 20 W. Such lasers are commercially available (e.g., Coherent Verdi G Series) and have been used to pump Ti:sapphire lasers. These pump lasers are more compact and less expensive than the diode-pumped single-mode Nd:YVO_4 lasers and show similar noise performance with an intensity noise specification below 0.02% rms in [10Hz, 100 MHz]. These optically pumped semiconductor lasers are operated with many longitudinal modes where the noise is reduced again [14Ver].

11.5 Some Words of Caution About Noise Characterization

11.5.1 General Remarks About Nonstationary Processes and Finite Measurement Durations

Theoretically, we can measure the noise of any random variable X, such as laser output power, timing jitter, etc., both in the time domain with an autocorrelation function $R_X(\tau) = \langle X(t)X(t+\tau)\rangle$, and in the frequency domain with the power spectral density $S_X(\omega)$, as defined in Tables 11.1 and 11.2 and discussed in Sect. 11.1.2. The final noise specification should then result in the same outcome because the power spectral density is the Fourier transform of the autocorrelation function. In reality, however, we can only measure the autocorrelation function over a limited time delay (see Sect. 11.5.2). In addition, we are typically dealing with nonstationary processes and we only measure the random variable over a limited time. Hence, we can only integrate the time-dependent measurements over a limited observation time T_o. We defined the finite-interval average for the mean value in (11.6). Similarly, this can be done for the autocorrelation function and variance.

For stationary noise, the mean value and the variance of any random variable does not change with time, so repeating the same measurement over and over again should result in the same mean and the same variance. Actually, in this case the variance cannot be reduced with repeated measurements because, for a perfectly stationary random variable, the noise is fully correlated. This means that successive measurements are statistically dependent. We can explain this as follows and refer the interested reader to the many more detailed textbooks (such as for example [87Dav]).

Consider a random variable X with mean value $\langle X \rangle$ and variance σ_X^2, as defined in Tables 11.1 and 11.2. Suppose we make N measurements of a random variable X, then define X_n to be the mean value of the n th measurement (or sample) for this variable. We can then define a sample mean M as the arithmetic mean of the N random variables X_n:

$$M \equiv \frac{1}{N} \sum_{n=1}^{N} X_n \,. \tag{11.60}$$

The statistical average of this sample mean M is the same as the mean value of the random variable under study, if all X_n values have the same statistical properties:

$$\langle X_n \rangle = \langle X \rangle \;, \;\; \forall n \implies \langle M \rangle \equiv \frac{1}{N} \sum_{n=1}^{N} \langle X_n \rangle = \langle X \rangle \,. \tag{11.61}$$

This is, for example, the case for a stationary process. We then need to study the convergence properties of the sample mean. We can therefore look at the variance of the sample mean:

$$\sigma^2(M) \equiv \frac{1}{N^2} \sum_{n=1}^{N} \sum_{m=1}^{N} \left[\langle X_n X_m \rangle - \langle X \rangle^2 \right] . \tag{11.62}$$

We now need to address the statistical relationship between the different samples. Assuming the successive experiments are statistically independent, the samples are statistically uncorrelated, which means:

$$\text{uncorrelated samples} \implies \langle X_n X_m \rangle = \langle X_n \rangle \langle X_m \rangle = \langle X \rangle^2 \;, \;\; \text{for}\, n \neq m \,. \tag{11.63}$$

In contrast, we can define correlated samples as follows:

$$\text{correlated samples} \implies \langle X_n X_m \rangle = \langle X^2 \rangle \,. \tag{11.64}$$

Therefore with (11.62), we obtain for the variance of the sample mean:

$$\boxed{\begin{aligned} \text{uncorrelated samples} &\implies \sigma^2(M) = \frac{1}{N} \sigma_X^2 \,, \\ \text{correlated samples} &\implies \sigma^2(M) = \sigma_X^2 \,. \end{aligned}} \tag{11.65}$$

This means that, if the statistical noise is highly correlated between the different measurements (i.e., samples), the variance of the sample mean is approximately equal to the variance of the random variable under study, whatever the number of samples. In this case then, a single measurement provides just as good (or bad) an estimate of the desired mean as does any other number of measurements.

In contrast, when the samples are fully uncorrelated, meaning that the successive measurements are statistically independent, then the variance of the sample mean becomes better with an increasing number of measurements and even goes to zero for $N \to \infty$.

However, it is possible and most likely the case that the value of the sample mean as obtained from a certain number of measurements will continue to be different, i.e., $\langle X_n \rangle \neq \langle X \rangle$. This is very often the case for our laser measurements, which are not necessarily stationary processes. For example, uncontrolled environmental influences strongly affect the low-frequency noise statistics over time, and the pulse arrival time typically has no restoring force in passively modelocked solid-state lasers. Normally, there is a limit on how much the variance of a random variable can be reduced with the number of measurements, because noise between consecutive measurements is not usually perfectly uncorrelated.

It is generally recommended to repeat a measurement many times, especially if the measurement duration is very short. Ideally, the mean value of each measurement should not change, i.e., $M = \langle X_n \rangle = \langle X \rangle$, and if the noise is uncorrelated, we benefit from a better SNR according to (11.65). If the mean value changes between measurements, i.e., $\langle X_n \rangle \neq \langle X \rangle$, we can then treat the mean value as a random variable with a variance that may have some level of statistical correlation between consecutive measurements. How to estimate the variance under these circumstances goes beyond the scope of this textbook. However, for the approximation of statistically independent samples, e.g., a set of many different consecutive measurements with uncorrelated noise, the probability distribution of the sample mean tends to become Gaussian with increasing sample numbers, as long as each sample has a finite mean and finite variance.

With this in mind, we now ask how nonstationary noise will affect our noise measurements. For a nonstationary random variable X, we have a mean and variance that change over time. This means that the variance and the mean from consecutive measurements also become random variables. Therefore, the relative noise power spectral density $\mathrm{RIN}(f) = P_{\mathrm{sb}}(f)/P_{\mathrm{c}}$ measured with a microwave spectrum analyzer is also a random variable (Figs. 11.7 and 11.8). We can see the random variation when we use a swept-tuned microwave spectrum analyzer (see Sect. 11.3.2) to gain some first insight into the noise statistics of the modelocked lasers. It is interesting to measure the noise sideband for different frequency spans with a reasonably large VBW setting, so that the sweep time is not too slow (11.43). Consecutive measurements of $\mathrm{RIN}(f)$ should be the same for stationary processes. The fluctuations in $\mathrm{RIN}(f)$ over many consecutive measurements then result in a Gaussian noise distribution with a certain width $\delta[\mathrm{RIN}(f)]$. This width determines a level of nonstationary noise statistics. It also becomes apparent that, for lower noise frequencies for which we need a smaller RBW and a smaller frequency span, the $\mathrm{RIN}(f)$ fluctuations become greater. It can become so severe that, during the full sweep time, the fluctuations are so large that the noise sidebands become asymmetric. This asymmetry is not possible for a stationary process, because the autocorrelation function is symmetric, i.e., $R_{\mathrm{X}}(\tau) = R_{\mathrm{X}}(-\tau)$, and therefore the noise sidebands need to be symmetric as well. Very often the noise is even highly uncorrelated, and then many successive noise measurements can generate an averaged RIN(f) with a Gaussian noise distribution $\delta[\text{RIN}(f)]$. In such a situation it is better to measure with a large VBW to obtain a faster sweep time and then average the different scans from consecutive measurements.

11.5.2 Noise Measurements in the Frequency Domain

Measuring a lot of noise is easy for modelocked lasers because the noise sidebands around the laser harmonics are well resolved above the system noise floor and are automatically normalized to a relative intensity noise or timing jitter in dBc/Hz (see Sect. 11.3). However, there are many potential problems with low-noise lasers, especially if a fully integrated commercial solution cannot be used. A comprehensive discussion about such measurement challenges is given in [01Sco].

No fully-integrated commercial solution is available for modelocked laser noise characterization at multi-gigahertz and higher pulse repetition rates. However, microwave spectrum analyzers alone can support a much higher frequency range, beyond 50 GHz and approaching 100 GHz. The system noise floor normally goes up with frequency and it becomes harder to obtain very high-dynamic range noise measurements.

11.5.3 Noise Measurements in the Time Domain

The time-domain measurements are in principle based on measurements of the autocorrelation $R_X(\tau) = \langle X(t)X(t+\tau)\rangle$. However, the maximum time delay τ_{max} required for a useful noise measurement is very long for modelocked lasers because the noise typically rolls off into shot noise at around 1 MHz. Thus, for a useful noise characterization, we typically need to integrate the noise well below 1 MHz. For example, for a lower noise frequency measurement starting at 1 Hz (resp. $\ll$ 1 MHz), we need to introduce a $\tau_{\text{max}} = 1$ s (resp. $\gg 1\ \mu$s). Using a simple time-of-flight interferometer, this corresponds to a path length difference in one arm of the interferometer of 3×10^8 m (resp. $\gg$ 300 m), which makes it practically impossible.

Sometimes, the laser to be characterized is used as the input to an interferometer to measure an autocorrelation at the output. The averaging time $T_0(\tau)$ of the autocorrelation signal for a fixed time delay τ is sometimes mistaken for the lower noise frequency f_l in the power spectral density, which is simply wrong. The lower noise frequency is only given by the maximum time delay in the autocorrelation measurement, viz.,

$$f_l = \frac{1}{\tau_{\text{max}}} \ll \frac{1}{T_0(\tau)}, \quad \forall \tau \le \tau_{\text{max}}. \tag{11.66}$$

According to the mathematical definitions in Tables 11.1 and 11.2, we see that the power spectral density is given by the Fourier transform of $R(\tau)$, independently of how long we average the measured $R(\tau)$ for a given time delay.

11.6 Signal-to-Noise Ratio (SNR) Optimization Techniques

11.6.1 Basic Principle of Pump–Probe Measurements

With pump–probe measurements, we can examine fast processes in time, and the time resolution is in principle only limited by the pulse duration of the ultrafast laser. We know from photography that we only obtain a sharp image of a skier in a ski race if we choose a short exposure time. Mechanical shutters for a camera can typically reduce the exposure time to take a picture to well below 0.1 ms. Flash photography can then reduce the exposure time to about 1 μs using an open mechanical camera shutter in a dark room and a short flash of light to take the picture. This can be repeated with different time delays so that a fast process can be observed similarly to a movie, e.g., a balloon bursting in Fig. 11.17.

Using short modelocked laser pulses instead of an incoherent flash of light leads to many more possibilities. In general, these techniques are called pump–probe measurements, because first a stronger pump pulse starts a fast process and after a time delay Δt a weaker probe pulse measures the changes caused by the pump pulse, e.g., through a change in transmission, absorption, polarization, diffraction, etc. (Fig. 11.18). The short pump pulse, for instance, starts a dissociation process in a molecule or an opto-electronic switching process. For example, the goal of measuring time-resolved structural dynamics with rearrangements of both the electronic charge distribution and the position of atoms in any material is currently a driving force behind the development of ever shorter laser pulses, towards the attosecond, and ever higher photon energies, reaching the hard X-ray regime. One can even envision expanding into the gamma ray regime to explore nuclear dynamics in the future. Once something can be observed, there may also be the potential to control the outcome of a dynamical process. The need for time-resolved measurements cannot be over-emphasized for both fundamental science and industrial applications. Making something visible that cannot normally be seen always opens up novel and unexpected possibilities. New discoveries and a better understanding of the underlying

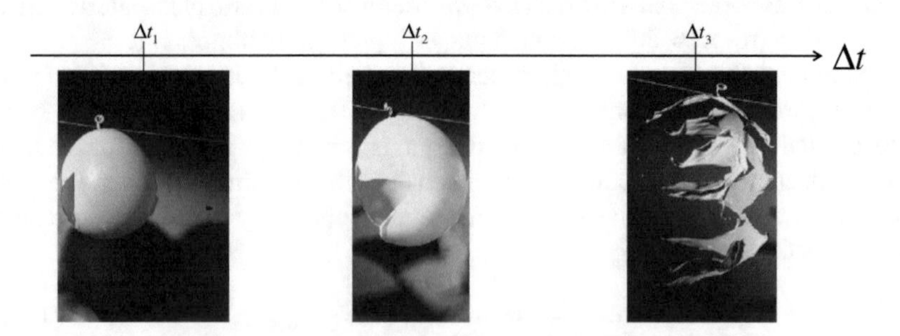

Fig. 11.17 Flash photography of a bursting balloon, taken when a bullet from a gun hits the balloon and a flash of light to record the picture is delayed by a certain time. Increasing this time delay from Δt_1 to Δt_2 and to Δt_3 shows the dynamical process. For each measurement, we assume that we can use an identical balloon under identical experimental conditions. Images from Harold Edgerton, MIT Museum

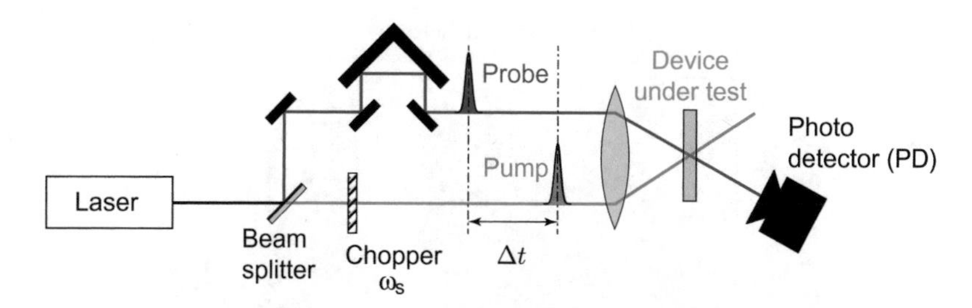

Fig. 11.18 Experimental setup for a laser pump–probe measurement. Both the pump and probe pulse come from the same pulsed laser source with no relative timing jitter. The pump pulse starts the fast process and the time-delayed probe pulse measures the dynamical process in the device under test (DUT). This can be transmission, diffraction, or any other interaction with the probe beam

physical, chemical, and biological processes has been made possible by the progress in ultrafast laser physics and technology. This is all enabled by the ever-improving laser performance with respect to all laser parameters, such as higher pulse energy, higher intensities, shorter pulses, broader and narrower optical spectra, higher and lower photon energies, higher pulse repetition rates, lower noise, etc. Many (but not all) ultrafast measurements are based on some form of pump–probe technique. However, a more detailed review of these techniques would go beyond the scope of this textbook.

The time delay between the pump and probe pulses can be established and controlled by a mechanical delay via the time of flight (Fig. 11.18). For example a 1 μm shift Δz in the mechanical delay line in the probe beam path introduces a time delay of $\Delta t = 2\Delta z/\upsilon_g = 2 \times 3.3$ fs, where we assume that the group velocity $\upsilon_g = c$, the speed of light. For each time delay Δt, we can average the probe signal after interacting with the DUT for as long as required. As long as no additional noise is generated in the relative optical beam path of the pump and probe pulses, we can neglect any timing jitter. The time resolution is then normally determined by the pulse duration or any other timing feature within the laser pulse that is relevant for the measurement.

Note that an extremely short attosecond pulse with a pulse duration of τ_p which has a very broad optical spectrum $\Delta\nu_p \propto 1/\tau_p$ can still resolve very narrow spectral absorption features, in transient absorption spectroscopy, for example. In transient absorption spectroscopy, the probe pulse measures the pump-induced change in the transmission of a DUT. The final spectral resolution for each time delay is given by a combination of the resolution of the spectrum analyzer, the averaging time at the fixed time delay, and the overall laser noise. Thus the spectral resolution $\delta\nu$ is typically much less than $\Delta\nu_p$. An example of this is shown in Fig. 11.19 for attosecond transient absorption.

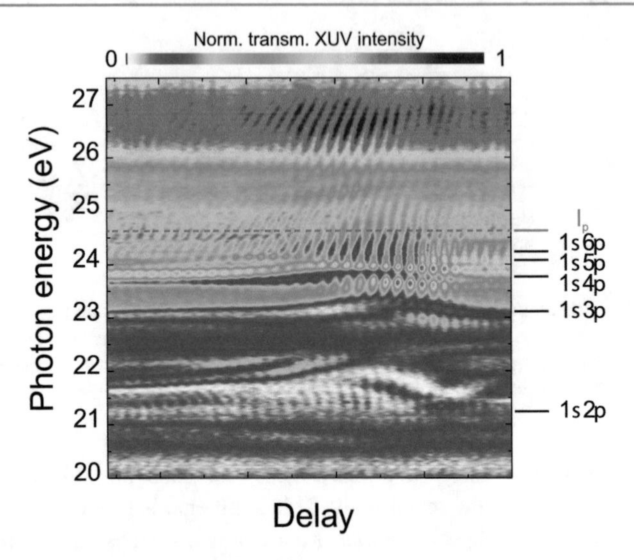

Fig. 11.19 Attosecond transient absorption spectroscopy in He with an infrared pump and attosecond XUV probe pulse over a time delay of about −20 fs to 20 fs [13Luc]

11.6.2 High Frequency Chopping with Lock-in Detection

The chopper in the arm of the pump pulse (Fig. 11.18) introduces an amplitude modulation at the chopper frequency into the signal to be measured, because the dynamical response in the DUT is initiated with this pump pulse. We refer to this frequency as the signal chopping frequency ω_S. This means that we shift the signal within the power spectral density $S_{\text{probe}}(\omega)$ of the probe laser beam as shown schematically in Fig. 11.20. Using a lock-in detection scheme, we can use a narrow detection bandwidth B. The signal-to-noise ratio is then given by the signal power P_s and the noise background within the detection bandwidth at the signal frequency ω_S. Within a small detection bandwidth, we can assume a constant power spectral density for the relative intensity noise of the probe pulse $S_{\text{N,probe}}(\omega_S)$, and the signal-to-noise ratio (SNR) is then given by

$$\text{SNR} = \frac{P_s}{2\int_{f_s-B/2}^{f_s+B/2} S_{\text{N,probe}}(f)\,\mathrm{d}f} \approx \frac{P_s}{2B S_{\text{N,probe}}(f_s)}\,, \tag{11.67}$$

because the signal does not depend on the detection bandwidth, whereas the spectrally integrated noise power $2\int_{f_s-B/2}^{f_s+B/2} S_{\text{N,probe}}(f_s)\mathrm{d}f$ within the detection bandwidth B increases approximately linearly for a small bandwidth. Here we use the two-side power spectral density of the relative intensity noise given by $S_N(f) = P_{sb}(f)/BP_c$, as defined in Sect. 11.3, shown in Fig. 11.8, and summarized in Table 11.4. Note that it is not possible to achieve a better SNR when we chop the probe beam, because the full noise of the probe beam will be shifted to the modulation frequency.

We have seen that the noise typically rolls off with increasing noise frequencies, whence the SNR can be increased for many dBs with high frequency chopping. With acousto-optic modulators, we can easily achieve megahertz chopping frequencies in the visible and infrared spectral region, where the noise sidebands of the ultrafast lasers typically roll off into the shot noise level. This means we can reach a shot-noise-limited SNR and we can neglect any excess laser noise in our measurement.

We have seen with many examples that the spectral noise power densities can have very strong noise peaks around relaxation oscillations, for example. These noise spikes need to be avoided in any case, so it is important to know the noise properties of the laser in our measurements or simply change the chopper frequency to find a better SNR.

According to (11.67), we can improve the SNR using narrowband filtering techniques to reduce the total noise power without reducing the signal level. However, there is a limit given by the bandwidth of the transient signal power $P_s(t)\exp(i\omega_s t)$. In addition, if the signal has significant harmonic content, a fundamental mixing approach, such as synchronous detection, cannot recover the harmonics. In this case, a fast averaging system that measures the periodic signal can be a solution (see Sect. 11.6.4).

11.6.3 Minimal Detectable Signal for Shot-Noise-Limited Detection

The noise rolls off with increasing frequency, as shown for example in Fig. 11.8. Figure 11.20 shows that the signal-to-noise ratio is much better, if the signal to be measured is modulated with a high frequency. For example, to achieve a signal-to-noise ratio of 10^6 in the measurement, one would have to have a signal modulation frequency of approximately 100 kHz for the laser noise in Fig. 11.8. Where does this come from? The double-logarithmic plot in Fig. 11.8 shows the power spectral density of the photocurrent and the photocurrent is proportional to the light intensity. Therefore, for a SNR, we have to use

$$20\log\frac{\mathrm{S}}{\mathrm{N}} = 20\log 10^{-6} = -120\ \mathrm{dBc}\,, \tag{11.68}$$

where SNR = S/N.

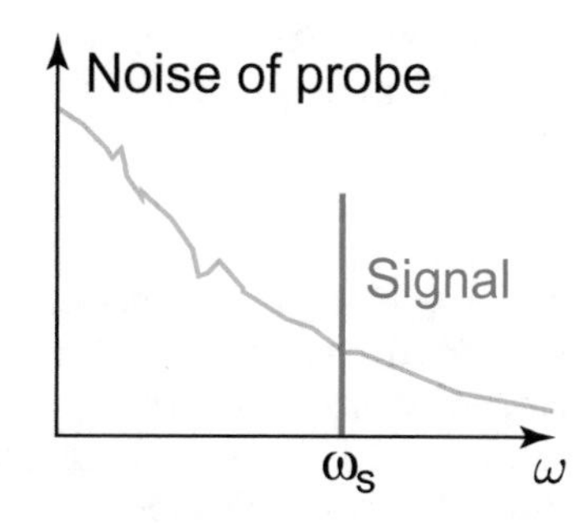

Fig. 11.20 High frequency chopping of the pump beam moves the signal to the noise frequency ω_s with respect to the spectral noise power density of the probe beam

Now we need to address the next question: How can we estimate the minimum detectable signal in a measurement? Ideally, the minimum detectable signal current J_s is equal to the shot noise current, i.e., $J_s = J_{SN}$. Using high-frequency chopping (Figs. 11.14 and 11.20) we can neglect any excess laser noise. Therefore, with a shot-noise-limited detector (see Sect. 11.2.1) and a lock-in detector bandwidth, the minimum detectable signal is determined for SNR = 1 [see (11.67)]. Two examples are discussed in the following.

Minimum Detectable Signal for Electro-optic Sampling

For electro-optic sampling of GaAs integrated circuits, the signal current J_s on a photodetector is given by [88Wei]

$$J_s = \frac{\pi J_0}{V_\pi} V(t) , \tag{11.69}$$

where J_0 is the average photocurrent, V_π is the half-wave voltage, i.e., 5 kV for GaAs, and $V(t)$ is the voltage to be measured in the DUT via the electro-optic effect. Assuming we can neglect any excess laser noise with a chopping frequency well above 1 MHz, the measurement can become shot-noise-limited. The minimal detectable voltage δV is obtained from the condition that $J_s(\delta V) = J_{SN}$, where J_{SN} is the shot noise current given in Table 11.3:

$$\frac{\delta V}{\sqrt{\mathrm{Hz}}} = \frac{V_\pi}{\pi}\sqrt{\frac{2q}{J_0}} \approx 30 \frac{\mu\mathrm{V}}{\sqrt{\mathrm{Hz}}} , \quad \text{for } J_0 = 1 \text{ mA} . \tag{11.70}$$

Assuming we observed some excess laser noise with our power spectral density measurements, such as, for example, 10 dB of excess laser noise, then the minimum detectable signal would increase by a factor of 3.16, i.e., (11.68), 10 dB = $20 \log(3.16)$, to about 90 $\mu\mathrm{V}/\sqrt{\mathrm{Hz}}$.

For a typical time-resolved measurement, we may want to use a 10 Hz scan rate with a 0.1 to 10 kHz measurement bandwidth in a lock-in detector. For example, a 10 kHz measurement bandwidth B increases the minimum detectable signal of (11.70) by an additional factor of $\sqrt{B} = 100$ to about 3 mV.

Minimum Detectable Signal for Optical Charge Sensing

For optical charge sensing in integrated silicon [86Hei] or GaAs [88Kel2] devices, we obtain a charge-induced signal current of

$$J_s = 2J_0 k n' N_s , \tag{11.71}$$

where J_0 is the average photocurrent, k is the vacuum wave vector, $\Delta n = n' N_s$ is the refractive index change for a free carrier density with n' the material dependent proportionality factor, and N_s is the sheet charge density, i.e., the integrated charge density along the probe beam path. For shot-noise-limited detection, we assume $J_s(\delta N_s) = J_{SN}$, which results in a minimum detectable sheet charge density δN_s :

$$\frac{\delta N_{\mathrm{s}}}{\sqrt{\mathrm{Hz}}} = \frac{1}{2kn'}\sqrt{\frac{2q}{J_0}} \approx 3 \times 10^7 \frac{1}{\mathrm{cm}^2\sqrt{\mathrm{Hz}}}\,, \quad \text{for } J_0 = 1\ \mathrm{mA}\,, \tag{11.72}$$

where the number is given for n-doped bulk GaAs with $n' = 6 \times 10^{-21}\ \mathrm{cm}^3$ at a probe wavelength of 1.32 μm [88Kel2]. Using a 10-kHz measurement bandwidth B once again, the minimum detectable signal given by (11.72) will be increased by an additional factor of $\sqrt{B} = 100$ to about 100 electrons/μm^2.

These two examples demonstrate how a good noise characterization allows for a quantitative prediction of the minimum detectable signal in an experiment. This gives clear guidelines for the required minimum chopping frequency and helps to optimize SNR in ultrafast laser measurements using lock-in detection schemes.

11.6.4 Short Measurement Durations for Lower SNR

(a) Basic Principle

The relevant rms intensity noise or the variance depends on the measurement duration (i.e., observation time) T_{o} and detection bandwidth f_2. This means that we have to integrate the power spectral density $S_{\mathrm{N}}(f)$ from $f_1 = 1/T_{\mathrm{o}}$ to f_2. We can then define a signal-to-noise ratio (SNR) given by

$$\mathrm{SNR} = \frac{P_{\mathrm{s}}}{2\displaystyle\int_{f_1}^{f_2} S_{\mathrm{N}}(f)\mathrm{d}f}\,, \tag{11.73}$$

where P_{s} is the signal power, and $S_{\mathrm{N}}(f) = P_{\mathrm{sb}}(f)/BP_{\mathrm{c}}$ is defined in Sect. 11.3, shown in Fig. 11.8, and summarized in Table 11.4. The factor of 2 arises because we use a two-sided PSD.

We have shown before in this chapter that the intensity noise in modelocked lasers typically decreases with increasing noise frequencies and reaches the shot-noise level at around 1 MHz. This means that we can obtain a significant improvement for the SNR when we keep the measurement duration as short as possible. For example, for a measurement duration of 1 ms, we only need to integrate the noise sideband from $f_1 = 1$ kHz. For a measurement duration of only 1 μs, the lower integration limit is $f_1 = 1$ MHz, which means that we start to become shot-noise-limited for measurement durations shorter than 1 μs for most ultrafast solid-state lasers.

Under certain conditions, we can further improve the SNR by repeating the same measurement many times, as discussed in Sect. 11.5.1. Normally, the scan rate in a mechanical delay line over a larger distance is limited. We can solve this problem as discussed in the next section.

(b) Equivalent Time Sampling or Asynchronous Optical Sampling

The time delay scan rate for pump–probe techniques is limited by the mechanical delay line (see Fig. 11.18). For mechanical delay lines, there is a trade-off between scan rate and maximum scan distance. Using for example a mechanical shaker from a loudspeaker will limit the maximum shaking rate to around 20 kHz. Much faster scan rates can be achieved with equivalent time sampling or asynchronous optical sampling, which in principle is the same. Within the more recent ultrafast community, i.e., more the physics than the electrical engineering community, it seems that asynchronous optical sampling (ASOPS) has become the preferred description, even though I personally do not like this (see my comments at the end of this section).

For equivalent time sampling a periodic signal on a fast timescale is converted to a much slower (equivalent) timescale which is easier to measure. This is similar to using a stroboscope light to stop or effectively slow down a fast repetitive motion, e.g., rotating fans, car wheels in movies (sampled by the frame rate of the film), etc. The time resolution is given by the flashlight duration, assuming there is no timing jitter between the periodic signal under test and the periodic repetition of the flashlight.

Equivalent time sampling (see also Sect. 10.1.6) was for example used for direct electro-optic sampling of the voltage inside an integrated GaAs circuit using a modelocked laser in the early 1980s. The pulse repetition rate of a modelocked laser (which replaces the repetitive flashlight mentioned above with much shorter laser pulses) was synchronized to an external microwave reference oscillator (Sect. 11.4.3) (Fig. 11.21) that also synchronizes a high-frequency synthesizer. This high-frequency synthesizer provides the pump signal for the integrated GaAs circuit, and is synchronized to the pulse repetition rate f_0 of the modelocked Nd:YAG probe laser with a frequency $Nf_0 + \Delta f$. For this pump–probe measurement, the time delay is scanned with a rate given by Δf. Initially, high-frequency megahertz chopping with lock-in detection was used to increase the sensitivity in the probed voltage close to the shot-noise limit (see Sect. 11.6.3) [88Wei]. However, much faster scan rates can be applied without lock-in detection, as demonstrated in [92Bla], where a 1024 point waveform is averaged with consecutive waveforms in a coherent fashion, such that N noisy waveforms yield a smooth averaged signal in less than a second.

Equivalent time sampling was also used with two synchronized modelocked lasers with slightly different pulse repetition rates f_{rep} and $f_{\text{rep}} + \Delta f$ to perform pump–probe measurements without a mechanical delay line, referred to as 'pump–probe spectroscopy by asynchronous optical sampling' [87Elz]. For the first demonstration [87Elz], a frequency-doubled modelocked Nd:YAG laser (the pump) was used to synchronously pump a dye laser (the probe) with a slightly detuned pulse repetition rate of 10 kHz. These two synchronized lasers were then used to measure the fluorescence lifetime of rhodamine B in methanol with a decay time of 2.3 ns. The 10 kHz beat frequency produces a repetitive relative phase walk-out of the pump and the probe pulses, which replaces the mechanical optical delay line used in Fig. 11.18 and is shown schematically in Fig. 11.22. Note that a traditional pump–probe measurement over a 10 ns time delay requires a mechanical displacement of 1.5 m, which is very difficult to achieve without mechanical pointing instabilities. In principle, today, such measurements can be done directly with high-speed optical photodetectors and oscilloscopes.

With the progress in active stabilization of two modelocked lasers, asynchronous optical sampling has been demonstrated more recently with two synchronized

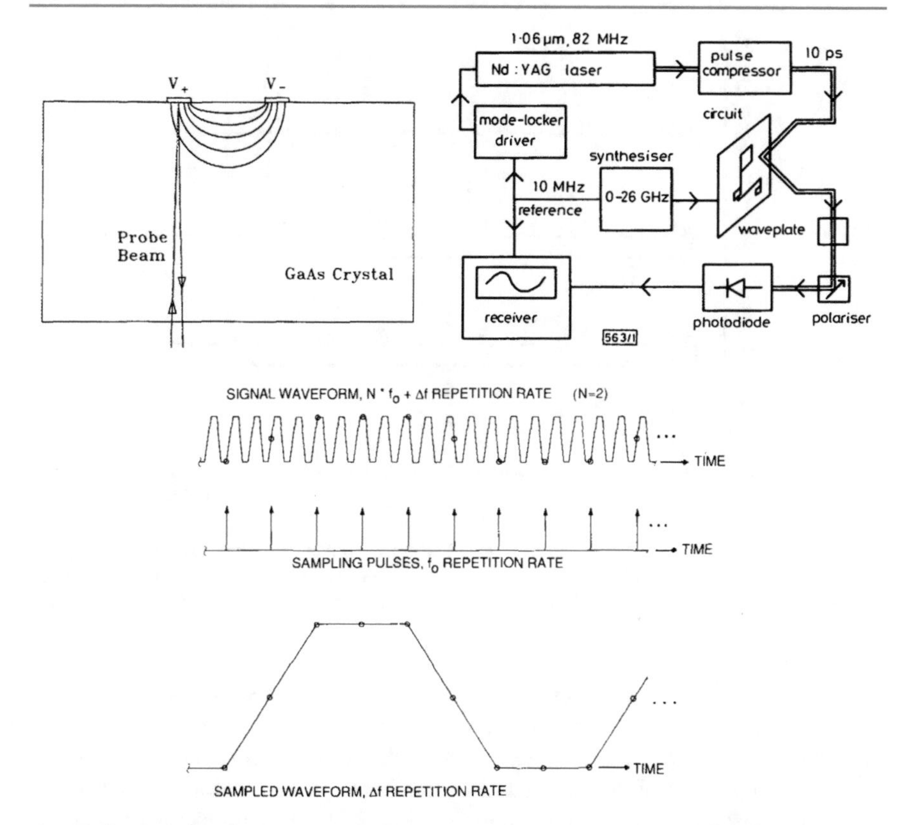

Fig. 11.21 Direct electro-optic sampling according to D. M. Bloom and coworkers in the early 1980s. *Top left*: Voltage measured using the electro-optic effect in the GaAs wafer [85Fre, Fig. 1]. *Top right*: Experimental setup [85Wei, Fig. 1] for equivalent time sampling shown at the *bottom* [88Wei]

femtosecond Ti:sapphire lasers [07Bar]. With the invention of dual-comb modelocking [15Lin] (see Sect. 9.10), two modelocked pulse trains with a tunable pulse repetition rate difference can be generated from a single laser without the need for any active stabilization. This substantially reduces the experimental effort involved in such measurements.

For a first demonstration of a dual-comb modelocked laser for asynchronous optical sampling [20Wil2], a diode-pumped dual-comb Yb:CaF_2 laser was used with 175 fs pulses at a center wavelength of 1050 nm and with two modelocked pulse trains around 137 MHz with a pulse repetition rate difference of 1 kHz. In this case, equivalent time sampling measurements were made on two different devices under test (DUT), without any further active stabilization of the relative timing jitter. The first DUT was a SESAM (see Chap. 7 for more on the DUT) and the nonlinear reflectivity was measured for different pump intensities. The second DUT was a VECSEL chip (see Sect. 9.8 for more on the DUT) for which the 1 μm pump pulse pumped the VECSEL gain via two-photon absorption, building up a population inversion within a few picoseconds, and within the same measurement scan, a

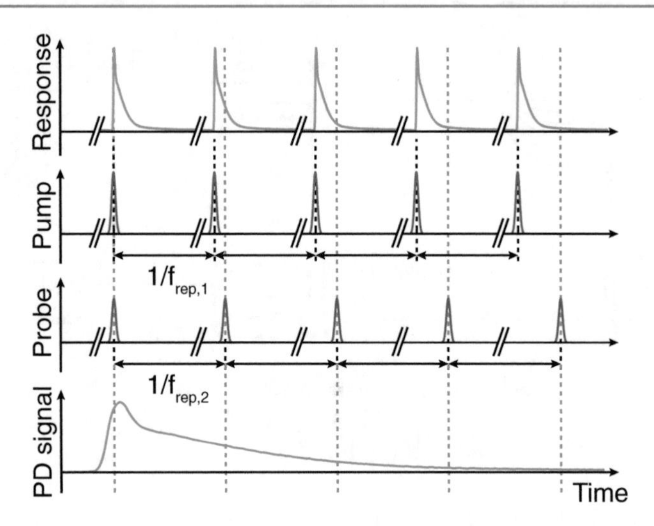

Fig. 11.22 Equivalent time sampling, resp. asynchronous optical sampling, where a pump pulse train with a pulse repetition rate $f_{\text{rep},1}$ starts a transient response in a device under test (DUT), which is then probed by a probe pulse train with $f_{\text{rep},2}$. The response is then measured in the equivalent time of $1/\Delta f$, with $\Delta f = |f_{\text{rep},1} - f_{\text{rep},2}|$

1.9 ns gain lifetime was also measured. These two measurements clearly showed the benefit of equivalent time sampling with (i) a very long scan delay, only limited by the pulse repetition period, i.e., $f_{\text{rep}} \approx 100$ MHz, which corresponds to 10 ns; (ii) very high temporal resolutions, i.e., autocorrelation of pulse durations around 100 fs; and (iii) a very short measurement duration, i.e., $1/\Delta f \leq 1$ ms, during which the relative timing jitter in dual-comb modelocking is negligible in comparison to the time resolution.

In conclusion, for both equivalent time sampling and asynchronous optical sampling, the probe scans the signal within an equivalent-time duration of $1/\Delta f$ (Figs. 11.21 and 11.22) and the real-time resolution is limited to the optical cross-correlation of the pump and probe pulse (which is a combination of the two pulse durations and their relative timing jitter). For example, a Δf of 1 kHz samples the signal within an equivalent time on the order of 1 ms, for transient signals that can occur in real time in the sub-picosecond regime. The higher the value of Δf, the faster the measurement and the better the SNR [see (11.73)]. It then becomes advantageous to average many such measurements for a better SNR, as shown before in [92Bla].

Why is asynchronous sampling a misnomer in my opinion? Note that 'asynchronous' is the opposite of 'synchronous.' However, in asynchronous sampling we still need to synchronize two repetitive signals with a fixed offset frequency, so they are in principle still synchronized. Moreover, in principle, two free running modelocked lasers with no synchronized pulse trains can still be used for pump–probe measurements when many single-shot pump–probe measurements are made in combination with a single-shot measurement for the relative time delay between pump and probe pulse. With a large data set and corresponding data analysis which puts them all in the correct order of increasing time delays, we can also obtain a full pump–probe signal. In this case there is no synchronization at all between the pump and the probe pulse.

12 Optical Frequency Comb from Modelocked Lasers

12.1 Introduction

The shortest pulses from modelocked solid-state lasers routinely reach pulse durations comparable to an optical cycle (see Sect. 9.10.1). For example, an optical cycle is 2.7 fs at a center wavelength of 800 nm. A 5 fs pulse then has only about two optical cycles underneath the pulse envelope. In this regime, a potentially noisy phase $\varphi(t)$ of the optical carrier $E(t) = A(t)\exp[\mathrm{i}\omega_\mathrm{c}t + \mathrm{i}\varphi(t)]$ with respect to the pulse envelope $A(t)$ substantially changes the peak electric field at a carrier frequency ω_c (Fig. 12.1). The exact position of the carrier, i.e., the electric field $E(t)$, below the pulse envelope $A(t)$, i.e., the carrier envelope offset (CEO), needs to be controlled. This becomes important in many nonlinear optical processes, and even becomes critical for highly nonlinear processes such as high-harmonic generation (HHG). The first step in HHG is based on tunnel ionization, which depends exponentially on the peak electric field amplitude. Attosecond pulse generation and attosecond science in general often critically depend on having a stable CEO phase.

Normally, no locking mechanisms exist between envelope and carrier inside a modelocked laser oscillator, and the relative phase offset between $E(t)$ and $A(t)$ experiences random fluctuations. In this chapter, we discuss the stabilization of this carrier-envelope offset (CEO) phase. The solution to this CEO problem [99Tel] not only solved the CEO phase problem for few-cycle pulses, but also enabled a fully stabilized frequency comb of modelocked lasers (which will be discussed in more detail in this chapter). This triggered a revolution in frequency metrology, where a complicated harmonic frequency chain was replaced by a single modelocked laser oscillator. This also brought together two research communities, which had rarely interacted previously. The frequency metrology community was interested in narrow-band single-frequency lasers, and the ultrafast laser science community in broadband ultrafast lasers. Both fields were relatively mature around the year 2000, with high-performance single-frequency and modelocked lasers, and with a well-established

© The Author(s), under exclusive licence to Springer Nature Switzerland AG 2021

U. Keller, *Ultrafast Lasers*, Graduate Texts in Physics,
https://doi.org/10.1007/978-3-030-82532-4_12

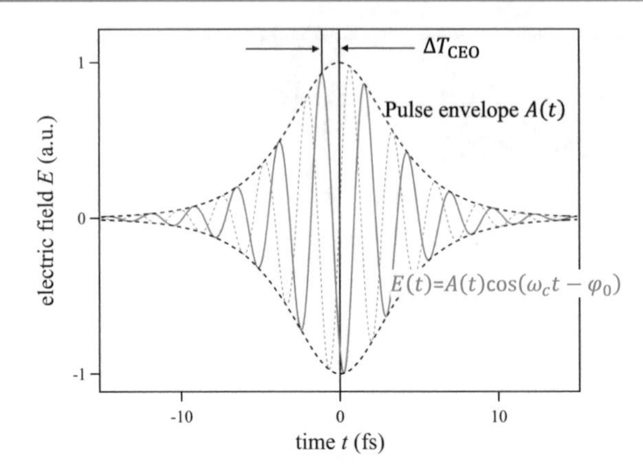

Fig. 12.1 Modelocked pulse in the time domain with a sech envelope function and for two different time-independent constant φ_0 values for the phase of the electric field (*red solid* and *dashed lines*) according to (12.4). The *black dashed line* $\pm|A(t)|$ shows the temporal boundary of the wavepacket. The carrier envelope offset (CEO) time ΔT_{CEO} is introduced [99Tel]

understanding in both the time and frequency domains in both communities. Properly referencing prior work was sometimes a challenging task, occasionally working against the goals of most scientists to be the first in their field. For example, see the foreword by Hall and Hänsch (Nobel Prize in Physics in 2005), published in 2005 [05Ye], where the prior work of [99Tel] was not referenced. Often, even earlier work on laser stability is neglected or overlooked. This textbook has tried to include at least some of these earlier references.

In this chapter, we first discuss the CEO phase noise, which means that we are interested in the optical phase and optical frequency noise. We also put the optical phase and frequency noise in the proper context, with the intensity noise and timing jitter discussed in Chap. 11, going back to the 1980s in the ultrafast community, when the modelocked optical frequency comb spacing was stabilized for the first time [86Rod,87Cot]. Since the typical modelocked pulse duration was much longer, the CEO frequency was essentially negligible at that time. As will be seen, the same mathematical concepts with regard to power spectral densities (PSDs) as introduced in Sect. 11.3 are applied to the frequency noise PSD $S_{\delta\nu}(f)$ and phase noise PSD $S_{\varphi}(f)$. We also show that the optical phase noise broadens the optical linewidth, which is given by the electric field PSD $S_E(\omega) = |\tilde{E}(\omega)|^2$. Many measurement applications greatly benefit from dual-comb generation, which will also be discussed at the end of this chapter.

The revolution in optical frequency comb (OFC) generation started around 2000 with three key publications [99Tel,00Apo,00Jon] and has remained a very active field. The goal in this textbook is to provide a basic introduction to assist in understanding or entering this field. The emphasis is on ultrafast solid-state lasers, which typically provide the best overall noise performance. A detailed review of all ongoing research goes beyond the scope of this textbook. Specific examples are shown to help explain the underlying physics.

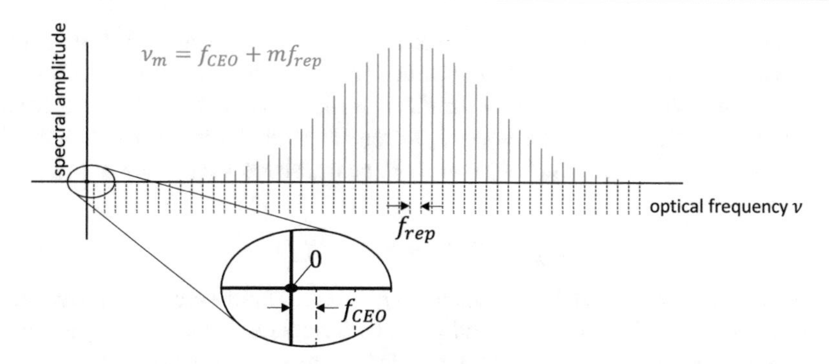

Fig. 12.2 Optical frequency comb (OFC) of a modelocked pulse train with a comb spacing given by the pulse repetition rate f_{rep} and a comb offset given by the carrier envelope offset frequency f_{CEO}. The optical comb lines (or modes) are given by $\nu_m = f_{\text{CEO}} + mf_{\text{rep}}$, where m is an integer number [99Tel]

For stable OFC generation we need a modelocked laser with low intensity and timing jitter noise as discussed in Chap. 11. The timing jitter determines the frequency comb spacing (i.e., f_{rep} in Fig. 12.2). In this chapter we learn how to measure and stabilize the frequency comb offset (i.e., f_{CEO} in Fig. 12.2). So far we have learned that Er:Yb:glass lasers are special with regard to their relaxation oscillations, which are strongly suppressed by the Yb to Er pump energy transfer, as discussed in Sect. 5.3 and shown in Fig. 5.17. This explains the extremely low-noise performance of SESAM modelocked diode-pumped Er:Yb:glass lasers [05Sch2, 16Sho], also shown in Fig. 11.16. We therefore use in this chapter a SESAM modelocked Er:Yb:glass laser as a tutorial example for OFC generation.

Some Words of Caution with Regard to Optical and Noise Frequencies

The difference between the optical spectrum and the frequency noise spectrum often causes some initial confusion. Note that $S_E(\omega) = |\tilde{E}(\omega)|^2$, or $S_E(\nu)$ with $\omega = 2\pi\nu$, determines the optical laser spectrum, whereas $S_{\delta\nu}(\omega)$, or $S_{\delta\nu}(f)$ with $\omega = 2\pi f$, determines the frequency noise spectrum, where ν is the optical frequency and f is the noise or Fourier frequency of the noise spectrum with $f \ll \nu$.

For example, an optical wavelength of 500 nm in vacuum has an optical frequency of $\nu = c/\lambda = 600$ THz, whereas typical laser noise rolls off at frequencies beyond a few megahertz (see Chap. 11). We will show here that this is also the case for the CEO phase noise. Noise frequencies are in the RF/microwave regime, i.e., below 10 MHz, and the optical frequencies in the 600 THz regime. For educational reasons, we also emphasize the difference here with different symbols, i.e., f for noise frequencies of the noise spectrum and ν for optical frequencies of the optical spectrum. However, we use the same symbol for the angular frequency $\omega = 2\pi f$ and $\omega = 2\pi\nu$. Note that, in other literature, this difference between ν and f is not applied consistently. For the CEO frequency, both symbols also make sense, i.e., ν_{CEO} or f_{CEO}, because the optical spectrum is given by $\nu_m = f_{\text{CEO}} + mf_{\text{rep}}$ (Fig. 12.2). In this chapter, we use f_{CEO}, as this has become more common in publications and because $f_{\text{CEO}} \leq f_{\text{rep}}/2$, i.e., in the RF/microwave frequency regime of the pulse repetition rate f_{rep}.

We continue to use the notation as introduced in Tables 11.1 and 11.2. Both two-sided and one-sided power spectral densities (PSD) are used [78Rut]. In Chap. 11, we always use the two-sided PSD for the intensity noise (Table 11.4) and timing jitter (Table 11.5), because the noise spectrum is normally measured around the modelocked laser harmonics. The results can be translated very easily with a factor of 2, i.e.,

$$2S_X^{\text{two-sided}}(f) = S_X^{\text{one-sided}}(f) , \tag{12.1}$$

where the one-sided PSD is only defined for $f \geq 0$. It is more common for frequency noise to show the PSD as a one-sided PSD. Unfortunately, this is not always carefully pointed out in other literature. Note also that the ultrafast community typically uses a two-sided PSD, and the frequency metrology and the electrical engineering community typically uses a one-sided PSD—but again not consistently.

Although ω is actually the angular frequency with units of rad/s, we normally also refer to angular frequency as frequency. Furthermore, we normally use the same function symbol S_X, even though the functions $S_X(\omega)$ and $S_X(f)$ are mathematically not identical, when we make a variable substitution of ω to $\omega = 2\pi f$.

In an optical spectrum analyzer, we measure the optical spectrum either as a function of frequency, i.e., $S_E(\nu) = |\tilde{E}(\nu)|^2$, or as a function of wavelength λ, i.e., $S_E(\lambda) = |\tilde{E}(\lambda)|^2$. In this case the relationship is a bit more complicated because $\omega = 2\pi c/\lambda$, and with $\mathrm{d}\omega = -2\pi c \mathrm{d}\lambda/\lambda^2$, we obtain

$$S_E(\lambda) = 2\pi \frac{c}{\lambda^2} S_E(\omega = 2\pi c/\lambda) . \tag{12.2}$$

This can be confirmed with the condition that the total power is the same if we integrate the wavelength or the frequency, i.e.,

$$\int S_E(\omega)\mathrm{d}\omega = \int S_E(\nu)\mathrm{d}\nu = \int S_E(\lambda)\mathrm{d}\lambda , \tag{12.3}$$

with the additional physical condition that the total power is a positive number and, according to the definition of the PSD in Table 11.1, $S_E(\omega) = |\tilde{E}(\omega)|^2$, $S_E(\nu) = |\tilde{E}(\nu)|^2$, and $S_E(\lambda) = |\tilde{E}(\lambda)|^2$. Again we use the same function symbol $\tilde{E}$.

12.2 Carrier Envelope Offset (CEO) Phase and Frequency

In this chapter we are interested in the optical phase noise $\varphi(t)$ of the electric field $E(t)$ as given in Chap. 2:

$$E(t) = A(t) \exp[\mathrm{i}\omega_{\mathrm{c}} t + \mathrm{i}\varphi(t)] , \tag{12.4}$$

where $E(t)$ is the electric field, ω_{c} is the (angular) carrier frequency, i.e., the center frequency or center of gravity of the modelocked laser spectrum, $A(t)$ is the pulse envelope (see Chap. 2), and $\varphi(t)$ is an additional phase that can be a slowly varying real function of time. Here we neglect any intensity noise and timing jitter, which is discussed in Chap. 11. Note that timing jitter is sometimes given in the form of phase noise. However, this phase noise of the arrival time of the pulse envelope should not be mixed up with the optical phase noise $\varphi(t)$, as discussed in Sect. 12.4.5.

An example of a pulse of the form given in (12.4) with two different time-independent phase values $\varphi(t) = \varphi_0$ is shown in Fig. 12.1, where we have introduced the carrier envelope offset (CEO) time ΔT_{CEO} which determines the time offset of the peak of the electric field in relation to the peak of the pulse envelope and defines a CEO phase ϕ_{CEO} [99Tel]:

$$\Delta T_{\mathrm{CEO}} = \phi_{\mathrm{CEO}}/\omega_{\mathrm{c}} \; . \tag{12.5}$$

We assume that the pulse envelope $A(t)$ does not contribute directly to the phase fluctuations and we neglect any intensity noise for now. We normally assume that the mean value of the phase noise has a zero time average:

$$\langle \varphi(t) \rangle = 0 \; . \tag{12.6}$$

The instantaneous angular frequency is given by

$$\omega(t) = \frac{\mathrm{d}}{\mathrm{d}t}[\omega_{\mathrm{c}} t + \varphi(t)] = \omega_{\mathrm{c}} + \dot{\varphi}(t) \equiv \omega_{\mathrm{c}} + \delta\omega(t) \; , \tag{12.7}$$

which then has a frequency fluctuation $\delta\omega(t) \equiv \dot{\varphi}(t)$. This clearly shows that the phase noise and the frequency noise are directly related and not independent of each other. From (12.4) and (12.7), it follows that $\varphi(t)$ is the instantaneous phase angle of the oscillator with respect to an ideal oscillator of frequency ω_{c}, and $\delta\omega(t) \equiv \dot{\varphi}(t)$ is the frequency departure away from ω_{c}.

In the previous chapters, the theoretical treatment for modelocking was based on differential equations in the complex pulse envelope $A(t)$ with the steady-state condition that the pulse envelope does not change after one cavity roundtrip [see, e.g., (6.50)]. We also used the slowly-varying envelope approximation (see Sect. 2.5.2), which means that no restoring force on the exact position of the electric field within the pulse envelope was introduced. Therefore we would expect $\varphi(t)$ in (12.4) to fluctuate and vary during the pulse propagation within the laser cavity. This generates a pulse train with random fluctuations in the CEO time ΔT_{CEO} (Fig. 12.1) and the CEO phase ϕ_{CEO} of (12.5).

As further progress was made with ever shorter modelocked pulses from KLM Ti:sapphire lasers at a carrier wavelength of around 800 nm, the pulse duration reached a regime comparable to the optical cycle of the carrier frequency, i.e., 2.7 fs at 800 nm (see Table 9.5). In the near-infrared spectral region for sub-10-femtosecond pulse durations, the exact position of the electric field within the pulse envelope (Fig. 12.1) starts to significantly affect the peak electric field amplitude within the laser pulse. The presence of arbitrary fluctuations in the CEO phase therefore introduces significant amplitude noise in the electric field. This noise will strongly affect nonlinear interactions and therefore requires CEO phase stabilization.

This challenge was recognized within the ultrafast community and was a hot topic during a Gordon Conference on nonlinear optics and lasers in 1995. Many of the leading ultrafast laser experts were present, but no solution was available at that time. Some experts even referred to this challenge as the 'holy grail' for ultrafast laser physics.

A first attempt at a solution was based on an optical Michelson interferometer which gives direct access to the phase [96Xu1]. In this joint publication between the groups led by Krausz and Hänsch, they proposed a Michelson-type interferometer to measure the interferometric cross-correlation of successive pulses from a KLM Ti:sapphire laser to give access to the phase noise $\varphi(t)$ in (12.4). However, this was not a solution for the 'holy grail' challenge because, within an optical interferometer, only the noise between very closely spaced pulses can be measured with sufficient mechanical stability. This was in principle well understood from the timing jitter measurements in the early 1980s (see Sect. 11.5.3). There is normally negligible phase noise between consecutive pulses in a modelocked pulse train, because, similarly to intensity noise and timing jitter, the noise is expected to roll off into a negligible level at a few megahertz. For noise frequencies much less than 1 MHz, we need to introduce a path length difference in one arm of the interferometer of well over 300 m (see Sect. 11.5.2), which cannot possibly be kept stable. A long fiber cable for the delay line would be more stable, but cannot be used because of the nonlinear interaction of the modelocked pulses during the propagation inside the fiber. The authors in [96Xu1] also realized this problem with their approach and could therefore only make some theoretical predictions for a potential low-frequency CEO phase noise from a nonlinear model of coupling between the measured pulse energy noise and the CEO phase noise. Such a model is highly complex (see Sect. 11.4.1) and therefore should be treated with some caution. Their theoretical prediction, however, confirmed our initial expectations that all the relevant CEO phase noise is expected to be below a few megahertz.

The solution for the 'holy grail' challenge was published in 1999 [99Tel]. In this case the low noise frequencies of the fluctuating CEO frequency $f_{\mathrm{CEO}}(t)$ can be directly measured in either the time or frequency domain, and gives access to the full CEO frequency noise power spectral density $S_{\mathrm{CEO}}(f)$. From timing jitter measurements in the 1980s (see Sect. 11.2.2), it was well known in the ultrafast laser community that the frequency noise can be measured with an electrical or optical discriminator. Such a discriminator transforms the frequency fluctuations into amplitude fluctuations, which can be analyzed for example via a Fourier transform in many commercial instruments. Two important new ideas were introduced to solve the 'holy grail' problem, providing a solution for the following two fundamental questions and demonstrating a first measured CEO frequency beat signal [99Tel]:

1. What is the experimental parameter that gives access to the full CEO phase noise spectrum? In the frequency domain the modelocked pulse train corresponds to an optical frequency comb (OFC) defined by two frequency parameters. For a modelocked laser, this OFC is given by the optical frequency $\nu_m = f_{\mathrm{CEO}} + m f_{\mathrm{rep}}$, where m is a large integer (Fig. 12.2). The comb spacing is equal to the pulse

repetition rate f_{rep}. The comb offset is given by the carrier envelope offset (CEO) frequency f_{CEO}, which is directly related to the time derivative of the CEO phase.
2. How can the CEO frequency be measured? Several methods have been proposed. The simplest approach is based on a self-referenced f-to-$2f$ interferometer, which requires an octave spanning optical spectrum and only one nonlinear process. Note that in this context an f-to-$2f$ interferometer is referred to as self-referenced when the fundamental and its second harmonic spectrum is used for the interference signal. This then results in a CEO frequency beat signal that can directly be measured with a microwave spectrum analyzer. The required microwave bandwidth is less than the pulse repetition rate f_{rep}.

With these fundamental ideas, the CEO frequency has become a measurable parameter, which is the time derivative of the CEO phase. The measured fluctuating CEO frequency $f_{\text{CEO}}(t)$ typically shows strong fluctuations in a free-running laser and can be used as an error signal for stabilizing the CEO frequency. Note that the CEO phase is generally not stabilized; it is the pulse-to-pulse CEO phase shift that is stabilized by locking f_{CEO}. If f_{CEO} is stabilized to $\mathrm{f_{rep}}/4$, it means that a similar pulse (same CEO phase) is retrieved every 4th pulse, but the CEO phase changes from pulse to pulse.

Therefore the problem is reduced to optical frequency or phase noise characterization and stabilization, a well understood problem for microwave and optical frequency metrology. There is strong nonlinear coupling between CEO noise, intensity noise, and timing jitter, so a modelocked laser had to be optimized for low-noise performance before a more stable CEO beat signal could be observed. In any case, we know from many experiments in physics and electronics that, once an error signal of a fluctuating variable has become measurable, there are always experimental parameters that can be found to control and stabilize such an error signal. The question then remains as to whether such an experimental control parameter has the necessary bandwidth to suppress all the relevant noise frequencies (see Sect. 12.6).

In Chap. 2, we discussed the fundamental concept that the electric field underneath the pulse envelope is propagating with the phase velocity υ_{p} and the envelope with the group velocity υ_{g}. Therefore, any control parameter that changes the phase and the group velocities by a different amount has the potential to control the CEO frequency. How effective such control parameters are can be investigated using transfer-function measurements, as introduced in Sect. 5.3 for pump power modulations and discussed with regards to intensity noise in Sect. 11.4.4. The stabilization guidelines will be discussed in more detail in Sect. 12.6.

Generally, $\upsilon_{\text{g}} \neq \upsilon_{\text{p}}$ because the group and phase velocities inside the cavity both depend on the material and other dispersion effects such as angular dispersion inside the cavity. When a pulse propagates through a dispersive medium of length L, its Fourier components experience a phase shift

$$\varphi = -\int_0^L \frac{n(\omega, x)\omega}{c} \mathrm{d}x \ , \tag{12.8}$$

where n is the refractive index and c the vacuum speed of light. We are interested in the fluctuations in the position of the electric field underneath the pulse envelope. We therefore want to determine the difference in the phase shift between the group velocities and the phase velocities, i.e., the group phase offset (GPO) $\Delta\varphi_{\mathrm{GPO}}$.

From the fundamental definitions of the phase velocity $\upsilon_{\mathrm{p}} = \omega/k$ and group velocity $\upsilon_{\mathrm{g}} = \mathrm{d}\omega/\mathrm{d}k$, we find that

$$\frac{1}{\upsilon_{\mathrm{g}}} = \frac{\mathrm{d}k}{\mathrm{d}\omega} = \frac{\mathrm{d}}{\mathrm{d}\omega}\left[n(\omega)\frac{\omega}{c}\right] = \frac{1}{c}\left[\frac{\mathrm{d}n(\omega)}{\mathrm{d}\omega}\omega + n(\omega)\right] = \frac{1}{\upsilon_{\mathrm{p}}} + \frac{\mathrm{d}n(\omega)}{\mathrm{d}\omega}\frac{\omega}{c} . \quad (12.9)$$

Combining (12.8) and (12.9), we can calculate the phase lag between the pulse envelope and the carrier by integrating along the beam path of one cavity roundtrip:

$$\Delta\varphi_{\mathrm{GPO}} = \omega \int_0^L \left(\frac{1}{\upsilon_{\mathrm{g}}} - \frac{1}{\upsilon_{\mathrm{p}}}\right) \mathrm{d}x = \int_0^L \frac{\omega^2}{c} \frac{\mathrm{d}n(\omega, x)}{\mathrm{d}\omega} \mathrm{d}x . \quad (12.10)$$

It is striking that $\Delta\varphi_{\mathrm{GPO}}$ only arises from the first-order dispersion of the material along the beam path. In a laser cavity, the GPO contributions of all intracavity components add up. For a typical Ti:sapphire laser cavity, the GPO contribution of the crystal alone is on the order of 1000 radians, with additional components from mirror dispersion and the air path. Moreover, geometrical dispersion, i.e., a spectrally dependent path length, may add to the cavity roundtrip $\Delta\varphi_{\mathrm{GPO}}$ if elements such as prisms or gratings are used.

In contrast to noise in cw lasers, parameters such as $\Delta\varphi_{\mathrm{GPO}}$ from a modelocked pulse train are not defined on a continuous time scale, but only on a discrete temporal grid. Equation (12.10) implies that the group-phase offset change by $\Delta\varphi_{\mathrm{GPO}}$ per roundtrip time T_{R} determines the GPO frequency f_{GPO} and therefore f_{CEO} [02Hel1, 02Hel2,03Hel]:

$$\boxed{2\pi f_{\mathrm{GPO}} \equiv \frac{\mathrm{d}}{\mathrm{d}t}\varphi_{\mathrm{GPO}} = \frac{\Delta\varphi_{\mathrm{GPO}}}{T_{\mathrm{R}}} = f_{\mathrm{rep}}\Delta\varphi_{\mathrm{GPO}} \Longrightarrow f_{\mathrm{CEO}} = \frac{f_{\mathrm{rep}}\phi_{\mathrm{CEO}}}{2\pi}} . \quad (12.11)$$

Here, we used the fact that the group phase offsets (GPO) of integer multiples of 2π do not change the electric field pattern of a laser pulse as shown in Fig. 12.1. The quantity of interest is

$$\phi_{\mathrm{CEO}} = \Delta\varphi_{\mathrm{GPO}} \ (\mathrm{mod}\ 2\pi) , \qquad f_{\mathrm{CEO}} = |f_{\mathrm{GPO}} - \xi f_{\mathrm{rep}}| , \quad (12.12)$$

where ξ is an integer chosen to yield the smallest possible value of f_{CEO}. As illustrated in Fig. 12.2, f_{CEO} is defined as the frequency of the line closest to zero with the additional requirement that $f_{\mathrm{CEO}} > 0$, when the comb spectrum is extrapolated to zero frequency. On a single pass through a 2 mm Ti:sapphire crystal, the group is typically delayed by 140 fs relative to the phase front. This delay corresponds to more than 50 optical cycles and leads to a typical value of $f_{\mathrm{GPO}} = 10$ GHz for a KLM laser with $f_{\mathrm{rep}} = 100$ MHz. Assuming the typical thermal and vibrational fluctuations reported for similar lasers, the resulting frequency fluctuations are expected to be small enough for an unambiguous choice of ξ.

The time scale of the CEO phase or frequency fluctuations is much slower (i.e., more than 1 μs and thus less than 1 MHz in frequency) than the pulse duration or the optical period, i.e., in the femtosecond regime. We can therefore treat the CEO frequency noise as a slowly varying random quantity and, using (12.4) and $\varphi(t) = \omega_{\mathrm{CEO}} t$, write the pulse train as follows [03Hel]:

$$E_{\mathrm{train}}(t) = A(t) \exp(\mathrm{i}\omega_{\mathrm{c}} t + \mathrm{i}\omega_{\mathrm{CEO}} t) * \sum \delta(t - mT_{\mathrm{R}}) \,. \tag{12.13}$$

Using the Fourier shift theorem, we can obtain the expression for the frequency domain:

$$\boxed{\tilde{E}_{\mathrm{train}}(\nu) = \tilde{E}(\nu) \sum \delta(\nu - m f_{\mathrm{rep}} - f_{\mathrm{CEO}}) = \tilde{A}(\nu - \nu_{\mathrm{c}}) \sum \delta(\nu - m f_{\mathrm{rep}} - f_{\mathrm{CEO}})} \,. \tag{12.14}$$

Here we have used

$$E(t) = A(t) \exp(\mathrm{i}\omega_{\mathrm{c}} t) \;\Rightarrow\; \tilde{E}(\nu) = F\{E(t)\} = \int E(t) \mathrm{e}^{-\mathrm{i}2\pi\nu t} \mathrm{d}t \,, \quad \tilde{E}(\nu) = \tilde{A}(\nu - \nu_{\mathrm{c}}) \,, \tag{12.15}$$

from Sect. 2.4, where the optical spectrum of the pulse envelope $\tilde{A}(\nu)$ is centered around zero frequency and is equal to the optical spectrum shifted by the center frequency (i.e., the carrier frequency) ν_{c}.

The whole equidistant frequency comb of (12.13) is shifted by f_{CEO} due to the per roundtrip carrier-envelope offset phase shift of $\omega_{\mathrm{CEO}} t$, where $\omega_{\mathrm{CEO}} = 2\pi f_{\mathrm{CEO}}$. Figure 12.2 depicts the spectrum of such a pulse train, where we observe an equidistant frequency comb with an offset. All frequencies of the comb lines can be calculated from the simple equation

$$\nu_m = f_{\mathrm{CEO}} + m f_{\mathrm{rep}} \,, \tag{12.16}$$

where m is a large integer. The frequency comb spacing is equidistant and given by the roundtrip propagation time of the pulse envelope, i.e., the inverse of the group velocity and not the inverse of the phase velocity. This means that the axial modes of a modelocked laser are not the same as those of a continuous wave (cw) laser for which the phase velocity determines the axial mode. Timing jitter in the arrival time of the pulses (see Chap. 11) produces a 'breathing' of the otherwise fully equidistant frequency comb. On the other hand, the timing fluctuations of the CEO result in a translation of the full frequency comb.

For $\upsilon_{\mathrm{g}} = \upsilon_{\mathrm{p}}$ the CEO does not change with time, so $\Delta\varphi_{\mathrm{GPO}} = 0$ and $\varphi(t) =$ constant. The pulse train from (12.13) and (12.14) is given in this case by

$$\tilde{E}_{\mathrm{train}}(\nu) = \tilde{E}(\nu) \sum \delta(\nu - m f_{\mathrm{rep}}) = \tilde{A}(\nu - \nu_{\mathrm{c}}) \sum \delta(\nu - m f_{\mathrm{rep}}) \,, \tag{12.17}$$

which is a frequency comb with $f_{\mathrm{CEO}} = 0$.

We are then interested in the noise properties of $\varphi(t) = 2\pi f_{\mathrm{CEO}} t$ and f_{CEO}. Even tiny variations in the intracavity material properties can give rise to variations in f_{CEO} due to the large value of $\Delta\varphi_{\mathrm{GPO}}$, according to (12.11) and (12.12). Note that the f_{CEO} fluctuations of two adjacent pulses in the modelocked pulse train are extremely small, as they correspond to noise fluctuations at very high noise frequencies at the pulse repetition rate. However, for much larger pulse separations τ within the pulse train, the relative variations in $f_{\mathrm{CEO}}(t)$ and $f_{\mathrm{CEO}}(t+\tau)$ start to become significant and correspond to much lower noise frequencies than f_{rep}.

12.3 Measurement of the CEO Frequency

12.3.1 Basic Principle with Heterodyne (Interference) Signals

While both the direct measurement of f_{rep} with an RF counter and of m with a wave-meter are straightforward in the gigahertz regime, the determination of f_{CEO} was initially the main challenge. Several new methods for the phase-coherent determination of f_{CEO} were proposed in 1999 [99Tel] using nonlinear optical processes, including second harmonic generation (SHG), sum frequency generation (SFG), and difference frequency generation (DFG).

The general idea of these f_{CEO} measurement methods is to generate mixing products from different phase-linked combs generated via nonlinear coherent processes. The frequencies of the fundamental optical frequency comb modes contain a unit of f_{CEO} (12.16), as shown in Fig. 12.2. Different parts of the other phase-linked combs contain frequencies $p f_{\mathrm{CEO}}$, with p an integer and $p \neq 1$. Consequently, heterodyning the mixing product of nearby comb lines results in a beat signal oscillating at $p-1$ times f_{CEO}. The complexity of the setup required to establish a coherent link to f_{CEO} depends on the phase-coherent frequency span of the optical frequency comb.

Heterodyning means that one frequency range is shifted into another. In electronics, two input frequencies f_1 and f_2 are combined in a nonlinear device usually called a mixer, for which we obtain the mixing products

$$2\cos(\omega_1 t)\cos(\omega_2 t) = \cos[(\omega_1 - \omega_2)t] + \cos[(\omega_1 + \omega_2)t] \,. \tag{12.18}$$

The low-frequency mixing signal proportional to $|\omega_1 - \omega_2| t$ is selected with a low-pass filter for further signal processing. In optics, we obtain the mixing product of two frequencies with optical interference and a photodetector as follows:

$$E(t) = E_1(t) + E_2(t) = E_{10} e^{i\omega_1 t} + E_{20} e^{i\omega_2 t} \,. \tag{12.19}$$

where ω_0 is the average (or center) frequency and $\Delta\omega$ the difference frequency, defined by

$$\omega_0 = \frac{\omega_1 + \omega_2}{2} \,, \qquad \Delta\omega = |\omega_2 - \omega_1| \,. \tag{12.20}$$

For simplicity we assume equal amplitudes, i.e., $E_{10} = E_{20} = E_0$. We then obtain

$$E(t) = E_0(e^{i\omega_1 t} + e^{i\omega_2 t}) = E_0 e^{i\omega_0 t}\left[\exp\left(-i\frac{\Delta\omega}{2}t\right) + \exp\left(i\frac{\Delta\omega}{2}t\right)\right]$$
$$= 2E_0 e^{i\omega_0 t}\cos\left(\frac{\Delta\omega}{2}t\right),$$

We can then write the interference signal as

$$E(t) = A(t)e^{i\omega_0 t}, \quad A(t) = 2E_0\cos\left(\frac{\Delta\omega}{2}t\right), \tag{12.21}$$

which is an amplitude modulated signal at a new center frequency, i.e., the average of the two frequencies, and an envelope function that is modulated at half of the difference frequency. The photodetector measures the intensity $I(t)$ of the interference signal, and with (12.20), we obtain

$$I(t) = 4I_0\cos^2\left(\frac{\Delta\omega}{2}t\right) = 2I_0[1 + \cos(\Delta\omega t)]. \tag{12.22}$$

This is referred to as the beat signal at the difference frequency $\Delta\omega$ of the two signals. The bandwidth of the photodetector needs to be large enough to resolve the time-dependent beat signal, otherwise the intensity is averaged in time:

$$\frac{1}{T}\int_0^T I(t)\mathrm{d}t = 2I_0, \tag{12.23}$$

which would just be the incoherent superposition of the two intensities.

Additional single-frequency transfer oscillators can be used when the optical frequency comb has very low power, so that one or several nonlinear processes become practically impossible. Such transfer oscillators are single-frequency cw lasers with an optical frequency ν_{trans} that is required to be phase-locked to one of the comb lines within the frequency comb. Note that a phase-locked loop (PLL) is a control system that keeps the two frequencies phase-locked, which also means that their frequencies are the same (see Sect. 12.6). The higher power in these transfer oscillators then allows for more efficient nonlinear frequency conversion and stronger mixing signals. More than one single-frequency oscillator can further reduce the requirement for the frequency span of an OFC for the CEO frequency measurement.

Generally, a wider comb span $\Delta\nu_{\mathrm{span}} = \nu_{\mathrm{high}} - \nu_{\mathrm{low}}$ requires fewer intermediate transfer oscillators, nonlinear processes, and mixing steps to determine f_{CEO} (see Table 12.1). All these methods were introduced in [99Tel] and can be summarized as follows. Note that we distinguish here between the comb span $\Delta\nu_{\mathrm{span}}$ as defined in Table 12.1 and the comb width, which is typically given as FWHM, i.e., $\Delta\nu_{\mathrm{p}} = \Delta\nu_{\mathrm{FWHM}}$. The high and low frequencies are approximately given by a measurable signal $|\tilde{E}(\nu_{\mathrm{high}})|^2$ and $|\tilde{E}(\nu_{\mathrm{low}})|^2$, which can be many 10 dBs below that of the carrier frequency $|\tilde{E}(\nu_{\mathrm{c}})|^2$. The frequency span is typically much larger than the optical spectrum of the modelocked pulses, i.e., $\Delta\nu_{\mathrm{span}} \gg \Delta\nu_{\mathrm{FWHM}}$.

Table 12.1 Comparison of different methods to measure the CEO frequency f_{CEO} in an optical frequency comb (OFC) $\nu_m = f_{CEO} + m f_{rep}$ as shown in Fig. 12.2. Methods are compared in terms of the required frequency span $\Delta\nu_{span} = \nu_{high} - \nu_{low}$ of the OFC, numbers of additional oscillators, and nonlinear conversion steps [99Tel]. Note that there is no exact definition of ν_{high} and ν_{low} in the community. This is the frequency for a spectral intensity in the range of −10 dB or even −20 dB below the peak value

Methods	Equation	ν_{trans}	Nonlinear conversion steps	ν_{high}/ν_{low}	Figure
1a. f-to-$2f$ interferometer: self-referenced direct SHG (or direct DFG)	(12.26)	0	1 SHG (or DFG)	2	Fig. 12.3
1b. f-to-$2f$ interferometer with transfer oscillator $\nu_{trans} = \nu_m$	(12.27)	1	1 $2\nu_{trans}$	2	Fig. 12.5
2. Frequency-doubled transfer oscillator $\nu_{trans} = n f_{rep}$	(12.28)	1	2 SFG and $2\nu_{trans}$	1.5	Fig. 12.9
3. Frequency-tripled transfer oscillator $\nu_{trans} = n f_{rep}$	(12.33)	1	3 SHG and THG (SHG+SFG)	1.33	
4. Two auxiliary oscillators or $2f$-to-$3f$ interferometer	(12.37) (12.38)	2	3 SHG and THG (SHG+SFG)	1.5	

12.3.2 Method 1: f-to-2f Interferometer

The simplest method for the measurement of f_{CEO} requires one nonlinear optical process such as second harmonic generation (SHG) with a coherent octave spanning frequency comb spectrum. The modelocked OFC (Fig. 12.2) is frequency doubled to generate a second harmonic OFC (i.e., SH-OFC) (Fig. 12.3). SHG is a coherent process and therefore the two optical frequency combs provide a phase-stable link between the optical and the microwave beat frequencies, such as the f_{CEO} beat signal.

The basic principle is very easy to understand. Assuming there is an octave-spanning OFC, then there is an overlap of the high-frequency part of the OFC with the low-frequency part of the SH-OFC (Fig. 12.3). As shown in Fig. 12.3, the pulse repetition rate does not change for the frequency-doubled pulse train. For a partial overlap of the fundamental OFC and the SH-OFC, we need at least a fundamental OFC spectrum of one octave, i.e., $\nu_{high}/\nu_{low} \geq 2$, as shown in Table 12.1. In the OFC/SH-OFC overlap regime, the measured beat signal between the OFC and the SH-OFC is f_{CEO}, which can be measured with interference on a photodetector and a spectrum analyzer, as shown in Fig. 12.4.

The phase noise is not lost in these beat-signal measurements. Following the derivation in (12.19) to (12.22), we can simply add the phase noise $\varphi(t)$. The interference is then given by

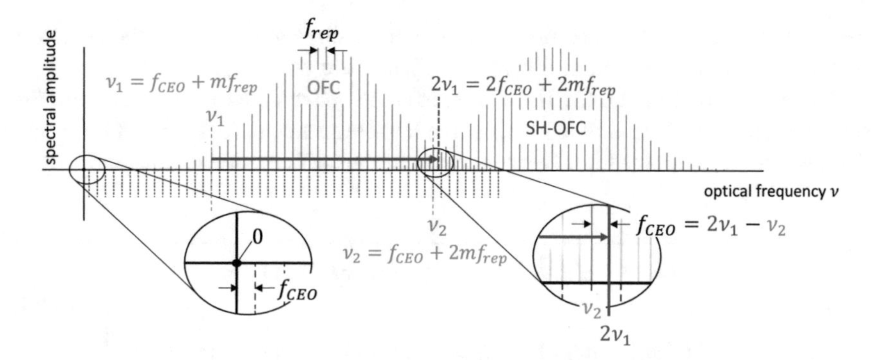

Fig. 12.3 Method 1: Self-referenced f-to-$2f$ interferometer for the f_{CEO} measurement according to (12.26). Note that in this context an f-to-$2f$ interferometer is referred to as self-referenced when the fundamental and its second harmonic spectrum are used for the interference signal. OFC: optical frequency comb of the modelocked laser, SH-OFC: second harmonic of OFC. The beat signal in the spectral overlap of OFC and SH-OFC gives the value of f_{CEO}. The total frequency span of the OFC needs to be more than one octave

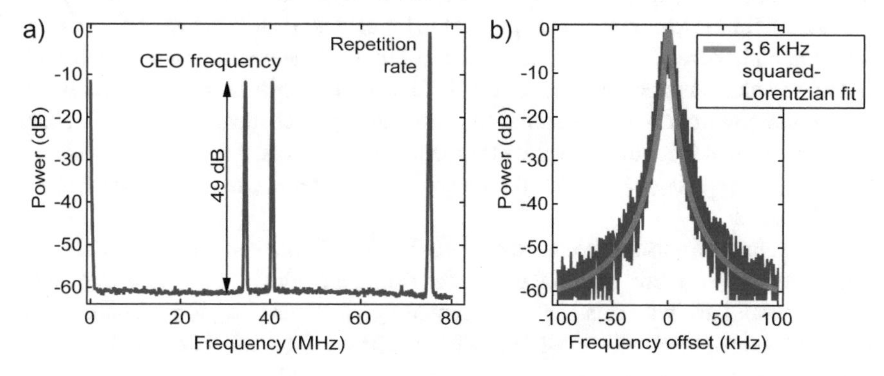

Fig. 12.4 Example of an f_{CEO} beat signal using Method 1 (Table 12.1) from a free-running SESAM-modelocked Er:Yb:glass laser with a pulse repetition rate $f_{\text{rep}} = 75$ MHz measured with a microwave spectrum analyzer: **a** over an 80 MHz frequency span (RBW = 100 kHz, VBW = 1 kHz), and **b** magnified and centered left CEO beat signal with a squared Lorentzian fit of 3.6 kHz FWHM (RBW = 1 kHz, VBW = 300 Hz) [10Stu]. RBW: resolution bandwidth, VBW: video bandwidth

$$E(t) = E_0(e^{i\phi_1} + e^{i\phi_2}) , \quad \phi_i = \omega_i t + \varphi_i(t) , \tag{12.24}$$

where i is 1 or 2. According to (12.22), this results in a beat signal intensity

$$I(t) = 2I_0[1 + \cos(\Delta\omega t + \Delta\phi)] \overset{\text{SHG}}{\Longrightarrow} I(t) = 2I_0\big[1 + \cos\big(\omega_1 t + \varphi_1(t)\big)\big] . \tag{12.25}$$

Here we have used the fact that, for second harmonic generation (SHG) with $\omega_2 = 2\omega_1$ and $\varphi_2 = 2\varphi_1$, both the optical frequency and the phase noise are doubled, since

$$\cos^2\phi(t) = \frac{1}{2}\big[1 + \cos\big(2\phi(t)\big)\big] .$$

Because this heterodyne detection scheme is insensitive to the sign of the beat note frequency, two lines appear at frequencies symmetric with respect to $f_{rep}/2$.

This means that the OFC/SH-OFC interference beat signal is obtained with the second harmonic of the m th mode from the low-frequency part of the OFC and with the $2m$ th mode from the high-frequency part of the OFC, as shown schematically in Fig. 12.3 [99Tel]:

$$\begin{aligned} &2\nu_1 - \nu_2 = 2(f_{CEO} + mf_{rep}) - (f_{CEO} + 2mf_{rep}) = f_{CEO}\,, \\ &\text{or equivalently,} \\ &2\nu_n - \nu_m = 2(f_{CEO} + nf_{rep}) - (f_{CEO} + mf_{rep}) = f_{CEO}\,, \quad m = 2n\,. \end{aligned} \tag{12.26}$$

The comb lines have a different phase noise, as the phase fluctuations of f_{rep} also contribute to the phase noise in addition to the noise of f_{CEO}. But the noise of f_{rep} cancels out in this f_{CEO} noise measurement.

Note that normally the SHG bandwidth is limited well below an optical octave in a nonlinear crystal. However, this is not a problem because we only need an SHG bandwidth for the lower frequency part of the fundamental OFC, which then overlaps with the higher frequency part, as shown in Fig. 12.3. Therefore many comb lines within the phase-matching range of the SHG crystal contribute to the measurement signal, thereby improving the signal-to-noise ratio. The challenge with this method is, however, that it requires very wide frequency comb spans covering at least one octave (i.e., $\nu_{high}/\nu_{low} \geq 2$), and the power in the extreme parts of the spectrum is usually very low.

The signal-to-noise ratio of the f-to-$2f$ interferometer can potentially be improved by using one additional transfer oscillator which is a single-frequency cw laser oscillator with a lasing frequency at ν_{trans}. When ν_{trans} is phase-locked to the m th mode of the modelocked frequency comb (see Fig. 12.5), (12.26) still applies with

$$2\nu_{trans} - \nu_{2m} = 2(f_{CEO} + mf_{rep}) - (f_{CEO} + 2mf_{rep}) = f_{CEO}\,. \tag{12.27}$$

In this case, only one comb line is used, but the efficiency of the nonlinear optical process in generating $2\nu_{trans}$ can be much higher when a strong single-frequency external transfer laser oscillator is used. However, this method requires an additional phase lock and an additional single-frequency laser oscillator. Method 1a of Table 12.1 is therefore often referred to as 'self-referenced,' because no additional laser oscillator has to be used for the CEO frequency measurement.

The same information can be obtained by DFG between modes from the opposite ends of the comb and the subsequent beating of this mixing product with modes from the low-frequency wing. This alternative approach may be preferred if the quantum efficiency of the photodetector is higher in this spectral region.

Initially, passively modelocked Ti:sapphire lasers were used for the first f_{CEO} beat signal measurements and stabilization [99Tel,00Apo,00Jon]. At that time, the octave-spanning spectrum required for the f-to-$2f$ frequency beat signal was only

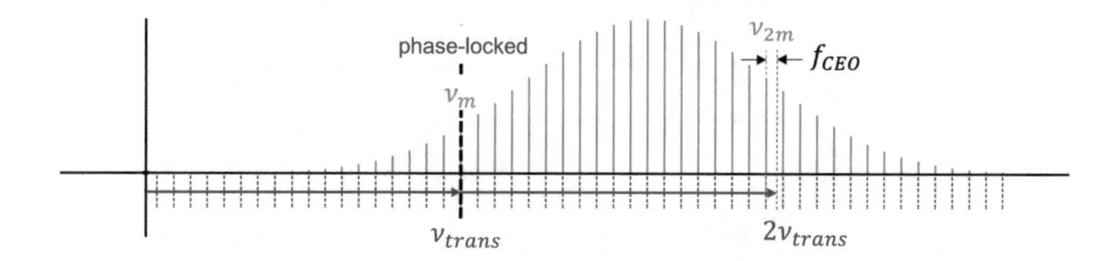

Fig. 12.5 Method 1b: f-to-$2f$ interferometer for f_{CEO} measurement of the OFC (*red line*) using a transfer oscillator phase-locked to $\nu_{\text{trans}} = \nu_m$, i.e., Method 1b in Table 12.1. Interference between $2\nu_{\text{trans}}$ and the ν_{2m} line in OFC gives a beat signal at f_{CEO}, as given by (12.27)

obtainable with additional nonlinear spectral broadening in an external fiber. However, such supercontinuum generation can add significant noise and needs to be optimized to generate a coherent spectrum [02Dud,06Dud]. A supercontinuum is coherent when ultrashort pulses can be supported with appropriate dispersion compensation. This was demonstrated, for example, with pulse compression of 15 fs Ti:sapphire laser pulses to 5.5 fs using supercontinuum generation in a very short (i.e., 5 mm long) microstructured fiber (see Fig. 4.7) [05Sch4]. Therefore, more research explored the possibility of generating an octave-spanning spectrum directly from a modelocked laser oscillator without further pulse compression and amplification. For example, this was achieved with improved intracavity dispersion management [08Cre].

In the following example, we shall discuss the experimental demonstration of a self-referenced f-to-$2f$ interferometer method to measure the f_{CEO} from a modelocked solid-state laser. Coherent supercontinuum generation was also used for the f_{CEO} beat signal measurement, using a SESAM-stabilized soliton modelocked Er:Yb:glass laser generating 170 fs pulses at a center wavelength of around 1.5 μm, at a 75 MHz pulse repetition rate with 110 mW average output power (see Sect. 9.5 for the passive modelocking technique) [05Spu1,08Stu,10Stu]. The modelocked optical spectrum has a FWHM of 14.7 nm for a sech^2 pulse shape, which is clearly not an octave-spanning spectrum. To minimize any additional noise, no external pulse compression or amplification was used. Furthermore, CEO phase noise was reduced by replacing an intracavity prism pair for dispersion compensation with GTI mirrors [03Hel]. The output pulses were directly launched into a polarization-maintaining highly nonlinear fiber (PM-HNLF) to generate a coherent octave-spanning frequency comb (Fig. 12.6).

Excellent coherence in supercontinuum generation was achieved by using a dispersion-flattened, polarization-maintaining, highly nonlinear fiber. Polarization maintenance (PM) is introduced by an elliptical core, while the nonlinearity is enhanced through Ge-doping. The fiber had an effective area of 12.7 μm^2 with an estimated nonlinear coefficient of 10.5 $(\text{W km})^{-1}$. A measured dispersion curve is plotted in Fig. 12.6. By optimizing the input polarization and choosing an appropriate fiber length of 1.3 m, the outer spectral peaks could be separated by exactly one octave (Fig. 12.6). The supercontinuum had an average power of 50 mW and it was polarized and stable.

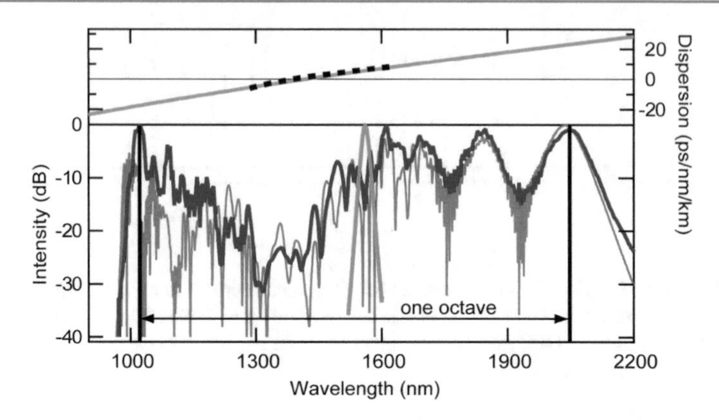

Fig. 12.6 Generated supercontinuum. *Blue line*: Measured spectrum of the output of the 1.3 m long PM-HNLF (i.e., polarization-maintaining, highly nonlinear fiber). *Red line*: Simulated spectrum [06Dud]. *Green line*: Unbroadened 14.7 nm laser spectrum. *Top panel*: Dispersion profile of the PM-HNLF used in the simulation (*grey line*) with measured dispersion profile (*black dots*) [10Stu]

After supercontinuum generation in the PM-HNLF (Fig. 12.6), the CEO frequency as shown in Fig. 12.4 was measured with the self-referenced f-to-$2f$ interferometer shown in Fig. 12.7. The latter's implementation is standard and should be very stable. Note that in this case the interferometer has two separate arms and is not a collinear interferometer with a common path. Therefore, noise can be introduced between the two arms. The supercontinuum is separated by a dichroic beamsplitter and combined again after the filtered long-wavelength (i.e., low-frequency ν_1 in Fig. 12.3) signal around 2050 nm has been frequency-doubled. At room temperature, a 4 mm long periodically poled lithium niobate (PPLN) crystal with a poling period of 31.1 μm generates a second harmonic signal with 8 nm bandwidth at 1025 nm. An average frequency-doubled power of up to 0.5 mW was achieved, corresponding to a 10% conversion efficiency of the available supercontinuum power at the long-wavelength part centered around 2050 nm. A high SHG efficiency is essential for a good signal-to-noise ratio and has mainly been achieved because of the high peak power from the outermost supercontinuum pulse structure (Fig. 12.6). The waveplates inside the interferometer set the correct polarization for the PPLN crystal and the combiner. After the combiner, an adjustable polarizer projects both fields into the same plane and also balances the power of both interferometer arms for optimal interference contrast. Additionally, the contrast is increased by the bandpass filter removing the background from the SHG bandwidth. The signal is finally detected with a fiber-coupled photodiode and monitored with a microwave spectrum analyzer (Fig. 12.4).

Both coherence and noise need to be optimized for stable supercontinuum generation. The signal-to-noise ratio and the linewidth of the CEO beat signal, as shown in Fig. 12.4, are very good for a free-running passively modelocked laser with a pulse repetition rate of 75 MHz. This CEO frequency linewidth can be further reduced with active stabilization, which will be discussed in Sect. 12.6. For now, we should address the potential issue that during supercontinuum generation additional noise

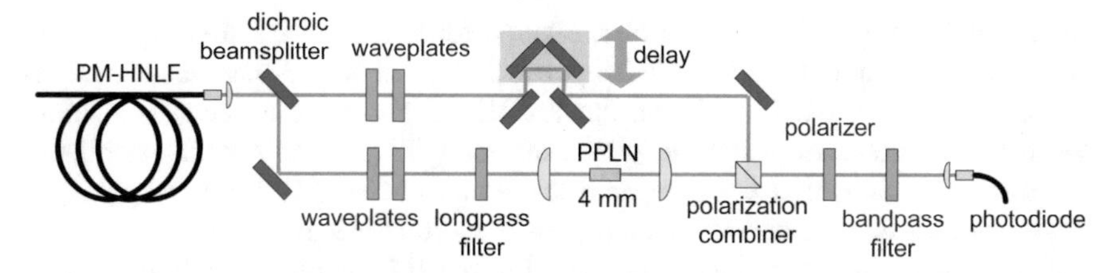

Fig. 12.7 Self-referenced f-to-$2f$ interferometer used for CEO frequency detection after octave-spanning supercontinuum generation in the PM-HNLF [10Stu] (Fig. 12.6). Note that in this case the interferometer has two separate arms and is not a collinear interferometer with a common path. Therefore, noise can be introduced between the two arms, which in principle should be prevented

can be generated, as shown in Fig. 12.8. The supercontinuum intensity noise can be directly measured and compared to the intensity noise of the modelocked solid-state laser using the methods introduced in Chap. 11. A direct measure of the coherence in experiments is difficult [05Sch4], but poor coherence is normally coupled with increased intensity noise in the supercontinuum generation [02Dud]. The measurements were carried out with a standard InGaAs photodiode with a peak sensitivity at the laser center wavelength and good sensitivity in the short wavelength range of the supercontinuum. Although RIN levels can vary considerably across the entire supercontinuum spectrum, it is very unlikely that instabilities would only occur in the long-wavelength spectral region above 1700 nm outside the sensitivity range of the photodiode. Three different fibers were tested for octave-spanning supercontinuum generation. A comparison reveals significant differences in their relative intensity

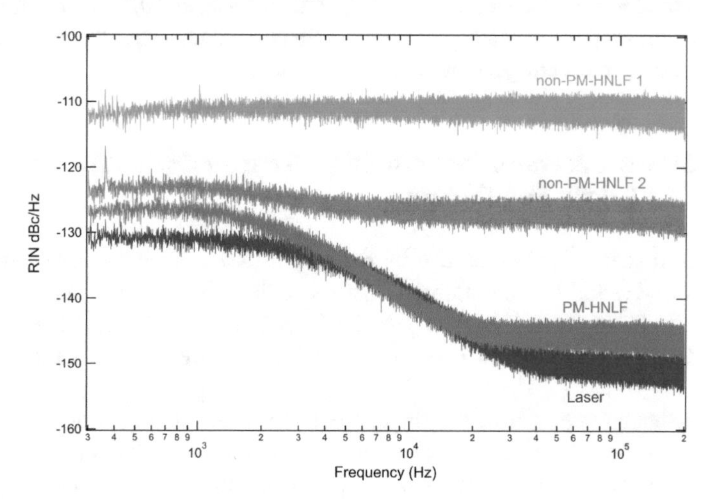

Fig. 12.8 Measured relative intensity noise (RIN) (see Chap. 11). RIN of the ultrafast Er:Yb:glass oscillator (*dark blue line*); the supercontinuum generated with a 10 m-long non-PM-HNLF 1 (*light blue line*), a 5 m-long non-PM-HNLF 2 (*red line*), and the 1.3 m-long PM-HNLF described in the text (*green line*) [10Stu]

noise. No CEO beat signal could be measured for a non-polarization-maintaining highly nonlinear fiber (non-PM-HNLF), even though an octave-spanning supercontinuum was generated. In addition, the PM-HNLF generated an octave-spanning spectrum at much shorter fiber lengths, even though it had very similar nonlinear and dispersive characteristics compared to the non-PM-HNLF. The fixed polarization obviously ensures stable interaction of the strong electric fields supporting a coherent, octave-spanning supercontinuum for longer input pulse durations.

Supercontinuum generation in waveguides on photonic chips support coherent supercontinuum generation with very low pulse energies [15Joh] and direct comparison with photonic crystal fibers (PCFs) resulted in better noise performance [14Kle3, 15May, 16Kle]. Standard PCFs for wavelengths above 1 μm often rely on the Raman effect to provide sufficient spectral broadening and this rapidly degrades coherence with increasing nonlinearity and longer pulses [16Kle]. The future vision is that the full f-to-$2f$ interferometer shown in Fig. 12.7 can be integrated on a photonic chip with much better overall stability [19Gae]. Silicon-based waveguides in particular provide material compatibility with existing complementary metal-oxide-semiconductor (CMOS) fabrication technology that allows for large-scale implementation of integrated chip-scale devices. Silicon nitride (Si_3N_4) is a CMOS-compatible material with a nonlinearity ($n_2 \approx 2.5 \times 10^{-15}$ cm^2/W) that is two times larger than that of high-index-doped silica and 10 times larger than that of silica [14Oh]. The combination of high nonlinearity with high-mode confinement (i.e., because of the high refractive index contrast between the Si_3N_4 core and silicon dioxide cladding), the large bandgap of Si_3N_4 (i.e., compared to either crystalline or amorphous silicon) for negligible two-photon absorption at near-infrared wavelengths, lithographically patterned waveguide structure to control dispersion, and the lack of significant Raman gain in single-pass waveguides makes Si_3N_4 an ideal candidate for low-noise coherent supercontinuum generation at 1 μm [13Mos].

12.3.3 Method 2: Frequency-Doubled Transfer Oscillator with Lower Bandwidth Requirement

The required span of the comb can be reduced if two nonlinear-optical processes (SFG and SHG) and one additional transfer oscillator at ν_{trans} are combined (see Fig. 12.9). In principle, it is still an f-to-$2f$ interferometer, but this time we phase lock the transfer oscillator to $\nu_{\text{trans}} = n f_{\text{ref}}$ with a self-referenced multiline SFG scheme.

The fundamental power of the single-frequency transfer oscillator at frequency ν_{trans} is used for a collective up-conversion of comb lines from the low-frequency to the high-frequency end of the comb. If $\nu_{\text{trans}} = n f_{\text{ref}}$, then $\nu_{m+n+j} = \nu_{m+j} + \nu_{\text{trans}}$ for all j leading to modes within the phase-matching range of SFG. With active control of ν_{trans}, the frequency shift can be forced to an integer number of modes n by tuning the frequency of the beat note between the modes of the comb-like SFG signal and the comb modes in the vicinity of ν_{m+n} to zero. Simultaneously, the second harmonic at $2\nu_{\text{trans}}$ is compared with the frequency $\nu_{2n} = f_{\text{CEO}} + 2n f_{\text{rep}}$ of

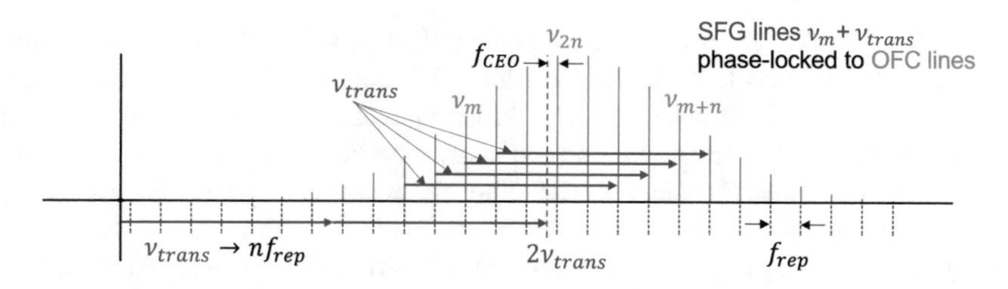

Fig. 12.9 Method 2: f-to-$2f$ interferometer for f_{CEO} measurement of the OFC (*red line*) using a transfer oscillator phase-locked to $\nu_{\mathrm{trans}} = nf_{\mathrm{ref}}$ with a self-referenced multiline sum frequency generation (SFG) scheme (12.28). This results in an f_{CEO} beat signal from interference of ν_{2n} with $2\nu_{\mathrm{trans}}$. The OFC bandwidth requirement $\nu_{\mathrm{high}}/\nu_{\mathrm{low}}$ is reduced to 1.5

the nearest comb mode:

$$\nu_{2n} - 2\nu_{\mathrm{trans}} = (f_{\mathrm{CEO}} + 2nf_{\mathrm{rep}}) - 2nf_{\mathrm{rep}} = f_{\mathrm{CEO}} \,. \tag{12.28}$$

In principle as shown in Fig. 12.9, ν_{trans} cannot exceed the frequency span of the comb, i.e., we have $\nu_{\mathrm{trans}} \leq \nu_{\mathrm{high}} - \nu_{\mathrm{low}}$. In addition, for a beat signal of $2\nu_{\mathrm{trans}}$ with the OFC, we need to have $2\nu_{\mathrm{trans}} \geq \nu_{\mathrm{low}}$. These two conditions then give a minimal bandwidth requirement of $\nu_{\mathrm{high}}/\nu_{\mathrm{low}} \geq 1.5$, as shown in Table 12.1 for Method 2. The trade-off for the lower OFC bandwidth requirement is that Method 2 requires an additional phase lock and an additional single-frequency laser oscillator. Note that Method 2 has the same additional requirements as Method 1b, but now with the benefit of a lower OFC bandwidth requirement going from 2 to 1.5 (see Table 12.1).

The bandwidth requirement $\nu_{\mathrm{high}}/\nu_{\mathrm{low}} \geq 1.5$ can be proven as follows. Assuming the full frequency span of the OFC is given by $\Delta\nu = \nu_{\mathrm{high}} - \nu_{\mathrm{low}}$, and the center frequency of the OFC is given by the carrier frequency ν_{c}, then we have $\nu_{\mathrm{high}} = \nu_{\mathrm{c}} + \Delta\nu/2$ and $\nu_{\mathrm{low}} = \nu_{\mathrm{c}} - \Delta\nu/2$. Therefore, with the condition that $2\nu_{\mathrm{trans}} \geq \nu_{\mathrm{low}}$ and $\nu_{\mathrm{trans}} \leq \Delta\nu$, we obtain

$$\nu_{\mathrm{low}} \leq 2\nu_{\mathrm{trans}} \leq 2\Delta\nu \,. \tag{12.29}$$

Using $\nu_{\mathrm{low}} = \nu_{\mathrm{c}} - \Delta\nu/2$, it then follows that

$$2\nu_{\mathrm{c}} - \Delta\nu \leq 4\Delta\nu \,, \tag{12.30}$$

and therefore,

$$2\nu_{\mathrm{c}} \leq 5\Delta\nu \,. \tag{12.31}$$

We then obtain

$$\frac{\nu_{\mathrm{high}}}{\nu_{\mathrm{low}}} = \frac{\nu_{\mathrm{c}} + \Delta\nu/2}{\nu_{\mathrm{c}} - \Delta\nu/2} = \frac{2\nu_{\mathrm{c}} + \Delta\nu}{2\nu_{\mathrm{c}} - \Delta\nu} \geq \frac{6\Delta\nu}{4\Delta\nu} = 1.5 \,. \tag{12.32}$$

This method looks very promising when the OFC bandwidth is limited. For example, in 1999 a few-cycle passively modelocked Ti:sapphire laser generated world-record short pulse durations and a very broad optical spectrum [99Sut]. The laser was pumped with 8 W from an argon-ion laser, generating 300 mW of average output power with pulse durations in the sub-6-femtosecond regime [99Gal]. In the experiments described here, slightly longer pulses have been used and the optical spectrum is shown in Fig. 12.10. For a sufficiently strong CEO beat signal (as shown in Fig. 12.4), a minimum photon flux per mode is required. Assuming a detection bandwidth of 100 kHz, 10^5 photons/s are needed for a shot-noise-limited signal-to-noise ratio of 1. To avoid cycle slips as described in [96Tel], the number of detected photons must be at least 100 times this value. Imperfect mode matching, limited quantum efficiency, and other losses may account for an additional factor of 100, which leads to a conservative requirement of 10^9 photons/s per mode.

The measured spectrum in Fig. 12.10 indicates that the spectral range from 650 to 1000 nm, corresponding to a width of more than 160 THz, satisfies this criterion. For most of the wavelength range, the spectrum is fitted by a Gaussian with a 55 THz FWHM, corresponding to a pulse duration of 8 fs, whereas the transform limit of the measured spectrum yields 6.6 fs.

Taking into account the availability of additional lasers at specific frequencies, Methods 2 to 4 (Table 12.1) appear to be the most promising candidates for the CEO

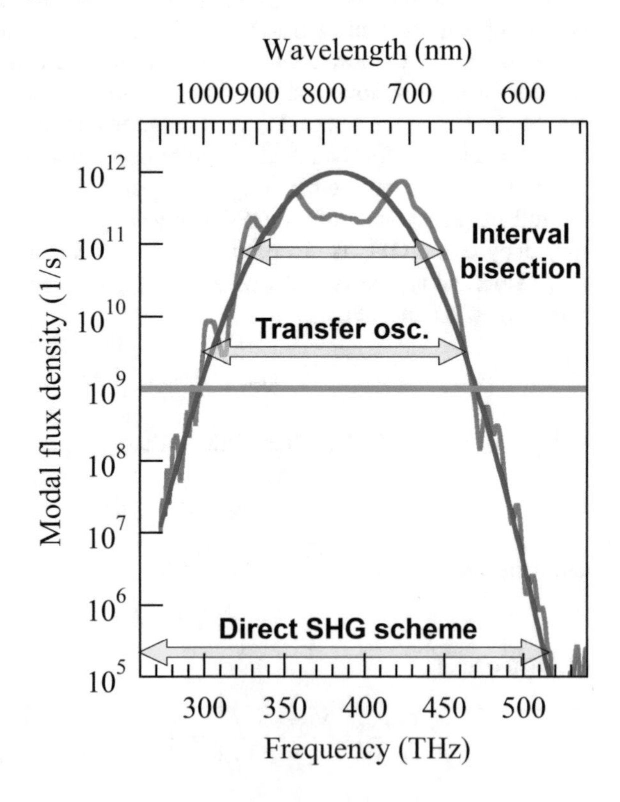

Fig. 12.10 Measured spectral flux density of the KLM Ti:sapphire laser [99Sut, 99Gal]. The spectrum is calibrated in units of photons/s and mode. TEM_{00} power is 100 mW. *Solid line*: Measured spectrum. The Fourier transform of this spectrum has a width of 6.6 fs. *Dashed line*: Gaussian fit with 55 THz FWHM, corresponding to an 8 fs duration of a bandwidth-limited pulse

phase control of this modelocked laser. The frequency-doubled transfer oscillator (Method 2, Fig. 12.9) requires a comb span of about 150 THz. Laser diodes in this 2 μm wavelength range were not readily available in 1999, and provided only small output power at large emission linewidths. The power in the spectral wings is insufficient to bridge broader frequency ranges in the 200 THz range, where convenient powerful lasers would have been readily available, i.e., 1.5 μm Er fiber lasers.

12.3.4 Method 3: Frequency-Tripled Transfer Oscillator

Given sufficient power, the high mixing efficiency of a combination of powerful diode-pumped solid-state lasers and quasi-phase-matched nonlinear crystals allows for third harmonic generation (THG). Because THG is usually accomplished by SHG and sum frequency mixing of the second harmonic with the fundamental, three nonlinear optical processes are required in total. Similarly to the frequency-doubling method (Method 2), the fundamental, which could be generated by a holmium laser emitting in the 2 μm range, is used to up-shift modes from the low-frequency wing of the spectrum, and its third harmonic is compared with the nearest comb mode at $\nu_{3n} = f_{\text{CEO}} + 3nf_{\text{rep}}$. This leads to

$$\nu_{3n} - 3\nu_{\text{trans}} = (f_{\text{CEO}} + 3nf_{\text{rep}}) - 3nf_{\text{rep}} = f_{\text{CEO}} \,. \tag{12.33}$$

The OFC bandwidth requirement in this case is then even lower, because this time $3\nu_{\text{trans}} \geq \nu_{\text{low}}$ and therefore, according to (12.29) to (12.32), we obtain

$$\nu_{\text{low}} \leq 3\nu_{\text{trans}} \leq 3\Delta\nu \,. \tag{12.34}$$

Therefore, $2\nu_{\text{c}} \leq 7\Delta\nu$ and

$$\frac{\nu_{\text{high}}}{\nu_{\text{low}}} = \frac{2\nu_{\text{c}} + \Delta\nu}{2\nu_{\text{c}} - \Delta\nu} \geq \frac{8\Delta\nu}{6\Delta\nu} = \frac{4}{3} = 1.33 \,. \tag{12.35}$$

12.3.5 Method 4: SHG and THG of Two Auxiliary Oscillators and Self-Referenced 2f-to-3f Interferometer

Using two additional lasers with frequencies ν_a and ν_b, the second and third harmonics of these two lasers can be phase-locked at $3\nu_a = 2\nu_b$ with a PLL control system. In a subsequent step, ν_a is phase-locked to the nearest mode ν_{2m}:

$$\nu_a = f_{\text{CEO}} + 2mf_{\text{rep}} \,, \tag{12.36}$$

and the beat note between ν_b and its nearest comb mode at ν_{3m} provides the desired information about f_{CEO}:

$$\nu_b - \nu_{3m} = \frac{3}{2}(f_{\text{CEO}} + 2mf_{\text{rep}}) - (f_{\text{CEO}} + 3mf_{\text{rep}}) = \frac{1}{2}f_{\text{CEO}} \,. \tag{12.37}$$

This method has the same efficiency constraints as the frequency doubling and tripling schemes.

A similar frequency span reduction can be achieved with higher order nonlinear frequency generation. For example, if we generate multiple harmonics of the fundamental OFC with SHG, THG, etc., we obtain SH-OFC, TH-OFC, etc. The frequency span requirement is reduced to 1.5, because the beat signal is then given by

$$3\nu_n - 2\nu_m = 3(f_{\mathrm{CEO}} + n f_{\mathrm{rep}}) - 2(f_{\mathrm{CEO}} + m f_{\mathrm{rep}}) = f_{\mathrm{CEO}}\ , \quad m = 3n/2\ . \qquad (12.38)$$

Normally, this is more challenging than using transfer oscillators because there is not much intensity in the ν_{high} and ν_{low} comb lines. Thus, the signal-to-noise ratio of the beat signal is typically degraded. However, a CEO beat signal was still measured for Ti:sapphire lasers (see, e.g., [01Mor]) and Er-doped fiber lasers (see, e.g., [15Hit]). In the latter case a dual-pitch periodically poled lithium niobate ridge waveguide (DP-PPLN) was used after supercontinuum generation in a highly nonlinear fiber to significantly reduce the complexity and efficiency of the $2f$-to-$3f$ interferometer. In this DP-PPLN, two different quasi-phase-matching pitch sizes were integrated, and this allowed for successful CEO beat signal stabilization [15Hit]. The complexity of such interferometers can be further reduced with simultaneous spectral broadening and nonlinear frequency conversion in a single waveguide PPLN device, where the pitch can be freely adjusted [07Lan]. Ultimately, for optimized signal-to-noise ratio of the beat signal, the spectral broadening and the nonlinear frequency conversion may still need to be done in two separate steps, as demonstrated in [15Hit].

12.3.6 Method 5: Frequency Interval Bisection

Frequency interval bisection can be employed to divide the required comb width. The simplest scheme of this type uses a one-octave interval divider stage that generates the frequency ν_a at the midpoint of ν_b and $2\nu_b$. This can be done by phase-locking the second harmonic of ν_a to the sum frequency of ν_b and $2\nu_b$, i.e., $2\nu_a = 3\nu_b$, to yield

$$\frac{\nu_a}{2\nu_b} = \frac{3\nu_b/2}{2\nu_b} = \frac{3}{4}\ . \qquad (12.39)$$

This results in $\nu_a = 3\nu_b/2$, at the midpoint of ν_b and $2\nu_b$, and all of them are phase locked. If ν_a is phase-locked to the nearest line ν_{3m} in the OFC,

$$\nu_a = f_{\mathrm{CEO}} + 3m f_{\mathrm{rep}}\ , \qquad (12.40)$$

and $2\nu_b$ with (12.39) is interfered with its nearest line ν_{4m} in the OFC. Then we obtain the following beat signal:

$$2\nu_b - \nu_{4m} = \frac{4}{3}(f_{\mathrm{CEO}} + 3m f_{\mathrm{rep}}) - (f_{\mathrm{CEO}} + 4m f_{\mathrm{rep}}) = \frac{1}{3} f_{\mathrm{CEO}}\ . \qquad (12.41)$$

For a few-cycle KLM Ti:sapphire laser spectrum, as shown in Fig. 12.10, we can benefit from Method 5. If we choose the strong 1.338 μm Nd:YAG line as the fundamental of the octave interval, we get the second harmonic at 669 nm and the midpoint frequency at 892 nm. This interval of 669 to 892 nm would then have to be bridged by the comb, and Fig. 12.10 shows that the laser provides sufficient power in this wavelength range.

A further advantage of this scheme is that several mW of the power required at 669 nm can be generated directly from the fundamental by SHG, thus eliminating the need for an additional laser. The field at 892 nm can be generated by narrow linewidth DBR laser diodes commercially available for spectroscopy of the cesium D_1 transition. Furthermore, all nonlinear processes can be carried out with highly efficient mixing crystals, including periodically poled KTP for SHG of 1338 nm and non-critically phase-matched $KNbO_3$ for SHG of 892 nm and SFG of 1338 and 669 nm.

For initial feasibility studies in 1999, two independent cw lasers were used to detect the beat notes with the argon-ion laser pumped KLM Ti:sapphire laser. Two extended cavity diode lasers (ECDL), emitting at 658 and 812 nm, were used as local oscillators in this initial experiment. Although these are not the exact wavelengths required for Method 5, it is reasonable to suggest that the lock at 669 nm will work if the beat at 658 nm is sufficiently strong and coherent. Additionally, because the single line power is similar at 812 and 892 nm, a measurement of the beat at 812 nm indicates the feasibility of the lock at 892 nm. Both beat frequencies were successfully measured with a microwave spectrum analyzer with a 16 dB signal-to-noise ratio (RBW = 100 kHz), but also showed strong frequency variations of about 10 MHz on a millisecond time scale [99Tel]. At that time no special efforts were made to package the laser and generally reduce both intensity noise and timing jitter.

High quality CEO beat signals as shown in Fig. 12.4 were only obtained later. Generally, the overall intensity noise and timing jitter of the laser has to be improved before high-quality CEO stabilization can be effectively implemented. For Ti:sapphire lasers, argon-ion pump lasers were replaced because of the large excess pump noise, as discussed in Sect. 11.4.5. In addition, intracavity prisms for dispersion compensation were also identified as an additional noise source [02Hel2,03Hel].

12.4 Phase and Frequency Noise

12.4.1 Spectrally Resolved CEO Frequency Noise

The measured linewidth of the CEO beat as shown in Fig. 12.4 is a single parameter that gives only partial information about the frequency noise of the CEO beat. As for intensity noise and timing jitter, we need to measure the CEO frequency noise power spectral density (PSD) $S_{CEO}(f) \equiv S_{\delta\nu}(f)$ of the CEO beat signal in (12.26):

$$f_{CEO}(t) = f_{CEO,0} + \delta\nu(t) , \tag{12.42}$$

where we assume for simplicity that there is no drift of the center frequency $f_{\mathrm{CEO},0}$ and $\langle \delta\nu(t) \rangle = 0$. The average CEO frequency $f_{\mathrm{CEO},0} = \langle f_{\mathrm{CEO}}(t) \rangle$ in a laser can be adjusted using intracavity dispersive elements. For example, a slightly wedged thin plate translated perpendicular to the beam propagation will change the group-phase offset and therefore $f_{\mathrm{CEO},0}$. A frequency drift results in a divergence of the PSD at low noise frequencies, as discussed in Sect. 11.1.2. We can accommodate for this by limiting the lower frequency and therefore the measurement time for free-running lasers.

Using the same laser as for Fig. 12.4, the measured spectrally resolved CEO frequency noise $S_{\mathrm{CEO}}(f) \equiv S_{\delta\nu}(f)$ is shown in Fig. 12.11 [11Sch2]. To measure the frequency noise PSD, we need to have a suitable frequency-to-voltage converter (also referred to as a frequency discriminator) which converts frequency fluctuations of the CEO beat $\delta\nu(t)$ (12.42) into fluctuations of an electrical signal, i.e., a voltage $\delta V(t)$ that can be spectrally analyzed (Fig. 12.12). The spectral analysis of these voltage fluctuations $\delta V(t)$ can be a fast Fourier transform (FFT) with some commercial signal analyzers, i.e., $\delta\tilde{V}(f) = F\{\delta V(t)\}$. We then obtain the voltage noise power spectral density $S_V(f) = |\delta\tilde{V}(f)|^2$.

As discussed in Sect. 11.2.2, there are optical and microwave/RF discriminators. Figure 12.12 shows the schematic of an analog phase-locked loop (PLL) discriminator [11Sch3]. The basic principle of this electrical discriminator is to phase-lock a voltage-controlled oscillator (VCO) to the beat signal of the frequency noise to be analyzed, using a high-bandwidth PLL that adjusts the VCO to follow any frequency fluctuations $\delta\nu(t)$ of the input CEO beat signal. The output signal in Fig. 12.12 $V(t) = V_0 + \delta V(t)$ reflects the frequency fluctuations of the input signal, which

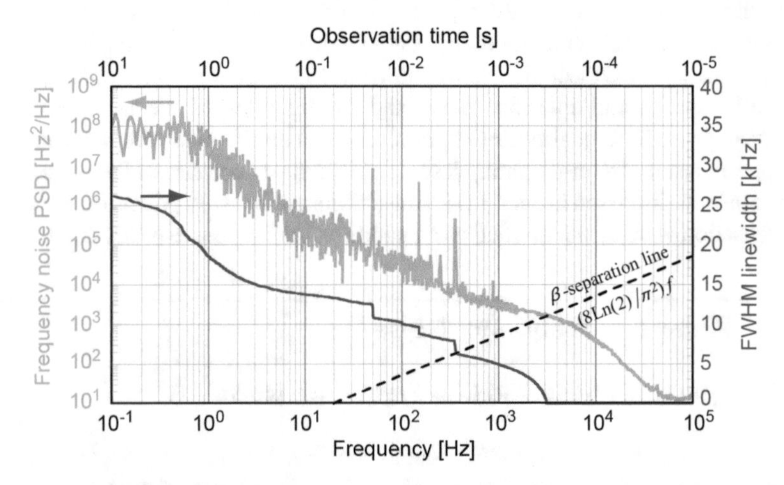

Fig. 12.11 One-sided CEO frequency noise power spectral density (PSD) $S_{\mathrm{CEO}}(f) \equiv S_{\delta\nu}(f)$ [compare with (12.42)]: spectrally resolved CEO frequency noise from a SESAM-modelocked Er:Yb:glass laser as used in Fig. 12.4. The calculated FWHM linewidth from the measured CEO frequency noise PSD is shown *on the right*, using the approximation given in Sect. 12.5.6. According to [11Sch2]

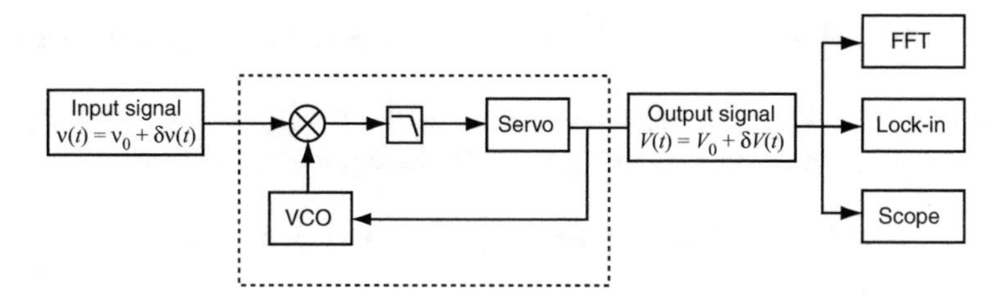

Fig. 12.12 Example of an electrical frequency discriminator based on an analog phase-locked loop discriminator according to [11Sch3]. This discriminator converts frequency fluctuations $\delta\nu(t)$ [compare equation (12.42)] into voltage fluctuations $\delta V(t)$ which can be analyzed in different ways, e.g., Scope, a lock-in amplifier, or an FFT analyzer. FFT: fast Fourier transform, VCO: voltage controlled oscillator

can then be analyzed to determine the $S_V(f)$ and therefore the $S_{\text{CEO}}(f) \equiv S_{\delta\nu}(f)$, according to (12.43).

A frequency discriminator that converts frequency fluctuations $\delta\nu(t)$ into voltage fluctuations $\delta V(t)$ is characterized by its discriminator slope D_ν in units of V/Hz. The frequency noise PSD as shown in Fig. 12.11 is then retrieved from the voltage noise PSD $S_V(f) = |\delta\tilde{V}(f)|^2$:

$$S_{\text{CEO}}(f) \equiv S_{\delta\nu}(f) = S_V(f)/D_\nu^2 . \tag{12.43}$$

Figure 12.11 shows the frequency noise power spectral density (PSD) of the free-running CEO beat (on the left scale) defined by (12.42) and (12.43) and measured with an electrical frequency discriminator according to Fig. 12.12. On the right scale of Fig. 12.11, the FWHM linewidth of the free-running CEO frequency noise spectrum is shown. This is calculated using the approximation described in Sect. 12.5.6, taking the measured PSD $S_{\text{CEO}}(f) \equiv S_{\delta\nu}(f)$ as a function of the observation time T_o, which sets the lower integration limit $f_l = 1/T_o$. Note that this calculated CEO linewidth is in reasonably good agreement with the previously reported value extracted from the microwave beat signal spectrum, e.g., a 3.6 kHz FWHM linewidth is measured with a 1 kHz RBW in Fig. 12.4b.

At any noise frequency f, the CEO frequency noise power spectral density $S_{\delta\nu}(f)$ only contributes to the linewidth if its value lies above the β-separation line [10Dom], i.e., if $S_{\delta\nu}(f) > (8\ln 2/\pi^2)f$. We will discuss the β-separation line in more detail in Sect. 12.5.5. From the measured spectrum shown in Fig. 12.11, this corresponds only to noise frequencies smaller than 3 kHz. This implies that the CEO linewidth can be significantly narrowed using a feedback bandwidth of just a few kilohertz.

Before we continue, we need to review the basic understanding of frequency and phase noise (see Sect. 12.4.2), following a broadly accepted international standard (see, e.g., [78Rut]). For example, it may be initially confusing that the units of a power spectral density are Hz^2 per hertz, as shown in Fig. 12.11. Moreover, we have to make sure we do not mix up the phase noise of the timing jitter with the phase noise of an optical frequency. And finally, it will be important to distinguish clearly between the frequency noise spectrum and the optical laser spectrum (see Sect. 12.5).

12.4.2 Basic Mathematical Principle of Phase and Frequency Noise

The basic principles are accepted standards in the field (see, e.g., [78Rut]). We consider a laser field with a certain amount of phase noise as follows:

$$E(t) = E_0 \exp\left[\mathrm{i}\big(\omega_0 t + \varphi(t)\big)\right] = A(t)\exp(\mathrm{i}\omega_0 t)\ . \tag{12.44}$$

Here we neglect any amplitude noise and assume

$$A(t) = E_0 \exp\big(\mathrm{i}\varphi(t)\big)\ , \quad |E_0|^2 = I_0 = \text{constant}\ , \tag{12.45}$$

where, for reasons of simplicity, we neglect any proper normalizations to obtain the correct units. The frequency noise is given by $\delta\omega(t) \equiv \dot{\varphi}(t)$ as follows [see (12.7)]:

$$\Delta\omega(t) = \frac{\mathrm{d}}{\mathrm{d}t}\big[\omega_0 t + \varphi(t)\big] = \omega_0 + \dot{\varphi}(t) \equiv \omega_0 + \delta\omega(t)\ . \tag{12.46}$$

As summarized in Chap. 11, Tables 11.1, and 11.2, we can define the autocorrelation $R_\varphi(\tau)$ and the power spectral density (PSD) $S_\varphi(\omega)$ of the phase noise $\varphi(t)$, and equivalently $R_{\delta\omega}(\tau)$ and $S_{\delta\omega}(\omega)$ of the frequency noise $\delta\omega(t)$. Because $\delta\omega(t)$ is the time derivative of the phase, i.e., $\delta\omega(t) \equiv \dot{\varphi}(t)$, and in the Fourier transform, the derivative in the time domain corresponds to multiplication by $\mathrm{i}\omega$, we obtain a multiplication by ω^2 in terms of PSDs. Thus,

$$\boxed{S_{\delta\omega}(\omega) = \omega^2 S_\varphi(\omega)}\ , \tag{12.47}$$

and also

$$\boxed{S_{\delta\nu}(f) = f^2 S_\varphi(f)}\ . \tag{12.48}$$

The detailed derivation of (12.47) is as follows, using the definition of the Fourier transform in Appendix A, and taking into account all the fundamental definitions given in Tables 11.1 and 11.2 with $\delta\omega(t) = \dot{\varphi}(t)$ (12.46):

$$S_{\delta\omega}(\omega) = |F\{\delta\omega(t)\}|^2 = \left|F\left\{\frac{\mathrm{d}}{\mathrm{d}t}\varphi(t)\right\}\right|^2\ ,$$

where

$$\begin{aligned}\frac{\mathrm{d}}{\mathrm{d}t}\varphi(t) &= \frac{\mathrm{d}}{\mathrm{d}t}F^{-1}\{\tilde{\varphi}(\omega)\} = \frac{\mathrm{d}}{\mathrm{d}t}\frac{1}{2\pi}\int \tilde{\varphi}(\omega)\mathrm{e}^{\mathrm{i}\omega t}\,\mathrm{d}\omega \\ &= \frac{1}{2\pi}\int \tilde{\varphi}(\omega)\mathrm{i}\omega\mathrm{e}^{\mathrm{i}\omega t}\,\mathrm{d}\omega = F^{-1}\{\mathrm{i}\omega\tilde{\varphi}(\omega)\}\ .\end{aligned}$$

Hence,

$$F\left\{\frac{\mathrm{d}}{\mathrm{d}t}\varphi(t)\right\} = FF^{-1}\{\mathrm{i}\omega\tilde{\varphi}(\omega)\} = \mathrm{i}\omega\tilde{\varphi}(\omega)\ ,$$

and

$$S_{\delta\omega}(\omega) = \omega^2 |\tilde{\varphi}(\omega)|^2 \ .$$

Since $S_\varphi(\omega) = |\tilde{\varphi}(\omega)|^2$, we therefore obtain $S_{\delta\omega}(\omega) = \omega^2 S_\varphi(\omega)$, as claimed. Similarly, (12.48) can be shown with (12.46):

$$\delta\nu(t) = \frac{1}{2\pi}\dot{\varphi}(t) = \frac{1}{2\pi}\frac{\mathrm{d}}{\mathrm{d}t}\varphi(t) \ ,$$

$$S_{\delta\nu}(f) = |F\{\delta\nu(t)\}|^2 = \frac{1}{4\pi^2}\left|F\left\{\frac{\mathrm{d}}{\mathrm{d}t}\varphi(t)\right\}\right|^2 \ ,$$

and

$$F\left\{\frac{\mathrm{d}}{\mathrm{d}t}\varphi(t)\right\} = FF^{-1}\{\mathrm{i}2\pi f\tilde{\varphi}(f)\} = \mathrm{i}2\pi f\tilde{\varphi}(f) \ .$$

With $S_\varphi(f) = |\tilde{\varphi}(f)|^2$ we then obtain (12.48).

We summarize all the relevant noise functions in Table 12.2, following the basic definitions in Sect. 11.1.2.

Dealing with laser and electrical signals, we typically determine the power spectral density (PSD) $S_{\mathrm{P}}(\omega)$ in units of watt/hertz. Note that with regard to the PSD for phase and frequency noise, i.e., $S_\varphi(\omega)$ and $S_{\delta\omega}(\omega)$, we still speak of 'power' spectral densities, even though there is no power involved. The units for $S_\varphi(\omega) = |\tilde{\varphi}(\omega)|^2$ are radians squared per hertz, and for $S_{\delta\nu}(f) = |F\{\delta\nu(t)\}|^2$ they are Hz^2 per hertz. This sometimes causes confusion.

As shown in Table 12.1 (and discussed in Sect. 11.1.2) the two-sided PSD and the autocorrelation are a Fourier transform pair of even functions, for which the cosine transform is the same as the Fourier transform:

$$S_\varphi(\omega) = F\{R_\varphi(\tau)\} = \int R_\varphi(\tau)\mathrm{e}^{-\mathrm{i}\omega\tau}\mathrm{d}\tau = 2\int_0^\infty R_\varphi(\tau)\cos(\omega\tau)\mathrm{d}\tau \ , \tag{12.49}$$

$$R_\varphi(\tau) = F^{-1}\{S_\varphi(\omega)\} = \frac{1}{2\pi}\int S_\varphi(\omega)\mathrm{e}^{\mathrm{i}\omega\tau}\mathrm{d}\omega = \frac{1}{\pi}\int_0^\infty S_\varphi^{\text{two-sided}}(\omega)\cos(\omega\tau)\mathrm{d}\omega \ , \tag{12.50}$$

where $F^{-1}\{\cdot\}$ denotes the inverse Fourier transform. In a similar way, $S_{\delta\omega}(\omega)$ and $R_{\delta\omega}(\tau)$ (i.e., defined in Table 12.1) form a Fourier transform pair, which are identical in form to $S_\varphi(\omega)$ and $R_\varphi(\tau)$. Note that, for (12.49) and (12.50), we have used the two-sided PSDs, which results in a factor of 2 in the cosine transform. We stress this fact by adding the superscript 'two-sided' on the symbol.

When we measure the frequency or phase noise PSD with a microwave spectrum analyzer, we obtain them as a function of a noise frequency f in units of Hz and not as a function of an angular frequency ω in units of radians per second. Normally, we also refer to angular frequency as frequency. However, we need to be careful about different factors of 2π here and there. The Fourier transform over f does not include the additional factor $1/2\pi$ as shown in Table 11.2. We normally use the same function symbol S_φ, even though the functions $S_\varphi(\omega)$ and $S_\varphi(f)$ are mathematically not identical, when we make a variable substitution of ω to $\omega = 2\pi f$.

Table 12.2 Summary of relevant equations for the frequency $\delta\omega(t)$ and phase $\varphi(t)$ noise, following the general notation for a random variable $X(t) = \delta\omega(t)$ or $X(t) = \varphi(t)$ in Tables 11.1 and 11.2. The symbol $\langle\cdot\rangle$ denotes the time average, i.e., $\langle X(t)\rangle = \overline{X(t)} = \lim_{T_0\to\infty}\int_{-T_0/2}^{T_0/2} X(t)\mathrm{d}t$, and $F\{\cdot\}$ denotes the Fourier transform. Here we use two-sided PSDs

	Phase noise	Frequency noise
Random variable (assuming real functions)	$\varphi(t)$ (12.44)	$\delta\omega(t) = \dot{\varphi}(t)$ (12.46)
Fourier transform $F\{\cdot\}$	$\tilde{\varphi} \equiv \tilde{\varphi}(\omega) = F\{\varphi(t)\}$	$\delta\tilde{\omega} \equiv \delta\tilde{\omega}(\omega) = F\{\delta\omega(t)\}$
Mean value	$\langle\varphi(t)\rangle = 0$	$\langle\delta\omega(t)\rangle = 0$ only for $\langle\omega(t)\rangle = 0$, i.e., no drift
Autocorrelation	$R_\varphi(\tau) = \langle\varphi(t)\varphi(t+\tau)\rangle$	$R_{\delta\omega}(\tau) = \langle\delta\omega(t)\delta\omega(t+\tau)\rangle$
Power spectral density (PSD)	$S_\varphi(\omega) = F\{R_\varphi(\tau)\}$ $S_\varphi(\omega) = \lvert\tilde{\varphi}(\omega)\rvert^2$	$S_{\delta\omega}(\omega) = F\{R_{\delta\omega}(\tau)\}$ $S_{\delta\omega}(\omega) = \lvert\delta\tilde{\omega}\rvert^2$
Equation (12.47)	$S_\varphi(\omega) = S_{\delta\omega}(\omega)/\omega^2$	$S_{\delta\omega}(\omega) = \omega^2 S_\varphi(\omega)$
Variance	$\sigma_\varphi^2 = \langle\varphi^2(t)\rangle = R_\varphi(0)$ $\sigma_\varphi^2 = \frac{1}{2\pi}\int S_\varphi(\omega)\mathrm{d}\omega$	$\sigma_{\delta\omega}^2 = \langle\delta\omega^2(t)\rangle = R_{\delta\omega}(0)$ $\sigma_{\delta\omega}^2 = \frac{1}{2\pi}\int S_{\delta\omega}(\omega)\mathrm{d}\omega$

The PSD $S_{\delta\nu}(f)$ determines the frequency noise spectrum and should not be confused with the optical spectrum (see the discussion in Sect. 12.5). Note that, when we change the PSD with respect to $\delta\omega$ (i.e., $S_{\delta\omega}$) versus $\delta\nu$ (i.e., $S_{\delta\nu}$), we should expect a change with $\omega = 2\pi f$ and therefore $\delta\omega = 2\pi\,\delta\nu$ as follows:

$$\boxed{S_{\delta\omega}(f) = 4\pi^2 S_{\delta\nu}(f)}\,. \tag{12.51}$$

Derivation of (12.51)
We have

$$\begin{aligned} S_{\delta\omega}(f) &= \lvert\delta\tilde{\omega}(\omega)\rvert^2 = \lvert F\{\delta\omega(t)\}\rvert^2 \\ &= 4\pi^2\lvert F\{\delta\nu(t)\}\rvert^2 = 4\pi^2 S_{\delta\nu}(f)\,, \end{aligned}$$

using $\delta\omega(t) = 2\pi\,\delta\nu(t)$ to start the second line.

12.4.3 Integrated Phase and Frequency Noise

The rms frequency fluctuation $\delta\nu_{\rm rms}$ relative to a mean value $\langle\delta\nu(t)\rangle = 0$ is given by the standard deviation [see Table 11.2 and (11.5)]:

$$\delta\nu_{\rm rms} \equiv \sigma_{\delta\nu} = \sqrt{\int S_{\delta\nu}(f)\mathrm{d}f}\ . \tag{12.52}$$

The noise can also be given in FWHM assuming Gaussian noise statistics [see (11.55)–(11.57) for the timing jitter noise]. In this case the FWHM frequency noise is equal to $2.355\delta\nu_{\rm rms}$.

Note that $\delta\nu_{\rm rms}$ determines the rms spectral width of the noise spectrum and should not be confused with the optical linewidth. The phase noise broadens the optical linewidth of a monochromatic laser and this linewidth broadening will be discussed in more detail in Sect. 12.5.

Very often we also define the rms frequency noise for a certain observation time $T_{\rm o}$, which sets the lower integration limit to $f_1 = 1/T_{\rm o}$ (as shown in Fig. 12.11):

$$\delta\nu_{\rm rms}[f_1,\infty] \equiv \sigma_{\delta\nu}[f_1,\infty] = \sqrt{\int_{f_1}^{\infty} S_{\delta\nu}^{\text{one-sided}}(f)\mathrm{d}f}\ , \tag{12.53}$$

as before with the intensity noise and timing jitter (see Sect. 11.3), except that here we have used the one-sided PSD $S_{\delta\nu}^{\text{one-sided}}(f)$ (see, e.g., Fig. 12.11), which is more typically used for frequency noise measurements. Note that an additional factor of 2 is only introduced when two-sided PSDs are used (e.g., shown in Tables 11.4 and 11.5).

Note that, for the frequency noise PSD of the CEO beat signal, we replace the frequency $\omega_0/2\pi$ in (12.44) with the center CEO frequency $f_{(\mathrm{CEO},0)}$ of (12.42) and $\delta\omega = 2\pi\,\delta\nu$, with the stationary noise condition $\langle f_{\mathrm{CEO}}(t)\rangle = f_{(\mathrm{CEO},0)}$. Therefore, all the definitions and discussions in this section also apply to the CEO frequency and phase noise.

Similarly, for an rms phase noise $\varphi_{\rm rms}$ in a frequency bandwidth between f_1 and f_2 relative to a mean value $\langle\varphi(t)\rangle = 0$, we have

$$\varphi_{\rm rms}[f_1,f_2] \equiv \sigma_{\varphi}[f_1,f_2] = \sqrt{\int_{f_1}^{f_2} S_{\varphi}^{\text{one-sided}}(f)\mathrm{d}f}\ . \tag{12.54}$$

For example, in Figs. 12.11 and 12.13, the noise is given as a frequency noise PSD $S_{\delta\nu}(f)$, because an electrical frequency discriminator was used, as shown in Fig. 12.12. Therefore, it is useful to express the phase noise PSD as a function of the frequency noise PSD. With (12.48), we obtain

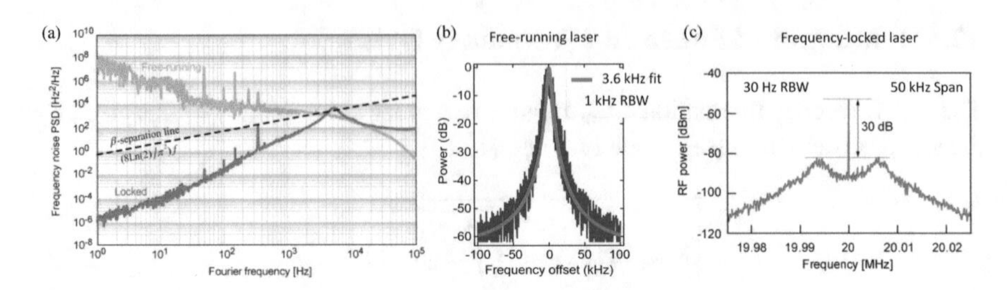

Fig. 12.13 (**a**) Reduction of the CEO frequency noise of a modelocked Er:Yb:glass laser obtained with a stabilization loop. The corresponding CEO frequency linewidth is shown for both the free-running (**b**) and locked CEO beat signal (**c**) [11Sch2]. RBW: resolution bandwidth of microwave spectrum analyzer

$$\boxed{\varphi_{\text{rms}}[f_1, f_2] \equiv \sigma_\varphi[f_1, f_2] = \sqrt{\int_{f_1}^{f_2} \frac{S_{\delta\nu}^{\text{one-sided}}(f)}{f^2}\mathrm{d}f}} \,. \tag{12.55}$$

Example SESAM-Modelocked Er:Yb:glass Laser

The same laser is used as before for Figs. 12.4, 12.11, and 12.13. The cavity design is shown later, in Fig. 12.15, but is not relevant for the following discussion. Here we shall use this laser example to demonstrate the effect on the PSD when the CEO frequency is stabilized and what impact it has on the optical linewidth and rms phase noise. How to stabilize a CEO beat signal will be discussed in more detail in Sect. 12.6.

Figure 12.13a shows the one-sided CEO frequency noise power spectral density $S_{\text{CEO}}(f) \equiv S_{\delta\nu}(f)$ of the free-running and stabilized laser [11Sch2]. The free-running CEO frequency linewidth shown in Fig. 12.4b is shown again in Fig. 12.13b for a direct comparison. The good stability of the free-running laser allows for an FWHM of 3.6 kHz. The reduced frequency noise PSD for the stabilized laser leads to a drastic reduction in the CEO linewidth, as shown in Fig. 12.13c, with a 30 dB signal-to-noise ratio at the 30 Hz resolution bandwidth (RBW), measured again with a microwave spectrum analyzer. The CEO frequency was locked to a 20 MHz reference oscillator (i.e., frequency-doubled 10 MHz signal from an H maser) using commercial locking electronics. The residual noise below the β-separation line contributes to the pedestal underneath the narrow CEO peak (see Sect. 12.5.5). The peak in the center of the CEO spectrum is a coherent peak with zero linewidth (i.e., only limited by the RBW of the spectrum analyzer).

Note that, for the CEO frequency linewidth, we can measure the 'optical' spectrum with an electronic spectrum analyzer (i.e., a microwave spectrum analyzer), because the CEO frequency is $f_{\text{CEO}} \leq f_{\text{rep}}/2$ in the RF resp. microwave regime. See also the warning in Sect. 12.1.

From the measured frequency noise PSD $S_{\delta\nu}(f)$, an integrated rms phase noise can be calculated according to (12.55). In (12.14), the rms phase noise is integrated within a frequency range of [0.1 Hz, $f_{\max}$], for which the lower frequency limit f_1 = 0.1 Hz sets the measurement/observation time to 10 s (i.e., $1/f_1$). In this case, a

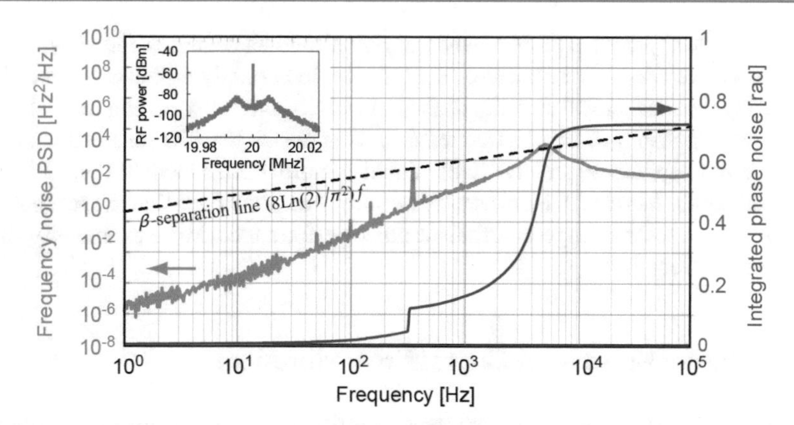

Fig. 12.14 One-sided CEO frequency noise power spectral density (PSD) $S_{\mathrm{CEO}}(f) \equiv S_{\delta\nu}(f)$ (12.42) of a stabilized CEO frequency beat signal from the same laser, as shown free-running in Fig. 12.11. The integrated rms phase noise $\varphi_{\mathrm{rms}}[0.1\ \mathrm{Hz},\ f]$ in units of rad is shown *on the right*. *Inset*: CEO frequency peak as shown in Fig. 12.13. According to [11Sch2]

one-sided PSD is used, so the integration from 0.1 Hz up to the cutoff frequency $f_{\max}$ is given by (12.55)

$$\varphi_{\mathrm{rms}}[0.1\ \mathrm{Hz},\ f_{\max}] = \sqrt{\int_{0.1\ \mathrm{Hz}}^{f_{\max}} \frac{S_{\delta\nu}(f)}{f^2} \mathrm{d}f}\ .$$

In order to integrate this into Fig. 12.14, $f_{\max}$ is set to 'Frequency'. The integrated rms phase noise $\varphi_{\mathrm{rms}}[0.1\ \mathrm{Hz},\ f_{\max}]$ increases with the frequency range until the phase noise at higher frequencies of over 10 kHz has no significant contributions. The fully integrated rms phase noise $\varphi_{\mathrm{rms}}[0.1\ \mathrm{Hz},\ 100\ \mathrm{kHz}]$ results in 0.72 rad rms. For Gaussian noise statistics, the FWHM is given by 2.355σ [see (11.59) in Sect. 11.3.3], so with $\varphi_{\mathrm{rms}} = \sigma_\varphi$, we obtain a $\varphi_{\mathrm{FWHM}} = 2.355\varphi_{\mathrm{rms}}$. An rms phase noise of 0.72 rad then results in an FWHM phase noise of 1.7 rad, i.e., close to 30% of an optical cycle, for an observation time of 10 s.

The major contribution to the integrated phase noise in the stabilized laser originates from the 5.5 kHz servo bump, as shown in Fig. 12.14. This is also the feedback bandwidth of the stabilization, mainly limited by the dynamics of the Er gain medium in the femtosecond laser [04Sch] (see also Sect. 11.4.4). An additional small technical contribution of about 10% results from a peak at 360 Hz which introduces a step-like increase in the integrated rms phase noise shown in Fig. 12.14. This noise peak is due to the mechanical resonance of the translational stage that holds the pump laser focusing lens in the laser resonator, which is excited by ambient acoustic noise. This acoustic sensitivity was confirmed by exciting a loudspeaker with a sine waveform of varying frequency in the vicinity of the unstabilized laser cavity and by demodulating the 20 MHz CEO beat using the frequency discriminator. The signal subsequently detected with a lock-in amplifier referenced to the speaker excitation frequency clearly showed a resonance at 360 Hz when scanning the speaker frequency. The residual phase noise of the laser could certainly be further reduced by improving the mechanical stability of the laser cavity and by extending the stabilization loop bandwidth to higher frequencies. This means that the noise background can be better suppressed and the 30 dB signal-to-noise ratio shown in Fig. 12.13 can be substantially increased. However, this example demonstrates that the mechanical stability of the laser is important for low noise performance. More examples will be discussed in Sect. 12.6.

12.4.4 Phase Noise from Cavity Mirror Vibrations

Mirror vibrations can make a significant contribution to the phase noise unless a very stable cavity setup is used. Following the derivation given in [07Pas], we first consider only cavity length changes, but no tilts (which would lead to misalignment). Such mirror vibrations ideally have no effect on the intensity noise, but a strong effect on the phase noise. A change δL in the roundtrip length (corresponding to twice the cavity length change in the case of a linear cavity) will change the optical phase by $\delta\varphi = k\delta L = 2\pi\delta L/\lambda$ (with $k = 2\pi/\lambda$) in each roundtrip. This means that the phase errors are integrated over subsequent roundtrips. This can be described by the differential equation

$$T_{\mathrm{R}}\frac{\mathrm{d}}{\mathrm{d}t}\delta\varphi = k\delta L\ , \tag{12.56}$$

where $\delta\varphi$ now describes the phase evolution over many cavity roundtrips. With a Fourier transform a time derivation goes into a multiplication by a factor of $\mathrm{i}\omega$. We then obtain

$$\delta\tilde{\varphi}(f) = \frac{k}{\mathrm{i}2\pi f T_{\mathrm{R}}}\delta\tilde{L}(f)\ . \tag{12.57}$$

From this we can conclude that random cavity length changes with a PSD $S_{\delta L}(f)$ lead to a phase noise PSD

$$S_{\varphi}(f) = \left(\frac{k}{2\pi f T_{\mathrm{R}}}\right)^2 S_{\delta L}(f)\ . \tag{12.58}$$

The f^{-2}-dependence of the coupling factor shows that the low-frequency phase noise is more sensitive to mirror vibrations. This effect is even more enhanced because, at high noise frequencies, above the mechanical resonance frequencies, $S_{\delta L}(f)$ is also strongly reduced.

The effect of mirror tilts is much more complicated to describe, because these lead to laser cavity misalignment. How exactly a laser will respond to that depends on the cavity design (determining how the resulting beam offset depends on the position in the cavity), as well as on factors like the pump distribution and thermal lensing in the gain medium. Both intensity and phase noise can be affected by such mirror tilts, while a general quantitative description becomes very difficult.

12.4.5 Intensity Noise, Timing Jitter, and CEO Frequency Noise

The timing jitter is sometimes given as a phase noise, which corresponds to the timing phase of the pulse envelope as shown in Figs. 11.1 and 11.2. We can therefore refer to this phase as the 'reprate' phase φ_{rep}, given by the timing error ΔT as defined in (11.44)

$$\frac{\varphi_{\text{rep}}}{2\pi} = \frac{\Delta T}{T} \implies \varphi_{\text{rep}} = 2\pi f_{\text{rep}} \Delta T \ , \tag{12.59}$$

whence the power spectral densities are related by

$$S_{\varphi_{\text{rep}}}(f) = (2\pi f_{\text{rep}})^2 S_{\Delta T}(f) \ . \tag{12.60}$$

Cavity length fluctuations affect the timing jitter ΔT and therefore the timing phase PSD $S_{\varphi_{\text{rep}}}(f)$, given in (12.60). This was, for example, used to stabilize the timing jitter in passively modelocked lasers (see Chap. 11). Therefore, with (12.60) and $f_{\text{CEO}} = f_{\text{rep}}\phi_{\text{CEO}}/2\pi$ from (12.11), we have a strong coupling between the CEO phase noise and timing jitter noise with cavity length fluctuations.

Generally, however, the coupling is nonlinear and significantly more complicated. In addition, the noise depends on the specific laser gain, saturable absorber, and cavity design [03Hel,06Pas]. We discussed the nonlinear coupling of noise in modelocked lasers in Sect. 11.4 with respect to intensity noise and timing jitter. For example, intracavity prisms for negative GDD should generally be avoided [02Hel1], because any pointing instabilities of the intracavity laser beam cause group-phase offset (GPO) noise, which strongly affects the CEO noise. Therefore, GTI mirrors for dispersion compensation result in lower CEO noise. In addition, for example, the Kerr effect couples intensity noise with the timing jitter and the CEO frequency noise [03Hel]. The Kerr effect affects the CEO noise more strongly than the timing jitter because a frequency-dependent Kerr effect [96DeS,91She] also induces an intensity-dependent group-phase offset.

Figure 12.15 demonstrates the coupling of the pulse repetition rate f_{rep} and the CEO frequency f_{CEO} with the cavity length and the intensity in a SESAM-modelocked diode-pumped Er:Yb:glass laser (see the top panel of Fig. 12.15). This is the same laser that was used for all the previous examples in this chapter. Soliton modelocking is used with GTIs for negative GDD and with the SESAM to start and stabilize modelocking (see Sect. 9.5). This laser generates 110 mW with 170 fs pulses at a center wavelength of about 1.55 μm at a pulse repetition rate of 75 MHz. The intracavity intensity is changed with the pump power, i.e., the pump current to the diodes and the cavity length can be changed with a piezo transducer (PZT) on the SESAM cavity end mirror.

A large, linear, and precise control of the repetition rate can be achieved with the PZT, with a tuning coefficient $\Delta f_{\text{rep}}/\Delta V_{\text{PZT}}$ around 30 Hz/V. Besides its direct effect on the repetition rate, a change in the cavity length also has a coupled effect on the 20 MHz CEO frequency, with a tuning coefficient that is four orders of magnitude larger ($\Delta f_{\text{CEO}}/\Delta V_{\text{PZT}} \approx 3 \times 10^5$ Hz/V). Note that a variation in the frequency of an optical comb line $\nu_m = f_{\text{CEO}} + m f_{\text{rep}}$ due to a variation in the cavity length is

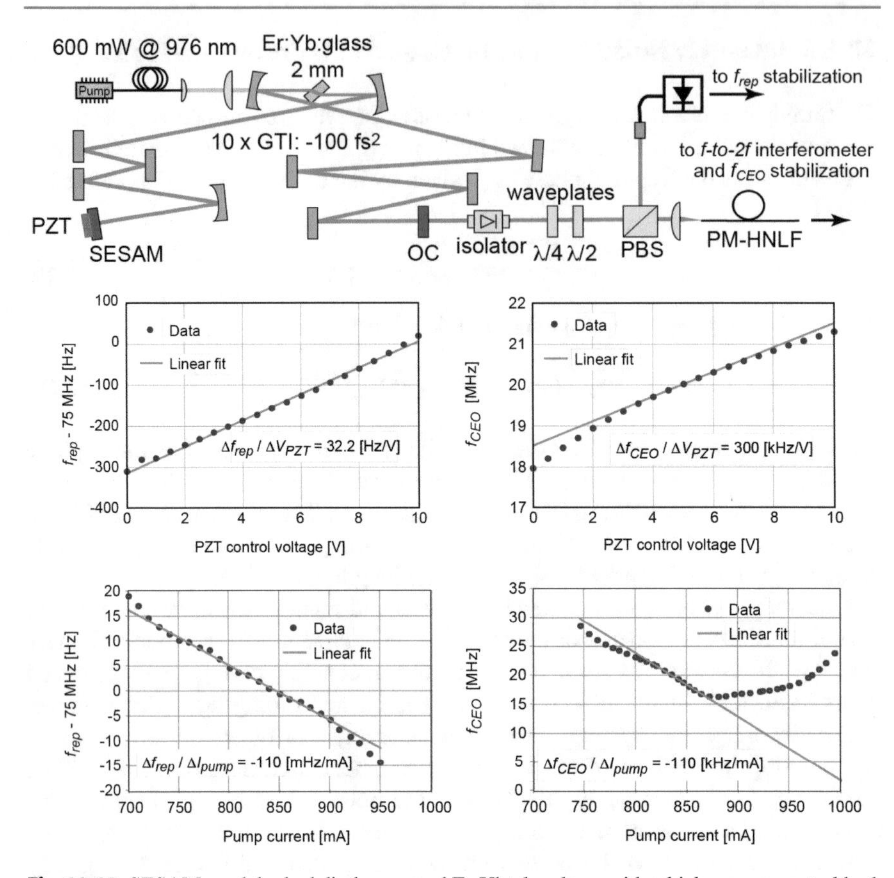

Fig. 12.15 SESAM-modelocked diode-pumped Er:Yb:glass laser with which we can control both the pulse repetition rate f_{rep} and the CEO frequency f_{CEO} with either the cavity length or the pump power. PZT: piezo transducer, OC: output coupler, PBS: polarizing beam splitter, PM-HNLF: polarization maintaining highly nonlinear fiber [12Sch]

strongly dominated by the contribution of the repetition rate compared to the CEO contribution, due to the large mode index ($m \approx 2.5 \times 10^6$ at $\lambda = 1.55$ μm) that multiplies f_{rep}.

The pump power has a direct influence on the pulse parameters, such as pulse duration and energy, which directly translates into a change in f_{CEO}. Therefore, the current of the pump laser can be used for fine tuning and stabilization of f_{CEO}, which is the standard stabilization method for diode-pumped femtosecond lasers. The dependence of f_{CEO} on the pump power may be nonlinear and even non-monotonic, as observed in Ti:sapphire combs [03Hol]. For this reason, an initial coarse adjustment is usually made with an intracavity element, such as for example a slightly wedged dispersive plate. The dependence of f_{CEO} on the pump current in the laser is also shown in Fig. 12.15 with a tuning coefficient in the range of $\Delta f_{\text{CEO}}/\Delta I_{\text{pump}} \approx 100$ kHz/mA around 20 MHz. This value can be changed by a significant amount and its sign can even be reversed, depending on the resonator configuration (precise adjustment of the intracavity elements).

In addition to its effect on f_{CEO}, the pump power also affects the repetition rate f_{rep} (Fig. 12.15) due to the resulting change in the resonator optical path length (combination of thermal and Kerr lensing as well as a slight redshift of the center wavelength at higher pump power due to the inversion-dependent gain profile). The tuning coefficient is about one million times smaller ($\Delta f_{\mathrm{rep}}/\Delta I_{\mathrm{pump}} \approx -100$ mHz/mA). However, this small tuning factor is not negligible, considering the effect of the pump current on an optical comb line. Due to the large scaling of the repetition rate by the mode index $m \approx 2.5 \times 10^6$, both the change in the CEO and the repetition rate for a variation of the pump current introduce similar contributions to the tuning of an optical comb line.

The cavity PZT and the pump laser current can be used as two actuators (i.e., control parameters) for the full stabilization of the modelocked optical frequency comb, even though they are not independent control parameters. Typically, the cavity PZT is used for the stabilization of the repetition rate and the pump laser current for the stabilization of the CEO frequency, which will be discussed in more detail in Sect. 12.6.

12.4.6 Phase-Time and Fractional (or Relative) Frequency Noise

The phase fluctuations $\varphi(t)$ and frequency fluctuations $\delta\omega(t) \equiv \dot{\varphi}(t)$ are often described by the following normalized form [78Rut,09Rub]:

$$x(t) \equiv \frac{\varphi(t)}{\omega_0} = \frac{\varphi(t)}{2\pi\,\nu_0} \tag{12.61}$$

and

$$y(t) = \dot{x}(t) = \frac{\dot{\varphi}(t)}{\omega_0} = \frac{\delta\omega(t)}{\omega_0} = \frac{\delta\nu(t)}{\nu_0}\,, \tag{12.62}$$

where $x(t)$ is the phase-time fluctuation and the time derivative $y(t) = \dot{x}(t)$ is the fractional (or relative) frequency fluctuation. These definitions are particularly useful in frequency multiplication or frequency synthesis schemes. For example, in second harmonic generation, the frequency-doubled light is given by

$$\left[\cos\varphi(t)\right]^2 = \frac{1}{2}\left[1 + \cos\left(2\varphi(t)\right)\right]\,, \tag{12.63}$$

which means that the phase noise in the time domain, i.e., $\varphi(t)$, is multiplied by a factor of 2. Note that the noise is most often described in terms of PSDs (in the frequency domain), which then results in a factor of $4 = 2^2$. Generally, under frequency multiplication by a factor of N, we have an increase in the phase noise $\varphi(t)$ given by

$$\omega_N = N\omega_0 \implies \varphi_N(t) = N\varphi(t)\,, \tag{12.64}$$

where $\varphi(t)$ is the phase noise at the fundamental frequency ω_0. The advantage with $x(t)$ and $y(t)$ is that they are invariant under frequency multiplications and synthesis.

Expressing the power spectral densities in terms of $S_\varphi(f)$, we then obtain

$$\boxed{\begin{aligned} &S_{\delta\nu}(f) = f^2 S_\varphi\,, \quad x \equiv \frac{\varphi}{\omega_0}\,, \quad y \equiv \frac{\delta\nu(t)}{\nu_0} \\ &\Longrightarrow \begin{cases} S_{\mathrm{x}}(f) = \dfrac{1}{\omega_0^2} S_\varphi(f)\,, \\ S_{\mathrm{y}}(f) = \dfrac{f^2}{\nu_0^2} S_\varphi(f)\,, \quad \text{resp.} \quad S_y(f) = \dfrac{1}{\nu_0^2} S_{\delta\nu}(f)\,. \end{cases} \end{aligned}} \tag{12.65}$$

The fractional or relative frequency noise PSD $S_{\mathrm{y}}(f)$ is also very useful for comparing the frequency noise levels of different lasers.

This also applies to the CEO frequency noise $f_{\mathrm{CEO}}(t) = f_{\mathrm{CEO},0} + \delta\nu(t)$ (12.42). When we want to compare different lasers with different pulse repetition rates and therefore typically very different CEO frequencies, it is better to compare them with the fractional CEO frequency PSD $S_{\mathrm{y,CEO}}(f)$:

$$S_{\mathrm{y,CEO}}(f) = \frac{1}{f_{\mathrm{CEO},0}^2} S_{\mathrm{CEO}}(f)\,, \tag{12.66}$$

for $S_{\mathrm{CEO}}(f) = S_{\delta\nu}(f)$. This is shown, for example, in Fig. 12.17, where an Er-fiber laser is compared with an Er:Yb:glass (ERGO) laser.

What happens if many frequency multiplications take place? According to (12.65), the phase/frequency noise PSD is multiplied by N^2 when the frequency is multiplied by N, and the integrated rms value is multiplied by N. If N becomes too large there are some issues.

For simplicity, we only assume one specific noise frequency Ω with a noise fluctuation of

$$\varphi(t) = \varphi_0 \cos(\Omega t)\,, \tag{12.67}$$

with $E(t)$ given by $E(t) = E_0 \exp\left[\mathrm{i}\big(\omega_0 t + \varphi_0 \cos(\Omega t)\big)\right]$. With frequency multiplication, we then obtain

$$E_N(t) = E_0 \exp\left\{\mathrm{i}\left[N\omega_0 t + N\varphi_0 \cos(\Omega t)\right]\right\} = E_0 \exp\left[\mathrm{i}(N\omega_0 t)\right] \sum_{n=-\infty}^{n=+\infty} J_n(N\varphi_0) \mathrm{e}^{\mathrm{i}n\Omega t}\,, \tag{12.68}$$

where J_n are the Bessel functions. As the factor N in (12.68) increases, the fluctuation power spreads to the $n > 1$ noise harmonics of Ω, and the $n = 0$ component eventually collapses when $N\varphi_0 = 2.4$, the first zero of J_0.

12.4.7 Polynomial Model: White, $1/f$ (Flicker), and $1/f^2$ Noise

Frequency-dependent trends in spectral power densities can be modelled by power law curves [78Rut]. The following polynomial form for the one-sided PSD is typically used:

$$\boxed{S_\varphi(f) = \sum_{i=-4\,\text{(or less)}}^{0} b_i f^i \,, \qquad S_{\delta\nu}(f) = f^2 S_\varphi(f)} \,. \tag{12.69}$$

Such power laws are useful models for limited noise frequency intervals. The noise is typically plotted on a double-logarithmic scale, where power laws appear as straight lines with different slopes. The physical meaning of the frequency and phase noise for the different slopes is summarized in Table 12.3.

White noise is a frequency-independent noise. For example, for white phase noise, we obtain a constant phase noise PSD according to Table 12.3:

$$S_\varphi(f) = b_0 = \text{constant} \,, \tag{12.70}$$

and an autocorrelation function

$$R_\varphi(\tau) = \int S_\varphi(f) \mathrm{e}^{\mathrm{i}2\pi f\tau} \mathrm{d}f = b_0 \delta(\tau) \,, \tag{12.71}$$

which is typical for uncorrelated noise.

For white frequency noise $S_{\delta\nu}(f) = \text{constant}$, we obtain a $1/f^2$-dependent phase noise, because $S_{\delta\nu}(f) = f^2 S_\varphi(f)$ and therefore, according to Table 12.3, for the two-sided PSD,

$$S_\varphi^{\text{two-sided}}(f) = \frac{1}{2}\frac{b_{-2}}{f^2} = \frac{a}{f^2} \,, \quad a \equiv \frac{1}{2} b_{-2} \,. \tag{12.72}$$

This corresponds to a random walk of the phase, which can be shown with the variance of the phase difference $\Delta\varphi(t, \tau) = \varphi(t + \tau) - \varphi(t)$, where τ is the time

Table 12.3 Most frequently encountered phase-noise processes for a one-sided PSD $S_\varphi^{\text{one-sided}}(f)$

Law	Slope	Noise process	Units of b_i
$b_0 f^0$	0	White phase noise	rad^2/Hz
$b_{-1} f^{-1}$	−1	Flicker phase noise	rad^2
$b_{-2} f^{-2}$	−2	White frequence noise (or random walk of phase)	rad^2Hz
$b_{-3} f^{-3}$	−3	Flicker frequency noise	rad^2Hz^2
$b_{-4} f^{-4}$	−4	Random walk of frequency	rad^2Hz^3

delay to calculate the autocorrelation function (e.g., experimentally, the time delay introduced in one arm of the interferometer). The variance of this phase difference is given in Sect. 12.5.2, and using (12.72), we obtain

$$\langle \Delta\varphi^2(t,\tau)\rangle = 8a\int_0^\infty \frac{\sin^2(\pi f\tau)}{f^2}\mathrm{d}f = 4\pi^2 a\tau \propto \tau \ , \tag{12.73}$$

where we have used $\int \mathrm{sinc}^2(x)\mathrm{d}x = 1$, with $\mathrm{sinc}(x) = \sin(\pi x)/\pi x$. Equation (12.73) implies that the variance of the phase difference increases linearly with the time delay τ, which is defined by a random walk. This shows that white frequency noise causes a random walk of the phase, as stated in Table 12.3.

The $1/f$ noise is typically referred to as flicker noise and has a $1/f$ power spectral density. However, there is also a generalized flicker noise definition $S_{\delta\nu}(f) = af^{-\alpha}$ with $1 \le \alpha \le 2$ [see, for example, (12.104)]. This more generalized flicker noise is typically observed in almost all electronic devices and lasers (see Chap. 11). Note that a $1/f$ flicker frequency noise then results in a $1/f^3$ phase noise because $S_{\delta\nu}(f) = f^2 S_\varphi(f)$, as stated in Table 12.3. In a double-logarithmic plot, $S_X(f) = 1/f$ noise is reduced by 10 dB for a noise frequency increase by a factor of 10, and for $S_X(f) = 1/f^2$, it is reduced by 20 dB.

12.4.8 Allan Variance

For a non-stationary noise quantity $X(t)$, the mean value $\langle |X(t)|\rangle$ does not exist, or is only defined for a specified time interval [see Sect. 11.1.2 and specifically (11.7)]. Therefore, a drift in the mean value over time produces a divergent power spectral density $S_X(f)$ at low noise frequencies, and the variance $\sigma_X^2 = \langle |X(t) - \langle |X(t)|\rangle|^2\rangle$ is also not defined. The solution for the mathematical description of such a noisy system was proposed by [66All].

It has become a broadly accepted standard for frequency fluctuations in the time domain [78Rut] to define an Allan variance $\sigma_y^2(\tau)$ for the relative frequency deviation $y(t) = \delta\nu(t)/\nu_0$ in (12.62):

$$\sigma_y^2(\tau) \equiv \frac{1}{2}\langle(\overline{y}_{k+1} - \overline{y}_k)^2\rangle \ , \tag{12.74}$$

with

$$\overline{y}_k = \frac{1}{\tau}\int_{t_k}^{t_k+\tau} y(t)\mathrm{d}t \ . \tag{12.75}$$

Note that it is generally better to use the term 'frequency noise' for the PSD in the frequency domain and to use the term 'fluctuations' or 'deviation' in the time domain, i.e., $y(t)$ is usually referred to as the relative (fractional) frequency deviation in the time-frequency community.

The relative frequency deviation is averaged over subsequent time intervals of width $\tau = t_{k+1} - t_k$, where the t_k are equidistant time values with integer index k and spacing τ. The Allan variance and its square root, referred to as the Allan (standard) deviation, are based on differences of adjacent, time-averaged frequency values, rather than on frequency differences from the mean value, as in the usual definition of the standard deviation. Plots usually show the Allan variance as a function of the averaging time τ, and the shape of such curves can be used to retrieve information on the noise processes. An Allan variance $\sigma_y^2(\tau)$ at measuring time τ suggests that there is a frequency instability between two measurements separated by τ, with a relative variance of $\sigma_y^2(\tau)$.

The basic idea for the Allan variance definition comes from the N-sample variance

$$\sigma_y^2 = \frac{1}{N-1}\sum_{k=1}^{N}\left(\overline{y}_k - \langle\overline{y}\rangle_N\right)^2 ,$$

for which $\langle\overline{y}\rangle_N$ grows with large N. The idea was then to replace $\langle\overline{y}\rangle_N$ by $\overline{y}_{k-1}$. This results in the Allan variance as defined in (12.74):

$$\sigma_y^2(\tau) = \frac{1}{2(M-1)}\sum_{k=1}^{M-1}\left(\overline{y}_{k+1} - \overline{y}_k\right)^2 . \tag{12.76}$$

The Allan (standard) deviation is then defined by

$$\sigma_y(\tau) = \sqrt{\sigma_y^2(\tau)} . \tag{12.77}$$

The Allan variance can also be expressed as the integral of the power spectral density, but filtered by a linear filter whose impulse response w_{A} is given in Fig. 12.16a. In linear systems theory, in the Fourier space, the filter function becomes multiplication by the transfer function $H_{\mathrm{A}}(f)$, which is the Fourier transform of the impulse response, i.e., $H_{\mathrm{A}}(f) = F\{w_{\mathrm{A}}\}$. For the power spectral density, it becomes multiplication by $|H_{\mathrm{A}}(f)|^2$:

$$H_{\mathrm{A}}(f) = F\{w_{\mathrm{A}}\} \Longrightarrow |H_{\mathrm{A}}(f)|^2 = 2\frac{\sin^4(\pi\tau f)}{(\pi\tau f)^2} . \tag{12.78}$$

The Allan variance then becomes, with the one-sided PSD $S_y^{\mathrm{I}}(f)$,

$$\sigma_y^2(\tau) = \int_0^{\infty} S_y^{\mathrm{I}}(f)|H_{\mathrm{A}}(f)|^2 \mathrm{d}f . \tag{12.79}$$

Equation (12.79) can be simplified in certain cases [see (12.69) and Table 12.3]:

- White frequency noise, i.e., $S_\varphi(f) = b_{-2}/f^2$, resp., $S_y(f) = h_0$, results in an Allan variance $\sigma_y^2(\tau) = h_0/2\tau$.
- Flicker frequency noise, i.e., $S_\varphi(f) = b_{-3}/f^3$, resp. $S_y(f) = h_{-1}/f$ leads to an Allan variance which is independent of the measuring time: $\sigma_y^2(\tau) = 2\ln(2)h_{-1}$.
- Random walk frequency noise, i.e., $S_\varphi(f) = b_{-4}/f^4$, resp. $S_y(f) = h_{-2}/f^2$ leads to an Allan variance $\sigma_y^2(\tau) = (2\pi)^2 h_{-2}\tau/6$.

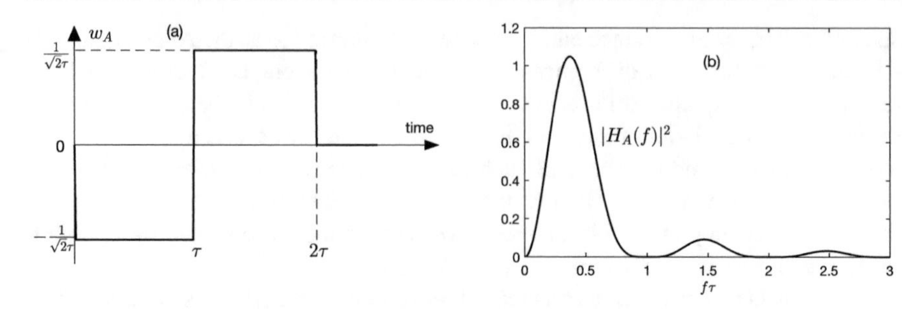

Fig. 12.16 The Allan variance as a function of the power spectral density $S_y(f)$ uses a filter in linear system theory given by (**a**) the impulse response w_A and (**b**) the transfer function $|H_A(f)|^2$

- A linear frequency drift $y(t) = bt$, which is typical for frequency standards, results in an Allan variance which increases with the squared measuring time: $\sigma_y^2(\tau) = (b\tau)^2/2$.

These are summarized in Table 12.4.

Example: SESAM-Modelocked Er:Yb:glass Laser

The following example for the Allan variance [11Sch2] is again based on the same laser as before (see Fig. 12.15), for which the CEO frequency was actively stabilized to 20 MHz, as shown in Figs. 12.13 and 12.14.

The CEO long-term stability was assessed by recording the stabilized CEO frequency with a 1 s gate time counter. Note that, as the Allan variance depends on the

Table 12.4 Noise types, power spectral densities, and Allan variance. f_H is the low-pass cutoff frequency needed for the noise power to remain finite

Noise type	$S_\varphi(f)$	$S_y(f)$	$S_\varphi \leftrightarrow S_y$	$\sigma_y^2(\tau)$
White phase noise	b_0	$h_2 f^2$	$h_2 = \dfrac{b_0}{\nu_0^2}$	$\dfrac{3 f_H h_2}{(2\pi)^2}\tau^{-2},\ 2\pi f_H \tau \gg 1$
Flicker phase noise	$b_{-1} f^{-1}$	$h_1 f$	$h_1 = \dfrac{b_{-1}}{\nu_0^2}$	$\left[1.038 + 3\ln(2\pi f_H \tau)\right]\dfrac{h_1}{(2\pi)^2}\tau^{-2}$
White frequency noise	$b_{-2} f^{-2}$	h_0	$h_0 = \dfrac{b_{-2}}{\nu_0^2}$	$\dfrac{1}{2} h_0 \tau^{-1}$
Flicker frequency noise	$b_{-3} f^{-3}$	$h_{-1} f^{-1}$	$h_{-1} = \dfrac{b_{-3}}{\nu_0^2}$	$2\ln(2) h_{-1}$
Random walk	$b_{-4} f^{-4}$	$h_{-2} f^{-2}$	$h_{-2} = \dfrac{b_{-4}}{\nu_0^2}$	$\dfrac{(2\pi)^2}{6} h_{-2}\tau$
Linear frequency drift $\dot{y}$				$\dfrac{1}{2}\dot{y}^2\tau^2$

counter filter function, it is therefore good practice to specify this filter function (e.g., in this case it is a Λ-type counter without dead time, as discussed in more detail in [11Sch2]).

Figure 12.17a shows a time series of more than 10 hours of continuous stable operation of the Er:Yb:glass laser comb. The Allan deviation (Fig. 12.17b) has a fractional frequency instability of $10^{-8}\tau^{-1}$ for the 20 MHz CEO frequency, which thus contributes only 10^{-15} to the optical carrier frequency instability ($\nu_N \approx 200$ THz) at 1 s integration time. For comparison, Fig. 12.17b also displays the same measurement performed with a commercial self-referenced Er fiber comb (FC1500-250 from Menlosystems, Germany, with 250 MHz repetition rate). The 20-fold improvement in the CEO fractional frequency stability observed in the Er:Yb:glass laser comb compared to a standard Er-fiber comb demonstrates the excellent CEO noise properties of the solid-state frequency comb. This results in part from the three times higher repetition rate of the fiber comb, but more significantly from the larger Q-factor of the Er:Yb:glass laser resonator and the suppression of the relaxation oscillations by the Yb to Er energy transfer (Sect. 5.3). These noise properties are attractive for the

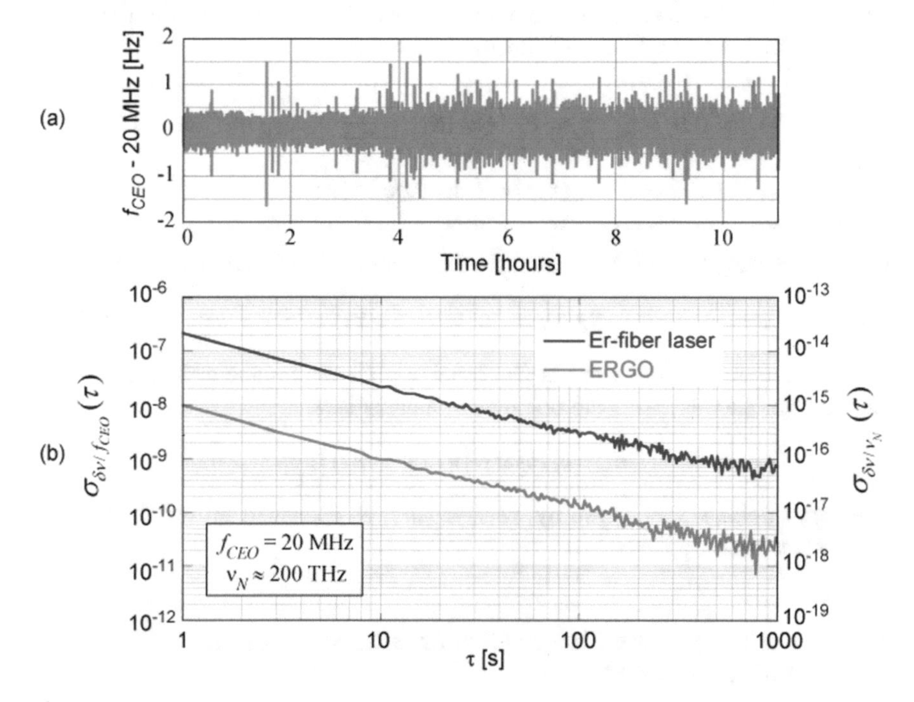

Fig. 12.17 CEO long-term stability of a SESAM-modelocked diode-pumped Er:Yb:glass laser (i.e., ERGO): **a** CEO frequency f_{CEO} stabilized to 20 MHz. Recorded time series of more than 10 hours of continuous stabilized operation of f_{CEO} acquired on a 1 s gate time counter. **b** Allan deviation of the CEO frequency, and for comparison a commercial self-referenced Er-fiber comb. The fractional stability is shown with respect to the CEO frequency $f_{\text{CEO}} = 20$ MHz on the *left scale* (i.e., $\sigma_{\delta\nu/f_{\text{CEO}}}$) and with respect to the optical frequency $\nu_N = 200$ THz on the *right scale* (i.e., $\sigma_{\delta\nu/\nu_N}$). According to [11Sch2]

future use of this comb as an optical-to-microwave frequency divider for all-optical ultra-low noise microwave generation (see Sect. 12.6.3).

12.5 Connection Between the Optical Laser Spectrum and the Phase Noise

In this section we will show that there is a direct link between the measured frequency noise or phase noise PSD and the optical laser spectrum.

12.5.1 Optical Laser Spectrum

The optical frequency power spectral density (PSD) $S_E(\omega) = |\tilde{E}(\omega)|^2$ determines the optical spectrum, the optical lineshape, and the optical linewidth. The spectral linewidth $\Delta\nu$ of the optical laser spectrum is typically given in FWHM of $S_E(\nu)$ (see Sect. 2.4). The autocorrelation function $R_E(\tau)$ and $S_E(\omega)$ form a Fourier transform pair (also known as the Wiener–Khinchin theorem):

$$\boxed{\begin{aligned} R_E(\tau) &= \langle E^*(t)E(t+\tau)\rangle = F^{-1}\{S_E(\omega)\} = \frac{1}{2\pi}\int S_E(\omega)\mathrm{e}^{\mathrm{i}\omega\tau}\mathrm{d}\omega \,, \\ S_E(\omega) &= |\tilde{E}(\omega)|^2 = F\{R_E(\tau)\} = \int R_E(\tau)\mathrm{e}^{-\mathrm{i}\omega\tau}\mathrm{d}\tau \,, \end{aligned}} \tag{12.80}$$

where, according to Table 11.1, the optical autocorrelation function is given by

$$R_E(\tau) = \langle E^*(t)E(t+\tau)\rangle \,. \tag{12.81}$$

To make the connection with the phase noise $\varphi(t)$, we assume a monochromatic electromagnetic wave with additional phase fluctuations:

$$E(t) = E_0 \exp\left\{\mathrm{i}\left[\omega_0 t + \varphi(t)\right]\right\} . \tag{12.82}$$

Frequency noise is then given by $\delta\omega(t) \equiv \dot{\varphi}(t)$, as discussed in Sect. 12.4.2. We neglect any intensity noise.

12.5.2 Variance of the Phase Noise Change and Optical Autocorrelation

This work goes back to 1960s, with a very detailed derivation also taking into account limited observation times (see, e.g., [66Cut]). Part of their work was published again in 1982 [82Ell]. The interference term in (12.81) with (12.82) can be written in the form

$$E^*(t)E(t+\tau) = E_0^* E_0 \mathrm{e}^{-\mathrm{i}\omega_0 t + \mathrm{i}\omega_0(t+\tau)}\mathrm{e}^{-\mathrm{i}\varphi(t)+\mathrm{i}\varphi(t+\tau)} = I_0 \mathrm{e}^{\mathrm{i}\omega_0\tau}\mathrm{e}^{\mathrm{i}\Delta\varphi(t,\tau)} \,, \tag{12.83}$$

where we can define a phase noise change as a function of time delay τ:

$$\Delta\varphi(t,\tau) \equiv \varphi(t+\tau) - \varphi(t) \; . \tag{12.84}$$

The optical autocorrelation function $R_E(\tau)$ is then given by

$$R_E(\tau) = \langle E^*(t)E(t+\tau)\rangle = I_0 \mathrm{e}^{\mathrm{i}\omega_0\tau}\langle \mathrm{e}^{\mathrm{i}\Delta\varphi(t,\tau)}\rangle \; . \tag{12.85}$$

We neglect any intensity noise, i.e., $I_0 =$ constant. In addition, we assume that there is no frequency drift, i.e., $\langle\omega_0\rangle = \omega_0$. The last term in (12.85) can be simplified using, for example, the Gaussian momentum theorem [82Ell]:

$$\langle \mathrm{e}^{\mathrm{i}\Delta\varphi(t,\tau)}\rangle = \exp\left[-\frac{1}{2}\langle\Delta\varphi^2(t,\tau)\rangle\right] \; . \tag{12.86}$$

This is a well-known relationship, which goes back to textbook knowledge in the 1960s (see, e.g., [65Row]). The physical meaning of $\langle\Delta\varphi^2(t,\tau)\rangle$ is the variance of $\Delta\varphi(t,\tau)$. The mean value $\langle\Delta\varphi(t,\tau)\rangle = 0$, because $\langle\varphi(t)\rangle = 0$ (see Table 12.2). As long as we assume stationary noise, we can also assume that $\langle\varphi(t+\tau)\rangle = 0$. We can rewrite $\langle\Delta\varphi^2(t,\tau)\rangle$ using (12.84):

$$\begin{aligned}\langle\Delta\varphi^2(t,\tau)\rangle &= \left\langle\left[\varphi(t+\tau) - \varphi(t)\right]^2\right\rangle \\ &= \langle\varphi^2(t)\rangle + \langle\varphi^2(t+\tau)\rangle - 2\langle\varphi(t)\varphi(t+\tau)\rangle \\ &= 2R_\varphi(0) - 2R_\varphi(\tau) \; ,\end{aligned} \tag{12.87}$$

where $R_\varphi(\tau)$ is the autocorrelation function of the phase noise (12.50) and we have assumed stationary noise, so that $R_\varphi(0) = \langle\varphi^2(t+\tau)\rangle = \langle\varphi^2(t)\rangle$ (see Table 12.2). Therefore,

$$-\frac{1}{2}\langle\Delta\varphi^2(t,\tau)\rangle = R_\varphi(\tau) - R_\varphi(0) \; . \tag{12.88}$$

The optical autocorrelation $R_E(\tau)$ is then given by

$$\begin{aligned}R_E(\tau) &= I_0 \mathrm{e}^{\mathrm{i}\omega_0\tau}\exp\left[-\frac{1}{2}\langle\Delta\varphi^2(t,\tau)\rangle\right] \\ &= I_0 \mathrm{e}^{\mathrm{i}\omega_0\tau}\exp\left[R_\varphi(\tau) - R_\varphi(0)\right] \; .\end{aligned} \tag{12.89}$$

This gives a direct link between the phase noise autocorrelation $R_\varphi(\tau)$ presented in Sect. 12.4.2 and the optical autocorrelation $R_E(\tau)$ of Sect. 12.5.1. Next we use the cosine transform in (12.50) to express the autocorrelation function $R_\varphi(\tau)$ in terms of the phase noise PSD $S_\varphi(\omega)$, i.e., by (12.50),

$$R_\varphi(\tau) = \frac{1}{\pi}\int_0^\infty S_\varphi^{\text{two-sided}}(\omega)\cos(\omega\tau)\mathrm{d}\omega \; .$$

Note with this definition of the cosine transform we use the two-sided PSD. With the identity $1 - \cos(\omega\tau) = 2\sin^2(\omega\tau/2)$, we then obtain the variance of the phase noise change as a function of the phase noise PSD:

$$\begin{aligned}\langle\Delta\varphi^2(t,\tau)\rangle &= 2\big[R_\varphi(0)-R_\varphi(\tau)\big] = \frac{2}{\pi}\int_0^\infty S_\varphi^{\text{two-sided}}(\omega)\big[1-\cos(\omega\tau)\big]\mathrm{d}\omega \\ &= \frac{4}{\pi}\int_0^\infty S_\varphi^{\text{two-sided}}(\omega)\sin^2(\omega\tau/2)\mathrm{d}\omega = 8\int_0^\infty S_\varphi^{\text{two-sided}}(f)\sin^2(\pi\tau f)\mathrm{d}f\,.\end{aligned} \tag{12.90}$$

Note that, in (12.90), the two-sided PSD is used, i.e., $2S_\varphi^{\text{two-sided}}(f) = S_\varphi^{\text{one-sided}}(f)$. This means that (12.90) is reduced by a factor of 2 when a one-sided PSD is used.

12.5.3 Optical Spectral Linewidth from Frequency and Phase Noise

As we will show in the following, the frequency noise PSD $S_{\delta\omega}(f)$ or the phase noise PSD $S_\varphi(f)$ determine the autocorrelation function $R_E(\tau)$ by (see derivation below):

$$\boxed{\begin{aligned}R_E(\tau) &= I_0\exp(\mathrm{i}2\pi\nu_0\tau)\exp\left[-2\int_0^\infty S_{\delta\nu}^{\text{one-sided}}(f)\frac{\sin^2(\pi f\tau)}{f^2}\mathrm{d}f\right] \\ &= I_0\exp(\mathrm{i}2\pi\nu_0\tau)\exp\left[-2\int_0^\infty S_\varphi^{\text{one-sided}}(f)\sin^2(\pi f\tau)\mathrm{d}f\right].\end{aligned}} \tag{12.91}$$

The optical spectrum is then given by (12.80)

$$S_E(\omega) = |\tilde{E}(\omega)|^2 = F\{R_E(\tau)\}\,.$$

Note that we have used the two-sided PSDs in the derivation, i.e., $2S_{\delta\nu}^{\text{two-sided}}(f) = S_{\delta\nu}^{\text{one-sided}}(f)$. Given that the one-sided PSD is typically used for frequency noise (e.g., Figs. 12.11 and 12.14), we modified (12.91) for the one-sided PSD.

Equation (12.91) gives a direct link between the frequency noise spectrum and the optical spectrum: the measured frequency noise PSD $S_{\delta\nu}(f)$ or phase noise PSD $S_\varphi(f)$ determine the autocorrelation $R_E(\tau)$ (12.91) and therefore the optical spectrum with the PSD $S_E(\omega) = |\tilde{E}(\omega)|^2 = F\{R_E(\tau)\}$, using (12.80). $S_E(\omega)$ determines the optical lineshape resp. the optical linewidth, as shown schematically in Fig. 12.18. This shows that the phase noise $\varphi(t)$, respectively, the related frequency noise $\delta\omega(t) = \dot{\varphi}(t)$, introduces a spectral broadening of the optical laser spectrum $|\tilde{E}(\omega)|^2$, centered around ω_0.

The optical laser spectrum and optical linewidth can therefore be determined directly using a frequency discriminator that converts, for example, the frequency noise directly to an amplitude signal (see Sect. 12.4.1). This amplitude signal can then be measured in the time domain and a simple Fourier transform determines the noise power spectral density (PSD). Either the frequency or phase noise PSD directly determines the optical laser spectrum. Normally, (12.91) has to be calculated numerically for a measured PSD. Only for the trivial case of white frequency noise

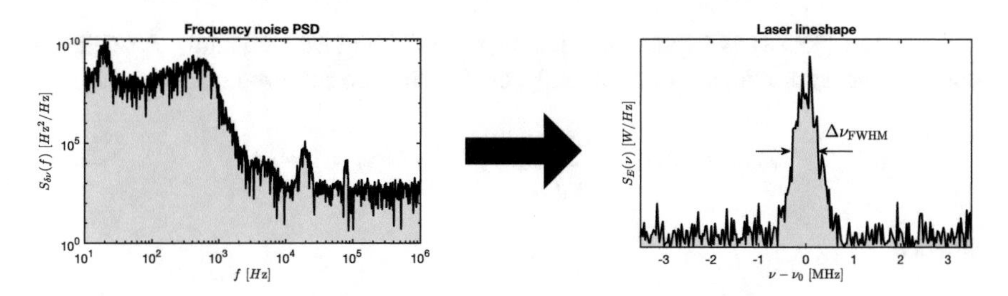

Fig. 12.18 Laser lineshape $S_E(\omega) = |\tilde{E}(\omega)|^2 = F\{R_E(\tau)\}$ and optical FWHM linewidth $\Delta\nu = \Delta\nu_{\mathrm{FWHM}}$ calculated from the frequency noise PSD $S_{\delta\nu}(f)$ or the phase noise PSD $S_\varphi(f)$ according to (12.91)

can an analytical solution be found, and this results in a Lorentzian-shaped spectrum (see Sect. 12.5.4). For the case of $1/f$ frequency noise, we can obtain an approximate solution, which results in a Gaussian lineshape (see Sect. 12.5.6).

The result in (12.91) implies that the spectral linewidth is only determined by the phase or frequency noise. However, we neglected any amplitude noise, i.e., $I_0 =$ constant. In most cases this is fully justified because the intensity noise of ultrafast solid-state lasers is very low.

The optical spectral linewidth can be reduced in a single-frequency laser, e.g., by cavity length stabilization, because there is a direct coupling between cavity length fluctuations and phase noise, as shown in Sect. 12.4.4. The ultimate linewidth limit is then given by quantum noise (see Sect. 12.5.7), if no squeezed light is used.

Derivation of (12.91)

Using (12.89),

$$R_E(\tau) = I_0 e^{i\omega_0\tau} \exp\left[R_\varphi(\tau) - R_\varphi(0)\right] ,$$

and since $1 - \cos(\omega\tau) = 2\sin^2(\omega\tau/2)$ and we are using the two-sided PSD,

$$R_\varphi(\tau) = \frac{1}{\pi}\int_0^\infty S_\varphi(\omega)\cos(\omega\tau)\mathrm{d}\omega ,$$

we obtain

$$\begin{aligned} R_E(\tau) &= I_0 e^{i\omega_0\tau} \exp\left[-\frac{2}{\pi}\int_0^\infty S_\varphi(\omega)\sin^2\frac{\omega\tau}{2}\mathrm{d}\omega\right] \\ &= I_0 e^{i\omega_0\tau} \exp\left[-\frac{2}{\pi}\int_0^\infty S_{\delta\omega}(\omega)\frac{\sin^2\omega\tau/2}{\omega^2}\mathrm{d}\omega\right] . \end{aligned}$$

Here, we have also used (12.47), which gives a simple relationship between the phase and the frequency noise PSDs, namely $S_\varphi(\omega) = S_{\delta\omega}(\omega)/\omega^2$.

Equation (12.91) is expressed with respect to the noise frequency f, which is what we measure with a spectrum analyzer. This can be achieved using

$$\omega = 2\pi f \implies \frac{\mathrm{d}\omega}{\mathrm{d}f} = 2\pi \implies \mathrm{d}\omega = 2\pi \mathrm{d}f ,$$

whence

$$R_E(\tau) = I_0 \mathrm{e}^{\mathrm{i}\omega_0\tau} \exp\left[-4\int_0^\infty S_\varphi(f)\sin^2(\pi f\tau)\mathrm{d}f\right] ,$$

which is equivalent to (12.91). Since $\omega = 2\pi f$, we obtain

$$R_E(\tau) = I_0 \mathrm{e}^{\mathrm{i}\omega_0\tau} \exp\left[-\int_0^\infty S_{\delta\omega}(f)\frac{\sin^2(\pi f\tau)}{\pi^2 f^2}\mathrm{d}f\right] .$$

Using $S_{\delta\omega}(f) = 4\pi^2 S_{\delta\nu}(f)$ from (12.51), we then obtain

$$R_E(\tau) = I_0 \mathrm{e}^{\mathrm{i}\omega_0\tau} \exp\left[-4\int_0^\infty S_{\delta\nu}(f)\frac{\sin^2(\pi f\tau)}{f^2}\mathrm{d}f\right] ,$$

which is equivalent to (12.91), assuming two-sided PSDs in this derivation.

12.5.4 Optical Linewidth for (Low-Pass Filtered) White Frequency Noise

White Frequency Noise

We have seen in Sect. 12.4.7 that, for a white frequency noise two-sided PSD $S_{\delta\nu}(f) = a$, we obtain a $1/f^2$ phase noise PSD $S_\varphi(f) = a/f^2$ as in (12.72). Here we want to show that white frequency noise, or $1/f^2$ phase noise, results in a Lorentzian lineshape.

This can be done as follows. From (12.91), we obtain for two-sided PSDs

$$\begin{aligned} R_E(\tau) &= I_0 \exp(\mathrm{i}2\pi\nu_0\tau)\exp\left[-4\int_0^\infty S_{\delta\nu}(f)\frac{\sin^2(\pi f\tau)}{f^2}\mathrm{d}f\right] \\ &= I_0 \exp(\mathrm{i}2\pi\nu_0\tau)\exp\left[-4a\int_0^\infty \frac{\sin^2(\pi f\tau)}{f^2}\mathrm{d}f\right] \\ &= I_0 \exp(\mathrm{i}2\pi\nu_0\tau)\exp(-2\pi^2 a\tau) . \end{aligned} \qquad (12.92)$$

Here, we have used $\int \mathrm{sinc}^2(x)\mathrm{d}x = 1$, from the definition $\mathrm{sinc}(x) = \sin(\pi x)/\pi x$, and hence $\int \mathrm{sinc}^2(ax)\mathrm{d}x = 1/a$. The lineshape is given by

$$S_E(\omega) = |\tilde{E}(\omega)|^2 = F\{R_E(\tau)\} ,$$

so we obtain

$$S_E(\omega) = F\{R_E(\tau)\} = F\{I_0 \exp(\mathrm{i}\omega_0\tau)\exp[-|\tau|/T]\} \ .$$

We have introduced $|\tau|$ because the autocorrelation function must be symmetric, i.e., $R_E(-\tau) = R_E(\tau)$ and $T \equiv 1/2\pi^2 a$. Using the Fourier transform table in Appendix A, we then obtain

$$F\{\exp(-|\tau|/T)\} = \frac{2T}{1+(\omega T)^2} \ ,$$

and with the shift theorem we obtain

$$F\{\exp(\mathrm{i}\omega_0\tau)\exp(-|\tau|/T)\} = \frac{2T}{1+[(\omega-\omega_0)T]^2} \ .$$

The optical power spectrum for the white frequency noise two-sided PSD $S_{\delta\nu}(f) = a$ is therefore a Lorentzian-shaped optical spectrum centered around ν_0 :

$$S_E(\nu) = |\tilde{E}(\nu)|^2 = F\{R_E(\tau)\} = I_0 \frac{a}{\pi^2 a^2 + (\nu-\nu_0)^2} \ , \tag{12.93}$$

with an FWHM linewidth

$$\Delta\nu_{\mathrm{FWHM}} = \frac{1}{\pi T} = 2\pi a \ . \tag{12.94}$$

Low-Pass Filtered White Frequency Noise

Next we shall investigate the lineshape for low-pass filtered white frequency noise with a cutoff frequency f_{LP} and frequency noise two-sided PSD given by

$$S_{\delta\nu}(f) = \mathrm{rect}_{f_{\mathrm{LP}}}(f) \equiv \begin{cases} a \ , & |f| \le f_{\mathrm{LP}} \ , \\ 0 \ , & |f| > f_{\mathrm{LP}} \ . \end{cases} \tag{12.95}$$

For the autocorrelation function, we then obtain, using (12.91),

$$\begin{aligned} R_E(\tau) &= I_0 \exp(\mathrm{i}2\pi\nu_0\tau)\exp\left[-4\int_0^\infty S_{\delta\nu}(f)\frac{\sin^2(\pi f\tau)}{f^2}\mathrm{d}f\right] \\ &= I_0 \exp(\mathrm{i}2\pi\nu_0\tau)\exp\left[-4a\int_0^{f_{\mathrm{LP}}} \frac{\sin^2(\pi f\tau)}{f^2}\mathrm{d}f\right] \ . \end{aligned} \tag{12.96}$$

This integral can no longer be solved analytically. In the limit $f_{\mathrm{LP}} \to \infty$, we obtain the white noise result with the Lorentzian lineshape. In the limit $f_{\mathrm{LP}} \to 0$, it helps to use the sine integral $\mathrm{Si}(x)$ defined by [10Dom]

$$\mathrm{Si}(x) \equiv \int_0^x \frac{\sin u}{u}\mathrm{d}u \ .$$

Using the identity $1 - \cos(x) = 2\sin^2(x/2)$ and integration by parts for

$$\frac{1}{x}\frac{\cos(x)}{x}\,,$$

and inserting $\alpha \equiv \pi\tau$, (12.96) implies

$$-\int_0^{f_{\mathrm{LP}}} \frac{\sin^2(\alpha f)}{f^2}\mathrm{d}f = -\left[\frac{\sin^2(\alpha f)}{f}\right]_0^{f_{\mathrm{LP}}} + \Big[\alpha\,\mathrm{Si}(2\alpha f)\Big]_0^{f_{\mathrm{LP}}}\,.$$

In the limit $f_{\mathrm{LP}} \to 0$, this equation simplifies since $\lim_{x\to 0}\mathrm{Si}(x) = 0$ and $\sin^2 x \approx x^2$, for $x \ll 1$. Using (12.96), we obtain

$$R_E(\tau) = I_0 \mathrm{e}^{\mathrm{i}\omega_0\tau}\exp\left(-4\pi^2 a f_{\mathrm{LP}}\tau^2\right)\,, \tag{12.97}$$

which is a Gaussian function. The optical lineshape is then also a Gaussian function

$$S_E(\nu) = F\{R_E(\tau)\} = \frac{I_0/2}{\sqrt{\pi a f_{\mathrm{LP}}}}\exp\left[-\frac{(\nu-\nu_0)^2}{4 a f_{\mathrm{LP}}}\right]\,, \tag{12.98}$$

with an FWHM linewidth in the limit $f_{\mathrm{LP}} \to 0$ given by

$$\Delta\nu_{\mathrm{FWHM}} = 4\sqrt{(\ln 2) a f_{\mathrm{LP}}}\,. \tag{12.99}$$

Note that, in contrast to the solution in [10Dom], we have used a two-sided PSD in (12.95), whereas one-sided PSDs were used in [10Dom]. The results are identical, given the condition that $2S_{\delta\nu}^{\text{two-sided}}(f) = S_{\delta\nu}^{\text{one-sided}}(f)$ and therefore $h_0 = 2a$.

As shown in Fig. 12.18, the FWHM laser linewidth $\Delta\nu$ is calculated numerically using (12.91). So far, we have shown that the optical lineshape is Gaussian for $f_{\mathrm{LP}} \ll a$, in agreement with the analytical solution with $f_{\mathrm{LP}} \to 0$ in (12.98). Furthermore, for $f_{\mathrm{LP}} \gg a$, the optical lineshape is Lorentzian, again in agreement with the analytical solution for $f_{\mathrm{LP}} \to \infty$. Numerical calculations confirm these different lineshapes over a larger cutoff frequency range of the low-pass filter.

Such numerical calculations also demonstrate that the optical linewidth above a critical value for $f_{\mathrm{LP}} \geq f_{\mathrm{LP}}^* \approx 3.5a$ remains constant. More detailed analysis of the numerical calculations showed that all the additional noise above the low-pass cutoff frequency $f_{\mathrm{LP}} > f_{\mathrm{LP}}^*$ only contributes to the wings of the optical lineshape (Fig. 12.19). The critical cutoff frequency is given by [10Dom]

$$\frac{f_{\mathrm{LP}}^*}{2a} = \frac{\pi^2}{8\ln 2} \approx 1.78\,. \tag{12.100}$$

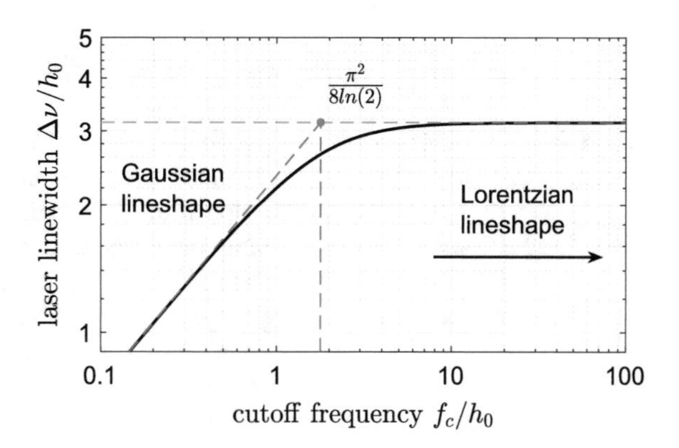

Fig. 12.19 Numerically calculated FWHM optical laser linewidth $\Delta\nu$, which is the FWHM of the optical spectrum $S_E(\nu) = |\tilde{E}(\nu)|^2 = F\{R_E(\tau)\}$ and (12.91) with the low-pass filtered white frequency noise PSD given in (12.100) and $h_0 = 2a$ [10Dom]

12.5.5 β-Separation Line

The β-separation line was introduced by the group of Pierre Thomann [10Dom]. The example of the low-pass filtered white frequency noise (Sect. 12.5.4) shows that only the lower noise frequencies contribute to the final linewidth (Fig. 12.19). This result can be generalized to flicker noise (see next section). Thomann et al. then introduced a β-separation line defined by the one-sided PSD:

$$S_{\delta\nu}(f) = \frac{8\ln(2)f}{\pi^2} . \tag{12.101}$$

For flicker noise, the low noise frequencies support a Gaussian lineshape and with the geometrical definition of an area A, i.e., the region of area A_1 above the β-separation line (Fig. 12.20), the FWHM linewidth is given by

$$\Delta\nu_{\text{FWHM}} = \sqrt{8\ln(2)A} . \tag{12.102}$$

The area A is the overall area under the portions of $S_{\delta\nu}(f)$ that exceed the β-separation line [10Dom]:

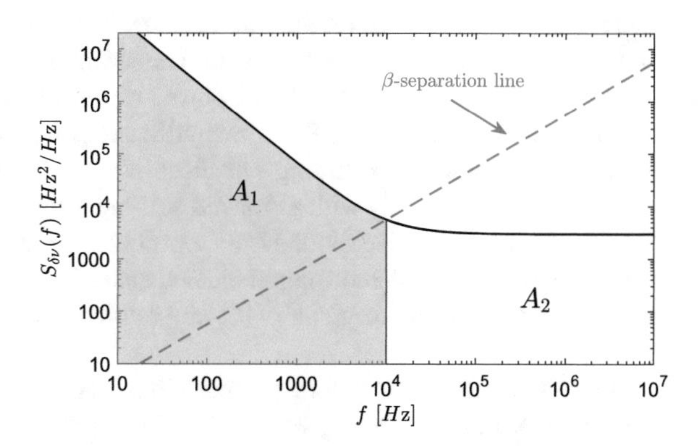

Fig. 12.20 The *dashed line*, i.e., the β-separation line, separates the frequency noise spectrum into two regions whose contribution to the optical lineshape is very different: the noise frequency regime A_1 contributes to the linewidth and A_2 contributes only to the wings of the lineshape [10Dom]

$$A = \int_{1/T_{\mathrm{obs}}}^{\infty} H\big(S_{\delta\nu}(f) - 8\ln(2) f/\pi^2\big) S_{\delta\nu}(f)\mathrm{d}f \,, \tag{12.103}$$

where $H(x)$ is the Heaviside unit step function with $H(x) = 1$ for $x \geq 0$, and $H(x) = 0$ for $x < 0$. The integration does not start at zero frequency but at a lower frequency $1/T_{\mathrm{obs}}$, where T_{obs} is the observation time. Theoretically, the flicker noise has a diverging PSD with $f \to 0$. In reality the measurement is always finite.

The β-separation line is a very useful concept. For example, the noise frequency, where the measured free-running frequency noise PSD is intersecting with the β-separation line, gives a quick guideline for the required stabilization feedback loop. It is very hard to make an excellent high-bandwidth stabilization and it helps to first optimize the laser parameters for a lower intersection frequency. Furthermore, we can use (12.102) to determine the linewidth for the flicker noise, as will be discussed in the next section.

12.5.6 Optical Linewidth from Flicker Noise

The more generalized flicker noise is given by the following frequency noise PSD:

$$S_{\delta\nu}(f) = af^{-\alpha} \,, \quad 1 \leq \alpha \leq 2 \,. \tag{12.104}$$

Equation (12.91) with this PSD can no longer be solved analytically. Different approximations have been proposed.

For example, [91Mer] gave an analytical approximation for $1/f$ frequency noise, i.e., $\alpha = 1$ in (12.104) with a Gaussian lineshape and an empirical expression for the linewidth. Note that, in this theoretical treatment, a two-sided PSD was used with the $1/f$ noise given by $S_{\delta\omega}(\omega) = k/|\omega|$. The approximate FWHM linewidth is given by [91Mer]:

$$S_{\delta\omega}^{\text{two-sided}}(\omega) = \frac{k}{|\omega|} \implies \Delta\nu_{\mathrm{FWHM}} \approx \sqrt{4\ln(2)\frac{k}{2\pi^3}\left[1 + \ln\left(T_{\mathrm{obs}}\sqrt{k/2\pi^3}\right)\right]} \,, \tag{12.105}$$

where T_{obs} is the observation time in seconds. This approximation is only valid for $\sqrt{k/2\pi^3} \gg 1$. He also showed that the Gaussian approximation is good near the line center, where about 90% of the power of the lineshape is contained. Only in the wings of the spectrum is there a significant deviation from the Gaussian lineshape approximation, confirming our discussion in the last section. This was also confirmed experimentally with a delayed self-heterodyne linewidth measurement of monochromatic low-noise diode lasers, optimized for coherent optical communication (see Sect. 12.5.8). A more sophisticated analytical model without any restriction for the observation time is given in [05Ste], which will not be discussed any further here.

The β-separation line offers another simple solution for the FWHM Gaussian linewidth from flicker noise given in (12.104) [10Dom]. The noise frequency f_{m} is introduced, for which the flicker noise PSD and the β-separation line intersect as follows:

$$S_{\delta\nu}(f_{\mathrm{m}}) = af_{\mathrm{m}}^{-\alpha} = \frac{8\ln(2)f_{\mathrm{m}}}{\pi^2} \Longrightarrow a = \frac{8\ln(2)}{\pi^2}f_{\mathrm{m}}^{\alpha+1} . \qquad (12.106)$$

This allows for a dimensionless representation of the flicker noise:

$$\frac{S_{\delta\nu}(f)}{f_{\mathrm{m}}} = \frac{8\ln(2)}{\pi^2}\left(\frac{f}{f_{\mathrm{m}}}\right)^{-\alpha} . \qquad (12.107)$$

Using (12.102) and (12.103) and assuming $T_{\mathrm{obs}}f_{\mathrm{m}} > 5$, we then obtain an FWHM linewidth for $\alpha = 1$,

$$\mathrm{FWHM} = f_{\mathrm{m}}\frac{8\ln(2)}{\pi}\left[\ln(f_{\mathrm{m}}T_{\mathrm{obs}})\right]^{1/2} , \qquad (12.108)$$

and for $\alpha > 1$,

$$\mathrm{FWHM} = f_{\mathrm{m}}\frac{8\ln(2)}{\pi}\left[\frac{(f_{\mathrm{m}}T_{\mathrm{obs}})^{\alpha-1} - 1}{\alpha - 1}\right]^{1/2} . \qquad (12.109)$$

In this case the linewidth is given by a simple parameter f_{m}, which can be determined from the measured frequency noise PSD (as shown, for example, in Figs. 12.11 and 12.14).

Normally, we have both flicker and white frequency noise components. Assuming uncorrelated noise contributions, i.e., statistically independent fluctuations, the total optical lineshape is a convolution between the Lorentzian spectrum (i.e., from white noise) and the Gaussian spectrum (i.e., from flicker noise). In this case we obtain the Voigt profile [91Mer,05Ste]. For uncorrelated noise, the variance of the total phase change $\langle\Delta\varphi^2(t,\tau)\rangle$ is then given by the sum of the two noise contributions. However, the white frequency noise normally comes from the higher noise frequencies, which typically do not significantly contribute to the integrated rms phase noise (see Fig. 12.14). Therefore, the simple Gaussian approximation is generally a good first approximation.

As we have seen before, there is strong coupling between all noise quantities. A linewidth enhancement factor [82Hen] was introduced for semiconductor edge-emitting diode lasers. These lasers have a strong coupling from fluctuations of the carrier density, which affect the refractive index and thus cause additional phase noise

$$\delta\varphi = \frac{1}{2}\alpha\delta g , \qquad (12.110)$$

for a fluctuation δg of the intensity gain. This increases the optical linewidth by a factor $1+\alpha^2$, as derived in [82Hen]. For semiconductor lasers, $1+\alpha^2$ can easily be between 10 and 100. For solid-state lasers based on ion-doped crystals or glasses, the $(1+\alpha^2)$-factor is usually neglected, although this may not always be fully justified for three-level lasers [07Pas].

12.5.7 Quantum Noise Limit

In Chap. 11, we introduced the shot noise (Table 11.3), which comes from the detection process with the random arrival of photons and results in a lower limit of the intensity noise, assuming no squeezed light is used. For a single-frequency laser, the quantum-limited finite linewidth is given by the Schawlow–Townes formula [58Sch] and shown here in a slightly different form [07Pas]:

$$\Delta\nu_{\text{laser}} = \frac{h\nu l_{\text{tot}} T_{\text{oc}}}{4\pi T_{\text{R}}^2 P_{\text{out}}} = \frac{h\nu l_{\text{tot}}}{4\pi T_{\text{R}}^2 P_{\text{int}}} , \tag{12.111}$$

where $\Delta\nu_{\text{laser}}$ is the full width at half maximum (FWHM) laser linewidth, $h\nu$ is the photon energy, l_{tot} denotes the total power losses per cavity roundtrip, T_{oc} the output coupler transmission, T_{R} the cavity roundtrip time, P_{out} the laser output power, and P_{int} the intracavity power. This formula is exactly equivalent to the original Schawlow–Townes formula if there are no parasitic losses, i.e., $l_{\text{tot}} = T_{\text{oc}}$, assuming $l_{\text{tot}} \ll 1$.

For a given output power, the linewidth scales with the square of the output coupler transmission (assuming that the power loss coefficient l_{tot} is dominated by the output coupler), since a higher transmission means both higher laser gain (introducing stronger fluctuations) and a lower intracavity power (making quantum effects due to spontaneous emission stronger). The linewidth also scales with the inverse square of the roundtrip time, which means that long laser cavities are potentially better in terms of linewidth—provided that classical noise can be well suppressed.

Derivation of the Schawlow–Townes Linewidth (12.111)

We follow the derivation given in [07Pas]. We assume a four-level single-frequency laser without amplitude/phase coupling. The output coupler transmission is T_{oc}, and there can be additional (parasitic) cavity losses l_{par}, so that the total cavity losses are $l_{\text{tot}} = T_{\text{oc}} + l_{\text{par}}$. At steady state, on average the gain must balance the losses, so $g = l_{\text{tot}}$. We describe the circulating field with a complex amplitude A, normalized so that the intracavity power is $P_{\text{int}} = |A|^2$.

During each roundtrip, the gain medium adds a fluctuating amplitude ΔA, where each of the quadrature components has the variance

$$\sigma^2 = \frac{h\nu}{4T_{\text{R}}} g ,$$

with T_{R} the roundtrip time. This amount of noise is consistent, e.g., with the results of [82Cav] and with the 3 dB noise figure of a high-gain laser amplifier. Its effect is that the optical phase changes by

$$\Delta\varphi = \frac{\Delta A_{\text{q}}}{A} ,$$

where ΔA_{q} is the quadrature component perpendicular to A in the complex plane. The cavity losses contribute another noise amplitude proportional to l_{tot}, where $g = l_{\text{tot}}$

because the gain must balance the cavity losses. As each roundtrip contributes statistically independent phase fluctuations, the variance of the phase change $\langle \Delta\varphi^2(t,\tau) \rangle$ grows with time (see Sect. 12.5.1):

$$\langle \Delta\varphi^2(t,\tau) \rangle = \frac{h\nu}{2T_{\mathrm{R}} P_{\mathrm{int}}} l_{\mathrm{tot}} \frac{\tau}{T_{\mathrm{R}}} = \frac{h\nu l_{\mathrm{tot}}}{2T_{\mathrm{R}}^2 P_{\mathrm{int}}} \tau \, ,$$

which is consistent with a two-sided phase noise PSD [see (12.72) and (12.73)]

$$S_{\varphi}(f) = \frac{C}{f^2} \, ,$$

where

$$C = \frac{1}{(2\pi)^2} \frac{h\nu l_{\mathrm{tot}}}{2T_{\mathrm{R}}^2 P_{\mathrm{int}}} \, .$$

We have shown that such a phase noise PSD corresponds to white frequency noise $S_{\delta\nu}(f) = C$ with a Lorentzian field spectrum (Sect. 12.5.4). By (12.93), this can be written as

$$L(\Delta f) = \frac{C}{(\pi C)^2 + (\Delta f)^2} \, .$$

It is normalized to give a unit integral and leads to the FWHM field linewidth (12.94), viz.,

$$\Delta\nu = 2\pi C = \frac{h\nu l_{\mathrm{tot}}}{4\pi T_{\mathrm{R}}^2 P_{\mathrm{int}}} = \frac{h\nu l_{\mathrm{tot}} T_{\mathrm{oc}}}{4\pi T_{\mathrm{R}}^2 P_{\mathrm{out}}} \, .$$

For $l_{\mathrm{tot}} = T_{\mathrm{oc}}$, i.e., without parasitic cavity losses, this result is exactly equivalent to the original Schawlow–Townes formula [58Sch]

$$\Delta\nu = \frac{4\pi h\nu (\Delta\nu_{\mathrm{c}})^2}{P_{\mathrm{out}}} \, ,$$

where

$$\Delta\nu_{\mathrm{c}} = \frac{1}{4\pi} \frac{l_{\mathrm{tot}}}{T_{\mathrm{R}}}$$

is the *half-width* of the resonances of the cavity (with unpumped gain medium).

12.5.8 Optical Interferometers, Coherence, and Phase Noise

In optical coherence theory, the optical autocorrelation function $R_E(\tau)$ given in (12.81) is also referred to as the temporal coherence function. The (instantaneous) intensity is given by $I(t) = |E(t)|^2$ and the average intensity $I_{\mathrm{av}} = \langle |E(t)|^2 \rangle = R_E(0)$. The coherence time τ_{c} is typically defined by the normalized autocorrelation function $g(\tau)$ as follows [91Sal2]:

$$g(\tau) \equiv \frac{R_E(\tau)}{R_E(0)} = \frac{\langle E^*(t)E(t+\tau)\rangle}{\langle E^*(t)E(t)\rangle}, \quad g(\tau_c) = 1/e . \tag{12.112}$$

The definition of the coherence time τ_c is somewhat arbitrary and different definitions are used in the literature.

Physically, the coherence time τ_c gives an approximate time difference up to which two components of the laser field can stably interfere. This is typically measured in an optical interferometer. The output of an optical interferometer, where one interferometer arm introduces a time difference of τ, is given by

$$E_{out}(t) = E(t) + E(t+\tau) . \tag{12.113}$$

Here we assume that $E(t)$ is given by a single-frequency laser with phase noise $\varphi(t)$, i.e., $E(t) = E_0 \exp\left[i\big(\omega_0 t + \varphi(t)\big)\right]$ by (12.82). The output intensity $I_{out}(t,\tau)$ is then given by

$$I_{out}(t,\tau) = |E_{out}(t)|^2 = |E(t)|^2 + |E(t+\tau)|^2 + 2\mathrm{Re}\{E^*(t)E(t+\tau)\} . \tag{12.114}$$

The average output intensity $\langle I_{out}(t,\tau)\rangle$ follows using (12.85):

$$\langle I_{out}(t,\tau)\rangle = 2\langle |E(t)|^2\rangle + 2\mathrm{Re}\{R_E(\tau)\} = 2I_{av}\big[1 + |g(\tau)|\cos(\omega_0\tau)\big] , \tag{12.115}$$

where $R_E(\tau)$ is the autocorrelation function of $E(t)$, $I_{av} = \langle |E(t)|^2\rangle = R_E(0)$ and $|g(\tau)|$ given by (12.112). Equations (12.85) and (12.86) imply

$$|g(\tau)| \equiv \left|\frac{R_E(\tau)}{R_E(0)}\right| = \big|\langle \exp[i\Delta\varphi(t,\tau)]\rangle\big| = \exp\left[-\frac{1}{2}\langle\Delta\varphi^2(t,\tau)\rangle\right] . \tag{12.116}$$

Defining the coherence time as in (12.112), we then obtain

$$\langle\Delta\varphi^2(t,\tau_c)\rangle = 2 . \tag{12.117}$$

Using (12.90), we obtain a direct link between the coherence time τ_c and the phase noise two-sided power spectral density $S_\varphi(f)$:

$$\begin{aligned}\frac{1}{\pi}\int_0^\infty S_\varphi(\omega)\big[1-\cos(\omega\tau_c)\big]d\omega &= \frac{2}{\pi}\int_0^\infty S_\varphi(\omega)\sin^2\frac{\omega\tau_c}{2}d\omega \\ &= 4\int_0^\infty S_\varphi(f)\sin^2(\pi f\tau_c)df = 1 .\end{aligned} \tag{12.118}$$

For example, white frequency noise introduces a two-sided phase noise PSD of $S_\varphi(f) = a/f^2$, and with (12.73) and (12.118), a coherence time

$$\langle\Delta\varphi^2(t,\tau_c)\rangle = 4\pi^2 a\tau_c = 2 \implies \tau_c = \frac{1}{2\pi^2 a} . \tag{12.119}$$

The relation between the coherence time τ_c and the spectral linewidth is typically given in FWHM and depends on the lineshape. A narrow linewidth then corresponds to a long coherence time. Examples with different lineshape functions are summarized in Table 12.5. Therefore, assuming a Lorentzian lineshape for white frequency noise (see Sect. 12.5.4), we obtain

$$\Delta\nu_{\mathrm{FWHM}} = \frac{1}{\pi\tau_c} = 2\pi a = \pi b_{-2} , \tag{12.120}$$

where b_{-2} is the coefficient for the one-sided PSD [see (12.72) and Table 12.4].

Example Linewidth Measurements of Monochromatic Semiconductor Diode Lasers

With coherent optical communication, the phase noise of monochromatic semiconductor diode lasers has become an important parameter. There have been different measurement techniques:

- Noise power spectral density measurements. The frequency noise measurement with an ideal optical discriminator determines the power spectral density, and with (12.91), the optical linewidth.
- Heterodyne measurements with an ultrastable single-frequency reference laser. The linewidth of a laser can be measured with a single-frequency reference laser using the heterodyne technique (see Sect. 12.3.1), which determines the linewidth of the laser under test (assuming the reference laser linewidth is known). However, such a reference laser is often unavailable.
- Self-heterodyne linewidth measurement [80Oko,91Mer]. This technique is easy to set up and the laser may operate at any optical frequency. This has been in widespread use for cw semiconductor lasers. Note that, with flicker noise, the linewidth increases very fast with increasing delay τ.

Table 12.5 Relation between spectral width $\Delta\nu_{\mathrm{FWHM}}$ and coherence time τ_c

Normalized autocorrelation $g(\tau)$ $\lvert g(\tau_c)\rvert \equiv \mathrm{e}^{-1}$	Normalized lineshape $S_E(\omega) = F\{g(\tau)\}$	$\Delta\nu_{\mathrm{FWHM}} = \dfrac{\Delta\omega_{\mathrm{FWHM}}}{2\pi}$
$\exp\left(-\dfrac{\lvert\tau\rvert}{\tau_c}\right)$ double-sided exponential function	$\dfrac{2\tau_c}{1+(\omega\tau_c)^2}$ Lorentzian	$\dfrac{1}{\pi\tau_c} \approx \dfrac{0.32}{\tau_c}$
$\mathrm{rect}_{\tau_c}(t) \equiv \begin{cases} 1 , & \lvert t\rvert \le \tau_c , \\ 0 , & \lvert t\rvert > \tau_c . \end{cases}$ rectangular function	$\dfrac{2\sin(\tau_c\omega)}{\omega}$ sinc function	$\dfrac{1}{\tau_c}$
$\exp\left[-\left(\dfrac{t}{\tau_c}\right)^2\right]$ Gaussian	$\sqrt{\pi}\tau_c \exp\left(-\dfrac{1}{4}\tau_c^2\omega^2\right)$ Gaussian	$\dfrac{\sqrt{2\ln(2)/\pi}}{\tau_c} \approx \dfrac{0.66}{\tau_c}$

Typically, for monochromatic semiconductor lasers in optical communication, the phase noise has been characterized with optical linewidth measurements. The required long time delay in the optical interferometer for self-heterodyne linewidth measurements is obtained with a very long fiber cable (e.g., longer than 10 km). The intensity in the long fiber cable needs to be low to avoid any additional measurement-induced noise and a lock-in detection scheme has to be applied. There are two regimes of self-heterodyne linewidth measurements with delay lines larger [80Oko,91Mer] and smaller than the coherence time.

The self-heterodyne technique cannot be transferred to ultrafast laser sources because of the required long delay ($\gg$1 km) in the optical interferometer, which cannot be done in a long fiber cable without adding any significant noise. In principle, noise power spectral density measurements are better for phase noise characterization because the linewidth measurement depends on the specific fiber cable length [see, for example, (12.105)].

12.6 Stabilized Optical Frequency Combs

12.6.1 Basic Principle of Active Stabilization

Figure 12.21 shows the basic principle for active stabilization with a feedback system, which can be used to stabilize different output parameters of a laser. This can be used to reduce intensity noise, timing jitter, CEO frequency noise, and spectral linewidth, for example. There is a lot of jargon in this field and the details of such a feedback system go beyond the scope of this textbook. Many commercial systems are available and can be adapted to a wide range of specific laser parameters to be stabilized.

The basic principle is to use a laser for which the output is compared to a reference signal. This results in an error signal. The error signal is conditioned to generate a control signal (i.e., the feedback), which reduces the error signal in the laser output. We have a closed loop in the simplified block diagram in Fig. 12.21 and therefore such a feedback system is also referred to as a feedback loop.

This feedback system has a certain frequency bandwidth and a certain frequency response, where the frequency corresponds to the modulation frequency of a control parameter (e.g., modulation frequency of pump current, or shaking frequency of cavity end mirror position). Generally, it is easier to obtain better noise suppression for a lower control loop bandwidth. The bandwidth depends on the control parameters, which are transducers (or actuators) that modify the output signal of the laser from a control input signal. The actuators are considered 'fast' when they can support a higher frequency bandwidth. Some commonly used transducers are laser temperature (very slow), piezo transducer (PZT) to control, for example, mirror positions (slow with a heavy cavity mirror), laser injection current for diode-pumped lasers (fast), and intracavity acousto-optic or electro-optic modulator (fast).

The response of such control parameters can be characterized independently from the electronic feedback system with transfer function measurements shown in Fig. 5.13, introduced in Sect. 5.3. For such transfer function measurements, the

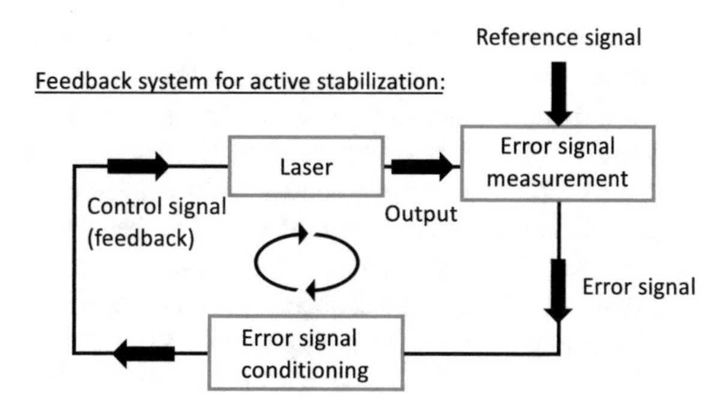

Fig. 12.21 Simplified block diagram of a feedback system for active stabilization. Typical control parameters (i.e., transducers or actuators) in the laser are pump power and cavity length adjustments

output signal is monitored as a function of the control signal modulation. In principle, more than one actuator can be used, such as for example a PZT for slow corrections and the laser pump current for fast corrections. The SESAM is normally used as one of the end mirrors in an ultrafast solid-state laser and has much less weight than a normal dielectric cavity mirror. Thus, mounting the SESAM on the PZT supports a much higher control bandwidth.

The error signal is conditioned for an optimized control signal, which includes different filters and some amplification of the error signal. The goal of the error signal conditioning electronics and their transfer function for a correction signal is to provide enough gain to damp the laser noise across a sufficiently broad noise frequency bandwidth. Ideally, the gain should be as high as possible, just limited by the stability of the feedback loop. For this, the phase response of the electronic transfer function also becomes important. The control signal is phase shifted from the original error signal. At some frequencies, the correction signal will be 180° out-of-phase compared to the fluctuations, which results in positive feedback with potential instabilities. To avoid such instabilities a loop gain of less than 0 dB (unit gain) is required in the frequency range where the loop phase shift approaches 180°. Therefore, the final goal of the error signal conditioning is a high enough gain at low frequency while keeping the unit gain frequency low enough to keep the system stable. The trade-off for a high gain at low frequency is a high unit gain frequency and therefore a large bandwidth.

A more detailed optimization strategy, however, would go beyond the scope of this textbook.

12.6.2 CEO Frequency Stabilization with Laser Feedback Control

Normally, stabilization of the CEO frequency is more challenging than timing jitter stabilization alone. The basic principle of CEO frequency stabilization with laser feedback control is shown in Fig. 12.21, for which the reference signal is, for example, an external stable microwave oscillator, e.g., a quartz crystal oscillator, a Cs-clock controlled microwave oscillator, a hydrogen maser, and more recently a sapphire whispering gallery mode resonator [05Gio]. The error signal in this case can be a phase difference with respect to this reference frequency, which needs to be locked to the reference with a feedback system. Such an active feedback system is typically referred to as a phase-locked loop (PLL), which also locks the two frequencies because the frequency is just the time derivative of the phase. The β-separation line (see Sect. 12.5.5) is a very good tool to determine the required feedback system bandwidth for the stabilization of the CEO frequency linewidth, because the noise only needs to be suppressed below this β-separation line.

Figure 12.22 shows the change in the CEO frequency bandwidth with active stabilization going from a free running to a weakly stabilized (i.e., weak lock) and finally to a fully stabilized CEO frequency (i.e., tight lock):

- For the free-running laser, the CEO frequency is very noisy and a kHz linewidth can only be observed for a very limited time, less than 10 ms, corresponding to a narrow noise frequency bandwidth (e.g., 100 Hz to 1 kHz). Averaging over a longer time typically washes out the strongly fluctuating CEO peak.
- For the weak lock, not all the noise above the β-separation line is stabilized. Note that noise spikes on a log scale do not look like much, but they do add significant noise (see, e.g., the 360 Hz noise peak in Fig. 12.14).
- Finally, for a tight lock, the noise is more strongly reduced below the β-separation line, even though the feedback control system adds more noise around the loop bandwidth. The noise below the β-separation line adds more noise in the wings of the CEO frequency. Thus, for a higher signal-to-noise of the CEO beat signal above the background noise, better noise suppression below the β-separation line is also required.

The error signal is measured with either optical or electronic discriminators (e.g., the electronic frequency discriminator in Fig. 12.12). The discriminator can be either

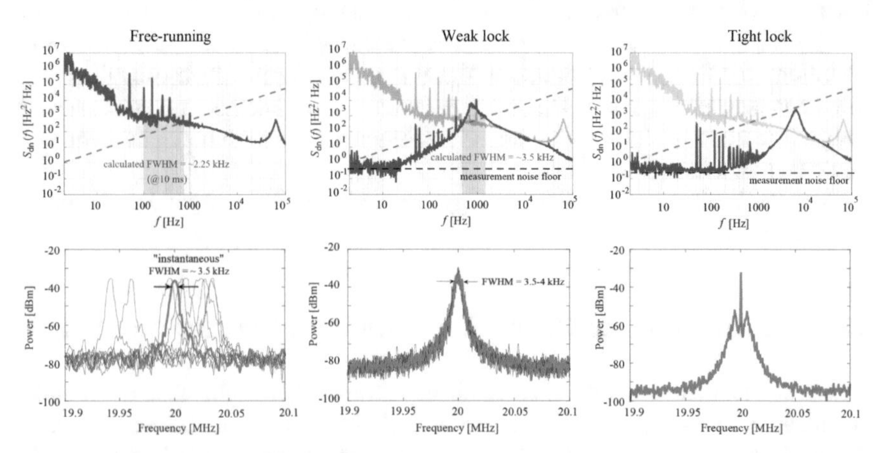

Fig. 12.22 Illustration of the different regimes of CEO frequency stabilization of a diode-pumped SESAM modelocked Er:Yb:glass laser, as shown in Fig. 12.15. The *red dashed line* in *top panels* is the β-separation line. One-sided frequency noise PSDs are shown in all cases [15Ste]

a frequency or a phase discriminator, which for a narrow frequency range has a strong frequency-dependent and, ideally, linear output. For instance, a microwave discriminator is a mixer or a phase detector and the signal can be either analog or digital. A digital phase detector can have a very broad linear range, e.g., $>\pm 30 \times 2\pi$, two orders of magnitude greater than a typical mixer, which is required for large fluctuations. However, they typically have significant nonlinearities. The noise performance of electronic discriminators is important and different solutions have been compared in [11Sch3].

The data in Fig. 12.22 is shown for the same SESAM-modelocked Er:Yb:glass laser as discussed before. The CEO frequency was stabilized to a 20 MHz reference, which in this case was a frequency-doubled 10 MHz H-maser. The control was the diode current of the diode-pumped laser, which has a sufficiently good response function (see Fig. 12.15). This response function critically depends on the specific laser and needs to be known in order to design a stable feedback system, as illustrated in the following example.

Example SESAM-Modelocked Er:Yb:glass Laser

The CEO frequency response as a function of pump power can have a rather complex behavior (see, e.g., [03Hol, 05New, 12Sch]). The response of f_{CEO} to a change in the pump power is thus not simple and generally leads to nonlinear behavior, which may impact the CEO stabilization. In a Ti:sapphire comb, the existence of a sign reversal in the dependence of f_{CEO} on the pump power has been shown [03Hol], with important consequences for CEO stabilization. A similar reversal point was also observed in a diode-pumped ultrafast Er:Yb:glass laser, the same laser we used earlier as an example [12Buc]. For this example, a static tuning of f_{CEO} with the pump current is nonlinear and even non-monotonic (Fig. 12.23). In principle, CEO stabilization should be possible on both sides of the reversal point, but this was not the case. The explanation comes from detailed transfer function measurements as a function of pump current modulation frequencies for different pump current operation points, as shown in Fig. 12.24. Figure 12.24 (left panels) shows that the CEO frequency transfer function depends on the position of the pump current operating point relative to the reversal point shown in Fig. 12.23. This plays a major role in the feasibility of phase-stabilizing f_{CEO}, because this results in a very different overall loop gain transfer function of the total stabilization feedback system (Fig. 12.24, right panels). As discussed in the introduction, the overall feedback loop system becomes unstable when the loop gain is greater than 0 dB (unit gain) in the frequency range where the loop phase shift approaches $\pm 180°$. As shown in Fig. 12.24 (right panels) for a pump current I_{high}, the phase is already below $-180°$ at 1 Hz modulation frequency, and this prevents CEO stabilization.

This example shows that detailed response functions and transfer function measurements may become necessary for successful active stabilization. Ultimately, the total transfer function for the loop amplitude and phase gain is important for a good stabilization, which according to Fig. 12.21 includes the discriminator response (i.e., error signal measurement), the error signal conditioning system, and the actuator (e.g., pump current) response for the specific laser.

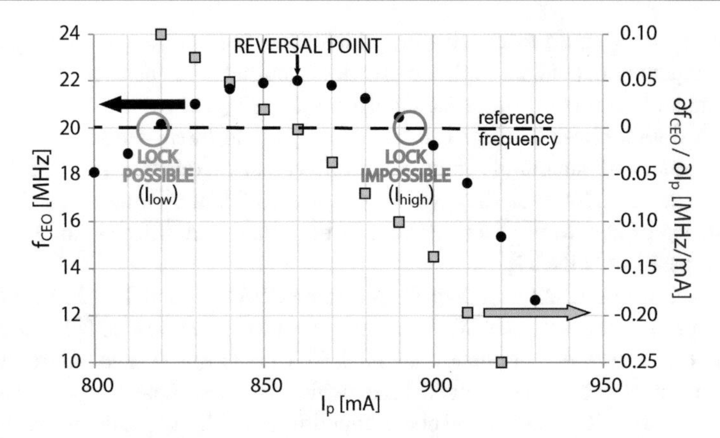

Fig. 12.23 Static tuning curve of the CEO frequency f_{CEO} with respect to the pump current, showing a reversal point at the pump current $I_p \approx 860$ mA. The derivative is indicated on the right scale in units of MHz/mA, and shows a sign change at the reversal point. Stable f_{CEO} locking was only possible for $\partial f_{\mathrm{CEO}}/\partial I_p > 0$ [12Buc]

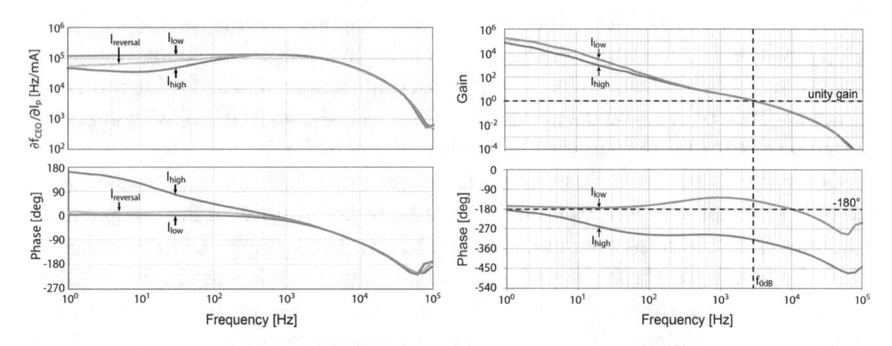

Fig. 12.24 Transfer function measurements as a function of pump current frequency modulation at different pump current operation points. *Left*: CEO frequency f_{CEO} response. *Right*: Overall CEO stabilization loop gain. In both cases, the response is shown for amplitude and phase. I_{low} is the pump current below the reversal point and I_{high} the pump current above the reversal point, as shown in Fig. 12.23 [12Buc]

12.6.3 Full Optical Frequency Comb Stabilization

The optical frequency comb from a modelocked laser has two degrees of freedom: the CEO frequency f_{CEO} and the pulse repetition rate f_{rep}. Both of them need to be stabilized. Timing jitter stabilization has been well established since the 1980s (see Chap. 11). Typically, the cavity length is the actuator for timing jitter stabilization with passively modelocked lasers. However, this also typically affects the CEO frequency (see, e.g., Fig. 12.15). In return, the control signal for the CEO frequency stabilization (e.g., the pump current) can potentially introduce additional timing jitter and intensity noise. Very often, however, even with only CEO stabilization, we obtain lower timing jitter and lower intensity noise (see, e.g., [14Kle2, Fig. 6]), even though the error signal only controls the CEO frequency with a reference frequency. However, this depends on the specific laser parameters and needs to be carefully evaluated. Complex nonlinear coupling between all the different noise quantities can make such a full stabilization more difficult.

For an ideal stabilization of the modelocked laser comb to an ultra-stable laser, the fractional frequency stability of the laser is transferred to the comb repetition rate f_{rep}. Moreover, the resulting phase noise PSD of the repetition rate should correspond to the laser phase noise divided by the large squared mode number N^2 (with $N \approx 2\,570\,000$ in the case of the 1.55 μm Er:Yb:glass laser comb), because the comb acts as a frequency divider. This property allows, for example, for ultra-low noise microwave generation (see, e.g., [05Bar]).

The noise performance of stabilized ultrafast solid-state laser frequency combs has become so good that such OFCs can provide <1 Hz linewidth in the visible and near-infrared regime with a relative frequency linewidth $\Delta\nu/\nu$ of a few 10^{-15} in each comb line of the OFC. This offers the possibility of transferring this relative stability of the laser back to the microwave to make an all-optical microwave oscillator, with better performance than some of the best microwave oscillators (e.g., the H-maser). In this case the optical-to-microwave transfer is made with an optical frequency comb from a modelocked laser. So this is the classic chicken or egg causality question: which came first, the chicken or the egg? However, a relative frequency stability $\Delta\nu/\nu = 10^{-15}$ does not come for free. For example, for a cavity length $L = 10$ cm and a center wavelength of 1.55 μm, a cavity length control of less than the diameter of a proton over the cavity length is required, i.e., $\Delta\nu/\nu = \Delta L/L$. Such a mechanical stability can only be achieved with a great deal of effort using an active anti-vibration system, optimized cavity support points, high temperature insulation, and a low pressure vacuum chamber. Time will tell what ultra-stable microwave oscillator technology will achieve the best performance for the best return of investment.

12.6.4 External CEO Frequency Stabilization

The CEO frequency f_{CEO} of a free-running modelocked laser output can be stabilized and controlled without any active feedback control of the laser. Such a technique can be based on the optical frequency shift of the diffracted laser beam from a travelling wave acousto-optic frequency shifter (AOFS), and is also referred to as the feed-forward method [10Kok]. This resulted in the record-low noise performance for a commercial version of the Er:Yb:glass laser discussed in this chapter [19Lem]. The long-term f_{CEO} stability was measured over 8 hours with 0.16 Hz rms, better than the result shown in Fig. 12.17a (which did not use commercial laser packaging).

Similar to an acousto-optic modulator (see Fig. 6.6), the AOFS consists of an acousto-optic substrate and a transducer that launches an acoustic wave into the substrate. The transducer generates a longitudinal acoustic wave of frequency ω_{ac} and the diffracted laser beam can be considered to be ‘reflected’ at the acoustic wave front and experience a Doppler shift. The acoustic transducer can generate gigahertz frequencies in fully transparent substrates [90Kel1] and can therefore be used to control the CEO frequency, which is less than $f_{rep}/2$ in modelocked lasers. In this way, it is possible to adjust the comb offset directly by replacing f_{CEO} with a fixed frequency and to stabilize the CEO frequency independently of the modelocked laser. The control bandwidth is limited by the travel time of the acoustic wave from the transducer to the interaction area inside the substrate. The trade-off is additional positive GDD from the propagation through the acoustic substrate and an angular chirp on the diffracted beam.

12.7 Laser Technology for Optical Frequency Combs

Free-running modelocked lasers can generate optical frequency combs (OFCs) with reasonable stability, making possible additional stabilization with very high performance. Many examples both in Chap. 11 and in this chapter have shown that the laser to be stabilized needs to be built for very high thermal and mechanical stability, and packaged in a housing to limit any airflow. Furthermore, any pump noise needs to be as low as possible. In this way, a fully monolithic SESAM-modelocked 1 GHz Er:Yb:glass laser has demonstrated record-low noise performance using low-noise 980 nm pump diodes (i.e., indirectly benefiting from major commercial efforts in optical communication) [16Sho].

Generally, diode-pumped passively modelocked solid-state lasers exhibit superior low-noise performance compared to all other lasers. Modelocking of such lasers has been discussed in Chap. 9. In particular, Yb- or Er-based solid-state lasers pumped by low-noise 980 nm laser diodes provide very stable modelocked frequency combs with pulse repetition rates typically ranging from around 100 MHz to 10 GHz. Gigahertz pulse repetition rates are especially interesting because of the small physical footprint, higher mechanical stability, and greater power per comb line. Such lasers are discussed and summarized in Sect. 9.10.3. Cascaded quadratic nonlinearities (CQNs) from phase-mismatched second-harmonic generation resolved the problems with SESAM damage due to self-Q-switching instabilities at high pulse repetition rates (around 10 GHz) [17May]. The frequency-dependent loss of these CQNs causes a power-dependent self-frequency shift (SFS) and allows for a better control of the CEO frequency [21Kru]. The Ti:sapphire laser is currently the best laser that can generate octave-spanning spectra directly inside the laser oscillator [01Ell,03For]. No additional supercontinuum generation is required, which generally degrades the coherence of the optical frequency comb (see, e.g., Sect. 12.3.2). However, the trade-off for Ti:sapphire lasers is the high-power green pump laser.

Ultrafast fiber lasers have an intrinsic higher phase noise because of their large cavity loss (e.g., typically above 50% compared to less than 5% for ion-doped solid-state lasers), large nonlinearities, limited intracavity pulse energies, and relatively large dispersion per cavity roundtrip. This was also confirmed by a direct comparison with Er-based solid-state and fiber lasers (Fig. 12.17, [11Sch2]). However, with more sophisticated active stabilization efforts, excellent noise performance can still be achieved, as for example for Er:fiber lasers (see, e.g., [17Leo]) and for Yb:fiber lasers (see, e.g., [20May]).

Modelocked semiconductor lasers, based on edge-emitting devices, have even more significant phase noise in comparison to fiber lasers, and for similar reasons. In addition, there is a complex coupling between fluctuations of the carrier density in the gain, which introduces significant phase noise and linewidth broadening (12.110). This is reduced significantly with diode-pumped vertical emitting semiconductor lasers such as VECSELs and MIXSELs (see Sect. 9.8), where the laser beam propagates perpendicularly to the quantum well or quantum dot gain layers, strongly limiting the interaction with the free carriers. Ultrafast diode-pumped MIXSELs have very low noise performance compared to edge-emitting semiconductor lasers because of the vertical emission geometry which also results in lower gain, lower output couplers, and therefore higher-Q cavities. The MIXSEL is passively modelocked with excellent noise properties even without any further active stabilization [14Man2]. This near-infrared free-running MIXSEL comb had a frequency comb spacing of 2 GHz, an average output power of more than 600 mW, and a timing jitter of 127 fs integrated over a frequency range of 100 Hz to 100 MHz. Moreover, a self-referenced CEO frequency stabilization has become possible without any further external pulse amplification [19Wal], which normally adds noise and complexity to the OFC.

There is a new group of optical frequency combs which are not based on passive modelocking in the traditional sense, typically referred to as Kerr combs, microresonator combs, or simply microcombs [19Gae]. These microcombs are high-Q cavities pumped with a single-frequency cw laser, and nonlinear intracavity interaction (i.e., cascaded four-wave mixing) generates a phase-locked frequency comb which under certain conditions can also support femtosecond pulses. Chip-based microcomb frequency spacing has been in the range of 20 GHz to more than 100 GHz. A gigahertz comb spacing requires long microresonators with few losses, and this is very challenging. The clear advantage of microcombs is the very high Q factor (above 10^8), corresponding to a linewidth at 1550 nm of around 1–10 MHz and an octave-wide comb span. The average power, however, is usually very low, typically in the range of a few μW to 1 mW total optical power, with much lower power per comb line (e.g., the nanowatt level).

Active modulators create coherent modulation sidebands which can be scaled up to many more modulation sidebands in a passive enhancement cavity and in cascaded configurations. Such a cascaded modulation scheme, however, increases the integrated phase noise in proportion to the number of sidebands, and the slope of this increase depends on the phase noise of the external microwave synthesizer [13Ish]. This is in strong contrast to active modelocking with an optical modulator inside a laser for which the nonlinear laser interaction leads to a very high intrinsic comb stability, and even better noise performance is obtained for femtosecond pulses with passive modelocking.

Optical frequency comb generation is still a very active field of research and different applications will benefit from all the different comb technologies. The frequency comb revolution [99Tel, 00Jon, 00Apo] and dual-comb spectroscopy [02Sch1, 16Cod] have motivated dual-comb measurement applications with two pulse trains with different pulse repetition rates for equivalent time sampling [85Wei, 87Elz, 88Wei]. Initially, such dual-comb measurements were based on two separate frequency comb sources that had to be stabilized to each other with four stabilization loops for their comb spacing and offset. Dual-comb modelocking from a single laser cavity with different pulse repetition rates was first demonstrated with polarization duplexing [15Lin], and in ring lasers with circular direction duplexing [16Ide, 16Meh] (see also Sect. 9.9). This resulted in a paradigm shift for dual-comb spectroscopy using only a free-running dual-comb modelocked laser [17Lin1]. Dual-comb measurements have become a hot topic as demonstrated with recent review articles (e.g., [16Cod, 19For, 19Gae, 20Lia]), and many further applications with free running dual-comb lasers have been demonstrated such as precise optical ranging [21Nur], high-speed pump–probe measurements via equivalent time sampling [20Wil2], and picosecond ultrasonics [21Pup], to mention only a few of the most recent publications from my group.

Fourier Transform

A

The Fourier transform of a function $f(t)$ is defined by

$$F(\omega) = F\{f(t)\} \equiv \int_{-\infty}^{+\infty} f(t)\mathrm{e}^{-\mathrm{i}\omega t}\mathrm{d}t \tag{A.1}$$

and the inverse Fourier transform by

$$f(t) = F^{-1}\{F(\omega)\} \equiv \frac{1}{2\pi}\int_{-\infty}^{+\infty} F(\omega)\mathrm{e}^{\mathrm{i}\omega t}\mathrm{d}\omega\ . \tag{A.2}$$

Table A.1 shows some standard results and Table A.2 some Fourier transforms of standard functions. The convolution is denoted by an asterisk and defined by

$$f(t) * g(t) \equiv \int_{-\infty}^{+\infty} f(\tau)g(t-\tau)\mathrm{d}\tau\ , \tag{A.3}$$

and it is commutative and distributive, i.e.,

$$f(t) * g(t) = g(t) * f(t)\ , \quad f * (g+h) = f * g + f * h\ . \tag{A.4}$$

Parseval's theorem states that

$$\int_{-\infty}^{+\infty} |f(t)|^2\mathrm{d}t = \frac{1}{2\pi}\int_{-\infty}^{+\infty} |F(\omega)|^2\mathrm{d}\omega\ . \tag{A.5}$$

The step function is defined by

$$s(t) = \begin{cases} 1\ , & t \geq 0\ , \\ 0\ , & t < 0\ . \end{cases} \tag{A.6}$$

© The Editor(s) (if applicable) and The Author(s), under exclusive licence to Springer Nature Switzerland AG 2021

U. Keller, *Ultrafast Lasers*, Graduate Texts in Physics,
https://doi.org/10.1007/978-3-030-82532-4

Table A.1 Standard results for the Fourier transform

	Time domain $f(t)$	Frequency domain $F(\omega)$
Addition	$af(t)+bg(t)$	$aF(\omega)+bG(\omega)$, for $a, b \in \mathbb{C}$
Similarity	$f(\alpha t)$	$\frac{1}{\lvert\alpha\rvert}F(\omega/\alpha)$, for $\alpha \in \mathbb{R}$
Time shift	$f(t-t_0)$	$F(\omega)\mathrm{e}^{-\mathrm{i}\omega t_0}$, for $t_0 \in \mathbb{R}$
Frequency shift	$f(t)\mathrm{e}^{\mathrm{i}\alpha t}$	$F(\omega-\alpha)$, for $\alpha \in \mathbb{R}$
Differentiation (nth derivative)	$f^{(n)}(t)$	$(\mathrm{i}\omega)^n F(\omega)\,,\quad \text{if } f(\pm\infty)=0$
	$t^n f(t)$	$\mathrm{i}^n F^{(n)}(\omega)$
Integration	$\int_{-\infty}^{t} f(\tau)\mathrm{d}\tau$	$\frac{1}{\mathrm{i}\omega}F(\omega)+\pi F(0)\delta(\omega)$
Convolution	$f(t) * g(t)$	$F(\omega)G(\omega)$
	$f(t)g(t)$	$\frac{1}{2\pi}F(\omega) * G(\omega)$
Average	$\frac{1}{2T}\int_{t-T}^{t+T} f(\tau)\mathrm{d}\tau$	$F(\omega)\frac{\sin(\omega T)}{\omega T}$, for $T \in \mathbb{R}$
Modulation	$f(t)\cos(\omega_0 t)$	$\frac{1}{2}\left[F(\omega-\omega_0)+F(\omega+\omega_0)\right]$, for $\omega_0 \in \mathbb{R}$
	$f(t)\sin(\omega_0 t)$	$\frac{1}{2\mathrm{i}}\left[F(\omega-\omega_0)-F(\omega+\omega_0)\right]$
Interchange (duality)	$F(t)$	$2\pi f(-\omega)$

The sign function is defined by

$$\operatorname{sgn}(t) = \begin{cases} -1\,, & t<0\,, \\ 1\,, & t \geq 0\,, \end{cases} \tag{A.7}$$

whence

$$\operatorname{sgn}(t) = 2s(t) - 1\,. \tag{A.8}$$

Table A.2 Fourier transforms of standard functions ($t_0, \omega_0, T \in \mathbb{R}$)

Time domain	Frequency domain
$f(t)$	$F(\omega) = F\{f(t)\}$
$\delta(t)$	1
$\delta(t - t_0)$	$e^{-i\omega t_0}$
1	$2\pi\,\delta(\omega)$
$e^{i\omega_0 t}$	$2\pi\,\delta(\omega - \omega_0)$
Step function $s(t)$	$\frac{1}{i\omega} + \pi\,\delta(\omega)$
Sign function $\operatorname{sgn}(t)$	$\frac{2}{i\omega}$
$\operatorname{rect}_a(t) \equiv \begin{cases} 1\,, & \|t\| \le a\,, \\ 0\,, & \|t\| > a\,, \end{cases}$	$2\frac{\sin(a\omega)}{\omega}$
$e^{-t/T}s(t)$	$\frac{T}{1 + i\omega T}$
$e^{-\|t\|/T}$	$\frac{2T}{1 + (\omega T)^2}$
$e^{-\|t\|/T}\operatorname{sgn}(t)$	$-2i\frac{\omega T^2}{1 + (\omega T)^2}$
$e^{-(t/T)^2/2}$	$\sqrt{2\pi}\,T e^{-(T\omega)^2/2}$
$e^{-t/T}\sin(\omega_0 t)s(t)$	$\frac{\omega_0}{\left(\frac{1}{T} + i\omega\right)^2 + \omega_0^2}$
$e^{-t/T}\cos(\omega_0 t)s(t)$	$\frac{1/T + i\omega}{\left(\frac{1}{T} + i\omega\right)^2 + \omega_0^2}$
$\frac{1}{t^2 + 1}$	$\pi e^{-\|\omega\|}$
$\frac{t}{t^2 + 1}$	$\begin{cases} i\pi e^{-\|\omega\|}\,, & \omega < 0\,, \\ 0\,, & \omega = 0\,, \\ -i\pi e^{-\|\omega\|}\,, & \omega > 0\,. \end{cases}$
$\sum_{n=-\infty}^{\infty} \delta(t - nT)$	$\omega_0 \sum_{n=-\infty}^{\infty} \delta(\omega - n\omega_0)\,, \quad \omega_0 = 2\pi/T$

Dispersion for Quantum Mechanical Particles

B

The discussion on laser pulse propagation also applies to any quantum mechanical wave packet such as free electrons, electrons in semiconductor materials, and phonons. For these cases, we can predict wave packet spreading and wave packet propagation in a very similar way as discussed for light pulses in Chap. 2.

The reason why the dispersion relation is typically described by $\omega(k)$ relies on the wave–particle duality of the e.m. wave in quantum mechanics. For a photon, the dispersion relation $\omega(k)$ together with the de Broglie relation results in an expression for the energy $E = \hbar\omega$ as a function of the momentum $p = \hbar k$:

$$\text{Photon in vacuum} \quad E = cp \iff \omega(k) = ck \,, \tag{B.1}$$

$$\text{Photon in medium} \quad E = c_n p_n \iff \omega(k) = c_n k_n \,. \tag{B.2}$$

The energy–momentum relation $E = cp$ also follows directly from the special theory of relativity for a massless particle moving at the speed of light.

We have shown that the propagation velocity of a photon is given by the group velocity υ_g :

$$\upsilon_g = \begin{cases} \dfrac{d\omega}{dk} & \text{in vacuum}\,, \\[2ex] \dfrac{d\omega}{dk_n} & \text{in a dispersive medium}\,, \end{cases} \tag{B.3}$$

which is not the same as the phase velocity υ_p :

$$\upsilon_p = \begin{cases} \dfrac{\omega}{k} & \text{in vacuum}\,, \\[2ex] \dfrac{\omega}{k_n} & \text{in a dispersive medium}\,. \end{cases} \tag{B.4}$$

© The Editor(s) (if applicable) and The Author(s), under exclusive licence to Springer Nature Switzerland AG 2021

U. Keller, *Ultrafast Lasers*, Graduate Texts in Physics,
https://doi.org/10.1007/978-3-030-82532-4

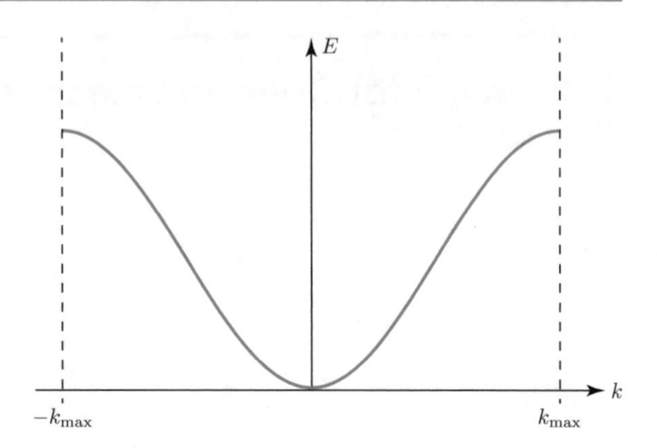

Fig. B.1 Band diagram for an electron in a semiconductor in the tight-binding approximation

In this book we focus on dispersion in the context of e.m. waves, but the discussion is valid much more generally. These group and phase velocities also apply for other quantum mechanical particles, except that the dispersion function for other quantum mechanical particles will in general be different from that for a photon.

For example, for a free non-relativistic electron with mass m and momentum p, we have the dispersion relation

$$\text{Free electron} \quad E = \frac{p^2}{2m} \iff \omega(k) = \frac{\hbar}{2m}k^2 \,. \tag{B.5}$$

For a free electron, we then obtain $\upsilon_\text{p} = \omega/k = \hbar k/2m$ for the phase velocity according to (B.4) and $\upsilon_\text{g} = \text{d}\omega/\text{d}k = \hbar k/m$ for the group velocity according to (B.3), i.e., υ_g is twice as big as υ_p. The group velocity can be transformed with the result $\upsilon_\text{g} = \hbar k/m = p/m = m\upsilon_\text{class}/m = \upsilon_\text{class}$, which corresponds to the classical velocity of a particle.

For another example, an electron in the conduction band of a semiconductor in the tight-binding approximation has the following dispersion:

$$E(k) \approx \tilde{E}_\text{at} - 2E_\text{s} \cos ka \iff \omega(k) \approx \tilde{\omega}_\text{at} - 2\omega_\text{s} \cos ka \,, \tag{B.6}$$

where $\tilde{E}_\text{at}$ is approximately the atomic energy state and E_s has to do with the overlap of the neighboring wave functions. Thus, an electron moves inside the conduction band of a semiconductor with different velocities for different momenta. For $\pm k_\text{max} = \pm\pi/a$ the group velocity is even zero, because the electron then satisfies the Bragg condition, and is therefore reflected by the periodic potential (Fig. B.1).

For a phonon in a one-dimensional grating with identical atoms of mass M, grating period a, and spring constant β, we obtain the dispersion

$$\text{Phonon} \quad \omega(k) = 2\sqrt{\frac{\beta}{M}} \sin\left(\frac{1}{2}|k|a\right) \,. \tag{B.7}$$

Delta-Function

C

The δ-function is defined by

$$\boxed{\delta(x) = 0\,, \quad x \neq 0\,, \quad \int_{-\infty}^{+\infty} \delta(x)\mathrm{d}x = 1}\,. \tag{C.1}$$

The δ-function can be expressed as the limit of a function series:

$$\boxed{\delta(x) = \lim_{\tau \to 0} \tau^{-1} \Pi\left(\frac{x}{\tau}\right)}\,, \tag{C.2}$$

where the function $\Pi(x)$ is the rectangle function:

$$\Pi(x) \equiv \begin{cases} 1\,, & |x| < 1/2\,, \\ 0\,, & |x| > 1/2\,, \end{cases} \tag{C.3}$$

and

$$\Pi\left(\frac{x}{\tau}\right) = \begin{cases} 1\,, & |x| < \tau/2\,, \\ 0\,, & |x| > \tau/2\,. \end{cases} \tag{C.4}$$

Thus, (C.2) means that the δ-function can be approximated by a rectangle which gets continuously narrower and higher, but still keeps a constant area of 1:

$$\int_{-\infty}^{+\infty} \delta(x)\mathrm{d}x = \lim_{\tau \to 0} \int_{-\infty}^{+\infty} \tau^{-1} \Pi\left(\frac{x}{\tau}\right) \mathrm{d}x = 1$$

and

$$\lim_{\tau \to 0} \tau^{-1} \Pi\left(\frac{x}{\tau}\right) = 0\,, \quad x \neq 0\,.$$

© The Editor(s) (if applicable) and The Author(s), under exclusive licence to Springer Nature Switzerland AG 2021

U. Keller, *Ultrafast Lasers*, Graduate Texts in Physics,
https://doi.org/10.1007/978-3-030-82532-4

Therefore, in the limiting case $\tau \to 0$, the function $\tau^{-1}\Pi(x/\tau)$ satisfies the definition (C.1).

Another function series which gives the δ-function in the limiting case is

$$\boxed{\delta(x) = \lim_{\tau\to 0} \frac{1}{\sqrt{\pi}\,\tau} \mathrm{e}^{-x^2/\tau^2}} \,. \tag{C.5}$$

As for the rectangle function, the definition in (C.1) is also satisfied for this limiting case, since

$$\int_{-\infty}^{+\infty} \mathrm{e}^{-a^2x^2}\mathrm{d}x = \frac{\sqrt{\pi}}{a} \,.$$

Thus, in this case the δ-function is the limit of a Gaussian function with a constant area 1 getting continuously narrower and higher.

We can also show that the δ-function $\delta(x)$ is the first derivative of the step function $\theta(x)$:

$$\boxed{\delta(x) = \frac{\mathrm{d}}{\mathrm{d}x}\theta(x)} \,, \tag{C.6}$$

where the step function is defined by

$$\theta(x) \equiv \begin{cases} 1\,, & x > 0\,, \\ 0\,, & x < 0\,. \end{cases} \tag{C.7}$$

Equation (C.6) is proven as follows. From the definition (C.1) of the δ-function, it follows that

$$\int_{-\infty}^{x} \delta\left(x'\right)\mathrm{d}x' = \begin{cases} 1\,, & x > 0\,, \\ 0\,, & x < 0\,. \end{cases}$$

Equation (C.7) then implies

$$\int_{-\infty}^{x} \delta\left(x'\right)\mathrm{d}x' = \theta(x)\,,$$

and (C.6) follows.

It can be shown that

$$\boxed{\int_{-\infty}^{+\infty} \delta(x) f(x)\mathrm{d}x = f(0)\,, \quad \int_{-\infty}^{+\infty} \delta(x-a) f(x)\mathrm{d}x = f(a)} \,. \tag{C.8}$$

This is proven as follows. Equation (C.2) implies

$$\int_{-\infty}^{+\infty} \delta(x) f(x)\mathrm{d}x = \lim_{\tau\to 0} \int_{-\infty}^{+\infty} \tau^{-1}\Pi\left(\frac{x}{\tau}\right) f(x)\mathrm{d}x \,.$$

In the limit as $\tau \to 0$, the height of the rectangle in the integral is given by $f(0)$, and we obtain

$$\lim_{\tau\to 0}\int_{-\infty}^{+\infty}\tau^{-1}\Pi\left(\frac{x}{\tau}\right)f(x)\mathrm{d}x = \lim_{\tau\to 0}\int_{-\tau/2}^{+\tau/2}\tau^{-1}f(0)\mathrm{d}x = f(0)\ .$$

For the convolution with a δ-function, we have

$$\boxed{f(x) * \delta(x) = \delta(x) * f(x) = f(x)}\ . \tag{C.9}$$

This is proven as follows. Equation (C.8) implies

$$f(x) * \delta(x) = \int_{-\infty}^{+\infty} f\left(x'\right)\delta\left(x - x'\right)\mathrm{d}x' = f(x)\ ,$$

and then (C.2) yields (C.9).

Directly from the definition (C.1) of the δ-function, it follows that

$$\boxed{\delta(-x) = \delta(x)}\ , \tag{C.10}$$

and thus

$$\boxed{\delta(ax) = \frac{1}{|a|}\delta(x)}\ . \tag{C.11}$$

This is proven as follows. Setting $y = ax$ and using (C.10), we have

$$\int_{-\infty}^{+\infty}\delta(ax)\mathrm{d}x = \frac{1}{|a|}\int_{-\infty}^{+\infty}\delta(y)\,\mathrm{d}y = \frac{1}{|a|}\int_{-\infty}^{+\infty}\delta(x)\mathrm{d}x\ .$$

The Fourier transform of the δ-function is given by

$$\boxed{F\{\delta(t)\} = \int_{-\infty}^{+\infty}\delta(t)\mathrm{e}^{-\mathrm{i}\omega t}\mathrm{d}t = 1}\ . \tag{C.12}$$

This is proven as follows. Equation (C.8) implies that

$$\tilde{\delta}(\omega) = F\{\delta(t)\} = \int_{-\infty}^{+\infty}\delta(t)\mathrm{e}^{-\mathrm{i}\omega t}\mathrm{d}t = \mathrm{e}^{-\mathrm{i}0} = \cos 0 - \mathrm{i}\sin 0 = 1\ .$$

The Fourier transform of 1 is given by

$$\boxed{F\{1\} = 2\pi\delta(\omega)}\ . \tag{C.13}$$

This is proven by noting that

$$1 = \int_{-\infty}^{+\infty} \delta(\omega) \mathrm{e}^{+\mathrm{i}\omega t} \mathrm{d}\omega = 2\pi F^{-1} \{\delta(\omega)\}$$

$$\Longrightarrow F\{1\} = 2\pi F F^{-1} \{\delta(\omega)\} = 2\pi \delta(\omega) \ .$$

Furthermore, by the shift theorem for the Fourier transform, we have

$$\boxed{F\left\{\mathrm{e}^{\mathrm{i}\omega_0 t}\right\} = 2\pi \delta(\omega - \omega_0)} \ . \tag{C.14}$$

The shift theorem yields

$$F\{f(t)\} = \tilde{f}(\omega) \Longrightarrow F\left\{\mathrm{e}^{\mathrm{i}\omega_0 t} f(t)\right\} = \tilde{f}(\omega - \omega_0) \ . \tag{C.15}$$

This follows directly from the definition of the Fourier transform:

$$F\left\{\mathrm{e}^{\mathrm{i}\omega_0 t} f(t)\right\} = \int_{-\infty}^{+\infty} f(t) \mathrm{e}^{\mathrm{i}\omega_0 t} \mathrm{e}^{-\mathrm{i}\omega t} \mathrm{d}t = \int_{-\infty}^{+\infty} f(t) \mathrm{e}^{-\mathrm{i}(\omega - \omega_0)t} \mathrm{d}t = \tilde{f}(\omega - \omega_0) \ .$$

Then (C.14) follows from (C.13) and (C.15).

Delta-Comb

D

The δ-comb is defined as a sum of periodically shifted δ-functions:

$$\text{III}(x) = \sum_{n=-\infty}^{+\infty} \delta(x-n) \; . \tag{D.1}$$

From Appendix C, we may deduce the following useful properties of the δ-comb:

$$\begin{aligned}
&\text{III}(x) = 0 \, , \quad x \neq n \, , \\
&\text{III}(-x) = \text{III}(x) \, , \\
&\text{III}(x+n) = \text{III}(x) \, , \quad n = \text{integer} \, , \\
&\text{III}(ax) = \tfrac{1}{|a|} \textstyle\sum_{n=-\infty}^{+\infty} \delta\left(x - \tfrac{n}{a}\right) \, , \\
&\textstyle\int_{n-1/2}^{n+1/2} \text{III}(x) \mathrm{d}x = 1 \; .
\end{aligned} \tag{D.2}$$

We have the sampling property:

$$\text{III}(x) f(x) = \sum_{n=-\infty}^{+\infty} f(n) \, \delta(x-n) \; . \tag{D.3}$$

This follows directly from (C.8). Another property is

$$\text{III}(x) * f(x) = \sum_{n=-\infty}^{+\infty} f(x-n) \; . \tag{D.4}$$

© The Editor(s) (if applicable) and The Author(s), under exclusive licence to Springer Nature Switzerland AG 2021

U. Keller, *Ultrafast Lasers*, Graduate Texts in Physics,
https://doi.org/10.1007/978-3-030-82532-4

This is proven by noting that

$$\begin{aligned}
\text{III}(x) * f(x) &= \int_{-\infty}^{+\infty} \text{III}\left(x'\right) f\left(x - x'\right) \mathrm{d}x' \\
&= \int_{-\infty}^{+\infty} \sum_{n=-\infty}^{+\infty} \delta\left(x' - n\right) f\left(x - x'\right) \mathrm{d}x' \\
&= \sum_{n=-\infty}^{+\infty} \int_{-\infty}^{+\infty} \delta\left(x' - n\right) f\left(x - x'\right) \mathrm{d}x' ,
\end{aligned}$$

then using (C.8) to see that

$$\int_{-\infty}^{+\infty} \delta\left(x' - n\right) f\left(x - x'\right) \mathrm{d}x' = f(x - n) ,$$

which yields (D.4).

The Fourier transform of the δ-comb is again a δ-comb:

$$F\left\{\text{III}(t)\right\} = 2\pi \,\text{III}\,(\omega) \;, \quad F\left\{\sum_{n=-\infty}^{+\infty} \delta(t - n)\right\} = 2\pi \sum_{n=-\infty}^{+\infty} \delta\left(\omega - n2\pi\right) , \tag{D.5}$$

or

$$F\left\{\sum_{n=-\infty}^{+\infty} \delta\left(t - nT\right)\right\} = 2\pi \frac{1}{T} \sum_{n=-\infty}^{+\infty} \delta\left(\omega - n\frac{2\pi}{T}\right) . \tag{D.6}$$

This is proven as follows. Due to the fact that the δ-comb is a periodic function with period T, it can be decomposed into a Fourier series with period $\omega_T = 2\pi/T$ and coefficients F_n:

$$\sum_{n=-\infty}^{+\infty} \delta(t - nT) = \sum_{n=-\infty}^{+\infty} F_n \mathrm{e}^{\mathrm{i}n\omega_T t} , \quad \omega_T = \frac{2\pi}{T} .$$

According to (C.12) the coefficients are then given by

$$F_n = \frac{1}{T} \int_{-T/2}^{T/2} \delta(t) \mathrm{e}^{-\mathrm{i}n\omega_T t} \mathrm{d}t = \frac{1}{T} ,$$

whence

$$\sum_{n=-\infty}^{+\infty} \delta(t - nT) = \frac{1}{T} \sum_{n=-\infty}^{+\infty} \mathrm{e}^{\mathrm{i}n\omega_T t} , \quad \omega_T = \frac{2\pi}{T} .$$

Equation (C.14) shows that

$$F\left\{\mathrm{e}^{\mathrm{i}n\omega_T t}\right\} = 2\pi \,\delta\left(\omega - n\omega_T\right) ,$$

which implies

$$F\left\{\sum_{n=-\infty}^{+\infty}\delta(t-nT)\right\}=2\pi\frac{1}{T}\sum_{n=-\infty}^{+\infty}\delta\left(\omega-n\omega_T\right)\ ,\quad \omega_T=\frac{2\pi}{T}\ .$$

We also have the following useful identity for the δ-comb, which we derived in the proof of (D.6):

$$\sum_{n=-\infty}^{+\infty}\delta(t-nT)=\frac{1}{T}\sum_{n=-\infty}^{+\infty}\mathrm{e}^{in\omega_T t}\ ,\quad \omega_T=\frac{2\pi}{T}\ . \tag{D.7}$$

Soliton Algebra

E

As derived in Chap. 4 of this book, the shape of a soliton pulse is described by the hyperbolic secant function $\operatorname{sech} x$. This means that in working with solitons it is useful to have a reference of hyperbolic identities. This appendix is intended to serve as such a reference.

The hyperbolic functions are defined as follows:

$$\text{hyperbolic cosine} \quad \cosh x \equiv \frac{1}{2}\left(e^{x}+e^{-x}\right) , \tag{E.1}$$

$$\text{hyperbolic sine} \quad \sinh x \equiv \frac{1}{2}\left(e^{x}-e^{-x}\right) , \tag{E.2}$$

$$\text{hyperbolic secant} \quad \operatorname{sech} x \equiv \frac{1}{\cosh x} = \frac{2}{e^{x}+e^{-x}} , \tag{E.3}$$

$$\text{hyperbolic tangent} \quad \tanh x \equiv \frac{\sinh x}{\cosh x} = \frac{e^{x}-e^{-x}}{e^{x}+e^{-x}} , \tag{E.4}$$

from which it follows that (Fig. E.1)

$$\cosh^2(x) - \sinh^2(x) = 1 , \tag{E.5}$$

$$\begin{aligned} \sinh(-x) &= -\sinh(x) , \quad \tanh(-x) = -\tanh(x) , \\ \cosh(-x) &= \cosh(x) , \quad \operatorname{sech}(-x) = \operatorname{sech}(x) , \end{aligned} \tag{E.6}$$

$$\operatorname{sech}^2(x) = 1 - \tanh^2(x) , \tag{E.7}$$

$$\sinh^2(x) = \frac{1-\operatorname{sech}^2(x)}{\operatorname{sech}^2(x)} . \tag{E.8}$$

© The Editor(s) (if applicable) and The Author(s), under exclusive licence to Springer Nature Switzerland AG 2021

U. Keller, *Ultrafast Lasers*, Graduate Texts in Physics,
https://doi.org/10.1007/978-3-030-82532-4

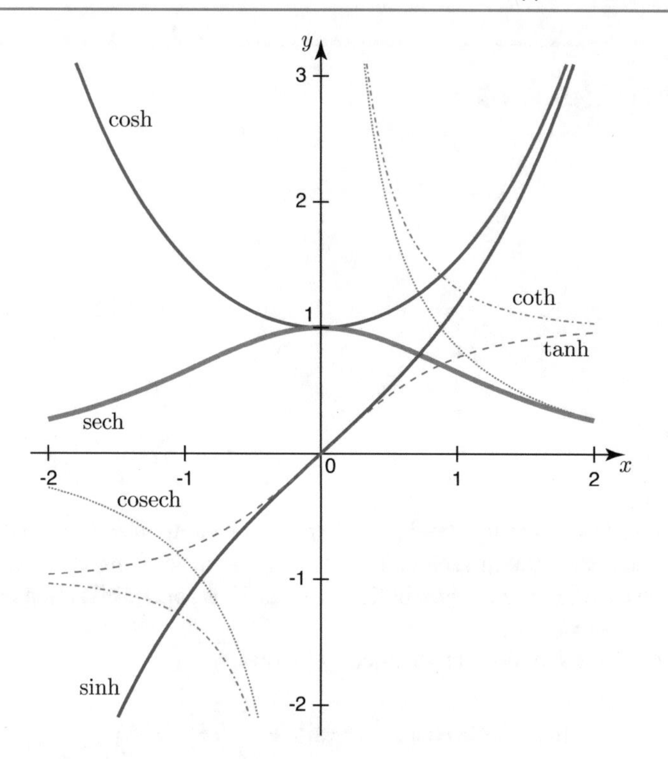

Fig. E.1 The hyperbolic functions (after Bronstein)

It is easy to see that

$$\frac{\mathrm{d}}{\mathrm{d}x}\sinh(x) = \cosh(x) \;, \tag{E.9}$$

$$\frac{\mathrm{d}}{\mathrm{d}x}\cosh(x) = \sinh(x) \;, \tag{E.10}$$

and

$$\frac{\mathrm{d}}{\mathrm{d}x}\tanh(x) = \frac{1}{\cosh^2(x)} = \operatorname{sech}^2(x) \;. \tag{E.11}$$

The first two derivatives of $\operatorname{sech}(x)$ can be calculated as follows:

$$\begin{aligned} \frac{\mathrm{d}}{\mathrm{d}x}\operatorname{sech}(x) &= \frac{\mathrm{d}}{\mathrm{d}x}\left(\frac{2}{\mathrm{e}^x + \mathrm{e}^{-x}}\right) \\ &= 2(-1)\left(\frac{1}{\mathrm{e}^x + \mathrm{e}^{-x}}\right)^2 \left(\mathrm{e}^x - \mathrm{e}^{-x}\right) \\ &= -\tanh(x)\operatorname{sech}(x) \;, \end{aligned} \tag{E.12}$$

$$\begin{aligned}
\frac{d^2}{dx^2}\operatorname{sech}(x) &= \frac{d}{dx}\big[-\tanh(x)\operatorname{sech}(x)\big] \\
&= -\left[\frac{d}{dx}\tanh(x)\right]\operatorname{sech}(x) - \tanh(x)\frac{d}{dx}\operatorname{sech}(x) \\
&= -\frac{1}{\cosh^2(x)}\operatorname{sech}(x) - \tanh(x)\big[-\tanh(x)\operatorname{sech}(x)\big] \\
&= -\operatorname{sech}^3(x) + \tanh^2(x)\operatorname{sech}(x) \\
&= \operatorname{sech}(x)\left[\tanh^2(x) - \operatorname{sech}^2(x)\right] .
\end{aligned} \tag{E.13}$$

The integral of $\operatorname{sech}(x)$ can be found as follows. First note that

$$\int_{-\infty}^{+\infty}\operatorname{sech}(ax)dx = \int_{-\infty}^{+\infty}\frac{2}{e^{ax}+e^{-ax}}dx = \int_{-\infty}^{+\infty}\frac{2e^{ax}}{e^{2ax}+1}dx .$$

Setting $y = e^{ax}$ so that $dy = e^{ax}a dx$, it follows that

$$\begin{aligned}
\int_{-\infty}^{+\infty}\operatorname{sech}(ax)dx &= \frac{2}{a}\int_{-\infty}^{+\infty}\frac{1}{y^2+1}dy \\
&= \frac{2}{a}\arctan(y) \implies \int_{-\infty}^{+\infty}\operatorname{sech}(ax)dx = \frac{2}{a}\arctan\left(e^{ax}\right)
\end{aligned} \tag{E.14}$$

whence also

$$\int_{-\infty}^{+\infty}\operatorname{sech}(x)dx = \pi . \tag{E.15}$$

The integral of $\operatorname{sech}^2(x)$ is found as follows. First note that

$$\operatorname{sech}^2(ax) = 1 - \tanh^2(ax)$$

and

$$\int_{-\infty}^{+\infty}\tanh^2(ax)dx = x - \frac{\tanh(ax)}{a} ,$$

whence

$$\begin{aligned}
\int_{-\infty}^{+\infty}\operatorname{sech}^2(ax)dx &= \int_{-\infty}^{+\infty}1\,dx - \int_{-\infty}^{+\infty}\tanh^2(ax)dx \\
&= x - x + \frac{\tanh(ax)}{a} \\
&= \frac{\tanh(ax)}{a} .
\end{aligned} \tag{E.16}$$

Table E.1 Fourier transforms of hyperbolic functions

Function $f(t)$	Fourier transform $\tilde{f}(\omega) = \int_{-\infty}^{+\infty} f(t)\mathrm{e}^{-\mathrm{i}\omega t}\,\mathrm{d}t$
$\mathrm{sech}(t)$	$\pi\,\mathrm{sech}\left(\frac{\pi}{2}\omega\right)$
$\mathrm{sech}^2(t)$	$\frac{\pi\omega}{\sinh(\pi\omega/2)}$
$\mathrm{sech}^3(t)$	$\frac{1}{2}\left(1+\omega^2\right)\pi\,\mathrm{sech}\left(\frac{\pi}{2}\omega\right)$
$\mathrm{sech}^5(t)$	$\frac{1}{24}\left(\omega^4+10\omega^2+9\right)\pi\,\mathrm{sech}\left(\frac{\pi}{2}\omega\right)$
$\tanh(t)\,\mathrm{sech}(t)$	$-\mathrm{i}\pi\omega\,\mathrm{sech}\left(\frac{\pi}{2}\omega\right)$
$\tanh^2(t)\,\mathrm{sech}(t)$	$\frac{1}{2}\left(1-\omega^2\right)\pi\,\mathrm{sech}\left(\frac{\pi}{2}\omega\right)$
$\tanh^3(t)\,\mathrm{sech}(t)$	$-\mathrm{i}\frac{\omega}{6}\left(5-\omega^2\right)\pi\,\mathrm{sech}\left(\frac{\pi}{2}\omega\right)$
$\tanh(t)\,\mathrm{sech}^3(t)$	$-\mathrm{i}\frac{\omega}{6}\left(1+\omega^2\right)\pi\,\mathrm{sech}\left(\frac{\pi}{2}\omega\right)$
$\tanh^2(t)\,\mathrm{sech}^3(t)$	$\frac{1}{2}\left(1+\omega^2\right)\pi\,\mathrm{sech}\left(\frac{\pi}{2}\omega\right)-\frac{1}{24}\left(\omega^4+10\omega^2+9\right)\pi\,\mathrm{sech}\left(\frac{\pi}{2}\omega\right)$
$t\tanh(t)\,\mathrm{sech}(t)$	$\pi\,\mathrm{sech}\left(\frac{\pi}{2}\omega\right)-\omega\frac{\pi^2}{2}\tanh\left(\frac{\pi}{2}\omega\right)\mathrm{sech}\left(\frac{\pi}{2}\omega\right)$
$t\tanh^2(t)\,\mathrm{sech}(t)$	$-\mathrm{i}\omega\pi\,\mathrm{sech}\left(\frac{\pi}{2}\omega\right)-\mathrm{i}\frac{\pi^2}{4}\left(1-\omega^2\right)\tanh\left(\frac{\pi}{2}\omega\right)\mathrm{sech}\left(\frac{\pi}{2}\omega\right)$
$t\tanh^3(t)\,\mathrm{sech}(t)$	$\frac{1}{6}\left(5-3\omega^2\right)\pi\,\mathrm{sech}\left(\frac{\pi}{2}\omega\right)-\frac{\pi^2}{2}\frac{\omega}{6}\left(5-\omega^2\right)\tanh\left(\frac{\pi}{2}\omega\right)\mathrm{sech}\left(\frac{\pi}{2}\omega\right)$
$t\tanh(t)\,\mathrm{sech}^3(t)$	$\frac{1}{6}\left(1+3\omega^2\right)\pi\,\mathrm{sech}\left(\frac{\pi}{2}\omega\right)-\frac{\pi^2}{2}\frac{\omega}{6}\left(1+\omega^2\right)\tanh\left(\frac{\pi}{2}\omega\right)\mathrm{sech}\left(\frac{\pi}{2}\omega\right)$
$t\,\mathrm{sech}(t)$	$-\mathrm{i}\frac{\pi^2}{2}\tanh\left(\frac{\pi}{2}\omega\right)\mathrm{sech}\left(\frac{\pi}{2}\omega\right)$
$t\,\mathrm{sech}^3(t)$	$\mathrm{i}\pi\omega\,\mathrm{sech}\left(\frac{\pi}{2}\omega\right)-\mathrm{i}\frac{\pi^2}{4}\left(1+\omega^2\right)\tanh\left(\frac{\pi}{2}\omega\right)\mathrm{sech}\left(\frac{\pi}{2}\omega\right)$

Then also (Table E.1)

$$\int_{-\infty}^{+\infty} \mathrm{sech}^2(x)\mathrm{d}x = 2 \tag{E.17}$$

and

$$\int_{-\infty}^{+\infty} x^2\,\mathrm{sech}^2(x)\mathrm{d}x = \frac{\pi^2}{6}\,. \tag{E.18}$$

Linearized Operators for the Master Equation

F

We use the Haus master equation approach [75Hau2, 84Hau] with linearized operators to derive a differential equation for the pulse envelope $A(t)$ in the time domain. In the following, the changes ΔA in the pulse envelope that occur during a resonator roundtrip due to gain, loss, SPM, and amplitude modulation will be determined. In this case we apply an approximation for small changes ΔA. The results are summarized in Table F.4.

Material Dispersion

The linearized operators for first and second order dispersion were derived in Chap. 4 and are summarized in Table F.1. For the second order dispersion we introduced a dispersion parameter D (4.39) which is half of the intracavity roundtrip GDD. According to Table 2.3 the group delay dispersion (GDD) is given by the second order dispersion with $\phi = k_n L_\mathrm{d}$:

$$\mathrm{GDD} = \frac{\mathrm{d}^2\phi}{\mathrm{d}\omega^2} = \frac{\mathrm{d}^2 k_n}{\mathrm{d}\omega^2} L_\mathrm{d} = 2D\ .$$

Here we discuss the third derivative k'''. For the propagation of very short pulses in the region of 10 fs, third order dispersion becomes important (see Chaps. 2 and 3).

Table F.1 Summary of the linearized dispersion operators

First order dispersion (4.32)	$1 - k_n' L_\mathrm{d} \frac{\partial}{\partial t}$
Second order dispersion (4.40)	$1 + \mathrm{i}\frac{1}{2} k_n'' L_\mathrm{d} \approx 1 + \mathrm{i} D \frac{\partial^2}{\partial t^2}$
Third order dispersion (Appendix F)	$1 + \frac{1}{6} k_n''' L_\mathrm{d} \frac{\partial^3}{\partial t^3}$

© The Editor(s) (if applicable) and The Author(s), under exclusive licence to Springer Nature Switzerland AG 2021

U. Keller, *Ultrafast Lasers*, Graduate Texts in Physics,
https://doi.org/10.1007/978-3-030-82532-4

We thus continue the Taylor expansion one step further:

$$k_n(\omega) \approx k_n(\omega_0) + k_n' \Delta\omega + \frac{1}{2} k_n'' \Delta\omega^2 + \frac{1}{6} k_n''' \Delta\omega^3 , \tag{F.1}$$

where

$$\Delta\omega = \omega - \omega_0 , \quad k_n' = \left. \frac{\mathrm{d}k_n}{\mathrm{d}\omega} \right|_{\omega_0} , \quad k_n'' = \left. \frac{\mathrm{d}^2 k_n}{\mathrm{d}\omega^2} \right|_{\omega_0} , \quad k_n''' = \left. \frac{\mathrm{d}^3 k_n}{\mathrm{d}\omega^3} \right|_{\omega_0} .$$

In full analogy with the first and second order dispersion (Chap. 4), with $k_n''' \Delta\omega^3 L_\mathrm{d} \ll 1$, we make a linearized approximation:

$$\tilde{A}(L_\mathrm{d}, \Delta\omega) = \exp\left(-\mathrm{i} \frac{1}{6} k_n''' \Delta\omega^3 L_\mathrm{d} \right) \tilde{A}(0, \Delta\omega) \approx \left(1 - \mathrm{i} \frac{1}{6} k_n''' \Delta\omega^3 L_\mathrm{d} \right) A(0, \Delta\omega) , \tag{F.2}$$

where L_d is the length of the dispersive material. Using the relation

$$F^{-1}\left\{ \Delta\omega^3 \tilde{A}(z, \Delta\omega) \right\} = \mathrm{i} \frac{\partial^3}{\partial t^3} A(z, t) , \tag{F.3}$$

we find

$$\boxed{A(L_\mathrm{d}, t) \approx \left(1 + \frac{1}{6} k_n''' L_\mathrm{d} \frac{\partial^3}{\partial t^3} \right) A(0, t)} , \tag{F.4}$$

for the linearized form in the time domain.

Constant Loss

The loss is found from

$$A_\mathrm{out}(t) = \mathrm{e}^{-l} A_\mathrm{in}(t) . \tag{F.5}$$

For small losses, the exponential function can be expanded $\mathrm{e}^x \approx 1 + x$ for $x \ll 1$:

$$A_\mathrm{out}(t) = \mathrm{e}^{-l} A_\mathrm{in}(t) \approx (1 - l) A_\mathrm{in}(t) ,$$

whence

$$\Delta A = A_\mathrm{out}(t) - A_\mathrm{in}(t) \approx -l A_\mathrm{in}(T, t) .$$

This implies

$$\boxed{\Delta A \approx -l A(T, t)} . \tag{F.6}$$

Self-Phase Modulation (SPM)

SPM was introduced in Sect. 4.1 and the linearized operator as shown in Table F.2 was derived in (4.7).

Table F.2 Linearized SPM operator, where δ is the SPM coefficient, n_2 is the nonlinear refractive index, and L_K is the length of the Kerr material

SPM (4.7)	$1 - i\delta|A(t)|^2\ ,\quad \delta = kn_2L_K$

Finite Gain Bandwidth and Gain Dispersion

The homogeneous gain of a laser typically has a Lorentzian line shape:

$$g(\omega) = \frac{g(z)L_g}{1 + \left[2(\omega - \omega_0)/\Delta\omega_g\right]^2} = \frac{g}{1 + \left[(\omega - \omega_0)/\Omega_g\right]^2}\ , \tag{F.7}$$

where ω_0 is the center frequency, $\Delta\omega_g$ is the full-width at half-maximum (FWHM) of the gain profile, and g is the spatially integrated saturated gain coefficient of the laser material of length L_g, i.e., $g = g(z)L_g$, with $g(z)$ the gain coefficient. Later it will also be useful to introduce the notation Ω_g (half-width at half-maximum, HWHM) to get rid of the factor of 2:

$$\Omega_g \equiv \frac{\Delta\omega_g}{2}\ . \tag{F.8}$$

The Fourier transform of the e.m. wave of a pulse $\tilde{A}(\omega)$ experiences the following change due to the gain material:

$$\exp\left[g(\omega)\right]\tilde{A}(\omega)\ . \tag{F.9}$$

In the case $g \ll 1$ and $\left[(\omega - \omega_0)/\Omega_g\right]^2 \ll 1$, one can make a parabolic approximation. With $e^x \approx 1 + x$, and $(1+x)^{-1} \approx 1 - x$, (F.7) and (F.9) imply

$$\exp\left[g(\omega)\right]\tilde{A}(\omega) \approx \left[1 + g\left(1 - \frac{\Delta\omega^2}{\Omega_g^2}\right)\right]\tilde{A}(\omega) = \left[1 + g - \frac{g}{\Omega_g^2}\Delta\omega^2\right]\tilde{A}(\omega)\ , \tag{F.10}$$

where $\Delta\omega = \omega - \omega_0$. Figure F.1 compares the parabolic approximation (F.10) with the exact gain profile. We obtain the correct gain curvature around the center wavelength, which makes an important contribution to the pulse shaping process and is discussed in Chap. 6.

As before for the material dispersion operators, the Fourier transform into the time domain is greatly simplified by the parabolic approximation:

$$A\left(L_g, t\right) \approx \left[1 + g\left(1 + \frac{1}{\Omega_g^2}\frac{\partial^2}{\partial t^2}\right)\right]A(0,t) \equiv \left[1 + g + D_g\frac{\partial^2}{\partial t^2}\right]A(0,t)\ , \tag{F.11}$$

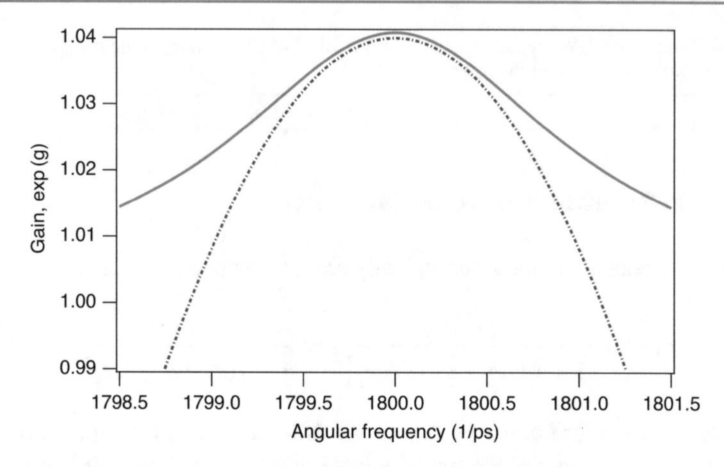

Fig. F.1 Lorentzian gain profile replaced by a parabolic approximation as a function of ω in units of THz. Example for a Nd:YLF laser with $\Omega_{\mathrm{g}} = 1.12$ THz, i.e. FWHM $\Delta\lambda_{\mathrm{g}} = 1.3$ nm, $\lambda_0 = 1.047\ \mu\mathrm{m}$, $\omega_0 = 1800$ THz, and $g = 0.04$ with (F.7) and (F.10)

Table F.3 Linearized gain dispersion operator

Gain and gain dispersion	$1 + g + D_{\mathrm{g}} \dfrac{\partial^2}{\partial t^2}\,, \quad D_{\mathrm{g}} \equiv \dfrac{g}{\Omega_{\mathrm{g}}^2}$

where L_{g} is the length of the gain material. By analogy with the material dispersion, we can define a parameter D_{g} for the gain dispersion by

$$D_{\mathrm{g}} \equiv \frac{g}{\Omega_{\mathrm{g}}^2}\,. \tag{F.12}$$

Both the frequency-dependent propagation constant $k_n(\omega)$ and the gain $g(\omega)$ will influence the pulse length during propagation. In materials with normal dispersion, the curvature leads to a broadening of the pulse, because D_{g} describes the curvature of the gain line, by analogy with the material dispersion parameter D. The greater the curvature the greater is the influence on the linear pulse propagation (Tables F.3 and F.4).

Table F.4 Summary of linearized operators that change the pulse envelope $A(t)$ with the following notation: D_g gain dispersion, g saturated amplitude gain coefficient, Ω_g HWHM of gain bandwidth in radians/second, $\Delta\nu_g = \Omega_g/\pi$, ω_m loss modulation frequency in radians/second, $2M$ peak-to-peak modulation depth for amplitude loss coefficient, l amplitude loss coefficient, ψ phase shift, γ_A fast absorber coefficient, $I_{sat,A}$ saturation intensity, A_A laser mode area in saturable absorber, D dispersion parameter (half of the total group delay dispersion per cavity roundtrip), δ SPM coefficient (4.5), n_2 nonlinear refractive index, A_L laser mode area inside the laser material. The length of the Kerr medium L_K is twice the length of the laser gain crystal in a standing wave cavity, when we assume that the dominant SPM comes from the laser gain crystal

Laser cavity element	Equation	Linearized operator	New constants
Dispersion			
First order	(4.32)	$\Delta A \approx -k_n' L_d \frac{\partial}{\partial t} A$	$k_n' = \frac{dk_n}{d\omega}$
Second order	(4.40)	$\Delta A \approx iD \frac{\partial^2}{\partial t^2} A$	$D \equiv \frac{1}{2} k_n'' z\,, \quad k_n'' = \frac{d^2 k_n}{d\omega^2}$
Third order	(F.4)	$\Delta A \approx \frac{1}{6} k_n''' L_d \frac{\partial^3}{\partial t^3} A$	$k_n''' = \frac{d^3 k_n}{d\omega^3}$
Gain	Table F.3	$\Delta A \approx \left(g + D_g \frac{\partial^2}{\partial t^2} \right) A$	$D_g \equiv \frac{g}{\Omega_g^2}\,, \quad 2\Omega_g = 2\pi \Delta\nu_g$
With dynamic gain saturation	(9.19)	$\Delta A \approx \left[g(t) + \frac{g_0}{\Omega_g^2} \frac{\partial^2}{\partial t^2} \right] A$	$g(t)$ given by (9.20)
Constant loss	(6.49), (F.6)	$\Delta A \approx -lA$	
Constant phase shifts	(6.74)	$\Delta A \approx i\psi A$	
Loss modulator	(6.47) and (6.54)	$\Delta A \approx -M_S t^2 A$	$M_S \equiv \frac{M\omega_m^2}{2}$
Fast saturable absorber	Sect. 9.4	$\Delta A \approx \bar{\gamma}_A \lvert A\rvert^2 A$	$\bar{\gamma}_A \equiv \frac{q_0}{I_{sat,A} A_A}\,,$ with $\lvert A\rvert^2 = P$ the pulse power
SPM	(4.7)	$\Delta A \approx -i\delta \lvert A\rvert^2 A$	$\delta = kn_2 L_K\,,$ with $\lvert A\rvert^2 = I$ the pulse intensity
	Sect. 9.4	$\Delta A \approx -i\bar{\delta} \lvert A\rvert^2 A$	$\bar{\delta} = \frac{\delta}{A_L}\,,$ with $\lvert A\rvert^2 = P$ the pulse power

References

[54Pie] J. R. Pierce, "Coupling of modes of propagation," J. Appl. Phys., vol. 25, pp. 179–183, 1954

[58Sch] A. L. Schawlow and C. H. Townes, "Infrared and optical masers," Phys. Rev., vol. 112, pp. 1940–1949, 1958

[60Kla] J. R. Klauder, A. C. Price, S. Darlington, W. J. Albersheim, "The theory and design of chirp radars," Bell Sys. Tech. J., vol. 39, pp. 745–808, 1960

[61Hel] R. W. Hellwarth, "Control of Fluorescent Pulsations," Advances in Quantum Electronics, pp. 334–341, ed. by J. R. Singer (Columbia University Press, New York, 1961)

[62Col] R. J. Collins, P. Kisliuk, "Control of Population Inversion in Pulsed Optical Masers by Feedback Modulation," J. Appl. Phys., vol. 33, pp. 2009–2011, 1962

[62McC] F. J. McClung, R. W. Hellwarth, "Giant Optical Pulsations from Ruby," J. Appl. Phys., vol. 33, pp. 828–829, 1962

[63Ell] R. J. Elliot, in *Polarons and Excitons*, C. G. Kuper and G. D. Whitefield Eds. New York: Plenum, 1963

[63Fra] L. M. Frantz, J. S. Nodvik, "Theory of pulse propagation in a laser amplifier," J. Appl. Phys., vol. 34, pp. 2346–2349, 1963

[63Joh] L. F. Johnson, R. E. Dietz, H. J. Guggenheim, "Optical maser oscillation from Ni^{2+} in MgF_2 involving simultaneous emmission of phonons," Phys. Rev. Lett., vol. 11, pp. 318–320, 1963

[63Vuy] A. A. Vuylsteke, "Theory of Laser Regeneration Switching," J. Appl. Phys., vol. 34, pp. 1615–1622, 1963

[63Wag] W. G.Wagner, B. A. Lengyel, "Evolution of the Giant Pulse in a Laser," J. Appl. Phys., vol. 34, pp. 2040–2046, 1963

[64Geu] J. E. Geusic, H. M. Marcos, L. G. van Uitert, "Laser oscillations in Nd-doped yttrium aluminum, yttrium gallium, and gadolinium garnets," Appl. Phys. Lett., vol. 4, pp. 182–184, 1964

[64Gir] F. Gires, P. Tournois, "Interféromètre utilisable pour la compression d'impulsions lumineuses modulées en fréquence," C. R. Acad. Sci. Paris, vol. 258, pp. 6112–6115, 1964

[64Wit] J. P. Wittke, P. J. Warter, "Pulse propagation in a laser amplifier," J. Appl. Phys., vol. 35, pp. 1668–1672, 1964

© The Editor(s) (if applicable) and The Author(s), under exclusive licence to Springer Nature Switzerland AG 2021

U. Keller, *Ultrafast Lasers*, Graduate Texts in Physics,
https://doi.org/10.1007/978-3-030-82532-4

[64Har] L. E. Hargrove, R. L. Fork, M. A. Pollack, "Locking of HeNe laser modes induced by synchronous intracavity modulation," Appl. Phys. Lett., vol. 5, pp. 4–5, 1964

[64Her] D. Herriott, H. Kogelnik, R. Kompfner, "Off-Axis Paths in Spherical Mirror Interferometers," Appl. Opt., vol. 3, pp. 523–526, 1964

[65Mal] H. Malitson, "Interspecimen comparison of the refractive index of fused silica," J. Opt. Soc. Am., vol. 55, pp. 1205–1209, 1965

[65Moc] H. W. Mocker, R. J. Collins, "Mode competition and self-locking effects in a Q-switched ruby laser," Appl. Phys. Lett., vol. 7, pp. 270–273, 1965

[65Row] H. E. Rowe, "Signal and Noise in Communication Systems". Princeton, NJ: Van Nostrand, 1965

[65Sni] E. Snitzer, R. Woodcock, "$Yb^{3+}Er^{3+}$ Glass Laser," Appl. Phys. Lett., vol. 6, pp. 45–46, 1965

[65Sza] A. Szabo, R. A. Stein, "Theory of Laser Giant Pulsing by a Saturable Absorber," J. of Appl. Phys., vol. 36, pp. 1562–1566, 1965

[66All] D. W. Allan, "Statistics of Atomic Frequency Standards," Proceedings of the IEEE, vol. 54, No. 2, pp. 221–230 (1966)

[66Cut] L. S. Cutler, C. L. Searle, "Some aspects of the theory and measurement of frequency fluctuations in frequency standards," Proceedings of the IEEE, vol. 54, No. 2, pp. 136–154, 1966

[66DeM] A. J. De Maria, D. A. Stetser, H. Heynau, "Self mode-locking of lasers with saturable absorbers," Appl. Phys. Lett., vol. 8, pp. 174–176, 1966

[66Eri] L. E. Erickson, A. Szabo, "Effects of Saturable Absorber Lifetime on the Performance of Giant-Pulse Lasers," J. Appl. Phys., vol. 37, pp. 4953–4961, 1966

[66Sch] F. P. Schäfer, F. P. W. Schmidth, J. Volze, "Organic Dye Solution Laser," Appl. Phys. Lett., vol. 9, pp. 306–309, 1966

[66Sor] P. P. Sorokin, J. R. Lankard, "Stimulated emission observed form an organic dye, chloro-aluminum phthalocyanine," IBM J. Res. Dev., vol. 10, pp. 162–163, 1966

[67Arm] J. A. Armstrong, "Measurement of picosecond laser pulse widths," Appl. Phys. Lett., vol. 10, pp. 16–18, 1967

[67DeM] F. DeMartini, C. H. Townes, T. K. Gustafson, P. L. Kelley, "Self-Steepening of Light Pulses," Phys. Rev., vol. 164, pp. 312–323, 1967

[67Eri] L. E. Erickson, A. Szabo, "Behavior of Saturable-Absorber Giant-Pulse Lasers in the Limit of Large Absorber Cross Section," J. Appl. Phys., vol. 38, pp. 2540–2542, 1967

[67Jon] R. J. Joenk, R. Landauer, "Laser pulse distortion in a nonlinear dielectric," Phys. Lett., vol. 24A, pp. 228–229, 1967

[67Ost] L. A. Ostrovskii, "Propagation of wave packets and space-time self-focusing in nonlinear medium," Sov. Phys. JETP, vol. 24, pp. 797–800, 1967

[67Shi] F. Shimizu, "Frequency broadening in liquids by a short light pulse," Phys. Rev. Lett., vol. 19, pp. 1097–1100, 1967

[67Web] H. P. Weber, "Method for pulsewidth measurement of ultrashort light pulses generated by phase-locked lasers using nonlinear optics," J. Appl. Phys., vol. 38, pp. 2231–2234, 1967

[68Lau] J. P. Laussade and A. Yariv, "Modelocking and ultrashort laser pulses by anisotropic molecular liquids," Appl. Phys. Lett., vol. 13, pp. 65–66, 1968

[69Gus] T. K. Gustafson, J. P. Taran, H. A. Haus, J. R. Lifsitz, P. L. Kelley, "Self- modulation, self-steepening, and spectral development of light in small-scale trapped filaments," Phys. Rev., vol. 177, pp. 306–313, 1969

[69Har] A. L. Harmer, A. Linz, D. R. Gabbe, "Fluorescence of Nd^{3+}In Lithium Yttrium Fluoride," J. Phys. Chem. Solids, vol. 30, pp. 1483–1491, 1969

[69Tre] E. B. Treacy, "Optical Pulse Compression with Diffraction Gratings," IEEE J. Quantum Electron., vol. 5, pp. 454-458, 1969

[70Alf] R. R. Alfano, S. L. Shapiro, "Observation of self-phase modulation and small- scale filaments in crystals and glasses," Phys. Rev. Lett., vol. 24, pp. 592–594, 1970; also vol. 24, p. 1217, 1970

[70Cub] R. Cubeddu, R. Polloni, C. A. Sacchi, O. Svelto, "Self-phase modulation and 'rocking' of molecules in trapped filaments of light with picosecond pulses," Phys. Rev. A, vol. 2, pp. 1955–1963, 1970

[70Kui1] D. J. Kuizenga, A. E. Siegman, "FM und AM mode locking of the homogeneous laser - Part I: Theory, Part II: Experimental results," IEEE J. Quantum Electron., vol. 6, pp. 694–708, 1970

[70Kui2] D. J. Kuizenga, A. E. Siegman, "FM and AM Mode Locking of the Homogeneous Laser – Part II: Experimental Results in a Nd:YAG Laser with Internal FM Modulation," IEEE J. Quantum Electron., vol. 6, pp. 709–715, 1970

[71Hal] D. Halford, in *Proceedings of the Frequency Standards and Metrology Seminar* (Laval University, Quebec, Canada 1971), p. 431

[71She] Y. R. Shen, M. M. T. Loy, "Theoretical interpretation of small-scale filaments of light originating from moving focal spots," Phys. Rev. A, vol. 3, pp. 2099–2105, 1971

[72Bec] M. F. Becker, K. J. Kuizenga, A. E. Siegman, "Harmonic mode locking of the Nd:YAG laser," IEEE J. Quantum Electron., vol. QE-8, pp. 687–693, 1972

[72Dah] L. Dahlström, "Passive modelocking and Q-switching of high power lasers by means of the optical Kerr effect," Optics Comm., vol. 5, pp. 157–162, 1972

[72Ipp] E. P. Ippen, C. V. Shank, A. Dienes, "Passive modelocking of the cw dye laser," Appl. Phys. Lett., vol. 21, pp. 348–350, 1972

[72New] G. H. C. New, "Modelocking of quasi-continuous lasers," Opt. Commun., vol. 6, pp. 188–192, 1972

[73Aul] B. A. Auld, Acoustic Fields and Waves in Solids (Wiley, New York, 1973), vol. 1

[73Gri] D. Grischkowsky, E. Courtens, J. A. Armstrong, "Observation of self-steepening of optical pulses with possible shock formation," Phys. Rev. Lett., vol. 31, pp. 422–425, 1973

[73Has] A. Hasegawa, F. Tappert, "Transmission of stationary nonlinear optical pulses in dispersive dielectric fibers. I. Anomalous dispersion," Appl. Phys. Lett., vol. 23, pp. 142–144, 1973

[73Sto] H. Stolen, E. P. Ippen, "Raman gain in glass optical waveguides," Appl. Phys. Lett., vol. 22, pp. 276–278, 1973

[74Gib] A. F. Gibson, M. F. Kimmitt, B. Norris, "Generation of bandwidth-limited pulses from a TEA CO_2 laser using p-type germanium," Appl. Phys. Lett., vol. 24, pp. 306–307, 1974

[74New] G. H. C. New, "Pulse evolution in mode-locked quasi-continuous lasers," IEEE J. Quantum Electron., vol. 10, pp. 115–124, 1974

[74Sha] C. V. Shank, E. P. Ippen, "Subpicosecond kilowatt pulses from a modelocked cw dye laser," Appl. Phys. Lett., vol. 24, pp. 373–375, 1974

[74Smi] P. W. Smith, M. A. Duguay, and E. P. Ippen, "Mode-Locking of Lasers," Prog. in Quant. Electron., vol. 3, pp. 107–229, 1974

[75All] L. Allen, J. H. Eberly, "Optical Resonance and Two-Level Atoms". New York: Dover, 1975

[75Chu] L. O. Chua, P. Lin, Computer aided analysis of electronic circuits: algorithms and computational techniques (Prentice-Hall, 1975, Englewood-Cliffs, N.J., 1975)

[75Hau1] H. A. Haus, "Theory of mode locking with a fast saturable absorber," J. Appl. Phys., vol. 46, pp. 3049–3058, 1975

[75Hau2] H. A. Haus, "A Theory of Forced Mode Locking," IEEE J. Quantum Electron., vol. 11, pp. 323–330, 1975

[75Hau3] H. A. Haus, "Theory of Mode Locking with a Slow Saturable Absorber," IEEE J. Quantum Electron., vol. 11, pp. 736–746, 1975

[75Mar] J. H. Marburger, "Self-focusing: theory," Prog. Quantum. Electron., vol. 4, pp. 35–110 (1975)

[75Mat] M. Matsuhara, K. O. Hill, A. Watanabe: "Optical-waveguide filters: synthesis," J. Opt. Soc. Am., vol. 65, pp. 804–809 (1975)

[76Bet] J. R. Bettis, R. A. House II, A. H. Guenther, "Spot size and pulse duration dependence of laser-induced damage," in *Laser Induced Damage in Optical Materials* (NBS Spec. Pub. 462, Washington DC, 1976) pp. 338–345

[76Hau] H. A. Haus, "Parameter ranges for cw passive modelocking," IEEE J. Quantum Electron., vol. 12, pp. 169–176, 1976

[76Rud] I. S. Ruddock, D. J. Bradley, "Bandwidth-limited subpicosecond pulse generation in modelocked cw dye lasers," Appl. Phys. Lett., vol. 29, pp. 296–297, 1976

[77Jai] R. K. Jain and J. P. Heritage, "Generation of synchronized cw trains of picosecond pulses at two independently tunable wavelengths," Appl. Phys. Lett., vol. 32, pp. 41–43, (1978)

[77Mil] D. Milam, M. J. Weber, A. J. Glass, "Nonlinear refractive index of fluoride crystals," App. Phys. Lett., vol. 31, pp. 822–825, 1977

[77Sal] K. Sala, M. C. Richardson, and N. R. Isenor, "Passive modelocking of lasers with the optical Kerr effect modulator," IEEE J. Quantum Electron., vol. 13, pp. 915–924, 1977

[78Bra] R. N. Bracewell, "The Fourier transform and its application," Second Edition, McGraw-Hill Book Company, 1978

[78Die] J. C. Diels, E. W. V. Stryland, G. Benedict, "Generation and measurement of pulses of 0.2 ps duration," Opt. Comm., vol. 25, p. 93, 1978

[78Rut] J. Rutman, "Characterization of phase and frequency instabilities in precision frequency source: fifteen years of progress," Proc. IEEE, vol. 66, pp. 1048–1075, 1978

[78Sch] Schott Glass Technologies, "Glass for Laser Applications, Data Sheets," 1978

[78Sto] R. H. Stolen, C. Lin, "Self-phase-modulation in silica optical fibers," Phys. Rev. A, vol. 17, pp. 1448–1453, 1978

[79Wal] J. C. Walling, H. P. Jenssen, R. C. Morris, E. W. O'Dell, O. G. Peterson, "Tunable laser performance in $BeAl_2O_4Cr^{3+}$," Opt. Lett., vol. 4, pp. 182–183, 1979

[80Ipp] E. P. Ippen, D. J. Eichenberger, R. W. Dixon, "Picosecond pulse generation by passive modelocking of diode lasers," Appl. Phys. Lett., vol. 37, pp. 267–269, 1980

[80Mol] L. F. Mollenauer, R. H. Stolen, J. P. Gordon, "Experimental Observation of Picosecond Pulse Narrowing and Solitons in Optical Fibers," Phys. Rev. Lett., vol. 45, pp. 1095–1098, 1980

[80Oko] T. Okoshi, K. Kikuchi, A. Nakayama, "Novel method for high resolution measurement of laser outpt spectrum," Electron. Lett., vol. 16, No. 16, pp. 630–631, 1980

[80Sal] K. L. Sala, G. A. Kenney-Wallace, G. E. Hall, "CW Autocorrelation Measurements of Picosecond Laser Pulses," IEEE J. Quantum Electron., vol. 16, pp. 990–996, 1980

[81For] R. L. Fork, B. I. Greene, C. V. Shank, "Generation of optical pulses shorter than 0.1 psec by colliding pulse modelocking," Appl. Phys. Lett., vol. 38, pp. 671–672, 1981

[81Tzo] N. Tzoar, M. Jain, "Self-phase modulation in long-geometry optical waveguides," Phys. Rev. A, vol. 23, pp. 1266–1270, 1981

[81Van] J. P. van der Ziel, R. A. Logan, R. M. Mikulyak, "Generation of subpicosecond pulses from an actively modelocked GaAs laser in an external ring cavity," Appl. Phys. Lett., vol. 39, pp. 867–869, 1981

[82Cav] C. M. Caves, "Quantum limits on noise in linear amplifiers," Phys. Rev. D, vol. 26, p. 1817, 1982.

[82Ell] D. S. Elliott, R. Roy, S. J. Smith, "Extracavity laser band-shape and bandwidth modification," Phys. Rev. A, vol. 26, Nr. 1, pp. 12–18, 1982

[82Gri] D. Grischkowsky and A. C. Balant, "Optical pulse compression based on enhanced frequency chirping," Appl. Phys. Lett., vol. 41, pp. 1–3, 1982

[82Hen] C. H. Henry, "Theory of the linewidth of semiconductor lasers," IEEE J. Quantum Electron., vol. 18, pp. 259–264, 1982

[82Mou] P. F. Moulton, "Ti-doped sapphire: tunable solid-state laser," Optics News, vol. 11, p. 9, 1982

[82Tak] M. Takeda, H. Ina, S. Kobayashi, "Fourier-transform method of fringe-pattern analysis for computer-based topography and interferometry," J. Opt. Soc. Am. B, vol. 72, pp. 156–160, 1982

[83And] D. Anderson, M. Lisak, "Nonlinear asymmetric self-phase modulation and self-steepening of pulses in long optical waveguides," Phys. Rev. A, vol. 27, pp. 1393–1398, 1983

[83Ben] R. A. Bendorius, E. K. Maldutis, "Nonlinear absorption of laser radiation in InP and GaAs," Sov. Phys. Coll, vol. 23, p. 69, 1983

[83For] R. L. Fork, C. V. Shank, C. Hirlimann, R. Yen, W. J. Tomlinson, "Femtosecond white-light continuum pulses," Opt. Lett., vol. 8, pp. 1–3, 1983

[83Joh] A. M. Johnson, W. M. Simpson, "Continuous-wave modelocked Nd:YAG-pumped subpicosecond dye lasers," Opt. Lett., vol. 8, pp. 554–556, 1983

[83Min] T. Mindl, P. Hefferle, S. Schneider, F. Dörr, "Characterisation of a Train of Subpicosecond Laser Pulses by Fringe Resolved Autocorrelation Measurements," Appl. Phys. B, vol. 31, pp. 201–207, 1983

[83Nik] B. Nikolaus, D. Grischkowsky, "12 × pulse compression using optical fibers," Appl. Phys. Lett., vol. 42, p. 3, 1983

[83Sch] R. Schimpe, "Intensity noise associated with the lasing mode of a (GaAl)As diode laser," IEEE J. Quant. Electron., vol. 19, pp. 895–897, 1983

[83Wei] A. M. Weiner, "Effect of Group Velocity Mismatch on the Measurement of Ultrashort Optical Pulses via Second Harmonic Generation," IEEE J. Quantum Electronics, vol. 19, pp. 1276–1283, 1983

[84Eck] R. C. Eckardt, J. Reintjes, "Phase matching limitations of high efficiency second harmonic generation," IEEE J. Quantum Electronics, vol. QE-20, pp. 1178–1187, 1984

[84For] R. L. Fork, O. E. Martinez, J. P. Gordon, "Negative dispersion using pairs of prisms," Opt. Lett., vol. 9, pp. 150–152, 1984

[84Hau] H. A. Haus: *Waves and Fields in Optoelectronics*, Prentice-Hall, Englewood Cliffs (1984)

[84Kaf] J. D. Kafka, B. H. Kolner, T. Baer, D. M. Bloom, "Compression of pulses from a continuous-wave modelocked Nd:YAG laser," Opt. Lett., vol. 9, pp. 505–506, 1984

[84Klu] J. Kluge, D. Wiechert, D. von der Linde, "Fluctuations in synchronously modelocked dye lasers," Opt. Commun., vol. 51, Nr. 4, pp. 271–277, 1984

[84Mar] O. E. Martinez, R. L. Fork, J. P. Gordon, "Theory of passively modelocked lasers including self-phase modulation and group-velocity dispersion," Opt. Lett., vol. 9, pp. 156–158 (1984)

[84Mar2] O. E. Martinez, J. P. Gordon, R. L. Fork, "Negative group-velocity dispersion using refraction," J. Opt. Soc. Am. A **1** (10), 1003–1006 (1984)

[84Mol] L. F. Mollenauer, R. H. Stolen, "The soliton laser," Opt. Lett., vol. 9, pp. 13–15, 1984

[84Pes] E. V. Pestryakov, V. I. Trunov, V. N. Matrosov, V. N. Razvalyaev, "Generation of ultrashort pulses that are tunable in the 0.7–0.8 μm range in a laser based on alexandrite," Bulletin of the Academy of Sciences of the USSR, vol. 48, pp. 94–98, 1984

[84Tom] W. J. Tomlinson, R. H. Stolen, C. V. Shank, "Compression of optical pulses chirped by self-phase modulation in fibers," J. Opt. Soc. Am. B, vol. 1, pp. 139–149, 1984

[84Yan] G. Yang, Y. R. Shen, "Spectral broadening of ultrashort pulses medium," Opt. Lett., vol. 9, pp. 510–512, 1984

[84Yar] A. Yariv, P. Yeh, *Optical Waves in Crystals* (John Wiley & Sons, New York, 1984)

[85Fre] J. L. Freeman, S. K. Diamond, H. Fong, D. M. Bloom, "Electro-optic sampling of planar digital GaAs integrated circuits," Appl. Phys. Lett., vol. 47, pp. 1083–1084, 1985

[85Har] D. J. Harter, Y. B. Band, E. I. Ippen, "Theory of mode-locked lasers containing a reverse saturable absorber," IEEE J. Quantum Electron., vol. 21, p. 1219, 1985

[85Joh] A. M. Johnson, W. M. Simpson, "Tunable femtosecond dye laser synchronously pumped by the compressed second harmonic of Nd:YAG," J. Opt. Soc. Am. B, vol. 2, pp. 619–625, 1985

[85Kan] T. J. Kane, R. L. Byer, "Monolithic, unidirectional single-mode Nd:YAG ring laser," Opt. Lett., vol. 10, p. 65, 1985

[85Mac] H. A. Macleod, *Thin-film optical filters*, Opt. Lett., vol. 10, pp. 65–67, 1985 (Adam Hilger, Bristol, 1985)

[85Mar] O. E. Martinez, R. L. Fork, J. P. Gordon, "Theory of passively mode-locked lasers for the case of a nonlinear complex-propagation coefficient," J. Opt. Soc. Am. B, vol. 2, pp. 753–760, 1985

[85Oud] J. L. Oudar, D. Hulin, A. Migus, A. Antonetti, F. Alexandre, "Subpicosecond spectral hole burning due to nonthermalized photoexcited carriers in GaAs," Phys. Rev. Lett., vol. 55, pp. 2074–2077, 1985

[85Smi] P. W. Smith, Y. Silberberg, and D. A. B. Miller, "Mode-locking of semiconductor diode lasers using saturable excitonic nonlinearities," J. Opt. Soc. Am. B, vol. 2, pp. 1228–1236, 1985

[85Str] D. Strickland and G. Mourou, "Compression of amplified chirped optical pulses," Opt. Commun., vol. 56, p. 219, 1985

[85Str2] E. W. V. Stryland, M. A. Woodall, H. Vanherzeele, M. J. Soileau, "Energy band-gap dependence of two-photon absorption," Opt. Lett., vol. 10, pp. 490–492, 1985

[85Val] J. A. Valdmanis, R. L. Fork, J. P. Gordon, "Generation of optical pulses as short as 27 fs directly from a laser balancing self-phase modulation, group-velocity dispersion, saturable absorption, and saturable gain," Opt. Lett., vol. 10, pp. 131–133, 1985

[85Van] E. W. Van Stryland, M. A. Woodall, H. Vanherzeele, M. J. Soileau, "Energy band-gap dependence of two-photon absorption," Opt. Lett., vol. 10, p. 490, 1985

[85Wei] K. J. Weingarten, M. J. W. Rodwell, H. K. Heinrich, B. H. Kolner, D. M. Bloom, "Direct electro-optic sampling of GaAs integrated circuits," Electron. Lett., vol. 21, pp. 765–766, 1985

[86Bae] T. Baer, "Large-amplitude fluctuations due to longitudinal mode coupling in diode-pumped intracavity-doubled Nd:YAG lasers," J. Opt. Soc. Am. B, vol. 3, p. 1175, 1986

[86Dod] M. J. Dodge, "Refractive Index" in Handbook of Laser Science and Technology, Volume IV, Optical Materials: Part 2, CRC Press, Boca Raton, 1986, p. 30

[86Gol] E. A. Golovchenko, E. M. Dianov, A. M. Prokhorov, V. N. Serkin, "Self-effect femtosecond optical wave packets," Sov. Phys. Dokl., vol. 31, pp. 494–497, 1986

[86Gor] J. P. Gordon, "Theory of the soliton self-frequency shift," Opt. Lett., vol. 11, pp. 662–664, 1986

[86Gou] P. L. Gourley and T. J. Drummond, "Single crystal, epitaxial multilayers of AlAs, GaAs, and $Al_xGa_{1-x}As$ for use as optical interferometric elements," Appl. Phys. Lett., vol. 49, pp. 489–491, 1986

[86Hau] H. A. Haus, Y. Silberberg, "Laser mode locking with addition of nonlinear index," IEEE J. Quantum Electron., vol. 22, pp. 325–331, 1986

[86Hei] H. K. Heinrich, D. M. Bloom, R. R. Hemenway, "Noninvasive sheet charge density probe for integrated silicon devices," Appl. Phys. Lett., vol. 48, pp. 1066–1068, 1986

[86Kno] W. H. Knox, C. Hirlimann, D. A. B. Miller, J. Shah, D. S. Chemla, and C. V. Shank, "Femtosecond excitation of nonthermal carrier populations in GaAs quantum wells," Phys. Rev. Lett., vol. 56, pp. 1191–1193, 1986

[86Lin] D. von der Linde, "Characterization of the noise in continuously operating mode-locked lasers," Appl. Phys. B, vol. 39, pp. 201–217, 1986
[86Man] J. T. Manassah, M. A. Mustafa, R. A. Alfano, P. P. Ho, "Spectral extent and pule shape of the supercontinuum for ultrashort laser pulse," IEEE J. Quantum Electron., vol. 22, p. 197, 1986
[86Mit] F. M. Mitschke, L. F. Mollenauer, "Discovery of the soliton self-frequency shift," Opt. Lett., vol. 11, pp. 659–661, 1986
[86Mou] P. F. Moulton, "Spectroscopic and laser characteristics of Ti:Al_2O_3," J. Opt. Soc. Am. B, vol. 3, pp. 125–132, 1986
[86Mys] A. Mysyrowicz, D. Hulin, A. Antonetti, A. Migus, "Dressed excitons in a multiple-quantum-well structure—evidence for an optical Stark effect with femtosecond response time," Phys. Rev. Lett., vol. 56, pp. 2748–2751, 1986
[86Oue] F. Ouelette and M. Piché, "Pulse shaping and passive mode-locking with a nonlinear Michelson interferometer," Opt. Commun., vol. 60, pp. 99–103, 1986
[86Pin] J. F. Pinto, E. Georgiou, C. R. Pollock, "Stable color-center laser in OH-doped NaCl operating in the 1.41 to 1.81 μm region," Opt. Lett., vol. 11, pp. 519–521, 1986
[86Rod] M. J. W. Rodwell, K. J. Weingarten, D. M. Bloom, T. Baer, and B. H. Kolner, "Reduction of timing fluctuations in a modelocked Nd:YAG laser by electronic feedback," Opt. Lett., vol. 11, pp. 638–640, 1986
[86Ros] H. Roskos, T. Robl, A. Seilmeier, "Pulse shortening to 25 ps in a cw mode-locked Nd:YAG laser by introducing an intracavity etalon," Appl. Phys. B, vol. 40, pp. 59–65, 1986
[86Sie] A. E. Siegman, *Lasers* (University Science Books, Mill Valley, California, 1986)
[86Val] J. A. Valdmanis, R. L. Fork, "Design Considerations for a Femtosecond Pulse Laser Balancing Self Phase Modulation, Group Velocity Dispersion, Saturable Absorption, and Saturable Gain," IEEE J. Quantum Electron., vol. 22, pp. 112–118, 1986
[86Vas] P. P. Vasil'ev, V. N. Morzov, Y. M. Popov, A. B. Sergeev, "Subpicosecond pulse generation by tandem-type AlGaAs DH laser with colliding pulse modelocking," IEEE J. Quantum Electron., vol. 22, pp. 149–151, 1986
[86vdL] D. von der Linde, "Characterization of the noise in continuously operating mode-locked lasers," Appl. Phys. B, vol. 39, pp. 201–217, 1986
[86Wal] A. C. Walker, A. K. Kar, Wei Ji, U. Keller, and S. D. Smith, "All-optical power limiting of CO_2 laser pulses using cascaded optical bistable elements," Appl. Phys. Lett., vol. 48, pp. 683–685, 1986
[86Yan] L. Yan, J. D. Ling, P.-T. Ho, C. H. Lee, "Picosecond-pulse generation from a continuous-wave neodymium:phosphate glass laser," Opt. Lett., vol. 11, pp. 502–503, 1986
[87Ada] S. Adachi, "Band gaps and refractive indices of AlGaAsSb, GaInAsSb, and InPAsSb: Key properties for a variety of the 2–4-μm optoelectronic device applications," J. Appl. Phys., vol. 61, pp. 4869–4876, 1987
[87Ada2] R. Adair, L. L. Chase, S. A. Payne, "Nonlinear refractive-index measurements of glasses using three-wave frequency mixing," J. Opt. Soc. Am. B, vol. 4, pp. 875–881, 1987
[87Bad] P. Bado, M. Bouvier, J. S. Coe, "Nd:YLF mode-locked oscillator and regenerative amplifier," Opt. Lett., vol. 12, pp. 319–321, 1987
[87Cot] D. Cotter, U.S. Patent Nr. 4,665,524, 12. May 1987 (filed 24 May 1985)
[87Dav] W. B. Davenport, W. L. Root, "An introduction to the theory of random signals and noise," IEEE press, 1987
[87Elz] P. A. Elzinga, R. J. Kneisler, F. E. Lytle, Y. Jiang, G. B. King, and N. M. Laurendeau, "Pump/probe method for fast analysis of visible spectral signatures utilizing asynchronous optical sampling," Appl. Opt., vol. 26, pp. 4303–4309, 1987
[87Fan] T. Y. Fan and R. L. Byer, "Modelling and CW Operation of a Quasi-Three-Level 946 nm Nd:YAG Laser," IEEE J. Quantum Electron., vol. 23, pp. 605–612, 1987

[87For] R. L. Fork, C. H. B. Cruz, P. C. Becker, C. V. Shank, "Compression of optical pulses to six femtoseconds by using cubic phase compensation," Opt. Lett., vol. 12, pp. 483–485, 1987

[87Gru] A. B. Grudinin, E. M. Dianov, D. V. Korobkin, A. M. Prokhorov, V. N. Serkin, D. V. Khaidarov, "Decay of femtosecond pulses in single-mode optical-fibers," JETP Lett., vol. 46, pp. 221–225, 1987

[87Hod] W. Hodel, H. P. Weber, "Decay of femtosecond higher-order solitons in an optical fiber induced by Raman self-pumping," Opt. Lett., vol. 12, pp. 924–926, 1987

[87Kin] S. Kino, Acoustic Waves: Devices, Imaging and Analog Signal Processing (Prentice-Hall, Englewood Cliffs, N.J., 1987)

[87Kod] Y. Kodama, K. Nozaki, "Soliton interaction in optical fibers," Opt. Lett., vol. 12, pp. 1038–1040, 1987

[87Kod2] Y. Kodama, A. Hasegawa, "Nonlinear pulse propagation in monomode dielectric guide," IEEE J. Quantum Electron., vol. QE-23, pp. 510–524, 1987

[87McP] A. McPherson, G. Gibson, H. Jara, U. Johann, T. S. Luk, I. McIntyre, K. Boyer, C. K. Rhodes, "Studies of multiphoton production of vacuum-ultraviolet radiation in the rare gases," J. Opt. Soc. Am. B, vol. 4, p. 595, 1987

[87Mit] F. M. Mitschke, L. F. Mollenauer, "Ultrashort Pulses from the Soliton Laser," Opt. Lett., vol. 12, pp. 407–409, 1987

[87Mol] L. F. Mollenauer, and J. C. White (eds.), Tunable Lasers, Topics Appl. Phys., vol. 59, Chapter 6 (Springer, Berlin, Heidelberg 1987)

[87Ohk] K. Ohkuma, Y. H. Ichikawa, Y. Abe, "Soliton propagation along optical fibers," Opt. Lett., vol. 12, pp. 516–518, 1987

[87Sch] R. W. Schoenlein, W. Z. Lin, E. P. Ippen, and J. G. Fujimoto, "Femtosecond hot-carrier energy relaxation in GaAs," Appl. Phys. Lett., vol. 51, pp. 1442–1444, 1987

[88Ang] N. B. Angert, N. I. Borodin, V. M. Garmash, V. A. Zhitnyuk, V. G. Okhrimchuk, O. G. Siyuchenko, A. V. Shestakov, "Lasing due to impurity color centers in yttrium aluminum garnet crystals at wavelengths in the range 1.35–1.45 μm," Sov. J. Quantum Electron., vol. 18, pp. 73–74, 1988

[88Bas] S. Basu, R. L. Byer, "Continuous-wave mode-locked Nd:glass laser pumped by a laser diode," Opt. Lett., vol. 13, pp. 458–460, 1988

[88Blo1] K. J. Blow and B. P. Nelson, "Improved mode-locking of an F-center laser with a nonlinear nonsoliton external cavity," Opt. Lett., vol. 13, pp. 1026–1028, 1988

[88Blo2] K. J. Blow and D. Wood, "Mode-locked lasers with nonlinear external cavities," J. Opt. Soc. Am B, vol. 5, pp. 629–632, 1988

[88Fer] M. Ferray, A. L'Huillier, X. F. Li, L. A. Lompré, G. Mainfray, C. Manus, "Multiple-harmonic conversion of 1064 nm radiation in rare gases," J. Phys. B: At. Mol. Opt. Phys., vol. 21, pp. L31–L35, 1988

[88Kai] C. V. Shank, "Generation of ultrashort optical pulses." In: W. Kaiser (Ed), Ultrashort Laser Pulses and Applications (Springer-Verlag, Berlin, Heidelberg 1988)

[88Kel] U. Keller, J. A. Valdmanis, M. C. Nuss, A. M. Johnson, "53 ps pulses at 1.32 μm from a harmonic mode-locked Nd:YAG laser," IEEE J. Quantum Electron., vol. 24, pp. 427–430, 1988

[88Kel2] U. Keller, S. K. Diamond, B. A. Auld, and D. M. Bloom, "A noninvasive optical probe of free charge and applied voltage in GaAs devices," Appl. Phys. Lett., vol. 53, pp. 388–390, 1988

[88Pay] S. A. Payne, L. L. Chase, H. W. Newkirk, L. K. Smith, W. F. Krupke, "Cr:LiCAF: a promising new solid-state laser material," IEEE J. Quantum Electron., vol. 24, pp. 2243–2252, 1988

[88Pet] V. Petricevic, S. K. Gayen, R. R. Alfano, "Laser action in chromium-doped forsterite," Appl. Phys. Lett., vol. 52, pp. 1040–1042, 1988

[88Pet2] K. Petermann, "Laser diode modulation and noise," Kluwer Academic Publishers, 1988, ISBN-13: 978-0-7923-1204-8

[88Ris] W. P. Risk, "Modelling of longitudinally pumped solid-state lasers exhibiting reabsorption losses," J. Opt. Soc. Am. B, vol. 5, pp. 1412–1423, 1988

[88Sha] C. V. Shank, "Generation of Ultrashort Optical Pulses," in *Ultrashort Laser Pulses and Applications*, W. Kaiser, Ed. (Springer Verlag, Heidelberg, 1988) Chap. 2
[88Sil] P. Silverberg, P. Omling, and L. Samuelson, "Hole photoionization cross sections of EL2 in GaAs," Appl. Phys. Lett., vol. 52, pp. 1689–91, 1988
[88Smi] F. W. Smith, A. R. Calawa, C.-L. Chen, M. J. Manfra, and L. J. Mahoney, "New MBE buffer used to eliminate backgating in GaAs MESFETs," IEEE Electron. Device Lett., vol. 9, pp. 77–80, 1988
[88Sta] K. A. Stankov, "A Mirror with an Intensity-Dependent Reflection Coefficient," Appl. Phys. B, vol. 45, pp. 191–195, 1988
[88Wei] K. J. Weingarten, M. J. W. Rodwell, D. M. Bloom, "Picosecond optical sampling of GaAs integrated circuits," IEEE J. Quantum Electron., vol. 24, pp. 198–220, 1988
[89Ada] R. Adair, L. L. Chase, S. A. Payne, "Nonlinear refractive index of optical crystals," Phys. Rev. B, vol. 39, pp. 3337–3350, 1989
[89Agr] G. P. Agrawal, N. A. Olsson, "Self-Phase Modulation and Spectral Broadening of Optical Pulses in Semiconductor Laser Amplifiers," IEEE J. Quantum Electron., vol. 25, pp. 2297–2306, 1989
[89Fre] P. M. W. French, J. A. R. Williams, and J. R. Taylor, "Femtosecond pulse generation from a titanium-doped sapphire laser using nonlinear external cavity feedback," Opt. Lett., vol. 14, pp. 686–688, 1989
[89Goo] J. Goodberlet, J. Wang, J. G. Fujimoto, P. A. Schulz, "Femtosecond passively mode-locked Ti:Al_2O_3 laser with a nonlinear external cavity," Opt. Lett., vol. 14, pp. 1125–1127, 1989
[89Ipp] E. P. Ippen, H. A. Haus, L. Y. Liu, "Additive Pulse Mode locking," J. Opt. Soc. Am. B, vol. 6, pp. 1736–1745, 1989
[89Isl] M. N. Islam, E. R. Sunderman, C. E. Soccolich, I. Bar-Joseph, N. Sauer, T. Y. Chang, B. I. Miller, "Color Center Lasers Passively Mode Locked by Quantum Wells," IEEE J. Quantum Electronics, vol. 25, pp. 2454–2463, 1989
[89Joh] M. B. Johnson, T. C. McGill, and N. G. Paulter, "Carrier lifetimes in ion-damaged GaAs," Appl. Phys. Lett., vol. 54, pp. 2424–2426, 1989
[89Kea] P. N. Kean, X. Zhu, D. W. Crust, R. S. Grant, N. Landford, W. Sibbett, "Enhanced mode locking of color center lasers," Opt. Lett., vol. 14, pp. 39–41, 1989
[89Kim] B. G. Kim, E. Garmire, S. G. Hummel, P. D. Dapkus, "Nonlinear Bragg reflector based on saturable absorption," Appl. Phys. Lett., vol. 54, pp. 1095–1097, 1989
[89Kel] U. Keller, K. D. Li, M. J. W. Rodwell, and D. M. Bloom, "Noise Characterization of Femtosecond Fiber Raman Soliton Lasers," IEEE J. Quantum Electron., vol. 25, pp. 280–288, 1989
[89Mak1] G. T. Maker and A. I. Ferguson, "Frequency-modulation mode locking of diode-pumped Nd:YAG laser," Opt. Lett., vol. 14, pp. 788–790, 1989
[89Mak2] G. T. Maker and A. I. Ferguson, "Modelocking and Q-switching of a diode laser pumped neodymium-doped yttrium lithium fluoride laser," Appl. Phys. Lett., vol. 54, pp. 403–405, 1989
[89Mar] J. Mark, L. Y. Liu, K. L. Hall, H. A. Haus, and E. P. Ippen, "Femtosecond pulse generation in a laser with a nonlinear external resonator," Opt. Lett., vol. 14, pp. 48–50, 1989
[89Mor] M. Morin, and M. Piché, "Interferential mode locking: Gaussian pulse analysis," Opt. Lett., vol. 15, pp. 1119–1121, 1989
[89Pay] S. A. Payne, L. L. Chase, L. K. Smith, W. L. Kway, H. Newkirk, "Laser performance of $LiSrAlF_6:Cr^{3+}$," J. Appl. Phys., vol. 66, pp. 1051–1056, 1989
[89Rod] M. J. W. Rodwell, D. M. Bloom, K. J. Weingarten, "Subpicosecond laser timing stabilization," IEEE J. Quantum Electron., vol. 25, pp. 817–827, 1989
[89Sam] R. C. Sam, "Alexandrite Lasers," in *Handbook of Solid-State Lasers*, P. K. Cheo, Ed. (Marcel Dekker, New York, 1989) pp. 349–455
[89Sas] M. W. Sasnett, "Propagation of Multimode Laser Beams - The M^2 Factor," in *The Physics and Technology of Laser Resonators*, D. R. Hall, P. E. Jackson, Eds. NY (1989) pp. 132–142

[89Vas] P. P. Vasilev and A. B. Sergeev, "Generation of Bandwidth-Limited 2 ps Pulses With 100 Ghz Repetition Rate From Multisegmented Injection-Laser," Electr. Lett., vol. 25, pp. 1049–1050, 1989

[89Zay] J. J. Zayhowski, A. Mooradian, "Single-frequency microchip Nd lasers," Opt. Lett., vol. 14, pp. 24–26, 1989

[89Zie] J. F. Ziegler, J. P. Biersack, and U. Littmark, *The Stopping and Range of Ions in Solids*, vol. 1 (New York: Pergamon, 1989)

[90Die] J.-C. Diels, "Femtosecond dye lasers," in *Dye Laser Principles: With Applications*, F. J. Duarte and L. W. Hillman, Eds. (Academic Press, Boston, 1990) pp. 41–132

[90Fan] T. Y. Fan, A. Sanchez, "Pump Source Requirements for End-Pumped Lasers," IEEE J.Quantum Electron., vol. 26, pp. 311–316, 1990

[90Fin] A. Finch, X. Zhu, P. N. Kean, W. Sibbett, "Noise characterization of mode-locked color-center laser sources," IEEE J. Quantum Electron., vol 26, pp. 1115–1123, 1990

[90Fon] C. Fontaine, P. Requena, and A. Munoz-Yagüe, "Generalization of Bragg reflector geometry: Application to (Ga,Al)As–(Ca,Sr)F_2 reflectors," J. Appl. Phys., vol. 68, pp. 5366–5368, 1990

[90Fre] P. M. W. French, S. M. J. Kelly, and J. R. Taylor, "Mode locking of a continuous-wave titanium-doped sapphire laser using a linear external cavity," Opt. Lett., vol. 15, pp. 378–380, 1990

[90Gob] E. O. Göbel, "Ultrafast spectroscopy of semiconductors," in *Advances in Solid State Physics*, vol. 30, U. Rössler, Ed. Braunschweig / Wiesbaden: Friedrich Vieweg & Sohn, 1990, pp. 269–294

[90Goo] J. Goodberlet, J. Jacobson, J. G. Fujimoto, P. A. Schulz, T. Y. Fan, "Self-starting additive pulse modelocked diode-pumped Nd:YAG laser," Opt. Lett., vol. 15, pp. 504–506, 1990

[90Ipp] E. P. Ippen, L. Y. Liu, and H. A. Haus, "Self-starting condition for additive-pulse mode-locked lasers," Opt. Lett., vol. 15, pp. 183–185, 1990

[90Juh] T. Juhasz, S. T. Lai, M. A. Pessot, "Efficient short-pulse generation from a diode-pumped Nd:YLF laser with a piezoelectrically induced diffraction modulator," Opt. Lett., vol. 15, pp. 1458–1460, 1990

[90Kel1] U. Keller, K. J. Weingarten, K. D. Li, D. C. Gerstenberger, P. T. Khuri-Yakub, D. M. Bloom, "High-frequency acousto-optic modelocker for picosecond pulse generation," Opt. Lett., vol. 15, pp. 45–47, 1990

[90Kel2] U. Keller, W. H. Knox, and H. Roskos, "Coupled-Cavity Resonant Passive Mode-locked Ti:Sapphire Laser," Opt. Lett., vol. 15, pp. 1377–1379, 1990

[90Kel3] U. Keller, C. E. Soccolich, G. Sucha, M. N. Islam, M. Wegener, "Noise characterization of femtosecond color center lasers," Opt. Lett., vol. 15, pp. 974–976, 1990

[90Liu1] L. Y. Liu, J. M. Huxley, E. P. Ippen, H. A. Haus, "Self-starting additive pulse modelocking of a Nd:YAG laser," Opt. Lett., vol. 15, pp. 553–555, 1990

[90Liu2] J. M. Liu, J. K. Chee, "Passive modelocking of a cw Nd:YLF laser with a nonlinear external coupled cavity," Opt. Lett., vol. 15, pp. 685–687, 1990

[90Mal] G. P. A. Malcom, P. F. Curley, A.I. Ferguson, "Additive pulse modelocking of a diode-pumped Nd:YLF laser," Opt. Lett., vol. 15, pp. 1303-1305, 1990

[90Mel] M. R. Melloch, N. Otsuka, J. M. Woodall, A. C. Warren, and J. L. Freeouf, "Formation of arsenic precipitates in GaAs buffer layers grown by molecular beam epitaxy at low substrate temperatures," Appl. Phys. Lett., vol. 57, p. 1531, 1990

[90Nag] K. Naganuma, K. Mogi, and H. Yamada, "Group-delay measurement using the Fourier transform of an interferometric cross correlation generated by white light," Opt. Lett., vol. 15, pp. 393–395, 1990

[90Nus] M. C. Nuss, U. Keller, G. T. Harvey, M. S. Heutmaker, P. R. Smith, "Amplitude noise reduction of 50 dB in colliding-pulse-modelocking dye lasers," Opt. Lett., vol. 15, pp. 1026–1028, 1990

[90Tel] H. R. Telle, D. Meschede, T. W. Hänsch, "Realization of a new concept for visible frequency division: phase locking of harmonic and sum frequencies," Opt. Lett., vol. 15, pp. 532–534, 1990

[90Wal] S. J. Walker, H. Avramopoulos, T. S. II, "Compact mode-locked solid-state lasers at 0.5- and 1-GHz repetition rates," Opt. Lett., vol. 15, pp. 1070–1072, 1990

[90Wei] K. J. Weingarten, D. C. Shannon, R. W. Wallace, U. Keller, "Two gigahertz repetition rate, diode-pumped, mode-locked, Nd:yttrium lithium fluoride (YLF) laser," Opt. Lett., vol. 15, pp. 962–964, 1990

[90Wu] M. C. Wu, Y. K. Chen, T. Tanbun-Ek, R. A. Logan, M. A. Chin, G. Raybon, "Transform-limited 1.4 ps optical pulses from a monolithic colliding-pulse mode-locked quantum well laser," Appl. Phys. Lett., vol. 57, pp. 759–761, 1990

[90Zay1] J. J. Zayhowski, "Limits imposed by spatial hole burning on the single-mode operation of standing-wave laser cavities," Opt. Lett., vol. 15, pp. 431–433, 1990

[90Zay2] J. J. Zayhowski, "Thermal Guiding in Microchip Lasers," in *Advanced Solid State Lasers*, G. Dube, Ed., Vol. 6 of OSA Proceedings Series (Optical Society of America, 1990), paper DPL3

[90Zha] X. C. Zhang, B. B. Hu, J. T. Darrow, D. H. Auston, "Generation of femtosecond electromagnetic pulses from semiconductor surfaces," Appl. Phys. Lett., vol. 56, pp. 1011–1013, 1990

[91Ben] H. Benisty, C. M. Sotomayor-Torres, C. Weisbuch, "Intrinsic mechanism for the poor luminescence properties of quantum box systems," Phys. Rev. B, vol. 44, pp. 1094510948, 1991

[91Bra] T. Brabec, C. Spielmann, F. Krausz, "Mode locking in solitary lasers," Opt. Lett., vol. 16, pp. 1961–1963, 1991

[91Che] Y. K. Chen, M. C. Wu, T. Tanbun-Ek, R. A. Logan, M. A. Chin, "Subpicosecond monolithic colliding-pulse modelocked multiple quantum well lasers," Appl. Phys. Lett., vol. 58, pp. 1253–1255, 1991

[91Chi] J. L. A. Chilla, O. E. Martinez, "Direct determination of the amplitude and the phase of femtosecond light pulses," Opt. Lett., vol. 16, pp. 39–41, 1991

[91Cur] P. F. Curley, A. I. Ferguson, "Actively modelocked Ti:Sapphire laser producing transform-limited pulses of 150 fs duration," Opt. Lett., vol. 16, pp. 1016–1018, 1991

[91God] A. A. Godil, A. S. Hou, B. A. Auld, D. M. Bloom, "Harmonic mode locking of a Nd:BEL laser using a 20-GHz dielectric resonator/optical modulator," Opt. Lett., vol. 16, pp. 1765–1767, 1991

[91Gup] S. Gupta, M. Y. Frankel, J. A. Valdmanis, J. F. Whitaker, G. A. Mourou, "Subpicosecond carrier lifetime in GaAs grown by molecular beam epitaxy at low temperatures," Appl. Phys. Lett., vol. 59, pp. 3276–3278, 1991

[91Har] G. T. Harvey, M. S. Heutmaker, P. R. Smith, M. C. Nuss, U. Keller, and J. A. Valdmanis, "Timing jitter and pump-induced amplitude modulation in the colliding-pulse mode-locked (CPM) laser," IEEE J. Quantum Electron., vol. 27, pp. 295–301, 1991

[91Hau] H. A. Haus, J. G. Fujimoto, E. P. Ippen, "Structures for additive pulse mode locking," J. Opt. Soc. Am. B, vol. 8, pp. 2068–2076, 1991

[91Hau2] H. A. Haus, U. Keller, and W. H. Knox, "Theory of Coupled-Cavity Mode Locking with a Resonant Nonlinearity," J. Opt. Soc. Am. B, vol. 8, pp. 1252–1258, 1991

[91Hug] D. W. Hughes, J. R. M. Barr, D. C. Hanna, "Mode locking of a diode-laser-pumped Nd:glass laser by frequency modulation," Opt. Lett., vol. 16, pp. 147–149, 1991

[91Kel] U. Keller, G. W. 'tHooft, W. H. Knox, J. E. Cunningham, "Femtosecond Pulses from a Continuously Self-Starting Passively Mode-Locked Ti:Sapphire Laser," Opt. Lett., vol. 16, pp. 1022–1024, 1991

[91Kel2] U. Keller, T. K. Woodward, D. L. Sivco, and A. Y. Cho, "Coupled-Cavity Resonant Passive Mode-locked Nd:Yttrium Lithium Fluoride Laser," Opt. Lett., vol. 16, pp. 390–392, 1991

[91Kut] S. A. Kutovoi, V. V. Laptev, S. Y. Matsnev, "Lanthanum scandoborate as a new highly efficient active medium of solid state lasers," Sov. J. Quantum Electr., vol. 21, pp. 131–132, 1991

[91Lac] P. Lacovara, H. K. Choi, C. A. Wang, R. L. Aggarwal, T. Y. Fan, "Room-temperature diode-pumped Yb:YAG laser," Opt. Lett., vol. 16, no. 14, pp. 1089–1091, 1991

[91Lam] M. Lambsdorff, J. Kuhl, J. Rosenzweig, A. Axmann, and J. Schneider, "Subpicosecond carrier lifetimes in radiation-damaged GaAs," Appl. Phys. Lett., vol. 58, pp. 1881–1883, 1991

[91Lap] P. Laporta, S. D. Silvestri, V. Magni, and O. Svelto, "Diode-pumped cw bulk Er:Yb:glass laser," Opt. Lett., vol. 16, pp. 1952–1954, 1991

[91Li] K. D. Li, J. A. Sheridan, D. M. Bloom, "Picosecond pulse generation in Nd:BEL with a high-frequency acousto-optic mode locker," Opt. Lett., vol. 16, pp. 1505–1507, 1991

[91McC] M. J. McCarthy, G. T. Maker, D. C. Hanna, "Efficient frequency doubling of a self-starting additive-pulse mode-locked diode-pumped Nd:YAG laser," Opt. Comm., vol. 82, pp. 327–332, 1991

[91Mer] L. B. Mercer, "$1/f$ frequency noise effects on self-heterodyne linewidth measurements," J. Lightwave Technol., vol. 9, pp. 485–493, 1991

[91Mur] M. M. Murnane, H. C. Kapteyn, M. D. Rosen, R. W. Falcone, "Ultrafast X-Ray Pulses from Laser-Produced Plasmas," Science, vol. 251, pp. 531–536, 1991

[91Neg] D. K. Negus, L. Spinelli, N. Goldblatt, G. Feugnet, "Sub-100 femtosecond pulse generation by Kerr lens modelocking in Ti:Al_2O_3," in *Advanced Solid-State Lasers*, G. Dubé, L. Chase, Eds. (Optical Society of America, Washington, D.C., 1991), vol. 10, pp. 120–124

[91Rod] M. Rodwell, M. Kamegawa, R. Yu, M. Case, E. Carman, K. Giboney, "GaAs Nonlinear Transmission Lines for Picosecond Pulse Generation and Millimeter-Wave Sampling." IEEE Transactions on Microwave Theory and Techniques, vol. 39, No. 7, pp. 1194–1204, 1991

[91Sal1] F. Salin, J. Squier, M. Piché, "Mode locking of TiAl_2O_3 lasers and self-focusing: a Gaussian approximation," Opt. Lett., vol. 16, pp. 1674–1676, 1991

[91Sal2] B. E. A. Saleh, M. C. Teich, *Fundamentals of Photonics* (John Wiley & Sons, Inc., 1991)

[91Sar] N. Sarukura, Y. Ishida, H. Nakano, "Generation of 50 fsec pulses from a pulse compressed, cw, passively mode-locked Ti:sapphire laser," Opt. Lett., vol. 16, pp. 153–155, 1991

[91Sch1] R. Scheps, J. F. Myers, H. B. Serreze, A. Rosenberg, R. C. Morris, M. Long, "Diode-pumped Cr:$LiSrAlF_6$ laser," Opt. Lett., vol. 16, pp. 820–822, 1991

[91Sch2] P. A. Schulz and S. R. Henion, "5-GHz mode locking of a Nd:YLF laser," Opt. Lett., vol. 16, pp. 1502–1504, 1991

[91She] M. Sheik-Bahae, D. C. Hutchings, D. J. Hagan, E. W. V. Stryland, "Dispersion of bound electronic nonlinear refraction in solids," IEEE J. Quantum Electron., vol. 27, pp. 1296–1309, 1991

[91She2] M. Sheik-Bahae, A. A. Said, D. J. Hagan, M. J. Soileau, E. W. Van Stryland, "Nonlinear refraction and optical limiting in thick media," Optical Engineering, vol. 30, pp. 1228–1235, 1991

[91Spe] D. E. Spence, P. N. Kean, W. Sibbett, "60-fsec pulse generation from a self-mode-locked Ti:sapphire laser," Opt. Lett., vol. 16, pp. 42–44, 1991

[91Spi] C. Spielmann, F. Krausz, T. Brabec, E. Wintner, A. J. Schmidt, "Femtosecond passive modelocking of a solid-state laser by dispersively balanced nonlinear interferometer," Appl. Phys. Lett., vol. 58, pp. 2470–2472, 1991

[91Sta] K. A. K. Stankov, V. Hamal, K., "Mode locking of a Nd:$YAlO_3$ laser at 1.08 and 1.34 μm wavelengths using a single $LiIO_3$ crystal," IEEE J. Quantum Electron., vol. 27, pp. 2135–2141, 1991

[91Tai] T. Taira, A. Mukai, Y. Nozawa, T. Kobayashi, "Single-mode oscillation of laser-diode-pumped Nd:YVO_4 microchip lasers," Opt. Lett., vol. 16, pp. 1955–1957, 1991

[91Tsa] H. K. Tsang, R. V. Penty, I. H. White, R. S. Grant, W. Sibbett, J. D. Soole, H. P. LeBlanc, N. C. Andreadakis, R. Bhat, M. A. Koza, "Two-photon absorption and self-phase modulation in InGaAsP/InP multi-quantum-well waveguides," J. Appl. Phys., vol. 70, p. 3992, 1991

[91Zay] J. J. Zayhowski, P. L. Kelley, "Optimization of Q-switched Lasers," IEEE J. Quantum Electronics, vol. 27, pp. 2220–2225, 1991

[91Zay2] J. J. Zayhowski, "Q-switched operation of microchip lasers," Opt. Lett., vol. 16, pp. 575–577, 1991

[91Zho] F. Zhou, G. P. A. Malcolm, A. I. Ferguson, "1-GHz repetition-rate frequency-modulation mode-locked neodymium lasers at 1.3 μm," Opt. Lett., vol. 16, pp. 1101–1103, 1991

[91Zir] M. Zirngibl, L. W. Stulz, J. Stone, J. Hugi, D. DiGiovanni, P. B. Hansen, "1.2 ps pulses from passively modelocked laser diode pumped Er-doped fiber ring lasers," Electronics Lett., vol. 27, pp. 1734–1735, 1991

[92Bea] Y. Beaudoin, C. Y. Chien, J. S. Coe, J. L. Tapié, G. Mourou, "Ultrahigh-contrast Ti:sapphire/Nd:glass terawatt laser system," Opt. Lett., vol. 17, pp. 865–867, 1992

[92Bla] A. Black, R. B. Apte, D. M. Bloom, "High-speed signal averaging system for periodic signals," Rev. Sci. Instrum., vol. 63, pp. 3191–3195, 1992

[92Cha] B. H. T. Chai, J.-L. Lefaucheur, M. Stalder, M. Bass, "Cr:$LiSr_{0.8}Ca_{0.2}AlF_6$ tunable laser," Opt. Lett., vol. 17, pp. 1584–1586, 1992

[92Che] Y.-K. Chen, M. C. Wu, "Monolithic Colliding Pulse Modelocked Quantum Well Lasers," IEEE J. Quantum Electron., vol. 28, pp. 2176–2185, 1992

[92Del] P. J. Delfyett, L. T. Florez, N. Stoffel, T. Gmitter, N. C. Andreadakis, Y. Silberberg, J. P. Heritage, G. A. Alphonse, "High-Power Ultrafast Laser Diodes," IEEE J. Quantum Electron., vol. 28, pp. 2203–2219, 1992

[92Dub] A. Dubietis, G. Jonusauskas, and A. Piskarskas, "Powerful femtosecond pulse generation by chirped and stretched pulse parametric amplification in BBO crystal," Opt. Communications, vol. 88, pp. 437–440, 1992

[92Eva] J. M. Evans, D. E. Spence, W. Sibbett, B. H. T. Chai, A. Miller, "50-fs pulse generation from a self-mode-locked Cr:$LiSrAlF_6$ laser," Opt. Lett., vol. 17, pp. 1447–1449, 1992

[92Gup] S. Gupta, J. F. Whitaker, and G. A. Mourou, "Ultrafast carrier dynamics in III–V semiconductors grown by molecular-beam epitaxy at very low substrate temperatures," IEEE J. Quantum Electron., vol. 28, pp. 2464–2472, 1992

[92Hau] H. A. Haus, J. G. Fujimoto, E. P. Ippen, "Analytic Theory of Additive Pulse and Kerr Lens Mode Locking," IEEE J. Quantum Electron., vol. 28, pp. 2086–2096, 1992

[92Hof] M. Hofer, M. H. Ober, F. Haberl, M. E. Fermann, "Characterization of Ultrashort Pulse Formation in Passively Mode-Locked Fiber Lasers," IEEE J. Quantum Electron., vol. 28, pp. 720–728, 1992

[92Hua1] C.-P. Huang, H. C. Kapteyn, J. W. McIntosh, M. M. Murnane, "Generation of transform-limited 32-fs pulses from a self-mode-locked Ti:sapphire laser," Opt. Lett., vol. 17, pp. 139–141, 1992

[92Hua2] C.-P. Huang, M. T. Asaki, S. Backus, M. M. Murnane, H. C. Kapteyn, H. Nathel, "17-fs pulses from a self-mode-locked Ti:sapphire laser," Opt. Lett., vol. 17, pp. 1289–1291, 1992

[92Hug] D. W. Hughes, M. W. Phillips, J. R. M. Barr, D. C. Hanna, "A Laser-Diode-Pumped Nd: Glass Laser: Mode-Locked, High Power, and Single Frequency Performance," IEEE J. Quantum Electron., vol. 28, pp. 1010–1017, 1992

[92Kaf] J. D. Kafka, M. L. Watts, J.-W. J. Pieterse, "Picosecond and Femtosecond Pulse Generation in a Regeneratively Mode-locked Ti:Sapphire Laser," IEEE J. Quantum Electron., vol. 28, pp. 2151–2162, 1992

[92Kel1] U. Keller, and T. H. Chiu, "Resonant Passive Mode-locked Nd:YLF Laser," IEEE J. Quantum Electron., vol. 28, pp. 1710–1721, 1992

[92Kel2] U. Keller, D. A. B. Miller, G. D. Boyd, T. H. Chiu, J. F. Ferguson, and M. T. Asom, "Solid-state low-loss intracavity saturable absorber for Nd:YLF lasers: an antiresonant semiconductor Fabry–Perot saturable absorber," Opt. Lett., vol. 17, pp. 505–507, 1992

[92Kel3] U. Keller, W. H. Knox, and G. W. 'tHooft, "Ultrafast Solid State Modelocked Lasers Using Resonant Nonlinearities," IEEE J. Quantum Electron., vol. 28, pp. 2123–2133, 1992

[92Kel4] S. M. J. Kelly, "Characteristic sideband instability of periodically amplified average soliton," Electron. Lett., vol. 28, pp. 806–807, 1992

[92Kod] Y. Kodama, A. Hasegawa, "Generation of asymptotically stable optical solitons and suppression of the Gordon–Haus effect," Opt. Lett., vol. 17, pp. 31–33, 1992

[92Koe] W. Koechner, Solid-State Laser Engineering (Springer-Verlag, Berlin, 1992)

[92Kra1] F. Krausz, C. Spielmann, T. Brabec, E. Wintner, A. J. Schmidt, "Generation of 33 fs optical pulses from a solid-state laser," Opt. Lett., vol. 17, pp. 204–206, 1992

[92Kra2] F. Krausz, M. E. Fermann, T. Brabec, P. F. Curley, M. Hofer, M. H. Ober, C. Spielmann, E. Wintner, A. J. Schmidt, "Femtosecond Solid-State Lasers," IEEE J. Quantum Electron., vol. 28, pp. 2097–2122, 1992

[92Lem] B. E. Lemoff, C. P. J. Barty, "Generation of high-peak-power 20-fs pulses from a regeneratively initiated, self-mode-locked Ti:sapphire laser," Opt. Lett., vol. 17, pp. 1367–1369, 1992

[92LiK] P. LiKamWa, B. H. T. Chai, A. Miller, "Self-mode-locked Cr^{3+}:$LiCaAlF_6$ laser," Opt. Lett., vol. 17, pp. 1438–1440, 1992

[92Liu] K. X. Liu, C. J. Flood, D. R. Walker, H. M. v. Driel, "Kerr lens mode locking of a diode-pumped Nd:YAG laser," Opt. Lett., vol. 17, pp. 1361–1363, 1992

[92Mar] D. Marcuse, "Simulations to demonstrate reduction of the Gordon-Haus effect," Opt. Lett., vol. 17, p. 34, 1992

[92Mil] A. Miller, P. LiKamWa, B. H. T. Chai, E. W. V. Stryland, "Generation of 150-fs tunable pulses in Cr:LiSAF," Opt. Lett., vol. 17, pp. 195–197, 1992

[92Nor] T. B. Norris, "Femtosecond pulse amplification at 250 kHz with a Ti:sapphire regenerative amplifier and application to continuum generation," Opt. Lett., vol. 17, pp. 1009–1011, 1992

[92Riz1] N. H. Rizvi, P. M. W. French, J. R. Taylor, "Continuously self-mode-locked Ti:Sapphire laser that produces sub-50-fs pulses," Opt. Lett., vol. 17, pp. 279–281, 1992

[92Riz2] N. H. Rizvi, P. M. W. French, J. R. Taylor, "50-fs pulse generation from a self-starting cw passively mode-locked Cr:$LiSrAlF_6$ laser," Opt. Lett., vol. 17, pp. 877–879, 1992

[92Riz3] N. H. Rizvi, P. M. W. French, J. R. Taylor, "Generation of 33-fs pulses from a passively mode-locked Cr^{3+}:$LiSrAlF_6$ laser," Opt. Lett., vol. 17, pp. 1605–1607, 1992

[92Smi] L. K. Smith, S. A. Payne, W. L. Kway, L. L. Chase, B. H. T. Chai, "Investigation of the Laser Properties of Cr^{3+}:$LiSrGaF_6$," IEEE J. Quantum Electron., vol. 28, pp. 2612–2618, 1992

[92Tap] J.-L. Tapié, G. Mourou, "Shaping of clean, femtosecond pulses at 1.053 μm for chirped-pulse amplification," Opt. Lett., vol. 17, pp. 136–138, 1992

[92Wei] K. J. Weingarten, A. A. Godil, M. Gifford, "FM modelocking at 2.85 GHz using a microwave resonant optical modulator," IEEE Photonics Technol. Lett., vol. 4, pp. 1106–1109, 1992

[92Wei2] A. M. Weiner, D. E. Leaird, J. S. Patel, J. R. Wullert, "Programmable shaping of femtosecond optical pulses by use of 128-element liquid crystal phase modulator," IEEE J. Quantum Electron., vol. 28, pp. 908–920, 1992

[92Wit] G. L. Witt, R. Calawa, U. Mishra, E. Weber, Eds., Low Temperature (LT) GaAs and Related Materials, vol. 241, Pittsburgh (1992)

[92Zay] J. J. Zayhowski and C. Dill, "Diode-pumped microchip lasers electro-optically Q-switched at high pulse repetition rates," Opt. Lett., vol. 17, pp. 1201–1203, 1992

[93Asa] M. T. Asaki, C.-P. Huang, D. Garvey, J. Zhou, H. C. Kapteyn, M. N. Murnane, "Generation of 11-fs pulses from a self-mode-locked Ti:sapphire laser," Opt. Lett. vol. 18, pp. 977–979, 1993

[93Bec] M. Beck, M. G. Raymer, I. A. Walmsley, and V. Wong, "Chronocyclic tomography for measuring the amplitude and phase structure of optical pulses," Opt. Lett., vol. 18, pp. 2041–2043, 1993

[93Car] T. J. Carrig, C. R. Pollock, "Performance of a cw Forsterite Laser with Krypton Ion, Ti:sapphire, and Nd:YAG Pump Lasers," IEEE J. Quantum Electron., vol. 29, pp. 2835–2844, 1993

[93Cur] P. F. Curley, C. Spielmann, T. Brabec, F. Krausz, E. Wintner, A. J. Schmidt, "Operation of a femtosecond Ti:sapphire solitary laser in the vicinity of zero group-delay dispersion," Opt. Lett., vol. 18, pp. 54–56, 1993

[93Eva] J. M. Evans, D. E. Spence, D. Burns, W. Sibbett, "Dual-wavelength self-mode-locked Ti:sapphire laser," Opt. Lett., vol. 18, pp. 1074–1076, 1993

[93Fan] T. Y. Fan, "Heat Generation in Nd:YAG and Yb:YAG," IEEE J. Quantum Electron., vol. 29, pp. 1457–1459, 1993

[93Fer] M. E. Fermann, M. J. Andrejco, M. L. Stock, Y. Silberberg, A. M. Weiner, "Passive modelocking in erbium fiber lasers with negative group delay," Appl. Phys. Lett., vol. 62, pp. 910–912, 1993

[93Fre] P. M. W. French, R. Mellish, J. R. Taylor, P. J. Delfyett, L. T. Florez, "Modelocked all-solid-state diode-pumped Cr:LiSAF laser," Opt. Lett., vol. 18, pp. 1934–1936, 1993

[93Hau] H. A. Haus and A. Mecozzi, "Noise of mode-locked lasers," IEEE J. Quantum Electron., vol. 29, pp. 983–996, 1993

[93Jia] W. Jiang, M. Shimizu, R. P. Mirin, T. E. Reynolds, and J. E. Bowers, "Electrically pumped mode-locked vertical-cavity semiconductor lasers," Opt. Lett., vol. 18, pp. 1937–1939, 1993

[93Kan] D. J. Kane, R. Trebino, "Characterization of Arbitrary Femtosecond Pulses Using Frequency-Resolved Optical Gating," IEEE J. Quantum Electron., vol. 29, pp. 571–578, 1993

[93Kel] U. Keller, T. H. Chiu, and J. F. Ferguson, "Self-starting and self-Q-switching dynamics of passively modelocked Nd:YLF and Nd:YAG lasers," Opt. Lett., vol. 18, pp. 217–219, 1993

[93Kwo] K. F. Kwong, D. Yankelevich, K. C. Chu, J. P. Heritage, A. Dienes, "400-Hz mechanical scanning optical delay line," Opt. Lett., vol. 18, pp. 558–560, 1993

[93Obe] M. H. Ober, M. Hofer, U. Keller, T. H. Chiu, "Self-starting, diode-pumped femtosecond Nd fiber laser," Opt. Lett., vol. 18, pp. 1532–1534, 1993

[93Pic] M. Piché, F. Salin, "Self-mode locking of solid-state lasers without apertures," Opt. Lett., vol. 18, pp. 1041–1043, 1993

[93Pla] H. Plaessmann, K. S. Yamada, C. E. Rich, W. M. Grossman, "Subnanosecond pulse generation from diode-pumped acousto-optically Q-switched solid-state lasers," Appl. Opt., vol. 32, pp. 6616–6619, 1993

[93Ram1] M. Ramaswamy, A. S. Gouveia-Neto, D. K. Negus, J. A. Izatt, J. G. Fujimoto, "2.3-ps pulses from a Kerr-lens mode-locked lamp-pumped Nd:YLF laser with a microdot mirror," Opt. Lett., vol. 18, pp. 1825–1827, 1993

[93Ram2] M. Ramaswamy, M. Ulman, J.Paye, and J. G. Fujimoto, "Cavity-dumped femtosecond Kerr-lens mode-locked Ti:Al_2O_3 laser," Opt. Lett., vol. 18, pp. 1822–1824, 1993

[93Riz] N. H. Rizvi, P. M. W. French, J. R. Taylor, P. J. Delfyett, L. T. Florez, "Generation of pulses as short as 93 fs from a self-starting femtosecond Cr:LiSrAlF_6 lasers by exploiting multiple-quantum-well absorbers," Opt. Lett., vol. 18, pp. 983–985, 1993

[93Sea] A. Seas, V. Petricevic, R. R. Alfano, "Self-mode-locked chromium-doped forsterite laser generates 50 fs pulses," Opt. Lett., vol. 18, pp. 891–893, 1993

[93Sch] R. Schiek, "Nonlinear refraction caused by cascaded second-order nonlinearity in optical waveguide structures," J. Opt. Soc. Am. B, vol. 10, pp. 1848–1855, 1993

[93Sen] A. Sennaroglu, C. R. Pollock, H. Nathel, "Generation of 48-fs pulses and measurement of crystal dispersion by using a regeneratively initiated self-mode-locked chromium-doped forsterite laser," Opt. Lett., vol. 18, pp. 826–828, 1993

[93Wei] K. J. Weingarten, U. Keller, T. H. Chiu, J. F. Ferguson, "Passively mode-locked diode-pumped solid-state lasers using an antiresonant Fabry–Perot saturable absorber," Opt. Lett., vol. 18, pp. 640–642, 1993

[93Wit] G. L. Witt, "LTMBE GaAs: present status and perspectives," Mater. Sci. Eng., vol. B22, p. 9, 1993

[93Yan] V. Yanovsky, Y. Pang, F. Wise, B. I. Minkov, "Generation of 25-fs from a self-modelocked Cr:forsterite laser with optimized group-delay dispersion," Opt. Lett., vol. 18, pp. 1541–1543, 1993

[94Bei] B. Beier, J.-P. Meyn, R. Knappe, K.-J. Boller, G. Huber, R. Wallenstein, "A 180-mW Nd:$LaSc_3(BO_3)_4$ single-frequency TEM00 microchip laser pumped by an injection-locked diode-laser array," Appl. Phys. B, vol. 58, 381–388 (1994)

[94Cer] G. Cerullo, S. De Silvestri, P. Laporta, S. Longhi, V. Magni, S. Taccheo, O. Svelto, "Continuous-wave mode locking of a bulk erbium–ytterbium glass laser," Opt. Lett., vol. 19, pp. 272–274, 1994

[94Con] P. J. Conlon, Y. P. Tong, P. M. W. French, J. R. Taylor, A. V. Shestakov, "Passive modelocking and dispersion measurement of a sub-100-fs Cr^{4+}:YAG laser," Opt. Lett., vol. 19, pp. 1468–1470, 1994

[94Del] L. D. DeLoach, S. A. Payne, L. K. Smith, W. L. Kway, and W. F. Krupke, "Laser and spectroscopic properties of $Sr_5(PO_4)_3F$:Yb," J. Opt Soc. Am. B, vol. 11, pp. 269–276, 1994

[94Den] M.L. Dennis, I.N. Duling, "Experimental study of sideband generation in femtosecond fiber lasers," IEEE J. Quantum Electron., vol. 30, pp. 1469–1477, 1994

[94Dym] M. J. P. Dymott, A. I. Ferguson, "Self-mode-locked diode-pumped Cr:LiSAF laser," Opt. Lett., vol. 19, pp. 1988–1990, 1994

[94Fer] M. E. Fermann, "Ultrashort-Pulse Sources Based on Single-Mode Rare-Earth-Doped Fibers," Appl. Phys. B, vol. 58, pp. 197–209, 1994

[94Gie] A. Giesen, H. Hügel, A. Voss, K. Wittig, U. Brauch, H. Opower, "Scalable Concept for Diode-Pumped High-Power Solid-State Lasers," Appl. Phys. B, vol. 58, pp. 363–372, 1994

[94Her] J. Herrmann, "Theory of Kerr-lens mode locking: role of self-focusing and radially varying gain," J. Opt. Soc. Am. B, vol. 11, pp. 498–512, 1994

[94Ipp] E. P. Ippen, "Principles of Passive Mode Locking," Appl. Phys. B, vol. 58, pp. 159–170, 1994

[94Ish] Y. Ishida, K. Naganuma, "Characteristics of femtosecond pulses near 1.5 μm in a self-mode-locked Cr^{4+}:YAG laser," Opt. Lett., vol. 19, pp. 2003–2005, 1994

[94Jen] T. Jensen, V. G. Ostroumov, J.-P. Meyn, G. Huber, A. I. Zagumennyi, and I. A. Shcherbakov, "Spectroscopic characterization and laser performance of diode-laser-pumped Nd: $GdVO_4$," Appl. Phys. B, vol. 58, pp. 373–379, 1994

[94Kel] U. Keller, "Ultrafast all-solid-state laser technology," Appl. Phys. B, vol. 58, pp. 347–363, 1994

[94Kop1] D. Kopf, K. J. Weingarten, L. R. Brovelli, M. Kamp, U. Keller, "Diode-pumped 100-fs passively mode-locked Cr:LiSAF laser with an antiresonant Fabry–Perot saturable absorber," Opt. Lett., vol. 19, pp. 2143–2145, 1994

[94Kop2] D. Kopf, F. X. Kärtner, K. J. Weingarten, U. Keller, "Pulse shortening in a Nd:glass laser by gain reshaping and soliton formation," Opt. Lett.., vol. 19, pp. 2146–2148, 1994

[94Kra] T. D. Krauss, F. W. Wise, "Femtosecond measurements of nonlinear absorption and refraction in CdS, ZnSe, and ZnS," Appl. Phys. Lett., vol. 65, pp. 1739–1741, 1994

[94Lew] M. Lewenstein, P. Balcou, M. Y. Ivanov, A. L'Huillier, P. B. Corkum, "Theory of high-harmonic generation by low-frequency laser fields," Phys. Rev. A, vol. 49, pp. 2117–2132, 1994

[94Lin1] J. R. Lincoln, M. J. P. Dymott, A. I. Ferguson, "Femtosecond pulses from an all-solid-state Kerr-lens mode-locked Cr:LiSAF laser," Opt. Lett., vol. 19, pp. 1210–1212, 1994

[94Lin2] J. R. Lincoln, A. I. Ferguson, "All-solid-state self-mode locking of a Nd:YLF laser," Opt. Lett., vol. 19, pp. 2119–2121, 1994

[94Liu] X. Liu, A. Prasad, W. M. Chen, A. Kurpiewski, A. Stoschek, Z. Liliental-Weber, and E. R. Weber, "Mechanism responsible for the semi-insulating properties of low-temperature-grown GaAs," Appl. Phys. Lett., vol. 65, pp. 3002–3004, 1994

[94Lon] S. Longhi, P. Laporta, S. Taccheo, O. Svelto, "Third-order-harmonic mode locking of a bulk erbium:ytterbium:glass at a 2.5-GHz repetition rate," Opt. Lett., vol. 19, pp. 1985–1987, 1994

[94Lon2] S. Longhi, "Theory of transverse modes in end-pumped microchip lasers," JOSA B, vol. 11, pp. 1098–1107, 1994

[94Mel] P. M. Mellish, P. M. W. French, J. R. Taylor, P. J. Delfyett, L. T. Florez, "All-solid-state femtosecond diode-pumped Cr:LiSAF laser," Electron. Lett., vol. 30, pp. 223–224, 1994

[94Mey] J.-P. Meyn, T. Jensen, G. Huber, "Spectroscopic Properties and Efficient Diode-Pumped Laser Operation of Neodymium Doped Lanthanum Scandium Borate," IEEE J. Quantum Electron., vol. 30, pp. 913–917, 1994

[94Psh] M. S. Pshenichnikov, W. P. de Boeij, and D. A. Wiersma, "Generation of 13-fs, 5-MW pulses from a cavity-dumped Ti:sapphire laser," Opt. Lett., vol. 19, pp. 572–574, 1994

[94Ram] M. Ramaswamy-Paye, J. G. Fujimoto, "Compact dispersion-compensating geometry for Kerr-lens mode-locked femtosecond lasers," Opt. Lett., vol. 19, pp. 1756–1758, 1994

[94Sen] A. Sennaroglu, C. R. Pollock, H. Nathel, "Continuous-wave self-mode-locked operation of a femtosecond Cr^{4+}:YAG laser, " Opt. Lett., vol. 19, pp. 390–392, 1994

[94Sti] A. Stingl, C. Spielmann, F. Krausz, "Generation of 11-fs pulses from a Ti:sapphire laser without the use of prisms," Opt. Lett., vol. 19, pp. 204–206, 1994

[94Szi] R. Szipöcs, K. Ferencz, C. Spielmann, F. Krausz, "Chirped multilayer coatings for broadband dispersion control in femtosecond lasers," Opt. Lett., vol. 19, pp. 201–203, 1994

[94Wan] H. S. Wang, P. L. K. Wa, J. L. Lefaucheur, B. H. T. Chai, A. Miller, "Cw and self-mode-locking performance of a red pumped Cr:LiSCAF laser," Opt. Comm., vol. 110, pp. 679–688, 1994

[94Wei] K. J. Weingarten, B. Braun, U. Keller, "In-situ small-signal gain of solid-state lasers determined from relaxation oscillation frequency measurements," Opt. Lett., vol. 19, pp. 1140–1142, 1994

[94Zay] J. J. Zayhowski and C. Dill, "Diode-pumped passively Q-switched picosecond microchip lasers," Opt. Lett., vol. 19, pp. 1427–1429, 1994

[94Zho] J. Zhou, G. Taft, C.-P. Huang, M. M. Murnane, H. C. Kapteyn, I. P. Christov, "Pulse evolution in a broad-bandwidth Ti:sapphire laser," Opt. Lett., vol. 19, pp. 1149–1151, 1994

[95Agr] G. A. Agrawal, Nonlinear Fiber Optics 2nd ed., Academic Press, ISBN 978-0123743022, San Diego CA, Chaps. 1 and 2

[95Bra1] B. Braun and U. Keller, "Single frequency Q-switched ring laser with an A-FPSA," Opt. Lett., vol. 20, pp. 1020–1022, 1995

[95Bra2] B. Braun, K. J. Weingarten, F. X. Kärtner, U. Keller, "Continuous-wave mode-locked solid-state lasers with enhanced spatial hole-burning, Part I: Experiments," Appl. Phys. B, vol. 61, pp. 429–437, 1995

[95Bra3] A. Braun, G. Korn, X. Liu, D. Du, J. Squier, G. Mourou, "Self-channeling of high-peak-power femtosecond laser pulses in air" Opt. Lett., vol. 20, 73–75, 1995

[95Bro1] L. R. Brovelli, U. Keller, and T. H. Chiu, "Design and Operation of Antiresonant Fabry–Perot Saturable Semiconductor Absorbers for Mode-Locked Solid-State Lasers," J. Opt. Soc. Am. B, vol. 12, pp. 311–322, 1995

[95Bro2] L. R. Brovelli, I. D. Jung, D. Kopf, M. Kamp, M. Moser, F. X. Kärtner and U. Keller, "Self-starting soliton modelocked Ti:sapphire laser using a thin semiconductor saturable absorber," Electronics Lett., vol. 31, pp. 287–289, 1995

[95Bro3] L. R. Brovelli, M. Lanker, U. Keller, K. W. Goossen, J. A. Walker, J. E. Cunningham, "Antiresonant Fabry–Perot quantum well modulator to actively modelock and synchronize solid-state lasers," Electronics Lett., vol. 31, pp. 381–382, 1995

[95Bro4] L. R. Brovelli, U. Keller, "Simple analytical expressions for the reflectivity and the penetration depth of a Bragg mirror between arbitrary media," Opt. Commun., vol. 116, pp. 343–350, 1995

[95Cas] Casix, Crystals & Materials (1995)

[95Cer] G. Cerullo, S. D. Silvestri, A. Monguzzi, D. Segala, and V. Magni, "Self-starting mode locking of a cw Nd:YAG laser using cascaded second-order nonlinearities," Opt. Lett., vol. 20, pp. 746–748, 1995

[95Chu1] J. Chung, A. E. Siegman, "Optical-Kerr-Enhanced Mode Locking of a Lamp-pumped Nd:YAG Laser," IEEE J. Quant. Electron., vol. 31, pp. 582–590, 1995.

[95Chu2] K. C. Chu, J. P. Heritage, R. S. Grant, K. X. Liu, A. Dienes, W. E. White, A. Sullivan, "Direct measurement of the spectral phase of femtosecond pulses," Opt. Lett., vol. 20, pp. 904–906, 1995

[95Cor] P. Corkum, "Breaking the attosecond barrier," Optics and Photonics, pp. 18–22, May 1995

[95Deg] J. J. Degnan, "Optimization of Passively Q-switched Lasers," IEEE J. Quantum Electronics, vol. 31, pp. 1890–1901, 1995

[95Del] P. J. Delfyett, "High power ultrafast semiconductor injection diode lasers," in *Compact Sources of Ultrashort Pulses*, I. I. N. Duling, Ed. (Cambridge University Press, New York, 1995) pp. 274–328

[95DiM] L. F. DiMauro, P. Agostini, "Ionization dynamics in strong laser fields," in *Advances in Atomic, Molecular and Optical Physics*, vol. 35, B. Bederson, H. Walther (eds.), San Diego, Academics Press, ISBN 0-12-003835-8, 1995, pp. 79–120

[95Dul] I. N. Duling, "Modelocking of all-fiber lasers," in *Compact Sources of Ultrashort Pulses*, I. N. Duling, Ed. (Cambridge University Press, New York, 1995) pp. 140–178

[95Dvo] M. D. Dvorak, B. L. Justus, "Z-scan studies of nonlinear absorption and refraction in bulk, undoped InP," Opt. Commun., vol. 114, pp. 147–150, 1995

[95Dym] M. J. P. Dymott, A. I. Ferguson, "Self-mode-locked diode-pumped Cr:LiSAF laser producing 34-fs pulses at 42 mW average output power," Opt. Lett., vol. 20, pp. 1157–1159, 1995

[95Fal] F. Falcoz, F. Balembois, P. Georges, A. Brun, "Self-starting self-mode-locked femtosecond diode-pumped Cr:LiSAF laser," Opt. Lett., vol. 20, pp. 1874–1876, 1995

[95Fer] M. E. Fermann, "Nonlinear polarization evolution in passively modelocked fiber lasers," in *Compact Sources of Ultrashort Pulses* (Cambridge University Press, New York, 1995), pp. 179–207

[95Flo] C. J. Flood, D. R. Walker, H. M. van Driel, "CW diode-pumped and FM modelocking of a Nd:KGW laser," Appl. Phys. B, vol. 60, pp. 309–312, 1995

[95Gal] G. M. Gale, M. Cavallari, T. J. Driscoll, F. Hache, "Sub-20 fs tunable pulses in the visible from an 82 MHz optical parametric oscillator," Opt. Lett., vol. 20, pp. 1562–1564, 1995

[95Gan] F. Ganikhanov, G.-R. Lin, W.-C. Chen, C.-S. Chang, and C.-L. Pan, "Subpicosecond carrier lifetimes in arsenic-implanted GaAs," Appl. Phys. Lett., vol. 67, pp. 3465–3467, 1995

[95Hau1] H. A. Haus, K. Tamura, L. E. Nelson, E. P. Ippen, "Stretched-Pulse Additive Pulse Mode-Locking in Fiber Ring Lasers: Theory and Experiment," IEEE J. Quantum Electron., vol. 31, pp. 591–598, 1995

[95Hau2] H. A. Haus, "Short pulse generation," in *Compact Sources of Ultrashort Pulses*, I. I. N. Duling, Ed. (Cambridge University Press 1995, New York, 1995) pp. 1–56

[95Hon] C. Hönninger, G. Zhang, U. Keller, A. Giesen, "Femtosecond Yb:YAG laser using semiconductor saturable absorbers," Opt. Lett., vol. 20, pp. 2402–2404, 1995

[95Iva] M. Ivanov, P. B. Corkum, T. Zuo, A. Bandrauk, "Routes to Control of Intense-Field Atomic Polarizability," Phys. Rev. Lett., vol. 74, pp. 2933–2936, 1995

[95Jac] M. Jacquemet, C. Jacquemet, N. Janel, F. Druon, F. Balembois, P. Georges, J. Petit, B. Viana, D. Vivien, and B. Ferrand, "Efficient laser action of Yb:LSO and Yb:YSO oxyorthosilicates crystals under high-power diode-pumping," Appl. Phys. B, vol. 80, pp. 171–176, 2005

[95Jia] W. Jiang, J. Bowers, "Ultrafast vertical cavity semiconductor lasers," in *Compact Sources of Ultrashort Pulses*, I. I. N. Duling, Eds. (Cambridge University Press, New York, 1995) pp. 208–273

[95Jun1] I. D. Jung, L. R. Brovelli, M. Kamp, U. Keller and M. Moser, "Scaling of the A-FPSA design toward a thin saturable absorber," Opt. Lett., vol. 20, pp. 1559–1561, 1995

[95Jun2] I. D. Jung, F. X. Kärtner, L. R. Brovelli, M. Kamp, U. Keller, "Experimental verification of soliton mode locking using only a slow saturable absorber," Opt. Lett., vol. 20, pp. 1892–1894, 1995

[95Kar1] F. X. Kärtner, U. Keller, "Stabilization of solitonlike pulses with a slow saturable absorber," Opt. Lett., vol. 20, pp. 16–18, 1995

[95Kar2] F. X. Kärtner, D. Kopf, U. Keller, "Solitary pulse stabilization and shortening in actively mode-locked lasers," JOSA B, vol. 12, pp. 486–496, 1995

[95Kar3] F. X. Kärtner, B. Braun, U. Keller, "Continuous-wave-mode-locked solid-state lasers with enhanced spatial hole-burning, Part II: Theory," Appl. Phys. B, vol. 61, pp. 569–579, 1995

[95Kar4] F. X. Kärtner, L. R. Brovelli, D. Kopf, M. Kamp, I. Calasso, and U. Keller, "Control of solid state laser dynamics by semiconductor devices," Opt. Eng., vol. 34, pp. 2024–2036, 1995

[95Kel] U. Keller, D. Kopf, "Optical component for generating pulsed laser radiation," European Patent Nr. EP 0 826 164 B1, priority date 19 May 1995

[95Kno] W. H. Knox, "Saturable Bragg Reflector," US Patent Nr. 5,627,854, priority date 15 March 1995

[95Kop1] D. Kopf, F. X. Kärtner, K. J. Weingarten, U. Keller, "Diode-pumped modelocked Nd:glass lasers using an A-FPSA," Opt. Lett., vol. 20, pp. 1169–1171, 1995

[95Kop2] D. Kopf, K. J. Weingarten, L. R. Brovelli, M. Kamp, U. Keller, "Sub-50-fs diode-pumped mode-locked Cr:LiSAF with an A-FPSA," Conference on Lasers and Electro-Optics (CLEO), 1995, paper CWM2

[95Kop3] D. Kopf, J. Aus der Au, U. Keller, G. L. Bona, P. Roentgen, "400-mW continuous-wave diode-pumped Cr:LiSAF laser based on a power-scalable concept" Opt. Lett., vol. 20, pp. 1782–1784, 1995

[95Kov] A. P. Kovacs, K. Osvay, Z. Bor, R. Szipöcs: "Group-delay measurement on laser mirrors by spectrally resolved white-light interferometry," Opt. Lett., vol. 20, pp. 788–790, 1995

[95Kro] A. Krotkus, S. Marcinkevicius, J. Jasinski, M. Kaminska, H. H. Tan, and C. Jagadish, "Picosecond carrier lifetime in GaAs implanted with high doses of As ions: An alternative material to low-temperature GaAs for optoelectronic applications," Appl. Phys. Lett., vol. 66, pp. 3304–3306, 1995

[95Kuc] S. Kück, K. Petermann, U. Pohlmann, G. Huber, "Near-infrared emission of Cr^{4+}-doped garnets: Lifetimes, quantum efficiencies, and emission cross sections," Phys. Rev. B, vol. 51, pp. 17323–17331, 1995

[95Lap] P. Laporta, S. Longhi, M. Marchesi, S. Taccheo, O. Svelto, "2.5 GHz and 5GHz Harmonic Mode-Locking of a Diode-Pumped Bulk Erbium–Ytterbium Glass Laser at 1.5 Microns," IEEE Photon. Technol. Lett., vol. 7, pp. 155–157, 1995

[95Leit] A. Leitenstorfer, C. Fürst, A. Laubereau, "Widely tunable two-color mode-locked Ti:sapphire laser with pulse jitter of less than 2 fs," Opt. Lett., vol. 20, pp. 916–918, 1995

[95Liu] X. Liu, A. Prasad, J. Nishio, E. R. Weber, Z. Liliental-Weber, and W. Walukiewicz, "Native point defects in low-temperature grown GaAs," Appl. Phys. Lett., vol. 67, p. 279, 1995

[95Mag] V. Magni, G. Cerullo, S. d. Silvestri, A. Monguzzi, "Astigmatism in Gaussian-beam self-focussing and in resonators for Kerr-lens mode-locking," JOSA B, vol. 12, pp. 476–485, 1995

[95Mel] R. Mellish, N. P. Barry, S. C. W. Hyde, R. Jones, P. M. W. French, J. R. Taylor, C. J. v. d. Poel, A. Valster, "Diode-pumped Cr:LiSAF all-solid-state femtosecond oscillator and regenerative amplifier," Opt. Lett., vol. 20, pp. 2312–2314, 1995

[95Ser] P. C. Sercel, "Multiphonon-assisted tunneling through deep levels: A rapid energy-relaxation mechanism in nonideal quantum-dot heterostructures," Phys. Rev. B, vol. 51, 14532, 1995

[95Ski] J. A. Skidmore, M. A. Emanuel, R. J. Beach, W. J. Benett, B. L. Freitas, N. W. Carlson, R. W. Solarz, "High-power continous wave 690 nm AlGaInP laser-diode arrays," Appl. Phys. Lett., vol. 66, pp. 1163–1165, 1995

[95Sol] G. S. Solomon, J. A. Trezza, and J. J. S. Harris, "Effects of monolayer coverage, flux ratio, and growth rate on the island density of InAs islands on GaAs," Appl. Phys. Lett., vol. 66, pp. 3161–3163, 1995

[95Sti] A. Stingl, M. Lenzner, Ch. Spielmann, F. Krausz, R. Szipöcs, "Sub-10-fs mirror-dispersion-controlled Ti:sapphire laser," Opt. Lett., vol. 20, pp. 602–604, 1995

[95Tam] K. Tamura, E. P. Ippen, H. A. Haus, "Pulse dynamics in stretched-pulse fiber lasers," Appl. Phys. Lett., vol. 67, pp. 158–160, 1995

[95Tsu] S. Tsuda, W. H. Knox, E. A. de Souza, W. Y. Jan, J. E. Cunningham, "Low-loss intracavity AlAs/AlGaAs saturable Bragg reflector for femtosecond mode locking in solid-state lasers," Opt. Lett., vol. 20, pp. 1406–1408, 1995

[95Vai] P. Vail'ev, *Ultrafast Diode Lasers* (Artech House, Inc., Boston, 1995)

[95Wan] P. Wang, S.-H. Zhou, K. K. Lee, and Y. C. Chen, "Picosecond laser pulse generation in a monolithic self-Q-switched solid-state laser," Opt. Comm., vol. 114, pp. 439–441, 1995

[95Yan] V. P. Yanovsky, F. W. Wise, A. Cassanho, and H. P. Jenssen, "Kerr-lens mode-locked diode-pumped Cr:LiSGAF laser," Opt. Lett., vol. 20, pp. 1304–1306, 1995

[95Zay] J. J. Zayhowski, C. Dill, "Coupled cavity electro-optically Q-switched Nd:YVO_4 microchip lasers," Opt. Lett., vol. 20, pp. 716–718, 1995

[96Ant] P. Antoine, A. L'Huillier, M. Lewenstein, "Attosecond pulse trains using high-order harmonics," Phys. Rev. Lett., vol. 77, pp. 1234–1237, 1996

[96Bra1] B. Braun, F. X. Kärtner, U. Keller, J.-P. Meyn, and G. Huber, "Passively Q-switched 180 ps Nd:LSB microchip laser," Opt. Lett., vol. 21, pp. 405–407, 1996

[96Bra2] B. Braun, C. Hönninger, G. Zhang, U. Keller, F. Heine, T. Kellner, G. Huber, "Efficient intracavity frequency doubling of a passively modelocked diode-pumped Nd:LSB laser," Opt. Lett., vol. 21, pp. 1567–1569, 1996

[96Col] B. C. Collings, J. B. Stark, S. Tsuda, W. H. Knox, J. E. Cunningham, W. Y. Jan, R. Pathak, K. Bergman, "Saturable Bragg reflector self-starting passive mode locking of a Cr^{4+}:YAG laser pumped with a diode-pumped Nd:YVO_4 laser," Opt. Lett., vol. 21, pp. 1171–1173, 1996

[96DeS] R. De Salvo, A. A. Said, D. J. Hagan, E. W. Van Stryland, M. Sheik-Bahae, "Infrared to ultraviolet measurements of two-photon absorption and n_2 in wide bandgap solids," IEEE J. Quantum Electron., vol. 32, pp. 1324–1333, 1996

[96Did] S. Diddams and J.-C. Diels, "Dispersion measurements with white-light interferometry," J. Opt. Soc. Am. B, vol. 13, pp. 1120–1129, 1996

[96Die] J.-C. Diels, W. Rudolph, *Ultrashort Laser Pulse Phenomena* (Academic Press, San Diego 1996)

[96Dob] J. A. Dobrowolski, A. V. Tikhonravov, M. K. Trubetskov, B. T. Sullivan, P. G. Verly, "Optimal single-band normal-incidence antireflection coatings," Appl. Opt., vol. 35, pp. 644–658, 1996

[96Flu1] R. Fluck, I. D. Jung, G. Zhang, F. X. Kärtner, U. Keller, "Broadband saturable absorber for 10-fs pulse generation," Opt. Lett., vol. 21, 743–745, 1996

[96Flu2] R. Fluck, G. Zhang, U. Keller, K. J. Weingarten, M. Moser, "Diode-pumped passively mode-locked 1.3 μm Nd:YVO_4 and Nd:YLF lasers by use of semiconductor saturable absorbers," Opt. Lett., vol. 21, pp. 1378–1380, 1996

[96Guy] O. Guy, V. Kubecek, and A. Barthelemy, "Mode-locked diode-pumped Nd:YAP laser," Opt. Comm., vol. 130, pp. 41–43, 1996

[96Gib] G. N. Gibson, R. Klnak, F. Gibson, and B. E. Bouma, "Electro-optically cavity-dumped ultrashort-pulse Ti:sapphire oscillator," Opt. Lett., vol. 21, pp. 1055–1057, 1996

[96Jag] C. Jagadish, H. H. Tan, A. Krotkus, S. Marcinkevicius, K. P. Korona, and M. Kaminska,"Ultrafast carrier trapping in high energy ion implanted gallium arsenide," Appl. Phys. Lett., vol. 68, no. 16, pp. 2225–7, 1996

[96Kar] F. X. Kärtner, I. D. Jung, U. Keller, "Soliton Mode-locking with Saturable Absorbers," Special Issue on Ultrafast Electronics, Photonics and Optoelectronics, IEEE J. Sel. Topics in Quantum Electronics (JSTQE), vol. 2, pp. 540–556, 1996

[96Kel] U. Keller, K. J. Weingarten, F. X. Kärtner, D. Kopf, B. Braun, I. D. Jung, R. Fluck, C. Hönninger, N. Matuschek, J. Aus der Au, "Semiconductor saturable absorber mirrors (SESAM's) for femtosecond to nanosecond pulse generation in solid-state lasers," IEEE J. Sel. Top. Quantum Electron., vol. 2, pp. 435–453, 1996

[96Kig] Kigre Inc., QX Laser Glasses, Data Sheet (1996)

[96Koe] W. Koechner, *Solid-State Laser Engineering*, A. L. Schawlow, A. E. Siegman, T. Tamir, H. K. V. Lotsch, Eds., Springer Series in Optical Sciences (Springer-Verlag, Heidelberg, Germany, 1996), vol. 1.

[96Kop1] D. Kopf, G. J. Spühler, K. J. Weingarten, U. Keller, "Mode-locked laser cavities with a single prism for dispersion compensation," Appl. Opt., vol. 35, pp. 912–915, 1996

[96Kop2] D. Kopf, G. Zhang, R. Fluck, M. Moser and U. Keller, "All-in-one dispersion-compensating saturable absorber mirror for compact femtosecond laser sources," Opt. Lett., vol. 21, pp. 486–488, 1996

[96Let] M. Lettenberger and K. Wolfrum, "Optimized Kerr lens mode-locking of a pulsed Nd:KGW laser," Opt. Comm., vol. 131, pp. 295–300, 1996

[96Lin] L.Y. Lin, M.C. Wu, T. Itoh, T.A. Vang, R.E. Muller, D.J. Sivco, and A.Y. Cho, "Velocity-matched distributed photodetectors with high saturation power and large bandwidth," IEEE Photon. Technol. Lett., Vol. 8, 1376–1378, 1996

[96Nel] L. E. Nelson, S. B. Fleischer, G. Lenz, E. P. Ippen, "Efficient frequency doubling of a femtosecond fiber laser," Opt. Lett., vol. 21, pp. 1759–1761, 1996

[96Rhe] J.-K. Rhee, T. S. Sosnowski, A.-C. Tien, T. B. Norris, "Real-time dispersion analyzer of femtosecond laser pulses with use of a spectrally and temporally resolved upconversion technique," J. Opt. Soc. Am. B, vol. 13, pp. 1780–1785, 1996

[96Sch] K. I. Schaffers, L. D. DeLoach, and S. A. Payne, "Crystal growth, frequency doubling, and infrared laser performance of Yb^{3+}:$BaCaBO_3F$," IEEE J. Quantum Electron., vol. 32, pp. 741–748, 1996

[96Sha1] R. C. Sharp, D. E. Spock, N. Pan, J. Elliot, "190-fs passively modelocked thulium fiber laser with a low threshold," Opt. Lett., 21, 881 (1996)

[96Shi] Z. Shi, H. Zogg, P. Müller, I. D. Jung, and U. Keller, "Wide Bandwidth (100) GaAs-Fluorides Quarter-Wavelength Bragg Reflectors Grown by Molecular Beam Epitaxy," Appl. Phys. Lett., vol. 69, pp. 3474–3476, 1996

[96Sie] U. Siegner, R. Fluck, G. Zhang, and U. Keller, "Ultrafast high-intensity nonlinear absorption dynamics in low-temperature grown gallium arsenide," Appl. Phys. Lett., vol. 69, pp. 2566–2568, 1996

[96Sor1] I. T. Sorokina, E. Sorokin, E. Wintner, A. Cassanho, H. P. Jenssen, M. A. Noginov, "Efficient cw TEM_{00} and femtosecond Kerr-lens modelocked Cr:LiSrGaF laser," Opt. Lett., vol. 21, pp. 204–206, 1996

[96Sor2] I. T. Sorokina, E. Sorokin, E. Wintner, A. Cassanho, H. P. Jenssen, R. Szipöcs, "Prismless passively mode-locked femtosecond Cr:LiSGaF laser," Opt. Lett., vol. 21, pp. 1165–1167, 1996

[96Ste] G. I. Stegeman, D. J. Hagan, and L. Torner, "$\chi(2)$ cascading phenomena and their applications to all-optical signal processing, mode-locking, pulse compression and solitons," Opt. Quantum Electron., vol. 28, pp.1691–1740, 1996

[96Taf] G. Taft, A. Rundquist, M. M. Murnane, I. P. Christov, H. C. Kapteyn, K. W. DeLong, D. N. Fittinghoff, M. A. Krumbügel, J. N. Sweetser, R. Trebino, "Measurement of 10-fs laser pulses," IEEE J. Sel. Top. Quantum Electron., vol. 2, pp. 575–585, 1996

[96Tel] H. R. Telle, in *Frequency Control of Semiconductor Lasers*, ed. by M. Ohtsu (Wiley, New York 1996), p. 137

[96Ton] Y. P. Tong, J. M. Sutherland, P. M. W. French, J. R. Taylor, A. V. Shestakov, B. H. T. Chai, "Self-starting Kerr-lens mode-locked femtosecond Cr^{4+}:YAG and picosecond Pr^{3+}:YLF solid-state lasers," Opt. Lett., vol. 21, pp. 644–646, 1996

[96Tsu] S. Tsuda, W. H. Knox, S. T. Cundiff, "High efficiency diode pumping of a saturable Bragg reflector-mode-locked Cr:LiSAF femtosecond laser," Appl. Phys. Lett., vol. 69, pp. 1538–1540, 1996

[96Xu1] L. Xu, C. Spielmann, A. Poppe, T. Brabec, F. Krausz, and T. W. Haensch, "Route to phase control of ultrashort light pulses," Opt. Lett., vol. 21, pp. 2008–2010, 1996

[96Xu2] L. Xu, C. Spielmann, F. Krausz, R. Szipöcs, "Ultrabroadband ring oscillator for sub-10-fs pulse generation," Opt. Lett., vol. 21, pp. 1259–1261, 1996

[96Zay] J. J. Zayhowski, "Ultraviolet generation with passively Q-switched microchip lasers," Opt. Lett., vol. 21, pp. 588–590, 1996

[97Agn] A. Agnesi, C. Pennacchio, G. C. Reali, V. Kubecek, "High-power diode-pumped picosecond Nd:YVO_4 laser," Opt. Lett., vol. 22, pp. 1645–1647, 1997

[97Aus] J. Aus der Au, D. Kopf, F. Morier-Genoud, M. Moser, U. Keller, "60-fs pulses from a diode-pumped Nd:glass laser," Opt. Lett., vol. 22, pp. 307–309, 1997

[97Bal] A. Baltuska, Z. Wei, M. S. Pshenichnikov, D. A. Wiersma, R. Szipöcs, "All-solid-state cavity dumped sub-5-fs laser," Appl. Phys. B, vol. 65, pp. 175–188, 1997

[97Bra] B. Braun, F. X. Kärtner, M. Moser, G. Zhang, and U. Keller, "56 ps passively Q-switched diode-pumped microchip laser," Opt. Lett., vol. 22, pp. 381–383, 1997

[97Bra2] T. Brabec, F. Krausz, "Nonlinear pulse propagation in single-cycle regime," Phys. Rev. Lett., vol. 78, pp. 3282–3285, 1997

[97Chr] I. P. Christov, M. M. Murnane, H. C. Kapteyn, "High-Harmonic Generation of Attosecond Pulses in the 'Single-Cycle' Regime," Phys. Rev. Lett., vol. 78, pp. 1251–1254, 1997

[97Col] B. C. Collings, K. Bergman, W. H. Knox, "True fundamental solitons in a passively mode-locked short cavity Cr^{4+}:YAG laser," Opt. Lett., vol. 22, pp. 1098–1100, 1997

[97Dym] M. J. P. Dymott, A. I. Ferguson, "Pulse duration limitations in a diode-pumped femtosecond Kerr-lens modelocked Cr:LiSAF laser," Appl. Phys. B., vol. 65, pp. 227–234, 1997

[97Fer] M. E. Fermann, A. Galvanauskas, G. Sucha, D. Harter, "Fiber-lasers for ultrafast optics," Appl. Phys. B, vol. 65, pp. 259–275, 1997

[97Flu] R. Fluck, B. Braun, E. Gini, H. Melchior, and U. Keller, "Passively Q-switched 1.34 μm Nd:YVO_4 microchip laser using semiconductor saturable-absorber mirrors," Opt. Lett., vol. 22, pp. 991–993, 1997

[97Gab] A. Gaber, H. Zillgen, P. Ehrhart, P. Partyka, R. S. Averback, "Lattice parameter changes and point defect reactions in low temperature electron irradiated AlAs," J. Appl. Physics, vol. 82, pp. 5348–5351, 1997

[97Hen] B. Henrich, R. Beigang, "Self-starting Kerr-lens mode locking of a Nd:YAG laser," Opt. Com., vol. 135, pp. 300–304, 1997

[97Jun1] I. D. Jung, F. X. Kärtner, N. Matuschek, D. H. Sutter, F. Morier-Genoud, Z. Shi, V. Scheuer, M. Tilsch, T. Tschudi, U. Keller, "Semiconductor saturable absorber mirrors supporting sub-10 fs pulses," Appl. Phys. B, vol. 65, pp. 137–150, 1997

[97Jun2] I. D. Jung, F. X. Kärtner, J. Henkmann, G. Zhang, U. Keller, "High-dynamic-range characterization of ultrashort pulses," Appl. Phys. B, vol. 65, pp. 307–310, 1997.

[97Jun3] I. D. Jung, F. X. Kärtner, N. Matuschek, D. H. Sutter, F. Morier-Genoud, G. Zhang, U. Keller, V. Scheuer, M. Tilsch, T. Tschudi, "Self-starting 6.5 fs pulses from a Ti:sapphire laser," Opt. Lett., vol. 22, pp. 1009–1011, 1997

[97Kal] V. L. Kalashnikov, V. P. Kalosha, I. G. Poloiko, V. P. Mikhailov, "Frequency mechanism of USP formation in solid-state lasers with slow saturable absorber," Opt. Spectrosc., vol. 82, pp. 469–472, 1997

[97Kar] F. X. Kärtner, N. Matuschek, T. Schibli, U. Keller, H. A. Haus, C. Heine, R. Morf, V. Scheuer, M. Tilsch, T. Tschudi, "Design and fabrication of double-chirped mirrors," Opt. Lett., vol. 22, pp. 831–833, 1997

[97Kon] M. Kondow, T. Kitatani, S. Nakatsuka, M. C. Larson, K. Nakahara, Y. Yazawa, M. Okai, K. Uomi, "GaInNAs: a novel material for long-wavelength semiconductor lasers," IEEE J. Sel. Top. Quantum Electron., vol. 3, pp. 719–730, 1997

[97Kop1] D. Kopf, U. Keller, M. A. Emanuel, R. J. Beach, J. A. Skidmore, "1.1-W cw Cr:LiSAF laser pumped by a 1-cm diode-array," Opt. Lett., vol. 22, pp. 99–101, 1997

[97Kop2] D. Kopf, A. Prasad, G. Zhang, M. Moser, U. Keller, "Broadly tunable femtosecond Cr:LiSAF laser," Opt. Lett., vol. 22, pp. 621–623, 1997

[97Kop3] D. Kopf, K. J. Weingarten, G. Zhang, M. Moser, M. A. Emanuel, R. J. Beach, J. A. Skidmore, U. Keller, "High-average-power diode-pumped femtosecond Cr:LiSAF lasers," Appl. Phys. B, vol. 65, pp. 235–243, 1997

[97Kul1] N. V. Kuleshov, A. A. Lagatsky, A. V. Podlipensky, V. P. Mikhailov, G. Huber, "Pulsed laser operation of Yb-doped $KY(WO_4)_2$ and $KGd(WO_4)_2$," Opt. Lett., vol. 22, pp. 1317–1319, 1997

[97Kul2] N. V. Kuleshov, A. A. Lagatsky, V. G. Shcherbitsky, V. P. Mikhailov, E. Heumann, T. Jensen, A. Diening, G. Huber, "CW laser performance of Yb and Er,Yb doped tungstates," Appl. Phys. B, vol. 64, pp. 409–413, 1997

[97Kuz] M. Kuznetsov, F. Hakimi, R. Sprague, and A. Mooradian, "High-Power (>0.5-W CW) Diode-Pumped Vertical-External-Cavity Surface-Emitting Semiconductor Lasers with Circular TEM_{00} Beams," IEEE Photon. Technol. Lett., vol. 9, pp. 1063–1065, 1997

[97Lin] D. v. d. Linde, K. Sokolowski-Tinten, and J. Bialkowski, "Laser-solid interactions in the femtosecond time regime," Appl. Surface Science, vol. 109/110, pp. 1–10, 1997

[97Liu] X. Liu, D. Du, and G. Mourou, "Laser ablation and micromachining with ultrashort laser pulses," IEEE J. Quantum Electron., vol. 33, pp. 1706–1716, 1997

[97Loe] F. H. Loesel, C. Horvath, F. Grasbon, M. Jost, M. H. Niemz, "Selfstarting femtosecond operation and transient dynamics of a diode-pumped Cr:LiSGaF laser with a semiconductor saturable absorber mirror," Appl. Phys. B, vol. 65, pp. 783–787, 1997

[97Mal] S. Malik, C. Roberts, R. Murray, and M. Pate, "Tuning self-assembled InAs quantum dots by rapid thermal annealing," Appl. Phys. Lett., vol. 71, pp. 1987–1989, 1997

[97Mat] N. Matuschek, F. X. Kärtner, U. Keller, "Exact Coupled-Mode Theories for Multilayer Interference Coatings with Arbitrary Strong Index Modulations," IEEE J. Quantum Electronics, vol. 33, pp. 295–302, 1997

[97McI] K. A. McIntosh, K. B. Nichols, S. Verghese, and E. R. Brown, "Investigation of ultrashort photocarrier relaxation times in low-temperature-grown GaAs," Appl. Phys. Lett., vol. 70, no. 3, pp. 354–356, 1997

[97Nib] E. T. J. Nibbering, G. Grillon, M. A. Franco, B. S. Prade, and A. Mysyrowicz, "Determination of the inertial contribution to the nonlinear refractive index of air, N_2, and O_2 by use of unfocused high-intensity femtosecond laser pulses," J. Opt. Soc. Am. B, vol. 14, no. 3, pp. 650–660, 1997

[97Nis] M. Nisoli, S. Stagira, S. D. Silvestri, O. Svelto, S. Sartania, Z. Cheng, M. Lenzner, C. Spielmann, F. Krausz, "A novel high-energy pulse compression system: generation of multigigawatt sub-5-fs pulses," Appl. Phys. B, vol. 65, pp. 189–196, 1997

[97Nis2] M. Nisoli, S. De Silvestri, O. Svelto, R. Szipöcs, K. Ferenz, C. Spielmann, S. Sartania, F. Krausz, "Compression of high energy laser pulses below 5 fs," Opt. Lett., vol. 22, pp. 522–524, 1997

[97Pag] R. H. Page, K. I. Schaffers, L. D. DeLoach, G. D. Wilke, F. D. Patel, J. B. Tassano, S. A. Payne, W. F. Krupke, K.-T. Chen, and A. Burger, "Cr^{2+}-doped zinc chalcogenides as efficient, widely tunable mid-infrared lasers," IEEE J. Quant. Electron., vol. 33, pp. 609–619, 1997

[97Ruf] B. Ruffing, A. Nebel, R. Wallenstein, "A 20-W KTA-OPO synchronously pumped by a cw modelocked Nd:YVO_4 oscillator–amplifier system," Conference on Lasers and Electro-Optics 1997, paper CWB2, p. 199

[97Sar] S. Sartania, Z. Cheng, M. Lenzner, G. Tempea, C. Spielmann, F. Krausz, K. Ferencz, "Generation of 0.1-TW 5-fs optical pulses at a 1-kHz repetition rate," Opt. Lett., vol. 22, pp. 1562–1564, 1997

[97Sor] I. T. Sorokina, E. Sorokin, E. Wintner, A. Cassanho, H. P. Jenssen, R. Szipöcs, "14-fs pulse generation in Kerr-lens modelocked prismless Cr:LiSGaF and Cr:LiSAF lasers observation of pulse self-frequency shift," Opt. Lett., vol. 22, pp. 1716–1718, 1997

[97Spa] S. Spälter, M. Böhm, M. Burk, B. Mikulla, R. Fluck, I. D. Jung, G. Zhang, U. Keller, A. Sizmann, G. Leuchs, "Self-starting soliton-modelocked femtosecond $Cr^{(4+)}$:YAG laser using an antiresonant Fabry–Perot saturable absorber," Appl. Phys. B, vol. 65, pp. 335–338, 1997

[97Spe] P. Specht, S. Jeong, H. Sohn, M. Luysberg, A. Prasad, J. Gebauer, R. Krause-Rehberg, E. R. Weber, "Defect control in As-rich GaAs," Mater. Sci. Forum, vol. 258–263, p. 951, 1997

[97Ton] Y. P. Tong, P. M. W. French, J. R. Taylor, J. O. Fujimoto, "All-solid-state femtosecond sources in the near infrared," Opt. Com., vol. 136, pp. 235–238, 1997

[97Tou] P. Tournois, "Acousto-optic programmable dispersive filter for adaptice compensation of group delay time dispersion in laser systems," Opt. Commun., vol. 140, pp. 245–249, 1997

[97Tre] R. Trebino, K. W. DeLong, D. N. Fittinghoff, J. Sweetser, M. A. Krumbügel, B. Richman, "Measuring ultrashort laser pulses in the time-frequency domain using frequency-resolved optical gating," Rev. Sci. Instrum., vol. 68, pp. 3277–3295, 1997

[97Ye] J. Ye, L.-S. Ma, T. Daly, J. L. Hall, "Highly selective terahertz optical frequency comb generator," Opt. Lett., vol. 22, pp. 301–303, 1997

[97Zay] J. J. Zayhowski, J. Harrison, "Miniature Solid-State Lasers," in *Handbook of Photonics*, M. C. Gupta, Eds. (CRC Press, New York, 1997) pp. 326–392

[97Zha1] Z. Zhang, K. Torizuka, T. Itatani, K. Kobayashi, T. Sugaya, T. Nakagawa, "Self-starting mode-locked femtosecond forsterite laser with a semiconductor saturable-absorber mirror," Opt. Lett., vol. 22, pp. 1006–1008, 1997

[97Zha2] Z. Zhang, K. Torizuka, T. Itatani, K. Kobayashi, T. Sugaya, T. Nakagawa, "Femtosecond Cr:forsterite Laser with Modelocking initiated by a quantum well saturable absorber," IEEE J. Quantum Electron., vol. 33, pp. 1975–1981, 1997

[98Akh] N. N. Akhmediev, A. Ankiewicz, M. J. Lederer, and B. Luther-Davies, "Ultrashort pulses generated by mode-locked lasers with either a slow or a fast saturable-absorber reponse," Opt. Lett., vol. 23, pp. 280–282, 1998

[98Aus] J. Aus der Au, F. H. Loesel, F. Morier-Genoud, M. Moser, U. Keller, "Femtosecond diode-pumped Nd:glass laser with more than 1-W average output power," Opt. Lett., vol. 23, pp. 271–273, 1998

[98Bal] A. Baltuska, M. S. Pshenichnikov, D. A. Wiersma, "Amplitude and phase characterization of 4.5-fs pulses by frequency-resolved optical gating," Opt. Lett., vol. 23, pp. 1474–1476, 1998

[98Cha] Y. Chang, R. Maciejko, R. Leonelli, A. Thorpe, "Self-starting passively mode-locked tunable Cr^{4+}:yttrium–aluminium–garnet laser with a single prism for dispersion compensation," Appl. Phys. Lett., vol. 73, pp. 2098–2100, 1998

[98deB] A. de Bohan, P. Antoine, D. B. Milosevic, B. Piraux, "Phase-dependent Harmonic Emission with Ultrashort Laser Pulses," Phys. Rev. Lett., vol. 81, pp. 1837–1840, 1998

[98Efi] A. Efimov, M. D. Moores, N. M. Beach, J. L. Krause, D. H. Reitze, "Adaptive control of pulse phase in a chirped-pulse amplifier," Opt. Lett., vol. 23, pp. 1915–1917, 1998

[98Flu] R. Fluck, R. Häring, R. Paschotta, E. Gini, H. Melchior, U. Keller, "Eyesafe pulsed microchip laser using semiconductor saturable absorber mirrors," Appl. Phys. Lett., vol. 72, pp. 3273–3275, 1998

[98For] L. Fornasiero, S. Kück, T. Jensen, G. Huber, and B. H. T. Chai, "Excited state absorption and stimulated emission of Nd^{3+} in crystals. Part2: YVO_4, $GdVO_4$, and $Sr_5(PO_4)_3F$," Appl. Phys. B, vol. 67, pp. 549–553, 1998

[98Gab] K. M. Gabel, P. Russbuldt, R. Lebert, and A. Valster, "Diode-pumped Cr^{3+}:LiCAF fs-laser," Opt. Commun., vol. 157, pp. 327–334, 1998

[98Hon] C. Hönninger, F. Morier-Genoud, M. Moser, U. Keller, L. R. Brovelli, C. Harder, "Efficient and tunable diode-pumped femtosecond Yb:glass lasers," Opt. Lett., vol. 23, pp. 126–128, 1998

[98Hop] J. M. Hopkins, G. J. Valentine, W. Sibbett, J. A. d. Au, F. Morier-Genoud, U. Keller, and A. Valster, "Efficient, low-noise, SESAM-based femtosecond Cr^{3+}: $LiSrAlF_6$ laser," Opt. Comm., vol. 154, pp. 54–58, 1998

[98Iac] C. Iaconis, I. A. Walmsley, "Spectral Phase Interferometry for Direct Electric Field Reconstruction of Ultrashort Optical Pulses," Opt. Lett., vol. 23, pp. 792–794, 1998

[98Iva] E. N. Ivanov, M. E. Tobar, and R. A. Woode, "Applications of interferometric signal processing to phase-noise reduction in microwave oscillators," IEEE Trans. Microwave Theory and Technol., vol. 46, pp. 1537–1545, 1998

[98Kal] V. L. Kalashnikov, V. P. Kalosha, I. G. Poloiko, and V. P. Mikhailov, "Frequency mechanism of USP formation in solid-state lasers with slow saturable absorber," Opt. Spectrosc., vol. 82, pp. 469–472, 1997

[98Kar] F. X. Kärtner, J. A. der Au, U. Keller, "Modelocking with slow and fast saturable absorbers - What's the difference?," IEEE J. Sel. Top. Quantum Electron., vol. 4, pp. 159–168, 1998

[98Kel] T. Kellner, F. Heine, G. Huber, C. Hönninger, B. Braun, F. Morier-Genoud, U. Keller, "Soliton mode-locked Nd:$YAlO_3$ laser at 930 nm," J. Opt. Soc. Am. B, vol. 15, pp. 1663–1666, 1998

[98Kel2] U. Keller, "Semiconductor nonlinearities for solid-state laser modelocking and Q-switching," in *Nonlinear Optics in Semiconductors*, A. Kost and E. Garmire, Eds. (Academic, Boston, Mass., 1998), vol. 59, chap. 4, pp. 211–285

[98Liu] X. Liu, L. Qian, F. Wise, Z. Zhang, T. Itatani, T. Sugaya, T. Nakagawa, K. Torizuka, "Femtosecond Cr:forsterite laser diode pumped by a double-clad fiber," Opt. Lett., vol. 23, pp. 129–131, 1998.

[98Lon] S. Longhi, G. Sorbello, S. Taccheo, P. Laporta, "5-GHz repetition-rate dual-wavelength pulse-train generation from an intracavity frequency-modulated Er-Yb:glass laser," Opt. Lett., vol. 23, pp. 1547–1549, 1998

[98Mel] R. Mellish, S. V. Chernikov, P. M. W. French, J. R. Taylor, "All-solid-state compact high repetition rate modelocked Cr^{4+}:YAG laser," Electron. Lett., vol. 34, pp. 552–553, 1998

[98Mil] D. Milam, "Review and assessment of measured values of the nonlinear refractive-index coefficient of fused silica," Appl. Opt., vol. 37, pp. 546–550, 1998

[98Pet] V. Petrov, V. Shcheslavskiy, T. Mirtchev, F. Noack, T. Itatani, T. Sugaya, T. Nakagawa, "High-power self-starting femtosecond Cr:forsterite laser," Electron. Lett., vol. 34, pp. 559–561, 1998

[98Pow] P. E. Powers, T. J. Kulp, S. E. Bisson, "Continuous tuning of a continuous-wave periodically poled lithium niobate optical parametric oscillator by use of a fan-out grating design," Opt. Lett., vol. 23, 159–161, 1998

[98Shi] Z. Shi, H. Zogg, and U. Keller, "Thick crack-free CaF_2 epitaxial layer on GaAs (100) substrate by molecular beam epitaxy," J. Electron. Mat., vol. 27, pp. 55–58, 1998

[98Sor] I. T. Sorokina, E. Sorokin, and E. Wintner, "On the way towards the pulse-duration limits in prismless KLM Cr:LiSGaF and Cr:LiSAF lasers," Laser Phys., vol. 8, pp. 607–611, 1998

[98Sut] D. H. Sutter, I. D. Jung, F. X. Kärtner, N. Matuschek, F. Morier-Genoud, V. Scheuer, M. Tilsch, T. Tschudi, U. Keller, "Self-starting 6.5-fs pulses from a Ti:sapphire laser using a semiconductor saturable absorber and double-chirped mirrors," IEEE J. of Sel. Topics in Quantum Electronics, vol. 4, pp. 169–178, 1998

[98Sve] O. Svelto, *Principles of Lasers*, Plenum Press, New York, 1998

[98Tan] C. Z. Tan, "Determination of refractive index of silica glass for infrared wavelengths by IR spectroscopy," J. Non-Cryst. Solids, vol. 223, pp. 158–163, 1998

[98Taw] T. Tawara, M. Arita, K. Uesugi, and I. Suemune, "MOVPE growth of ZnSe/ZnS distributed Bragg reflectors on GaAs(100) and (311)B substrates," J. Cryst. Growth, vol. 184/185, p. 777, 1998

[98Tho] M. E. Thomas, S. K. Andersson, R. M. Sova, R. I. Joseph, "Frequency and temperature dependence of the refractive index of sapphire," Infrared Phys. Technol., vol. 39, pp. 235–249, 1998

[98Xu] L. Xu, G. Tempea, C. Spielmann, F. Krausz, "Continuous-wave mode-locked Ti:sapphire laser focusable to 5×10^{13} W/cm^2," Opt. Lett., vol. 23, pp. 789791, 1998

[98Zav] M. Zavelani-Rossi, G. Cerullo, V. Magni, "Mode locking by cascading of second-order nonlinearities," IEEE J. Quantum Electron., vol. 34, pp. 61–70, 1998

[99Agn] A. Agnesi, S. Dell'Acqua, and G. Reali, "Nonlinear mirror mode-locking of efficiently diode-pumped pulsed neodymium lasers," J. Opt. Soc. Am. B, vol. 16, pp. 1236–1242, 1999

[99Att] D. Attwood, *Soft X-rays and extreme ultraviolet radiation: principles and applications*, Cambridge University Press, 1999

[99Aus] J. Aus der Au, S. F. Schaer, R. Paschotta, C. Hönninger, U. Keller, M. Moser, "High-power diode-pumped passively modelocked Yb:YAG lasers," Opt. Lett., vol. 24, pp. 1281–1283, 1999

[99Bar] A. Bartels, T. Dekorsky, H. Kurz, "Femtosecond Ti:sapphire ring laser with 2-GHz repetition rate and its application in time-resolved spectroscopy," Opt. Lett., vol. 24, pp. 996–998, 1999

[99Bed] T. Beddard, W. Sibbett, D. T. Reid, J. Garduno-Mejia, N. Jamasbi, M. Mohebi, "High-average-power, 1-MW peak-power self-mode-locked Ti:sapphire oscillator," Opt. Lett., vol. 24, pp. 163–165, 1999

[99Bil] I. P. Bilinsky, J. G. Fujimoto, J. N. Walpole, L. J. Missaggia, "InAs-doped silica films for saturable absorber applications," Appl. Phys. Lett., vol. 74, pp. 2411–2413, 1999

[99Che] Z. Cheng, A. Fürbach, S. Sartania, M. Lenzner, C. Spielmann, F. Krausz, "Amplitude and chirp characterization of high-power laser pulses in the 5-fs regime," Opt. Lett., vol. 24, pp. 247–249, 1999

[99Cho] S. H. Cho, B. E. Bouma, E. P. Ippen, J. G. Fujimoto, "Low-repetition-rate high-peak-power Kerr-lens mode-locked Ti:Al_2O_3 laser with a multiple-pass cell," Opt. Lett., vol. 24, pp. 417–419, 1999

[99Cid] P. E. Ciddor, R. J. Hill: "Refractive index of air. 2. Group index," Appl. Opt., vol. 38, pp. 1663–1667, 1999

[99Cou] V. Couderc, F. Louradour, A. Barthelemy, "2.8 ps pulses from a modelocked diode pumped Nd:YVO_4 laser using quadratic polarization switching," Opt. Com., vol. 166, pp. 103–111, 1999

[99Dur] C. G. Durfee, S. Backus, H. C. Kapteyn, M. M. Murnane, "Intense 8-fs pulse generation in the deep ultraviolet," Opt. Lett., vol. 24, pp. 697–699, 1999

[99Fit] D. N. Fittinghoff, A. C. Millard, J. A. Squier, and M. Müller, "Frequency-Resolved Optical Gating Measurement of Ultrashort Pulses Passing Through a High Numerical Aperture Objective," IEEE J. Quantum Electron., vol. 35, pp. 479–486, 1999

[99Gal] L. Gallmann, D. H. Sutter, N. Matuschek, G. Steinmeyer, U. Keller, C. Iaconis, and I. A. Walmsley, "Characterization of sub-6-fs optical pulses with spectral phase interferometry for direct electric-field reconstruction," Opt. Lett., vol. 24, pp. 1314–1316, 1999

[99Gra] Th. Graf, A. I. Ferguson, E. Bente, D. Burns, M. D. Dawson, "Multi-watt Nd:YVO_4 laser, mode locked by a semiconductor saturable absorber mirror and side-pumped by a diode-laser bar," Optics Comm., vol. 159, pp. 84–87, 1999

[99Hai1] M. Haiml, U. Siegner, F. Morier-Genoud, U. Keller, M. Luysberg, P. Specht, and E. R. Weber, "Femtosecond response times and high optical nonlinearity in beryllium doped low-temperature grown GaAs," Appl. Phys. Lett., vol. 74, pp. 1269–1271, 1999

[99Hai2] M. Haiml, U. Siegner, F. Morier-Genoud, U. Keller, M. Luysberg, R. C. Lutz, P. Specht, and E. R. Weber, "Optical nonlinearity in low-temperature grown GaAs: microscopic limitations and optimization strategies," Appl. Phys. Lett., vol. 74, pp. 3134–3136, 1999

[99Hol] M. A. Holms, P. Cusumano, D. Burns, A. I. Ferguson, M. D. Dawson, "Mode-locked operation of a diode-pumped, external-cavity GaAs/AlGaAs surface emitting laser," *Technical Digest. Summaries of papers presented at the Conference on Lasers and Electro-Optics. Postconference Edition. CLEO '99*, 1999, pp. 153–154

[99Hon1] C. Hönninger, R. Paschotta, F. Morier-Genoud, M. Moser, U. Keller, "Q-switching stability limits of continuous-wave passive modelocking," J. Opt. Soc. Am. B, vol. 16, pp. 46–56, 1999

[99Hon2] C. Hönninger, R. Paschotta, M. Graf, F. Morier-Genoud, G. Zhang, M. Moser, S. Biswal, J. Nees, A. Braun, G. A. Mourou, I. Johannsen, A. Giesen, W. Seeber, U. Keller, "Ultrafast ytterbium-doped bulk lasers and laser amplifiers," Appl. Phys. B, vol. 69, pp. 3–17, 1999

[99Iac] C. Iaconis, I. A. Walmsley, "Self-Referencing Spectral Interferometry for Measuring Ultrashort Optical Pulses," IEEE J. Quantum Electron., vol. 35, pp. 501–509, 1999

[99Jeo] T. M. Jeong, E. C. Kang, C. H. Nam, "Temporal and spectral characteristics of an additive-pulse mode-locked Nd:YLF laser with Michelson-type configuration," Opt. Com., vol. 166, pp. 95–102, 1999

[99Kai] R. A. Kaindl, F. Eickemeyer, M. Woerner, T. Elsaesser, "Broadband phase-matched difference frequency mixing of femtosecond pulses in GaSe: Experiment and theory," Appl. Phys. Lett., vol. 75, pp. 1060–1062, 1999

[99Kan] D. J. Kane, "Recent Progress Toward Real-Time Measurement of Ultrashort Laser Pulses," IEEE J. Quantum Electron., vol. 35, pp. 421–431, 1999

[99Kar] F. X. Kärtner, D. M. Zumbühl, N. Matuschek, "Turbulence in mode-locked lasers," Phys. Rev. Lett., vol. 82, pp. 4428–4431, 1999

[99Kel] U. Keller, "Semiconductor nonlinearities for solid-state laser modelocking and Q-switching," in Chap. 4 in *Nonlinear Optics in Semiconductors*, E. Garmire, A. Kost, Eds. (Academic Press, Inc., Boston, 1998), vol. 59, pp. 211–286

[99Kra1] L. Krainer, R. Paschotta, J. Aus der Au, C. Hönninger, U. Keller, M. Moser, D. Kopf, K. J. Weingarten, "Passively mode-locked Nd:YVO_4 laser with up to 13 GHz repetition rate," Appl. Phys. B, vol. 69, pp. 245–247, 1999

[99Kra2] L. Krainer, R. Paschotta, G. Spühler, M. Moser, U. Keller, "29 GHz modelocked miniature Nd:YVO_4 laser," Electronics Lett., vol. 35, pp. 1160–1161, 1999

[99Kub] V. Kubecek, V. Couderc, B. Bourliaguet, F. Louradour, A. Barthelemy, "4-W and 23-ps pulses from a lamp-pumped Nd:YAG laser passively mode-locked by polarization switching in a KTP crystal," Appl. Phys. B, vol. 69, pp. 99–102, 1999

[99Kuz] M. Kuznetsov, F. Hakimi, R. Sprague, and A. Mooradian, "Design and Characteristics of High-Power (>0.5-W CW) Diode-Pumped Vertical-External-Cavity Surface-Emitting Semiconductor Lasers with Circular TEM_{00} Beams," IEEE J. Sel. Top. Quantum Electron., vol. 5, pp. 561–573, 1999

[99Lap] P. Laporta, S. Taccheo, S. Longhi, O. Svelto, and C. Svelto, "Erbium–ytterbium microlasers: optical properties and lasing characteristics," Opt. Materials, vol. 11, pp. 269–288, 1999

[99Led] M. J. Lederer, B. Luther-Davies, H. H. Tan, C. Jagadish, M. Haiml, U. Siegner, and U. Keller, "Nonlinear optical absorption and temporal response of arsenic- and oxygen-implanted GaAs," Appl. Phys. Lett., vol. 74, pp. 1993–1995, 1999

[99Lef] L. Lefort, K. Puech, S. D. Butterworth, Y. P. Svirko, D. C. Hanna, "Generation of femtosecond pulses from order-of-magnitude pulse compression in a synchronously pumped optical parametric oscillator based on periodically poled lithium niobate," Opt. Lett, vol. 24, pp. 28–30, 1999

[99Liu] Z. Liu, S. Izumida, S. Ono, H. Ohtake, and N. Sarukura, "High-repetition-rate, high-average-power, mode-locked Ti:sapphire laser with an intracavity continuous-wave amplification scheme," Appl. Phys. Lett., vol. 74, pp. 3622–3623, 1999

[99Liu2] X. Liu, L. Qian, and F. Wise, "High-energy pulse compression by use of negative phase shifts produced by the cascade $\chi(2)$: $\chi(2)$ nonlinearity," Opt. Lett. vol. 24, 1777–1779, 1999

[99Mat] N. Matuschek, F. X. Kärtner, U. Keller, "Analytical Design of Double-Chirped Mirrors with Custom-Tailored Dispersion Characteristics," IEEE J. Quantum Electron., vol. 35, pp. 129–137, 1999

[99Mat2] N. Matuschek: "Theory and design of double-chirped mirrors," Ph.D. Thesis, Swiss Federal Institute of Technology Zurich (1999), Hartung-Gorre Verlag, ISBN 3-89649-501-1

[99Mor1] U. Morgner, F. X. Kärtner, S. H. Cho, Y. Chen, H. A. Haus, J. G. Fujimoto, E. P. Ippen, V. Scheuer, G. Angelow, T. Tschudi, "Sub-two-cycle pulses from a Kerr-lens mode-locked Ti:sapphire laser," Opt. Lett., vol. 24, pp. 411–413, 1999

[99Mor2] U. Morgner, F. X. Kärtner, S. H. Cho, Y. Chen, H. A. Haus, J. G. Fujimoto, E. P. Ippen, V. Scheuer, G. Angelow, T. Tschudi, "Sub-two-cycle pulses from a Kerr-lens mode-locked Ti:sapphire laser: addenda," Opt. Lett., vol. 24, p. 920, 1999

[99Mou] F. Mougel, K. Dardenne, G. Aka, A. Kahn-Harari, D. Vivien, "Ytterbium-doped $Ca_4GdO(BO_3)_3$: an efficient infrared laser and self-frequency doubling crystal," J. Opt. Soc. Am. B, vol. 16, pp. 164–172, 1999

[99Pas] R. Paschotta, G. J. Spühler, D. H. Sutter, N. Matuschek, U. Keller, M. Moser, R. Hövel, V. Scheuer, G. Angelow, T. Tschudi, "Double-chirped semiconductor mirror for dispersion compensation in femtosecond lasers," Appl. Phys. Lett., vol. 75, pp. 2166–2168, 1999

[99Pas2] R. Paschotta, J. A. d. Au, U. Keller, "Strongly enhanced negative dispersion from thermal lensing or other focussing elements in femtosecond laser cavities," J. Opt. Soc. Am. B, vol. 17, pp. 646–651, 1999

[99Pre] G. Pretzler, A. Kasper, K. J. Witte, "Angular chirp and tilted pulses in CPA lasers," Appl. Phys. B, vol. 70, pp. 1–9, 2000

[99Qia] L. Qian, X. Liu, and F. Wise, "Femtosecond Kerr-lens mode locking with negative nonlinear phase shifts," Opt. Lett., vol. 24, pp. 166–168, 1999

[99Rei] J. Reichert, R. Holzwarth, T. Udem, and T. W. Hänsch, "Measuring the frequency of light with mode-locked lasers," Opt. Commun., vol. 172, pp. 59–68, 1999

[99Rop] A. Robertson, U. Ernst, R. Knappe, R. Wallenstein, V. Scheuer, T. Tschudi, D. Burns, M. D. Dawson, and A. I. Ferguson, "Prismless diode-pumped mode-locked femtosecond Cr:LiSAF laser," Opt. Comm., vol. 163, pp. 38–43, 1999

[99Sal] P. Salieres, A. L'Huillier, P. Antoine, and M. Lewenstein, "Study of the spatial and temporal coherence of high-order harmonics," in *Advances in Atomic, Molecular, and Optical Physics*, Academic Press ISBN 0-12-003841-2, p. 83, 1999

[99Sch] Schott Glass Technologies, Glass for Laser Applications, Data Sheets 1999

[99Sch2] S. Schön, H. Zogg, and U. Keller, "Growth of novel broadband high reflection mirrors by molecular beam epitaxy," J. Cryst. Growth, vol. 201/202, pp. 1020–1023, 1999

[99Sha] J. Shah, *Ultrafast Spectroscopy of Semiconductors and Semiconductor Nanostructures*, 2nd ed. (Springer-Verlag, Berlin 1999)

[99Shi] A. Shirakawa, I. Sakane, M. Takasaka, T. Kobayashi, "Sub-5-fs visible pulse generation by pulse-front-matched noncollinear optical parametric amplification," Appl. Phys. Lett., vol. 74, pp. 2268–2270, 1999

[99Sid] C. W. Siders, A. Cavalleri, K. Sokolowski-Tinten, C. Toth, T. Guo, M. Kammler, M. H. v. Hoegen, K. R. Wilson, D. v. d. Linde, C. P. J. Barty, "Detection of Nonthermal Melting by Ultrafast X-ray Diffraction," Science, vol. 286, pp. 1340–1342, 1999

[99Sie] U. Siegner, M. Haiml, F. Morier-Genoud, R. C. Lutz, P. Specht, E. R. Weber, U. Keller, "Femtosecond nonlinear optics of low-temperature grown semiconductors," Physica B: Condensed Matter, vol. 273–274, pp. 733–736, 1999

[99Spu1] G. J. Spühler, R. Paschotta, R. Fluck, B. Braun, M. Moser, G. Zhang, E. Gini, and U. Keller, "Experimentally confirmed design guidelines for passively Q-switched microchip lasers using semiconductor saturable absorbers," J. Opt. Soc. Am. B, vol. 16, pp. 376–388, 1999; Erratum J. Opt. Soc. Am. B, vol. 18, 886, 2001

[99Spu2] G. J. Spühler, L. Gallmann, R. Fluck, G. Zhang, L. R. Brovelli, C. Harder, P. Laporta, U. Keller, "Passively modelocked diode-pumped erbium–ytterbium glass laser using a semiconductor saturable absorber mirror," Electron. Lett., vol. 35, pp. 567–568, 1999

[99Spu3] G. J. Spühler, R. Paschotta, U. Keller, M. Moser, M. J. P. Dymott, D. Kopf, J. Meyer, K. J. Weingarten, J. D. Kmetec, J. Alexander, G. Truong, "Diode-pumped passively mode-locked Nd:YAG laser with 10-W average power in diffraction-limited beam," Opt. Lett., vol. 24, pp. 528–530, 1999

[99Ste] G. Steinmeyer, D. H. Sutter, L. Gallmann, N. Matuschek, U. Keller, "Frontiers in Ultrashort Pulse Generation: Pushing the Limits in Linear and Nonlinear Optics," Science, vol. 286, pp. 1507–1512, 1999

[99Sut] D. H. Sutter, G. Steinmeyer, L. Gallmann, N. Matuschek, F. Morier-Genoud, U. Keller, V. Scheuer, G. Angelow, T. Tschudi, "Semiconductor saturable-absorber mirror-assisted Kerr-lens mode-locked Ti:sapphire laser producing pulses in the two-cycle regime," Opt. Lett., vol. 24, pp. 631–633, 1999

[99Tan] H. H. Tan, C. Jagadish, "Role of implantation-induced defects on the response time of semiconductor saturable absorbers," Appl. Phys. Lett., vol. 75, pp. 1437–1439, 1999

[99Tel] H. R. Telle, G. Steinmeyer, A. E. Dunlop, J. Stenger, D. H. Sutter, U. Keller, "Carrier-envelope offset phase control: A novel concept for asolute optical frequency measurement and ultrashort pulse generation," Appl. Phys. B, vol. 69, pp. 327–332, 1999

[99Tho] E. R. Thoen, E. M. Koontz. M. Joschko, P. Langlois, T. R. Schibli, F. X. Kärtner, E. P. Ippen, and L. A. Kolodziejski, "Two-photon absorption in semiconductor saturable absorber mirrors," Appl. Phys. Lett., vol. 74, 3927–3929, 1999

[99Tsu] H. Tsuchida, "Pulse timing stabilization of a mode-locked Cr:LiSAF laser," Opt. Lett., vol. 24, pp. 1641–1643, 1999

[99Tsu2] H. Tsuchida, "Time-interval analysis of laser-pulse-timing fluctuations," Opt. Lett., vol. 24, pp. 1434–1436, 1999

[99Ude] T. Udem, J. Reichert, R. Holzwarth, and T. W. Hänsch, "Accurate measurement of large optical frequency differences with a mode-locked laser," Opt. Lett., vol. 24, pp. 881–883, 1999

[99Ude2] Th. Udem, J. Reichert, R. Holzwarth, T. W. Hänsch, "Absolute Optical Frequency Measurement of the Cesium D_1 Line with a Mode-Locked Laser," Phys. Rev. Lett., vol. 82, pp. 3568–3571, 1999

[99Uem] S. Uemura, K. Torizuka, "Generation of 12-fs pulses from a diode-pumped Kerr-lens mode-locked Cr:LiSAF laser," Opt. Lett., vol. 24, pp. 780–782, 1999

[99Zee] E. Zeek, K. Maginnis, S. Backus, U. Russek, M. Murnane, G. Mourou, H. Kapteyn, "Pulse compression by use of deformable mirrors," Opt. Lett., vol. 24, pp. 493–495, 1999

[99Zha] Z. Zhang, T. Nakagawa, K. Torizuka, T. Sugaya, K. Kobayashi, "Self-starting modelocked Cr^{4+}:YAG laser with a low-loss broadband semiconductor saturable-absorber mirror," Opt. Lett., vol. 24, pp. 1768–1770, 1999

[99Zha2] Zhang, Q., B. Ozygus, H. Weber, "Degeneration effects in laser cavities," Eur. Phys. J. AP, vol. 6, pp. 293–298, 1999

[00Apo] A. Apolonski, A. Poppe, G. Tempea, C. Spielmann, T. Udem, R. Holzwarth, T. W. Hänsch, and F. Krausz, "Controlling the phase evolution of few-cycle light pulses," Phys. Rev. Lett., vol. 85, p. 740, 2000

[00Aus] J. Aus der Au, G. J. Spühler, T. Südmeyer, R. Paschotta, R. Hövel, M. Moser, S. Erhard, M. Karszewski, A. Giesen, U. Keller, "16.2 W average power from a diode-pumped femtosecond Yb:YAG thin disk laser," Opt. Lett., vol. 25, pp. 859–861, 2000

[00Avr] E. A. Avrutin, J. H. Marsh, and E. L. Portnoi, "Monolithic and multi-gigahertz mode-locked semiconductor lasers: Constructions, experiments, models and applications," IEE Proc. Optoelectron., vol. 147, pp. 251–278, 2000

[00Bra] T. Brabec and F. Krausz, "Intense few-cycle laser fields: Frontiers of nonlinear optics," Rev. Mod. Phys., vol. 72, pp. 545–591, 2000

[00Bru] F. Brunner, G. J. Spühler, J. Aus der Au, L. Krainer, F. Morier-Genoud, R. Paschotta, N. Lichtenstein, S. Weiss, C. Harder, A. A. Lagatsky, A. Abdolvand, N. V. Kuleshov, U. Keller, "Diode-pumped femtosecond Yb:KGd$(WO_4)_2$ laser with 1.1-W average power," Opt. Lett., vol. 25, pp. 1119–1121, 2000

[00Bur] D. Burns, M. Hetterich, A. I. Ferguson, E. Bente, M. D. Dawson, J. I. Davies, S. W. Bland, "High-average-power (>20 W) Nd:YVO_4 lasers mode locked by strain-compensated saturable Bragg reflectors," J. Opt. Soc. Am. B, vol. 17, pp. 919–926, 2000

[00Car] T. J. Carrig, G. J. Wagner, A. Sennaroglu, J. Y. Jeong, and C. R. Pollock, "Mode-locked Cr^{2+}:ZnSe laser," Opt. Lett., vol. 25, pp. 168–170, 2000

[00Dem] A. A. Demidovich, A. N. Kuzmin, G. I. Ryabtsev, M. B. Danailov, W. Strek, A. N. Titov, "Influence of Yb concentration on Yb:KYW laser properties," J. Alloys and Compounds, vol. 300-301, pp. 238–241, 2000

[00Dor] C. Dorrer, N. Belabas, J. P. Likforman, and M. Joffre, "Spectral resolution and sampling issues in Fourier-transform spectral interferometry," J. Opt. Soc. Am. B, vol. 17, pp. 1795–1802, 2000

[00Dru] F. Druon, F. Balembois, P. Georges, A. Brun, A. Courjaud, C. Hönninger, F. Salin, A. Aron, F. Mougel, G. Aka, D. Vivien, "Generation of 90-fs pulses from a modelocked diode-pumped Yb:$Ca_4Gdo(Bo_3)_3$ laser," Opt. Lett., vol. 25, pp. 423–425, 2000

[00Fib] G. Fibich, A. L. Gaeta, "Critical power for self-focusing in bulk media and in hollow waveguides," Opt. Lett., vol. 25, pp. 335–337, 2000

[00Gal1] L. Gallmann, G. Steinmeyer, D. H. Sutter, N. Matuschek, and U. Keller, "Collinear type-II second-harmonic-generation frequency-resolved optical gating for the characterization of sub-10-fs optical pulses," Opt. Lett., vol. 25, pp. 269–271, 2000

[00Gal2] L. Gallmann, D. H. Sutter, N. Matuschek, G. Steinmeyer, and U. Keller, "Techniques for the Characterization of sub-10-fs Optical Pulses: A Comparison," Appl. Phys. B, vol. 70, pp. S67–S75, 2000

[00Hau] H. A. Haus, "Mode-Locking of Lasers," IEEE Journal on Sel. Top. in Quantum Electron., vol. 6, pp. 1173–1185, 2000

[00Hoo] S. Hoogland, S. Dhanjal, J. S. Roberts, A. C. Tropper, R. Häring, R. Paschotta, F. Morier-Genoud, U. Keller, "Passively modelocked diode-pumped surface-emitting semiconductor laser," IEEE Photonics Tech. Lett., vol. 12, pp. 1135–1137, 2000

[00Hub] R. Huber, A. Brodschelm, F. Tauser, A. Leitenstorfer, "Generation and field-resolved detection of femtosecond electromagnetic pulses tunable up to 41 THz," Appl. Phys. Lett., vol. 76, pp. 3191–3193, 2000

[00Ime] G. Imeshev, M. A. Arbore, S. Kasriel, and M. M. Fejer, "Pulse shaping and compression by second-harmonic generation with quasi-phase-matching gratings in the presence of arbitrary dispersion," J. Opt. Soc. Am. B, vol. 17, pp. 1420–1437, 2000

[00Jon] D. J. Jones, S. A. Diddams, J. K. Ranka, A. Stentz, R. S. Windeler, J. L. Hall, and S. T. Cundiff, "Carrier-envelope phase control of femtosecond mode-locked lasers and direct optical frequency synthesis," Science, vol. 288, pp. 635–639, 2000

[00Kra1] L. Krainer, R. Paschotta, M. Moser, U. Keller, "Passively modelocked picosecond lasers with up to 59 GHz repetition rate," Appl. Phys. Lett., vol. 77, pp. 2104–2105, 2000

[00Kra2] L. Krainer, R. Paschotta, M. Moser, U. Keller, "77 GHz soliton modelocked Nd:YVO_4 laser," Electron. Lett., vol. 36, pp. 1846–1848, 2000

[00Lar] M. A. Larotonda, A. A. Hnilo, and F. P. Diodati, "Diode-pumped self-starting Kerr-lens mode locking Nd: YAG laser," Opt. Commun., vol. 183, pp. 485–491, 2000

[00Li] C. Li, J. Song, D. Shen, N. S. Kim, J. Lu, K. Ueda, "Diode-pumped passively Q-switched ND:$GdVO_4$ lasers operating at 1.06 μm wavelength," App. Phys. B, vol. 70, p. 471, 2000

[00Mat] N. Matuschek, L. Gallmann, D. H. Sutter, G. Steinmeyer, U. Keller, "Back-side coated chirped mirror with ultra-smooth broadband dispersion characteristics," Appl. Phys. B, vol. 71, pp. 509–522, 2000

[00Meh] M. Mehendale, S. A. Mitchell, J. P. Likforman, D. M. Villeneuve, P. B. Corkum, "Method for single-shot measurement of the carrier envelope phase of a few-cycle laser pulse," Opt. Lett., vol. 25, pp. 1672–1674, 2000

[00Pai] M. Paillard, X. Marie, E. Vanelle, T. Amand, V. K. Kalevich, A. R. Kovsh, A. E. Zhukov, and V. M. Ustinov, "Time-resolved photoluminescence in self- assembled InAs/GaAs quantum dots under strictly resonant excitation," Appl. Phys. Lett., vol. 76, pp. 76–78, 2000

[00Pas] R. Paschotta, J. Aus der Au, G. J. Spühler, F. Morier-Genoud, R. Hövel, M. Moser, S. Erhard, M. Karszewski, A. Giesen, U. Keller, "Diode-pumped passively mode-locked lasers with high average power," Appl. Phys. B, vol. 70, pp. S25–S31, 2000

[00Pas2] R. Paschotta, J. A. der Au, U. Keller, "Thermal effects in high power end-pumped lasers with elliptical mode geometry," J. Sel. Topics in Quantum Electron., vol. 6, pp. 636–642, 2000

[00Pop] A. Poppe, R. Holzwarth, A. Apolonski, G. Tempea, C. Spielmann, T. W. Hänsch, F. Krausz, "Few-cycle optical waveform synthesis," Appl. Phys. B, vol. 72, pp. 373–376, 2000

[00Pre] G. Pretzler, A. Kasper, K. J. Witte, "Angular chirp and tilted pulses in CPA lasers," Appl. Phys. B, vol. 70, pp. 1–9, 2000

[00Ran] J. K. Ranka, R. S. Windeler, A. J. Stentz, "Visible continuum generation in air–silica microstructure optical fibers with anomalous dispersion at 800 nm," Opt. Lett., vol. 25, pp. 25–27, 2000

[00Rub] E. Rubiola and V. Giordano, "Correlation-based phase noise measurements," Rev. Sci. Instrum., vol. 71, pp. 3085–3091, 2000

[00Sch] S. Schön, M. Haiml, U. Keller, "Ultrabroadband AlGaAs/CaF_2 semiconductor saturable absorber mirrors," Appl. Phys. Lett., vol. 77, p. 782, 2000

[00Sch2] T. R. Schibli, E. R. Thoen, F. X. Kärtner, E. P. Ippen, "Suppression of Q-switched mode locking and break-up into multiple pulses by inverse saturable absorption," Appl. Phys. B, vol. 70, pp. S41–S49, 2000

[00Sie] U. Siegner, U. Keller, "Nonlinear optical processes for ultrashort pulse generation," in *Handbook of Optics*, Vol. III, M. Bass, E. W. Stryland, D. R. Williams, W. L. Wolfe, Eds. (McGraw-Hill, Inc., New York, 2000)

[00Spu] G. J. Spühler, T. Südmeyer, R. Paschotta, M. Moser, K. J. Weingarten, U. Keller, "Passively mode-locked high-power Nd:YAG lasers with multiple laser heads," Appl. Phys. B, vol. 71, pp. 19-25, 2000

[00Sut] D. H. Sutter, L. Gallmann, N. Matuschek, F. Morier-Genoud, V. Scheuer, G. Angelow, T. Tschudi, G. Steinmeyer, U. Keller, "Sub-6-fs pulses from a SESAM-assisted Kerr-lens modelocked Ti:sapphire laser: At the frontiers of ultrashort pulse generation," Appl. Phys. B, vol. 70, pp. S5–S12, 2000

[00Sut2] D. Sutter, "New Frontiers in Ultrashort Pulse Generation," Doktorarbeit, 28. Jan. 2000 (Hartung- Gore Verlag, Konstanz, 2000, ISBN 3-89649-537-2)

[00Sve] O. Svelto, *Principles of Lasers*, 4th edn (Plenum Press, New York)

[00Szi] R. Szipocs, A. Koházi-Kis, S. Lako, P. Apai, A. P. Kovács, G. DeBell, L. Mott, A. W. Louderback, A. V. Tikhonravov, and M. K. Trubetskov, "Negative Dispersion Mirrors for Dispersion Control in Femtosecond Lasers: Chirped Dielectric Mirrors and Multi-cavity Gires–Tournois Interferometers," Appl. Phys. B, vol. 70 (S1), pp. S51–S57, 2000

[00Tom] T. Tomaru, H. Petek, "Femtosecond Cr^{4+}:YAG laser with an L-fold cavity operating at a 1.2 GHz repetition rate," Opt. Lett., vol. 25, pp. 584–586, 2000

[00Tre] R. Trebino, *Frequency-Resolved Optical Gating: The Measurement of Ultrashort Laser Pulses*, Kluwer Academic Publishers, Boston, London 2000, ISBN 1-4020-7066-7

[00Ver1] F. Verluise, V. Laude, Z. Cheng, Ch. Spielmann, P. Tournois, "Amplitude and phase control of ultrashort pulses by use of an acousto-optic programmable dispersive filter: pulse compression and shaping," Opt. Lett., vol. 25, pp. 575–577, 2000

[00Ver2] F. Verluise, V. Laude, J-P. Huignard, P. Tournois, A. Migus, "Arbitrary dispersion control of ultrashort optical pulses with acoustic waves," J. Opt. Soc. Am. B, vol. 17, pp. 138–145, 2000

[00Wan] P. Wang, J. M. Dawes, P. Dekker, and J. A. Piper, "Highly efficient diode-pumped ytterbium-doped yttrium aluminum berate laser," Opt. Comm., vol. 174, pp. 467–470, 2000

[00Wei] A. M. Weiner, "Femtosecond pulse shaping using spatial light modulators," Rev. Sci. Instrum., vol. 71, no. 5, pp. 1929–1960, 2000

[00Zha] Z. G. Zhang, T. Nakagawa, H. Takada, K. Torizuka, T. Sugaya, T. Mirua, and K. Kobayashi, "Low-loss broadband semiconductor saturable absorber mirror for modelocked Ti:sapphire lasers," Opt. Commun., vol. 176, pp. 171–175, 2000

[01Agr] G. P. Agrawal, *Nonlinear Fiber Optics*, 3rd edn. (San Diego, CA: Academic Press, 2001)

[01Bau] M. Bauer, C. Lei, K. Read, R. Tobey, J. Gland, M. M. Murnane, H. C. Kapteyn, "Direct observation of surface chemistry using ultrafast soft-X-ray pulses," Phys. Rev. Lett., vol. 87, p. 025501, 2001

[01Ber] T. W. Berg, S. Bischoff, I. Magnusdottir, and J. Mork, "Ultrafast gain recovery and modulation limitations in self-assembled quantum-dot devices," IEEE Photon. Technol. Lett., vol. 13, pp. 541–543, 2001

[01Bru] F. Brunner, R. Paschotta, J. Aus der Au, G. J. Spühler, F. Morier-Genoud, R. Hövel, M. Moser, S. Erhard, M. Karszewski, A. Giesen, U. Keller, "Widely tunable pulse durations from a passively mode-locked thin-disk Yb:YAG laser," Opt. Lett., vol. 26, pp. 379–381, 2001

[01Che] Y. F. Chen, S. W. Tsai, Y. P. Lan, S. C. Wang, and K. F. Huang, "Diode-end-pumped passively mode-locked high-power Nd: YVO_4 laser with a relaxed saturable Bragg reflector," Opt. Lett., vol. 26, pp. 199–201, 2001

[01Che2] Y. F. Chen, K. f. Huang, S. W. Tsai, Y. P. Lan, S. C. Wang, J. Chen, "Simultaneous modelocking in a diode-pumped passively Q-switched Nd:YVO_4 laser with a GaAs saturable absorber," Appl. Opt., vol. 40, pp. 6038–6041, 2001

[01Chu] C. Chudoba, J. G. Fujimoto, E. P. Ippen, H. A. Haus, U. Morgner, F. X. Kärtner, V. Scheuer, G. Angelow, T. Tschudi, "All-solid-state Cr:forsterite laser generating 14-fs pulses at 1.3 μm," Opt. Lett., vol. 26, pp. 292–294, 2001

[01Cun] S. T. Cundiff, J. Ye, and J. L. Hall, "Optical frequency synthesis based on mode-locked lasers," Rev. Sci. Instrum., vol. 27, pp. 3750–3771, 2001

[01Dai] J. M. Dai, W. L. Zhang, L. Z. Zhang, L. Chai, Y. Wang, Z. G. Zhang, Q. R. Xing, C. Y. Wang, K. Torizuka, T. Nakagawa, and T. Sugaya, "A diode-pumped, self-starting, all-solid-state self-mode-locked Cr: LiSGAF laser," Optics and Laser Technology, vol. 33, pp. 71–73, 2001

[01Dre] M. Drescher, M. Hentschel, R. Kienberger, G. Tempea, C. Spielmann, G. A. Reider, P. B. Corkum, and F. Krausz, "X-ray pulses approaching the attosecond frontier," Science, vol. 291, pp. 1923–1927, 2001

[01Ell] R. Ell, U. Morgner, F. X. Kärtner, J. G. Fujimoto, E. P. Ippen, V. Scheuer, G. Angelow, T. Tschudi, M. J. Lederer, A. Boiko, B. Luther-Davis, "Generation of 5-fs pulses and octave-spanning spectra directly from a Ti:sapphire laser," Opt. Lett., vol. 26, pp. 373–375, 2001

[01Gae] A. L. Gaeta, "Nonlinear propagation and continuum generation in microstructure optical fibers," Opt. Lett., vol. 27, pp. 924–926, 2001

[01Gal] L. Gallmann, G. Steinmeyer, D. H. Sutter, T. Rupp, C. Iaconis, I. A. Walmsley, U. Keller, "Spatially resolved amplitude and phase characterization of femtosecond optical pulses," Opt. Lett., vol. 26, pp. 96–98, 2001

[01Gar] A. Garnache, S. Hoogland, A. C. Tropper, J. M. Gerard, V. Thierry-Mieg, J. S. Roberts, "Pico-second passively mode locked surface-emitting laser with self-assembled semiconductor quantum dot absorber" CLEO/Europe-EQEC, pp. Post-deadline Paper C-PSL 163, 2001

[01Han] B. T. Hansson and A. T. Friberg, "Eye-safe Q-switched microchip laser with an electro-absorbing semiconductor modulator," Opt. Lett., vol. 26, pp. 1057–1059, 2001

[01Har1] R. Häring, R. Paschotta, R. Fluck, E. Gini, H. Melchior, U. Keller, "Passively Q-switched microchip laser at 1.5 μm," J. Opt. Soc. Am. B, vol. 18, pp. 1805–1812, 2001

[01Har2] R. Häring, R. Paschotta, E. Gini, F. Morier-Genoud, H. Melchior, D. Martin, U. Keller, "Picosecond surface-emitting semiconductor laser with > 200 W average power," Electron. Lett., vol. 37, pp. 766–767, 2001

[01Har3] R. Häring, R. Paschotta, R. Fluck, E. Gini, H. Melchior, and U. Keller, "Passively Q-switched Microchip Laser at 1.5 μm," J. Opt. Soc. Am. B, vol. 18, pp. 1805–1812, 2001

[01Hel] F. W. Helbing, G. Steinmeyer, U. Keller, R. S. Windeler, J. Stenger, H. R. Telle, "Carrier-envelope offset dynamics of mode-locked lasers," Opt. Lett., vol. 27, pp. 194–196, 2001

[01Hen] M. Hentschel, R. Kienberger, Ch. Spielmann, G.A. Reider, N. Milosevic, T. Brabec, P. Corkum, U. Heinzmann, M. Drescher, F. Krausz, "Attosecond metrology," Nature, vol. 414, pp. 511–515, 2001

[01Hol] R. Holzwarth, M. Zimmermann, T. Udem, and T. W. Hänsch, "Optical clockworks and the measurement of laser frequencies with a mode-locked frequency comb," IEEE J. Quantum Electron., vol. 37, pp. 1493–1501, 2001

[01Juo] P. W. Juodawlkis, J. C. Twichell, J. L. Wasserman, G. E. Betts, and R. C. Williamson, "Measurement of mode-locked laser timing jitter by use of phase-encoded optical sampling," Opt. Lett, vol. 26, pp. 289–291, 2001

[01Kar] F. X. Kärtner, U. Morgner, R. Ell, T. Schibli, J. G. Fujimoto, E. P. Ippen, V. Scheuer, G. Angelow, T. Tschudi, "Ultrabroadband double-chirped mirror pairs for generation of octave spectra," JOSA B, vol. 18, pp. 882–885, 2001

[01Kem] A. J. Kemp, B. Stormont, B. Agate, C. T. A. Brown, U. Keller and W. Sibbett, "Gigahertz repetition-rates from a directly diode-pumped femtosecond Cr:LiSAF laser," Electron. Lett., vol. 37, pp. 1457–1458, 2001

[01Led] M. J. Lederer, V. Kolev, B. Luther-Davies, H. H. Tan, and C. Jagadish, "Ion-implanted InGaAs single quantum well semiconductor saturable absorber mirrors for passive mode-locking," J. Phys. D – Appl. Phys., vol. 34, pp. 2455–2464, 2001

[01Li] H. P. Li, C. H. Kam, Y. L. Lam, and W. Ji, "Femtosecond Z-scan measurements of nonlinear refraction in nonlinear optical crystals," Optical Materials, vol. 15, no. 4, pp. 237–242, 2001

[01Liu] H. Liu, J. Nees, G. Mourou, "Diode-pumped Kerr-lens mode-locked Yb:KY$(WO_4)_2$ laser," Opt. Lett., vol. 26, pp. 1723–1725, 2001

[01Liu2] C. Liu, Z. Dutton, C. H. Behroozi, L. V. Hau, "Observation of coherent optical information storage in an atomic medium using halted light pulses," Nature, vol. 409, pp. 490–493, 2001

[01Lu] W. Lu, L. Yan, and C. R. Menyuk, "Kerr-lens mode-locking of Nd:glass laser," Opt. Comm., vol. 200, pp. 159–163, 2001

[01Mor] U. Morgner, R. Ell, G. Metzler, T. R. Schibli, F. X. Kärtner, J. G. Fujimoto, H. A. Haus, E. P. Ippen, "Nonlinear optics with phase-controlled pulses in the sub-two-cycle regime," Phys. Rev. Lett., vol. 86, pp. 5462–5465, 2001

[01Pas] R. Paschotta, U. Keller, "Passive mode locking with slow saturable absorbers," Appl. Phys. B, vol. 73, pp. 653–662, 2001

[01Pas2] R. Paschotta, J. Aus der Au, G. J. Spühler, S. Erhard, A. Giesen, U. Keller, "Passive mode locking of thin disk lasers: effects of spatial hole burning," Appl. Phys. B, vol. 72, pp. 267–278, 2001

[01Pau] G. G. Paulus, F. Grasborn, H. Walther, P. Villoresi, M. Nisoli, S. Stagira, E. Priori, and S. D. Silvestri, "Absolute-phase phenomena in photoionization with few-cycle laser pulses," Nature, vol. 414, pp. 182–184, 2001

[01Pau2] P. M. Paul, E. S. Toma, P. Breger, G. Mullot, F. Augé, P. Balcou, H. G. Muller, P. Agostini, "Observation of a Train of Attosecond Pulses from High Harmonic generation," Science, vol. 292, pp. 1689–1692, 2001

[01Rou] A. Rousse, C. Rischel, S. Fourmaux, I. Uschmann, S. Sebban, G. Grillon, P. Balcou, E. Förster, J. P. Geindre, P. Audebert, J. C. Gauthier, D. Hulin, "Non-thermal melting in semiconductors measured at femtosecond resolution," Nature, vol. 410, pp. 65–68, 2001

[01Sag] I. Sagnes, G. L. Roux, C. Meriadec, A. Mereuta, G. Saint-Girons, and M. Bensoussan, "MOCVD InP/AlGaInAs distributed Bragg reflector for 1.55 μm VCSELs," Electr. Lett., vol. 37, pp. 500–501, 2001

[01Sch] T. R. Schibli, T. Kremp, U. Morgner, F. X. Kärtner, R. Butendeich, J. Schwarz, H. Schweizer, F. Scholz, J. Hetzler, and M. Wegener, "Continuous-wave operation and Q-switched mode locking of Cr^{4+}:YAG microchip lasers," Opt. Lett., vol. 26, pp. 941–943, 2001

[01Sco] R. P. Scott, C. Langrock, B. H. Kolner, "High-dynamic-range laser amplitude and phase noise measurement techniques," IEEE J. Sel. Topics Quantum Electron., vol. 7, no. 4, pp. 641–655, 2001

[01Spu1] G. J. Spühler, S. Reffert, M. Haiml, M. Moser, U. Keller, "Output-coupling semiconductor saturable absorber mirror," Appl. Phys. Lett., vol. 78, pp. 2733–2735, 2001

[01Spu2] G. J. Spühler, R. Paschotta, M. P. Kullberg, M. Graf, M. Moser, E. Mix, G. Huber, C. Harder, and U. Keller, "A passively Q-switched Yb:YAG microchip laser," Appl. Phys. B, vol. 72, pp. 285–287, 2001

[01Sud] T. Südmeyer, J. Aus der Au, R. Paschotta, U. Keller, P. G. R. Smith, G. W. Ross, D. C. Hanna, "Novel ultrafast parametric systems: high repetition rate single-pass OPG and fiber-feedback OPO," Journal of Physics D: Appl. Phys., vol. 34, pp. 2433–2439, 2001

[01Tem] G. Tempea, "Tilted-front-interface chirped mirrors," J. Opt. Soc. Am. B, vol. 18, pp. 1747–1750, 2001

[01Tom] T. Tomaru, "Two-element-cavity femtosecond Cr^{4+}:YAG laser operating at a 2.6-GHz repetition rate," Opt. Lett., vol. 26, pp. 1439–1441, 2001

[01Wal] I. A. Walmsley, L. Waxer, C. Dorrer, "The role of dispersion in optics," Rev. Sci. Instrum., vol. 72, 1, 2001

[01Was] G. Wasik, F. W. Helbing, F. König, A. Sizmann, G. Leuchs, "Bulk Er:Yb:glass soliton femtosecond laser," Conference on Lasers and Electro-Optics (CLEO) 2001, paper CMA4

[02Aga] B. Agate, B. Stormont, A. J. Kemp, C. T. A. Brown, U. Keller, W. Sibbett, "Simplified cavity designs for efficient and compact femtosecond Cr:LiSAF lasers," Opt. Commun., vol. 205, pp. 207–213, 2002

[02And] M. E. Anderson, L. E. E. de Araujo, E. M. Kosik, and I. A. Walmsley, "The effects of noise on ultrashort-optical-pulse measurement using SPIDER," Appl. Phys. B, vol. 70, pp. 85–93, 2000

[02Ash] S. Ashihara, J. Nishina, T. Shimura, and K. Kuroda, "Soliton compression of femtosecond pulses in quadratic media," J. Opt. Soc. Am. B, vol. 19, pp. 2505–2510, 2002

[02Bal] A. Baltuska, T. Fuji, T. Kobayashi, "Visible pulse compression to 4 fs by optical parametric amplification and programmable dispersion control," Opt. Lett., vol. 27, pp. 306–308, 2002

[02Bal2] A. Baltuska, T. Fuji, T. Kobayashi, "Controlling the carrier-envelope phase of ultrashort light pulses with optical parametric amplifiers," Phys. Rev. Lett., p. 133901, 2002

[02Bar] A. Bartels and H. Kurz, "Generation of a broadband continuum by a Ti:sapphire femtosecond oscillator with a 1-GHz repetition rate," Opt. Lett., vol. 27, pp. 1839–1841, 2002

[02Bau] A. Bauch and H. R. Telle, "Frequency standards and frequency measurement," Rep. Prog. Phys, vol. 65, pp. 789–843, 2002

[02Bru] F. Brunner, T. Südmeyer, E. Innerhofer, F. Morier-Genoud, R. Paschotta, V. E. Kisel, V. G. Shcherbitsky, N. V. Kuleshov, J. Gao, K. Contag, A. Giesen, U. Keller, "240-fs pulses with 22-W average power from a mode-locked thin-disk Yb:KY$(WO_4)_2$ laser," Opt. Lett., vol. 27, pp. 1162–1164, 2002

[02Con] C. Conti, S. Trillo, P. Di Trapani, J. Kilius, A. Bramati, S. Minardi, W. Chinaglia, and G. Valiulis, "Effective lensing effects in parametric frequency conversion," J. Opt. Soc. Am. B, vol. 19, p. 852, 2002

[02Cun] S. T. Cundiff, "Phase stabilization of ultrashort optical pulses," J. Phys. D: Appl. Phys., vol. 35, pp. R43–R59, 2002

[02Dor1] C. Dorrer, I. A. Walmsley, "Accuracy criterion for ultrashort pulse characterization techniques: application to spectral phase interferometry for direct electric field reconstruction," J. Opt. Soc. Am. B, vol. 19, pp. 1019–1029, 2002

[02Dor2] C. Dorrer, I. A. Walmsley, "Precision and consistency criteria in spectral phase interferometry for direct electric-field reconstruction," J. Opt. Soc. Am. B, vol. 19, pp. 1030–1038, 2002

[02Dre] M. Drescher, M. Hentschel, R. Kienberger, M. Uiberacker, V. Yakovlev, A. Scrinzi, Th. Westerwalbesloh, U. Kleineberg, U. Heinzmann, F. Krausz, "Time-resolved atomic inner-shell spectroscopy," Nature, vol. 419, pp. 803–807, 2002

[02Dru1] F. Druon, S. Chenais, P. Raybaut, F. Balembois, P. Georges, R. Gaume, G. Aka, B. Viana, S. Mohr, D. Kopf, "Diode-pumped Yb:$Sr_3Y(BO_3)_3$ femtosecond laser," Opt. Lett., vol. 27, pp. 197–199, 2002

[02Dru2] F. Druon, S. Chenais, F. Balembois, P. Georges, A. Brun, A. Courjaud, C. Honninger, F. Salin, M. Zavelani-Rossi, F. Auge, J. P. Chambaret, A. Aron, F. Mougel, G. Aka, and D. Vivien, "High-power diode-pumped Yb:GdCOB laser: from continuous-wave to femtosecond regime," Opt. Materials, vol. 19, pp. 73–80, 2002

[02Dud] J. M. Dudley and S. Coen, "Coherence properties of supercontinuum spectra generated in photonic crystal and tapered optical fibers," Opt. Lett., vol. 27, pp. 1180–1182, 2002

[02For1] T. M. Fortier, J. Ye, S. T. Cundiff, "Nonlinear phase noise generated in air–silica microstructure fiber and its effect on carrier-envelope phase," Opt. Lett., vol. 27, pp. 445–447, 2002

[02For2] T. M. Fortier, D. J. Jones, J. Ye, S. T. Cundiff, R. S. Windeler, "Long-term carrier-envelope phase coherence," Opt. Lett., vol. 27, pp. 1436–1438, 2002

[02Gae] A. L. Gaeta, "Nonlinear propagation and continuum generation in microstructured optical fibers," Opt. Lett., vol. 27, pp. 924–926, 2002

[02Gal] L. Gallmann, G. Steinmeyer, G. Imeshev, J.-P. Meyn, M. M. Fejer, U. Keller, "Sub-6-fs blue pulses generated by quasi phase-matching second-harmonic generation pulse compression," Appl. Phys. B, vol. 74, S237–S243, 2002

[02Gal2] L. Gallmann, *Generation and Characterization of few-femtosecond optical pulses*, Series in Quantum Electronics, vol. 28, Reprint of ETH Thesis No. 14475, ISBN 3-89649-715-4

[02Gar] A. Garnache, S. Hoogland, A. C. Tropper, I. Sagnes, G. Saint-Girons, and J. S. Roberts, "Sub-500-fs soliton pulse in a passively mode-locked broadband surface-emitting laser with 100-mW average power," Appl. Phys. Lett., vol. 80, pp. 3892–3894, 2002

[02Han] S. Han, W. Lu, B. Y. Sheh, L. Yan, M. Wraback, H. Shen, J. Pamulapati, and P. G. Newman, "Generation of sub-40 fs pulses from a mode-locked dual-gain-media Nd: glass laser," Appl. Phys. B, vol. 74, pp. S177–S179, 2002

[02Har] J. S. Harris, Jr., "GaInNAs long-wavelength lasers: progress and challenges," Semicond. Sci. Technol., vol. 17, pp. 880–891, 2002

[02Har2] R. Häring. R. Paschotta, A. Aschwanden, E. Gini, F. Morier-Genoud, U. Keller, "High power passively modelocked semiconductor laser," IEEE J. Quantum Electron., vol. 38, pp. 1268–1275, 2002

[02Hel1] F. W. Helbing, G. Steinmeyer, J. Stenger, H. R. Telle, U. Keller, "Carrier-envelope offset dynamics and stabilization of femtosecond pulses," Appl. Phys. B, vol. 74, S35–S42, 2002

[02Hel2] F. W. Helbing, G. Steinmeyer, U. Keller, R. S. Windeler, J. Stenger, and H. R. Telle, "Carrier-envelope offset dynamics of mode-locked lasers," Opt. Lett., vol. 27, pp. 194–196, 2002

[02Hop] J. M. Hopkins, G. J. Valentine, B. Agate, A. J. Kemp, U. Keller, and W. Sibbett, "Highly compact and efficient femtosecond Cr: LiSAF lasers," IEEE J. Quantum Electron., vol. 38, pp. 360–368, 2002

[02Kak] M. Kakehata, Y. Fujihira, H. Takada, Y. Kobayashi, K. Torizuka, T. Homma, H. Takahashi, "Measurements of carrier-envelope phase changes of 100-Hz amplified laser pulses," Appl. Phys. B, vol. 74, pp. S43–S50, 2002

[02Klo] P. Klopp, V. Petrov, U. Griebner, and G. Erbert, "Passively mode-locked Yb: KYW laser pumped by a tapered diode laser," Opt. Express, vol. 10, pp. 108–113, 2002

[02Kra1] L. Krainer, R. Paschotta, G. J. Spühler, I. Klimov, C. Teisset, K. J. Weingarten, U. Keller, "Tunable picosecond pulse-generating laser with a repetition rate exceeding 10 GHz," Electron. Lett., vol. 38, pp. 225–226, 2002

[02Kra2] L. Krainer, R. Paschotta, S. Lecomte, M. Moser, K. J. Weingarten, U. Keller, "Compact Nd:YVO_4 lasers with pulse repetition rates up to 160 GHz," IEEE J. Quantum Electron., vol. 38, pp. 1331–1338, 2002

[02Lec] S. Lecomte, L. Krainer, R. Paschotta, M. J. P. Dymott, K. J. Weingarten, and U. Keller, "Optical parametric oscillator with 10 GHz repetition rate and 100 mW average output power in the 1.5-μm spectral region," Opt. Lett., vol. 27, pp. 1714–1717, 2002

[02Led] M. J. Lederer, M. Hildebrandt, V. Z. Kolev, B. Luther-Davies, B. Taylor, J. Dawes, P. Dekker, J. Piper, H. H. Tan, and C. Jagadish, "Passive mode locking of a self-frequency-doubling Yb:$YAl_3(BO_3)_4$ laser," Opt. Lett., vol. 27, pp. 436–438, 2002

[02Leu] J. Leuthold, G. Raybon, Y. Su, R. Essiambre, S. Cabot, J. Jaques, M. Kauer, "40 Gbit/s transmission and cascaded all-optical wavelength conversion over 1 000 000 km," Electron. Lett. vol. 38, pp. 890–892, 2002

[02Maj1] A. Major, N. Langford, T. Graf, A. I. Ferguson, "Addititve-pulse mode locking of a diode-pumped Nd:KGd$(WO_4)_2$ laser," Appl. Phys. B, vol. 75, pp. 467–469, 2002

[02Maj2] A. Major, N. Langford, T. Graf, D. Burns, A. I. Ferguson, "Diode-pumped passively mode-locked Nd:KGd$(WO_4)_2$ laser with 1-W average output power," Opt. Lett., vol. 27, pp. 1478–1480, 2002

[02Maj3] A. Major, L. Giniunas, N. Langford, A. I. Ferguson, D. Burns, E. Bente, and R. Danielius, "Saturable Bragg reflector-based continuous-wave mode locking of Yb: KGd$(WO_4)_2$ laser," J. Mod. Opt., vol. 49, pp. 787–793, 2002

[02Pas] R. Paschotta, R. Häring, U. Keller, A. Garnache, S. Hoogland, and A. C. Tropper, "Soliton-like pulse formation mechanism in passively mode-locked surface-emitting semiconductor lasers," Appl. Phys. B, vol. 75, pp. 445–451, 2002

[02Pra] R. P. Prasankumar, C. Chudoba, J. G. Fujimoto, P. Mak, M. F. Ruane, "Self-starting mode locking in a Cr:forsterite laser by use of non-epitaxially-grown semiconductor-doped silica films," Opt. Lett., vol. 27, pp. 1564–1566, 2002

[02Puj] M. C. Pujol, M. A. Bursukova, F. Güell, X. Mateos, R. Sole, Jna. Gavalda, M. Aguilo, J. Massons, F. Diaz, P. Klopp, U. Griebner and V. Petrov, "Growth, optical characterization, and laser operation of a stoichiometric crystal KYb$(WO_4)_2$," Phys. Rev. B, vol. 65, p. 165121, 2002

[02Quo] F. Quochi, M. Dinu, N. H. Bonadeo, J. Shah, L. N. Pfeiffer, K. W. West, P. M. Platzman, "Ultrafast carrier dynamics of resonantly excited 1.3-μm InAs/GaAs self-assembled quantum dots," Physica B: Condens. Matter, vol. 314, pp. 263–267, 2002

[02Ram] T. M. Ramond, S. A. Diddams, L. Hollberg, A. Bartels, "Phase-coherent link from optical to microwave frequencies by means of the broadband continuum from a 1-GHz Ti:sapphire femtosecond oscillator," Opt. Lett., vol. 27, pp. 1842–1844, 2002

[02Rie] H. Riechert, A. Ramakrishnan, G. Steinle, "Development of InGaAsN-based 1.3 μm VCSELs," Semicond. Sci. Technol., vol. 17, pp. 892–897, 2002

[02Rip1] D. J. Ripin, C. Chudoba, J. T. Gopinath, J. G. Fujimoto, E. P. Ippen, U. Morgner, F. X. Kärtner, V. Scheuer, G. Angelow, T. Tschudi, "Generation of 20-fs pulses by a prismless Cr:YAG laser," Opt. Lett., vol. 27, pp. 61–63, 2002

[02Rip2] D. J. Ripin, J. T. Gopinath, H. M. Shen, A. A. Erchak, G. S. Petrich, L. A. Kolodziejski, F. X. Kärtner, and E. P. Ippen, "Oxidized GaAs/AlAs mirror with a quantum-well saturable absorber for ultrashort-pulse Cr^{4+}:YAG laser," Opt. Comm., vol. 214, pp. 285–289, 2002

[02Rot] U. Roth and J. E. Balmer, "Neodymium: YLF lasers at 1053 nm passively mode locked with a saturable Bragg reflector," Appl. Optics, vol. 41, pp. 459–463, 2002

[02Sch1] S. Schön, M. Haiml, L. Gallmann, U. Keller, "Fluoride semiconductor saturable-absorber mirror for ultrashort pulse generation," Opt. Lett., vol. 27, pp. 1845–1847, 2002

[02Sch2] O. Schmidt, M. Bauer, C. Wiemann, R. Porath, M. Scharte, O. Andreyev, G. Schönhense, M. Aeschlimann, "Time-resolved two-photon photoemission electron microscopy," Appl. Phys. B, vol. 74, pp. 223–227, 2002

[02She] D. Y. Shen, D. Y. Tang, and K. Ueda, "Continuous wave and Q-Switched mode-locking of a Nd: YVO_4 laser with a single crystal GaAs wafer," Jpn. J. Appl. Phys. Part 2, vol. 41, pp. L1224–L1227, 2002

[02She2] R. K. Shelton, S. M. Foreman, L.-S. Ma, J. L. Hall, H. C. Kapteyn, M. M. Murnane, M. Notcutt, J. Ye, "Subfemtosecond timing jiter between two independent, actively synchronized, mode-locked lasers," Opt. Lett., vol. 27, pp. 312–314, 2002

[02Spu] G. J. Spühler, M. Dymott, I. Klimov, G. Luntz, L. Baraldi, I. Kilburn, P. Crosby, S. Thomas, O. Zehnder, C. Y. Teisset, M. Brownell, K. J. Weingarten, R. Dangel, B. J. Offrein, G. L. Bona, O. Buccafusca, Y. Kaneko, L. Krainer, R. Paschotta, U. Keller, "40 GHz pulse generating source with sub-picosecond timing jitter," Electron. Lett., vol. 38, pp. 1031–1033, 2002

[02Sun] H. D. Sun, G. J. Valentine, R. Macaluso, S. Calvez, D. Burns, M. D. Dawson, T. Jouhti, and M. Pessa, "Low-loss 1.3 μm GaInNAs saturable Bragg reflector for high-power picosecond neodymium lasers," Opt. Lett., vol. 27, pp. 2124–2126, 2002

[02Ude] T. Udem, R. Holzwarth, T. W. Hänsch, "Optical frequency metrology," Nature, vol. 416, pp. 233–237, 2002

[02Wag] P. C. Wagenblast, U. Morgner, F. Grawert, T. R. Schibli, F. X. Kärtner, V. Scheuer, G. Angelow, M. J. Lederer, "Generation of sub-10-fs pulses from a Kerr-lens mode-locked Cr:LiCAF laser oscillator by use of third-order dispersion-compensating double-chirped mirrors," Opt. Lett., vol. 27, pp. 1726–1728, 2002

[02Zav] M. Zavelani-Rossi, D. Polli, G. Cerullo, S. De Silvestri, L. Gallmann, G. Steinmeyer, U. Keller, "Few-optical-cycle laser pulses by OPA: broadband chirped mirror compression and SPIDER characterization," Appl. Phys. B, vol. 74, S245–S251, 2002

[03Del] P. J. Delfyett, "Ultrafast single- and multiwavelength modelocked semiconductor lasers: physics and applications," in *Ultrafast Lasers: Technology and Applications*, eds. M. E. Fermann, A. Galvanauskas, G. Sucha (Marcel Dekker Inc., New York, 2003) pp. 219–321

[03Don] J. Dong, M. Bass, Y. Mao, P. Deng, and F. Gan, "Dependence of the Yb^{3+} emission cross section and lifetime on temperature and concentration in yttrium aluminum garnet," J. Opt. Soc. Am. B vol. 20, pp. 1975–1979, 2003

[03Dru] F. Druon, F. Balembois, and P. Georges, "Laser crystals for the production of ultra-short laser pulses," Ann. Chim. (Paris), vol. 28, pp. 47–72, 2003

[03Fer] M. E. Fermann, "Ultrafast fiber oscillators," in *Ultrafast Lasers: Technology and Applications*, eds. M. E. Fermann, A. Galvanauskas, G. Sucha (Marcel Dekker Inc., New York, 2003) pp. 89–154

[03For] T. M. Fortier, D. J. Jones, and S. T. Cundiff, "Phase stabilization of an octave-spanning Ti:sapphire laser," Opt. Lett., vol. 28, no. 22, pp. 2198–2200, 2003

[03Gal] A. Galvanauskas, "Ultrashort-pulse fiber amplifiers," in *Ultrafast Lasers: Technology and Applications*, eds. M. E. Fermann, A. Galvanauskas, G. Sucha (Marcel Dekker Inc., New York, 2003) pp. 155–217

[03Gar] J. Garduno-Mejia, E. Ramsay, A. Greenaway, and D. T. Reid, "Real time femtosecond optical pulse measurement using a video-rate frequency-resolved optical gating system," Rev. Sci. Instr., vol. 74, pp. 3624–3627, 2003

[03Har] A. Härkönen, T. Jouhti, N. V. Tkachenko, H. Lemmetyinen, B. Ryvkin, O. G. Okhotnikov, T. Sajavaara, and J. Keinonen, "Dynamics of photoluminescence in GaInNAs saturable absorber mirrors," Appl. Phys. A, vol. 77, pp. 861–863, 2003

[03He] J. L. He, C. K. Lee, J. Y. J. Huang, S. C. Wang, C. L. Pan, and K. F. Huang, "Diode-pumped passively mode-locked multiwatt Nd:$GdVO_4$ laser with a saturable Bragg reflector," Appl. Optics, vol. 42, pp. 5496-5499, 2003

[03Hel] F. W. Helbing, G. Steinmeyer, U. Keller, "Carrier-envelope offset phase-locking with attosecond timing jitter," IEEE J. Sel. Top. Quantum Electron., vol. 9, pp. 1030–1040, 2003

[03Hol] K.W. Holman, R.J. Jones, A. Marian, S.T. Cundiff, Y. Je, "Detailed studies and control of intensity-related dynamics of femtosecond frequency combs from mode-locked Ti:sapphire lasers," IEEE J. of Sel. Top. in Quantum Electron., vol. 9, Nr. 4, pp. 1018–1024, 2003

[03Hoo] S. Hoogland, A. Garnache, I. Sagnes, B. Paldus, K. J. Weingarten, R. Grange, M. Haiml, R. Paschotta, U. Keller, A. C. Tropper, "Picosecond pulse generation with a 1.5-μm passively mode-locked surface-emitting semiconductor laser," Electron. Lett., vol. 39, no. 11, pp. 846–847, 2003

[03Inn] E. Innerhofer, T. Südmeyer, F. Brunner, R. Häring, A. Aschwanden, R. Paschotta, C. Hönninger, M. Kumkar, U. Keller, "60-W average power in 810-fs pulses from a thin-disk Yb:YAG laser," Opt. Lett., vol. 28, pp. 367–369, 2003

[03Iva] E. N. Ivanov, S. A. Diddams, L. Hollberg, "Experimental study of noise properties of a Ti:sapphire femtosecond laser," IEEE Trans. Ultrason., Ferroelectr., Freq. Control, vol. 45, pp. 355–360, 2003

[03Jas] K. Jasim, Q. Zhang, A. V. Nurmikko, A. Mooradian, G. Carey, W. Ha, and E. Ippen, "Passively modelocked vertical extended cavity surface emitting diode laser," Electron. Lett., vol. 39, pp. 373–375, 2003

[03Kan] I. Kang, C. Dorrer, and F. Quochi, "Implementation of electro-optic spectral shearing interferometry for ultrashort pulse characterization," Opt. Lett., vol. 28, pp. 2264–2266, 2003

[03Kel] U. Keller, "Recent developments in compact ultrafast lasers," Nature, vol. 424, pp. 831–838, 2003

[03Kol] V. Z. Kolev, M. J. Lederer, B. Luther-Davies, and A. V. Rode, "Passive mode locking of a Nd:YVO_4 laser with an extra-long optical resonator," Opt. Lett., vol. 28, pp. 1275–1277, 2003

[03Kor] W. Kornelis, J. Biegert, J. W. G. Tisch, M. Nisoli, G. Sansone, S. De Silvestri, U. Keller, "Single-shot kilohertz characterization of ultrashort pulses by spectral phase interferometry for direct electric-field reconstruction," Opt. Lett., vol. 28, No. 4, pp. 281–283, 2003

[03Kow] A. M. Kowalevicz, A. T. Zare, F. X. Kartner, J. G. Fujimoto, S. Dewald, U. Morgner, V. Scheuer, and G. Angelow, "Generation of 150-nJ pulses from a multiple-pass cavity Kerr-lens mode-locked Ti:Al_2O_3 oscillator," Opt. Lett., vol. 28, pp. 1597–1599, 2003

[03Kul] Private communication with N. V. Kuleshov

[03Lag1] A. A. Lagatsky, E. U. Rafailov, C. G. Leburn, C. T. A. Brown, N. Xiang, O. G. Okhotnikov, and W. Sibbett, "Highly efficient femtosecond Yb:KYW laser pumped by single narrow-stripe laser diode," Electron. Lett., vol. 39, pp. 1108–1110, 2003

[03Lag2] A. A. Lagatsky, C. G. Leburn, C. T. A. Brown, W. Sibbett, and W. H. Knox, "Compact self-starting femtosecond Cr^{4+}:YAG laser diode pumped by a Yb-fiber laser," Opt. Comm., vol. 217, pp. 363–367, 2003

[03Lim] J. Limpert, T. Clausnitzer, A. Liem, T. Schreiber, H.-J. Fuchs, H. Zellmer, E.-B. Kley, and A. Tünnermann, "High-average-power femtosecond fiber chirped-pulse amplification system," Opt. Lett, vol. 28, pp. 1984–1986, 2003

[03Maj] A. Major, N. Langford, S. T. Lee, and A. I. Ferguson, "Additive-pulse mode locking of a thin-disk Yb:YAG laser," Appl. Phys. B, vol. 76, pp. 505–508, 2003

[03McI] J. G. McInerney, A. Mooradian, A. Lewis, A. V. Shchegrov, E. M. Strzelecka, D. Lee, J. P. Watson, M. Liebman, G. P. Carey, B. D. Cantos, W. R. Hitchens, and D. Heald, "High-power surface emitting semiconductor laser with extended vertical compound cavity," Electron. Lett., vol. 39, pp. 523–525, 2003

[03Nau] S. Naumov, E. Sorokin, V. L. Kalashnikov, G. Tempea, and I. T. Sorokina, "Self-starting five optical cycle pulse generation in Cr^{4+}:YAG laser," Appl. Phys. B, vol. 76, pp. 1–11, 2003

[03Okh] O. G. Okhotnikov, T. Jouhti, J. Konttinen, S. Karirinne, and M. Pessa, "1.5 μm monolithic GaInNAs semiconductor saturable-absorber mode locking of an erbium fiber laser," Opt. Lett., vol. 28, pp. 364–366, 2003

[03Pap] D. N. Papadopoulos, S. Forget, M. Delaigue, F. Druon, F. Balembois, and P. Georges, "Passively mode-locked diode-pumped Nd: YVO_4 oscillator operating at an ultralow repetition rate," Opt. Lett., vol. 28, pp. 1838–1840, 2003

[03Pas] R. Paschotta and U. Keller, "Ultrafast solid-state lasers," in *Ultrafast Lasers: Technology and Applications*, eds. M. E. Fermann, A. Galvanauskas, G. Sucha (Marcel Dekker Inc., New York, 2003) pp. 1–60

[03Pra] R. P. Prasankumar, Y. Hirakawa, A. M. Kowalevicz, F. X. Kaertner, J. G. Fujimoto, and W. H. Knox, "An extended cavity femtosecond Cr:LiSAF laser pumped by low cost diode lasers," Opt. Express, vol. 11, pp. 1265–1269, 2003

[03Qin] L. Qin, X. Meng, C. Du, L. Zhu, B. Xu, Z. Shao, Z. Liu, Q. Fang, R. Cheng, "Growth and properties of mixed crystal Nd:YGdVO$_4$," J. Alloys Compd., vol. 354, pp. 259–262, 2003
[03Rus] P. Russell, "Photonic Crystal Fibers," Science, vol. 299, pp. 358–362 (2003)
[03Saa] F. Saadallah, N. Yacoubi, F. Genty, and C. Alibert, "Photothermal investigations of thermal and optical properties of GaAlAsSb and AlAsSb thin layers," J. of Appl. Phys., vol. 94, pp. 5041–5048, 2003
[03Sch] B. Schenkel, J. Biegert, U. Keller, C. Vozzi, M. Nisoli, G. Sansone, S. Stagira, S. D. Silvestri, and O. Svelto, "Generation of 3.8-fs pulses from adaptive compression of a cascaded hollow fiber supercontinuum," Opt. Lett., vol. 28, pp. 1987–1989, 2003
[03Sch2] T. R. Schibli, J. Kim, O. Kuzucu, J. T. Gopinath, S. N. Tandon, G. S. Petrich, L. A. Kolodziejski, J. G. Fujimoto, E. P. Ippen, and F. X. Kaertner, "Attosecond active synchronization of passively mode-locked lasers by balanced cross correlation," Opt. Lett., vol. 28, pp. 947–949, 2003
[03Sei] W. Seitz, R. Ell, U. Morgner, and F. X. Kaertner, "All-optical synchronization and mode locking of solid-state lasers with nonlinear semiconductor Fabry–Perot mirrors," IEEE J. Sel. Top. Quan. Electron., vol. 9, pp. 1093–1101, 2003
[03Spu] G. J. Spühler, P. S. Golding, L. Krainer, I. J. Kilburn, P. A. Crosby, M. Brownell, K. J. Weingarten, R. Paschotta, M. Haiml, R. Grange, U. Keller, "Novel multi-wavelength source with 25 GHz channel spacing tunable over the C-band," Electron. Lett., vol. 39, pp. 778–780, 2003
[03Sud] T. Südmeyer, F. Brunner, E. Innerhofer, R. Paschotta, K. Furusawa, J. C. Baggett, T. M. Monro, D. J. Richardson, and U. Keller, "Nonlinear femtosecond pulse compression at high average power levels using a large mode area holey fiber," Opt. Lett., vol. 28, pp. 951–953, 2003
[03Ste] G. Steinmeyer, "Brewster-angled chirped mirrors for high-fidelity dispersion compensation and bandwidths exceeding one optical octave," Opt. Express, vol. 11, pp. 2385–2396, 2003
[03Sto] B. Stormont, I. G. Cormack, M. Mazilu, C. T. A. Brown, D. Burns, and W. Sibbett, "Low-threshold, multi-gigahertz repetition-rate femtosecond Ti:sapphire laser," Electron. Lett., vol. 39, pp. 1820–1822, 2003
[03Tom] T. Tomaru, "Mode-locking operating points of a three-element-cavity femtosecond Cr^{4+}:YAG laser," Opt. Comm., vol. 225, pp. 163–175, 2003
[03Uem] S. Uemura, K. Torizuka, "Development of a diode-pumped Kerr-lens mode-locked Cr:LiSAF laser," IEEE J. Quant. Electron., vol. 39, pp. 68–73, 2003
[03Wag] P. Wagenblast, R. Ell, U. Morgner, F. Grawert, and F. X. Kartner, "Diode-pumped 10-fs Cr^{3+}:LiCAF laser," Opt. Lett., vol. 28, pp. 1713–1715, 2003
[03Yam] K. Yamane, Z. Zhang, K. Oka, R. Morita, and M. Yamashita, "Optical pulse compression to 3.4 fs in the monocycle region by feedback phase compensation," Opt. Lett., vol. 28, pp. 2258–2260, 2004
[03Zay] J. J. Zayhowski and A. L. Wilson, "Pump-induced bleaching of the saturable absorber in short-pulse ND:YAG/Cr^{4+}:YAG passively Q-switched microchip lasers," IEEE J. Quant. Electron., vol. 39, pp. 1588–1593, 2003
[03Zel] S. C. Zeller, L. Krainer, G. J. Spühler, K. J. Weingarten, R. Paschotta, and U. Keller, "Passively modelocked 40-GHz Er:Yb:glass laser," Appl. Phys. B, vol. 76, pp. 1181–1182, 2003
[03Zha] B. Zhang, L. Gang, M. Chen, Z. Zhang, and Y. Wang, "Passive mode locking of a diode-end-pumped Nd:GdVO$_4$ laser with a semiconductor saturable absorber mirror," Opt. Lett., vol. 28, pp. 1829–1831, 2003
[04Agi] Agilent Spectrum Analysis Basics, Application Note 150, 2004
[04Ago] P. Agostini and L. F. DiMauro, "The physics of attosecond light pulses," Rep. Prog. Phys., vol. 67, pp. 813–855, 2004
[04Bal] N. Balkan,"The physics and technology of dilute nitrides," Special Issue in J. Phys. Condens. Matter, vol. 16, No. 31, 2004

[04Bru] F. Brunner, E. Innerhofer, S. V. Marchese, T. Südmeyer, R. Paschotta, T. Usami, H. Ito, S. Kurimura, K. Kitamura, G. Arisholm, and U. Keller, "Powerful red-green-blue laser source pumped with a mode-locked thin disk laser," Opt. Lett., vol. 29, pp. 1921–1923, 2004

[04Bur] O. A. Buryy, S. B. Ubiszkii, S. S. Melnyk, and A. O. Matkovskii, "The Q-switched Nd:YAG and Yb:YAG microchip lasers optimization and comparative analysis," Appl. Phys. B, vol. 78, pp. 291–297, 2004

[04Buy] I. A. Buyanova, W. M. Chen, C. W. Tu, "Defects in dilute nitrides," J. Phys.: Condens. Matter, vol. 16, pp. S3027–S3035, 2004

[04Chi] J. Chilla, S. Butterworth, A. Zeitschel, J. Charles, A. Caprara, M. Reed, and L. Spinelli, "High Power Optically Pumped Semiconductor Lasers," presented at Photonics West 2004, Solid State Lasers XIII: Technology and Devices, in Proc. SPIE 5332, 2004

[04Dru] F. Druon, F. Balembois, and P. Georges, "Ultra-short-pulsed and highly-efficient diode-pumped Yb:SYS mode-locked oscillators," Opt. Express, vol. 12, pp. 5005–5012, 2004

[04Ern] C. Erny, G. J. Spühler, L. Krainer, R. Paschotta, K. J. Weingarten, and U. Keller, "Simple repetition rate tunable picosecond pulse-generating 10 GHz laser," Electron. Lett., vol. 40, pp. 877–878, 2004

[04Fer] A. Fernandez, T. Fuji, A. Poppe, A. Fürbach, F. Krausz, and A. Apolonski, "Chirped-pulse oscillators: a route to high-power femtosecond pulses without external amplification," Opt. Lett., vol. 29, pp. 1366–1368, 2004

[04Gra] R. Grange, O. Ostinelli, M. Haiml, L. Krainer, G. J. Spühler, S. Schön, M. Ebnöther, E. Gini, and U. Keller, "Antimonide semiconductor saturable absorber for 1.5-μm," Electron. Lett., vol. 40, pp. 1414–1415, 2004

[04Gri] U. Griebner, V. Petrov, K. Petermann, and V. Peters, "Passively mode-locked Yb:Lu_2O_3 laser," Opt. Express, vol. 12, pp. 3125–3130, 2004

[04Hai] M. Haiml, R. Grange, and U. Keller, "Optical characterization of semiconductor saturable absorbers," Appl. Phys. B, vol. 79, pp. 331–339, 2004

[04Hau] C. P. Hauri, W. Kornelis, F. W. Helbing, A. Heinrich, A. Courairon, A. Mysyrowicz, J. Biegert, U. Keller, "Generation of intense, carrier-envelope phase-locked few-cycle laser pulses through filamentation," Appl. Phys. B, vol. 79, pp. 673–677, 2004

[04Hau2] C. P. Hauri, P. Schlup, G. Arisholm, J. Biegert, U. Keller, "Phase-preserving chirped-pulse optical parametric amplification to 17.3 fs directly from a Ti:sapphire oscillator," Opt. Lett., vol. 29, pp. 1369–1371, 2004

[04He] J.-L. He, Y.-X. Fan, J. Du, Y.-G. Wang, S. Liu, H.-T. Wang, L.-H. Zhang, and Y. Hang, "4-ps passively mode-locked Nd:$Gd_{0.5}Y_{0.5}YVO_4$ laser with a semiconductor saturable-absorber mirror," Opt. Lett, vol. 29, pp. 2803–2805, 2004

[04Ild] F. O. Ilday, K. Beckwitt, Y.-F. Chen, H. Lim, and F. W. Wise, "Controllable Raman-like nonlinearities from nonstationary, cascaded quadratic processes," J. Opt. Soc. Am. B, vol. 21, pp. 376–383, 2004

[04Jac] M. Jacquemet, F. Balembois, S. Chenais, F. Druon, P. Georges, R. Gaume, and B. Ferrand, "First diode-pumped Yb-doped solid-state laser continuously tunable between 1000 and 1010 nm," Appl. Phys. B, vol. 78, pp. 13–18, 2004

[04Jas] K. Jasim, Q. Zhang, A. V. Nurmikko, E. Ippen, A. Mooradian, G. Carey, and W. Ha, "Picosecond pulse generation from passively modelocked vertical cavity diode laser at up to 15 GHz pulse repetition rate," Electron. Lett., vol. 40, pp. 34–35, 2004

[04Jen] S. Jensen and M. E. Anderson, "Measuring ultrashort optical pulses in the presence of noise: an empirical study of the performance of spectral phase interferometry for direct electric field reconstruction," Appl. Opt., vol. 43, pp. 883–893, 2004

[04Kel] U. Keller, "Ultrafast solid-state lasers," Prog. Opt., vol. 46, pp. 1–115, 2004 (ISBN 0 444 51468 6, Elsevier, 2004)

[04Kil] A. Killi, U. Morgner, M. J. Lederer, D. Kopf, "Diode-pumped femtosecond laser oscillator with cavity dumping," Opt. Lett., vol. 29, pp. 1288–1290, 2004

[04Kis] V. E. Kisel, A. E. Troshin, N. A. Tolstik, V. G. Shcherbitsky, N. V. Kuleshov, V. N. Matrosov, T. A. Matrosova, and M. I. Kupchenko, "Spectroscopy and continuous-wave diode-pumped laser action of Yb^{3+}:YVO_4," Opt. Lett., vol. 29, pp. 2491–2493, 2004

[04Kol] M. Kolesik, J. V. Moloney, "Nonlinear optical pulse propagation simulation: from Maxwell's to unidirectional equations," Phys. Rev. E, vol. 70, 036604, 2004

[04Kon] J. Kong, D. Y. Tang, S. P. Ng, B. Zhao, L. J. Qin, and X. L. Meng, "Diode-pumped passively mode-locked Nd:$GdVO_4$ laser with a GaAs saturable absorber mirror," Appl. Phys. B, vol. 79, pp. 203–206, 2004

[04Kra] L. Krainer, D. Nodop, G. J. Spühler, S. Lecomte, M. Golling, R. Paschotta, D. Ebling, T. Ohgoh, T. Hayakawa, K. J. Weingarten, and U. Keller, "Compact 10-GHz Nd:$GdVO_4$ laser with 0.5-W average output power and low timing jitter," Opt. Lett., vol. 29, pp. 2629–2631, 2004

[04Kra2] C. Kränkel, D. Fagundes-Peters, S. T. Friedrich, J. Johannsen, M. Mond, G. Huber, M. Bernhagen, and R. Uecker, "Continuous wave laser operation of Yb^{3+}:YVO_4," App. Phys. B, vol. 79, pp. 543–546, 2004

[04Lag1] A. A. Lagatsky, C. T. A. Brown, and W. Sibbett, "Highly efficient and low threshold diode-pumped Kerr-lens mode-locked Yb:KYW laser," Opt. Express, vol. 12, pp. 3928–3933, 2004

[04Lag2] A. A. Lagatsky, C. G. Leburn, C. T. A. Brown, W. Sibbett, A. M. Malyarevich, V. G. Savitski, K. V. Yumashev, E. L. Raaben, and A. A. Zhilin, "Passive mode locking of a $Cr^{(4+)}$: YAG laser by PbS quantum-dot-doped glass saturable absorber," Opt. Comm., vol. 241, pp. 449–454, 2004

[04Lec] S. Lecomte, R. Paschotta, M. Golling, D. Ebling, and U. Keller, "Synchronously pumped optical parametric oscillators in the 1.5-μm spectral region with a repetition rate of 10 GHz," J. Opt. Soc. Am. B, vol. 21, pp. 844–850, 2004

[04Liv] V. Liverini, S. Schön, R. Grange, M. Haiml, S. C. Zeller, and U. Keller, "A low-loss GaInNAs SESAM mode-locking a 1.3-μm solid-state laser," Appl. Phys. Lett., vol. 84, pp. 4002–4004, 2004

[04Lor] D. Lorenser, H. J. Unold, D. J. H. C. Maas, A. Aschwanden, R. Grange, R. Paschotta, D. Ebling, E. Gini, and U. Keller, "Towards Wafer-Scale Integration of High Repetition Rate Passively Mode-Locked Surface-Emitting Semiconductor Lasers," Appl. Phys. B, vol. 79, pp. 927–932, 2004

[04Luc] A. Lucca, M. Jacquemet, F. Druon, F. Balembois, P. Georges, P. Camy, J. L. Doualan, and R. Moncorge, "High-power tunable diode-pumped Yb^{3+}:CaF_2 laser," Opt. Lett., vol. 29, pp. 1879–1881, 2004

[04Lut] B. Luther-Davies, V. Z. Kolev, M. J. Lederer, N. R. Madsen, A. V. Rode, J. Giesekus, K. M. Du, and M. Duering, "Table-top 50-W laser system for ultra-fast laser ablation," Appl. Phys. A, vol. 79, pp. 1051–1055, 2004

[04Maj] A. Major, F. Yoshino, I. Nikolakakos, J. S. Aitchison, P. W. E. Smith, "Dispersion of the nonlinear refractive index in sapphire," Opt. Lett., vol. 29, pp. 602–604, 2004

[04Nau] S. Naumov, E. Sorokin, and I. T. Sorokina, "Directly diode-pumped Kerr-lens mode-locked Cr^{4+}:YAG laser," Opt. Lett., vol. 29, pp. 1276–1278, 2004

[04Osb] S. W. Osborne, P. Blood, P. M. Smowton, Y. C. Xin, A. Stintz, D. Huffaker, and L. F. Lester, "Optical absorption cross section of quantum dots," J. Phys.: Condens. Matter, vol. 16, pp. S3749–S3756, 2004

[04Ost] O. Ostinelli, G. Almuneau, M. Ebnöther, E. Gini, M. Haiml, and W. Bächtold, "MOVPE growth of long wavelength AlGaAsSb/InP Bragg mirrors," Electron. Lett., vol. 40, pp. 940–942, 2004

[04Pas] R. Paschotta, L. Krainer, S. Lecomte, G. J. Spühler, S. C. Zeller, A. Aschwanden, D. Lorenser, H. J. Unold, K. J. Weingarten, and U. Keller, "Picosecond pulse sources with multi-GHz repetition rates and high output power," New J. Phys., vol. 6, p. 174, 2004

[04Pau] G. Paunescu, J. Hein, and R. Sauerbrey, "100-fs diode-pumped Yb:KGW mode-locked laser," Appl. Phys. B, vol. 79, pp. 555–558, 2004

[04Pet] V. Petit, J. L. Doualan, P. Camy, V. Menard, and R. Moncorge, "CW and tunable laser operation of Yb^{3+} doped CaF_2," Appl. Phys. B, vol. 78, pp. 681–684, 2004

[04Pra] R. P. Prasankumar, I. Hartl, J. T. Gopinath, E. P. Ippen, J. G. Fujimoto, P. Mak, and M. E. Ruane, "Design and characterization of semiconductor-doped silica film saturable absorbers," J. Opt. Soc. Am. B, vol. 21, pp. 851–857, 2004

[04Raf] E. U. Rafailov, S. J. White, A. A. Lagatsky, A. Miller, W. Sibbett, D. A. Livshits, A. E. Zhukov, and V. M. Ustinov, "Fast quantum-dot saturable absorber for passive mode-locking of solid-state lasers," IEEE Photon. Technol. Lett. vol. 16, pp. 2439–2441, 2004

[04San] G. Sansone, S. Stagira, M. Nisoli, S. DeSilvestri, C. Vozzi, B. Schenkel, J. Biegert, A. Gosteva, K. Starke, D. Ristau, G. Steinmeyer, U. Keller, "Mirror dispersion control of a hollow fiber supercontinuum," Appl. Phys. B, vol. 78, pp. 551–555, 2004

[04Sch] A. Schlatter, S. C. Zeller, R. Grange, R. Paschotta, and U. Keller, "Pulse energy dynamics of passively mode-locked solid-state lasers above the Q-switching threshold," J. Opt. Soc. Am. B, vol. 21, pp. 1469–1478, 2004

[04Sch2] S. Schön, A. Rutz, V. Liverini, R. Grange, M. Haiml, S. C. Zeller, U. Keller, "Dilute nitride absorbers in passive devices for mode locking of solid-state lasers," J. Cryst. Growth, vol. 278, pp. 239–243, 2004

[04Smi] S. A. Smith, J. M. Hopkins, J. E. Hastie, D. Burns, S. Calvez, M. D. Dawson, T. Jouhti, J. Kontinnen, and M. Pessa, "Diamond-microchip GaInNAs vertical external-cavity surface-emitting laser operating CW at 1315 nm," Electr. Lett., vol. 40, pp. 935–937, 2004

[04Sor] I. T. Sorokina, "Cr^{2+}-doped II–VI materials for lasers and nonlinar optics," Opt. Mat., vol. 26, pp. 395–412, 2004

[04Tan] S. N. Tandon, J. T. Gopinath, H. M. Shen, G. S. Petrich, L. A. Kolodziejski, F. X. Kartner, and E. P. Ippen, "Large-area broadband saturable Bragg reflectors by use of oxidized AlAs," Opt. Lett., vol. 29, pp. 2551–2553, 2004.

[04Vig] D. Vignaud, J. F. Lampin, F. Mollot, "Two-photon absorption in InP substrates in the 1.55 μm range," Appl. Phys. Lett., vol. 85, p. 239, 2004

[04Yam] K. Yamane, T. Kito, R. Morita, and M. Yamashita, "Experimental and theoretical demonstration of validity and limitations in fringe-resolved autocorrelation measurements for pulses of few optical cycles," Opt. Express, vol. 12, pp. 2762–2773, 2004

[04Zel] S. C. Zeller, L. Krainer, G. J. Spühler, R. Paschotta, M. Golling, D. Ebling, K. J. Weingarten, and U. Keller, "Passively mode-locked 50-GHz Er:Yb:glass laser," Electron. Lett., vol. 40, pp. 875–876, 2004

[04Zha1] S. Zhang, E. Wu, H. F. Pan, and H. P. Zeng, "Passive mode locking in a diode-pumped $Nd:GdVO_4$ laser with a semiconductor saturable absorber mirror," IEEE J. Quantum Electron., vol. 40, pp. 505–508, 2004

[04Zha2] Q. A. Zhang, K. Jasim, A. V. Nurmikko, A. Mooradian, G. Carey, W. Ha, and E. Ippen, "Operation of a passively mode-locked extended-cavity surface-emitting diode laser in multi-GHz regime," IEEE Phot. Tech. Lett., vol. 16, pp. 885–887, 2004

[05Agn] A. Agnesi, F. Pirzio, A. Tomaselli, G. Reali, and C. Braggio, "Multi-GHz tunable-repetition-rate mode-locked $Nd:GdVO_4$ laser," Opt. Express, vol. 13, pp. 5302–5307, 2005

[05Agn2] A. Agnesi, A. Guandalini, G. Reali, "Self-stabilized and dispersion-compensated passively mode-locked Yb:yttrium aluminum garnet laser," Appl. Phys. Lett., vol. 86, p. 171105, 2005

[05Asc1] A. Aschwanden, D. Lorenser, H. J. Unold, R. Paschotta, E. Gini, and U. Keller, "2.1-W picosecond passively mode-locked external-cavity semiconductor laser," Opt. Lett., vol. 30, pp. 272–274, 2005

[05Asc2] A. Aschwanden, D. Lorenser, H. J. Unold, R. Paschotta, E. Gini, and U. Keller, "10-GHz passively mode-locked surface emitting semiconductor laser with 1.4-W average output power," Appl. Phys. Lett., vol. 86, p. 131102, 2005

[05Bar] A. Bartels, S.A. Diddams, C.W. Oates, G. Wilpers, J.C. Bergquist, W.H. Oskay, L. Hollberg, "Femtosecond-laser-based synthesis of ultrastable microwave signals from optical frequency references," Opt. Lett., vol. 30, Nr. 6, pp. 667–669, 2005

[05Cas] O. Casel, D. Woll, M. A. Tremont, H. Fuchs, R. Wallenstein, E. Gerster, P. Unger, M. Zorn, and M. Weyers, "Blue 489-nm picosecond pulses generated by intracavity frequency doubling in a passively mode-locked optically pumped semiconductor disk laser," Appl. Phys. B, vol. 81, pp. 443–446, 2005

[05Cha] W. Chao, B. D. Harteneck, J. A. Liddle, E. H. Anderson, and D. T. Attwood, "Soft-X-ray microscopy at a spatial resolution better than 15 nm," Nature, vol. 435, pp. 1210–1213, 2005

[05Dat] P. K. Datta, S. Mukhopadhyay, G. K. Samanta, S. K. Das, and A. Agnesi, "Realization of inverse saturable absorption by intracavity third-harmonic generation for efficient nonlinear mirror mode-locking," Appl. Phys. Lett., vol. 86, pp. 151105, 2005

[05Fan] Y. X. Fan, J. L. He, Y. G. Wang, S. Liu, H. T. Wang, and X. Y. Ma, "2-ps passively mode-locked Nd:YVO_4 laser using an output-coupling-type semiconductor saturable absorber mirror," Appl. Phys. Lett., vol. 86, pp. 101103, 2005

[05Gio] V. Giordano, P. Y. Bourgeois, Y. Gruson, N. Boubekeur, R. Boudot, E. Rubiola, N. Bazin, Y. Kersale, "New advances in ultra–stable microwave oscillators," Eur. Phys. J. Appl. Phys., vol. 32, pp. 133–141, 2005

[05Gos] A. Gosteva, M. Haiml, R. Paschotta, U. Keller, "Noise-related resolution limit of dispersion measurements with white light interferometers," J. Opt. Soc. Am. B, vol. 22, Issue 9, pp. 1868–1874, 2005

[05Gra1] F. J. Grawert, J. T. Gopinath, F. O. Ilday, H. M. Shen, E. P. Ippen, F. X. Kartner, S. Akiyama, J. Liu, K. Wada, and L. C. Kimerling, "220-fs erbium–ytterbium:glass laser mode locked by a broadband low-loss silicon/germanium saturable absorber," Opt. Lett., vol. 30, pp. 329–331, 2005

[05Gra2] R. Grange, M. Haiml, R. Paschotta, G. J. Spühler, L. Krainer, O. Ostinelli, M. Golling, and U. Keller, "New regime of inverse saturable absorption for self-stabilizing passively mode-locked lasers," Appl. Phys. B, vol. 80, pp. 151–158, 2005

[05Gra3] R. Grange, A. Rutz, V. Liverini, M. Haiml, S. Schön, and U. Keller, "Nonlinear absorption edge properties of 1.3 μm GaInNAs saturable absorbers," Appl. Phys. Lett, vol. 87, pp. 132103, 2005

[05Guo] L. Guo, W. Hou, H. B. Zhang, Z. P. Sun, D. F. Cui, Z. Y. Xu, Y. G. Wang, and X. Y. Ma, "Diode-end-pumped passively mode-locked ceramic Nd:YAG Laser with a semiconductor saturable mirror," Opt. Express, vol. 13, pp. 4085–4089, 2005

[05Has] J. E. Hastie, L. G. Morton, S. Calvez, M. D. Dawson, T. Leinonen, M. Pessa, G. Gibson, and M. J. Padgett, "Red microchip VECSEL array," Opt. Exp., vol. 13, pp. 7209–7214, 2005

[05Hau] C. P. Hauri, A. Guandalini, P. Eckle, W. Kornelis, J. Biegert, U. Keller, "Generation of intense few-cycle laser pulses through filamentation – parameter dependence," Opt. Express, vol. 13, No. 19, pp. 7541–7547, 2005

[05Hol] S. Holmgren, V. Pasiskevicius, and F. Laurell, "Generation of 2.8 ps pulses by mode-locking a Nd:$GdVO_4$ laser with defocusing cascaded Kerr lensing in periodically poled KTP," Opt. Express, vol. 13, pp. 5270–5278, 2005

[05Hoo] S. Hoogland, A. Garnache, I. Sagnes, J. S. Roberts, and A. C. Tropper, "10-GHz Train of Sub-500-fs Optical Soliton-Like Pulses From a Surface-Emitting Semiconductor Laser," IEEE Phot. Tech. Lett., vol. 17, pp. 267–269, 2005

[05Inn] E. Innerhofer, B. Brunner, S. V. Marchese, R. Paschotta, U. Keller, K. Furusawa, J. C. Baggett, T. M. Monro, D. J. Richardson, "32 W of average power in 24-fs pulses

from a passively modelocked thin disk laser with nonlinear fiber compression" – Talk TuA3, Advanced Solid-State Photonics (ASSP '05), Vienna, Austria, Feb. 6–9, 2005
[05Jac] M. Jacquemet, C. Jacquemet, N. Janel, F. Druon, F. Balembois, P. Georges, J. Petit, B. Viana, D. Vivien, and B. Ferrand, "Efficient laser action of Yb: LSO and Yb:YSO oxyorthosilicates crystals under high-power diode-pumping," Appl. Phys. B, vol. 80, pp. 171–176, 2005
[05Kil1] A. Killi, A. Steinmann, J. Dörring, U. Morgner, M. J. Lederer, D. Kopf, and C. Fallnich, "High-peak-power pulses from a cavity-dumped Yb:KY$(WO_4)_2$ oscillator," Opt. Lett., vol. 30, pp. 1891–1893, 2005
[05Kil2] A. Killi, J. Dorring, U. Morgner, M. J. Lederer, J. Frei, and D. Kopf, "High speed electro-optical cavity dumping of mode-locked laser oscillators," Opt. Exp., vol. 13, pp. 1916–1922, 2005
[05Kim] K. Kim, S. Lee, and P. J. Delfyett, "1.4 kW high peak power generation from an all semiconductor mode-locked master oscillator power amplifier system based on eXtreme Chirped Pulse Amplification (X-CPA)," Opt. Express, vol. 13, pp. 4600–4606, 2005
[05Kis] V. E. Kisel, A. E. Troshin, V. G. Shcherbitsky, N. V. Kuleshov, V. N. Matrosov, T. A. Matrosova, M. I. Kupchenko, F. Brunner, R. Paschotta, F. Morier-Genoud, and U. Keller, "Femtosecond pulse generation with a diode-pumped Yb^{3+}:YVO_4 laser," Opt. Lett., vol. 30, pp. 1150–1152, 2005
[05Lag] A. A. Lagatsky, E. U. Rafailov, W. Sibbett, D. A. Livshits, A. E. Zhukov, and V. M. Ustinov, "Quantum-dot-based saturable absorber with p-n junction for mode-locking of solid-state lasers," IEEE Phot. Tech. Lett., vol. 17, pp. 294–296, 2005
[05Lec1] S. Lecomte, M. Kalisch, L. Krainer, G. J. Spühler, R. Paschotta, L. Krainer, M. Golling, D. Ebling, T. Ohgoh, T. Hayakawa, S. Pawlik, B. Schmidt, and U. Keller, "Diode-pumped passively mode-locked Nd:YVO_4 lasers with 40-GHz repetition rate," IEEE J. Quantum Electron., vol. 41, pp. 45–52, 2005
[05Lec2] S. Lecomte, R. Paschotta, S. Pawlik, B. Schmidt, K. Furusawa, A. Malinowski, D. J. Richardson, and U. Keller, "Synchronously pumped optical parametric oscillator with a repetition rate of 81.8 GHz," IEEE Phot. Tech. Lett., vol. 17, pp. 483–485, 2005
[05Lin1] J. H. Lin, W. H. Yang, W. F. Hsieh, and K. H. Lin, "Low threshold and high power output of a diode-pumped nonlinear mirror mode-locked Nd:$GdVO_4$ laser," Opt. Express, vol. 13, pp. 6323–6329, 2005
[05Lin2] H. Lindberg, M. Sadeghi, M. Westlund, S. Wang, A. Larsson, M. Strassner, and S. Marcinkevicius, "Mode locking a 1550 nm semiconductor disk laser by using a GaInNAs saturable absorber," Opt. Lett., vol. 30, pp. 2793–2795, 2005
[05Liu] J. Liu, X. Mateos, H. Zhang, J. Wang, M. Jiang, U. Griebner, and V. Petrov, "Continuous-wave laser operation of Yb:$LuVO_4$," Opt. Lett., vol. 30, pp. 3162–3164, 2005
[05Mai] Y. Mairesse and F. Quere, "Frequency-resolved optical gating for complete reconstruction of attosecond bursts," Phys. Rev. A, vol. 71, p. 011401, 2005
[05Man] A. V. Mandrik, A. E. Troshin, V. E. Kisel, A. S. Yasukevich, G. N. Klavsut1, N. V. Kuleshov, and A. A. Pavlyuk, "CW and Q-switched diode-pumped laser operation of Yb^{3+}:NaLa$(MoO_4)_2$," Appl. Phys. B, vol. 81, pp. 1119–1121, 2005
[05Mar] S. V. Marchese, E. Innerhofer, R. Paschotta, S. Kurimura, K. Kitamura, G. Arisholm, U. Keller, "Room temperature femtosecond optical parametric generation in MgO-doped stoichiometric $LiTaO_3$" Appl. Phys. B, vol. 239, pp. 1049–1052, 2005
[05Maz] Y. I. Mazur, Z. M. Wang, G. G. Tarasov, M. Xiao, G. J. Salamo, J. W. Tomm, V. Talalaev, and H. Kissel, "Interdot carrier transfer in asymmetric bilayer InAs/GaAs quantum dot structures," Appl. Phys. Lett., vol. 86, 063102–3, 2005
[05Nau] S. Naumov, A. Fernandez, R. Graf, P. Dombi, F. Krausz, and A. Apolonski, "Approaching the microjoule frontier with femtosecond laser oscillators," New J. Phys., vol. 7, p. 216, 2005

[05New] N. Newbury and B. Washburn, "Theory of the frequency comb output from a femtosecond fiber laser," IEEE J. Quantum Electron., vol. 41, Nr. 11, p. 1388, 2005

[05Nob] See http://nobelprize.org/physics/laureates/2005/adv.html to download text with "Advanced Information"

[05Pas] R. Paschotta, B. Budin, A. Schlatter, G. J. Spühler, L. Krainer, S. C. Zeller, N. Haverkamp, H. R. Telle, U. Keller, "Relative timing jitter measurements with an indirect phase comparison method," Appl. Phys. B, vol. 80, pp. 185–192, 2005

[05Pet] J. Petit, P. Goldner, and B. Viana, "Laser emission with low quantum defect in Yb:$CaGdAlO_4$," Opt. Lett., vol. 30, pp. 1345–1347, 2005

[05Rom] J. J. Romero, J. Johannsen, M. Mond, K. Petermann, G. Huber, and E. Heumann, "Continuous-wave laser action of Yb^{3+}-doped lanthanum scandium borate," Appl. Phys. B, vol. 80, pp. 159–163, 2005

[05Ros] F. Röser, J. Rothhard, B. Ortac, A. Liem, O. Schmidt, T. Schreiber, J. Limpert, and A. Tünnermann, "131 W 220 fs fiber laser system," Opt. Lett., vol. 30, pp. 2754–2756, 2005

[05Riv] S. Rivier, X. Mateos, V. Petrov, U. Griebner, A. Aznar, O. Silvestre, R. Sole, M. Aguilo, F. Diaz, M. Zorn, and M. Weyers, "Mode-locked laser operation of epitaxially grown Yb: $KLu(WO_4)_2$ composites," Opt. Lett., vol. 30, pp. 2484–2486, 2005

[05Rut] A. Rutz, R. Grange, V. Liverini, M. Haiml, S. Schön, and U. Keller, "1.5 μm GaInNAs semiconductor saturable absorber for passively modelocked solid-state lasers," Electr. Lett., vol. 41, pp. 321–323, 2005

[05Sch1] A. Schlatter, L. Krainer, M. Golling, R. Paschotta, D. Ebling, and U. Keller, "Passively mode-locked 914-nm Nd:YVO_4 laser," Opt. Lett., vol. 30, pp. 44–46, 2005

[05Sch2] A. Schlatter, B. Rudin, S. C. Zeller, R. Paschotta, G. J. Spühler, L. Krainer, N. Haverkamp, H. R. Telle, and U. Keller, "Nearly quantum-noise-limited timing jitter from miniature Er:Yb:glass lasers," Opt. Lett., vol. 30, pp. 1536–1538, 2005

[05Sch3] T. Schibli, K. Minoshima, H. Kataura, E. Itoga, N. Minami, S. Kazaoui, K. Miyashita, M. Tokumoto, and Y. Sakakibara, "Ultrashort pulse-generation by saturable absorber mirrors based on polymer-embedded carbon nanotubes," Opt. Exp., vol. 13, pp. 8025–8031, 2005

[05Sch4] B. Schenkel, R. Paschotta, U. Keller, "Pulse compression with supercontinuum generation in microstructure fibers," J. Opt. Soc. Am. B, vol. 22, pp. 687–693, 2005

[05Sor] E. Sorokin, S. Naumov, and I. T. Sorokina, "Ultrabroadband infrared solid-state lasers," IEEE J. Sel. Top. Quan. Electron., vol. 11, pp. 690–712, 2005

[05Spu1] G. J. Spühler, L. Krainer, E. Innerhofer, R. Paschotta, K. J. Weingarten, and U. Keller, "Soliton mode-locked Er:Yb: glass laser," Opt. Lett., vol. 30, pp. 263–265, 2005

[05Spu2] G. J. Spühler, L. Krainer, V. Liverini, S. Schön, R. Grange, M. Haiml, A. Schlatter, S. Pawlik, B. Schmidt, U. Keller, "Passively mode-locked multi-GHz 1.3-μm Nd:YVO_4 lasers with low timing jitter," IEEE Phot. Tech. Lett., vol. 17, pp. 1319–1321, 2005

[05Spu3] G. J. Spühler, K. J. Weingarten, R. Grange, L. Krainer, M. Haiml, V. Liverini, M. Golling, S. Schon, and U. Keller, "Semiconductor saturable absorber mirror structures with low saturation fluence," Appl. Phys. B, vol. 81, pp. 27–32, 2005

[05Ste] G. M. Stéphan, T. T. Tam, S. Blin, P. Besnard, and M. Têtu, "Laser line shape and spectral density of frequency noise," Phys. Rev. A , vol. 71, 043809, 2005

[05Su] K. W. Su, H. C. Lai, A. Li, Y. F. Chen, and K. E. Huang, "InAs/GaAs quantum-dot saturable absorber for a diode-pumped passively mode-locked Nd:YVO_4 laser at 1342 nm," Opt. Lett., vol. 30, pp. 1482–1484, 2005

[05Uem] S. Uemura and K. Torizuka, "Center-wavelength-shifted passively mode-locked diode-pumped ytterbium(Yb):yttrium aluminum garnet (YAG) laser," Jpn. J. Appl. Phys. Part 2, vol. 44, pp. L361–L363, 2005

[05Voi] S. V. Voitikov, A. A. Demidovich, L. E. Batay, A. N. Kuzmin, M. B. Danailov, "Subnanosecond pulse dynamics of Nd: LSB microchip laser passively Q-switched by Cr:YAG saturable absorber," Opt. Commun., vol. 251, pp. 154–164, 2005

[05Wan] Y. B. Wang, X. Y. Ma, Y. X. Fan, and H. T. Wang, "Passively mode-locking Nd: $Gd_{0.5}Y_{0.5}VO_4$ laser with an $In_{0.25}Ga_{0.75}As$ absorber grown at low temperature," Appl. Optics, vol. 44, pp. 4384–4387, 2005

[05Wit] S. Witte, R. T. Zinkstok, W. Hogervorst, and K. W. E. Eikema, "Generation of few-cycle terawatt light pulses using optical parametric chirped pulse amplification," Opt. Express, vol. 13, pp. 4903–4908, 2005

[05Yam] S. Yamashita, Y. Inoue, K. Hsu, T. Kotake, H. Yaguchi, D. Tanaka, M. Jablonski, S. Y. Set, "5-GHz pulsed fiber Fabry–Perot laser mode-locked using carbon nanotubes," IEEE Photon. Technol. Lett., vol. 17, pp. 750–752, 2005

[05Ye] J. Ye, S. T. Cundiff, Femtosecond optical frequency comb: principle, operation, and applications, Springer 2005, ISBN 0-387-23790-9

[05Zha1] B. Y. Zhang, G. Li, M. Chen, H. J. Yu, Y. G. Wang, X. Y. Ma, "Passive mode locking of diode-end-pumped Nd:$GdVO_4$ laser with an $In_{0.25}Ga_{0.75}As$ output coupler," Opt. Comm., vol. 244, pp. 311–314, 2005

[05Zha2] Q. Zhang, K. Jasim, A. V. Nurmikko, E. Ippen, A. Mooradian, G. Carey, and W. Ha, "Characteristics of a high-speed passively mode-locked surface-emitting semiconductor InGaAs laser diode," IEEE Phot. Tech. Lett., vol. 17, pp. 525–527, 2005

[06Cas] C. Cascales, M. D. Serrano, F. Esteban-Betegón, C. Zaldo, R. Peters, K. Petermann, G. Huber, L. Ackermann, D. Rytz, C. Dupré, M. Rico, J. Liu, U. Griebner, and V. Petrov, "Structural, spectroscopic, and tunable laser properties of Yb^{3+}-doped $NaGd(WO_4)_2$," Phys. Rev. B, vol. 74, pp. 174114–174115, 2006

[06Cou] A. Courairon, J. Biegert, C. P. Hauri, W. Kornelis, F. W. Helbing, U. Keller, A. Mysyrowicz, "Self-compression of ultrashort laser pulses down to one optical cycle by filamentation," J. Mod. Opt., vol. 53, pp. 75–85, 2006

[06Dew] S. Dewald, T. Lang, C. D. Schroter, R. Moshammer, J. Ullrich, M. Siegel, and U. Morgner, "Ionization of noble gases with pulses directly from a laser oscillator," Opt. Lett., vol. 31, pp. 2072–2074, 2006

[06Dud] J. M. Dudley, G. Genty, S. Coen, "Supercontinuum generation in photonic crystal fiber," Rev. Mod. Phys., vol. 78, no. 4, pp. 1135–1184, 2006

[06For] T. M. Fortier, A. Bartels, and S. A. Diddams, "Octave-spanning Ti:sapphire laser with a repetition rate >1 GHz for optical frequency measurements and comparisons," Opt. Lett., vol. 31, pp. 1011–1013, 2006

[06Gau] D. M. Gaudiosi, E. Gagnon, A. L. Lytle, J. L. Fiore, E. A. Gibson, S. Kane, J. Squier, M. M. Murnane, H. C. Kapteyn, R. Jimenez, and S. Backus, "Multi-kilohertz repetition rate Ti:sapphire amplifier based on down-chirped pulse amplification," Opt. Express, vol. 14, pp. 9277–9283, 2006

[06Gra] R. Grange, S. C. Zeller, S. Schön, M. Haiml, O. Ostinelli, M. Ebnöther, E. Gini, and U. Keller, "Antimonide semiconductor saturable absorber for passive modelocking of a 1.5-μm Er:Yb:glass laser at 10 GHz," IEEE Phot. Tech. Lett., vol. 18, pp. 805–807, 2006

[06Gua] A. Guandalini, P. Eckle, M. P. Anscombe, P. Schlup, J. Biegert, U. Keller, "5.1 fs pulses generated by filamentation and carrier envelope phase stability analysis," J. Phys. B: At. Mol. Opt. Phys., vol. 39, pp. S257–S264, 2006

[06Hol] G. R. Holtom, "Mode-locked Yb:KGW laser longitudinally pumped by polarization-coupled diode bars," Opt. Lett., vol. 31, pp. 2719–2721, 2006

[06Hon] K. H. Hong, T. J. Yu, S. Kostritsa, J. H. Sung, I. W. Choi, Y. C. Noh, D. K. Ko, J. Lee, "Development of 100 kHz femtosecond high-power laser using down-chirped regenerative amplification," Laser Phys., vol. 16, pp. 673–677, 2006

[06Hop] J. M. Hopkins, S. A. Smith, C. W. Jeon, H. D. Sun, D. Burns, S. Calvez, M. D. Dawson, T. Jouhti, and M. Pessa, "0.6 W CW GaInNAs vertical external-cavity surface emitting laser operating at 1.32 μm," Electr. Lett., vol. 40, pp. 30–31, 2004

[06Inn] E. Innerhofer, F. Brunner, S. V. Marchese, R. Paschotta, G. Arisholm, S. Kurimura, K. Kitamura, T. Usami, H. Ito, and U. Keller, "Analysis of nonlinear wavelength conversion system for a red-green-blue laser-projection source," J. Opt. Soc. Am. B, vol. 23, pp. 265–275, 2006

[06Kel] U. Keller and A. C. Tropper, "Passively modelocked surface-emitting semiconductor lasers," Physics Report, vol. 429, pp. 67–120, 2006

[06Kor] J. D. Koralek, J. F. Douglas, N. C. Plumb, Z. Sun, A. V. Fedorov, M. M. Murnane, H. C. Kapteyn, S. T. Cundiff, Y. Aiura, K. Oka, H. Eisaki, D. S. Dessau, "Laser based angle-resolved photoemission, the sudden approximation, and quasiparticle-like spectral peaks in $Bi_2Sr_2CaCu_2O_{8+\delta}$," Phys. Rev. Lett., vol. 96, p. 017005, 2006

[06Lor] D. Lorenser, D. J. H. C. Maas, H. J. Unold, A.-R. Bellancourt, B. Rudin, E. Gini, D. Ebling, and U. Keller, "50-GHz passively mode-locked surface-emitting semiconductor laser with 100 mW average output power," IEEE J. Quant. Electron., vol. 42, pp. 838–847, 2006

[06Maj] A. Major, R. Cisek, and V. Barzda, "Femtosecond Yb:KGd$(WO_4)_2$ laser oscillator pumped by a high power fiber-coupled diode laser module," Opt. Exp., vol. 14, pp. 12163–12168, 2006

[06Mar] S. V. Marchese, T. Südmeyer, M. Golling, R. Grange, and U. Keller, "Pulse energy scaling to 5 μJ from a femtosecond thin disk laser," Opt. Lett., vol. 31, pp. 2728–2730, 2006

[06Ost] O. Ostinelli, M. Haiml, R. Grange, G. Almuneau, M. Ebnöther, E. Gini, E. Müller, U. Keller, and W. Bächtold, "Highly reflective AlGaAsSb/InP Bragg reflectors at 1.55 μm grown by MOVPE," J. Cryst. Growth, vol. 286, pp. 247–254, 2006

[06Pas] R. Paschotta, A. Schlatter, S. C. Zeller, U. Keller, "Optical phase noise and carrier-envelope offset noise of modelocked lasers," Appl. Phys. B, vol. 82, pp. 265–273, 2006

[06Pla] J. J. Plant, J. T. Gopinath, B. Chann, D. J. Ripin, R. K. Huang, and P. W. Juodawlkis, "250 mW, 1.5 μm monolithic passively modelocked slab-coupled optical waveguide laser," Opt. Lett., vol. 31, pp. 223–225, 2006

[06Riv] S. Rivier, X. Mateos, J. Liu, V. Petrov, U. Griebner, M. Zorn, M. Weyers, H. Zhang, J. Wang, and M. Jiang, "Passively mode-locked Yb:$LuVO_4$ oscillator," Opt. Exp., vol. 14, pp. 11668–11671, 2006

[06Rut] A. Rutz, V. Liverini, D. J. H. C. Maas, B. Rudin, A.-R. Bellancourt, S. Schön, U. Keller, "Passively modelocked GaInNAs VECSEL at centre wavelength around 1.3 μm," Electron. Lett., vol. 42, p. 926, 2006

[06Sti] G. Stibenz, N. Zhavoronkov, and G. Steinmeyer, "Self-compression of millijoule pulses to 7.8 fs duration in a white-light filament," Opt. Lett., vol. 21, no. 2, pp. 274–276, 2006

[06Thi] F. Thibault, D. Pelenc, F. Druon, Y. Zaouter, M. Jacquemet, and P. Georges, "Efficient diode-pumped Yb^{3+}:Y_2SiO_5 and Yb^{3+}:Lu_2SiO_5 high-power femtosecond laser operation," Opt. Lett., vol. 31, pp. 1555–1557, 2006

[06Xue] Y. Xue, C. Wang, Q. Liu, Y. Li, L. Chai, C. Yan, G. Zhao, L. Su, X. Xu, and J. Xu, "Characterization of diode-pumped laser operation of a novel Yb:GSO crystal," IEEE J. Quantum Electron., vol. 42, pp. 517–521, 2006

[06Zao] Y. Zaouter, J. Didierjean, F. Balembois, G. L. Leclin, F. Druon, P. Georges, J. Petit, P. Goldner, and B. Viana, "47-fs diode-pumped Yb^{3+}:$CaGdAlO_4$ laser," Opt. Lett., vol. 31, pp. 119–121, 2006

[06Zel] S. C. Zeller, R. Grange, V. Liverini, A. Rutz, S. Schön, M. Haiml, U. Keller, S. Pawlik, and B. Schmidt, "A low-loss buried resonant GaInNAs SESAM for 1.3-μm Nd:YLF laser at 1.4 GHz," Advanced Solid-State Photonics (ASSP '06), 2006

[06Zho] X. B. Zhou, H. Kapteyn, and M. Murnane, "Positive-dispersion cavity-dumped Ti: sapphire laser oscillator and its application to white light generation," Opt. Exp., vol. 14, pp. 9750–9757, 2006

[07Bar] A. Bartels, R. Cerna, C. Kistner, A. Thoma, F. Hudert, C. Janke, and T. Dekorsy, "Ultrafast time-domain spectroscopy based on high-speed asynchronous optical sampling," Rev. Sci. Instrum., vol. 78, 035107, 2007

[07Bar2] A. Bartels, R. Gebs, M. S. Kirchner, S. A. Diddams, "Spectrally resolved optical frequency comb from a self-referenced 5 GHz femtosecond laser," Opt. Lett., vol. 32, pp. 2553–2555, 2007

[07Ber] L. Bergé, S. Skupin, R. Nuter, J. Kasparian, J.-P. Wolf, "Ultrashort filaments of light in weakly ionized optically transparent media," Rep. Prog. Phys., vol. 70, pp. 1633–1713, 2007

[07Bou] J. Boudeile, F. Druon, M. Hanna, P. Georges, Y. Zaouter, E. Cormier, J. Petit, P. Goldner, B. Viana, "Continuous-wave and femtosecond laser operation of Yb:$CaGdAlO_4$ under high-power diode pumping," Opt. Lett., vol. 32, pp. 1962–1964, 2007

[07Cou] A. Couairon, A. Mysyrowicz, "Femtosecond filamentation in transparent media," Physics Reports, vol. 441, pp. 47–189, 2007

[07Fon] K. H. Fong, K. Kikuchi, C. S. Goh, S. Y. Set, R. Grange, M. Haiml, A. Schlatter, and U. Keller, "Solid-state Er:Yb:glass laser mode-locked by using single-wall carbon nanotube thin film," Opt. Lett., vol. 32, pp. 38–40, 2007

[07Gar] A. Garcia-Cortes, J. M. Cano-Torres, M. D. Serrano, C. Cascales, C. Zaldo, S. Rivier, X. Mateos, U. Griebner, and V. Petrov, "Spectroscopy and Lasing of Yb-doped $NaY(WO_4)_2$: Tunable and Femtosecond Mode-Locked Laser Operation," IEEE J. Quantum Electron., vol. 43, no. 9, pp. 758–764, 2007

[07Gie] A. Giesen and J. Speiser, "Fifteen Years of Work on Thin-Disk Lasers: Results and Scaling Laws," IEEE J. Sel. Top. Quant., vol. 13, pp. 598–609, 2007

[07Kel] U. Keller, "Ultrafast solid-state lasers," Landolt-Börnstein, Group VIII/1B1, *Laser Physics and Applications. Subvolume B: Laser Systems*. Part 1. Edited by G. Herziger, H. Weber, R. Proprawe (Springer-Verlag, Berlin, Heidelberg, New York, October, 2007), pp. 33–167, ISBN 978-3-540-26033-2

[07Lan] C. Langrock, M. M. Fejer, I. Hartl, M. E. Fermann, "Generation of octave-spanning spectra inside reverse-proton-exchanged periodically poled lithium niobate waveguides," Opt. Lett., vol. 32, pp. 2478–2480, 2007

[07Lar] M. Larionov, F. Butze, D. Nickel, and A. Giesen, "High-repetition-rate regenerative thin-disk amplifier with 116 μJ pulse energy and 250 fs pulse duration," Opt. Lett., vol. 32, pp. 494–496, 2007

[07Li] W. Li, Q. Hao, H. Zhai, H. Zeng, W. Lu, G. Zhao, L. Zheng, L. Su, and J. Xu, "Diode-pumped Yb:GSO femtosecond laser," Opt. Exp., vol. 15, pp. 2354–2359, 2007

[07Maa] D. J. H. C. Maas, A.-R. Bellancourt, B. Rudin, M. Golling, H. J. Unold, T. Südmeyer, and U. Keller, "Vertical integration of ultrafast semiconductor lasers," Appl. Phys. B, vol. 88, pp. 493–497, 2007

[07Mar] S. V. Marchese, S. Hashimoto, C. R. E. Baer, M. S. Ruosch, R. Grange, M. Golling, T. Südmeyer, U. Keller, G. Lépine, G. Gingras, and B. Witzel, "Passively mode-locked thin disk lasers reach 10 microjoules pulse energy at megahertz repetition rate and drive high field physics experiments," Paper CF3-2-MON, presented at Conference on Lasers and Electro-Optics (Europe), Munich, Germany, 2007

[07Mar2] S. V. Marchese, C. R. E. Baer, R. Peters, C. Kränkel, A. G. Engqvist, M. Golling, D. J. H. C. Maas, K. Petermann, T. Südmeyer, G. Huber, and U. Keller, "Efficient femtosecond high power Yb:Lu_2O_3 thin disk laser," Opt. Express, vol. 15, 16966–16971, 2007

[07Nod] D. Nodop, J. Limpert, R. Hohmuth, W. Richter, M. Guina, A. Tünnermann, "High-pulse-energy passively Q-switched quasi-monolithic microchip lasers operating in the sub-100-ps pulse regime," Opt. Lett., vol. 32, pp. 2115–2117, 2007

[07Pal] G. Palmer, M. Siegel, A. Steinmann, and U. Morgner, "Microjoule pulses from a passively mode-locked Yb:$KY(WO_4)_2$ thin-disk oscillator with cavity dumping," Opt. Lett., vol. 32, pp. 1593–1595, 2007

[07Pas] R. Paschotta, H. Telle, and U. Keller, "Noise of solid-state lasers," Chap. 12 in *Solid-State Lasers and Applications*, ed. by A. Sennaroglu, CRC Press, Taylor and Francis Group, LLC, pp. 473–510, 2007, ISBN 0-8493-3589-2

[07Riv] S. Rivier, A. Schmidt, C. Kränkel, R. Peters, K. Petermann, G. Huber, M. Zorn, M. Weyers, A. Klehr, G. Erbert, V. Petrov, U. Griebner, "Ultrashort pulse $Yb:LaSc_3(BO_3)_4$ modelocked oscillator," Opt. Express, vol. 15, pp. 15539–15544, 2007

[07Ros] F. Röser, D. Schimpf, O. Schmidt, B. Ortac, K. Rademaker, J. Limpert, and A. Tünnermann, "90 W average power 100 μJ energy femtosecond fiber chirped-pulse amplification system," Opt. Lett., vol. 32, pp. 2230–2232, 2007

[07Saa] E. J. Saarinen, A. Harkonen, R. Herda, S. Suomalainen, L. Orsila, T. Hakulinen, M. Guina, and O. G. Okhotnikov, "Harmonically mode-locked VECSELs for multi-GHz pulse train generation," Opt. Exp., vol. 15, pp. 955–964, 2007

[07Sch] A. Schlatter, S. C. Zeller, R. Paschotta, U. Keller, "Simultaneous measurement of the phase noise on all optical modes of a mode-locked laser," Appl. Phys. B, vol. 88, pp. 385–391, 2007

[07Sor] I. T. Sorokina, E. Sorokin, "Chirped-mirror dispersion controlled femtosecond Cr:ZnSe laser," paper WA7, in *Advanced Solid-State Photonics*, OSA Technical Digest, Jan. 28–31, Vancouver, Canada (2007)

[07Yos] M. Yoshida, K. Kasai, M. Nakazawa, "Mode-hop-free, optical frequency tunable 40-GHz mode-locked fiber laser," IEEE J. Quantum Electron., vol. 43, no. 8, pp. 704–708, 2007

[07Zai] A. Zair, A. Guandalini, F. Schapper, M. Holler, J. Biegert, L. Gallmann, A. Couairon, M. Franco, A. Mysyrowicz, U. Keller, "Spatio-temporal characterization of few-cycle pulses obtained by filamentation," Opt. Exp., vol. 15, pp. 5394–5404, 2007

[07Zel] S. C. Zeller, T. Südmeyer, K. J. Weingarten, and U. Keller, "Passively modelocked 77 GHz Er:Yb:glass laser," Electron. Lett., vol. 43, pp. 32–33, 2007

[08Bar] A. Bartels, D. Heinecke, and S. A. Diddams, "Passively mode-locked 10 GHz femtosecond Ti:sapphire laser," Opt. Lett., vol. 33, pp. 1905–1907, 2008

[08Boy] R. W. Boyd, Nonlinear Optics (Academic, 2008, 3rd edition)

[08Cre] H. M. Crespo, J. R. Birge, E. L. Falcão-Filho, M. Y. Sander, A. Benedick, and F. X. Kärtner, "Nonintrusive phase stabilization of sub-two-cycle pulses from a prismless octave-spanning Ti:sapphire laser," Opt. Lett., vol. 33, pp. 833–835, 2008

[08Ero] A. Erol, Dilute III–V Nitride Semiconductors and Material Systems, Physics and Technology, Springer Series in Materials Science (Springer, 2008)

[08Maa1] D. J. H. C. Maas, "MIXSELs—a new class of ultrafast semiconductor lasers," PhD thesis No. 18121, ETH Zurich

[08Maa2] D. J. H. C. Maas, A. R. Bellancourt, M. Hoffmann, B. Rudin, Y. Barbarin, M. Golling, T. Südmeyer, U. Keller, "Growth parameter optimization for fast quantum dot SESAMs," Opt. Express, vol. 16, No. 23, pp. 18646–18656, 2008

[08Maa3] D. J. H. Maas, B. Rudin, A.-R. Bellancourt, D. Iwaniuk, S. V. Marchese, T. Südmeyer, U. Keller, "High precision optical characterization of semiconductor saturable absorber mirrors," Opt. Exp., vol. 16, pp. 7571–7579, 2008

[08Mar] S. V. Marchese, C. R. E. Baer, A. G. Engqvist, S. Hashimoto, D. J. H. C. Maas, M. Golling, T. Südmeyer, U. Keller, "Femtosecond thin disk laser oscillator with pulse energy beyond the 10-microjoule level," Opt. Express, vol. 16, pp. 6397–6407, 2008

[08Men] J. Mende, J. Speiser, G. Spindler, W. L. Bohn, A. Giesen, "Mode dynamics and thermal lens effects of thin-disk lasers," Proceedings of the Society of Photo-Optical Instrumentation Engineers (SPIE), vol. 6871, pp. M8710, 2008

[08Neu] J. Neuhaus, J. Kleinbauer, A. Killi, S. Weiler, D. Sutter, T. Dekorsy, "Passively mode-locked Yb:YAG thin-disk laser with pulse energies exceeding 13 μJ by use of an active multipass geometry," Opt. Lett., vol. 33, no. 7, pp. 726–728, 2008

[08Neu2] J. Neuhaus, D. Bauer, J. Zhang, A. Killi, J. Kleinbauer, M. Kumkar, S. Weiler, M. Guina, D. H. Sutter, T. Dekorsky, "Subpicosecond thin-disk laser oscillator with pulse energies of up to 25.9 microjoules by use of an active multipass geometry," Opt. Express, vol. 16, no. 25, pp. 20530–20539, 2008

[08Oeh] A. E. H. Oehler, T. Südmeyer, K. J. Weingarten, U. Keller, "100-GHz passively modelocked Er:Yb:glass laser at 1.5 μm with 1.6-ps pulses," Opt. Exp., vol. 16, No. 26, pp. 21930–21935, 2008

[08Pal] G. Palmer, M. Schultze, M. Siegel, M. Emons, U. Bünting, U. Morgner, "Passively mode-locked Yb:KLu$(WO_4)_2$ thin-disk oscillator operated in the positive and negative dispersion regime," Opt. Lett., vol. 33, pp. 1608–1610, 2008

[08Pet] R. Peters, C. Kränkel, K. Petermann, G. Huber, "Power scaling potential of Yb:NGW in thin disk laser configuration," Appl. Phys. B, vol. 91, pp. 25–28, 2008

[08Qui] F. Quinlan, C. Williams, S. Ozharar, S. Gee, P. J. Delfyett, "Self-stabilization of the optical frequencies and the pulse repetition rate in a coupled optoelectronic oscillator," J. Lightwave Technol., vol. 26, no. 15, pp 2571–2577, 2008

[08Stu] M. C. Stumpf, S. C. Zeller, A. Schlatter, T. Okuno, T. Südmeyer, U. Keller, "Compact Er:Yb:glass-laser-based supercontinuum source for high-resolution optical coherence tomography," Opt. Exp., vol. 16, Nr. 14, pp. 10572–10579, 2008

[08Sud] T. Südmeyer, S. V. Marchese, S. Hashimoto, C. R. E. Baer, G. Gingras, B. Witzel, U. Keller, "Femtosecond laser oscillators for high-field science," Nature Photonics, vol. 2, pp. 599–604, 2008

[09Bae] C. R. E. Baer, C. Kränkel, O. H. Heckl, M. Golling, T. Südmeyer, R. Peters, K. Petermann, G. Huber, U. Keller, "227-fs pulses from a modelocked Yb:$LuScO_3$ thin disk laser," Opt. Express, vol. 17, No. 13, 10725–10730, 2009

[09Bae2] C. R. E. Baer, C. Kränkel, C. J. Saraceno, O. H. Heckl, M. Golling, T. Südmeyer, R. Peters, K. Petermann, G. Huber, U. Keller, "Femtosecond Yb:Lu_2O_3 thin disk laser with 63 W of average power," Opt. Lett., vol. 34, pp. 2823–2825, 2009

[09Bel] A. R. Bellancourt, D. J. H. C. Maas, B. Rudin, M. Golling, T. Südmeyer, U. Keller, "Modelocked integrated external-cavity surface emitting laser (MIXSEL)," IET Optoelectronics, vol. 3, pp. 61–72, 2009

[09Kil] A. Killi, C. Stolzenburg, I. Zawischa, D. Sutter, J. Kleinbauer, S. Schad, R. Brockmann, S. Weiler, J. Neuhaus, S. Kalfhues, E. Mehner, D. Bauer, H. Schlueter, C. Schmitz, "The broad applicability of the disk laser principle: from CW to ps," Proc. SPIE 7193, 71931T (2009)

[09Kra] G. Krauss, S. Lohss, T. Hanke, A. Sell, S. Eggert, R. Huber, A. Leitenstorfer, "Synthesis of single cycle of light with compact erbium-doped fibre technology," Nature Photonics, vol. 4, pp. 33–36, 2010

[09Rub] E. Rubiola, Phase Noise and Frequency Stability in Oscillators, Cambridge Univ. Press (2009), ISBN 978-0-521-88677-2

[09Rus] P. Russbueldt, T. Mans, G. Rotarius, J. Weitenberg, H. D. Hoffmann, and R. Poprawe, "400 W Yb:YAG Innoslab fs-amplifier," Opt. Express, vol. 17, pp. 12230–12245, 2009

[09Ste] A. Steinmetz, D. Nodop, J. Limpert, R. Hohmuth, W. Richter, A. Tünnermann, "22 MHz repetition rate, 200 ps pulse duration from a monolithic, passively Q-switched microchip laser," Appl. Phys. B, vol. 97, pp. 317–320, 2009

[09Suc] H. Suchowski, V. Prabhudesai, D. Oron, A. Arie, Y. Silberberg, "Robust adiabatic sum frequency conversion," Opt. Express, vol. 17, pp. 12731–12740, 2009

[09Was] P. Wasylczyk, P. Wnuk, and C. Radzewicz, "Passively modelocked, diode-pumped Yb:KYW femtosecond oscillator with 1 GHz repetition rate," Opt. Express, vol. 17, pp. 5630–5635, 2009

[10Bae] C. R. E. Baer, C. Kränkel, C. J. Saraceno, O. H. Heckl, M. Golling, R. Peters, K. Petermann, T. Südmeyer, G. Huber, U. Keller, "Femtosecond thin-disk laser with 141 W of average power," Opt. Lett., vol. 35, No. 13, pp. 2302–2304, 2010

[10Chr] D. N. Christodoulides, I. C. Khoo, G. J. Salamo, G. I. Stegeman, and E. W. Van Stryland, "Nonlinear refraction and absorption: mechanisms and magnitudes," Adv. Opt. Photon., vol.2, pp. 60–200, 2010

[10Dom] G. D. Domenico, S. Schilt, P. Thomann, "Simple approach to the relation between laser frequency noise and laser line shape," Appl. Optics, vol. 49, No. 25, pp. 4801–4807, 2010

[10Eid] T. Eidam, S. Hanf, E. Seise, T. V. Andersen, T. Gabler, C. Wirth, T. Schreiber, J. Limpert, and A. Tünnermann, "Femtosecond fiber CPA system emitting 830 W average output power," Opt. Lett., vol. 35, pp. 94–96, 2010

[10Har] A. Härkönen, J. Paajaste, S. Suomalainen, J.-P. Alanko, C. Grebing, R. Koskinen, G. Steinmeyer, M. Guina, "Picosecond passively mode-locked GaSb-based semiconductor disk laser operating at 2 μm," Opt. Lett., vol. 35, pp. 4090–4092, 2010

[10Hec] O. H. Heckl, C. Kränkel, C. R. E. Baer, C. J. Saraceno, T. Südmeyer, K. Petermann, G. Huber, U. Keller, "Continuous wave and mode-locked Yb:YCOB thin disk laser: first demonstration and future prospects," Opt. Express, vol. 18, No. 18, pp. 19201–19208, 2010

[10Hua] Z. Huang, Gaoming Li, Yishen Qiu, "Power calculation of wavelength tunable Yb^{3+}:LSO laser," Opt. Express., vol. 18, pp. 20979–20987, 2010

[10Ip] E. M. Ip, J. M. Kahn, "Fiber Impairment Compensation Using Coherent Detection and Digital Signal Processing," J. Lightwave Technol., vol. 28, pp. 502–519, 2010

[10Kel] U. Keller, "Ultrafast solid-state laser oscillators: a success story for the last 20 years with no end in sight," Appl. Phys. B, vol. 100, pp. 15–28, 2010

[10Kok] S. Koke, C. Grebing, H. Frei, A. Anderson, A. Assion, G. Steinmeyer, "Direct frequency comb synthesis with arbitrary offset and shot-noise limited phase noise," Nat. Photonics, vol. 4, pp. 462–465, 2010

[10Lev] A. Leven, N. Kaneda, S. Corteselli, "Real-time implementation of digital signal processing for coherent optical digital communication systems," IEEE J. Sel. Top. Quantum Electron., vol. 16, no. 5, pp. 1227–1234, 2010

[10Li] D. Li, U. Demirbas, J. R. Birge, G. S. Petrich, L. A. Kolodziejski, A. Sennaroglu, F. X. Kärtner, and J. G. Fujimoto, "Diode-pumped passively mode-locked GHz femtosecond Cr:LiSAF laser with kW peak power," Opt. Lett. 35, 1446–1448 (2010)

[10Oeh] A. E. H. Oehler, M. C. Stumpf, S. Pekarek, T. Südmeyer, K. J. Weingarten, U. Keller, "Picosecond diode-pumped 1.5 μm Er,Yb:glass lasers operating at 10–100 GHz repetition rate," Appl. Phys. B, vol. 99, 53–62, 2010

[10Pek] S. Pekarek, C. Fiebig, M. C. Stumpf, A. E. H. Oehler, K. Paschke, G. Erbert, T. Südmeyer, and U. Keller, "Diode-pumped gigahertz femtosecond Yb:KGW laser with a peak power of 3.9 kW," Opt. Express, vol. 18, pp 16320–16326, 2010

[10Qua] A. H. Quarterman, A. Perevedentsev, K. G. Wilcox, V. Apostolopoulos, H. E. Beere, I. Farrer, D. A. Ritchie, A. C. Tropper, "Passively harmonically mode-locked vertical-external-cavity surface-emitting laser emitting 1.1 ps pulses at 147 GHz repetition rate," Appl. Phys. Lett., vol. 97, no. 25, pp. 251101–3, 2010

[10Rud] B. Rudin, V. J. Wittwer, D. J. H. C. Maas, M. Hoffmann, O. D. Sieber, Y. Barbarin, M. Golling, T. Südmeyer, U. Keller, "Novel ultrafast semiconductor laser with 6.4 W average output power," Opt. Exp., vol. 18, pp. 27582–27588, 2010

[10Rus] P. Russbueldt, T. Mans, J. Weitenberg, H. D. Hoffmann, and R. Poprawe, "Compact diode-pumped 1.1 kW Yb:YAG Innoslab femtosecond amplifier," Opt. Lett., vol. 35, pp. 4169–4171, 2010

[10Sav] S. J. Savory, "Digital Coherent Optical Receivers: Algorithms and Subsystems," IEEE J. Sel. Top. Quantum Electron., vol. 16, no. 5, pp. 1164–1179, 2010

[10Spi] B. Spinnler, "Equalizer Design and Complexity for Digital Coherent Receivers," IEEE J. Sel. Top. Quantum Electron., vol. 16, no. 5, pp. 1180–1192, 2010

[10Stu] M. C. Stumpf, S. Pekarek, A. E. H. Oehler, T. Südmeyer, J. M. Dudley, U. Keller, "Self-referencable frequency comb from a 170-fs, 1.5-μm solid-state laser oscillator," Appl. Phys. B, vol. 99, Issue 3, pp. 401–408, 2010

[10Yam] S. Yamazoe, M. Katou, T. Adachi, T. Kasamatsu, "Palm-top-size, 1.5 kW peak power, and femtosecond (160 fs) diode-pumped mode-locked Yb^{+3}:$KY(WO_4)_2$ solid-state laser with a semiconductor saturable absorber mirror," Opt. Lett., vol. 35, pp. 748–750, 2010

[11Bao] Q. Bao, H. Zhang, Z. Ni, Y. Wang, L. Polavarapu, Z. Shen, Q-H. Xu, D. Tang, K. P. Loh, "Monolayer graphene as a saturable absorber in a mode-locked laser," Nano Research 4, pp. 297–307, 2011

[11Bau] D. Bauer, F. Schättiger, J. Kleinbauer, D. H. Sutter, A. Killi, T. Dekorsy, "Energies above 30 μJ and average power beyond 100 W directly from a mode-locked thin-disk oscillator," in Advanced Solid-State Photonics, OSA Technical Digest (CD) (Optical Society of America, 2011), paper ATuC2

[11Ben] F. Benabid and P. J. Roberts, "Linear and nonlinear optical properties of hollow core photonic crystal fiber," J. Mod. Opt., vol. 58, no. 2, pp. 87–124, 2011

[11Che] Y. F. Chen, Y. C. Lee, H. C. Liang, K. Y. Lin, K. W. Su, and K. F. Huang, "Femtosecond high-power spontaneous mode-locked operation in vertical-external cavity surface-emitting laser with gigahertz oscillation," Opt. Lett., vol. 36, no. 23, pp. 4581–4583, 2011

[11Har] A. Harkonen, C. Grebing, J. Paajaste, R. Koskinen, J. P. Alanko, S. Suomalainen, G. Steinmeyer, M. Guina, "Modelocked GaSb disk laser producing 384 fs pulses at 2 μm wavelength," Electron. Lett., vol. 47, pp. 454–456, 2011

[11Hec] O. H. Heckl, C. J. Saraceno, C. R. E. Baer, T. Südmeyer, Y. Y. Wang, Y. Cheng, F. Benabid, U. Keller, "Temporal pulse compression in a xenon-filled Kagome-type hollow-core photonic crystal fiber at high average power," Opt. Express, vol. 19, no. 20, pp. 19142–19149, 2011

[11Hei] D. C. Heinecke, A. Bartels, S. A. Diddams, "Offset frequency dynamics and phase noise properties of a self-referenced 10 GHz Ti:sapphire frequency comb," Opt. Express, vol. 19, pp. 18440–18451, 2011

[11Hil] D. Hillerkuss, R. Schmogrow, T. Schellinger, M. Jordan, M. Winter, G. Huber, T. Vallaitis, R. Bonk, P. Kleinow, F. Frey, M. Roeger, S. Koenig, A. Ludwig, A. Marculescu, J. Li, M. Hoh, M. Dreschmann, J. Meyer, S. Ben Ezra, N. Narkiss, B. Nebendahl, F. Parmigiani, P. Petropoulos, B. Resan, A. Oehler, K. Weingarten, T. Ellermeyer, J. Lutz, M. Moeller, M. Huebner, J. Becker, C. Koos, W. Freude, J. Leuthold, "26 Tbit s^{-1} line-rate super-channel transmission utilizing all-optical fast Fourier transform processing," Nature Photon., vol. 5, no. 6, pp. 364–371, 2011.

[11Hof] M. Hoffmann, O. D. Sieber, V. J. Wittwer, I. L. Krestnikov, D. A. Livshits, T. Südmeyer, U. Keller, "Femtosecond high-power quantum dot vertical external cavity surface emitting laser," Opt. Exp., vol. 19, No. 9, pp. 8108–8116, 2011

[11Ili] H. Iliev, I. Buchvarov, S. Kurimura, V. Petrov, "1.34-μm Nd:YVO_4 laser mode-locked by SHG-lens formation in periodically-poled stoichiometric lithium tantalate," Opt. Express, vol. 19, pp. 21754–21759, 2011

[11Mar] A. Martinez, S. Yamashita, "Multi-gigahertz repetition rate passively modelocked fiber lasers using carbon nanotubes," Opt. Express, vol. 19, 6155–6163, 2011

[11Pek] S. Pekarek, T. Südmeyer, S. Lecomte, S. Kundermann, J. M. Dudley, and U. Keller, "Self-referencable frequency comb from a gigahertz diode-pumped solid state laser," Opt. Express, vol. 19, pp. 16491–16497, 2011

[11Phi] C. R. Phillips, C. Langrock, J. S. Pelc, M. M. Fejer, I. Hartl, M. E. Fermann, "Supercontinuum generation in quasi-phasematched waveguides," Opt. Express, vol. 19, pp. 18754–18773, 2011

[11Pro] O. Pronin, J. Brons, C. Grasse, V. Pervak, G. Boehm, M.-C. Amann, V. L. Kalashnikov, A. Apolonski, F. Krausz, "High-power 200 fs Kerr-lens mode-locked Yb:YAG thin-disk oscillator," Opt. Lett., vol. 36, pp. 4746–4748, 2011

[11Sar] C. J. Saraceno, O. H. Heckl, C. R. E. Baer, M. Golling, T. Südmeyer, K. Beil, C. Kränkel, K. Petermann, G. Huber, U. Keller, "SESAMs for high-power femtosecond modelocking: power scaling of an Yb:$LuScO_3$ thin disk laser to 23 W and 235 fs," Opt. Exp., vol. 19, No. 21, pp. 20288–20300, 2011

[11Sch] C. Schäfer, C. Fries, C. Theobald, J. A. L'Huillier, "Parametric Kerr lens mode-locked, 888 nm pumped Nd:YVO_4 laser," Opt. Lett., vol. 36, pp. 2674–2676, 2011

[11Sch2] S. Schilt, N. Bucalovic, V. Dolgovskiy, C. Schori, M. C. Stumpf, G. Di Domenico, S. Pekarek, A. E. H. Oehler, T. Südmeyer, U. Keller, P. Thomann, "Fully stabilized optical frequency comb with sub-radian CEO phase noise from a SESAM-modelocked 1.5-μm solid-state laser," Opt. Exp., vol. 19, pp. 24171–24181, 2011

[11Sch3] S. Schilt, N. Bucalovic, L. Tombez, V. Dolgovskiy, C. Schori, G. Di Domenico, M. Zaffalon, P. Thomann, "Frequency discriminators for the characterization of narrow-spectrum heterodyne beat signals: Application to the measurement of sub-hertz carrier-envelope-offset beat in an optical frequency comb," Rev. Sci. Instruments, vol. 82, p. 123116, 2011

[12Bau] D. Bauer, I. Zawischa, D. H. Sutter, A. Killi, T, Dekorsy, "Mode-locked Yb:YAG thin-disk oscillator with 41 μJ pulse energy at 145 W average infrared power and high power frequency conversion," Opt. Express, vol. 20, pp. 9698–9704, 2012

[12Buc] N. Bucalovic, V. Dolgovskiy, M. C. Stumpf, C. Schori, G. Di Domenico, U. Keller, S. Schilt, T. Südmeyer, "Effect of the carrier-envelope-offset dynamics on the stabilization of a diode-pumped solid-state frequency comb," Opt. Lett., vol. 37, No. 21, pp. 4428–4430, 2012

[12But] A. C. Butler, D. J. Spence, D. W. Coutts, "Scaling Q-switched microchip lasers for shortest pulses," Appl. Phys. B, vol. 109, pp. 81–88, 2012

[12Che] H.-W. Chen, G. Chang, S. Xu, Z. Yang, F. X. Kärtner, "3 GHz, fundamentally mode-locked, femtosecond Yb-fiber laser," Opt. Lett., vol. 37, 3522–3524, 2012

[12End] M. Endo, A. Ozawa, and Y. Kobayashi, "Kerr-lens mode-locked Yb:KYW laser at 4.6-GHz repetition rate," Opt. Express, vol. 20, pp. 12191–12196, 2012

[12Hei] B. Heinen, T.-L. Wang, M. Sparenberg, A. Weber, B. Kunert, J. Hader, S. Koch, J. V. Moloney, M. Koch, W. Stolz, "106 W continuous-wave output power from vertical-external-cavity surface-emitting laser," Electron. Lett., vol. 48, no. 9, pp. 516–517, 2012

[12Joc] C. Jocher, T. Eidam, S. Hendrich, J. Limpert, and A. Tünnermann, "Sub 25 fs pulses from solid-core nonlinear compression stage at 250 W of average power," Opt. Lett., vol. 37, no. 21, pp. 4407–4409, 2012

[12Kri] V. Kristijonas, A. Markus-Christian, "Room-temperature 3.73 μm GaSb-based type-I quantum-well lasers with quinternary barriers," Semicond. Sci. Techn., vol. 27, 032001, 2012

[12Mar] A. Martinez, S. Yamashita, "10-GHz fundamental mode fiber laser using a graphene saturable absorber," Appl. Phys. Lett., vol. 101, 041118, 2012

[12Nak] H. Nakao, A. Shirakawa, K. Ueda, H. Yagi, T. Yanagitani, "CW and mode-locked operation of Yb^{3+}-doped $Lu_3Al_5O_{12}$ ceramic laser," Opt. Express, vol. 20, pp. 15385–15391, 2012

[12Nov] K. S. Novoselov, V. I. Fal'ko, L. Colombo, Pr. R. Gellert, M. G. Schwab. K. Kim, "A roadmap for graphene," Nature, vol. 490, pp. 192–200, 2012

[12Pek] S. Pekarek, A. Klenner, T. Südmeyer, C. Fiebig, K. Paschke, G. Erbert, and U. Keller, "Femtosecond diode-pumped solid-state laser with a repetition rate of 4.8 GHz," Opt. Express, vol. 20, pp. 4248–4253, 2012

[12Per] V. Pervak, O. Pronin, O. Razskazovskaya, J. Brons, I. B. Angelov, M. K. Trubetskov, A. V. Tikhonravov, and F. Krausz, "High-dispersive mirrors for high power applications," Opt. Exp., vol. 20, pp. 4503–4508, 2012

[12Sar] C. J. Saraceno, F. Emaury, O. H. Heckl, C. R. E. Baer, M. Hoffmann, C. Schriber, M. Golling, T. Südmeyer, U. Keller, "275 W average output power from a femtosecond thin disk oscillator operated in a vaccum environment," Opt. Exp., vol. 20, No. 21, pp. 23535–23541, 2012

[12Sar2] C. J. Saraceno, O. H. Heckl, C. R. E. Baer, C. Schriber, M. Golling, K. Beil, C. Kränkel, T. Südmeyer, G. Huber, U. Keller, "Sub-100 femtosecond pulses from a SESAM modelocked thin disk laser," Appl. Phys. B, vol. 163, pp. 559– 562, 2012

[12Sar3] C. J. Saraceno, S. Pekarek, O. H. Heckl, C. R. E. Baer, C. Schriber, M. Golling, K. Beil, C. Kränkel, G. Huber, U. Keller, T. Südmeyer, "Self-referencable frequency comb from an ultrafast thin disk laser," Opt. Exp., vol. 20, No. 9, pp. 9650–9656, 2012

[12Sar4] C. J. Saraceno, C. Schriber, M. Mangold, M. Hoffmann, O. H. Heckl, C. R. E. Baer, M. Golling, T. Südmeyer, U. Keller, "SESAMs for high-power oscillators: design guidelines and damage thresholds," IEEE J. Sel. Top. Quantum Electron., vol. 18, No. 1, pp. 29–41, 2012

[12Sch] S. Schilt, V. Dolgovskiy, N. Bucalovic, C. Schori, M. C. Stumpf, G. Di Domenico, S. Pekarek, A. E. H. Oehler, T. Südmeyer, U. Keller, P. Thomann, "Noise properties of an optical frequency comb from SESAM-modelocked 1.5-μm solid-state laser stabilized to the 10^{-13} level," Appl. Phys. B, vol. 109, pp. 391–402, 2012

[12Sch2] T. C. Schratwieser, C. G. Leburn, D. T. Reid, "Highly efficient 1 GHz repetition-frequency femtosecond Yb^3:$KY(WO_4)_2$ laser," Opt. Lett., vol. 37, pp. 1133–1135, 2012

[12Sun] Z. Sun, T. Hasan, A. C. Ferrari, "Ultrafast lasers mode-locked by nanotubes and graphene," Physica E, vol. 44, pp. 1082–1091, 2012

[12Tok] M. Tokurakawa, A. Shirakawa, K. Ueda, H. Yagi, T. Yanagitani, A. A. Kaminskii, K. Beil, C. Kränkel, G. Huber, "Continuous wave and mode-locked Yb^{3+}:Y_2O_3 ceramic thin disk laser," Opt. Exp., vol. 20, pp. 10847–10852, 2012

[12Wen] K. S. Wentsch, L. Zheng, J. Xu, M. A. Ahmed, T. Graf, "Passively mode-locked Yb^{3+}:Sc_2SiO_5 thin-disk laser," Opt. Lett., vol. 37, pp. 4750–4752, 2012

[12Wil] K. G. Wilcox , A. H. Quarterman, V. Apostolopoulos, H. E. Beere, I. Farrer, D. A. Ritchie, A. C. Tropper "175 GHz, 400-fs-pulse harmonically mode-locked surface emitting semiconductor laser," Opt. Exp., vol. 20, no. 7, pp. 7040–7045, 2012

[12Zau] C. A. Zaugg, M. Hoffmann, W. P. Pallmann, O. D. Sieber, V. J. Wittwer, M. Mangold, M. Golling, K. J. Weingarten, B. W. Tilma, T. Südmeyer, U. Keller, "Low repetition rate SESAM modelocked VECSEL using an extendable active multipass-cavity approach," Opt. Exp., vol. 20, No. 25, pp. 27915–27921, 2012

[12Zho] B. Zhou, A. Chong, F. Wise, M. Bache, "Ultrafast and octave spanning optical nonlinearities from strongly phase-mismatched quadratic nonlinearities," Phys. Rev. Lett. vol. 109, pp. 043902, 2012

[13Bot] R. C. Botha, H. J. Strauss, C. Bollig, W. Koen, O. Collett, N. V. Kuleshov, M. J. D. Esser, W. L. Combrinck, H. M. von Bergmann, "High average power 1314 nm Nd:YLF laser, passively Q-switched with V:YAG," Opt. Lett., vol. 38, pp. 980–982, 2013

[13End] M. Endo, A. Ozawa, and Y. Kobayashi, "6-GHz, Kerr-lens mode-locked Yb:Lu_2O_3 ceramic laser for comb-resolved broadband spectroscopy," Opt. Lett., vol. 38, 4502–4505, 2013

[13Fer] M. E. Fermann, I. Hartl, "Ultrafast fiber lasers," Nature Photonics, vol. 7, pp. 868–874, 2013

[13Fle] R. Fleischhaker, N. Krauss, F. Schättiger, T. Dekorsy, "Consistent characterization of semiconductor saturable absorber mirrors with single-pulse and pump–probe spectroscopy," Opt. Express, vol. 21, p. 6764, 2013

[13Ish] A. Ishizawa, T. Nishikawa, A. Mizutori, H. Takara, A. Takada, T. Sogawa, M. Koga, "Phase-noise characteristics of a 25-GHz-spaced optical frequency comb based on a phase- and intensity-modulated laser," Opt. Express, vol. 21, no. 24, pp. 29186–29194, 2013

[13Kle] A. Klenner, M. Golling, U. Keller, "A gigahertz multimode-diode-pumped Yb:KGW enables a strong frequency comb offset beat signal," Opt. Express, vol. 21, pp. 10351–10356, 2013

[13Luc] M. Lucchini, J. Herrmann, A. Ludwig, R. Locher, M. Sabbar, L. Gallmann, U. Keller, "Role of electron wavepacket interference in the optical response of helium atoms," New J. Phys., vol. 15, 103010, 2013

[13Man] M. Mangold, V. J. Wittwer, C. A. Zaugg, S. M. Link, M. Golling, B. W. Tilma, U. Keller, "Femtosecond pulses from a modelocked integrated external- cavity surface emitting laser (MIXSEL)," Opt. Express., vol. 21, pp. 24904–24911, 2013

[13Mos] D. J. Moss, R. Morandotti, A. L. Gaeta, M. Lipson, "New CMOS-compatible platforms based on silicon nitride and Hydex for nonlinear optics," Nature Photonics, vol. 7, pp. 597–607, 2013

[13Pen] Y. H. Peng, Y. X. Lim, J. Cheng, Y. Guo, Y. Y. Cheah, K. S. Lai, "Near fundamental mode 1.1 kW Yb:YAG thin-disk laser," Opt. Lett., vol. 38, pp. 1709–1711, 2013

[13Phi] C. R. Phillips, C. Langrock, D. Chang, Y. W. Lin, L. Gallmann, M. M. Fejer, "Apodization of chirped quasi-phasematching devices," J. Opt. Soc. Am. B, vol. 30, pp. 1551–1568, 2013

[13Sch] M. Schultze, E. M. Bothschafter, A. Sommer, S. Holzner, W. Schweinberger, M. Fiess, M. Hofstetter, R. Kienberger, V. Apalkov, V. S. Yakovlev, M. I. Stockman F. Krausz, "Controlling dielectrics with the electric field of light," Nature, vol. 493, no. 7430, pp. 75–78, 2013

[13Seg] K. Seger, N. Meiser, S. Y. Choi, B. H. Jung, D-I Yeom, F. Rotermund, O. Okhotnikov, F. Laurell, V. Pasiskevicius, "Carbon nanotube mode-locked optically-pumped semiconductor disk laser," Opt. Express., vol. 21, no. 15, pp. 17806–17813, 2013

[13Tei] C. Y. Teisset, M. Schultze, R. Bessing, M. Häfner, S. Prinz, D. Sutter, and T. Metzger, "300 W Picosecond Thin-Disk Regenerative Amplifier at 10 kHz Repetition Rate," in Advanced Solid-State Lasers Congress Postdeadline, G. Huber and P. Moulton, eds., OSA Postdeadline Paper Digest (online) (Optical Society of America, 2013), paper JTh5A.1

[13Zau] C. A. Zaugg, Z. Sun, V. J. Wittwer, D. Popa, S. Milana, T. Kulmala, R. S. Sundaram, M. Mangold, O. D. Sieber, M. Golling, Y. Lee, J. H. Ahn, A. C. Ferrari, U. Keller, "Ultrafast and widely tuneable vertical-external-cavity surface-emitting laser, mode-locked by a graphene-integrated distributed Bragg reflector," Opt. Exp., vol. 21, No. 25, pp. 31548–31559, 2013

[14Bro] J. Brons, V. Pervak, E. Fedulova, D. Bauer, D. Sutter, V. Kalashnikov, A. Apolonsky, O. Pronin, F. Krausz, "Energy scaling of Kerr-lens mode-locked thin-disk oscillators," Opt. Lett., vol. 39, pp. 6442–6445, 2014

[14Dan] B. Dannecker, X. Delen, K. S. Wentsch, B. Weichelt, C. Hönninger, A. Voss, M. A. Ahmed, T. Graf, "Passively mode-locked Yb:CaF_2 thin-disk laser," Opt. Express, vol. 22, pp. 22278–22284 (2014)

[14Ema] F. Emaury, C. J. Saraceno, B. Debord, D. Ghosh, A. Diebold, F. Gerome, T. Südmeyer, F. Benabid, U. Keller, "Efficient spectral broadening in the 100-W average power regime using gas-filled kagome HC-PCF and pulse compression," Opt. Lett., vol. 39, no. 24, pp. 6843–6846, 2014

[14Kle] A. Klenke, S. Hädrich, T. Eidam, J. Rothhardt, M. Kienel, S. Demmler, T. Gottschall, J. Limpert, and A. Tünnermann, "22 GW peak-power fiber chirped-pulse-amplification system," Opt. Lett., vol. 39, pp. 6875–6878, 2014

[14Kle2] A. Klenner, M. Golling, and U. Keller, "High peak power gigahertz Yb:CALGO laser," Opt. Express, vol. 22, No. 10, pp. 11884–11891, 2014

[14Kle3] A. Klenner, S. Schilt, T. Südmeyer, U. Keller, "Gigahertz frequency comb from a diode-pumped solid-state laser," Opt. Exp., vol. 22, No. 25, pp. 31008–31019, 2014

[14Man] M. Mangold, C. A. Zaugg, S. M. Link, M. Golling, B. W. Tilma, U. Keller, "Pulse repetition rate scaling from 5 to 100 GHz with a high-power semiconductor disk laser," Opt. Exp., vol. 22, No. 5, pp. 6099–6107, 2014

[14Man2] M. Mangold, S. M. Link, A. Klenner, C. A. Zaugg, M. Golling, B. W. Tilma, U. Keller, "Amplitude noise and timing jitter characterization of a high-power mode-locked integrated external-cavity surface emitting laser," IEEE Photon. J., vol. 6, pp.1–9, 2014

[14Mei] N. Meiser, K. Seger, V. Pasiskevicius, A. Zukauskas, C. Canalias, F. Laurell, "Cascaded mode-locking of a spectrally controlled Yb:KYW laser," Appl. Phys. B, vol. 116, pp. 493–499, 2014

[14Oh] D. Y. Oh, D. Sell, H. Lee, K. Y. Yang, S. A. Diddams, K. J. Vahala, "Supercontinuum generation in an on-chip silica waveguide," Opt. Lett., vol. 39, pp. 1046–1048, 2014

[14Phi] C. R. Phillips, A. S. Mayer, A. Klenner, U. Keller, "SESAM mode-locked Yb:$CaGdAlO_4$ laser in the soliton modelocking regime with positive intracavity dispersion," Opt. Express, vol. 22, pp. 6060–6077, 2014

[14Sar] C. J. Saraceno, F. Emaury, C. Schriber, M. Hoffmann, M. Golling, T. Südmeyer, U. Keller, "Ultrafast thin disk laser with 80 μJ pulse energy and 240 W of average power," Opt. Lett., vol. 39, No. 1, pp. 9–12, 2014

[14Sch] C. Schriber, F. Emaury, A. Diebold, S. M. Link, M. Golling, K. Beil, C. Kränkel, C. J. Saraceno, T. Südmeyer, U. Keller, "Dual-gain SESAM modelocked thin disk laser based on Yb:Lu_2O_3 and Yb:Sc_2O_3," Opt. Exp., vol. 22, No. 16, pp. 18979–18986, 2014

[14Sut] D. V. Sutyrin, N. Poli, N. Beverini, G. M. Tino, "Carrier-envelope offset frequency noise analysis in Ti:sapphire frequency combs," Opt. Engineering, vol. 53, Nr. 12, 122603, 2014

[14Ver] A. Vernaleken, B. Schmidt, T. W. Hänsch, R. Holzwarth, P. Hommelhoff, "Carrier-envelope frequency stabilization of a Ti:sapphire oscillator using different pump lasers: part II," Appl. Phys. B, vol. 117, pp. 33–39, 2014

[14Zau] C. A. Zaugg, S. Gronenborn, H. Moench, M. Mangold, M. Miller, U. Weichmann, W. P. Pallmann, M. Golling, B. W. Tilma, U. Keller, "Absorber and gain chip optimization to improve performance from a passively modelocked electrically pumped vertical external cavity surface emitting laser," Appl. Phys. Lett., vol. 104, no. 12, p. 121115, 2014

[15Bal] T. Balciunas, C. Fourdade-Dutin, G. Fan, T. Witting, A. A. Voronin, A. M. Zheltikov, G. Gerome, G. G. Paulus, A. Baltuska, F. Benabid, "A strong-field driver in the single-cycle regime based on self-compression in a kagome fibre," Nat. Commun., vol. 6, p. 6117, 2015

[15Bor] B. Borchers, C. Schäfer, C. Fries, M. Larionov, and R. Knappe, "Nonlinear polarization rotation mode-locking via phase-mismatched type I SHG of a thin disk femtosecond laser," in Advanced Solid State Lasers, OSA Technical Digest (online) (Optical Society of America, 2015), paper ATh4A.9

[15End] M. Endo, I. Ito, and Y. Kobayashi, "Direct 15-GHz mode-spacing optical frequency comb with a Kerr-lens mode-locked Yb:Y_2O_3 ceramic laser," Opt. Express, vol. 23, pp. 1276–1282, 2015

[15Hit] K. Hitachi, A. Ishizawa, O. Tadanaga, T. Nishikawa, H. Mashiko, T. Sogawa, H. Gotoh, "Frequency stabilization of an Er-doped fiber laser with a collinear $2f$-to-$3f$ self-referencing interferometer," Appl. Phys. Lett., vol. 106, no. 23, p. 231106, 2015

[15Hol] P. Holl, M. Rattunde, S. Adler, S. Kaspar, W. Bronner, A. Bächle, R. Aidam, J. Wagner, "Recent Advances in Power Scaling of GaSb-Based Semiconductor Disk Lasers," IEEE J. Sel. Top. Quantum Electron., vol. 21, pp. 324–335, 2015

[15Joh] A. R. Johnson, A. S. Mayer, A. Klenner, K. Luke, E. S. Lamb, M. R. E. Lamont, C. Joshi, Y. Okawachi, F. W. Wise, M. Lipson, U. Keller, A. L. Gaeta, "Octave-spanning coherent supercontinuum generation in a silicon nitride waveguide," Opt. Lett., vol. 40, No. 21, pp. 5117–5120, 2015

[15Kle] A. Klenner, U. Keller, "All-optical Q-switching limiter for high-power gigahertz modelocked diode-pumped solid-state lasers," Opt. Exp., vol. 23, No. 7, pp. 8532–8544, 2015

[15Kuh] V. Kuhn, T. Gottwald, C. Stolzenburg, S.-S. Schad, A. Killi, T. Ryba, "Latest advances in high brightness disk lasers," Proc. SPIE 9342, 93420Y (2015)

[15Lin] S. M. Link, A. Klenner, M. Mangold, C. A. Zaugg, M. Golling, B. W. Tilma, U. Keller, "Dual-comb modelocked laser," Opt. Exp., vol. 23, No. 5, pp. 5521–5531, 2015

[15Loc] R. Locher, L. Castiglioni, M. Lucchini, M. Greif, L. Gallmann, J. Osterwalder, M. Hengsberger, U. Keller, "Energy-dependent photoemission delays from noble metal surfaces by attosecond interferometry," Optica, vol. 2, No. 5, pp. 405–410, 2015

[15May] A. S. Mayer, A. Klenner, A. R. Johnson, K. Luke, M. R. E. Lamont, Y. Okawachi, M. Lipson, A. L. Gaeta, U. Keller, "Frequency comb offset detection using supercontinuum generation in silicon nitride waveguides," Opt. Exp., vol. 23, No. 12, pp. 15440–15451, 2015

[15Phi] C. R. Phillips, A. S. Mayer, A. Klenner, U. Keller, "Femtosecond mode locking based on adiabatic excitation of quadratic solitons," Optica, vol. 2, No. 8, pp. 667–674, 2015

[15Sar] C. J. Saraceno, F. Emaury, C. Schriber, A. Diebold, M. Hoffmann, M. Golling, T. Südmeyer, U. Keller, "Toward millijoule-level high-power ultrafast thin-disk oscillators," IEEE J. Sel. Top. Quantum Electron., vol. 21, No. 1, 1100318, 2015

[15Sor] I. T. Sorokina, E. Sorokin, "Femtosecond Cr^{2+}-based lasers," IEEE J. Sel. Top. Quantum Electron., vol. 21, pp. 273–291, 2015

[15Ste] S. Schilt, T. Südmeyer, "Carrier-envelope offset stabilized ultrafast diode-pumped solid-state lasers," Appl. Sci., vol. 5, pp. 787–816, 2015

[15Til] B. W. Tilma, M. Mangold, C. A. Zaugg, S. M. Link, D. Waldburger, A. Klenner, A. S. Mayer, E. Gini, M. Golling, U. Keller, "Recent advances in ultrafast semiconductor disk lasers," Light: Science & Applications (2015), vol. 4, e310

[15Zha] Z. Zhao, Y. Kobayashi, "Ytterbium fiber-based, 270 fs, 100 W chirped pulse amplification laser system with 1 MHz repetition rate," Appl. Phys. Express, vol. 9, 012701, 2015

[16Alf] C. G. E. Alfieri, A. Diebold, F. Emaury, E. Gini, C. J. Saraceno, U. Keller, "Improved SESAMs for femtosecond pulse generation approaching the kW average power regime," Opt. Exp., vol. 24, No. 24, pp. 27587–27599, 2016

[16Bro] J. Brons, V. Pervak, D. Bauer, D. Sutter, O. Pronin, F. Krausz, "Powerful 100-fs-scale Kerr-lens mode-locked thin-disk oscillator," Opt. Lett., vol. 41, pp. 3567–3570, 2016

[16Cod] I. Coddington, N. R. Newbury, W. C. Swann, "Dual-comb spectroscopy," Optica, vol. 3, pp. 414–426, 2016

[16Dan] B. Dannecker, M. A. Ahmed, T. Graf, "SESAM-modelocked Yb:CaF_2 thin-disk-laser generating 285 fs pulses with 1.78 μJ of pulse energy," Laser Phys. Lett., vol. 13, 055801, 2016

[16Die] A. Diebold, T. Zengerle, C. G. E. Alfieri, C. Schriber, F. Emaury, M. Mangold, M. Hoffmann, C. J. Saraceno, M. Golling, D. Follman, G. D. Cole, M. Aspelmeyer, T. Südmeyer, U. Keller, "Optimized SESAMs for kilowatt-level ultrafast lasers," Opt. Exp., vol. 24, No. 10, pp. 10512–10526, 2016

[16Dre] F. Dreisow, S. Döring, A. Ancona, J. König, S. Nolte, Ultrashort Pulse Laser Technology: Laser Sources and Applications. S. Nolte et al., eds. Springer Series in Optical Sciences, p. 195, 2016

[16Gaa] M. A. Gaafar, A. Rahimi-Iman, K. A. Fedorova, W. Stolz, E. U. Rafailov, M. Koch, "Mode-locked semiconductor disk lasers," Adv Opt Photonics, vol. 8, no. 3, pp. 370–400, 2016

[16Had] S. Hädrich, M. Kienel, M. Müller, A. Klenke, J. Rothhardt, R. Klas, T. Gottschall, T. Eidam, A. Drozdy, P. Jójárt, Z. Várallyay, E. Cormier, K. Osvay, A. Tünnermann, J. Limpert,"Energetic sub-2-cycle laser with 216 W average power," Opt. Lett., vol. 41, no. 18, pp. 4332–4335, 2016

[16Ide] T. Ideguchi, T. Nakamura, Y. Kobayashi, K. Goda, "Kerr-lens mode-locked bidirectional dual-comb ring laser for broadband dual-comb spectroscopy," Optica, vol. 3, pp. 748–753, 2016

[16Kah] H. Kahle, C. M. N. Mateo, U. Brauch, P. Tatar-Mathes, R. Bek, M. Jetter, T. Graf, P. Michler, "Semiconductor membrane external-cavity surface-emitting laser (MECSEL)," Optica, vol. 3, no. 12, pp. 1506–1512, 2016

[16Kie] M. Kienel, M. Müller, A. Klenke, J. Limpert, A. Tünnermann, "12 mJ kW-class ultrafast fiber laser system using multidimensional coherent pulse addition," Opt. Lett., vol. 41, pp. 3343–3346, 2016

[16Kle] A. Klenner, A. S. Mayer, A. R. Johnson, K. Luke, M. R. E. Lamont, Y. Okawachi, M. Lipson, A. L. Gaeta, U. Keller, "Gigahertz frequency comb offset stabilization based on supercontinuum generation in silicon nitride waveguides," Opt. Exp., vol. 24, No. 10, pp. 11043–11053, 2016

[16Lau] A. Laurain, D. Marah, R. Rockmore, J. McInerney, J. Hader, A. R. Perez, W. Stolz, J. V. Moloney, "Colliding pulse mode locking of vertical-external cavity surface-emitting laser," Optica, vol. 3, pp. 781–784, 2016

[16Lin] S. M. Link, A. Klenner, U. Keller, "Dual-comb modelocked lasers: semiconductor saturable absorber mirror decouples noise stabilization," Opt. Exp., vol. 24, No. 3, pp. 1889–1902, 2016

[16Luc] M. Lucchini, S. A. Sato, A. Ludwig, J. Herrmann, M. Volkov, L. Kasmi, Y. Shinohara, K. Yabana, L. Gallmann, U. Keller, "Attosecond dynamical Franz–Keldysh effect in polycrystalline diamond," Science, vol. 353, Issue 6302, pp. 916–919, 2016

[16Meh] S. Mehravar, R. A. Norwood, N. Peyghambarian, K. Q. Kieu, "Real-time dual-comb spectroscopy with a free-running bidirectionally mode-locked fiber laser," Appl. Phys. Lett., vol. 108, 231104, 2016

[16Mul] M. Müller, M. Kienel, A. Klenke, T. Gottschall, E. Shestaev, M. Plötner, J. Limpert, A. Tünnermann, "1 kW 1 mJ eight-channel ultrafast fiber laser," Opt. Lett., vol. 41, pp. 3439–3442, 2016

[16Res] B. Resan, S. Kurmulis, Z. Y. Zhang, A. E. H. Oehler, V. Markovic, M. Mangold, T. Südmeyer, U. Keller, R. A. Hogg, K. J. Weingarten, "10 GHz pulse repetition rate Er:Yb:glass laser modelocked with quantum dot semiconductor saturable absorber mirror," Appl. Optics, vol. 55, No. 14, pp. 3776–3780, 2016

[16Sch] J. Schulte, T. Sartorius, J. Weitenberg, A. Vernaleken, and P. Russbueldt, "Nonlinear pulse compression in a multi-pass cell," Opt. Lett., vol. 41, no. 19, pp. 4511–4514, 2016

[16Sch2] M. Schultze, C. Wandt, S. Klingebiel, C. Y. Teisset, M. Häfner, R. Bessing, T. Herzig, S. Prinz, S. Stark, K. Michel, and T. Metzger, "Toward Kilowatt-Level Ultrafast Regenerative Thin-Disk Amplifiers," in Lasers Congress 2016 (ASSL, LSC, LAC), OSA Technical Digest (online) (Optical Society of America, 2016), paper ATu4A.4.

[16Sei1] M. Seidel, G. Arisholm, J. Brons, V. Pervak, and O. Pronin, "All solid-state spectral broadening: an average and peak power scalable method for compression of ultrashort pulses," Opt. Express., vol. 24, no. 9, pp. 9412–9428, May 2, 2016

[16Sei2] M. Seidel, J. Brons, G. Arisholm, K. Fritsch, V. Pervak, and O. Pronin, "Efficient High-Power Pulse Compression in Self-Defocusing Bulk Media," Europhoton Conference 2016, PD1-2, 2016

[16Sha] E. A. Shaw, A. H. Quarterman, A. P. Turnbull, T. C. Sverre, C. R. Head, A. C. Tropper, K. G. Wilcox, "Nonlinear Lensing in an unpumped antiresonant semiconductor disk laser gain structure," IEEE Phot. Tech. Lett., vol. 28, No. 13, p. 1395, 2016

[16Sho] T. D. Shoji, W. Xie, K. L. Silverman, A. Feldman, T. Harvey, R. P. Mirin, T. R. Schibli, "Ultra-low-noise monolithic mode-locked solid-state laser," Optica, vol. 3, No. 9, pp. 995–998, 2016

[16Wal] D. Waldburger, S. M. Link, M. Mangold, C. G. E. Alfieri, E. Gini, M. Golling, B. W. Tilma, U. Keller, "High-power 100-femtosecond semiconductor disk lasers," Optica, vol. 3, No. 8, pp. 844–852, 2016

[17Alf] C. G. E. Alfieri, D. Waldburger, S. M. Link, E. Gini, M. Golling, G. Eisenstein, U. Keller, "Optical efficiency and gain dynamics of modelocked semiconductor disk lasers," Opt. Exp., vol. 25, No. 6, pp. 6402–6420, 2017

[17Gra] I. J. Graumann, A. Diebold, C. G. E. Alfieri, F. Emaury, B. Deppe, M. Golling, D. Bauer, D. Sutter, C. Kränkel, C. J. Saraceno, C. R. Phillips, U. Keller, "Peak-power scaling of femtosecond Yb:Lu_2O_3 thin-disk lasers," Opt. Exp., vol. 25, No. 19, pp. 22519–22536, 2017

[17Gui] M. Guina, A. Rantamäki, and A. Härkönen, "Optically pumped VECSELs: review of technology and progress," J. Phys. D: Appl. Phys., vol. 50, no. 38, p. 383001, 2017

[17Leo] H. Leopardi, J. D. Rodriguez, F. Quinlan, J. Olson, J. A. Sherman, S. A. Diddams, T. M. Fortier, "Single-branch Er:fiber frequency comb for precision optical metrology with 10^{-18} fractional instability," Optica, vol. 4, no. 8, pp. 879–885, 2017

[17Lin1] S. M. Link, D. J. H. C. Maas, D. Waldburger, U. Keller, "Dual-comb spectroscopy of water-vapor with a free-running semiconductor disk laser," Science, vol. 356, pp. 1164–1168, 2017

[17Lin2] S. M. Link, D. Waldburger, C. G. E. Alfieri, M. Golling, U. Keller, "Coherent beam combining and noise analysis of a colliding pulse modelocked VECSEL," Opt. Exp., vol. 25, No. 16, pp. 19281–19290, 2017

[17May] A. S. Mayer, C. R. Phillips, U. Keller, "Watt-level 10-gigahertz solid-state laser enabled by self-defocusing nonlinearities," Nat. Commun., vol. 8, p. 1673, 2017

[17Nub] T. Nubbemeyer, M. Kaumanns, M. Ueffing, M. Gorjan, A. Alismail, H. Fattahi, J. Brons, O. Pronin, H. G. Barros, Z. Major, T. Metzger, D. Sutter, and F. Krausz, "1 kW, 200 mJ picosecond thin-disk laser system," Opt. Lett., vol. 42, pp. 1381–1384, 2017

[17Sal] F. Saltarelli, A. Diebold, I. J. Graumann, C. R. Phillips, U. Keller, "Modelocking of thin-disk laser with the frequency-doubling nonlinear-mirror technique," Opt. Exp., vol. 25, No. 19, pp. 23254–23266, 2017

[17Vas] S. Vasilyev, I. Moskalev, M. Mirov, V. Smolski, S. Mirov, V. Gapontsev, "Ultrafast middle-IR lasers and amplifiers based on polycrystalline Cr:ZnS and Cr:ZnSe," Optical Materials Express, vol. 7, p. 2636, 2017

[17Wei] J. Weitenberg, A. Vernaleken, J. Schulte, A. Ozawa, T. Sartorius, V. Pervak, H. Hoffmann, T. Udem, P. Russbüldt, T. W. Hänsch, "Multi-pass-cell-based nonlinear pulse compression to 115 fs at 7.5 μJ pulse energy and 300 W average power," Opt. Express, vol. 25, no. 17, pp. 20502–20510, 2017

[17Wei2] J. Weitenberg, T. Saule, J. Schulte, P. Rüsbüldt, "Nonlinear pulse compression to sub-40 fs at 4.5 μJ pulse energy by multi-pass-cell spectral broadening," IEEE J. of Quantum Electron., vol. 53, p. 8600204, 2017

[17Win] T. Winzer, M. Mittendorff, S. Winnerl, H. Mittenzwey, R. Jago, M. Helm, E. Malic, A. Knorr, "Unconventional double-bended saturation of carrier occupation in optically excited graphene due to many-particle interactions," Nat. Commun., vol. 8, 15042, 2017

[18Alf] C. G. E. Alfieri, D. Waldburger, M. Golling, U. Keller, "High-power sub-300-femtosecond optically pumped quantum dot semiconductor disk lasers," IEEE Photonics Tech. Lett., vol. 30, No. 6, pp. 525–528, 2018

[18Die] A. Diebold, F. Saltarelli, I. J. Graumann, C. J. Saraceno, C. R. Phillips, U. Keller, "Gas-lens effect in kW class thin-disk lasers," Opt. Exp., vol. 26, No. 10, pp. 12648–12659, 2018

[18Guo] X. Guo, S. Tokita, J. Kawanaka, "12 mJ Yb:YAG/Cr:YAG microchip laser," Opt. Lett., vol. 43, pp. 459–461, 2018

[18Jia] M. Jiang, H. Xiao, S. Peng, L. Qiao, G. Yang, Z. Liu, X. Zu, "First-principles study of point defects in GaAs/AlAs superlattice: the phase stability and the effects on the band structure and carrier mobility," Nanoscale Res. Lett., vol. 13, 301, 2018

[18Mod] N. Modsching, C. Paradis, F. Labaye, M. Gaponenko, I. J. Graumann, A. Diebold, F. Emaury, V. J. Wittwer, T. Südmeyer, "Kerr lens mode-locked Yb:CALGO thin-disk laser," Opt. Lett., vol. 43, pp. 879–882, 2018

[18Mul] M. Müller, A. Klenke, A. Steinkopff, H. Stark, A. Tünnermann, J. Limpert, "3.5 kW coherently combined ultrafast fiber laser," Opt. Lett., vol. 43, pp. 6037–6040, 2018

[18Sal] F. Saltarelli, A. Diebold, I. J. Graumann, C. R. Phillips, U. Keller, "Self-phase modulation cancellation in a high-power ultrafast thin-disk laser oscillator," Optica, vol. 5, No. 12, pp. 1603–1606, 2018

[18Wal] D. Waldburger, C. G. E. Alfieri, S. M. Link, S. Meinecke, L. C. Jaurique, K. Lüdge, U. Keller, "Multipulse instabilities of a femtosecond SESAM- modelocked VECSEL," Opt. Exp., vol. 26, No. 17, pp. 21872–21886, 2018

[19Alf] C. G. E. Alfieri, D. Waldburger, J. Nürnberg, M. Golling, U. Keller, "Sub-150-fs pulses from a broadband optically pumped MIXSEL," Opt. Lett., vol. 44, No. 1, pp. 25–28, 2019

[19Fel] J. Fellinger, A. S. Mayer, G. Winkler, W. Grosinger, G.-W. Truong, S. Droste, C. Li, C. M. Heyl, I. Hartl, O. H. Heckl, "Tunable dual-comb from an all-polarization-maintaining single-cavity dual-color Yb:fiber laser," Opt. Express, vol. 27, pp. 28062–28074, 2019

[19For] T. Fortier, E. Baumann, "20 years of developments in optical frequency comb technology and applications," Communications Physics, vol. 2, 153, 2019

[19Gae] A. L. Gaeta, M. Lipson, T. J. Kippenber, "Photonic-chip-based frequency combs," Nature Photonics, vol. 13, pp. 158–169, 2019

[19Gra] I. J. Graumann, F. Saltarelli, L. Lang, V. J. Wittwer, T. Südmeyer, C. R. Phillips, U. Keller, "Power-scaling of nonlinear-mirror modelocked thin-disk lasers," Opt. Exp., vol. 27, No. 26, pp. 37349–37363, 2019

[19Kim] S. Kimura, S. Tani, Y. Kobayashi, "Kerr-lens mode locking above 20 GHz repetition rate," Optica, vol. 6, pp. 532–533, 2019

[19Lem] R. Lemons, W. Liu, I. Fernandez de Fuentes, S. Droste, G. Steinmeyer, C. G. Durfee, S. Carbajo, "Carrier-envelope phase stabilization of an Er:Yb:glass laser via a feed-forward technique," Opt. Lett., vol. 44, No. 22, pp. 5610–5613, 2019

[19Nur] J. Nürnberg, C. G. E. Alfieri, Z. Chen, D. Waldurger, N. Picque, U. Keller, "An unstabilized femtosecond semiconductor laser for dual-comb spectroscopy of acetylene," Opt. Exp., vol. 27, No. 3, pp. 3190–3199, 2019

[19Poe] M. Poetzlberger, J. Zhang, S. Gröbmeyer, D. Bauer, D. Sutter, J. Brons, O. Pronin, "Kerr-lens mode-locked thin-disk oscillator with 50% output coupling rate," Opt. Lett., vol. 44, pp. 4227–4230, 2019

[19Rus] P. Russbueldt, J. Weitenberg, J. Schulte, R. Meyer, C. Meinhardt, H. D. Hoffmann, R. Poprawe, "Scalable 30 fs laser source with 530 W average power," Opt. Lett., vol. 44, pp. 5222–5225, 2019

[19Sal] F. Saltarelli, I. J. Graumann, L. Lang, D. Bauer, C. R. Phillips, U. Keller, "Power scaling of ultrafast oscillators: 350-W average-power sub-picosecond thin-disk laser," Opt. Exp., vol. 27, No. 22, pp. 31465–31474, 2019

[19Sta] H. Stark, J. Buldt, M. Müller, A. Klenke, A. Tünnermann, J. Limpert, "23 mJ high-power fiber CPA system using electro-optically controlled divided-pulse amplification," Opt. Lett., vol. 44, pp. 5529–5532, 2019

[19Tsa] C.-L. Tsai, F. Meyer, A. Omar, Y. Wang, A.-Y. Liang, C.-H. Lu, M. Hoffmann, S.-D. Yang, C. J. Saraceno, "Efficient nonlinear compression of a mode-locked thin-disk oscillator to 27 fs at 98 W average power," Opt. Lett., vol. 44, pp. 4115–4118, 2019

[19Wal] D. Waldburger, A. S. Mayer, C. G. E. Alfieri, A. R. Johnson, X. Ji, A. Klenner, Y. Okawachi, M. Lipson, A. L. Gaeta, U. Keller, "Tightly locked optical frequency comb from a semiconductor disk laser," Opt. Exp., vol. 27, No. 3, pp. 1786–1797, 2019

[20Die] T. Dietz, M. Jenne, D. Bauer, M. Scharun, D. Sutter, and A. Killi, "Ultrafast thin-disk multi-pass amplifier system providing 1.9 kW of average output power and pulse energies in the 10 mJ range at 1 ps of pulse duration for glass-cleaving applications," Opt. Express, vol. 28, pp. 11415–11423, 2020

[20Fin] T. Finke, J. Nürnberg, V. Sichkovskyi, M. Golling, U. Keller, J. P. Reithmaier, "Temperature resistant fast $In_xGa_{1-x}As/GaAs$ quantum dot saturable absorber for epitaxial integration into semiconductor surface emitting lasers," Opt. Exp., vol. 28, No. 14, pp. 20954–20965, 2020

[20For] T. Fortier, E. Baumann, "20 years of developments in optical frequency comb technology and applications," Commun. Phys., vol. 2, 153, 2019

[20Gre] C. Grebing, M. Müller, J. Stark, J. Limpert, "Kilowatt-average-power compression of millijoule pulses in a gas-filled multi-pass cell," Opt. Lett., vol. 45, pp. 6250–6253, 2020

[20Kru] L. M. Krüger, A. S. Mayer, Y. Okawachi, X. Ji, A. Klenner, A. R. Johnson, C. Langrock, M. M. Fejer, M. Lipson, A. L. Gaeta, V. J. Wittwer, T. Südmeyer, C. R. Phillips, U. Keller, "Performance scaling of a 10-GHz solid-state laser enabling self-referenced CEO frequency detection without amplification," Opt. Exp., vol. 28, No. 9, pp. 12755–12770, 2020

[20Lia] R. Liao, H. Tian, W. Liu, R. Li, Y. Son, M. Hu, "Dual-comb generation from a single laser source: principles and spectroscopic applications towards mid-IR – a review," J. Phys. Photonics, vol. 2, 042006, 2020

[20May] A. S. Mayer, W. Grosinger, J. Fellinger, G. Winkler, L. W. Perner, S. Droste, S. H. Salman, C. Li, C. M. Heyl, I. Hartl, O. H. Heckl, "Flexible all-PM NALM Yb:fiber laser design for frequency comb applications: operation regimes and their noise properties," Opt. Express, vol. 28, 13, pp. 18946–18968, 2020

[20Pup] J. Pupeikis, P.-A. Chevreuil, N. Bigler, L. Gallmann, C. R. Phillips, U. Keller, "Water window soft x-ray source enabled by 25 W few-cycle 2.2 μm OPCPA at 100 kHz," Optica, vol. 7, No. 2, pp. 168–171, 2020

[20Sal] F. Saltarelli, ETH Zurich PhD thesis, Aug. 2020

[20Ste] A. Steinkoff, C. Jauregui, C. Aleshire, A. Klenke, J. Limpert, "Impact of thermo-optical effects in coherently combined multicore fiber amplifiers," Opt. Exp., vol. 28, pp. 38093–38105, 2020

[20Wil] B. Willenberg, F. Brunner, C. R. Phillips, U. Keller, "High-power picosecond deep-UV source via group velocity matched frequency conversion," Optica, vol. 7, No. 5, pp. 485–491, 2020

[20Wil2] B. Willenberg, J. Pupeikis, L. M. Krüger, F. Koch, C. R. Phillips, U. Keller, "Femtosecond dual-comb Yb:CaF_2 laser from a single free-running polarization-multiplexed cavity for optical sampling applications," Opt. Express, vol. 28, No. 20, pp. 30275–30288, 2020

[21Bar] A. Barh, J. Heidrich, B. O. Alaydin, M. Gaulke, M. Golling, C. R. Phillips, U. Keller, "Watt-level and sub-100-fs self-starting modelocking Cr:ZnS oscillator enabled by GaSb-SESAMs," Opt. Exp., vol. 29, No. 4, pp. 5934–5946, 2021

[21Hei] J. Heidrich, M. Gaulke, B. O. Aladin, M. Golling, A. Barh, U. Keller, "Full optical SESAM characterization methods in the 1.9 to 3-μm wavelength regime," Opt. Exp., vol. 29, No. 5, pp. 6647–6656, 2021

[21Hil] J. Hillbrand, L. M. Krüger, S. D. Cin, H. Knötig, J. Heidrich, A. M. Andrews, G. Strasser, U. Keller, B. Schwarz, "High-speed quantum cascade detector characterized with a mid-infrared femtosecond oscillator," Opt. Exp., vol. 29, No. 4, pp. 5774–5781, 2021

[21Kru] L. M. Krüger, S. L. Camenzind, C. R. Phillips, U. Keller, "Carrier-envelope offset frequency dynamics of a 10-GHz modelocked laser based on cascaded quadratic nonlinearities," Opt. Exp., vol. 29, No. 22, pp. 36915–36925, 2021

[21Lan] L. Lang, F. Saltarelli, G. Lacaille, S. Rowan, J. Hough, I. J. Graumann, C. R. Phillips, U. Keller, "Silicate bonding of sapphire to SESAMs: adjustable thermal lensing for high-power lasers," Opt. Exp., vol. 29, No. 12, pp. 18059–18069, 2021

[21Mul] M. Müller, J. Buldt, H. Stark, C. Grebing, J. Limpert, "Multipass cell for high-power few-cycle compression," Opt. Lett., vol. 46, pp. 2678–2681, 2021

[21Nur] J. Nürnberg, B. Willenberg, C. R. Phillips, U. Keller, "Dual-comb ranging with frequency combs from single cavity free-running laser oscillators," Opt. Exp., vol. 29, No. 16, pp. 24910–24918, 2021

[21Nus] A. Nussbaum-Lapping, C. R. Phillips, B. Willenberg, J. Pupeikis, U. Keller, "Absolute SESAM characterization via polarization-resolved non-collinear equivalent time sampling" Appl. Phys. B, submitted (invited paper should be published in Dec. 2021 by Springer Verlag)

[21Pup] J. Pupeikis, B. Willenberg, F. Bruno, M. Hettich, A Nussbaum-Lapping, M. Golling, C. P. Bauer, S. L. Camenzind, A. Benayad, P. Camy, B. Audoin, C. R. Phillips, U. Keller "Picosecond ultrasonics with a free-running dual-comb laser" Opt. Exp., vol. 29, No. 22, pp. 35735–35754, 2021

[21Sta] H. Stark, J. Buldt, M. Müller, A. Klenke, J. Limpert, "1 kW, 10 mJ, 120 fs coherently combined fiber CPA laser system," Opt. Lett., vol. 46, pp. 969–972, 2021

Index

© The Editor(s) (if applicable) and The Author(s), under exclusive licence to Springer Nature Switzerland AG 2021

U. Keller, *Ultrafast Lasers*, Graduate Texts in Physics,
https://doi.org/10.1007/978-3-030-82532-4

Y

Z

Printed in the United States
by Baker & Taylor Publisher Services